BASIC CURVES

$$y = mx + b$$

$$y = ax^2 + bx + c \quad (a > 0)$$

$$y = ax^2 + bx + c \quad (a < 0)$$

$$x^2 + y^2 = a^2$$

$$y^2 = 4px \quad (p > 0)$$

$$x^2 = 4py \quad (p > 0)$$

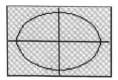

$$\frac{x^2}{a^2} + \frac{y^2}{b^2} = 1$$

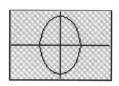

$$\frac{y^2}{a^2} + \frac{x^2}{b^2} = 1$$

$$\frac{x^2}{a^2} - \frac{y^2}{b^2} = 1$$

$$\frac{y^2}{a^2} - \frac{x^2}{b^2} = 1$$

$$xy = a \quad (a > 0)$$

$$y = a\sqrt{x} \quad (a > 0)$$

$$y = ax^3 \quad (a > 0)$$

$$y = ax^4 \quad (a > 0)$$

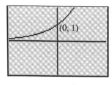

$$y = b^x \quad (b > 1)$$

$$y = b^{-x} \quad (b > 1)$$

$$y = \log_b x$$

$$y = a \sin(bx + c)$$
$$(a > 0, c > 0)$$

$$y = a \cos(bx + c)$$
$$(a > 0, c > 0)$$

$$y = a \tan x \quad (a > 0)$$

Basic Technical Mathematics

NINTH EDITION

Basic Technical Mathematics

Allyn J. Washington
Dutchess Community College

PEARSON

Prentice
Hall

Upper Saddle River, New Jersey
Columbus, Ohio

Library of Congress Cataloging-in-Publication Data

Washington, Allyn J.
  Basic technical mathematics / Allyn J. Washington.
    p. cm.
  ISBN 978-0-13-814225-4 (alk. paper)
  1. Mathematics. I. Title.
  QA39.3.W365 2009
  512'.1—dc22

                              2008020788

EDITOR IN CHIEF: Vernon Anthony
SENIOR ACQUISITIONS EDITOR: Gary Bauer
DEVELOPMENT EDITOR: Linda Cupp
EDITORIAL ASSISTANT: Megan Heintz
PRODUCTION COORDINATION: Greg Hubit, Bookworks
PROJECT MANAGER: Rex Davidson
SENIOR OPERATIONS SUPERVISOR: Pat Tonneman
ART DIRECTOR: Diane Ernsberger

INTERIOR DESIGN: Susan Schmidler
COVER DESIGNER: Mike Fruhbeis
COVER IMAGE: Getty Images
MEDIA PRODUCER: Carl Cottrell
DIRECTOR OF MARKETING: David Gesell
MARKETING MANAGER: Leigh Ann Sims
MARKETING ASSISTANT: Les Roberts
CHAPTER-OPENER ILLUSTRATIONS: Jim Bryant

About the Cover: The solar-electric Helios Prototype flying wing flies over the Hawaiian islands of Nilhau and Lehua during the first solar-powered test flight July 14, 2001, from the U.S. Navy's Pacific Missile Range Facility on Kauai, HI. The 18-hour flight was a functional checkout of the aircraft's systems and performance in preparation for an attempt to reach sustained flight at 100,000 feet altitude later in the summer. (Photo courtesy of NASA/Getty Images)

This book was set in Times Roman by Pre-PressPMG and was printed and bound by R. R. Donnelley & Sons. The cover was printed by Phoenix Color Corp.

Pearson Prentice Hall™ is a trademark of Pearson Education, Inc.
Pearson® is a registered trademark of Pearson plc
Prentice Hall® is a registered trademark of Pearson Education, Inc.

Pearson Education Ltd., London
Pearson Education Singapore Pte. Ltd.
Pearson Education Canada, Inc.
Pearson Education—Japan

Pearson Education Australia Pty. Limited
Pearson Education North Asia Ltd., Hong Kong
Pearson Educación de Mexico, S.A. de C.V.
Pearson Education Malaysia Pte. Ltd.

10 9 8 7

ISBN-13: 978-0-13-814225-4
ISBN-10:   0-13-814225-4

In memory of my loving wife, Millie

Contents

x   Contents

Preface

Scope of the Book

Basic Technical Mathematics, Ninth Edition, is intended primarily for students in technical and pre-engineering technology programs or other programs for which coverage of basic mathematics is required.

Chapters 1 through 20 provide the necessary background for further study with an integrated treatment of algebra and trigonometry. Chapter 21 covers the basic topics of analytic geometry, and Chapter 22 gives an introduction to statistics. In the examples and exercises, numerous applications from many fields of technology are included, primarily to indicate where and how mathematical techniques are used. However, it is not necessary that the student have a specific knowledge of the technical area from which any given problem is taken.

Most students using this text will have a background that includes some algebra and geometry. However, the material is presented in adequate detail for those who may need more study in these areas. The material presented here is sufficient for two to three semesters.

One of the principal reasons for the arrangement of topics in this text is to present material in an order that allows a student to take courses concurrently in allied technical areas, such as physics and electricity. These allied courses normally require a student to know certain mathematical topics by certain definite times; yet the traditional order of topics in mathematics courses makes it difficult to attain this coverage without loss of continuity. However, the material in this book can be rearranged to fit any appropriate sequence of topics. Another feature of this text is that certain topics traditionally included for mathematical completeness have been covered only briefly or have been omitted.

The approach used in this text is not unduly rigorous mathematically, although all appropriate terms and concepts are introduced as needed and given an intuitive or algebraic foundation. The aim is to help the student develop an understanding of mathematical methods without simply providing a collection of formulas. The text material is developed recognizing that it is essential for the student to have a sound background in algebra and trigonometry in order to understand and succeed in any subsequent work in mathematics.

New Features

The ninth edition of *Basic Technical Mathematics* includes all the basic features of the earlier editions. Many sections include revised explanatory material. A number of the examples have been rewritten, and new exercises have been included in all sections. Specifically, among the new features of this edition are the following:

NEW AND REVISED COVERAGE

The topics covered as Supplementary Topics in the eighth edition have now been integrated into the regular chapters. The sections on "Gaussian Elimination" and

"Higher-Order Determinants" are now included in Chapter 16, which has been retitled as "Matrices; Systems of Linear Equations." The section on "Rotation of Axes" is now included in Chapter 21, "Plane Analytic Geometry."

THE GRAPHING CALCULATOR

The graphing calculator is used throughout the text in examples and exercises to help reinforce and develop many topics. There are more than 190 graphing calculator screens, over 20 of which are new in this edition. The coverage starts in Section 1.3, where it is used for calculational purposes, and its use for graphing starts in Section 3.5. Additional coverage on the use of calculators is found in Appendix C.

Appendix C also includes 14 graphing calculator programs that show how the calculator's use can be expanded. Also, a *Graphing Calculator Manual* showing detailed procedures for several models is available as a supplement.

CHAPTER AND SECTION CONTENTS

A listing of the section titles for each chapter is now given on the introductory page of the chapter. Also, a listing of the key topics of each section is given below the section number and title on the first page of the section. This gives the student and instructor a quick preview of the chapter and section contents.

EXAMPLE DESCRIPTIONS

A brief descriptive title is given with each example number. This gives an easy reference for the example, particularly when reviewing the contents of the section.

PRACTICE EXERCISES

Throughout the text there are *practice exercises* in the margin. Most sections have at least one (up to as many as four) of these basic exercises. They are included so that a student is more actively involved in the learning process and can check his or her understanding of the material to that point in the section. They can also be used for classroom exercises. The answers to these exercises are given at the end of the exercise set for the section.

NEW EXERCISES

There are over 1500 new exercises, including over 250 practice exercises and over 270 that illustrate technical applications. There is a total of about 8700 exercises in the ninth edition.

Many of the new exercises included after the basic exercises of each set are directed more specifically at the student's understanding of the concepts presented. These are generally given before the exercises that show applications in the various technical areas.

Continuing Features

PAGE LAYOUT

Special attention has been given to the page layout. Nearly all examples are started and completed on the same page (there are only two exceptions, and each of these is presented on facing pages). Also, all figures are shown immediately adjacent to the material in which they are discussed.

CHAPTER INTRODUCTIONS

Each chapter introduction illustrates specific examples of how the development of technology has been related to the development of mathematics. In these introductions, it is shown that past discoveries in technology led to some of the methods in mathematics, whereas in other cases mathematical topics already known were later very useful in bringing about advances in technology.

SPECIAL EXPLANATORY COMMENTS

Throughout the book, special explanatory comments in color have been used in the examples to emphasize and clarify certain important points. Arrows are often used to indicate clearly the part of the example to which reference is made.

PROBLEM-SOLVING TECHNIQUES

Techniques and procedures that summarize the approaches in solving many types of problems have been clearly outlined in color-shaded boxes.

IMPORTANT FORMULAS

Throughout the book, important formulas are set off and displayed so that they can be easily referenced for use.

SUBHEADS AND KEY TERMS

Many sections include subheads to indicate where the discussion of a new topic starts within the section. Other key terms are noted in the margin for emphasis and easy reference.

SPECIAL CAUTION AND NOTE INDICATORS

CAUTION ▶
NOTE ▶

Two special margin indicators (as shown at the left) are used. The caution indicator identifies errors students commonly make or places where they frequently have difficulty. The note indicator points out material that is of particular importance in developing or understanding the topic under discussion.

EXERCISES DIRECTLY REFERENCED TO TEXT EXAMPLES

The first few exercises in most of the text sections are referenced directly to a specific example of the section. These exercises are worded so that it is necessary for the student to refer to the example in order to complete the required solution. In this way, the student should be able to better review and understand the text material before attempting to solve the exercises that follow.

WRITING EXERCISES

One specific writing exercise is included at the end of each chapter. These exercises give the student practice in explaining their solutions. Also, there are over 260 additional exercises throughout the book (at least eight in each chapter) that require at least a sentence or two of explanation as part of the answer. These are noted by the (W) symbol next to the exercise number. A special Index of Writing Exercises is included at the back of the book.

WORD PROBLEMS

There are over 70 examples throughout the text that show the complete solutions of word problems. There are also over 750 exercises, of which about 100 are new, in which word problems are to be solved.

CHAPTER EQUATIONS, REVIEW EXERCISES, AND PRACTICE TESTS

At the end of each chapter, all important equations are listed together for easy reference. Each chapter is also followed by a set of review exercises that covers all the material in the chapter. Following the chapter equations and review exercises is a chapter practice test that students can use to check their understanding of the material. Solutions to all practice test problems are given in the back of the book.

APPLICATIONS AND UNITS OF MEASUREMENT

Examples and exercises illustrate the application of mathematics in all fields of technology. Many relate to modern technology such as computer design, electronics, solar energy, lasers, fiber optics, the environment, and space technology. A special Index of Applications is included near the end of the book.

EXAMPLES

There are over 1000 worked examples in this text. Of these, over 200 illustrate technical applications.

FIGURES

There are over 1000 figures in the text, about 80 of which are new to this edition.

MARGIN NOTES

Throughout the text, some margin notes briefly point out relevant historical events in mathematics and technology. Other margin notes are used to make specific comments related to the text material. Also, where appropriate, equations from earlier material are shown for reference in the margin.

ANSWERS TO EXERCISES

The answers to all odd-numbered exercises (except the end-of-chapter writing exercises) are given at the back of the book. The *Student's Solution Manual* contains solutions for every other odd-numbered exercise and the *Instructor's Solution Manual* contains solutions for all section exercises.

FLEXIBILITY OF MATERIAL COVERAGE

The order of material coverage can be changed in many places and certain sections may be omitted without loss of continuity of coverage. Users of earlier editions have indicated the successful use of numerous variations in coverage. Any changes will depend on the type of course and completeness required.

Supplements

SUPPLEMENTS FOR THE INSTRUCTOR

To access supplementary materials online, instructors need to request an instructor access code. Go to **www.pearsonhighered.com/irc**, where you can register for an instructor access code. Within 48 hours after registering, you will receive a confirming email including an instructor access code. Once you have received your code, go to the site and log on for full instructions on downloading the materials you wish to use.

Instructor's Solutions Manual

The *Instructor's Solution Manual* by Bob Martin contains detailed solutions to every section exercise, including review exercises.

TestGen with Algorithmically Generated Questions

Instructors can easily create tests from textbook section objectives. Algorithmically generated questions allow unlimited versions. Instructors can edit problems or create their own using the built-in question editor to generate graphs, import graphics, and insert math notation, insert variable numbers, or text. Tests can be printed or administered online via the Web or other network.

MyMathLab® with MathXL

Powered by CourseCompass™ (Pearson's online teaching and learning environment) and MathXL® (Pearson's online homework, tutorial, and assessment system), MyMathLab gives instructors the tools needed to deliver all or a portion of their courses online, whether students are in a lab setting or working from home. MyMathLab provides a rich and flexible set of course materials, featuring free-response exercises that are algorithmically generated for unlimited practice and mastery. Students can also use online tools, such as video lectures and a multimedia textbook to independently improve their understanding and performance. Instructors can use MyMathLab's homework and test managers to select and assign online exercises correlated directly to the textbook and create and assign their own online exercises. TestGen tests can also be imported for added flexibility. MyMathLab's online grade book automatically tracks students' homework and test results and gives the instructor control over how to calculate final grades. Instructors can also add offline (paper-and-pencil) grades to the grade book. MyMath-Lab includes student access to the Pearson Math Tutor Center. MyMathLab is available to qualified adopters. Pearson's sales representatives can provide more information.

MathXL® Online

MathXL® is a powerful online homework, tutorial, and assessment system. With MathXL, instructors can create, edit, and assign online homework and tests using algorithmically generated exercises correlated at the outcome level to the textbook. Instructors can also create and assign their own online exercises and import TestGen tests for added flexibility. All student work is tracked in MathXL's online grade book. Students can take chapter tests in MathXL and receive personalized study plans based on their test results. The study plan diagnoses weaknesses and links students directly to tutorial exercises for the objectives they need to study and retest. For more information, please contact your local sales representative.

SUPPLEMENTS FOR THE STUDENT

Student's Solutions Manual (ISBN: 0-13-814563-6)
The *Student's Solutions Manual* by Bob Martin includes detailed solutions for every other odd-numbered section exercise.

Graphing Calculator Manual (ISBN 0-321-19740-2)
This manual by Robert Seaver presents detailed instructions on the use of various graphing calculators in this course.

Acknowledgments

Special thanks go to Bob Martin and John Garlow, both of Tarrant County College, Arlington, TX, for preparing the *Answer Book,* the *Student's Solutions Manual*, and the *Instructor's Solutions Manual*. Thanks also to Suzanne Garlow of Bedford, TX, for the keyboarding and typesetting of these supplements. Special thanks also to Robert Seaver of Lorain County Community College for preparing the *Graphing Calculator Manual*. Also, I again wish to thank Thomas Stark of Cincinnati State Technical and Community College for the *RISERS* approach to solving word problems in Appendix A.

My thanks and gratitude go to Jim Bryant who drew all of the chapter-opener illustrations; to Charles Cox and Carole Quandt for assisting in the tedious job of reading proof; and to Roxanne Barrows, Doug Ewert, Ellena Reda, and Sue Schroeder for checking the accuracy of answers and examples.

I gratefully acknowledge the cooperation and support of my editors, Gary Bauer and Linda Cupp, at Prentice Hall. I am especially grateful to the production editor, Greg Hubit, who has provided valuable assistance for the past seven editions. Also, I wish to acknowledge the very fine work of Tracy Duff at Pre-PressPMG, who set all the type for this edition.

Also of great assistance during the production of this edition were Rex Davidson, Megan Heintz, Alicia Anderson, Ruth Berry, Carl Cottrell, Eileen Moore, Mary Durnwald, and Marty Wright.

The author gratefully acknowledges the contributions of the following reviewers. Their detailed comments and many suggestions were of great assistance in preparing this ninth edition.

Imad Abouzahr
Oklahoma State University

Laurie Bishop
McKicken College

Ruth Dalrymple
St. Philips College

Joe Jordan
John Tyler Community College

Rosa Kavanaugh
Ozarks Technical Community College

Kellie Knox
Southwest Wisconsin Technical College

Nestor Komar
Niagara College

Larry Mason
Fairmont State University

Todd Mattson
DeVry University, DuPage Campus

Scott Randby
University of Akron

Hank Regis
Valencia Community College

Janet Schachtner
San Jacinto College

Elizabeth Tregoning
Southern Illinois College

Richard Watkins
Tidewater Community College

Finally, very special mention is due my wife Millie, who passed away during the production of this edition. I will never forget her loving patience, understanding, support, and keen sense of humor over our more than fifty years together.

A.J.W.

Basic Technical Mathematics

Basic Algebraic Operations

1

Interest in things such as the land on which they lived, the structures they built, and the motion of the planets led people in early civilizations to keep records and to create methods of counting and measuring. In turn, some of the early ideas of arithmetic, geometry, and trigonometry were developed. From such beginnings, mathematics has played a key role in the great advances in science and technology.

Often, mathematical methods were developed from studies made in sciences, such as astronomy and physics, to better describe and understand the subject being studied. Some of these methods resulted from the needs in a particular area of application.

Many people were interested in the math itself, and added to what was then known. Although this additional mathematical knowledge may not have been related to applications at the time it was developed, it often later became useful in applied areas.

In the chapter introductions that follow, examples of the interaction of technology and mathematics are given. From these examples and the text material, it is hoped you will better understand the important role that math has had and still has in technology. In this text, there are applications from technologies including (but not limited to) aeronautical, business, communications, electricity, electronics, engineering, environmental, heat and air conditioning, mechanical, medical, meteorology, petroleum, product design, solar, and space. To solve the applied problems in this text will require a knowledge of the mathematics presented but will *not* require prior knowledge of the field of application.

We begin by reviewing the concepts that deal with numbers and symbols. This will enable us to develop topics in algebra, an understanding of which is essential for progress in other areas such as geometry, trigonometry, and calculus.

The Great Pyramid at Giza in Egypt was built about 2560 BC or about 4,500 years ago.

In the 1500s, 1600s, and 1700s, discoveries in astronomy and the need for more accurate maps and instruments in navigation were very important in leading scientists and mathematicians to develop useful new ideas and methods in mathematics.

Late in the 1800s, scientists were studying the nature of light. This led to a mathematical prediction of the existence of radio waves, now used in many types of communication. Also, in the 1900s and 2000s, mathematics has been vital to the development of electronics and space travel.

1

■ Irrational numbers were discussed by the Greek mathematician Pythagoras in about 540 B.C.E.

■ A notation that is often used for repeating decimals is to place a bar over the digits that repeat. Using this notation we can write $\frac{1121}{1665} = 0.6\overline{732}$ and $\frac{2}{3} = 0.\overline{6}$.

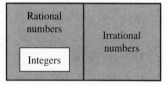

Real Numbers

Fig. 1.1

■ Real numbers and imaginary numbers are both included in the *complex number system.* See Exercise 37.

1.1 Numbers

In technology and science, as well as in everyday life, we use the very familiar **counting numbers** 1, 2, 3, and so on. They are also called **natural numbers** or **positive integers.** The **negative integers** $-1, -2, -3$, and so on are also very necessary and useful in mathematics and its applications. *The* **integers** *include the positive integers and the negative integers and* **zero,** *which is neither positive nor negative.* This means the integers are the numbers $\ldots, -3, -2, -1, 0, 1, 2, 3$, and so on.

To specify parts of a quantity, *rational numbers* are used. *A* **rational number** *is any number that can be represented by the division of one integer by another nonzero integer.* Another type of number, an **irrational number,** *cannot be written as the division of one integer by another.*

EXAMPLE 1 Identifying rational numbers and irrational numbers

The numbers 5 and -19 are integers. They are also rational numbers since they can be written as $\frac{5}{1}$ and $\frac{-19}{1}$, respectively. Normally, we do not write the 1's in the denominators.

The numbers $\frac{5}{8}$ and $\frac{-11}{3}$ are rational numbers because the numerator and the denominator of each are integers.

The numbers $\sqrt{2}$ and π are irrational numbers. It is not possible to find two integers, one divided by the other, to represent either of these numbers. It can be shown that square roots (and other roots) that cannot be expressed exactly in decimal form are irrational. Also, $\frac{22}{7}$ is sometimes used as an *approximation* for π, but it is not equal *exactly* to π. We must remember that $\frac{22}{7}$ is rational and π is irrational.

The decimal number 1.5 is rational since it can be written as $\frac{3}{2}$. Any such *terminating decimal* is rational. The number $0.6666\ldots$, where the 6's continue on indefinitely, is rational since we may write it as $\frac{2}{3}$. In fact, any *repeating decimal* (in decimal form, a specific sequence of digits is repeated indefinitely) is rational. The decimal number $0.6732732732\ldots$ is a repeating decimal where the sequence of digits 732 is repeated indefinitely ($0.6732732732\ldots = \frac{1121}{1665}$). ■

The integers, the rational numbers, and the irrational numbers, including all such numbers that are positive, negative, or zero, make up the **real number system** (see Fig. 1.1). There are times we will encounter an **imaginary number,** *the name given to the square root of a negative number.* Imaginary numbers are not real numbers and will be discussed in Chapter 12. However, unless specifically noted, we will use real numbers. Until Chapter 12, it will be necessary to only *recognize* imaginary numbers when they occur.

Also in Chapter 12, we will consider **complex numbers,** which include both the real numbers and imaginary numbers. See Exercise 37 of this section.

EXAMPLE 2 Identifying real numbers and imaginary numbers

The number 7 is an integer. It is also rational since $7 = \frac{7}{1}$, and it is a real number since the real numbers include all the rational numbers.

The number 3π is irrational, and it is real since the real numbers include all the irrational numbers.

The numbers $\sqrt{-10}$ and $-\sqrt{-7}$ are imaginary numbers.

The number $\frac{-3}{7}$ is rational and real. The number $-\sqrt{7}$ is irrational and real.

The number $\frac{\pi}{6}$ is irrational and real. The number $\frac{\sqrt{-3}}{2}$ is imaginary. ■

Fractions were used by early Egyptians and Babylonians. They were used for calculations that involved parts of measurements, property, and possessions.

A **fraction** *may contain any number or symbol representing a number in its numerator or in its denominator.* The fraction indicates the division of the numerator by the denominator, as we previously indicated in writing rational numbers. Therefore, a fraction may be a number that is rational, irrational, or imaginary.

EXAMPLE 3 Fractions

The numbers $\frac{2}{7}$ and $\frac{-3}{2}$ are fractions, and they are rational.

The numbers $\frac{\sqrt{2}}{9}$ and $\frac{6}{\pi}$ are fractions, but they are not rational numbers. It is not possible to express either as one integer divided by another integer.

The number $\frac{\sqrt{-5}}{6}$ is a fraction, and it is an imaginary number. ∎

The Number Line

Real numbers may be represented by points on a line. We draw a horizontal line and designate some point on it by O, which we call the **origin** (see Fig. 1.2). The integer *zero* is located at this point. Equal intervals are marked to the right of the origin, and the positive integers are placed at these positions. The other positive rational numbers are located between the integers. The points that cannot be defined as rational numbers represent irrational numbers. We cannot tell whether a given point represents a rational number or an irrational number unless it is specifically marked to indicate its value.

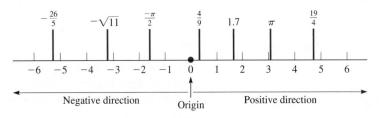

Fig. 1.2

The negative numbers are located on the number line by starting at the origin and marking off equal intervals *to the left, which is the* **negative direction.** As shown in Fig. 1.2, *the positive numbers are to the right of the origin and the negative numbers are to the left of the origin.* Representing numbers in this way is especially useful for graphical methods.

We next define another important concept of a number. *The* **absolute value** *of a positive number is the number itself, and the absolute value of a negative number is the corresponding positive number.* On the number line, we may interpret the absolute value of a number as the distance (which is always positive) between the origin and the number. Absolute value is denoted by writing the number between vertical lines, as shown in the following example.

EXAMPLE 4 Absolute value

Practice Exercises

1. $|-4.2| = ?$ 2. $-\left|-\frac{3}{4}\right| = ?$

The absolute value of 6 is 6, and the absolute value of -7 is 7. We write these as $|6| = 6$ and $|-7| = 7$. See Fig. 1.3.

Fig. 1.3

Other examples are $\left|\frac{7}{5}\right| = \frac{7}{5}$, $|-\sqrt{2}| = \sqrt{2}$, $|0| = 0$, $-|\pi| = -\pi$, $|-5.29| = 5.29$, $-|-9| = -9$ since $|-9| = 9$. ∎

■ The symbols =, <, and > were introduced by English mathematicians in the late 1500s.

On the number line, *if a first number is to the right of a second number, then the first number is said to be* **greater than** *the second. If the first number is to the left of the second, it is* **less than** *the second number.* The symbol > designates "is greater than," and the symbol < designates "is less than." These are called **signs of inequality.** See Fig. 1.4.

EXAMPLE 5   Signs of inequality

Fig. 1.4   ■

Practice Exercises

Place the correct sign of inequality (< or >) between the given numbers.

3. −5   4   **4.** 0   −3

Every number, except zero, has a **reciprocal.** *The reciprocal of a number is 1 divided by the number.*

EXAMPLE 6   Reciprocal

The reciprocal of 7 is $\frac{1}{7}$. The reciprocal of $\frac{2}{3}$ is

$$\frac{1}{\frac{2}{3}} = 1 \times \frac{3}{2} = \frac{3}{2} \qquad \text{invert denominator and multiply (from arithmetic)}$$

The reciprocal of 0.5 is $\frac{1}{0.5} = 2$. The reciprocal of $-\pi$ is $-\frac{1}{\pi}$. Note that the negative sign is retained in the reciprocal of a negative number.

We showed the multiplication of 1 and $\frac{3}{2}$ as $1 \times \frac{3}{2}$. We could also show it as $1 \cdot \frac{3}{2}$ or $1\left(\frac{3}{2}\right)$. We will often find the form with parentheses is preferable.   ■

In applications, *numbers that represent a measurement and are written with units of measurement are called* **denominate numbers.** The next example illustrates the use of units and the symbols that represent them.

EXAMPLE 7   Denominate numbers

■ For reference, see Appendix B for units of measurement and the symbols used for them.

To show that a certain HDTV set weights 62 pounds, we write the weight as 62 lb.

To show that a giant redwood tree is 330 feet high, we write the height as 300 ft.

To show that the speed of a rocket is 1500 meters per second, we write the speed as 1500 m/s. (Note the use of s for second. We use s rather than sec.)

To show that the area of a computer chip is 0.75 square inch, we write the area as 0.75 in.^2. (We will not use sq in.)

To show that the volume of water in a glass tube is 25 cubic centimeters, we write the volume as 25 cm^3. (We will not use cu cm nor cc.)   ■

Literal Numbers

It is usually more convenient to state definitions and operations on numbers in a general form. *To do this, we represent the numbers by letters, called* **literal numbers.** For example, if we want to say "If a first number is to the right of a second number on the number line, then the first number is greater than the second number," we can write "If a is to the right of b on the number line, then $a > b$." Another example of using a literal number is "The reciprocal of n is $1/n$."

Certain literal numbers may take on any allowable value, whereas other literal numbers represent the same value throughout the discussion. *Those literal numbers that may vary in a given problem are called* **variables,** and *those literal numbers that are held fixed are called* **constants.**

EXAMPLE 8 Variables and constants

(a) The resistance of an electric resistor is R. The current I in the resistor equals the voltage V divided by R, written as $I = V/R$. For this resistor, I and V may take on various values, and R is fixed. This means I and V are variables and R is a constant. For a *different* resistor, the value of R may differ.

(b) The fixed cost for a calculator manufacturer to operate a certain plant is b dollars per day, and it costs a dollars to produce each calculator. The total daily cost C to produce n calculators is

$$C = an + b$$

Here, C and n are variables, and a and b are constants, and the product of a and n is shown as an. For *another* plant, the values of a and b would probably differ.

If specific numerical values of a and b are known, say $a = \$7$ per calculator and $b = \$3000$, then $C = 7a + 3000$. Thus, constants may be numerical or literal. ∎

EXERCISES 1.1

In Exercises 1–4, make the given changes in the indicated examples of this section, and then answer the given questions.

1. In the first line of Example 1, change the 5 to -3 and the -19 to 14. What other changes must then be made in the first paragraph?

2. In Example 4, change the 6 to -6. What other changes must then be made in the first paragraph?

3. In the left figure of Example 5, change the 2 to -6. What other changes must then be made?

4. In Example 6, change the $\frac{2}{3}$ to $\frac{3}{2}$. What other changes must then be made?

In Exercises 5 and 6, designate each of the given numbers as being an integer, rational, irrational, real, or imaginary. (More than one designation may be correct.)

5. 3, $\sqrt{-4}$, $-\dfrac{\pi}{6}$

6. $-\sqrt{-6}$, -2.33, $\dfrac{\sqrt{7}}{3}$

In Exercises 7 and 8, find the absolute value of each number.

7. 3, -4, $-\dfrac{\pi}{2}$

8. -0.857, $\sqrt{2}$, $-\dfrac{19}{4}$

In Exercises 9–16, insert the correct sign of inequality ($>$ or $<$) between the given numbers.

9. 6   8
10. 7   5
11. π -3.2
12. -4 0
13. -4 $-|-3|$
14. $-\sqrt{2}$ -1.42
15. $-\dfrac{1}{3}$ $-\dfrac{1}{2}$
16. -0.6 0.2

In Exercises 17 and 18, find the reciprocal of each number.

17. 3, $-\dfrac{4}{\sqrt{3}}$, $\dfrac{y}{b}$

18. $-\dfrac{1}{3}$,   0.25, x

In Exercises 19 and 20, locate each number on a number line as in Fig. 1.2.

19. 2.5, $-\dfrac{12}{5}$, $\sqrt{3}$

20. $-\dfrac{\sqrt{2}}{2}$, 2π, $\dfrac{123}{19}$

In Exercises 21–44, solve the given problems. Refer to Appendix B for units of measurement and their symbols.

(W) 21. Is an absolute value always positive? Explain.

(W) 22. Is 2.17 rational? Explain.

23. What is the reciprocal of the reciprocal of any positive or negative number?

24. Find a rational number between -0.9 and -1.0 that can be written with a denominator of 11 and an integer in the numerator.

25. Find a rational number between 0.13 and 0.14 that can be written with a numerator of 3 and an integer in the denominator.

26. If $b > a$ and $a > 0$, is $|b - a| < |b| - |a|$?

27. List the following numbers in numerical order, starting with the smallest: $-1, 9, \pi, \sqrt{5}, |-8|, -|-3|, -3.1$.

28. List the following numbers in numerical order, starting with the smallest: $\frac{1}{5}, -\sqrt{10}, -|-6|, -4, 0.25, |-\pi|$.

29. If a and b are positive integers and $b > a$, what type of number is represented by the following?

    (a) $b - a$ (b) $a - b$ (c) $\dfrac{b - a}{b + a}$

30. If a and b represent positive integers, what kind of number is represented by (a) $a + b$, (b) a/b, and (c) $a \times b$?

31. For any positive or negative integer: (a) Is its absolute value always an integer? (b) Is its reciprocal always a rational number?

32. For any positive or negative rational number: (a) Is its absolute value always a rational number? (b) Is its reciprocal always a rational number?

(W) 33. Describe the location of a number x on the number line when (a) $x > 0$ and (b) $x < -4$.

(W) **34.** Describe the location of a number x on the number line when (a) $|x| < 1$ and (b) $|x| > 2$.

(W) **35.** For a number $x > 1$, describe the location on the number line of the reciprocal of x.

(W) **36.** For a number $x < 0$, describe the location on the number line of the number with a value of $|x|$.

37. A *complex number* is defined as $a + bj$, where a and b are real numbers and $j = \sqrt{-1}$. For what values of a and b is the complex number $a + bj$ a real number? (All real numbers and all imaginary numbers are also complex numbers.)

38. A sensitive gauge measures the total weight w of a container and the water that forms in it as vapor condenses. It is found that $w = c\sqrt{0.1t + 1}$, where c is the weight of the container and t is the time of condensation. Identify the variables and constants.

39. In an electric circuit, the reciprocal of the total capacitance of two capacitors in series is the sum of the reciprocals of the capacitances. Find the total capacitance of two capacitances of 0.0040 F and 0.0010 F connected in series.

40. Alternating-current (ac) voltages change rapidly between positive and negative values. If a voltage of 100 V changes to -200 V, which is greater in absolute value?

41. The memory of a certain computer has a bits in each byte. Express the number N of bits in n kilobytes in an equation. (A *bit* is a single digit, and bits are grouped in *bytes* in order to represent special characters. Generally, there are 8 bits per byte. If necessary, see Appendix B for the meaning of *kilo*.)

42. The computer design of the base of a truss is x ft. long. Later it is redesigned and shortened by y in. Give an equation for the length L, in inches, of the base in the second design.

(W) **43.** In a laboratory report, a student wrote "$-20°C > -30°C$." Is this statement correct? Explain.

44. After 5 s, the pressure on a valve is less than 60 lb/in.^{2} (pounds per square inch). Using t to represent time and p to represent pressure, this statement can be written "for $t > 5$ s, $p < 60$ lb/in.^{2}." In this way, write the statement "when the current I in a circuit is less than 4 A, the voltage V is greater than 12 V."

Answers to Practice Exercises

1. 4.2   **2.** $-\dfrac{3}{4}$ **3.** $<$ **4.** $>$

1.2   Fundamental Operations of Algebra

Fundamental Laws of Algebra • Operations on Positive and Negative Numbers • Order of Operations • Operations with Zero

The Commutative and Associative Laws

If two numbers are added, it does not matter in which order they are added. (For example, $5 + 3 = 8$ and $3 + 5 = 8$, or $5 + 3 = 3 + 5$.) This statement, generalized and accepted as being correct for all possible combinations of numbers being added, is called the **commutative law** for addition. It states that *the sum of two numbers is the same, regardless of the order in which they are added*. We make no attempt to prove this law in general, but accept that it is true.

In the same way, we have the **associative law** for addition, which states that *the sum of three or more numbers is the same, regardless of the way in which they are grouped for addition*. For example, $3 + (5 + 6) = (3 + 5) + 6$.

The laws just stated for addition are also true for multiplication. Therefore, *the product of two numbers is the same, regardless of the order in which they are multiplied*, and *the product of three or more numbers is the same, regardless of the way in which they are grouped for multiplication*. For example, $2 \times 5 = 5 \times 2$, and $5 \times (4 \times 2) = (5 \times 4) \times 2$.

The Distributive Law

Another very important law is the **distributive law**. It states that *the product of one number and the sum of two or more other numbers is equal to the sum of the products of the first number and each of the other numbers of the sum*. For example,

■ Note carefully the difference:
associative law: $5 \times (4 \times 2)$
distributive law: $5 \times (4 + 2)$

$$5(4 + 2) = 5 \times 4 + 5 \times 2$$

In this case, it can be seen that the total is 30 on each side.

In practice, these **fundamental laws of algebra** are used naturally without thinking about them, except perhaps for the distributive law.

Not all operations are commutative and associative. For example, division is not commutative, since the order of division of two numbers does matter. For instance, $\frac{6}{5} \neq \frac{5}{6}$ ($\neq$ is read "does not equal)". (Also, see Exercise 50.)

Using literal numbers, the fundamental laws of algebra are as follows:

Commutative law of addition: $a + b = b + a$

Associative law of addition: $a + (b + c) = (a + b) + c$

Commutative law of multiplication: $ab = ba$

Associative law of multiplication: $a(bc) = (ab)c$

Distributive law: $a(b + c) = ab + ac$

Each of these laws is an example of an *identity,* in that the expression to the left of the $=$ sign equals the expression to the right for any value of each of a, b, and c.

OPERATIONS ON POSITIVE AND NEGATIVE NUMBERS

When using the basic operations (addition, subtraction, multiplication, division) on positive and negative numbers, we determine the result to be either positive or negative according to the following rules.

Addition of two numbers of the same sign *Add their absolute values and assign the sum their common sign.*

EXAMPLE 1 Adding numbers of the same sign

(a) $2 + 6 = 8$ the sum of two positive numbers is positive

(b) $-2 + (-6) = -(2 + 6) = -8$ the sum of two negative numbers is negative

■ From Section 1.1, we recall that a positive number is preceded by no sign. Therefore, in using these rules, we show the "sign" of a positive number by simply writing the number itself.

The negative number -6 is placed in parentheses since it is also preceded by a plus sign showing addition. It is not necessary to place the -2 in parentheses. ■

Addition of two numbers of different signs *Subtract the number of smaller absolute value from the number of larger absolute value and assign to the result the sign of the number of larger absolute value.*

EXAMPLE 2 Adding numbers of different signs

(a) $2 + (-6) = -(6 - 2) = -4$ the negative 6 has the larger absolute value

(b) $-6 + 2 = -(6 - 2) = -4$

(c) $6 + (-2) = 6 - 2 = 4$ the positive 6 has the larger absolute value

(d) $-2 + 6 = 6 - 2 = 4$

the subtraction of absolute values ■

Subtraction of one number from another *Change the sign of the number being subtracted and change the subtraction to addition. Perform the addition.*

EXAMPLE 3 Subtracting positive and negative numbers

(a) $2 - 6 = 2 + (-6) = -(6 - 2) = -4$

Note that after changing the subtraction to addition, and changing the sign of 6 to make it -6, we have precisely the same illustration as Example 2(a).

(b) $-2 - 6 = -2 + (-6) = -(2 + 6) = -8$

Note that after changing the subtraction to addition, and changing the sign of 6 to make it -6, we have precisely the same illustration as Example 1(b).

(c) $-a - (-a) = -a + a = 0$

Subtraction of a Negative Number

This shows that subtracting a number from itself results in zero, even if the number is negative. Therefore, *subtracting a negative number is equivalent to adding a positive number of the same absolute value.* ■

Multiplication and division of two numbers *The product (or quotient) of two numbers of the same sign is positive. The product (or quotient) of two numbers of different signs is negative.*

EXAMPLE 4 Multiplying and dividing positive and negative numbers

(a) $3(12) = 3 \times 12 = 36$ $\dfrac{12}{3} = 4$ result is positive if both numbers are positive

(b) $-3(-12) = 3 \times 12 = 36$ $\dfrac{-12}{-3} = 4$ result is positive if both numbers are negative

(c) $3(-12) = -(3 \times 12) = -36$ $\dfrac{-12}{3} = -\dfrac{12}{3} = -4$ result is negative if one number is positive and the other is negative

(d) $-3(12) = -(3 \times 12) = -36$ $\dfrac{12}{-3} = -\dfrac{12}{3} = -4$ ∎

ORDER OF OPERATIONS

Often, how we are to combine numbers is clear by grouping the numbers using symbols such as **parentheses, ( )**, the **bar, _____** , between the numerator and denominator of a fraction, and **vertical lines** for absolute value. Otherwise, for an expression in which there are several operations, we use the following order of operations.

ORDER OF OPERATIONS

1. *Operations within specific groupings are done first.*

2. *Perform multiplications and divisions (from left to right).*

3. *Then perform additions and subtractions (from left to right).*

EXAMPLE 5 Order of operations

■ Note that $20 \div (2 + 3) = \frac{20}{2+3}$, whereas $20 \div 2 + 3 = \frac{20}{2} + 3$.

(a) $20 \div (2 + 3)$ is evaluated by first adding $2 + 3$ and then dividing. The grouping of $2 + 3$ is clearly shown by the parentheses. Therefore, $20 \div (2 + 3) = 20 \div 5 = 4$.

(b) $20 \div 2 + 3$ is evaluated by first dividing 20 by 2 and then adding. No specific grouping is shown, and therefore the division is done before the addition. This means $20 \div 2 + 3 = 10 + 3 = 13$.

CAUTION ▶ **(c)** $16 - 2 \times 3$ is evaluated by *first multiplying* 2 *by* 3 and then subtracting. We *do not first subtract* 2 *from* 16. Therefore, $16 - 2 \times 3 = 16 - 6 = 10$.

(d) $16 \div 2 \times 4$ is evaluated by first dividing 16 by 2 and then multiplying. From left to right, the division occurs first. Therefore, $16 \div 2 \times 4 = 8 \times 4 = 32$.

(e) $|3 - 5| - |-3 - 6|$ is evaluated by first performing the subtractions within the absolute value vertical bars, then evaluating the absolute values, and then subtracting. This means that $|3 - 5| - |-3 - 6| = |-2| - |-9| = 2 - 9 = -7$.   ∎

Practice Exercises

Evaluate: **1.** $12 - 6 \div 2$
2. $16 \div (2 \times 4)$

When evaluating expressions, it is generally more convenient to change the operations and numbers so that the result is found by the addition and subtraction of positive numbers. When this is done, we must remember that

$$a + (-b) = a - b \qquad\qquad (1.1)$$

$$a - (-b) = a + b \qquad\qquad (1.2)$$

Practice Exercises

Evaluate: **3.** $2(-3) - \dfrac{4-8}{2}$

4. $\dfrac{|5-15|}{2} - \dfrac{-9}{3}$

EXAMPLE 6 Evaluating numerical expressions

(a) $7 + (-3) - 6 = 7 - 3 - 6 = 4 - 6 = -2$ using Eq. (1.1)

(b) $\dfrac{18}{-6} + 5 - (-2)(3) = -3 + 5 - (-6) = 2 + 6 = 8$ using Eq. (1.2)

(c) $\dfrac{|3-15|}{-2} - \dfrac{8}{4-6} = \dfrac{12}{-2} - \dfrac{8}{-2} = -6 - (-4) = -6 + 4 = -2$

(d) $\dfrac{-12}{2-8} + \dfrac{5-1}{2(-1)} = \dfrac{-12}{-6} + \dfrac{4}{-2} = 2 + (-2) = 2 - 2 = 0$

In illustration (b), we see that the division and multiplication were done before the addition and subtraction. In (c) and (d), we see that the groupings were evaluated first. Then we did the divisions, and finally the subtraction and addition. ∎

EXAMPLE 7 Evaluating in an application

A 3000-lb van going at 40 mi/h ran head-on into a 2000-lb car going at 20 mi/h. An insurance investigator determined the velocity of the vehicles immediately after the collision from the following calculation. See Fig. 1.5.

$$\frac{3000(40) + (2000)(-20)}{3000 + 2000} = \frac{120{,}000 + (-40{,}000)}{3000 + 2000} = \frac{120{,}000 - 40{,}000}{5000}$$

$$= \frac{80{,}000}{5000} = 16 \text{ mi/h}$$

3000 lb
40 mi/h

2000 lb
20 mi/h

→ 16 mi/h

Fig. 1.5

The numerator and the denominator must be evaluated before the division is performed. The multiplications in the numerator are performed first, followed by the addition in the denominator and the subtraction in the numerator. ∎

OPERATIONS WITH ZERO

Since operations with zero tend to cause some difficulty, we will show them here.

If a is a real number, the operations of addition, subtraction, multiplication, and division with zero are as follows:

$$a + 0 = a$$
$$a - 0 = a \qquad 0 - a = -a$$
$$a \times 0 = 0$$
$$0 \div a = \frac{0}{a} = 0 \qquad (\text{if } a \neq 0) \qquad (\neq \text{ means "is not equal to"})$$

EXAMPLE 8 Operations with zero

(a) $5 + 0 = 5$ **(b)** $-6 - 0 = -6$ **(c)** $0 - 4 = -4$

(d) $\dfrac{0}{6} = 0$ **(e)** $\dfrac{0}{-3} = 0$ **(f)** $\dfrac{5 \times 0}{7} = \dfrac{0}{7} = 0$ ∎

Note that there is no result defined for division by zero. To understand the reason for this, consider the results for $\frac{6}{2}$ and $\frac{6}{0}$.

$$\frac{6}{2} = 3 \quad \text{since} \quad 2 \times 3 = 6$$

If $\frac{6}{0} = b$, then $0 \times b = 6$. This cannot be true because $0 \times b = 0$ for any value of b. Thus,

NOTE ▶ *division by zero is undefined*

(The special case of $\frac{0}{0}$ is termed *indeterminate*. If $\frac{0}{0} = b$, then $0 = 0 \times b$, which is true for any value of b. Therefore, no specific value of b can be determined.)

EXAMPLE 9   Division by zero is undefined

see bottom of page 9

$$\frac{2}{5} \div 0 \text{ is undefined} \qquad \frac{8}{0} \text{ is undefined} \qquad \frac{7 \times 0}{0 \times 6} \text{ is indeterminate} \qquad ■$$

The operations with zero will not cause any difficulty if we remember to

CAUTION ▶   *never divide by zero*

Division by zero is the only undefined basic operation. All the other operations with zero may be performed as for any other number.

EXERCISES 1.2

In Exercises 1–4, make the given changes in the indicated examples of this section, and then solve the resulting problems.

1. In Example 5(c), change 3 to (-3) and then evaluate.

2. In Example 6(b), change 18 to -18 and then evaluate.

3. In Example 6(d), interchange the 2 and 8 in the first denominator and then evaluate.

4. In the rightmost illustration in Example 9, interchange the 6 and the 0 above the 6. Is any other change needed?

In Exercises 5–36, evaluate each of the given expressions by performing the indicated operations.

5. $8 + (-4)$

6. $-4 + (-7)$

7. $-3 + 9$

8. $18 - 21$

9. $-19 - (-16)$

10. $8 - (-4)$

11. $8(-3)$

12. $-9(3)$

13. $-7(-5)$

14. $\dfrac{-9}{3}$

15. $\dfrac{-6(20 - 10)}{-3}$

16. $\dfrac{28}{-7(6 - 5)}$

17. $-2(4)(-5)$

18. $3(-4)(6)$

19. $2(2 - 7) \div 10$

20. $\dfrac{-64}{-2|4 - 8|}$

21. $-9 - |2 - 10|$

22. $(7 - 7) \div (5 - 7)$

23. $\dfrac{17 - 7}{7 - 7}$

24. $\dfrac{7 - 7}{7 - 7}$

25. $8 - 3(-4)$

26. $20 + 8 \div 4$

27. $-2(-6) + \left|\dfrac{8}{-2}\right|$

28. $-10 - (-6)(-8)$

29. $30(-6)(-2) \div (0 - 40)$

30. $\dfrac{7 - |-5|}{-1(-2)}$

31. $\dfrac{24}{3 + (-5)} - 4(-9)$

32. $\dfrac{-18}{3} - \dfrac{4 - 6}{-1}$

33. $-7 - \dfrac{|-14|}{2(2 - 3)} - 3|6 - 8|$

34. $-7(-3) + \dfrac{6}{-3} - (-9)$

35. $\dfrac{3(-9) - 2(-3)}{3 - 10}$

36. $\dfrac{20(-12) - 40(-15)}{98 - |-98|}$

In Exercises 37–44, determine which of the fundamental laws of algebra is demonstrated.

37. $6(7) = 7(6)$

38. $6 + 8 = 8 + 6$

39. $6(3 + 1) = 6(3) + 6(1)$

40. $4(5 \times \pi) = (4 \times 5)(\pi)$

41. $3 + (5 + 9) = (3 + 5) + 9$

42. $8(3 - 2) = 8(3) - 8(2)$

43. $(\sqrt{5} \times 3) \times 9 = \sqrt{5} \times (3 \times 9)$

44. $(3 \times 6) \times 7 = 7 \times (3 \times 6)$

In Exercises 45–48, for numbers a and b, determine which of the following expressions equals the given expression.
(a) $a + b$ (b) $a - b$ (c) $b - a$ (d) $-a - b$

45. $-a + (-b)$

46. $b - (-a)$

47. $-b - (-a)$

48. $-a - (-b)$

In Exercises 49–64, answer the given questions. Refer to Appendix B for units of measurement and their symbols.

49. (a) What is the sign of the product of an even number of negative numbers? (b) What is the sign of the product of an odd number of negative numbers?

(W) 50. Is subtraction commutative? Explain.

(W) 51. Explain why the following definition of the absolute value of a real number x is either correct or incorrect (the symbol $\geq$ means "is equal to or greater than": If $x \geq 0$, then $|x| = x$; if $x < 0$, then $|x| = -x$).

(W) 52. Explain what is the error if the expression $24 - 6 \div 2 \cdot 3$ is evaluated as 27. What is the correct value?

(W) 53. Describe the values of x and y for which (a) $-xy = 1$ and (b) $\dfrac{x - y}{x - y} = 1$.

(W) 54. Describe the values of x and y for which (a) $|x + y| = |x| + |y|$ and (b) $|x - y| = |x| + |y|$.

55. Some solar energy systems are used to supplement the utility power company power supplied to a home such that the meter runs backward if the solar energy being generated is greater than the energy being used. With such a system, if the solar power averages 1.5 kW for a 3.0-h period and only 2.1 kW · h is used during this period, what will be the change in the meter reading for this period?

(W) 56. A baseball player's batting average (total number of hits divided by total number of at-bats) is expressed in decimal form from 0.000 (no hits for all at-bats) to 1.000 (one hit for each at-bat). A player's batting average is often shown as 0.000 before the first at-bat of the season. Is this a correct batting average? Explain.

57. The daily high temperatures (in °C) in the Falkland Islands in the southern Atlantic Ocean during the first week in July were recorded as 7, 3, −2, −3, −1, 4, and 6. What was the average daily temperature for the week? (Divide the algebraic sum of the readings by the number of readings.)

58. A flare is shot up from the top of a tower. Distances above the flare gun are positive and those below it are negative. After 5 s the vertical distance (in ft) of the flare from the flare gun is found by evaluating $(70)(5) + (−16)(25)$. Find this distance.

59. Find the sum of the voltages of the batteries shown in Fig. 1.6. Note the directions in which they are connected.

6 V   −2 V   8 V   −5 V   3 V

Fig. 1.6

60. The electric current was measured in a given ac circuit at equal intervals as 0.7 mA, −0.2 mA, −0.9 mA, and −0.6 mA. What was the change in the current between (a) the first two readings, (b) the middle two readings, and (c) the last two readings?

61. One oil-well drilling rig drills 100 m deep the first day and 200 m deeper the second day. A second rig drills 200 m deep the first day and 100 m deeper the second day. In showing that the total depth drilled by each rig was the same, state what fundamental law of algebra is illustrated.

62. A water tank leaks 12 gal each hour for 7 h, and a second tank leaks 7 gal each hour for 12 h. In showing that the total amount leaked is the same for the two tanks, what fundamental law of algebra is illustrated?

63. Each of four persons spends 8 min browsing one website and 6 min browsing a second website. Set up the expression for the total time these persons spent browsing these websites. What fundamental law of algebra is illustrated?

64. A jet travels 600 mi/h relative to the air. The wind is blowing at 50 mi/h. If the jet travels with the wind for 3 h, set up the expression for the distance traveled. What fundamental law of algebra is illustrated?

Answers to Practice Exercises

1. 9   **2.** 2   **3.** −4   **4.** 8

1.3 Calculators and Approximate Numbers

Graphing Calculators • Approximate Numbers • Significant Digits • Accuracy and Precision • Rounding Off • Operations with Approximate Numbers • Estimating Results

■ The calculator screens shown with text material are for a TI-83 or TI-84. They are intended only as an illustration of a calculator screen for the particular operation. Screens for other models may differ.

You will be doing many of your calculations on a calculator, and a *graphing calculator* can be used for these calculations and many other operations. In this text, we will restrict our coverage of calculator use to graphing calculators because a *scientific calculator* cannot perform many of the required operations we will cover.

A brief discussion of the graphing calculator appears in Appendix C, and sample calculator screens appear throughout the book. Since there are many models of graphing calculators, *the notation and screen appearance for many operations will differ from one model to another.* You should *practice using your calculator* and **review its manual** to be sure how it is used. Following is an example of a basic calculation done on a graphing calculator.

> **EXAMPLE 1 Calculating on a graphing calculator**

38.3−12.9*-3.58
                84.482

Fig. 1.7

In order to calculate the value of $38.3 − 12.9(−3.58)$, the numbers are entered as follows. The calculator will perform the multiplication first, following the order of operations shown in Section 1.2. The sign of −3.58 is entered using the $\boxed{(-)}$ key, before 3.58 is entered. The display on the calculator screen is shown in Fig. 1.7.

$$38.3 \;\boxed{-}\; 12.9 \;\boxed{\times}\; \boxed{(-)} \; 3.58 \;\boxed{\text{ENTER}} \qquad \text{keystrokes}$$

This means that $38.3 − 12.9(−3.58) = 84.482$.

Note in the display that the negative sign of −3.58 is smaller and a little higher to distinguish it from the minus sign for subtraction. Also note the * shown for multiplication; the asterisk is the standard computer symbol for multiplication. ■

■ Some calculator keys on different models are labeled differently. For example, on some models, the EXE key is equivalent to the ENTER key.

Looking back into Section 1.2, we see that *the minus sign is used in two different ways:* (1) to indicate subtraction and (2) to designate a negative number. This is clearly shown on a graphing calculator because there is a key for each purpose. The $\boxed{-}$ key is used for subtraction, and the $\boxed{(-)}$ key is used before a number to make it negative.

■ Calculator keystrokes will generally not be shown, except as they appear in the display screens. They may vary from one model to another.

We will first use a graphing calculator for the purpose of graphing in Section 3.5. Before then, we will show some calculational uses of a graphing calculator.

APPROXIMATE NUMBERS AND SIGNIFICANT DIGITS

Most numbers in technical and scientific work are **approximate numbers,** having been determined by some *measurement*. Certain other numbers are **exact numbers,** having been determined by a *definition* or *counting process*.

EXAMPLE 2   Approximate numbers and exact numbers

If a voltage on a voltmeter is read as 116 V, the 116 is approximate. Another voltmeter might show the voltage as 115.7 V. However, the voltage cannot be determined *exactly*.

If a computer prints out the number of names on a list of 97, this 97 is exact. We know it is not 96 or 98. Since 97 was found from precise counting, it is exact.

By definition, 60 s = 1 min, and the 60 and the 1 are exact.   ■

An approximate number may have to include some zeros to properly locate the decimal point. *Except for these zeros, all other digits are called* **significant digits.**

EXAMPLE 3   Significant digits

All numbers in this example are assumed to be approximate.

34.7 has three significant digits.

0.039 has two significant digits. The zeros properly locate the decimal point.

706.1 has four significant digits. The zero is not used for the location of the decimal point. It shows the number of tens in 706.1.

CAUTION ▶

5.90 has three significant digits. ***The zero is not necessary as a placeholder*** and should not be written unless it is significant.

1400 has two significant digits, unless information is known about the number that makes either or both zeros significant. (A temperature shown as 1400°C has two significant digits. If a price list gives all costs in dollars, a price shown as $1400 has four significant digits.) Without such information, we assume that the zeros are placeholders for proper location of the decimal point.

■ To show that zeros at the end of a whole number are significant, a notation that can be used is to place a bar over the last significant zero. Using this notation, $78{,}0\overline{0}0$ is shown to have four significant digits.

Other approximate numbers with the number of significant digits are 0.0005 (one), 960,000 (two), 0.0709 (three), 1.070 (four), and 700.00 (five).   ■

From Example 3, we see that *all nonzero digits are significant. Also, zeros not used as placeholders (for location of the decimal point) are significant.*

In calculations with approximate numbers, the number of significant digits and the position of the decimal point are important. *The* **accuracy** *of a number refers to the number of significant digits it has,* whereas *the* **precision** *of a number refers to the decimal position of the last significant digit.*

EXAMPLE 4   Accuracy and precision

An electric current is measured as 0.31 A on one ammeter and as 0.312 A on another ammeter. Here, 0.312 is more precise since its last digit represents thousandths and 0.31 is expressed only to hundredths. Also, 0.312 is more accurate since it has three significant digits and 0.31 has only two.

A concrete driveway is 230 ft long and 0.4 ft thick. Here, 230 is more accurate (two significant digits) and 0.4 is more precise (expressed to tenths).   ■

The last significant digit of an approximate number is not exact. It has usually been determined by estimating or *rounding off*. However, it is not off by more than one-half of a unit in its place value.

EXAMPLE 5 Meaning of the last digit of an approximate number

When we write the voltage in Example 2 as 115.7 V, we are saying that the voltage is at least 115.65 V and no more than 115.75 V. Any value between these two, rounded off to tenths, would be expressed as 115.7 V.

In changing the fraction $\frac{2}{3}$ to the approximate decimal value 0.667, we are saying that the value is between 0.6665 and 0.6675.   ■

■ On graphing calculators, it is possible to set the number of decimal places (to the right of the decimal point) to which results will be rounded off.

*To **round off** a number to a specified number of significant digits, discard all digits to the right of the last significant digit (replace them with zeros if needed to properly place the decimal point). If the first digit discarded is 5 or more, increase the last significant digit by 1 (round up). If the first digit discarded is less than 5, do not change the last significant digit (round down).*

EXAMPLE 6 Rounding off

70,360 rounded off to three significant digits is 70,400. Here, 3 is the third significant digit and the next digit is 6. Since $6 > 5$, we add 1 to 3 and the result, 4, becomes the third significant digit of the approximation. The 6 is then replaced with a zero in order to keep the decimal point in the proper position.

70,430 rounded off to three significant digits, or to the nearest hundred, is 70,400. Here the 3 is replaced with a zero.

187.35 rounded off to four significant digits, or to tenths, is 187.4.

187.349 rounded off to four significant digits is 187.3. *We do not round up the 4 and then round up the 3.*

CAUTION ▶

35.003 rounded off to four significant digits is 35.00. *We do not discard the zeros since they are significant* and are not used only to properly place the decimal point.   ■

Practice Exercises

Round off each number to three significant digits.

1. 2015   **2.** 0.3004

OPERATIONS WITH APPROXIMATE NUMBERS

NOTE ▶

When performing operations on approximate numbers, *we must not express the result to an accuracy or precision that is not valid.* Consider the following examples.

EXAMPLE 7 Application of precision

	smallest values	largest values
16.3 ft	16.25 ft	16.35 ft
0.927 ft	0.9265 ft	0.9275 ft
17.227 ft	17.1765 ft	17.2775 ft

A pipe is made in two sections. One is measured as 16.3 ft long and the other as 0.927 ft long. What is the total length of the two sections together?

It may appear that we simply add the numbers as shown at the left. However, both numbers are approximate, and adding the smallest possible values and the largest possible values, the result differs by 0.1 (17.2 and 17.3) when rounded off to tenths. Rounded off to hundredths (17.18 and 17.28), they do not agree at all since the tenths digit is different. Thus, we get a good approximation for the total length if it is rounded off to *tenths,* the precision of the least precise length, and it is written as 17.2 ft.   ■

EXAMPLE 8 Application of accuracy

We find the area of the rectangular piece of land in Fig. 1.8 by multiplying the length, 207.54 ft, by the width, 81.4 ft. Using a calculator, we find that $(207.54)(81.4) = 16,893.756$. This apparently means the area is 16,893.756 ft^2.

However, the area should not be expressed with this accuracy. Since the length and width are both approximate, we have

$$(207.535 \text{ ft})(81.35 \text{ ft}) = 16,882.97225 \text{ ft}^2 \quad \text{least possible area}$$
$$(207.545 \text{ ft})(81.45 \text{ ft}) = 16,904.54025 \text{ ft}^2 \quad \text{greatest possible area}$$

These values agree when rounded off to three significant digits (16,900 ft^2) but do not agree when rounded off to a greater accuracy. Thus, we conclude that the result is accurate only to *three* significant digits, the accuracy of the least accurate measurement, and that the area is written as 16,900 ft^2.   ■

Fig. 1.8

Following are the rules used in expressing the result when we perform basic operations on approximate numbers. They are based on reasoning similar to that shown in Examples 7 and 8.

OPERATIONS WITH APPROXIMATE NUMBERS

1. *When approximate numbers are added or subtracted, the result is expressed with the precision of the least precise number.*
2. *When approximate numbers are multiplied or divided, the result is expressed with the accuracy of the least accurate number.*
3. *When the root of an approximate number is found, the result is expressed with the accuracy of the number.*
4. *When approximate numbers and exact numbers are involved, the accuracy of the result is limited only by the approximate numbers.*

CAUTION ▶   *Always express the result of a calculation with the proper accuracy or precision. When using a calculator, if additional digits are displayed, round off the **final** result* (do not round off in any of the intermediate steps).

EXAMPLE 9 Adding approximate numbers

■ When rounding off a number, it may seem difficult to discard the extra digits. However, if you keep those digits, you show a number with too great an accuracy, and it is incorrect to do so.

Find the sum of the approximate numbers 73.2, 8.0627, and 93.57.
    Showing the addition in the standard way and using a calculator, we have

$$
\begin{array}{l}
73.2 \quad \longleftarrow \text{ least precise number (expressed to tenths)}\\
8.0627\\
\underline{93.57}\\
174.8327 \quad \longleftarrow \text{ final display must be rounded to tenths}
\end{array}
$$

Therefore, the sum of these approximate numbers is 174.8. ■

EXAMPLE 10 Combined operations

In finding the product of the approximate numbers 2.4832 and 30.5 on a calculator, the final display shows 75.7376. However, since 30.5 has only three significant digits, the product is 75.7.

    In Example 1, we calculated that $38.3 - 12.9(-3.58) = 84.482$. We know that $38.3 - 12.9(-3.58) = 38.3 + 46.182 = 84.482$. If these numbers are approximate, we must round off the result to tenths, which means the sum is 84.5. We see that *where there is a combination of operations, the final operation determines how the final result is to be rounded off.* ■

Practice Exercise

Evaluate using a calculator.

3. $40.5 + \dfrac{3275}{-60.041}$ (Numbers are approximate.)

EXAMPLE 11 Operations with exact numbers and approximate numbers

Using the exact number 600 and the approximate number 2.7, we express the result to tenths if the numbers are added or subtracted. If they are multiplied or divided, we express the result to two significant digits. Since 600 is exact, the accuracy of the result depends only on the approximate number 2.7.

$$
\begin{array}{ll}
600 + 2.7 = 602.7 & 600 - 2.7 = 597.3\\
600 \times 2.7 = 1600 & 600 \div 2.7 = 220
\end{array}
$$ ■

NOTE ▶   A note regarding the equal sign ($=$) is in order. We will use it for its defined meaning of "equals exactly" and when the result is an approximate number that has been properly rounded off. Although $\sqrt{27.8} \approx 5.27$, where $\approx$ means "equals approximately," we write $\sqrt{27.8} = 5.27$, since 5.27 has been properly rounded off.

You should *make a rough estimate* of the result when using a calculator. An estimation may prevent accepting an incorrect result after using an incorrect calculator sequence, particularly if the calculator result is far from the estimated value.

EXAMPLE 12 Estimating results

In Example 1, we found that

$$38.3 - 12.9(-3.58) = 84.482 \qquad \text{using exact numbers}$$

When using the calculator, if we forgot to make 3.58 negative, the display would be -7.882, or if we incorrectly entered 38.3 as 83.3, the display would be 129.482.

However, if we estimate the result as

$$40 - 10(-4) = 80$$

we know that a result of -7.882 or 129.482 cannot be correct.

When estimating, we can often use one-significant-digit approximations. If the calculator result is far from the estimate, we should do the calculation again. ∎

EXERCISES 1.3

In Exercises 1–4, make the given changes in the indicated examples of this section, and then solve the given problems.

1. In Example 3, change 0.039 (the second number discussed) to 0.390. Is there any change in the conclusion?

2. In the last paragraph of Example 6, change 35.003 to 35.303 and then find the result.

3. In the first paragraph of Example 10, change 2.4832 to 2.483 and then find the result.

4. In Example 12, change 12.9 to 21.9 and then find the estimated value.

In Exercises 5–8, determine whether the given numbers are approximate or exact.

5. A car with 8 cylinders travels at 55 mi/h.

6. A computer chip 0.002 mm thick is priced at $7.50.

7. In 24 h there are 1440 min.

8. A calculator has 50 keys, and its battery lasted for 50 h of use.

In Exercises 9–14, determine the number of significant digits in each of the given approximate numbers.

9. 107; 3004   10. 3600; 730   11. 6.80; 6.08

12. 0.8735; 0.0075   13. 3000; 3000.1   14. 1.00; 0.01

In Exercises 15–20, determine which of the pair of approximate numbers is (a) more precise and (b) more accurate.

15. 30.8; 0.01   16. 0.041; 7.673   17. 0.1; 78.0

18. 7040; 0.004   19. 7000; 0.004   20. 50.060; 8.914

In Exercises 21–28, round off the given approximate numbers (a) to three significant digits and (b) to two significant digits.

21. 4.936   22. 80.53   23. 50,893   24. 7.004

25. 9549   26. 30.96   27. 0.9449   28. 0.9999

In Exercises 29–36, assume that all numbers are approximate. (a) Estimate the result and (b) perform the indicated operations on a calculator and compare with the estimate.

29. $12.78 + 1.0495 - 1.633$

30. $3.64(17.06)$

31. $0.0350 - \dfrac{0.0450}{1.909}$

32. $\dfrac{0.3275}{1.096 \times 0.50085}$

33. $\dfrac{23.962 \times 0.01537}{10.965 - 8.249}$

34. $\dfrac{0.69378 + 0.04997}{257.4 \times 3.216}$

35. $\dfrac{3872}{503.1} - \dfrac{2.056 \times 309.6}{395.2}$

36. $\dfrac{1}{0.5926} + \dfrac{3.6957}{2.935 - 1.054}$

In Exercises 37–40, perform the indicated operations. The first number is approximate, and the second number is exact.

37. $0.9788 + 14.9$

38. $17.311 - 22.98$

39. $3.142(65)$

40. $8.62 \div 1728$

In Exercises 41–44, answer the given questions. Refer to Appendix B for units of measurement and their symbols.

41. The manual for a heart monitor lists the frequency of the ultrasound wave as 2.75 MHz. What are the least possible and the greatest possible frequencies?

42. A car manufacturer states that the engine displacement for a certain model is 2400 cm³. What should be the least possible and greatest possible displacements?

(W)43. A flash of lightning struck a tower 3.25 mi from a person. The thunder was heard 15 s later. The person calculated the speed of sound and reported it as 1144 ft/s. What is wrong with this conclusion?

(W)44. A technician records 4.4 s as the time for a robot arm to swing from the extreme left to the extreme right, 2.72 s as the time for the return swing, and 1.68 s as the difference in these times. What is wrong with this conclusion?

In Exercises 45–60, perform the calculations on a calculator.

45. Evaluate: (a) $2.2 + 3.8 \times 4.5$ (b) $(2.2 + 3.8) \times 4.5$

46. Evaluate: (a) $6.03 \div 2.25 + 1.77$ (b) $6.03 \div (2.25 + 1.77)$

47. Evaluate: (a) $2 + 0$ (b) $2 - 0$ (c) $0 - 2$ (d) 2×0
(e) $2 \div 0$ Compare with operations with zero on page 9.

(W) 48. Evaluate: (a) $2 \div 0.0001$ and $2 \div 0$ (b) $0.0001 \div 0.0001$ and $0 \div 0$ (c) Explain why the displays differ.

49. Enter a positive integer x (five or six digits is suggested) and then rearrange the same digits to form another integer y. Evaluate $(x - y) \div 9$. What type of number is the result?

50. Enter the digits in the order 9, 8, 7, 6, 5, 4, 3, 2, 1, 0, using between them any of the operations $(+, -, \times, \div)$ that will lead to a result of 100.

51. Show that π is not equal exactly to (a) 3.1416, or (b) 22/7.

52. At some point in the decimal equivalent of a rational number, some sequence of digits will start repeating endlessly. An irrational number never has an endlessly repeating sequence of digits. Find the decimal equivalents of (a) 8/33 and (b) π. Note the repetition for 8/33 and that no such repetition occurs for π.

53. Following Exercise 52, show that the decimal equivalents of the following fractions indicate they are rational: (a) 1/3 (b) 5/11 (c) 2/5. What is the repeating part of the decimal in (c)?

(W) 54. Following Exercise 52, show that the decimal equivalent of the fraction 124/990 indicates that it is rational. Why is the last digit different?

55. In 3 successive days, a home solar system produced 32.4 MJ, 26.704 MJ, and 36.23 MJ of energy. What was the total energy produced in these 3 days?

56. Two jets flew at 938 km/h and 1450 km/h, respectively. How much faster was the second jet?

57. If 1 K of computer memory has 1024 bytes, how many bytes are there in 256 K of memory? (All numbers are exact.)

58. Find the voltage in a certain electric circuit by multiplying the sum of the resistances 15.2 Ω, 5.64 Ω, and 101.23 Ω by the current 3.55 A.

59. The percent of alcohol in a certain car engine coolant is found by performing the calculation $\dfrac{100(40.63 + 52.96)}{105.30 + 52.96}$. Find this percent of alcohol. The number 100 is exact.

60. The tension (in N) in a pulley cable lifting a certain crate was found by calculating the value of $\dfrac{50.45(9.80)}{1 + 100.9 \div 23}$, where the 1 is exact. Calculate the tension.

Answers to Practice Exercises

1. 2020 **2.** 0.300 **3.** -14.0

1.4 Exponents

Positive Integer Exponents • Zero and Negative Exponents • Order of Operations • Evaluating Algebraic Expressions

In mathematics and its applications, we often have a number multiplied by itself several times. To show this type of product, we use the notation a^n, where a is the number and n is the number of times it appears. *In the expression a^n, the number a is called the* **base,** *and n is called the* **exponent;** *in words, a^n is read as "the **n**th power of a."*

EXAMPLE 1 Meaning of exponents

(a) $4 \times 4 \times 4 \times 4 \times 4 = 4^5$ the fifth power of 4

(b) $(-2)(-2)(-2)(-2) = (-2)^4$ the fourth power of -2

(c) $a \times a = a^2$ the second power of a, called "a squared"

(d) $\left(\dfrac{1}{5}\right)\left(\dfrac{1}{5}\right)\left(\dfrac{1}{5}\right) = \left(\dfrac{1}{5}\right)^3$ the third power of $\frac{1}{5}$, called "$\frac{1}{5}$ cubed" ■

We now state the basic operations with exponents using positive integers as exponents. Therefore, with m and n as positive integers, we have the following operations:

■ Two forms are shown for Eqs. (1.4) in order that the resulting exponent is a positive integer. We consider negative and zero exponents after the next three examples.

$$a^m \times a^n = a^{m+n} \tag{1.3}$$

$$\frac{a^m}{a^n} = a^{m-n} \ (m > n, a \neq 0) \qquad \frac{a^m}{a^n} = \frac{1}{a^{n-m}} \ (m < n, a \neq 0) \tag{1.4}$$

$$(a^m)^n = a^{mn} \tag{1.5}$$

$$(ab)^n = a^n b^n \qquad \left(\frac{a}{b}\right)^n = \frac{a^n}{b^n} \ (b \neq 0) \tag{1.6}$$

■ In a^3, which equals $a \times a \times a$, each a is called a factor. A more general definition of factor is given in Section 1.7.

■ Here we are using the fact that a (not zero) divided by itself equals 1, or $a/a = 1$.

EXAMPLE 2 Illustrating Eqs. (1.3) and (1.4)

Using Eq. (1.3):

add exponents

$$a^3 \times a^5 = a^{3+5} = a^8$$

Using the meaning of exponents:

8 factors of a

$$a^3 \times a^5 = (a \times a \times a)(a \times a \times a \times a \times a) = a^8$$

(3 factors of a)(5 factors of a)

Using first form Eq. (1.4):

$5 > 3$

$$\frac{a^5}{a^3} = a^{5-3} = a^2$$

Using the meaning of exponents:

$$\frac{a^5}{a^3} = \frac{\overset{1}{\cancel{a}} \times \overset{1}{\cancel{a}} \times \overset{1}{\cancel{a}} \times a \times a}{\underset{1}{\cancel{a}} \times \underset{1}{\cancel{a}} \times \underset{1}{\cancel{a}}} = a^2$$

Using second form Eq. (1.4):

$$\frac{a^3}{a^5} = \frac{1}{a^{5-3}} = \frac{1}{a^2}$$

$5 > 3$

Using the meaning of exponents:

$$\frac{a^3}{a^5} = \frac{\overset{1}{\cancel{a}} \times \overset{1}{\cancel{a}} \times \overset{1}{\cancel{a}}}{\underset{1}{\cancel{a}} \times \underset{1}{\cancel{a}} \times \underset{1}{\cancel{a}} \times a \times a} = \frac{1}{a^2}$$

■

EXAMPLE 3 Illustrating Eqs. (1.5) and (1.6)

Using Eq. (1.5):

multiply exponents

$$(a^5)^3 = a^{5(3)} = a^{15}$$

Using the meaning of exponents:

$$(a^5)^3 = (a^5)(a^5)(a^5) = a^{5+5+5} = a^{15}$$

Using first form Eq. (1.6):

$$(ab)^3 = a^3 b^3$$

Using the meaning of exponents:

$$(ab)^3 = (ab)(ab)(ab) = a^3 b^3$$

Using second form Eq. (1.6):

$$\left(\frac{a}{b}\right)^3 = \frac{a^3}{b^3}$$

Using the meaning of exponents:

$$\left(\frac{a}{b}\right)^3 = \left(\frac{a}{b}\right)\left(\frac{a}{b}\right)\left(\frac{a}{b}\right) = \frac{a^3}{b^3}$$

■

CAUTION ▶ When an expression involves a product or a quotient of different bases, ***only exponents of the same base may be combined.*** Consider the following example.

EXAMPLE 4 Other illustrations of exponents

Practice Exercises

Use Eqs. (1.3) – (1.6) to simplify the given expressions.

1. $ax^3(-ax)^2$ **2.** $\dfrac{(2c)^5}{(3cd)^2}$

(a) $(-x^2)^3 = [(-1)x^2]^3 = (-1)^3(x^2)^3 = -x^6$

exponent of 1   add exponents of a

(b) $ax^2(ax)^3 = ax^2(a^3x^3) = a^4x^5$ ◀ add exponents of x

(c) $\dfrac{(3 \times 2)^4}{(3 \times 5)^3} = \dfrac{3^4 2^4}{3^3 5^3} = \dfrac{3 \times 2^4}{5^3}$

(d) $\dfrac{(ry^3)^2}{r(y^2)^4} = \dfrac{r^2 y^6}{ry^8} = \dfrac{r}{y^2}$

CAUTION ▶ In illustration (b), note that ***ax^2 means a times the square of x and does not mean a^2x^2,*** whereas $(ax)^3$ *does* mean a^3x^3. ■

EXAMPLE 5   Exponents in an application

In the analysis of the deflection of a beam (the amount it bends), the expression that follows is simplified as shown.

$$\frac{1}{2}\left(\frac{PL}{4EI}\right)\left(\frac{2}{3}\right)\left(\frac{L}{2}\right)^2 = \frac{1}{2}\left(\frac{PL}{4EI}\right)\left(\frac{2}{3}\right)\left(\frac{L^2}{2^2}\right)$$

$$= \frac{\overset{1}{\cancel{2}}PL(L^2)}{\underset{1}{\cancel{2}}(3)(4)(4)EI} = \frac{PL^3}{48EI}$$

L is the length of the beam, and P is the force applied to it. E and I are constants related to the beam. In *simplifying* this expression, we combined exponents of L and divided out the 2 that was in the numerator and in the denominator.   ∎

ZERO AND NEGATIVE EXPONENTS

If we let $n = m$ in Eqs. (1.4), we would have $a^m/a^m = a^{m-m} = a^0$. Also, $a^m/a^m = 1$, since any nonzero quantity divided by itself equals 1. Therefore, for Eqs. (1.4) to hold, when $m = n$, we have

$$a^0 = 1 \qquad (a \neq 0) \tag{1.7}$$

Equation (1.7) states that *any nonzero expression raised to the zero power is 1.* Zero exponents can be used with any of the operations for exponents.

EXAMPLE 6   Zero as an exponent

(a) $5^0 = 1$ **(b)** $(-3)^0 = 1$ **(c)** $-(-3)^0 = -1$ **(d)** $(2x)^0 = 1$

(e) $(ax + b)^0 = 1$ **(f)** $(a^2 b^0 c)^2 = a^4 c^2$ **(g)** $2t^0 = 2(1) = 2$

$b^0 = 1$

We note in illustration (g) that *only t is raised to the zero power.* If the quantity $2t$ were raised to the zero power, it would be written as $(2t)^0$.   ∎

If we apply the first form of Eqs. (1.4) to the case where $n > m$, the resulting exponent is negative. This leads to the definition of a negative exponent.

EXAMPLE 7   Basis for negative exponents

Applying both forms of Eqs. (1.4) to a^2/a^7, we have

$$\frac{a^2}{a^7} = a^{2-7} = a^{-5} \quad \text{and} \quad \frac{a^2}{a^7} = \frac{1}{a^{7-2}} = \frac{1}{a^5}$$

If these results are to be consistent, then $a^{-5} = \dfrac{1}{a^5}$.   ∎

Following the reasoning of Example 7, if we define

$$a^{-n} = \frac{1}{a^n} \qquad (a \neq 0) \tag{1.8}$$

then all of the laws of exponents will hold for negative integers.

■ Although positive exponents are generally preferred in a final result, there are some cases in which zero or negative exponents are to be used. Also, negative exponents are very useful in some operations that we will use later.

Practice Exercises

Simplify: **3.** $\dfrac{-7^0}{c^{-3}}$ **4.** $\dfrac{(3x)^{-1}}{2a^{-2}}$

EXAMPLE 8 Negative exponents

(a) $3^{-1} = \dfrac{1}{3}$ **(b)** $4^{-2} = \dfrac{1}{4^2} = \dfrac{1}{16}$ **(c)** $\dfrac{1}{a^{-3}} = a^3$ change signs of exponents

(d) $\left(\dfrac{a^3 t}{b^2 x}\right)^{-2} = \dfrac{(a^3 t)^{-2}}{(b^2 x)^{-2}} = \dfrac{(b^2 x)^2}{(a^3 t)^2} = \dfrac{b^4 x^2}{a^6 t^2}$ **(e)** $3x^{-1} = 3\left(\dfrac{1}{x}\right) = \dfrac{3}{x}$

CAUTION ▶   From illustration (a), where $3^{-1} = 1/3$, we see that the reciprocal of a number x (not 0) is x^{-1}. In illustration (e), *only the x has the exponent of* -1. ∎

ORDER OF OPERATIONS

In Section 1.2, we saw that it is necessary to follow a particular order of operations when performing the basic operations on numbers. Since raising a number to a power is a form of multiplication, this operation is performed before additions and subtractions. In fact, it is performed before multiplications and divisions.

> **ORDER OF OPERATIONS**
>
> **1.** *Operations within specific groupings*
> **2.** *Powers*
> **3.** *Multiplications and divisions (from left to right)*
> **4.** *Additions and subtractions (from left to right)*

■ The use of exponents is taken up in more detail in Chapter 11.

EXAMPLE 9 Order of operations

$$8 - (-1)^2 - 2(-3)^2 = 8 - 1 - 2(9)$$
$$= 8 - 1 - 18 = -11$$

CAUTION ▶   Since there were no specific groupings, we first squared -1 and -3. Next, we found the product $2(9)$ in the last term. Finally, the subtractions were performed. Note carefully that *we did not change the sign of* -1 *before we squared it.* ∎

EVALUATING ALGEBRAIC EXPRESSIONS

■ On many calculators, there is a specific key or key sequence to evaluate x^3.

An algebraic expression is **evaluated** *by* **substituting** *given values of the literal numbers in the expression and calculating the result.* On a calculator, the $\boxed{x^2}$ key is used to square numbers, and the $\boxed{\wedge}$ or $\boxed{x^y}$ key is used for other powers.

EXAMPLE 10 Evaluating using exponents

```
16*4.2²
            282.24
```

Fig. 1.9

■ When less than half of a calculator screen is needed, the figure for that calculator screen will show only a partial screen.

The distance (in ft) that an object falls in 4.2 s is found by substituting 4.2 for t in the expression $16.0t^2$. We show this as

$$t = 4.2 \text{ s} \longleftarrow \text{substituting}$$
$$\downarrow$$
$$16.0(4.2)^2 = 280 \text{ ft} \qquad \text{estimation} \longrightarrow 20(4)^2 = 320$$

The result is rounded off to two significant digits (the accuracy of t). The calculator will square 4.2 before multiplying. See Fig. 1.9. ∎

As stated in Section 1.3, in evaluating an expression on a calculator, we should also estimate its value as in Example 10. For an estimate, note that *a negative number raised to an even power gives a positive value and a negative number raised to an odd power gives a negative value.*

EXAMPLE 11 Power of a number

Using the meaning of a power of a number, we have

$$(-2)^2 = (-2)(-2) = 4 \qquad (-2)^3 = (-2)(-2)(-2) = -8$$
$$(-2)^4 = 16 \qquad (-2)^5 = -32 \qquad (-2)^6 = 64 \qquad (-2)^7 = -128 \qquad \blacksquare$$

EXAMPLE 12 Evaluating in an application

A wire made of a special alloy has an electric resistance R (in Ω) given by $R = a + 0.0115T^3$, where T (in °C) is the temperature (between -4°C and 4°C). Find R for $a = 0.838\ \Omega$ and $T = -2.87$°C.

Substituting these values, we have

$$R = 0.838 + 0.0115(-2.87)^3$$

estimation:
$$0.8 + 0.01(-3)^3 = 0.8 + 0.01(-27) = 0.53$$

$$= 0.566\ \Omega$$

Note in the estimation that $(-3)^3 = -27$. $\qquad \blacksquare$

Graphing calculators generally use computer symbols in the display for some of the operations to be performed. These symbols are as follows:

Multiplication: $*$ Division: $/$ Powers: $\wedge$

Therefore, to calculate the value of $20 \times 6 + 200/5 - 3^4$, we use the key sequence

20 ⨯ 6 + 200 ÷ 5 − 3 ∧ 4

```
20*6+200/5-3^4
              79
```

Fig. 1.10

CAUTION ▶ with the result of 79 shown in the display of Fig. 1.10. ***Note carefully that 200 is divided only by 5.*** If it were divided by $5 - 3^4$, then we would use parentheses and show the expression to be evaluated as $20 \times 6 + 200/(5 - 3^4)$.

EXERCISES 1.4

In Exercises 1–4, make the given changes in the indicated examples of this section, and then simplify the resulting expression.

1. In Example 4(a), change $(-x^2)^3$ to $(-x^3)^2$.

2. In Example 6(d), change $(2x)^0$ to $2x^0$.

3. In Example 8(d), interchange the a^3 and b^2.

4. In Example 9, change $(-1)^2$ to $(-1)^3$.

In Exercises 5–48, simplify the given expressions. Express results with positive exponents only.

5. x^3x^4 **6.** y^2y^7 **7.** $2b^4b^2$ **8.** $3k^5k$

9. $\dfrac{m^5}{m^3}$ **10.** $\dfrac{2x^6}{x}$ **11.** $\dfrac{n^5}{7n^9}$ **12.** $\dfrac{3s}{s^4}$

13. $(P^2)^4$ **14.** $(x^8)^3$ **15.** $(2\pi)^3$ **16.** $(ax)^5$

17. $(aT^2)^{30}$ **18.** $(3r^2)^3$ **19.** $\left(\dfrac{2}{b}\right)^3$ **20.** $\left(\dfrac{F}{t}\right)^{20}$

21. $\left(\dfrac{x^2}{2}\right)^4$ **22.** $\left(\dfrac{3}{n^3}\right)^3$ **23.** $(8a)^0$ **24.** $6v^0$

25. $-3x^0$ **26.** $-(-2)^0$ **27.** 6^{-1} **28.** $-w^{-5}$

29. $\dfrac{1}{R^{-2}}$ **30.** $\dfrac{1}{-t^{-48}}$ **31.** $(-t^2)^7$ **32.** $(-y^3)^5$

33. $(2v^2)^{-6}$ **34.** $-(-c^4)^{-4}$ **35.** $-\dfrac{L^{-3}}{L^{-5}}$ **36.** $2i^{40}i^{-70}$

37. $\dfrac{2v^4}{(2v)^4}$ **38.** $\dfrac{x^2x^3}{(x^2)^3}$ **39.** $\dfrac{(n^2)^4}{(n^4)^2}$ **40.** $\dfrac{(3t)^{-1}}{3t^{-1}}$

41. $(\pi^0 x^2 a^{-1})^{-1}$ **42.** $(3m^{-2}n^4)^{-2}$ **43.** $(-8g^{-1}s^3)^2$

44. $ax^{-2}(-a^2x)^3$ **45.** $\left(\dfrac{4x^{-1}}{a^{-1}}\right)^{-3}$ **46.** $\left(\dfrac{2b^2}{y^5}\right)^{-2}$

47. $\dfrac{15n^2T^5}{3n^{-1}T^6}$ **48.** $\dfrac{(nRT^{-2})^{32}}{R^{-2}T^{32}}$

In Exercises 49–56, evaluate the given expressions. In Exercises 55–60, all numbers are approximate.

49. $7(-4) - (-5)^2$ **50.** $6 + (-2)^5 - (-2)(8)$

51. $-(-26.5)^2 - (-9.85)^3$ **52.** $-0.711^2 - (-0.809)^6$

53. $\dfrac{3.07(-1.86)}{(-1.86)^4 + 1.596}$ **54.** $\dfrac{15.66^2 - (-4.017)^4}{1.044(-3.68)}$

55. $2.38(-60.7)^2 - 2540/1.17^3 + 0.806^5(26.1^3 - 9.88^4)$

56. $0.513(-2.778) - (-3.67)^3 + 0.889^4/(1.89 - 1.09^2)$

In Exercises 57–68, perform the indicated operations.

57. Does $\left(\dfrac{1}{x^{-1}}\right)^{-1}$ represent the reciprocal of x?

(W) 58. Does $\left(\dfrac{0.2 - 5^{-1}}{10^{-2}}\right)^{0}$ equal 1? Explain.

59. If $a^3 = 5$, then what does a^{12} equal?

(W) 60. Is $a^{-2} < a^{-1}$ for any negative value of a? Explain.

61. If a is a positive integer, simplify $(x^a \cdot x^{-a})^5$.

62. If a and b are positive integers, simplify $(y^{a-b} \cdot y^{a+b})^2$.

63. In developing the "big bang" theory of the origin of the universe, the expression $(kT/(hc))^3 (GkThc)^2 c$ arises. Simplify this expression.

64. In studying planetary motion, the expression $(GmM)(mr)^{-1}(r^{-2})$ arises. Simplify this expression.

65. In designing a cam for a pump, the expression $\pi\left(\dfrac{r}{2}\right)^3 \left(\dfrac{4}{3\pi r^2}\right)$ is used. Simplify this expression.

66. For a certain integrated electric circuit, it is necessary to simplify the expression $\dfrac{gM}{2\pi f C (2\pi f M)^2}$. Perform this simplification.

67. If \$2500 is invested at 4.2% interest, compounded quarterly, the amount in the account after 6 years is $2500(1 + 0.042/4)^{24}$. Calculate this amount (the 1 is exact).

68. In designing a building, it was determined that the forces acting on an I beam would deflect the beam an amount (in cm), given by $\dfrac{x(1000 - 20x^2 + x^3)}{1850}$, where x is the distance (in m) from one end of the beam. Find the deflection for $x = 6.85$ m. (The 1000 and 20 are exact.)

Answers to Practice Exercises

1. $a^3 x^5$ **2.** $\dfrac{2^5 c^3}{3^2 d^2} = \dfrac{32 c^3}{9 d^2}$ **3.** $-c^3$ **4.** $\dfrac{a^2}{6x}$

1.5 Scientific Notation

Meaning of Scientific Notation • Changing Numbers to and from Scientific Notation • Scientific Notation on a Calculator

■ Television was invented in the 1920s and first used commercially in the 1940s.

The use of fiber optics was developed in the 1950s.

X-rays were discovered by Roentgen in 1895.

In technical and scientific work, we often encounter numbers that are either very large or very small. Examples of such numbers are: (a) Radio and television signals travel at about 30,000,000,000 cm/s; (b) The mass of the earth is about 6,600,000,000,000,000,000,000 tons; (c) A typical individual fiber in a fiber-optic communications cable has a diameter of 0.000005 m; (d) some X-rays have a wavelength of about 0.000000095 cm.

As we can see, writing numbers such as these is inconvenient in ordinary notation. Although calculators and computers can often handle such numbers, there is a more efficient way of expressing numbers, in order to work more easily with them. A convenient and useful notation, called *scientific notation*, is used to represent these or any other numbers.

A number in **scientific notation** *is expressed as the product of a number greater than or equal to 1 and less than* 10, *and a power of* 10, *and is written as*

$$P \times 10^k$$

where $1 \le P < 10$ and k is an integer. (The symbol $\le$ means "is less than or equal to.")

EXAMPLE 1 Scientific notation

(a) $34{,}000 = 3.4 \times 10{,}000 = 3.4 \times 10^4$ **(b)** $6.82 = 6.82 \times 1 = 6.82 \times 10^0$

between 1 and 10

(c) $0.00503 = \dfrac{5.03}{1000} = \dfrac{5.03}{10^3} = 5.03 \times 10^{-3}$ ■

From Example 1, we see how a number is changed from ordinary notation to scientific notation. *The decimal point is moved so that only one nonzero digit is to its left. The number of places moved is the power of* 10 *(k), which is positive if the decimal point is moved to the left and negative if moved to the right.*

EXAMPLE 2 Changing numbers to scientific notation

(a) $34{,}000 = 3.4 \times 10^4$ **(b)** $6.82 = 6.82 \times 10^0$ **(c)** $0.00503 = 5.03 \times 10^{-3}$

4 places to left   0 places   3 places to right   ■

To change a number from scientific notation to ordinary notation, we reverse the procedure used in Example 2.

EXAMPLE 3   Changing numbers to ordinary notation

(a) To change 5.83×10^6 to ordinary notation, we move the decimal point six places to the right, including additional zeros to properly locate the decimal point.

$$5.83 \times 10^6 = 5,830,000$$

6 places to right

(b) To change 8.06×10^{-3} to ordinary notation, we must move the decimal point three places to the left, again including additional zeros to properly locate the decimal point.

$$8.06 \times 10^{-3} = 0.00806$$

3 places to left ∎

Scientific notation provides a practical way to handle calculations with very large or very small numbers, particularly products, quotients, and powers. First, all numbers are expressed in scientific notation. Then the calculation can be performed on numbers between 1 and 10, using the laws of exponents to find the power of 10 in the result. In this way, scientific notation gives an important application of the use of exponents.

EXAMPLE 4   Scientific notation in calculations

In designing a computer, it was determined that it would be able to process 803,000 bits of data in 0.00000525 s. (See Exercise 41 of Exercises 1.1 for a brief note on computer data.) The rate of processing the data is

$$5 - (-6) = 11$$

$$\frac{803,000}{0.00000525} = \frac{8.03 \times 10^5}{5.25 \times 10^{-6}} = \left(\frac{8.03}{5.25}\right) \times 10^{11} = 1.53 \times 10^{11} \text{ bits/s}$$

■ Another important use of scientific notation is shown by the metric system use of prefixes as shown in Appendix B.

As shown, it is proper to leave the result *(rounded off)* in scientific notation. This method is useful when using a calculator and then estimating the result. In this case, the estimate is $(8 \times 10^5) \div (5 \times 10^{-6}) = 1.6 \times 10^{11}$. ∎

Another advantage of scientific notation is that the precise number of significant digits of a number can be shown directly, even when the final significant digit is 0.

EXAMPLE 5   Scientific notation and significant digits

In evaluating $750,000,000,000^2$, if we know that $750,000,000,000$ has *three* significant digits, we can show the significant digits by writing

$$750,000,000,000^2 = (7.50 \times 10^{11})^2 = 7.50^2 \times 10^{2 \times 11} = 56.3 \times 10^{22}$$

If the answer is to be written in scientific notation, we should write it as the product of a number between 1 and 10 and a power of 10. This means we should rewrite it as

$$56.3 \times 10^{22} = (5.63 \times 10)(10^{22}) = 5.63 \times 10^{23}$$ ∎

■ The number 7.50×10^{11} cannot be entered on a scientific calculator in the form 750000000000. It could be entered in this form on a graphing calculator.

We can enter numbers in scientific notation on a calculator, as well as have the calculator give results automatically in scientific notation. See the next example.

EXAMPLE 6   Scientific notation on a calculator

The wavelength λ (in m) of the light in a red laser beam can be found from the following calculation. Note the significant digits in the numerator.

```
3E6/4.74E12
      6.329113924E-7
```

Fig. 1.11

$$\lambda = \frac{3,000,000}{4,740,000,000,000} = \frac{3.00 \times 10^6}{4.74 \times 10^{12}} = 6.33 \times 10^{-7} \text{ m}$$

The key sequence is 3 [EE] 6 [÷] 4.74 [EE] 12 [ENTER]. See Fig. 1.11. ∎

EXERCISES 1.5

In Exercises 1 and 2, make the given changes in the indicated examples of this section, and then rewrite the number as directed.

1. In Example 3(b), change the exponent -3 to 3 and then write the number in ordinary notation.

2. In Example 5, change the exponent 2 to -1 and then write the result in scientific notation.

In Exercises 3–10, change the numbers from scientific notation to ordinary notation.

3. 4.5×10^4 **4.** 6.8×10^7 **5.** 2.01×10^{-3} **6.** 9.61×10^{-5}

7. 3.23×10^0 **8.** 8×10^0 **9.** 1.86×10 **10.** 1×10^{-1}

In Exercises 11–20, change the numbers from ordinary notation to scientific notation.

11. 4000 **12.** 56,000 **13.** 0.0087 **14.** 0.7

15. 6.09 **16.** 100 **17.** 0.063 **18.** 0.0000908

19. 1 **20.** 10

In Exercises 21–24, perform the indicated calculations using a calculator and by first expressing all numbers in scientific notation.

21. $28,000(2,000,000,000)$ **22.** $50,000(0.006)$

23. $\dfrac{88,000}{0.0004}$ **24.** $\dfrac{0.00003}{6,000,000}$

In Exercises 25–28, perform the indicated calculations and then check the result using a calculator. Assume that all numbers are exact.

25. $2 \times 10^{-35} + 3 \times 10^{-34}$ **26.** $5.3 \times 10^{12} - 3.7 \times 10^{10}$

27. $(1.2 \times 10^{29})^3$ **28.** $(2 \times 10^{-16})^{-5}$

In Exercises 29–36, perform the indicated calculations using a calculator. All numbers are approximate.

29. $1280(865,000)(43.8)$ **30.** $0.0000569(3,190,000)$

31. $\dfrac{0.0732(6710)}{0.00134(0.0231)}$ **32.** $\dfrac{0.00452}{2430(97,100)}$

33. $(3.642 \times 10^{-8})(2.736 \times 10^5)$ **34.** $\dfrac{(7.309 \times 10^{-1})^2}{5.9843(2.5036 \times 10^{-20})}$

35. $\dfrac{(3.69 \times 10^{-7})(4.61 \times 10^{21})}{0.0504}$

36. $\dfrac{(9.907 \times 10^7)(1.08 \times 10^{12})^2}{(3.603 \times 10^{-5})(2054)}$

In Exercises 37–44, change numbers in ordinary notation to scientific notation or change numbers in scientific notation to ordinary notation. See Appendix B for an explanation of symbols used.

37. The power plant at Grand Coulee Dam produces 6,500,000 kW of power.

38. A certain computer has 85,000,000,000 bytes (85 gigabytes) of memory.

39. A fiber-optic system requires 0.000003 W of power.

40. A red blood cell measures 0.0075 mm across.

41. The frequency of a certain TV signal is 2,000,000,000 Hz.

42. A *parsec*, a unit used in astronomy, is about 3.086×10^{16} m.

43. The power of the signal of a laser beam probe is 1.6×10^{-12} W.

44. The electrical force between two electrons is about 2.4×10^{-43} times the gravitational force between them.

In Exercises 45–48, solve the given problems.

45. *Engineering notation* expresses a number as $P \times 10^k$, where P has one, two, or three digits to the left of the decimal point, and k is an integral multiple of 3. Write the following numbers in engineering notation. (a) 2300 (b) 0.23 (c) 23

46. See Exercise 45. Write the following numbers in engineering notation. (a) 8,090,000 (b) 809,000 (c) 0.0809

47. A *googol* is defined as 1 followed by 100 zeros. (a) Write this number in scientific notaion. (b) A *googolplex* is defined as 10 to the googol power. Write this number using powers of 10, and not the word *googol*. (Note the name of the Internet company.)

48. The number of electrons in the universe has been estimated at 10^{79}. How many times greater is a googol than the estimated number of electrons in the universe? (See Exercise 47.)

In Exercises 49–52, perform the indicated calculations.

49. A computer can do an addition in 7.5×10^{-15} s. How long does it take to perform 5.6×10^6 additions?

50. Uranium is used in nuclear reactors to generate electricity. About 0.000000039% of the uranium disintegrates each day. How much of 0.085 mg of uranium disintegrates in a day?

51. A TV signal travels at 1.86×10^5 mi/s for a total of 4.57×10^4 mi from the station transmitter to a satellite and then to a receiver dish. How long does it take the signal to go from the transmitter to the dish?

52. (a) Determine the number of seconds in a day in scientific notation. (b) Using the result of part (a), determine the number of seconds in a century (assume 365.24 days/year).

In Exercises 53–56, perform the indicated calculations by first expressing all numbers in scientific notation.

53. One *atomic mass unit* (amu) is 1.66×10^{-27} kg. If one oxygen atom has 16 amu (an exact number), what is the mass of 125,000,000 oxygen atoms?

54. The rate of energy radiation (in W) from an object is found by evaluating the expression kT^4, where T is the thermodynamic temperature. Find this value for the human body, for which $k = 0.000000057$ W/K^4 and $T = 303$ K.

55. In a microwave receiver circuit, the resistance R of a wire 1 m long is given by $R = k/d^2$, where d is the diameter of the wire. Find R if $k = 0.00000002196\ \Omega \cdot$m^2 and $d = 0.00007998$ m.

56. The average distance between the sun and earth is about 149,600,000 km, and this distance is called an *astronomical unit* (AU). It takes light about 499.0 s to travel 1 AU. What is the speed of light? Compare this with the speed of the TV signal in Exercise 51.

Answers to Practice Exercises

1. 0.00235 **2.** 2.35×10^2

1.6 Roots and Radicals

Principal *n*th Root • Simplifying Radicals • Using a Calculator • Imaginary Numbers

■ Unless we state otherwise, when we refer to the root of a number, it is the principal root.

At times, we have to find the *square root* of a number, or maybe some other root of a number, such as a *cube root*. This means we must find a number that when squared, or cubed, and so on equals some given number. For example, to find the square root of 9, we must find a number that when squared equals 9. In this case, either 3 or −3 is an answer. Therefore, *either 3 or −3 is a square root of 9 since* $3^2 = 9$ and $(-3)^2 = 9$.

To have a general notation for the square root and have it represent *one* number, *we define the* **principal square root** *of a to be positive if a is positive and represent it by* $\sqrt{a}$. This means $\sqrt{9} = 3$ and not −3.

The general notation for the **principal *n*th root** *of a is* $\sqrt[n]{a}$. (When $n = 2$, do not write the 2 for *n*.) *The* $\sqrt{\ }$ *sign is called a* **radical sign.**

EXAMPLE 1 Roots of numbers

(a) $\sqrt{2}$ (the square root of 2) (b) $\sqrt[3]{2}$ (the cube root of 2)

(c) $\sqrt[4]{2}$ (the fourth root of 2) (d) $\sqrt[7]{6}$ (the seventh root of 6) ■

To have a single defined value for all roots (not just square roots) and to consider only real-number roots, *we define the* **principal *n*th root** *of a to be positive if a is positive and to be negative if a is negative and n is odd.* (If *a* is negative and *n* is even, the roots are not real.)

EXAMPLE 2 Principal *n*th root

(a) $\sqrt{169} = 13$ ($\sqrt{169} \neq -13$) (b) $-\sqrt{64} = -8$

(c) $\sqrt[3]{27} = 3$ since $3^3 = 27$ (d) $\sqrt{0.04} = 0.2$ since $0.2^2 = 0.04$

(e) $-\sqrt[4]{256} = -4$ (f) $\sqrt[3]{-27} = -3$ (g) $-\sqrt[3]{27} = -(+3) = -3$ ■

Another property of square roots is developed by noting illustrations such as $\sqrt{36} = \sqrt{4 \times 9} = \sqrt{4} \times \sqrt{9} = 2 \times 3 = 6$. In general, this property states that *the square root of a product of positive numbers is the product of their square roots.*

$$\sqrt{ab} = \sqrt{a}\sqrt{b} \quad (a \text{ and } b \text{ positive real numbers})$$ (1.9)

This property is used in simplifying radicals. It is most useful if either *a* or *b* is a **perfect square,** *which is the square of a rational number.*

EXAMPLE 3 Simplifying square roots

(a) $\sqrt{8} = \sqrt{(4)(2)} = \sqrt{4}\sqrt{2} = 2\sqrt{2}$

(b) $\sqrt{75} = \sqrt{(25)(3)} = \sqrt{25}\sqrt{3} = 5\sqrt{3}$

(c) $\sqrt{4 \times 10^2} = \sqrt{4}\sqrt{10^2} = 2(10) = 20$

(Note that the square root of the square of a positive number is that number.) ■

■ Try this one on your calculator:
$\sqrt{12345678987654321}$

.25√(1260)
        8.874119675

Fig. 1.12

In order to represent the square root of a number *exactly*, use Eq. (1.9) to write it in simplest form. However, a decimal *approximation* is often acceptable, and we use the $\boxed{\sqrt{x}}$ key on a calculator. (We will show another way of finding the root of a number in Chapter 11.)

EXAMPLE 4 Using a calculator in an application

After reaching its greatest height, the time (in s) for a rocket to fall h ft is found by evaluating $0.25\sqrt{h}$. Find the time for the rocket to fall 1260 ft.

The calculator evaluation is shown in Fig. 1.12. From the display, we see that the rocket takes 8.9 s to fall 1260 ft. The result is rounded off to two significant digits, the accuracy of 0.25 (an approximate number). ■

In simplifying a radical, *all operations under a radical sign must be done before finding the root*. The horizontal bar groups the numbers under it.

EXAMPLE 5 More on simplifying square roots

(a) $\sqrt{16 + 9} = \sqrt{25}$ first perform the addition 16 + 9

$\qquad\qquad\quad = 5$

CAUTION ▶   However, $\sqrt{16 + 9}$ is *not* $\sqrt{16} + \sqrt{9} = 4 + 3 = 7$.

(b) $\sqrt{2^2 + 6^2} = \sqrt{4 + 36} = \sqrt{40} = \sqrt{4}\sqrt{10} = 2\sqrt{10}$, but
$\sqrt{2^2 + 6^2}$ is *not* $\sqrt{2^2} + \sqrt{6^2} = 2 + 6 = 8$. ■

Practice Exercises
Simplify:
1. $\sqrt{12}$ 2. $\sqrt{36 + 144}$

In defining the principal square root, we did not define the square root of a negative number. However, in Section 1.1, we defined the square root of a negative number to be an **imaginary number.** More generally, *the even root of a negative number is an imaginary number, and the odd root of a negative number is a negative real number*.

EXAMPLE 6 Imaginary roots and real roots

even   even   odd
↓   ↓   ↓
$\sqrt{-64}$ is imaginary $\sqrt[4]{-243}$ is imaginary $\sqrt[3]{-64} = -4$ (a real number) ■

A much more detailed coverage of imaginary numbers, roots, and radicals is taken up in Chapters 11 and 12.

EXERCISES 1.6

In Exercises 1–4, make the given changes in the indicated examples of this section and then solve the given problems.

1. In Example 2(b), change the square root to a cube root and then evaluate.

(W) 2. In Example 3(b), change $\sqrt{(25)(3)}$ to $\sqrt{(15)(5)}$ and explain whether or not this would be a better expression to use.

3. In Example 5(a), change the $+$ to $\times$ and then evaluate.

4. In the first illustration of Example 6, place a $-$ sign before the radical. Is there any other change in the statement?

In Exercises 5–36, simplify the given expressions. In each of 5–9 and 12–21, the result is an integer.

5. $\sqrt{81}$ 6. $\sqrt{225}$ 7. $-\sqrt{121}$ 8. $-\sqrt{36}$
9. $-\sqrt{49}$ 10. $\sqrt{0.25}$ 11. $\sqrt{0.09}$ 12. $-\sqrt{900}$
13. $\sqrt[3]{125}$ 14. $\sqrt[4]{16}$ 15. $\sqrt[3]{-216}$ 16. $\sqrt[5]{-32}$

17. $\left(\sqrt{5}\right)^2$ 18. $\left(\sqrt[3]{31}\right)^3$ 19. $\left(-\sqrt[3]{-47}\right)^3$ 20. $\left(\sqrt[5]{-23}\right)^5$
21. $\left(-\sqrt[4]{53}\right)^4$ 22. $-\sqrt{32}$ 23. $\sqrt{1200}$ 24. $\sqrt{50}$
25. $2\sqrt{84}$ 26. $\dfrac{\sqrt{108}}{2}$ 27. $\sqrt{\dfrac{80}{7-3}}$ 28. $\sqrt{81 \times 10^2}$
29. $\sqrt[3]{8^2}$ 30. $\sqrt[4]{9^2}$ 31. $\dfrac{7^2\sqrt{81}}{3^2\sqrt{49}}$ 32. $\dfrac{2^5\sqrt[5]{243}}{3\sqrt{144}}$

33. $\sqrt{36 + 64}$ 34. $\sqrt{25 + 144}$
35. $\sqrt{3^2 + 9^2}$ 36. $\sqrt{8^2 - 4^2}$

In Exercises 37–44, find the value of each square root by use of a calculator. Each number is approximate.

37. $\sqrt{85.4}$ 38. $\sqrt{3762}$ 39. $\sqrt{0.4729}$ 40. $\sqrt{0.0627}$
41. (a) $\sqrt{1296 + 2304}$ (b) $\sqrt{1296} + \sqrt{2304}$
42. (a) $\sqrt{10.6276 + 2.1609}$ (b) $\sqrt{10.6276} + \sqrt{2.1609}$

43. (a) $\sqrt{0.0429^2 - 0.0183^2}$ (b) $\sqrt{0.0429^2} - \sqrt{0.0183^2}$

44. (a) $\sqrt{3.625^2 + 0.614^2}$ (b) $\sqrt{3.625^2} + \sqrt{0.614^2}$

In Exercises 45–56, solve the given problems.

45. The speed (in mi/h) of a car that skids to a stop on dry pavement is often estimated by $\sqrt{24s}$, where s is the length (in ft) of the skid marks. Estimate the speed if $s = 150$ ft.

46. The resistance in an amplifier circuit is found by evaluating $\sqrt{Z^2 - X^2}$. Find the resistance for $Z = 5.362\ \Omega$ and $X = 2.875\ \Omega$.

47. The speed (in m/s) of sound in seawater is found by evaluating $\sqrt{B/d}$ for $B = 2.18 \times 10^9$ Pa and $d = 1.03 \times 10^3\,\text{kg/m}^3$. Find this speed, which is important in locating underwater objects using sonar.

48. The terminal speed (in m/s) of a skydiver can be approximated by $\sqrt{40m}$, where m is the mass (in kg) of the skydiver. Calculate the terminal speed (after reaching this speed, the skydiver's speed remains fairly constant before opening the parachute) of a 75-kg skydiver.

49. An HDTV screen is 36.6 in. wide and 20.5 in. high. The length of a diagonal (the dimension used to describe it—from one corner to the opposite corner) is found by evaluating $\sqrt{w^2 + h^2}$, where w is the width and h is the height. Find the diagonal.

50. A car costs $22,000 new and is worth $15,000 2 years later. The annual rate of depreciation is found by evaluating $100\left(1 - \sqrt{V/C}\right)$, where C is the cost and V is the value after 2 years. At what rate did the car depreciate? (100 and 1 are exact.)

(W) 51. Is it always true that $\sqrt{a^2} = a$? Explain.

52. For what values of x is (a) $x > \sqrt{x}$, (b) $x = \sqrt{x}$, and (c) $x < \sqrt{x}$?

53. A graphing calculator has a specific key sequence to find cube roots. Using a calculator, find $\sqrt[3]{2140}$ and $\sqrt[3]{-0.214}$.

54. A graphing calculator has a specific key sequence to find nth roots. Using a calculator, find $\sqrt[7]{0.382}$ and $\sqrt[7]{-382}$.

55. A graphing calculator can determine if a number is real or imaginary. Use the $a + bi$ setting for numbers (see Exercise 37 on page 6), where i represents $\sqrt{-1}$ on a calculator. Use a calculator to determine if the following numbers are real or imaginary: (a) $\sqrt{-64}$ (b) $\sqrt[3]{-64}$

56. Use a calculator to determine if the following numbers are real or imaginary (see Exercise 55): (a) $\sqrt[5]{-32}$ (b) $\sqrt[4]{-64}$

Answers to Practice Exercises

1. $2\sqrt{3}$ 2. $6\sqrt{5}$

1.7 Addition and Subtraction of Algebraic Expressions

Algebraic Expressions • Terms • Factors • Polynomials • Similar Terms • Simplifying • Symbols of Grouping

Since we use letters to represent numbers, we can see that all operations that can be used on numbers can also be used on literal numbers. In this section, we discuss the methods for adding and subtracting literal numbers.

Addition, subtraction, multiplication, division, and taking of roots are known as **algebraic operations.** *Any combination of numbers and literal symbols that results from algebraic operations is known as an* **algebraic expression.**

When an algebraic expression consists of several parts connected by plus signs and minus signs, each part (along with its sign) is known as a **term** *of the expression. If a given expression is made up of the product of a number of quantities, each of these quantities, or any product of them, is called a* **factor** *of the expression.*

CAUTION ▶ It is very important to **distinguish clearly between terms and factors,** because some operations that are valid for terms are not valid for factors, and conversely. Some of the common errors in handling algebraic expressions occur because these operations are not handled properly.

EXAMPLE 1 Terms and factors

In the study of the motion of a rocket, the following algebraic expression may be used.

$$\text{terms} \quad gt^2 - 2vt + 2s$$
$$\text{factors}$$

This expression has three terms: gt^2, $-2vt$, and $2s$. The first term, gt^2, has a factor of g and two factors of t. Any product of these factors is also a factor of gt^2. This means other factors are gt, t^2, and gt^2 itself. ∎

EXAMPLE 2 Terms and factors

$7a(x^2 + 2y) - 6x(5 + x - 3y)$ is an expression with terms $7(x^2 + 2y)$ and $-6x(5 + x - 3y)$.

The term $7a(x^2 + 2y)$ has individual factors of 7, a, and $(x^2 + 2y)$, as well as products of these factors. The factor $x^2 + 2y$ has two terms, x^2 and $2y$.

The term $-6x(5 + x - 3y)$ has factors 2, 3, x, and $(5 + x - 3y)$. The negative sign in front can be treated as a factor of -1. The factor $5 + x - 3y$ has three terms, 5, x, and $-3y$. ∎

■ In Chapter 11, we will see that roots are equivalent to noninteger exponents.

A **polynomial** *is an algebraic expression with only nonnegative integer exponents on one or more variables, and has no variable in a denominator. The* **degree of a term** *is the sum of the exponents of the variables of that term, and the* **degree of the polynomial** *is the degree of the term of highest degree.*

A **multinomial** *is any algebraic expression of more than one term.* Terms like $1/x$ and $\sqrt{x}$ can be included in a multinomial, but not in a polynomial. (Since $1/x = x^{-1}$, the exponent is negative.

EXAMPLE 3 Polynomials

Some examples of polynomials are as follows:

(a) $4x^2 - 5x + 3$ (degree 2)   **(b)** $2x^6 - x$ (degree 6)   **(c)** $3x$ (degree 1)

(d) $xy^3 + 7x - 3$ (degree 4) (add exponents of x and y)

(e) -6 (degree 0) $\left(-6 = -6x^0\right)$

From (c), note that a single term can be a polynomial, and from (e), note that a constant can be a polynomial. The expressions in (a), (b), and (d) are also multinomials.

The expression $x^2 + \sqrt{y + 2} - 8$ is a multinomial, but not a polynomial because of the square root term. ∎

A *polynomial with one term is called a* **monomial.** *A polynomial with two terms is called a* **binomial,** *and one with three terms is called a* **trinomial.** *The numerical factor is called the* **numerical coefficient** *(or simply* **coefficient***) of the term. All terms that differ at most in their numerical coefficients are known as* **similar** *or* **like** *terms.* That is, similar terms have the same variables with the same exponents.

■ In practice, the numerical coefficient is usually called the coefficient. However, a more general definition is: *any factor of a product is the coefficient of the other factors.*

EXAMPLE 4 Monomial, binomial, trinomial

(a) $7x^4$ is a monomial. The numerical coefficient is 7.

(b) $3ab - 6a$ is a binomial. The numerical coefficient of the first term is 3, and the numerical coefficient of the second term is -6. Note that the sign is attached to the coefficient.

(c) $8cx^3 - x + 2$ is a trinomial. The coefficients of the first two terms are 8 and -1.

(d) $x^2y^2 - 2x + 3y - 9$ is a polynomial with four terms (no special name). ∎

EXAMPLE 5 Similar terms

(a) $8b - 6ab + 81b$ is a trinomial. The first and third terms are similar since they differ only in their numerical coefficients. The middle term is not similar to the others since it has a factor of a.

(b) $4x^2 - 3x$ is a binomial. The terms are not similar since the first term has two factors of x, and the second term has only one factor of x.

(c) $3x^2y^3 - 5y^3x^2 + x^2 - 2y^3$ is a polynomial. The commutative law tells us that $x^2y^3 = y^3x^2$, which means the first two terms are similar. ∎

In adding and subtracting algebraic expressions, we combine similar terms into a single term. The ***simplified*** expression will contain only terms that are not similar.

■ Some calculators can display algebraic expressions and perform algebraic operations.

EXAMPLE 6 Simplifying expressions

(a) $3x + 2x - 5y = 5x - 5y$ Add similar terms—result has unlike terms

(b) $6a^2 - 7a + 8ax$ cannot be simplified since there are no like terms.

(c) $6a + 5c + 2a - c = 6a + 2a + 5c - c$ commutative law

$$= 8a + 4c$$ add like terms ■

To group terms in an algebraic expression, we use **symbols of grouping.** In this text, we use **parentheses, ( ), brackets, [ ],** and **braces, { }.** The **bar,** which is used with radicals and fractions, also groups terms. In earlier sections, we used parentheses and the bar.

When adding and subtracting algebraic expressions, it may be necessary to remove symbols of grouping. To do so, we must *change the sign of **every term*** *within the symbols if the grouping is preceded by a minus sign. If the symbols of grouping are preceded by a plus sign, each term within the symbols retains its original sign.* This is a result of the distributive law, $a(b + c) = ab + ac$.

CAUTION ▶

EXAMPLE 7 Symbols of grouping

(a) $2(a + 2x) = 2a + 2(2x)$ use distributive law

$$= 2a + 4x$$

(b) $-(+a - 3c) = (-1)(+a - 3c)$ treat − sign as −1

$$= (-1)(+a) + (-1)(-3c)$$

$$= -a + 3c$$ note change of signs

Normally, $+a$ would be written simply as a. ■

In Example 7(a), we see that $2(a + 2x) = 2a + 4x$. This is true for any values of a and x, and is therefore an *identity* (see p. 7). In fact, an expression and any proper form to which it may be changed form an identity when they are shown to be equal.

EXAMPLE 8 Simplifying: signs before parentheses

+ sign before parentheses

(a) $3c + (2b - c) = 3c + 2b - c = 2b + 2c$ use distributive law

$2b = +2b$ signs retained

− sign before parentheses

(b) $3c - (2b - c) = 3c - 2b + c = -2b + 4c$ use distributive law

$2b = +2b$ signs changed

(c) $3c - (-2b + c) = 3c + 2b - c = 2b + 2c$ use distributive law

signs changed ■

EXAMPLE 9 Simplifying in an application

In designing a certain machine part, it is necessary to perform the following simplification.

$$16(8 - x) - 2(8x - x^2) - (64 - 16x + x^2) = 128 - 16x - 16x + 2x^2 - 64 + 16x - x^2$$

$$= 64 - 16x + x^2$$ ■

NOTE ▶ At times, we have expressions in which more than one symbol of grouping is to be removed in the simplification. Normally, *when several symbols of grouping are to be removed, it is more convenient to* **remove the innermost symbols first.**

CAUTION ▶ One of the most common errors made is changing the sign of only the first term when removing symbols of grouping preceded by a minus sign. *Remember,* **if the symbols are preceded by a minus sign, we must change the sign of every term.**

EXAMPLE 10 Several symbols of grouping

(a) $3ax - [ax - (5s - 2ax)] = 3ax - [ax - 5s + 2ax]$ ◀┐
$$= 3ax - ax + 5s - 2ax \qquad \text{remove parentheses}$$
$$= 5s \qquad \text{remove brackets}$$

(b) $3a^2b - \{[a - (2a^2b - a)] + 2b\} = 3a^2b - \{[a - 2a^2b + a] + 2b\}$ ◀┐
$$= 3a^2b - \{a - 2a^2b + a + 2b\} \qquad \text{remove parentheses}$$
$$= 3a^2b - a + 2a^2b - a - 2b \qquad \text{remove brackets}$$
$$= 5a^2b - 2a - 2b \qquad \text{remove braces} \qquad ■$$

Calculators use only parentheses for grouping symbols, and we often need to use one set of parentheses within another set. These are called **nested parentheses.** In the next example, note that the innermost parentheses are removed first.

EXAMPLE 11 Nested parentheses

$$2 - (3x - 2(5 - (7 - x))) = 2 - (3x - 2(5 - 7 + x))$$
$$= 2 - (3x - 10 + 14 - 2x)$$
$$= 2 - 3x + 10 - 14 + 2x = -x - 2 \qquad ■$$

EXERCISES 1.7

In Exercises 1–4, make the given changes in the indicated examples of this section, and then solve the resulting problems.

1. In Example 6(a), change $2x$ to $2y$.

2. In Example 8(a), change the sign before $(2b - c)$ from + to −.

3. In Example 10(a), change $[ax - (5s - 2ax)]$ to $[(ax - 5s) - 2ax]$.

4. In Example 10(b), change $\{[a - (2a^2b - a)] + 2b\}$ to $\{a - [2a^2b - (a + 2b)]\}$.

In Exercises 5–54, simplify the given algebraic expressions.

5. $5x + 7x - 4x$

6. $6t - 3t - 4t$

7. $2y - y + 4x$

8. $4C + L - 6C$

9. $2F - 2T - 2 + 3F - T$

10. $x - 2y + 3x - y + z$

11. $a^2b - a^2b^2 - 2a^2b$

12. $xy^2 - 3x^2y^2 + 2xy^2$

13. $s + (3s - 4 - s)$

14. $5 + (3 - 4n + p)$

15. $v - (4 - 5x + 2v)$

16. $2a - (b - a)$

17. $2 - 3 - (4 - 5a)$

18. $\sqrt{A} + (h - 2\sqrt{A}) - 3\sqrt{A}$

19. $(a - 3) + (5 - 6a)$

20. $(4x - y) - (-2x - 4y)$

21. $-(t - 2u) + (3u - t)$

22. $2(x - 2y) + (5x - y)$

23. $3(2r + s) - (-5s - r)$

24. $3(a - b) - 2(a - 2b)$

25. $-7(6 - 3j) - 2(j + 4)$

26. $-(5t + a^2) - 2(3a^2 - 2st)$

27. $-[(6 - n) - (2n - 3)]$

28. $-[(A - B) - (B - A)]$

29. $2[4 - (t^2 - 5)]$

30. $3[-3 - (a - 4)]$

31. $-2[-x - 2a - (a - x)]$

32. $-2[-3(x - 2y) + 4y]$

33. $aZ - [3 - (aZ + 4)]$

34. $9v - [6 - (v - 4) + 4v]$

35. $8c - \{5 - [2 - (3 + 4c)]\}$

36. $7y - \{y - [2y - (x - y)]\}$

37. $5p - (q - 2p) - [3q - (p - q)]$

38. $-(4 - \sqrt{LC}) - [(5\sqrt{LC} - 7) - (6\sqrt{LC} + 2)]$

39. $-2\{(4 - x^2) - [3 + (4 - x^2)]\}$

40. $-\{-[-(x - 2a) - b] - (a - x)\}$

41. $5V^2 - (6 - (2V^2 + 3))$

42. $-2F + 2((2F - 1) - 5)$

43. $-(3t - (7 + 2t - (5t - 6)))$

44. $a^2 - 2(x - 5 - (7 - 2(a^2 - 2x) - 3x))$

45. $-4[4R - 2.5(Z - 2R) - 1.5(2R - Z)]$

46. $3\{2.1e - 1.3[f - 2(e - 5f)]\}$

47. In determining the size of a V belt to be used with an engine, the expression $3D - (D - d)$ is used. Simplify this expression.

48. When finding the current in a transistor circuit, the expression $i_1 - (2 - 3i_2) + i_2$ is used. Simplify this expression. (The numbers below the i's are *subscripts*. Different subscripts denote different variables.)

49. Research on a plastic building material leads to
$$\left[\left(B + \tfrac{4}{3}\alpha\right) + 2\left(B - \tfrac{2}{3}\alpha\right)\right] - \left[\left(B + \tfrac{4}{3}\alpha\right) - \left(B - \tfrac{2}{3}\alpha\right)\right].$$
Simplify this expression.

50. One car goes 30 km/h for $t - 1$ hours, and a second car goes 40 km/h for $t + 2$ hours. Find the expression for the sum of the distances traveled by the two cars.

51. A shipment contains x film cartridges for 15 exposures each and $x + 10$ cartridges for 25 exposures each. What is the total number of photographs that can be taken with the film from this shipment?

52. Each of two stores has $2n + 1$ mouse pads costing \$3 each and $n - 2$ mouse pads costing \$2 each. How much more is the total value of the \$3 mouse pads than the \$2 mouse pads in the two stores?

53. For the expressions $2x^2 - y + 2a$ and $3y - x^2 - b$ find (a) the sum, and (b) the difference if the second is subtracted from the first.

54. For the following expressions, subtract the third from the sum of the first two: $3a^2 + b - c^3, 2c^3 - 2b - a^2, 4c^3 - 4b + 3$.

In Exercises 55 and 56, answer the given questions.

(W)**55.** For any real numbers a and b, is it true that $|a - b| = |b - a|$? Explain.

(W)**56.** Is subtraction associative? That is, in general, does $(a - b) - c$ equal $a - (b - c)$? Explain.

Answers to Practice Exercises

1. $6a + 3y$ **2.** $6r - 3s$ **3.** $5x - 12y$

1.8 Multiplication of Algebraic Expressions

Multiplying Monomials • Products of Monomials and Polynomials • Powers of Polynomials

To find the product of two or more monomials, we use the laws of exponents as given in Section 1.4 and the laws for multiplying signed numbers as stated in Section 1.2. We first multiply the numerical coefficients to find the numerical coefficient of the product. Then we multiply the literal numbers, remembering that *the exponents may be combined only if the base is the same.*

EXAMPLE 1 Multiplying monomials

(a) $3c^5(-4c^2) = -12c^7$ multiply numerical coefficients and add exponents of c

(b) $(-2b^2y^3)(-9aby^5) = 18ab^3y^8$ add exponents of same base

(c) $2xy(-6cx^2)(3xcy^2) = -36c^2x^4y^3$ ∎

If a product contains a monomial that is raised to a power, *we must first raise it to the indicated power* before proceeding with the multiplication.

EXAMPLE 2 Product containing power of a monomial

(a) $3(2a^2x)^3(-ax) = 3(8a^6x^3)(-ax) = -24a^7x^4$

(b) $2s^3(-st^4)^2(4s^2t) = 2s^3(s^2t^8)(4s^2t) = 8s^7t^9$ ∎

We find the product of a monomial and a polynomial by using the distributive law, which states that we *multiply each term of the polynomial by the monomial.* In doing so, we must be careful to give the correct sign to each term of the product.

EXAMPLE 3 Product of monomial and polynomial

(a) $2ax(3ax^2 - 4yz) = 2ax(3ax^2) + (2ax)(-4yz) = 6a^2x^3 - 8axyz$

(b) $5cy^2(-7cx - ac) = (5cy^2)(-7cx) + (5cy^2)(-ac) = -35c^2xy^2 - 5ac^2y^2$ ∎

It is generally not necessary to write out the middle step as it appears in the preceding example. We write the answer directly. For instance, Example 3(a) would appear as $2ax(3ax^2 - 4yz) = 6a^2x^3 - 8axyz$.

Practice Exercises

Perform the indicated multiplications.

1. $2a^3b(-6ab^2)$ **2.** $-5x^2y^3(2xy - y^4)$

We find the product of two polynomials by using the distributive law. The result is that *we multiply each term of one polynomial by each term of the other and add the results.* Again we must be careful to give each term of the product its correct sign.

EXAMPLE 4 Product of polynomials

■ Note that, using the distributive law,
$(x - 2)(x + 3) = (x - 2)(x) + (x - 2)(3)$
leads to the same result.

$$(x - 2)(x + 3) = x(x) + x(3) + (-2)(x) + (-2)(3)$$

$$= x^2 + 3x - 2x - 6 = x^2 + x - 6$$ ■

Finding the power of a polynomial is equivalent to using the polynomial as a factor the number of times indicated by the exponent. It is sometimes convenient to write the power of a polynomial in this form before multiplying.

EXAMPLE 5 Power of a polynomial

two factors

(a) $(x + 5)^2 = (x + 5)(x + 5) = x^2 + 5x + 5x + 25$

$$= x^2 + 10x + 25$$

Practice Exercises

Perform the indicated multiplications.

3. $(2s - 5t)(s + 4t)$ **4.** $(3u + 2v)^2$

(b) $(2a - b)^3 = (2a - b)(2a - b)(2a - b)$ the exponent 3 indicates three factors

$$= (2a - b)(4a^2 - 2ab - 2ab + b^2)$$

$$= (2a - b)(4a^2 - 4ab + b^2)$$

$$= 8a^3 - 8a^2b + 2ab^2 - 4a^2b + 4ab^2 - b^3$$

$$= 8a^3 - 12a^2b + 6ab^2 - b^3$$

We should note in illustration (a) that

CAUTION ▶ $(x + 5)^2$ *is __not equal to__* $x^2 + 25$

since the term $10x$ is not included. We must follow the proper procedure and not simply square each of the terms within the parentheses. ■

EXAMPLE 6 Products in an application

An expression used with a lens of a certain telescope is simplified as shown.

$$a(a + b)^2 + a^3 - (a + b)(2a^2 - s^2)$$

$$= a(a + b)(a + b) + a^3 - (2a^3 - as^2 + 2a^2b - bs^2)$$

$$= a(a^2 + ab + ab + b^2) + a^3 - 2a^3 + as^2 - 2a^2b + bs^2$$

$$= a^3 + a^2b + a^2b + ab^2 - a^3 + as^2 - 2a^2b + bs^2$$

$$= ab^2 + as^2 + bs^2$$ ■

EXERCISES 1.8

In Exercises 1–4, make the given changes in the indicated examples of this section, and then solve the resulting problems.

1. In Example 2(b), change the factor $(-st^4)^2$ to $(-st^4)^3$.

2. In Example 3(a), change the factor $2ax$ to $-2ax$.

3. In Example 4, change the factor $(x + 3)$ to $(x - 3)$.

4. In Example 5(b), change the exponent 3 to 2.

In Exercises 5–64, perform the indicated multiplications.

5. $(a^2)(ax)$ **6.** $(2xy)(x^2y^3)$ **7.** $-ac^2(acx^3)$

8. $-2cs^2(-4cs)^2$ **9.** $(2ax^2)^2(-2ax)$ **10.** $6pq^3(3pq^2)^2$

11. $i^2(R + 2r)$ **12.** $2x(p - q)$ **13.** $-3s(s^2 - 5t)$

14. $-3b(2b^2 - b)$ **15.** $5m(m^2n + 3mn)$ **16.** $a^2bc(2ac - 3b^2c)$

17. $3M(-M - N + 2)$ **18.** $-4c^2(-9gc - 2c + g^2)$

19. $ax(cx^2)(x + y^3)$ **20.** $-2(-3st^3)(3s - 4t)$

21. $(x - 3)(x + 5)$ **22.** $(a + 7)(a + 1)$

23. $(x + 5)(2x - 1)$ **24.** $(4t_1 + t_2)(2t_1 - 3t_2)$

25. $(2a - b)(3a - 2b)$ **26.** $(4w^2 - 3)(3w^2 - 1)$

27. $(2s + 7t)(3s - 5t)$ **28.** $(5p - 2q)(p + 8q)$

29. $(x^2 - 1)(2x + 5)$ **30.** $(3y^2 + 2)(2y - 9)$

31. $(x - 2y - 4)(x - 2y + 4)$

32. $(2a + 3b + 1)(2a + 3b - 1)$

33. $2(a + 1)(a - 9)$ **34.** $-5(y - 3)(y + 6)$

35. $-3(3 - 2T)(3T + 2)$ **36.** $2n(5 - n)(6n + 5)$

37. $2L(L + 1)(4 - L)$ **38.** $ax(x + 4)(7 - x^2)$

39. $(2x - 5)^2$ **40.** $(x - 3y)^2$ **41.** $(x_1 + 3x_2)^2$

42. $(7m + 1)^2$ **43.** $(xyz - 2)^2$ **44.** $(b - 6x^2)^2$

45. $2(x + 8)^2$ **46.** $3(3R + 4)^2$

47. $(2 + x)(3 - x)(x - 1)$ **48.** $(3x - c^2)^3$

49. $3T(T + 2)(2T - 1)$ **50.** $[(x - 2)^2(x + 2)]^2$

51. Let $x = 3$ and $y = 4$ to show that (a) $(x + y)^2 \neq x^2 + y^2$ and (b) $(x - y)^2 \neq x^2 - y^2$. ($\neq$ means "does not equal")

52. Evaluate the product $(98)(102)$ by expressing it as $(100 - 2)(100 + 2)$.

(W) **53.** Square an integer between 1 and 9 and subtract 1 from the result. Explain why the result is the product of the integer before and the integer after the one you chose.

(W) **54.** Explain how, by appropriate grouping, the product $(x - 2)(x + 3)(x + 2)(x - 3)$ is easier to find. Find the product.

55. By multiplication, show that $(x + y)^3$ is not equal to $x^3 + y^3$.

56. By multiplication, show that $(x + y)(x^2 - xy + y^2) = x^3 + y^3$.

57. In finding the value of a certain savings account, the expression $P(1 + 0.01r)^2$ is used. Multiply out this expression.

58. The force due to the liquid in a parabolic container leads to the expression $w(1 - x)(4 - x^2)$. Multiply out this expression.

59. In using aircraft radar, the expression $(2R - X)^2 - (R^2 + X^2)$ arises. Simplify this expression.

60. In calculating the temperature variation of an industrial area, the expression $(2T^3 + 3)(T^2 - T - 3)$ arises. Perform the indicated multiplication.

61. In a particular computer design containing n circuit elements, n^2 switches are needed. Find the expression for the number of switches needed for $n + 100$ circuit elements.

62. Simplify the expression $(T^2 - 100)(T - 10)(T + 10)$, which arises when analyzing the energy radiation from an object.

63. In finding the maximum power in part of a microwave transmitter circuit, the expression $(R_1 + R_2)^2 - 2R_2(R_1 + R_2)$ is used. Multiply and simplify.

64. In determining the deflection of a certain steel beam, the expression $27x^2 - 24(x - 6)^2 - (x - 12)^3$ is used. Multiply and simplify.

Answers to Practice Exercises

1. $-12a^4b^3$ **2.** $-10x^3y^4 + 5x^2y^7$

3. $2s^2 + 3st - 20t^2$ **4.** $9u^2 + 12uv + 4v^2$

1.9 Division of Algebraic Expressions

Dividing Monomials • Dividing by a Monomial • Dividing One Polynomial by Another

To find the quotient of one monomial divided by another, we use the laws of exponents and the laws for dividing signed numbers. Again, the *exponents may be combined only if the base is the same.*

EXAMPLE 1 Dividing monomials

(a) $\dfrac{3c^7}{c^2} = 3c^{7-2} = 3c^5$

(b) $\dfrac{16x^3y^5}{4xy^2} = \dfrac{16}{4}(x^{3-1})(y^{5-2}) = 4x^2y^3$

                                                               divide   subtract

(c) $\dfrac{-6a^2xy^2}{2axy^4} = -\left(\dfrac{6}{2}\right)\dfrac{a^{2-1}x^{1-1}}{y^{4-2}} = -\dfrac{3a}{y^2}$

                                                               coefficients exponents

As shown in illustration (c), we use only positive exponents in the final result unless there are specific instructions otherwise. ∎

From arithmetic, we may show how a multinomial is to be divided by a monomial. When adding fractions $\left(\text{say } \tfrac{2}{7} \text{ and } \tfrac{3}{7}\right)$, we have $\dfrac{2}{7} + \dfrac{3}{7} = \dfrac{2 + 3}{7}$.

Looking at this from *right to left*, we see that *the quotient of a multinomial divided by a monomial is found by dividing each term of the multinomial by the monomial and adding the results.* This can be shown as

■ This is an identity and is valid for all values of a and b, and all values of c except zero (which would make it undefined).

$$\frac{a + b}{c} = \frac{a}{c} + \frac{b}{c}$$

CAUTION ▶   Be careful: Although $\dfrac{a + b}{c} = \dfrac{a}{c} + \dfrac{b}{c}$, we must note that $\dfrac{c}{a + b}$ *is not* $\dfrac{c}{a} + \dfrac{c}{b}$.

EXAMPLE 2 Dividing by a monomial

(a) $\dfrac{4a^2 + 8a}{2a} = \dfrac{4a^2}{2a} + \dfrac{8a}{2a} = 2a + 4$

each term of
numerator divided
by denominator

(b) $\dfrac{4x^3y - 8x^3y^2 + 2x^2y}{2x^2y} = \dfrac{4x^3y}{2x^2y} - \dfrac{8x^3y^2}{2x^2y} + \dfrac{2x^2y}{2x^2y}$

$$= 2x - 4xy + 1$$

Practice Exercise

Divide: **1.** $\dfrac{4ax^2 - 6a^2x}{2ax}$

■ Until you are familiar with the method, it is recommended that you do write out the middle steps.

We usually do not write out the middle step as shown in these illustrations. The divisions of the terms of the numerator by the denominator are usually done by inspection (mentally), and the result is shown as it appears in the next example. ■

EXAMPLE 3 Application of dividing by a monomial

The expression $\dfrac{2p + v^2d + 2ydg}{2dg}$ is used when analyzing the operation of an irrigation pump. Performing the indicated division, we have

$$\frac{2p + v^2d + 2ydg}{2dg} = \frac{p}{dg} + \frac{v^2}{2g} + y$$ ■

DIVISION OF ONE POLYNOMIAL BY ANOTHER

To divide one polynomial by another, use the following steps. (1) Arrange the dividend (the polynomial to be divided) and the divisor in descending powers of the variable. (2) Divide the first term of the dividend by the first term of the divisor. The result is the first term of the quotient. (3) Multiply the entire divisor by the first term of the quotient and *subtract* the product from the dividend. (4) Divide the first term of this difference by the first term of the divisor. This gives the second term of the quotient. (5) Multiply this term by the entire divisor and *subtract* the product from the first difference. (6) Repeat this process until the remainder is zero or a term of lower degree than the divisor.

■ This is similar to long division of numbers.

EXAMPLE 4 Dividing one polynomial by another

Perform the division $(6x^2 + x - 2) \div (2x - 1)$.
$\left(\text{This division can also be indicated in the fractional form } \dfrac{6x^2 + x - 2}{2x - 1}. \right)$

We set up the division as we would for long division in arithmetic. Then, following the procedure outlined above, we have the following:

$$
\begin{array}{r}
3x + 2 \\
2x - 1 \overline{\smash{\big)}\, 6x^2 + x - 2} \\
\end{array}
$$

$\dfrac{6x^2}{2x}$ divide first term of dividend by first term of divisor

CAUTION ▶

subtract

$6x^2 - 6x^2 = 0$

$x - (-3x) = 4x$

$6x^2 - 3x$ ◀—— $3x(2x - 1)$

$4x - 2$

$4x - 2$ subtract

0

Practice Exercise

Divide: **2.** $(6x^2 + 7x - 3) \div (3x - 1)$

The remainder is zero and the quotient is $3x + 2$. Note that when we subtracted $-3x$ from x, we obtained $4x$. ■

EXAMPLE 5 Quotient with a remainder

Perform the division $\dfrac{8x^3 + 4x^2 + 3}{4x^2 - 1}$. Since there is no x-term in the dividend, we should leave space for any x-terms that might arise (which we will show as $0x$).

$$
\begin{array}{r}
2x + 1 \longleftarrow \\
4x^2 - 1 \overline{)\,8x^3 + 4x^2 + 0x + 3} \\
\end{array}
$$

divisor $\quad 4x^2 - 1 \overline{)\,8x^3 + 4x^2 + 0x + 3}$ $\quad$ dividend $\qquad \dfrac{8x^3}{4x^2} = 2x$

$\qquad\qquad\qquad \underline{8x^3 \qquad\quad - 2x}$ $\qquad$ subtract

$\qquad\qquad\qquad\qquad\quad 4x^2 + 2x + 3$ $\qquad\qquad\qquad \dfrac{4x^2}{4x^2} = 1$

$0x - (-2x) = 2x$

$\qquad\qquad\qquad\qquad\quad \underline{4x^2 \qquad\quad - 1}$ $\qquad$ subtract

$\qquad\qquad\qquad\qquad\qquad\qquad 2x + 4$ $\qquad$ remainder

Since the degree of the remainder $2x + 4$ is less than that of the divisor, we now show the quotient in this case as $2x + 1 + \dfrac{2x + 4}{4x^2 - 1}$. ∎

EXERCISES 1.9

In Exercises 1–4, make the given changes in the indicated examples of this section and then perform the indicated divisions.

1. In Example 1(c), change the denominator to $-2a^2xy^5$.

2. In Example 2(b), change the denominator to $2xy^2$.

3. In Example 4, change the dividend to $6x^2 - 7x + 2$.

4. In Example 5, change the sign of the middle term of the numerator from $+$ to $-$.

In Exercises 5–24, perform the indicated divisions.

5. $\dfrac{8x^3y^2}{-2xy}$ **6.** $\dfrac{-18b^7c^3}{bc^2}$ **7.** $\dfrac{-16r^3t^5}{-4r^5t}$ **8.** $\dfrac{51mn^5}{17m^2n^2}$

9. $\dfrac{(15x^2)(4bx)(2y)}{30bxy}$ **10.** $\dfrac{(5sT)(8s^2T^3)}{10s^3T^2}$ **11.** $\dfrac{6(ax)^2}{-ax^2}$

12. $\dfrac{12a^2b}{(3ab^2)^2}$ **13.** $\dfrac{3a^2x + 6xy}{3x}$ **14.** $\dfrac{2m^2n - 6mn}{2m}$

15. $\dfrac{3rst - 6r^2st^2}{3rs}$ **16.** $\dfrac{-5a^2n - 10an^2}{5an}$

17. $\dfrac{4pq^3 + 8p^2q^2 - 16pq^5}{4pq^2}$ **18.** $\dfrac{a^2x_1x_2^2 + ax_1^3 - ax_1}{ax_1}$

19. $\dfrac{2\pi fL - \pi fR^2}{\pi fR}$ **20.** $\dfrac{9(aB)^4 - 6aB^4}{3aB^3}$

21. $\dfrac{3ab^2 - 6ab^3 + 9a^2b^2}{9a^2b^2}$ **22.** $\dfrac{2x^{n+2} + 4ax^n}{2x^n}$

23. $\dfrac{6y^{2n} - 4ay^{n+1}}{2y^n}$ **24.** $\dfrac{3a(F + T)b^2 - (F + T)}{a(F + T)}$

In Exercises 25–42, perform the indicated divisions. Express the answer as shown in Example 5 when applicable.

25. $(2x^2 + 7x + 3) \div (x + 3)$ **26.** $(3t^2 - 7t + 4) \div (t - 1)$

27. $(x^2 - 3x + 2) \div (x - 2)$ **28.** $(2x^2 - 5x - 7) \div (x + 1)$

29. $(x - 14x^2 + 8x^3) \div (2x - 3)$

30. $(6 + 7y + 6y^2) \div (2y + 1)$

31. $(4Z^2 - 5Z - 7) \div (4Z + 3)$

32. $(6x^2 - 5x - 9) \div (3x - 4)$

33. $\dfrac{x^3 + 3x^2 - 4x - 12}{x + 2}$ **34.** $\dfrac{3x^3 + 19x^2 + 13x - 20}{3x - 2}$

35. $\dfrac{2a^4 + 4a^2 - 16}{a^2 - 2}$ **36.** $\dfrac{6T^3 + T^2 + 2}{3T^2 - T + 2}$

37. $\dfrac{x^3 + 8}{x + 2}$ **38.** $\dfrac{D^3 - 1}{D - 1}$

39. $\dfrac{x^2 - 2xy + y^2}{x - y}$ **40.** $\dfrac{3r^2 - 5rR + 2R^2}{r - 3R}$

41. $\dfrac{x^2 - y^2 + 2yz - z^2}{x + y - z}$ **42.** $\dfrac{a^4 + b^4}{a^2 - 2ab + 2b^2}$

In Exercises 43–52, perform the indicated divisions.

43. When $2x^2 - 9x - 5$ is divided by $x + c$, the quotient is $2x + 1$. Find c.

44. When $6x^2 - x + k$ is divided by $3x + 4$, the remainder is zero. Find k.

45. By division show that $\dfrac{x^4 + 1}{x + 1}$ is not equal to x^3.

46. By division show that $\dfrac{x^3 + y^3}{x + y}$ is not equal to $x^2 + y^2$.

47. In the optical theory dealing with lasers, the following expression arises: $\dfrac{8A^5 + 4A^3\mu^2E^2 - A\mu^4E^4}{8A^4}$. ($\mu$ is the Greek letter mu.)

48. In finding the total resistance of the resistors shown in Fig. 1.13, the following expression is used.

$$\dfrac{6R_1 + 6R_2 + R_1R_2}{6R_1R_2}$$

Fig. 1.13

49. When analyzing the potential energy associated with gravitational forces, the expression $\dfrac{GMm[(R + r) - (R - r)]}{2rR}$ arises. Perform the indicated division.

50. A computer model shows that the temperature change T in a certain freezing unit is found by using the expression $\dfrac{3T^3 - 8T^2 + 8}{T - 2}$. Perform the indicated division.

51. In analyzing the displacement of a certain valve, the expression $\dfrac{s^2 - 2s - 2}{s^4 + 4}$ is used. Find the reciprocal of this expression and then perform the indicated division.

52. The voltage and resistance in a certain electric circuit vary with time such that the current is given by the expression $\dfrac{2t^3 + 94t^2 - 290t + 500}{2t + 100}$. By performing the indicated division, find the expression for the current.

Answers to Practice Exercises

1. $2x - 3a$ **2.** $2x + 3$

1.10 Solving Equations

Types of Equations • Solving Basic Types of Equations • Checking the Solution • First Steps • Ratio and Proportion

In this section, we show how algebraic operations are used in solving equations. In the following sections, we show some of the important applications of equations.

An **equation** *is an algebraic statement that two algebraic expressions are equal.* Any value of the literal number representing the **unknown** that produces equality when **substituted** in the equation is said to **satisfy** the equation.

EXAMPLE 1 Valid values for equations

The equation $3x - 5 = x + 1$ is true only if $x = 3$. Substituting 3 for x in the equation, we have $3(3) - 5 = 3 + 1$, or $4 = 4$; substituting $x = 2$, we have $1 = 3$, which is not correct.

This equation is valid for only one value of the unknown. *An equation valid only for certain values of the unknown is a* **conditional equation.** In this section, nearly all equations we solve will be conditional equations that are satisfied by only one value of the unknown. ∎

EXAMPLE 2 Identity and contradiction

(a) The equation $x^2 - 4 = (x - 2)(x + 2)$ is true for all values of x. For example, substituting $x = 3$ in the equation, we have $3^2 - 4 = (3 - 2)(3 + 2)$, or $5 = 5$. Substituting $x = -1$, we have $(-1)^2 - 4 = (-1 - 2)(-1 + 2)$, or $-3 = -3$. *An equation valid for all values of the unknown is an* **identity.**

■ We have previously noted *identities* on pages 7 and 28.

(b) The equation $x + 5 = x + 1$ is not true for any value of x. For any value of x we try, we find that the left side is 4 greater than the right side. *Such an equation is called a* **contradiction.** ∎

To **solve** *an equation, we find the values of the unknown that satisfy it.* There is one basic rule to follow when solving an equation:

■ Equations can be solved on most graphing calculators. An estimate (or guess) of the answer may be required to find the solution. See Exercises 43 and 44.

Perform the same operation on both sides of the equation.

We do this to isolate the unknown and thus to find its value.

By performing the same operation on both sides of an equation, the two sides remain equal. Thus,

■ Although we may multiply both sides of an equation by zero, this produces $0 = 0$, which is not useful in finding the solution.

we may add the same number to both sides, subtract the same number from both sides, multiply both sides by the same number, or divide both sides by the same number (not zero).

EXAMPLE 3 Basic operations used in solving

In each of the following equations, we may isolate x, and thereby solve the equation, by performing the indicated operation.

$x - 3 = 12$	$x + 3 = 12$	$\dfrac{x}{3} = 12$	$3x = 12$
add 3 to both sides	subtract 3 from both sides	multiply both sides by 3	divide both sides by 3
$x - 3 + 3 = 12 + 3$	$x + 3 - 3 = 12 - 3$	$3\left(\dfrac{x}{3}\right) = 3(12)$	$\dfrac{3x}{3} = \dfrac{12}{3}$
$x = 15$	$x = 9$	$x = 36$	$x = 4$

NOTE ▶ Each solution should be checked by substitution in the original equation. ■

■ The word *algebra* comes from Arabic and means "a restoration." It refers to the fact that when a number has been added to one side of an equation, the same number must be added to the other side to maintain equality.

EXAMPLE 4 Operations used for solution; checking

Solve the equation $2t - 7 = 9$.

We are to perform basic operations to both sides of the equation to finally isolate t on one side. The steps to be followed are suggested by the form of the equation.

$$2t - 7 = 9 \qquad \text{original equation}$$
$$2t - 7 + 7 = 9 + 7 \qquad \text{add 7 to both sides}$$
$$2t = 16 \qquad \text{combine like terms}$$
$$\frac{2t}{2} = \frac{16}{2} \qquad \text{divide both sides by 2}$$
$$t = 8 \qquad \text{simplify}$$

■ Note that the solution generally requires a combination of basic operations.

NOTE ▶ Therefore, we conclude that $t = 8$. Checking *in the original equation*, we have

$$2(8) - 7 \stackrel{?}{=} 9, \qquad 16 - 7 \stackrel{?}{=} 9, \qquad 9 = 9$$

The solution checks. ■

EXAMPLE 5 First remove parentheses

■ With simpler numbers, many basic steps are done by inspection and not actually written down.

Solve the equation $x - 7 = 3x - (6x - 8)$.

$x - 7 = 3x - 6x + 8$	parentheses removed
$x - 7 = -3x + 8$	*x*-terms combined on right
$4x - 7 = 8$	3x added to both sides —by inspection
$4x = 15$	7 added to both sides —by inspection
$x = \frac{15}{4}$	both sides divided by 4 —by inspection

Checking in the original equation, we obtain (after simplifying) $-\frac{13}{4} = -\frac{13}{4}$. ■

Note that we **always check in the original** equation. This is done since errors may have been made in finding the later equations.

From these examples, we see that the following steps are used in solving the basic equations of this section.

Practice Exercises

Solve for x.

1. $3x + 4 = x - 6$
2. $2(5 - x) = x - 8$

■ Many other types of equations require more advanced methods for solving. These are considered in later chapters.

PROCEDURE FOR SOLVING EQUATIONS

1. *Remove grouping symbols (distributive law).*
2. *Combine any like terms on each side (also after step 3).*
3. *Perform the same operations on both sides until $x = $ result is obtained.*
4. *Check the solution in the original equation.*

If an equation contains numbers not easily combined by inspection, the best procedure is to *first solve for the unknown and then perform the calculation.*

NOTE ▶

EXAMPLE 6 First solve for unknown – then calculate

When finding the current i (in A) in a certain radio circuit, the following equation and solution are used.

$$0.0595 - 0.525i - 8.85(i + 0.00316) = 0$$
$$0.0595 - 0.525i - 8.85i - 8.85(0.00316) = 0 \qquad \text{note how the above}$$
$$(-0.525 - 8.85)i = 8.85(0.00316) - 0.0595 \qquad \text{procedure is followed}$$
$$i = \frac{8.85(0.00316) - 0.0595}{-0.525 - 8.85} \qquad \text{evaluate}$$
$$= 0.00336 \text{ A}$$

The calculator solution of this equation, using the *Solver* feature, is shown in Fig. 1.14. When doing the calculation indicated above in the solution for i, be careful to group the numbers in the denominator for the division. Also, be sure to round off the result as shown above, but do not round off values before the final calculation. ■

Fig. 1.14

Ratio and Proportion

The quotient a/b is also called the **ratio** *of a to b. An equation stating that two ratios are equal is called a* **proportion.** Since a proportion is an equation, if one of the numbers is unknown, we can solve for its value as with any equation. Usually, this is done by noting the denominators and multiplying each side by a number that will clear the fractions.

EXAMPLE 7 Ratio

If the ratio of x to 8 equals the ratio of 3 to 4, we have the proportion

$$\frac{x}{8} = \frac{3}{4}$$

We can solve this equation by multiplying both sides by 8. This gives

$$8\left(\frac{x}{8}\right) = 8\left(\frac{3}{4}\right), \quad \text{or} \quad x = 6$$

Substituting $x = 6$ into the original proportion gives the proportion $\frac{6}{8} = \frac{3}{4}$. Since these ratios are equal, the solution checks. ■

Practice Exercise

3. If the ratio of 2 to 5 equals the ratio of x to 30, find x.

EXAMPLE 8 Applied proportion

The ratio of electric current I (in A) to the voltage V across a resistor is constant. If $I = 1.52$ A for $V = 60.0$ V, find I for $V = 82.0$ V.

Since the ratio of I and V is constant, we have the following solution.

$$\frac{1.52}{60.0} = \frac{I}{82.0}$$
$$82.0\left(\frac{1.52}{60.0}\right) = 82.0\left(\frac{I}{82.0}\right)$$
$$2.08 \text{ A} = I$$
$$I = 2.08 \text{ A}$$

■ Generally, units of measurement will not be shown in intermediate steps. The proper units will be shown with the data and final result.

Checking in the original equation, we have $0.0253 \text{ A/V} = 0.0254 \text{ A/V}$. Since the value of I is rounded up, the solution checks. ■

We will use the terms *ratio* and *proportion* (particularly ratio) when studying trigonometry in Chapter 4. A detailed discussion of ratio and proportion is found in Chapter 18. A general method of solving equations involving fractions is given in Chapter 6.

EXERCISES 1.10

In Exercises 1–4, make the given changes in the indicated examples of this section and then solve the resulting problems.

1. In Example 3, change 12 to -12 in each of the four illustrations and then solve.

2. In Example 4, change $2t - 7$ to $7 - 2t$ and then solve.

3. In Example 5, change $(6x - 8)$ to $(8 - 6x)$ and then solve.

4. In Example 8, for the same given values of I and V, find I for $V = 48.0$ V.

In Exercises 5–40, solve the given equations.

5. $x - 2 = 7$
6. $x - 4 = -1$
7. $x + 5 = 4$

8. $s + 6 = -3$
9. $\dfrac{t}{2} = -5$
10. $\dfrac{x}{-4} = 2$

11. $4E = -20$
12. $2x = 12$
13. $3t + 5 = -4$

14. $5D - 2 = 13$
15. $5 - 2y = -3$
16. $8 - 5t = 18$

17. $3x + 7 = x$
18. $6 + 4L = 5 - 3L$

19. $2(s - 4) = s$
20. $3(4 - n) = -n$

21. $6 - (r - 4) = 2r$
22. $5 - (x + 2) = 5x$

23. $2(x - 3) = -x$
24. $4(7 - F) = -7$

25. $0.1x - 0.5(x - 2) = 2$
26. $1.5x - 0.3(x - 4) = 6$

27. $7 - 3(1 - 2p) = 4 + 2p$
28. $3 - 6(2 - 3t) = t - 5$

29. $\dfrac{4x - 2(x - 4)}{3} = 8$
30. $2x = \dfrac{3 - 5(7 - 3x)}{4}$

31. $|x| - 1 = 8$
32. $2 - |x| = 4$

In Exercises 33–40, all numbers are approximate.

33. $5.8 - 0.3(x - 6.0) = 0.5x$
34. $1.9t = 0.5(4.0 - t) - 0.8$

35. $-0.24(C - 0.50) = 0.63$
36. $27.5(5.17 - 1.44x) = 73.4$

37. $\dfrac{x}{2.0} = \dfrac{17}{6.0}$
38. $\dfrac{3.0}{7.0} = \dfrac{R}{42}$

39. $\dfrac{165}{223} = \dfrac{13V}{15}$
40. $\dfrac{276x}{17.0} = \dfrac{1360}{46.4}$

In Exercises 41–52, solve the given problems.

41. Identify each of the following equations as a conditional equation, an identity, or a contradiction.
    (a) $2x + 3 = 3 + 2x$ (b) $2x - 3 = 3 - 2x$

42. For what values of a is the equation $2x + a = 2x$ a conditional equation? Explain.

43. Solve the equation of Example 5 by using the Equation Solver of a graphing calculator.

44. Solve the equation of Example 6 by using the Equation Solver of a graphing calculator.

45. In finding the speed v (in mi/h) at which a certain person was traveling, the equation $2.0v + 40 = 2.5(v + 5.0)$ is to be solved. Find the speed.

46. In finding the rate v (in km/h) at which a polluted stream is flowing, the equation $15(5.5 + v) = 24(5.5 - v)$ is used. Find v.

47. In finding the maximum operating temperature T (in °C) for a computer integrated circuit, the equation $1.1 = (T - 76)/40$ is used. Find the temperature.

48. To find the voltage V in a circuit in a TV remote-control unit, the equation $1.12V - 0.67(10.5 - V) = 0$ is used. Find V.

49. In blending two gasolines of different octanes, in order to find the number n of gallons of one octane needed, the equation $0.14n + 0.06(2000 - n) = 0.09(2000)$ is used. Find n, given that 0.06 and 0.09 are exact and the first zero of 2000 is significant.

50. In order to find the distance x such that the weights are balanced on the lever shown in Fig. 1.15, the equation $210(3x) = 55.3x + 38.5(8.25 - 3x)$ must be solved. Find x. (3 is exact.)

Fig. 1.15

51. The manufacturer of a certain car powered partly by gas and partly by batteries (a hybrid car) claims it can go 1250 mi on one full tank of 15 gal. How far can the car go on 5.5 gal?

52. A person 1.8 m tall is photographed with a 35-mm camera, and the film image is 20 mm. Under the same conditions, how tall is a person whose film image is 16 mm?

Answers to Practice Exercises

1. -5 2. 6 3. 12

1.11 Formulas and Literal Equations

Formulas • Literal Equations • Subscripts • Solve for Symbol before Substituting Numerical Values

■ Einstein published his first paper on relativity in 1905.

An important application of equations is in the use of *formulas* that are found in geometry and nearly all fields of science and technology. *A* **formula** *is an equation that expresses the relationship between two or more related quantities.* For example, Einstein's famous formula $E = mc^2$ shows the equivalence of energy E to the mass m of an object and the speed of light c.

NOTE ▶ We can solve a formula for a particular symbol just as we solve any equation. That is, we *isolate the required symbol by using algebraic operations on literal numbers.*

EXAMPLE 1 Solving for symbol in formula

In Einstein's formula $E = mc^2$, solve for m.

$$\frac{E}{c^2} = m \qquad \text{divide both sides by } c^2$$

$$m = \frac{E}{c^2} \qquad \text{switch sides to place } m \text{ at left}$$

The required symbol is usually placed on the left, as shown. ■

EXAMPLE 2 Symbol with subscript in formula

A formula relating acceleration a, velocity v, initial velocity v_0, and time is $v = v_0 + at$. Solve for t.

$$v - v_0 = at \qquad v_0 \text{ subtracted from both sides}$$

$$t = \frac{v - v_0}{a} \qquad \text{both sides divided by } a \text{ and then sides switched} \quad ■$$

■ The subscript $_0$ makes v_0 a different literal symbol from v. (We have used subscripts in a few of the earlier exercises.)

EXAMPLE 3 Symbol in capital and in lower case

In the study of the forces on a certain beam, the equation $W = \dfrac{L(wL + 2P)}{8}$ is used. Solve for P.

$$8W = \frac{8L(wL + 2P)}{8} \qquad \text{multiply both sides by 8}$$

$$8W = L(wL + 2P) \qquad \text{simplify right side}$$

$$8W = wL^2 + 2LP \qquad \text{remove parentheses}$$

$$8W - wL^2 = 2LP \qquad \text{subtract } wL^2 \text{ from both sides}$$

$$P = \frac{8W - wL^2}{2L} \qquad \text{divide both sides by } 2L \text{ and switch sides} \quad ■$$

■ Be careful! Just as subscripts can indicate different literal numbers, a capital letter and the same letter in lowercase are different literal numbers. In this example, W and w are different. This is shown in several of the exercises for this section.

EXAMPLE 4 Formula with groupings

The effect of temperature upon measurements is important when measurements must be made with great accuracy. The volume V of a special precision container at temperature T in terms of the volume V_0 at temperature T_0 is given by $V = V_0[1 + b(T - T_0)]$, where b depends on the material of which the container is made. Solve for T.

Since we are to solve for T, we must isolate the term containing T. This can be done by first removing the grouping symbols.

$$V = V_0[1 + b(T - T_0)] \qquad \text{original equation}$$

$$V = V_0[1 + bT - bT_0] \qquad \text{remove parentheses}$$

$$V = V_0 + bTV_0 - bT_0V_0 \qquad \text{remove brackets}$$

$$V - V_0 + bT_0V_0 = bTV_0 \qquad \text{subtract } V_0 \text{ and add } bT_0V_0 \text{ to both sides}$$

$$T = \frac{V - V_0 + bT_0V_0}{bV_0} \qquad \text{divide both sides by } bV_0 \text{ and switch sides} \quad ■$$

Practice Exercises

Solve for the indicated letter. Each comes from the indicated area of study.

1. $\theta = kA + \lambda$, for λ (robotics)

2. $P = n(p - c)$, for p (economics)

NOTE ▶ To determine the values of any literal number in an expression for which we know values of the other literal numbers, we should *first solve for the required symbol and then evaluate.*

EXAMPLE 5   Solve for symbol before substituting

The electric resistance R (in Ω) of a resistor changes with the temperature T (in °C) according to $R = R_0 + R_0\alpha T$, where R_0 is the resistance at 0°C. For a given resistor, $R_0 = 712\ \Omega$ and $\alpha = 0.00455/°C$. Find the value of T for $R = 825\ \Omega$.

We first solve for T and then substitute the given values.

$$R = R_0 + R_0\alpha T$$
$$R - R_0 = R_0\alpha T$$
$$T = \frac{R - R_0}{\alpha R_0}$$

Now substituting, we have

$$T = \frac{825 - 712}{(0.00455)(712)}$$

estimation:

$$\frac{800 - 700}{0.005(700)} = \frac{1}{0.035} = 30$$

$$= 34.9°C \quad \text{rounded off}$$

■

EXERCISES 1.11

In Exercises 1–4, solve for the given letter from the indicated example of this section.

1. For the formula in Example 2, solve for a.

2. For the formula in Example 3, solve for w.

3. For the formula in Example 4, solve for T_0.

4. For the formula in Example 5, solve for α. (Do not evaluate.)

In Exercises 5–40, each of the given formulas arises in the technical or scientific area of study shown. Solve for the indicated letter.

5. $E = IR$, for R (electricity)

6. $PV = nRT$, for T (chemistry)

7. $rL = g_2 - g_1$, for g_1 (surveying)

8. $W = S_d T - Q$, for Q (air conditioning)

9. $Q = SLd^2$, for L (machine design)

10. $P = 2\pi T f$, for T (mechanics)

11. $p = p_a + dgh$, for h (hydrodynamics)

12. $2Q = 2I + A + S$, for I (nuclear physics)

13. $A = \dfrac{Rt}{PV}$, for t (jet engine design)

14. $u = -\dfrac{eL}{2m}$, for L (spectroscopy)

15. $ct^2 = 0.3t - ac$, for a (medical technology)

16. $2p + dv^2 = 2d(C - W)$, for C (fluid flow)

17. $T = \dfrac{c + d}{v}$, for d (traffic flow)

18. $L = \dfrac{N\Phi}{i}$, for Φ (electricity)

19. $\dfrac{K_1}{K_2} = \dfrac{m_1 + m_2}{m_1}$, for m_2 (kinetic energy)

20. $f = \dfrac{F}{d - F}$, for d (photography)

21. $a = \dfrac{2mg}{M + 2m}$, for M (pulleys)

22. $v = \dfrac{V(m + M)}{m}$, for M (ballistics)

23. $C_0^2 = C_1^2(1 + 2V)$, for V (electronics)

24. $A_1 = A(M + 1)$, for M (photography)

25. $N = r(A - s)$, for s (engineering)

26. $T = 3(T_2 - T_1)$, for T_1 (oil drilling)

27. $T_2 = T_1 - \dfrac{h}{100}$, for h (air temperature)

28. $p_2 = p_1 + rp_1(1 - p_1)$, for r (population growth)

29. $Q_1 = P(Q_2 - Q_1)$, for Q_2 (refrigeration)

30. $p - p_a = dg(y_2 - y_1)$, for y_2 (pressure gauges)

31. $N = N_1 T - N_2(1 - T)$, for N_1 (machine design)

32. $t_a = t_c + (1 - h)t_m$, for h (computer access time)

33. $L = \pi(r_1 + r_2) + 2x_1 + x_2$, for r_1 (pulleys)

34. $I = \dfrac{VR_2 + VR_1(1 + \mu)}{R_1 R_2}$, for μ (electronics)

35. $P = \dfrac{V_1(V_2 - V_1)}{gJ}$, for V_2 (jet engine power)

36. $W = T(S_1 - S_2) - Q$, for S_2 (refrigeration)

37. $C = \dfrac{2eAk_1 k_2}{d(k_1 + k_2)}$, for e (electronics)

38. $d = \dfrac{3LPx^2 - Px^3}{6EI}$, for L (beam deflection)

39. $V = C\left(1 - \dfrac{n}{N}\right)$, for n (property deprecation)

40. $\dfrac{p}{P} = \dfrac{AI}{B + AI}$, for B (atomic theory)

In Exercises 41–46, find the indicated values.

41. The efficiency e (in %) of a certain engine is given by the formula $e = T_2/(T_1 + T_2)$. Find T_1 if $e = 45\%$ and $T_2 = 850$ K.

42. A formula used in determining the total transmitted power P_t in an AM radio signal is $P_t = P_c(1 + 0.500\, m^2)$. Find P_c if $P_t = 680$ W and $m = 0.925$.

43. A formula relating the Fahrenheit temperature F and the Celsius temperature C is $F = \frac{9}{5}C + 32$. Find the Celsius temperature that corresponds to $90.2°$F.

44. In forestry, a formula used to determine the volume V of a log is $V = \frac{1}{2}L(B + b)$, where L is the length of the log and B and b are the areas of the ends. Find b (in ft^2) if $V = 38.6$ ft^3, $L = 16.1$ ft, and $B = 2.63$ ft^2. See Fig. 1.16.

Fig. 1.16

45. The voltage V_1 across resistance R_1 is $V_1 = \dfrac{VR_1}{R_1 + R_2}$, where V is the voltage across resistances R_1 and R_2. See Fig. 1.17. Find R_2 (in Ω) if $R_1 = 3.56\ \Omega$, $V_1 = 6.30$ V, and $V = 12.0$ V.

Fig. 1.17

46. The efficiency E of a computer multiprocessor compilation is given by $E = \dfrac{1}{q + p(1 - q)}$, where p is the number of processors and q is the fraction of the compilation that can be performed by the available parallel processors. Find p for $E = 0.66$ and $q = 0.83$.

In Exercises 47 and 48, set up the required formula and solve for the indicated letter.

47. One missile travels at a speed of v_2 mi/h for 4 h, and another missile travels at a speed of v_1 for $t + 2$ hours. If they travel a total of d mi, solve the resulting formula for t.

48. A microwave transmitter can handle x telephone communications, and 15 separate cables can handle y connections each. If the combined system can handle C connections, solve for y.

Answers to Practice Exercises

1. $\theta - kA$ **2.** $\dfrac{P + nc}{n}$

1.12 Applied Word Problems

Procedure for Solving Word Problems • Identifying the Unknown Quantities • Setting Up the Proper Equation • Examples of Solving Word Problems

Many applied problems are at first word problems, and we must put them into mathematical terms for solution. Usually, the most difficult part in solving a word problem is identifying the information needed for setting up the equation that leads to the solution. To do this, you must read the problem carefully to be sure that you understand all of the terms and expressions used. Following is an approach you should use.

■ See Appendix A, page A-1, for a variation to the method outlined in these steps. You might find it helpful.

CAUTION ▶

> **PROCEDURE FOR SOLVING WORD PROBLEMS**
>
> 1. *Read the statement of the problem.* First, read it quickly for a general overview. Then *reread* slowly and carefully, *listing the information* given.
> 2. *Clearly identify the unknown quantities* and then *assign an appropriate letter to represent one of them,* stating this choice clearly.
> 3. *Specify the other unknown quantities* in terms of the one in step 2.
> 4. *If possible, make a sketch* using the known and unknown quantities.
> 5. *Analyze the statement* of the problem and *write the necessary equation.* This is often the most difficult step because *some of the information may be implied and not explicitly stated.* Again, a very careful reading of the statement is necessary.
> 6. *Solve the equation,* clearly stating the solution.
> 7. *Check the solution with the original statement* of the problem.

Read the following examples very carefully and note just how the outlined procedure is followed.

EXAMPLE 1 Sum of forces on a beam

A 17-lb beam is supported at each end. The supporting force at one end is 3 lb more than at the other end. Find the forces.

Since the force at each end is required, we write

$$\text{let } F = \text{ the smaller force (in lb)} \qquad \text{step 2}$$

as a way of establishing the unknown for the equation. Any appropriate letter could be used, and we could have let it represent the larger force.

Also, since the other force is 3 lb more, we write

$$F + 3 = \text{ the larger force (in lb)} \qquad \text{step 3}$$

We now draw the sketch in Fig. 1.18.   step 4

Since the forces at each end of the beam support the weight of the beam, we have the equation

$$F + (F + 3) = 17 \qquad \text{step 5}$$

This equation can now be solved: $2F = 14$

$$F = 7 \text{ lb} \qquad \text{step 6}$$

Thus, the smaller force is 7 lb, and the larger force is 10 lb. This checks with the original statement of the problem.   step 7   ■

EXAMPLE 2 Total resistance of electric resistors

In designing an electric circuit, it is found that 34 resistors with a total resistance of 56 Ω are required. Two different resistances, 1.5 Ω and 2.0 Ω, are used. How many of each are in the circuit?

Since we want to find the number of each resistance, we

$$\text{let } x = \text{ number of 1.5-}\Omega \text{ resistors}$$

Also, since there are 34 resistors in all,

$$34 - x = \text{ number of 2.0-}\Omega \text{ resistors}$$

We also know that the total resistance of all resistors is 56 Ω. This means that

1.5 Ω number
each ┐ │
      ↓ ↓
     $1.5x$ 2.0 Ω
                ┌ each ┌ number
                ↓   ↓
        $+ 2.0(34 - x) = 56$ ◄——— total resistance of all resistors

total resistance   total resistance
of 1.5-Ω resistors   of 2.0-Ω resistors

$$1.5x + 68 - 2.0x = 56$$
$$-0.5x = -12$$
$$x = 24$$

Therefore, there are 24 1.5-Ω resistors and 10 2.0-Ω resistors. The total resistance of these is $24(1.5) + 10(2.0) = 36 + 20 = 56$ Ω. We see that this checks with the statement of the problem.   ■

■ Be sure to carefully identify your choice for the unknown. In most problems, there is really a choice. Using the word *let* clearly shows that a specific choice has been made.

■ The statement after "let x (or some other appropriate letter) =" should be clear. It should completely define the chosen unknown.

F 17 lb $F + 3$

Fig. 1.18

■ "Let x = 1.5-Ω resistors" is incomplete. We want to find out how many of them there are.

■ See Appendix A, page A-1, for a "sketch" that might be used with this example.

EXAMPLE 3 Number of medical slides

A medical researcher finds that a given sample of an experimental drug can be divided into 4 more slides with 5 mg each than with 6 mg each. How many slides with 5 mg each can be made up?

We are asked to find the number of slides with 5 mg, and therefore we

$$\text{let } x = \text{number of slides with 5 mg}$$

Since the sample may be divided into 4 more slides with 5 mg each than of 6 mg each, we know that

$$x - 4 = \text{number of slides with 6 mg}$$

Since *it is the same sample that is to be divided,* the total mass of the drug on each type is the same. This means

5 mg   number   6 mg   number
each ┐   │   each ──────┐   │
     ▼   ▼   ▼   ▼
$$5x \qquad = \qquad 6(x - 4)$$

total mass   total mass
5-mg slides   6-mg slides

$$5x = 6x - 24$$
$$-x = -24 \quad \text{or} \quad x = 24$$

Therefore, the sample can be divided into 24 slides with 5 mg each, or 20 slides with 6 mg each. Since the total mass, 120 mg, is the same for each set of slides, the solution checks with the statement of the problem. ∎

Practice Exercise

1. Solve the problem in Example 3 by letting $y =$ number of slides with 6 mg.

EXAMPLE 4 Distance traveled – space travel

A space shuttle maneuvers so that it may "capture" an already orbiting satellite that is 6000 km ahead. If the satellite is moving at 27,000 km/h and the shuttle is moving at 29,500 km/h, how long will it take the shuttle to reach the satellite? (All digits shown are significant.)

First, we let $t =$ the time for the shuttle to reach the satellite. Then, using the fact that the shuttle must go 6000 km farther in the same time, we draw the sketch in Fig. 1.19. Next, we use the formula *distance = rate × time* ($d = rt$). This leads to the following equation and solution.

speed of   speed of
shuttle ──┐ ┌ time   satellite ┐ ┌ time
          ▼ ▼   ▼ ▼
$$29{,}500t \qquad = \qquad 6000 \qquad + \qquad 27{,}000t$$

distance traveled   distance between   distance traveled
by shuttle   at beginning   by satellite

$$2500t = 6000$$
$$t = 2.400 \text{ h}$$

This means that it will take the shuttle 2.400 h to reach the satellite. In 2.400 h, the shuttle will travel 70,800 km, and the satellite will travel 64,800 km. We see that the solution checks with the statement of the problem. ∎

Shuttle

6000 km

Satellite

Fig. 1.19

■ A maneuver similar to the one in this example was used to repair the Hubble space telescope in 1994 and 2001.

EXAMPLE 5 Mixture – gasoline and methanol

A refinery has 7250 L of a gasoline-methanol blend that is 6.00% methanol. How much pure methanol must be added so that the resulting blend is 10.0% methanol?

First, let x = the number of liters of methanol to be added. The total volume of methanol in the final blend is the volume in the original blend plus that which is added. This total volume is to be 10.0% of the final blend. See Fig. 1.20.

■ "Let x = methanol" is incomplete. We want to find out the volume (in L) that is to be added.

$$0.0600(7250) + x = 0.100(7250 + x)$$

methanol in original mixture   methanol added   methanol in final mixture

$$435 + x = 725 + 0.100x$$
$$0.900x = 290, \quad x = 322\,\text{L} \quad \text{to be added}$$

Checking (to three sig. digits), there would be 757 L of methanol of a total 7570 L. ■

7600
5.00%

$+$

x
100%

$=$

7600 + x
10.0%

Liters of methanol

Fig. 1.20

EXERCISES 1.12

In Exercises 1–4, make the given changes in the indicated examples of this section and then solve the resulting problems.

1. In Example 2, in the second line, change 2.0 Ω to 2.5 Ω.

2. In Example 3, in the second line, change "4 more slides" to "3 more slides."

3. In Example 4, in the second line, change 27,000 km/h to 27,500 km/h.

4. In Example 5, change "pure methanol" to "of a blend with 50.0% methanol."

In Exercises 5–32, set up an appropriate equation and solve. Data are accurate to two sig. digits unless greater accuracy is given.

5. A certain new car costs $5000 more than the same model new car cost 6 years ago. Together a new model today and six years ago cost $49,000. What was the cost of each?

6. The flow of one stream into a lake is 1700 ft³/s more than the flow of a second stream. In 1 h, 1.98×10^7 ft³ flow into the lake from the two streams. What is the flow rate of each?

7. Approximately 4.5 million wrecked cars are recycled in two consecutive years. There were 700,000 more recycled the second year than the first year. How many are recycled each year?

8. A business website was accessed as often on the first day of a promotion as on the next two days combined. The first day hits were four times those on the third. The second day hits were 4000 more than the third. How many were there each day?

9. Petroleum rights to 140 acres of land are leased for $37,000. Part of the land leases for $200 per acre, and the reminder for $300 per acre. How much is leased at each price?

10. A vial contains 2000 mg, which is to be used for two dosages. One patient is to be administered 660 mg more than another. How much should be administered to each?

11. After installing a pollution control device, a car's exhaust contained the same amount of pollutant after 5.0 h as it had in 3.0 h. Before the installation the exhaust contained 150 ppm/h (parts per million per hour) of the pollutant. By how much did the device reduce the emission?

12. Three meshed spur gears have a total of 107 teeth. If the second gear has 13 more teeth than the first and the third has 15 more teeth than the second, how many teeth does each have?

13. In the design of a bridge, an engineer determines that four fewer 18-m girders are needed for the span length than 15-m girders. How many 18-m girders are needed?

14. A fuel oil storage depot had an 8-week supply on hand. However, cold weather caused the supply to be used in 6 weeks when 5000 gal extra were used each week. How many gallons were in the original supply?

15. The sum of three electric currents that come together at a point in a circuit is zero. If the second current is twice the first and the third current is 9.2 μA more than the first, what are the currents? (The sign indicates the direction of flow.)

16. A delivery firm uses one fleet of trucks on daily routes of 8 h. A second fleet, with five more trucks than the first, is used on daily routes of 6 h. Budget allotments allow for 198 h of daily delivery time. How many trucks are in each fleet?

17. A natural gas pipeline feeds into three smaller pipelines, each of which is 2.6 km longer than the main pipeline. The total length of the four pipelines is 35.4 km. How long is each section?

18. At 100% efficiency two generators would produce 750 MW of power. At efficiencies of 65% and 75%, they produce 530 MW. At 100% efficiency, what power would each produce?

19. A person has a total of 54 CDs and DVDs, which cost a total of $876. Each CD cost $15 and each DVD cost $18. How many of each does the person have?

20. A person won a state lottery prize of $20,000, from which 25% was deducted for taxes. The remainder was invested, partly for a 40% gain, and the rest for a 10% loss. How much was each investment if there was a $2000 net investment gain?

21. A ski lift takes a skier up a slope at 50 m/min. The skier then skis down the slope at 150 m/min. If a round trip takes 24 min, how long is the slope?

22. Before being put out of service, the supersonic jet Concorde made a trip averaging 120 mi/h less than the speed of sound for 0.1 h, and 410 mi/h more than the speed of sound for 3.0 h. If the trip covered 3990 mi, what is the speed of sound?

23. Trains at each end of the 50.0-km-long Eurotunnel under the English Channel start at the same time into the tunnel. Find their speeds if the train from France travels 8.0 km/h faster than the train from England and they pass in 17.0 min. See Fig. 1.21.

England   English Channel   France

17.0 min   50.0 km

Fig. 1.21

24. An executive would arrive 10.0 min early for an appointment if traveling at 60.0 mi/h, or 5.0 min early if traveling at 45.0 mi/h. How much time is there until the appointment?

25. One lap at the Indianapolis Speedway is 2.50 mi. In a race, a car stalls and then starts 30.0 s after a second car. The first car travels at 260 ft/s, and the second car travels at 240 ft/s. How long does it take the first car to overtake the second, and which car will be ahead after eight laps?

26. A computer chip manufacturer produces two types of chips. In testing a total of 6100 chips of both types, 0.50% of one type and 0.80% of the other type were defective. If a total of 38 defective chips were found, how many of each type were tested?

27. Two gasoline distributors, A and B, are 228 mi apart on Interstate 80. A charges $3.40/gal and B charges $3.20/gal. Each charges 0.2¢/gal per mile for delivery. Where on Interstate 80 is the cost to the customer the same?

28. An outboard engine uses a gasoline-oil fuel mixture in the ratio of 15 to 1. How much gasoline must be mixed with a gasoline-oil mixture, which is 75% gasoline, to make 8.0 L of the mixture for the outboard engine?

29. A car's radiator contains 12 L of antifreeze at a 25% concentration. How many liters must be drained and then replaced by pure antifreeze to bring the concentration to 50% (the manufacturer's "safe" level)?

30. How much sand must be added to 250 lb of a cement mixture that is 22% sand to have a mixture that is 25% sand?

31. To pass a 20-m long semitrailer traveling at 70 km/h in 10 s, how fast must a 5.0-m long car go?

32. An earthquake emits primary waves moving at 8.0 km/s and secondary waves moving at 5.0 km/s. How far from the epicenter of the earthquake is the seismic station if the two waves arrive at the station 2.0 min apart?

CHAPTER 1   EQUATIONS

Commutative law of addition: $a + b = b + a$

Commutative law of multiplication: $ab = ba$

Associative law of addition: $a + (b + c) = (a + b) + c$

Associative law of multiplication: $a(bc) = (ab)c$

Distributive law: $a(b + c) = ab + ac$

$a + (-b) = a - b$ **(1.1)**

$a - (-b) = a + b$ **(1.2)**

$a^m \times a^n = a^{m+n}$ **(1.3)**

$\dfrac{a^m}{a^n} = a^{m-n} \quad (m > n, a \neq 0),$ **(1.4)**

$\dfrac{a^m}{a^n} = \dfrac{1}{a^{n-m}} \quad (m < n, a \neq 0)$

$(a^m)^n = a^{mn}$ **(1.5)**

$(ab)^n = a^n b^n, \quad \left(\dfrac{a}{b}\right)^n = \dfrac{a^n}{b^n} \quad (b \neq 0)$ **(1.6)**

$a^0 = 1 \quad (a \neq 0)$ **(1.7)**

$a^{-n} = \dfrac{1}{a^n} \quad (a \neq 0)$ **(1.8)**

$\sqrt{ab} = \sqrt{a}\sqrt{b} \quad (a \text{ and } b \text{ positive real numbers})$ **(1.9)**

CHAPTER 1   REVIEW EXERCISES

In Exercises 1–12, evaluate the given expressions.

1. $(-2) + (-5) - 3$

2. $6 - 8 - (-4)$

3. $\dfrac{(-5)(6)(-4)}{(-2)(3)}$

4. $\dfrac{(-9)(-12)(-4)}{24}$

5. $-5 - |2(-6)| + \dfrac{-15}{3}$

6. $3 - 5|-3 - 2| - \dfrac{12}{-4}$

7. $\dfrac{18}{3 - 5} - (-4)^2$

8. $-(-3)^2 - \dfrac{-8}{(-2) - |-4|}$

9. $\sqrt{16} - \sqrt{64}$

10. $-\sqrt{81} + 144$

11. $(\sqrt{7})^2 - \sqrt[3]{8}$

12. $-\sqrt[4]{16} + (\sqrt{6})^2$

In Exercises 13–20, simplify the given expressions. Where appropriate, express results with positive exponents only.

13. $(-2rt^2)^2$

14. $(3a^0 b^{-2})^3$

15. $-3mn^{-5}t(8m^{-3}n^4)$

16. $\dfrac{15p^4 q^2 r}{5pq^5 r}$

17. $\dfrac{-16N^{-2}(NT^2)}{-2N^0 T^{-1}}$

18. $\dfrac{-35x^{-1}y(x^2 y)}{5xy^{-1}}$

19. $\sqrt{45}$

20. $\sqrt{9 + 36}$

In Exercises 21–24, for each number, (a) determine the number of significant digits and (b) round off each to two significant digits.

21. 8840

22. 21,450

23. 9.040

24. 0.700

In Exercises 25–28, evaluate the given expressions. All numbers are approximate.

25. $37.3 - 16.92(1.067)^2$

26. $\dfrac{8.896 \times 10^{-12}}{3.5954 + 6.0449}$

27. $\dfrac{\sqrt{0.1958 + 2.844}}{3.142(65)^2}$

28. $\dfrac{1}{0.03568} + \dfrac{37,466}{29.63^2}$

In Exercises 29–60, perform the indicated operations.

29. $a - 3ab - 2a + ab$

30. $xy - y - 5y - 4xy$

31. $6LC - (3 - LC)$

32. $-(2x - b) - 3(-x - 5b)$

33. $(2x - 1)(x + 5)$

34. $(C - 4D)(2C - D)$

35. $(x + 8)^2$

36. $(2r - 9s)^2$

37. $\dfrac{2h^3 k^2 - 6h^4 k^5}{2h^2 k}$

38. $\dfrac{4a^2 x^3 - 8ax^4}{-2ax^2}$

39. $4R - [2r - (3R - 4r)]$

40. $3b - [3a - (a - 3b)] + 4a$

41. $2xy - \{3z - [5xy - (7z - 6xy)]\}$

42. $x^2 + 3b + [(b - y) - 3(2b - y + z)]$

43. $(2x + 1)(x^2 - x - 3)$

44. $(x - 3)(2x^2 - 3x + 1)$

45. $-3y(x - 4y)^2$

46. $-s(4s - 3t)^2$

47. $3p[(q - p) - 2p(1 - 3q)]$

48. $3x[2y - r - 4(s - 2r)]$

49. $\dfrac{12p^3 q^2 - 4p^4 q + 6pq^5}{2p^4 q}$

50. $\dfrac{27s^3 t^2 - 18s^4 t + 9s^2 t}{9s^2 t}$

51. $(2x^2 + 7x - 30) \div (x + 6)$

52. $(4x^2 - 41) \div (2x + 7)$

53. $\dfrac{3x^3 - 7x^2 + 11x - 3}{3x - 1}$

54. $\dfrac{w^3 - 4w^2 + 7w - 12}{w - 3}$

55. $\dfrac{4x^4 + 10x^3 + 18x - 1}{x + 3}$

56. $\dfrac{8x^3 - 14x + 3}{2x + 3}$

57. $-3\{(r + s - t) - 2[(3r - 2s) - (t - 2s)]\}$

58. $(1 - 2x)(x - 3) - (x + 4)(4 - 3x)$

59. $\dfrac{2y^3 + 9y^2 - 7y + 5}{2y - 1}$

60. $\dfrac{6x^2 + 5xy - 4y^2}{2x - y}$

In Exercises 61–72, solve the given equations.

61. $3x + 1 = x - 8$

62. $4y - 3 = 5y + 7$

63. $\dfrac{5x}{7} = \dfrac{3}{2}$

64. $\dfrac{2(N - 4)}{3} = \dfrac{5}{4}$

65. $6x - 5 = 3(x - 4)$

66. $-2(-4 - y) = 3y$

67. $2s + 4(3 - s) = 6$

68. $2|x| - 1 = 3$

69. $3t - 2(7 - t) = 5(2t + 1)$

70. $-(8 - x) = x - 2(2 - x)$

71. $2.7 + 2.0(2.1x - 3.4) = 0.1$

72. $0.250(6.721 - 2.44x) = 2.08$

In Exercises 73–82, change numbers in ordinary notation to scientific notation or change numbers in scientific notation to ordinary notation. (See Appendix B for an explanation of the symbols that are used.)

73. A certain computer has 60,000,000,000 bytes of memory.

74. The escape velocity (the velocity required to leave the earth's gravitational field) is about 25,000 mi/h.

75. When pictures of the surface of Mars were transmitted to the earth from the Pathfinder mission in 1997, Mars was about 192,000,000 km from earth.

76. Police radar has a frequency of 1.02×10^9 Hz.

77. Among the stars nearest the earth, Centaurus A is about 2.53×10^{13} mi away.

78. Before its destruction in 2001, the World Trade Center had nearly 10^7 ft^2 of office space. (See Exercise 32, page 128.)

79. The faintest sound that can be heard has an intensity of about 10^{-12} W/m^2.

80. An optical coating on glass to reduce reflections is about 0.00000015 m thick.

81. The maximum safe level of radiation in the air of a home due to radon gas is 1.5×10^{-1} Bq/L. (Bq is the symbol for bequerel, the metric unit of radioactivity, where 1 Bq = 1 decay/s.)

82. A certain virus was measured to have a diameter of about 0.00000018 m.

In Exercises 83–96, solve for the indicated letter. Where noted, the given formula arises in the technical or scientific area of study.

83. $R = n^2Z$, for Z (electricity)

84. $R = \dfrac{2GM}{c^2}$, for G (astronomy: black holes)

85. $P = \dfrac{\pi^2EI}{L^2}$, for E (mechanics)

86. $f = p(c - 1) - c(p - 1)$, for p (thermodynamics)

87. $Pp + Qq = Rr$, for q (moments of forces)

88. $V = IR + Ir$, for R (electricity)

89. $d = (n - 1)A$, for n (optics)

90. $mu = (m + M)v$, for M (physics: momentum)

91. $N_1 = T(N_2 - N_3) + N_3$, for N_2 (mechanics: gears)

92. $q = \dfrac{KA(B - C)}{L}$, for B (solar heating)

93. $R = \dfrac{A(T_2 - T_1)}{H}$, for T_2 (thermal resistance)

94. $Z^2\left(1 - \dfrac{\lambda}{2a}\right) = k$, for λ (radar design)

95. $d = kx^2[3(a + b) - x]$, for a (mechanics: beams)

96. $V = V_0[1 + 3a(T_2 - T_1)]$, for T_2 (thermal expansion)

In Exercises 97–102, perform the indicated calculations.

97. A computer's memory is 5.25×10^{10} bytes, and that of a model 20 years older is 6.4×10^4 bytes. What is the ratio of the newer computer's memory to the older computer's memory?

98. The time (in s) for an object to fall h feet is given by the expression $0.25\sqrt{h}$. How long does it take a person to fall 66 ft from a sixth-floor window into a net while escaping a fire?

99. The CN Tower in Toronto is 0.553 km high. The Sears Tower in Chicago is 443 m high. How much higher is the CN Tower than the Sears Tower?

100. The time (in s) it takes a computer to check n memory cells is found by evaluating $(n/2650)^2$. Find the time to check 48 cells.

101. The combined electric resistance of two parallel resistors is found by evaluating the expression $\dfrac{R_1R_2}{R_1 + R_2}$. Evaluate this for $R_1 = 0.0275\ \Omega$ and $R_2 = 0.0590\ \Omega$.

102. The distance (in m) from the earth for which the gravitational force of the earth on a spacecraft equals the gravitational force of the sun on it is found by evaluating $1.5 \times 10^{11}\sqrt{m/M}$, where m and M are the masses of the earth and sun, respectively. Find this distance for $m = 5.98 \times 10^{24}$ kg and $M = 1.99 \times 10^{30}$ kg.

In Exercises 103–106, simplify the given expressions.

103. One transmitter antenna is $(x - 2a)$ ft long, and another is $(x + 2a)$ yd long. What is the sum, in feet, of their lengths?

104. In finding the value of an annuity, the expression $(Ai - R)(1 + i)^2$ is used. Multiply out this expression.

105. A computer analysis of the velocity of a link in an industrial robot leads to the expression $4(t + h) - 2(t + h)^2$. Simplify this expression.

106. When analyzing the motion of a communications satellite, the expression $\dfrac{k^2r - 2h^2k + h^2rv^2}{k^2r}$ is used. Perform the indicated division.

In Exercises 107–114 solve the given problems.

107. Does the value of $3 \times 18 \div (9 - 6)$ change if the parentheses are removed?

108. Does the value of $(3 \times 18) \div 9 - 6$ change if the parentheses are removed?

(W) 109. In solving the equation $x - (3 - x) = 2x - 3$, what conclusion can be made?

(W) 110. In solving the equation $7 - (2 - x) = x + 2$, what conclusion can be made?

111. Show that $(x - y)^3 = -(y - x)^3$.

112. Is division associative? That is, is it true (if $b \neq 0, c \neq 0$) that $(a \div b) \div c = a \div (b \div c)$?

113. What is the ratio of 8×10^{-3} to 2×10^4?

114. What is the ratio of $\sqrt{4 + 36}$ to $\sqrt{4}$?

In Exercises 115–128, solve the given problems. All data are accurate to two significant digits unless greater accuracy is given.

115. Two computer software programs cost $190 together. If one costs $72 more than the other, what is the cost of each?

116. A sponsor pays a total of $9500 to run a commercial on two different TV stations. One station charges $1100 more than the other. What does each charge to run the commercial?

117. Three chemical reactions each produce oxygen. If the first produces twice that of the second, the third produces twice that of the first, and the combined total is $560\ \text{cm}^3$, what volume is produced by each?

118. In testing the rate at which a polluted stream flows, a boat that travels at 5.5 mi/h in still water took 5.0 h to go downstream between two points, and it took 8.0 h to go upstream between the same two points. What is the rate of flow of the stream?

119. The voltage across a resistor equals the current times the resistance. In a microprocessor circuit, one resistor is 1200 Ω greater than another. The sum of the voltages across them is 12.0 mV. Find the resistances if the current is 2.4 μA in each.

120. An air sample contains 4.0 ppm (parts per million) of two pollutants. The concentration of one is four times the other. What are the concentrations?

121. One road crew constructs 450 m of road bed in 12 h. If another crew works at the same rate, how long will it take them to construct another 250 m of road bed?

122. The fuel for a two-cycle motorboat engine is a mixture of gasoline and oil in the ratio of 15 to 1. How many liters of each are in 6.6 L of mixture?

123. A ship enters the Red Sea from the Suez Canal, moving south at 17.4 mi/h. Two hours later, a second ship enters the Red Sea at the southern end, moving north at 21.8 mi/h. If the Red Sea is 1390 mi long, when will the ships pass?

124. A helicopter used in fighting a forest fire travels at 105 mi/h from the fire to a pond and 70 mi/h with water from the pond to the fire. If a round-trip takes 30 min, how long does it take from the pond to the fire? See Fig. 1.22.

105 mi/h   70 mi/h

30 min

Fig. 1.22

125. One grade of oil has 0.50% of an additive, and a higher grade has 0.75% of the additive. How many liters of each must be used to have 1000 L of a mixture with 0.65% of the additive?

126. How much water must be added to 20.0 mL of a 60.0% saline solution in order to make a 45.0% saline solution?

127. An architect plans to have 25% of the floor area of a house in ceramic tile. In all but the kitchen and entry, there are 2200 ft² of floor area, 15% of which is tile. What area can be planned for the kitchen and entry if each has an all-tile floor?

128. A *karat* equals 1/24 part of gold in an alloy (for example, 9-karat gold is 9/24 gold). How many grams of 9-karat gold must be mixed with 18-karat gold to get 200 g of 14-karat gold?

Writing Exercise

129. In calculating the simple interest earned by an investment, the equation $P = P_0 + P_0 rt$ is used, where P is the value after an initial principal P_0 is invested for t years at interest rate r. Solve for r, and then evaluate r for $P = \$7625$, $P_0 = \$6250$, and $t = 4.000$ years. Write a paragraph or two explaining (a) your method for solving for r, and (b) the calculator steps used to evaluate r, noting the use of parentheses.

CHAPTER 1   PRACTICE TEST

In Problems 1–5, evaluate the given expressions. In Problems 3 and 5, the numbers are approximate.

1. $\sqrt{9 + 16}$ **2.** $\dfrac{(7)(-3)(-2)}{(-6)(0)}$ **3.** $\dfrac{3.372 \times 10^{-3}}{7.526 \times 10^{12}}$

4. $\dfrac{(+6)(-2) - 3(-1)}{|2 - 5|}$ **5.** $\dfrac{346.4 - 23.5}{287.7} - \dfrac{0.944^3}{(3.46)(0.109)}$

In Problems 6–12, perform the indicated operations and simplify. When exponents are used, use only positive exponents in the result.

6. $(2a^0 b^{-2} c^3)^{-3}$ **7.** $(2x + 3)^2$ **8.** $3m^2(am - 2m^3)$

9. $\dfrac{8a^3 x^2 - 4a^2 x^4}{-2ax^2}$ **10.** $\dfrac{6x^2 - 13x + 7}{2x - 1}$

11. $(2x - 3)(x + 7)$ **12.** $3x - [4x - (3 - 2x)]$

13. Solve for y: $5y - 2(y - 4) = 7$

14. Solve for x: $3(x - 3) = x - (2 - 3d)$

15. Express 0.0000036 in scientific notation.

16. List the numbers -3, $|-4|$, $-\pi$, $\sqrt{2}$, and 0.3 in numerical order.

17. What fundamental law is illustrated by $3(5 + 8) = 3(5) + 3(8)$?

18. (a) How many significant digits are in the number 3.0450?

(b) Round it off to two significant digits.

19. If P dollars is deposited in a bank that compounds interest n times a year, the value of the account after t years is found by evaluating $P(1 + i/n)^{nt}$, where i is the annual interest rate. Find the value of an account for which $P = \$1000$, $i = 5\%$, $n = 2$, and $t = 3$ years (values are exact).

20. In finding the illuminance from a light source, the expression $8(100 - x)^2 + x^2$ is used. Simplify this expression.

21. The equation $L = L_0[1 + \alpha(t_2 - t_1)]$ is used when studying thermal expansion. Solve for t_2.

22. An alloy weighing 20 lb is 30% copper. How many pounds of another alloy, which is 80% copper, must be added for the final alloy to be 60% copper?

Geometry

When building the pyramids nearly 5000 years ago, and today when using MRI (magnetic resonance imaging) to detect a tumor in a human being, the size and shape of an object are measured. Since geometry deals with size and shape, the topics and methods of geometry are important in many of the applications of technology.

Many of the methods of measuring geometric objects were known in ancient times, and most of the geometry used in technology has been known for hundreds of years. In about 300 B.C.E., the Greek mathematician Euclid (who lived and taught in Alexandria, Egypt) organized what was known in geometry. He added many new ideas in a 13-volume set of writings known as the *Elements*. Centuries later it was translated into various languages, and today is second only to the Bible as the most published book in history.

The study of geometry includes the properties and measurements of angles, lines, and surfaces and the basic figures they form. In this chapter, we review the more important methods and formulas for calculating the important geometric measures, such as area and volume. Technical applications are included from areas such as architecture, construction, instrumentation, surveying and civil engineering, mechanical design, and product design of various types, as well as other areas of engineering.

Geometric figures and concepts are also basic to the development of many areas of mathematics, such as graphing and trigonometry. We will start our study of graphs in Chapter 3 and trigonometry in Chapter 4.

In Section 2.5, we see how to find an excellent approximation of the area of an irregular geometric figure, such as a lake. At the left is a satellite photograph of Lake Ontario, one of the Great Lakes between the United States and Canada.

2.1   Lines and Angles

Basic Angles • Parallel Lines and Perpendicular Lines • Supplementary Angles and Complementary Angles • Angles Between Intersecting Lines • Segments of Transversals

■ The use of 360 comes from the ancient Babylonians and their number system based on 60 rather than 10, as we use today. However, the specific reason for the choice of 360 is not known. (One theory is that 360 is divisible by many smaller numbers and is close to the number of days in a year.)

It is not possible to define every term we use. In geometry, *the meanings of* **point, line,** and **plane** *are accepted without being defined.* These terms give us a starting point for the definitions of other useful geometric terms.

*The amount of rotation of a **ray** (or **half-line**) about its endpoint is called an* **angle.** A ray is that part of a line (the word *line* means "straight line") to one side of a fixed point on the line. *The fixed point is the* **vertex** *of the angle. One complete rotation of a ray is an angle with a measure of* 360 **degrees,** *written as* 360°. Some special types of angles are as follows:

Name of angle	Measure of angle
Right angle	*90°*
Straight angle	*180°*
Acute angle	*Between 0° and 90°*
Obtuse angle	*Between 90° and 180°*

EXAMPLE 1   Basic angles

Figure 2.1(a) shows a right angle (marked as ∟). The vertex of the angle is point *B*, and the ray is the half-line *BA*. Figure 2.1(b) shows a straight angle. Figure 2.1(c) shows an acute angle, denoted as ∠*E* (or ∠*DEF* or ∠*FED*). In Fig. 2.1(d), ∠*G* is an obtuse angle.

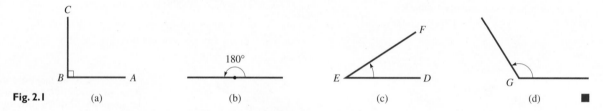

Fig. 2.1   (a)   (b)   (c)   (d)   ■

If two lines intersect such that the angle between them is a right angle, the lines are **perpendicular.** *Lines in the same plane that do not intersect are* **parallel.** These are illustrated in the following example.

EXAMPLE 2   Parallel lines and perpendicular lines

In Fig. 2.2(a), lines *AC* and *DE* are perpendicular (which is shown as *AC* ⊥ *DE*) since they meet in a right angle (again, shown as ∟) at *B*.

In Fig. 2.2(b), lines *AB* and *CD* are drawn so they do not meet, even if extended. Therefore, these lines are parallel (which can be shown as *AB* ∥ *CD*).

In Fig. 2.2(c), *AB* ⊥ *BC*, *DC* ⊥ *BC*, and *AB* ∥ *DC*.

Fig. 2.2   (a)   *E*   (b)   (c)

It will be important to recognize perpendicular sides and parallel sides in many of the geometric figures in later sections.   ■

■ Recognizing complementary angles is important in trigonometry.

If the sum of the measures of two angles is 180°, *then the angles are called* **supplementary angles.** Each angle is the **supplement** of the other. *If the sum of the measures of two angles is* 90°, *the angles are called* **complementary angles.** Each is the **complement** of the other.

EXAMPLE 3 Supplementary angles and complementary angles

(a) In Fig. 2.3(a), $\angle BAC = 55°$, and in Fig. 2.3(b), $\angle DEF = 125°$. Since $55° + 125° = 180°$, $\angle BAC$ and $\angle DEF$ are supplementary angles.

(b) In Fig. 2.4, we see that $\angle POQ$ is a right angle, or $\angle POQ = 90°$. Since $\angle POR + \angle ROQ = \angle POQ = 90°$, $\angle POR$ is the complement of $\angle ROQ$ (or $\angle ROQ$ is the complement of $\angle POR$).

Practice Exercise

1. What is the measure of the complement of $\angle A$ in Fig. 2.3(a)?

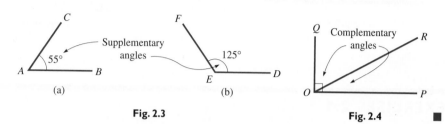

Fig. 2.3

Fig. 2.4 ■

It is often necessary to refer to certain special pairs of angles. *Two angles that have a common vertex and a side common between them are known as* **adjacent angles.** *If two lines cross to form equal angles on opposite sides of the point of intersection, which is the common vertex, these angles are called* **vertical angles.**

EXAMPLE 4 Adjacent angles and vertical angles

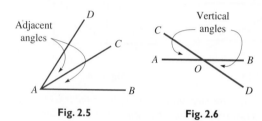

Fig. 2.5   Fig. 2.6

(a) In Fig. 2.5, $\angle BAC$ and $\angle CAD$ have a common vertex at A and the common side between them. This means that $\angle BAC$ and $\angle CAD$ are adjacent angles.

(b) In Fig. 2.6, lines AB and CD intersect at point O. Here, $\angle AOC$ and $\angle BOD$ are vertical angles, and they are equal. Also, $\angle BOC$ and $\angle AOD$ are vertical angles and are equal. ■

We should also be able to identify *the sides of an angle that are adjacent to the angle.* In Fig. 2.5, sides AB and AC are adjacent to $\angle BAC$, and in Fig. 2.6, sides OB and OD are adjacent to $\angle BOD$. Identifying sides adjacent and opposite an angle in a triangle is important in trigonometry.

In a plane, *if a line crosses two or more parallel or nonparallel lines, it is called a* **transversal.** In Fig. 2.7, $AB \parallel CD$, and the transversal of these two parallel lines is the line EF.

When a transversal crosses a pair of parallel lines, certain pairs of equal angles result. In Fig. 2.7, the **corresponding angles** are equal (that is, $\angle 1 = \angle 5$, $\angle 2 = \angle 6$, $\angle 3 = \angle 7$, and $\angle 4 = \angle 8$). Also, the **alternate-interior angles** are equal ($\angle 3 = \angle 6$ and $\angle 4 = \angle 5$), and the **alternate-exterior angles** are equal ($\angle 1 = \angle 8$ and $\angle 2 = \angle 7$).

When more than two parallel lines are crossed by *two* transversals, such as is shown in Fig. 2.8, *the segments of the transversals between the same two parallel lines are called* **corresponding segments.** A useful theorem is that *the ratios of corresponding segments of the transversals are equal.* In Fig. 2.8, this means that

Fig. 2.7

Practice Exercise

2. In Fig. 2.7, if $\angle 2 = 42°$, then $\angle 5 = ?$

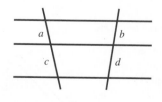

Fig. 2.8

$$\frac{a}{b} = \frac{c}{d}$$

(2.1)

Fig. 2.9

EXAMPLE 5 Segments of transversals

In Fig. 2.9, part of the beam structure within a building is shown. The vertical beams are parallel. From the distances between beams that are shown, determine the distance x between the middle and right vertical beams.

Using Eq. (2.1), we have

$$\frac{6.50}{5.65} = \frac{x}{7.75}$$

$$x = \frac{6.50(7.75)}{5.65} = 8.92 \text{ ft} \qquad \text{rounded off}$$ ∎

EXERCISES 2.1

In Exercises 1–4, answer the given questions about the indicated examples of this section.

1. In Example 2, what is the measure of $\angle ABE$ in Fig. 2.2(a)?

2. In Example 3(b), if $\angle POR = 32°$ in Fig. 2.4, what is the measure of $\angle QOR$?

3. In Example 4, how many different pairs of adjacent angles are there in Fig. 2.6?

4. In Example 5, if the segments of 6.50 ft and 7.75 ft are interchanged, (a) what is the answer, and (b) is the beam along which x is measured more nearly vertical or more nearly horizontal?

In Exercises 5–12, identify the indicated angles and sides in Fig. 2.10. In Exercises 9 and 10, also find the measures of the indicated angles.

5. Two acute angles

6. Two right angles

7. The straight angle

8. The obtuse angle

9. If $\angle CBD = 65°$, find its complement.

10. If $\angle CBD = 65°$, find its supplement.

11. The sides adjacent to $\angle DBC$

12. The acute angle adjacent to $\angle DBC$.

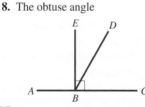

Fig. 2.10

In Exercises 13–15, use Fig. 2.11. In Exercises 16–18, use Fig. 2.12. Find the measures of the indicated angles.

13. $\angle AOB$ 14. $\angle AOC$ 15. $\angle BOD$

16. $\angle 3$ 17. $\angle 4$ 18. $\angle 5$

Fig. 2.11   **Fig. 2.12**

In Exercises 19–22, find the measures of the angles in Fig. 2.13.

19. $\angle 1$ 20. $\angle 2$ 21. $\angle 3$ 22. $\angle 4$

     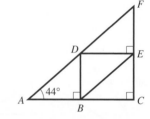

Fig. 2.13   **Fig. 2.14**

In Exercises 23–28, find the measures of the angles in the truss shown in Fig. 2.14. A truss is a rigid support structure that is used in the construction of buildings and bridges.

23. $\angle BDF$ 24. $\angle ABE$ 25. $\angle DEB$

26. $\angle DBE$ 27. $\angle DFE$ 28. $\angle ADE$

In Exercises 29–32, find the indicated distances between the straight irrigation ditches shown in Fig. 2.15. The vertical ditches are parallel.

29. a 30. b 31. c 32. d

Fig. 2.15

In Exercises 33–36, solve the given problems.

33. A steam pipe is connected in sections AB, BC, and CD as shown in Fig. 2.16. What is the angle between BC and CD if $AB \parallel CD$?

Fig. 2.16

34. Part of a laser beam striking a surface is reflected and the remainder passes through (see Fig. 2.17). Find the angle between the surface and the part that passes through.

Laser beam $AB \perp CO$

Fig. 2.17

35. Find the distance on Dundas St. W between Dufferin St. and Ossington Ave. in Toronto, as shown in Fig. 2.18. The north–south streets are parallel.

Fig. 2.18

36. An electric circuit board has equally spaced parallel wires with connections at points A, B, and C, as shown in Fig. 2.19. How far is A from C, if $BC = 2.15\,\text{cm}$?

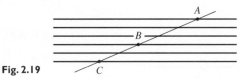

Fig. 2.19

In Exercises 37–40, solve the given problems related to Fig 2.20.

37. $\angle 1 + \angle 2 + \angle 3 = ?$

38. $\angle 4 + \angle 2 + \angle 5 = ?$

$AD \parallel BC$

(W) 39. Based on Exercise 38, what conclusion can be drawn about a closed geometric figure like the one with vertices at A, B, and D?

Fig. 2.20

(W) 40. Based on Exercises 38 and 39, what conclusion can be drawn about a closed geometric figure like the one with vertices A, B, C, and D?

Answers to Practice Exercises

1. $35°$ **2.** $138°$

2.2 Triangles

Types and Properties of Triangles • Perimeter and Area • Hero's Formula • Pythagorean Theorem • Similar Triangles

When part of a plane is bounded and closed by straight-line segments, it is called a **polygon,** *and it is named according to the number of sides it has. A* **triangle** *has three sides, a* **quadrilateral** *has four sides, a* **pentagon** *has five sides, a* **hexagon** *has six sides, and so on.* The most important polygons are the triangle, which we consider in this section, and the quadrilateral, which we study in the next section.

■ The properties of the triangle are important in the study of trigonometry, which we start in Chapter 4.

TYPES AND PROPERTIES OF TRIANGLES

In a **scalene triangle,** no two sides are equal in length. In an **isosceles triangle,** two of the sides are equal in length, and the two *base angles* (the angles opposite the equal sides) are equal. In an **equilateral triangle,** the three sides are equal in length, and each of the three angles is $60°$.

The most important triangle in technical applications is the **right triangle.** *In a right triangle, one of the angles is a right angle. The side opposite the right angle is the* **hypotenuse,** *and the other two sides are called* **legs.**

EXAMPLE 1 Types of triangles

Figure 2.21(a) shows a scalene triangle; each side is of a different length. Figure 2.21(b) shows an isosceles triangle with two equal sides of 2 in. and equal base angles of $40°$. Figure 2.21(c) shows an equilateral triangle with equal sides of 5 cm and equal angles of $60°$. Figure 2.21(d) shows a right triangle. The hypotenuse is side AB.

Fig. 2.21 (a) (b) (c) (d) ■

One very important property of a triangle is that

NOTE ▶ *the sum of the measures of the three angles of a triangle is* **180°**.

In the next example, we show this property by using material from Section 2.1.

Fig. 2.22

EXAMPLE 2 Sum of angles of triangle

In Fig. 2.22, since $\angle 1$, $\angle 2$, and $\angle 3$ constitute a straight angle,

$$\angle 1 + \angle 2 + \angle 3 = 180°$$

Also, by noting alternate interior angles, we see that $\angle 1 = \angle 4$ and $\angle 3 = \angle 5$. Therefore, by substitution, we have

$$\angle 4 + \angle 2 + \angle 5 = 180° \qquad \text{see Exercises 37–40 on page 53}$$

Therefore, if two of the angles of a triangle are known, the third may be found by subtracting the sum of the first two from 180°. ■

EXAMPLE 3 Application of sum of angles

An airplane is flying north and then makes a 90° turn to the west. Later, it makes another left turn of 150°. What is the angle of a third left turn that will cause the plane to again fly north? See Fig. 2.23.

From Fig. 2.23, we see that the interior angle of the triangle at A is the supplement of 150°, or 30°. Since the sum of the measures of the interior angles of the triangle is 180°, the interior angle at B is

$$\angle B = 180° - (90° + 30°) = 60°$$

The required angle is the supplement of 60°, which is 120°. ■

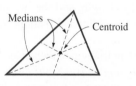

Fig. 2.23

A line segment drawn from a vertex of a triangle to the *midpoint* of the opposite side is called a **median** of the triangle. A basic property of a triangle is that *the three medians meet at a single point, called the* **centroid** *of the triangle.* See Fig. 2.24. Also, *the three* **angle bisectors** (lines from the vertices that divide the angles in half) *meet at a common point.* See Fig. 2.25.

Practice Exercise

1. If the triangle in Fig. 2.24 is isosceles and the vertex angle (at the left) is 30°, what are the base angles?

Fig. 2.24

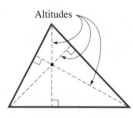

Fig. 2.25

Fig. 2.26

An **altitude** *(or* **height***) of a triangle is the line segment drawn from a vertex perpendicular to the opposite side (or its extension), which is called the* **base** *of the triangle. The three altitudes of a triangle meet at a common point.* See Fig. 2.26. The three common points of the medians, angle bisectors, and altitudes are generally not the same point for a given triangle.

PERIMETER AND AREA OF A TRIANGLE

We now consider two of the most basic measures of a plane geometric figure. The first of these is its **perimeter,** *which is the total distance around it.* In the following example, we find the perimeter of a triangle.

2.56 m 3.22 m

4.89 m

Fig. 2.27

EXAMPLE 4 Perimeter of triangle

Find the perimeter p of a triangle with sides 2.56 m, 3.22 m, and 4.89 m. See Fig. 2.27.
Using the definition of perimeter, for this triangle we have

$$p = 2.56 + 3.22 + 4.89 = 10.67 \text{ m}$$

Therefore, the distance around the triangle is 10.67 m. We express the results to hundredths since each side is given to hundredths. ■

The second important measure of a geometric figure is its **area.** Although the concept of area is primarily intuitive, it is easily defined and calculated for the basic geometric figures. *Area gives a measure of the surface of the figure,* just as perimeter gives the measure of the distance around it.

The area A of a triangle of base b and altitude h is

$$A = \tfrac{1}{2}bh \qquad (2.2)$$

The following example illustrates the use of Eq. (2.2).

EXAMPLE 5 Area of triangle

Find the areas of the triangles in Fig. 2.28(a) and Fig. 2.28(b).
Even though the triangles are of different shapes, the base b of each is 16.2 in., and the altitude h of each is 5.75 in. Therefore, the area of each triangle is

$$A = \tfrac{1}{2}bh = \tfrac{1}{2}(16.2)(5.75) = 46.6 \text{ in.}^2 \qquad ■$$

5.75 in.

16.2 in. 16.2 in.
(a) (b)

Fig. 2.28

Another formula for the area of a triangle that is particularly useful when we have *a triangle with three known sides and no right angle* is **Hero's formula,** which is

■ Named for Hero (or Heron), a first-century Greek mathematician.

$$A = \sqrt{s(s-a)(s-b)(s-c)},$$
$$\text{where } s = \tfrac{1}{2}(a+b+c) \qquad (2.3)$$

In Eq. (2.3), a, b, and c are the lengths of the sides, and s is one-half of the perimeter.

EXAMPLE 6 Application of Hero's formula

A surveyor measures the sides of a triangular parcel of land between two intersecting straight roads to be 206 ft, 293 ft, and 187 ft. Find the area of this parcel (see Fig. 2.29).
To use Eq. (2.3), we first find s:

$$s = \tfrac{1}{2}(206 + 293 + 187)$$
$$= \tfrac{1}{2}(686) = 343 \text{ ft}$$

Now, substituting in Eq. (2.3), we have

$$A = \sqrt{343(343-206)(343-293)(343-187)}$$
$$= 19{,}100 \text{ ft}^2$$

187 ft
293 ft
206 ft

Fig. 2.29

```
(206+293+187)/2→
X
              343
√(X(X-206)(X-293
)(X-187))
      19144.96801
```

Fig. 2.30

The calculator solution is shown in Fig. 2.30. Note that the value of s was stored in memory (as x) and then used to find A. Using the calculator, it is not necessary to write down anything but the final result, properly *rounded off.* ■

Practice Exercise

2. What is the perimeter of the triangle in Fig. 2.29?

■ Named for the Greek mathematician Pythagoras (sixth century B.C.E.)

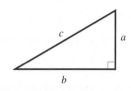

Fig. 2.31

THE PYTHAGOREAN THEOREM

As we have noted, one of the most important geometric figures in technical applications is the right triangle. A very important property of a right triangle is given by the **Pythagorean theorem,** which states that

*in a **right triangle,** the square of the length of the hypotenuse equals the sum of the squares of the lengths of the other two sides.*

If c is the length of the hypotenuse and a and b are the lengths of the other two sides (see Fig. 2.31), the Pythagorean theorem is

$$c^2 = a^2 + b^2 \qquad (2.4)$$

EXAMPLE 7 Application of Pythagorean theorem

A pole is perpendicular to the level ground around it. A guy wire is attached 3.20 m up the pole and at a point on the ground, 2.65 m from the pole. How long is the guy wire?

We sketch the pole and guy wire as shown in Fig. 2.32. Using the Pythagorean theorem, and then substituting, we have

$$AC^2 = AB^2 + BC^2$$
$$= 2.65^2 + 3.20^2$$
$$AC = \sqrt{2.65^2 + 3.20^2} = 4.15\,\text{m}$$

Fig. 2.32   **Fig. 2.33**

The guy wire is 4.15 m long. The calculator solution is shown in Fig 2.33. Note the use of parentheses to group $2.65^2 + 3.20^2$. ■

SIMILAR TRIANGLES

■ Calculators can be programmed to perform specific calculations. See Appendix C for a graphing calculator program PYTHAGTH. It can be used to find a side of a right triangle, given the other two sides.

The perimeter and area of a triangle are measures of its *size*. We now consider the shape of triangles.

Two triangles are **similar** *if they have the same shape (but not necessarily the same size).* There are two very important properties of similar triangles.

PROPERTIES OF SIMILAR TRIANGLES

1. *The corresponding angles of similar triangles are equal.*

2. *The corresponding sides of similar triangles are proportional.*

NOTE ▶ For two triangles that are similar, *if one property is true, then the other is also true.* In two similar triangles, the **corresponding sides** *are the sides, one in each triangle, that are between the same pair of equal corresponding angles.*

EXAMPLE 8 Similar triangles

In Fig. 2.34, a pair of similar triangles are shown. They are similar even though the corresponding parts are not in the same position relative to the page. Using standard symbols, we can write $\triangle ABC \sim \triangle A'B'C'$, where $\triangle$ means "triangle" and $\sim$ means "is similar to."

The pairs of corresponding angles are A and A', B and B', and C and C'. This means $A = A'$, $B = B'$, and $C = C'$.

The pairs of corresponding sides are AB and $A'B'$, BC and $B'C'$, and AC and $A'C'$. In order to show that these corresponding sides are proportional, we write

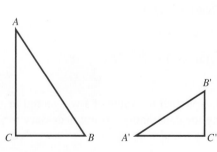

Fig. 2.34

$$\frac{AB}{A'B'} = \frac{BC}{B'C'} = \frac{AC}{A'C'} \quad \begin{array}{l} \leftarrow\text{sides of } \triangle ABC \\ \leftarrow\text{sides of } \triangle A'B'C' \end{array}$$

■

If we know that two triangles are similar, we can use the two basic properties of similar triangles to find the unknown parts of one triangle from the known parts of the other triangle. The next example illustrates this in a practical application.

EXAMPLE 9 Application of similar triangles

Fig. 2.35

On level ground, a silo casts a shadow 24 ft long. At the same time, a nearby vertical pole 4.0 ft high casts a shadow 3.0 ft long. How tall is the silo? See Fig. 2.35.

The rays of the sun are essentially parallel. The two triangles in Fig. 2.35 are similar since *each has a right angle and the angles at the tops are equal.* The other angles must be equal since the sum of the angles is 180°. The lengths of the hypotenuses are of no importance in this problem, so we use only the other sides in stating the ratios of corresponding sides. Denoting the height of the silo as h, we have

$$\frac{h}{4.0} = \frac{24}{3.0}, \qquad h = 32 \text{ ft}$$

We conclude that the silo is 32 ft high.   ■

Practice Exercise

3. In Fig. 2.35, knowing the value of h, what is the distance between the top of the silo and the end of its shadow?

One of the most practical uses of similar geometric figures is that of **scale drawings.** Maps, charts, architectural blueprints, engineering sketches, and most drawings that appear in books (including many that have already appeared, and will appear, in this text) are familiar examples of scale drawings.

In any scale drawing, all distances are drawn at a certain ratio of the distances that they represent, and all angles are drawn equal to the angles they represent. Note the distances and angles shown in Fig. 2.36 in the following example.

EXAMPLE 10 Scale drawing

In drawing a map of the area shown in Fig. 2.36, a scale of 1 cm = 200 km is used. In measuring the distance between Chicago and Toronto on the map, we find it to be 3.5 cm. The actual distance x between Chicago and Toronto is found from the proportion

If we did not have the scale but knew that the distance between Chicago and Toronto is 700 km, then by measuring distances on the map between Chicago and Toronto (3.5 cm) and between Toronto and Philadelphia (2.7 cm), we could find the distance between Toronto and Philadelphia. It is found from the following proportion, determined by use of similar triangles:

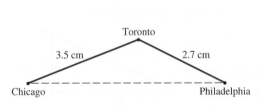

Fig. 2.36

$$\frac{700 \text{ km}}{3.5 \text{ cm}} = \frac{y}{2.7 \text{ cm}}$$

$$y = \frac{2.7(700)}{3.5} = 540 \text{ km}$$ ■

Similarity requires *equal* angles and *proportional* sides. *If the corresponding angles and the corresponding sides of two triangles are equal, the two triangles are* **congruent.** As a result of this definition, the areas and perimeters of congruent triangles are also equal. Informally, we can say that similar triangles have the same shape, whereas congruent triangles have the same shape and same size.

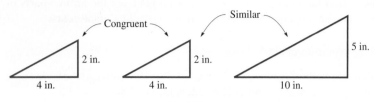

Fig. 2.37

EXAMPLE 11 Similar and congruent triangles

A right triangle with legs of 2 in. and 4 in. is congruent to any other right triangle with legs of 2 in. and 4 in. It is also similar to any right triangle with legs of 5 in. and 10 in., or any other right triangle that has legs in the ratio of 1 to 2, since the corresponding sides are proportional. See Fig. 2.37. ■

EXERCISES 2.2

In Exercises 1–4, answer the given questions about the indicated examples of this section.

1. In Example 2, if $\angle 1 = 70°$ and $\angle 5 = 45°$ in Fig. 2.22, what is the measure of $\angle 2$?

2. In Example 5, change 16.2 in. to 61.2 in. What is the answer?

3. In Example 7, change 2.65 m to 6.25 m. What is the answer?

4. In Example 9, interchange 4.0 ft and 3.0 ft. What is the answer?

In Exercises 5–8, determine $\angle A$ in the indicated figures.

5. Fig. 2.38(a)
6. Fig. 2.38(b)
7. Fig. 2.38(c)
8. Fig. 2.38(d)

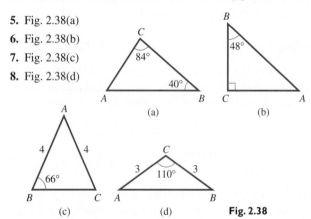

Fig. 2.38

In Exercises 9–16, find the area of each triangle.

9. Fig. 2.39(a)
10. Fig. 2.39(b)
11. Fig. 2.39(c)
12. Fig. 2.39(d)

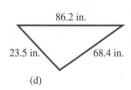

Fig. 2.39

13. Right triangle with legs 3.46 ft and 2.55 ft.
14. Right triangle with legs 234 mm and 342 mm.
15. Isosceles triangle, equal sides of 0.986 m, third side of 0.884 m
16. Equilateral triangle of sides 320 yd

In Exercises 17–20, find the perimeter of each triangle.

17. Fig. 2.39(c) 18. Fig. 2.39(d)
19. An equilateral triangle of sides 21.5 cm
20. Isosceles triangle, equal sides of 2.45 in., third side of 3.22 in.

In Exercises 21–24, find the third side of the right triangle shown in Fig. 2.40 for the given values.

21. $a = 13.8$ ft, $b = 22.7$ ft
22. $a = 2.48$ m, $b = 1.45$ m
23. $a = 175$ cm, $c = 551$ cm
24. $b = 0.474$ in., $c = 0.836$ in.

Fig. 2.40

In Exercises 25–28, use the right triangle in Fig. 2.41.

25. Find $\angle B$.
26. Find side c.
27. Find the perimeter.
28. Find the area.

Fig. 2.41

In Exercises 29–54, solve the given problems.

29. What is the angle between the bisectors of the acute angles of a right triangle?

W 30. If the midpoints of the sides of an isosceles triangle are joined, another triangle is formed. What do you conclude about this inner triangle?

31. For what type of triangle is the centroid the same as the intersection of altitudes and the intersection of angle bisectors?

W 32. Is it possible that the altitudes of a triangle meet, when extended, outside the triangle? Explain.

W **33.** The altitude to the hypotenuse of a right triangle divides the triangle into two smaller triangles. What do you conclude about these two triangles?

W **34.** The altitude to the hypotenuse of a right triangle divides the triangle into two smaller triangles. What do you conclude about the original triangle and the two new triangles? Explain.

35. In Fig. 2.42, show that $\triangle MKL \sim \triangle MNO$.

36. In Fig. 2.43, show that $\triangle ACB \sim \triangle ADC$.

Fig. 2.42   **Fig. 2.43**

37. In Fig. 2.42, if $KN = 15$, $MN = 9$, and $MO = 12$, find LM.

38. In Fig. 2.43, if $AD = 9$ and $AC = 12$, find AB.

39. The angle between the roof sections of an A-frame house is 50°. What is the angle between either roof section and a horizontal rafter?

40. A transmitting tower is supported by a wire that makes an angle of 52° with the level ground. What is the angle between the tower and the wire?

41. A wall pennant is in the shape of an isosceles triangle. If each equal side is 76.6 cm long and the third side is 30.6 cm, what is the area of the pennant?

42. The Bermuda Triangle is sometimes defined as an equilateral triangle 1600 km on a side, with vertices in Bermuda, Puerto Rico, and the Florida coast. Assuming it is flat, what is its approximate area?

43. The sail of a sailboat is in the shape of a right triangle with sides of 8.0 ft, 15 ft, and 17 ft. What is the area of the sail?

44. An observer is 550 m horizontally from the launch pad of a rocket. After the rocket has ascended 750 m, how far is it from the observer?

45. The base of a 20.0-ft ladder is 8.0 ft from a wall. How far up on the wall does the ladder reach?

46. The beach shade shown in Fig. 2.44 is made of 30°-60°-90° triangular sections. Find x. (In a 30°-60°-90° triangle, the side opposite the 30° angle is one-half the hypotenuse.)

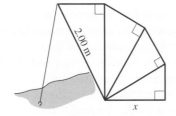

Fig. 2.44

47. A rectangular room is 18 ft long, 12 ft wide, and 8.0 ft high. What is the length of the longest diagonal from one corner to another corner of the room?

48. On a blueprint, a hallway is 45.6 cm long. The scale is 1.2 cm = 1.0 m. How long is the hallway?

49. Two parallel guy wires are attached to a vertical pole 4.5 m and 5.4 m above the ground. They are secured on the level ground at points 1.2 m apart. How long are the guy wires?

50. The two sections of a folding door, hinged in the middle, are at right angles. If each section is 2.5 ft wide, how far are the hinges from the far edge of the other section?

51. A 4.0-ft wall stands 2.0 ft from a building. The ends of a straight pole touch the building and the ground 6.0 ft from the wall. A point on the pole touches the top of the wall. How long is the pole? See Fig. 2.45.

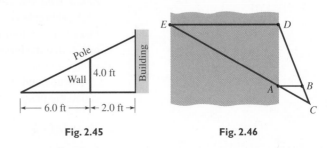

Fig. 2.45   **Fig. 2.46**

52. To find the width ED of a river, a surveyor places markers at A, B, C, and D, as shown in Fig. 2.46. The markers are placed such that $AB \parallel ED$, $BC = 50.0$ ft, $DC = 312$ ft, and $AB = 80.0$ ft. How wide is the river?

53. A water pumping station is to be built on a river at point P in order to deliver water to points A and B. See Fig. 2.47. The design requires that $\angle APD = \angle BPC$ so that the total length of piping that will be needed is a minimum. Find this minimum length of pipe.

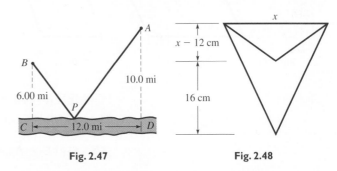

Fig. 2.47   **Fig. 2.48**

54. The cross section of a drainage trough has the shape of an isosceles triangle whose depth is 12 cm less than its width. If the depth is increased by 16 cm and the width remains the same, the area of the cross section is increased by 160 cm². Find the original depth and width. See Fig. 2.48.

Answers to Practice Exercises

1. 75°   **2.** 686 ft   **3.** 40 ft

2.3 Quadrilaterals

Types and Properties of Quadrilaterals • Perimeter and Area

| Fig. 2.49 | Fig. 2.50 |

A **quadrilateral** *is a closed plane figure with four sides,* and these four sides form four interior angles. A general quadrilateral is shown in Fig. 2.49.

A **diagonal** *of a polygon is a straight line segment joining any two nonadjacent vertices.* The dashed line is one of two diagonals of the quadrilateral in Fig. 2.50.

TYPES OF QUADRILATERALS

A **parallelogram** *is a quadrilateral in which opposite sides are parallel.* In a parallelogram, opposite sides are equal and opposite angles are equal. *A* **rhombus** *is a parallelogram with four equal sides.*

A **rectangle** *is a parallelogram in which intersecting sides are perpendicular,* which means that all four interior angles are right angles. In a rectangle, the longer side is usually called the **length,** and the shorter side is called the **width.** *A* **square** *is a rectangle with four equal sides.*

A **trapezoid** *is a quadrilateral in which two sides are parallel.* The parallel sides are called the **bases** of the trapezoid.

Practice Exercise

1. Develop a formula for the length d of a diagonal for the rectangle in Fig. 2.51(c).

EXAMPLE 1 Types of quadrilaterals

A parallelogram is shown in Fig. 2.51(a). Opposite sides a are equal in length, as are opposite sides b. A rhombus with equal sides s is shown in Fig. 2.51(b). A rectangle is shown in Fig. 2.51(c). The length is labeled l, and the width is labeled w. A square with equal sides s is shown in Fig. 2.51(d). A trapezoid with bases b_1 and b_2 is shown in Fig. 2.51(e).

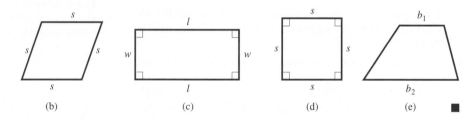

| Fig. 2.51 | (a) | (b) | (c) | (d) | (e) |

PERIMETER AND AREA OF A QUADRILATERAL

The perimeter of a quadrilateral is the sum of the lengths of the four sides.

EXAMPLE 2 Application of perimeter

An architect designs a room with a rectangular window 36 in. high and 21 in. wide, with another window above in the shape of an equilateral triangle, 21 in. on a side. See Fig. 2.52. How much molding is needed for these windows?

The length of molding is the sum of the perimeters of the windows. For the rectangular window, the opposite sides are equal, which means the perimeter is twice the length l plus twice the width w. For the equilateral triangle, the perimeter is three times the side s. Therefore, the length L of molding is

$$L = 2l + 2w + 3s$$
$$= 2(36) + 2(21) + 3(21)$$
$$= 177 \text{ in.} \qquad ■$$

21 in.

36 in.

21 in.

Fig. 2.52

We could write down formulas for the perimeters of the different kinds of triangles and quadrilaterals. However, if we *remember the meaning of perimeter as being the total distance around a geometric figure,* such formulas are not necessary.

Practice Exercise

2. If we did develop perimeter formulas, what is the formula for the perimeter p of a rhombus of side s?

For the areas of the square, rectangle, parallelogram, and trapezoid, we have the following formulas.

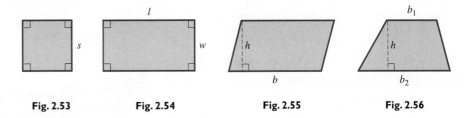

$A = s^2$	Square of side s (Fig. 2.53)	**(2.5)**
$A = lw$	Rectangle of length l and width w (Fig. 2.54)	**(2.6)**
$A = bh$	Parallelogram of base b and height h (Fig. 2.55)	**(2.7)**
$A = \frac{1}{2}h(b_1 + b_2)$	Trapezoid of bases b_1 and b_2 and height h (Fig. 2.56)	**(2.8)**

Fig. 2.53   **Fig. 2.54**   **Fig. 2.55**   **Fig. 2.56**

Since a rectangle, a square, and a rhombus are special types of parallelograms, the area of these figures can be found from Eq (2.7). The area of a trapezoid is of importance when we find areas of irregular geometric figures in Section 2.5.

EXAMPLE 3 Application of area

A city park is designed with lawn areas in the shape of a right triangle, a parallelogram, and a trapezoid, as shown in Fig. 2.57, with walkways between them. Find the area of each section of lawn and the total lawn area.

$$A_1 = \tfrac{1}{2}bh = \tfrac{1}{2}(72)(45) = 1600 \text{ ft}^2$$
$$A_2 = bh = (72)(45) = 3200 \text{ ft}^2$$
$$A_3 = \tfrac{1}{2}h(b_1 + b_2) = \tfrac{1}{2}(45)(72 + 35) = 2400 \text{ ft}^2$$

The total lawn area is about 7200 ft^2.

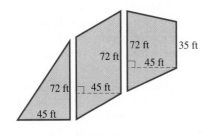

Fig. 2.57

■ The computer microprocessor chip was first commercially available in 1971.

Fig. 2.58

EXAMPLE 4 Perimeter in a word problem

The length of a rectangular computer chip is 2.0 mm longer than its width. Find the dimensions of the chip if its perimeter is 26.4 mm.

Since the dimensions, the length and the width, are required, let w = the width of the chip. Since the length is 2.0 mm more than the width, we know that $w + 2.0$ = the length of the chip. See Fig. 2.58.

Since the perimeter of a rectangle is twice the length plus twice the width, we have the equation

$$2(w + 2.0) + 2w = 26.4$$

since the perimeter is given as 26.4 mm. This is the equation we need.

Solving this equation, we have

$$2w + 4.0 + 2w = 26.4$$
$$4w = 22.4$$
$$w = 5.6 \text{ mm} \quad \text{and} \quad w + 2.0 = 7.6 \text{ mm}$$

Therefore, the length is 7.6 mm and the width is 5.6 mm. These values check with the statements of the original problem. ■

EXERCISES 2.3

In Exercises 1–4, make the given changes in the indicated examples of this section and then solve the given problems.

1. In Example 1, interchange the lengths of b_1 and b_2 in Fig. 2.51(e). What type of quadrilateral is the resulting figure?

2. In Example 2, change the equilateral triangle of side 21 in. to a square of side 21 in. and then find the length of molding.

3. In Example 3, change the dimension of 45 ft to 55 ft in each figure and then find the area.

4. In Example 4, change 2.0 mm to 3.0 mm, and then solve.

In Exercises 5–12, find the perimeter of each figure.

5. Square: side of 65 m
6. Rhombus: side of 2.46 ft
7. Rectangle: $l = 0.920$ in., $w = 0.742$ in.
8. Rectangle: $l = 142$ cm, $w = 126$ cm
9. Parallelogram in Fig. 2.59
10. Parallelogram in Fig. 2.60
11. Trapezoid in Fig. 2.61
12. Trapezoid in Fig. 2.62

Fig. 2.59   **Fig. 2.60**

Fig. 2.61   **Fig. 2.62**

In Exercises 13–20, find the area of each figure.

13. Square: $s = 2.7$ mm
14. Square: $s = 15.6$ ft
15. Rectangle: $l = 0.920$ in., $w = 0.742$ in.
16. Rectangle: $l = 142$ cm, $w = 126$ cm
17. Parallelogram in Fig. 2.59
18. Parallelogram in Fig. 2.60
19. Trapezoid in Fig. 2.61
20. Trapezoid in Fig. 2.62

In Exercises 21–24, set up a formula for the indicated perimeter or area. (Do not include dashed lines.)

21. The perimeter of the figure in Fig. 2.63 (a parallelogram and a square attached)

22. The perimeter of the figure in Fig. 2.64 (two trapezoids attached)

23. Area of figure in Fig. 2.63

24. Area of figure in Fig. 2.64

Fig. 2.63

Fig. 2.64

In Exercises 25–40, solve the given problems.

(W) 25. If the angle between adjacent sides of a parallelogram is 90°, what conclusion can you make about the parallelogram?

(W) 26. What conclusion can you make about the two triangles formed by the sides and diagonal of a parallelogram? Explain.

27. Find the area of a square whose diagonal is 24.0 cm.

28. In a trapezoid, find the angle between the bisectors of the two angles formed by the bases and one nonparallel side.

(W) 29. Noting how a diagonal of a rhombus divides an interior angle, explain why the automobile jack in Fig. 2.65 is in the shape of a rhombus.

Fig. 2.65   **Fig. 2.66**

30. Part of an electric circuit is wired in the configuration of a rhombus and one of its altitudes as shown in Fig. 2.66. What is the length of wire in this part of the circuit?

31. A walkway 3.0 m wide is constructed along the outside edge of a square courtyard. If the perimeter of the courtyard is 320 m, what is the perimeter of the square formed by the outer edge of the walkway?

32. An architect designs a rectangular window such that the width of the window is 18 in. less than the height. If the perimeter of the window is 180 in., what are its dimensions?

33. A designer plans the top of a rectangular workbench to be four times as long as it is wide and then determines that if the width is 2.5 ft greater and the length is 4.7 ft less, it would be a square. What are its dimensions?

34. A beam support in a building is in the shape of a parallelogram, as shown in Fig. 2.67. Find the area of the side of the beam shown.

Fig. 2.67   **Fig. 2.68**

35. Each of two walls (with rectangular windows) of an A-frame house has the shape of a trapezoid as shown in Fig. 2.68. If a gallon of paint covers 320 ft², how much paint is required to paint these walls? (All data are accurate to two significant digits.)

36. Six equal trapezoidal sections form a conference table in the shape of a hexagon, with a hexagonal opening in the middle. See Fig. 2.69. From the dimensions shown, find the area of the table top.

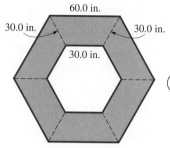

60.0 in.
30.0 in. 30.0 in.
30.0 in.

Fig. 2.69

37. A fenced section of a ranch is in the shape of a quadrilateral whose sides are 1.74 km, 1.46 km, 2.27 km, and 1.86 km, the last two sides being perpendicular to each other. Find the area of the section.

38. A rectangular security area is enclosed on one side by a wall, and the other sides are fenced. The length of the wall is twice the width of the area. The total cost of building the wall and fence is $13,200. If the wall costs $50.00/m and the fence costs $5.00/m, find the dimensions of the area.

W 39. What is the sum of the measures of the interior angles of a quadrilateral? Explain.

40. Find a formula for the area of a rhombus in terms of its diagonals d_1 and d_2. (See Exercise 29.)

Answers to Practice Exercises

1. $d = \sqrt{l^2 + w^2}$ **2.** $p = 4s$

2.4   Circles

Properties of Circles • Tangent and Secant Lines • Circumference and Area • Circular Arcs and Angles • Radian Measure of an Angle

The next geometric figure we consider is the circle. *All points on a* **circle** *are at the same distance from a fixed point, the* **center** *of the circle. The distance from the center to a point on the circle is the* **radius** *of the circle. The distance between two points on the circle on a line through the center is the* **diameter.** Therefore, the diameter d is twice the radius r, or $d = 2r$. See Fig. 2.70.

There are also certain types of lines associated with a circle. *A* **chord** *is a line segment having its endpoints on the circle. A* **tangent** *is a line that touches (does not pass through) the circle at one point. A* **secant** *is a line that passes through two points of the circle.* See Fig. 2.71.

An important property of a tangent is that *a tangent is perpendicular to the radius drawn to the point of contact.* This is illustrated in the following example.

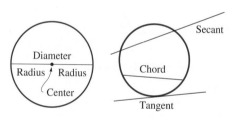

Diameter
Radius   Radius
Center

Secant
Chord
Tangent

Fig. 2.70   **Fig. 2.71**

EXAMPLE 1   Tangent line perpendicular to radius

In Fig. 2.72, O is the center of the circle, and AB is tangent at B. If $\angle OAB = 25°$, find $\angle AOB$.

Since the center is O, OB is a radius of the circle. A tangent is perpendicular to a radius at the point of tangency, which means $\angle ABO = 90°$ so that

$$\angle OAB + \angle OBA = 25° + 90° = 115°$$

Since the sum of the angles of a triangle is 180°, we have

$$\angle AOB = 180° - 115° = 65°   \blacksquare$$

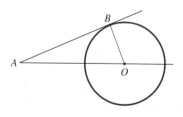

Fig. 2.72

CIRCUMFERENCE AND AREA OF A CIRCLE

The perimeter of a circle is called the **circumference.** The formulas for the circumference and area of a circle are as follows:

■ The symbol π (the Greek letter pi), which we use as a number, was first used in this way as a number in the 1700s.

■ On the TI-83 and TI-84, values are stored in memory using up to 14 digits with a two-digit exponent.

$c = 2\pi r$	Circumference of a circle of radius r	**(2.9)**
$A = \pi r^2$	Area of a circle of radius r	**(2.10)**

Here, π equals approximately 3.1416. In using a calculator, π can be entered to a much greater accuracy by using the $\boxed{\pi}$ key.

Fig. 2.73

EXAMPLE 2 Application of area

A circular oil spill has a diameter of 2.4 km. It is to be enclosed within special flexible tubing. What is the area of the spill, and how long must the tubing be? See Fig. 2.73.

Since $d = 2r$, $r = d/2 = 1.2$ km. Using Eq. (2.10), the area is

$$A = \pi r^2 = \pi(1.2)^2$$
$$= 4.5 \text{ km}^2$$

The length of tubing needed is the circumference of the circle. Therefore,

$$c = 2\pi r = 2\pi(1.2) \qquad \text{note that } c = \pi d$$
$$= 7.5 \text{ km} \qquad \text{rounded off}$$ ∎

Many applied problems involve a combination of geometric figures. The following example illustrates one such combination.

EXAMPLE 3 Application of perimeter and area

$s = 3.25$ in.

Fig. 2.74

A machine part is a square of side 3.25 in. with a quarter-circle removed (see Fig. 2.74). Find the perimeter and the area of one side of the part.

Setting up a formula for the perimeter, we add the two sides of length s to *one-fourth of the circumference of a circle with radius s*. For the area, we *subtract the area of one-fourth of a circle from the area of the square*. This gives

$$p = \underset{\substack{\text{bottom} \\ \text{and left}}}{2s} + \underset{\substack{\text{circular} \\ \text{section}}}{\frac{2\pi s}{4}} = 2s + \frac{\pi s}{2} \qquad A = \underset{\text{square}}{s^2} - \underset{\substack{\text{quarter} \\ \text{circle}}}{\frac{\pi s^2}{4}}$$

where s is the side of the square and the radius of the circle. Evaluating, we have

$$p = 2(3.25) + \frac{\pi(3.25)}{2} = 11.6 \text{ in.}$$

$$A = 3.25^2 - \frac{\pi(3.25)^2}{4} = 2.27 \text{ in.}^2$$

Fig. 2.75

The calculator solution for p and A is shown in Fig. 2.75. Note that s is stored (as x) in order to shorten the solution. ∎

CIRCULAR ARCS AND ANGLES

An arc is part of a circle, and *an angle formed at the center by two radii is a* **central angle.** The measure of an arc is the same as the central angle between the ends of the radii that define the arc. *A* **sector** *of a circle is the region bounded by two radii and the arc they intercept. A* **segment** *of a circle is the region bounded by a chord and its arc.* (There are two possible segments for a given chord. The smaller region is a *minor segment,* and the larger region is a *major segment.*) These are illustrated in the following example.

EXAMPLE 4 Sector and segment

In Fig. 2.76, a sector of the circle is between radii OA and OB and arc AB (denoted as $\overset{\frown}{AB}$). If the measure of the central angle at O between the radii is 70°, the measure of $\overset{\frown}{AB}$ is 70°.

A segment of the circle is the region between chord BC and arc BC ($\overset{\frown}{BC}$). ∎

An **inscribed angle** *of an arc is one for which the endpoints of the arc are points on the sides of the angle and for which the vertex is a point (not an endpoint) of the arc.* An important property of a circle is that *the measure of an inscribed angle is one-half of its intercepted arc.*

Practice Exercises

1. Find the circumference of a circle with a radius of 20.0 cm.
2. Find the area of the circle in Exercise 1.

Fig. 2.76

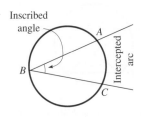

Inscribed angle

Intercepted arc

Fig. 2.77

Fig. 2.78

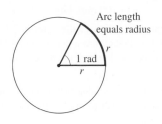

Arc length equals radius

1 rad

Fig. 2.79

EXAMPLE 5 Inscribed angle

(a) In the circle shown in Fig. 2.77, $\angle ABC$ is inscribed in $\overset{\frown}{ABC}$, and it intercepts $\overset{\frown}{AC}$. If $\overset{\frown}{AC} = 60°$, then $\angle ABC = 30°$.

(b) In the circle shown in Fig. 2.78, PQ is a diameter, and $\angle PRQ$ is inscribed in the semicircular $\overset{\frown}{PRQ}$. Since $\overset{\frown}{PSQ} = 180°$, $\angle PRQ = 90°$. From this we conclude that *an angle inscribed in a semicircle is a right angle.* ■

RADIAN MEASURE OF AN ANGLE

To this point, we have measured all angles in degrees. There is another measure of an angle, the *radian,* that is defined in terms of an arc of a circle. We will find it of importance when we study trigonometry.

If a central angle of a circle intercepts an arc equal in length to the radius of the circle, the measure of the central angle is defined as **1 radian.** See Fig. 2.79. The radius can be marked off along the circumference 2π times (about 6.283 times). Thus, 2π rad $= 360°$ (where rad is the symbol for radian), and the basic relationship between radians and degrees is

$$\boxed{\pi \text{ rad} = 180°} \qquad (2.11)$$

EXAMPLE 6 Radian measure of an angle

(a) If we divide each side of Eq. (2.11) by π, we get

$$1 \text{ rad} = 57.3°$$

where the result has been rounded off.

(b) To change an angle of 118.2° to radian measure, we have

$$118.2° = 118.2°\left(\frac{\pi \text{ rad}}{180°}\right) = 2.06 \text{ rad}$$

Multiplying 118.2° by π rad/180°, the unit that remains is rad, since degrees "cancel." We will review radian measure again when we study trigonometry. ■

Practice Exercise

3. Express the angle 85.0° in radian measure.

EXERCISES 2.4

In Exercises 1–4, answer the given questions about the indicated examples of this section.

1. In Example 1, if $\angle AOB = 72°$ in Fig. 2.72, then what is the measure of $\angle OAB$?

2. In the first line of Example 2, if "diameter" is changed to "radius," what are the results?

3. In Example 3, if the machine part is the unshaded part (rather than the shaded part) of Fig. 2.74, what are the results?

4. In Example 5(a), if $\angle ABC = 25°$ in Fig. 2.77, then what is the measure of $\overset{\frown}{AC}$?

In Exercises 5–8, refer to the circle with center at O in Fig. 2.80. Identify the following.

5. (a) A secant line
   (b) A tangent line

6. (a) Two chords
   (b) An inscribed angle

7. (a) Two perpendicular lines
   (b) An isosceles triangle

8. (a) A segment
   (b) A sector with an acute central angle

Fig. 2.80

In Exercises 9–12, find the circumference of the circle with the given radius or diameter.

9. $r = 275$ ft   **10.** $r = 0.563$ m

11. $d = 23.1$ mm   **12.** $d = 8.2$ in.

In Exercises 13–16, find the area of the circle with the given radius or diameter.

13. $r = 0.0952$ yd   **14.** $r = 45.8$ cm

15. $d = 2.33$ m   **16.** $d = 1256$ ft

In Exercises 17–20, refer to Fig. 2.81, where AB is a diameter, TB is a tangent line at B, and $\angle ABC = 65°$. Determine the indicated angles.

17. $\angle CBT$

18. $\angle BCT$

19. $\angle CAB$

20. $\angle BTC$

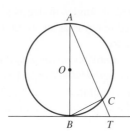

Fig. 2.81

In Exercises 21–24, refer to Fig. 2.82. Determine the indicated arcs and angles.

21. $\overset{\frown}{BC}$

22. $\overset{\frown}{AB}$

23. $\angle ABC$

24. $\angle ACB$

Fig. 2.82

In Exercises 25–28, change the given angles to radian measure.

25. $22.5°$ **26.** $60.0°$ **27.** $125.2°$ **28.** $323.0°$

In Exercises 29–32, find a formula for the indicated perimeter or area.

Fig. 2.83   **Fig. 2.84**

29. The perimeter of the quarter-circle in Fig. 2.83

30. The perimeter of the figure in Fig. 2.84. A quarter-circle is attached to a triangle.

31. The area of the segment of the quarter-circle in Fig. 2.83

32. The area of the figure in Fig. 2.84

In Exercises 33–48, solve the given problems.

(W) 33. Describe the location of the midpoints of a set of parallel chords of a circle.

34. The measure of $\overset{\frown}{AB}$ on a circle of radius r is $45°$. What is the length of the arc in terms of r and π?

35. In a circle, a chord connects the ends of two perpendicular radii of 6.00 in. What is the area of the minor segment?

(W) 36. In Fig. 2.85, chords AB and DE are parallel. What is the relation between $\triangle ABC$ and $\triangle CDE$? Explain.

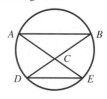

Fig. 2.85

37. The radius of the earth's equator is 3960 mi. What is the circumference?

38. As a ball bearing rolls along a straight track, it makes 11.0 revolutions while traveling a distance of 109 mm. Find its radius.

39. The rim on a basketball hoop has an inside diameter of 18.0 in. The largest cross section of a basketball has a diameter of 12.0 in. What is the ratio of the cross sectional area of the basketball to the area of the hoop?

40. With no change in the speed of flow, by what factor should the diameter of a pipe be increased in order to double the amount of water that flows through the pipe?

41. Using a tape measure, the circumference of a tree is found to be 112 in. What is the diameter of the tree (assuming a circular cross section)?

42. What is the area of the largest circle that can be cut from a rectangular plate 21.2 cm by 15.8 cm?

43. A window designed between semicircular regions is shown in Fig. 2.86. Find the area of the window.

Fig. 2.86   **Fig. 2.87**

44. The cross section of a large circular conduit has seven smaller equal circular conduits within it. The conduits are tangent to each other as shown in Fig. 2.87. What fraction of the large conduit is occupied by the seven smaller conduits?

45. Find the area of the room in the plan shown in Fig. 2.88.

Fig. 2.88

46. Find the length of the pulley belt shown in Fig. 2.89 if the belt crosses at right angles. The radius of each pulley wheel is 5.50 in.

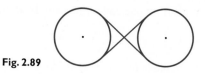

Fig. 2.89

W **47.** The velocity of an object moving in a circular path is directed tangent to the circle in which it is moving. A stone on a string moves in a vertical circle, and the string breaks after 5.5 revolutions. If the string was initially in a vertical position, in what direction does the stone move after the string breaks? Explain.

48. Part of a circular gear with 24 teeth is shown in Fig. 2.90. Find the indicated angle.

Fig. 2.90

Answers to Practice Exercises

1. 126 cm   **2.** 1260 cm^2 **3.** 1.48 rad

2.5 Measurement of Irregular Areas

Trapezoidal Rule • Simpson's Rule

To this point, the figures for which we have found areas are well defined, and the areas can be found by direct use of a specific formula. In practice, however, it may be necessary to find the area of a figure with an irregular perimeter or one for which there is no specific formula. In this section, we show two methods of finding a very good *approximation* of such an area. These methods are particularly useful in technical areas such as surveying, architecture, and mechanical design.

THE TRAPEZOIDAL RULE

The first method is based on dividing the required area into trapezoids with equal heights. Considering the area shown in Fig. 2.91, we draw parallel lines at n equal intervals between the edges of the area. We then join the ends of these parallel line segments to form adjacent trapezoids. The sum of the areas of the trapezoids gives a good approximation to the required area.

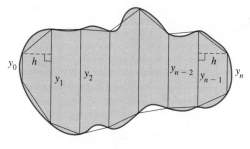

Fig. 2.91

Calling the lengths of the parallel lines $y_0, y_1, y_2, \ldots, y_n$ and the height of each trapezoid h (the distance between the parallel lines), the total area A is the sum of areas of all the trapezoids. This gives us

first trapezoid, second trapezoid, third trapezoid, next-to-last trapezoid, last trapezoid

$$A = \frac{h}{2}(y_0 + y_1) + \frac{h}{2}(y_1 + y_2) + \frac{h}{2}(y_2 + y_3) + \cdots + \frac{h}{2}(y_{n-2} + y_{n-1}) + \frac{h}{2}(y_{n-1} + y_n)$$

$$= \frac{h}{2}(y_0 + y_1 + y_1 + y_2 + y_2 + y_3 + \cdots + y_{n-2} + y_{n-1} + y_{n-1} + y_n)$$

Therefore, the approximate area is

■ Note carefully that the values of y_0 and y_n are *not* multiplied by 2.

$$A = \frac{h}{2}(y_0 + 2y_1 + 2y_2 + \cdots + 2y_{n-1} + y_n) \tag{2.12}$$

Equation (2.12) is known as the **trapezoidal rule.** The following examples illustrate its use.

Fig. 2.92

EXAMPLE 1 Application of trapezoidal rule

A plate cam for opening and closing a valve is shown in Fig. 2.92. Widths of the face of the cam are shown at 2.00-cm intervals from one end of the cam. Find the area of the face of the cam.

From the figure, we see that

$$y_0 = 2.56 \text{ cm} \qquad y_1 = 3.82 \text{ cm} \qquad y_2 = 3.25 \text{ cm}$$
$$y_3 = 2.95 \text{ cm} \qquad y_4 = 1.85 \text{ cm} \qquad y_5 = 0.00 \text{ cm}$$

(In making such measurements, often a y-value at one end—or both ends—is zero. In such a case, the end "trapezoid" is actually a triangle.) From the given information in this example, $h = 2.00$ cm. Therefore, using the trapezoidal rule, Eq. (2.12), we have

$$A = \frac{2.00}{2}[2.56 + 2(3.82) + 2(3.25) + 2(2.95) + 2(1.85) + 0.00]$$
$$= 26.3 \text{ cm}^2$$

The area of the face of the cam is approximately 26.3 cm^2. ∎

When approximating the area with trapezoids, we omit small parts of the area for some trapezoids and include small extra areas for other trapezoids. The omitted areas often approximate the extra areas, which makes the approximation better. Also, the use of smaller intervals improves the approximation since the total omitted area or total extra area is smaller.

EXAMPLE 2 Application of trapezoidal rule

■ See the chapter introduction.

From a satellite photograph of Lake Ontario (as shown on page 49), one of the Great Lakes between the United States and Canada, measurements of the width of the lake were made along its length, starting at the west end, at 26.0-km intervals. The widths are shown in Fig. 2.93 and are given in the following table.

Distance from West End (km)	0.0	26.0	52.0	78.0	104	130	156
Width (km)	0.0	46.7	52.1	59.2	60.4	65.7	73.9

Distance from West End (km)	182	208	234	260	286	312
Width (km)	87.0	75.5	66.4	86.1	77.0	0.0

Fig. 2.93

Here, we see that $y_0 = 0.0$ km, $y_1 = 46.7$ km, $y_2 = 52.1$ km, ..., and $y_n = 0.0$ km. Therefore, using the trapezoidal rule, the approximate area of Lake Ontario is found as follows:

$$A = \frac{26.0}{2}[0.0 + 2(46.7) + 2(52.1) + 2(59.2) + 2(60.4) + 2(65.7) + 2(73.9)$$
$$+ 2(87.0) + 2(75.5) + 2(66.4) + 2(86.1) + 2(77.0) + 0.0]$$
$$= 19,500 \text{ km}^2$$

The area of Lake Ontario is actually $19,477 \text{ km}^2$. ∎

Practice Exercise

1. In Example 2, use only the distances from west end of (in km) 0.0, 52.0, 104, 156, 208, 260, and 312. Calculate the area and compare with the answer in Example 2.

Parabola

Fig. 2.94

Fig. 2.95

■ ·Named for the English mathematician Thomas Simpson (1710–1761).

SIMPSON'S RULE

For the second method of measuring an irregular area, we also draw parallel lines at equal intervals between the edges of the area. We then join the ends of these parallel lines with curved *arcs*. This takes into account the fact that the perimeters of most figures are curved. The arcs used in this method are not arcs of a circle, but arcs of a *parabola*. A parabola is shown in Fig. 2.94 and is discussed in detail in Chapter 21. (Examples of parabolas are (1) the path of a ball that has been thrown and (2) the cross section of a microwave "dish.")

The development of this method requires advanced mathematics. Therefore, we will simply state the formula to be used. It might be noted that the form of the equation is similar to that of the trapezoidal rule.

The approximate area of the geometric figure shown in Fig. 2.95 is given by

$$A = \frac{h}{3}(y_0 + 4y_1 + 2y_2 + 4y_3 + \cdots + 2y_{n-2} + 4y_{n-1} + y_n)$$ (2.13)

Equation (2.13) is known as **Simpson's rule.** *In using Eq. (2.13), the number n of intervals of width h must be even.*

EXAMPLE 3 Application of Simpson's rule

A parking lot is proposed for a riverfront area in a town. The town engineer measured the widths of the area at 100-ft (three sig. digits) intervals, as shown in Fig. 2.96. Find the area available for parking.

First, we see that there are six intervals, which means Eq. (2.13) may be used. With $y_0 = 407$ ft, $y_1 = 483$ ft, ..., $y_6 = 495$ ft, and $h = 100$ ft, we have

$$A = \frac{100}{3}[407 + 4(483) + 2(382) + 4(378) + 2(285) + 4(384) + 495]$$

$$= 241,000 \text{ ft}^2$$ ■

For most areas, Simpson's rule gives a somewhat better approximation than the trapezoidal rule. The accuracy of Simpson's rule is also usually improved by using smaller intervals.

River

Street

407 ft · 483 ft · 382 ft · 378 ft · 285 ft · 384 ft · 495 ft

Fig. 2.96

EXAMPLE 4 Application of Simpson's rule

From an aerial photograph, a cartographer determines the widths of Easter Island (in the Pacific Ocean) at 1.50-km intervals as shown in Fig. 2.97. The widths found are as follows:

Distance from South End (km)	0	1.50	3.00	4.50	6.00	7.50	9.00	10.5	12.0	13.5	15.0
Width (km)	0	4.8	5.7	10.5	15.2	18.5	18.8	17.9	11.3	8.8	3.1

Since there are ten intervals, Simpson's rule may be used. From the table, we have the following values: $y_0 = 0$, $y_1 = 4.8$, $y_2 = 5.7$, ..., $y_9 = 8.8$, $y_{10} = 3.1$, and $h = 1.5$. Using Simpson's rule, the cartographer would approximate the area of Easter Island as follows:

$$A = \frac{1.50}{3}(0 + 4(4.8) + 2(5.7) + 4(10.5) + 2(15.2) + 4(18.5)$$

$$+ 2(18.8) + 4(17.9) + 2(11.3) + 4(8.8) + 3.1) = 174 \text{ km}^2$$ ■

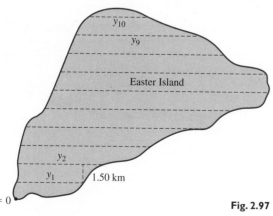

y_{10}

y_9

Easter Island

y_2

y_1 1.50 km

$y_0 = 0$

Fig. 2.97

EXERCISES 2.5

In Exercises 1 and 2, answer the given questions related to the indicated examples of this section.

(W) **1.** In Example 1, if widths of the face of the same cam were given at 1.00-cm intervals (five more widths would be included), from the methods of this section, what is probably the most accurate way of finding the area? Explain.

(W) **2.** In Example 4, if you use only the data from the south end of (in km) 0, 3.00, 6.00, 9.00, 12.0, and 15.0, would you choose the trapezoidal rule or Simpson's rule to calculate the area? Explain. Do not calculate the area for these data.

In Exercises 3 and 4, answer the given questions related to Fig. 2.98.

(W) **3.** Which should be more accurate for finding the area, the trapezoidal rule or Simpson's rule? Explain.

(W) **4.** If the trapezoidal rule is used to find the area, will the result probably be too high, about right, or too little? Explain.

Fig. 2.98

In Exercises 5–16, calculate the indicated areas. All data are accurate to at least two significant digits.

5. The widths of a kidney-shaped swimming pool were measured at 2.0-m intervals, as shown in Fig. 2.99. Calculate the surface area of the pool, using the trapezoidal rule.

Fig. 2.99

6. Calculate the surface area of the swimming pool in Fig. 2.99, using Simpson's rule.

7. The widths of a cross section of an airplane wing are measured at 1.00-ft intervals, as shown in Fig. 2.100. Calculate the area of the cross section, using Simpson's rule.

Fig. 2.100

8. Calculate the area of the cross section of the airplane wing in Fig. 2.100, using the trapezoidal rule.

9. Using aerial photography, the widths of an area burned by a forest fire were measured at 0.5-mi intervals, as shown in the following table:

Distance (mi)	0.0	0.5	1.0	1.5	2.0	2.5	3.0	3.5	4.0
Width (mi)	0.6	2.2	4.7	3.1	3.6	1.6	2.2	1.5	0.8

Determine the area burned by the fire by using the trapezoidal rule.

10. Find the area burned by the forest fire of Exercise 9, using Simpson's rule.

11. A cartographer measured the width of Kruger National Park (and game reserve) in South Africa at 6.0-km intervals on a map, as shown in Fig. 2.101. The widths are shown in the list that follows. Find the area of the park if the scale of the map is 1.0 mm = 6.0 km.

$y_0 = 7$ mm $y_1 = 15$ mm
$y_2 = 7$ mm $y_3 = 11$ mm
$y_4 = 13$ mm $y_5 = 10$ mm
$y_6 = 9$ mm $y_7 = 12$ mm
$y_8 = 8$ mm $y_9 = 3$ mm

Fig. 2.101

12. The widths of an oval-shaped floor were measured at 1.5-m intervals, as shown in the following table:

Distance (m)	0.0	1.5	3.0	4.5	6.0	7.5	9.0	10.5	12.0
Width (m)	0.0	5.0	7.2	8.3	8.6	8.3	7.2	5.0	0.0

Find the area of the floor by using Simpson's rule.

13. The widths of the baseball playing area in Boston's Fenway Park at 45-ft intervals are shown in Fig. 2.102. Find the playing area using the trapezoidal rule.

14. Find the playing area of Fenway Park (see Exercise 13) by Simpson's rule.

Fig. 2.102

15. Soundings taken across a river channel give the following depths with the corresponding distances from one shore.

Distance (ft)	0	50	100	150	200	250	300	350	400	450	500
Depth (ft)	5	12	17	21	22	25	26	16	10	8	0

Find the area of the cross section of the channel using Simpson's rule.

16. The widths of a bell crank are measured at 2.0-in. intervals, as shown in Fig. 2.103. Find the area of the bell crank if the two connector holes are each 2.50 in. in diameter.

3.5 in.
6.0 in.
7.6 in.
10.8 in.
16.2 in.
18.6 in.
19.0 in.
17.8 in. — 2.0 in.
12.5 in.
8.2 in.

Fig. 2.103

In Exercises 17–20, calculate the area of the circle by the indicated method.

The lengths of parallel chords of a circle that are 0.250 in. apart are given in the following table. The diameter of the circle is 2.000 in. The distance shown is the distance from one end of a diameter.

Distance (in.)	0.000	0.250	0.500	0.750	1.000	1.250	1.500	1.750	2.000
Length (in.)	0.000	1.323	1.732	1.936	2.000	1.936	1.732	1.323	0.000

Using the formula $A = \pi r^2$, the area of the circle is 3.14 in.^{2}.

W **17.** Find the area of the circle using the trapezoidal rule and only the values of distance of 0.000 in., 0.500 in., 1.000 in., 1.500 in., and 2.000 in. with the corresponding values of the chord lengths. Explain why the value found is less than 3.14 in.^{2}.

W **18.** Find the area of the circle using the trapezoidal rule and all values in the table. Explain why the value found is closer to 3.14 in.^{2} than the value found in Exercise 17.

W **19.** Find the area of the circle using Simpson's rule and the same table values as in Exercise 17. Explain why the value found is closer to 3.14 in.^{2} than the value found in Exercise 17.

W **20.** Find the area of the circle using Simpson's rule and all values in the table. Explain why the value found is closer to 3.14 in.^{2} than the value found in Exercise 19.

Answer to Practice Exercise

1. 18,100 km^2

2.6 Solid Geometric Figures

Rectangular Solid • Cylinder • Prism • Cone • Pyramid • Sphere • Frustum

We now review the formulas for the *volume* and *surface area* of some basic solid geometric figures. Just as area is a measure of the surface of a plane geometric figure, **volume** is a measure of the space occupied by a solid geometric figure.

One of the most common solid figures is the **rectangular solid.** This figure has six sides (**faces**), and opposite sides are rectangles. All intersecting sides are perpendicular to each other. The **bases** of the rectangular solid are the top and bottom faces. A **cube** is a rectangular solid with all six faces being equal squares.

A **right circular cylinder** is *generated* by rotating a rectangle about one of its sides. Each **base** is a circle, and the *cylindrical surface* is perpendicular to each of the bases. The **height** is one side of the rectangle, and the **radius** of the base is the other side.

A **right circular cone** is generated by rotating a right triangle about one of its legs. The **base** is a circle, and the **slant height** is the hypotenuse of the right triangle. The **height** is one leg of the right triangle, and the **radius** of the base is the other leg.

The bases of a **right prism** are equal and parallel polygons, and the sides are rectangles. The **height** of a prism is the perpendicular distance between bases. The base of a **pyramid** is a polygon, and the other faces, the **lateral faces,** are triangles that meet at a common point, the **vertex.** A **regular pyramid** has congruent triangles for its lateral faces.

A **sphere** is generated by rotating a circle about a diameter. The **radius** is a line segment joining the center and a point on the sphere. The **diameter** is a line segment through the center and having its endpoints on the sphere.

In the following formulas, V represents the *volume*, A represents the *total surface area*, S represents the *lateral surface area* (bases not included), B represents the *area of the base*, and p represents the *perimeter of the base*.

Fig. 2.104

Fig. 2.105

Fig. 2.106

Fig. 2.107

Fig. 2.108

Fig. 2.109

Fig. 2.110

Fig. 2.111

$V = lwh$	Rectangular solid (Fig. 2.104)	**(2.14)**
$A = 2lw + 2lh + 2wh$		**(2.15)**
$V = e^3$	Cube (Fig. 2.105)	**(2.16)**
$A = 6e^2$		**(2.17)**
$V = \pi r^2 h$	Right circular cylinder (Fig. 2.106)	**(2.18)**
$A = 2\pi r^2 + 2\pi rh$		**(2.19)**
$S = 2\pi rh$		**(2.20)**
$V = Bh$	Right prism (Fig. 2.107)	**(2.21)**
$S = ph$		**(2.22)**
$V = \frac{1}{3}\pi r^2 h$	Right circular cone (Fig. 2.108)	**(2.23)**
$A = \pi r^2 + \pi rs$		**(2.24)**
$S = \pi rs$		**(2.25)**
$V = \frac{1}{3}Bh$	Regular pyramid (Fig. 2.109)	**(2.26)**
$S = \frac{1}{2}ps$		**(2.27)**
$V = \frac{4}{3}\pi r^3$	Sphere (Fig. 2.110)	**(2.28)**
$A = 4\pi r^2$		**(2.29)**

Equation (2.21) is valid for any prism, and Eq. (2.26) is valid for any pyramid. There are other types of cylinders and cones, but we restrict our attention to right circular cylinders and right circular cones, and we will often use "cylinder" or "cone" when referring to them.

The **frustum** of a cone or pyramid is the solid figure that remains after the top is cut off by a plane parallel to the base. Figure 2.111 shows the frustum of a cone.

EXAMPLE 1 Volume of rectangular solid

What volume of concrete is needed for a driveway 25.0 m long, 2.75 m wide, and 0.100 m thick?

The driveway is a rectangular solid for which $l = 25.0$ m, $w = 2.75$ m, and $h = 0.100$ m. Using Eq. (2.14), we have

$$V = (25.0)(2.75)(0.100)$$
$$= 6.88 \text{ m}^3$$ ∎

Fig. 2.112

EXAMPLE 2 Total surface area of right circular cone

Calculate the total surface area of a right circular cone for which the radius $r = 11.9$ cm and the height is $h = 10.4$ cm. See Fig. 2.112.

To find the total surface area using Eq. (2.24), we need the radius and the slant height s of the cone. Therefore, we must first find s. The radius and height are legs of a right triangle, and the slant height is the hypotenuse. To find s, we *use the Pythagorean theorem:*

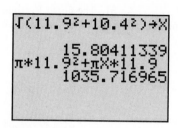

Fig. 2.113

$$s^2 = r^2 + h^2 \qquad \text{Pythagorean theorem}$$
$$s = \sqrt{r^2 + h^2} \qquad \text{solve for } s$$
$$= \sqrt{11.9^2 + 10.4^2} = 15.8 \text{ cm}$$

Now, calculating the total surface area, we have

$$A = \pi r^2 + \pi r s \qquad \text{Eq. (2.24)}$$
$$= \pi (11.9)^2 + \pi (11.9)(15.8) \qquad \text{substituting}$$
$$= 1040 \text{ cm}^2$$

Practice Exercise

1. Find the volume of the cone in Example 2.

Note in the calculator display in Fig. 2.113 that we stored s as x, and did not have to actually record its value. ∎

EXAMPLE 3 Volume of combination of solids

A grain storage building is in the shape of a cylinder surmounted by a hemisphere (*half a sphere*). See Fig. 2.114. Find the volume of grain that can be stored if the height of the cylinder is 122 ft and its radius is 40.0 ft.

The total volume of the structure is the volume of the cylinder plus the volume of the hemisphere. By the construction we see that the radius of the hemisphere is the same as the radius of the cylinder. Therefore,

$$V = \overset{\text{cylinder}}{\pi r^2 h} + \overset{\text{hemisphere}}{\frac{1}{2}\left(\frac{4}{3}\pi r^3\right)} = \pi r^2 h + \frac{2}{3}\pi r^3$$

$$= \pi (40.0)^2 (122) + \frac{2}{3}\pi (40.0)^3$$

$$= 747{,}000 \text{ ft}^3 \qquad ∎$$

Fig. 2.114

EXERCISES 2.6

In Exercises 1–4, answer the given questions about the indicated examples of this section.

1. In Example 1, if the length is doubled and the thickness is doubled, by what factor is the volume changed?

2. In Example 2, if the value of the slant height $s = 17.5$ cm is given instead of the height, what is the height?

3. In Example 2, if the radius is halved and the height is doubled, what is the volume?

4. In Example 3, if h is halved, what is the volume?

In Exercises 5–20, find the volume or area of each solid figure for the given values. See Figs. 2.104 to 2.110.

5. Volume of cube: $e = 7.15$ ft

6. Volume of right circular cylinder: $r = 23.5$ cm, $h = 48.4$ cm

7. Total surface area of right circular cylinder: $r = 689$ m, $h = 233$ m

8. Area of sphere: $r = 0.067$ in.

9. Volume of sphere: $r = 0.877$ yd

10. Volume of right circular cone: $r = 25.1$ mm, $h = 5.66$ mm

11. Lateral area of right circular cone: $r = 78.0$ cm, $s = 83.8$ cm

12. Lateral area of regular pyramid: $p = 345$ ft, $s = 272$ ft

13. Volume of regular pyramid: square base of side 76 in., $h = 130$ in.

14. Volume of right prism: square base of side 29.0 cm, $h = 11.2$ cm

15. Lateral area of regular prism: equilateral triangle base of side 1.092 m, $h = 1.025$ m

16. Lateral area of right circular cylinder: diameter = 250 ft, $h = 347$ ft

17. Volume of hemisphere: diameter = 0.83 yd

18. Volume of regular pyramid: square base of side 22.4 m, $s = 14.2$ m

19. Total surface area of right circular cone: $r = 3.39$ cm, $h = 0.274$ cm

20. Total surface area of pyramid: All faces and base are equilateral triangles of side 3.67 in.

In Exercises 21–40, solve the given problems.

21. Eq. (2.28) expresses the volume V of a sphere in terms of the radius r. Express V in terms of the diameter d.

22. Derive a formula for the total surface area A of a hemispherical volume of radius r (curved surface and flat surface).

23. The radius of a cylinder is twice as long as the radius of a cone, and the height of the cylinder is half as long as the height of the cone. What is the ratio of the volume of the cylinder to that of the cone?

24. The base area of a cone is one-fourth of the total area. Find the ratio of the radius to the slant height.

25. In designing a weather balloon, it is decided to double the diameter of the balloon so that it can carry a heavier instrument load. What is the ratio of the final surface area to the original surface area?

26. During a rainfall of 1.00 in., what weight of water falls on an area of 1.00 mi²? Each cubic foot of water weighs 62.4 lb.

27. A rectangular box is to be used to store radioactive materials. The inside of the box is 12.0 in. long, 9.50 in. wide, and 8.75 in. deep. What is the area of sheet lead that must be used to line the inside of the box?

28. A swimming pool is 50.0 ft wide, 78.0 ft long, 3.50 ft deep at one end, and 8.75 ft deep at the other end. How many cubic feet of water can it hold? (The slope on the bottom is constant.) See Fig. 2.115.

78.0 ft
50.0 ft
3.50 ft
8.75 ft

Fig. 2.115

29. The Alaskan oil pipeline is 750 mi long and has a diameter of 4.0 ft. What is the maximum volume of the pipeline?

30. A glass prism used in the study of optics has a right triangular base. The legs of the triangle are 3.00 cm and 4.00 cm. The prism is 8.50 cm high. What is the total surface area of the prism? See Fig. 2.116.

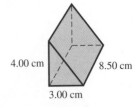

4.00 cm
8.50 cm
3.00 cm

Fig. 2.116

31. The Great Pyramid of Egypt has a square base approximately 250 yd on a side. The height of the pyramid is about 160 yd. What is its volume? See Fig. 2.117.

160 yd

Fig. 2.117

32. A paper cup is in the shape of a cone as shown in Fig. 2.118. What is the surface area of the cup?

3.60 in.
3.50 in.

Fig. 2.118

165 ft

Fig. 2.119

33. *Spaceship Earth* (shown in Fig. 2.119) at Epcot Center in Florida is a sphere of 165 ft in diameter. What is the volume of *Spaceship Earth*?

34. A propane tank is constructed in the shape of a cylinder with a hemisphere at each end, as shown in Fig. 2.120. Find the volume of the tank.

6.50 ft
4.00 ft

Fig. 2.120

35. A special wedge in the shape of a regular pyramid has a square base 16.0 mm on a side. The height of the wedge is 40.0 mm. What is the total surface area of the wedge?

36. A lawn roller is a cylinder 0.96.0 m long and 0.60 m in diameter. How many revolutions of the roller are needed to roll 76 m² of lawn?

37. The circumference of a basketball is about 29.8 in. What is its volume?

38. What is the area of a paper label that is to cover the lateral surface of a cylindrical can 3.00 in. in diameter and 4.25 in. high? The ends of the label will overlap 0.25 in. when the label is placed on the can.

39. The side view of a rivet is shown in Fig. 2.121. It is a conical part on a cylindrical part. Find the volume of the rivet.

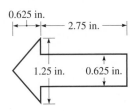

Fig. 2.121

40. A dipstick is made to measure the volume remaining in the conical container shown in Fig. 2.122. How far below the full mark (at the top of the container) on the stick should the mark for half-full be placed?

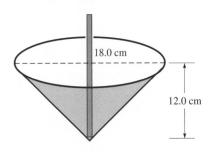

Fig. 2.122

Answer to Practice Exercise

1. 1540 cm³

CHAPTER 2   EQUATIONS

Line segments	Fig. 2.8	$\dfrac{a}{b} = \dfrac{c}{d}$	(2.1)
Triangle		$A = \frac{1}{2}bh$	(2.2)
Hero's formula		$A = \sqrt{s(s-a)(s-b)(s-c)}$, where $s = \frac{1}{2}(a+b+c)$	(2.3)
Pythagorean theorem	Fig. 2.31	$c^2 = a^2 + b^2$	(2.4)
Square	Fig. 2.53	$A = s^2$	(2.5)
Rectangle	Fig. 2.54	$A = lw$	(2.6)
Parallelogram	Fig. 2.55	$A = bh$	(2.7)
Trapezoid	Fig. 2.56	$A = \frac{1}{2}h(b_1 + b_2)$	(2.8)
Circle		$c = 2\pi r$	(2.9)
		$A = \pi r^2$	(2.10)
Radians	Fig. 2.79	$\pi \text{ rad} = 180°$	(2.11)
Trapezoidal rule	Fig. 2.91	$A = \dfrac{h}{2}(y_0 + 2y_1 + 2y_2 + \cdots + 2y_{n-1} + y_n)$	(2.12)
Simpson's rule	Fig. 2.95	$A = \dfrac{h}{3}(y_0 + 4y_1 + 2y_2 + 4y_3 + \cdots + 2y_{n-2} + 4y_{n-1} + y_n)$	(2.13)

Rectangular solid	Fig. 2.104	$V = lwh$	(2.14)
		$A = 2lw + 2lh + 2wh$	(2.15)
Cube	Fig. 2.105	$V = e^3$	(2.16)
		$A = 6e^2$	(2.17)
Right circular cylinder	Fig. 2.106	$V = \pi r^2 h$	(2.18)
		$A = 2\pi r^2 + 2\pi rh$	(2.19)
		$S = 2\pi rh$	(2.20)
Right prism	Fig. 2.107	$V = Bh$	(2.21)
		$S = ph$	(2.22)
Right circular cone	Fig. 2.108	$V = \frac{1}{3}\pi r^2 h$	(2.23)
		$A = \pi r^2 + \pi rs$	(2.24)
		$S = \pi rs$	(2.25)
Regular pyramid	Fig. 2.109	$V = \frac{1}{3}Bh$	(2.26)
		$S = \frac{1}{2}ps$	(2.27)
Sphere	Fig. 2.110	$V = \frac{4}{3}\pi r^3$	(2.28)
		$A = 4\pi r^2$	(2.29)

CHAPTER 2   REVIEW EXERCISES

In Exercises 1–4, use Fig. 2.123. Determine the indicated angles.

1. $\angle CGE$ 32°
2. $\angle EGF$ 58°
3. $\angle DGH$ 32°
4. $\angle EGI$

$AB \parallel CD$

Fig. 2.123

In Exercises 5–12, find the indicated sides of the right triangle shown in Fig. 2.124.

5. $a = 9, b = 40, c = ?$
6. $a = 14, b = 48, c = ?$
7. $a = 400, b = 580, c = ?$
8. $b = 56, c = 65, a = ?$
9. $a = 6.30, b = 3.80, c = ?$
10. $a = 126, b = 25.1, c = ?$
11. $b = 29.3, c = 36.1, a = ?$
12. $a = 0.782, c = 0.885, b = ?$ **Fig. 2.124**

In Exercises 13–20, find the perimeter or area of the indicated figure.

13. Perimeter: equilateral triangle of side 8.5 mm
14. Perimeter: rhombus of side 15.2 in.
15. Area: triangle, $b = 3.25$ ft, $h = 1.88$ ft
16. Area: triangle of sides 175 cm, 138 cm, 119 cm
17. Circumference of circle: $d = 98.4$ mm
18. Perimeter: rectangle, $l = 2980$ yd, $w = 1860$ yd
19. Area: trapezoid, $b_1 = 67.2$ in., $b_2 = 126.7$ in., $h = 34.2$ in.
20. Area: circle, $d = 32.8$ m

In Exercises 21–24, find the volume of the indicated solid geometric figure.

21. Prism: base is right triangle with legs 26.0 cm and 34.0 cm, height is 14.0 cm
22. Cylinder: base radius 36.0 in., height 2.40 in.
23. Pyramid: base area 3850 ft^2, height 125 ft
24. Sphere: diameter 22.1 mm

In Exercises 25–28, find the surface area of the indicated solid geometric figure.

25. Total area of cube of edge 0.520 m
26. Total area of cylinder: base diameter 1.20 ft, height 5.80 ft
27. Lateral area of cone: base radius 1.82 in., height 11.5 in.
28. Total area of sphere: $d = 12{,}760$ km

In Exercises 29–32, use Fig. 2.125. Line CT is tangent to the circle with center at O. Find the indicated angles.

29. $\angle BTA$
30. $\angle TAB$
31. $\angle BTC$
32. $\angle ABT$

Fig. 2.125

In Exercises 33–36, use Fig. 2.126. Given that $AB = 4$, $BC = 4$, $CD = 6$, and $\angle ADC = 53°$, find the indicated angle and lengths.

33. $\angle ABE$

34. AD

35. BE

36. AE

Fig. 2.126

In Exercises 37–40, find the formulas for the indicated perimeters and areas.

37. Perimeter of Fig. 2.127 (a right triangle and semicircle attached)

38. Perimeter of Fig. 2.128 (a square with a quarter circle at each end)

39. Area of Fig. 2.127   **40.** Area of Fig. 2.128

Fig. 2.127   **Fig. 2.128**

In Exercises 41–46, answer the given questions.

41. Is a square also a rectangle, a parallelogram, and a rhombus?

42. If the measures of two angles of one triangle equal the measures of two angles of a second triangle, are the two triangles similar?

(W) 43. If the dimensions of a plane geometric figure are each multiplied by n, by how much is the area multiplied? Explain, using a circle to illustrate.

(W) 44. If the dimensions of a solid geometric figure are each multiplied by n, by how much is the volume multiplied? Explain, using a cube to illustrate.

45. What is an equation relating chord segments a, b, c, and d shown in Fig. 2.129? The dashed chords are an aid in the solution.

Fig. 2.129

46. From a common point, two line segments are tangent to the same circle. If the angle between the lines segments is 36°, what is the angle between the two radii of the circle drawn from the points of tangency?

In Exercises 47–72, solve the given problems.

47. A tooth on a saw is in the shape of an isosceles triangle. If the angle at the point is 38°, find the two base angles.

48. A lead sphere 1.50 in. in diameter is flattened into a circular sheet 14.0 in. in diameter. How thick is the sheet?

49. A ramp for the disabled is designed so that it rises 1.2 m over a horizontal distance of 7.8 m. How long is the ramp?

50. An airplane is 2100 ft directly above one end of a 9500-ft runway. How far is the plane from the glide-slope indicator on the ground at the other end of the runway?

51. A machine part is in the shape of a square with equilateral triangles attached to two sides (see Fig. 2.130). Find the perimeter of the machine part.

Fig. 2.130   **Fig. 2.131**

52. A patio is designed with semicircular areas attached to a square, as shown in Fig. 2.131. Find the area of the patio.

53. A radio transmitting tower is supported by guy wires. The tower and three parallel guy wires are shown in Fig. 2.132. Find the distance AB along the tower.

Fig. 2.132   **Fig. 2.133**

54. Find the areas of lots A and B in Fig. 2.133. A has a frontage on Main St. of 140 ft, and B has a frontage on Main St. of 84 ft. The boundary between lots is 120 ft.

55. To find the height of a flagpole, a person places a mirror at M, as shown in Fig. 2.134. The person's eyes at E are 160 cm above the ground at A. From physics, it is known that $\angle AME = \angle BMF$. If $AM = 120$ cm and $MB = 4.5$ m, find the height BF of the flagpole.

Fig. 2.134   **Fig. 2.135**

56. The metal support in the form of $\triangle ABC$ shown in Fig. 2.135 is strengthened by brace DE, which is parallel to BC. How long is the brace if $AB = 24$ in., $AD = 16$ in., and $BC = 33$ in.?

57. A typical scale for an aerial photograph is 1/18450. In an 8.00-by 10.0-in. photograph with this scale, what is the longest distance (in mi) between two locations in the photograph?

58. For a hydraulic press, the mechanical advantage is the ratio of the large piston area to the small piston area. Find the mechanical advantage if the pistons have diameters of 3.10 cm and 2.25 cm.

59. The diameter of the earth is 7920 mi, and a satellite is in orbit at an altitude of 210 mi. How far does the satellite travel in one rotation about the earth?

60. The roof of the Louisiana Superdome in New Orleans is supported by a circular steel tension ring 651 m in circumference. Find the area covered by the roof.

61. A rectangular piece of wallboard with two holes cut out for heating ducts is shown in Fig. 2.136. What is the area of the remaining piece?

Fig. 2.136

62. The diameter of the sun is 1.38×10^6 km, the diameter of the earth is 1.27×10^4 km, and the distance from the earth to the sun (center to center) is 1.50×10^8 km. What is the distance from the center of the earth to the end of the shadow due to the rays from the sun?

63. Using aerial photography, the width of an oil spill is measured at 250-m intervals, as shown in Fig. 2.137. Using Simpson's rule, find the area of the oil spill.

Fig. 2.137

64. To build a highway, it is necessary to cut through a hill. A surveyor measured the cross-sectional areas at 250-ft intervals through the cut as shown in the following table. Using the trapezoidal rule, determine the volume of soil to be removed.

Dist. (ft)	0	250	500	750	1000	1250	1500	1750
Area (ft²)	560	1780	4650	6730	5600	6280	2260	230

65. The Hubble space telescope is within a cylinder 4.3 m in diameter and 13 m long. What is the volume within this cylinder?

66. A horizontal cross section of a concrete bridge pier is a regular hexagon (six sides, all equal in length, and all internal angles are equal), each side of which is 2.50 m long. If the height of the pier is 6.75 m, what is the volume of concrete in the pier?

67. A railroad track 1000.00 ft long expands 0.20 ft (2.4 in.) during the afternoon (due to an increase in temperature of about 30°F). Assuming that the track cannot move at either end and that the increase in length causes a bend straight up in the middle of the track, how high is the top of the bend?

68. Two persons are talking to each other on cellular phones. If the angle between their signals at the tower is 90°, as shown in Fig. 2.138, how far apart are they?

Fig. 2.138

69. A hot-water tank is in the shape of a right circular cylinder surmounted by a hemisphere as shown in Fig. 2.139. How many gallons does the tank hold? (1.00 ft³ contains 7.48 gal.)

Fig. 2.139

70. A tent is in the shape of a regular pyramid surmounted on a cube. If the edge of the cube is 2.50 m and the total height of the tent is 3.25 m, find the area of the material used in making the tent (not including any floor area).

71. The diagonal of a HDTV screen is 107 cm. If the ratio of the width of the screen to the height of the screen is 16 to 9, what are the width and height?

72. A satellite is 1590 km directly above the center of the eye of a circular (approximately) hurricane that has formed in the Atlantic Ocean. The distance from the satellite to the edge of the hurricane is 1620 km. What area does the hurricane cover? Neglect the curvature of the Earth and any possible depth of the hurricane.

Writing Exercise

73. The Pentagon, headquarters of the U.S. Department of Defense, is the world's largest office building. It is a regular pentagon (five sides, all equal in length, and all interior angles are equal) 921 ft on a side, with a diagonal of length 1490 ft. Using these data, draw a sketch and write one or two paragraphs to explain how to find the area covered within the outside perimeter of the Pentagon. (What is the area?)

CHAPTER 2   PRACTICE TEST

1. In Fig. 2.140, determine $\angle 1$.

2. In Fig. 2.140, determine $\angle 2$.

Fig. 2.140 $AB \parallel CD$

3. A tree is 8.0 ft high and casts a shadow 10.0 ft long. At the same time, a telephone pole casts a shadow 25.0 ft long. How tall is the pole?

4. Find the area of a triangle with sides of 2.46 cm, 3.65 cm, and 4.07 cm.

5. What is the diagonal distance between corners of a rectangular field 125 ft wide and 170 ft long?

6. What is the area of the trapezoid shown in Fig. 2.141?

Fig. 2.141

7. The edge of a cube is 4.50 cm. What are (a) the surface area and (b) the volume of the cube?

8. Find the surface area of a tennis ball whose circumference is 21.0 cm.

9. Find the volume of a right circular cone of radius 2.08 m and height 1.78 m.

10. In Fig. 2.142, find $\angle 1$.

11. In Fig. 2.142, find $\angle 2$.

Fig. 2.142   **Fig. 2.143**

12. In Fig. 2.143, find the perimeter of the figure shown. It is a square with a semicircle removed.

13. In Fig. 2.143, find the area of the figure shown.

14. The width of a marshy area is measured at 50-ft intervals, with the results shown in the following table. Using the trapezoidal rule, find the area of the marsh. (All data accurate to two or more significant digits.)

Distance (ft)	0	50	100	150	200	250	300
Width (ft)	0	90	145	260	205	110	20

3 Functions and Graphs

By noting, for example, that stones fall faster than leaves, the Greek philosopher Aristotle (about 350 B.C.E.) reasoned that heavier objects fall faster than lighter ones. For about 2000 years, this idea was generally accepted. Then in about 1600, the Italian scientist Galileo showed, by dropping objects from the Leaning Tower of Pisa, that the distance an object falls in a given time does not depend on its weight.

Galileo is generally credited with first using the *experimental method* by which controlled experiments are used to study natural phenomena. His aim was to find mathematical formulas that could be used to describe these phenomena. He realized that such formulas provided a way of showing a compact and precise relation between the variables.

In technology and science, determining how one quantity depends on others is a primary goal. A rule that shows such a relation is of great importance, and in mathematics such a rule is called a *function*. This chapter starts with a discussion of functions.

Examples of such relations in technology and science, as well as in everyday life, are numerous. Plant growth depends on sunlight and rainfall; traffic flow depends on roadway design; the sales tax on an item depends on the cost of an item; the time to access the Internet depends on how fast a computer processes data; distance traveled depends on time and speed of travel; electric voltage depends on the current and resistance. These are but a few of the innumerable possibilities.

A way of actually seeing how one quantity depends on another is by means of a *graph*. The basic method of graphing was devised by the French mathematicians Descartes and Fermat in the 1630s from their work to combine the methods of algebra and geometry. Their work was very influential in later developments in mathematics and technology, and we will discuss this more in later chapters. In this chapter, we take up these basic methods of drawing and using graphs. In doing so, we will also expand our use of the *graphing calculator*, one of the more recent innovations of technology. As noted in Chapter 1, they were first used in the 1980s.

The electric power produced in a circuit depends on the resistance R in the circuit. In Section 3.4, we draw a graph of $P = \dfrac{100R}{(0.50 + R)^2}$ to see this type of relationship.

3.1 Introduction to Functions

**Definition of Function •
Independent and Dependent Variables •
Functional Notation**

An important method of finding formulas such as those in Section 1.11 is through scientific observation and experimentation, like that of Galileo. For example, if we were to perform an experiment to determine the relationship between the distance an object falls and the time it falls, we should find (approximately, at least) that $s = 16t^2$, where s is the distance in feet and t is the time in seconds. Another example is that electrical measurements of voltage V and current I through a particular resistor would show that $V = kI$, where k is a constant.

Considerations such as these lead us to one of the basic concepts in mathematics, that of a *function*.

Definition of a Function

> *Whenever a relationship exists between two variables—such that for every value of the first, there is only one corresponding value of the second—we say that the second variable is a* **function** *of the first variable.*
>
> *The first variable is called the* **independent variable,** *and the second variable is called the* **dependent variable.**

The first variable is termed *independent* since permissible values can be assigned to it arbitrarily, and the second variable is termed *dependent* since its value is determined by the choice of the independent variable. *Values of the independent variable and dependent variable are to be real numbers.* Therefore, there may be restrictions on their possible values. This is discussed in the following section.

EXAMPLE 1 Examples of functions

(a) In the equation $y = 2x$, y is a function of x, since for each value of x there is only one value of y. For example, if $x = 3$, then $y = 6$ and no other value. By arbitrarily assigning values to x, and substituting, we see that the values of y we obtain *depend* on the values of x. Therefore, x is the independent variable, and y is the dependent variable.

(b) The power P developed in a certain resistor by a current I is given by $P = 4I^2$. Here, P is a function of I. The dependent variable is P and the independent variable is I. ■

EXAMPLE 2 Independent and dependent variables

Fig. 3.1

Figure 3.1 shows a cube of edge e. From Chapter 2, we know that the volume V in terms of the edge e is $V = e^3$. Here, V is a function of e, since for each value of e, there is only one value of V. The dependent variable is V, and the independent variable is e.

If the equation relating the volume and the edge of a cube is written as $e = \sqrt[3]{V}$—that is, if the edge is expressed in terms of the volume—e is a function of V. In this case, e is the dependent variable, and V is the independent variable. ■

There are many ways to express functions. Formulas, tables, charts, and graphs can also define functions. Functions will be of importance throughout the book, and we will use a number of different types of functions in later chapters.

FUNCTIONAL NOTATION

For convenience of notation, the phrase

"function of x" is written as $f(x)$.

This means that "y is a function of x" may be written as $y = f(x)$. Here, f denotes *dependence* and does not represent a quantity or a variable. Therefore, it also follows

CAUTION ▶ that $f(x)$ *does not mean f times x.*

EXAMPLE 3 Functional notation $f(x)$

If $y = 6x^3 - 5x$, we say that y is a function of x. This function is $6x^3 - 5x$. It is also common to write such a function as $f(x) = 6x^3 - 5x$. However, y and $f(x)$ represent the same expression, $6x^3 - 5x$. Using y, the quantities are shown, and using $f(x)$, the functional dependence is shown. ∎

One of the most important uses of functional notation is to designate the value of the function for a particular value of the independent variable. That is,

the value of the function $f(x)$ when $x = a$ is written as $f(a)$.

EXAMPLE 4 Meaning of $f(a)$

For a function $f(x)$, the value of $f(x)$ for $x = 2$ may be expressed as $f(2)$. Thus, substituting 2 for x in $f(x) = 3x - 7$, we have

$$f(2) = 3(2) - 7 = -1 \qquad \text{substitute 2 for } x$$

The value of $f(x)$ for $x = -1.4$ is

$$f(-1.4) = 3(-1.4) - 7 = -11.2 \qquad \text{substitute } -1.4 \text{ for } x \qquad ∎$$

■ A graphing calculator can be used to evaluate a function in several ways. One is to directly substitute the value into the function as shown in Fig. 3.2(a). A second is to enter the function as Y_1 and evaluate as shown in Fig. 3.2(b). A third way, which is very useful when many values are to be used, is to enter the function as Y_1 and use the *table* feature as shown in Fig. 3.2(c).

```
3*2-7
            -1
3* -1.4-7
          -11.2
```

(a)

```
Y₁(2)
            -1
Y₁(-1.4)
          -11.2
```

(b)

X	Y₁
2	-1
-1.4	-11.2

X=

(c)

Fig. 3.2

We must note that *whatever number a represents, to find $f(a)$, we substitute a for x in $f(x)$.* This is true even if a is a literal number, as in the following examples.

EXAMPLE 5 Meaning of $f(a)$

If $g(t) = \dfrac{t^2}{2t + 1}$, to find $g(a^3)$ we substitute a^3 for t in $g(t)$:

$$g(a^3) = \frac{(a^3)^2}{2a^3 + 1} = \frac{a^6}{2a^3 + 1}$$

For the same function, $g(3) = \dfrac{3^2}{2(3) + 1} = \dfrac{9}{7}$. In both cases, to obtain the value of $g(t)$, we simply substitute the value within the parentheses for t. ∎

Practice Exercise

1. For the function in Example 5, find $g(-1)$.

EXAMPLE 6 Application: $f(T)$

The electric resistance R of a certain resistor as a function of the temperature T (in °C) is given by $R = 10.0 + 0.01T + 0.001T^2$. If a given temperature T is increased by 10°C, what is the value of R for the increased temperature as a function of T?

We are to determine R for a temperature of $T + 10$. Since

$$f(T) = 10.0 + 0.10T + 0.001T^2$$

then

$$f(T + 10) = 10.0 + 0.10(T + 10) + 0.001(T + 10)^2 \qquad \text{substitute } T + 10 \text{ for } T$$
$$= 10.0 + 0.10T + 1.0 + 0.001T^2 + 0.02T + 0.1$$
$$= 11.1 + 0.12T + 0.001T^2 \qquad ∎$$

At times, we need to define more than one function. We then use different symbols, such as $f(x)$ and $g(x)$, to denote these functions.

EXAMPLE 7 $f(x)$ and $g(x)$

For the functions $f(x) = 5x - 3$ and $g(x) = ax^2 + x$, where a is a constant, we have

$$f(-4) = 5(-4) - 3 = -23 \qquad \text{substitute } -4 \text{ for } x \text{ in } f(x)$$

$$g(-4) = a(-4)^2 + (-4) = 16a - 4 \qquad \text{substitute } -4 \text{ for } x \text{ in } g(x) \qquad ■$$

Practice Exercise

2. For the function in Example 7, find $f(-a^2x)$ and $g(-a^2x)$.

A function may be looked upon as a set of instructions. These instructions tell us how to obtain the value of the dependent variable for a particular value of the independent variable, even if the instructions are expressed in literal symbols.

EXAMPLE 8 Function as a set of instructions

The function $f(x) = x^2 - 3x$ tells us to "square the value of the independent variable, multiply the value of the independent variable by 3, and subtract the second result from the first." An analogy would be a computer that was programmed so that when a number was entered into the program, it would square the number, then multiply the number by 3, and finally subtract the second result from the first. This is diagramed in Fig. 3.3.

Fig. 3.3

The functions $f(t) = t^2 - 3t$ and $f(n) = n^2 - 3n$ are the same as the function $f(x) = x^2 - 3x$, since the operations performed on the independent variable are

NOTE ▶ the same. ***Although different literal symbols appear, this does not change the function.*** ■

EXERCISES 3.1

In Exercises 1–4, solve the given problems related to the indicated examples of this section.

1. In Example 4, find the value of $f(-2)$.

2. In Example 5, evaluate $g(-a^2)$.

3. In Example 6, change "increased" to "decreased" in the second and third lines and then evaluate the function to find the proper expression.

4. In Example 8, change $x^2 - 3x$ to $x^3 + 4x$ and then determine the statements that should be placed in the three boxes in Fig. 3.3.

In Exercises 5–12, find the indicated functions.

5. Express the area A of a circle as a function of (a) its radius r and (b) its diameter d.

6. Express the circumference c of a circle as a function of (a) its radius r and (b) its diameter d.

7. Express the diameter d of a sphere as a function of its volume V.

8. Express the edge e of a cube as a function of its surface area A.

9. Express the area A of a square as a function of its side s; express the side s of a square as a function of its area A.

10. Express the perimeter p of a square as a function of its side s; express the side s of a square as a function of its perimeter p.

11. A circle is inscribed in a square (the circle is tangent to each side of the square). Express the area A of the four corner regions bounded by the circle and the square as a function of the radius r of the circle.

12. Express the area A of an equilateral triangle as a function of its side s.

In Exercises 13–24, evaluate the given functions.

13. $f(x) = 2x + 1$; find $f(1)$ and $f(-1)$.

14. $f(x) = 5x - 9$; find $f(2)$ and $f(-2)$.

15. $f(x) = 5$; find $f(-2)$ and $f(0.4)$.

16. $f(T) = 7.2 - 2.5|T|$; find $f(2.6)$ and $f(-4)$.

17. $\phi(x) = \dfrac{6 - x^2}{2x}$; find $\phi(2\pi)$ and $\phi(-2)$.

18. $H(q) = \dfrac{8}{q} + 2\sqrt{q}$; find $H(4)$ and $H(0.16)$.

19. $g(t) = at^2 - a^2t$; find $g\left(-\frac{1}{2}\right)$ and $g(a)$.

20. $s(y) = 6\sqrt{y + 11} - 3$; find $s(-2)$ and $s(a^2)$.

21. $K(s) = 3s^2 - s + 6$; find $K(-s)$ and $K(2s)$.

22. $T(t) = 5t + 7$; find $T(-2t)$ and $T(t + 1)$.

23. $f(x) = 2x + 4$; find $f(3x) - 3f(x)$.

24. $f(x) = 2x^2 + 1$; find $f(x + 2) - [f(x) + 2]$.

In Exercises 25–28, evaluate the given functions. The values of the independent variable are approximate.

25. Given $f(x) = 5x^2 - 3x$, find $f(3.86)$ and $f(-6.92)$.

26. Given $g(t) = \sqrt{t + 1.0604} - 6t^3$, find $g(0.9261)$.

27. Given $F(H) = \dfrac{2H^2}{H + 36.85}$, find $F(-84.466)$.

28. Given $f(x) = \dfrac{x^4 - 2.0965}{6x}$, find $f(1.9654)$.

(W) *In Exercises 29–36, state the instructions of the function in words as in Example 9.*

29. $f(x) = x^2 + 2$

30. $f(x) = 2x - 6$

31. $g(y) = 6y - y^3$

32. $\phi(s) = 8 - 5s + s^2$

33. $R(r) = 3(2r + 5) - 1$

34. $f(z) = \dfrac{4z}{5 - z}$

35. $p(t) = \dfrac{2t - 3}{t + 2}$

36. $Y(y) = 2 + \dfrac{5y}{2(y - 3)}$

In Exercises 37–40, write the equation as given by the statement. Then write the indicated function using functional notation.

37. The surface area A of a cubical open-top aquarium equals 5 times the square of an edge e of the aquarium.

38. A helicopter is at an altitude of 1000 m and is x m horizontally from a fire. Its distance d from the fire is the square root of the sum of 1000 squared and x squared.

39. The area A of the Bering Glacier in Alaska, given that its present area is 8430 km² and that it is melting at the rate of $140t$ km², where t is the time in centuries.

40. The electrical resistance R of a certain ammeter, in which the resistance of the coil is R_c, is the product of 10 and R_c divided by the sum of 10 and R_c.

In Exercises 41–48, solve the given problems.

41. A demolition ball is used to tear down a building. Its distance s (in m) above the ground as a function of time t (in s) after it is dropped is $s = 17.5 - 4.9t^2$. Since $s = f(t)$, find $f(1.2)$.

42. The length L (in ft) of a cable hanging between equal supports 100 ft apart is $L = 100(1 + 0.0003s^2)$, where s is the sag (in ft) in the middle of the cable. Since $L = f(s)$, find $f(15)$.

43. The stopping distance d (in ft) of a car going v mi/h is given by $d = v + 0.05v^2$. Since $d = f(v)$, find $f(30)$, $f(2v)$, and $f(60)$, using both $f(v)$ and $f(2v)$.

44. The electric power P (in W) dissipated in a resistor of resistance R (in Ω) is given by the function $P = \dfrac{200R}{(100 + R)^2}$. Since $P = f(R)$, find $f(R + 10)$.

45. A motorist travels at 55 km/h for t hours. Express the distance d traveled as a function of t.

46. The sales tax in a city is 7%. The price tag on an item shows D dollars. Express the total cost C of the item as a function of D.

(W) **47.** (a) Explain the meaning of $f[f(x)]$. (b) Find $f[f(x)]$ for $f(x) = 2x^2$.

48. If $f(x) = x$ and $g(x) = x^2$, find (a) $f[g(x)]$, and (b) $g[f(x)]$.

Answers to Practice Exercises

1. $f(-1) = -1$ **2.** $f(-a^2x) = -5a^2x - 3$; $g(-a^2x) = a^5x^2 - a^2x$

3.2 More about Functions

Domain and Range • Functions from Verbal Statements • Relation

For a given function, *the complete set of possible values of the independent variable is called the* **domain** *of the function, and the complete set of all possible resulting values of the dependent variable is called the* **range** *of the function.*

As noted earlier, *we will be using only* ***real numbers*** when using functions. This means there may be restrictions as to the values that may be used since

NOTE ▶   *values that lead to* **division by zero** *or to* **imaginary numbers may not be included** *in the domain or the range.*

EXAMPLE 1 Domain and range

The function $f(x) = x^2 + 2$ is defined for all real values of x. This means its domain is written as *all real numbers*. However, since x^2 is never negative, $x^2 + 2$ is never less than 2. We then write the range as *all real numbers $f(x) \geq 2$*, where the symbol $\geq$ means "is greater than or equal to."

The function $f(t) = \frac{1}{t + 2}$ is not defined for $t = -2$, for this value would require division by zero. Also, no matter how large t becomes, $f(t)$ will never exactly equal zero. Therefore, the domain of this function is *all real numbers except* -2, and the range is *all real numbers except* 0.   ■

EXAMPLE 2 Domain and range

The function $g(s) = \sqrt{3 - s}$ is not defined for real numbers greater than 3, since such values make $3 - s$ negative and would result in imaginary values for $g(s)$. This means that the domain of this function is *all real numbers $s \leq 3$*, where the symbol $\leq$ means "is less than or equal to."

Also, since $\sqrt{3 - s}$ means the principal square root of $3 - s$ (see Section 1.6), we know that $g(s)$ cannot be negative. This tells us that the range of the function is *all real numbers $g(s) \geq 0$*.   ■

In Examples 1 and 2, we found the domains by looking for values of the independent variable that cannot be used. The range was found through an inspection of the function. We generally use this procedure, although it may be necessary to use more advanced methods to find the range. Therefore, until we develop other methods, we will look only for the domain of some functions.

EXAMPLE 3 Find domain only

Find the domain of the function $f(x) = 16\sqrt{x} + \frac{1}{x}$.

From the term $16\sqrt{x}$, we see that x must be greater than or equal to zero in order to have real values. The term $\frac{1}{x}$ indicates that x cannot be zero, because of division by zero. Thus, putting these together, the domain is *all real numbers $x > 0$*.

As for the range, it is *all real numbers $f(x) \geq 12$*. More advanced methods are needed to determine this.   ■

We have seen that the domain may be restricted since we do not use imaginary numbers or divide by zero. The domain may also be restricted by the definition of the function, or by practical considerations in an application.

EXAMPLE 4 Restricted domain

A function defined as

$$f(x) = x^2 + 4 \qquad (\text{for } x > 2)$$

has a domain restricted to real numbers greater than 2 by definition. Thus, $f(5) = 29$, but $f(1)$ is not defined, since 1 is not in the domain. Also, the range is all real numbers greater than 8.

The height h (in m) of a certain projectile as a function of the time t (in s) is

$$h = 20t - 4.9t^2$$

Negative values of time have no real meaning in this case. This is generally true in applications. Therefore, the domain is $t \geq 0$. Also, since we know the projectile will not continue in flight indefinitely, there is some upper limit on the value of t. These restrictions are not usually stated unless it affects the solution.

There could be negative values of h if it is possible that the projectile is *below* the launching point at some time (such as a stone thrown from the top of a cliff).   ■

■ Note that $f(2) = 6$ and $f(-2) = 6$. This is all right. Two different values of x may give the same value of y, but there may not be two different values of y for one value of x.

Practice Exercise

1. Find the domain and range of the function $f(x) = \sqrt{x + 4}$.

■ Later in this chapter, we will see that the graphing calculator is useful in finding the range of a function.

The following example illustrates a function that is defined differently for different intervals of the domain.

EXAMPLE 5　Function defined for intervals of domain

In a certain electric circuit, the current i (in mA) is a function of the time t (in s), which means $i = f(t)$. The function is

$$f(t) = \begin{cases} 8 - 2t & \text{(for } 0 \le t \le 4 \text{ s)} \\ 0 & \text{(for } t > 4 \text{ s)} \end{cases}$$

Since negative values of t are not usually meaningful, $f(t)$ is not defined for $t < 0$. Find the current for $t = 3$ s, $t = 6$ s, and $t = -1$ s.

We are to find $f(3), f(6)$, and $f(-1)$, and we see that values of this function are determined differently depending on the value of t. Since 3 is between 0 and 4,

$$f(3) = 8 - 2(3) = 2 \quad \text{or} \quad i = 2 \text{ mA}$$

Since 6 is greater than 4, $f(6) = 0$, or $i = 0$ mA. We see that $i = 0$ mA for all values of t that are 4 or greater.

Since $f(t)$ is not defined for $t < 0$, $f(-1)$ is not defined. ■

FUNCTIONS FROM VERBAL STATEMENTS

To find a mathematical function from a verbal statement, we use methods like those for setting up equations in Chapter 1. This is shown in the following examples.

EXAMPLE 6　Function from verbal statement

The fixed cost for a company to operate a certain plant is $3000 per day. It also costs $4 for each unit produced in the plant. Express the daily cost C of operating the plant as a function of the number n of units produced.

The daily total cost C equals the fixed cost of $3000 plus the cost of producing n units. Since the cost of producing one unit is $4, the cost of producing n units is $4n$. Thus, the total cost C, where $C = f(n)$, is

$$C = 3000 + 4n$$

Here, we know that the domain is all values of $n \ge 0$, with some upper limit on n based on the production capacity of the plant. ■

EXAMPLE 7　Function from verbal statement

A metallurgist melts and mixes m grams (g) of solder that is 40% tin with n grams of another solder that is 20% tin to get a final solder mixture that contains 200 g of tin. Express n as a function of m. See Fig. 3.4.

The statement leads to the following equation:

tin in first solder		tin in second solder		total amount of tin
$0.40m$	$+$	$0.20n$	$=$	200

Since we want $n = f(m)$, we now solve for n:

$$0.20n = 200 - 0.40m$$
$$n = 1000 - 2m$$

Grams of tin

Fig. 3.4

This is the required function. Since neither m nor n can be negative, the domain is all values $0 \le m \le 500$ g, which means that m is greater than or equal to 0 g and less than or equal to 500 g. The range is all values $0 \le n \le 1000$ g. ■

$\frac{1}{2}(2\pi r)$

$2r + 10$

$2r$

Fig. 3.5

Practice Exercise

2. In Example 8, find p as a function of r if there is to be a semicircular section of the window at the bottom, as well as at the top.

EXAMPLE 8 Function from verbal statement

An architect designs a window in the shape of a rectangle with a semicircle on top, as shown in Fig. 3.5. The base of the window is 10 cm less than the height of the rectangular part. Express the perimeter p of the window as a function of the radius r of the circular part.

The perimeter is the distance around the window. Since the top part is a semicircle, the length of this top circular part is $\frac{1}{2}(2\pi r)$, and the base of the window is $2r$ because it is equal in length to the dashed line (the diameter of the circle). Finally, the base being 10 cm less than the height of the rectangular part tells us that each vertical side of the rectangle is $2r + 10$. Therefore, the perimeter p, where $p = f(r)$, is

$$p = \tfrac{1}{2}(2\pi r) + 2r + 2(2r + 10)$$
$$= \pi r + 2r + 4r + 20$$
$$= \pi r + 6r + 20$$

We see that the required function is $p = \pi r + 6r + 20$. Since the radius cannot be negative and there would be no window if $r = 0$, the domain of the function is all values $0 < r \le R$, where R is a maximum possible value of r determined by design considerations. ■

From the definition of a function, we know that any value of the independent variable must yield only one value of the dependent variable. *If a value of the independent variable yields one or more values of the dependent variable, the relationship is called a **relation**.* A function is a relation for which each value of the independent variable yields only one value of the dependent variable. Therefore, a function is a special type of relation. There are also relations that are not functions.

EXAMPLE 9 Relation

For $y^2 = 4x^2$, if $x = 2$, then y can be either 4 or -4. Since a value of x yields more than one value for y, we see that $y^2 = 4x^2$ is a relation, not a function. ■

EXERCISES 3.2

In Exercises 1–4, solve the given problems related to the indicated examples of this section.

1. In Example 1, in the first line, change x^2 to $-x^2$. What other changes must be made in the rest of the paragraph?

2. In Example 3, in the first line, change $\frac{1}{x}$ to $\frac{1}{x-1}$. What other changes must be made in the first paragraph?

3. In Example 5, find $f(2)$ and $f(5)$.

4. In Example 7, interchange 40% and 20% and then find the function.

In Exercises 5–14, find the domain and range of the given functions. In Exercises 11 and 12, explain your answers.

5. $f(x) = x + 5$

6. $g(u) = 3 - u^2$

7. $G(R) = \dfrac{3.2}{R}$

8. $F(r) = \sqrt{r + 4}$

9. $f(s) = \dfrac{2}{s^2}$

10. $T(t) = 2t^4 + t^2 - 1$

(W) 11. $H(h) = 2h + \sqrt{h} + 1$

(W) 12. $f(x) = \dfrac{-6}{\sqrt{2 - x}}$

13. $y = |x - 3|$

14. $y = x + |x|$

In Exercises 15–18, find the domain of the given functions.

15. $Y(y) = \dfrac{y + 1}{\sqrt{y - 2}}$

16. $f(n) = \dfrac{n}{6 - 2n}$

17. $f(D) = \dfrac{D}{D - 2} + \dfrac{4}{D + 4}$

18. $g(x) = \dfrac{\sqrt{x - 2}}{x - 3}$

In Exercises 19–22, evaluate the indicated functions.

$F(t) = 3t - t^2 \quad$ (for $t \le 2$)

$h(s) = \begin{cases} 2s & \text{(for } s < -1) \\ s + 1 & \text{(for } s \ge -1) \end{cases}$

$f(x) = \begin{cases} x + 1 & \text{(for } x < 1) \\ \sqrt{x + 3} & \text{(for } x \ge 1) \end{cases}$

$g(x) = \begin{cases} \dfrac{1}{x} & \text{(for } x \ne 0) \\ 0 & \text{(for } x = 0) \end{cases}$

19. Find $F(2)$ and $F(3)$.

20. Find $h(-8)$ and $h(-0.5)$.

21. Find $f(1)$ and $f(-0.25)$.

22. Find $g(0.2)$ and $g(0)$.

In Exercises 23–36, determine the appropriate functions.

23. A motorist travels at 40 mi/h for 2 h and then at 55 mi/h for t h. Express the distance d traveled as a function of t.

24. Express the cost C of insulating a cylindrical water tank of height 2 m as a function of its radius r, if the cost of insulation is \$3 per square meter.

25. A rocket burns up at the rate of 2 tons/min after falling out of orbit into the atmosphere. If the rocket weighed 5500 tons before reentry, express its weight w as a function of the time t, in minutes, of reentry.

26. A computer part costs \$3 to produce and distribute. Express the profit p made by selling 100 of these parts as a function of the price of c dollars each.

27. Upon ascending, a weather balloon ices up at the rate of 0.5 kg/m after reaching an altitude of 1000 m. If the mass of the balloon below 1000 m is 110 kg, express its mass m as a function of its altitude h if $h > 1000$ m.

28. A chemist adds x L of a solution that is 50% alcohol to 100 L of a solution that is 70% alcohol. Express the number n of liters of alcohol in the final solution as a function of x.

29. A company installs underground cable at a cost of \$500 for the first 50 ft (or up to 50 ft) and \$5 for each foot thereafter. Express the cost C as a function of the length l of underground cable if $l > 50$ ft.

30. The *mechanical advantage* of an inclined plane is the ratio of the length of the plane to its height. Express the mechanical advantage M of a plane of length 8 m as a function of its height h.

31. The capacities (in L) of two oil-storage tanks are x and y. The tanks are initially full; 1200 L is removed from them by taking 10% of the contents of the first tank and 40% of the contents of the second tank. (a) Express y as a function of x. (b) Find $f(400)$.

32. A city does not tax the first \$30,000 of a resident's income but taxes any amount over \$30,000 at 5%. (a) Find the tax T as a function of a resident's income I and (b) find $f(25,000)$ and $f(45,000)$.

33. A company finds that it earns a profit of \$15 on each cell phone and a profit of \$25 on each DVD player that it sells. If x cell phones and y DVD players are sold, the profit is \$2750. Express y as a function of x.

34. A satellite telephone is leased at a cost C of \$200 plus \$10 per minute t the phone is used. For $t = f(C)$, find $f(500)$.

35. For studying the electric current that is induced in wire rotating through a magnetic field, a piece of wire 60 cm long is cut into two pieces. One of these is bent into a circle and the other into a square. Express the total area A of the two figures as a function of the perimeter p of the square.

36. The cross section of an air-conditioning duct is in the shape of a square with semicircles on each side. See Fig. 3.6. Express the area A of this cross section as a function of the diameter d (in cm) of the circular part.

Fig. 3.6

In Exercises 37–50, solve the given problems.

37. A helicopter 120 m from a person takes off vertically. Express the distance d from the person to the helicopter as a function of the height h of the helicopter. What are the domain and the range of $d = f(h)$? See Fig. 3.7.

Fig. 3.7

38. A computer program displays a circular image of radius 6 in. If the radius is decreased by x in., express the area of the image as a function of x. What are the domain and range of $A = f(x)$?

39. A truck travels 300 km in t h. Express the average speed s of the truck as a function of t. What are the domain and range of $s = f(t)$?

40. A rectangular grazing range with an area of 8 mi² is to be fenced. Express the length l of the field as a function of its width w. What are the domain and range of $l = f(w)$?

W **41.** The resonant frequency f of a certain electric circuit as a function of the capacitance is $f = \dfrac{1}{2\pi\sqrt{C}}$. Describe the domain.

42. A jet is traveling directly between Calgary, Alberta, and Portland, Oregon, which are 550 mi apart. If the jet is x mi from Calgary and y mi from Portland, find the domain of $y = f(x)$.

43. Express the mass m of the weather balloon in Exercise 27 as a function of any height h in the same manner as the function in Example 5 (and Exercises 19–22) was expressed.

44. Express the cost C of installing any length l of the underground cable in Exercise 29 in the same manner as the function in Example 5 (and Exercises 19–22) was represented.

45. A rectangular piece of cardboard twice as long as wide is to be made into an open box by cutting 2-in. squares from each corner and bending up the sides. (a) Express the volume V of the box as a function of the width w of the piece of cardboard. (b) Find the domain of the function.

46. A spherical buoy 36 in. in diameter is floating in a lake and is more than half above the water. (a) Express the circumference c of the circle of intersection of the buoy and water as a function of the depth d to which the buoy sinks. (b) Find the domain and range of the function.

47. For $f(x - 1) = |x|$, find $f(0)$.

48. For $f(x) = x^2$, find $\dfrac{f(x + h) - f(x)}{h}$.

49. What is the range of the function $f(x) = |x| + |x - 2|$?

W **50.** For $f(x) = \sqrt{x - 1}$ and $g(x) = x^2$, find the domain of $g[f(x)]$. Explain.

Answers to Practice Exercises

1. Domain: all real numbers $x \geq -4$; Range: all real numbers $f(x) \geq 0$ **2.** $p = 2\pi r + 4r + 20$

3.3 Rectangular Coordinates

Rectangular Coordinate System

■ Rectangular (Cartesian) coordinates were developed by the French mathematician René Descartes (1596–1650).

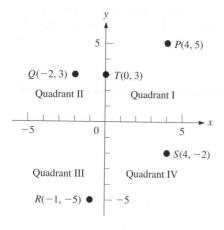

Fig. 3.8

One of the most valuable ways of representing a function is by graphical representation. By using graphs, we can obtain a "picture" of the function; by using this picture, we can learn a great deal about the function.

To make a graphical representation of a function, recall from Chapter 1 that numbers can be represented by points on a line. For a function, we have values of the independent variable as well as the corresponding values of the dependent variable. Therefore, it is necessary to use two different lines to represent the values from each of these sets of numbers. We do this by placing the lines perpendicular to each other.

Place one line horizontally and label it the x-axis. The values of the independent variable are normally placed on this axis. *The other line is placed vertically and labeled the y-axis.* Normally, the y-axis is used for values of the dependent variable. *The point of intersection is called the* **origin.** This is the **rectangular coordinate system.**

On the x-axis, positive values are to the right of the origin, and negative values are to the left of the origin. On the y-axis, positive values are above the origin, and negative values are below it. *The four parts into which the plane is divided are called* **quadrants,** which are numbered as in Fig. 3.8.

A point P in the plane is designated by the pair of numbers (x, y), where x is the value of the independent variable and y is the value of the dependent variable. *The x-value is the perpendicular distance of P from the y-axis, and the y-value is the perpendicular distance of P from the x-axis.* The values of x and y, written as (x, y), are the **coordinates** of the point P.

EXAMPLE 1 Locating points

(a) Locate the points $A(2, 1)$ and $B(-4, -3)$ on the rectangular coordinate system.

The coordinates $(2, 1)$ for A mean that the point is 2 units to the *right* of the y-axis and 1 unit *above* the x-axis, as shown in Fig. 3.9. The coordinates $(-4, -3)$ for B mean that the point is 4 units to the *left* of the y-axis and 3 units *below* the x-axis, as shown. The x-coordinate of A is 2, and the y-coordinate of A is 1. For point B, the x-coordinate is -4, and the y-coordinate is -3.

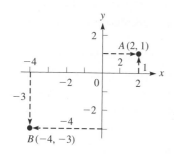

Fig. 3.9

(b) The positions of points $P(4, 5)$, $Q(-2, 3)$, $R(-1, -5)$, $S(4, -2)$, and $T(0, 3)$ are shown in Fig. 3.8. We see that this representation allows for *one point for any pair of values* (x, y). Also note that the point $T(0, 3)$ is on the y-axis. Any such point that is on either axis is not *in* any of the four quadrants. ■

NOTE ▶   Note very carefully that the **x-coordinate is always written first,** and the y-coordinate *is always written second.* It is very important to keep the proper order when writing the coordinates of a point. (The x-coordinate is also known as the *abscissa,* and the y-coordinate is also known as the *ordinate.*)

EXAMPLE 2 Coordinates of vertices of rectangle

Three vertices of the rectangle in Fig. 3.10 are $A(-3, -2)$, $B(4, -2)$, and $C(4, 1)$. What is the fourth vertex?

We use the fact that opposite sides of a rectangle are equal and parallel to find the solution. Since both vertices of the base AB of the rectangle have a y-coordinate of -2, the base is parallel to the x-axis. Therefore, the top of the rectangle must also be parallel to the x-axis. Thus, the vertices of the top must both have a y-coordinate of 1, since one of them has a y-coordinate of 1. In the same way, the x-coordinates of the left side must both be -3. Therefore, the fourth vertex is $D(-3, 1)$. ■

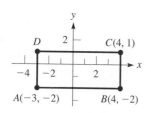

Fig. 3.10

Fig. 3.11

Fig. 3.12

Practice Exercise

1. Where are all points for which $x = 0$ and $y > 0$?

EXAMPLE 3   Locating sets of points

(a) Where are all the points whose y-coordinates are 2?

We can see that the question could be stated as: "Where are all the points for which $y = 2$?" Since all such points are two units above the x-axis, the answer could be stated as "on a line 2 units above the x-axis." See Fig. 3.11.

(b) Where are all points (x, y) for which $x < 0$ and $y < 0$?

Noting that $x < 0$ means "x is less than zero," or "x is negative," and that $y < 0$ means the same for y, we want to determine where both x and y are negative. Our answer is "in the third quadrant," since both coordinates are negative for all points in the third quadrant, and this is the only quadrant for which this is true. See Fig. 3.12.   ∎

EXERCISES 3.3

In Exercises 1 and 2, make the given changes in the indicated examples of this section, and then solve the resulting problems.

1. In Example 2, change $A(-3, -2)$ to $A(-1, -2)$ and then find the fourth vertex.

2. In Example 3(b), change $y < 0$ to $y > 0$ and then find the location of the points (x, y).

In Exercises 3 and 4, determine (at least approximately) the coordinates of the points shown in Fig. 3.13.

3. A, B, C 4. D, E, F

In Exercises 5 and 6, plot the given points.

5. $A(2, 7), B(-1, -2), C(-4, 2)$

6. $A\left(3, \frac{1}{2}\right), B(-6, 0), C\left(-\frac{5}{2}, -5\right)$

Fig. 3.13

In Exercises 7–10, plot the given points and then join these points, in the order given, by straight-line segments. Name the geometric figure formed.

7. $A(-1, 4), B(3, 4), C(1, -2)$

8. $A(0, 3), B(0, -1), C(4, -1)$

9. $A(-2, -1), B(3, -1), C(3, 5), D(-2, 5)$

10. $A(-5, -2), B(4, -2), C(6, 3), D(-3, 3)$

In Exercises 11–14, find the indicated coordinates.

11. Three vertices of a rectangle are $(5, 2), (-1, 2)$, and $(-1, 4)$. What are the coordinates of the fourth vertex?

12. Two vertices of an equilateral triangle are $(7, 1)$ and $(2, 1)$. What is the x-coordinate of the third vertex?

13. P is the point $(3, 2)$. Locate point Q such that the x-axis is the perpendicular bisector of the line segment joining P and Q.

14. P is the point $(-4, 1)$. Locate point Q such that the line segment joining P and Q is bisected by the origin.

In Exercises 15–34, answer the given questions.

15. Where are all points whose x-coordinates are 1?

16. Where are all points whose y-coordinates are -3?

17. Where are all points such that $y = 3$?

18. Where are all points such that $|x| = 2$?

19. Where are all points whose x-coordinates equal their y-coordinates?

20. Where are all points whose x-coordinates equal the negative of their y-coordinates?

21. What is the x-coordinate of all points on the y-axis?

22. What is the y-coordinate of all points on the x-axis?

23. Where are all points for which $x > 0$?

24. Where are all points for which $y < 0$?

25. Where are all points for which $x < -1$?

26. Where are all points for which $y > 4$?

27. Where are all points for which $xy > 0$?

28. Where are all points for which $y/x < 0$?

29. Where are all points for which $xy = 0$?

30. Where are all points for which $x < y$?

31. If the point (a, b) is in the second quadrant, in which quadrant is $(a, -b)$?

(W) 32. The points $(3, -1), (3, 0)$, and (x, y) are on the same straight line. Describe this line.

33. Find the distance (a) between $(3, -2)$ and $(-5, -2)$ and (b) between $(3, -2)$ and $(3, 4)$.

(W) 34. Find the distance between $(-5, -2)$ and $(3, 4)$. Explain your method. (*Hint*: See Exercise 33.)

Answer to Practice Exercise

1. On the positive y-axis

3.4   The Graph of a Function

Table of Values • Plotting Points • Be Careful about the Domain • Vertical-Line Test

Now that we have introduced the concepts of a function and the rectangular coordinate system, we are in a position to determine the graph of a function. In this way, we will obtain a visual representation of a function.

The graph of a function is the set of all points whose coordinates (x, y) satisfy the functional relationship $y = f(x)$. Since $y = f(x)$, we can write the coordinates of the points on the graph as $(x, f(x))$. Writing the coordinates in this manner tells us exactly how to find them. *We assume a certain value for x and then find the value of the function of x. These two numbers are the coordinates of the point.*

Since there is no limit to the possible number of points that can be chosen, we normally select a few values of x, obtain the corresponding values of the function, plot these points, and then join them. Therefore, we use the following basic procedure in plotting a graph.

■ As to just what values of x to choose and how many to choose, with a little experience you will usually be able to tell if you have enough points to plot an accurate graph.

> **PROCEDURE FOR PLOTTING THE GRAPH OF A FUNCTION**
>
> 1. *Let x take on several values and calculate the corresponding values of y.*
> 2. *Tabulate these values, arranging the table so that **values of x are increasing.***
> 3. *Plot the points and join them from left to right by a **smooth curve** (not short straight-line segments).*

EXAMPLE 1 Graphing a function by plotting points

Graph the function $f(x) = 3x - 5$.

■ We will show how a graph is displayed on a graphing calculator in the next section.

For purposes of graphing, let $y = f(x)$, or $y = 3x - 5$. Then, let x take on various values and determine the corresponding values of y. Note that once we choose a given value of x, we have no choice about the corresponding y-value, as it is determined by evaluating the function. If $x = 0$, we find that $y = -5$. This means that the point $(0, -5)$ is on the graph of the function $3x - 5$. Choosing another value of x—for example, 1—we find that $y = -2$. This means that the point $(1, -2)$ is on the graph of the function $3x - 5$. Continuing to choose a few other values of x, we tabulate the results, as shown in Fig. 3.14. It is best to arrange the table so that the values of x increase; then there is no doubt how they are to be connected, for they are then connected in the order shown. Finally, we connect the points in Fig. 3.14 and see that the graph of the function $3x - 5$ is a straight line.

■ The table of values for a graph can be seen on a graphing calculator using the *table* feature. Figure 3.15 shows a calculator display of the table for Example 1.

X	Y1
-1	-8
0	-5
1	-2
2	1
3	4

X=

Fig. 3.15

Select values of x and calculate corresponding values of y		Tabulate with values of x increasing		Plot points and join from left to right with smooth curve
$f(x) = 3x - 5$		x	y	
$f(-1) = 3(-1) - 5 = -8$		-1	-8	
$f(0) = 3(0) - 5 = -5$		0	-5	
$f(1) = 3(1) - 5 = -2$		1	-2	
$f(2) = 3(2) - 5 = 1$		2	1	
$f(3) = 3(3) - 5 = 4$		3	4	

Fig. 3.14

EXAMPLE 2 Be careful: negative numbers

Graph the function $f(x) = 2x^2 - 4$.

First, let $y = 2x^2 - 4$ and tabulate the values as shown in Fig. 3.16. In determining the values in the table, take particular care to obtain the correct values of y for negative values of x. ***Mistakes are relatively common when dealing with negative numbers.*** We must carefully use the laws for signed numbers. For example, if $x = -2$, we have $y = 2(-2)^2 - 4 = 2(4) - 4 = 8 - 4 = 4$. Once the values are obtained, plot and connect the points with a smooth curve, as shown.

CAUTION ◆

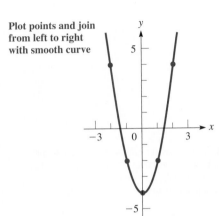

Select values of x and calculate corresponding values of y	Tabulate with values of x increasing	Plot points and join from left to right with smooth curve

	x	y
$f(x) = 2x^2 - 4$		
$f(-2) = 2(-2)^2 - 4 = 4$	-2	4
$f(-1) = 2(-1)^2 - 4 = -2$	-1	-2
$f(0) = 2(0)^2 - 4 = -4$	0	-4
$f(1) = 2(1)^2 - 4 = -2$	1	-2
$f(2) = 2(2)^2 - 4 = 4$	2	4

Fig. 3.16 ∎

When graphing a function, we must use extra care with certain parts of the graph. These include the following:

Special Notes on Graphing

1. Since the graphs of most common functions are smooth, *any place that the graph changes in a way that is not expected should be checked* with care. It usually helps to check values of x between which the question arises.

2. The domain of the function may not include all values of x. Remember, ***division by zero is not defined, and only real values of the variables may be used.***

3. In applications, we must *use only values of the variables that have meaning*. In particular, negative values of some variables, such as time, generally have no meaning.

The following examples illustrate these points.

EXAMPLE 3 Be careful: another point needed

Graph the function $y = x - x^2$.

First, we determine the values in the table, as shown with Fig. 3.17. Again, we must be careful when dealing with negative values of x. For the value $x = -1$, we have $y = (-1) - (-1)^2 = -1 - 1 = -2$. Once all values in the table have been found and plotted, note that $y = 0$ *for both $x = 0$ and $x = 1$*. The question arises—*what happens between these values?* Trying $x = \frac{1}{2}$, we find that $y = \frac{1}{4}$. Using this point completes the information needed to complete an accurate graph. ∎

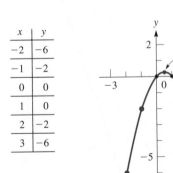

x	y
-2	-6
-1	-2
0	0
1	0
2	-2
3	-6

Fig. 3.17

Note that in plotting these graphs, we do not stop the graph with the points we found, but indicate that the curve continues by drawing the graph past these points. However, this is not true for all graphs. As we will see in later examples, the graph should not include points for values of x not in the domain, and therefore may not continue past a particular point.

EXAMPLE 4 **Be careful: division by zero**

Graph the function $y = 1 + \dfrac{1}{x}$.

In finding the points on this graph, as shown in Fig. 3.18, note that y is not defined for $x = 0$, due to division by zero. Thus, $x = 0$ is not in the domain, and *we must be careful not to have any part of the curve cross the y-axis* $(x = 0)$. Although we cannot let $x = 0$, we can choose other values for x between -1 and 1 that are close to zero. In doing so, we find that as x gets closer to zero, the points get closer to the y-axis, although they do not reach or touch it. In this case, the y-axis is called an **asymptote** of the curve.

We see that the curve is smooth, except when $x = 0$, where it is not defined. ∎

x	y		x	y
-4	3/4		1/3	4
-3	2/3		1/2	3
-2	1/2		1	2
-1	0		2	3/2
$-1/2$	-1		3	4/3
$-1/3$	-2		4	5/4

Fig. 3.18

EXAMPLE 5 **Be careful: imaginary values**

Graph the function $y = \sqrt{x + 1}$.

When finding the points for the graph, we may not let x take on any value less than -1, for *all such values would lead to imaginary values for y* and are not in the domain. Also, since we have the positive square root indicated, the range consists of all values of y that are positive or zero $(y \geq 0)$. See Fig. 3.19. Note that the graph starts at $(-1, 0)$.

Although not defined for all values of x, the curve is smooth. ∎

x	y		x	y
-1	0		4	2.2
0	1		5	2.4
1	1.4		6	2.6
2	1.7		7	2.8
3	2		8	3

Fig. 3.19

Practice Exercises

Determine for what value(s) of x there are no points on the graph.

1. $y = \dfrac{2x}{x + 5}$ **2.** $y = \sqrt{x - 3}$

Functions of a particular type have graphs with a certain basic shape, and many have been named. Three examples of this are the *straight line* (Example 1), the *parabola* (Examples 2 and 3), and the *hyperbola* (Example 4). We consider the straight line again in Chapter 5 and the parabola in Chapter 7. All of these graphs and others are studied in detail in Chapter 21. Other types of graphs are found in many of the later chapters.

The following example illustrates an application of graphing in which a different scale is used for each variable.

EXAMPLE 6 **See the chapter introduction**

The electric power P (in W) delivered by a certain fuel cell as a function of the resistance R (in Ω) in the circuit is given by $P = \dfrac{100R}{(0.50 + R)^2}$. Plot P as a function of R.

Since negative values for the resistance have no physical significance, we should not plot any values of P for negative values of R. The following table is obtained:

$R(\Omega)$	0	0.25	0.50	1.0	2.0	3.0	4.0	5.0	10.0
$P(W)$	0.0	44.4	50.0	44.4	32.0	24.5	19.8	16.5	9.1

■ The unit of power, the watt (W), is named for James Watt (1736–1819), a British engineer.

The R values of 0.25 and 0.50 are used since P is less for $R = 2\,\Omega$ than for $R = 1\,\Omega$. The sharp change in direction should be checked by using these additional points to get a smoother curve near $R = 1\,\Omega$. See Fig. 3.20.

Note that the scale on the P-axis differs from that on the R-axis. Different scales are normally used when the variables differ in magnitudes and ranges.

The graph gives us information about P and R. For example, the maximum power of 50 W occurs for $R = 0.50\,\Omega$. Also, P decreases as R increases beyond $0.50\,\Omega$. Reading information from a graph is discussed in the next section. ∎

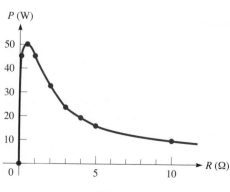

Fig. 3.20

EXAMPLE 7 **Function defined for intervals of domain**

Graph the function $f(x) = \begin{cases} 2x + 1 & (\text{for } x \leq 1) \\ 6 - x^2 & (\text{for } x > 1) \end{cases}$.

First, let $y = f(x)$ and then tabulate the necessary values. In evaluating $f(x)$, we must be careful to use the proper part of the definition. To see where to start the curve for $x > 1$, we evaluate $6 - x^2$ for $x = 1$, but we must realize that *the curve does not include this point* $(1, 5)$ and starts immediately to its right. To show that it is not part of the curve, draw it as an open circle. See Fig. 3.21.

A function such as this one, with a "break" in it, is called *discontinuous*. ∎

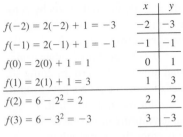

	x	y
$f(-2) = 2(-2) + 1 = -3$	-2	-3
$f(-1) = 2(-1) + 1 = -1$	-1	-1
$f(0) = 2(0) + 1 = 1$	0	1
$f(1) = 2(1) + 1 = 3$	1	3
$f(2) = 6 - 2^2 = 2$	2	2
$f(3) = 6 - 3^2 = -3$	3	-3

Fig. 3.21

A function is defined such that there is only one value of the dependent variable for each value of the independent variable, but a *relation* may have more than one such value of the dependent variable. The following example illustrates how to use a graph to see whether or not a relation is also a function by using the **vertical-line test.**

EXAMPLE 8 **Vertical-line test**

In Example 9 of Section 3.2, we noted that $y^2 = 4x^2$ is a relation, but not a function. Since $y = 4$ or $y = -4$, normally written as $y = \pm 4$, for $x = 2$, we have two values of y for $x = 2$. Therefore, it is not a function. Making the table as shown at the left, we then draw the graph in Fig. 3.22. When making the table, we also note that there are two values of y for every value of x, except $x = 0$.

If any *vertical line* that crosses the x-axis in the domain intersects the graph in more than one point, it is the graph of a relation that is not a function. Any such vertical line in any of the previous graphs of this section would intersect the graph in only one point. This shows that they are graphs of functions. ∎

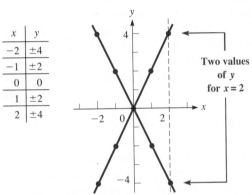

x	y
-2	± 4
-1	± 2
0	0
1	± 2
2	± 4

Two values of y for $x = 2$

Fig. 3.22

The use of graphs is extensive in mathematics and its applications. For example, in the next section, we use graphs to solve equations. For another example, the parabola has applications in microwave-dish design, suspension bridge design, and the path (trajectory) of a baseball. We will note many more applications throughout the book.

EXERCISES 3.4

In Exercises 1–4, make the given changes in the indicated examples of this section and then plot the graphs.

1. In Example 1, change the $-$ sign to $+$.

2. In Example 2, change $2x^2 - 4$ to $4 - 2x^2$.

3. In Example 4, change the x in the denominator to $x - 1$.

4. In Example 5, change the $+$ sign to $-$.

In Exercises 5–36, graph the given functions.

5. $y = 3x$

6. $y = -2x$

7. $y = 2x - 4$

8. $y = 3x + 5$

9. $s = 7 - 2t$

10. $y = -3$

11. $y = \frac{1}{2}x - 2$

12. $A = 6 - \frac{1}{3}r$

13. $y = x^2$

14. $y = -2x^2$

15. $y = 3 - x^2$

16. $y = x^2 - 3$

17. $y = \frac{1}{2}x^2 + 2$

18. $y = 2x^2 + 1$

19. $y = x^2 + 2x$

20. $h = 20t - 5t^2$

21. $y = x^2 - 3x + 1$

22. $y = 2 + 3x + x^2$

23. $V = e^3$

24. $y = -2x^3$

25. $y = x^3 - x^2$

26. $L = 3e - e^3$

27. $D = v^4 - 4v^2$

28. $y = x^3 - x^4$

29. $P = \dfrac{1}{V} + 1$

30. $y = \dfrac{2}{x + 2}$

31. $y = \dfrac{4}{x^2}$

32. $p = \dfrac{1}{n^2 + 0.5}$

33. $y = \sqrt{x}$

34. $y = \sqrt{4 - x}$

35. $v = \sqrt{16 - h^2}$

36. $y = \sqrt{x^2 - 16}$

In Exercises 37–60, graph the indicated functions.

37. In blending gasoline, the number of gallons n of 85-octane gas to be blended with m gal of 92-octane gas is given by the equation $n = 0.40m$. Plot n as a function of m.

38. The consumption of fuel c (in L/h) of a certain engine is determined as a function of the number r of r/min of the engine, to be $c = 0.011r + 40$. This formula is valid for $500\,\text{r/min}$ to $3000\,\text{r/min}$. Plot c as a function of r. (r is the symbol for revolution.)

39. For a certain model of truck, its resale value V (in dollars) as a function of its mileage m is $V = 50{,}000 - 0.2m$. Plot V as a function of m for $m \le 100{,}000$ mi.

40. The resistance R (in Ω) of a resistor as a function of the temperature T (in °C) is given by $R = 250(1 + 0.0032T)$. Plot R as a function of T.

41. The rate H (in W) at which heat is developed in the filament of an electric light bulb as a function of the electric current I (in A) is $H = 240I^2$. Plot H as a function of I.

42. The total annual fraction f of energy supplied by solar energy to a home as a function of the area A (in m^2) of the solar collector is $f = 0.065\sqrt{A}$. Plot f as a function of A.

43. The maximum speed v (in mi/h) at which a car can safely travel around a circular turn of radius r (in ft) is given by $r = 0.42v^2$. Plot r as a function of v.

44. The height h (in m) of a rocket as a function of the time t (in s) is given by the function $h = 1500t - 4.9t^2$. Plot h as a function of t, assuming level terrain.

45. The power P (in W/h) that a certain windmill generates is given by $P = 0.004v^3$, where v is the wind speed (in km/h). Plot the graph of P vs. v.

46. An astronaut weighs 750 N at sea level. The astronaut's weight at an altitude of x km above sea level is given by $w = 750\left(\dfrac{6400}{6400 + x}\right)^2$. Plot w as a function of x for $x = 0$ to $x = 8000$ km.

47. A formula used to determine the number N of board feet of lumber that can be cut from a 4-ft section of a log of diameter d (in in.) is $N = 0.22d^2 - 0.71d$. Plot N as a function of d for values of d from 10 in. to 40 in.

48. A copper electrode with a mass of 25.0 g is placed in a solution of copper sulfate. An electric current is passed through the solution, and 1.6 g of copper is deposited on the electrode each hour. Express the total mass m on the electrode as a function of the time t and plot the graph.

49. A land developer is considering several options of dividing a large tract into rectangular building lots, many of which would have perimeters of 200 m. For these, the minimum width would be 30 m and the maximum width would be 70 m. Express the areas A of these lots as a function of their widths w and plot the graph.

50. The distance p (in m) from a camera with a 50-mm lens to the object being photographed is a function of the magnification m of the camera, given by $p = \dfrac{0.05(1 + m)}{m}$. Plot the graph for positive values of m up to 0.50.

51. A measure of the light beam that can be passed through an optic fiber is its numerical aperture N. For a particular optic fiber, N is a function of the index of refraction n of the glass in the fiber, given by $N = \sqrt{n^2 - 1.69}$. Plot the graph for $n \le 2.00$.

52. $F = x^4 - 12x^3 + 46x^2 - 60x + 25$ is the force (in N) exerted by a cam on the arm of a robot, as shown in Fig. 3.23. Noting that x varies from 1 cm to 5 cm, plot the graph.

Fig. 3.23

53. The number of times S that a certain computer can perform a computation faster with a multiprocessor than with a uniprocessor is given by $S = \dfrac{5n}{4 + n}$, where n is the number of processors. Plot S as a function of n.

54. The voltage V across a capacitor in a certain electric circuit for a 2-s interval is $V = 2t$ during the first second and $V = 4 - 2t$ during the second second. Here, t is the time (in s). Plot V as a function of t.

(W) 55. Given that the point $(1, 2)$ is on the graph of $y = f(x)$, must it be true that $f(2) = 1$? Explain.

56. In Fig. 3.24, part of the graph of $y = 2x^2 + 0.5$ is shown. What is the area of the rectangle, if its vertex is on the curve, as shown?

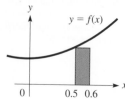

Fig. 3.24

(W) 57. Plot the graphs of $y = x$ and $y = |x|$ on the same coordinate system. Explain why the graphs differ.

(W) 58. Plot the graphs of $y = 2 - x$ and $y = |2 - x|$ on the same coordinate system. Explain why the graphs differ.

59. Plot the graph of $f(x) = \begin{cases} 3 - x & (\text{for } x < 1) \\ x^2 + 1 & (\text{for } x \geq 1) \end{cases}$.

60. Plot the graph of $f(x) = \begin{cases} \dfrac{1}{x - 1} & (\text{for } x < 0) \\ \sqrt{x + 1} & (\text{for } x \geq 0) \end{cases}$.

(W) 61. Plot the graphs of (a) $y = x + 2$ and (b) $y = \dfrac{x^2 - 4}{x - 2}$. Explain the difference between the graphs.

(W) 62. Plot the graphs of (a) $y = x^2 - x + 1$ and (b) $y = \dfrac{x^3 + 1}{x + 1}$. Explain the difference between the graphs.

_____

Answers to Practice Exercises

1. $x = -5$ 2. $x < 3$

In Exercises 63–66, determine whether or not the indicated graph is that of a function.

63. Fig. 3.25(a)

64. Fig. 3.25(b)

65. Fig. 3.25(c)

66. Fig. 3.25(d)

(a)   (b)

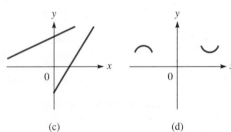

(c)   (d)

Fig. 3.25

3.5   Graphs on the Graphing Calculator

Calculator Setup for Graphing •
Solving Equations Graphically •
Finding the Range of a Function •
Shifting a Graph

In this section, we will see that a graphing calculator can display a graph quickly and easily. As we have pointed out, to use your calculator effectively, *you must know the sequence of keys* for any operation you intend to use. The manual for any particular model should be used for a detailed coverage of its features.

EXAMPLE 1   Entering the function: window settings

To graph the function $y = 2x + 8$, first display $Y_1 =$ and then enter the $2x + 8$. The display for this is shown in Fig. 3.26(a).

Next use the *window* (or *range*) feature to set the part of the domain and the range that will be seen in the *viewing window*. For this function, set

$$\text{Xmin} = -6, \text{Xmax} = 2, \text{Xscl} = 1, \text{Ymin} = -2, \text{Ymax} = 10, \text{Yscl} = 1$$

in order to get a good view of the graph. The display for the *window* settings is shown in Fig. 3.26(b).

Then display the graph, using the *graph* (or *exe*) key. The display showing the graph of $y = 2x + 8$ is shown in Fig. 3.26(c).

Practice Exercise

1. Use your calculator to determine good window settings for graphing $y = 6 - 3x$.

■ The specific detail shown in any nongraphic display depends on the model. A graph with the same *window* settings should appear about the same on all models.

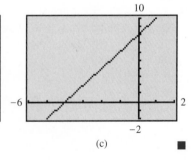

Fig. 3.26   (a)   (b)   (c)   ■

In Example 1, we noted that the *window* feature sets the intervals of the domain and range of the function, which is seen in the viewing window. Unless the settings are appropriate, you may not get a good view of the graph.

EXAMPLE 2 Choose window settings carefully

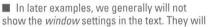

Fig. 3.27

In Example 1, the *window* settings were chosen to give a good view of the graph of $y = 2x + 8$. However, if we had chosen settings of $\text{Xmin} = -3, \text{Xmax} = 3$, $\text{Ymin} = -3$, and $\text{Ymax} = 3$, we would get the view shown in Fig. 3.27. We see that very little of the graph can be seen. With some settings, it is possible that no part of the graph can be seen. Therefore, it is necessary to be careful in choosing the settings. This includes the scale settings (Xscl and Yscl), which should be chosen so that several (not too many or too few) axis markers are used.

Since the settings are easily changed, to get the general location of the graph, it is usually best to first choose intervals between the min and max values that are greater than is probably necessary. They can be reduced as needed.

Also, note that the calculator makes the graph just as we have been doing—by plotting points (square dots called *pixels*). It just does it a lot faster. ∎

SOLVING EQUATIONS GRAPHICALLY

■ In later examples, we generally will not show the *window* settings in the text. They will be shown outside the display, as in Fig. 3.27.

■ See Appendix C for a list of graphing calculator features and examples of calculator use (with page references) that are included in this text.

An equation can be solved by use of a graph. Most of the time, the solution will be approximate, but with a graphing calculator, it is possible to get good accuracy in the result. The procedure used is as follows:

PROCEDURE FOR SOLVING AN EQUATION GRAPHICALLY

1. *Collect all terms on one side of the equal sign.* This gives us the equation $f(x) = 0$.
2. Set $y = f(x)$ *and graph this function.*
3. *Find the points where the graph crosses the x-axis. These points are called the **x-intercepts** of the graph.* At these points, $y = 0$.
4. *The values of x for which $y = 0$ are the solutions of the equation. (These values are called the **zeros** of the function $f(x)$.)*

The next example is done without a graphing calculator to illustrate the method.

EXAMPLE 3 Graphical solution of equation

Graphically solve the equation $x^2 - 2x = 1$.

Following the above procedure, first rewrite the equation as $x^2 - 2x - 1 = 0$ and then set $y = x^2 - 2x - 1$. Next, plot the graph of this function, as shown in Fig. 3.28.

For this curve, we see that it crosses the *x*-axis between $x = -1$ and $x = 0$ and between $x = 2$ and $x = 3$. This means that there are *two* solutions of the equation, and they can be estimated from the graph as

$$x = -0.5 \quad \text{and} \quad x = 2.5$$

For these values, $y = 0$, which means $x^2 - 2x - 1 = 0$, which in turn means $x^2 - 2x = 1$. ∎

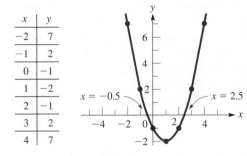

x	y
-2	7
-1	2
0	-1
1	-2
2	-1
3	2
4	7

Fig. 3.28

We now show the use of a graphing calculator in solving equations graphically. In the next example, we solve the equation of Example 3.

■ Most calculators have an *intersect* feature. The equation in Example 4 can be solved using this feature and finding points where the curves $y = x^2 - 2x$ and $y = 1$ cross.

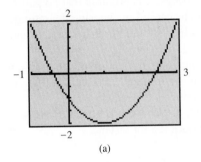

(a)

EXAMPLE 4   Graphing calculator solution of equation

Using a graphing calculator, solve the equation $x^2 - 2x = 1$.

Again, rewrite the equation as $x^2 - 2x - 1 = 0$ and set $y = x^2 - 2x - 1$. Then display this function on the calculator, as shown in Fig. 3.29(a).

To get values of x for which $y = 0$, we can use the *trace* and *zoom* features or the *zero* feature. Using the *trace* feature, we move the *cursor* (the blinking pixel) with the arrow keys to find the value of y is closest to zero (generally not *exactly* zero). We then have the display in Fig. 3.29(b), which shows solutions of about $x = -0.4$ and $x = 2.4$. More accurate values can be found by using the *zoom* feature with which any region of the screen can be magnified. In Fig. 3.29(c), we see that the solutions are about $x = -0.415$ and $x = 2.415$.

In using the *zero* feature, it is necessary to give the part of the domain where you want to find the zero, and make a reasonable guess as to its value. In Fig. 3.29(d), this has been done, and the zero is shown to be $x = -0.4142136$. The other zero is at $x = 2.4142136$.

Using the *value* feature, we can check these solutions. In all cases, we find that the value of y is very near zero, which means that these solutions check. Obviously, the values found using the zero feature are the most accurate.

-2

Fig. 3.29

(b)

(c)

-2

(d)   ■

Fig. 3.30

Fig. 3.31

EXAMPLE 5   Calculator solution of word problem

A rectangular box whose volume is 30 cm^3 (two significant digits) is made with a square base and a height that is 2.0 cm less than the length of a side of the base. Find the dimensions of the box by first setting up the necessary equation and then solving it graphically.

Let x = the length of the side of the square base (see Fig. 3.30); the height is then $x - 2.0$. This means that the volume is $(x)(x)(x - 2.0)$, or $x^3 - 2.0x^2$. Since the volume is 30 cm^3, we have the equation

$$x^3 - 2.0x^2 = 30$$

To solve it graphically, first rewrite the equation as $x^3 - 2.0x^2 - 30 = 0$ and then set $y = x^3 - 2.0x^2 - 30$. The graphing calculator view is shown in Fig. 3.31. We use only positive values of x since negative values have no meaning. Using the *trace, zero,* or *intersect* feature, we find that $x = 3.9$ cm is the approximate solution. Therefore, the dimensions are 3.9 cm, 3.9 cm, and 1.9 cm. Checking, these dimensions give a volume of 29 cm^3. The difference from 30 cm^3 is due to the use of the approximate value of x. A better check can be made by using the unrounded calculator value for x.   ■

FINDING THE RANGE OF A FUNCTION

In Section 3.2, we stated that advanced methods are often necessary to find the range of a function and that we would see that a graphing calculator is useful for this purpose. As with solving equations, we can get very good approximations of the range for most functions by using a graphing calculator.

The method is simply to graph the function and see for what values of y there is a point on the graph. These values of y give us the range of the function.

EXAMPLE 6 Graphically finding range of function

Graphically find the range of the function $y = 16\sqrt{x} + 1/x$.

If we start by using the default *window* settings of Xmin = −10, Xmax = 10, Ymin = −10, Ymax = 10, we see no part of the graph. Looking back to the function, note that x cannot be zero because that would require dividing by zero in the $1/x$ term. Also, $x > 0$ since $\sqrt{x}$ is not defined for $x < 0$. Thus, we should try Xmin = 0 and Ymin = 0.

We also can easily see that $y = 17$ for $x = 1$. This means we should try a value more than 17 for Ymax. Using Ymax = 20, after two or three *window* settings, we use the settings shown in Fig. 3.32. The display shows the range to be all real numbers greater than about 12. (Actually, the range is all real numbers $y \geq 12$.) ■

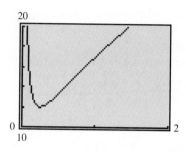

Fig. 3.32

SHIFTING A GRAPH

By adding a positive constant to the right side of the function $y = f(x)$, the graph of the function is *shifted* straight up. If a negative constant is added to the right side of the function, the graph is shifted straight down.

EXAMPLE 7 Shifting a graph vertically

If we add 2 to the right side of $y = x^2$, we get $y = x^2 + 2$. The graphs of these functions are displayed on a calculator as shown in Fig. 3.33. We see that the graph of $y = x^2$ has been shifted up 2 units.

If we add −3 to the right side of $y = x^2$, we get $y = x^2 - 3$. The graphs of these functions are displayed in Fig. 3.34. We see that the graph of $y = x^2$ has been shifted down 3 units.

Fig. 3.33

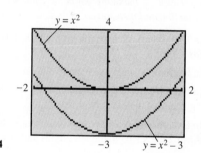

Fig. 3.34 ■

CAUTION ▶ By adding a constant to x in the function $y = f(x)$, the graph of the function is shifted to the right or to the left. *Adding a positive constant to x shifts the graph to the **left**, and adding a negative constant to x shifts the graph to the **right**.*

EXAMPLE 8 Shifting a graph horizontally

For the function $y = x^2$, if we add 2 to x, we get $y = (x + 2)^2$, or if we add −3 to x, we get $y = (x - 3)^2$. The graphs of these three functions are displayed in Fig. 3.35.

We see that the graph of $y = (x + 2)^2$ is 2 units to the left of $y = x^2$ and that the graph of $y = (x - 3)^2$ is 3 units to the right of $y = x^2$. ■

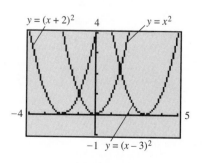

Fig. 3.35

Be very careful when shifting graphs horizontally. To check the direction and magnitude of a horizontal shift, find the value of x that makes the expression in parentheses equal to 0. For example, in Example 8, $y = (x - 3)^2$ is 3 units to the *right* of $y = x^2$ because $x = +3$ makes $x - 3$ equal to zero. The point $(3, 0)$ on $y = (x - 3)^2$ is equivalent to the point $(0, 0)$ on $y = x^2$. In the same way $x + 2 = 0$ for $x = -2$, and the graph of $y = (x + 2)^2$ is shifted 2 units to the *left* of $y = x^2$.

Summarizing how a graph is shifted, we have the following:

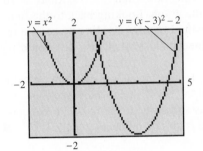

Fig. 3.36

Practice Exercise

2. Describe the graph of $y = (x + 2)^2 + 3$ relative to the graph of $y = x^2$.

> **SHIFTING A GRAPH**
>
> **Vertical shifts:** $y = f(x) + k$ shifts the graph of $y = f(x)$ up k units if $k > 0$ and down k units if $k < 0$.
>
> **Horizontal shifts:** $y = f(x + k)$ shifts the graph of $y = f(x)$ **left** k units if $k > 0$ and **right** k units if $k < 0$.

EXAMPLE 9 Shifting vertically and horizontally

The graph of a function can be shifted both vertically and horizontally. To shift the graph of $y = x^2$ to the right 3 units and down two units, add –3 to x and add –2 to the resulting function. In this way, we get $y = (x - 3)^2 - 2$. The graph of this function and the graph of $y = x^2$ are shown in Fig. 3.36.

Note that the point $(3, -2)$ on the graph of $y = (x - 3)^2 - 2$ is equivalent to the point $(0, 0)$ on the graph of $y = x^2$. Checking, by setting $x - 3 = 0$, we get $x = 3$. This means the graph of $y = x^2$ has been shifted 3 units to the *right*. The vertical shift of -2 is clear. ∎

EXERCISES 3.5

In Exercises 1–4, make the indicated changes in the given examples of this section and then solve.

1. In Example 3, change the sign on the left side of the equation from − to +.

2. In Example 5, change 2.0 cm to 3.0 cm.

3. In Example 6, change the denominator of the second term on the right side of the equation to x^2.

4. In Example 9, in the second line, change "to the right 3 units and down 2 units" to "to the left 2 units and down 3 units."

In Exercises 5–18, display the graphs of the given functions on a graphing calculator. Use appropriate window settings.

5. $y = 3x - 1$

6. $y = 4 - 0.5x$

7. $y = x^2 - 4x$

8. $y = 8 - 2x^2$

9. $y = 6 - x^3$

10. $y = x^4 - 6x^2$

11. $y = x^4 - 2x^3 - 5$

12. $y = x^2 - 3x^5 + 3$

13. $y = \dfrac{2x}{x - 2}$

14. $y = \dfrac{3}{x^2 - 4}$

15. $y = x + \sqrt{x + 3}$

16. $y = \sqrt[3]{2x + 1}$

17. $y = 3 + \dfrac{2}{x}$

18. $y = \dfrac{x^2}{\sqrt{3 - x}}$

In Exercises 19–28, use a graphing calculator to solve the given equations to the nearest 0.001.

19. $x^2 + x - 5 = 0$

20. $v^2 - 2v - 4 = 0$

21. $x^3 - 3 = 3x$

22. $x^4 - 2x = 0$

23. $s^3 - 4s^2 = 6$

24. $3x^2 - x^4 = 2 + x$

25. $\sqrt{5R + 2} = 3$

26. $\sqrt{x} + 3x = 7$

27. $\dfrac{1}{x^2 + 1} = 0$

28. $T - 2 = \dfrac{1}{T}$

In Exercises 29–36, use a graphing calculator to find the range of the given functions. (The functions of Exercises 33–36 are the same as the functions of Exercises 15–18 of Section 3.2.)

29. $y = \dfrac{4}{x^2 - 4}$

30. $y = \dfrac{x + 1}{x^2}$

31. $y = \dfrac{x^2}{x + 1}$

32. $y = \dfrac{x}{x^2 - 4}$

33. $Y(y) = \dfrac{y + 1}{\sqrt{y - 2}}$

34. $f(n) = \dfrac{n}{6 - 2n}$

35. $f(D) = \dfrac{D}{D - 2} + \dfrac{4}{D + 4}$

36. $g(x) = \dfrac{\sqrt{x - 2}}{x - 3}$

In Exercises 37–44, a function and how it is to be shifted is given. Find the shifted function and then display the given function and the shifted function on the same screen of a graphing calculator.

37. $y = 3x$, up 1

38. $y = x^3$, down 2

39. $y = \sqrt{x}$, right 3

40. $y = \dfrac{2}{x}$, left 4

41. $y = -2x^2$, down 3, left 2

42. $y = -4x$, up 4, right 3

43. $y = \sqrt{2x + 1}$, up 1, left 1

44. $y = \sqrt{x^2 + 4}$, down 2, right 2

In Exercises 45–52, solve the indicated equations graphically. Assume all data are accurate to two significant digits unless greater accuracy is given.

45. In an electric circuit, the current i (in A) as a function of voltage v is given by $i = 0.01v - 0.06$. Find v for $i = 0$.

46. For tax purposes, a corporation assumes that one of its computers depreciates according to the equation $V = 90{,}000 - 12{,}000t$, where V is the value (in dollars) of the computer after t years. According to this formula, when will the computer be fully depreciated (no value)?

47. Two cubical coolers together hold 40.0 L (40,000 cm³). If the inside edge of one is 5.00 cm greater than the inside edge of the other, what is the inside edge of each?

48. The height h (in ft) of a rocket as a function of time t (in s) of flight is given by $h = 50 + 280t - 16t^2$. Determine when the rocket is at ground level.

49. The length of a rectangular solar panel is 12 cm more than its width. If its area is 520 cm², find its dimensions.

50. A computer model shows that the cost (in dollars) to remove x percent of a pollutant from a lake is $C = \dfrac{8000x}{100 - x}$. What percent can be removed for $25,000?

51. In finding the illumination at a point x feet from one of two light sources that are 100 ft apart, it is necessary to solve the equation $9x^3 - 2400x^2 + 240{,}000x - 8{,}000{,}000 = 0$. Find x.

52. A rectangular storage bin is to be made from a rectangular piece of sheet metal 12 in. by 10 in., by cutting out equal corners of side x and bending up the sides. See Fig. 3.37. Find x if the storage bin is to hold 90 in.³.

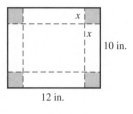

Fig. 3.37

In Exercises 53–60, solve the given problems.

53. In Example 5, assume the data are good to four significant digits, and then find the dimensions of the box.

(W)54. Explain how to show the graph of the *relation* $y^2 = 4x^2$ on a graphing calculator, and then display it on the calculator. See Example 8 on page 94.

55. The cutting speed s (in ft/min) of a saw in cutting a particular type of metal piece is given by $s = \sqrt{t - 4t^2}$, where t is the time in seconds. What is the maximum cutting speed in this operation (to two significant digits)? (*Hint*: Find the range.)

(W)56. Referring to Exercise 52, explain how to determine the maximum possible capacity for a storage bin constructed in this way. What is the maximum possible capacity (to three significant digits)?

57. A balloon is being blown up at a constant rate. (a) Sketch a reasonable graph of the radius of the balloon as a function of time. (b) Compare to a typical situation that can be described by $r = \sqrt[3]{3t}$, where r is the radius (in cm) and t is the time (in s).

58. A hot-water faucet is turned on. (a) Sketch a reasonable graph of the water temperature as a function of time. (b) Compare to a typical situation described by $T = \dfrac{t^3 + 80}{0.015t^3 + 4}$, where T is the water temperature (in °C) and t is the time (in s).

(W)59. Display the graph of $y = cx^3$ with $c = -2$ and with $c = 2$. Describe the effect of the value of c.

(W)60. Display the graph of $y = cx^4$, with $c = 4$ and with $c = 1/4$. Describe the effect of the value of c.

Answers to Practice Exercises

1. x: -1 to 3; y: -2 to 8 probably best

2. 2 units to left, 3 units up

3.6 Graphs of Functions Defined by Tables of Data

Straight-Line Segments Graph • Smooth Curve Graph • Reading a Graph • Linear Interpolation

As we noted in Section 3.1, there are ways other than formulas to show functions. One important way to show the relationship between variables is by means of a table of values found by observation or from an experiment.

Statistical data often give values that are taken for certain intervals or are averaged over various intervals, and there is no meaning to the intervals *between* the points. Such points should be connected by straight-line segments only to make them stand out better and make the graph easier to read. See Example 1.

EXAMPLE 1 Graph using straight-line segments

The electric energy usage (in kW · h) for a certain all-electric house for each month of a year is shown in the following table. Plot these data.

Month	Jan	Feb	Mar	Apr	May	Jun
Energy Usage	2626	3090	2542	1875	1207	892

Month	July	Aug	Sep	Oct	Nov	Dec
Energy Usage	637	722	825	1437	1825	2427

■ Using statistical plotting features, Fig. 3.39 shows a graphing calculator display of the data of Example 1.

Fig. 3.39

Fig. 3.38

We know that there is no meaning to the intervals between the months, since we have the total number of kilowatt-hours for each month. Therefore, we use straight-line segments, but only to make the points stand out better. See Fig. 3.38. ■

Data from experiments in science and technology often indicate that the variables could have a formula relating them, although the formula may not be known. In this case, when plotting the graph, the points should be connected by a smooth curve.

EXAMPLE 2 Graph using smooth curve

Steam in a boiler was heated to 150°C and then allowed to cool. Its temperature T (in °C) was recorded each minute, as shown in the following table. Plot the graph.

Time (min)	0.0	1.0	2.0	3.0	4.0	5.0
Temperature (°C)	150.0	142.8	138.5	135.2	132.7	130.8

■ The Celsius degree is named for the Swedish astronomer Andres Celsius (1701–1744). He designated 100° as the freezing point of water and 0° as the boiling point. These were later reversed.

Since the temperature changes in a continuous way, there is meaning to the values in the intervals between points. Therefore, these points are joined by a smooth curve, as in Fig. 3.40. Also, note that most of the vertical scale was used for the required values, with the indicated break in the scale between 0 and 130.

■ In Chapter 22, we see how to find an equation that approximates the function relating the variables in a table of values.

Fig. 3.40

If the graph relating two variables is known, values can be obtained directly from the graph. In finding such a value through the inspection of the graph, we are *reading the graph*.

EXAMPLE 3 Reading a graph

For the cooling steam in Example 2, we can estimate values of one variable for given values of the other variable.

If we want to know the temperature after 2.5 min, estimate 0.5 of the interval between 2 and 3 on the t-axis and mark this point. See Fig. 3.41. Then draw a vertical line from this point to the curve. From the point where it intersects the curve, draw a horizontal line to the T-axis. We now estimate that the line crosses at $T = 136.7°$C. Obviously, the number of tenths is a rough estimate.

In the same way, if we want to determine how long the steam took to cool down to 141.0°C, go from 141.0 marker on the T-axis to the curve and then to the t-axis. This crosses at about $t = 1.4$ min.

Note that we cannot read values between points on the graph in Example 1, since there is no real meaning to values between those listed. However, in this example we assume that the general trend shown by the graph is probably valid for values between those in the table. ∎

Fig. 3.41

LINEAR INTERPOLATION

In Example 3, we see that we can estimate values from a graph. However, unless a very accurate graph is drawn with expanded scales for both variables, only very approximate values can be found. There is a method, called *linear interpolation*, that uses the table itself to get more accurate results.

Linear interpolation *assumes that if a particular value of one variable lies between two of those listed in the table, then the corresponding value of the other variable is at the same proportional distance between the listed values.* On the graph, linear interpolation assumes that two points defined in the table are connected by a straight line. Although this is generally not correct, it is a good approximation if the values in the table are sufficiently close together.

EXAMPLE 4 Linear interpolation

For the cooling steam in Example 2, we can use interpolation to find its temperature after 1.4 min. Since 1.4 min is $\frac{4}{10}$ of the way from 1.0 min to 2.0 min, we will assume that the value of T we want is $\frac{4}{10}$ of the way between 142.8 and 138.5, the values of T for 1.0 min and 2.0 min, respectively. The difference between these values is 4.3, and $\frac{4}{10}$ of 4.3 is 1.7 (rounded off to tenths). Subtracting (the values of T are decreasing) 1.7 from 142.8, we obtain 141.1. Thus, the required value of T is about 141.1°C. (Note that this agrees well with the result in Example 3.)

Another method of indicating the interpolation is shown in Fig. 3.42. From the figure, we have the proportion

$$\frac{0.4}{1.0} = \frac{x}{4.3}$$

$$x = 1.7 \quad \text{rounded off}$$

Fig. 3.42

Therefore,

$$142.8 - 1.7 = 141.1°C$$

is the required value of T. If the values of T had been increasing, we would have added 1.7 to the value of T for 1.0 min. ∎

■ Before the use of electronic calculators, interpolation was used extensively in mathematics textbooks for finding values from mathematics tables. It is still of use when using scientific and technical tables.

Practice Exercise

1. In Example 4, interpolate to find T for $t = 4.2$ min.

EXERCISES 3.6

In Exercises 1–8, represent the data graphically.

1. The diesel fuel production (in 1000's of gallons) at a certain refinery during an 8-wk period was as follows:

Week	1	2	3	4	5	6	7	8
Production	765	780	840	850	880	840	760	820

2. The *exchange rate* for the number of Canadian dollars equal to one U.S. dollar for 1992–2008 is as follows:

Year	1992	1994	1996	1998	2000	2002	2004	2006
Can. Dol.	1.21	1.38	1.37	1.48	1.49	1.43	1.30	1.13

3. The amount of material necessary to make a cylindrical gallon container depends on the diameter, as shown in this table:

Diameter (in.)	3.0	4.0	5.0	6.0	7.0	8.0	9.0
Material (in.^{2})	322	256	224	211	209	216	230

4. An oil burner propels air that has been heated to 90°C. The temperature then drops as the distance from the burner increases, as shown in the following table:

Distance (m)	0.0	1.0	2.0	3.0	4.0	5.0	6.0
Temperature (°C)	90	84	76	66	54	46	41

5. A changing electric current in a coil of wire will induce a voltage in a nearby coil. Important in the design of transformers, the effect is called *mutual inductance*. For two coils, the mutual inductance (in H) as a function of the distance between them is given in the following table:

Distance (cm)	0.0	2.0	4.0	6.0	8.0	10.0	12.0
M. ind. (H)	0.77	0.75	0.61	0.49	0.38	0.25	0.17

6. The temperatures felt by the body as a result of the *wind-chill factor* for an outside temperature of 20°F (as determined by the National Weather Service) are given in the following table:

Wind speed (mi/h)	5	10	15	20	25	30	35	40
Temp. felt (°F)	13	9	6	4	3	1	0	−1

7. The time required for a sum of money to double in value, when compounded annually, is given as a function of the interest rate in the following table:

Rate (%)	4	5	6	7	8	9	10
Time (years)	17.7	14.2	11.9	10.2	9.0	8.0	7.3

8. The torque T of an engine, as a function of the frequency f of rotation, was measured as follows:

f (r/min)	500	1000	1500	2000	2500	3000	3500
T (ft · lb)	175	90	62	45	34	31	27

In Exercises 9 and 10, use the graph in Fig. 3.40, which relates the temperature of cooling steam and the time. Find the indicated values by reading the graph.

9. (a) For $t = 4.3$ min, find T. (b) For $T = 145.0$°C, find t.

10. (a) For $t = 1.8$ min, find T. (b) For $T = 133.5$°C, find t.

In Exercises 11 and 12, use the following table, which gives the valve lift L (in mm) of a certain cam as a function of the angle θ (in degrees) through which the cam is turned. Plot the values. Find the indicated values by reading the graph.

θ (°)	0	20	40	60	80	100	120	140
L (mm)	0	1.2	2.3	3.3	3.8	3.0	1.6	0

11. (a) For $\theta = 25°$, find L. (b) For $\theta = 96°$, find L.

12. For $L = 2.0$ mm, find θ.

In Exercises 13–16, find the indicated values by means of linear interpolation.

13. In Exercise 5, find the inductance for $d = 9.2$ cm.

14. In Exercise 6, find the temperature for $s = 12$ mi/h.

15. In Exercise 7, find the rate for $t = 10.0$ years.

16. In Exercise 8, find the torque for $t = 2300$ r/min.

In Exercises 17–20, use the following table that gives the rate R of discharge from a tank of water as a function of the height H of water in the tank. For Exercises 17 and 18, plot the graph and find the values from the graph. For Exercises 19 and 20, find the indicated values by linear interpolation.

Height (ft)	0	1.0	2.0	4.0	6.0	8.0	12
Rate (ft^3/s)	0	10	15	22	27	31	35

17. (a) for $R = 20$ ft^3/s, find H. (b) For $H = 2.5$ ft, find R.

18. (a) For $R = 34$ ft^3/s, find H. (b) For $H = 6.4$ ft, find R.

19. Find R for $H = 1.7$ ft.   **20.** Find H for $R = 25$ ft^3/s.

In Exercises 21–24, use the following table, which gives the fraction (as a decimal) of the total heating load of a certain system that will be supplied by a solar collector of area A (in m^2). Find the indicated values by linear interpolation.

f	0.22	0.30	0.37	0.44	0.50	0.56	0.61
A (m^2)	20	30	40	50	60	70	80

21. For $A = 36$ m^2, find f.   **22.** For $A = 52$ m^2, find f.

23. For $f = 0.59$, find A.   **24.** For $f = 0.27$, find A.

*In Exercises 25–28, a method of finding values beyond those given is considered. By using a straight-line segment to extend a graph beyond the last known point, we can estimate values from the extension of the graph. The method is known as **linear extrapolation.** Use this method to estimate the required values from the given graphs.*

25. Using Fig. 3.41, estimate T for $t = 5.3$ min.

26. Using the graph for Exercise 7, estimate T for $R = 10.4\%$.

27. Using the graph for Exer. 17–20, estimate R for $H = 13$ ft.

28. Using the graph for Exer. 17–20, estimate R for $H = 16$ ft.

Answer to Practice Exercise

1. $T = 132.3$°C

CHAPTER 3 REVIEW EXERCISES

In Exercises 1–4, determine the appropriate function.

1. The radius of a circular water wave increases at the rate of 2 m/s. Express the area of the circle as a function of the time t (in s).

2. A conical sheet-metal hood is to cover an area 6 m in diameter. Find the total surface area A of the hood as a function of its height h.

3. A person invests x dollars at 5% APR (annual percentage rate) and y dollars at 4% APR. If the total annual income is $2000, solve for y as a function of x.

4. Fencing around a rectangular storage depot area costs twice as much along the front as along the other three sides. The back costs $10 per foot. Express the cost C of the fencing as a function of the width w if the length (along the front) is 20 ft longer than the width.

In Exercises 5–12, evaluate the given functions.

5. $f(x) = 7x - 5$; find $f(3)$ and $f(-6)$.

6. $g(I) = 8 - 3I$; find $g\left(\frac{1}{6}\right)$ and $g(-4)$.

7. $H(h) = \sqrt{1 - 2h}$; find $H(-4)$ and $H(2h)$.

8. $\phi(v) = \dfrac{3v - 2}{|v + 1|}$; find $\phi(-2)$ and $\phi(v + 1)$.

9. $F(x) = x^3 + 2x^2 - 3x$; find $F(3 + h) - F(3)$.

10. $f(x) = 3x^2 - 2x + 4$; find $\dfrac{f(x + h) - f(x)}{h}$.

11. $f(x) = 3 - 2x$; find $f(2x) - 2f(x)$.

12. $f(x) = 1 - x^2$; find $[f(x)]^2 - f(x^2)$.

In Exercises 13–16, evaluate the given functions. Values of the independent variable are approximate.

13. $f(x) = 8.07 - 2x$; find $f(5.87)$ and $f(-4.29)$.

14. $g(x) = 7x - x^2$; find $g(45.81)$ and $g(-21.85)$.

15. $G(S) = \dfrac{S - 0.087629}{3.0125S}$; find $G(0.17427)$ and $G(0.053206)$.

16. $h(t) = \dfrac{t^2 - 4t}{t^3 + 564}$; find $h(8.91)$ and $h(-4.91)$.

In Exercises 17–22, determine the domain and the range of the given functions.

17. $f(x) = x^4 + 1$

18. $G(z) = \dfrac{4}{z^3}$

19. $g(t) = \dfrac{2}{\sqrt{t + 4}}$

20. $F(y) = 1 - 2\sqrt{y}$

21. $f(n) = 1 + \dfrac{2}{(n - 5)^2}$

22. $F(x) = 3 - |x|$

In Exercises 23–32, plot the graphs of the given functions. Check these graphs by using a graphing calculator.

23. $y = 4x + 2$

24. $y = 5x - 10$

25. $s = 4t - t^2$

26. $y = x^2 - 8x - 5$

27. $y = 3 - x - 2x^2$

28. $V = 3 - 0.5s^3$

29. $A = 2 - s^4$

30. $y = x^4 - 4x$

31. $y = \dfrac{x}{x + 1}$

32. $Z = \sqrt{25 - 2R^2}$

In Exercises 33–40, use a graphing calculator to solve the given equations to the nearest 0.1.

33. $7x - 3 = 0$

34. $3x + 11 = 0$

35. $x^2 + 1 = 6x$

36. $3t - 2 = t^2$

37. $x^3 - x^2 = 2 - x$

38. $5 - x^3 = 2x^2$

39. $\dfrac{1}{v} = 2v$

40. $\sqrt{x} = 2x - 1$

In Exercises 41–44, use a graphing calculator to find the range of the given function.

41. $y = x^4 - 5x^2$

42. $y = x\sqrt{4 - x^2}$

43. $A = w + \dfrac{2}{w}$

44. $y = 2x + \dfrac{3}{\sqrt{x}}$

In Exercises 45–60, solve the given problems.

W 45. Explain how $A(a, b)$ and $B(b, a)$ may be in different quadrants.

46. Determine the distance from the origin to the point (a, b).

47. Two vertices of an equilateral triangle are $(0, 0)$ and $(2, 0)$. What is the third vertex?

48. The points $(1, 2)$ and $(1, -3)$ are two adjacent vertices of a square. Find the other vertices.

49. Where are all points for which $\left|y/x\right| > 0$?

50. Describe the values of x and y for which $(1, -2)$, $(-1, -2)$, and (x, y) are on the same straight line.

51. Sketch the graph of a function for which the domain is $0 \le x \le 4$ and the range is $1 \le y \le 3$.

52. Sketch the graph of a function for which the domain is all values of x and the range is $-2 < y < 2$.

53. If the function $y = \sqrt{x - 1}$ is shifted left 2 and up 1, what is the resulting function?

54. If the function $y = 3 - 2x$ is shifted right 1 and down 3, what is the resulting function?

W 55. For $f(x) = 2x^3 - 3$, display the graphs of $f(x)$ and $f(-x)$ on a graphing calculator. Describe the graphs in relation to the y-axis.

W 56. For $f(x) = 2x^3 - 3$, display the graphs of $f(x)$ and $-f(x)$ on a graphing calculator. Describe the graphs in relation to the x-axis.

57. Is it possible that the points $(2, 3)$, $(5, -1)$, and $(-1, 3)$ are all on the graph of the same function? Explain.

58. For the functions $f(x) = |x|$ and $g(x) = \sqrt{x^2}$, use a graphing calculator to determine whether or not $f(x) = g(x)$ for all real x.

59. An equation used in electronics with a transformer antenna is $I = 12.5\sqrt{1 + 0.5m^2}$. For $I = f(m)$, find $f(0.55)$.

60. The percent p of wood lost in cutting it into boards 1.5 in. thick due to the thickness t (in in.) of the saw blade is $p = \dfrac{100t}{t + 1.5}$. Find p if $t = 0.4$ in. That is, since $p = f(t)$, find $f(0.4)$.

61. The angle A (in degrees) of a robot arm with the horizontal as a function of time t (for 0.0 s to 6.0 s) is given by $A = 8.0 + 12t^2 - 2.0t^3$. What is the greatest value of A to the nearest $0.1°$? See Fig. 3.43. (*Hint*: Find the range.)

Fig. 3.43

62. The electric power P (in W) produced by a certain battery is $P = \dfrac{24R}{R^2 + 1.40R + 0.49}$, where R is the resistance (in Ω) in the circuit. What is the maximum power produced? (*Hint*: Find the range.) See Example 6 on page 93.

In Exercises 63–78, plot the graphs of the indicated functions.

63. When El Niño, a Pacific Ocean current, moves east and warms the water off South America, weather patterns in many parts of the world change significantly. Special buoys along the equator in the Pacific Ocean send data via satellite to monitoring stations. If the temperature T (in °C) at one of these buoys is $T = 28.0 + 0.15t$, where t is the time in weeks between Jan. 1 and Aug. 1 (30 weeks), plot the graph of $T = f(t)$.

64. The change C (in in.) in the length of a 100-ft steel bridge girder from its length at 40°F as a function of the temperature T is given by $C = 0.014(T - 40)$. Plot the graph for $T = 0°$F to $T = 120°$F.

65. The profit P (in dollars) a retailer makes in selling 50 cell phones is given by $P = 50(p - 50)$, where p is the selling price. Plot P as a function of p for $p = \$30$ to $p = \$150$.

66. There are 500 L of oil in a tank that has the capacity of 100,000 L. It is filled at the rate of 7000 L/h. Determine the function relating the number of liters N and the time t while the tank is being filled. Plot N as a function of t.

67. The length L (in cm) of a pulley belt is 12 cm longer than the circumference of one of the pulley wheels. Express L as a function of the radius r of the wheel and plot L as a function of r.

68. The pressure loss P (in lb/in.^{2} per 100 ft) in a fire hose is given by $P = 0.00021Q^2 + 0.013Q$, where Q is the rate of flow (in gal/min). Plot the graph of P as a function of Q.

69. A *thermograph* measures the infrared radiation from each small area of a person's skin. Since the skin over a tumor radiates more than skin from nearby areas, a thermograph can help detect cancer cells. The emissivity e (in %) of radiation as a function of skin temperature T (in °K) is $e = f(T) = 100(T^4 - 307^4)/307^4$, if nearby skin is at 34°C (307°K). Find e for $T = 309°$K.

70. A company buys a new copier for $1000 and determines that it costs $10 per day to use it (for paper, toner, etc.). Plot the total cost C of the copier as a function of the number n of days of use.

71. For a certain laser device, the laser output power P (in mW) is negligible if the drive current i is less than 80 mA. From 80 mA to 140 mA, $P = 1.5 \times 10^{-6}i^3 - 0.77$. Plot the graph of $P = f(i)$.

72. It is determined that a good approximation for the cost C (in cents/mi) of operating a certain car at a constant speed v (in mi/h) is given by $C = 0.025v^2 - 1.4v + 35$. Plot C as a function of v for $v = 10$ mi/h to $v = 60$ mi/h.

73. A medical researcher exposed a virus culture to an experimental vaccine. It was observed that the number of live cells N in the culture as a function of the time t (in h) after exposure was given by $N = \dfrac{1000}{\sqrt{t + 1}}$. Plot the graph of $N = f(t)$.

74. The electric field E (in V/m) from a certain electric charge is given by $E = 25/r^2$, where r is the distance (in m) from the charge. Plot the graph of $E = f(r)$ for values of r up to 10 cm.

75. To draw the approximate shape of an irregular shoreline, a surveyor measured the distances d from a straight wall to the shoreline at 20-ft intervals along the wall, as shown in the following table. Plot the graph of distance d as a function of the distance D along the wall.

D (ft)	0	20	40	60	80	100	120	140	160
d (ft)	15	32	56	33	29	47	68	31	52

76. The percent p of a computer network that is in use during a particular loading cycle as a function of the time t (in s) is given in the following table. Plot the graph of $p = f(t)$.

t (s)	0.0	0.2	0.4	0.6	0.8	1.0	1.2	1.4	1.6
P (%)	0	45	85	90	85	85	60	10	0

77. The vertical sag s (in ft) at the middle of an 800-ft power line as a function of the temperature T (in °F) is given in the following table. See Fig. 3.44. For the function $s = f(T)$, find $f(47)$ by linear interpolation.

T (°F)	0	20	40	60
s (ft)	10.2	10.6	11.1	11.7

Fig. 3.44

78. In an experiment measuring the pressure p (in kPa) at a given depth d (in m) of seawater, the results in the following table were found. Plot the graph of $p = f(d)$ and from the graph determine $f(10)$.

d (m)	0.0	3.0	6.0	9.0	12	15
p (kPa)	101	131	161	193	225	256

In Exercises 79–86, solve the indicated equations graphically.

79. A person 250 mi from home starts toward home and travels at 60 mi/h for the first 2.0 h and then slows down to 40 mi/h for the rest of the trip. How long does it take the person to be 70 mi from home?

80. One industrial cleaner contains 30% of a certain solvent, and another contains 10% of the solvent. To get a mixture containing 50 gal of the solvent, 120 gal of the first cleaner is used. How much of the second must be used?

81. The solubility s (in kg/m^3 of water) of a certain type of fertilizer is given by $s = 135 + 4.9T + 0.19T^2$, where T is the temperature (in °C). Find T for $s = 500$ kg/m^3.

82. A 2.00-L (2000-cm^3) metal container is to be made in the shape of a right circular cylinder. Express the total area A of metal necessary as a function of the radius r of the base. Then find A for $r = 6.00$ cm, 7.00 cm, and 8.00 cm.

83. In an oil pipeline, the velocity v (in ft/s) of the oil as a function of the distance x (in ft) from the wall of the pipe is given by $v = 7.6x - 2.1x^2$. Find x for $v = 5.6$ ft/s. The diameter of the pipe is 3.50 ft.

84. One ball bearing is 1.00 mm more in radius and has twice the volume of another ball bearing. What is the radius of each?

85. A computer, using data from a refrigeration plant, estimates that in the event of a power failure, the temperature (in °C) in the freezers would be given by $T = \dfrac{4t^2}{t + 2} - 20$, where t is the number of hours after the power failure. How long would it take for the temperature to reach 0°C?

86. Two electrical resistors in parallel (see Fig. 3.45) have a combined resistance R_T given by $R_T = \dfrac{R_1 R_2}{R_1 + R_2}$. If $R_2 = R_1 + 2.0$, express R_T as a function of R_1 and find R_1 if $R_T = 6.0\ \Omega$.

Fig. 3.45

Writing Exercise

87. Find the inner surface area A of a cylindrical 250.0-cm^3 cup as a function of the radius r of the base. Then if $A = 175.0$ cm^2, solve for r. Write one or two paragraphs explaining your method for setting up the function, and how you used a graphing calculator to solve for r with the given value of A.

CHAPTER 3   PRACTICE TEST

1. Given $f(x) = 2x - x^2 + \dfrac{8}{x}$, find $f(-4)$ and $f(2.385)$.

2. A rocket has a mass of 2000 Mg at liftoff. If the first-stage engines burn fuel at the rate of 10 Mg/s, find the mass m of the rocket as a function of the time t (in s) while the first-stage engines operate.

3. Plot the graph of the function $f(x) = 4 - 2x$.

4. Use a graphing calculator to solve the equation $2x^2 - 3 = 3x$ to the nearest 0.1.

5. Plot the graph of the function $y = \sqrt{4 + 2x}$.

6. Locate all points (x, y) for which $x < 0$ and $y = 0$.

7. Find the domain and the range of the function $f(x) = \sqrt{6 - x}$.

8. If the function $y = 2x^2 - 3$ is shifted right 1 and up 3, what is the resulting function?

9. Use a graphing calculator to find the range of the function $y = \dfrac{x^2 + 2}{x + 2}$.

10. A window has the shape of a semicircle over a square, as shown in Fig. 3.46. Express the area of the window as a function of the radius of the circular part.

Fig. 3.46

11. The voltage V and current i (in mA) for a certain electrical experiment were measured as shown in the following table. Plot the graph of $i = f(V)$ and from the graph find $f(45.0)$.

Voltage (V)	10.0	20.0	30.0	40.0	50.0	60.0
Current (mA)	145	188	220	255	285	315

12. From the table in Problem 11, find the voltage for $i = 200$ mA.

4 The Trigonometric Functions

Triangles are often used in solving applied problems in technology. A short list of such problems includes those in air navigation, surveying, rocket motion, carpentry, structural design, electric circuits, and astronomy. In fact, it was because the Greek astronomer Hipparchus (about 150 B.C.E.) was interested in measuring distances such as that between the earth and the moon that he started the study of **trigonometry.** Using records of earlier works, he organized and developed the real beginnings of this very important field of mathematics in order to make a number of astronomical measurements.

In trigonometry, we develop methods for measuring the sides and angles of triangles, and this in turn allows us to solve related applied problems. Because trigonometry has a great number of applications in many areas of study, it is considered one of the most practical branches of mathematics.

In this chapter, we introduce the basic trigonometric functions and show many applications of right triangles from science and technology. In later chapters, we will discuss other types of triangles and their applications.

As mathematics developed, it became clear, particularly in the 1800s and 1900s, that the trigonometric functions used for solving problems involving triangles were also very valuable in applications in which a triangle is not involved. This important use of the trigonometric functions is now essential in areas such as electronics, mechanical vibrations, acoustics, and optics, and we will study it in later chapters.

Using trigonometry, it is often possible to calculate distances that may not be directly measured. In Section 4.5, we show how we can measure the height of Horseshoe Falls on the Canadian side of Niagara Falls.

4.1 Angles

**Positive and Negative Angles •
Coterminal Angles • Angle Conversions •
Standard Position of an Angle**

Fig. 4.1

Fig. 4.2

Fig. 4.3

In Chapter 2, we gave a basic definition of an *angle*. In this section, we extend this definition and also give some other important definitions related to angles.

An **angle** *is generated by rotating a ray about its fixed endpoint from an* **initial position** *to a* **terminal position.** *The initial position is called the* **initial side** *of the angle, the terminal position is called the* **terminal side,** *and the fixed endpoint is the* **vertex.** The angle itself is the amount of rotation from the initial side to the terminal side.

If the rotation of the terminal side from the initial side is **counterclockwise,** *the angle is defined as* **positive.** *If the rotation is* **clockwise,** *the angle is* **negative.** In Fig. 4.1, $\angle 1$ is positive and $\angle 2$ is negative.

Many symbols are used to designate angles. Among the most widely used are certain Greek letters such·as θ (theta), ϕ (phi), α (alpha), and β (beta). Capital letters representing the vertex (e.g., $\angle A$ or simply A) and other literal symbols, such as x and y, are also used commonly.

In Chapter 2, we introduced two measurements of an angle. These are the *degree* and the *radian.* Since degrees and radians are both used on calculators and computers, we will briefly review the relationship between them in this section. However, we will not make use of radians until Chapter 8.

From Section 2.1, we recall that a **degree** *is* $1/360$ *of one complete rotation.* In Fig. 4.2, $\angle\theta = 30°$, $\angle\phi = 140°$, $\angle\alpha = 240°$, and $\angle\beta = -120°$. Note that β is drawn in a clockwise direction to show that it is negative. The other angles are drawn in a counterclockwise direction to show that they are positive angles.

In Chapter 2, we used degrees and decimal parts of a degree. Most calculators use degrees in this decimal form. Another traditional way is to divide a degree into 60 equal parts called **minutes;** each minute is divided into 60 equal parts called **seconds.** The symbols ′ and ″ are used to designate minutes and seconds, respectively.

In Fig. 4.2, we note that angles α and β have the same initial and terminal sides. *Such angles are called* **coterminal angles.** An understanding of coterminal angles is important in certain concepts of trigonometry.

EXAMPLE 1 Coterminal angles

Determine the measures of two angles that are coterminal with an angle of 145.6°.

Since there are 360° in one complete rotation, we can find a coterminal angle by adding 360° to the given angle of 145.6° to get 505.6°. Another coterminal angle can be found by subtracting 360° from 145.6° to get −214.4°. See the angles in Fig. 4.3. We could continue to add 360°, or subtract 360°, as many times as needed to get as many additional coterminal angles as may be required. ∎

ANGLE CONVERSIONS

We will use only degrees as a measure of angles in this chapter. Therefore, when using a calculator, *be sure to use the* **mode** *feature to set the calculator for degrees.* In later chapters, we will use radians. We can change between degrees and radians by using a calculator feature or by the definition (see Section 2.4) of π rad $= 180°$.

NOTE ▶

Before the extensive use of calculators, it was common to use degrees and minutes in tables, whereas calculators use degrees and decimal parts of a degree. Changing from one form to another can be done directly on a calculator by use of the *dms (degree-minute-second)* feature. The following examples illustrate angle conversions by using the definitions and by using the appropriate calculator features.

Fig. 4.4

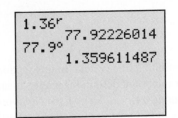

Fig. 4.5

EXAMPLE 2 Convert radians to degrees

Express 1.36 rad in degrees.

We know that π rad $= 180°$, which means 1 rad $= 180°/\pi$. Therefore,

$$1.36\,\text{rad} = 1.36\left(\frac{180°}{\pi}\right) = 77.9° \qquad \text{to nearest } 0.1°$$

This angle is shown in Fig. 4.4. We again note that degrees and radians are simply two different ways of measuring an angle.

In Fig. 4.5, a graphing calculator display shows the conversions of 1.36 rad to degrees (calculator in *degree* mode) and 77.9° to radians (calculator in *radian* mode). ∎

EXAMPLE 3 Degrees, minutes – decimal form

To change 17°53′ to decimal form, we use the fact that $1° = 60'$. Therefore, at this point we change the 53′ to degrees, which gives us $53' = \left(\frac{53}{60}\right)° = 0.88°$ (to nearest 0.01°). This means that 17°53′ = 17.88°. This angle is shown in Fig. 4.6.

To change 154.36° to an angle measured to the nearest minute, we have

$$0.36° = 0.36(60') = 22'$$

This means that 154.36° = 154°22′. See Fig. 4.7. ∎

Practice Exercise

1. Change 17°24′ to decimal form.

STANDARD POSITION OF AN ANGLE

If the initial side of the angle is the positive x-axis and the vertex is the origin, the angle is said to be in **standard position.** The angle is then determined by the position of the terminal side. *If the terminal side is in the first quadrant, the angle is called a* **first-quadrant angle.** Similar terms are used when the terminal side is in the other quadrants. *If the terminal side coincides with one of the axes, the angle is a* **quadrantal angle.** For an angle in standard position, the terminal side can be determined if we know any point, except the origin, on the terminal side.

EXAMPLE 4 Angles in standard position

A standard position angle of 60° is a first-quadrant angle with its terminal side 60° from the *x*-axis. See Fig. 4.8(a).

A second-quadrant angle of 130° is shown in Fig. 4.8(b).

A third-quadrant angle of 225° is shown in Fig. 4.8(c).

A fourth-quadrant angle of 340° is shown in Fig. 4.8(d).

A standard-position angle of −120° is shown in Fig. 4.8(e). Since the terminal side is in the third quadrant, it is a third-quadrant angle.

A standard-position angle of 90° is a quadrantal angle since its terminal side is the positive *y*-axis. See Fig. 4.8(f).

Practice Exercise

2. In Fig. 4.8, which terminal side is that of a standard position angle of 240°?

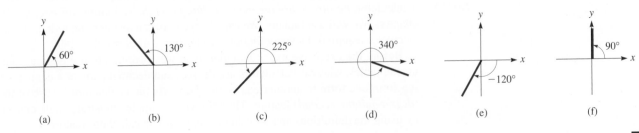

(a) (b) (c) (d) (e) (f)

Fig. 4.8 ∎

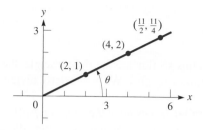

Fig. 4.9

EXAMPLE 5   Standard position – terminal side

In Fig. 4.9, θ is in standard position, and the terminal side is uniquely determined by knowing that it passes through the point $(2, 1)$. The same terminal side passes through the points $(4, 2)$ and $\left(\frac{11}{2}, \frac{11}{4}\right)$, among an unlimited number of other points. Knowing that the terminal side passes through any one of these points makes it possible to determine the terminal side of the angle. ∎

EXERCISES 4.1

In Exercises 1–4, find the indicated angles in the given examples of this section.

1. In Example 1, find another angle that is coterminal with the given angle.

2. In Example 3, change 53′ to 35′ and then find the decimal form.

3. In Example 4, find another standard-position angle that has the same terminal side as the angle in Fig. 4.8(c).

4. In Example 4, find another standard-position angle that has the same terminal side as the angle in Fig. 4.8(e).

In Exercises 5–8, draw the given angles.

5. $60°, 120°, -90°$
6. $330°, -150°, 450°$
7. $50°, -360°, -30°$
8. $45°, 245°, -250°$

In Exercises 9–16, determine one positive and one negative coterminal angle for each angle given.

9. $45°$
10. $73°$
11. $-150°$
12. $462°$
13. $70°30'$
14. $153°47'$
15. $278.1°$
16. $-197.6°$

In Exercises 17–20, by means of the definition of a radian, change the given angles in radians to equal angles expressed in degrees to the nearest 0.01°.

17. 0.265 rad
18. 0.838 rad
19. 1.447 rad
20. −3.642 rad

In Exercises 21–24, use a calculator conversion sequence to change the given angles in radians to equal angles expressed in degrees to the nearest 0.01°.

21. 0.329 rad
22. 2.089 rad
23. −4.110 rad
24. 6.705 rad

In Exercises 25–28, use a calculator conversion sequence to change the given angles to equal angles expressed in radians to three significant digits.

25. $56.0°$
26. $137.4°$
27. $384.8°$
28. $-17.5°$

In Exercises 29–32, change the given angles to equal angles expressed to the nearest minute.

29. $47.50°$
30. $715.80°$
31. $-5.62°$
32. $142.87°$

In Exercises 33–36, change the given angles to equal angles expressed in decimal form to the nearest 0.01°.

33. $15°12'$
34. $517°39'$
35. $301°16'$
36. $-4°47'$

In Exercises 37–44, draw angles in standard position such that the terminal side passes through the given point.

37. $(4, 2)$
38. $(-3, 8)$
39. $(-3, -5)$
40. $(6, -1)$
41. $(-7, 5)$
42. $(-4, -2)$
43. $(-2, 0)$
44. $(0, 6)$

In Exercises 45–52, the given angles are in standard position. Designate each angle by the quadrant in which the terminal side lies, or as a quadrantal angle.

45. $31°, 310°$
46. $180°, 92°$
47. $435°, -270°$
48. $-5°, 265°$
49. 1 rad, 2 rad
50. 3 rad, π rad
51. 4 rad, $\pi/3$ rad
52. 12 rad, −2 rad

In Exercises 53 and 54, change the given angles to equal angles expressed in decimal form to the nearest 0.001°. In Exercises 55 and 56, change the given angles to equal angles expressed to the nearest second.

53. $21°42'36''$
54. $7°16'23''$
55. $86.274°$
56. $257.019°$

Answers to Practice Exercises

1. $17.4°$ 2. (e)

4.2   Defining the Trigonometric Functions

Similar Triangles • Definitions of the
Trigonometric Functions • Evaluating the
Trigonometric Functions

Important to the definitions and development in this section are the right triangle, the Pythagorean theorem, and the properties of similar triangles. We now briefly review similar triangles and their properties.

As stated in Section 2.2, *two triangles are* **similar** *if they have the same shape (but not necessarily the same size)*. Similar triangles have the following important properties.

PROPERTIES OF SIMILAR TRIANGLES

1. *Corresponding angles are equal.*
2. *Corresponding sides are proportional.*

The **corresponding sides** *are the sides, one in each triangle, that are between the same pair of equal* **corresponding angles.**

EXAMPLE 1   Similar triangles

In Fig. 4.10, the triangles are similar and are lettered so that corresponding sides and angles have the same letters. That is, angles A_1 and A_2, angles B_1 and B_2, and angles C_1 and C_2 are pairs of corresponding angles. The pairs of corresponding sides are a_1 and a_2, b_1 and b_2, and c_1 and c_2. From the properties of similar triangles, we know that the corresponding angles are equal, or

$$\angle A_1 = \angle A_2 \quad \angle B_1 = \angle B_2 \quad \angle C_1 = \angle C_2$$

Also, the corresponding sides are proportional, which we can show as

$$\frac{a_1}{a_2} = \frac{b_1}{b_2} \quad \frac{a_1}{a_2} = \frac{c_1}{c_2} \quad \frac{b_1}{b_2} = \frac{c_1}{c_2}$$ ∎

Fig. 4.10

In Example 1, if we multiply both sides of $a_1/a_2 = b_1/b_2$ by a_2/b_1, we get

$$\frac{a_1}{a_2}\left(\frac{a_2}{b_1}\right) = \frac{b_1}{b_2}\left(\frac{a_2}{b_1}\right), \quad \text{or} \quad \frac{a_1}{b_1} = \frac{a_2}{b_2}$$

NOTE ▶   This shows us that *when two triangles are similar, the ratio of one side to another side in one triangle is the same as the ratio of the corresponding sides in the other triangle.* Using this, we now proceed to the definitions of the trigonometric functions.

We now place an angle θ in standard position and drop perpendiculars from points on the terminal side to the x-axis, as shown in Fig. 4.11. In doing this, we set up similar triangles, each with one vertex at the origin and one side along the x-axis.

EXAMPLE 2   Equality of ratios of corresponding sides

In Fig. 4.11, we can see that triangles ORP and OSQ are similar since their corresponding sides are equal (each has the same angle at O, a right angle, and therefore equal angles at P and Q). This means ratios of the lengths of corresponding sides are equal. For example,

$$\frac{RP}{OR} = \frac{SQ}{OS}, \quad \text{which is the same as} \quad \frac{y}{x} = \frac{b}{a}$$

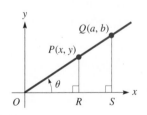

Fig. 4.11

For any position (except at the origin) of Q on the terminal side of θ, the ratio of b/a of the ordinate to the abscissa will equal y/x. ∎

For an angle in standard position, we may set up six different ratios of the values of x, y and r, as shown in Fig. 4.12. Because of the properties of similar triangles, for a given angle, any of these ratios has the same value for any point on the terminal side. This means the *values of the ratios depend on the size of the angle, and there is only one value for each ratio.* This means *the ratios are **functions** of the angle, and they are called the **trigonometric functions.*** We now define them, giving their names and the abbreviations that are used when using them.

Fig. 4.12

$$\text{sine of } \theta: \quad \sin \theta = \frac{y}{r} \qquad\qquad \text{cosine of } \theta: \quad \cos \theta = \frac{x}{r}$$

$$\text{tangent of } \theta: \quad \tan \theta = \frac{y}{x} \qquad \text{cotangent of } \theta: \quad \cot \theta = \frac{x}{y} \qquad (4.1)$$

$$\text{secant of } \theta: \quad \sec \theta = \frac{r}{x} \qquad \text{cosecant of } \theta: \quad \csc \theta = \frac{r}{y}$$

Here, *the distance r from the origin to the point is called the **radius vector,*** and it is assumed that $r > 0$ (if $r = 0$, there would be no terminal side and therefore no angle).

A given function is not defined when the denominator is zero, and if either $x = 0$ or $y = 0$, this does affect the domain. We will discuss the domains and ranges of these functions in Chapter 10, when we discuss their graphs.

In this chapter, we use the trigonometric functions of acute angles (angles between $0°$ and $90°$). However, *the definitions above are general and may be used with angles of any size.* We will discuss these functions in general in Chapters 8 and 20.

EVALUATING THE TRIGONOMETRIC FUNCTIONS

The definitions in Eqs. (4.1) are used in evaluating the trigonometric functions. Also, we often use the *Pythagorean theorem*, which is discussed in Section 2.2, and which we restate here for reference. For the right triangle in Fig. 4.13, with hypotenuse c and legs a and b,

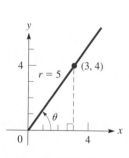

Fig. 4.13

$$c^2 = a^2 + b^2 \qquad (4.2)$$

EXAMPLE 3   Values of trigonometric functions

Find the values of the trigonometric functions of the standard-position angle θ with its terminal side passing through the point $(3, 4)$.

By placing the angle in standard position, as shown in Fig. 4.14, and drawing the terminal side through $(3, 4)$, we find by use of the Pythagorean theorem that

$$r = \sqrt{3^2 + 4^2} = \sqrt{25} = 5$$

Using the values $x = 3$, $y = 4$, and $r = 5$, we find that

$$\sin \theta = \frac{4}{5} \qquad \cos \theta = \frac{3}{5} \qquad \tan \theta = \frac{4}{3}$$

$$\cot \theta = \frac{3}{4} \qquad \sec \theta = \frac{5}{3} \qquad \csc \theta = \frac{5}{4}$$

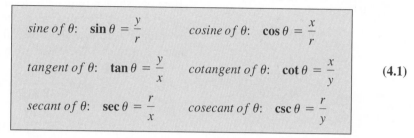

Fig. 4.14

We have left each of these results in the form of a fraction, which is considered to be an *exact form* in that there has been no approximation made. In writing decimal values, we find that $\tan \theta = 1.333$ and $\sec \theta = 1.667$, where these values have been rounded off and are therefore *approximate*. ■

Practice Exercise

1. In Example 3, change $(3, 4)$ to $(4, 3)$ and then find $\tan \theta$ and $\sec \theta$.

Fig. 4.15

Fig. 4.16

CAUTION ▶

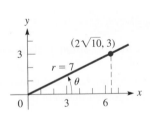

Fig. 4.17

Practice Exercise

2. In Example 5, change $\sin \theta = 3/7$ to $\cos \theta = 3/7$, and then find approximate values of $\sin \theta$ and $\cot \theta$.

EXAMPLE 4   Values of trigonometric functions

Find the values of the trigonometric functions of the standard-position angle whose terminal side passes through (7.27, 4.19). The coordinates are approximate.

We show the angle and the given point in Fig. 4.15. From the Pythagorean theorem, we have

$$r = \sqrt{7.27^2 + 4.49^2} = 8.545$$

(Here, we show a rounded-off value of r. It is not actually necessary to record the value of r since its value can be stored in the memory of a calculator. The reason for recording it here is to show the values used in the calculation of each of the trigonometric functions.) Therefore, we have the following values:

$$\sin \theta = \frac{4.49}{8.545} = 0.525 \qquad \cos \theta = \frac{7.27}{8.545} = 0.851$$

$$\tan \theta = \frac{4.49}{7.27} = 0.618 \qquad \cot \theta = \frac{7.27}{4.49} = 1.62$$

$$\sec \theta = \frac{8.545}{7.27} = 1.18 \qquad \csc \theta = \frac{8.545}{4.49} = 1.90$$

Since the coordinates are approximate, the results are rounded off.

The key sequence for a calculator solution is shown in Fig. 4.16. The result shown in line 2 is for r, which was then stored under x (for convenience of calculator use—it is not the x-coordinate). Storing x as shown is to emphasize the fact the r can be stored for later use (it can be stored simply by using $sto \blacktriangleright x$ in line 1). The value of $\sin \theta$ is shown in lines 5 and 6. The values of the other functions can be found by doing the appropriate divisions, using the stored value of r when needed. ■

In Example 4, we expressed the result as $\sin \theta = 0.525$. A common error is to omit the angle and give the value as $\sin = 0.525$. This is a meaningless expression, for *we must show the angle* for which we have the value of a function.

If one of the trigonometric functions is known, it is possible to find the values of the other functions. The following example illustrates the method.

EXAMPLE 5   Given one function – find others

If we know that $\sin \theta = 3/7$, and that θ is a first-quadrant angle, we know the ratio of the ordinate to the radius vector (y to r) is 3 to 7. Therefore, the point on the terminal side for which $y = 3$ can be found by use of the Pythagorean theorem. The x-value for this point is

$$x = \sqrt{7^2 - 3^2} = \sqrt{49 - 9} = \sqrt{40} = 2\sqrt{10}$$

Therefore, the point $(2\sqrt{10}, 3)$ is on the terminal side, as shown in Fig. 4.17.

Therefore, using the values $x = 2\sqrt{10}$, $y = 3$, and $r = 7$, we have the other trigonometric functions of θ. They are

$$\cos \theta = \frac{2\sqrt{10}}{7} \qquad \tan \theta = \frac{3}{2\sqrt{10}} \qquad \cot \theta = \frac{2\sqrt{10}}{3} \qquad \sec \theta = \frac{7}{2\sqrt{10}} \qquad \csc \theta = \frac{7}{3}$$

These values are *exact*. *Approximate* decimal values found on a calculator are

$$\cos \theta = 0.9035 \qquad \tan \theta = 0.4743 \qquad \cot \theta = 2.108$$

$$\sec \theta = 1.107 \qquad \csc \theta = 2.333$$

■

EXERCISES 4.2

In Exercises 1 and 2, answer the given questions about the indicated examples of this section.

1. In Example 3, if the point $(4, 3)$ replaces the point $(3, 4)$, what are the values?

2. In Example 5, if 4/7 replaces 3/7, what are the values?

In Exercises 3–16, find values of the trigonometric functions of the angle (in standard position) whose terminal side passes through the given points. For Exercises 3–14, give answers in exact form. For Exercises 15 and 16, the coordinates are approximate.

3. $(6, 8)$ **4.** $(5, 12)$ **5.** $(15, 8)$ **6.** $(240, 70)$

7. $(0.09, 0.40)$ **8.** $(3.2, 6.0)$ **9.** $(1, \sqrt{15})$ **10.** $(\sqrt{3}, 2)$

11. $(1, 1)$ **12.** $(6, 5)$ **13.** $(50, 20)$ **14.** $\left(1, \frac{1}{2}\right)$

15. $(0.687, 0.943)$ **16.** $(37.65, 21.87)$

In Exercises 17–24, find the values of the indicated functions. In Exercises 17–20, give answers in exact form. In Exercises 21–24, the values are approximate.

17. Given $\cos\theta = 12/13$, find $\sin\theta$ and $\cot\theta$.

18. Given $\sin\theta = 1/2$, find $\cos\theta$ and $\csc\theta$.

19. Given $\tan\theta = 2$, find $\sin\theta$ and $\sec\theta$.

20. Given $\sec\theta = \sqrt{5}/2$, find $\tan\theta$ and $\cos\theta$.

21. Given $\sin\theta = 0.750$, find $\cot\theta$ and $\csc\theta$.

22. Given $\cos\theta = 0.0326$, find $\sin\theta$ and $\tan\theta$.

23. Given $\cot\theta = 0.254$, find $\cos\theta$ and $\tan\theta$.

24. Given $\csc\theta = 1.20$, find $\sec\theta$ and $\cos\theta$.

In Exercises 25–28, each point listed is on the terminal side of an angle. Show that each of the indicated functions is the same for each of the points.

25. $(3, 4), (6, 8), (4.5, 6)$, $\sin\theta$ and $\tan\theta$

26. $(5, 12), (15, 36), (7.5, 18)$, $\cos\theta$ and $\cot\theta$

27. $(0.3, 0.1), (9, 3), (33, 11)$, $\tan\theta$ and $\sec\theta$

28. $(40, 30), (8, 6), (36, 27)$, $\csc\theta$ and $\cos\theta$

In Exercises 29–36, answer the given questions.

29. If $\tan\theta = 3/4$, what is the value of $\sin^2\theta + \cos^2\theta$? $[\sin^2\theta = (\sin\theta)^2]$

30. If $\sin\theta = 2/3$, what is the value of $\sec^2\theta - \tan^2\theta$?

31. If $y = \sin\theta$, what is $\cos\theta$ in terms of y?

32. If $x = \cos\theta$, what is $\tan\theta$ in terms of x?

33. What is x if $(x + 1, 4)$ and $(-2, 6)$ are on the same terminal side of a standard-position angle?

34. What is x if $(2, 5)$ and $(7, x)$ are on the same terminal side of a standard-position angle?

35. From the definitions of the trigonometric functions, it can be seen that $\csc\theta$ is the reciprocal of $\sin\theta$. What function is the reciprocal of $\cos\theta$?

W **36.** Refer to the definitions of the trigonometric functions in Eqs. (4.1). Is the quotient of one of the functions divided by $\cos\theta$ equal to $\tan\theta$? Explain.

Answers to Practice Exercises

1. $\tan\theta = 3/4$, $\sec\theta = 5/4$ **2.** $\sin\theta = 0.9035$, $\cot\theta = 0.4743$

4.3 Values of the Trigonometric Functions

**Function Values Using Geometry •
Function Values from Calculator •
Inverse Trigonometric Functions •
Accuracy of Trigonometric Functions •
Reciprocal Functions**

In practice, we often need the values of the trigonometric functions for angles measured in degrees. One way to find some of these values is to use certain basic facts from geometry to find the functions of particular angles. This is illustrated in the next two examples.

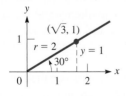

Fig. 4.18

EXAMPLE 1 Function values of 30° and 60°

From geometry, we find that the side opposite a 30° angle in a right triangle is one-half of the hypotenuse. Therefore, letting $y = 1$ and $r = 2$ (see Fig. 4.18), and using the Pythagorean theorem, we have $x = \sqrt{2^2 - 1^2} = \sqrt{3}$. Now, with $x = \sqrt{3}, y = 1$, and $r = 2$,

$$\sin 30° = \frac{1}{2} \qquad \cos 30° = \frac{\sqrt{3}}{2} \qquad \tan 30° = \frac{1}{\sqrt{3}}$$

Using this same method, we find the functions of 60° to be

$$\sin 60° = \frac{\sqrt{3}}{2} \qquad \cos 60° = \frac{1}{2} \qquad \tan 60° = \sqrt{3}$$

∎

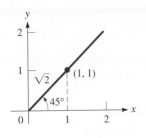

Fig. 4.19

EXAMPLE 2   Function values of 45°

Find sin 45°, cos 45°, and tan 45°.

If we place an isosceles right triangle with one of its 45° angles in standard position and hypotenuse along the radius vector (see Fig. 4.19), the terminal side passes through (1, 1), since the legs of the triangle are equal. Using this point, $x = 1$, $y = 1$, and $r = \sqrt{2}$. Thus,

$$\sin 45° = \frac{1}{\sqrt{2}} \qquad \cos 45° = \frac{1}{\sqrt{2}} \qquad \tan 45° = 1$$

In Examples 1 and 2, we have given *exact* values. Decimal approximations are also given in the following table that summarizes the results for 30°, 45°, and 60°.

Trigonometric Functions of 30°, 45°, and 60°

	(exact values)			(decimal approximations)		
θ	30°	45°	60°	30°	45°	60°
$\sin \theta$	$\dfrac{1}{2}$	$\dfrac{1}{\sqrt{2}}$	$\dfrac{\sqrt{3}}{2}$	0.500	0.707	0.866
$\cos \theta$	$\dfrac{\sqrt{3}}{2}$	$\dfrac{1}{\sqrt{2}}$	$\dfrac{1}{2}$	0.866	0.707	0.500
$\tan \theta$	$\dfrac{1}{\sqrt{3}}$	1	$\sqrt{3}$	0.577	1.000	1.732

■ It is helpful to be familiar with these values, as they are used in later sections.

Another way to find values of the functions is to use a scale drawing. Measure the angle with a protractor, then measure directly the values of x, y, and r for some point on the terminal side, and finally use the proper ratios to evaluate the functions. However, this method is only approximate, and geometric methods work only for a limited number of angles. As it turns out, it is possible to find these values to any required accuracy through more advanced methods (using calculus and what are known as *power series*).

Values of sin θ, cos θ, and tan θ are programmed into graphing calculators. For the remainder of this chapter, *be sure that your calculator is set for **degrees*** (not radians). The following examples illustrate using a calculator to find trigonometric values.

EXAMPLE 3   Values from calculator

Using a graphing calculator to find the value of tan 67.36°, first enter the function and then the angle, just as it is written. The resulting display is shown in Fig. 4.20.

Therefore, tan 67.36° = 2.397626383.

Fig. 4.20

Not only are we able to find values of the trigonometric functions if we know the angle, but we can also find the angle if we know that value of a function. In doing this, we are actually using another important type of mathematical function, an **inverse trigonometric function.** They are discussed in detail in Chapter 20. For the purpose of using a calculator at this point, it is sufficient to recognize and understand the notation that is used.

■ Another notation that is used for $\sin^{-1} x$ is arcsin x.

The notation for "the angle whose sine is x" is $\sin^{-1} x$. This is called the *inverse sine function.* Equivalent meanings are given to $\cos^{-1} x$ (the angle whose cosine is x) and $\tan^{-1} x$ (the angle whose tangent is x). Carefully note that *the −1 used with a* **CAUTION** ▸ *trigonometric function in this way shows an inverse trigonometric function and is **not** a negative exponent* ($\sin^{-1} x$ represents an *angle*, not a function of an angle). On a calculator, the $\sin^{-1}$ key is used to find the angle when the sine of that angle is known. The following example illustrates the use of the equivalent $\cos^{-1}$ key.   ■

EXAMPLE 4 Inverse function value from calculator

If $\cos \theta = 0.3527$, which means that $\theta = \cos^{-1} 0.3527$ (θ is the angle whose cosine is 0.3527), we can use a graphing calculator to find θ. The display is shown in Fig. 4.21.

Therefore, $\theta = 69.35°$ (rounded off). ∎

When using the trigonometric functions, the angle is often *approximate*. Angles of 2.3°, 92.3°, and 182.3° are angles with equal accuracy, which shows that *the accuracy of an angle does not depend on the number of digits shown*. The measurement of an angle and the accuracy of its trigonometric functions are shown in the following table:

Angles and Accuracy of Trigonometric Functions

Measurements of Angle to Nearest	Accuracy of Trigonometric Function
1°	2 significant digits
0.1° or 10′	3 significant digits
0.01° or 1′	4 significant digits

We rounded off the result in Example 4 according to this table.

Although we can usually set up the solution of a problem in terms of the sine, cosine, or tangent, there are times when a value of the cotangent, secant, or cosecant is used. We now show how values of these functions are found on a calculator.

Reciprocal Functions

From the definitions, $\csc \theta = r/y$ and $\sin \theta = y/r$. This means *the value of $\csc \theta$ is the reciprocal of the value of $\sin \theta$*. Again, using the definitions, we find that *the value of $\sec \theta$ is the reciprocal of $\cos \theta$* and *the value of $\cot \theta$ is the reciprocal of $\tan \theta$*. Since **NOTE ▶** the reciprocal of x equals x^{-1}, we use the x^{-1} key along with the sin, cos, and tan keys, to find the values of $\csc \theta$, $\sec \theta$, and $\cot \theta$.

EXAMPLE 5 Reciprocal function value from calculator

To find the value of $\sec 27.82°$, we use the fact that

$$\sec 27.82° = \frac{1}{\cos 27.82°} \quad \text{or} \quad \sec 27.82° = (\cos 27.82°)^{-1}$$

Therefore, we are to find the reciprocal of the value of $\cos 27.82°$. It can be found using either the first two lines, or the third and fourth lines, of the calculator display in Fig. 4.22. From this display, $\sec 27.82° = 1.131$, rounded off according to the above table.

Note that we calculated the reciprocal $(\cos 27.82°)^{-1}$, and not the angle that would be denoted by using the $\cos^{-1}$ notation. ∎

EXAMPLE 6 Given a function – find θ

To find the value of θ if $\cot \theta = 0.354$, we use the fact that

$$\tan \theta = \frac{1}{\cot \theta} = \frac{1}{0.354}$$

The graphing calculator value is shown in the fifth and sixth lines of Fig. 4.22. Therefore, $\theta = 70.5°$ (rounded off). ∎

EXAMPLE 7 Given one function – find another

Find $\sin \theta$ if $\sec \theta = 2.504$.

Since the value of $\sec \theta$ is known, we know that $\cos \theta = 2.504^{-1}$ (or 1/2.504). This in turn tells us that $\theta = \cos^{-1}(2.504^{-1})$. Since we are to find the value of $\sin \theta$, we can see that

$$\sin \theta = \sin\left(\cos^{-1}(2.504^{-1})\right)$$

Therefore, we have the calculator display shown in Fig. 4.23.

This means that $\sin \theta = 0.9168$ (rounded off). ∎

Fig. 4.21

Practice Exercises

1. Find the value of $\sin 12.5°$.
2. Find θ if $\tan \theta = 1.039$.

Fig. 4.22

Practice Exercises

3. Find the values of $\cot 56.4°$.
4. Find θ if $\csc \theta = 1.904$.

Fig. 4.23

Calculators have been in common use since the 1970s. Until then, the values of the trigonometric functions were generally found by the use of tables (where linear interpolation was used to find values to one more place than was shown in the table). Many standard sources with tables are still available with a precision to at least 10′ or 0.1° (these tables give values with a precision to about five decimal places). However, a calculator is much easier to use than a table, and it can give values to a much greater accuracy. Therefore, we will not use tables in this text.

The following example illustrates the use of the value of a trigonometric function in an applied problem. We consider various types of applications later in the chapter.

EXAMPLE 8 Applied problem

When a rocket is launched, its horizontal velocity v_x is related to the velocity v with which it is fired by the equation $v_x = v \cos \theta$ (which means $v(\cos \theta)$, but does *not* mean $\cos \theta\, v$, which is the same as $\cos (\theta\, v)$). Here, θ is the angle between the horizontal and the direction in which it is fired (see Fig. 4.24). Find v_x if $v = 1250$ m/s and $\theta = 36.0°$.

Substituting the given values of v and θ in $v_x = v \cos \theta$, we have

$$v_x = 1250 \cos 36.0°$$
$$= 1010 \text{ m/s}$$

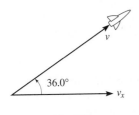

Fig. 4.24

Therefore, the horizontal velocity is 1010 m/s. ∎

EXERCISES 4.3

In Exercises 1–4, make the given changes in the indicated examples of this section and then find the indicated values.

1. In Example 4, change $\cos \theta$ to $\sin \theta$ and then find the angle.
2. In Example 5, change $\sec 27.82°$ to $\csc 27.82°$ and then find the value.
3. In Example 6, change 0.354 to 0.345 and then find the angle.
4. In Example 7, change $\sin \theta$ to $\tan \theta$ and then find the value.

In Exercises 5–8, use a protractor to draw the given angle. Measure off 10 units (centimeters are convenient) along the radius vector. Then measure the corresponding values of x and y. From these values, determine the trigonometric functions of the angle.

5. 40°   6. 75°   7. 15°   8. 53°

In Exercises 9–24, find the values of the trigonometric functions. Round off results according to the table following Example 4.

9. $\sin 22.4°$
10. $\cos 72.5°$
11. $\tan 57.6°$
12. $\sin 36.0°$
13. $\cos 15.71°$
14. $\tan 8.653°$
15. $\sin 89°$
16. $\cos 0.7°$
17. $\cot 67.78°$
18. $\csc 22.81°$
19. $\sec 50.4°$
20. $\cot 41.8°$
21. $\csc 0.49°$
22. $\sec 7.8°$
23. $\cot 85.96°$
24. $\csc 76.30°$

In Exercises 25–40, find θ for each of the given trigonometric functions. Round off results according to the table following Example 4.

25. $\cos \theta = 0.3261$
26. $\tan \theta = 2.470$
27. $\sin \theta = 0.9114$
28. $\cos \theta = 0.0427$
29. $\tan \theta = 0.207$
30. $\sin \theta = 0.109$
31. $\cos \theta = 0.65007$
32. $\tan \theta = 5.7706$
33. $\csc \theta = 1.245$
34. $\sec \theta = 2.045$
35. $\cot \theta = 0.0606$
36. $\csc \theta = 1.002$
37. $\sec \theta = 3.65$
38. $\cot \theta = 14.4$
39. $\csc \theta = 8.26$
40. $\cot \theta = 0.1519$

In Exercises 41–44, use a calculator to verify the given relationships or statements. $\left[\sin^2 \theta = (\sin \theta)^2 \right]$

41. $\dfrac{\sin 43.7°}{\cos 43.7°} = \tan 43.7°$

42. $\sin^2 77.5° + \cos^2 77.5° = 1$

43. $\tan 70° = \dfrac{\tan 30° + \tan 40°}{1 - (\tan 30°)(\tan 40°)}$

44. $\sin 78.4° = 2(\sin 39.2°)(\cos 39.2°)$

(W) *In Exercises 45–48, explain why the given statements are true for an acute angle θ.*

45. $\sin \theta$ is always between 0 and 1.
46. $\tan \theta$ can equal any positive real number.
47. $\cos \theta$ decreases in value from 0° to 90°.
48. The value of $\sec \theta$ is never less than 1.

In Exercises 49–52, find the values of the indicated trigonometric functions.

49. Find $\sin\theta$, given $\tan\theta = 1.936$.

50. Find $\cos\theta$, given $\sin\theta = 0.6725$.

51. Find $\tan\theta$, given $\sec\theta = 1.3698$.

52. Find $\csc\theta$, given $\cos\theta = 0.1063$.

In Exercises 53–58, solve the given problems.

53. Find the first quadrant angle between the line $y = 3x$ and the x-axis.

54. Find the angle in the first quadrant between the lines $y = 2x$ and $y = 4x$.

55. The sound produced by a jet engine was measured at a distance of 100 m in all directions. The loudness d of the sound (in decibels) was found to be $d = 70.0 + 30.0\cos\theta$, where the 0° line was directed in front of the engine. Calculate d for $\theta = 54.5°$.

56. A brace is used in the structure shown in Fig. 4.25. Its length is $l = a(\sec\theta + \csc\theta)$. Find l if $a = 28.0$ cm and $\theta = 34.5°$.

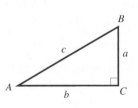

Fig. 4.25

57. The signal from an AM radio station with two antennas d meters apart has a wavelength λ (in m). The intensity of the signal depends on the angle θ as shown in Fig. 4.26. An angle of minimum intensity is given by $\sin\theta = 1.50\,\lambda/d$. Find θ if $\lambda = 200$ m and $d = 400$ m.

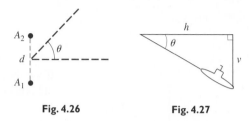

Fig. 4.26   **Fig. 4.27**

58. A submarine dives such that the horizontal distance h it moves and the vertical distance v it dives are related by $v = h\tan\theta$. Here, θ is the angle of the dive, as shown in Fig. 4.27. Find θ if $h = 2.35$ mi and $v = 1.52$ mi.

Answers to Practice Exercises

1. 0.216   **2.** 46.10°   **3.** 0.664   **4.** 31.68°

4.4 The Right Triangle

Solving a Triangle • Cofunctions • Procedure for Solving a Right Triangle • Calculator Solutions

We know that a triangle has three sides and three angles. If one side and any other two of these six parts are known, we can find the other three parts. One of the known parts must be a side, for if we know only the three angles, we know only that any triangle with these angles is similar to any other triangle with these angles.

EXAMPLE 1 Parts of a triangle

Assume that one side and two angles are known, such as the side of 5 and the angles of 35° and 90° in the triangle in Fig. 4.28. Then we may determine the third angle α by the fact that the sum of the angles of a triangle is always 180°. Of all possible similar triangles having the three angles of 35°, 90°, and 55° (which is α), we have the one with the particular side of 5 between angles of 35° and 90°. Only one triangle with these parts is possible (in the sense that all triangles with the given parts are *congruent* and have equal corresponding angles and sides). ■

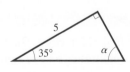

Fig. 4.28

To **solve a triangle** *means that, when we are given three parts of a triangle (at least one a side), we are to find the other three parts.* In this section, we are going to demonstrate the method of solving a right triangle. *Since one angle of the triangle is 90°, it is necessary to know one side and one other part.* Also, since the sum of the three angles is 180°, we know that *the sum of the other two angles is 90°, and they are acute angles.* It also means they are **complementary angles,** following the definition on page 50.

For consistency, when we are labeling the parts of the right triangle, *we will use the letters A and B to denote the acute angles and C to denote the right angle. The letters a, b, and c will denote the sides opposite these angles, respectively. Thus, side c is the hypotenuse of the right triangle.* See Fig. 4.29.

Fig. 4.29

Fig. 4.30

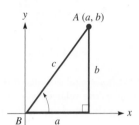
Fig. 4.31

In solving right triangles, we will find it convenient to express the trigonometric functions of the acute angles in terms of the sides. By placing the vertex of angle A at the origin and the vertex of right angle C on the positive x-axis, as shown in Fig. 4.30, we have the following ratios for angle A in terms of the sides of the triangle.

$$\sin A = \frac{a}{c} \qquad \cos A = \frac{b}{c} \qquad \tan A = \frac{a}{b}$$

$$\cot A = \frac{b}{a} \qquad \sec A = \frac{c}{b} \qquad \csc A = \frac{c}{a} \qquad \text{(4.3)}$$

If we should place the vertex of B at the origin, instead of the vertex of angle A, we would obtain the following ratios for the functions of angle B (see Fig. 4.31):

$$\sin B = \frac{b}{c} \qquad \cos B = \frac{a}{c} \qquad \tan B = \frac{b}{a}$$

$$\cot B = \frac{a}{b} \qquad \sec B = \frac{c}{a} \qquad \csc B = \frac{c}{b} \qquad \text{(4.4)}$$

Equations (4.3) and (4.4) show that we may generalize our definitions of the trigonometric functions of an acute angle of a right triangle (we have chosen $\angle A$ in Fig. 4.32) to be as follows:

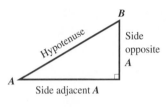
Fig. 4.32

$$\sin A = \frac{\text{side opposite } A}{\text{hypotenuse}} \qquad \cos A = \frac{\text{side adjacent } A}{\text{hypotenuse}}$$

$$\tan A = \frac{\text{side opposite } A}{\text{side adjacent } A} \qquad \cot A = \frac{\text{side adjacent } A}{\text{side opposite } A} \qquad \text{(4.5)}$$

$$\sec A = \frac{\text{hypotenuse}}{\text{side adjacent } A} \qquad \csc A = \frac{\text{hypotenuse}}{\text{side opposite } A}$$

■ Which side is adjacent or opposite depends on the angle being considered. In Fig. 4.31, the side opposite A is adjacent to B, and the side adjacent to A is opposite B.

Using the definitions in this form, we can solve right triangles without placing the angle in standard position. The angle need only be a part of any right triangle.

We note from the above discussion that $\sin A = \cos B$, $\tan A = \cot B$, and $\sec A = \csc B$. From this, we conclude that *cofunctions of acute complementary angles are equal.* The sine function and cosine function are cofunctions, the tangent function and cotangent function are cofunctions, and the secant function and cosecant function are cofunctions.

EXAMPLE 2 Cofunctions of complementary angles

Given $a = 4$, $b = 7$, and $c = \sqrt{65}$ (see Fig. 4.33), find $\sin A$, $\cos A$, $\tan A$, $\sin B$, $\cos B$, and $\tan B$ in exact form and in approximate decimal form (to three significant digits).

Fig. 4.33

$$\sin A = \frac{\text{side opposite angle } A}{\text{hypotenuse}} = \frac{4}{\sqrt{65}} = 0.496 \qquad \sin B = \frac{\text{side opposite angle } B}{\text{hypotenuse}} = \frac{7}{\sqrt{65}} = 0.868$$

$$\cos A = \frac{\text{side adjacent angle } A}{\text{hypotenuse}} = \frac{7}{\sqrt{65}} = 0.868 \qquad \cos B = \frac{\text{side adjacent angle } B}{\text{hypotenuse}} = \frac{4}{\sqrt{65}} = 0.496$$

$$\tan A = \frac{\text{side opposite angle } A}{\text{side adjacent angle } A} = \frac{4}{7} = 0.571 \qquad \tan B = \frac{\text{side opposite angle } B}{\text{side adjacent angle } B} = \frac{7}{4} = 1.75$$

We see that A and B are complementary angles. Comparing values of the functions of angles A and B, we see that $\sin A = \cos B$ and $\cos A = \sin B$. ∎

We are now ready to solve right triangles. See how the following procedure is used in the examples that follow.

Fig. 4.34

PROCEDURE FOR SOLVING A RIGHT TRIANGLE

1. *Sketch a right triangle and label the known and unknown sides and angles.*
2. *Express each of the three unknown parts in terms of the known parts and solve for the unknown parts.*
3. *Check the results.* The sum of the angles should be 180°. If only one side is given, check the computed side with the Pythagorean theorem. If two sides are given, check the angles and computed side by using appropriate trigonometric functions.

```
6.7tan(50)
          7.98474907
6.7/cos(50)
          10.42334964
90-50
               40
```

Fig. 4.35

EXAMPLE 3 Given angle and side – find other parts

Solve the right triangle with $A = 50.0°$ and $b = 6.70$.

We first sketch the right triangle shown in Fig. 4.34. (In making the sketch, we should be careful to follow proper labeling of the triangle as outlined on page 120.) We then express unknown side a in terms of known side b and known angle A and solve for a. We will then do the same for unknown side c and unknown angle B.

Finding side a, we know that $\tan A = \dfrac{a}{b}$, which means that $a = b \tan A$. Thus,

$$a = 6.70 \tan 50.0° = 7.98 \quad \text{lines 1 \& 2 of calculator display in Fig. 4.35}$$

Next, solving for side c, we have $\cos A = \dfrac{b}{c}$, which means $c = \dfrac{b}{\cos A}$.

$$c = \frac{6.70}{\cos 50.0°} = 10.4 \quad \text{lines 3 \& 4 of calculator display in Fig. 4.35}$$

Now, solving for B, we know that $A + B = 90°$, or

$$B = 90° - A$$
$$= 90° - 50.0° = 40.0° \quad \text{lines 5 \& 6 of calculator display in Fig. 4.35}$$

Therefore, $a = 7.98$, $c = 10.4$, and $B = 40.0°$.

Checking the angles: $A + B + C = 50.0° + 40.0° + 90° = 180°$

Checking the sides: $10.4^2 = 108.16$
$$7.98^2 + 6.70^2 = 108.57$$

Since the computed values were rounded off, the values 108.16 and 108.57 show that the values for sides a and c check.

As we calculate the values of the unknown parts, if we store each in the calculator memory, we can get a better check of the solution. In Fig. 4.36, we see the solution for each side, its storage in memory, and the check of these values. The lines above the display are those that are replaced as the solution proceeds. From this calculator display, we see that the values for sides a and c check very accurately. ∎

Practice Exercise

1. In a right triangle, find a if $B = 20.0°$ and $c = 8.50$.

```
6.7tan(50)
          7.98474907
Ans→A
          7.98474907
6.7/cos(50)
          10.42334964
Ans→C
          10.42334964
A²+6.7²
          108.6462177
C²
          108.6462177
```

Fig. 4.36

NOTE ▸ In finding the unknown parts, we first expressed them in terms of the known parts. We do this because *it is best to use given values in calculations.* If we use one computed value to find another computed value, any error in the first would be carried to the value of the second. For instance, in Example 3, if we were to find the value of c by using the value of a, any error in a would cause c to be in error as well.

■ See Appendix C for a graphing calculator program SLVRTTRI. It can be used to solve a right triangle, given the two legs.

We should also point out that, by inspection, we can make a rough check on the sides and angles of any triangle.

The longest side is always opposite the largest angle, and the shortest side is always opposite the smallest angle.

In a right triangle, *the hypotenuse is always the longest side.* We see that this is true for the sides and angles for the triangle in Example 3, where c is the longest side (opposite the 90° angle) and b is the shortest side and is opposite the angle of 40°.

EXAMPLE 4 Given two sides – find other parts

Solve the right triangle with $b = 56.82$ and $c = 79.55$.

We sketch the right triangle as shown in Fig. 4.37. Since two sides are given, we will use the Pythagorean theorem to find the third side a. Also, we will use the cosine to find $\angle A$.

Since $c^2 = a^2 + b^2, a^2 = c^2 - b^2$. Therefore,

$$a = \sqrt{c^2 - b^2} = \sqrt{79.55^2 - 56.82^2}$$
$$= 55.67 \qquad \text{lines 1 \& 2 shown in Fig. 4.38}$$

Since $\cos A = \dfrac{b}{c}$, we have

$$\cos A = \frac{56.82}{79.55}$$
$$A = \cos^{-1}\left(\frac{56.82}{79.55}\right) = 44.42° \qquad \text{lines 3–5 shown in Fig. 4.38}$$

It is not necessary to actually calculate the ratio 56.82/79.55. In the same way, we find the value of angle B:

$$\sin B = \frac{56.82}{79.55}$$
$$B = \sin^{-1}\left(\frac{56.82}{79.55}\right) = 45.58° \qquad \text{lines 6–8 shown in Fig. 4.38}$$

Although we used a different function, we did use exactly the same ratio to find B as we used to find A. Therefore, many texts would find B from the fact that $A + B = 90°$, or $B = 90° - A = 90° - 44.42° = 45.58°$. This is also acceptable since any possible error should be discovered when the solution is checked.

We have now found that

$$a = 55.67 \qquad A = 44.42° \qquad B = 45.58°$$

Checking the sides and angles, we first note that side a is the shortest side and is opposite the smallest angle, $\angle A$. Also, the hypotenuse is the longest side. Next, using the sine function (we could use the cosine or tangent) to check the sides, we have

$$\sin 44.42° = \frac{55.67}{79.55}, \text{ or } 0.6999 \approx 0.6998 \qquad \sin 45.58° = \frac{56.82}{79.55}, \text{ or } 0.7142 \approx 0.7143$$

This shows that the values check. As we noted at the end of Example 3, we would get a more accurate check if we save the calculator values as they are found and use them for the check. ■

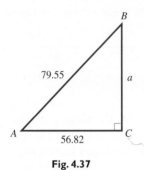

Fig. 4.37

Practice Exercise

2. In a right triangle, find B if $a = 20.0$ and $b = 28.0$.

```
√(79.55²-56.82²)
          55.67486057
cos⁻¹(56.82/79.55
)
          44.41677878
sin⁻¹(56.82/79.55
)
          45.58322122
```

Fig. 4.38

■ As we noted earlier, the symbol ≈ means "equals approximately."

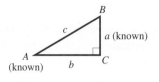

Fig. 4.39

EXAMPLE 5 Unknown parts in terms of known parts

If A and a are known, express the unknown parts of a right triangle in terms of A and a. We sketch a right triangle as in Fig. 4.39, and then set up the required expressions.

Since $\dfrac{a}{b} = \tan A$, we have $a = b \tan A$, or $b = \dfrac{a}{\tan A}$.

Since $\dfrac{a}{c} = \sin A$, we have $a = c \sin A$, or $c = \dfrac{a}{\sin A}$.

Since A is known, $B = 90° - A$. ∎

EXERCISES 4.4

In Exercises 1–4, make the given changes in the indicated examples of this section and then find the indicated values.

1. In Example 2, interchange a and b and then find the values.

2. In Example 3, change 6.70 to 7.60 and then solve the triangle.

3. In Example 4, change 56.82 to 65.82 and then solve the triangle.

4. In Example 5, in line 1, change a to b and then find the expressions for the unknown parts.

In Exercises 5–8, draw appropriate figures and verify through observation that only one triangle may contain the given parts (that is, any others which may be drawn will be congruent).

5. A 60° angle included between sides of 3 in. and 6 in.

6. A side of 4 in. included between angles of 40° and 50°

7. A right triangle with a hypotenuse of 5 cm and a leg of 3 cm

8. A right triangle with a 70° angle between the hypotenuse and a leg of 5 cm

In Exercises 9–32, solve the right triangles with the given parts. Round off results. Refer to Fig. 4.40.

9. $A = 77.8°, a = 6700$

10. $A = 18.4°, c = 0.0897$

11. $a = 150, c = 345$

12. $a = 932, c = 1240$

13. $B = 32.1°, c = 23.8$

14. $B = 64.3°, b = 0.652$

Fig. 4.40

15. $b = 82, c = 88$

16. $a = 5920, b = 4110$

17. $A = 32.10°, c = 56.85$

18. $B = 12.60°, c = 18.42$

19. $a = 56.73, b = 44.09$

20. $a = 9.908, c = 12.63$

21. $B = 37.5°, a = 0.862$

22. $A = 87.25°, b = 8.450$

23. $B = 74.18°, b = 1.849$

24. $A = 51.36°, a = 3692$

25. $a = 591.87, b = 264.93$

26. $b = 2.9507, c = 5.0864$

27. $A = 2.975°, b = 14.592$

28. $B = 84.942°, a = 7413.5$

29. $B = 9.56°, c = 0.0973$

30. $a = 1.28, b = 16.3$

31. $a = 35.0, C = 90.0°$

32. $A = 25.7°, B = 64.3°$

In Exercises 33–36, find the part of the triangle labeled either x or A in the indicated figure.

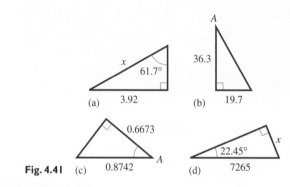

Fig. 4.41 (a)   (b)   (c)   (d)

33. Fig. 4.41(a)

34. Fig. 4.41(b)

35. Fig. 4.41(c)

36. Fig. 4.41(d)

In Exercises 37–40, find the indicated part of the right triangle that has the given parts.

37. One leg is 25.6, and the hypotenuse is 37.5. Find the smaller acute angle.

38. One leg is 8.50, and the angle opposite this leg is 52.3°. Find the other leg.

39. The hypotenuse is 827, and one angle is 17.6°. Find the longer leg.

40. The legs are 0.596 and 0.842. Find the larger acute angle.

In Exercises 41–44, refer to Fig. 4.40. In Exercises 41–43, the listed parts are assumed known. Express the other parts in terms of the known parts.

41. A, c

42. a, b

43. B, a

(W) 44. In Fig. 4.40, is there any combination of two given parts (not including $\angle C$) that does not give a unique solution of the triangle? Explain.

Answers to Practice Exercises

1. $a = 7.99$ 2. $54.5°$

Applications of Right Triangles

Angles of Elevation and Depression •
Indirect Measurements of Distances
and Angles

Many problems in science, technology, and everyday life can be solved by finding the missing parts of a right triangle. In this section, we illustrate a number of these in the examples and exercises.

EXAMPLE 1 Angle of elevation

Horseshoe Falls on the Canadian side of Niagara Falls can be seen from a small boat 2500 ft downstream. The **angle of elevation** *(the angle between the horizontal and the line of sight, when the object is above the horizontal)* from the observer to the top of Horseshoe Falls is 4.0°. How high are the Falls?

■ See the chapter introduction.

By drawing an appropriate figure, as shown in Fig. 4.42, we show the given information and what we are to find. Here, we let h be the height of the Falls and label the horizontal distance from the boat to the base of the Falls as 2500 ft. From the figure, we see that

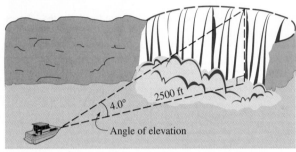

$$\frac{h}{2500} = \tan 4.0° \qquad \frac{\text{required opposite side}}{\text{given adjacent side}} = \text{tangent of given angle}$$

$$h = 2500 \tan 4.0°$$

$$= 170 \text{ ft}$$

We have rounded off the result since the data are good only to two significant digits. (Niagara Falls is on the border between the United States and Canada and is divided into the American Falls and the Horseshoe Falls. About 500,000 tons of water flow over the Falls each minute.) ■

Fig. 4.42

EXAMPLE 2 Angle of depression

The Goodyear blimp is 1850 ft above the ground and south of the Rose Bowl in California during a Super Bowl game. The **angle of depression** *(the angle between the horizontal and the line of sight, when the object is below the horizontal)* of the north goal line from the blimp is 58.5°. How far is the observer in the blimp from the goal line?

Again, we sketch a figure as shown in Fig. 4.43. Here, we let d be the distance between the blimp and the north goal line. From the figure, we see that

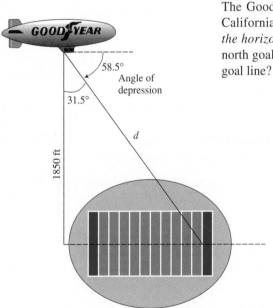

$$\frac{1850}{d} = \cos 31.5° \qquad \frac{\text{given adjacent side}}{\text{required hypotenuse}} = \text{cosine of known angle}$$

$$d = \frac{1850}{\cos 31.5°}$$

$$= 2170 \text{ ft}$$

In this case, we have rounded off the result to three significant digits, which is the accuracy of the given information. (The Super Bowl games in 1977, 1980, 1983, 1987, and 1993 were played in the Rose Bowl.) ■

Fig. 4.43

Carefully note the difference between the angle of elevation and the angle of depression. The angle of elevation is the angle through which an object is observed by *elevating* the line of sight above the horizontal. The angle of depression is the angle through which the object is observed by *depressing* (lowering) the line of sight below the horizontal.

EXAMPLE 3 Height of missile

A missile is launched at an angle of 26.55° with respect to the horizontal. If it travels in a straight line over level terrain for 2.000 min and its average speed is 6355 km/h, what is its altitude at this time?

Fig. 4.44

In Fig. 4.44, we let h represent the altitude of the missile after 2.000 min (altitude is measured on a perpendicular line). Also, we determine that in this time the missile has flown 211.8 km in a direct line from the launching site. This is found from the fact that it travels at 6355 km/h for $\frac{1}{30.00}$ h (2.000 min), and distance = speed × time. We therefore have $\left(6355\,\frac{\text{km}}{\text{h}}\right)\left(\frac{1}{30.00}\,\text{h}\right)$ = 211.8 km. This means

$$\frac{h}{211.8} = \sin 26.55° \qquad \frac{\text{required opposite side}}{\text{known hypotenuse}} = \text{sine of given angle}$$

$$h = 211.8(\sin 26.55°)$$

$$= 94.67 \text{ km} \qquad \blacksquare$$

EXAMPLE 4 Measurement of angle

A driver coming to an intersection sees the word STOP in the roadway. From the measurements shown in Fig. 4.45, find the angle θ that the letters make at the driver's eye.

From the figure, we know sides BS and BE in triangle BES and sides BT and BE in triangle BET. This means we can find $\angle TEB$ and $\angle SEB$ by use of the tangent. We then find θ from the fact that $\theta = \angle TEB - \angle SEB$.

$$\tan \angle TEB = \frac{18.0}{1.20}, \qquad \angle TEB = 86.2°$$

$$\tan \angle SEB = \frac{15.0}{1.20}, \qquad \angle SEB = 85.4°$$

$$\theta = 86.2° - 85.4° = 0.8° \qquad \blacksquare$$

Practice Exercise

1. Find θ if the letters in the road are 2.0 m long, rather than 3.0 m long.

Fig. 4.45

EXAMPLE 5 Surveyor – indirect measurement

Using lasers, a surveyor makes the measurements shown in Fig. 4.46, where points B and C are in a marsh. Find the distance between B and C.

Since the distance $BC = AC - AB$, BC is found by finding AC and AB and subtracting:

$$\frac{AB}{265.74} = \tan 21.66°$$

$$AB = 265.74 \tan 21.66°$$

$$\frac{AC}{265.74} = \tan (21.66° + 8.85°)$$

$$AC = 265.74 \tan 30.51°$$

$$BC = AC - AB = 265.74 \tan 30.51° - 265.74 \tan 21.66°$$

$$= 51.06 \text{ ft} \qquad \blacksquare$$

Fig. 4.46

■ Lasers were first produced in the late 1950s.

The indirect measurement of distances, such as the one illustrated in Example 5, has been one of the most useful applications of trigonometry. As we mentioned in the chapter introduction, the Greek astronomer Hipparchus used some of the methods of trigonometry for measurements in astronomy. Since the time of Hipparchus, methods of indirect measurement have also been used in many fields, such as surveying and navigation.

EXERCISES 4.5

In Exercises 1 and 2, make the given changes in the indicated examples of this section and then find the indicated values.

1. In Example 2, in line 4, change 58.5° to 62.1° and then find the distance.

2. In Example 3, in line 2, change 2.000 min to 3.000 min and then find the altitude.

In Exercises 3–40, solve the given problems. Sketch an appropriate figure, unless the figure is given.

3. A straight 120-ft culvert is built down a hillside that makes an angle of 54.0° with the horizontal. Find the height of the hill.

4. In 2000, about 70 metric tons of soil were removed from under the Leaning Tower of Pisa, and the angle the tower made with the ground was increased by about 0.5°. Before that, a point near the top of the tower was 50.5 m from a point at the base (measured along the tower), and this top point was directly above a point on the ground 4.25 m from the same base point. See Fig. 4.47. How much did the point on the ground move toward the base point?

50.5 m
4.25 m
Fig. 4.47

5. A tree has a shadow 22.8 ft long when the angle of elevation of the sun is 62.6°. How tall is the tree?

6. The straight arm of a robot is 1.25 m long and makes an angle of 13.0° above a horizontal conveyor belt. How high above the belt is the end of the arm? See Fig. 4.48.

1.25 m
13.0°
Fig. 4.48
Belt

7. The headlights of an automobile are set such that the beam drops 2.00 in. for each 25.0 ft in front of the car. What is the angle between the beam and the road?

8. A bullet was fired such that it just grazed the top of a table. It entered a wall, which is 12.60 ft from the graze point in the table, at a point 4.63 ft above the tabletop. At what angle was the bullet fired above the horizontal? See Fig. 4.49.

4.63 ft
12.60 ft
Fig. 4.49

9. A robot is on the surface of Mars. The angle of depression from a camera in the robot to a rock on the surface of Mars is 13.33°. The camera is 196.0 cm above the surface. How far from the camera is the rock?

10. The Sears Tower in Chicago can be seen from a point on the ground known to be 5200 ft from the base of the tower. The angle of elevation from the observer to the top of the tower is 16°. How high is the Sears Tower? See Fig. 4.50.

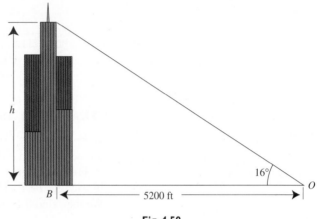
h
B
5200 ft
16°
O
Fig. 4.50

11. In the design of a new building, a doorway is 2.65 ft above the ground. A ramp for the disabled, at an angle of 6.0° with the ground, is to be built to the doorway. How long will the ramp be?

12. On a test flight, during the landing of the space shuttle, the ship was 325 ft above the end of the landing strip. It then came in on a constant angle of 7.5° with the landing strip. How far from the end of the landing strip did it first touch ground?

13. From a point on the South Rim of the Grand Canyon, it is found that the angle of elevation of a point on the North Rim is 1.2°. If the horizontal distance between the points is 9.8 mi, how much higher is the point on the North Rim?

14. What is the steepest angle between the surface of a board 3.50 cm thick and a nail 5.00 cm long if the nail is hammered into the board such that it does not go through?

15. A rectangular piece of plywood 4.00 ft by 8.00 ft is cut from one corner to an opposite corner. What are the angles between edges of the resulting pieces?

16. A guardrail is to be constructed around the top of a circular observation tower. The diameter of the observation area is 12.3 m. If the railing is constructed with 30 equal straight sections, what should be the length of each section?

17. The angle of inclination of a road is often expressed as *percent grade*, which is the vertical rise divided by the horizontal run (expressed as a percent). See Fig. 4.51. A 6.0% grade corresponds to a road that rises 6.0 ft for every 100 ft along the horizontal. Find the angle of inclination that corresponds to a 6.0% grade.

Fig. 4.51   Horizontal run   Rise

18. A tabletop is in the shape of a regular octagon (eight sides). What is the greatest distance across the table if one side of the octagon is 0.750 m?

19. To get a good view of a person in front of a teller's window, it is determined that a surveillance camera at a bank should be directed at a point 15.5 ft to the right and 6.75 ft below the camera. See Fig. 4.52. At what angle of depression should the camera be directed?

Fig. 4.52   **Fig. 4.53**

20. A street light is designed as shown in Fig. 4.53. How high above the street is the light?

21. A straight driveway is 85.0 ft long, and the top is 12.0 ft above the bottom. What angle does it make with the horizontal?

22. Part of the Tower Bridge in London is a drawbridge. This part of the bridge is 76.0 m long. When each half is raised, the distance between them is 8.0 m. What angle does each half make with the horizontal? See Fig. 4.54.

Fig. 4.54

23. A square wire loop is rotating in the magnetic field between two poles of a magnet in order to induce an electric current. The axis of rotation passes through the center of the loop and is midway between the poles, as shown in the side view in Fig. 4.55. How far is the edge of the loop from either pole if the side of the square is 7.30 cm and the poles are 7.66 cm apart when the angle between the loop and the vertical is 78.0°?

Fig. 4.55

24. From a space probe circling Io, one of Jupiter's moons, at an altitude of 552 km, it was observed that the angle of depression of the horizon was 39.7°. What is the radius of Io?

25. A manufacturing plant is designed to be in the shape of a regular pentagon with 92.5 ft on each side. A security fence surrounds the building to form a circle, and each corner of the building is to be 25.0 ft from the closest point on the fence. How much fencing is required?

26. A surveyor on the New York City side of the Hudson River wishes to find the height of a cliff *(palisade)* on the New Jersey side. Figure 4.56 shows the measurements that were made. How high is the cliff? (In the figure, the triangle containing the height *h* is vertical and perpendicular to the river.)

Fig. 4.56   **Fig. 4.57**

27. Find the angle θ in the taper shown in Fig. 4.57. (The front face is an isosceles trapezoid.)

28. What is the circumference of the Arctic Circle (latitude 66°32′ N)? The radius of the earth is 3960 mi.

29. A draftsman sets the legs of a pair of dividers so that the angle between them is 32.0°. If each leg is 4.50 in. long, what is the distance between their ends where they touch the paper?

30. The ratio of the width to the height of an HDTV screen is 16 to 9. What is the angle between the width and a diagonal of the screen?

31. A stairway 1.0 m wide goes from the bottom of a cylindrical storage tank to the top at a point halfway around the tank. The handrail on the outside of the stairway makes an angle of 31.8° with the horizontal, and the radius of the tank is 11.8 m. Find the length of the handrail. See Fig. 4.58.

Fig. 4.58

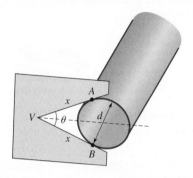

Fig. 4.59

32. An antenna was on the top of the World Trade Center (before it was destroyed in 2001). From a point on the river 7800 ft from the Center, the angles of elevation of the top and bottom of the antenna were 12.1° and 9.9°, respectively. How tall was the antenna? (Disregard the small part of the antenna near the base that could not be seen.) The former World Trade Center is shown in Fig. 4.59. *(This problem is included in memory of those who suffered and died as a result of the terrorist attack of September 11, 2001.)*

33. Some of the streets of San Francisco are shown in Fig. 4.60. The distances between intersections *A* and *B*, *A* and *D*, and *C* and *E* are shown. How far is it between intersections *C* and *D*?

Fig. 4.60

34. A supporting girder structure is shown in Fig. 4.61. Find the length *x*.

Fig. 4.61

35. The diameter *d* of a pipe can be determined by noting the distance *x* on the V-gauge shown in Fig. 4.62. Points *A* and *B* indicate where the pipe touches the gauge, and *x* equals either *AV* or *VB*. Find a formula for *d* in terms of *x* and θ.

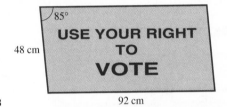

Fig. 4.62

36. The political banner shown in Fig. 4.63 is in the shape of a parallelogram. Find its area.

Fig. 4.63

37. Find a formula for the area of the trapezoidal aqueduct cross section shown in Fig. 4.64.

Fig. 4.64

38. A communications satellite is in orbit 35,300 km directly above the earth's equator. What is the greatest latitude from which a signal can travel from the earth's surface to the satellite in a straight line? The radius of the earth is 6400 km.

39. What is the angle between the base of a cubical glass paperweight and a diagonal of the cube (from one corner to the opposite corner)?

40. Find the angle of view θ of the camera lens (see Fig. 4.65), given the measurements shown in the figure.

Fig. 4.65

Answer to Practice Exercise

1. $\theta = 0.6°$

CHAPTER 4 EQUATIONS

$$\sin\theta = \frac{y}{r} \qquad \cos\theta = \frac{x}{r}$$

$$\tan\theta = \frac{y}{x} \qquad \cot\theta = \frac{x}{y}$$

$$\sec\theta = \frac{r}{x} \qquad \csc\theta = \frac{r}{y}$$

(4.1)

Pythagorean theorem

$$c^2 = a^2 + b^2$$

(4.2)

$$\sin A = \frac{\text{side opposite } A}{\text{hypotenuse}} \qquad \cos A = \frac{\text{side adjacent } A}{\text{hypotenuse}}$$

$$\tan A = \frac{\text{side opposite } A}{\text{side adjacent } A} \qquad \cos A = \frac{\text{side adjacent } A}{\text{side opposite } A}$$

$$\sec A = \frac{\text{hypotenuse}}{\text{side adjacent } A} \qquad \csc A = \frac{\text{hypotenuse}}{\text{side opposite } A}$$

(4.5)

CHAPTER 4 REVIEW EXERCISES

In Exercises 1–4, find the smallest positive angle and the smallest negative angle (numerically) coterminal with but not equal to the given angle.

1. 17.0° **2.** 248.3° **3.** −217.5° **4.** −7.6°

In Exercises 5–8, express the given angles in decimal form.

5. 31°54′ **6.** 574°45′ **7.** −38°6′ **8.** 321°27′

In Exercises 9–12, express the given angles to the nearest minute.

9. 17.5° **10.** −65.4° **11.** 749.7° **12.** 126.25°

In Exercises 13–16, determine the trigonometric functions of the angles (in standard position) whose terminal side passes through the given points. Give answers in exact form.

13. $(24, 7)$ **14.** $(5, 4)$ **15.** $(48, 48)$ **16.** $(1.2, 0.5)$

In Exercises 17–20, find the indicated trigonometric functions. Give answers in decimal form, rounded off to three significant digits.

17. Given $\sin\theta = \frac{5}{13}$, find $\cos\theta$ and $\cot\theta$.
18. Given $\cos\theta = \frac{3}{8}$, find $\sin\theta$ and $\tan\theta$.

19. Given $\tan\theta = 2$, find $\cos\theta$ and $\csc\theta$.
20. Given $\cot\theta = 40$, find $\sin\theta$ and $\sec\theta$.

In Exercises 21–28, find the values of the trigonometric functions. Round off results.

21. $\sin 72.1°$ **22.** $\cos 40.3°$
23. $\tan 88.64°$ **24.** $\sin 0.91°$
25. $\sec 18.4°$ **26.** $\csc 82.4°$
27. $(\cot 7.06°)(\sin 7.06°) - \cos 7.06°$
28. $(\sec 79.36°)(\sin 79.36°) - \tan 79.36°$

In Exercises 29–40, find θ for each of the given trigonometric functions. Round off results.

29. $\cos\theta = 0.950$ **30.** $\sin\theta = 0.63052$
31. $\tan\theta = 1.574$ **32.** $\cos\theta = 0.0135$
33. $\csc\theta = 4.713$ **34.** $\cot\theta = 0.7561$
35. $\sec\theta = 25.4$ **36.** $\csc\theta = 1.92$
37. $\cot\theta = 7.117$ **38.** $\sec\theta = 1.006$
39. $\sin\theta = 0.9998$ **40.** $\cos\theta = 1.402$

In Exercises 41–52, solve the right triangles with the given parts. Refer to Fig. 4.66.

41. $A = 17.0°, b = 6.00$

42. $B = 68.1°, a = 1080$

43. $a = 81.0, b = 64.5$

44. $a = 106, c = 382$

45. $A = 37.5°, a = 12.0$

46. $B = 15.7°, c = 126$

Fig. 4.66

47. $b = 6.508, c = 7.642$ **48.** $a = 0.721, b = 0.144$

49. $A = 49.67°, c = 0.8253$ **50.** $B = 4.38°, b = 5682$

51. $a = 11.652, c = 15.483$ **52.** $a = 724.39, b = 852.44$

In Exercises 53–94, solve the given problems.

53. Find the value of x for the triangle shown in Fig. 4.67.

Fig. 4.67

W **54.** Explain three ways in which the value of x can be found for the triangle shown in Fig. 4.68. Which of these methods is the easiest?

Fig. 4.68

55. Find the perimeter of a regular octagon (eight equal sides with equal interior angles) that is inscribed in a circle (all vertices of the octagon touch the circle) of radius 10.

W **56.** Explain why values of $\sin \theta$ increase as θ increases from 0° to 90°.

57. What is x if $(3, 2)$ and $(x, 7)$ are on the same terminal side of an acute angle?

58. Two legs of a right triangle are 2.607 and 4.517. What is the smaller acute angle?

59. Show that the side c of any triangle ABC is related to the perpendicular h from C to side AB by the equation

$$c = h \cot A + h \cot B.$$

60. For the isosceles triangle shown in Fig. 4.69, show that $c = 2a \sin \frac{A}{2}$.

Fig. 4.69

61. If $x = \tan \theta$, what is $\csc \theta$ in terms of x?

62. Find the angle between the line passing through the origin and $(3, 2)$, and the line passing through the origin and $(2, 3)$.

63. A sloped cathedral ceiling is between walls that are 7.50 ft high and 12.0 ft high. If the walls are 15.0 ft apart, at what angle does the ceiling rise?

64. A pendulum 1.25 m long swings through an angle of 5.6°. What is the distance between the extreme positions of the pendulum?

65. The voltage e at any instant in a coil of wire that is turning in a magnetic field is given by $e = E \cos \alpha$, where E is the maximum voltage and α is the angle the coil makes with the field. Find the acute angle α if $e = 56.9$ V and $E = 339$ V.

66. A formula for the area of a quadrilateral is $A = \frac{1}{2} d_1 d_2 \sin \theta$, where d_1 and d_2 are the lengths of the diagonals and θ is the angle between them. Find the area of a four-sided carpet remnant with diagonals 3.25 ft and 4.38 ft and $\theta = 72.0°$.

67. For a car rounding a curve, the road should be banked at an angle θ according to the equation $\tan \theta = \dfrac{v^2}{gr}$. Here, v is the speed of the car and r is the radius of the curve in the road. See Fig. 4.70. Find θ for $v = 80.7$ ft/s (55.0 mi/h), $g = 32.2$ ft/s², and $r = 950$ ft.

Fig. 4.70

68. The *apparent power S* in an electric circuit in which the power is P and the impedance phase angle is θ is given by $S = P \sec \theta$. Given $P = 12.0$ V·A and $\theta = 29.4°$, find S.

69. A surveyor measures two sides and the included angle of a triangular tract of land to be $a = 31.96$ m, $b = 47.25$ m, and $C = 64.09°$. (a) Show that a formula for the area A of the tract is $A = \frac{1}{2} ab \sin C$. (b) Find the area of the tract.

70. A water channel has the cross section of an isosceles trapezoid. See Fig. 4.71. (a) Show that a formula for the area of the cross section is $A = bh + h^2 \cot \theta$. (b) Find A if $b = 12.6$ ft, $h = 4.75$ ft, and $\theta = 37.2°$.

Fig. 4.71

71. In tracking an airplane on radar, it is found that the plane is 27.5 km on a direct line from the control tower, with an angle of elevation of 10.3°. What is the altitude of the plane?

72. A straight emergency chute for an airplane is 16.0 ft long. In being tested, the top of the chute is 8.5 ft above the ground. What angle does the chute make with the ground?

73. The windshield on an automobile is inclined 42.5° with respect to the horizontal. Assuming that the windshield is flat and rectangular, what is its area if it is 4.80 ft wide and the bottom is 1.50 ft in front of the top?

74. A water slide at an amusement park is 85 ft long and is inclined at an angle of 52° with the horizontal. How high is the top of the slide above the water level?

75. Find the area of the patio shown in Fig. 4.72.

Fig. 4.72

76. The cross section (a regular trapezoid) of a levee to be built along a river is shown in Fig. 4.73. What is the volume of rock and soil that will be needed for a one-mile length of the levee?

Fig. 4.73

77. The vertical cross section of an attic room in a house is shown in Fig. 4.74. Find the distance d across the floor.

Fig. 4.74

78. The impedance Z and resistance R in an ac circuit may be represented by letting the impedance be the hypotenuse of a right triangle and the resistance be the side adjacent to the phase angle θ. If $R = 1750\ \Omega$ and $\theta = 17.38°$, find Z.

79. A typical aqueduct built by the Romans dropped on average at an angle of about 0.03° to allow gravity to move the water from the source to the city. For such an aqueduct of 65 km in length, how much higher was the source than the city?

80. The distance from the ground level to the underside of a cloud is called the *ceiling*. See Fig. 4.75. A ground observer 950 m from a searchlight aimed vertically notes that the angle of elevation of the spot of light on a cloud is 76°. What is the ceiling?

Fig. 4.75

81. The window of a house is shaded as shown in Fig. 4.76. What percent of the window is shaded when the angle of elevation θ of the sun is 65°?

Fig. 4.76

82. A person standing on a level plain hears the sound of a plane, looks in the direction of the sound, but the plane is not there (familiar?). When the sound was heard, it was coming from a point at an angle of elevation of 25°, and the plane was traveling at 450 mi/h (660 ft/s) at a constant altitude of 2800 ft along a straight line. If the plane later passes directly over the person, at what angle of elevation should the person have looked directly to see the plane when the sound was heard? (The speed of sound is 1130 ft/s.) See Fig. 4.77.

Fig. 4.77

83. In the structural support shown in Fig. 4.78, find x.

Fig. 4.78

84. A Coast Guard boat 2.75 km from a straight beach can travel at 37.5 km/h. By traveling along a line that is at 69.0° with the beach, how long will it take it to reach the beach? See Fig. 4.79.

Fig. 4.79

85. The main span of the Mackinac Bridge (see Fig. 4.80) in northern Michigan is 1160 m long. The angle *subtended* by the span at the eye of an observer in a helicopter is 2.2°. Show that the distance calculated from the helicopter to the span is about the same if the line of sight is perpendicular to the end or to the middle of the span.

Fig. 4.80

86. Each side piece of the trellis shown in Fig. 4.81 makes an angle of 80.0° with the ground. Find the length of each side piece and the area covered by the trellis.

Fig. 4.81

87. A laser beam is transmitted with a "width" of 0.00200°. What is the diameter of a spot of the beam on an object 52,500 km distant? See Fig. 4.82.

Fig. 4.82

88. Find the gear angle θ in Fig. 4.83 if $t = 0.180$ in.

Fig. 4.83

89. The surface of a soccer ball consists of 20 regular hexagons (six sides) interlocked around 12 regular pentagons (five sides). See Fig. 4.84. (a) If the side of each hexagon and pentagon is 45.0 mm, what is the surface area of the soccer ball? (b) Find the surface area, given that the diameter of the ball is 222 mm, (c) Assuming that the given values are accurate, account for the difference in the values found in parts (a) and (b).

Fig. 4.84   **Fig. 4.85**

90. Through what angle θ must the crate shown in Fig. 4.85 be tipped in order that its center of gravity C is directly above the pivot point P?

91. A hang glider is directly above the shore of a lake. An observer on a hill is 375 m along a straight line from the shore. From the observer, the angle of elevation of the hang glider is 42.0°, and the angle of depression of the shore is 25.0°. How far above the shore is the hang glider?

92. A ground observer sights a weather balloon to the east at an angle of elevation of 15.0°. A second observer 2.35 mi to the east of the first also sights the balloon to the east at an angle of elevation of 24.0°. How high is the balloon? See Fig. 4.86.

Fig. 4.86

93. A uniform strip of wood 5.0 cm wide frames a trapezoidal window, as shown in Fig. 4.87. Find the left dimension l of the outside of the frame.

Fig. 4.87

94. A crop-dusting plane flies over a level field at a height of 25 ft. If the dust leaves the plane through a 30° angle and hits the

ground after the plane travels 75 ft, how wide a strip is dusted? See Fig. 4.88.

Fig. 4.88

Writing Exercise

95. A patio is designed in the shape of an isosceles trapezoid with bases 5.0 m and 7.0 m. The other sides are 6.0 m each. Write one or two paragraphs explaining how to use (a) the sine and (b) the cosine to find the internal angles of the patio, and (c) the tangent in finding the area of the patio.

CHAPTER 4   PRACTICE TEST

1. Express 37°39′ in decimal form.

2. Find the value of tan 73.8°.

3. Find θ if cos $\theta = 0.3726$.

4. A ship's captain, desiring to travel due south, discovers due to an improperly functioning instrument, the ship has gone 22.62 km in a direction 4.05° east of south. How far from its course (to the east) is the ship?

5. Find tan θ in fractional form if sin $\theta = \dfrac{2}{3}$.

6. Find csc θ if tan $\theta = 1.294$.

7. Solve the right triangle in Fig. 4.89 if $A = 37.4°$ and $b = 52.8$.

8. Solve the right triangle in Fig. 4.89 if $a = 2.49$ and $c = 3.88$.

Fig. 4.89

9. The equal sides of an isosceles triangle are each 12.0, and each base angle is 42.0°. What is the length of the third side?

10. If tan $\theta = 9/40$, find values of sin θ and cos θ. Then evaluate sin θ/cos θ.

11. In finding the wavelength λ (the Greek letter lambda) of light, the equation $\lambda = d \sin \theta$ is used. Find λ if $d = 30.05 \, \mu$m and $\theta = 1.167°$. (μ is the prefix for 10^{-6}.)

12. Determine the trigonometric functions of an angle in standard position if its terminal side passes through $(5, 2)$. Give answers in exact and decimal forms.

13. The loading ramp at the back of a truck is 9.5 ft long. What angle does it make with the ground if the top of the ramp is 2.5 ft above the ground?

14. A surveyor sights two points directly ahead. Both are at an elevation 18.525 m lower than the observation point. How far apart are the points if the angles of depression are 13.500° and 21.375°, respectively? See Fig. 4.90.

Fig. 4.90

5

Systems of Linear Equations; Determinants

As knowledge about electric circuits was first developing, in 1848 the German physicist Gustav Kirchhoff formulated what are now known as *Kirchhoff's current law* and *Kirchhoff's voltage law*. These laws are still widely used today, and in using them, more than one equation is usually set up. To find the needed information about the circuit, it is necessary to find solutions that satisfy all equations at the same time. We will show the solution of circuits using Kirchhoff's laws in some of the exercises later in the chapter. Although best known for these laws of electric circuits, Kirchhoff is also credited in the study of optics as a founder of the modern chemical process known as *spectrum analysis.*

Methods of solving such *systems* of equations were well known to Kirchhoff, and this allowed the study of electricity to progress rapidly. In fact, 100 years earlier a book by the English mathematician Colin Maclaurin was published (2 years after his death) in which many well organized topics in algebra were covered, including a general method of solving systems of equations. This method is now called *Cramer's rule* (named for the Swiss mathematician Gabriel Cramer, who popularized it in a book he wrote in 1750.) We will explain some of the methods of solution, including Cramer's rule, later in the chapter.

Two or more equations that relate variables are found in many fields of science and technology. These include aeronautics, business, transportation, the analysis of forces on a structure, medical doses, and robotics, as well as electric circuits. These applications often require solutions that satisfy all equations at the same time.

In this chapter, we restrict our attention to *linear* equations (variables occur only to the first power). We will consider systems of two equations with two unknowns and systems of three equations with three unknowns. Systems with other kinds of equations and systems with more unknowns are taken up later in the book.

The correct amounts of acid solutions are needed to make a solution of a certain strength. In Section 5.7, we see how a system of equations is used to find these amounts in a certain case.

5.1 Linear Equations

Linear Equation in One Unknown •
Linear Equation in Two Unknowns •
System of Linear Equations

Many of the equations we have encountered are examples of a very important type of equation, the *linear equation*. In general, *an equation is termed* **linear** *in a given set of variables if each term contains only one variable, to the first power, or is a constant.*

EXAMPLE 1   Meaning of linear

The equation $5x - t + 6 = 0$ is linear in x and t, but $5x^2 - t + 6 = 0$ is not linear, due to the presence of x^2.

The equation $4x + y = 8$ is linear in x and y, but $4xy + y = 8$ is not, due to the presence of xy.

The equation $x - 6y + z - 4w = 7$ is linear in x, y, z, and w, but the equation $x - \frac{6}{y} + z - 4w = 7$ is not, due to the presence of $\frac{6}{y}$, where y appears in the denominator. ∎

An equation that can be written in the form

$$ax + b = 0 \tag{5.1}$$

is known as a **linear equation in one unknown.** Here, a and b are constants. We discussed this type of equation in Section 1.10. *The* **solution,** *or* **root,** *of the equation is* $x = -b/a$. We can also see that the solution is the same as the *zero* of the **linear function** $f(x) = ax + b$.

EXAMPLE 2   Linear equation in one unknown

The equation $2x + 7 = 0$ is a linear equation of the form of Eq. (5.1), with $a = 2$ and $b = 7$.

The solution to the linear equation $2x + 7 = 0$, or the zero of the linear function $f(x) = 2x + 7$, is $-7/2$. From Chapter 3, recall that the *zero of a function* $f(x)$ is the value of x for which $f(x) = 0$. ∎

In the next example, two specific illustrations are given of applied problems that involve more than one unknown.

EXAMPLE 3   Applications of linear equations

(a) A basic law of direct-current electricity, known as *Kirchhoff's current law*, may be stated as "The algebraic sum of the currents entering any junction in a circuit is zero." If three wires are joined at a junction as in Fig. 5.1, this law leads to the linear equation.

$$i_1 + i_2 + i_3 = 0$$

where i_1, i_2, and i_3 are the currents in each of the wires. (Either one or two of these currents must have a negative sign, showing that it is actually leaving the junction.)

(b) Two forces, F_1 and F_2, acting on a beam might be related by the equation

$$2F_1 + 4F_2 = 200$$

∎

An equation that can be written in the form

$$ax + by = c \tag{5.2}$$

is a **linear equation in two unknowns.** For such equations, in Chapter 3, we found that for each value of x there is a corresponding value of y. Each of these pairs of numbers is a *solution* to the equation. *A* **solution** *is any set of numbers, one for each variable, that satisfies the equation.*

Practice Exercise

1. Find the zero of the function $f(x) = 3x + 9$.

∎ See the chapter introduction.

Fig. 5.1

NOTE ▶

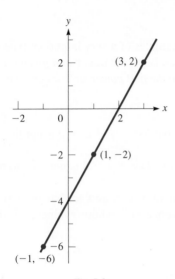

Fig. 5.2

2. Is $x = -3, y = 2$ a solution to the equation of Example 4?

Fig. 5.3

EXAMPLE 4   Linear equation in two unknowns

The equation $2x - y - 4 = 0$ is a linear equation in two unknowns, x and y, since we can write it in the form of Eq. (5.2) as $2x - y = 4$. To graph this equation, we can write it as $y = 2x - 4$. From Fig. 5.2, we see that the graph is a straight line.

The coordinates of any point on the line give us a solution of this equation. For example, the point $(1, -2)$ is on the line. This means that $x = 1, y = -2$ is a solution of the equation. We can show that these values are a solution by substituting in the equation $2x - y - 4 = 0$. This gives us

$$2(1) - (-2) - 4 = 0, \qquad 2 + 2 - 4 = 0, \qquad 0 = 0$$

Since we have equality, $x = 1, y = -2$ is a solution. In the same way, we can show that $x = 3, y = 2$ is a solution.

If we are given the value of one variable, we find the value of the other variable, which gives a solution by substitution. For example, for the equation $2x - y - 4 = 0$, if $x = -1$, we have

$$2(-1) - y - 4 = 0$$
$$-2 - y - 4 = 0$$
$$y = -6$$

Therefore, $x = -1, y = -6$ is a solution, and the point $(-1, -6)$ is on the graph.   ∎

Two linear equations, each containing the same two unknowns,

$$\boxed{\begin{aligned} a_1x + b_1y &= c_1 \\ a_2x + b_2y &= c_2 \end{aligned}} \tag{5.3}$$

are said to form a **system of simultaneous linear equations.** *A solution of the system is any pair of values* (x, y) *that satisfies both equations.* Methods of finding the solutions to such systems are the principal concern of this chapter.

EXAMPLE 5   System of equations – application

The perimeter of the page on which this problem is printed (also see Fig. 5.3) is 98.2 cm. The length is 6.1 cm more than the width.

From the two statements, we can set up two equations in the two unknown quantities, the length and the width. Letting l = the length and w = the width, we have

$$2l + 2w = 98.2$$
$$l - w = 6.1$$

as a system of simultaneous linear equations. The solution of this system is $l = 27.6$ cm and $w = 21.5$ cm. These values satisfy both equations since

$$2(27.6) + 2(21.5) = 98.2 \quad \text{and} \quad 27.6 - 21.5 = 6.1$$

This is the only pair of values that satisfies *both* equations. Methods for finding such solutions are taken up in later sections of this chapter.   ∎

EXERCISES 5.1

In Exercises 1–4, answer the given questions about the indicated examples of this section.

1. In the second line of the third paragraph of Example 1, if $\frac{6}{y}$ is replaced by its reciprocal, is there any change in the conclusion drawn?

2. In the first line of Example 2, if $+$ is changed to $-$, what changes result in the rest of the example?

3. In Example 4, what is the solution if $x = 4$?

4. In the second line of Example 5, if 6.1 cm is replaced with 3.7 cm, what is the solution?

In Exercises 5–8, determine whether or not the given pairs of values are solutions of the given linear equations in two unknowns.

5. $2x + 3y = 9$; $(3, 1), \left(5, \frac{1}{3}\right)$
6. $5x + 2y = 1$; $(0.2, -1); (1, -2)$
7. $-3x + 5y = 13$; $(-1, 2), (4, 5)$
8. $x - 4y = 10$; $(2, -2), (2, 2)$

In Exercises 9–14, for each given value of x, determine the value of y that gives a solution to the given linear equations in two unknowns.

9. $3x - 2y = 12$; $x = 2, x = -3$
10. $-5x + 6y = 60$; $x = -10, x = 8$
11. $x - 4y = 2$; $x = 3, x = -0.4$
12. $3x - 2y = 9$; $x = \frac{2}{3}, x = -3$
13. $24x - 9y = 16$; $x = 2/3, x = -1/2$
14. $2.4y - 4.5x = -3.0$; $x = -0.4, x = 2.0$

In Exercises 15–24, determine whether or not the given pair of values is a solution of the given system of simultaneous linear equations.

15. $x - y = 5$ $x = 4, y = -1$
    $2x + y = 7$
16. $2x + y = 8$ $x = -1, y = 10$
    $3x - y = -13$
17. $A + 5B = -7$ $A = -2, B = 1$
    $3A - 4B = -4$
18. $-30x + 5y = 1$ $x = \frac{1}{3}, y = 2$
    $6x - 3y = -4$
19. $2x - 5y = 0$ $x = \frac{1}{2}, y = -\frac{1}{5}$
    $4x + 10y = 4$
20. $6i_1 + i_2 = 5$ $i_1 = 1, i_2 = -1$
    $3i_1 - 4i_2 = -1$
21. $3x - 2y = 2.2$ $x = 0.6, y = -0.2$
    $5x + y = 2.8$
22. $s - 7t = -3.2$ $s = -1.1, t = 0.3$
    $2s + t = 2.5$

23. $30x - 21y = -675$ $x = -12, y = 15$
    $42x + 15y = -279$
24. $3.8a + 7.5b = -26.5$ $a = 2.5, b = -4.8$
    $2.6a - 12.6b = -51.1$

In Exercises 25–30, solve the given problems.

25. If $x = -2$ is a root of the equation $3x + b = 0$, find b.

26. If $x = 2, y = -3$ is a solution of the system of equations
    $$ax + y = 1$$
    $$x - ay = c$$
    find a and c.

27. In planning a search pattern from an aircraft carrier, a pilot plans to fly at p mi/h relative to a wind that is blowing at w mi/h. Traveling with the wind, the ground speed would be 300 mi/h, and against the wind the ground speed would be 220 mi/h. This leads to two equations:
    $$p + w = 300$$
    $$p - w = 220$$
    Are the speeds 260 mi/h and 40 mi/h?

28. The electric resistance R of a certain resistor is a function of the temperature T given by the equation $R = aT + b$, where a and b are constants. If $R = 1200\ \Omega$ when $T = 10.0°C$ and $R = 1280\ \Omega$ when $T = 50.0°C$, we can find the constants a and b by substituting and obtaining the equations
    $$1200 = 10.0a + b$$
    $$1280 = 50.0a + b$$
    Are the constants $a = 4.00\ \Omega/°C$ and $b = 1160\ \Omega$?

29. The forces acting on part of a structure are shown in Fig. 5.4. An analysis of the forces leads to the equations
    $$0.80F_1 + 0.50F_2 = 50$$
    $$0.60F_1 - 0.87F_2 = 12$$
    Are the forces 45 N and 28 N?

Fig. 5.4

30. A student earned $4000 during the summer and decided to put half into an IRA (Individual Retirement Account). If the IRA was invested in two accounts earning 4.0% and 5.0%, the total income for the first year is $92. The equations to determine the amounts of x and y are
    $$x + y = 2000$$
    $$0.040x + 0.050y = 92$$
    Are the amounts $x = \$1200$ and $y = \$800$?

Answers to Practice Exercises

1. $x = -3$ 2. No

5.2   Graphs of Linear Functions

Slope • Slope-Intercept Form •
Sketching Lines by Intercepts •
Displaying Lines on Calculator

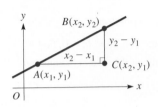

Fig. 5.5

Slope

The first method of solving a system of equations that will be studied is a graphical method. Before taking up this method in the next section, we will develop additional ways of graphing a linear equation. We will then be able to analyze and check the graph of a linear equation quickly, often by inspection.

Consider the line that passes through points A, with coordinates (x_1, y_1), and B, with coordinates (x_2, y_2), in Fig. 5.5. Point C is horizontal from A and vertical from B. Thus, C has coordinates (x_2, y_1), and there is a right angle at C. One way of measuring the steepness of this line is to find the ratio of the vertical distance to the horizontal distance between two points. Therefore, *we define the* **slope** *of the line through two points as the difference in the y-coordinates divided by the difference in the x-coordinates.* For points A and B, the slope m is

$$m = \frac{y_2 - y_1}{x_2 - x_1} \tag{5.4}$$

The slope is often referred to as the *rise* (vertical change) over the *run* (horizontal change). Note that the slope of a vertical line, for which $x_2 = x_1$, is undefined (for $x_2 = x_1$, the denominator of Eq. (5.4) is zero).

EXAMPLE 1   Slope of line through two points

Find the slope of the line through the points $(2, -3)$ and $(5, 3)$.

In Fig. 5.6, we draw the line through the two given points. By taking $(5, 3)$ as (x_2, y_2), then (x_1, y_1) is $(2, -3)$. We may choose either point as (x_2, y_2), but *once the choice is made the order must be maintained.* Using Eq. (5.4), the slope is

$$m = \frac{3 - (-3)}{5 - 2}$$

$$= \frac{6}{3} = 2$$

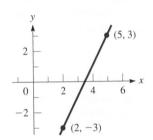

Fig. 5.6

The rise is 2 units for each unit (of run) in going from left to right.   ∎

EXAMPLE 2   Slope of line through two points

Find the slope of the line through $(-1, 2)$ and $(3, -1)$.

In Fig. 5.7, we draw the line through these two points. By taking (x_2, y_2) as $(3, -1)$ and (x_1, y_1) as $(-1, 2)$, the slope is

$$m = \frac{-1 - 2}{3 - (-1)}$$

$$= \frac{-3}{3 + 1} = -\frac{3}{4}$$

Fig. 5.7

The line *falls* 3 units for each 4 units in going from left to right.   ∎

Practice Exercise

1. Find the slope of the line through
$(-2, 4)$ and $(-5, -6)$.

We note in Example 1 that as *x increases, y increases and that slope is* **positive.** In Example 2, *as x increases, y decreases and that slope is* **negative.** Also, *the larger the absolute value of the slope, the steeper is the line.*

EXAMPLE 3 Slope and steepness of line

For each of the following lines shown in Fig. 5.8, we show the difference in the y-coordinates and in the x-coordinates between two points.

In Fig. 5.8(a), a line with a slope of 5 is shown. It rises sharply.

In Fig. 5.8(b), a line with a slope of $\frac{1}{2}$ is shown. It rises slowly.

In Fig. 5.8(c), a line with a slope of -5 is shown. It falls sharply.

In Fig. 5.8(d), a line with a slope of $-\frac{1}{2}$ is shown. It falls slowly.

Fig. 5.8   ■

SLOPE-INTERCEPT FORM OF THE EQUATION OF A STRAIGHT LINE

We now show how the slope is related to the equation of a straight line. In Fig. 5.9, if we have two points, $(0, b)$ and a general point (x, y), the slope is

$$m = \frac{y - b}{x - 0}$$

Simplifying this, we have $mx = y - b$, or

$$\boxed{y = mx + b} \tag{5.5}$$

Fig. 5.9

NOTE ▶

In Eq. (5.5), m is the slope, and b is the y-coordinate of the point where it crosses the y-axis. *This point is the **y-intercept** of the line,* and its coordinates are $(0, b)$. *Equation (5.5) is the **slope-intercept** form of the equation of a straight line. The coefficient of x is the slope, and the constant is the ordinate of the y-intercept.* The point $(0, b)$ and simply b are both referred to as the y-intercept.

EXAMPLE 4 Slope-intercept form of equation

Find the slope and the y-intercept of the line $y = \frac{3}{2}x - 3$. Since the coefficient of x is $\frac{3}{2}$, this means that the slope is $\frac{3}{2}$. Also, because we can write

$$y = \frac{3}{2}x + (-3)$$

we see that the constant is -3, which means the y-intercept is the point $(0, -3)$. The line is shown in Fig. 5.10.   ■

Fig. 5.10

EXAMPLE 5 Slope-intercept form of equation

Find the slope and the y-intercept of the line $2x + 3y = 4$.

We must first write the equation in slope-intercept form. Solving for y, we have

$$y = -\frac{2}{3}x + \frac{4}{3}$$

Therefore, the slope is $-\frac{2}{3}$, and the y-intercept is the point $\left(0, \frac{4}{3}\right)$. See Fig. 5.11.   ■

Fig. 5.11

Fig. 5.12

Practice Exercise

2. Write the equation $3x - 5y - 15 = 0$ in slope-intercept form.

NOTE ▶

Practice Exercise

3. Find the intercepts of the line $3x - 5y - 15 = 0$.

■ Further details and discussion of slope and the graphs of linear equations are found in Chapter 21.

EXAMPLE 6 Application of slope-intercept form

In analyzing an electric current, two of the currents i_1 and i_2, were related by $i_2 = 2i_1 + 1$. Graph this equation using the slope-intercept form.

Since the equation is solved for i_2, we treat i_1 as the independent variable and i_2 as the dependent variable. This means that the slope of the line is 2. Also, since the b-term is 1, we can see that the intercept is $(0, 1)$.

We can use this information to sketch the line, as shown in Fig. 5.12. Since the slope is 2, we know that i_2 increases 2 units for each unit increase of i_1. Thus, starting at the i_2-intercept $(0, 1)$, if i_1 increases by 1, i_2 increases by 2, and we are at the point $(1, 3)$. The line must pass through $(1, 3)$, as well as $(0, 1)$. Therefore, we draw the line through these points. (Negative values may be used for electric currents since the sign shows the direction of flow.) ■

SKETCHING LINES BY INTERCEPTS

Another way of sketching the graph of a straight line is to find two points on the line and then draw the line through these points. Two points that are easily determined are those where the line crosses the y-axis and the x-axis. We already know that the point where it crosses the y-axis is the y-intercept. In the same way, *the point where it crosses the x-axis is called the **x-intercept**,* and the coordinates of the x-intercept are $(a, 0)$.

The intercepts are easily found because one of the coordinates is zero. By setting $x = 0$ and $y = 0$, in turn, and finding the value of the other variable, we get the coordinates of the intercepts. This method works except when both intercepts are at the origin, and we must find one other point, or use the slope-intercept method. In using the intercept method, a third point should be found as a check. The next example shows how a line is sketched by finding its intercepts.

EXAMPLE 7 Sketching line by intercepts

Sketch the graph of the line $2x - 3y = 6$ by finding its intercepts and one check point. See Fig. 5.13.

First, we let $x = 0$. This gives us $-3y = 6$, or $y = -2$. This gives us the y-intercept, which is the point $(0, -2)$. Next, we let $y = 0$, which gives us $2x = 6$, or $x = 3$. This means the x-intercept is the point $(3, 0)$.

The intercepts are enough to sketch the line as shown in Fig. 5.13. To find a check point, we can use any value for x other than 3 or any value of y other than -2. Choosing $x = 1$, we find that $y = -\frac{4}{3}$. This means that the point $\left(1, -\frac{4}{3}\right)$ should be on the line. In Fig. 5.13, we can see that it is on the line.

Fig. 5.13

■

Fig. 5.14

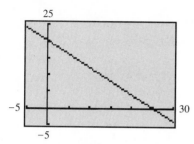

Fig. 5.15

DISPLAYING LINES ON A GRAPHING CALCULATOR

We can display the graph of a straight line on a graphing calculator, just as in Section 3.5. In order to enter the equation into the calculator, we must first solve for y. In doing so, the equation will often be in slope-intercept form or be easily written in this form.

Good choices for the *window* settings can be made by using the y-intercept and the slope. The y-intercept tells us a value of y that we usually want displayed, and the slope tells us how the line is directed from that point. Using the y-intercept along with the x-intercept should also lead to good choices for the settings.

EXAMPLE 8 Displaying line on calculator

(a) Use a graphing calculator to display the graph of the equation in Example 7.

First, solve the equation $2x - 3y = 6$ for y. This gives us $y = \frac{2}{3}x - 2$. Therefore, on the calculator we enter $y_1 = 2x/3 - 2$ and use the *window* settings shown in Fig. 5.14. Note that the graph is the same as that in Fig. 5.13.

(b) Use a graphing calculator to display the graph of the equation $4x + 5y = 100$.

Solving for y, we get $y = -\frac{4}{5}x + 20$, and therefore in the calculator we enter the function $y_1 = -4x/5 + 20$. We also note immediately that the y-intercept is $(0, 20)$, which means we must be careful in choosing the *window* settings. We can also determine that the x-intercept is $(25, 0)$. Therefore, we have the settings and graph shown in Fig. 5.15. ■

EXERCISES 5.2

In Exercises 1–4, answer the given questions about the indicated examples of this section.

1. In Example 2, if the first y-coordinate is changed to -2, what changes result in the example?

2. In the first line of Example 5, if $+$ is changed to $-$, what changes result in the example?

3. In the first line of Example 7, if $-$ is changed to $+$, what changes occur in the graph?

4. In the first line of Example 8(b), if $4x + 5y$ is changed to $5x - 4y$, what changes occur in the graph?

In Exercises 5–12, find the slope of the line that passes through the given points.

5. $(1, 0), (3, 8)$

6. $(3, 1), (2, -7)$

7. $(-1, 2), (-4, 17)$

8. $(-1, -2), (2, 10)$

9. $(5, -3), (-2, -5)$

10. $(-3, 4), (-7, -4)$

11. $(0.4, 0.5), (-0.2, 0.2)$

12. $(-2.8, 3.4), (1.2, 4.2)$

In Exercises 13–20, sketch the line with the given slope and y-intercept.

13. $m = 2, (0, -1)$

14. $m = 3, (0, 1)$

15. $m = 0, (0, 2)$

16. $m = -4, (0, -2)$

17. $m = \frac{1}{2}, (0, 0)$

18. $m = \frac{2}{3}, (0, -1)$

19. $m = -9, (0, 20)$

20. $m = -0.3, (0, -1.4)$

In Exercises 21–28, find the slope and the y-intercept of the line with the given equation and sketch the graph using the slope and the y-intercept. A graphing calculator can be used to check your graph.

21. $y = -2x + 1$

22. $y = -4x$

23. $y = x - 4$

24. $y = \frac{4}{5}x + 2$

25. $5x - 2y = 40$

26. $-2y = 7$

27. $24x + 40y = 15$

28. $1.5x - 2.4y = 3.0$

In Exercises 29–36, find the x-intercept and the y-intercept of the line with the given equation. Sketch the line using the intercepts. A graphing calculator can be used to check the graph.

29. $x + 2y = 4$

30. $3x + y = 3$

31. $4x - 3y = 12$

32. $5y - x = 5$

33. $y = 3x + 6$

34. $y = -2x - 4$

35. $y = -12x + 30$

36. $y = 0.25x + 4.5$

In Exercises 37 and 38, solve for k.

37. Find the value of k such that the line $kx - 2y = 9$ has a slope of 3.

38. Find the value of k such that the line through $(0, k)$ and $(-k, -1)$ has a slope of 1/2.

In Exercises 39–42, sketch the indicated lines.

39. The diameter of the large end, d (in in.), of a certain type of machine tool can be found from the equation $d = 0.2l + 1.2$, where l is the length of the tool. Sketch d as a function of l, for values of l to 10 in. See Fig. 5.16.

1.2 in.

d

l

Fig. 5.16

40. In mixing 85-octane gasoline and 93-octane gasoline to produce 91-octane gasoline, the equation $0.85x + 0.93y = 910$ is used. Sketch the graph.

41. Two electric currents, I_1 and I_2 (in mA), in part of a circuit in a computer are related by the equation $4I_1 - 5I_2 = 2$. Sketch I_2 as a function of I_1. These currents can be negative.

42. In 2000, about 9.7×10^9 email messages were sent, and in 2004 about 30.1×10^9 email messages were sent. Assuming that the growth in email messages is linear, set up a function relating the number N of email messages and the time t (in years). Let $t = 0$ be the year 2000. Sketch the graph.

Answers to Practice Exercises

1. $10/3$ **2.** $y = \frac{3}{5}x - 3$ **3.** $(5, 0)\ (0, -3)$

5.3 Solving Systems of Two Linear Equations in Two Unknowns Graphically

**Determine the Point of Intersection •
Using a Graphing Calculator •
Inconsistent and Dependent Systems**

Since a solution of a system of simultaneous linear equations in two unknowns is any pair of values (x, y) that satisfies *both* equations, graphically *the solution would be the coordinates of the point of intersection of the two lines.* This must be the case, for the coordinates of this point constitute the only pair of values to satisfy *both* equations. (In some special cases, there may be no solution; in others, there may be many solutions. See Examples 5 and 6.)

Therefore, when we solve two simultaneous linear equations in two unknowns graphically, *we must graph each line and determine the point of intersection.* This may, of course, lead to *approximate results* if the lines cross at points not used to determine the graph.

EXAMPLE 1 Determine the point of intersection

Solve the system of equations

$$y = x - 3$$
$$y = -2x + 1$$

Since each of the equations is in slope-intercept form, note that $m = 1$ and $b = -3$ for the first line and that $m = -2$ and $b = 1$ for the second line. Using these values, we sketch the lines, as shown in Fig. 5.17.

From the figure, it can be seen that *the lines cross at about the point* $(1.3, -1.7)$. This means that the solution is approximately

$$x = 1.3 \qquad y = -1.7$$

$\left(\text{The exact solution is } x = \frac{4}{3}, y = -\frac{5}{3}.\right)$

NOTE ▶   In checking the solution, be careful to substitute the values in *both* equations. Making these substitutions gives us

$$-1.7 \overset{?}{=} 1.3 - 3 \quad \text{and} \quad -1.7 \overset{?}{=} -2(1.3) + 1$$
$$= -1.7 \qquad\qquad\qquad \approx -1.6$$

These values show that the solution checks. (The point $(1.3, -1.7)$ is *on* the first line and *almost on* the second line. The difference in values when checking the values for the second line is due to the fact that the solution is *approximate*.) ∎

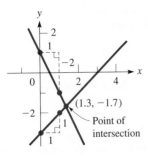

(1.3, −1.7)

Point of intersection

Fig. 5.17

Fig. 5.18

Practice Exercise

1. In Example 2, change the 6 to 3, and then solve. (Can be done using Fig. 5.18.)

EXAMPLE 2 Determine the point of intersection

Solve the system of equations

$$2x + 5y = 10$$
$$3x - y = 6$$

We could write each equation in slope-intercept form in order to sketch the lines. Also, we could use the form in which they are written to find the intercepts. Choosing to find the intercepts and draw lines through them, let $y = 0$; then $x = 0$. Therefore, the intercepts of the first line are the points $(5, 0)$ and $(0, 2)$. A third point is $\left(-1, \frac{12}{5}\right)$. The intercepts of the second line are $(2, 0)$ and $(0, -6)$. A third point is $(1, -3)$. Plotting these points and drawing the proper straight lines, we see that the lines cross at about $(2.3, 1.1)$. $\left[\text{The exact values are } \left(\frac{40}{17}, \frac{18}{17}\right).\right]$ The solution of the system of equations is approximately $x = 2.3$, $y = 1.1$ (see Fig. 5.18).

Checking, we have

$$2(2.3) + 5(1.1) \stackrel{?}{=} 10 \quad \text{and} \quad 3(2.3) - 1.1 \stackrel{?}{=} 6$$
$$10.1 \approx 10 \qquad\qquad\qquad 5.8 \approx 6$$

This shows the solution is correct to the accuracy we can get from the graph. ■

SOLVING SYSTEMS OF EQUATIONS USING A GRAPHING CALCULATOR

A graphing calculator can be used to find the point of intersection with much greater accuracy than is possible by hand-sketching the lines. Once we locate the point of intersection, the features of the calculator allow us to get the accuracy we need.

EXAMPLE 3 Using a graphing calculator

Using a graphing calculator, we now solve the systems of equations in Examples 1 and 2.

Solving the system for Example 1, let $y_1 = x - 3$ and $y_2 = -2x + 1$. Then display the lines as shown in Fig. 5.19(a). Then using the *intersect* feature of a graphing calculator, we find (rounded to the nearest 0.001) that the solution is $x = 1.333$, $y = -1.667$.

Solving the system for Example 2, let $y_1 = -2x/5 + 2$ and $y_2 = 3x - 6$. Then display the lines as shown in Fig. 5.19(b). Then using the *intersect* feature, we find (rounded to the nearest 0.001) that the solution is $x = 2.333$, $y = 1.059$.

■ We could use the *trace* and *zoom* features, but the *intersect* feature is meant for this type of problem, and is more accurate.

Fig. 5.19   (a)   (b)   ■

WORD PROBLEMS INVOLVING TWO LINEAR EQUATIONS

Linear equations in two unknowns are often useful in solving word problems. Just as in Section 1.12, *we must read the statement carefully in order to identify the unknowns and the information for setting up the equations.*

(a)

(b)

Fig. 5.20

Fig. 5.21

EXAMPLE 4 Application of solving a system

A driver traveled for 1.5 h at a constant speed along a highway. Then, through a construction zone, the driver reduced the car's speed by 20 mi/h for 30 min. If 100 mi were covered in the 2.0 h, what were the two speeds?

First, let v_h = the highway speed and v_c = the speed in the construction zone. Two equations are found by using

1. distance = rate × time $\left[\text{for units, mi} = \left(\tfrac{\text{mi}}{\text{h}}\right)\text{h}\right]$, and

2. the fact that "the driver reduced the car's speed by 20 mi/h for 30 min."

$$
\underset{\substack{\text{distance} \\ \text{on highway}}}{\underset{\substack{\text{time on} \\ \text{highway}}}{\downarrow}\,\underset{\substack{\text{highway} \\ \text{speed}}}{\downarrow}}\!\!\!\!\!\!\!\!\! \text{First equation:} \quad 1.5\,v_h \;+\; \underset{\substack{\text{distance in} \\ \text{construction}}}{\underset{\substack{30\text{ min} \\ =0.5\text{ h}}}{\downarrow}\,\underset{\substack{\text{speed in} \\ \text{construction zone}}}{\downarrow}}\!\!\!\!\!\!\!\!\! 0.5\,v_c \;=\; 100 \quad \text{total distance}
$$

Second equation: $\qquad\qquad v_c = v_h - 20$

speed reduced by 20 mi/h

Using v_h as the independent variable and v_c as the dependent variable, the sketch is shown in Fig. 5.20(a). With $v_c = y$ and $v_h = x$, we have the calculator display shown in Fig. 5.20(b) with the indicated settings, and $y_1 = -3x + 200$, $y_2 = x - 20$.

We see that the point of intersection is $(55, 35)$, which means that the solution is $v_h = 55$ mi/h and $v_c = 35$ mi/h. Checking *in the statement of the problem*, we have $(1.5\,\text{h})(55\,\text{mi/h}) + (0.5\,\text{h})(35\,\text{mi/h}) = 100\,\text{mi}$. ∎

INCONSISTENT AND DEPENDENT SYSTEMS

The lines of each system in the previous examples intersect in a single point, and each system has *one* solution. Such systems are called *consistent* and *independent*. Most systems that we will encounter have just one solution. However, as we now show, *not all systems have just one such solution.* The following examples illustrate a system that has *no solution* and a system that has an *unlimited number of solutions.*

EXAMPLE 5 Inconsistent system

Solve the system of equations

$$
x = 2y + 6
$$
$$
6y = 3x - 6
$$

Writing each of these equations in slope-intercept form (Eq. 5.5), we have for the first equation

$$
y = \tfrac{1}{2}x - 3
$$

For the second equation, we have

$$
y = \tfrac{1}{2}x - 1
$$

From these, we see that each line has a slope of $\tfrac{1}{2}$ and that the y-intercepts are $(0, -3)$ and $(0, -1)$. Therefore, we know that the y-intercepts are different, but the *slopes are the same.* Since the slope indicates that each line rises $\tfrac{1}{2}$ unit for y for each unit x increases, *the lines are parallel and do not intersect,* as shown in Fig. 5.21. This means that ***there are no solutions*** for this system of equations. *Such a system is called **inconsistent.*** ∎

EXAMPLE 6 Dependent system

Solve the system of equations

$$x - 3y = 9$$
$$-2x + 6y = -18$$

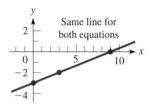

Same line for both equations

Fig. 5.22

We find that the intercepts and a third point for the first line are $(9, 0)$, $(0, -3)$, and $(3, -2)$. For the second line, we then find that the intercepts are the same as for the first line. We also find that the check point $(3, -2)$ also satisfies the equation of the second line. This means that the two lines are really the same line.

Another check is to write each equation in slope-intercept form. This gives us the equation $y = \frac{1}{3}x - 3$ for each line. See Fig. 5.22.

Since the lines are the same, *the coordinates of any point on this common line constitute a solution of the system*. Since ***no unique solution can be determined***, *the system is called* **dependent**. ∎

Later in the chapter, we will show algebraic ways of finding out whether a given system is consistent, inconsistent, or dependent.

EXERCISES 5.3

In Exercises 1 and 2, answer the given questions about the indicated examples of this section.

1. In the second equation of Example 2, if $-$ is replaced with $+$, what is the solution?

2. In Example 6, by changing what one number in the first equation does the system become (a) inconsistent? (b) consistent?

In Exercises 3–20, solve each system of equations by sketching the graphs. Use the slope and the y-intercept or both intercepts. Estimate each result to the nearest 0.1 if necessary.

3. $y = -x + 4$
$y = x - 2$

4. $y = \frac{1}{2}x - 1$
$y = -x + 8$

5. $y = 2x - 6$
$y = -\frac{1}{3}x + 1$

6. $y = \frac{1}{2}x - 4$
$y = 2x + 2$

7. $3x + 2y = 6$
$x - 3y = 3$

8. $4R - 3V = -8$
$6R + V = 6$

9. $2x - 5y = 10$
$3x + 4y = -12$

10. $-5x + 3y = 15$
$2x + 7y = 14$

11. $s - 4t = 8$
$2s = t + 4$

12. $y = 4x - 6$
$y = 2x + 4$

13. $y = -x + 3$
$y = -2x + 3$

14. $p - 6 = 6v$
$v = 3 - 3p$

15. $x - 4y = 6$
$2y = x + 4$

16. $x + y = 3$
$3x - 2y = 14$

17. $-2r_1 + 2r_2 = 7$
$4r_1 - 2r_2 = 1$

18. $2x - 3y = -5$
$3x + 2y = 12$

19. $15y = 20x - 48$
$32x + 21y = 60$

20. $7A = 8B + 17$
$5B = 12A - 22$

In Exercises 21–32, solve each system of equations to the nearest 0.001 for each variable by using a graphing calculator.

21. $x = 4y + 2$
$3y = 2x + 3$

22. $1.2x - 2.4y = 4.8$
$3.0x = -2.0y + 7.2$

23. $4.0x - 3.5y = 1.5$
$0.7y + 0.1x = 0.7$

24. $5F - 2T = 7$
$3F + 4T = 8$

25. $x - 5y = 10$
$2x - 10y = 20$

26. $18x - 3y = 7$
$2y = 1 + 12x$

27. $1.9v = 3.2t$
$1.2t - 2.6v = 6$

28. $3y = 14x - 9$
$12x + 23y = 0$

29. $5x = y + 3$
$4x = 2y - 3$

30. $0.75u + 0.67v = 5.9$
$2.1u - 3.9v = 4.8$

31. $7R = 18V + 13$
$-1.4R + 3.6V = 2.6$

32. $y = 6x + 2$
$12x - 2y = -4$

In Exercises 33–38, graphically solve the given problems.

33. Chains support a crate, as shown in Fig. 5.23. The equations relating tensions T_1 and T_2 are given below. Determine the tensions to the nearest 1 N from the graph.

$$0.8T_1 - 0.6T_2 = 12$$
$$0.6T_1 + 0.8T_2 = 68$$

Fig. 5.23

34. The equations relating the currents i_1 and i_2 shown in Fig. 5.24 are given below. Find the currents to the nearest 0.1 A.

$$2i_1 + 6(i_1 + i_2) = 12$$
$$4i_2 + 6(i_1 + i_2) = 12$$

Fig. 5.24

35. An architect designing a parking lot has a row 202 ft wide to divide into spaces for compact cars and full-size cars. The architect determines that 16 compact car spaces and 6 full-size car spaces use the width, or that 12 compact car spaces and 9 full-size car spaces use all but 1 ft of the width. What are the widths (to the nearest 0.1 ft) of the spaces being planned? See Fig. 5.25.

Fig. 5.25

36. An airplane flies into a headwind with an effective ground speed of 70 mi/h. On the return trip it flies with the tailwind and has an effective ground speed of 120 mi/h. Find the speed p of the plane in still air, and the speed w of the wind.

37. One alloy is 70% lead and 30% zinc, and another alloy is 40% lead and 60% zinc. Find the amount of each alloy needed to make 120 kg of an alloy that is 50% lead and 50% zinc.

38. A certain car gets 21 mi/gal in city driving and 28 mi/gal in highway driving. If 18 gal of gas are used in traveling 448 mi, how many miles were driven in the city, and how many were driven on the highway (assuming that only the given rates of usage were actually used)?

Answer to Practice Exercise

1. $x = 1.5, y = 1.4$

5.4   Solving Systems of Two Linear Equations in Two Unknowns Algebraically

Solution by Substitution • Solution by Addition or Subtraction

The graphical method of solving two linear equations in two unknowns is good for getting a "picture" of the solution. One problem is that graphical methods usually give *approximate* results, although the accuracy obtained when using a graphing calculator is excellent. If *exact* solutions are required, we turn to other methods. In this section, we present two algebraic methods of solution.

SOLUTION BY SUBSTITUTION

The first method involves the *elimination of one variable by* **substitution.** The basic idea of this method is to *solve one of the equations for one of the unknowns and substitute this into the other equation.* The result is an equation with only one unknown, and this equation can then be solved for the unknown. Following is the basic procedure to be used.

> **SOLUTION OF TWO LINEAR EQUATIONS BY SUBSTITUTION**
>
> 1. *Solve one equation for one of the unknowns.*
> 2. *Substitute this solution into the **other** equation.* At this point, we have a linear equation in one unknown.
> 3. *Solve the resulting equation for the value of the unknown it contains.*
> 4. *Substitute this value into the equation of step 1 and solve for the other unknown.*
> 5. *Check the values in **both original equations.***

In following this procedure, you must first choose an unknown for which to solve. Often it makes little difference, but if it is easier to solve a particular equation for one of the unknowns, that is the one to use.

EXAMPLE 1 Solution by substitution

Solve the following system of equations by substitution.

$$x - 3y = 6$$
$$2x + 3y = 3$$

Here, it is easiest to solve the first equation for x:

step 1
$$x = 3y + 6 \qquad\qquad\text{(A1)}$$

_____ in second equation, x replaced by $3y + 6$

step 2
$$2(3y + 6) + 3y = 3 \qquad\qquad \text{substituting}$$

step 3
$$6y + 12 + 3y = 3 \qquad\text{solving for } y$$
$$9y = -9$$
$$y = -1$$

Now, put the value $y = -1$ into the first of the original equations. Since this equation is already solved for x in terms of y, Eq. (A1), we obtain

step 4
$$x = 3(-1) + 6 = 3 \qquad\text{solving for } x$$

Therefore, the solution of the system is $x = 3, y = -1$. As a check, substitute these
step 5 values into each of the original equations. This gives us $3 - 3(-1) = 6$ and $2(3) + 3(-1) = 3$, which verifies the solution. ∎

EXAMPLE 2 Solution by substitution

Solve the following system of equations by substitution.

$$-5x + 2y = -4$$
$$10x + 6y = \;\; 3$$

It makes little difference which equation or which unknown is chosen. Therefore,

$$2y = 5x - 4 \qquad\text{solving first equation for } y$$
$$y = \frac{5x - 4}{2} \qquad\qquad\text{(B1)}$$

_____ in second equation, y replaced by $\frac{5x - 4}{2}$

$$10x + 6\left(\frac{5x - 4}{2}\right) = 3 \qquad\qquad\text{substituting}$$

$$10x + 3(5x - 4) = 3 \qquad\text{solving for } x$$
$$10x + 15x - 12 = 3$$
$$25x = 15$$
$$x = \frac{3}{5}$$

Practice Exercise

1. Solve the system in Example 2 by first solving for x and then substituting.

Substituting this value into the expression for y, Eq. (B1), we obtain

$$y = \frac{5(3/5) - 4}{2} = \frac{3 - 4}{2} = -\frac{1}{2} \qquad\text{solving for } y$$

Therefore, the solution of this system is $x = \frac{3}{5}, y = -\frac{1}{2}$. Substituting these values in both original equations shows that the solution checks. ∎

SOLUTION BY ADDITION OR SUBTRACTION

The method of substitution is useful if one of the equations can easily be solved for one of the unknowns. However, often a fraction results (such as in Example 2), and this leads to additional algebraic steps to find the solution. Even worse, if the coefficients are themselves decimals or fractions, the algebraic steps are even more involved.

Therefore, we now present another algebraic method of solving a system of equations, *elimination of a variable by* **addition** *or* **subtraction**. Following is the basic procedure to be used.

■ For reference, Eqs. (5.3) are

$$a_1x + b_1y = c_1$$
$$a_2x + b_2y = c_2$$

SOLUTION OF TWO LINEAR EQUATIONS BY ADDITION OR SUBTRACTION

1. *Write the equations in the form of Eqs. (5.3), if they are not already in this form.*
2. *If necessary, multiply all terms of each equation by a constant chosen so that the coefficients of one unknown will be numerically the same in both equations. (They can have the same or different signs.)*
3. (a) *If the numerically equal coefficients have **different** signs, **add** the terms on each side of the resulting equations.*
   (b) *If the numerically equal coefficients have the **same** sign, **subtract** the terms on each side of one equation from the terms of the other equation.*
4. *Solve the resulting linear equation in the other unknown.*
5. *Substitute this value into one of the **original** equations to find the value of the other unknown.*
6. *Check by substituting both values into both original equations.*

■ Some prefer always to use addition of the terms of the resulting equations. This method is appropriate because it avoids possible errors that may be caused when subtracting. However, it may require that all signs of one equation be changed.

EXAMPLE 3 Solution by addition

Use the method of elimination by addition or subtraction to solve the system of equations.

$$x - 3y = 6 \qquad \text{already in the form of Eqs. (5.3)}$$
$$2x + 3y = 3$$

■ Some graphing calculators have a specific feature for solving simultaneous linear equations. It is necessary only to enter the coefficients and constants to get the solution.

We look at the coefficients to determine the best way to eliminate one of the unknowns. Since the coefficients of the *y*-terms are *numerically the same and opposite in sign*, we may immediately *add* terms of the two equations together to eliminate *y*. Adding the terms of the left sides and adding terms of the right sides, we obtain

$$x + 2x - 3y + 3y = 6 + 3$$
$$3x = 9$$
$$x = 3$$

Substituting this value into the first equation, we obtain

$$3 - 3y = 6$$
$$-3y = 3$$
$$y = -1$$

Practice Exercise

2. Solve the system in Example 3 by first multiplying the terms of the first equation by 2, and then subtracting.

The solution $x = 3, y = -1$ agrees with the results obtained for the same problem illustrated in Example 1.   ■

EXAMPLE 4 Solution by subtraction

Use the method of addition or subtraction to solve the following system of equations.

$$3x - 2y = 4$$
$$x + 3y = 2$$

Looking at the coefficients of x and y, we see that we must multiply the second equation by 3 to make the coefficients of x the same. To make the coefficients of y numerically the same, we must multiply the first equation by 3 and the second equation by 2. Thus, the best method is to multiply the second equation by 3 and eliminate x.

CAUTION ▶ Doing this (be careful to multiply the terms on *both* sides: *a common error is to forget to multiply the value on the* ***right***), the coefficients of x have the *same sign*. Therefore, we *subtract* terms of the second equation from those of the first equation:

$$3x - 2y = 4$$
$$\underline{3x + 9y = 6} \qquad \text{each term of second equation multiplied by 3}$$

$3x - 3x = 0 \longrightarrow -11y = -2 \qquad \text{subtract}$

$-2y - (+9y) = -11y \longrightarrow y = \dfrac{2}{11} \qquad 4 - 6 = -2$

In order to find the value of x, substitute $y = \frac{2}{11}$ into one of the original equations. Choosing the second equation (its form is somewhat simpler), we have

$$x + 3\left(\frac{2}{11}\right) = 2$$

$$11x + 6 = 22 \qquad \text{multiply each term by 11}$$

$$x = \frac{16}{11}$$

Therefore, the solution is $x = \frac{16}{11}, y = \frac{2}{11}$. Substituting these values into both of the original equations shows that the solution checks. ∎

EXAMPLE 5 System of Example 4 solved by addition

As noted in Example 4, we can solve that system of equations by first multiplying the terms of the first equation by 3 and those of the second equation by 2, and thereby eliminating y. Doing this, we have

$$9x - 6y = 12 \qquad \text{each term of first equation multiplied by 3}$$
$$\underline{2x + 6y = 4} \qquad \text{each term of second equation multiplied by 2}$$

$11x = 16 \qquad \text{add}$

$9x + 2x = 11x \longrightarrow$

$-6y + 6y = 0 \qquad x = \dfrac{16}{11} \qquad 12 + 4 = 16$

$$3\left(\tfrac{16}{11}\right) - 2y = 4 \qquad \text{substituting } x = 16/11 \text{ in the first original equation}$$
$$48 - 22y = 44 \qquad \text{multiply each term by 11}$$
$$-22y = -4$$

$$y = \frac{2}{11}$$

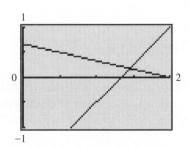

Fig. 5.26

Therefore, the solution is $x = \frac{16}{11}, y = \frac{2}{11}$, as found in Example 4.

A calculator solution of this system is shown in Fig. 5.26, where $y_1 = 3x/2 - 2$ and $y_2 = -x/3 + 2/3$. The point of intersection is $(1.455, 0.182)$. This solution is the same as the algebraic solutions since $\frac{16}{11} = 1.455$ and $\frac{2}{11} = 0.182$. ∎

■ The second equation is based on the amount of copper. We could have used the amount of zinc, which would have led to the equation $0.30A + 0.60B = 0.40(300)$. Since we need only two equations, we may use any two of these three equations to find the solution.

EXAMPLE 6   Application – solution by subtraction

By weight, one alloy is 70% copper and 30% zinc. Another alloy is 40% copper and 60% zinc. How many grams of each are required to make 300 g of an alloy that is 60% copper and 40% zinc?

Let A = the required number of grams of the first alloy and B = the required number of grams of the second alloy. Our equations are determined from:

1. The total weight of the final alloy is 300 g: $A + B = 300$.

2. The final alloy will have 180 g of copper (60% of 300 g), and this comes from 70% of A ($0.70A$) and 40% of B ($0.40B$): $0.70A + 0.40B = 180$.

These two equations can now be solved simultaneously:

$$A + \quad B = 300 \quad \text{sum of weights is 300 g}$$

copper → $0.70A + 0.40B = 180$ ← 60% of 300 g

70% weight 40% weight of
of first   second
alloy   alloy

$$4A + 4B = 1200 \quad \text{multiply each term of first equation by 4}$$
$$\underline{7A + 4B = 1800} \quad \text{multiply each term of second equation by 10}$$
$$3A \quad\quad = 600 \quad \text{subtract first equation from second equation}$$
$$A = 200 \text{ g}$$
$$B = 100 \text{ g} \quad \text{by substituting into first equation}$$

Checking with the statement of the problem, using the percentages of zinc, we have $0.30(200) + 0.60(100) = 0.40(300)$, or $60 \text{ g} + 60 \text{ g} = 120 \text{ g}$. We use zinc here since we used the percentages of copper for the equation. ■

EXAMPLE 7   Inconsistent system

In solving the system of equations

$$4x = 2y + 3$$
$$-y + 2x - 2 = 0$$

first note that the equations are not in the correct form. Therefore, writing them in the form of Eqs. (5.3), we have

$$4x - 2y = 3$$
$$2x - \ y = 2$$

Now, multiply the second equation by 2 and subtract to get

$$4x - 2y = 3$$
$$\underline{4x - 2y = 4}$$

$$4x - 4x = 0 \longrightarrow 0 = -1 \longleftarrow 3 - 4 = -1$$
$$-2y - (-2y) = 0 \nearrow$$

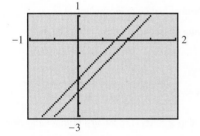

1

−1   2

−3

Fig. 5.27

NOTE ▶ Since 0 does not equal -1, we conclude that there is no solution. When the result is $0 = a \ (a \neq 0)$, the system is *inconsistent*. From Section 5.3, we know this means the lines representing the equations are parallel. See Fig. 5.27, where $y_1 = 2x - 3/2$, $y_2 = 2x - 2$.

NOTE ▶ If the result of solving a system is $0 = 0$, the system is *dependent*. As shown in Section 5.3, this means there is an unlimited number of solutions, and the lines that represent the equations are really the same line. ■

EXERCISES 5.4

In Exercises 1–4, make the given changes in the indicated examples of this section and then solve the resulting problems.

1. In Example 1, change the $+$ to $-$ in the second equation and then solve the system of equations.

2. In Example 3, change the $+$ to $-$ in the second equation and then solve the system of equations.

3. In Example 4, change x to $2x$ in the second equation and then solve the system of equations.

4. In Example 7, change 3 to 4 in the first equation and then find if there is any change in the conclusion that is drawn.

In Exercises 5–14, solve the given systems of equations by the method of elimination by substitution.

5. $x = y + 3$
   $x - 2y = 5$

6. $x = 2y + 1$
   $2x - 3y = 4$

7. $p = V - 4$
   $V + p = 10$

8. $y = 2x + 10$
   $2x + y = -2$

9. $x + y = -5$
   $2x - y = 2$

10. $3x + y = 1$
    $3x - 2y = 16$

11. $2x + 3y = 7$
    $6x - y = 1$

12. $2s + 2t = 1$
    $4s - 2t = 17$

13. $33x + 2y = 34$
    $40y = 9x + 11$

14. $3A + 3B = -1$
    $5A = -6B - 1$

In Exercises 15–24, solve the given systems of equations by the method of elimination by addition or subtraction.

15. $x + 2y = 5$
    $x - 2y = 1$

16. $x + 3y = 7$
    $2x + 3y = 5$

17. $2x - 3y = 4$
    $2x + y = -4$

18. $R - 4r = 17$
    $3R + 4r = 3$

19. $12t + 9y = 14$
    $6t = 7y - 16$

20. $3x - y = 3$
    $4x = 3y + 14$

21. $v + 2t = 7$
    $2v + 4t = 9$

22. $3x - y = 5$
    $-9x + 3y = -15$

23. $2x - 3y - 4 = 0$
    $3x + 2 = 2y$

24. $3i_1 + 5 = -4i_2$
    $3i_2 = 5i_1 - 2$

In Exercises 25–36, solve the given systems of equations by either method of this section.

25. $2x - y = 5$
    $6x + 2y = -5$

26. $3x + 2y = 4$
    $6x - 6y = 13$

27. $6x + 3y + 4 = 0$
    $5y = -9x - 6$

28. $1 + 6q = 5p$
    $3p - 4q = 7$

29. $15x + 10y = 11$
    $20x - 25y = 7$

30. $2x + 6y = -3$
    $-6x - 18y = 5$

31. $12V + 108 = -84C$
    $36C + 48V + 132 = 0$

32. $66x + 66y = -77$
    $33x - 132y = 143$

33. $44A = 1 - 15B$
    $5B = 22 + 7A$

34. $3.0x - 2.0y = 4.0$
    $2.9x - 2.0y = 8.0$

35. $2b = 6a - 16$
    $33a = 4b + 39$

36. $30P = 55 - Q$
    $19P + 14Q + 32 = 0$

In Exercises 37–40, in order to make the coefficients easier to work with, first multiply each term of the equation or divide each term of the equation by a number selected by inspection. Then proceed with the solution of the system by an appropriate algebraic method.

37. $0.3x - 0.7y = 0.4$
    $0.2x + 0.5y = 0.7$

38. $250R + 225Z = 400$
    $375R - 675Z = 325$

39. $40s - 30t = 60$
    $20s - 40t = -50$

40. $0.060x + 0.048y = -0.084$
    $0.013x - 0.065y = -0.078$

In Exercises 41–44, solve the given systems of equations by an appropriate algebraic method.

41. Find the voltages V_1 and V_2 of the batteries shown in Fig. 5.28. The terminals are aligned in the same direction in Fig. 5.28(a) and in the opposite directions in Fig. 5.28(b).

Fig. 5.28

42. A spring of length L is stretched x cm for each newton of weight hung from it. Weights of 3 N and then 5 N are hung from the spring, leading to the equations

    $L + 3x = 18$
    $L + 5x = 22$

    Solve for L and x.

43. Two grades of gasoline are mixed to make a blend with 1.50% of a special additive. Combining x liters of a grade with 1.80% of the additive to y liters of a grade with 1.00% of the additive gives 10,000 L of the blend. The equations relating x and y are

    $x + y = 10,000$
    $0.0180x + 0.0100y = 0.0150(10,000)$

    Find x and y (to three significant digits).

44. A 6.0% solution and a 15.0% solution of a drug are added to 200 mL of a 20.0% solution to make 1200 mL of a 12.0% solution for a proper dosage. The equations relating the number of milliliters of the added solutions are

    $x + y + 200 = 1200$
    $0.060x + 0.150y + 0.200(200) = 0.120(1200)$

    Find x and y (to three significant digits).

In Exercises 45–56, set up appropriate systems of two linear equations and solve the systems algebraically. All data are accurate to at least two significant digits.

45. A person's email for a week contained a total of 79 messages. The number of spam messages was four more than twice the number of other messages. How many were spam?

46. A 150-m cable is cut into two pieces such that one piece is four times as long as the other. How long is each piece?

47. The weight W_f supported by the front wheels of a certain car and the weight W_r supported by the rear wheels together equal the weight of the car, 17,700 N. See Fig. 5.29. Also, the ratio of W_r to W_f is 0.847. What are the weights supported by each set of wheels?

Fig. 5.29 W_r 17,700 N W_f

48. A sprinkler system is used to water two areas. If the total water flow is 980 L/h and the flow through one sprinkler is 65% as much as the other, what is the flow in each?

49. In a test of a heat-seeking rocket, a first rocket is launched at 2000 ft/s, and the heat-seeking rocket is launched along the same flight path 12 s later at a speed of 3200 ft/s. Find the times t_1 and t_2 of flight of the rockets until the heat-seeking rocket destroys the first rocket.

50. The *torque* of a force is the product of the force and the perpendicular distance from a specified point. If a lever is supported at only one point and is in balance, the sum of the torques (about the support) of forces acting on one side of the support must equal the sum of the torques of the forces acting on the other side. Find the forces F_1 and F_2 that are in the positions shown in Fig. 5.30(a) and then move to the positions in Fig. 5.30(b). The lever weighs 20 N and is in balance in each case.

Fig. 5.30

51. There are two types of offices in an office building, and a total of 54 offices. One type rents for \$900/month and the other type rents for \$1250/month. If all offices are occupied and the total rental income is \$55,600/month, how many of each type are there?

52. An underwater (but near the surface) explosion is detected by sonar on a ship 30 s before it is heard on the deck. If sound travels at 5000 ft/s in water and 1100 ft/s in air, how far is the ship from the explosion? See Fig. 5.31.

Fig. 5.31

53. A small isolated farm uses a windmill and a gas generator for power. During a 10-day period, they produced 3010 kW·h of power, with the windmill operating at 45.0% of capacity and the generator at capacity. During the following 10-day period, they produced 2900 kW·h with the windmill at 72.0% of capacity and the generator down 60 h for repairs (at capacity otherwise). What is the capacity (in kW) of each?

54. In mixing a weed-killing chemical, a 40% solution of the chemical is mixed with an 85% solution to get 20 L of a 60% solution. How much of each solution is needed?

(W)55. What conclusion can you draw from a sales report that states that "sales this month were \$8000 more than last month, which means that total sales for both months are \$4000 more than twice the sales last month"?

(W)56. For an electric circuit, a report stated that current i_1 is twice current i_2 and that twice the sum of the two currents less 6 times i_2 is 6 mA. Explain your conclusion about the values of the currents found from this report.

In Exercises 57–60, answer the given questions.

57. What condition(s) must be placed on the constants of the system of equations

$$ax + y = c$$
$$bx + y = d$$

such that there is a unique solution for x and y?

58. What conditions must be placed on the constants of the system of equations in Exercise 57 such that the system is (a) inconsistent? (b) Dependent?

59. For the dependent system of Example 6 on page 145, both equations can be written as $y = \frac{1}{3}x - 3$. The solution for the system can then be shown in terms of a general point on the graph as $(x, \frac{1}{3}x - 3)$. This is referred to as the solution with arbitrary x. Find the solutions in this form for this system for $x = -3$, and $x = 9$.

60. For the dependent system of Example 6 on page 145, write the form for the solution with arbitrary y. See Exercise 59.

Answers to Practice Exercises

1. $x = 3/5, y = -1/2$ **2.** $x = 3, y = -1$

5.5 Solving Systems of Two Linear Equations in Two Unknowns by Determinants

Determinant of the Second Order • Cramer's Rule • Solving Systems of Equations by Determinants

Consider two linear equations in two unknowns, as given in Eqs. (5.3):

$$
\begin{aligned}
a_1 x + b_1 y &= c_1 \\
a_2 x + b_2 y &= c_2
\end{aligned}
\tag{5.3}
$$

If we multiply the first of these equations by b_2 and the second by b_1, we obtain

$$
\begin{aligned}
a_1 b_2 x + b_1 b_2 y &= c_1 b_2 \\
a_2 b_1 x + b_2 b_1 y &= c_2 b_1
\end{aligned}
\tag{5.6}
$$

We see that the coefficients of y are the same. Thus, subtracting the second equation from the first, we can solve for x. The solution can be shown to be

$$
x = \frac{c_1 b_2 - c_2 b_1}{a_1 b_2 - a_2 b_1}
\tag{5.7}
$$

In the same manner, we may show that

$$
y = \frac{a_1 c_2 - a_2 c_1}{a_1 b_2 - a_2 b_1}
\tag{5.8}
$$

The expression $a_1 b_2 - a_2 b_1$, which appears in each of the denominators of Eqs. (5.7) and (5.8), is an example of a special kind of expression called a *determinant of the second order.* The determinant $a_1 b_2 - a_2 b_1$ is denoted by

$$
\begin{vmatrix} a_1 & b_1 \\ a_2 & b_2 \end{vmatrix}
$$

Second-order Determinant

■ Determinants were invented by the German mathematician Gottfried Wilhelm Leibniz (1646–1716).

Therefore, by definition, *a* **determinant of the second order** *is*

$$
\begin{vmatrix} a_1 & b_1 \\ a_2 & b_2 \end{vmatrix} = a_1 b_2 - a_2 b_1
\tag{5.9}
$$

The numbers a_1 and b_1 are called the **elements** *of the first* **row** *of the determinant. The numbers a_1 and a_2 are the elements of the first* **column** *of the determinant. In the same manner, the numbers a_2 and b_2 are the elements of the second row, and the numbers b_1 and b_2 are the elements of the second column. The numbers a_1 and b_2 are the elements of the* **principal diagonal,** *and the numbers a_2 and b_1 are the elements of the* **secondary diagonal.** Thus, one way of stating the definition indicated in Eq. (5.9) is that *the value of a determinant of the second order is found by taking the product of the elements of the principal diagonal and subtracting the product of the elements of the secondary diagonal.*

A diagram that is often helpful for remembering the expansion of a second-order determinant is shown in Fig. 5.32. The following examples illustrate how we carry out the evaluation of determinants.

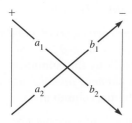

Fig. 5.32

EXAMPLE 1 Evaluating a second-order determinant

$$\begin{vmatrix} -5 & 8 \\ 3 & 7 \end{vmatrix} = (-5)(7) - 3(8) = -35 - 24 = -59$$

■

EXAMPLE 2 Evaluating second-order determinants

(a) $\begin{vmatrix} 4 & 6 \\ 3 & 17 \end{vmatrix} = 4(17) - (3)(6) = 68 - 18 = 50$

(b) $\begin{vmatrix} 4 & 6 \\ -3 & 17 \end{vmatrix} = 4(17) - (-3)(6) = 68 + 18 = 86$

(c) $\begin{vmatrix} 3.6 & 6.1 \\ -3.2 & -17.2 \end{vmatrix} = 3.6(-17.2) - (-3.2)(6.1) = -42.4$

Practice Exercise

1. Evaluate the determinant $\begin{vmatrix} 2 & -4 \\ 3 & 5 \end{vmatrix}$

Note the signs of the terms being combined.   ■

We note that the numerators and denominators of Eqs. (5.7) and (5.8) may be written as determinants. The numerators of the equations are

$$\begin{vmatrix} c_1 & b_1 \\ c_2 & b_2 \end{vmatrix} \quad \text{and} \quad \begin{vmatrix} a_1 & c_1 \\ a_2 & c_2 \end{vmatrix}$$

Therefore, the solutions for x and y of the system of equations

$$\begin{aligned} a_1 x + b_1 y &= c_1 \\ a_2 x + b_2 y &= c_2 \end{aligned} \tag{5.3}$$

may be written directly in terms of determinants, without algebraic operations, as

Note Carefully the Location of c_1 and c_2

$$x = \frac{\begin{vmatrix} c_1 & b_1 \\ c_2 & b_2 \end{vmatrix}}{\begin{vmatrix} a_1 & b_1 \\ a_2 & b_2 \end{vmatrix}} \quad \text{and} \quad y = \frac{\begin{vmatrix} a_1 & c_1 \\ a_2 & c_2 \end{vmatrix}}{\begin{vmatrix} a_1 & b_1 \\ a_2 & b_2 \end{vmatrix}} \tag{5.10}$$

For this reason, determinants provide a quick and easy method of solution of systems of equations. Again, *the denominator of each of Eqs. (5.10) is the same.*

The determinant of the denominator is made up of the coefficients of x and y. Also, the determinant of the numerator of the solution for x is obtained from the determinant of the denominator by **replacing the column of a's by the column of c's.** *The determinant of the numerator of the solution for y is obtained from the determinant of the denominator by* **replacing the column of b's by the column of c's.**

■ Named for the Swiss mathematician Gabriel Cramer (1704–1752).

This result is referred to as **Cramer's rule.** In using Cramer's rule, we must be sure that the equations are written in the form of Eqs. (5.3) before setting up the determinants.

The following examples illustrate the method of solving systems of equations by determinants.

EXAMPLE 3 Solving a system using Cramer's rule

Solve the following system of equations by determinants:

$$2x + y = 1$$
$$5x - 2y = -11$$

First, note that the equations are in the proper form of Eqs. (5.3) for solution by determinants. Next, set up the determinant for the denominator, which consists of the four coefficients in the system, written as shown. It is

$$x\text{-coefficients} \underbrace{\begin{vmatrix} 2 & 1 \\ 5 & -2 \end{vmatrix}}_{} y\text{-coefficients}$$

For finding x, the determinant in the numerator is obtained from this determinant by replacing the first column by the constants that appear on the right sides of the equations. Thus, *the numerator for the solution for x* is

$$\begin{vmatrix} 1 & 1 \\ -11 & -2 \end{vmatrix} \quad \text{replace } x\text{-coefficients with the constants}$$

For finding y, the determinant in the numerator is obtained from the determinant of the denominator by replacing the second column by the constants that appear on the right sides of the equations. Thus, *the numerator for the solution for y* is

$$\begin{vmatrix} 2 & 1 \\ 5 & -11 \end{vmatrix} \quad \text{replace } y\text{-coefficients with the constants}$$

Now, set up the solutions for x and y using the determinants above:

$$x = \frac{\begin{vmatrix} 1 & 1 \\ -11 & -2 \end{vmatrix}}{\begin{vmatrix} 2 & 1 \\ 5 & -2 \end{vmatrix}} = \frac{1(-2) - (-11)(1)}{2(-2) - (5)(1)} = \frac{-2 + 11}{-4 - 5} = \frac{9}{-9} = -1$$

$$y = \frac{\begin{vmatrix} 2 & 1 \\ 5 & -11 \end{vmatrix}}{\begin{vmatrix} 2 & 1 \\ 5 & -2 \end{vmatrix}} = \frac{2(-11) - (5)(1)}{-9} = \frac{-22 - 5}{-9} = 3$$

Therefore, the solution to the system of equations is $x = -1, y = 3$.

Substituting these values into the equations, we have

$$2(-1) + 3 \stackrel{?}{=} 1 \quad \text{and} \quad 5(-1) - 2(3) \stackrel{?}{=} -11$$
$$1 = 1 \qquad\qquad\qquad -11 = -11$$

which shows that they check.

NOTE ▸   Since the same determinant appears in each denominator, *it needs to be evaluated only once.* This means that three determinants are to be evaluated in order to solve the system.   ■

Practice Exercise

2. Solve the following system by determinants.

$$x + 2y = 4$$
$$3x - y = -9$$

■ Determinants can be evaluated on a graphing calculator. This is shown on page 164.

EXAMPLE 4   Solving a system using Cramer's rule

Solve the following system of equations by determinants. Numbers are approximate.

$$5.3x + 7.2y = 4.5$$
$$3.2x - 6.9y = 5.7$$

constants

$$x = \frac{\begin{vmatrix} 4.5 & 7.2 \\ 5.7 & -6.9 \end{vmatrix}}{\begin{vmatrix} 5.3 & 7.2 \\ 3.2 & -6.9 \end{vmatrix}} = \frac{4.5(-6.9) - 5.7(7.2)}{5.3(-6.9) - 3.2(7.2)} = \frac{-72.09}{-59.61} = 1.2$$

coefficients

constants

$$y = \frac{\begin{vmatrix} 5.3 & 4.5 \\ 3.2 & 5.7 \end{vmatrix}}{\begin{vmatrix} 5.3 & 7.2 \\ 3.2 & -6.9 \end{vmatrix}} = \frac{5.3(5.7) - 3.2(4.5)}{-59.61} = \frac{15.81}{-59.61} = -0.27$$

coefficients

The calculations can all be done on a calculator, using the following procedure: (1) Evaluate the denominator and store this value; (2) divide the value of each numerator by the value of the denominator, storing the values of x and y; (3) use the stored values of x and y to check the solution; (4) round off results (if the numbers are approximate).

For this system, we store the value of the denominator, $D = -59.61$, in order to calculate x and y, stored as X and Y. The check is shown in the calculator display in Fig. 5.33. ∎

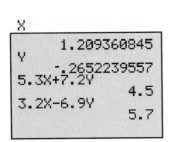

X
　　　1.209360845
Y
　　　-.2652239557
5.3X+7.2Y
　　　　　4.5
3.2X-6.9Y
　　　　　5.7

■ The line with X may be displaced off the screen.

Fig. 5.33

EXAMPLE 5   Application – solving a system

Two investments totaling $18,000 yield an annual income of $700. If the first investment has an interest rate of 5.5% and the second a rate of 3.0%, what is the value of each?

Let x = the value of the first investment and y = the value of the second investment. We know that the total of the two investments is $18,000. This leads to the equation $x + y = \$18,000$. The first investment yields $0.055x$ dollars annually, and the second yields $0.030y$ dollars annually. This leads to the equation $0.055x + 0.030y = 700$. These two equations are then solved simultaneously:

$$x + \quad y = 18,000 \quad \text{sum of investments}$$
$$0.055x + 0.030y = \quad 700 \leftarrow \text{income}$$

5.5% value   3.0% value

$$x = \frac{\begin{vmatrix} 18,000 & 1 \\ 700 & 0.030 \end{vmatrix}}{\begin{vmatrix} 1 & 1 \\ 0.055 & 0.030 \end{vmatrix}} = \frac{540 - 700}{0.030 - 0.055} = \frac{-160}{-0.025} = 6400$$

The value of y can be found most easily by substituting this value of x into the first equation, $y = 18,000 - x = 18,000 - 6400 = 11,600$.

Therefore, the values invested are $6400 and $11,600. Checking, the total income is $6400(0.055) + \$11,600(0.030) = \$700$, which agrees with the statement of the problem. ∎

CAUTION ▶ *The equations must be in the form of Eqs. (5.3) before the determinants are set up.* The specific positions of the values in the determinants are based on that form of writing the system. If either unknown is missing from an equation, a zero must be placed in the proper position. Also, from Example 4, we see that determinants are easier to use than other algebraic methods when the coefficients are decimals.

NOTE ▶   If the determinant of the denominator is zero, we do not have a unique solution since this would require division by zero. If the determinant of the denominator is zero and that of the numerator is not zero, the system is *inconsistent*. If the determinants of both numerator and denominator are zero, the system is *dependent*.

EXERCISES 5.5

In Exercises 1–4, make the given changes in the indicated examples of this section and then solve the resulting problems.

1. In Example 2(a), change the 6 to -6 and then evaluate.

2. In Example 2(a), change the 4 to -4 and the 6 to -6 and then evaluate.

3. In Example 3, change the $+$ to $-$ in the first equation and then solve the system of equations.

4. In Example 5, change \$700 to \$830 and then solve for the values of the investments.

In Exercises 5–16, evaluate the given determinants.

5. $\begin{vmatrix} 2 & 4 \\ 3 & 1 \end{vmatrix}$

6. $\begin{vmatrix} -1 & 3 \\ 2 & 6 \end{vmatrix}$

7. $\begin{vmatrix} 3 & -5 \\ 7 & -2 \end{vmatrix}$

8. $\begin{vmatrix} -4 & 7 \\ 1 & -3 \end{vmatrix}$

9. $\begin{vmatrix} 8 & -10 \\ 0 & 4 \end{vmatrix}$

10. $\begin{vmatrix} -4 & -3 \\ -8 & -6 \end{vmatrix}$

11. $\begin{vmatrix} -20 & 110 \\ -70 & -80 \end{vmatrix}$

12. $\begin{vmatrix} -6.5 & 12.2 \\ -15.5 & 34.6 \end{vmatrix}$

13. $\begin{vmatrix} 0.75 & -1.32 \\ 0.15 & 1.18 \end{vmatrix}$

14. $\begin{vmatrix} 0.20 & -0.05 \\ 0.28 & 0.09 \end{vmatrix}$

15. $\begin{vmatrix} 16 & -8 \\ 42 & -15 \end{vmatrix}$

16. $\begin{vmatrix} 43 & -7 \\ -81 & 16 \end{vmatrix}$

In Exercises 17–26, solve the given systems of equations by determinants. (These are the same as those for Exercises 15–24 of Section 5.4.)

17. $x + 2y = 5$
    $x - 2y = 1$

18. $x + 3y = 7$
    $2x + 3y = 5$

19. $2x - 3y = 4$
    $2x + y = -4$

20. $R - 4r = 17$
    $3R + 4r = 3$

21. $12t + 9y = 14$
    $6t = 7y - 16$

22. $2x + y = 4$
    $19x = 80 - 10y$

23. $v + 2t = 7$
    $2v + 4t = 9$

24. $3x - y = 5$
    $-9x + 3y = -15$

25. $2x - 3y - 4 = 0$
    $3x + 2 = 2y$

26. $3i_1 + 5 = -4i_2$
    $3i_2 = 5i_1 - 2$

In Exercises 27–34, solve the given systems of equations by determinants. All numbers are approximate. (Exercises 27–30 are the same as Exercises 37–40 of Section 5.4.)

27. $0.3x - 0.7y = 0.4$
    $0.2x + 0.5y = 0.7$

28. $250R + 225Z = 400$
    $375R - 675Z = 325$

29. $40s - 30t = 60$
    $20s - 40t = -50$

30. $0.060x + 0.048y = -0.084$
    $0.013x - 0.065y = -0.078$

31. $301x - 529y = 1520$
    $385x - 741y = 2540$

32. $0.25d + 0.63n = -0.37$
    $-0.61d - 1.80n = 0.55$

33. $1.2y + 10.8 = -8.4x$
    $3.5x + 4.8y + 12.9 = 0$

34. $6541x + 4397y = -7732$
    $3309x - 8755y = 7622$

In Exercises 35–38, answer the given questions about the determinant to the right. $\begin{vmatrix} a & b \\ c & d \end{vmatrix}$

35. What is the value of the determinant if $c = d = 0$?

36. What change in value occurs if rows and columns are interchanged (b and c are interchanged)?

37. What is the value of the determinant if $a = kb$ and $c = kd$?

38. How does the value change if a and c are doubled?

In Exercises 39–42, solve the given systems of equations by determinants. All numbers are accurate to at least two significant digits.

39. The forces acting on a link of an industrial robot are shown in Fig. 5.34. The equations for finding forces F_1 and F_2 are

$F_1 + F_2 = 21$
$2F_1 = 5F_2$

Find F_1 and F_2.

Fig. 5.34

40. The area of a quadrilateral is

$$A = \frac{1}{2}\left(\begin{vmatrix} x_0 & x_1 \\ y_0 & y_1 \end{vmatrix} + \begin{vmatrix} x_1 & x_2 \\ y_1 & y_2 \end{vmatrix} + \begin{vmatrix} x_2 & x_3 \\ y_2 & y_3 \end{vmatrix} + \begin{vmatrix} x_3 & x_0 \\ y_3 & y_0 \end{vmatrix} \right)$$

where $(x_0, y_0), (x_1, y_1), (x_2, y_2),$ and (x_3, y_3) are the rectangular coordinates of the vertices of the quadrilateral, listed counterclockwise. (This *surveyor's formula* can be generalized to find the area of any polygon.)

A surveyor records the locations of the vertices of a quadrilateral building lot on a rectangular coordinate system as $(12.79, 0.00)$, $(67.21, 12.30)$, $(53.05, 47.12)$, and $(10.09, 53.11)$, where distances are in meters. Find the area of the lot.

41. An airplane begins a flight with a total of 36.0 gal of fuel stored in two separate wing tanks. During the flight, 25.0% of the fuel in one tank is used, and in the other tank 37.5% of the fuel is used. If the total fuel used is 11.2 gal, the amounts x and y used from each tank can be found by solving the system of equations

$x + y = 36.0$
$0.250x + 0.375y = 11.2$

Find x and y.

42. In applying Kirchhoff's laws (see the chapter introduction; for the equations, see, for example, Beiser, *Modern Technical Physics,* 6th ed., p. 550) to the electric circuit shown in Fig. 5.35, the following equations are found. Find the indicated currents I_1 and I_2 (in A).

$$52I_1 - 27I_2 = -420$$
$$-27I_1 + 76I_2 = 210$$

Fig. 5.35

In Exercises 43–52, set up appropriate systems of two linear equations in two unknowns and then solve the systems by determinants. All numbers are accurate to at least two significant digits.

43. A developer built twice as many three-bedroom homes as four-bedroom homes in a new development. The profit from each three-bedroom home was $25,000, and the profit from each four-bedroom home was $35,000. If the total profit was $6,800,000, how many of each were built?

44. Two joggers are 2.0 mi apart. If they jog toward each other, they will meet in 12 min. If they jog in the same direction, the faster one will overtake the slower one in 2.0 h. At what rate does each jog?

45. A shipment of 320 cell phones and radar detectors was destroyed due to a truck accident. On the insurance claim, the shipper stated that each phone was worth $110, each detector was worth $160, and their total value was $40,700. How many of each were in the shipment?

46. Two types of electromechanical carburetors are being assembled and tested. Each of the first type requires 15 min of assembly time and 2 min of testing time. Each of the second type requires 12 min of assembly time and 3 min of testing time. If 222 min of assembly time and 45 min of testing time are available, how many of each type can be assembled and tested, if all the time is used?

47. A machinery sales representative receives a fixed salary plus a sales commission each month. If $6200 is earned on sales of $70,000 in one month and $4700 is earned on sales of $45,000 in the following month, what are the fixed salary and the commission percent?

48. A moving walkway at an airport is 65.0 m long. A child running at a constant speed takes 20.0 s to run along the walkway in the direction it is moving, and then 52.0 s to run all the way back. What are the speed of the walkway and the speed of the child?

49. A boat carrying illegal drugs leaves a port and travels at 42 mi/h. A Coast Guard cutter leaves the port 24 min later and travels at 50 mi/h in pursuit of the boat. Find the times each has traveled when the cutter overtakes the boat with drugs. See Fig. 5.36.

Fig. 5.36

50. Sterling silver is 92.5% silver and 7.5% copper. One silver-copper alloy is 94.0% silver, and a second silver-copper alloy is 85.0% silver. How much of each should be used in order to make 100 g of sterling silver?

51. In an experiment, a variable voltage V is in a circuit with a fixed voltage V_0 and a resistance R. The voltage V is related to the current i in the circuit by $V = Ri - V_0$. If $V = 5.8$ V for $i = 2.0$ A and $V = 24.7$ V for $i = 6.2$ A, find V as a function of i.

52. The velocity of sound in steel is 15,900 ft/s faster than the velocity of sound in air. One end of a long steel bar is struck, and an instrument at the other end measures the time it takes for the sound to reach it. The sound in the bar takes 0.0120 s, and the sound in the air takes 0.180 s. What are the velocities of sound in air and in steel?

Answers to Practice Exercises

1. 22 **2.** $x = -2, y = 3$

5.6 **Solving Systems of Three Linear Equations in Three Unknowns Algebraically**

Algebraic Method Using Addition or Subtraction

Many technical problems involve systems of linear equations with more than two unknowns. In this section, we solve systems with three unknowns, and later we will show how systems with even more unknowns are solved.

Solving such systems is very similar to solving systems in two unknowns. In this section, we will show the algebraic method, and in the next section we will show how determinants are used. Graphical solutions are not used since a linear equation in three unknowns represents a plane in space. We will, however, briefly show graphical interpretations of systems of three linear equations at the end of this section.

A system of three linear equations in three unknowns written in the form

$$\begin{aligned} a_1x + b_1y + c_1z &= d_1 \\ a_2x + b_2y + c_2z &= d_2 \\ a_3x + b_3y + c_3z &= d_3 \end{aligned}$$

(5.11)

has as its solution the set of values x, y, and z that satisfy all three equations simultaneously. The method of solution involves multiplying *two* of the equations by the proper numbers to eliminate **one** of the unknowns between these equations. We then repeat this process, using a *different pair* of the original equations, being sure that we eliminate the same unknown as we did between the first pair of equations. At this point we have two linear equations in two unknowns that can be solved by any of the methods previously discussed.

EXAMPLE 1 Algebraically solving a system

Solve the following system of equations:

■ We can choose to first eliminate any one of the three unknowns. We have chosen y.

(1) $\qquad 4x + y + 3z = 1$
(2) $\qquad 2x - 2y + 6z = 11$
(3) $\quad -6x + 3y + 12z = -4$

(4) $\qquad 8x + 2y + 6z = 2$ (1) multiplied by 2
$\qquad\quad 2x - 2y + 6z = 11$ (2)
(5) $\qquad 10x \quad\quad + 12z = 13$ adding

■ At this point we could have used Eqs. (1) and (3) or Eqs. (2) and (3), but we must set them up to eliminate y.

(6) $\qquad 12x + 3y + 9z = 3$ (1) multiplied by 3
$\quad -6x + 3y + 12z = -4$ (3)
(7) $\qquad 18x \quad\quad - 3z = 7$ subtracting

$\qquad\quad 10x + 12z = 13$ (5)

Practice Exercise

1. Solve the system in Example 1 by first eliminating x.

(8) $\qquad 72x - 12z = 28$ (7) multiplied by 4
(9) $\qquad 82x \quad\quad = 41$ adding
(10) $\qquad\qquad x = \frac{1}{2}$
(11) $\qquad 18\left(\frac{1}{2}\right) - 3z = 7$ substituting (10) in (7)
(12) $\qquad\qquad -3z = -2$

■ Some graphing calculators have a specific feature for solving simultaneous linear equations. It is necessary only to enter the coefficients and constants to get the solution.

(13) $\qquad\qquad z = \frac{2}{3}$
(14) $\quad 4\left(\frac{1}{2}\right) + y + 3\left(\frac{2}{3}\right) = 1$ substituting (13) and (10) in (1)
(15) $\qquad 2 + y + 2 = 1$
(16) $\qquad\qquad y = -3$

Thus, the solution is $x = \frac{1}{2}$, $y = -3$, $z = \frac{2}{3}$. Substituting in the equations, we have

$$4\left(\tfrac{1}{2}\right) + (-3) + 3\left(\tfrac{2}{3}\right) \stackrel{?}{=} 1 \qquad 2\left(\tfrac{1}{2}\right) - 2(-3) + 6\left(\tfrac{2}{3}\right) \stackrel{?}{=} 11 \qquad -6\left(\tfrac{1}{2}\right) + 3(-3) + 12\left(\tfrac{2}{3}\right) \stackrel{?}{=} -4$$
$$1 = 1 \qquad\qquad\qquad 11 = 11 \qquad\qquad\qquad -4 = -4$$

We see that the solution checks. If we had first eliminated x or z, we would have completed the solution in a similar manner. ■

Fig. 5.37

EXAMPLE 2 Solving a system – application

An industrial robot and the forces acting on its main link are shown in Fig. 5.37. An analysis of the forces leads to the following equations relating the forces. Determine the forces.

(1)	$A + 60 = 0.8T$	
(2)	$B = 0.6T$	⌐ choose 5 to ↓ clear decimals
(3)	$8A + 6B + 80 = 5T$	
(4)	$5A - 4T = -300$	(1) multiplied by 5 and rewritten
(5)	$5B - 3T = 0$	(2) multiplied by 5 and rewritten
(6)	$8A + 6B - 5T = -80$	(3) rewritten
(7)	$40A - 32T = -2400$	(4) multiplied by 8
(8)	$40A + 30B - 25T = -400$	(6) multiplied by 5
(9)	$-30B - 7T = -2000$	subtracting
(10)	$30B - 18T = 0$	(5) multiplied by 6
(11)	$-25T = -2000$	adding
(12)	$T = 80\,\text{N}$	
(13)	$A + 60 = 64$	(12) substituted in (1)
(14)	$A = 4\,\text{N}$	
(15)	$B = 48\,\text{N}$	(12) substituted in (2)

Therefore, the forces are $A = 4\,\text{N}, B = 48\,\text{N}$, and $T = 80\,\text{N}$. The solution checks when substituted into the original equations. Note that we could have started the solution by substituting Eq. (2) into Eq. (3), thereby first eliminating B. ∎

EXAMPLE 3 Setting up and solving a system

Three voltages, e_1, e_2, and e_3, where e_3 is three times e_1, are in series with the same polarity (see Fig. 5.38(a)) and have a total voltage of 85 mV. If e_2 is reversed in polarity (see Fig. 5.38(b)), the voltage is 35 mV. Find the voltages.

Since the voltages are in series with the same polarity, $e_1 + e_2 + e_3 = 85$. Then, since e_3 is three times e_1, we have $e_3 = 3e_1$. Then, with the reversed polarity of e_2, we have $e_1 - e_2 + e_3 = 35$. Writing these equations in standard form, we have the following solution:

(a)

(b)

Fig. 5.38

(1)	$e_1 + e_2 + e_3 = 85$	
(2)	$3e_1 - e_3 = 0$	rewriting second equation
(3)	$e_1 - e_2 + e_3 = 35$	
(4)	$2e_1 + 2e_3 = 120$	adding (1) and (3)
(5)	$6e_1 - 2e_3 = 0$	(2) multiplied by 2
(6)	$8e_1 = 120$	adding
(7)	$e_1 = 15\,\text{mV}$	
(8)	$3(15) - e_3 = 0$	substituting (7) in (2)
(9)	$e_3 = 45\,\text{mV}$	
(10)	$15 + e_2 + 45 = 85$	substituting (7) and (9) in (1)
(11)	$e_2 = 25\,\text{mV}$	

Therefore, the three voltages are 15 mV, 25 mV, and 45 mV.

Checking the solution, the sum of the three voltages is 85 mV, e_3 is three times e_1, and the sum of e_1 and e_3 less e_2 is 35 mV. ∎

(a)

(b)

Fig. 5.39

Systems with four or more unknowns are solved in a manner similar to that used for three unknowns. With four unknowns, one is eliminated between three different pairs of equations, and the resulting three equations are then solved.

Linear systems with more than two unknowns may have an unlimited number of solutions or be inconsistent. After eliminating unknowns, if we have $0 = 0$, there is an unlimited number of solutions. If we have $0 = a(a \neq 0)$, the system is inconsistent, and there is no solution. (See Exercises 29–32.)

As noted earlier, a linear equation in three unknowns represents a plane in space. For three linear equations in three unknowns, if the planes intersect at a point, there is a unique solution (Fig. 5.39(a)); if they intersect in a line, there is an unlimited number of solutions (Fig. 5.39(b)). If the planes do not have a common intersection, the system is inconsistent. The planes can be parallel (Fig. 5.40(a)), two can be parallel (Fig. 5.40(b)), or they can intersect in three parallel lines (Fig. 5.40(c)). If one plane is coincident with another plane, the system has an unlimited number of solutions if they intersect, or it is inconsistent.

  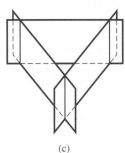

Fig. 5.40   (a)   (b)   (c)

EXERCISES 5.6

In Exercises 1 and 2, make the given changes in Example 1 of this section and then solve the resulting system of equations.

1. In the second equation, change the constant to the right of the $=$ sign from 11 to 12, and in the third equation, change the constant to the right of the $=$ sign from -4 to -14.

2. Change the second equation to $8x + 9z = 10$ (no y-term).

In Exercises 3–16, solve the given systems of equations.

3. $x + y + z = 2$
$x - z = 1$
$x + y = 1$

4. $x + y - z = -3$
$x + z = 2$
$2x - y + 2z = 3$

5. $2x + 3y + z = 2$
$-x + 2y + 3z = -1$
$-3x - 3y + z = 0$

6. $2x + y - z = 4$
$4x - 3y - 2z = -2$
$8x - 2y - 3z = 3$

7. $5l + 6w - 3h = 6$
$4l - 7w - 2h = -3$
$3l + w - 7h = 1$

8. $3r + s - t = 2$
$r - 2s + t = 0$
$4r - s + t = 3$

9. $2x - 2y + 3z = 5$
$2x + y - 2z = -1$
$4x - y - 3z = 0$

10. $2u + 2v + 3w = 0$
$3u + v + 4w = 21$
$-u - 3v + 7w = 15$

11. $3x - 7y + 3z = 6$
$3x + 3y + 6z = 1$
$5x - 5y + 2z = 5$

12. $18x + 24y + 4z = 46$
$63x + 6y - 15z = -75$
$-90x + 30y - 20z = -55$

13. $10x + 15y - 25z = 35$
$40x - 30y - 20z = 10$
$16x - 2y + 8z = 6$

14. $2i_1 - 4i_2 - 4i_3 = 3$
$3i_1 + 8i_2 + 2i_3 = -11$
$4i_1 + 6i_2 - i_3 = -8$

15. $r - s - 3t - u = 1$
$2r + 4s - 2u = 2$
$3r + 4s - 2t = 0$
$r + 2t - 3u = 3$

16. $3x + 2y - 4z + 2t = 3$
$5x - 3y - 5z + 6t = 8$
$2x - y + 3z - 2t = 1$
$-2x + 3y + 2z - 3t = -2$

In Exercises 17 and 18, three solutions of the equation $Ax + By + Cz = D$ are given, along with the value of D. Find the constants A, B, and C, and write the equation.

17. $x = 2, y = 4, z = 4$
$x = 3, y = -2, z = 8$
$x = -1, y = 8, z = 6$
$D = 12$

18. $x = 4, y = 1, z = 1$
$x = -1, y = 2, z = 3$
$x = 2, y = -1, z = 0$
$D = 5$

In Exercises 19–28, solve the systems of equations. In Exercises 23–28, it is necessary to set up the appropriate equations. All numbers are accurate to at least three significant digits.

19. A medical supply company has 1150 worker-hours for production, maintenance, and inspection. Using this and other factors, the number of hours used for each operation, P, M, and I, respectively, is found by solving the following system of equations:

$$P + M + I = 1150$$
$$P = 4I - 100$$
$$P = 6M + 50$$

20. Three oil pumps fill three different tanks. The pumping rates of the pumps (in L/h) are r_1, r_2, and r_3, respectively. Because of malfunctions, they do not operate at capacity each time. Their rates can be found by solving the following system of equations:

$$r_1 + r_2 + r_3 = 14{,}000$$
$$r_1 + 2r_2 = 13{,}000$$
$$3r_1 + 3r_2 + 2r_3 = 36{,}000$$

21. The forces acting on a certain girder, as shown in Fig. 5.41, can be found by solving the following system of equations:

$$0.707F_1 - 0.800F_2 = 0$$
$$0.707F_1 + 0.600F_2 - F_3 = 10.0$$
$$3.00\ F_2 - 3.00\ F_3 = 20.0$$

Find the forces, in newtons.

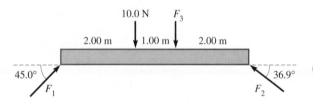

Fig. 5.41

22. Using Kirchhoff's laws (see the chapter introduction; for the equations, see, for example, Beiser, *Modern Technical Physics,* 6th ed., p. 550) with the electric circuit shown in Fig. 5.42, the following equations are found. Find the indicated currents (in A) I_1, I_2, and I_3.

$$1.0I_1 + 3.0(I_1 - I_3) = 12$$
$$2.0I_2 + 4.0(I_2 + I_3) = 12$$
$$1.0I_1 - 2.0I_2 + 3.0I_3 = 0$$

Fig. 5.42

23. Find angles A, B, and C in the roof truss shown in Fig. 5.43.

Fig. 5.43

24. Under certain conditions, the cost per mile C of operating a car is a function of the speed v (in mi/h) of the car, given by $C = av^2 + bv + c$. If $C = 28$ ¢/mi for $v = 10$ mi/h, $C = 22$ ¢/mi for $v = 30$ mi/h, and $C = 24$ ¢/mi for $v = 50$ mi/h, find C as a function of v.

25. The angle θ between two links of a robot arm is given by $\theta = at^3 + bt^2 + ct$, where t is the time during an 11.8-s cycle. If $\theta = 19.0°$ for $t = 1.00$ s, $\theta = 30.9°$ for $t = 3.00$ s, and $\theta = 19.8°$ for $t = 5.00$ s, find the equation $\theta = f(t)$. See Fig. 5.44.

Fig. 5.44

26. The computer systems at three weather bureaus have a combined hard-disk memory capacity of 8.0 TB (terabytes). The memory capacity of systems A and C have 0.2 TB more memory than twice that of system B, and twice the sum of the memory capacities of systems A and B is three times that of system C. What are the memory capacities of each of these computer systems?

27. By weight, one fertilizer is 20% potassium, 30% nitrogen, and 50% phosphorus. A second fertilizer has percents of 10, 20, and 70, respectively, and a third fertilizer has percents of 0, 30, and 70, respectively. How much of each must be mixed to get 200 lb of fertilizer with percents of 12, 25, and 63, respectively?

(W) 28. The average traffic flow (number of vehicles) from noon until 1 P.M. in a certain section of one-way streets in a city is shown in Fig. 5.45. Explain why an analysis of the flow through these intersections is not sufficient to obtain unique values for x, y, and z.

Fig. 5.45

In Exercises 29–32, show that the given systems of equations have either an unlimited number of solutions or no solution. If there is an unlimited number of solutions, find one of them.

29.
$$x - 2y - 3z = 2$$
$$x - 4y - 13z = 14$$
$$-3x + 5y + 4z = 0$$

30.
$$x - 2y - 3z = 2$$
$$x - 4y - 13z = 14$$
$$-3x + 5y + 4z = 2$$

31.
$$3x + 3y - 2z = 2$$
$$2x - y + z = 1$$
$$x - 5y + 4z = -3$$

32.
$$3x + y - z = -3$$
$$x + y - 3z = -5$$
$$-5x - 2y + 3z = -7$$

Answer to Practice Exercise

1. $x = 1/2, y = -3, z = 2/3$

Solving Systems of Three Linear Equations in Three Unknowns by Determinants

Determinant of the Third Order • Cramer's Rule • Solving Systems of Equations by Determinants • Determinants on the Calculator

Just as systems of two linear equations in two unknowns can be solved by determinants, so can systems of three linear equations in three unknowns. The system

$$\begin{array}{c} a_1x + b_1y + c_1z = d_1 \\ a_2x + b_2y + c_2z = d_2 \\ a_3x + b_3y + c_3z = d_3 \end{array} \qquad (5.11)$$

can be solved in general terms by the method of elimination by addition or subtraction. This leads to the following solutions for x, y, and z.

$$x = \frac{d_1b_2c_2 + d_3b_1c_2 + d_2b_3c_1 - d_3b_2c_1 - d_1b_3c_2 - d_2b_1c_3}{a_1b_2c_3 + a_3b_1c_2 + a_2b_3c_1 - a_3b_2c_1 - a_1b_3c_2 - a_2b_1c_3}$$

$$y = \frac{a_1d_2c_3 + a_3d_1c_2 + a_2d_3c_1 - a_3d_2c_1 - a_1d_3c_2 - a_2d_1c_3}{a_1b_2c_3 + a_3b_1c_2 + a_2b_3c_1 - a_3b_2c_1 - a_1b_3c_2 - a_2b_1c_3} \qquad (5.12)$$

$$z = \frac{a_1b_2d_3 + a_3b_1d_2 + a_2b_3d_1 - a_3b_2d_1 - a_1b_3d_2 - a_2b_1d_3}{a_1b_2c_3 + a_3b_1c_2 + a_2b_3c_1 - a_3b_2c_1 - a_1b_3c_2 - a_2b_1c_3}$$

The expressions that appear in the numerators and denominators of Eqs. (5.12) are examples of a **determinant of the third order.** *This determinant is defined by*

$$\begin{vmatrix} a_1 & b_1 & c_1 \\ a_2 & b_2 & c_2 \\ a_3 & b_3 & c_3 \end{vmatrix} = a_1b_2c_3 + a_3b_1c_2 + a_2b_3c_1 - a_3b_2c_1 - a_1b_3c_2 - a_2b_1c_3 \qquad (5.13)$$

The elements, rows, columns, and diagonals of a third-order determinant are defined just as are those of a second-order determinant. For example, the principal diagonal is made up of the elements a_1, b_2, and c_3.

Probably the easiest way of remembering the method of finding the value of a third-order determinant is as follows: *Rewrite the first two columns to the right of the determinant. The products of the elements of the principal diagonal and the two parallel diagonals to the right of it are then added. The products of the elements of the secondary diagonal and the two parallel diagonals to the right of it are subtracted from the first sum. The algebraic sum of these six products gives the value of the determinant.* These products are indicated in Fig. 5.46.

■ This method is used only for third-order determinants. It does not work for determinants of order higher than three.

Fig. 5.46

Examples 1 and 2 illustrate this method of evaluating third-order determinants.

EXAMPLE 1   Evaluating a third-order determinant

$$\begin{vmatrix} 1 & 5 & 4 \\ -2 & 3 & -1 \\ 2 & -1 & 5 \end{vmatrix} \begin{matrix} 1 & 5 \\ -2 & 3 \\ 2 & -1 \end{matrix} = 15 + (-10) + (+8) - (24) - (1) - (-50) = 38$$

$p_1 \quad p_2 \quad p_3$

$p_4 \quad p_5 \quad p_6$

∎

EXAMPLE 2   Evaluating a determinant on calculator

$$\begin{vmatrix} 3 & -2 & 8 \\ -5 & 5 & 0 \\ 4 & 9 & -6 \end{vmatrix} \begin{matrix} 3 & -2 \\ -5 & 5 \\ 4 & 9 \end{matrix} = (-90) + 0 + (-360) - 160 - 0 - (-60) = -550$$

∎

■ Determinants can be evaluated on a calculator using the *matrix* feature. Figure 5.47 shows two windows for the determinant in Example 2. The first shows the window for entering the numbers in the *matrix* (the array of numbers—see page 436). The first four lines of the second window show the matrix displayed, and the last two lines show the evaluation of the determinant.

Fig. 5.47

Practice Exercise

1. Using a calculator, evaluate the determinant in Example 1.

Inspection of Eqs. (5.12) reveals that the numerators of these solutions may also be written in terms of determinants. Thus, we may write the general solution to a system of three equations in three unknowns as

$$x = \frac{\begin{vmatrix} d_1 & b_1 & c_1 \\ d_2 & b_2 & c_2 \\ d_3 & b_3 & c_3 \end{vmatrix}}{\begin{vmatrix} a_1 & b_1 & c_1 \\ a_2 & b_2 & c_2 \\ a_3 & b_3 & c_3 \end{vmatrix}} \quad y = \frac{\begin{vmatrix} a_1 & d_1 & c_1 \\ a_2 & d_2 & c_2 \\ a_3 & d_3 & c_3 \end{vmatrix}}{\begin{vmatrix} a_1 & b_1 & c_1 \\ a_2 & b_2 & c_2 \\ a_3 & b_3 & c_3 \end{vmatrix}} \quad z = \frac{\begin{vmatrix} a_1 & b_1 & d_1 \\ a_2 & b_2 & d_2 \\ a_3 & b_3 & d_3 \end{vmatrix}}{\begin{vmatrix} a_1 & b_1 & c_1 \\ a_2 & b_2 & c_2 \\ a_3 & b_3 & c_3 \end{vmatrix}} \quad (5.14)$$

If the determinant of the denominator is not zero, there is a unique solution to the system of equations. (If all determinants are zero, there is an *unlimited number of solutions.* If the determinant of the denominator is zero and any of the determinants of the numerators is not zero, the system is *inconsistent,* and there is *no solution.*)

An analysis of Eqs. (5.14) shows that the situation is precisely the same as it was when we were using determinants to solve systems of two linear equations. That is, the determinants in the denominators in the expressions for x, y, and z are the same. They consist of elements that are the coefficients of the unknowns. The determinant of the numerator of the solution for x is the same as that of the denominator, except that the column of d's replaces the column of a's. The determinant in the numerator of the solution for y is the same as that of the denominator, except that the column of d's replaces the column of b's. The determinant of the numerator of the solution for z is the same as the determinant of the denominator, except that the column of d's replaces the column of c's. To summarize:

Cramer's Rule

NOTE ▶

The determinant in the denominator of each is made up of the coefficients of x, y, and z. The determinants in the numerators are the same as that in the denominator, except that the column of d's replaces the column of coefficients of the unknown for which we are solving.

This, again, is **Cramer's rule.** Remember, the equations must be written in the standard form shown in Eqs. (5.11) before the determinants are formed.

EXAMPLE 3 Solving a system using Cramer's rule

Solve the following system by determinants.

$$3x + 2y - 5z = -1$$
$$2x - 3y - z = 11$$
$$5x - 2y + 7x = 9$$

■ Figure 5.48(a) shows the calculator display for the solutions for x and y in Example 3. Here, det[A] and det[B] are the determinants for the numerators for x and y, respectively, and det[D] is the determinant for the denominator.

Figure 5.48(b) shows the solution for z and the use of the *fraction* feature in changing the result for z into a fraction.

(a)

(b)

Fig. 5.48

constants ——→

coefficients ——→

$$x = \cfrac{\begin{vmatrix} -1 & 2 & -5 \\ 11 & -3 & -1 \\ 9 & -2 & 7 \end{vmatrix} \begin{matrix} -1 & 2 \\ 11 & -3 \\ 9 & -2 \end{matrix}}{\begin{vmatrix} 3 & 2 & -5 \\ 2 & -3 & -1 \\ 5 & -2 & 7 \end{vmatrix} \begin{matrix} 3 & 2 \\ 2 & -3 \\ 5 & -2 \end{matrix}}$$

evaluate the denominator first; if it is zero, there is no need to evaluate the numerator

$$= \frac{21 - 18 + 110 - 135 + 2 - 154}{-63 - 10 + 20 - 75 - 6 - 28} = \frac{-174}{-162} = \frac{29}{27}$$

constants ——→

$$y = \cfrac{\begin{vmatrix} 3 & -1 & -5 \\ 2 & 11 & -1 \\ 5 & 9 & 7 \end{vmatrix} \begin{matrix} 3 & -1 \\ 2 & 11 \\ 5 & 9 \end{matrix}}{-162} = \frac{231 + 5 - 90 + 275 + 27 + 14}{-162}$$

denominator $= -162$ from solution for x

$$= \frac{462}{-162} = -\frac{77}{27}$$

constants ——→

$$z = \cfrac{\begin{vmatrix} 3 & 2 & -1 \\ 2 & -3 & 11 \\ 5 & -2 & 9 \end{vmatrix} \begin{matrix} 3 & 2 \\ 2 & -3 \\ 5 & -2 \end{matrix}}{-162} = \frac{-81 + 110 + 4 - 15 + 66 - 36}{-162}$$

denominator $= -162$ from solution for x

$$= \frac{48}{-162} = -\frac{8}{27}$$

Substituting in each of the original equations shows that the solution checks:

$$3\left(\frac{29}{27}\right) + 2\left(-\frac{77}{27}\right) - 5\left(-\frac{8}{27}\right) = \frac{87 - 154 + 40}{27} = \frac{-27}{27} = -1$$

$$2\left(\frac{29}{27}\right) - 3\left(-\frac{77}{27}\right) - \left(-\frac{8}{27}\right) = \frac{58 + 231 + 8}{27} = \frac{297}{27} = 11$$

$$5\left(\frac{29}{27}\right) - 2\left(-\frac{77}{27}\right) + 7\left(-\frac{8}{27}\right) = \frac{145 + 154 - 56}{27} = \frac{243}{27} = 9$$

After the values of x and y were determined, we could have evaluated z by substituting the values of x and y into one of the original equations. ■

EXAMPLE 4 Setting up and solving a system

An 8.0% solution, an 11% solution, and an 18% solution of nitric acid are to be mixed to get 150 mL of a 12% solution. If the volume of acid from the 8.0% solution equals half the volume of acid from the other two solutions, how much of each is needed?

Let x = volume of 8.0% solution needed, y = volume of 11% solution needed, and z = volume of 18% solution needed.

■ See the chapter introduction.

The fact that the sum of the volumes of the three solutions is 150 mL leads to the equation $x + y + z = 150$. Since there are $0.080x$ mL of pure acid from the first solution, $0.11y$ mL from the second solution, and $0.18z$ mL from the third solution, and $0.12(150)$ mL in the final solution, we are led to the equation $0.080x + 0.11y + 0.18z = 18$. Finally, using the last stated condition, we have the equation $0.080x = 0.5(0.11y + 0.18z)$. These equations are then written in the form of Eqs. (5.11) and solved.

$$
\begin{aligned}
x + \quad y + \quad z &= 150 \qquad &&\text{sum of volumes}\\
0.080x + 0.11y + 0.18z &= 18 \qquad &&\text{volumes of pure acid}\\
\text{acid in 8.0\% solution} \quad 0.080x \qquad\qquad\qquad &= 0.055y + 0.090z \qquad &&\text{one-half of acid in others}
\end{aligned}
$$

$$
\begin{aligned}
x + \quad y + \quad z &= 150 \qquad &&\text{standard form of}\\
0.080x + 0.11y + 0.18z &= 18 \qquad &&\text{Eqs. (5.11) by rewriting}\\
0.080x - 0.055y - 0.090z &= 0 \qquad &&\text{third equation}
\end{aligned}
$$

$$
x = \frac{\begin{vmatrix} 150 & 1 & 1 \\ 18 & 0.11 & 0.18 \\ 0 & -0.055 & -0.090 \end{vmatrix}\begin{matrix} 150 & 1 \\ 18 & 0.11 \\ 0 & -0.055 \end{matrix}}{\begin{vmatrix} 1 & 1 & 1 \\ 0.080 & 0.11 & 0.18 \\ 0.080 & -0.055 & -0.090 \end{vmatrix}\begin{matrix} 1 & 1 \\ 0.080 & 0.11 \\ 0.080 & -0.055 \end{matrix}}
$$

$$
= \frac{-1.485 + 0 - 0.990 - 0 + 1.485 + 1.620}{-0.0099 + 0.0144 - 0.0044 - 0.0088 + 0.0099 + 0.0072} = \frac{0.630}{0.0084} = 75
$$

$$
y = \frac{\begin{vmatrix} 1 & 150 & 1 \\ 0.080 & 18 & 0.18 \\ 0.080 & 0 & -0.090 \end{vmatrix}\begin{matrix} 1 & 150 \\ 0.080 & 18 \\ 0.080 & 0 \end{matrix}}{0.0084}
$$

$$
= \frac{-1.620 + 2.160 + 0 - 1.440 - 0 + 1.080}{0.0084} = \frac{0.180}{0.0084} = 21
$$

$$
z = \frac{\begin{vmatrix} 1 & 1 & 150 \\ 0.080 & 0.11 & 18 \\ 0.080 & -0.055 & 0 \end{vmatrix}\begin{matrix} 1 & 1 \\ 0.080 & 0.11 \\ 0.080 & -0.055 \end{matrix}}{0.0084}
$$

the value of z can also be found by substituting $x = 75$ and $y = 21$ into the first equation

$$
= \frac{0 + 1.440 - 0.660 - 1.320 + 0.990 - 0}{0.0084} = \frac{0.450}{0.0084} = 54
$$

Therefore, 75 mL of the 8.0% solution, 21 mL of the 11% solution, and 54 mL of the 18% solution are required to make the 12% solution. Results have been rounded off to two significant digits, the accuracy of the data. Checking with the statement of the problem, we see that these volumes total 150 mL. ■

As we previously noted, the calculations can be done completely on the calculator using either the *matrix* feature (as illustrated with Examples 2 and 3) or using the method outlined after Example 4 on page 156. Briefly, using this method for three equations is: (1) Evaluate and store the value of the denominator; (2) divide the value of each numerator by the value of the denominator, storing the values of x, y, and z; (3) use the stored values of x, y, and z to check the solution.

EXERCISES 5.7

In Exercises 1 and 2, make the given change in the indicated examples of this section and then solve the resulting problems.

1. In Example 1, interchange the first and second rows of the determinant and then evaluate it.

2. In Example 3, change the constant to the right of the $=$ sign in the first equation from -1 to -3, change the constant to the right of the $=$ sign in the third equation from 9 to 11, and then solve the resulting system of equations.

In Exercises 3–14, evaluate the given third-order determinants.

3. $\begin{vmatrix} 5 & 4 & -1 \\ -2 & -6 & 8 \\ 7 & 1 & 1 \end{vmatrix}$

4. $\begin{vmatrix} -7 & 0 & 0 \\ 2 & 4 & 5 \\ 1 & 4 & 2 \end{vmatrix}$

5. $\begin{vmatrix} 8 & 9 & -6 \\ -3 & 7 & 2 \\ 4 & -2 & 5 \end{vmatrix}$

6. $\begin{vmatrix} -2 & 4 & -1 \\ 5 & -10 & 4 \\ 4 & -8 & 2 \end{vmatrix}$

7. $\begin{vmatrix} -3 & -4 & -8 \\ 5 & -1 & 0 \\ 2 & 10 & -1 \end{vmatrix}$

8. $\begin{vmatrix} 10 & 2 & -7 \\ -2 & -3 & 6 \\ 6 & 5 & -2 \end{vmatrix}$

9. $\begin{vmatrix} 4 & -3 & -11 \\ -9 & 2 & -2 \\ 0 & 1 & -5 \end{vmatrix}$

10. $\begin{vmatrix} 9 & -2 & 0 \\ -1 & 3 & -6 \\ -4 & -6 & -2 \end{vmatrix}$

11. $\begin{vmatrix} 25 & 18 & -50 \\ -30 & 40 & -12 \\ -20 & 55 & -22 \end{vmatrix}$

12. $\begin{vmatrix} 20 & 0 & -15 \\ -4 & 30 & 1 \\ 6 & -1 & 40 \end{vmatrix}$

13. $\begin{vmatrix} 0.1 & -0.2 & 0 \\ -0.5 & 1 & 0.4 \\ -2 & 0.8 & 2 \end{vmatrix}$

14. $\begin{vmatrix} 0.25 & -0.54 & -0.42 \\ 1.20 & 0.35 & 0.28 \\ -0.50 & 0.12 & -0.44 \end{vmatrix}$

In Exercises 15–30, solve the given systems of equations by use of determinants. (Exercises 17–26 are the same as Exercises 3–12 of Section 5.6.)

15. $2x + 3y + z = 4$
$3x - z = -3$
$x - 2y + 2z = -5$

16. $4x + y + z = 2$
$2x - y - z = 4$
$3y + z = 2$

17. $x + y + z = 2$
$x - z = 1$
$x + y = 1$

18. $x + y - z = -3$
$x + z = 2$
$2x - y + 2z = 3$

19. $2x + 3y + z = 2$
$-x + 2y + 3z = -1$
$-3x - 3y + z = 0$

20. $2x + y - z = 4$
$4x - 3y - 2z = -2$
$8x - 2y - 3z = 3$

21. $5l + 6w - 3h = 6$
$4l - 7w - 2h = -3$
$3l + w - 7h = 1$

22. $3r + s - t = 2$
$r - 2s + t = 0$
$4r - s + t = 3$

23. $2x - 2y + 3z = 5$
$2x + y - 2z = -1$
$4x - y - 3z = 0$

24. $2u + 2v + 3w = 0$
$3u + v + 4w = 21$
$-u - 3v + 7w = 15$

25. $3x - 7y + 3z = 6$
$3x + 3y + 6z = 1$
$5x - 5y + 2z = 5$

26. $18x + 24y + 4z = 46$
$63x + 6y - 15z = -75$
$-90x + 30y - 20z = -55$

27. $p + 2q + 2r = 0$
$2p + 6q - 3r = -1$
$4p - 3q + 6r = -8$

28. $9x + 12y + 2z = 23$
$21x + 2y - 5z = -25$
$-18x + 6y - 4z = -11$

29. $3.0x + 4.5y - 7.5z = 10.5$
$4.8x - 3.6y - 2.4z = 1.2$
$4.0x - 0.5y + 2.0z = 1.5$

30. $26L - 52M - 52N = 39$
$45L + 96M + 40N = -80$
$55L + 62M - 11N = -48$

In Exercises 31–34, use the determinant at the right. Answer the questions about the determinant for the changes given in each exercise. $\begin{vmatrix} 4 & 2 & 1 \\ 3 & 6 & 5 \\ 7 & 9 & 8 \end{vmatrix} = 19$

31. How does the value change if the first two rows are interchanged?

32. What is the value if the second row is replaced with the first row (the first row remains unchanged—the first and second rows are the same)?

33. How does the value change if the elements of the first row are added to the corresponding elements of the second row (the first row remains unchanged)?

34. How does value change if each element of the first row is multiplied by 2?

In Exercises 35–44, solve the given problems by determinants. In Exercises 38–44, set up appropriate systems of equations. All numbers are accurate to at least two significant digits.

35. In analyzing the forces on the bell-crank mechanism shown in Fig. 5.49, the following equations are found. Find the forces.

$A - 0.60F = 80$

$B - 0.80F = 0$

$6.0A - 10F = 0$

Fig. 5.49   80 N

36. Using Kirchhoff's laws (see the chapter introduction; for the equations, see, for example, Beiser, *Modern Technical Physics*, 6th ed., p. 550) with the electric circuit shown in Fig. 5.50, the following equations are found. Find the indicated currents (in A) I_1, I_2, and I_3.

$19I_1 - 12I_2 \qquad = 60$

$12I_1 - 18I_2 + 6.0I_3 = 0$

$\qquad 6.0I_2 - 18I_3 = 0$

Fig. 5.50

37. In a laboratory experiment to measure the acceleration of an object, the distances traveled by the object were recorded for three different time intervals. These data led to the following equations:

$s_0 + 2v_0 + \;\;2a = 20$

$s_0 + 4v_0 + \;\;8a = 54$

$s_0 + 6v_0 + 18a = 104$

Here, s_0 is the initial displacement (in ft), v_0 is the initial velocity (in ft/s), and a is the acceleration (in ft/s^2). Find s_0, v_0, and a.

38. A certain 18-hole golf course has par-3, par-4, and par-5 holes, and there are twice as many par-4 holes as par-5 holes. How many holes of each type are there if a golfer has par on every hole for a score of 70?

39. The voltage V across part of an electric circuit is a function of the temperature T (in °C) given by $V = a + bT + cT^2$, where a, b, and c are constants. By evaluating a, b, and c, find $V = f(T)$ if $V = 6.4$ V for $T = 2.0$°C, $V = 8.6$ V for $T = 4.0$°C, and $V = 11.6$ V for $T = 6.0$°C.

40. A mail-order company charges $4 for shipping orders of less than $50, $6 for orders from $50 to $200, and $8 for orders over $200. One day the total shipping charges were $2160 for 384 orders. Find the number of orders shipped at each rate if the number of orders under $50 was 12 more than twice the number of orders over $200.

41. An alloy used in electrical transformers contains nickel (Ni), iron (Fe), and molybdenum (Mo). The percent of Ni is 1% less than five times the percent of Fe. The percent of Fe is 1% more than three times the percent of Mo. Find the percent of each in the alloy.

42. A person invests $20,000, partly at 5.00%, partly at 6.00%, and the remainder at 6.50%. The total annual interest is $1170. Three times the amount invested at 6.00% equals the amount invested at 5.00% and 6.50% combined. How much is invested at each rate?

43. A person spent 1.10 h in a car going to an airport, 1.95 h flying in a jet, and 0.520 h in a taxi to reach the final destination. The jet's speed averaged 12.0 times that of the car, which averaged 15.0 mi/h more than the taxi. What was the average speed of each if the trip covered 1140 mi?

44. An intravenous aqueous solution is made from three mixtures to get 500 mL with 6.0% of one medication, 8.0% of a second medication, and 86% water. The percents in the mixtures are, respectively, 5.0, 20, 75 (first), 0, 5.0, 95 (second), and 10, 5.0, 85 (third). How much of each is used?

Answer to Practice Exercise

1. 38

CHAPTER 5 **EQUATIONS**

Linear equation in one unknown $\qquad ax + b = 0$ $\qquad$ (5.1)

Linear equation in two unknowns $\qquad ax + by = c$ $\qquad$ (5.2)

System of two linear equations $\qquad a_1x + b_1y = c_1$ $\qquad$ (5.3)

$\qquad a_2x + b_2y = c_2$

Definition of slope $\qquad m = \dfrac{y_2 - y_1}{x_2 - x_1}$ $\qquad$ (5.4)

Slope-intercept form $\qquad y = mx + b$ $\qquad$ (5.5)

Second-order determinant
$$\begin{vmatrix} a_1 & b_1 \\ a_2 & b_2 \end{vmatrix} = a_1 b_2 - a_2 b_1 \tag{5.9}$$

Cramer's rule
$$x = \frac{\begin{vmatrix} c_1 & b_1 \\ c_2 & b_2 \end{vmatrix}}{\begin{vmatrix} a_1 & b_1 \\ a_2 & b_2 \end{vmatrix}} \quad \text{and} \quad y = \frac{\begin{vmatrix} a_1 & c_1 \\ a_2 & c_2 \end{vmatrix}}{\begin{vmatrix} a_1 & b_1 \\ a_2 & b_2 \end{vmatrix}} \tag{5.10}$$

System of three linear equations
$$\begin{aligned} a_1 x + b_1 y + c_1 z &= d_1 \\ a_2 x + b_2 y + c_2 z &= d_2 \\ a_3 x + b_3 y + c_3 z &= d_3 \end{aligned} \tag{5.11}$$

Third-order determinant
$$\begin{vmatrix} a_1 & b_1 & c_1 \\ a_2 & b_2 & c_2 \\ a_3 & b_3 & c_3 \end{vmatrix} = a_1 b_2 c_3 + a_3 b_1 c_2 + a_2 b_3 c_1 - a_3 b_2 c_1 - a_1 b_3 c_2 - a_2 b_1 c_3 \tag{5.13}$$

Cramer's rule
$$x = \frac{\begin{vmatrix} d_1 & b_1 & c_1 \\ d_2 & b_2 & c_2 \\ d_3 & b_3 & c_3 \end{vmatrix}}{\begin{vmatrix} a_1 & b_1 & c_1 \\ a_2 & b_2 & c_2 \\ a_3 & b_3 & c_3 \end{vmatrix}} \quad y = \frac{\begin{vmatrix} a_1 & d_1 & c_1 \\ a_2 & d_2 & c_2 \\ a_3 & d_3 & c_3 \end{vmatrix}}{\begin{vmatrix} a_1 & b_1 & c_1 \\ a_2 & b_2 & c_2 \\ a_3 & b_3 & c_3 \end{vmatrix}} \quad z = \frac{\begin{vmatrix} a_1 & b_1 & d_1 \\ a_2 & b_2 & d_2 \\ a_3 & b_3 & d_3 \end{vmatrix}}{\begin{vmatrix} a_1 & b_1 & c_1 \\ a_2 & b_2 & c_2 \\ a_3 & b_3 & c_3 \end{vmatrix}} \tag{5.14}$$

CHAPTER 5   REVIEW EXERCISES

In Exercises 1–4, evaluate the given determinants.

1. $\begin{vmatrix} -2 & 5 \\ 3 & 1 \end{vmatrix}$
2. $\begin{vmatrix} 40 & 10 \\ -20 & -60 \end{vmatrix}$
3. $\begin{vmatrix} -18 & -33 \\ -21 & 44 \end{vmatrix}$
4. $\begin{vmatrix} 0.91 & -1.2 \\ 0.73 & -5.0 \end{vmatrix}$

In Exercises 5–8, find the slopes of the lines that pass through the given points.

5. $(2, 0), (4, -8)$
6. $(-1, -5), (-4, 4)$
7. $(40, -20), (-30, -40)$
8. $\left(-6, \frac{1}{2}\right), \left(1, -\frac{7}{2}\right)$

In Exercises 9–12, find the slope and the y-intercepts of the lines with the given equations. Sketch the graphs.

9. $y = -2x + 4$
10. $2y = \frac{2}{3}x - 3$
11. $8x - 2y = 5$
12. $3x = 8 + 3y$

In Exercises 13–20, solve the given systems of equations graphically.

13. $y = 2x - 4$
    $y = -\frac{3}{2}x + 3$
14. $y = -3x + 3$
    $y = 2x - 6$
15. $4A - B = 6$
    $3A + 2B = 12$
16. $2x - 5y = 10$
    $3x + y = 6$
17. $7x = 2y + 14$
    $y = -4x + 4$
18. $5v = 15 - 3t$
    $t = 6v - 12$
19. $3M + 4N = 6$
    $2M - 3N = 2$
20. $5x + 2y = 5$
    $2x - 4y = 3$

In Exercises 21–30, solve the given systems of equations by using an appropriate algebraic method.

21. $x + 2y = 5$
    $x + 3y = 7$
22. $2x - y = 7$
    $x + y = 2$
23. $4x + 3y = -4$
    $y = 2x - 3$
24. $r = -3s - 2$
    $-2r - 9s = 2$
25. $10i - 27v = 29$
    $40i + 33v = 69$
26. $3x - 6y = 5$
    $7x + 2y = 4$
27. $7x = 2y - 6$
    $7y = 12 - 4x$
28. $3R = 8 - 5I$
    $6I = 8R + 11$
29. $90x - 110y = 40$
    $30x - 15y = 25$
30. $0.42x - 0.56y = 1.26$
    $0.98x - 1.40y = -0.28$

In Exercises 31–40, solve the given systems of equations by determinants. (These are the same as for Exercises 21–30.)

31. $x + 2y = 5$
    $x + 3y = 7$
32. $2x - y = 7$
    $x + y = 2$

33. $4x + 3y = -4$
$y = 2x - 3$

34. $r = -3s - 2$
$-2r - 9s = 2$

35. $10i - 27v = 29$
$40i + 33v = 69$

36. $3x - 6y = 5$
$7x + 2y = 4$

37. $7x = 2y - 6$
$7y = 12 - 4x$

38. $3R = 8 - 5I$
$6I = 8R + 11$

39. $90x - 110y = 40$
$30x - 15y = 25$

40. $0.42x - 0.56y = 1.26$
$0.98x - 1.40y = -0.28$

(W) *In Exercises 41–44, explain your answers. In Exercises 41–43, choose an exercise from among Exercises 31–40 that you think is most easily solved by the indicated method.*

41. Substitution **42.** Addition or subtraction **43.** Determinants

44. Considering the slope m and the y-intercept b, explain how to determine if there is (a) a unique solution, (b) an inconsistent solution, or (c) a dependent solution.

In Exercises 45–48, evaluate the given determinants.

45. $\begin{vmatrix} 4 & -1 & 8 \\ -1 & 6 & -2 \\ 2 & 1 & -1 \end{vmatrix}$

46. $\begin{vmatrix} -500 & 0 & -500 \\ 250 & 300 & -100 \\ -300 & 200 & 200 \end{vmatrix}$

47. $\begin{vmatrix} -2.2 & -4.1 & 7.0 \\ 1.2 & 6.4 & -3.5 \\ -7.2 & 2.4 & -1.0 \end{vmatrix}$

48. $\begin{vmatrix} 30 & 22 & -12 \\ 0 & -34 & 44 \\ 35 & -41 & -27 \end{vmatrix}$

In Exercises 49–54, solve the given systems of equations algebraically. In Exercises 53 and 54, the numbers are approximate.

49. $2x + y + z = 4$
$x - 2y - z = 3$
$3x + 3y - 2z = 1$

50. $x + 2y + z = 2$
$3x - 6y + 2z = 2$
$2x - z = 8$

51. $2r + s + 2t = 8$
$3r - 2s - 4t = 5$
$-2r + 3s + 4t = -3$

52. $4u + 4v - 2w = -4$
$20u - 15v + 10w = -10$
$24u - 12v - 9w = 39$

53. $3.6x + 5.2y - z = -2.2$
$3.2x - 4.8y + 3.9z = 8.1$
$6.4x + 4.1y + 2.3z = 5.1$

54. $32t + 24u + 63v = 32$
$42t - 31u + 19v = 132$
$48t + 12u + 11v = 0$

In Exercises 55–60, solve the given systems of equations by determinants. In Exercises 59 and 60, the numbers are approximate. (These systems are the same as for Exercises 49–54.)

55. $2x + y + z = 4$
$x - 2y - z = 3$
$3x + 3y - 2z = 1$

56. $x + 2y + z = 2$
$3x - 6y + 2z = 2$
$2x - z = 8$

57. $2r + s + 2t = 8$
$3r - 2s - 4t = 5$
$-2r + 3s + 4t = -3$

58. $4u + 4v - 2w = -4$
$20u - 15v + 10w = -10$
$24u - 12v - 9w = 39$

59. $3.6x + 5.2y - z = -2.2$
$3.2x - 4.8y + 3.9z = 8.1$
$6.4x + 4.1y + 2.3z = 5.1$

60. $32t + 24u + 63v = 32$
$42t - 31u + 19v = 132$
$48t + 12u + 11v = 0$

In Exercises 61–64, solve for x. Here, we see that we can solve an equation in which the unknown is an element of a determinant.

61. $\begin{vmatrix} 2 & 5 \\ 1 & x \end{vmatrix} = 3$

62. $\begin{vmatrix} -1 & x \\ 3 & 4 \end{vmatrix} = 7$

63. $\begin{vmatrix} x & 1 & 2 \\ 0 & -1 & 3 \\ -2 & 2 & 1 \end{vmatrix} = 5$

64. $\begin{vmatrix} 1 & 2 & -1 \\ -2 & 3 & x \\ -1 & 2 & -2 \end{vmatrix} = -3$

In Exercises 65–68, let $1/x = u$ and $1/y = v$. Solve for u and v and then solve for x and y. In this way, we see how to solve systems of equations involving reciprocals.

65. $\dfrac{1}{x} - \dfrac{1}{y} = \dfrac{1}{2}$
$\dfrac{1}{x} + \dfrac{1}{y} = \dfrac{1}{4}$

66. $\dfrac{1}{x} + \dfrac{1}{y} = 3$
$\dfrac{2}{x} + \dfrac{1}{y} = 1$

67. $\dfrac{2}{x} + \dfrac{3}{y} = 3$
$\dfrac{5}{x} - \dfrac{6}{y} = 3$

68. $\dfrac{3}{x} - \dfrac{2}{y} = 4$
$\dfrac{2}{x} + \dfrac{4}{y} = 1$

In Exercises 69 and 70, determine the value of k that makes the system dependent. In Exercises 71 and 72, determine the value of k that makes the system inconsistent.

69. $3x - ky = 6$
$x + 2y = 2$

70. $5x + 20y = 15$
$2x + ky = 6$

71. $kx - 2y = 5$
$4x + 6y = 1$

72. $2x - 5y = 7$
$kx + 10y = 2$

In Exercises 73 and 74, solve the given systems of equations by any appropriate method. All numbers in 73 are accurate to two significant digits, and in 74 they are accurate to three significant digits.

73. A 20-ft crane arm with a supporting cable and with a 100-lb box suspended from its end has forces acting on it, as shown in Fig. 5.51. Find the forces (in lb) from the following equations.

$F_1 + 2.0F_2 = 280$
$0.87F_1 - F_3 = 0$
$3.0F_1 - 4.0F_2 = 600$

Fig. 5.51

74. In applying Kirchhoff's laws (see Exercise 22 of Section 5.6) to the electric circuit shown in Fig. 5.52, the following equations result. Find the indicated currents (in A).

$i_1 + i_2 + i_3 = 0$
$5.20i_1 - 3.25i_2 = 8.33 - 6.45$
$3.25i_2 - 2.62i_3 = 6.45 - 9.80$

Fig. 5.52

In Exercises 75–94, set up systems of equations and solve by any appropriate method. All numbers are accurate to at least two significant digits.

75. A study found that the fuel consumption for transportation contributes a percent p_1 that is 16% less than the percent p_2 of all other sources combined. Find p_1 and p_2.

76. A certain amount of a fuel contains 150,000 Btu of potential heat. Part is burned at 80% efficiency, and the rest is burned at 70% efficiency, such that the total amount of heat actually delivered is 114,000 Btu. Find the amounts burned at each efficiency.

77. The sales representatives of a company have a choice of being paid 10% of their sales in a month, or $2400 plus 4% of their sales in the month. For what monthly sales is the income the same, and what is that income?

78. A person invested a total of $20,900 into two bonds, one with an annual interest rate of 6.00% and the other with an annual interest rate of 5.00% per year. If the total annual interest from the bonds is $1170, how much is invested in each bond?

79. A total of 42 tons of two types of ore is to be loaded into a smelter. The first type contains 6.0% copper, and the second contains 2.4% copper. Find the necessary amounts of each ore (to the nearest 1 ton) to produce 2 tons of copper.

80. As a 40-ft pulley belt makes one revolution, one of the two pulley wheels makes one more revolution than the other. Another wheel of half the radius replaces the smaller wheel and makes six more revolutions than the larger wheel for one revolution of the belt. Find the circumferences of the wheels. See Fig. 5.53.

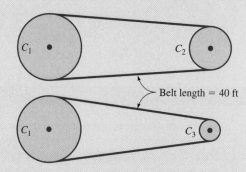

Belt length = 40 ft

Fig. 5.53

81. A computer analysis showed that the temperature T of the ocean water within 1000 m of a nuclear-plant discharge pipe was given by $T = \dfrac{a}{x + 100} + b$, where x is the distance from the pipe and a and b are constants. If $T = 14°C$ for $x = 0$ and $T = 10°C$ for $x = 900$ m, find a and b.

(W) 82. In measuring the angles of a triangular parcel, a surveyor noted that one of the angles equaled the sum of the other two angles. What conclusion can be drawn?

83. A satellite is to be launched from a space shuttle. It is calculated that the satellite's speed will be 24,200 km/h if launched directly ahead of the shuttle or 21,400 km/h if launched directly to the rear of the shuttle. What is the speed of the shuttle and the launching speed of the satellite relative to the shuttle? See Fig. 5.54.

21,400 km/h   24,200 km/h

Fig. 5.54

84. The velocity v of sound is a function of the temperature T according to the function $v = aT + b$, where a and b are constants. If $v = 337.5$ m/s for $T = 10.0°C$ and $v = 346.6$ m/s for $T = 25.0°C$, find v as a function of T.

85. The power (in W) dissipated in an electric resistance (in Ω) equals the resistance times the square of the current (in A). If 1.0 A flows through resistance R_1 and 3.0 A flows through resistance R_2, the total power dissipated is 14.0 W. If 3.0 A flows through R_1 and 1.0 A flows through R_2, the total power dissipated is 6.0 W. Find R_1 and R_2.

86. Twelve equal rectangular ceiling panels are placed as shown in Fig. 5.55. If each panel is 6.0 in. longer than it is wide and a total of 132.0 ft of edge and middle strips is used, what are the dimensions of the room?

Width

Length

Fig. 5.55

87. The weight of a lever may be considered to be at its center. A 20-ft lever of weight w is balanced on a fulcrum 8 ft from one end by a load L at that end. If a load of $4L$ is placed at that end, it requires a 20-lb weight at the other end to balance the lever. What are the initial load L and the weight w of the lever? (See Exercise 50 of Section 5.4.) See Fig. 5.56.

Fig. 5.56

88. Two fuel mixtures, one of 2.0% oil and 98.0% gasoline and another of 8.0% oil and 92.0% gasoline, are to be used to make 10.0 L of a fuel that is 4.0% oil and 96.0% gasoline for use in a chain saw. How much of each mixture is needed?

89. For the triangular truss shown in Fig. 5.57, ∠A is 55° less than twice ∠B, and ∠C is 25° less than ∠B. Find the measures of the three angles.

Fig. 5.57

90. A manufacturer produces three models of DVD players in a year. Four times as many of model A are produced as model C, and 7000 more of model B than model C. If the total production for the year is 97,000 units, how many of each are produced?

91. Three computer programs A, B, and C, require a total of 140 MB (megabytes) of hard-disk memory. If three other programs, two requiring the same memory as B and one the same as C, are added to a disk with A, B, and C, a total of 236 MB are required. If three other programs, one requiring the same memory as A and two the same memory as C, are added to a disk with A, B, and C, a total of 304 MB are required. How much memory is required for each of A, B, and C?

92. One ampere of electric current is passed through a solution of sulfuric acid, silver nitrate, and cupric sulfate, releasing hydrogen gas, silver, and copper. A total mass of 1.750 g is released. The mass of silver deposited is 3.40 times the mass of copper deposited, and the mass of copper and 70.0 times the mass of hydrogen combined equals the mass of silver deposited less 0.037 g. How much of each is released?

93. Gold loses about 5.3%, and silver about 10%, of its weight when immersed in water. If a gold-silver alloy weighs 6.0 N in air and 5.6 N in water, find the weight in air of the gold and the silver in the alloy.

94. A person engaged a workman for 48 days. For each day that he labored, he received 24 cents, and for each day he was idle, he paid 12 cents for his board. At the end of the 48 days, the account was settled, when the laborer received 504 cents. *Required the number of working days, and the number of days he was idle.* (Copied from Davies, Charles, *Elementary Algebra*, New York: A. S. Barnes & Burr, 1852.)

Writing Exercise

95. Write one or two paragraphs giving reasons for choosing a particular method of solving the following problem. If a first pump is used for 2.2 h and a second pump is used for 2.7 h, 1100 ft³ can be removed from a wastewater-holding tank. If the first pump is used for 1.4 h and the second for 2.5 h, 840 ft³ can be removed. How much can each pump remove in 1.0 h? (What is the result to two significant digits?)

CHAPTER 5 **PRACTICE TEST**

1. Find the slope of the line through $(2, -5)$ and $(-1, 4)$.

2. Is $x = -2, y = 3$ a solution to the system of equations

$2x + 5y = 11$
$y - 5x = 12$?

Explain.

3. Evaluate: $\begin{vmatrix} -1 & 3 & -2 \\ 4 & -3 & 0 \\ 5 & -4 & 2 \end{vmatrix}$

4. Solve by substitution:

$x + 2y = 5$
$4y = 3 - 2x$

5. Solve by determinants:

$3x - 2y = 4$
$2x + 5y = -1$

6. By finding the slope and y-intercept, sketch the graph of $2x + y = 4$.

7. The perimeter of a rectangular ranch is 24 km, and the length is 6.0 km more than the width. Set up equations relating the length l and the width w and then solve for l and w.

8. Solve by addition or subtraction:

$6N - 2P = 13$
$4N + 3P = -13$

9. In testing an anticholesterol drug, it was found that each milligram of drug administered reduced a person's blood cholesterol level by 2 units. Set up the function relating the cholesterol level C as a function of the dosage d for a person whose cholesterol level is 310 before taking the drug. Sketch the graph.

10. Solve the following system of equations graphically. Determine the values of x and y to the nearest 0.1.

$2x - 3y = 6$
$4x + y = 4$

11. By volume, one alloy is 60% copper, 30% zinc, and 10% nickel. A second alloy has percents of 50, 30, and 20, respectively. A third alloy is 30% copper and 70% nickel. How much of each alloy is needed to make 100 cm³ of a resulting alloy with percents of 40, 15, and 45, respectively?

12. Solve for y by determinants:

$3x + 2y - z = 4$
$2x - y + 3z = -2$
$x + 4z = 5$

Factoring and Fractions

6

The algebraic methods reviewed in Chapter 1 have been enough for our use to now. However, in later chapters, additional algebraic methods are needed. In this chapter, we further develop operations with products, quotients, and fractions and show how they are used in solving equations.

The development of the symbols now used in algebra in itself led to advances in mathematics and science. Until about 1500, most problems and their solutions were stated in words, which made them very lengthy and often difficult to follow. For example, to solve the equation $3x + 7 = 8(2x - 5)$, the problem would be stated something like "find a number such that seven added to three times the number is equal to the product of 8 and the quantity of five subtracted from twice the number." Then the solution would be written out, word by word.

As time went on, writers abbreviated some of the words in a problem, but they still essentially wrote out the solution in words. Some symbols did start to come into use. For example, the $+$ and $-$ signs first appeared in a published book

in the late 1400s, the square root symbol $\sqrt{}$ was first used in 1525, and the $=$ sign was introduced in 1557. Then, in the late 1500s, the French lawyer François Viète wrote several articles in which he used symbols, including letters to represent numbers. He so improved the symbolism of algebra that he is often called "the father of algebra." By the mid-1600s, the notation being used was reasonably similar to that we use today.

The use of symbols in algebra made it more useful in all fields of mathematics. It was important in the development of calculus in the 1600s and 1700s. In turn, this gave scientists a powerful tool that allowed for much greater advancement in all areas of science and technology.

Although the primary purpose of this chapter is to develop additional algebraic methods, many technical applications of these operations will be shown. The important applications in optics are noted by the picture of the Hubble telescope above. Other areas of application include electronics and mechanical design.

Great advances in our knowledge of the universe have been made through the use of the Hubble space telescope (shown at the left) since the mid-1990s. In Section 6.8, we illustrate the use of algebraic operations in the study of planetary motion and in the design of lenses and reflectors in telescopes.

6.1 Special Products

Special Algebraic Products •
Special Products Involving Cubes

Certain types of algebraic products are used so often that we should be very familiar with them. These are shown in the following equations.

$$a(x + y) = ax + ay \tag{6.1}$$

$$(x + y)(x - y) = x^2 - y^2 \tag{6.2}$$

$$(x + y)^2 = x^2 + 2xy + y^2 \tag{6.3}$$

$$(x - y)^2 = x^2 - 2xy + y^2 \tag{6.4}$$

$$(x + a)(x + b) = x^2 + (a + b)x + ab \tag{6.5}$$

$$(ax + b)(cx + d) = acx^2 + (ad + bc)x + bd \tag{6.6}$$

We see that Eq. (6.1) is the very important *distributive law.* Equations (6.2) to (6.6) are found by using the distributive law along with other basic operations that were developed in Chapter 1.

■ Equations (6.1) to (6.6) are identities. (See page 35.) Each is valid for all sets of values of the literal numbers.

Equations (6.1) to (6.6) are important **special products** *and should be known* **thoroughly.** By carefully studying the products, you will *see how they are formed,* and this allows them to be used quickly. Equation (6.6) looks complicated, but when you see how each of the terms on the right side is formed, you will not have to actually memorize it. In using these equations, we must realize that any of the literal numbers may represent any symbol or expression that represents a number.

EXAMPLE 1 Using special products 6.1 and 6.2

(a) Using Eq. (6.1) in the following product, we have

$$6(3r + 2s) = 6(3r) + 6(2s) = 18r + 12s$$

(b) Using Eq. (6.2), we have

Difference of Squares

$$(3r + 2s)(3r - 2s) = (3r)^2 - (2s)^2 = 9r^2 - 4s^2$$

sum of   difference   difference
3r and 2s   of 3r   of squares
          and 2s

Using Eq. (6.1) in (a), we have $a = 6$. In (a) and (b), $3r = x$ and $2s = y$. ∎

EXAMPLE 2 Using special products 6.3 and 6.4

Using Eqs. (6.3) and (6.4) in the following products, we have

■ $(5a + 2)^2 = (5a + 2)(5a + 2)$

(a) $(5a + 2)^2 = (5a)^2 + 2(5a)(2) + 2^2 = 25a^2 + 20a + 4$ Eq. (6.3)

square   twice   square
          product

(b) $(5a - 2)^2 = (5a)^2 - 2(5a)(2) + 2^2 = 25a^2 - 20a + 4$ Eq. (6.4)

In these illustrations, we let $x = 5a$ and $y = 2$. *It should be emphasized that*

■ Avoid a very common error.

CAUTION ▶ $(5a + 2)^2$ is **not** $(5a)^2 + 2^2$, or $25a^2 + 4$

CAUTION ▶ We must very carefully follow the forms of Eqs. (6.3) and (6.4) and be certain to *include the middle term,* 20a. (See Example 5 on page 31.) ∎

EXAMPLE 3   Using special products 6.5 and 6.6

Using Eqs. (6.5) and (6.6) in the following products, we have

(a) $(x + 5)(x - 3) = x^2 + [5 + (-3)]x + (5)(-3) = x^2 + 2x - 15$ Eq. (6.5)

(b) $(4x + 5)(2x - 3) = (4x)(2x) + [(4)(-3) + (5)(2)]x + (5)(-3)$ Eq. (6.6)
$$= 8x^2 - 2x - 15$$ ∎

Generally, when we use these special products, we find the middle term mentally and write down the result directly, as shown in the next example.

EXAMPLE 4   Middle term often found mentally

(a) $(y - 5)(y + 5) = y^2 - 25$ no middle term—Eq. (6.2)

(b) $(3x - 2)^2 = 9x^2 - 12x + 4$ middle term $= 2(3x)(-2)$—Eq. (6.4)

(c) $(x^2 - 4)(x^2 + 7) = x^4 + 3x^2 - 28$ middle term $= (7 - 4)x^2$—Eq. (6.5)
$(x^2)^2 = x^4$ ∎

At times, it is necessary to use more than one of the special products to simplify an algebraic expression. When this happens, it may be necessary to write down an intermediate step. This is illustrated in the following examples.

EXAMPLE 5   Applications of special products

(a) When analyzing the forces on a certain type of beam, the expression $Fa(L - a)(L + a)$ occurs. In expanding this expression, we first multiply $L - a$ by $L + a$ by use of Eq. (6.2). The expansion is completed by using Eq. (6.1), the distributive law.

$$Fa(L - a)(L + a) = Fa(L^2 - a^2)$$ Eq. (6.2)
$$= FaL^2 - Fa^3$$ Eq. (6.1)

(b) The electrical power delivered to the resistor R in Fig. (6.1) is $R(i_1 + i_2)^2$. Here, i_1 and i_2 are electric currents. To expand this expression, we first perform the square by use of Eq. (6.3) and then complete the expansion by use of Eq. (6.1).

$$R(i_1 + i_2)^2 = R(i_1^2 + 2i_1i_2 + i_2^2)$$ Eq. (6.3)
$$= Ri_1^2 + 2Ri_1i_2 + Ri_2^2$$ Eq. (6.1)   ∎

EXAMPLE 6   Group terms then use special products

In determining the product $(x + y - 2)^2$, we may group the quantity $(x + y)$ in an intermediate step. This leads to

$$(x + y - 2)^2 = [(x + y) - 2]^2$$
$$= (x + y)^2 - 2(x + y)(2) + 2^2$$
$$= x^2 + 2xy + y^2 - 4x - 4y + 4$$

In this example, we used Eqs. (6.3) and (6.4).   ∎

Practice Exercises

Find the indicated products

1. $(3x + 4)(3x - 4)$
2. $(3x + 4)^2$
3. $(3x + 4)(x - 2)$

Fig. 6.1

SPECIAL PRODUCTS INVOLVING CUBES

Four other special products occur less frequently. However, they are sufficiently important that they should be readily recognized. They are shown in Eqs. (6.7) to (6.10).

$$(x + y)^3 = x^3 + 3x^2y + 3xy^2 + y^3 \tag{6.7}$$
$$(x - y)^3 = x^3 - 3x^2y + 3xy^2 - y^3 \tag{6.8}$$
$$(x + y)(x^2 - xy + y^2) = x^3 + y^3 \tag{6.9}$$
$$(x - y)(x^2 + xy + y^2) = x^3 - y^3 \tag{6.10}$$

EXAMPLE 7 Using special products 6.7 and 6.8

(a) $(x + 4)^3 = x^3 + 3(x^2)(4) + 3(x)(4^2) + 4^3$ Eq. (6.7)
$$= x^3 + 12x^2 + 48x + 64$$

(b) $(2x - 5)^3 = (2x)^3 - 3(2x)^2(5) + 3(2x)(5^2) - 5^3$ Eq. (6.8)
$$= 8x^3 - 60x^2 + 150x - 125$$ ∎

EXAMPLE 8 Using special products 6.9 and 6.10

(a) $(x + 3)(x^2 - 3x + 9) = x^3 + 3^3$ Eq. (6.9)
$$= x^3 + 27$$

(b) $(x - 2)(x^2 + 2x + 4) = x^3 - 2^3$ Eq. (6.10)
$$= x^3 - 8$$ ∎

EXERCISES 6.1

In Exercises 1–4, make the given changes in the indicated examples of this section and then solve the resulting problems.

1. In Example 1(b), interchange the + and − signs and then find the product.

2. In Example 3(b), interchange the + and − signs and then find the product.

3. In Example 6, group $y - 2$ instead of $x + y$ and then find the product.

4. In Example 7(a), change + to − in the original expression and then find the product.

In Exercises 5–36, find the indicated products directly by inspection. It should not be necessary to write down intermediate steps [except possibly when using Eq. (6.6)].

5. $40(x - y)$

6. $2x(a - 3)$

7. $2x^2(x - 4)$

8. $3a^2(2a + 7)$

9. $(T + 6)(T - 6)$

10. $(s + 2t)(s - 2t)$

11. $(3v - 2)(3v + 2)$

12. $(ab - c)(ab + c)$

13. $(4x - 5y)(4x + 5y)$

14. $(7s + 2t)(7s - 2t)$

15. $(12 + 5ab)(12 - 5ab)$

16. $(2xy - 11)(2xy + 11)$

17. $(5f + 4)^2$

18. $(i_1 + 3)^2$

19. $(2x + 17)^2$

20. $(9a + 8b)^2$

21. $(L^2 - 1)^2$

22. $(b^2 - 6)^2$

23. $(4a + 7xy)^2$

24. $(3A + 10z)^2$

25. $(0.6s - t)^2$

26. $(0.3p - 0.4q)^2$

27. $(x + 1)(x + 5)$

28. $(y - 8)(y + 5)$

29. $(3 + C^2)(6 + C^2)$

30. $(1 - e^3)(7 - e^3)$

31. $(4x - 5)(5x + 1)$

32. $(2y - 1)(3y - 1)$

33. $(10v - 3)(4v + 15)$

34. $(7s + 12)(8s + 5)$

35. $(30x + 7y)(20x - 9y)$

36. $(24x - y)(15x + 4y)$

Use the special products of this section to determine the products of Exercises 37–62. You may need to write down one or two intermediate steps.

37. $2(x - 2)(x + 2)$

38. $25(V - 5)(V + 5)$

39. $2a(2a - 1)(2a + 1)$

40. $4c(2c - 3)(2c + 3)$

41. $6a(x + 2b)^2$

42. $7r(5r + 2b)^2$

43. $5n^2(2n + 5)^2$

44. $8T(T - 7)^2$

45. $[(2R + 3r)(2R - 3r)]^2$

46. $[(6t - y)(6t + y)]^2$

47. $(x + y + 1)^2$

48. $(x + 2 + 3y)^2$

49. $(3 - x - y)^2$

50. $2(x - y + 1)^2$

51. $(5 - t)^3$

52. $(2s + 3)^3$

53. $(3L + 7R)^3$

54. $(2A - 5B)^3$

55. $(w + h - 1)(w + h + 1)$

56. $(2a - c + 2)(2a - c - 2)$

57. $(x + 2)(x^2 - 2x + 4)$

58. $(s - 3)(s^2 + 3s + 9)$

59. $(4 - 3x)(16 + 12x + 9x^2)$

60. $(2x + 3a)(4x^2 - 6ax + 9a^2)$

61. $(x + y)^2(x - y)^2$

62. $(x - y)(x + y)(x^2 + y^2)$

In Exercises 63–72, use the special products of this section to determine the products. Each comes from the technical area indicated.

63. $P_1(P_0c + G)$ (computers)

64. $akT(t_2 - t_1)$ (heat conduction)

65. $4(p + DA)^2$ (photography)

66. $(2J + 3)(2J - 1)$ (lasers)

67. $\frac{1}{2}\pi(R + r)(R - r)$ (architecture)

68. $K(T - T_0)(T + T_0)(T^2 + T_0^2)$ (radiation)

69. $\frac{L}{6}(x - a)^3$ (mechanics: beams)

70. $(1 - z)^2(1 + z)$ (motion: gyroscope)

71. $L_0[1 + a(T - T_0)]$ (thermal expansion)

72. $(s + 1 + j)(s + 1 - j)$ (electricity)

In Exercises 73–80, solve the given problems.

73. Multiply 49 by 51 by writing $(49)(51) = (50 - 1)(50 + 1)$ and using one of the special products.

74. Multiply 82 by 78 by writing $(82)(78) = (80 + 2)(80 - 2)$ and using one of the special products.

75. The length of a piece of rectangular floor tile is 3 cm more than twice the side x of a second square piece of tile. The width of the rectangular piece is 3 cm less than twice the side of the square piece. Find the area of the rectangular piece in terms of x (in expanded form).

76. The radius of a circular oil spill is r. It then increases in radius by 40 m before being contained. Find the area of the oil spill at the time it is contained in terms of r (in expanded form). See Fig. 6.2.

Fig. 6.2

Fig. 6.3

77. Referring to Fig. 6.3, find (a) an expression for the area of the large square (the entire figure) as the power of a binomial; (b) the sum of the areas of the four figures within the large square as a trinomial. Noting that the results of (a) and (b) should be equal, what special product is shown geometrically?

78. If $(x + k)(x - 1) = x^2 + k(x + 2) - (x - 3)$, find the value of k.

79. Find the product $(2x - y - 2)(2x + y + 2)$ by first grouping terms as in Example 6. (*Hint:* This requires grouping the last two terms of each factor.)

(W) 80. Verify Eqs. (6.8) and (6.9) by multiplication. Then explain why we may refer to Eqs. (6.1) to (6.10) as *identities*.

Answers to Practice Exercises

1. $9x^2 - 16$ 2. $9x^2 + 24x + 16$ 3. $3x^2 - 2x - 8$

6.2 Factoring: Common Factor and Difference of Squares

Factoring • Prime Factors • Common Monomial Factors • Factoring Difference of Two Squares • Complete Factoring • Factoring by Grouping

At times, we want to determine which expressions can be multiplied together to equal a given algebraic expression. We know from Section 1.7 that when an algebraic expression is the product of two or more quantities, each of these quantities is a ***factor*** of the expression. Therefore, *determining these factors, which is essentially reversing the process of finding a product, is called* **factoring.**

In our work on factoring, we will consider only the factoring of polynomials (see Section 1.7) that have integers as coefficients for all terms. Also, all factors will have integral coefficients. *A polynomial or a factor is called* **prime** *if it contains no factors other than* $+1$ *or* -1 *and plus or minus itself.* Also, we say that an *expression is* **factored completely** *if it is expressed as a product of its prime factors.*

EXAMPLE 1   Factoring completely

When we factor the expression $12x + 6x^2$ as

$$12x + 6x^2 = 2(6x + 3x^2)$$

we see that it has not been factored completely. The factor $6x + 3x^2$ is not prime, because it may be factored as

$$6x + 3x^2 = 3x(2 + x)$$

Therefore, the expression $12x + 6x^2$ is factored completely as

$$12x + 6x^2 = 6x(2 + x)$$

The factors x and $(2 + x)$ are prime. We can factor the numerical coefficient, 6, as $2(3)$, but it is standard not to write numerical coefficients in factored form.   ■

To factor expressions easily, we must know how to do algebraic multiplication and really know the special products of the previous section.

NOTE ▶   *The ability to factor algebraic expressions depends heavily on the proper recognition of the special products.*

The special products also give us a way of checking answers and deciding whether a given factor is prime.

COMMON MONOMIAL FACTORS

Often an expression contains a monomial that is common to each term of the expression. Therefore, *the first step in factoring any expression should be to factor out any* **common monomial factor** *that may exist.* To do this, we note the common factor by inspection and then use the reverse of the distributive law, Eq. (6.1), to show the factored form. The following examples illustrate factoring a common monomial factor out of an expression.

■ For reference, Eq. (6.1) is $a(x + y) = ax + ay$.

EXAMPLE 2   Common monomial factor

In factoring $6x - 2y$, we note each term contains a factor of 2:

$$6x - 2y = 2(3x) - 2y = 2(3x - y)$$

Here, 2 is the common monomial factor, and $2(3x - y)$ is the required factored form of $6x - 2y$. Once the common factor has been identified, it is not actually necessary to write a term like $6x$ as $2(3x)$. The result can be written directly.

NOTE ▶   *We check the result by multiplication.* In this case,

$$2(3x - y) = 6x - 2y$$

Since the result of the multiplication gives the original expression, the factored form is correct.   ■

In Example 2, we determined the common factor of 2 by inspection. This is normally the way in which a common factor is found. Once the common factor has been found, the other factor can be determined by dividing the original expression by the common factor.

The next example illustrates the case where the common factor is the same as one of the terms, and special care must be taken to complete the factoring correctly.

EXAMPLE 3 Common factor same as term

Factor: $4ax^2 + 2ax$.

The numerical factor 2 and the literal factors a and x are common to each term. Therefore, the common monomial factor of $4ax^2 + 2ax$ is $2ax$. This means that

$$4ax^2 + 2ax = 2ax(2x) + 2ax(1) = 2ax(2x + 1)$$

Note the presence of the 1 in the factored form. When we divide $4ax^2 + 2ax$ by $2ax$, we get

$$\frac{4ax^2 + 2ax}{2ax} = \frac{4ax^2}{2ax} + \frac{2ax}{2ax}$$

$$= 2x + 1 \ \hookleftarrow$$

Practice Exercises

Factor: **1.** $3cx^3 - 9cx$ **2.** $9cx^3 - 3cx$

NOTE ▶ noting very carefully that $2ax$ divided by $2ax$ is 1 and *does not simply cancel out leaving nothing*. In cases like this, where the common factor is the same as one of the terms, it is a common error to omit the 1. However,

CAUTION ▶ *we must include the 1 in the factor $2x + 1$.*

Without the 1, when the factored form is multiplied out, we would not obtain the original expression.

Usually, the division shown in this example is done by inspection. However, we show it here to emphasize the actual operation that is being performed when we factor out a common factor. ■

EXAMPLE 4 Common factor by inspection

Factor: $6a^5x^2 - 9a^3x^3 + 3a^3x^2$.

After inspecting each term, we determine that each contains a factor of 3, a^3, and x^2. Thus, the common monomial factor is $3a^3x^2$. This means that

$$6a^5x^2 - 9a^3x^3 + 3a^3x^2 = 3a^3x^2(2a^2 - 3x + 1)$$ ■

In these examples, note that factoring an expression does not actually change the expression, although it does change the *form* of the expression. In equating the expression to its factored form, we write an *identity*.

It is often necessary to use factoring when solving an equation. This is illustrated in the following example.

EXAMPLE 5 Using factoring in solving an equation

An equation used in the analysis of FM reception is $R_F = \alpha(2R_A + R_F)$. Solve for R_F.

The steps in the solution are as follows:

$$R_F = \alpha(2R_A + R_F) \qquad \text{original equation}$$
$$R_F = 2\alpha R_A + \alpha R_F \qquad \text{use distributive law}$$
$$R_F - \alpha R_F = 2\alpha R_A \qquad \text{subtract } \alpha R_F \text{ from both sides}$$
$$R_F(1 - \alpha) = 2\alpha R_A \qquad \text{factor out } R_F \text{ on left}$$
$$R_F = \frac{2\alpha R_A}{1 - \alpha} \qquad \text{divide both sides by } 1 - \alpha$$

■ FM radio was developed in the early 1930s.

We see that we collected both terms containing R_F on the left so that we could factor and thereby solve for R_F. ■

FACTORING THE DIFFERENCE OF TWO SQUARES

In Eq. (6.2), we see that the product of the sum and the difference of two numbers results in the difference between the squares of two numbers. Therefore, *factoring the difference of two squares gives factors that are the sum and the difference of the numbers.*

■ For reference, Eq. (6.2) is
$(x + y)(x - y) = x^2 - y^2$.

EXAMPLE 6 Factoring difference of two squares

In factoring $x^2 - 16$, note that x^2 is the square of x and 16 is the square of 4. Therefore,

■ Usually, in factoring an expression of this type, where it is very clear what numbers are squared, we do not actually write out the middle step as shown. However, if in doubt, write it out.

$$\underset{\text{difference}}{x^2} - 16 = x^2 - \overset{\text{squares}}{4^2} = (x + 4)(x - 4)$$

difference   sum   difference   ■

EXAMPLE 7 Factoring difference of two squares

(a) Since $4x^2$ is the square of $2x$ and 9 is the square of 3, we may factor $4x^2 - 9$ as

$$4x^2 - 9 = (2x)^2 - 3^2 = (2x + 3)(2x - 3)$$

(b) In the same way,

$$(y - 3)^2 - 16x^4 = [(y - 3) + 4x^2][(y - 3) - 4x^2]$$
$$= (y - 3 + 4x^2)(y - 3 - 4x^2)$$

where we note that $16x^4 = (4x^2)^2$.   ■

COMPLETE FACTORING

NOTE ▶

CAUTION ▶

As noted before, *a common monomial factor should be factored out first.* However, we must be careful to *see if the other factor can itself be factored.* It is possible, for example, that the other factor is the difference of squares. This means that **complete factoring** *often requires more than one step.* Be sure to include only prime factors in the final result.

EXAMPLE 8 Complete factoring

(a) In factoring $20x^2 - 45$, note a common factor of 5 in each term. Therefore, $20x^2 - 45 = 5(4x^2 - 9)$. However, the factor $4x^2 - 9$ itself is the difference of squares. Therefore, $20x^2 - 45$ is completely factored as

Practice Exercises
Factor: **3.** $9c^2 - 64$ **4.** $18c^2 - 128$

$$20x^2 - 45 = \overset{\text{common factor}}{5(4x^2 - 9)} = 5(2x + 3)(2x - 3)$$

difference of squares

(b) In factoring $x^4 - y^4$, note that we have the difference of two squares. Therefore, $x^4 - y^4 = (x^2 + y^2)(x^2 - y^2)$. However, the factor $x^2 - y^2$ is also the difference of squares. This means that

$$x^4 - y^4 = (x^2 + y^2)(x^2 - y^2) = (x^2 + y^2)(x + y)(x - y)$$

CAUTION ▶   *The factor $x^2 + y^2$ is prime.* It is **not** equal to $(x + y)^2$. (See Example 2 of Section 6.1.)   ■

FACTORING BY GROUPING

The terms in a polynomial can sometimes be grouped such that the polynomial can then be factored by the methods of this section. The following example illustrates this method of *factoring by grouping.*

EXAMPLE 9 Factoring by grouping

Factor: $2x - 2y + ax - ay$.

We see that there is no common factor to all four terms, but that each of the first two terms contains a factor of 2, and each of the third and fourth terms contains a factor of a. Grouping terms this way and then factoring each group, we have

$$2x - 2y + ax - ay = (2x - 2y) + (ax - ay)$$

NOTE ▸
$$= 2(x - y) + a(x - y) \quad \text{now note the common factor of } (x - y)$$
$$= (x - y)(2 + a) \qquad ■$$

The general method of factoring by grouping can be used with several types of groupings. We will discuss another type in the following section.

EXERCISES 6.2

In Exercises 1–4, make the given changes in the indicated examples of this section and then solve the indicated problems.

1. In Example 3, change the $+$ sign to $-$ and then factor.

2. In Example 3, set the given expression equal to B and then solve for a.

3. In Example 8(a), change the coefficient of the first term from 20 to 5 and then factor.

4. In Example 9, change both $-$ signs to $+$ and then factor.

In Exercises 5–44, factor the given expressions completely.

5. $6x + 6y$ **6.** $3a - 3b$ **7.** $5a - 5$

8. $2x^2 + 2$ **9.** $3x^2 - 9x$ **10.** $4s^2 + 20s$

11. $7b^2h - 28b$ **12.** $5a^2 - 20ax$

13. $288n^2 + 24n$ **14.** $90p^3 - 15p^2$

15. $2x + 4y - 8z$ **16.** $23a - 46b + 69c$

17. $3ab^2 - 6ab + 12ab^3$ **18.** $4pq - 14q^2 - 16pq^2$

19. $12pq^2 - 8pq - 28pq^3$ **20.** $27a^2b - 24ab - 9a$

21. $2a^2 - 2b^2 + 4c^2 - 6d^2$ **22.** $5a + 10ax - 5ay - 20az$

23. $x^2 - 4$ **24.** $r^2 - 25$ **25.** $100 - 9A^2$

26. $49 - Z^4$ **27.** $36a^4 + 1$ **28.** $324z^2 - 4$

29. $162s^2 - 50t^2$ **30.** $36s^2 - 121t^2$

31. $144n^2 - 169p^4$ **32.** $36a^2b^2 + 169c^2$

33. $(x + y)^2 - 9$ **34.** $(a - b)^2 - 1$

35. $2x^2 - 8$ **36.** $5a^2 - 125$

37. $300x^2 - 2700z^2$ **38.** $28x^2 - 700y^2$

39. $2(I - 3)^2 - 8$ **40.** $a(x + 2)^2 - ay^2$

41. $x^4 - 16$ **42.** $y^4 - 81$

43. $x^8 - 1$ **44.** $2x^4 - 8y^4$

In Exercises 45–50, solve for the indicated letter.

45. $2a - b = ab + 3$, for a **46.** $n(x + 1) = 5 - x$, for x

47. $3 - 2s = 2(3 - st)$, for s

48. $k(2 - y) = y(2k - 1)$, for y

49. $(x + 2k)(x - 2) = x^2 + 3x - 4k$, for k

50. $(2x - 3k)(x + 1) = 2x^2 - x - 3$, for k

In Exercises 51–58, factor the given expressions by grouping as illustrated in Example 9.

51. $3x - 3y + bx - by$ **52.** $am + an + cn + cm$

53. $a^2 + ax - ab - bx$ **54.** $2y - y^2 - 6y^4 + 12y^3$

55. $x^3 + 3x^2 - 4x - 12$ **56.** $S^3 - 5S^2 - S + 5$

57. $x^2 - y^2 + x - y$ **58.** $4p^2 - q^2 + 2p + q$

In Exercises 59 and 60, evaluate the given expressions by using factoring. The results may be checked with a calculator.

59. $\dfrac{8^9 - 8^8}{7}$ **60.** $\dfrac{5^9 - 5^7}{7^2 - 5^2}$

Ⓦ *In Exercises 61 and 62, give the required explanations.*

61. Factor $n^2 + n$, and then explain why it represents a positive even integer if n is a positive integer.

62. Factor $n^3 - n$, and then explain why it represents a multiple of 6 if n is an integer greater than 1.

In Exercises 63–70, factor the expressions completely. In Exercises 69 and 70, it is necessary to set up the proper expression. Each expression comes from the technical area indicated.

63. $2\pi rh + 2\pi r^2$ (container surface area)

64. $4d^2D^2 - 4d^3D - d^4$ (machine design)

65. $Rv + Rv^2 + Rv^3$ (business)

66. $PbL^2 - Pb^3$ (architecture)

67. $rR^2 - r^3$ (pipeline flow)

68. $p_1R^2 - p_1r^2 - p_2R^2 + p_2r^2$ (fluid flow)

69. As large as possible square is cut from a circular metal plate of radius r. Express in factored form the area of the metal pieces that are left.

70. A pipe of outside diameter d is inserted into a pipe of inside radius r. Express in factored form the cross-sectional area within the larger pipe that is outside the smaller pipe.

In Exercises 71–76, solve for the indicated letter. Each equation comes from the technical area indicated.

71. $i_1 R_1 = (i_2 - i_1)R_2$, for i_1 (electricity: ammeter)

72. $nV + n_1 v = n_1 V$, for n_1 (acoustics)

73. $3BY + 5Y = 9BS$, for B (physics: elasticity)

74. $Sq + Sp = Spq + p$, for q (computer design)

75. $ER = AtT_0 - AtT_1$, for t (energy conservation)

76. $R = kT_2^4 - kT_1^4$, for k (factor resulting denominator) (radiation)

Answers to Practice Exercises

1. $3cx(x^2 - 3)$ **2.** $3cx(3x^2 - 1)$

3. $(3c - 8)(3c + 8)$ **4.** $2(3c - 8)(3c + 8)$

6.3 Factoring Trinomials

When Coefficient of Square Term is 1 •
Importance of the Middle Term •
Factoring General Trinomials •
Factoring by Grouping

In the previous section, we considered factoring based on the special products of Eqs. (6.1) and (6.2). The trinomials formed from Eqs. (6.3) to (6.6) are important expressions to be factored, and this section is devoted to them.

When factoring is based on Eq. (6.5), we start with the expression on the right and then find the factors on the left. By writing Eq. (6.5) with sides reversed, we have

■ For reference, Eq. (6.5) is
$(x + a)(x + b) = x^2 + (a + b)x + ab$.

$$\text{coefficient} = 1 \longrightarrow x^2 + (a + b)x + ab = (x + a)(x + b)$$

■ Factoring a trinomial for which the coefficient of x^2 is 1.

We are to find integers a and b, and they are found by noting that

1. *the coefficient of x^2 is 1,*

2. *the final constant is the product of the constants a and b in the factors, and*

CAUTION ▶   3. *the coefficient of x is the sum of a and b.*

As in Section 6.2, we consider only factors in which all terms have integral coefficients.

> **EXAMPLE 1 Factoring trinomial $x^2 + (a + b)x + ab$**
>
> In factoring $x^2 + 3x + 2$, we set it up as
>
> $$x^2 + 3x + \ 2\ = (x \boxed{})\ (x \boxed{})$$
>
> The constant 2 tells us that the product of the required integers is 2. Thus, the only possibilities are 2 and 1 (or 1 and 2). The $+$ sign before the 2 indicates that the sign before the 1 and 2 in the factors must be the same. The $+$ sign before the 3, the sum of the integers, tells us that both signs are positive. Therefore,
>
> $$x^2 + 3x + 2 = (x + 2)(x + 1)$$
>
> In factoring $x^2 - 3x + 2$, the analysis is the same until we note that the middle term is negative. This tells us that both signs are negative in this case. Therefore,
>
> $$x^2 - 3x + 2 = (x - 2)(x - 1)$$
>
> For a trinomial with first term x^2 and constant $+2$ to be factorable, the middle term must be $+3x$ or $-3x$. No other middle terms are possible. This means, for example, the expressions $x^2 + 4x + 2$ and $x^2 - x + 2$ cannot be factored.   ∎

EXAMPLE 2 Factoring trinomials $x^2 + (a + b)x + ab$

(a) In order to factor $x^2 + 7x - 8$, *we must find two integers whose product is −8 and whose sum is +7*. The possible factors of −8 are

$$-8 \text{ and } +1 \qquad +8 \text{ and } -1 \qquad -4 \text{ and } +2 \qquad +4 \text{ and } -2$$

Inspecting these, we see that only +8 and −1 have the sum of +7. Therefore,

$$x^2 + 7x - 8 = (x + 8)(x - 1)$$

In choosing the correct values for the integers, it is usually fairly easy to find a pair for which the product is the final term. However, choosing the pair of integers that correctly fits the middle term is the step that often is not done properly. *Special attention must be given to choosing the integers so that the expansion of the resulting factors has the correct **middle term** of the original expression.*

■ Always multiply the factors together to check to see that you get the **correct middle term.**

(b) In the same way, we have

$$x^2 - x - 12 = (x - 4)(x + 3)$$

since −4 and +3 is the only pair of integers whose product is −12 and whose sum is −1.

(c) Also,

$$x^2 - 5xy + 6y^2 = (x - 3y)(x - 2y)$$

since −3 and −2 is the only pair of integers whose product is +6 and whose sum is −5. Here, we find second terms of each factor with a product of $6y^2$ and sum of $-5xy$, which means that each second term must have a factor of y, as we have shown above.   ■

■ For reference, Eqs. (6.3) and (6.4) are $(x + y)^2 = x^2 + 2xy + y^2$ and $(x - y)^2 = x^2 - 2xy + y^2$.

In factoring a trinomial in which the first and third terms are perfect squares, we may find that the expression fits the form of Eq. (6.3) or Eq. (6.4), as well as the form of Eq. (6.5). The following example illustrates this case.

EXAMPLE 3 Factoring perfect square trinomials

(a) To factor $x^2 + 10x + 25$, we must find two integers whose product is +25 and whose sum is +10. Since $5^2 = 25$, we note that this expression may fit the form of Eq. (6.3). This can be the case only if the first and third terms are perfect squares. Since the sum of +5 and +5 is +10, we have

$$x^2 + 10x + 25 = (x + 5)(x + 5)$$

or

$$x^2 + 10x + 25 = (x + 5)^2$$

Practice Exercises

Factor: **1.** $x^2 - x - 2$ **2.** $x^2 - 4x + 4$

(b) To factor $A^4 - 20A^2 + 100$, we note that $A^4 = (A^2)^2$ and $100 = 10^2$. This means that this expression may fit the form of Eq. (6.4). Since the sum of −10 and −10 is −20, we have

$$A^4 - 20A^2 + 100 = (A^2 - 10)(A^2 - 10)$$
$$= (A^2 - 10)^2$$

(c) Just because the first and third terms are perfect squares, we must realize that the expression may be factored but does *not* fit the form of either Eq. (6.3) or (6.4). One example, using x^2 and 100 for the first and third terms, is

$$x^2 + 29x + 100 = (x + 25)(x + 4)$$ ■

FACTORING GENERAL TRINOMIALS

■ For reference, Eq. (6.6) is
$(ax + b)(cx + d) =$
$acx^2 + (ad + bc)x + bd$.

Factoring based on the special product of Eq. (6.6) often requires some trial and error. However, the amount of trial and error can be kept to a minimum by carefully noting the coefficients of x^2 and the constant. Rewriting Eq. (6.6) with sides reversed, we have

$$acx^2 + (ad + bc)x + bd = (ax + b)(cx + d)$$

product ← coefficients

product ← integers

$$acx^2 + (ad + bc)x + bd = (ax + b)(cx + d)$$

outer product

inner product

■ Factoring a trinomial for which the coefficient of x^2 is any positive integer.

This diagram shows us that

1. *the coefficient of x^2 is the product of the coefficients a and c in the factors,*

2. *the final constant is the product of the constants b and d in the factors, and*

CAUTION ▶ **3.** *the coefficient of x is the sum of the inner and outer products.*

In finding the factors, we must try possible combinations of a, b, c, and d that give

CAUTION ▶ *the proper inner and outer products for the middle term*

of the given expression.

EXAMPLE 4 Factoring general trinomial

To factor $2x^2 + 11x + 5$, we take the factors of 2 to be $+2$ and $+1$ (we use only positive coefficients a and c when the coefficient of x^2 is positive). We set up the factoring as

$$2x^2 + 11x + 5 = (2x\boxed{})(x\boxed{})$$

NOTE ▶

Since the product of the integers to be found is $+5$, only integers of the same sign need to be considered. Also since *the sum of the outer and inner products is $+11$,* the integers are positive. The factors of $+5$ are $+1$ and $+5$, and -1 and -5, which means that $+1$ and $+5$ is the only possible pair. Now, trying the factors

$$(2x + 5)(x + 1)$$

$+5x$

$+2x$

$+2x + 5x = +7x$

we see that $7x$ is not the correct middle term. Therefore, we now try

$$(2x + 1)(x + 5)$$

$+x$

$+10x$

$+x + 10x = +11x$

and we have the correct sum of $+11x$. Therefore,

$$2x^2 + 11x + 5 = (2x + 1)(x + 5)$$

For a trinomial with a first term $2x^2$ and a constant $+5$ to be factorable, we can now see that the middle term must be either $+11x$ or $+7x$. This means that $2x^2 + 7x + 5 = (2x + 5)(x + 1)$, but a trinomial such as $2x^2 + 8x + 5$ is not factorable. ■

EXAMPLE 5 Factoring general trinomial

NOTE ▶

In factoring $4x^2 + 4x - 3$, the coefficient 4 in $4x^2$ shows that the possible coefficients of x in the factors are 4 and 1, or 2 and 2. The 3 shows that the only possible constants in the factors are 1 and 3, and *the minus sign with the 3 tells us that these integers have different signs.* This gives us the following possible combinations of factors, along with the resulting sum of the outer and inner products:

$$(4x + 3)(x - 1): \quad -4x + 3x = -x$$
$$(4x + 1)(x - 3): \quad -12x + x = -11x$$
$$(4x - 3)(x + 1): \quad 4x - 3x = +x$$
$$(4x - 1)(x + 3): \quad 12x - x = +11x$$
$$(2x + 3)(2x - 1): \quad -2x + 6x = +4x$$
$$(2x - 3)(2x + 1): \quad 2x - 6x = -4x$$

We see that the factors that have the correct middle term of $+4x$ are $(2x + 3)(2x - 1)$. This means that

$$4x^2 + 4x - 3 = (2x + 3)(2x - 1)$$

$$+6x$$
$$-2x$$
$$-2x + 6x = +4x$$

Expressing the result with the factors reversed is an equally correct answer.

NOTE ▶

Another hint given by the coefficients of the original expression is that *the plus sign with the 4x tells us that the larger of the outer and inner products must be positive.* ■

EXAMPLE 6 Factoring general trinomial

$$6s^2 + 19st - 20t^2 = (6s - 5t)(s + 4t)$$

$$+24st - 5st = +19st$$

CAUTION ▶

There are numerous possibilities for the combinations of 6 and 20. However, we must remember to **check carefully that the middle term** *of the expression is the **proper result of the factors*** we have chosen. ■

EXAMPLE 7 Check carefully the middle term

(a) In factoring $9x^2 - 6x + 1$, note that $9x^2$ is the square of $3x$ and 1 is the square of 1. Therefore, we recognize that this expression might fit the perfect square form of Eq. (6.4). This leads us to factor it tentatively as

$$9x^2 - 6x + 1 = (3x - 1)^2 \quad \text{check middle term: } 2(3x)(-1) = -6x$$

However, before we can be certain that this is correct, we must check to see if the middle term of the expansion of $(3x - 1)^2$ is $-6x$, which is what it must be to fit the form of Eq. (6.4). When we expand $(3x - 1)^2$, we find that the middle term is $-6x$ and therefore that the factorization is correct.

(b) In the same way, we have

$$36x^2 + 84xy + 49y^2 = (6x + 7y)^2 \quad \text{check middle term: } 2(6x)(7y) = 84xy$$

NOTE ▶

since the middle term of the expansion of $(6x + 7y)^2$ is $84xy$. *We must be careful to include the factors of y in the second terms of the factors.* ■

Practice Exercise

Factor: **3.** $4x^2 + x - 5$

Common Monomial Factors

As noted before, we must be careful to factor an expression completely. *We first look for common monomial factors* and *then check each resulting factor to see if it can be factored* when we complete each step.

EXAMPLE 8   First, look for common monomial factor

When factoring $2x^2 + 6x - 8$, first note the common monomial factor of 2. This leads to

$$2x^2 + 6x - 8 = 2(x^2 + 3x - 4)$$

Now, notice that $x^2 + 3x - 4$ is also factorable. Therefore,

$$2x^2 + 6x - 8 = 2(x + 4)(x - 1)$$

Having noted the common factor of 2 prevents our having to check factors of 2 and 8. If we had not noted the common factor, we might have arrived at

$$(2x + 8)(x - 1) \quad \text{or} \quad (2x - 2)(x + 4)$$

CAUTION ▶
(possibly after a number of trials). Each is correct as far as it goes, but also each of these is ***not complete.*** Since $2x + 8 = 2(x + 4)$, or $2x - 2 = 2(x - 1)$, we can arrive at the proper result shown above. In this way, we would have

$$2x^2 + 6x - 8 = (2x + 8)(x - 1)$$
$$= 2(x + 4)(x - 1)$$

Practice Exercise

Factor: **4.** $6x^2 + 9x - 6$

or
$$2x^2 + 6x - 8 = (2x - 2)(x + 4)$$
$$= 2(x - 1)(x + 4)$$

Although these factorizations are correct, *it is better to factor out the common factor first.* By doing so, the number of possible factoring combinations is greatly reduced. ∎

EXAMPLE 9   Factoring a trinomial – application

A study of the path of a certain rocket leads to the expression $16t^2 + 240t - 1600$, where t is the time of flight. Factor this expression.

An inspection shows that there is a common factor of 16. (This might be found by noting successive factors of 2 or 4.) Factoring out 16 leads to

$$16t^2 + 240t - 1600 = 16(t^2 + 15t - 100)$$
$$= 16(t + 20)(t - 5)$$

■ Liquid-fuel rockets were designed in the United States in the 1920s but were developed by German engineers. They were first used in the 1940s during World War II.

Here, factors of 100 need to be checked for sums equal to 15. This might take some time, but it is much simpler than looking for factors of 16 and 1600 with sums of 240. ∎

FACTORING BY GROUPING

EXAMPLE 10   Factoring by grouping

Factor the trinomial $6x^2 + 7x - 20$ by grouping.

To factor the trinomial $ax^2 + bx + c$ by grouping, we use the following procedure.

1. Find two numbers whose product is ac and whose sum is b.

2. Write the trinomial with two x-terms having these numbers as coefficients.

3. Complete the factorization by grouping.

For $6x^2 + 7x - 20$, this means we want two numbers with a product of -120 and a sum of 7. Trying products of numbers *with different signs* and a sum of 7, we find the numbers that work are -8 and 15. We then complete steps 2 and 3 of the above procedure as follows:

$$6x^2 + 7x - 20 = 6x^2 - 8x + 15x - 20 \qquad \text{\small $7x = -8x + 15x$}$$
$$= (6x^2 - 8x) + (15x - 20) \qquad \text{\small group first two terms and last two terms}$$
$$= 2x(3x - 4) + 5(3x - 4) \qquad \text{\small find common factor of each group and}$$
$$= (3x - 4)(2x + 5) \qquad \text{\small note common factor of $3x - 4$}$$

Multiplication verifies that these are the correct factors. ∎

EXAMPLE 11 Factoring by grouping

Factor: $x^2 - 4xy + 4y^2 - 9$.

The first three terms represent $(x - 2y)^2$. Thus, grouping these terms, we have

$$x^2 - 4xy + 4y^2 - 9 = (x^2 - 4xy + 4y^2) - 9 \qquad \text{group terms}$$
$$= (x - 2y)^2 - 9 \qquad \text{factor grouping (note difference of squares)}$$
$$= [(x - 2y) + 3][(x - 2y) - 3] \qquad \text{factor difference of squares}$$
$$= (x - 2y + 3)(x - 2y - 3)$$

Not all groupings work. Noting the factorable combination $4y^2 - 9$, and then grouping the first two terms and the last two terms, would not have led to the factorization. ∎

EXERCISES 6.3

In Exercises 1–6, make the given changes in the indicated examples of this section and then factor.

1. In Example 1, change the 3 to 4 and the 2 to 3.
2. In Example 2(a), change the + before $7x$ to −.
3. In Example 4, change the + before $11x$ to −.
4. In Example 6, change the coefficient 19 to 7.
5. In Example 8, change the 8 to 36.
6. In Example 10, change the + before $7x$ to −.

In Exercises 7–56, factor the given expressions completely.

7. $x^2 + 5x + 4$
8. $x^2 - 5x - 6$
9. $s^2 - s - 42$
10. $a^2 + 14a - 32$
11. $t^2 + 5t - 24$
12. $r^3 - 11r^2 + 18r$
13. $x^2 + 2x + 1$
14. $D^2 + 8D + 16$
15. $L^2 - 4LK + 4K^2$
16. $b^2 - 12bc + 36c^2$
17. $3x^2 - 5x - 2$
18. $2n^2 - 13n - 7$
19. $12y^2 - 32y - 12$
20. $25x^2 + 45x - 10$
21. $2s^2 + 13s + 11$
22. $7y^2 - 12y + 5$
23. $3f^4 - 16f^2 + 5$
24. $5R^4 - 3R^2 - 2$
25. $2t^2 + 7t - 15$
26. $3n^2 - 20n + 20$
27. $3t^2 - 7tu + 4u^2$
28. $3x^2 + xy - 14y^2$
29. $4x^2 - 3x - 7$
30. $2z^2 + 13z - 5$
31. $9x^2 + 7xy - 2y^2$
32. $4r^2 + 11rs - 3s^2$
33. $4m^2 + 20m + 25$
34. $16q^2 + 24q + 9$
35. $8x^2 - 24x + 18$
36. $3a^2c^2 - 6ac + 3$
37. $9t^2 - 15t + 4$
38. $6t^4 + t^2 - 12$
39. $8b^6 + 31b^3 - 4$
40. $12n^4 + 8n^2 - 15$
41. $4p^2 - 25pq + 6q^2$
42. $12x^2 + 4xy - 5y^2$
43. $12x^2 + 47xy - 4y^2$
44. $8r^2 - 14rs - 9s^2$
45. $2x^2 - 14x + 12$
46. $6y^2 - 33y - 18$
47. $4x^5 + 14x^3 - 8x$
48. $12B^2 + 22BH - 4H^2$
49. $ax^3 + 4a^2x^2 - 12a^3x$
50. $6x^4 - 13x^3 + 5x^2$
51. $a^2 + 2ab + b^2 - 4$
52. $x^2 - 6xy + 9y^2 - 4z^2$
53. $25a^2 - 25x^2 - 10xy - y^2$
54. $r^2 - s^2 + 2st - t^2$
55. $4x^{2n} + 13x^n - 12$
56. $12B^{2n} + 19B^nH - 10H^2$

In Exercises 57–68, factor the given expressions completely. Each is from the technical area indicated.

57. $16t^2 - 80t + 64$ (projectile motion)
58. $9x^2 - 33Lx + 30L^2$ (civil engineering)
59. $4s^2 + 16s + 12$ (electricity)
60. $3e^2 + 18e - 1560$ (fuel efficiency)
61. $200n^2 - 2100n - 3600$ (biology)
62. $bT^2 - 40bT + 400b$ (thermodynamics)
63. $V^2 - 2nBV + n^2B^2$ (chemistry)
64. $a^4 + 8a^2\pi^2f^2 + 16\pi^4f^4$ (periodic motion: energy)
65. $wx^4 - 5wLx^3 + 6wL^2x^2$ (beam design)
66. $1 - 2r^2 + r^4$ (lasers)
67. $3Adu^2 - 4Aduv + Adv^2$ (water power)
68. $k^2A^2 + 2k\lambda A + \lambda^2 - \alpha^2$ (robotics)

In Exercises 69–74, solve the given problems.

69. Find the integral value of k that makes $4x^2 + 4x - k$ a perfect square trinomial, and express the result in factored form.
70. Find the integral value of k that makes $49x^2 - 70x + k$ a perfect square trinomial, and express the result in factored form.
71. Often, the sum of squares cannot be factored, but $x^4 + 4$ can be factored. Add and subtract $4x^2$ and then use factoring by grouping.
72. Using the method of Exercise 71, factor $x^4 + x^2 + 1$. (First add and subtract x^2.)
73. If sum of squares $36x^2 + 9$ can be factored, factor it.
(W) 74. Explain why most students would find $24x^2 - 23x - 12$ more difficult to factor than $23x^2 - 18x - 5$.

Answers to Practice Exercises

1. $(x - 2)(x + 1)$ 2. $(x - 2)^2$
3. $(4x + 5)(x - 1)$ 4. $3(2x - 1)(x + 2)$

6.4 The Sum and Difference of Cubes

**Factoring the Sum of Cubes •
Factoring the Difference of Cubes •
Summary of Methods of Factoring**

We have seen that the difference of squares can be factored, but that the sum of squares often cannot be factored. We now turn our attention to the sum and difference of cubes, both of which can be factored. Writing Eqs. (6.9) and (6.10) with sides reversed,

■ Note carefully the positions of the + and − signs.

$$x^3 + y^3 = (x + y)(x^2 - xy + y^2) \tag{6.9}$$
$$x^3 - y^3 = (x - y)(x^2 + xy + y^2) \tag{6.10}$$

In these equations, the second factors are prime.

EXAMPLE 1 Factoring sum and difference of cubes

Table of Cubes

$1^3 =$	1
$2^3 =$	8
$3^3 =$	27
$4^3 =$	64
$5^3 =$	125
$6^3 =$	216

(a) $x^3 + 8 = x^3 + 2^3$

$= (x + 2)[(x)^2 - 2x + 2^2]$
$= (x + 2)(x^2 - 2x + 4)$

(b) $x^3 - 1 = x^3 - 1^3$

$= (x - 1)[(x)^2 + (1)(x) + 1^2]$
$= (x - 1)(x^2 + x + 1)$

(c) $8 - 27x^3 = 2^3 - (3x)^3$ $8 = 2^3$ and $27x^3 = (3x)^3$

$= (2 - 3x)[2^2 + 2(3x) + (3x)^2]$
$= (2 - 3x)(4 + 6x + 9x^2)$ ■

EXAMPLE 2 First, factor out the common factor

In factoring $ax^5 - ax^2$, we first note that each term has a common factor of ax^2. This is factored out to get $ax^2(x^3 - 1)$. However, the expression is not completely factored since $1 = 1^3$, which means that $x^3 - 1$ is the difference of cubes. We complete the factoring by the use of Eq. (6.10). Therefore,

$$ax^5 - ax^2 = ax^2(x^3 - 1)$$
$$= ax^2(x - 1)(x^2 + x + 1)$$ ■

Practice Exercises

Factor: **1.** $x^3 + 216$ **2.** $x^3 - 216$

EXAMPLE 3 Factoring difference of cubes – application

The volume of material used to make a steel bearing with a hollow core is given by $\frac{4}{3}\pi R^3 - \frac{4}{3}\pi r^3$. Factor this expression.

$$\frac{4}{3}\pi R^3 - \frac{4}{3}\pi r^3 = \frac{4}{3}\pi(R^3 - r^3)$$ common factor of $\frac{4}{3}\pi$

$$= \frac{4}{3}\pi(R - r)(R^2 + Rr + r^2)$$ using Eq. (6.10) ■

SUMMARY OF METHODS OF FACTORING

1. *Common monomial factor*   **Always check for this first.**
2. *Difference of squares*
3. *Factorable trinomial*
4. *Sum or difference of cubes*
5. *Factorable by grouping*

CAUTION ▶   *Be sure the expression is factored **completely**.*

EXERCISES 6.4

In Exercises 1 and 2, make the given changes in the indicated examples of this section and then factor.

1. In Example 1(a), change the $+$ before the 8 to $-$.

2. In Example 2, change $-ax^2$ to $+a^4x^2$.

In Exercises 3–26, factor the given expressions completely.

3. $x^3 + 1$ **4.** $R^3 + 27$ **5.** $8 - t^3$ **6.** $8r^3 - 1$

7. $27x^3 - 8a^6$ **8.** $64x^4 + 125x$ **9.** $4x^3 + 32$

10. $3y^3 - 81$ **11.** $6A^6 - 6A^3$ **12.** $8s^9 - 64$

13. $54x^3y - 6x^3y^4$ **14.** $12a^3 + 96a^3b^3$ **15.** $x^6y^3 + x^3y^6$

16. $16r^3 - 432$ **17.** $3a^6 - 3a^2$ **18.** $x^6 - 81y^2$

19. $0.001R^3 - 0.064r^3$ **20.** $0.027x^3 + 0.125$

21. $27L^6 + 216L^3$ **22.** $a^3s^5 - 8000a^3s^2$

23. $(a + b)^3 + 64$ **24.** $125 + (2x + y)^3$

25. $64 - x^6$ **26.** $a^6 - 27b^6$

In Exercises 27–32, factor the given expressions completely. Each is from the technical area indicated.

27. $2x^3 + 250$ (computer image)

28. $kT^3 - kT_0^3$ (thermodynamics)

29. $D^4 - d^3D$ (machine design)

30. $(h + 2t)^3 - h^3$ (container design)

31. $QH^4 + Q^4H$ (thermodynamics)

32. $\left(\dfrac{s}{r}\right)^{12} - \left(\dfrac{s}{r}\right)^6$ (molecular interaction)

In Exercises 33 and 34, perform the indicated operations.

33. Perform the division $(x^5 - y^5) \div (x - y)$. Noting the result, determine the quotient $(x^7 - y^7) \div (x - y)$, without dividing. From these results, factor $x^5 - y^5$ and $x^7 - y^7$.

34. Perform the division $(x^5 + y^5) \div (x + y)$. Noting the result, determine the quotient $(x^7 + y^7) \div (x + y)$, without dividing. From these results, factor $x^5 + y^5$ and $x^7 + y^7$.

In Exercises 35–38, solve the given problems.

35. Factor $x^6 - y^6$ as the difference of cubes.

(W) 36. Factor $x^6 - y^6$ as the difference of squares. Then explain how the result of Exercise 35 can be shown to be the same as the result in this exercise. (*Hint:* See Exercise 72 on page 187.)

(W) 37. By factoring $(n^3 + 1)$, explain why this expression represents a number that is not prime if n is an integer greater than one.

(W) 38. Explain how Eq. (6.9) can be used to factor the *difference* of cubes.

Answers to Practice Exercises

1. $(x + 6)(x^2 - 6x + 36)$ **2.** $(x - 6)(x^2 + 6x + 36)$

6.5 Equivalent Fractions

Fundamental Principle of Fractions • Equivalent Fractions • Simplest Form, or Lowest Terms, of a Fraction • Factors That Differ Only in Sign

When we deal with algebraic expressions, we must be able to work effectively with fractions. Since algebraic expressions are representations of numbers, the basic operations on fractions from arithmetic form the basis of our algebraic operations. In this section, we demonstrate a very important property of fractions, and in the following two sections, we establish the basic algebraic operations with fractions.

Fundamental Principle of Fractions

This important property of fractions, often referred to as the **fundamental principle of fractions,** is that *the value of a fraction is unchanged if both numerator and denominator are multiplied or divided by the same number, provided this number is not zero.* Two fractions are said to be **equivalent** if one can be obtained from the other by use of the fundamental principle.

EXAMPLE 1 Equivalent arithmetic fractions

If we multiply the numerator and the denominator of the fraction $\frac{6}{8}$ by 2, we obtain the equivalent fraction $\frac{12}{16}$. If we divide the numerator and the denominator of $\frac{6}{8}$ by 2, we obtain the equivalent fraction $\frac{3}{4}$. This means that the fractions $\frac{6}{8}, \frac{3}{4}$, and $\frac{12}{16}$ are equivalent. Obviously, there is an unlimited number of other fractions that are equivalent to these fractions. ∎

EXAMPLE 2 Equivalent algebraic fractions

We may write

$$\frac{ax}{2} = \frac{3a^2x}{6a}$$

since we get the fraction on the right by multiplying the numerator and the denominator of the fraction on the left by $3a$. This means that the fractions are equivalent. ∎

SIMPLEST FORM, OR LOWEST TERMS, OF A FRACTION

One of the most important operations with a fraction is that of *reducing* it to its **simplest form,** or **lowest terms.**

> *A fraction is said to be in its simplest form if the numerator and the denominator have no common integral factors other than $+1$ or -1.*

NOTE ▶ In *reducing* a fraction to its simplest form, use the fundamental principle of fractions by *dividing* both the numerator and the denominator by all *factors* that are common to each. It will be assumed throughout this text that *if any of the literal symbols were to be evaluated, numerical values would be such that none of the denominators would be equal to zero.* Therefore, all operations are done properly since the undefined operation of division by zero is avoided.

EXAMPLE 3 Reducing fraction to lowest terms

In order to reduce the fraction

$$\frac{16ab^3c^2}{24ab^2c^5}$$

to its lowest terms, note that both the numerator and the denominator contain the factor $8ab^2c^2$. Therefore,

$$\frac{16ab^3c^2}{24ab^2c^5} = \frac{2b(8ab^2c^2)}{3c^3(8ab^2c^2)} = \frac{2b}{3c^3} \quad \text{— common factor}$$

Here, we divided out the common factor of $8ab^2c^2$. The resulting fraction is in its lowest terms, since there are no common factors in the numerator and the denominator other than $+1$ or -1. In performing this division, we also assume that neither a, b, nor c is equal to zero so as to avoid the possibility of dividing by zero. ∎

Cancellation In simplifying a fraction, be very careful in performing the basic step of reducing the fraction to its simplest form. That is,

> *divide* both the numerator and the denominator **by the common factor.**

CAUTION ▶ *This process is called* **cancellation.** However, it is a common error to remove any expression that appears in both the numerator and the denominator. If a *term* is removed in this way, it is an incorrect application of the cancellation process. We must always *cancel only factors* that are in both the numerator and the denominator. (Remember—*terms* are separated by $+$ and $-$ signs.) In the following example, we illustrate this very common error in the simplification of fractions. Follow it very carefully.

EXAMPLE 4 Cancel factors only

When simplifying the expression

$$\frac{x^2(x-2)}{x^2-4}$$ — a term, but not a factor, of the denominator

CAUTION ▶ many students would "cancel" the x^2 from the numerator and the denominator. This is incorrect, since x^2 **is a term only** of the denominator.

In order to simplify the above fraction properly, we should factor the denominator. We get

$$\frac{x^2\overset{1}{\cancel{(x-2)}}}{\underset{1}{\cancel{(x-2)}}(x+2)} = \frac{x^2}{x+2}$$

Here, the common **factor** $x-2$ has been divided out. ■

The following examples illustrate the proper simplification of fractions.

EXAMPLE 5 Distinguishing between term and factor

(a) $\dfrac{2a}{2ax} = \dfrac{1}{x}$ — $2a$ is a factor of the numerator and the denominator

We divide out the common factor of $2a$.

(b) $\dfrac{2a}{2a+x}$ — $2a$ is a term, but not a factor, of the denominator

CAUTION ▶ **This cannot be reduced,** since *there are no common **factors** in the numerator and the denominator.* ■

EXAMPLE 6 Remaining factor of 1 in denominator

$$\frac{2x^2+8x}{x+4} = \frac{2x\overset{1}{\cancel{(x+4)}}}{\underset{1}{\cancel{(x+4)}}} = \frac{2x}{1}$$
$$= 2x$$

The numerator and the denominator were each divided by $x+4$ after factoring the numerator. The only remaining factor in the denominator is 1, and it is generally not written in the final result. Another way of writing the denominator is $1(x+4)$, which shows the **factor** of 1 more clearly. ■

EXAMPLE 7 Cancel factors only

$$\frac{x^2-4x+4}{x^2-4} = \frac{(x-2)\overset{1}{\cancel{(x-2)}}}{(x+2)\underset{1}{\cancel{(x-2)}}}$$

$$= \frac{x-2}{x+2}$$ — x is a term but not a factor

CAUTION ▶ Here, the numerator and the denominator have each been *factored first and then the common factor $x-2$ has been divided out.* In the final form, neither the x's nor the 2's may be canceled, since they are not common *factors.* ■

Practice Exercise

2. Reduce to lowest terms: $\dfrac{4a}{4a-2x}$

Practice Exercise

3. Reduce to lowest terms: $\dfrac{x^2-x-2}{x^2+3x+2}$

EXAMPLE 8 **Reducing fraction – application**

In the mathematical analysis of the vibrations in a certain mechanical system, the following expression and simplification are used:

$$\frac{8s + 12}{4s^2 + 26s + 30} = \frac{4(2s + 3)}{2(2s^2 + 13s + 15)} = \frac{\overset{2}{\cancel{4}}\overset{1}{\cancel{(2s + 3)}}}{\underset{1}{\cancel{2}}\underset{1}{\cancel{(2s + 3)}}(s + 5)}$$
$$= \frac{2}{s + 5}$$

In the third fraction, note that the factors common to both the numerator and the denominator are 2 and $(2s + 3)$. ∎

FACTORS THAT DIFFER ONLY IN SIGN

In simplifying fractions, we must be able to distinguish between factors that differ only in *sign.* Since $-(y - x) = -y + x = x - y$, we have

$$\boxed{x - y = -(y - x)} \tag{6.11}$$

CAUTION ▶ Here, the *factors $x - y$ and $y - x$ differ only in sign.* The following examples illustrate the simplification of fractions where a change of signs is necessary.

EXAMPLE 9 **Factors differ only in sign**

$$\frac{x^2 - 1}{1 - x} = \frac{\cancel{(x - 1)}(x + 1)}{-\cancel{(x - 1)}} = \frac{x + 1}{-1} = -(x + 1)$$

CAUTION ▶ In the second fraction, *we replaced $1 - x$ with the equal expression $-(x - 1)$.* In the third fraction, the common factor $x - 1$ was divided out. Finally, we expressed the result in the more convenient form by dividing $x + 1$ by -1. Replacing $1 - x$ with $-(x - 1)$ is the same as factoring -1 from the terms of $1 - x$.

It should be clear that the factors $1 - x$ and $x - 1$ differ only in sign. However, in simplifying fractions, the fact that they can provide a cancellation often goes unnoticed, and a simplification may be incomplete. ∎

Practice Exercise

4. Reduce to lowest terms: $\dfrac{y - 2}{4 - y^2}$

EXAMPLE 10 **Factors differ only in sign**

$$\frac{2x^4 - 128x}{20 + 7x - 3x^2} = \frac{2x(x^3 - 64)}{(4 - x)(5 + 3x)} = \frac{2x\cancel{(x - 4)}(x^2 + 4x + 16)}{-\cancel{(x - 4)}(3x + 5)}$$
$$= -\frac{2x(x^2 + 4x + 16)}{3x + 5}$$

Again, the factor $4 - x$ has been replaced by the equal expression $-(x - 4)$. This allows us to recognize the common factor of $x - 4$.

Also, note that the order of the terms of the factor $5 + 3x$ was changed in writing the third fraction. This was done only to write the terms in the more standard form with the x-term first. However, since both terms are *positive,* it is simply an application of the commutative law of addition, and the factor itself is not actually changed. ∎

EXERCISES 6.5

In Exercises 1–4, make the given changes in the indicated examples of this section and then solve the resulting problems.

1. In Example 3, change the numerator to $18abc^6$, and then reduce the fraction to lowest terms.

2. In Example 4, change the $-$ sign in the numerator to $+$ and then simplify.

3. In Example 7, change the $-$ sign in the numerator to $+$ and then simplify.

4. In Example 10, change the numerator to $2x^4 - 32x^2$ and then simplify.

In Exercises 5–12, multiply the numerator and the denominator of each fraction by the given factor and obtain an equivalent fraction.

5. $\dfrac{2}{3}$ (by 7)

6. $\dfrac{7}{5}$ (by 9)

7. $\dfrac{ax}{y}$ (by $2x$)

8. $\dfrac{2x^2y}{3n}$ (by $2xn^2$)

9. $\dfrac{2}{x+3}$ (by $x-2$)

10. $\dfrac{7}{a-1}$ (by $a+2$)

11. $\dfrac{a(x-y)}{x-2y}$ (by $x+y$)

12. $\dfrac{B-1}{B+1}$ (by $B-1$)

In Exercises 13–20, divide the numerator and the denominator of each fraction by the given factor and obtain an equivalent fraction.

13. $\dfrac{28}{44}$ (by 4)

14. $\dfrac{25}{65}$ (by 5)

15. $\dfrac{4x^2y}{8xy^2}$ (by $2x$)

16. $\dfrac{6a^3b^2}{9a^5b^4}$ (by $3a^2b^2$)

17. $\dfrac{2(R-1)}{(R-1)(R+1)}$ (by $R-1$)

18. $\dfrac{(x+5)(x-3)}{3(x+5)}$ (by $x+5$)

19. $\dfrac{s^2-3s-10}{2s^2+3s-2}$ (by $s+2$)

20. $\dfrac{6x^2+13x-5}{6x^3-2x^2}$ (by $3x-1$)

In Exercises 21–28, replace the A with the proper expression such that the fractions are equivalent.

21. $\dfrac{3x}{2y} = \dfrac{A}{6y^2}$

22. $\dfrac{2R}{R+T} = \dfrac{2R^2T}{A}$

23. $\dfrac{7}{a+5} = \dfrac{7a-35}{A}$

24. $\dfrac{a+1}{5a^2c} = \dfrac{A}{5a^3c-5a^2c}$

25. $\dfrac{2x^3+2x}{x^4-1} = \dfrac{A}{x^2-1}$

26. $\dfrac{n^2-1}{n^3+1} = \dfrac{A}{n^2-n+1}$

27. $\dfrac{x^2+3bx-4b^2}{x-b} = \dfrac{x+4b}{A}$

28. $\dfrac{4y^2-1}{4y^2+6y-4} = \dfrac{A}{2y+4}$

In Exercises 29–64, reduce each fraction to simplest form.

29. $\dfrac{2a}{8a}$

30. $\dfrac{6x}{15x}$

31. $\dfrac{18x^2y}{24xy}$

32. $\dfrac{2a^2xy}{6axyz^2}$

33. $\dfrac{a+b}{5a^2+5ab}$

34. $\dfrac{t-a}{t^2-a^2}$

35. $\dfrac{4a-4b}{4a-2b}$

36. $\dfrac{20s-5r}{10r-5s}$

37. $\dfrac{4x^2+1}{4x^2-1}$

38. $\dfrac{x^2-y^2}{x^2+y^2}$

39. $\dfrac{3x^2-6x}{x-2}$

40. $\dfrac{10T^2+15T}{2T+3}$

41. $\dfrac{3+2y}{4y^3+6y^2}$

42. $\dfrac{6-3t}{4t^3-8t^2}$

43. $\dfrac{x^2-8x+16}{x^2-16}$

44. $\dfrac{4a^2+12ab+9b^2}{4a^2+6ab}$

45. $\dfrac{2w^4+5w^2-3}{w^4+11w^2+24}$

46. $\dfrac{3y^3+7y^2+4y}{y^2+5y+4}$

47. $\dfrac{5x^2-6x-8}{x^3+x^2-6x}$

48. $\dfrac{5s^2+8rs-4s^2}{6r^2-17rs+5s^2}$

49. $\dfrac{N^4-16}{8N-16}$

50. $\dfrac{3+x(4+x)}{3+x}$

51. $\dfrac{t+4}{(2t+9)t+4}$

52. $\dfrac{2A^3+8A^4+8A^5}{4A+2}$

53. $\dfrac{(x-1)(3+x)}{(3-x)(1-x)}$

54. $\dfrac{(2x-1)(x+6)}{(x-3)(1-2x)}$

55. $\dfrac{y^2-x^2}{2x-2y}$

56. $\dfrac{x^2-y^2-4x+4y}{x^2-y^2+4x-4y}$

57. $\dfrac{x^3+x^2-x-1}{x^3-x^2-x+1}$

58. $\dfrac{3a^2-13a-10}{5+4a-a^2}$

59. $\dfrac{(x+5)(x-2)(x+2)(3-x)}{(2-x)(5-x)(3+x)(2+x)}$

60. $\dfrac{(2x-3)(3-x)(x-7)(3x+1)}{(3x+2)(3-2x)(x-3)(7+x)}$

61. $\dfrac{x^3+y^3}{2x+2y}$

62. $\dfrac{w^3-8}{w^2+2w+4}$

63. $\dfrac{6x^2+2x}{27x^3+1}$

64. $\dfrac{24-3a^3}{a^2-4a+4}$

(W) In Exercises 65–68, after finding the simplest form of each fraction, explain why it cannot be simplified more.

65. (a) $\dfrac{x^2(x + 2)}{x^2 + 4}$ (b) $\dfrac{x^4 + 4x^2}{x^4 - 16}$

66. (a) $\dfrac{2x + 3}{2x + 6}$ (b) $\dfrac{2(x + 6)}{2x + 6}$

67. (a) $\dfrac{x^2 - x - 2}{x^2 - x}$ (b) $\dfrac{x^2 - x - 2}{x^2 + x}$

68. (a) $\dfrac{x^3 - x}{1 - x}$ (b) $\dfrac{2x^2 + 4x}{2x^2 + 4}$

In Exercises 69–72, reduce each fraction to simplest form. Each is from the indicated area of application.

69. $\dfrac{mu^2 - mv^2}{mu - mv}$ (nuclear energy)

70. $\dfrac{16(t^2 - 2tt_0 + t_0^2)(t - t_0 - 3)}{3t - 3t_0}$ (rocket motion)

71. $\dfrac{E^2R^2 - E^2r^2}{(R^2 + 2Rr + r^2)^2}$ (electricity)

72. $\dfrac{r_0^3 - r_i^3}{r_0^2 - r_i^2}$ (machine design)

Answers to Practice Exercises

1. $\dfrac{3y^3}{5x^2}$ **2.** $\dfrac{2a}{2a - x}$ **3.** $\dfrac{x - 2}{x + 2}$ **4.** $-\dfrac{1}{y + 2}$

6.6 Multiplication and Division of Fractions

Multiplication of Fractions • Division of Fractions • Using Calculator for Checking Answers

From arithmetic, recall that *the product of two fractions is a fraction whose numerator is the product of the numerators and whose denominator is the product of the denominators of the given fractions.* Also, *the quotient of two fractions is found by inverting the denominator and proceeding as in multiplication.* Symbolically, these operations are shown as

Multiplication of Fractions

$$\frac{a}{b} \times \frac{c}{d} = \frac{ac}{bd}$$

Division of Fractions

$$\frac{a}{b} \div \frac{c}{d} = \frac{\dfrac{a}{b}}{\dfrac{c}{d}} = \frac{a}{b} \times \frac{d}{c} = \frac{ad}{bc}$$

The rule for division may be verified by use of the fundamental principle of fractions. By multiplying the numerator and the denominator of the fraction

$$\frac{\dfrac{a}{b}}{\dfrac{c}{d}} \quad \text{by} \quad \frac{d}{c} \quad \text{we obtain} \quad \frac{\dfrac{a}{b} \times \dfrac{d}{c}}{\dfrac{c}{d} \times \dfrac{d}{c}} = \frac{\dfrac{ad}{bc}}{1} = \frac{ad}{bc}$$

The following three examples illustrate the multiplication of fractions.

EXAMPLE 1 Multiplying basic fractions

(a) $\dfrac{3}{5} \times \dfrac{2}{7} = \dfrac{(3)(2)}{(5)(7)} = \dfrac{6}{35}$ ← multiply numerators
← multiply denominators

(b) $\dfrac{3a}{5b} \times \dfrac{15b^2}{a} = \dfrac{(3a)(15b^2)}{(5b)(a)}$
$= \dfrac{45ab^2}{5ab} = \dfrac{9b}{1}$
$= 9b$

(c) $(6x)\left(\dfrac{2y}{3x^2}\right) = \left(\dfrac{6x}{1}\right)\left(\dfrac{2y}{3x^2}\right)$
$= \dfrac{12xy}{3x^2} = \dfrac{4y}{x}$

In (b), we divided out the common factor $5ab$ to reduce the resulting fraction to its lowest terms. In (c), note that we treated $6x$ as $\dfrac{6x}{1}$. ∎

When multiplying fractions, we usually want to express the final result in simplest form, which is generally its most useful form. Since all factors in the numerators and all factors in the denominators are to be multiplied, we should

NOTE ▶ *first only* **indicate** *the multiplication, but not actually perform it and then factor the numerator and the denominator of the result.*

By *indicate,* we mean to write the factors of the numerator side by side and do the same for the factors of the denominator. If we were to multiply out the numerator and the denominator before factoring, it is very possible that we would not see how to factor the result and therefore would not be able to simplify it. The following example illustrates this point.

EXAMPLE 2   First indicate the multiplications

In performing the multiplication

$$\frac{3(x-y)}{(x-y)^2} \times \frac{(x^2-y^2)}{6x+9y}$$

if we multiplied out the numerators and the denominators before performing any factoring, we would have to simplify the fraction

$$\frac{3x^3-3x^2y-3xy^2+3y^3}{6x^3-3x^2y-12xy^2+9y^3}$$

NOTE ▶ It is possible to factor the resulting numerator and denominator, but finding any common factors this way is very difficult. However, as stated above, we should *first indicate the multiplications* and not actually do them, and *the solution is much easier.* Then we factor the resulting numerator and denominator, divide out (cancel) any common factors, and the solution is then complete. Doing this, we have

$$\frac{3(x-y)}{(x-y)^2} \times \frac{(x^2-y^2)}{6x+9y} = \frac{3(x-y)(x^2-y^2)}{(x-y)^2(6x+9y)}$$

$$= \frac{3(x-y)(x+y)(x-y)}{(x-y)^2(3)(2x+3y)}$$

$$= \frac{3(x-y)^2(x+y)}{3(x-y)^2(2x+3y)}$$

$$= \frac{x+y}{2x+3y}$$

The common factor $3(x-y)^2$ is readily recognized using this procedure.   ∎

EXAMPLE 3   Multiplying algebraic fractions

$$\left(\frac{2x-4}{4x+12}\right)\left(\frac{2x^2+x-15}{3x-1}\right) = \frac{2(x-2)(2x-5)(x+3)}{4(x+3)(3x-1)} \quad \begin{array}{l}\leftarrow \text{multiplications}\\ \leftarrow \text{indicated}\end{array}$$

$$= \frac{(x-2)(2x-5)}{2(3x-1)}$$

Here, the common factor is $2(x+3)$. It is permissible to multiply out the final form of the numerator and the denominator, but it is often preferable to leave the numerator and the denominator in factored form, as indicated.   ∎

Practice Exercise

1. Multiply: $\dfrac{2x}{4x+8} \times \dfrac{x+2}{xy-x}$

■ It is possible to factor and indicate the product of the factors, showing only a single step, as we have done here.

The following examples illustrate the division of fractions.

EXAMPLE 4 Dividing basic fractions

(a) $\dfrac{6x}{7} \div \dfrac{5}{3} = \dfrac{6x}{7} \times \dfrac{3}{5} = \dfrac{18x}{35}$
multiply ↓
invert

(b) $\dfrac{\dfrac{3a^2}{5c}}{\dfrac{2c^2}{a}} = \dfrac{3a^2}{5c} \times \dfrac{a}{2c^2} = \dfrac{3a^3}{10c^3}$
multiply ↓
invert ∎

EXAMPLE 5 Division by a fraction – application

When finding the center of gravity (c.g.) of a uniform flat semicircular metal plate, the equation $X = \dfrac{4\pi r^3}{3} \div \left(\dfrac{\pi r^2}{2} \times 2\pi \right)$ is derived. Simplify the right side of this equation to find X as a function of r in simplest form. See Fig. 6.4.

The parentheses indicate that we should perform the multiplication first:

$$X = \frac{4\pi r^3}{3} \div \left(\frac{\pi r^2}{2} \times 2\pi \right) = \frac{4\pi r^3}{3} \div \left(\frac{2\pi^2 r^2}{2} \right) \qquad 2\pi = \frac{2\pi}{1}$$

$$= \frac{4\pi r^3}{3} \div (\pi^2 r^2) = \frac{4\pi r^3}{3} \times \frac{1}{\pi^2 r^2}$$

$$= \frac{4\pi r^3}{3\pi^2 r^2} = \frac{4r}{3\pi} \qquad \text{divide out the common factor of } \pi r^2$$

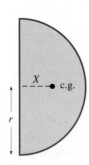

Fig. 6.4

This is the exact solution. Approximately, $X = 0.424r$. ∎

EXAMPLE 6 Dividing algebraic fractions

$$(x + y) \div \frac{2x + 2y}{6x + 15y} = \frac{x + y}{1} \times \frac{6x + 15y}{2x + 2y} = \frac{\cancel{(x + y)}(3)(2x + 5y)}{2\cancel{(x + y)}} \qquad \text{indicate multiplication}$$

invert

$$= \frac{3(2x + 5y)}{2} \qquad \text{simplify} \qquad ∎$$

Practice Exercise

2. Divide: $\dfrac{3x}{a + 1} \div \dfrac{x^2 + 2x}{a^2 + a}$

EXAMPLE 7 Dividing algebraic fractions

invert

$$\frac{\dfrac{4 - x^2}{x^2 - 3x + 2}}{\dfrac{x + 2}{x^2 - 9}} = \frac{4 - x^2}{x^2 - 3x + 2} \times \frac{x^2 - 9}{x + 2} = \frac{(2 - x)(2 + x)(x - 3)(x + 3)}{(x - 2)(x - 1)(x + 2)} \qquad \text{factor and indicate multiplications}$$

$$= \frac{-\cancel{(x - 2)}\cancel{(x + 2)}(x - 3)(x + 3)}{\cancel{(x - 2)}(x - 1)\cancel{(x + 2)}} \qquad \begin{array}{l}\text{replace } (2 - x) \text{ with } -(x - 2) \\ \text{and } (2 + x) \text{ with } (x + 2)\end{array}$$

$$= -\frac{(x - 3)(x + 3)}{x - 1} \quad \text{or} \quad \frac{(x - 3)(x + 3)}{1 - x} \qquad \text{simplify}$$

Note the use of Eq. (6.11) when the factor $(2 - x)$ was replaced by $-(x - 2)$ to get the first form of the answer. As shown, it can also be used to get the alternate form of the answer, although it is not necessary to give this form. ∎

USE OF A GRAPHING CALCULATOR FOR CHECKING ANSWERS

A calculator can be used to check algebraic results. Different numbers can be assigned to each variable to see if the values of the initial expression and the resulting expression are the same. Care must be taken in choosing values (such as not choosing 0 or 1), but in any case, the check is valid only for the numbers used.

A graphing calculator gives us a more general way of checking an algebraic result if only one variable is involved. (With more than one variable, values must be assigned to all but one, and the check is less general.) This is shown in the next example.

EXAMPLE 8 Using calculator to check answer

To graphically check the result of Example 7, we first let y_1 equal the original expression and y_2 equal the final result. In this case,

$$y_1 = ((4 - x^2)/(x^2 - 3x + 2))/((x + 2)/(x^2 - 9))$$
$$y_2 = -(x - 3)(x + 3)/(x - 1)$$

By graphing y_2 with a heavier curve than y_1 and watching carefully, the graph of y_2 simply goes over the graph of y_1 (see Fig. 6.5(a)). Since the graphs are the same, we can consider this as a *check,* but not as a *proof* that the expressions are equal.

The calculator shows a check of the solution from $x = -8$ to $x = 8$, except $x = 1$. Actually, using values of -3, -2, 1, 2, or 3 is incorrect in that each of these indicates division by zero in the original expression. The calculator does show this if we use Xmin and Xmax values such that these values come up *exactly*. When we use the *trace* feature for y_1, no y-values show, and there are missing pixels for these x-values. This is easier to see with *axes off* as in Fig. 6.5(b). ∎

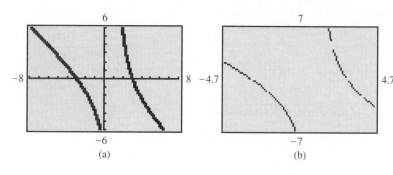

(a)   (b)

Fig. 6.5

EXERCISES 6.6

In Exercises 1–4, make the given changes in the indicated examples of this section and then solve the resulting problems.

1. In Example 2, change the first numerator to $4x + 6y$ and do the multiplication.

2. In Example 3, change the first denominator to $4x - 10$ and do the multiplication.

3. In Example 6, change the denominator $6x + 15y$ to $6x + 12y$ and then do the division.

4. In Example 7, change the numerator $x + 2$ of the divisor to $x + 3$ and then simplify.

In Exercises 5–40, simplify the given expressions involving the indicated multiplications and divisions.

5. $\dfrac{3}{8} \times \dfrac{2}{7}$

6. $11 \times \dfrac{13}{33}$

7. $\dfrac{4x}{3y} \times \dfrac{9y^2}{2}$

8. $\dfrac{18sy^3}{ax^2} \times \dfrac{(ax)^2}{3s}$

9. $\dfrac{2}{9} \div \dfrac{4}{7}$

10. $\dfrac{5}{16} \div \dfrac{25}{13}$

11. $\dfrac{xy}{az} \div \dfrac{bz}{ay}$

12. $\dfrac{sr^2}{2t} \div \dfrac{st}{4}$

13. $\dfrac{4x + 12}{5} \times \dfrac{15t}{3x + 9}$

14. $\dfrac{2y^2 + 6y}{6z} \times \dfrac{z^3}{y^2 - 9}$

15. $\dfrac{u^2 - v^2}{u + 2v}(3u + 6v)$

16. $(x - y)\dfrac{x + 2y}{x^2 - y^2}$

17. $\dfrac{2a + 8}{15} \div \dfrac{a^2 + 8a + 16}{125}$

18. $\dfrac{a^2 - a}{3a + 9} \div \dfrac{a^2 - 2a + 1}{18 - 2a^2}$

19. $\dfrac{x^4 - 9}{x^2} \div (x^2 + 3)^2$

20. $\dfrac{9B^2 - 16}{B + 1} \div (4 - 3B)$

21. $\dfrac{3ax^2 - 9ax}{10x^2 + 5x} \times \dfrac{2x^2 + x}{a^2x - 3a^2}$

22. $\dfrac{4R^2 - 36}{R^3 - 25R} \times \dfrac{7R - 35}{3R^2 + 9R}$

23. $\left(\dfrac{x^4 - 1}{8x + 16}\right)\left(\dfrac{2x^2 - 8x}{x^3 + x}\right)$

24. $\left(\dfrac{2x^2 - 4x - 6}{x^2 - 3x}\right)\left(\dfrac{x^3 - 4x^2}{4x^2 - 4x - 8}\right)$

25. $\dfrac{\dfrac{x^2 + ax}{2b - cx}}{\dfrac{a^2 + 2ax + x^2}{2bx - cx^2}}$

26. $\dfrac{\dfrac{x^4 - 11x^2 + 28}{2x^2 + 6}}{\dfrac{4 - x^2}{2x^2 + 3}}$

27. $\dfrac{35a + 25}{12a + 33} \div \dfrac{28a + 20}{36a + 99}$

28. $\dfrac{2a^3 + a^2}{2b^3 + b^2} \div \dfrac{2ab + a}{2ab + b}$

29. $\dfrac{x^2 - 6x + 5}{4x^2 - 17x - 15} \times \dfrac{6x + 21}{2x^2 + 5x - 7}$

30. $\dfrac{n^2 + 5n}{3n^2 + 8n + 4} \times \dfrac{2n^2 - 8}{n^3 + 3n^2 - 10n}$

31. $\dfrac{\dfrac{6T^2 - NT - N^2}{2V^2 - 9V - 35}}{\dfrac{8T^2 - 2NT - N^2}{20V^2 + 26V - 60}}$

32. $\dfrac{\dfrac{4L^3 - 9L}{8L^2 + 10L - 3}}{2L^3 - 3L^2}$

33. $\dfrac{7x^2}{3a} \div \left(\dfrac{a}{x} \times \dfrac{a^2x}{x^2} \right)$

34. $\left(\dfrac{3u}{8v^2} \div \dfrac{9u^2}{2w^2} \right) \times \dfrac{2u^4}{15vw}$

35. $\left(\dfrac{4t^2 - 1}{t - 5} \div \dfrac{2t + 1}{2t} \right) \times \dfrac{2t^2 - 50}{4t^2 + 4t + 1}$

36. $\dfrac{2x^2 - 5x - 3}{x - 4} \div \left(\dfrac{x - 3}{x^2 - 16} \times \dfrac{1}{3 - x} \right)$

37. $\dfrac{x^3 - y^3}{2x^2 - 2y^2} \times \dfrac{y^2 + 2xy + x^2}{x^2 + xy + y^2}$

38. $\dfrac{2M^2 + 4M + 2}{6M - 6} \div \dfrac{5M + 5}{M^2 - 1}$

39. $\left(\dfrac{ax + bx + ay + by}{p - q} \right)\left(\dfrac{3p^2 + 4pq - 7q^2}{a + b} \right)$

40. $\dfrac{x^4 + x^5 - 1 - x}{x - 1} \div \dfrac{x - 1}{x}$

In Exercises 41–44, simplify the given expressions and then check your answers with a graphing calculator as in Example 8.

41. $\dfrac{x}{2x + 4} \times \dfrac{x^2 - 4}{3x^2}$

42. $\dfrac{4x^2 - 25}{4x^2} \div \dfrac{4x + 10}{8}$

43. $\dfrac{2x^2 + 3x - 2}{2 + 3x - 2x^2} \div \dfrac{5x + 10}{4x + 2}$

44. $\dfrac{16x^2 - 8x + 1}{9x} \times \dfrac{12x + 3}{1 - 16x^2}$

In Exercises 45–48, simplify the given expressions. The technical application of each is indicated.

45. $\dfrac{d}{2} \div \dfrac{v_1 d + v_2 d}{4v_1 v_2}$ (average velocity)

46. $\dfrac{c\lambda^2 - c\lambda_0^2}{\lambda_0^2} \div \dfrac{\lambda^2 + \lambda_0^2}{\lambda_0^2}$ (cosmology)

47. $\dfrac{2\pi}{\lambda}\left(\dfrac{a + b}{2ab} \right)\left(\dfrac{ab\lambda}{2a + 2b} \right)$ (optics)

48. $(p_1 - p_2) \div \left(\dfrac{\pi a^4 p_1 - \pi a^4 p_2}{81u} \right)$ (hydrodynamics)

Answers to Practice Exercises

1. $\dfrac{1}{2(y - 1)}$ **2.** $\dfrac{3a}{x + 2}$

6.7 Addition and Subtraction of Fractions

Lowest Common Denominator •
Addition and Subtraction of Fractions •
Complex Fractions

From arithmetic, recall that *the sum of a set of fractions that all have the same denominator is the sum of the numerators divided by the common denominator.* Since algebraic expressions represent numbers, this fact is also true in algebra. Addition and subtraction of such fractions are illustrated in the following example.

EXAMPLE 1 Combining basic fractions

(a) $\dfrac{5}{9} + \dfrac{2}{9} - \dfrac{4}{9} = \dfrac{5 + 2 - 4}{9}$ ← sum of numerators
← same denominators

$= \dfrac{3}{9} = \dfrac{1}{3}$ ← final result in lowest terms

CAUTION ▶

use parentheses to show subtraction of both terms

(b) $\dfrac{b}{ax} + \dfrac{1}{ax} - \dfrac{2b - 1}{ax} = \dfrac{b + 1 - (2b - 1)}{ax} = \dfrac{b + 1 - 2b + 1}{ax}$

$= \dfrac{2 - b}{ax}$ ∎

LOWEST COMMON DENOMINATOR

NOTE ▶

If the fractions to be combined do not all have the same denominator, we must first change each to an equivalent fraction so that the resulting fractions do have the same denominator. Normally, the denominator that is most convenient and useful is the **lowest common denominator** (abbreviated as **LCD**). *This is the product of all the prime factors that appear in the denominators, with each factor raised to the **highest power** to which it appears **in any one** of the denominators.* This means that the lowest common denominator is the *simplest* algebraic expression into which all given denominators will divide exactly. Following is the procedure for finding the lowest common denominator of a set of fractions.

PROCEDURE FOR FINDING THE LOWEST COMMON DENOMINATOR

1. *Factor each denominator into its prime factors.*
2. *For each different prime factor that appears, note the highest power to which it is raised in any one of the denominators.*
3. *Form the product of all the different prime factors, each raised to the power found in step 2. This product is the lowest common denominator.*

The two examples that follow illustrate the method of finding the LCD.

EXAMPLE 2 Lowest common denominator (LCD)

Find the LCD of the fractions

$$\frac{3}{4a^2b} \qquad \frac{5}{6ab^3} \qquad \frac{1}{4ab^2}$$

We now express each denominator in terms of powers of its prime factors:

highest powers   already seen to be highest power of 2

$$4a^2b = 2^2a^2b \qquad 6ab^3 = 2 \times 3 \times ab^3 \qquad 4ab^2 = 2^2ab^2$$

The prime factors to be considered are 2, 3, a, and b. The largest exponent of 2 that appears is 2. Therefore, 2^2 is a factor of the LCD.

CAUTION ▶

*What matters is that **the highest power of 2 that appears is 2,** not the fact that 2 appears in all three denominators with a total of five factors.*

The largest exponent of 3 that appears is 1 (understood in the second denominator). Therefore, 3 is a factor of the LCD. The largest exponent of a that appears is 2, and the largest exponent of b that appears is 3. Thus, a^2 and b^3 are factors of the LCD. Therefore, the LCD of the fractions is

$$2^2 \times 3 \times a^2b^3 = 12a^2b^3$$

This is the simplest expression into which ***each*** of the denominators above will divide exactly. ∎

Finding the LCD of a set of fractions can be a source of difficulty. Remember, it is necessary to find *all of the prime factors* that may be in any of the denominators, and then find *the highest power to which each is raised in any of the denominators.* The following example illustrates this when factoring the denominators is the first step.

EXAMPLE 3 Lowest common denominator

Find the LCD of the following fractions:

$$\frac{x-4}{x^2-2x+1} \qquad \frac{1}{x^2-1} \qquad \frac{x+3}{x^2-x}$$

Factoring each of the denominators, we find that the fractions are

$$\frac{x-4}{(x-1)^2} \qquad \frac{1}{(x-1)(x+1)} \qquad \frac{x+3}{x(x-1)}$$

The factor $(x-1)$ appears in all the denominators. It is squared in the first fraction and appears only to the first power in the other two fractions. Thus, we must have $(x-1)^2$ as a factor in the LCD. We do not need a higher power of $(x-1)$ since, as far as this factor is concerned, each denominator will divide into it evenly. Next, the second denominator has a factor of $(x+1)$. Therefore, the LCD must also have a factor of $(x+1)$; otherwise, the second denominator would not divide into it exactly. Finally, the third denominator shows that a factor of x is also needed. The LCD is therefore $x(x+1)(x-1)^2$. All three denominators will divide exactly into this expression, and there is no simpler expression for which this is true. ∎

Practice Exercise

1. Find the LCD of the following fractions:
$$\frac{x}{2x^2+2x}, \quad \frac{1}{x^2+2x+1}, \quad \frac{2x+4}{4x^2}$$

ADDITION AND SUBTRACTION OF FRACTIONS

Once we have found the LCD for the fractions, we multiply the numerator and the denominator of each fraction by the proper quantity to make the resulting denominator in each case the lowest common denominator. After this step, it is necessary only to add the numerators, place this result over the common denominator, and simplify.

EXAMPLE 4 Finding LCD – combining fractions

Combine: $\dfrac{2}{3r^2} + \dfrac{4}{rs^3} - \dfrac{5}{3s}$.

By looking at the denominators, notice that the factors necessary in the LCD are 3, r, and s. The 3 appears only to the first power, the largest exponent of r is 2, and the largest exponent of s is 3. Therefore, the LCD is $3r^2s^3$. Now, write each fraction with this quantity as the denominator. Since the denominator of the first fraction already contains factors of 3 and r^2, **it is necessary to introduce the factor of s^3.** In other words, we must multiply the numerator and the denominator of this fraction by s^3. For similar reasons, we must multiply the numerators and the denominators of the second and third fractions by $3r$ and r^2s^2, respectively. This leads to

NOTE ▶

$$\frac{2}{3r^2} + \frac{4}{rs^3} - \frac{5}{3s} = \frac{2(s^3)}{(3r^2)(s^3)} + \frac{4(3r)}{(rs^3)(3r)} - \frac{5(r^2s^2)}{(3s)(r^2s^2)} \qquad \text{change to equivalent fractions with LCD}$$

factors needed in each

$$= \frac{2s^3}{3r^2s^3} + \frac{12r}{3r^2s^3} - \frac{5r^2s^2}{3r^2s^3}$$

$$= \frac{2s^3 + 12r - 5r^2s^2}{3r^2s^3} \qquad \text{combine numerators over LCD}$$

Note that the minus sign for the third term is used in the numerator. ∎

EXAMPLE 5 Find LCD – adding fractions

$$\frac{a}{x - 1} + \frac{a}{x + 1} = \frac{a(x + 1)}{(x - 1)(x + 1)} + \frac{a(x - 1)}{(x + 1)(x - 1)}$$ change to equivalent fractions with LCD

factors needed

$$= \frac{ax + a + ax - a}{(x + 1)(x - 1)}$$ combine numerators over LCD

$$= \frac{2ax}{(x + 1)(x - 1)}$$ simplify

When we multiply each fraction by the quantity required to obtain the proper denominator, we do not actually have to write the common denominator under each numerator. Placing all the products that appear in the numerators over the common denominator is sufficient. Hence, the illustration in this example would appear as

$$\frac{a}{x - 1} + \frac{a}{x + 1} = \frac{a(x + 1) + a(x - 1)}{(x - 1)(x + 1)} = \frac{ax + a + ax - a}{(x - 1)(x + 1)}$$

$$= \frac{2ax}{(x - 1)(x + 1)}$$ ∎

EXAMPLE 6 Combining fractions – application

The following expression is found in the analysis of the dynamics of missile firing. The indicated addition is performed as shown.

$$\frac{1}{s} - \frac{1}{s + 4} + \frac{8}{s^2 + 8s + 16} = \frac{1}{s} - \frac{1}{s + 4} + \frac{8}{(s + 4)^2}$$ factor third denominator

NOTE ▶

$$= \frac{1(s + 4)^2 - 1(s)(s + 4) + 8s}{s(s + 4)^2}$$ LCD has one factor of s and two factors of $(s + 4)$

$$= \frac{s^2 + 8s + 16 - s^2 - 4s + 8s}{s(s + 4)^2}$$ expand terms of the numerator

Practice Exercise

2. Combine:

$$\frac{3}{2x + 2} - \frac{2}{x^2 + 2x + 1} + \frac{1}{2x^3 + 2x^2}$$

$$= \frac{12s + 16}{s(s + 4)^2} = \frac{4(3s + 4)}{s(s + 4)^2}$$ simplify/factor

We factored the numerator in the final result to see whether or not there were any factors common to the numerator and the denominator. Since there are none, either form of the result is acceptable. ∎

EXAMPLE 7 Combining fractions

$$\frac{3x}{x^2 - x - 12} - \frac{x - 1}{x^2 - 8x + 16} - \frac{6 - x}{2x - 8} = \frac{3x}{(x - 4)(x + 3)} - \frac{x - 1}{(x - 4)^2} - \frac{6 - x}{2(x - 4)}$$ factor denominators

$$= \frac{3x(2)(x - 4) - (x - 1)(2)(x + 3) - (6 - x)(x - 4)(x + 3)}{2(x - 4)^2(x + 3)}$$ change to equivalent fraction with LCD

$$= \frac{[6x^2 - 24x] - [2x^2 + 4x - 6] - [-x^3 + 7x^2 + 6x - 72]}{2(x - 4)^2(x + 3)}$$ expand in numerator

$$= \frac{6x^2 - 24x - 2x^2 - 4x + 6 + x^3 - 7x^2 - 6x + 72}{2(x - 4)^2(x + 3)} = \frac{x^3 - 3x^2 - 34x + 78}{2(x - 4)^2(x + 3)}$$ simplify

In doing this kind of problem, many errors may arise in the use of the minus sign. **CAUTION ▶** Remember, *if a minus sign precedes an expression, the **signs of all terms must be changed** before they can be combined with other terms.* ∎

COMPLEX FRACTIONS

A **complex fraction** *is one in which the numerator, the denominator, or both the numerator and the denominator contain fractions.* The following examples illustrate the simplification of complex fractions.

EXAMPLE 8 Complex fraction

■ The original complex fraction can be written as a division as follows:

$$\frac{2}{x} \div \left(1 - \frac{4}{x}\right)$$

$$\frac{\dfrac{2}{x}}{1 - \dfrac{4}{x}} = \frac{\dfrac{2}{x}}{\dfrac{x-4}{x}} \qquad \text{first, perform subtraction in denominator}$$

$$= \frac{2}{x} \times \frac{x}{x-4} = \frac{2\cancel{x}}{\cancel{x}(x-4)} \qquad \text{invert divisor and multiply}$$

$$= \frac{2}{x-4} \qquad \text{simplify} \qquad\blacksquare$$

EXAMPLE 9 Complex fraction – calculator check

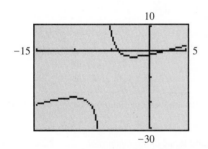

$$\frac{1 - \dfrac{2}{x}}{\dfrac{1}{x} + \dfrac{2}{x^2 + 4x}} = \frac{1 - \dfrac{2}{x}}{\dfrac{1}{x} + \dfrac{2}{x(x+4)}} = \frac{\dfrac{x-2}{x}}{\dfrac{x+4+2}{x(x+4)}} \qquad \text{perform subtraction and addition}$$

$$= \frac{x-2}{x} \times \frac{x(x+4)}{x+6} \qquad \text{invert divisor and multiply}$$

$$= \frac{(x-2)(\cancel{x})(x+4)}{\cancel{x}(x+6)} \qquad \text{indicate multiplication}$$

$$= \frac{(x-2)(x+4)}{x+6} \qquad \text{simplify}$$

Fig. 6.6

With $y_1 = (1 - 2/x)/(1/x + 2/(x^2 + 4x))$ and $y_2 = (x - 2)(x + 4)/(x + 6)$, a graphing calculator check of the result (see Example 8 on page 197) is shown in Fig. 6.6. ■

EXERCISES 6.7

In Exercises 1–4, make the given changes in the indicated examples of this section and then solve the resulting problems.

1. In Example 2, change the denominator of the third fraction to $4a^2b^2$ and then find the LCD.

2. In Example 5, add the fraction $\dfrac{2}{x^2 - 1}$ to those being added and then find the result.

3. In Example 7, change the denominator of the third fraction to $2x + 6$ and then find the result.

4. In Example 8, change the fraction in the numerator to $2/x^2$ and then simplify.

In Exercises 5–44, perform the indicated operations and simplify. For Exercises 33, 34, 39, and 40, check the solution with a graphing calculator.

5. $\dfrac{3}{5} + \dfrac{6}{5}$

6. $\dfrac{2}{13} + \dfrac{6}{13}$

7. $\dfrac{1}{x} + \dfrac{7}{x}$

8. $\dfrac{2}{a} + \dfrac{3}{a}$

9. $\dfrac{1}{2} + \dfrac{3}{4}$

10. $\dfrac{5}{9} - \dfrac{1}{3}$

11. $\dfrac{3}{4x} + \dfrac{7a}{4} + 2$

12. $\dfrac{t-3}{a} - \dfrac{t}{2a}$

13. $\dfrac{a}{x} - \dfrac{b}{x^2}$

14. $\dfrac{3}{2s^2} + \dfrac{5}{4s}$

15. $\dfrac{6}{5x^3} + \dfrac{a}{25x}$

16. $\dfrac{a}{6y} - \dfrac{2b}{3y^4}$

17. $\dfrac{2}{5a} + \dfrac{1}{a} - \dfrac{a}{10}$

18. $\dfrac{1}{2A} - \dfrac{6}{B} - \dfrac{9}{4C}$

19. $\dfrac{x+1}{2x} - \dfrac{y-3}{4y} - \dfrac{2-x}{xy}$

20. $5 + \dfrac{1-x}{2y} - \dfrac{3+x}{4y}$

21. $\dfrac{3}{2x-1} + \dfrac{1}{4x-2}$

22. $\dfrac{5}{6y+3} - \dfrac{a}{8y+4}$

23. $\dfrac{4}{x(x+1)} - \dfrac{3}{2x}$

24. $\dfrac{3}{ax+ay} - \dfrac{1}{a^2}$

25. $\dfrac{s}{2s-6} + \dfrac{1}{4} - \dfrac{3s}{4s-12}$

26. $\dfrac{2}{x+2} - \dfrac{3-x}{x^2+2x} + \dfrac{1}{x}$

27. $\dfrac{3R}{R^2 - 9} - \dfrac{2}{3R + 9}$

28. $\dfrac{2}{n^2 + 4n + 4} - \dfrac{3}{4 + 2n}$

29. $\dfrac{3}{x^2 - 8x + 16} - \dfrac{2}{4 - x}$

30. $\dfrac{2a - b}{c - 3d} - \dfrac{b - 2a}{3d - c}$

31. $\dfrac{v + 4}{v^2 + 5v + 4} - \dfrac{v - 2}{v^2 - 5v + 6}$

32. $\dfrac{N - 1}{2N^3 - 4N^2} - \dfrac{5}{2 - N}$

33. $\dfrac{x - 1}{3x^2 - 13x + 4} - \dfrac{3x + 1}{4 - x}$

34. $\dfrac{x}{4x^2 - 12x + 5} + \dfrac{2x - 1}{4x^2 - 4x - 15}$

35. $\dfrac{t}{t^2 - t - 6} - \dfrac{2t}{t^2 + 6t + 9} + \dfrac{t}{9 - t^2}$

36. $\dfrac{5}{2x^3 - 3x^2 + x} - \dfrac{x}{x^4 - x^2} + \dfrac{2 - x}{2x^2 + x - 1}$

37. $\dfrac{1}{w^3 + 1} + \dfrac{1}{w + 1} - 2$

38. $\dfrac{2}{8 - x^3} + \dfrac{1}{x^2 - x - 2}$

39. $\dfrac{\dfrac{1}{x}}{1 - \dfrac{1}{x}}$

40. $\dfrac{x - \dfrac{1}{x}}{1 - \dfrac{1}{x}}$

41. $\dfrac{\dfrac{x}{y} - \dfrac{y}{x}}{1 + \dfrac{y}{x}}$

42. $\dfrac{\dfrac{V^2 - 9}{V}}{\dfrac{1}{V} - \dfrac{1}{3}}$

43. $\dfrac{\dfrac{3}{x} + \dfrac{1}{x^2 + x}}{\dfrac{1}{x + 1} - \dfrac{1}{x - 1}}$

44. $\dfrac{\dfrac{1}{u - v} + \dfrac{1}{2u + 2v}}{\dfrac{2u}{2u^2 - 3uv + v^2} + \dfrac{2}{2u - v}}$

The expression $f(x + h) - f(x)$ is frequently used in the study of calculus. (If necessary, refer to Section 3.1 for a review of functional notation.) In Exercises 45–48, determine and then simplify this expression for the given functions.

45. $f(x) = \dfrac{x}{x + 1}$

46. $f(x) = \dfrac{3}{2x - 1}$

47. $f(x) = \dfrac{1}{x^2}$

48. $f(x) = \dfrac{2}{x^2 + 4}$

In Exercises 49–57, simplify the given expressions. In Exercise 58, answer the given question.

49. Using the definitions of the trigonometric functions given in Section 4.2, find an expression that is equivalent to $(\tan \theta)(\cot \theta) + (\sin \theta)^2 - \cos \theta$, in terms of x, y, and r.

50. Using the definitions of the trigonometric functions given in Section 4.2, find an expression that is equivalent to $\sec \theta - (\cot \theta)^2 + \csc \theta$, in terms of x, y, and r.

51. If $f(x) = 2x - x^2$, find $f\left(\dfrac{1}{a}\right)$.

52. If $f(x) = x^2 + x$, find $f\left(a + \dfrac{1}{a}\right)$.

53. If $f(x) = x - \dfrac{2}{x}$, find $f(a + 1)$.

54. If $f(x) = 2x - 3$, find $f[1/f(x)]$.

55. The sum of two numbers a and b is divided by the sum of their reciprocals. Simplify the expression for this quotient.

56. Find A and B if $\dfrac{2x - 9}{x^2 - x - 6} = \dfrac{A}{x - 3} + \dfrac{B}{x + 2}$.

57. If $x = \dfrac{mn}{m + n}$ and $y = \dfrac{mn}{m - n}$, show that $\dfrac{y^2 - x^2}{y^2 + x^2} = \dfrac{2mn}{m^2 + n^2}$.

(W) 58. When adding fractions, explain why it is better to find the lowest common denominator rather than any denominator that is common to the fractions.

In Exercises 59–68, perform the indicated operations. Each expression occurs in the indicated area of application.

59. $\dfrac{3}{4\pi} - \dfrac{3H_0}{4\pi H}$ (transistor theory)

60. $1 + \dfrac{9}{128T} - \dfrac{27P}{64T^3}$ (thermodynamics)

61. $\dfrac{2n^2 - n - 4}{2n^2 + 2n - 4} + \dfrac{1}{n - 1}$ (optics)

62. $\dfrac{b}{x^2 + y^2} - \dfrac{2bx^2}{x^4 + 2x^2y^2 + y^4}$ (magnetic field)

63. $\left(\dfrac{3Px}{2L^2}\right)^2 + \left(\dfrac{P}{2L}\right)^2$ (force of a weld)

64. $\dfrac{a}{b^2 h} + \dfrac{c}{bh^2} - \dfrac{1}{6bh}$ (strength of materials)

65. $\dfrac{\dfrac{L}{C} + \dfrac{R}{sC}}{sL + R + \dfrac{1}{sC}}$ (electricity)

66. $\dfrac{\dfrac{m}{c}}{1 - \dfrac{p^2}{c^2}}$ (airfoil design)

67. $\dfrac{1}{R^2} + \left(\omega c - \dfrac{1}{\omega L}\right)^2$ (electricity)

68. $\dfrac{\dfrac{x}{h_1} + \dfrac{x - L}{h_2}}{1 + \dfrac{x(L - x)}{h_1 h_2}}$ (optics)

Answers to Practice Exercises

1. $4x^2(x + 1)^2$ **2.** $\dfrac{3x^3 - x^2 + x + 1}{2x^2(x + 1)^2}$

6.8 Equations Involving Fractions

Multiply Each Term by the LCD •
Solving Equations • Applications •
Extraneous Solutions

Many important equations in science and technology have fractions in them. Although the solution of these equations will still involve the use of the basic operations stated in Section 1.10, an additional procedure can be used to eliminate the fractions and thereby help lead to the solution. The method is to

NOTE ▶   *multiply each term of the equation by the LCD.*

The resulting equation will not involve fractions and can be solved by methods previously discussed. The following examples illustrate how to solve equations involving fractions.

EXAMPLE 1 Multiply each term by LCD

Solve for x: $\dfrac{x}{12} - \dfrac{1}{8} = \dfrac{x + 2}{6}$.

First, note that the LCD of the terms of the equation is 24. Therefore, multiply each term by 24. This gives

$$\frac{24(x)}{12} - \frac{24(1)}{8} = \frac{24(x + 2)}{6} \qquad \text{each term multiplied by LCD}$$

■ Note carefully that we are multiplying both sides of the equation by the LCD and not combining terms over a common denominator as when adding fractions.

Reduce each term to its lowest terms and solve the resulting equation:

$$2x - 3 = 4(x + 2) \qquad \text{each term reduced}$$
$$2x - 3 = 4x + 8$$
$$-2x = 11$$
$$x = -\frac{11}{2}$$

Fig. 6.7

When we check this solution in the original equation, we obtain $-7/12$ on each side of the equal sign. Therefore, the solution is correct.

A graphing calculator solution (see page 97) using $y_1 = x/12 - 1/8 - (x + 2)/6$ is shown in Fig. 6.7. The zero of -5.5 agrees with solution $-11/2$.   ■

EXAMPLE 2 Solving a literal equation

Solve for x: $\dfrac{x}{2} - \dfrac{1}{b^2} = \dfrac{x}{2b}$.

First, determine that the LCD of the terms of the equation is $2b^2$. Then multiply each term by $2b^2$ and continue with the solution:

$$\frac{2b^2(x)}{2} - \frac{2b^2(1)}{b^2} = \frac{2b^2(x)}{2b} \qquad \text{each term multiplied by LCD}$$
$$b^2x - 2 = bx \qquad \text{each term reduced}$$
$$b^2x - bx = 2$$
$$x(b^2 - b) = 2 \qquad \text{factor}$$
$$x = \frac{2}{b^2 - b}$$

■ The procedure for solving equations on page 36 can now be stated as:

1. If the equation has any fractions, first remove the fractions.
2. Combine like terms on each side.
3. Remove grouping symbols.
4. Perform the same operations on both sides, and factor if necessary, until $x = $ result is obtained.
5. Check the solution in the original equation.

Note the use of factoring in arriving at the final result. Checking shows that each side of the original equation is equal to $\dfrac{1}{b^2(b - 1)}$.   ■

EXAMPLE 3 Solving an equation – application

An equation relating the focal length f of a lens with the object distance p and the image distance q is given below. See Fig. 6.8. Solve for q.

$$f = \frac{pq}{p + q} \qquad \text{given equation}$$

Since the only denominator is $p + q$, the LCD is also $p + q$. By first multiplying each term by $p + q$, the solution is completed as follows:

$$f(p + q) = \frac{pq(p + q)}{p + q} \qquad \text{each term multiplied by LCD}$$

$$fp + fq = pq \qquad \text{reduce term on right}$$

$$fq - pq = -fp$$

$$q(f - p) = -fp \qquad \text{factor}$$

$$q = \frac{-fp}{f - p} \qquad \text{divide by } f - p$$

$$= -\frac{fp}{-(p - f)} \qquad \text{use Eq. (6.11)}$$

$$= \frac{fp}{p - f}$$

The last form is preferred since there is no minus sign before the fraction. However, either form of the result is correct. ■

EXAMPLE 4 Solving an equation – application

In astronomy, when developing the equations that describe the motion of the planets, the equation

$$\frac{1}{2}v^2 - \frac{GM}{r} = -\frac{GM}{2a}$$

is found. Solve for M.

First, determine that the LCD of the terms of the equation is $2ar$. Multiplying each term by $2ar$ and proceeding, we have

$$\frac{2ar(v^2)}{2} - \frac{2ar(GM)}{r} = -\frac{2ar(GM)}{2a} \qquad \text{each term multiplied by LCD}$$

$$arv^2 - 2aGM = -rGM \qquad \text{each term reduced}$$

$$rGM - 2aGM = -arv^2$$

$$M(rG - 2aG) = -arv^2 \qquad \text{factor}$$

$$M = -\frac{arv^2}{rG - 2aG}$$

$$= \frac{arv^2}{2aG - rG}$$

The second form of the result is obtained by using Eq. (6.11). Again, note the use of factoring to arrive at the final result. ■

■ See the chapter introduction.

■ The first telescope was invented by Lippershay, a Dutch lens maker, in about 1608.

Object

Focal point

Image

f

p

q

Fig. 6.8

■ For reference, Eq. (6.11) is $x - y = -(y - x)$.

■ See the chapter introduction.

■ In 1609, the Italian scientist Galileo (1564–1642) learned of the invention of the telescope and developed it for astronomical observations. Among his first discoveries were the four largest moons of the planet Jupiter.

Practice Exercise

1. Solve for x: $\dfrac{x - 1}{3a} = \dfrac{5}{6} + \dfrac{x}{a^2}$

EXAMPLE 5 Extraneous solution

Solve for x: $\dfrac{2}{x+1} - \dfrac{1}{x} = -\dfrac{2}{x^2+x}$.

The solution is as follows:

$$\frac{2(x)(x+1)}{x+1} - \frac{x(x+1)}{x} = -\frac{2x(x+1)}{x(x+1)} \qquad \text{multiplying each side by the LCD of } x(x+1)$$

$$2x - (x+1) = -2 \qquad \text{simplifying each fraction}$$

$$2x - x - 1 = -2 \qquad \text{completing the solution}$$

$$x = -1$$

CAUTION ▶

Extraneous Solutions

Checking this solution *in the original equation,* notice the zero in the denominators of the first and third terms of the equation. Since division by zero is undefined (see Section 1.2), $x = -1$ cannot be a solution. *Thus, there is* **no solution** *to this equation.* This example points out clearly why it is necessary to check solutions in the original equation. It also shows that *whenever we multiply each term by a common denominator that* **contains the unknown,** *it is possible to obtain a value that is not a solution of the original equation. Such a value is termed an* **extraneous solution.** Only certain equations will lead to extraneous solutions, but we must be careful to identify them when they occur. ■

A number of stated problems give rise to equations involving fractions. The following examples illustrate the solution of such problems.

EXAMPLE 6 Solving a word problem

■ The first large-scale electronic computer was the ENIAC. It was constructed at the Univ. of Pennsylvania in the mid-1940s and used until 1955. It had 18,000 vacuum tubes and occupied 15,000 ft² of floor space.

■ A military programmable computer called *Colossus* was used to break the German codes in World War II. It was developed, in secret, before ENIAC.

An industrial firm uses a computer system that processes and prints out its data for an average day in 20 h. To process the data more rapidly and to handle increased future computer needs, the firm plans to add new components to the system. One set of new components can process the data in 12 h, without the present system. How long would it take the new system, a combination of the present system and the new components, to process the data?

First, let x = the number of hours for the new system to process the data. Next, we know that it takes the present system 20 h to do it. This means that it processes $\frac{1}{20}$ of the data in 1 h, or $\frac{1}{20}x$ of the data in x h. In the same way, the new components can process $\frac{1}{12}x$ of the data in x h. When x h have passed, the new system will have processed all of the data. Therefore,

<table>
<tr><td align="center">part of data processed
by present system</td><td></td><td align="center">part of data processed
by new components</td><td></td><td></td></tr>
<tr><td></td><td></td><td></td><td></td><td align="center">one complete processing</td></tr>
<tr><td align="center">$\dfrac{x}{20}$</td><td align="center">$+$</td><td align="center">$\dfrac{x}{12}$</td><td align="center">$=\quad 1$</td><td align="center">(all of data)</td></tr>
</table>

$$\frac{60x}{20} + \frac{60x}{12} = 60(1) \qquad \text{each term multiplied by LCD of 60}$$

$$3x + 5x = 60 \qquad \text{each term reduced}$$

$$8x = 60$$

$$x = \frac{60}{8} = 7.5 \text{ h}$$

Therefore, the new system should take about 7.5 h to process the data. ■

EXAMPLE 7 Solving a word problem

A bus averaging 80.0 km/h takes 1.50 h longer to travel from city A to city B than a train that averages 96.0 km/h. How far apart are the cities?

Let d = the distance from city A to city B. Since distance = rate $\times$ time, then distance $\div$ rate = time. Therefore, the time the train takes is $d/96.0$ h, and the time the bus takes is $d/80.0$ h. This means

$$\frac{d}{80.0} - \frac{d}{96.0} = 1.50 \qquad \begin{array}{l} 80 = 16(5) = (2^4)(5) \\ 96 = 32(3) = (2^5)(3) \\ \text{LCD} = (2^5)(3)(5) = 480 \end{array}$$

$$\frac{480d}{80.0} - \frac{480d}{96.0} = 480(1.50)$$

$$6.00d - 5.00d = 720, \qquad d = 720 \text{ km} \qquad \blacksquare$$

EXERCISES 6.8

In Exercises 1–4, make the given changes in the indicated examples of this section and then solve for the indicated variable.

1. In Example 2, change the second denominator from b^2 to b and then solve for x.

2. In Example 3, solve for p.

3. In Example 4, solve for G.

4. In Example 5, change the numerator on the right to 1 and then solve for x.

In Exercises 5–32, solve the given equations and check the results.

5. $\dfrac{x}{2} + 6 = 2x$

6. $\dfrac{x}{5} + 2 = \dfrac{15 + x}{10}$

7. $\dfrac{x}{6} - \dfrac{1}{2} = \dfrac{x}{3}$

8. $\dfrac{3N}{8} - \dfrac{3}{4} = \dfrac{N - 4}{2}$

9. $1 - \dfrac{t - 5}{6} = \dfrac{3}{4}$

10. $\dfrac{2x - 7}{3} + 5 = \dfrac{1}{5}$

11. $\dfrac{3x}{7} - \dfrac{5}{21} = \dfrac{2 - x}{14}$

12. $\dfrac{F - 3}{12} - \dfrac{2}{3} = \dfrac{1 - 3F}{2}$

13. $\dfrac{3}{T} + 2 = \dfrac{5}{3}$

14. $\dfrac{1}{2y} - \dfrac{1}{2} = 4$

15. $3 - \dfrac{x - 2}{5x} = \dfrac{1}{5}$

16. $\dfrac{1}{2R} - \dfrac{1}{3} = \dfrac{2}{3R}$

17. $\dfrac{2y}{y - 1} = 5$

18. $\dfrac{x}{2x - 3} = 4$

19. $\dfrac{2}{s} = \dfrac{3}{s - 1}$

20. $\dfrac{5}{n + 2} = \dfrac{3}{2n}$

21. $\dfrac{5}{2x + 4} + \dfrac{3}{6x + 12} = 2$

22. $\dfrac{3}{4x - 6} + \dfrac{1}{4} = \dfrac{5}{3 - 2x}$

23. $\dfrac{2}{Z - 5} - \dfrac{3}{10 - 2Z} = 3$

24. $\dfrac{4}{4 - x} + 2 - \dfrac{2}{12 - 3x} = \dfrac{1}{3}$

25. $\dfrac{1}{4x} + \dfrac{3}{2x} = \dfrac{2}{x + 1}$

26. $\dfrac{3}{t + 3} - \dfrac{1}{t} = \dfrac{5}{6 + 2t}$

27. $\dfrac{7}{y} = \dfrac{3}{y - 4} + \dfrac{7}{2y^2 - 8y}$

28. $\dfrac{1}{2x + 3} = \dfrac{5}{2x} - \dfrac{4}{2x^2 + 3x}$

29. $\dfrac{1}{x^2 - x} - \dfrac{1}{x} = \dfrac{1}{x - 1}$

30. $\dfrac{2}{x^2 - 1} - \dfrac{2}{x + 1} = \dfrac{1}{x - 1}$

31. $\dfrac{2}{B^2 - 4} - \dfrac{1}{B - 2} = \dfrac{1}{2B + 4}$

32. $\dfrac{2}{2x^2 + 5x - 3} - \dfrac{1}{4x - 2} + \dfrac{3}{2x + 6} = 0$

In Exercises 33–48, solve for the indicated letter. In Exercises 37–48, each of the given formulas arises in the technical or scientific area of study listed.

33. $2 - \dfrac{1}{b} + \dfrac{3}{c} = 0$, for c

34. $\dfrac{2}{3} - \dfrac{h}{x} = \dfrac{1}{6x}$, for x

35. $\dfrac{t - 3}{b} - \dfrac{t}{2b - 1} = \dfrac{1}{2}$, for t

36. $\dfrac{1}{a^2 + 2a} - \dfrac{y}{2a} = \dfrac{2y}{a + 2}$, for y

37. $\dfrac{s - s_0}{t} = \dfrac{v + v_0}{2}$, for v (velocity of object)

38. $S = \dfrac{P}{A} + \dfrac{Mc}{I}$, for A (machine design)

39. $V = 1.2\left(5.0 + \dfrac{8.0R}{8.0 + R}\right)$, for R

(electric resistance in Fig. 6.9)

Fig. 6.9

40. $C = \dfrac{7p}{100 - p}$, for p (environmental pollution)

41. $z = \dfrac{1}{g_m} - \dfrac{jX}{g_m R}$, for R (FM transmission)

42. $A = \dfrac{1}{2} wp - \dfrac{1}{2} w^2 - \dfrac{\pi}{8} w^2$, for p (architecture)

43. $P = \dfrac{RT}{V - b} - \dfrac{a}{V^2}$, for T (thermodynamics)

44. $\dfrac{1}{x} + \dfrac{1}{nx} = \dfrac{1}{f}$, for f (photography)

45. $\dfrac{1}{C} = \dfrac{1}{C_2} + \dfrac{1}{C_1 + C_3}$, for C_1 (electricity: capacitors)

46. $D = \dfrac{wx^4}{24EI} - \dfrac{wLx^3}{6EI} + \dfrac{wL^2x^2}{4EI}$, for w (beam design)

47. $f(n - 1) = \dfrac{1}{\dfrac{1}{R_1} + \dfrac{1}{R_2}}$, for R_1 (optics)

48. $P = \dfrac{\dfrac{1}{1 + i}}{1 - \dfrac{1}{1 + i}}$, for i (business)

In Exercises 49–58, set up appropriate equations and solve the given stated problems. All numbers are accurate to at least two significant digits.

49. One pump can empty an oil tanker in 5.0 h, and a second pump can empty the tanker in 8.0 h. How long would it take the two pumps working together to empty the tanker?

50. One company determines that it will take its crew 450 h to clean up a chemical dump site, and a second company determines that it will take its crew 600 h to clean up the site. How long will it take the two crews working together?

51. One automatic packaging machine can package 100 boxes of machine parts in 12 min, and a second machine can do it in 10 min. A newer model machine can do it in 8.0 min. How long will it take the three machines working together?

52. A painting crew can paint a structure in 12 h, or the crew can paint it in 7.2 h when working with a second crew. How long would it take the second crew to do the job if working alone?

53. An elevator traveled from the first floor to the top floor of a building at an average speed of 2.0 m/s and returned to the first floor at 2.2 m/s. If it was on the top floor for 90 s and the total elapsed time was 5.0 min, how far above the first floor is the top floor?

54. A commuter rapid transit train travels 24 km farther between stops A and B than between stops B and C. If it averages 60 km/h from A to B and 30 km/h between B and C, and an express averages 50 km/h between A and C (not stopping at B), how far apart are stops A and C?

55. A jet takes the same time to travel 2580 km with the wind as it does to travel 1800 km against the wind. If its speed relative to the air is 450 km/h, what is the speed of the wind?

56. An engineer travels from Aberdeen, Scotland, to the Montrose oil field in the North Sea on a ship that averages 28 km/h. After spending 6.0 h at the field, the engineer returns to Aberdeen in a helicopter that averages 140 km/h. If the total trip takes 15.0 h, how far is the Montrose oil field from Aberdeen?

57. The current through each of the resistances R_1 and R_2 in Fig. 6.10 equals the voltage V divided by the resistance. The sum of the currents equals the current i in the rest of the circuit. Find the voltage if $i = 1.2$ A, $R_1 = 2.7\ \Omega$, and $R_2 = 6.0\ \Omega$.

Fig. 6.10

58. A fox, pursued by a greyhound, has a start of 60 leaps. He makes 9 leaps while the greyhound makes but 6; but, 3 leaps of the greyhound are equivalent to 7 of the fox. How many leaps must the greyhound make to overcome the fox? (Copied from Davies, Charles, *Elementary Algebra,* New York: A. S. Barnes & Burr, 1852.) (*Hint:* Let the unit of distance be one fox leap.)

In Exercises 59 and 60, find constants A and B such that the equation is true.

59. $\dfrac{x - 12}{x^2 + x - 6} = \dfrac{A}{x + 3} + \dfrac{B}{x - 2}$

60. $\dfrac{23 - x}{2x^2 + 7x - 4} = \dfrac{A}{2x - 1} - \dfrac{B}{x + 4}$

Answer to Practice Exercise

1. $x = \dfrac{5a^2 + 2a}{2a - 6}$

CHAPTER 6   EQUATIONS

$$a(x + y) = ax + ay \tag{6.1}$$
$$(x + y)(x - y) = x^2 - y^2 \tag{6.2}$$
$$(x + y)^2 = x^2 + 2xy + y^2 \tag{6.3}$$
$$(x - y)^2 = x^2 - 2xy + y^2 \tag{6.4}$$

$$(x + a)(x + b) = x^2 + (a + b)x + ab \tag{6.5}$$

$$(ax + b)(cx + d) = acx^2 + (ad + bc)x + bd \tag{6.6}$$

$$(x + y)^3 = x^3 + 3x^2y + 3xy^2 + y^3 \tag{6.7}$$

$$(x - y)^3 = x^3 - 3x^2y + 3xy^2 - y^3 \tag{6.8}$$

$$(x + y)(x^2 - xy + y^2) = x^3 + y^3 \tag{6.9}$$

$$(x - y)(x^2 + xy + y^2) = x^3 - y^3 \tag{6.10}$$

$$x - y = -(y - x) \tag{6.11}$$

CHAPTER 6 REVIEW EXERCISES

In Exercises 1–12, find the products by inspection. No intermediate steps should be necessary.

1. $3a(4x + 5a)$ **2.** $-7xy(4x^2 - 7y)$

3. $(2a + 7b)(2a - 7b)$ **4.** $(x - 4z)(x + 4z)$

5. $(2a + 1)^2$ **6.** $(5C - 2D)^2$

7. $(b - 4)(b + 7)$ **8.** $(y - 5)(y - 7)$

9. $(2x + 5)(x - 9)$ **10.** $(4ax - 3)(5ax + 7)$

11. $(2c^a + d)(2c^a - d)$ **12.** $(3s^n - 2t)(3s^n + 2t)$

In Exercises 13–44, factor the given expressions completely.

13. $3s + 9t$ **14.** $7x - 28y$

15. $a^2x^2 + a^2$ **16.** $3ax - 6ax^4 - 9a$

17. $W^2b^{x+2} - 144b^x$ **18.** $900n^a - n^{a+4}$

19. $16(x + 2)^2 - t^4$ **20.** $25s^4 - 36t^2$

21. $36t^2 - 24t + 4$ **22.** $4x^2 - 12x + 9$

23. $25t^2 + 10t + 1$ **24.** $4c^2 + 36cd + 81d^2$

25. $x^2 + x - 56$ **26.** $x^2 - 4x - 45$

27. $t^4 - 5t^2 - 36$ **28.** $3N^4 - 33N^2 + 30$

29. $2k^2 - k - 36$ **30.** $5x^2 + 2x - 3$

31. $8x^2 - 8x - 70$ **32.** $27F^3 + 21F^2 - 48F$

33. $10b^2 + 23b - 5$ **34.** $12x^2 - 7xy - 12y^2$

35. $4x^2 - 64y^2$ **36.** $4a^2x^2 + 26a^2x + 36a^2$

37. $250 - 16y^6$ **38.** $8a^4 + 64a$

39. $8x^3 + 27$ **40.** $R^3 - 125r^3$

41. $ab^2 - 3b^2 + a - 3$ **42.** $axy - ay + ax - a$

43. $nx + 5n - x^2 + 25$ **44.** $ty - 4t + y^2 - 16$

In Exercises 45–68, perform the indicated operations and express results in simplest form.

45. $\dfrac{48ax^3y^6}{9a^3xy^6}$ **46.** $\dfrac{-39r^2s^4t^8}{52rs^5t}$

47. $\dfrac{6x^2 - 7x - 3}{4x^2 - 8x + 3}$ **48.** $\dfrac{p^4 - 4p^2 - 4}{p^4 - p^2 - 12}$

49. $\dfrac{4x + 4y}{35x^2} \times \dfrac{28x}{x^2 - y^2}$ **50.** $\left(\dfrac{6x - 3}{x^2}\right)\left(\dfrac{4x^2 - 12x}{6 - 12x}\right)$

51. $\dfrac{18 - 6L}{L^2 - 6L + 9} \div \dfrac{L^2 - 2L - 15}{L^2 - 9}$

52. $\dfrac{6x^2 - xy - y^2}{2x^2 + xy - y^2} \div \dfrac{16y^2 - 4x^2}{2x^2 + 6xy + 4y^2}$

53. $\dfrac{\dfrac{3x}{7x^2 + 13x - 2}}{\dfrac{6x^2}{x^2 + 4x + 4}}$ **54.** $\dfrac{\dfrac{3x - 3y}{2x^2 + 3xy - 2y^2}}{\dfrac{3x^2 - 3y^2}{x^2 + 4xy + 4y^2}}$

55. $\dfrac{x + \dfrac{1}{x} + 1}{x^2 - \dfrac{1}{x}}$ **56.** $\dfrac{\dfrac{4}{y} - 4y}{2 - \dfrac{2}{y}}$

57. $\dfrac{4}{9x} - \dfrac{5}{12x^2}$ **58.** $\dfrac{3}{10a^2} + \dfrac{1}{4a^3}$

59. $\dfrac{6}{x} - \dfrac{7}{2x} + \dfrac{3}{xy}$ **60.** $\dfrac{T}{T^2 + 2} - \dfrac{1}{2T + T^3}$

61. $\dfrac{a + 1}{a + 2} - \dfrac{a + 3}{a}$ **62.** $\dfrac{y}{y + 2} - \dfrac{1}{y^2 + 2y}$

63. $\dfrac{2x}{x^2 + 2x - 3} - \dfrac{1}{6x + 2x^2}$ **64.** $\dfrac{x}{4x^2 + 4x - 3} - \dfrac{3}{4x^2 - 9}$

65. $\dfrac{3x}{2x^2 - 2} - \dfrac{2}{4x^2 - 5x + 1}$ **66.** $\dfrac{2n - 1}{4 - n} + \dfrac{n + 2}{5n - 20}$

67. $\dfrac{3x}{x^2 + 2x - 3} - \dfrac{2}{x^2 + 3x} + \dfrac{x}{x - 1}$

68. $\dfrac{3}{y^4 - 2y^3 - 8y^2} + \dfrac{y - 1}{y^2 + 2y} - \dfrac{y - 3}{y^2 - 4y}$

In Exercises 69–72, graphically check the results for the indicated exercises of this set on a graphing calculator.

69. Exercise 47   **70.** Exercise 50

71. Exercise 55   **72.** Exercise 64

In Exercises 73–76, factor the given expressions. In Exercises 73 and 74, disregard the restriction that coefficients and exponents must be integers. The expressions in Exercises 75 and 76 can be factored if they are first rewritten in a different form.

73. $x^2 - 5$

74. $x - y$

75. $x + y$ (one of two factors is to be x)

76. $x + y$ (one of two factors is to be a)

In Exercises 77–84, solve the given equations.

77. $\dfrac{x}{2} - 3 = \dfrac{x - 10}{4}$

78. $\dfrac{2x}{c} - \dfrac{1}{2c} = \dfrac{3}{c} - x$, for x

79. $\dfrac{2}{t} - \dfrac{1}{at} = 2 + \dfrac{a}{t}$, for t

80. $\dfrac{3}{a^2 y} - \dfrac{1}{ay} = \dfrac{9}{a}$, for y

81. $\dfrac{2x}{2x^2 - 5x} - \dfrac{3}{x} = \dfrac{1}{4x - 10}$

82. $\dfrac{3}{x^2 + 3x} - \dfrac{1}{x} = \dfrac{1}{3 + x}$

83. Given $f(x) = \dfrac{1}{x + 2}$, solve for x if $f(x + 2) = 2f(x)$.

84. Given $f(x) = \dfrac{x}{x + 1}$, solve for x if $3f(x) + f\left(\dfrac{1}{x}\right) = 2$.

In Exercises 85–102, solve the given problems. Where indicated, the expression is found in the stated technical area.

85. In an algebraic fraction, what effect on the result is made if the sign is changed of; (a) an odd number of factors?; (b) an even number of factors?

86. Algebraically show that the reciprocal of the reciprocal of the number x is x.

87. Show that $xy = \dfrac{1}{4}[(x + y)^2 - (x - y)^2]$.

88. Show that $x^2 + y^2 = \dfrac{1}{2}[(x + y)^2 + (x - y)^2]$.

89. Multiply: $2zS(S + 1)$ (solid-state physics)

90. Expand: $kr(R - r)$ (blood flow)

91. Expand: $[2b + (n - 1)\lambda]^2$ (optics)

92. Factor: $\pi r_1^2 l - \pi r_2^2 l$ (jet plane fuel supply)

93. Factor: $cT_2 - cT_1 + RT_2 - RT_1$ (pipeline flow)

94. Factor: $9600t + 8400t^2 - 1200t^3$ (solar energy)

95. Simplify and express in factored form: $(2R - r)^2 - (r^2 + R^2)$ (aircraft radar)

96. Express in factored form: $2R(R + r) - (R + r)^2$ (electricity: power)

97. Expand and simplify: $(n + 1)^3(2n + 1)^3$ (fluid flow in pipes)

98. Expand and simplify: $2(e_1 - e_2)^2 + 2(e_2 - e_3)^2$ (mechanical design)

99. Expand and simplify: $10a(T - t) + a(T - t)^2$ (instrumentation)

100. Expand the third term and then factor by grouping: $pa^2 + (1 - p)b^2 - [pa + (1 - p)b]^2$ (nuclear physics)

101. A metal cube of edge x is heated and each edge increases by 4 mm. Express the increase in volume in factored form.

102. Express the difference in volumes of two ball bearings of radii r mm and 3 mm in factored form. ($r > 3$ mm)

In Exercises 103–114, perform the indicated operations and simplify the given expressions. Each expression is from the indicated technical area of application.

103. $\left(\dfrac{2wtv^2}{Dg}\right)\left(\dfrac{b\pi^2 D^2}{n^2}\right)\left(\dfrac{6}{bt^2}\right)$ (machine design)

104. $\dfrac{m}{c} \div \left[1 - \left(\dfrac{p}{c}\right)^2\right]$ (airfoil design)

105. $\dfrac{\frac{\pi ka}{2}(R^4 - r^4)}{\pi ka(R^2 - r^2)}$ (flywheel rotation)

106. $\dfrac{V}{kp} - \dfrac{RT}{k^2 p^2}$ (electric motors)

107. $1 - \dfrac{d^2}{2} + \dfrac{d^4}{24} - \dfrac{d^6}{120}$ (aircraft emergency locator transmitter)

108. $\dfrac{wx^2}{2T_0} + \dfrac{kx^4}{12T_0}$ (bridge design)

109. $\dfrac{4k - 1}{4k - 4} + \dfrac{1}{2k}$ (spring stress)

110. $\dfrac{Am}{k} - \dfrac{g}{2}\left(\dfrac{m}{k}\right)^2 + \dfrac{AML}{k}$ (rocket fuel)

111. $1 - \dfrac{3a}{4r} - \dfrac{a^3}{4r^3}$ (hydrodynamics)

112. $\dfrac{1}{F} + \dfrac{1}{f} - \dfrac{d}{fF}$ (optics)

113. $\dfrac{\frac{u^2}{2g} - x}{\frac{1}{2gc^2} - \frac{u^2}{2g} + x}$ (mechanism design)

114. $\dfrac{V}{\frac{1}{2R} + \frac{1}{2R + 2}}$ (electricity)

In Exercises 115–124, solve for the indicated letter. Each equation is from the indicated technical area of application.

115. $W = mgh_2 - mgh_1$, for m (work done on object)

116. $R_1(V_2 - V_1) + R_2 V_2 = 3R_1 R_2$, for V_2 (electricity)

117. $R = \dfrac{wL}{H(w + L)}$, for L (architecture)

118. $\dfrac{J}{T} = \dfrac{t}{\omega_1 - \omega_2}$, for ω_2 (rotational torque)

119. $E = V_0 + \dfrac{(m + M)V^2}{2} + \dfrac{p^2}{2I}$, for I (nuclear physics)

120. $\dfrac{q_2 - q_1}{d} = \dfrac{f + q_1}{D}$, for q_1 (photography)

121. $s^2 + \dfrac{cs}{m} + \dfrac{kL^2}{mb^2} = 0$, for c (mechanical vibrations)

122. $I = \dfrac{A}{x^2} + \dfrac{B}{(10 - x)^2}$, for A (optics)

123. $V = \dfrac{i}{sC} + \dfrac{V_0}{s}$, for C (electricity)

124. $C = \dfrac{k}{n(1 - k)}$, for k (reinforced concrete design)

In Exercises 125–132, set up appropriate equations and solve the given stated problems. All numbers are accurate to at least two significant digits.

125. If a certain car's lights are left on, the battery will be dead in 4.0 h. If only the radio is left on, the battery will be dead in 24 h. How long will the battery last if both the lights and the radio are left on?

126. Two pumps are being used to fight a fire. One pumps 5000 gal in 20 min, and the other pumps 5000 gal in 25 min. How long will it take the two pumps together to pump 5000 gal?

127. A number m is the *harmonic mean* of number x and y if $1/m$ equals the average of $1/x$ and $1/y$ (divide $1/x + 1/y$ by 2). Find the harmonic mean of the musical notes that have frequencies of 400 Hz and 1200 Hz.

128. An auto mechanic can do a certain motor job in 3.0 h, and with an assistant he can do it in 2.1 h. How long would it take the assistant to do the job alone?

129. The *relative density* of an object may be defined as its weight in air w_a, divided by the difference of its weight in air and its weight when submerged in water, w_w. For a lead weight, $w_a = 1.097\, w_w$. Find the relative density of lead.

130. A car travels halfway to its destination at 80.0 km/h and the remainder of the distance at 60.0 km/h. What is the average speed of the car for the trip?

131. For electric resistors in parallel, the reciprocal of the combined resistance equals the sum of the reciprocals of the individual resistances. For three resistors of 12 Ω, R ohms, and $2R$ ohms, in parallel, the combined resistance is 6.0 Ω. Find R.

132. An ambulance averaged 36 mi/h going to an accident and 48 mi/h on its return to the hospital. If the total time for the round-trip was 40 min, including 5 min at the accident scene, how far from the hospital was the accident?

Writing Exercise

133. In analyzing a certain electric circuit, the expression

$$\dfrac{\left(1 + \dfrac{1}{s}\right)\left(1 + \dfrac{1}{s/2}\right)}{3 + \dfrac{1}{s} + \dfrac{1}{s/2}}$$ is used. Simplify this expression, and

write a paragraph describing your procedure. When you "cancel," explain what basic operation is being performed.

CHAPTER 6 PRACTICE TEST

1. Find the product: $2x(2x - 3)^2$.

2. The following equation is used in elecricity.

Solve for R_1: $\dfrac{1}{R} = \dfrac{1}{R_1 + r} + \dfrac{1}{R_2}$

3. Reduce to simplest form: $\dfrac{2x^2 + 5x - 3}{2x^2 + 12x + 18}$

4. Factor: $4x^2 - 16y^2$

5. Factor: $pb^3 + 8a^3p$ (business application)

6. Factor: $2a - 4T - ba + 2bT$

7. Factor: $36x^2 + 14x - 16$

In problems 8–10, perform the indicated operations and simplify.

8. $\dfrac{3}{4x^2} - \dfrac{2}{x^2 - x} - \dfrac{x}{2x - 2}$

9. $\dfrac{x^2 + x}{2 - x} \div \dfrac{x^2}{x^2 - 4x + 4}$

10. $\dfrac{1 - \dfrac{3}{2x + 2}}{\dfrac{x}{5} - \dfrac{1}{2}}$

11. If one riveter can do a job in 12 days, and a second riveter can do it in 16 days, how long would it take for them to do it together?

12. Solve for x: $\dfrac{3}{2x^2 - 3x} + \dfrac{1}{x} = \dfrac{3}{2x - 3}$

Quadratic Equations

We showed earlier how to solve some basic equations, and in this chapter we develop methods of solving the important *quadratic equation*. In the first three sections, we present algebraic methods of solution, and in Section 7.4 we discuss graphical solutions, including the use of the graphing calculator.

We will see that the graph of the *quadratic function* is a *parabola*, which has many modern applications in science and technology (Examples 2 and 3 on page 92 illustrate the graph of a parabola). We will discuss the parabola more in this chapter and in detail in Chapter 21.

One of the applications of a parabola is shown by a TV signal from a space satellite to a microwave dish at a home. The signal is reflected off the dish to a receiver at the *focus* of the dish, from which the signal goes to the TV set in the home. The surface of the dish is parabolic, and the fact that parabolic surfaces have this property of reflection was first shown by the Greek mathematician Diocles in about 200 B.C.E. He wanted to find a mirror to reflect the rays of the Sun to a point and cause burning. He proved that this was true for a mirror with a parabolic surface.

Another important application of a parabola (and thereby of quadratic functions) is that of a projectile, examples of which are a baseball, an artillery shell, and a rocket. When Galileo showed that the distance an object falls does not depend on its weight, he discovered that the distance fallen depends on the square of the time of fall. This, in turn, was shown to mean that the path of a projectile is parabolic (not considering air resistance). To find the location of a projectile for any given time of flight requires the solution of a quadratic equation.

The Babylonians developed methods of solving quadratic equations nearly 2000 years ago. However, their use in applied situations was limited, as they had little real-world use for them. Today, in addition to those mentioned above, quadratic equations have applications in architecture, electric circuits, mechanical systems, forces on structures, and product design.

In Section 7.3, we see how the design of a swimming pool area involves the solution of a quadratic equation.

7.1 Quadratic Equations; Solution by Factoring

General Quadratic Equation •
Solutions of a Quadratic Equation •
Procedure for Solving by Factoring •
Equations with Fractions

■ It is usually easier to have integral coefficients and $a > 0$ in a quadratic equation. This can be done by multiplying each term by the LCD if there are fractions, and to change the sign of all terms if $a < 0$.

Given that a, b, and c are constants $(a \neq 0)$, the equation

$$ax^2 + bx + c = 0 \qquad\qquad (7.1)$$

is called the **general quadratic equation in x.** The left side of Eq. (7.1) is a polynomial function of degree 2. *This function, $f(x) = ax^2 + bx + c$, is known as the* **quadratic function.** Any equation that can be simplified and then written in the form of Eq. (7.1) is a quadratic equation in one unknown.

Among the applications of quadratic equations and functions are the following examples: In finding the time t of flight of a projectile, we have the equation $s_0 + v_0 t - 16t^2 = 0$; in analyzing the electric current i in a circuit, the function $f(i) = Ei - Ri^2$ is found; and in determining the forces at a distance x along a beam, the function $f(x) = ax^2 + bLx + cL^2$ is used.

Since it is the x^2-term in Eq. (7.1) that distinguishes the quadratic equation from other types of equations, the equation is not quadratic if $a = 0$. However, either b or c (or both) may be zero, and the equation is still quadratic. No power of x higher than the second may be present in a quadratic equation. Also, we should be able to properly identify a quadratic equation even when it does not initially appear in the form of Eq. (7.1). The following two examples illustrate how we may recognize quadratic equations.

EXAMPLE 1 Examples of quadratic equations

The following are quadratic equations.

■ If either (or both) the first-power term or the constant is missing, the quadratic is called *incomplete.*

$$x^2 - 4x - 5 = 0$$
$$\uparrow \qquad \uparrow \qquad \uparrow$$
$$a = 1 \quad b = -4 \quad c = -5$$

To show this equation in the form of Eq. (7.1), it can be written as $1x^2 + (-4)x + (-5) = 0$.

$$3x^2 - 6 = 0$$
$$\uparrow \qquad \uparrow$$
$$a = 3 \quad c = -6$$

Since there is no x-term, $b = 0$.

$$2x^2 + 7x = 0$$
$$\uparrow \qquad \uparrow$$
$$a = 2 \quad b = 7$$

Since no constant appears, $c = 0$.

$$(m - 3)x^2 - mx + 7 = 0$$

The constants in Eq. (7.1) may include literal expressions. In this case, $m - 3$ takes the place of a, $-m$ takes the place of b, and $c = 7$.

$$4x^2 - 2x = x^2$$

After all nonzero terms have been collected on the left side, the equation becomes $3x^2 - 2x = 0$.

$$(x + 1)^2 = 4$$

Expanding the left side and collecting all nonzero terms on the left, we have $x^2 + 2x - 3 = 0$. ■

EXAMPLE 2 Examples of equations not quadratic

The following are not quadratic equations.

$$bx - 6 = 0$$

There is no x^2-term.

$$x^3 - x^2 - 5 = 0$$

There should be no term of degree higher than 2. Thus, there can be no x^3-term in a quadratic equation.

$$x^2 + x - 7 = x^2$$

When terms are collected, there will be no x^2-term. ■

SOLUTIONS OF A QUADRATIC EQUATION

Recall that *the* **solution** *of an equation consists of all numbers* (**roots**) *which, when substituted in the equation, give equality.* There are *two* roots for a quadratic equation. At times, these roots are equal (see Example 3), and only one number is actually a solution. Also, the roots can be imaginary, and if this happens, all we wish to do at this point is to recognize that they are imaginary.

EXAMPLE 3 Solutions (roots) of a quadratic equation

(a) The quadratic equation $3x^2 - 7x + 2 = 0$ has roots $x = 1/3$ and $x = 2$. This is seen by substituting these numbers in the equation.

$$3\left(\tfrac{1}{3}\right)^2 - 7\left(\tfrac{1}{3}\right) + 2 = 3\left(\tfrac{1}{9}\right) - \tfrac{7}{3} + 2 = \tfrac{1}{3} - \tfrac{7}{3} + 2 = \tfrac{0}{3} = 0$$

$$3(2)^2 - 7(2) + 2 = 3(4) - 14 + 2 = 14 - 14 = 0$$

■ Until the 1600s, most mathematicians did not accept negative, irrational, or imaginary roots of an equation. It was also generally accepted that an equation had only one root.

(b) The quadratic equation $4x^2 - 4x + 1 = 0$ has a **double root** (*both roots are the same*) of $x = 1/2$. Showing that this number is a solution, we have

$$4\left(\tfrac{1}{2}\right)^2 - 4\left(\tfrac{1}{2}\right) + 1 = 4\left(\tfrac{1}{4}\right) - 2 + 1 = 1 - 2 + 1 = 0$$

(c) The quadratic equation $x^2 + 9 = 0$ has the imaginary roots $x = 3j$ and $x = -3j$, which means $x = 3\sqrt{-1}$ and $x = -3\sqrt{-1}$. ■

This section deals only with quadratic equations whose quadratic expression is factorable. Therefore, all roots will be rational. Using the fact that

NOTE ▶

a product is zero if any of its factors is zero

we have the following steps in solving a quadratic equation.

PROCEDURE FOR SOLVING A QUADRATIC EQUATION BY FACTORING

1. *Collect all terms on the left and simplify* (to the form of Eq. (7.1)).
2. *Factor the quadratic expression.*
3. *Set each factor equal to zero.*
4. *Solve the resulting linear equations. These numbers are the roots of the quadratic equation.*
5. *Check the solutions in the original equation.*

EXAMPLE 4 Solving quadratic equation by factoring

$$x^2 - x - 12 = 0$$
$$(x - 4)(x + 3) = 0 \qquad \text{factor}$$

$$x - 4 = 0 \qquad x + 3 = 0 \qquad \text{set each factor equal to zero}$$
$$x = 4 \qquad x = -3 \qquad \text{solve}$$

The roots are $x = 4$ and $x = -3$. We can check them in the original equation by substitution. Therefore,

$$(4)^2 - (4) - 12 \overset{?}{=} 0 \qquad (-3)^2 - (-3) - 12 \overset{?}{=} 0$$
$$0 = 0 \qquad\qquad\qquad 0 = 0$$

Both roots satisfy the original equation. ■

Practice Exercise

1. Solve for x: $x^2 + 4x - 21 = 0$

EXAMPLE 5 Solving quadratic equation by factoring

$$2N^2 + 7N - 4 = 0$$

$$(2N - 1)(N + 4) = 0 \qquad \text{factor}$$

$$2N - 1 = 0, \qquad N = \frac{1}{2} \qquad \text{see each factor to zero and solve}$$

$$N + 4 = 0, \qquad N = -4$$

Therefore, the roots are $N = \frac{1}{2}$ and $N = -4$. These roots can be checked by the same procedure used in Example 4. ∎

EXAMPLE 6 First put equation in proper form

$$x^2 + 4 = 4x \qquad \text{equation not in form of Eq. (7.1)}$$

$$x^2 - 4x + 4 = 0 \qquad \text{subtract } 4x \text{ from both sides}$$

$$(x - 2)^2 = 0 \qquad \text{factor}$$

$$x - 2 = 0, \qquad x = 2 \qquad \text{solve}$$

Practice Exercise

2. Solve for x: $9x^2 + 1 = 6x$

Since $(x - 2)^2 = (x - 2)(x - 2)$, both factors are the same. This means there is a double root of $x = 2$. Substitution shows that $x = 2$ satisfies the original equation. ∎

NOTE ▶   *It is essential for the quadratic expression on the left to be equal to zero* (on the right), because if a product equals a nonzero number, it is probable that neither factor will give a correct root. Again, the first step must be to write the equation in the form of Eq. (7.1).

A number of equations involving fractions lead to quadratic equations after the fractions are eliminated. The following two examples, the second being a stated problem, **Equations with Fractions**   illustrate the process of solving such equations with fractions.

EXAMPLE 7 Fractional equation solved as quadratic

Solve for x: $\dfrac{1}{x} + 3 = \dfrac{2}{x + 2}$.

$$\frac{x(x + 2)}{x} + 3x(x + 2) = \frac{2x(x + 2)}{x + 2} \qquad \begin{array}{l}\text{multiply each term by}\\ \text{the LCD, } x(x + 2)\end{array}$$

$$x + 2 + 3x^2 + 6x = 2x \qquad \text{reduce each term}$$

$$3x^2 + 5x + 2 = 0 \qquad \text{collect terms on left}$$

$$(3x + 2)(x + 1) = 0 \qquad \text{factor}$$

$$3x + 2 = 0, \qquad x = -\frac{2}{3} \qquad \begin{array}{l}\text{set each factor equal}\\ \text{to zero and solve}\end{array}$$

$$x + 1 = 0, \qquad x = -1$$

Checking in the original equation, we have

$$\frac{1}{-\frac{2}{3}} + 3 \overset{?}{=} \frac{2}{-\frac{2}{3} + 2} \qquad\qquad \frac{1}{-1} + 3 \overset{?}{=} \frac{2}{-1 + 2}$$

$$-\frac{3}{2} + 3 \overset{?}{=} \frac{2}{\frac{4}{3}} \qquad\qquad -1 + 3 \overset{?}{=} \frac{2}{1}$$

$$\frac{3}{2} = \frac{6}{4} \qquad\qquad\qquad 2 = 2$$

We see that the roots check. Remember, if either value gives division by zero, the root is extraneous, and must be excluded from the solution. ∎

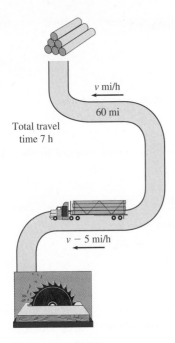

v mi/h

60 mi

Total travel
time 7 h

$v - 5$ mi/h

Fig. 7.1

EXAMPLE 8 Solving a word problem

A lumber truck travels 60 mi from a sawmill to a lumber camp and then back in 7 h travel time. If the truck averages 5 mi/h less on the return trip than on the trip to the camp, find its average speed to the camp. See Fig. 7.1.

Let v = the average speed (in mi/h) of the truck going to the camp. This means that the average speed of the return trip was $(v - 5)$ mi/h.

We also know that $d = vt$ (distance equals speed times time), which tells us that $t = d/v$. Thus, the time for each part of the trip is the distance divided by the speed.

$$\underset{\substack{\text{time to} \\ \text{camp}}}{\frac{60}{v}} + \underset{\substack{\text{time from} \\ \text{camp}}}{\frac{60}{v - 5}} = \underset{\substack{\text{total} \\ \text{time}}}{7}$$

$$60(v - 5) + 60v = 7v(v - 5) \qquad \text{multiply each term by } v(v - 5)$$

$$7v^2 - 155v + 300 = 0 \qquad \text{collect terms on the left}$$

$$(7v - 15)(v - 20) = 0 \qquad \text{factor}$$

$$7v - 15 = 0, \qquad v = \frac{15}{7} \qquad \text{set each factor equal to zero and solve}$$

$$v - 20 = 0, \qquad v = 20$$

The value $v = 15/7$ mi/h cannot be a solution since the return speed of 5 mi//h less would be negative. Thus, the solution is $v = 20$ mi/h, which means the return speed was 15 mi//h. The trip to the camp took 3 h, and the return took 4 h, which shows the solution checks. ∎

EXERCISES 7.1

In Exercises 1 and 2, make the given changes in the indicated examples of this section and then solve the resulting quadratic equations.

1. In Example 5, change the + sign before $7N$ to − and then solve.

2. In Example 7, change the numerator of the first term to 2 and the numerator of the term on the right to 1 and then solve.

In Exercises 3–8, determine whether or not the given equations are quadratic. If the resulting form is quadratic, identify a, b, and c, with $a > 0$. Otherwise, explain why the resulting form is not quadratic.

3. $x(x - 2) = 4$

4. $(3x - 2)^2 = 2$

5. $x^2 = (x + 2)^2$

6. $x(2x^2 + 5) = 7 + 2x^2$

7. $n(n^2 + n - 1) = n^3$

8. $(T - 7)^2 = (2T + 3)^2$

In Exercises 9–44, solve the given quadratic equations by factoring.

9. $x^2 - 4 = 0$

10. $B^2 - 400 = 0$

11. $4x^2 = 9$

12. $x^2 = 0.16$

13. $x^2 - 8x - 9 = 0$

14. $x^2 + x - 6 = 0$

15. $R^2 + 12 = 7R$

16. $x^2 + 30 = 11$

17. $40x - 16x^2 = 0$

18. $15L = 20L^2$

19. $27m^2 = 3$

20. $a^2x^2 = 9$

21. $3x^2 - 13x + 4 = 0$

22. $A^2 + 8A + 16 = 0$

23. $7x^2 + 3x = 4$

24. $4x^2 + 25 = 20x$

25. $6x^2 = 13x - 6$

26. $6z^2 = 6 + 5x$

27. $4x(x + 1) = 3$

28. $9t^2 = 9 - t(43 + t)$

29. $6y^2 + by = 2b^2$

30. $2x^2 - 7ax + 4a^2 = a^2$

31. $8s^2 + 16s = 90$

32. $18t^2 = 48t - 32$

33. $(x + 2)^3 = x^3 + 8$

34. $V(V^2 - 4) = V^2(V - 1)$

35. $(x + a)^2 - b^2 = 0$

36. $bx^2 - b = x - b^2x$

37. $x^2 + 2ax = b^2 - a^2$

38. $x^2(a^2 + 2ab + b^2) = x(a + b)$

39. In Eq. (7.1), for $a = 2$, $b = -7$, and $c = 3$, show that the sum of the roots is $-b/a$.

40. For the equation of Exercise 39, show that the product of the roots is c/a.

41. The voltage V across a semiconductor in a computer is given by $V = \alpha I + \beta I^2$, where I is the current (in A). If a 6-V battery is conducted across the semiconductor, find the current if $\alpha = 2\ \Omega$ and $\beta = 0.5\ \Omega/A$.

42. The mass m (in Mg) of the fuel supply in the first-stage booster of a rocket is $m = 135 - 6t - t^2$, where t is the time (in s) after launch. When does the booster run out of fuel?

43. The power P (in MW) produced between midnight and noon by a nuclear power plant is $P = 4h^2 - 48h + 744$, where h is the hour of the day. At what time is the power 664 MW?

44. In determining the speed s (in mi/h) of a car while studying its fuel economy, the equation $s^2 - 16s = 3072$ is used. Find s.

In Exercises 45 and 46, although the equations are not quadratic, factoring will lead to one quadratic factor and the solution can be completed by factoring as with a quadratic equation. Find the three roots of each equation.

45. $x^3 - x = 0$ **46.** $x^3 - 4x^2 - x + 4 = 0$

In Exercises 47–50, solve the given equations involving fractions.

47. $\dfrac{1}{x-3} + \dfrac{4}{x} = 2$ **48.** $2 - \dfrac{1}{x} = \dfrac{3}{x+2}$

49. $\dfrac{1}{2x} - \dfrac{3}{4} = \dfrac{1}{2x+3}$ **50.** $\dfrac{x}{2} + \dfrac{1}{x-3} = 3$

In Exercises 51–54, set up the appropriate quadratic equations and solve.

51. The spring constant k is the force F divided by the amount x the spring stretches $(k = F/x)$. See Fig. 7.2(a). For two springs in series (see Fig. 7.2(b)), the reciprocal of the spring constant k_c for the combination equals the sum of the reciprocals of the individual spring constants. Find the spring constants for each of two springs in series if $k_c = 2\,\text{N/cm}$ and one spring constant is 3 N/cm more than the other.

Fig. 7.2

52. The reciprocal of the combined resistance R of two resistances R_1 and R_2 connected in parallel (see Fig. 7.3(a)) is equal to the sum of the reciprocals of the individual resistances. If the two resistances are connected in series (see Fig. 7.3(b)), their combined resistance is the sum of their individual resistances. If two resistances connected in parallel have a combined resistance of 3.0 Ω and the same two resistances have a combined resistance of 16 Ω when connected in series, what are the resistances?

Fig. 7.3   (a)   (b)

53. A hydrofoil made the round-trip of 120 km between two islands in 3.5 h of travel time. If the average speed going was 10 km/h less than the average speed returning, find these speeds.

54. A rectangular solar panel is 20 cm by 30 cm. By adding the same amount to each dimension, the area is doubled. How much is added?

Answers to Practice Exercises

1. $x = -7$, $x = 3$ **2.** $x = 1/3$, $x = 1/3$

7.2 Completing the Square

Solving a Quadratic Equation by Completing the Square

Most quadratic equations that arise in applications cannot be solved by factoring. Therefore, we now develop a method, called **completing the square,** that can be used to solve any quadratic equation. In the next section, this method is used to develop a general formula that can be used to solve any quadratic equation.

In the first example, we show the solution of a type of quadratic equation that arises while using the method of completing the square. In the examples that follow it, the method itself is used and described.

EXAMPLE 1 Quadratic equation solutions

In solving $x^2 = 16$, we may write $x^2 - 16 = 0$ and complete the solution by factoring. This gives us $x = 4$ and $x = -4$. Therefore, the principal square root of 16 and its negative both satisfy the original equation. Thus, we may solve $x^2 = 16$ by equating x to the principal square root of 16 and to its negative. The roots of $x^2 = 16$ are 4 and -4.

■ The method of completing the square is used later in Section 7.4 and also in Chapter 21.

We may solve $(x - 3)^2 = 16$ in a similar way by equating $x - 3$ to 4 and to -4:

$$x - 3 = 4 \quad \text{or} \quad x - 3 = -4$$

Solving these equations, we obtain the roots 7 and -1.

We may solve $(x - 3)^2 = 17$ in the same way. Thus,

$$x - 3 = \sqrt{17} \quad \text{or} \quad x - 3 = -\sqrt{17}$$

The roots are therefore $3 + \sqrt{17}$ and $3 - \sqrt{17}$. Decimal approximations of these roots are 7.123 and -1.123. ■

EXAMPLE 2 Method of completing the square

To find the roots of the quadratic equation

$$x^2 - 6x - 8 = 0$$

first note that the left side is not factorable. However,

NOTE ▶ $x^2 - 6x$ *is part of the special product* $(x - 3)^2 = x^2 - 6x + 9$

and this special product is a *perfect square*. By adding 9 to $x^2 - 6x$, we have $(x - 3)^2$. Therefore, we rewrite the original equation as

$$x^2 - 6x = 8$$

NOTE ▶
$$x^2 - 6x + 9 = 17 \qquad \text{add 9 to both sides}$$
$$(x - 3)^2 = 17 \qquad \text{rewriting the left side}$$
$$x - 3 = \pm\sqrt{17} \qquad \text{as in third illustration of Example 1}$$

The $\pm$ sign means that $x - 3 = \sqrt{17}$ or $x - 3 = -\sqrt{17}$.
    By adding 3 to each side, we obtain

$$x = 3 \pm \sqrt{17}$$

which means $x = 3 + \sqrt{17}$ and $x = 3 - \sqrt{17}$ are the two roots of the equation.
    Therefore, by creating an expression that is a perfect square and then using the principal square root and its negative, we were finally able to solve the equation as two linear equations. ■

How we determine the number to be added to complete the square is based on the special products in Eqs. (6.3) and (6.4). We rewrite these as

$$(x + a)^2 = x^2 + 2ax + a^2 \qquad \textbf{(7.2)}$$
$$(x - a)^2 = x^2 - 2ax + a^2 \qquad \textbf{(7.3)}$$

We must be certain that the coefficient of the x^2-term is 1 before we start to complete the square. The coefficient of x in each case is numerically $2a$, and the number added
CAUTION ▶ to complete the square is a^2. Thus, *if we* ***take half the coefficient of the x-term and square this result,*** *we have the number that completes the square.* In our example, the numerical coefficient of the x-term was 6, and 9 was added to complete the square. Therefore, the procedure for solving a quadratic equation by completing the square is as follows:

Practice Exercise

1. Solve by completing the square:
   $x^2 + 4x - 12 = 0$

SOLVING A QUADRATIC EQUATION BY COMPLETING THE SQUARE

1. *Divide each side by a* (the coefficient of x^2).

2. *Rewrite the equation with the constant on the right side.*

3. *Complete the square: Add the square of one-half of the coefficient of x to both sides.*

4. *Write the left side as a square and simplify the right side.*

5. *Equate the square root of the left side to the principal square root of the right side and to its negative*

6. *Solve the two resulting linear equations.*

EXAMPLE 3 Method of completing the square

Solve $2x^2 + 16x - 9 = 0$ by completing the square.

$$x^2 + 8x - \frac{9}{2} = 0$$

1. Divide each term by 2 to make coefficient of x^2 equal to 1.

$$x^2 + 8x = \frac{9}{2}$$

2. Put constant on right by adding 9/2 to each side.

$$\frac{1}{2}(8) = 4; \qquad 4^2 = 16$$

$$x^2 + 8x + 16 = \frac{9}{2} + 16 = \frac{41}{2}$$

3. Divide coefficient 8 of x by 2, square the 4, and add to both sides.

$$(x + 4)^2 = \frac{41}{2}$$

4. Write left side as $(x + 4)^2$

$$x + 4 = \pm\sqrt{\frac{41}{2}}$$

5. Take the square root of each side and equate the $x + 4$ to the principal square root of 41/2 and its negative.

$$x = -4 \pm\sqrt{\frac{41}{2}}$$

6. Solve for x

CAUTION ▶

(a)   (b)

Fig. 7.4

Therefore, the roots are $-4 + \sqrt{\frac{41}{2}}$ and $-4 - \sqrt{\frac{41}{2}}$. The calculator solution, using the *Solver* feature, is shown in Figs. 7.4(a) and 7.4(b). We use the *Solver* feature twice, since we can find only one root at a time. ∎

EXERCISES 7.2

In Exercises 1 and 2, make the given changes in the indicated examples of this section and then solve the resulting quadratic equations by completing the square.

1. In Example 2, change the $-$ sign before $6x$ to $+$.

2. In Example 3, change the coefficient of the second term from 16 to 12.

In Exercises 3–10, solve the given quadratic equations by finding appropriate square roots as in Example 1.

3. $x^2 = 25$

4. $x^2 = 100$

5. $x^2 = 7$

6. $x^2 = 15$

7. $(x - 2)^2 = 25$

8. $(x + 2)^2 = 10$

9. $(x + 3)^2 = 7$

10. $\left(x - \frac{5}{2}\right)^2 = 100$

In Exercises 11–30, solve the given quadratic equations by completing the square. Exercises 11–14 and 17–20 may be checked by factoring.

11. $x^2 + 2x - 8 = 0$

12. $x^2 - 8x - 20 = 0$

13. $D^2 + 3D + 2 = 0$

14. $t^2 + 5t - 6 = 0$

15. $n^2 = 4n - 2$

16. $(R + 9)(R + 1) = 13$

17. $v(v + 2) = 15$

18. $Z^2 + 12 = 8Z$

19. $2s^2 + 5s = 3$

20. $4x^2 + x = 3$

21. $3y^2 = 3y + 2$

22. $3x^2 = 3 - 4x$

23. $2y^2 - y - 2 = 0$

24. $9v^2 - 6v - 2 = 0$

25. $5T^2 - 10T + 4 = 0$

26. $4V^2 + 9 = 12V$

27. $9x^2 + 6x + 1 = 0$

28. $2x^2 - 3x + 2a = 0$

29. $x^2 + 2bx + c = 0$

30. $px^2 + qx + r = 0$

In Exercises 31–34, use completing the square to solve the given problems.

31. The voltage V across a certain electronic device is related to the temperature T (in °C) by $V = 4.0T - 0.2T^2$. For what temperature(s) is $V = 15\,V$?

32. A flare is shot vertically into the air such that its distance s (in ft) above the ground is given by $s = 64t - 16t^2$, where t is the time (in s) after it was fired. Find t for $s = 48$ ft.

33. A surveillance camera is 34 ft on a direct line from an ATM. The camera is 14 ft more to the right of the ATM than it is above the ATM. How far above the ATM is the camera?

34. A rectangular storage area is 8.0 m longer than it is wide. If the area is 28 m², what are its dimensions?

Answer to Practice Exercise

1. $x = -6, \quad x = 2$

7.3 The Quadratic Formula

The Quadratic Formula • Character of the Roots of a Quadratic Equation

We now use the method of completing the square to derive a general formula that may be used for the solution of any quadratic equation.

Consider Eq. (7.1), the general quadratic equation:

$$ax^2 + bx + c = 0 \quad (a \neq 0)$$

When we divide through by a, we obtain

$$x^2 + \frac{b}{a}x + \frac{c}{a} = 0$$

Subtracting c/a from each side, we have

$$x^2 + \frac{b}{a}x = -\frac{c}{a}$$

Half of b/a is $b/2a$, which squared is $b^2/4a^2$. Adding $b^2/4a^2$ to each side gives us

$$x^2 + \frac{b}{a}x + \frac{b^2}{4a^2} = -\frac{c}{a} + \frac{b^2}{4a^2}$$

Writing the left side as a perfect square and combining fractions on the right side, we have

$$\left(x + \frac{b}{2a}\right)^2 = \frac{b^2 - 4ac}{4a^2}$$

Equating $x + \dfrac{b}{2a}$ to the principal square root of the right side and its negative,

$$x + \frac{b}{2a} = \frac{\pm\sqrt{b^2 - 4ac}}{2a}$$

■ The quadratic formula, with the $\pm$ sign, means that the solutions to the quadratic equation

$$ax^2 + bx + c = 0$$

are

$$x = \frac{-b + \sqrt{b^2 - 4ac}}{2a}$$

and

$$x = \frac{-b - \sqrt{b^2 - 4ac}}{2a}$$

When we subtract $b/2a$ from each side and simplify the resulting expression, we obtain the **quadratic formula:**

$$x = \frac{-b \pm \sqrt{b^2 - 4ac}}{2a} \tag{7.4}$$

The quadratic formula gives us a quick general way of solving any quadratic equation. We need only write the equation in the standard form of Eq. (7.1), substitute these numbers into the formula, and simplify.

EXAMPLE 1 Quadratic formula – rational roots

$$\text{Solve: } \underset{\substack{\uparrow \\ a=1}}{x^2} - \underset{\substack{\uparrow \\ b=-5}}{5x} + \underset{\substack{\uparrow \\ c=6}}{6} = 0.$$

Here, using the indicated values of a, b, and c in the quadratic formula, we have

$$x = \frac{-(-5) \pm \sqrt{(-5)^2 - 4(1)(6)}}{2(1)} = \frac{5 \pm \sqrt{25 - 24}}{2} = \frac{5 \pm 1}{2}$$

$$x = \frac{5 + 1}{2} = 3 \quad \text{or} \quad x = \frac{5 - 1}{2} = 2$$

The roots $x = 3$ and $x = 2$ check when substituted in the original equation. ■

CAUTION ▶

It must be emphasized that, in using the quadratic formula, the entire expression $-b \pm \sqrt{b^2 - 4ac}$ is divided by $2a$. *It is a relatively common error to divide only the radical* $\sqrt{b^2 - 4ac}$.

EXAMPLE 2   Quadratic formula – irrational roots

Solve: $2x^2 - 7x - 5 = 0.$

$$\begin{array}{ccc} \uparrow & \uparrow & \uparrow \\ a=2 & b=-7 & c=-5 \end{array}$$

Substituting the values for a, b, and c in the quadratic formula, we have

$$x = \frac{-(-7) \pm \sqrt{(-7)^2 - 4(2)(-5)}}{2(2)} = \frac{7 \pm \sqrt{49 + 40}}{4} = \frac{7 \pm \sqrt{89}}{4}$$

$$x = \frac{7 + \sqrt{89}}{4} = 4.108 \quad \text{or} \quad x = \frac{7 - \sqrt{89}}{4} = -0.6085$$

Practice Exercise

1. Solve using the quadratic formula:
$3x^2 + x - 5 = 0$

The exact roots are $x = \dfrac{7 \pm \sqrt{89}}{4}$ (this form is often used when the roots are irrational). Approximate decimal values are $x = 4.108$ and $x = -0.6085$. ■

■ See Appendix C for a graphing calculator program QUADFORM. It can be used to find the roots of a quadratic equation.

EXAMPLE 3   Quadratic formula – double root

Solve: $9x^2 + 24x + 16 = 0.$

In this example, $a = 9$, $b = 24$, and $c = 16$. Thus,

$$x = \frac{-24 \pm \sqrt{24^2 - 4(9)(16)}}{2(9)} = \frac{-24 \pm \sqrt{576 - 576}}{18} = \frac{-24 \pm 0}{18} = -\frac{4}{3}$$

Here, both roots are $-\frac{4}{3}$, and we write the result as $x = -\frac{4}{3}$ and $x = -\frac{4}{3}$. We will get a double root when $b^2 = 4ac$, as in this case. ■

EXAMPLE 4   Quadratic formula – imaginary roots

Solve: $3x^2 - 5x + 4 = 0.$

In this example, $a = 3$, $b = -5$, and $c = 4$. Therefore,

$$x = \frac{-(-5) \pm \sqrt{(-5)^2 - 4(3)(4)}}{2(3)} = \frac{5 \pm \sqrt{25 - 48}}{6} = \frac{5 \pm \sqrt{-23}}{6}$$

■ Many calculators have an equation-solving feature. Also, some calculators have a specific feature for solving higher-degree equations.

These roots contain imaginary numbers. This happens if $b^2 < 4ac$. ■

Examples 1–4 illustrate the *character of the roots* of a quadratic equation. If a, b, and c are rational numbers (see Section 1.1), by noting the value of $b^2 - 4ac$ (called the **discriminant**), we have the following:

■ If $b^2 - 4ac$ is positive and a perfect square, $ax^2 + bx + c$ is factorable.

> **CHARACTER OF THE ROOTS OF A QUADRATIC EQUATION**
>
> **1.** If $b^2 - 4ac$ is positive and a perfect square (see Section 1.6), the roots are real, rational, and unequal. (See Example 1, where $b^2 - 4ac = 1$.)
>
> **2.** If $b^2 - 4ac$ is positive but not a perfect square, the roots are real, irrational, and unequal. (See Example 2, where $b^2 - 4ac = 89$.)
>
> **3.** If $b^2 - 4ac = 0$, the roots are real, rational, and equal. (See Example 3, where $b^2 - 4ac = 0$.)
>
> **4.** If $b^2 - 4ac < 0$, the roots contain imaginary numbers and are unequal. (See Example 4, where $b^2 - 4ac = -23$.)

We can use the value of $b^2 - 4ac$ to help in checking the roots or in finding the character of the roots without having to solve the equation completely.

g is the acceleration due to gravity t is the time

Fig. 7.5

EXAMPLE 5   Quadratic formula – literal numbers

The equation $s = s_0 + v_0 t - \frac{1}{2} g t^2$ is used in the analysis of projectile motion (see Fig. 7.5). Solve for t.

$$gt^2 - 2v_0 t - 2(s_0 - s) = 0 \qquad \text{multiply by } -2, \text{ put in form of Eq. (7.1)}$$

In this form, we see that $a = g$, $b = -2v_0$, and $c = -2(s_0 - s)$:

$$t = \frac{-(-2v_0) \pm \sqrt{(-2v_0)^2 - 4g(-2)(s_0 - s)}}{2g} = \frac{2v_0 \pm \sqrt{4(v_0^2 + 2gs_0 - 2gs)}}{2g}$$

$$= \frac{2v_0 \pm 2\sqrt{v_0^2 + 2gs_0 - 2gs}}{2g} = \frac{v_0 \pm \sqrt{v_0^2 + 2gs_0 - 2gs}}{g} \qquad ∎$$

EXAMPLE 6   Quadratic formula – word problem

A rectangular area 17.0 m long and 12.0 m wide is to be used for a patio with a rectangular pool. One end and one side of the patio area around the pool (the chairs, sunning, etc.) are to be the same width. The other end with the diving board is to be twice as wide, and the other side is to be three times as wide as the narrow side. The pool area is to be 96.5 m². What are the widths of the patio ends and sides, and the dimensions of the pool? See Fig. 7.6.

First, let x = the width of the narrow end and side of the patio. The other end is then $2x$ in width, and the other side is $3x$ in width. Since the pool area is 96.5 m², we have

$$\overset{\text{pool length}}{(17.0 - 3x)} \, \overset{\text{pool width}}{(12.0 - 4x)} = \overset{\text{pool area}}{96.5}$$

$$204 - 68.0x - 36.0x + 12x^2 = 96.5$$

$$12x^2 - 104.0x + 107.5 = 0$$

$$x = \frac{-(-104.0) \pm \sqrt{(-104.0)^2 - 4(12)(107.5)}}{2(12)} = \frac{104.0 \pm \sqrt{5656}}{24}$$

Evaluating, we get $x = 7.5$ m and $x = 1.2$ m. The value 7.5 m cannot be the required result since the width of the patio would be greater than the width of the entire area. For $x = 1.2$ m, the pool would have a length 13.4 m and width 7.2 m. These give an area of 96.5 m², which checks. The widths of the patio area are then 1.2 m, 1.2 m, 2.4 m, and 3.6 m. ∎

■ See the chapter introduction.

$2x$

Patio

Pool

17.0 m

96.5 m²

x $3x$

x

12.0 m

Fig. 7.6

EXERCISES 7.3

In Exercises 1–4, make the given changes in the indicated examples of this section and then solve the resulting equations by the quadratic formula.

1. In Example 1, change the $-$ sign before $5x$ to $+$.

2. In Example 2, change the coefficient of x^2 from 2 to 3.

3. In Example 3, change the $+$ sign before $24x$ to $-$.

4. In Example 4, change 4 to 3.

In Exercises 5–36, solve the given quadratic equations, using the quadratic formula. Exercises 5–8 are the same as Exercises 11–14 of Section 7.2.

5. $x^2 + 2x - 8 = 0$ **6.** $x^2 - 8x - 20 = 0$

7. $D^2 + 3D + 2 = 0$ **8.** $t^2 + 5t - 6 = 0$

9. $x^2 - 4x + 2 = 0$

10. $x^2 + 10x - 4 = 0$

11. $v^2 = 15 - 2v$

12. $8V - 12 = V^2$

13. $2s^2 + 5s = 3$

14. $4x^2 + x = 3$

15. $3y^2 = 3y + 2$

16. $3x^2 = 3 - 4x$

17. $y + 2 = 2y^2$

18. $2 + 6v = 9v^2$

19. $30y^2 + 23y - 40 = 0$

20. $40x^2 - 62x - 63 = 0$

21. $8t^2 + 61t = -120$

22. $2d(d - 2) = -7$

23. $s^2 = 9 + s(1 - 2s)$

24. $20r^2 = 20r + 1$

25. $25y^2 = 121$

26. $37T = T^2$

27. $15 + 4z = 32z^2$

28. $4x^2 - 12x = 7$

29. $x^2 - 0.20x - 0.40 = 0$

30. $3.2x^2 = 2.5x + 7.6$

31. $0.29Z^2 - 0.18 = 0.63Z$

32. $12.5x^2 + 13.2x = 15.5$

33. $x^2 + 2cx - 1 = 0$

34. $x^2 - 7x + (6 + a) = 0$

35. $b^2x^2 + 1 - a = (b + 1)x$

36. $c^2x^2 - x - 1 = x^2$

In Exercises 37–40, without solving the given equations, determine the character of the roots.

37. $2x^2 - 7x = -8$

38. $3x^2 + 19x = 14$

39. $3.6t^2 + 2.1 = 7.7t$

40. $0.45s^2 + 0.33 = 0.12s$

In Exercises 41–58, solve the given problems. All numbers are accurate to at least two significant digits.

41. Find k if the equation $x^2 + 4x + k = 0$ has a real double root.

42. Find the smallest positive integral value of k if the equation $x^2 + 3x + k = 0$ has roots with imaginary numbers.

43. Solve the equation $x^4 - 5x^2 + 4 = 0$ for x. (*Hint:* The equation can be written as $(x^2)^2 - 5(x^2) + 4 = 0$. First solve for x^2.)

 44. Without drawing the graph or completely solving the equation, explain how to find the number of x-intercepts of a quadratic function.

45. In machine design, in finding the outside diameter D_0 of a hollow shaft, the equation $D_0^2 - DD_0 - 0.25D^2 = 0$ is used. Solve for D_0 if $D = 3.625$ cm.

46. A missile is fired vertically into the air. The distance s (in ft) above the ground as a function of time t (in s) is given by $s = 300 + 500t - 16t^2$. (a) When will the missile hit the ground? (b) When will the missile be 1000 ft above the ground?

47. For a rectangle, if the ratio of the length to the width equals the ratio of the length plus the width to the length, the ratio is called the *golden ratio*. Find the value of the golden ratio, which the ancient Greeks thought had the most pleasing properties to look at.

48. When focusing a camera, the distance r the lens must move from the infinity setting is given by $r = f^2/(p - f)$, where p is the distance from the object to the lens, and f is the focal length of the lens. Solve for f.

49. In calculating the current in an electric circuit with an inductance L, a resistance R, and a capacitance C, it is necessary to solve the equation $Lm^2 + Rm + 1/C = 0$. Solve for m in the terms of L, R, and C. See Fig. 7.7.

Fig. 7.7

50. In finding the radius r of a circular arch of height h and span b, an architect used the following formula. Solve for h.

$$r = \frac{b^2 + 4h^2}{8h}$$

51. A flat-screen computer monitor is 37 cm wide and 31 cm high with a uniform edge around the viewing screen. If the edge covers 20% of the monitor front, what is the width of the edge?

52. An investment of $2000 is deposited at a certain annual interest rate. One year later, $3000 is deposited in another account at the same rate. At the end of the second year, the accounts have a total value of $5319.05. What is the interest rate?

53. The length of a tennis court is 12.8 m more than its width. If the area of the tennis court is 262 m², what are its dimensions? See Fig. 7.8.

Fig. 7.8

54. Two circular oil spills are tangent to each other. If the distance between centers is 800 m and they cover a combined area of 1.02×10^6 m², what is the radius of each? See Fig. 7.9.

Fig. 7.9

55. In remodeling a house, an architect finds that by adding the same amount to each dimension of a 12-ft by 16-ft rectangular room, the area would be increased by 80 ft². How much must be added to each dimension?

56. Two pipes together drain a wastewater-holding tank in 6.00 h. If used alone to empty the tank, one takes 2.00 h longer than the other. How long does each take to empty the tank if used alone?

57. On some California highways, a car can legally travel 15 mi/h faster than a truck. Traveling at maximum legal speeds, a car can travel 77 mi in 18 min less than a truck. What are the maximum legal speeds for cars and for trucks?

58. For electric capacitors connected in series, the sum of the reciprocals of the capacitances equals the reciprocal of the combined capacitance. If one capacitor has 5.0 μF more capacitance than another capacitor and they are connected in series, what are their capacitances if their combined capacitance is 4.0 μF?

Answer to Practice Exercise

1. $x = \dfrac{-1 \pm \sqrt{61}}{6}$

7.4 The Graph of the Quadratic Function

The Parabola • Vertex and *y*-intercept • Solving Quadratic Equations Graphically

In this section, we discuss the graph of the quadratic function $ax^2 + bx + c$ and show the graphical solution of a quadratic equation. By letting $y = ax^2 + bx + c$, we can graph this function, as in Chapter 3. The next example shows the graph of the quadratic function.

EXAMPLE 1 Graphing a quadratic function

Graph the function $f(x) = x^2 + 2x - 3$.

First, let $y = x^2 + 2x - 3$. We can now set up a table of values and graph the function as shown in Fig. 7.10, or display the graph on a graphing calculator as shown in Fig. 7.11.

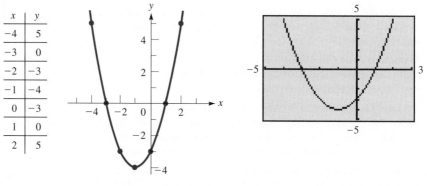

x	y
−4	5
−3	0
−2	−3
−1	−4
0	−3
1	0
2	5

Fig. 7.10   **Fig. 7.11**

The graph of any quadratic function shown in Figs. 7.10 and 7.11 is that of a **parabola,** and *the graph of any quadratic function $y = ax^2 + bx + c$ will have the same basic shape.* (In Section 3.4, we briefly noted that graphs in Examples 2 and 3 were parabolas.) A parabola can open upward (as in Example 1) or downward. The location of the parabola and how it opens depends on the values of a, b, and c, as we will show after Example 2.

NOTE ▶ In Example 1, the parabola has a minimum point at $(-1, -4)$, and the curve opens upward. *All parabolas have an* **extreme point** *of this type. For $y = ax^2 + bx + c$, if $a > 0$, the parabola has a* **minimum point** *and opens* **upward.** *If $a < 0$, the parabola has a* **maximum point** *and opens* **downward.** *The extreme point of the parabola is known as its* **vertex.**

(a)

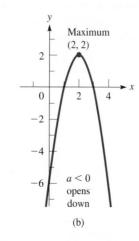

(b)

Fig. 7.12

EXAMPLE 2 Parabola – extreme points

The graph of $y = 2x^2 - 8x + 6$ is shown in Fig. 7.12(a). For this parabola, $a = 2$ $(a > 0)$ and it opens upward. The vertex (a minimum point) is $(2, -2)$.

The graphs of $y = -2x^2 + 8x - 6$ is shown in Fig. 7.12(b). For this parabola, $a = -2$ $(a < 0)$ and it opens downward. The vertex (a maximum point) is $(2, 2)$. ∎

We can sketch the graphs of parabolas like these by using its basic shape and knowing two or three points, including the vertex. Even when using a graphing calculator, we can get a check on the graph by knowing the vertex and how the parabola opens.

In order to find the coordinates of the extreme point, start with the quadratic function

$$y = ax^2 + bx + c$$

Then factor a from the two terms containing x, obtaining

$$y = a\left(x^2 + \frac{b}{a}x\right) + c$$

Now, completing the square of the terms within parentheses, we have

$$y = a\left(x^2 + \frac{b}{a}x + \frac{b^2}{4a^2}\right) + c - \frac{b^2}{4a}$$

$$= a\left(x + \frac{b}{2a}\right)^2 + c - \frac{b^2}{4a}$$

Now, look at the factor $\left(x + \frac{b}{2a}\right)^2$. If $x = -b/2a$, the term is zero. If x is any other value, $\left(x + \frac{b}{2a}\right)^2$ is positive. Thus, if $a > 0$, the value of y increases from the minimum

NOTE ▶ value, where $x = -b/2a$, and if $a < 0$, the value of y decreases from the maximum value, where $x = -b/2a$. *This means that $x = -b/2a$ is the x-coordinate of the vertex (extreme point). The y-coordinate is found by substituting this x-value in the function.*

Another easily found point is the y-intercept. As with a linear equation, we find the y-intercept where $x = 0$. For $y = ax^2 + bx + c$, if $x = 0$, then $y = c$. *This means that the point $(0, c)$ is the y-intercept.*

EXAMPLE 3   Graphing a parabola – vertex – y-intercept

For the graph of the function $y = 2x^2 - 8x + 6$, find the vertex and y-intercept and sketch the graph. (This function is also used in Example 2.)

First, $a = 2$ and $b = -8$. This means that the x-coordinate of the vertex is

$$\frac{-b}{2a} = \frac{-(-8)}{2(2)} = \frac{8}{4} = 2$$

and the y-coordinate is

$$y = 2(2^2) - 8(2) + 6 = -2$$

Thus, the vertex is $(2, -2)$. Since $a > 0$, it is a minimum point.

Since $c = 6$, the y-intercept is $(0,6)$.

We can use the minimum point $(2, -2)$ and the y-intercept $(0, 6)$, along with the fact that the graph is a parabola, to get an approximate sketch of the graph. Noting that a parabola increases (or decreases) away from the vertex in the same way on each side of it (it is *symmetric* to a vertical line through the vertex), we sketch the graph in Fig. 7.13. It is the same graph as that shown in Fig. 7.12(a). ∎

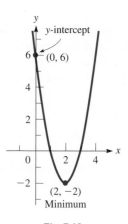

Fig. 7.13

If we are not using a graphing calculator, we may need one or two additional points to get a good sketch of a parabola. This would be true if the y-intercept is close to the vertex. Two points we can find are the x-intercepts, if the parabola crosses the x-axis (one point if the vertex is on the x-axis). They are found by setting $y = 0$ and solving the quadratic equation $ax^2 + bx + c = 0$. Also, we may simply find one or two points other than the vertex and the y-intercept. Sketching a parabola in this way is shown in the following two examples.

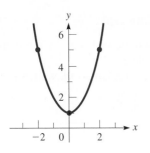

Fig. 7.14

Practice Exercise

1. (a) Find the vertex and y-intercept for the graph of the function $y = 3x^2 + 12x - 4$.
   (b) Is the vertex a minimum or a maximum point?

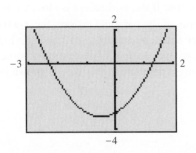

Fig. 7.15

Fig. 7.16

EXAMPLE 4 Graphing a parabola – vertex – y-intercept

Sketch the graph of $y = -x^2 + x + 6$.

We first note that $a = -1$ and $b = 1$. Therefore, the x-coordinate of the maximum point $(a < 0)$ is $-\frac{1}{2(-1)} = \frac{1}{2}$. The y-coordinate is $-\left(\frac{1}{2}\right)^2 + \frac{1}{2} + 6 = \frac{25}{4}$. This means that the maximum point is $\left(\frac{1}{2}, \frac{25}{4}\right)$.

The y-intercept is $(0,6)$.

Using these points in Fig. 7.14, since they are close together, they do not give a good idea of how wide the parabola opens. Therefore, setting $y = 0$, we solve the equation

$$-x^2 + x + 6 = 0 \qquad \text{multiply each term by } -1$$

or

$$x^2 - x - 6 = 0$$

This equation is factorable. Thus,

$$(x - 3)(x + 2) = 0$$
$$x = 3, -2$$

This means that the x-intercepts are $(3, 0)$ and $(-2, 0)$, as shown in Fig. 7.14.

Also, rather than finding the x-intercepts, we can let $x = 2$ (or some value to the right of the vertex) and then use the point $(2, 4)$. ∎

EXAMPLE 5 Graphing a parabola – no x-term

Sketch the graph of $y = x^2 + 1$.

Since there is no x-term, $b = 0$. This means that the x-coordinate of the minimum point $(a > 0)$ is 0 and that the minimum point and the y-intercept are both $(0, 1)$. We know that the graph opens upward, since $a > 0$, which in turn means that it does not cross the x-axis. Now, letting $x = 2$ and $x = -2$, find the points $(2, 5)$ and $(-2, 5)$ on the graph, which is shown in Fig. 7.15. ∎

We can see that the domain of the quadratic function is all x. Knowing that the graph has either a maximum point or a minimum point, the range of the quadratic function must be restricted to certain values of y. In Example 3, the range is $f(x) \geq -2$; in Example 4, the range is $f(x) \leq \frac{25}{4}$; and in Example 5, the range is $f(x) \geq 1$.

SOLVING QUADRATIC EQUATIONS GRAPHICALLY

In Chapter 3, we showed how an equation can be solved graphically. Following that method, to solve the equation $ax^2 + bx + c = 0$, let $y = ax^2 + bx + c$ and graph the function. The roots of the equation are the x-coordinates of the points for which $y = 0$ (the x-intercepts). The following examples illustrate solving quadratic equations graphically.

EXAMPLE 6 Solving quadratic equation graphically

Solve the equation $3x = x(2 - x) + 3$ graphically.

First, collect all terms on the left of the equal sign. This gives $x^2 + x - 3 = 0$. We then let $y = x^2 + x - 3$ and graph this function. The minimum point is $\left(-\frac{1}{2}, -\frac{13}{4}\right)$ and the y-intercept is $(0, -3)$.

Using the graphing calculator, these points help us choose the *window* settings shown for the graph of the function in Fig. 7.16. Now, using the *zero* feature (or the *trace* and *zoom* features), we find the roots to be

$$x = -2.30 \quad \text{and} \quad x = 1.30$$

These values (or the stored calculator values) check when substituted in the original equation. ∎

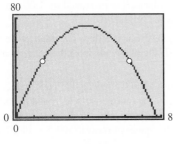

Fig. 7.17

EXAMPLE 7 Solving equation graphically – application

A projectile is fired vertically upward from the ground with a velocity of 38 m/s. Its distance above the ground is given by $s = -4.9t^2 + 38t$, where s is the distance (in m) and t is the time (in s). Graph the function and from the graph determine (1) when the projectile will hit the ground and (2) how long it takes to reach 45 m above the ground.

Since $c = 0$, the s-intercept is at the origin. Using $-b/2a$, we find the maximum point at about (3.9, 74). These points help us choose (using x for t, and y for s, on the calculator) the *window* settings for the graph shown in Fig. 7.17.

Using the *trace* and *zoom* features (or the *intersect* feature), we can now determine the required answers.

1. The projectile will hit the ground for $s = 0$, which is shown at the right t-intercept. Thus, it will take about 7.8 s to hit the ground.

2. We find the time for the projectile to get 45 m above the ground by finding the values of t for $s = 45$ m. We see that there are *two* points for which $s = 45$ m. The values of t at these points show that the times are $t = 1.5$ s and $t = 6.3$ s.

From the maximum point, we can also conclude that the maximum height reached by the projectile is about 74 m. This can also be seen on the graph. ∎

EXERCISES 7.4

In Exercises 1 and 2, make the given changes in the indicated examples of this section and then solve the resulting problems.

1. In Example 3, change the − sign before $8x$ to + and then sketch the graph.

2. In Example 6, change the coefficient of x on the left from 3 to 5 and then find the solution graphically.

In Exercises 3–8, sketch the graph of each parabola by using only the vertex and the y-intercept. Check the graph using a graphing calculator.

3. $y = x^2 - 6x + 5$
4. $y = -x^2 - 4x - 3$
5. $y = -3x^2 + 10x - 4$
6. $s = 2t^2 + 8t - 5$
7. $R = v^2 - 4v$
8. $y = -2x^2 - 5x$

In Exercises 9–12, sketch the graph of each parabola by using the vertex, the y-intercept, and the x-intercepts. Check the graph using a graphing calculator.

9. $y = x^2 - 4$
10. $y = x^2 + 3x$
11. $y = -2x^2 - 6x + 8$
12. $u = -3v^2 + 12v - 5$

In Exercises 13–16, sketch the graph of each parabola by using the vertex, the y-intercept, and two other points, not including the x-intercepts. Check the graph using a graphing calculator.

13. $y = 2x^2 + 3$
14. $y = x^2 + 2x + 2$
15. $y = -2x^2 - 2x - 6$
16. $y = -3x^2 - x$

In Exercises 17–24, use a graphing calculator to solve the given equations. If there are no real roots, state this as the answer.

17. $2x^2 - 3 = 0$
18. $5 - x^2 = 0$
19. $-3x^2 + 11x - 5 = 0$
20. $2t^2 = 7t + 4$
21. $x(2x - 1) = -3$
22. $2w - 5 = w^2$
23. $6R^2 = 18 - 7R$
24. $3x^2 - 25 = 20x$

(W) *In Exercises 25–30, use a graphing calculator to graph all three parabolas on the same coordinate system. In Exercises 25–28, describe (a) the shifts (see page 99) of $y = x^2$ that occur and (b) how each parabola opens. In Exercises 29 and 30, describe (a) the shifts and (b) the stretching and shrinking.*

25. (a) $y = x^2$ (b) $y = x^2 + 3$ (c) $y = x^2 - 3$
26. (a) $y = x^2$ (b) $y = (x - 3)^2$ (c) $y = (x + 3)^2$
27. (a) $y = x^2$ (b) $y = (x - 2)^2 + 3$ (c) $y = (x + 2)^2 - 3$
28. (a) $y = x^2$ (b) $y = -x^2$ (c) $y = -(x - 2)^2$
29. (a) $y = x^2$ (b) $y = 3x^2$ (c) $y = \frac{1}{3}x^2$
30. (a) $y = x^2$ (b) $y = -3(x - 2)^2$ (c) $y = \frac{1}{3}(x + 2)^2$

In Exercises 31–46, solve the given applied problem.

31. A quadratic equation $f(x) = 0$ has a solution $x = -1$. Its graph has its vertex at (3, 4). What is the other solution?

32. Find c such that $y = x^2 - 12x + c$ has exactly one real root.

33. Find the smallest integral value of c such that $y = 2x^2 - 4x - c$ has two real roots.

34. Find the smallest integral value of c such that $y = 3x^2 - 12x + c$ has no real roots.

35. The vertical distance d (in cm) of the end of a robot arm above a conveyor belt in its 8-s cycle is given by $d = 2t^2 - 16t + 47$. Sketch the graph of $d = f(t)$.

36. When mineral deposits form a uniform coating 1 mm thick on the inside of a pipe of radius r (in mm), the cross-sectional area A through which water can flow is $A = \pi(r^2 - 2r + 1)$. Sketch $A = f(r)$.

37. An equipment company determines that the area A (in m²) covered by a rectangular tarpaulin is given by $A = w(8 - w)$, where w is the width of the tarpaulin and its perimeter is 16 m. Sketch the graph of A as a function of w.

38. Under specified conditions, the pressure loss L (in lb/in.² per 100 ft), in the flow of water through a water line in which the flow is q gal/min, is given by $L = 0.0002q^2 + 0.005q$. Sketch the graph of L as a function of q, for $q < 100$ gal/min.

39. When analyzing the power P (in W) dissipated in an electric circuit, the equation $P = 50i - 3i^2$ results. Here, i is the current (in A). Sketch the graph of $P = f(i)$.

40. Tests show that the power P rpm (in hp) of an automobile engine as a function of r (in r/min) is given by
$P = -5.0 \times 10^{-6}r^2 + 0.050r - 45(1500 < r < 6000 \text{ r/min})$.
Sketch the graph of P vs. r and find the maximum power that is produced.

41. A missile is fired vertically upward such that its distance s (in ft) above the ground is given by $s = 150 + 250t - 16t^2$, where t is the time (in s). Sketch the graph and then determine from the graph (a) when the missile will hit the ground (to 0.1 s), (b) how high it will go (to three significant digits), and (c) how long it will take to reach a height of 800 ft (to 0.1 s).

42. In a certain electric circuit, the resistance R (in Ω) that gives resonance is found by solving the equation $25R = 3(R^2 + 4)$. Solve this equation graphically (to 0.1 Ω).

43. If the radius of a circular solar cell is increased by 1.00 cm, its area is 96.0 cm². What was the original radius?

44. The diagonal of a rectangular floor is 3.00 ft less than twice the length of one of the sides. If the other side is 15.0 ft long, what is the area of the floor?

45. A security fence is to be built around a rectangular parking area of 20,000 ft². If the front side of the fence costs \$20/ft and the other three sides cost \$10/ft, solve graphically for the dimensions (to 1 ft) of the parking area if the fence is to cost \$7500. See Fig. 7.18.

Fig. 7.18

46. An airplane pilot could decrease the time t (in h) needed to travel the 630 mi from Ottawa to Milwaukee by 20 min if the plane's speed v is increased by 40 mi/h. Set up the appropriate equation and solve graphically for v (to two significant digits).

Answers to Practice Exercise

1. (a) $(-2, -16)$, $(0, -4)$ **(b)** minimum

CHAPTER 7 EQUATIONS

Quadratic equation	$ax^2 + bx + c = 0$	(7.1)
Quadratic formula	$x = \dfrac{-b \pm \sqrt{b^2 - 4ac}}{2a}$	(7.4)

CHAPTER 7 REVIEW EXERCISES

In Exercises 1–12, solve the given quadratic equations by factoring.

1. $x^2 + 3x - 4 = 0$ **2.** $x^2 + 3x - 10 = 0$

3. $x^2 - 10x + 16 = 0$ **4.** $P^2 - 27 = 6P$

5. $3x^2 + 11x = 4$ **6.** $6y^2 = 11y - 3$

7. $6t^2 = 13t - 5$ **8.** $3x^2 + 5x + 2 = 0$

9. $6s^2 = 25s$ **10.** $23n + 35 = 6n^2$

11. $4B^2 = 8B + 21$ **12.** $6x^2 = 8 - 47x$

In Exercises 13–24, solve the given quadratic equations by using the quadratic formula.

13. $x^2 - x - 110 = 0$ **14.** $x^2 + 3x - 18 = 0$

15. $m^2 + 2m = 6$ **16.** $1 + 7D = D^2$

17. $2x^2 - x = 36$ **18.** $3x^2 = 14 - x$

19. $3s + 2 = 4s^2$ **20.** $5x^2 + 7x - 2 = 0$

21. $2.1x^2 + 2.3x + 5.5 = 0$ **22.** $0.30R^2 - 0.42R = 0.15$

23. $6x^2 = 9 - 4x$ **24.** $24x^2 = 25x + 20$

In Exercises 25 – 36, solve the given quadratic equations by any appropriate algebraic method.

25. $x^2 + 4x - 4 = 0$ **26.** $x^2 + 3x + 1 = 0$

27. $3x^2 + 8x + 2 = 0$ **28.** $3p^2 = 28 - 5p$

29. $4v^2 = v + 5$ **30.** $n - 2 = 6n^2$

31. $7 + 3C = -2C^2$ **32.** $4y^2 - 5y = 8$

33. $a^2x^2 + 2ax + 2 = 0$ **34.** $16r^2 = 8r - 1$

35. $ay^2 = a - 3y$ **36.** $2bx = x^2 - 3b$

In Exercises 37–40, solve the given quadratic equations by completing the square.

37. $x^2 - x - 30 = 0$ **38.** $x^2 = 2x + 5$

39. $2t^2 = t + 4$ **40.** $4x^2 - 8x = 3$

In Exercises 41 – 44, solve the given equations.

41. $\dfrac{x - 4}{x - 1} = \dfrac{2}{x}$ **42.** $\dfrac{V - 1}{3} = \dfrac{5}{V} + 1$

43. $\dfrac{x^2 - 3x}{x - 3} = \dfrac{x^2}{x + 2}$ **44.** $\dfrac{x - 2}{x - 5} = \dfrac{15}{x^2 - 5x}$

In Exercises 45 – 48, sketch the graphs of the given functions by using the vertex, the y-intercept, and one or two other points.

45. $y = 2x^2 - x - 1$ **46.** $y = -4x^2 - 1$

47. $y = x - 3x^2$ **48.** $y = 2x^2 + 8x - 10$

In Exercises 49 – 52, solve the given equations by using a graphing calculator. If there are no real roots, state this as the answer.

49. $2x^2 + x - 4 = 0$ **50.** $-4x^2 - x - 1 = 0$

51. $3x^2 = -x - 2$ **52.** $x(15x - 12) = 8$

In Exercises 53 and 54, solve the given problems.

53. A quadratic equation $f(x) = 0$ has a solution $x = 2$. Its graph has its vertex at $(-1, 6)$. What is the other solution?

54. Find c such that $y = 2x^2 + 16x + c$ has exactly one real root.

In Exercises 55 – 68, solve the given quadratic equations by any appropriate method. All numbers are accurate to at least two significant digits.

55. The bending moment M of a simply supported beam of length L with a uniform load of w kg/m at a distance x from one end is $M = 0.5wLx - 0.5wx^2$. For what values of x is $M = 0$?

56. A nuclear power plant supplies a fixed power level at a constant voltage. The current I (in A) is found by solving the equation $I^2 - 17I - 12 = 0$. Solve for $I > 0$.

57. A computer analysis shows that the cost C (in dollars) for a company to make x units of a certain product is given by $C = 0.1x^2 + 0.8x + 7$. How many units can be made for 50?

58. For laminar flow of fluids, the coefficient K used to calculate energy loss due to sudden enlargements is given by $K = 1.00 - 2.67R + R^2$, where R is the ratio of cross-sectional areas. If $K = 0.500$, what is the value of R?

59. At an altitude h (in ft) above sea level, the boiling point of water is lower by $T°$F than the boiling point at sea level, which is

212°F. The difference can be approximated by solving the equation $T^2 + 520T - h = 0$. What is the boiling point in Boulder, Colorado (altitude 5300 ft)?

60. In a natural gas pipeline, the velocity v (in m/s) of the gas as a function of the distance x (in cm) from the wall of the pipe is given by $v = 5.2x - x^2$. Determine x for $v = 4.8$ m/s.

61. The height h of an object ejected at an angle θ from a vehicle moving with velocity v is given by $h = vt \sin \theta - 16t^2$, where t is the time of flight. Find t (to 0.1 s) if $v = 44$ ft/s, $\theta = 65°$, and $h = 18$ ft.

62. In studying the emission of light, in order to determine the angle at which the intensity is a given value, the equation $\sin^2 A - 4 \sin A + 1 = 0$ must be solved. Find angle A (to 0.1°). ($\sin^2 A = (\sin A)^2$.)

63. A computer analysis shows that the number n of electronic components a company should produce for supply to equal demand is found by solving.

$$\frac{n^2}{500,000} = 144 - \frac{n}{500}. \text{ Find } n.$$

64. To determine the resistances of two resistors that are to be in parallel in an electric circuit, it is necessary to solve the equation

$$\frac{20}{R} + \frac{20}{R + 10} = \frac{1}{5}. \text{ Find } R \text{ (to nearest 1 } \Omega).$$

65. In designing a cylindrical container, the formula $A = 2\pi r^2 + 2\pi rh$ (Eq. (2.19)) is used. Solve for r.

66. In determining the number of bytes b that can be stored on a hard disk, the equation $b = kr(R - r)$ is used. Solve for r.

67. In the study of population growth, the equation $p_2 = p_1 + rp_1(1 - p_1)$ occurs. Solve for p_1.

68. In the study of the velocities of deep-water waves, the equation $v^2 = k^2\left(\dfrac{L}{C} + \dfrac{C}{L}\right)$ occurs. Solve for L.

In Exercises 69 – 82, set up the necessary equation where appropriate and solve the given problems. All numbers are accurate to at least two significant digits.

69. In testing the effects of a drug, the percent of the drug in the blood was given by $p = 0.090t - 0.015t^2$, where t is the time (in h) after the drug was administered. Sketch the graph of $p = f(t)$.

70. In an electric circuit, the voltage V as a function of the time t (in min) is given by $V = 9.8 - 9.2t + 2.3t^2$. Sketch the graph of $V = f(t)$, for $t \le 5$ min.

71. By adding the same amount to its length and its width, a developer increased the area of a rectangular lot by 3000 m^2 to make it 80 m by 100 m. What were the original dimensions of the lot? See Fig. 7.19.

Fig. 7.19

72. A machinery pedestal is made of two concrete cubes, one on top of the other. The pedestal is 8.00 ft high and contains 152 ft³ of concrete. Find the edge of each cube.

73. Concrete contracts as it dries. If the volume of a cubical concrete block is 29 cm³ less and each edge is 0.10 cm less after drying, what was the original length of an edge of the block?

74. A jet flew 1200 mi with a tailwind of 50 mi/h. The tailwind then changed to 20 mi/h for the remaining 570 mi of the flight. If the total time of the flight was 3.0 h, find the speed of the jet relative to the air.

75. The length of one rectangular field is 400 m more than the side of a square field. The width is 100 m more than the side of the square field. If the rectangular field has twice the area of the square field, what are the dimensions of each field?

76. A military jet flies directly over and at right angles to the straight course of a commercial jet. The military jet is flying at 200 mi/h faster than four times the speed of the commercial jet. How fast is each going if they are 2050 mi apart (on a direct line) after 1 h?

77. The width of a rectangular TV screen is 5.7 in. more than the height. If the diagonal is 27.0 in., find the dimensions of the screen. See Fig. 7.20.

Fig. 7.20

(W) 78. Find the exact value of x that is defined in terms of the *continuing fraction* at the right (the pattern continues endlessly). Explain your method of solution.

$$x = 2 + \cfrac{1}{2 + \cfrac{1}{2 + \cfrac{1}{2 + \cdots}}}$$

79. An electric utility company is placing utility poles along a road. It is determined that five fewer poles per kilometer would be necessary if the distance between poles were increased by 10 m. How many poles are being placed each kilometer?

80. An architect is designing a Norman window (a semicircular part over a rectangular part) as shown in Fig. 7.21. If the area of the window is to be 16.0 ft² and the height of the rectangular part is 4.00 ft, find the radius of the circular part.

Fig. 7.21

81. A testing station found p parts per million (ppm) of sulfur dioxide in the air as a function of the hour h of the day to be $p = 0.00174(10 + 24h - h^2)$. Sketch the graph of $p = f(h)$ and, from the graph, find the time when $p = 0.205$ ppm.

82. A compact disc (CD) is made such that it is 53.0 mm from the edge of the center hole to the edge of the disc. Find the radius of the hole if 1.36% of the disc is removed in making the hole. See Fig. 7.22

Fig. 7.22

Writing Exercise

83. An electronics student is asked to solve the equation $\frac{1}{2} = \frac{1}{R} + \frac{1}{R + 1}$ for R. Write one or two paragraphs explaining your procedure for the solution, including a discussion of what methods of the chapter may be used in completing the solution.

CHAPTER 7 **PRACTICE TEST**

1. Solve by factoring: $2x^2 + 5x = 12$

2. Solve by using the quadratic formula: $x^2 = 3x + 5$

3. Solve graphically using a calculator: $4x^2 - 5x - 3 = 0$

4. Solve algebraically: $2x^2 - x = 6 - 2x(3 - x)$

5. Solve algebraically: $\frac{3}{x} - \frac{2}{x + 2} = 1$

6. Sketch the graph of $y = 2x^2 + 8x + 5$ using the extreme point and the y-intercept.

7. In electricity, the formula $P = EI - RI^2$ is used. Solve for I in terms of E, P, and R.

8. Solve by completing the square: $x^2 - 6x - 9 = 0$.

9. The perimeter of a rectangular window is 8.4 m, and its area is 3.8 m². Find its dimensions.

10. If $y = x^2 - 8x + 8$, sketch the graph using the vertex and any other useful points and find graphically the values of x for which $y = 0$.

Trigonometric Functions of Any Angle

8

When we introduced the trigonometric functions in Chapter 4, we defined them in general but used them only with acute angles. In this chapter, we show how these functions are used with angles of any size and with angles measured in radians.

By the mid 1700s, the trigonometric functions had been used for many years as ratios, as in Chapter 4. It was also known that they are useful in describing periodic functions (functions for which values repeat at specific intervals) without reference to triangles. In about 1750, this led the Swiss mathematician Leonhard Euler to include, for the first time in a textbook, the trigonometric functions of numbers (not angles). As we will see in this chapter, this is equivalent to using these functions of angles of any size, measured in radians.

Euler wrote over 70 volumes in mathematics and applied subjects such as astronomy, mechanics, and music. (Many of these volumes were dictated, as he was blind for the last 17 years of his life.) Although he is noted as one of the great mathematicians of all time, his interest in applied subjects often led him to study and develop topics in mathematics that were used in these applications.

Today, trigonometric functions of numbers are of importance in many areas of application such as electric circuits, mechanical vibrations, and rotational motion. Although electronics were unknown in the 1700s, the trigonometric functions of numbers developed at that time became of great importance to the development of electronics in the 1900s. We will show some of these applications in this chapter and in Chapter 10.

Cruise ships generally measure distances traveled in nautical miles. In Section 8.4, we will see how a nautical mile is defined as a distance along an arc of the earth's surface.

8.1 Signs of the Trigonometric Functions

Signs of the Trigonometric Functions in Each of the Four Quadrants • Evaluating Functions Knowing Point on Terminal Side

Recall the definitions of the trigonometric functions that were given in Section 4.2. *Here, the point (x, y) is a point on the terminal side of angle θ, and r is the radius vector.* See Fig. 8.1.

$$\sin\theta = \frac{y}{r} \qquad \cos\theta = \frac{x}{r} \qquad \tan\theta = \frac{y}{x}$$

$$\cot\theta = \frac{x}{y} \qquad \sec\theta = \frac{r}{x} \qquad \csc\theta = \frac{r}{y} \qquad\qquad (8.1)$$

Fig. 8.1

As noted before, these definitions are valid for a standard-position angle of any size. In this section, we determine the *sign* of each function of angles with terminal sides in each of the four quadrants.

We can find the values of the functions if we know the coordinates (x, y) on the terminal side and the radius vector r. *Since r is always taken to be positive, the functions will vary in sign, depending on the values of x and y.* If either x or y is zero in the denominator, the function is undefined. We will consider this in the next section.

Since $\sin\theta = y/r$, *the sign of $\sin\theta$ depends on the sign of y.* Since $y > 0$ in the first and second quadrants and $y < 0$ in the third and fourth quadrants, $\sin\theta$ is positive if the terminal side is in the first or second quadrant, and $\sin\theta$ is negative if the terminal side is in the third or fourth quadrant. See Fig. 8.2.

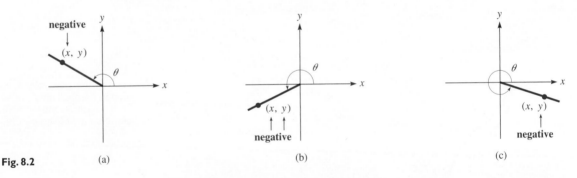

Fig. 8.2   (a)   (b)   (c)

EXAMPLE 1 Sign of $\sin\theta$ in each quadrant

The value of $\sin 20°$ is positive, because the terminal side of $20°$ is in the first quadrant. The value of $\sin 160°$ is positive, because the terminal side of $160°$ is in the second quadrant. The values of $\sin 200°$ and $\sin 340°$ are negative, because the terminal sides are in the third and fourth quadrants, respectively. ∎

Since $\tan\theta = y/x$, *the sign of $\tan\theta$ depends on the ratio of y to x.* Since x and y are both positive in the first quadrant, both negative in the third quadrant, and have different signs in the second and fourth quadrants, $\tan\theta$ is positive if the terminal side is in the first or third quadrant, and $\tan\theta$ is negative if the terminal side is in the second or fourth quadrant. See Fig. 8.2.

EXAMPLE 2 Sign of $\tan\theta$ in each quadrant

The values of $\tan 20°$ and $\tan 200°$ are positive, because the terminal sides of these angles are in the first and third quadrants, respectively. The values of $\tan 160°$ and $\tan 340°$ are negative, because the terminal sides of these angles are in the second and fourth quadrants, respectively. ∎

Sin θ   All
Csc θ

Tan θ   Cos θ
Cot θ   Sec θ

Positive functions

Fig. 8.3

Since $\cos \theta = x/r$, *the sign of* $\cos \theta$ *depends on the sign of x.* Since $x > 0$ in the first and fourth quadrants and $x < 0$ in the second and third quadrants, $\cos \theta$ *is positive if the terminal side is in the first or fourth quadrant,* and $\cos \theta$ *is negative if the terminal side is in the second or third quadrant.* See Fig. 8.2.

EXAMPLE 3 Sign of cos θ in each quadrant

The values of $\cos 20°$ and $\cos 340°$ are positive, because the terminal sides of these angles are in the first and fourth quadrants, respectively. The values of $\cos 160°$ and $\cos 200°$ are negative, because the terminal sides of these angles are in the second and third quadrants, respectively. ∎

Since $\csc \theta$ is defined in terms of y and r, as in $\sin \theta$, $\csc \theta$ has the same sign as $\sin \theta$. For similar reasons, $\cot \theta$ has the same sign as $\tan \theta$, and $\sec \theta$ has the same sign as $\cos \theta$. Therefore, as shown in Fig. 8.3,

All functions of first-quadrant angles are positive. Sin θ and csc θ are positive for second-quadrant angles. Tan θ and cot θ are positive for third-quadrant angles. Cos θ and sec θ are positive for fourth-quadrant angles. All others are negative.

This discussion does not include the *quadrantal angles,* those angles with terminal sides on one of the axes. They will be discussed in the next section.

EXAMPLE 4 Positive and negative functions

(a) The following are *positive:*

$$\sin 150° \quad \cos 290° \quad \tan 190° \quad \cot 260° \quad \sec 350° \quad \csc 100°$$

(b) The following are *negative:*

$$\sin 300° \quad \cos 150° \quad \tan 100° \quad \cot 300° \quad \sec 200° \quad \csc 250°$$ ∎

EXAMPLE 5 Sign of function related to terminal side

(a) If $\sin \theta > 0$, then the terminal side of θ is in either the first or second quadrant.

(b) If $\sec \theta < 0$, then the terminal side of θ is in either the second or third quadrant.

(c) If $\cos \theta > 0$, and $\tan \theta < 0$, then the terminal side of θ is in the fourth quadrant. Only in the fourth quadrant are both signs correct. ∎

EXAMPLE 6 Evaluating functions

Determine the trigonometric functions of θ if the terminal side of θ passes through $(-1, \sqrt{3})$. See Fig. 8.4.

We know that $x = -1$, $y = \sqrt{3}$, and from the Pythagorean theorem, we find that $r = 2$. Therefore, the trigonometric functions of θ are

$$\sin \theta = \frac{\sqrt{3}}{2} = 0.8660 \qquad \cos \theta = -\frac{1}{2} = -0.5000 \qquad \tan \theta = -\sqrt{3} = -1.732$$

$$\cot \theta = -\frac{1}{\sqrt{3}} = -0.5774 \qquad \sec \theta = -2 = -2.000 \qquad \csc \theta = \frac{2}{\sqrt{3}} = 1.155$$

The point $(-1, \sqrt{3})$ is in the second quadrant, and the signs of the functions of θ are those of a second-quadrant angle. ∎

When dealing with negative angles, or angles greater than 360°, the sign of the function is still determined by the location of the terminal side of the angle. For example, $\sin(-120°) < 0$ since the terminal side of $-120°$ is in the third quadrant. Also, for the same reason, $\sin 600° < 0$ (same terminal side as $-120°$ or 240°).

Practice Exercise

1. Determine the sign of the given functions:
(a) $\sin 140°$ (b) $\tan 255°$ (c) $\sec 175°$

Practice Exercise

2. Determine the value of $\cos \theta$ if the terminal side of θ passes through $(-1, 4)$.

EXERCISES 8.1

In Exercises 1 and 2, answer the given questions about the indicated examples of this section.

1. In Example 4, if 90° is added to each angle, what is the sign of each resulting function?
2. In Example 6, if the point $(-1, \sqrt{3})$ is replaced with the point $(1, -\sqrt{3})$, what are the resulting values?

In Exercises 3–14, determine the sign of the given functions.

3. sin 36°, cos 120°
4. tan 320°, sec 185°
5. csc 98°, cot 82°
6. cos 260°, csc 290°
7. sec 150°, tan 220°
8. sin 335°, cot 265°
9. cos 348°, csc 238°
10. cot 110°, sec 309°
11. tan 460°, sin(−110°)
12. csc(−200°), cos 550°
13. cot(−2°), cos 710°
14. sin 539°, tan(−480°)

In Exercises 15–22, find the trigonometric functions of θ if the terminal side of θ passes through the given point.

15. $(2, 1)$
16. $(-1, 1)$
17. $(-2, -3)$
18. $(4, -3)$
19. $(-0.5, 1.2)$
20. $(-9, -12)$
21. $(20, -8)$
22. $(0.9, 4)$

In Exercises 23–28, for the given values, determine the quadrant(s) in which the terminal side of the angle lies.

23. $\sin \theta = 0.5000$
24. $\cos \theta = 0.8666$
25. $\tan \theta = 1.500$
26. $\sin \theta = -0.8666$
27. $\cos \theta = -0.5000$
28. $\tan \theta = -0.7500$

In Exercises 29–38, determine the quadrant in which the terminal side of θ lies, subject to both given conditions.

29. $\sin \theta > 0, \cos \theta < 0$
30. $\tan \theta > 0, \cos \theta < 0$
31. $\sec \theta < 0, \cot \theta < 0$
32. $\cos \theta > 0, \csc \theta < 0$
33. $\csc \theta < 0, \tan \theta < 0$
34. $\tan \theta < 0, \cos \theta > 0$
35. $\sin \theta < 0, \tan \theta > 0$
36. $\sec \theta > 0, \csc \theta < 0$
37. $\sin \theta > 0, \cot \theta < 0$
38. $\tan \theta > 0, \csc \theta < 0$

In Exercises 39–42, with (x, y) in the given quadrant, determine whether the given ratio is positive or negative.

39. III, $\dfrac{x}{r}$
40. II, $\dfrac{y}{r}$
41. IV, $\dfrac{y}{x}$
42. III, $\dfrac{x}{y}$

Answers to Practice Exercises

1. (a) + (b) + (c) −   2. $-1/\sqrt{17}$

8.2 Trigonometric Functions of Any Angle

Reference Angle • Evaluating Functions • Evaluations on Calculator • Quadrantal Angles • Negative Angles

Any negative angle or angle greater than 360° is coterminal with a positive angle less than 360°. Since the terminal sides of coterminal angles are the same, the trigonometric functions of coterminal angles are equal. Therefore, we can evaluate the functions of an angle of any size if we can find the trigonometric functions of 0° and positive angles less than 360°. Such values will be needed for vectors, oblique triangles, and graphs in Chapters 9 and 10.

EXAMPLE 1 Coterminal angles

The following pairs of angles are coterminal.

$$390° \text{ and } 30° \qquad 900° \text{ and } 180° \qquad -150° \text{ and } 210° \text{ (see Fig. 8.5)}$$

The trigonometric functions of both angles in each pair are equal. For example,

$$\sin 390° = \sin 30° \quad \text{and} \quad \tan(-150°) = \tan 210° \qquad \blacksquare$$

Fig. 8.5

The values of the functions depend on the values of x, y, and r. Therefore, the absolute value of a function of a second-quadrant angle equals the value of the same function of a first-quadrant angle. For example, in Fig. 8.6, $\tan \theta_2 = -4/3$, or $|\tan \theta_2| = 4/3$, and $\tan \theta_1 = 4/3$. This means $|\tan \theta_2| = \tan \theta_1$. The triangles with θ_1 and α are congruent, which means $\theta_1 = \alpha$. If F represents any of the trigonometric functions, knowing that the absolute value of a function of θ_2 equals the value of the same function of θ_1 means that

$$|F(\theta_2)| = |F(\theta_1)| = |F(\alpha)| \qquad (8.2)$$

Fig. 8.6

The angle labeled α is called the reference angle. The reference angle of a given angle is the acute angle formed by the terminal side of the angle and the x-axis.

Using Eq. (8.2) and the fact that $\alpha = 180° - \theta_2$, we may conclude that the value of any trigonometric function of any second-quadrant angle is found from

$$F(\theta_2) = \pm F(180° - \theta_2) = \pm F(\alpha) \tag{8.3}$$

NOTE ▶ The *sign* used depends on the *sign of the function* in the second quadrant.

EXAMPLE 2 Quadrant II reference angles

In Fig. 8.6, the trigonometric functions of θ are as follows:

$$\sin\theta_2 = \sin(180° - \theta_2) = \sin\alpha = \sin\theta_1 = \tfrac{4}{5} = 0.8000$$

$$\cos\theta_2 = -\cos\theta_1 = -\tfrac{3}{5} = -0.6000 \qquad \tan\theta_2 = -\tfrac{4}{3} = -1.333$$

$$\cot\theta_2 = -\tfrac{3}{4} = -0.7500 \qquad \sec\theta_2 = -\tfrac{5}{3} = -1.667 \qquad \csc\theta_2 = \tfrac{5}{4} = 1.250 \qquad ■$$

In the same way, we derive the formulas for trigonometric functions of any third- or fourth-quadrant angle. In Fig. 8.7, the reference angle α is found by subtracting 180° from θ_3, and the functions of α and θ_1 are numerically equal. In Fig. 8.8, the reference angle α is found by subtracting θ_4 from 360°.

$$F(\theta_3) = \pm F(\theta_3 - 180°) = \pm F(\alpha) \tag{8.4}$$

$$F(\theta_4) = \pm F(360° - \theta_4) = \pm F(\alpha) \tag{8.5}$$

Fig. 8.6

Fig. 8.7

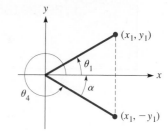

Fig. 8.8

EXAMPLE 3 Quadrant III reference angles

The trigonometric functions of $\theta_3 = 210°$ can be expressed in terms of the reference angle of 30° (see Fig. 8.9), and then evaluated, as follows:

Fig. 8.9

same function   reference angle

$$\sin 210° = -\sin(210° - 180°) = -\sin 30° = -\frac{1}{2} = -0.5000$$

quadrant III $\sin\theta, \csc\theta$ − in quadrant III

$$\csc 210° = -\csc 30° = -2.000$$

$$\cos 210° = -\cos 30° = -0.8660 \qquad \sec 210° = -\sec 30° = -1.155$$

$\cos\theta, \sec\theta$ − in quadrant III

$$\tan 210° = +\tan 30° = +0.5774 \qquad \cot 210° = +\cot 30° = +1.732$$

$\tan\theta, \cot\theta$ + in quadrant III   ■

Fig. 8.10

Practice Exercise

1. Express each of the functions of 225°
   in terms of the reference angle.

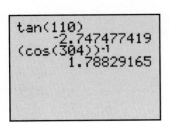

Fig. 8.11

EXAMPLE 4 Quadrant IV reference angles

If $\theta_4 = 315°$, the functions of θ_4 are found by using Eq. (8.5) as follows. See Fig. 8.10.

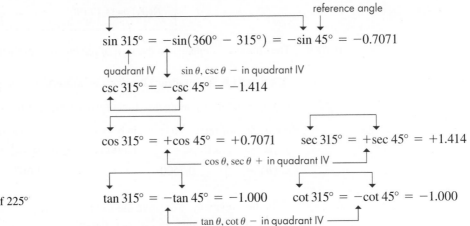

EXAMPLE 5 Evaluating using reference angles

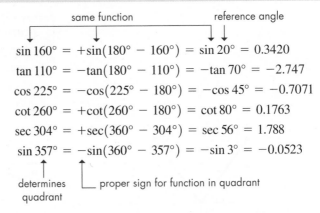

A calculator will give values, with the proper signs, of functions like those in Examples 3, 4, and 5. The reciprocal key is used for $\cot \theta$, $\sec \theta$, and $\csc \theta$, as shown on page 117. The display for $\tan 110°$ and $\sec 304°$ of Example 5 is shown in Fig. 8.11.

NOTE ▶ In most examples, we will round off values to four significant digits (as in Examples 3, 4, and 5). However, *if the angle is approximate, use the guidelines in Section 4.3 for rounding off values.*

EXAMPLE 6 Quadrant II angle – application

A formula for finding the area of a triangle, knowing sides a and b and the included $\angle C$, is $A = \frac{1}{2} ab \sin C$. A surveyor uses this formula to find the area of a triangular tract of land for which $a = 173.2$ m, $b = 156.3$ m, and $C = 112.51°$. See Fig. 8.12.

$$A = \tfrac{1}{2} (173.2)(156.3) \sin 112.51°$$
$$= 12{,}500 \text{ m}^2 \quad \text{rounded to four significant digits}$$

The calculator automatically uses a positive value for $\sin 112.51°$. ■

Knowing how to use the reference angle is important when using a calculator, because *for a given value of a function of an angle, when finding the angle*

CAUTION ▶ *the calculator will not necessarily give us directly the required angle.*

It will give an angle we can use, but whether or not it is the required angle for the problem will depend on the problem being solved.

Fig. 8.12

■ The reason that the calculator displays these angles is shown in Chapter 20, when the inverse trigonometric functions are discussed in detail.

When a value of a trigonometric function is entered into a calculator, it is programmed to give the angle as follows:

For values of sin θ, *the calculator displays angles from* −90° *to* 90°:

$$[\sin(-90°) = -1, \sin 0° = 0, \sin 90° = 1]$$

For values of cos θ, *the calculator displays angles from* 0° *to* 180°:

$$[\cos 0° = 1, \cos 90° = 0, \cos 180° = -1]$$

For values of tan θ, *the calculator displays angles* **between** −90° *and* 90°:

$$[\tan 0° = 0, \textit{if } \tan θ < 0, θ \textit{ is between } -90° \text{ and } 0°]$$

This means that if sin θ or tan θ is negative, the displayed angle is negative. If cos θ is negative, the displayed angle is greater than 90° and no greater than 180°.

EXAMPLE 7   Finding angles given sin θ

If sin θ = 0.2250, we see from the first two lines of the calculator display shown in Fig. 8.13 that θ = 13.00° (rounded off).

This result is correct, but remember that

$$\sin(180° − 13.00°) = \sin 167.00°$$
$$= 0.2250$$

also. If we need only an acute angle, θ = 13.00° is correct. However, if a second-quadrant angle is required, we see that θ = 167.00° is the angle (see Fig. 8.14). These can be checked by finding the values of sin 13.00° and sin 167.00°.   ■

Fig. 8.13

Fig. 8.15

EXAMPLE 8   Finding angles given sec θ

For sec θ = −2.722 and 0° ≤ θ < 360° (this means θ may equal 0° or be between 0° and 360°), see from the third and fourth lines of the calculator display shown in Fig. 8.13 that θ = 111.55° (rounded off).

The angle 111.55° is the second-quadrant angle, but sec θ < 0 in the third quadrant as well. The reference angle is α = 180° − 111.55° = 68.45°, and the third-quadrant angle is 180° + 68.45° = 248.45°. Therefore, the two angles between 0° and 360° for which sec θ = −2.722 are 111.55° and 248.45° (see Fig. 8.15). These angles can be checked by finding sec 111.55° and sec 248.45°.   ■

EXAMPLE 9   Finding angles given tan θ

Given that tan θ = 2.050 and cos θ < 0, find θ for 0° ≤ θ < 360°.

Since tan θ is positive and cos θ is negative, θ must be a third-quadrant angle. A calculator will display an angle of 64.00° (rounded off) for tan θ = 2.050. However, since we need a third-quadrant angle, *we must add* 64.00° to 180°. Thus, the required angle is 244.00° (see Fig. 8.16). Check by finding tan 244.00°.

If tan θ = −2.050 and cos θ < 0, the calculator will display an angle of −64.00° for tan θ = −2.050. We would then have to *recognize that the reference angle is 64.00° and subtract it from 180° to get 116.00°, the required second-quadrant angle. This can be checked by finding tan 116.00°.   ■

Practice Exercise

2. If cos θ = 0.5736, find θ for 0° ≤ θ < 360°.

NOTE ▶

The calculator gives the reference angle (disregarding any minus signs) in all cases except when cos θ is negative. To avoid confusion from the angle displayed by the calculator, *a good procedure is to find the reference angle first.* Then it can be used to determine the angle required by the problem.

We can *find the reference angle by entering the absolute value of the function. The displayed angle will be the reference angle.* The required angle θ is found by using the reference angle as described earlier and as shown in Eqs. (8.6).

$$
\begin{aligned}
\theta &= \alpha && \text{(first quadrant)} \\
\theta &= 180° - \alpha && \text{(second quadrant)} \\
\theta &= 180° + \alpha && \text{(third quadrant)} \\
\theta &= 360° - \alpha && \text{(fourth quadrant)}
\end{aligned}
$$

(8.6)

EXAMPLE 10 Using reference angle

Given that $\cos \theta = -0.1298$, find θ for $0° \le \theta < 360°$.

Since $\cos \theta$ is negative, θ is either a second-quadrant angle or a third-quadrant angle. Using 0.1298, the calculator tells us that the reference angle is 82.54°.

For the second-quadrant angle, we subtract 82.54° from 180° to get 97.46°. For the third-quadrant angle, we add 82.54° to 180° to get 262.54°. See Fig. 8.17(a).

Using −0.1298, the calculator displays the second-quadrant angle of 97.46°. For the third-quadrant angle, we must subtract 97.46° from 180° to get the reference angle 82.54°, which is then added to 180° to get 262.54°. See Fig. 8.17(b). ■

(a)

(b)

Fig. 8.17

Quadrantal Angles We use Eqs. (8.3), (8.4), and (8.5) when the terminal side is in one of the quadrants. A **quadrantal angle** *has its terminal side along one of the axes.* Using the definitions of the functions (recalling that $r > 0$), we get the following values for quadrantal angles. These values may be verified by referring to Fig. 8.18.

θ	$\sin \theta$	$\cos \theta$	$\tan \theta$	$\cot \theta$	$\sec \theta$	$\csc \theta$
0°	0.000	1.000	0.000	undef.	1.000	undef.
90°	1.000	0.000	undef.	0.000	undef.	1.000
180°	0.000	−1.000	0.000	undef.	−1.000	undef.
270°	−1.000	0.000	undef.	0.000	undef.	−1.000
360°	Same as the functions of 0° (same terminal side)					

Fig. 8.18   (a)   (b)   (c)   (d)

EXAMPLE 11 Quadrantal angles

Since $\sin \theta = y/r$, by looking at Fig. 8.18(a), we can see that $\sin 0° = 0/r = 0$.

Since $\tan \theta = y/x$, from Fig. 8.18(b), we see that $\tan 90° = r/0$, which is undefined due to the division by zero. Using a calculator to find $\tan 90°$, the display would indicate an error (due to division by zero).

Since $\cos \theta = x/r$, from Fig. 8.18(c), we see that $\cos 180° = -r/r = -1$.

Since $\cot \theta = x/y$, from Fig. 8.18(d), we see that $\cot 270° = 0/-r = 0$. ■

Negative Angles

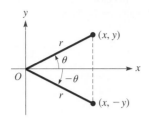

Fig. 8.19

To evaluate functions of negative angles, we can use functions of corresponding positive angles, if we use the correct *sign*. In Fig. 8.19, note that $\sin \theta = y/r$, and $\sin(-\theta) = -y/r$, which means $\sin(-\theta) = -\sin \theta$. In the same way, we can get all of the relations between functions of $-\theta$ and the functions of θ. Therefore, using the definitions, we have

$$
\boxed{
\begin{array}{lll}
\sin(-\theta) = -\sin\theta & \cos(-\theta) = \cos\theta & \tan(-\theta) = -\tan\theta \\
\csc(-\theta) = -\csc\theta & \sec(-\theta) = \sec\theta & \cot(-\theta) = -\cot\theta
\end{array}
}
\tag{8.7}
$$

EXAMPLE 12 Negative angles

$$\sin(-60°) = -\sin 60° = -0.8660 \qquad \cos(-60°) = \cos 60° = 0.5000$$
$$\tan(-60°) = -\tan 60° = -1.732 \qquad \cot(-60°) = -\cot 60° = -0.5774$$
$$\sec(-60°) = \sec 60° = 2.000 \qquad \csc(-60°) = -\csc 60° = -1.155 \qquad ■$$

EXERCISES 8.2

In Exercises 1–4, make the given changes in the indicated examples of this section and then solve the resulting problems.

1. In Example 5, add 40° to each angle, express in terms of the same function of a positive acute angle, and then evaluate.

2. In Example 7, make the value of the function negative and then find θ between 0° and 360°.

3. In Example 9, change 2.050 to -2.050 and $<$ to $>$, then find θ.

4. In Example 10, change the $-$ to $+$ and then find θ.

In Exercises 5–10, express the given trigonometric function in terms of the same function of a positive acute angle.

5. $\sin 160°$, $\cos 220°$
6. $\tan 91°$, $\sec 345°$
7. $\tan 105°$, $\csc 302°$
8. $\cos 190°$, $\cot 290°$
9. $\cos 400°$, $\tan(-400°)$
10. $\tan 920°$, $\csc(-550°)$

In Exercises 11–44, the given angles are approximate. In Exercises 11–18, find the values of the given trigonometric functions by finding the reference angle and attaching the proper sign.

11. $\sin 195°$
12. $\tan 311°$
13. $\cos 106.3°$
14. $\sin 103.4°$
15. $\sec 328.33°$
16. $\cot 516.53°$
17. $\tan(-31.5°)$
18. $\csc(-108.4°)$

In Exercises 19–26, find the values of the given trigonometric functions directly from a calculator.

19. $\tan 152.4°$
20. $\cos 341.4°$
21. $\sin 310.36°$
22. $\tan 242.68°$
23. $\csc 194.82°$
24. $\sec 441.08°$
25. $\cos(-72.61°)$
26. $\sin(-215.5°)$

In Exercises 27–40, find θ for $0° \leq \theta < 360°$.

27. $\sin \theta = -0.8480$
28. $\tan \theta = -1.830$
29. $\cos \theta = 0.4003$
30. $\sin \theta = 0.6374$
31. $\cot \theta = -0.012$
32. $\csc \theta = -8.09$
33. $\sin \theta = 0.870$, $\cos \theta < 0$
34. $\tan \theta = 0.932$, $\sin \theta < 0$
35. $\cos \theta = -0.12$, $\tan \theta > 0$
36. $\sin \theta = -0.192$, $\tan \theta < 0$

37. $\tan \theta = -1.366$, $\cos \theta > 0$
38. $\cos \theta = 0.0726$, $\sin \theta < 0$
39. $\sec \theta = 2.047$, $\cot \theta < 0$
40. $\cot \theta = -0.3256$, $\csc \theta > 0$

In Exercises 41–44, determine the function that satisfies the given conditions.

41. Find $\tan \theta$ when $\sin \theta = -0.5736$ and $\cos \theta > 0$.
42. Find $\sin \theta$ when $\cos \theta = 0.422$ and $\tan \theta < 0$.
43. Find $\cos \theta$ when $\tan \theta = -0.809$ and $\csc \theta > 0$.
44. Find $\cot \theta$ when $\sec \theta = 6.122$ and $\sin \theta < 0$.

(W) *In Exercises 45–48, insert the proper sign, $>$ or $<$ or $=$, between the given expressions. Explain your answers.*

45. $\sin 90°$ $2 \sin 45°$
46. $\cos 360°$ $2 \cos 180°$
47. $\tan 180°$ $\tan 0°$
48. $\sin 270°$ $3 \sin 90°$

In Exercises 49–52, solve the given problems. In Exercises 51 and 52, assume $0° < \theta < 90°$. (Hint: review cofunctions on page 120.)

49. Using the fact that $\sin 75° = 0.9659$, evaluate $\cos 195°$.
50. Using the fact that $\cot 20° = 2.747$, evaluate $\tan 290°$.
51. Express $\tan(270° - \theta)$ in terms of the cot θ.
52. Express $\cos(90° + \theta)$ in terms of $\sin \theta$.

In Exercises 53–56, evaluate the given expressions.

53. The current i in an alternating-current circuit is given by $i = i_m \sin \theta$, where i_m is the maximum current in the circuit. Find i if $i_m = 0.0259$ A and $\theta = 495.2°$.

54. The force F that a rope exerts on a crate is related to force F_x directed along the x-axis by $F = F_x \sec \theta$, where θ is the standard-position angle for F. See Fig. 8.20. Find F if $F_x = -365$ N and $\theta = 127.0°$.

Fig. 8.20

55. For the slider mechanism shown in Fig. 8.21, $y \sin \alpha = x \sin \beta$. Find y if $x = 6.78$ in., $\alpha = 31.3°$, and $\beta = 104.7°$.

Fig. 8.21

56. A laser follows the path shown in Fig. 8.22. The angle θ is related to the distances a, b, and c by $2ab \cos \theta = a^2 + b^2 - c^2$. Find θ if $a = 12.9$ cm, $b = 15.3$ cm, and $c = 24.5$ cm.

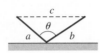

Fig. 8.22

Answers to Practice Exercises

1. $\sin 225° = -\sin 45°$ $\cos 225° = -\cos 45°$
$\tan 225° = \tan 45°$ $\cot 225° = \cot 45°$
$\sec 225° = -\sec 45°$ $\csc 225° = -\csc 45°$
2. $55.0°, 305.0°$

8.3   Radians

Definition of Radian • Procedure for Converting Angle Measurements • Evaluations on Calculator

Fig. 8.23

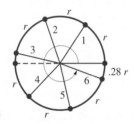

Fig. 8.24

For many problems in which trigonometric functions are used, particularly those involving the solution of triangles, degree measurements of angles are convenient and quite sufficient. However, division of a circle into 360 equal parts is by definition, and it is arbitrary and artificial (see the margin comment on page 50).

In numerous other types of applications and in more theoretical discussions, the *radian* is a more meaningful measure of an angle. We defined the radian in Chapter 2 and reviewed it briefly in Chapter 4. In this section, we discuss the radian in detail and start by reviewing its definition.

A **radian** *is the measure of an angle with its vertex at the center of a circle and with an intercepted arc on the circle equal in length to the radius of the circle.* See Fig. 8.23.

Since the circumference of any circle in terms of its radius is given by $c = 2\pi r$, the ratio of the circumference to the radius is 2π. This means that the radius may be laid off 2π (about 6.28) times along the circumference, regardless of the length of the radius. Therefore, note that radian measure is independent of the radius of the circle. The definition of a radian is based on an important property of a circle and is therefore a more natural measure of an angle. In Fig. 8.24, the numbers on each of the radii indicate the number of radians in the angle measured in standard position. The circular arrow shows an angle of 6 radians.

Since the radius may be laid off 2π times along the circumference, it follows that there are 2π radians in one complete rotation. Also, there are 360° in one complete rotation. Therefore, 360° is *equivalent* to 2π radians. It then follows that the relation between degrees and radians is 2π rad = 360°, or

Converting Angles

$$\pi \text{ rad} = 180° \tag{8.8}$$

Degrees to Radians

$$1° = \frac{\pi}{180} \text{ rad} = 0.01745 \text{ rad} \tag{8.9}$$

Radians to Degrees

$$1 \text{ rad} = \frac{180°}{\pi} = 57.30° \tag{8.10}$$

From Eqs. (8.8), (8.9), and (8.10), note that we convert angle measurements from degrees to radians or radians to degrees using the following procedure:

> **PROCEDURE FOR CONVERTING ANGLE MEASUREMENTS**
> 1. *To convert an angle measured in degrees to the same angle measured in radians*, **multiply the number of degrees by $\pi/180°$.**
> 2. *To convert an angle measured in radians to the same angle measured in degrees*, **multiply the number of radians by $180°/\pi$.**

EXAMPLE 1   Converting to and from radians

converting degrees to radians

(a) $18.0° = \left(\dfrac{\pi}{180°}\right)(18.0°) = \dfrac{\pi}{10.0} = 0.314 \text{ rad}$

degrees cancel   (See Fig. 8.25.)

converting radians to degrees

(b) $2.00 \text{ rad} = \left(\dfrac{180°}{\pi}\right)(2.00) = \dfrac{360°}{\pi} = 114.6°$ (See Fig. 8.25.)

Multiplying by $\pi/180°$ or $180°/\pi$ is actually multiplying by 1, because π rad $= 180°$. The unit of measurement is different, but *the angle is the same.* ∎

Because of the definition of the radian, it is common to express radians in terms of π, particularly for angles whose degree measure is a fraction of $180°$.

EXAMPLE 2   Radians in terms of π

(a) Converting $30°$ to radian measure, we have

$$30° = \left(\frac{\pi}{180°}\right)(30°) = \frac{\pi}{6} \text{ rad}$$ (See Fig. 8.26)

(b) Converting $3\pi/4$ rad to degrees, we have

$$\frac{3\pi}{4} \text{ rad} = \left(\frac{180°}{\pi}\right)\left(\frac{3\pi}{4}\right) = 135°$$ (See Fig. 8.26) ∎

We wish now to make a very important point. Since π is a number (a little greater than 3) that is the ratio of the circumference of a circle to its diameter, it is the ratio of one length to another. This means *radians have no units,* and *radian measure amounts to measuring angles in terms of real numbers.* It is this property that makes radians useful in many applications. Therefore,

CAUTION ▶ *when the angle is measured in radians, it is customary that no units are shown. The radian is understood to be the unit of measurement.*

If the symbol rad is used, it is only to emphasize that the angle is in radians.

EXAMPLE 3   No angle units indicates radians

(a) $60° = \left(\dfrac{\pi}{180°}\right)(60.0°) = \dfrac{\pi}{3.00} = 1.05$ $1.05 = 1.05 \text{ rad}$

no units indicates radian measure

(b) $3.80 = \left(\dfrac{180°}{\pi}\right)(3.80) = 218°$

Because no units are shown for 1.05 and 3.80, they are known to be measured in radians. See Fig. 8.27. ∎

Fig. 8.25

2.00 rad = 114.6°

18.0° = 0.314 rad

■ See Appendix B on unit conversions.

Fig. 8.26

$\frac{3\pi}{4}$ rad = 135°

$30° = \frac{\pi}{6}$ rad

Practice Exercises

1. Convert $36°$ to radians in terms of π.
2. Convert $7\pi/9$ to degrees.

3.80 = 218°

Fig. 8.27

Fig. 8.28

Using the *angle* feature, a calculator can be used directly to change an angle expressed in degrees to an angle expressed in radians, or from an angle expressed in radians to an angle expressed in degrees (as pointed out in Chapter 4 on page 110). In Fig. 8.28, we show the calculator display for changing $\pi/2$ rad to degrees (calculator in degree mode) and for changing 60° to radians (calculator in radian mode). How these operations are done depends on the calculator model.

We can use a calculator to find the value of a function of an angle in radians. If the calculator is in radian mode, it then uses values in radians directly and will *consider any angle entered to be in radians*. The mode can be changed as needed, but

CAUTION ▶   *always be careful to have your calculator in the proper mode.*

Check the setting in the *mode* feature. If you are working in degrees, use the degree mode, but if you are working in radians, use the radian mode.

EXAMPLE 4 Calculator evaluations

(a) To find the value of sin 0.7538, put the calculator in radian mode (note that no units are shown with 0.7538), and the value is found as shown in the first two lines of the calculator display in Fig. 8.29. Therefore,

$$\sin 0.7538 = 0.6844$$

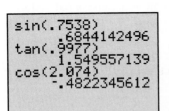

no units indicates radian measure

Fig. 8.29

(b) From the third and fourth lines of the display in Fig. 8.29, we see that

$$\tan 0.9977 = 1.550$$

(c) From the last two lines of the display in Fig. 8.29, we see that

$$\cos 2.074 = -0.4822$$

In each case, the calculator was in radian mode.   ■

Practice Exercise

3. Find the value of sin 3.56.

In the following application, the resulting angle is a unitless number, and it is therefore in radian measure.

EXAMPLE 5 Application of radian measure

The velocity v of an object undergoing simple harmonic motion at the end of a spring is given by

$$v = A\sqrt{\frac{k}{m}}\cos\sqrt{\frac{k}{m}}\,t \qquad \text{the angle is } \left(\sqrt{\frac{k}{m}}\right)(t)$$

Here, m is the mass of the object (in g), k is a constant depending on the spring, A is the maximum distance the object moves, and t is the time (in s). Find the velocity (in cm/s) after 0.100 s of a 36.0-g object at the end of a spring for which $k = 400\,\text{g/s}^2$, if $A = 5.00$ cm.

Substituting, we have

$$v = 5.00\sqrt{\frac{400}{36.0}}\cos\sqrt{\frac{400}{36.0}}(0.100)$$

Using calculator memory for $\sqrt{\frac{400}{36.0}}$, and with the calculator in radian mode, we have

$$v = 15.7\,\text{cm/s}$$ ■

If we need a reference angle in radians, recall that $\pi/2 = 90°$, $\pi = 180°$, $3\pi/2 = 270°$, and $2\pi = 360°$ (see Fig. 8.30). These and their decimal approximations are shown in Table 8.1.

Table 8.1 Quadrantal Angles

Degrees	Radians	Radians (decimal)
90°	$\frac{1}{2}\pi$	1.571
180°	π	3.142
270°	$\frac{3}{2}\pi$	4.712
360°	2π	6.283

Fig. 8.30

EXAMPLE 6 Angles of any size in radians

An angle of 3.402 is greater than 3.142 but less than 4.712. Thus, it is a third-quadrant angle, and the reference angle is $3.402 - \pi = 0.260$. The calculator $\boxed{\pi}$ key can be used. See Fig. 8.31.

An angle of 5.210 is between 4.712 and 6.283. Therefore, it is in the fourth quadrant and the reference angle is $2\pi - 5.210 = 1.073$. ∎

Fig. 8.31

EXAMPLE 7 Find angle in radians for given function

Express θ in radians, such that $\cos\theta = 0.8829$ and $0 \le \theta < 2\pi$.

Since $\cos\theta$ is positive and θ between 0 and 2π, we want a first-quadrant angle and a fourth-quadrant angle. With the calculator in radian mode (see Fig. 8.32),

$\cos^{-1} 0.8829 = 0.4888$ first-quadrant angle

$2\pi - 0.4888 = 5.794$ fourth-quadrant angle

$\theta = 0.4888$ or $\theta = 5.794$ see Fig. 8.33 ∎

Fig. 8.33

```
cos⁻¹(.8829)→X
        .4887935563
2π-X
        5.794391751
```

Fig. 8.32

When one first encounters radian measure, *expressions such as* **sin 1** *and* **sin θ = 1** *are often confused.* The first is equivalent to $\sin 57.30°$ (since 1 radian = 57.30°). The second means θ is the angle for which the sine is 1. Since $\sin 90° = 1$, we can say that $\theta = 90°$ or $\theta = \pi/2$. The following example gives additional illustrations.

Practice Exercise

4. If $\sin\theta = 0.4235$, find the smallest positive θ (in radians).

EXAMPLE 8 Angle in radians – value of function

(a) $\sin \pi/3 = \sqrt{3}/2$ **(b)** $\cos\theta = 0.5$, $\theta = 60° = \pi/3$ (smallest positive θ)

(c) $\tan 2 = \tan 114.6°$ **(d)** $\tan\theta = 2$, $\theta = 1.107$ (smallest positive θ)

EXERCISES 8.3

In Exercises 1–4, make the given changes in the indicated examples of this section and then solve the resulting problems.

1. In Example 3(b), change 3.80 to 2.80.

2. In Example 4(a), change sin to cos.

3. In Example 7, change cos to sin.

4. In Example 8(a), change $\frac{\pi}{3}$ to $\frac{\pi}{4}$.

In Exercises 5–12, express the given angle measurements in radian measure in terms of π.

5. $15°, 150°$ 6. $12°, 225°$ 7. $75°, 330°$ 8. $36°, 315°$

9. $210°, 27°$ 10. $5°, 300°$ 11. $720°, -9°$ 12. $66°, 540°$

In Exercises 13–20, the given numbers express angle measure. Express the measure of each angle in terms of degrees.

13. $\dfrac{2\pi}{5}, \dfrac{3\pi}{2}$ 14. $\dfrac{3\pi}{10}, \dfrac{5\pi}{6}$ 15. $\dfrac{\pi}{18}, \dfrac{7\pi}{4}$ 16. $\dfrac{8\pi}{15}, \dfrac{4\pi}{3}$

17. $\dfrac{7\pi}{18}, \dfrac{5\pi}{3}$ 18. $\dfrac{\pi}{40}, \dfrac{5\pi}{4}$ 19. $-\dfrac{\pi}{9}, \dfrac{3\pi}{20}$ 20. $\dfrac{9\pi}{2}, \dfrac{4\pi}{15}$

In Exercises 21–28, express the given angles in radian measure. Round off results to the number of significant digits in the given angle.

21. $23.0°$ 22. $54.3°$ 23. $252°$ 24. $104°$

25. $333.5°$ 26. $168.7°$ 27. $478.5°$ 28. $-86.1°$

In Exercises 29–36, the given numbers express the angle measure. Express the measure of each angle in terms of degrees, with the same accuracy as the given value.

29. 0.750   **30.** 0.240   **31.** 3.407   **32.** 1.703

33. 12.4   **34.** 34.4   **35.** −16.42   **36.** 100.0

In Exercises 37–44, evaluate the given trigonometric functions by first changing the radian measure to degree measure. Round off results to four significant digits.

37. $\sin\dfrac{\pi}{4}$ **38.** $\cos\dfrac{\pi}{6}$ **39.** $\tan\dfrac{5\pi}{12}$

40. $\sin\dfrac{7\pi}{36}$ **41.** $\cos\dfrac{5\pi}{6}$ **42.** $\tan\dfrac{7\pi}{3}$

43. $\sec 4.5920$ **44.** $\cot 3.2732$

In Exercises 45–52, evaluate the given trigonometric functions directly, without first changing to degree measure.

45. $\tan 0.7359$ **46.** $\cos 0.9308$ **47.** $\sin 4.24$

48. $\tan 3.47$ **49.** $\sec 2.07$ **50.** $\sin(-2.34)$

51. $\cot(-4.86)$ **52.** $\csc 6.19$

In Exercises 53–60, find θ to four significant digits for $0 \le \theta < 2\pi$.

53. $\sin\theta = 0.3090$ **54.** $\cos\theta = -0.9135$

55. $\tan\theta = -0.2126$ **56.** $\sin\theta = -0.0436$

57. $\cos\theta = 0.6742$ **58.** $\cot\theta = 1.860$

59. $\sec\theta = -1.307$ **60.** $\csc\theta = 3.940$

In Exercises 61–66, solve the given problems. (Hint: For problems 63–66, review cofunctions on page 120.)

61. Find the radian measure of an angle at the center of a circle of radius 12 cm that intercepts an arc of 15 cm on the circle.

62. Find the length of arc of a circle of radius 10 in. that is intercepted from the center of the circle by an angle of 3 radians.

63. Using the fact that $\sin\dfrac{\pi}{8} = 0.3827$, find the value of $\cos\dfrac{5\pi}{8}$. (A calculator should be used only to check the result.)

64. Using the fact that $\tan\dfrac{\pi}{6} = 0.5774$, find the value of $\cot\dfrac{5\pi}{3}$. (A calculator should be used only to check the result.)

65. Express $\tan\!\left(\dfrac{\pi}{2} + \theta\right)$ in terms of $\cot\theta$. $\left(0 < \theta < \dfrac{\pi}{2}\right)$

66. Express $\cos\!\left(\dfrac{3\pi}{2} + \theta\right)$ in terms of $\sin\theta$. $\left(0 < \theta < \dfrac{\pi}{2}\right)$

In Exercises 67–74, evaluate the given problems.

67. A unit of angle measurement used in artillery is the *mil*, which is defined as a central angle of a circle that intercepts an arc equal in length to $1/6400$ of the circumference. How many mils are in a central angle of 34.4°?

68. Through how many radians does the minute hand of a clock move in 25 min?

69. After the brake was applied, a bicycle wheel went through 1.75 rotations. Through how many radians did a spoke rotate?

70. Through how many radians does a Ferris wheel with 18 seats move when loading passengers on the first 12 seats, assuming the loading process started at seat 1?

71. A flat plate of weight W oscillates as shown in Fig. 8.34. Its potential energy V is given by $V = \frac{1}{2}Wb\theta^2$, where θ is measured in radians. Find V if $W = 8.75$ lb, $b = 0.75$ ft, and $\theta = 5.5°$.

Fig. 8.34

W **72.** The charge q (in C) on a capacitor as a function of time is $q = A\sin\omega t$. If t is measured in seconds, in what units is ω measured? Explain.

73. The height h of a rocket launched 1200 m from an observer is found to be $h = 1200\tan\dfrac{5t}{3t + 10}$ for $t < 10$ s, where t is the time after launch. Find h for $t = 8.0$ s.

74. The electric intensity I (in W/m²) from the two radio antennas in Fig. 8.35 is a function of θ given by $I = 0.023\cos^2(\pi\sin\theta)$. Find I for $\theta = 40.0°$. $(\cos^2\alpha = (\cos\alpha)^2.)$

Fig. 8.35

Answers to Practice Exercises

1. $\pi/5$ **2.** 140°   **3.** −0.4063   **4.** 0.4373

8.4 Applications of Radian Measure

Arc Length • Area of a Sector of a Circle • Angular Velocity

Radian measure has numerous applications in mathematics and technology, some of which were illustrated in the last eight exercises of the previous section. In this section, several more applications are shown.

ARC LENGTH

From geometry, we know that *the length of an arc on a circle is proportional to the central angle* formed by the radii that intercept the arc. The length of arc of a complete circle is the circumference. Letting s represent the length of arc, we may state that $s = 2\pi r$ for a complete circle. Since 2π is the central angle (in radians) of the complete circle, *we have for the length of arc*

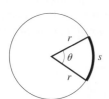

Fig. 8.36

$$s = \theta r \quad (\theta \text{ in radians}) \tag{8.11}$$

for any circular arc with central angle θ. Therefore, if we know the central angle θ *in radians* and the radius of a circle, we can find the length of a circular arc directly by using Eq. (8.11). See Fig. 8.36.

EXAMPLE 1 Arc length

Find the length of arc on a circle of radius $r = 3.00$ in., for which the central angle $\theta = \pi/6$. See Fig. 8.37.

$$\overset{\theta \text{ in radians}}{s = \left(\frac{\pi}{6}\right)(3.00) = \frac{\pi}{2.00}}$$
$$= 1.57 \text{ in.}$$

Fig. 8.37

Therefore, the length of arc s is 1.57 in. ∎

Among the important applications of arc length are distances on the earth's surface. For most purposes, the earth may be regarded as a sphere (the diameter at the equator is slightly greater than the distance between the poles). A *great circle* of the earth (or any sphere) is the circle of intersection of the surface of the sphere and a plane that passes through the center.

The equator is a great circle and is designated as 0° *latitude*. Other *parallels of latitude* are parallel to the equator with diameters decreasing to zero at the poles, which are 90° N and 90° S. See Fig. 8.38.

Meridians of longitude are half great circles between the poles. The *prime meridian* through Greenwich, England, is designated as 0°, with meridians to 180° measured east and west from Greenwich. Positions on the surface of the earth are designated by longitude and latitude.

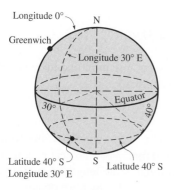

Fig. 8.38

EXAMPLE 2 Arc length – nautical mile

The traditional definition of a *nautical mile* is the length of arc along a great circle of the earth for a central angle of 1'. The modern international definition is a distance of 1852 m. What measurement of the earth's radius does this definition use?

Here, $\theta = 1' = (1/60)°$, and $s = 1852$ m. Solving for r, we have

$$r = \frac{s}{\theta} = \frac{1852}{\left(\frac{1}{60}\right)°\left(\frac{\pi}{180°}\right)} = 6.367 \times 10^6 \text{ m}$$
$$= 6367 \text{ km}$$

■ See the chapter introduction.

Practice Exercise

1. Find θ in degrees if $s = 2.50$ m and $r = 1.75$ m.

Historically, the fact that the earth is not a perfect sphere has led to many variations in the distance used for a nautical mile. ∎

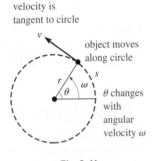

Fig. 8.39

AREA OF A SECTOR OF A CIRCLE

Another application of radians is finding the area of a sector of a circle (see Fig. 8.39). Recall from geometry that areas of sectors of circles are proportional to their central angles. The area of a circle is $A = \pi r^2$, which can be written as $A = \frac{1}{2}(2\pi)r^2$. Since the angle for a complete circle is 2π, *the area of any sector of a circle in terms of the radius and central angle (in radians) is*

$$A = \frac{1}{2}\theta r^2 \qquad (\theta \text{ in radians}) \tag{8.12}$$

EXAMPLE 3 Area of sector of circle

(a) The area of a sector of a circle with central angle 218° and a radius of 5.25 in. (see Fig. 8.40(a)) is

$$A = \frac{1}{2}(218)\overbrace{\left(\frac{\pi}{180}\right)}^{\theta \text{ in radians}}(5.25)^2 = 52.4 \text{ in.}^2$$

(b) Given that the area of a sector is 75.5 ft² and the radius is 12.2 ft (see Fig. 8.40(b)), we find the central angle by solving for θ and then substituting:

$$\theta = \frac{2A}{r^2} = \frac{2(75.5)}{(12.2)^2} = 1.01 \quad \overset{\text{no units indicates radian measure}}{\downarrow}$$

This means that the central angle is 1.01 rad, or 57.9°. ∎

(a) (b)

Fig. 8.40

Practice Exercise

2. Find A if $r = 17.5$ in. and $\theta = 125°$.

CAUTION ▶ We should note again that the equations in this section require that the angle θ *is expressed in radians.* A common error is to use θ in degrees.

ANGULAR VELOCITY

The average velocity of a moving object is defined by $v = s/t$, where v is the average velocity, s is the distance traveled, and t is the elapsed time. For an object moving in a circular path with constant speed, the distance traveled is the length of arc through which it moves. Therefore, if we divide both sides of Eq. (8.11) by t, we obtain

$$\frac{s}{t} = \frac{\theta r}{t} = \frac{\theta}{t}r$$

where θ/t is called the *angular velocity* and is designated by ω. Therefore,

$$v = \omega r \tag{8.13}$$

velocity is tangent to circle

v

object moves along circle

r s

θ ω

θ changes with angular velocity ω

Fig. 8.41

■ Some of the typical units used for angular velocity are

rad/s   rad/min   rad/h

°/s   °/min

r/s   r/min

(r represents revolutions.)

(r/min is the same as rpm. This text does not use rpm.)

Equation (8.13) expresses the relationship between the **linear velocity v** *and the* **angular velocity** ω *of an object moving around a circle of radius r.* See Fig. 8.41. In the figure, v is shown directed tangent to the circle, for that is its direction for the position shown. The direction of v changes constantly.

The units for ω are radians per unit of time. In this way, the formula can be used directly. However, in practice, ω is often given in revolutions per minute or in some similar unit. In these cases, it is necessary to convert the units of ω to radians per unit of time before substituting in Eq. (8.13).

EXAMPLE 4   Angular velocity – application

A person on a hang glider is moving in a horizontal circular arc of radius 90.0 m with an angular velocity of 0.125 rad/s. The person's linear velocity is

$$v = (0.125\,\text{rad/s})(90.0\,\text{m}) = 11.3\,\text{m/s}$$

(Remember that radians are numbers and are not included in the final set of units.) This means that the person is moving along the circumference of the arc at 11.3 m/s (40.7 km/h). ∎

EXAMPLE 5   Angular velocity – application

A communications satellite remains at an altitude of 22,320 mi above a point on the equator. If the radius of the earth is 3960 mi, what is the velocity of the satellite?

■ The first U.S. communications satellite was launched in July 1962.

In order for the satellite to remain over a point on the equator, it must rotate exactly once each day around the center of the earth (and it must remain at an altitude of 22,320 mi). Since there are 2π radians in each revolution, the angular velocity is

$$\omega = \frac{1\,\text{r}}{1\,\text{day}} = \frac{2\pi\,\text{rad}}{24\,\text{h}} = 0.2618\,\text{rad/h}$$

The radius of the circle through which the satellite moves is its altitude plus the radius of the earth, or $22{,}320 + 3960 = 26{,}280$ mi. Thus, the velocity is

$$v = 0.2618(26{,}280) = 6880\,\text{mi/h}$$ ∎

Practice Exercise

3. Find r if $v = 25.0$ ft/s and the angular velocity is $6.0°/\text{min}$.

EXAMPLE 6   Angular velocity – application

A pulley belt 10.0 ft long takes 2.00 s to make one complete revolution. The radius of the pulley is 6.00 in. What is the angular velocity (in revolutions per minute) of a point on the rim of the pulley? See Fig. 8.42.

Since the linear velocity of a point on the rim of the pulley is the same as the velocity of the belt, $v = 10.0/2.00 = 5.00$ ft/s. The radius of the pulley is $r = 6.00$ in. $= 0.500$ ft, and we can find ω by substituting into Eq. (8.13). This gives us

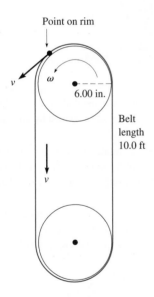

Point on rim

6.00 in.

Belt
length
10.0 ft

v

Fig. 8.42

$$v = \omega r$$
$$5.00 = \omega(0.500)$$
$$\omega = 10.0\,\text{rad/s} \qquad \text{multiply by 60 s/1 min}$$
$$= 600\,\text{rad/min} \qquad \text{multiply by 1 r/}2\pi\text{ rad}$$
$$= 95.5\,\text{r/min} \qquad \text{r is the symbol for revolution}$$

As shown in Appendix B, the change of units can be handled algebraically as

$$10.0\,\frac{\text{rad}}{\text{s}} \times 60\,\frac{\text{s}}{\text{min}} = 600\,\frac{\text{rad}}{\text{min}}$$

$$\frac{600\,\text{rad/min}}{2\pi\,\text{rad/r}} = 600\,\frac{\text{rad}}{\text{min}} \times \frac{1}{2\pi}\,\frac{\text{r}}{\text{rad}} = 95.5\,\text{r/min}$$ ∎

Many types of problems use radian measure. Following is one involving electric current.

EXAMPLE 7   Application to electric current

The current at any time in a certain alternating-current electric circuit is given by $i = I\sin 120\pi t$, where I is the maximum current and t is the time in seconds. Given that $I = 0.0685$ A, find i for $t = 0.00500$ s.

Substituting, with the calculator in radian mode, we get

$$i = 0.0685\sin[(120\pi)(0.00500)]$$
$$= 0.0651\,\text{A}$$

$(120\pi)(0.00500)$ is a pure number, and therefore is an angle in radians. ∎

EXERCISES 8.4

In Exercises 1–4, make the given changes in the indicated examples of this section, and then solve the resulting problems.

1. In Example 1, change $\pi/6$ to $\pi/4$.
2. In Example 3(a), change 218° to 258°.
3. In Example 4, change 90.0 m to 115 m.
4. In Example 6, change 2.00 s to 2.50 s.

In Exercises 5–16, for an arc length s, area of sector A, and central angle θ of a circle of radius r, find the indicated quantity for the given values.

5. $r = 3.30$ in., $\theta = \pi/3$, $s = ?$
6. $r = 21.2$ cm, $\theta = 2.65$, $s = ?$
7. $s = 1010$ mm, $\theta = 136.0°$, $r = ?$
8. $s = 0.3456$ ft, $\theta = 73.61°$, $A = ?$
9. $s = 0.3913$ mi, $r = 0.9449$ mi, $A = ?$
10. $s = 3.19$ m, $r = 2.29$ m, $\theta = ?$
11. $r = 4.9$ cm, $\theta = 5.6$, $A = ?$
12. $r = 46.3$ in., $\theta = 2\pi/5$, $A = ?$
13. $A = 0.0119$ ft^2, $\theta = 326.0°$, $r = ?$
14. $A = 1200$ mm^2, $\theta = 17°$, $s = ?$
15. $A = 16.5$ m^2, $r = 4.02$ m, $s = ?$
16. $A = 67.8$ mi^2, $r = 67.8$ mi, $\theta = ?$

In Exercises 17–58, solve the given problems.

17. In traveling three-fourths of the way around a traffic circle a car travels 0.203 mi. What is the radius of the traffic circle? See Fig. 8.43.

Fig. 8.43

18. The latitude of Miami is 26° N, and the latitude of the north end of the Panama Canal is 9° N. Both are at a longitude of 80° W. What is the distance between Miami and the Canal? Explain how the angle used in the solution is found. The radius of the earth is 3960 mi.

19. Of the estimated natural gas reserves in North America, 4.59×10^9 m^3 are in the United States, 2.66×10^9 m^3 are in Canada, and 1.99×10^9 m^3 are in Mexico. In making a *circle graph* (circular sectors represent percentages of the whole—a *pie chart*) with a radius of 4.00 cm for these data, what are the central angle and area of the sector that represents Canada's reserves?

20. A section of sidewalk is a circular sector of radius 3.00 ft and central angle 50.6°. What is the area of this section of sidewalk?

21. When between 12:00 noon and 1:00 P.M. are the minute and hour hands of a clock 180° apart?

22. A cam is in the shape of a circular sector, as shown in Fig. 8.44. What is the perimeter of the cam?

Fig. 8.44   165.58°   1.875 in.

23. A lawn sprinkler can water up to a distance of 65.0 ft. It turns through an angle of 115.0°. What area can it water?

24. A spotlight beam sweeps through a horizontal angle of 75.0°. If the range of the spotlight is 250 ft, what area can it cover?

25. If a car makes a U-turn in 6.0 s, what is its average angular velocity in the turn?

26. The roller on a computer printer makes 2200 r/min. What is its angular velocity?

27. What is the floor area of the hallway shown in Fig. 8.45? The outside and inside of the hallway are circular arcs.

Fig. 8.45   3.755 m   7.535 m   8.250 m

28. The arm of a car windshield wiper is 12.75 in. long and is attached at the middle of a 15.00-in. blade. (Assume that the arm and blade are in line.) What area of the windshield is cleaned by the wiper if it swings through 110.0° arcs?

29. Part of a railroad track follows a circular arc with a central angle of 28.0°. If the radius of the arc of the inner rail is 93.67 ft and the rails are 4.71 ft apart, how much longer is the outer rail than the inner rail?

30. A wrecking ball is dropped as shown in Fig. 8.46. Its velocity at the bottom of its swing is $v = \sqrt{2gh}$, where g is the acceleration due to gravity. What is its angular velocity at the bottom if $g = 9.80$ m/s^2 and $h = 4.80$ m?

Fig. 8.46   13.8 m   h

31. Part of a security fence is built 2.50 m from a cylindrical storage tank 11.2 m in diameter. What is the area between the tank and this part of the fence if the central angle of the fence is 75.5°? See Fig. 8.47.

Fig. 8.47

32. Through what angle does the drum in Fig. 8.48 turn in order to lower the crate 18.5 ft?

Fig. 8.48

33. A section of road follows a circular arc with a central angle of 15.6°. The radius of the inside of the curve is 285.0 m, and the road is 15.2 m wide. What is the volume of the concrete in the road if it is 0.305 m thick?

34. The propeller of the motor on a motorboat is rotating at 130 rad/s. What is the linear velocity of a point on the tip of a blade if it is 22.5 cm long?

35. A storm causes a pilot to follow a circular-arc route, with a central angle of 12.8°, from city A to city B rather than the straight-line route of 185.0 km. How much farther does the plane fly due to the storm?

36. An interstate route exit is a circular arc 330 m long with a central angle of 79.4°. What is the radius of curvature of the exit?

37. The paddles of a riverboat have a radius of 8.50 ft and revolve at 20.0 r/min. What is the speed of a tip of one of the paddles?

38. The sweep second hand of a watch is 15.0 mm long. What is the linear velocity of the tip?

39. A DVD has a diameter of 4.75 in. and rotates at 360.0 r/min. What is the linear velocity of a point on the outer edge?

40. A Ferris wheel 18.0 m in diameter makes one revolution in 3.00 min. Find the speed of a seat on the rim.

41. In 2005 a planet was found revolving about a distant sunlike star in 2.88 days. The radius of its orbit is 3,500,000 mi. What is its linear velocity (in mi/h) about its star?

42. Assume that earth rotates around the sun in a circular orbit of radius 93,000,000 mi. (which is approximately correct). What is earth's linear velocity (in mi/h)?

43. The sprocket assembly for a 28.0-in. bike is shown in Fig. 8.49. How fast (in r/min) does the rider have to pedal in order to go 15.0 mi/h on level ground?

Fig. 8.49

44. The flywheel of a car engine is 0.36 m in diameter. If it is revolving at 750 r/min, through what distance does a point on the rim move in 2.00 s?

45. Two streets meet at an angle of 82.0°. What is the length of the piece of curved curbing at the intersection if it is constructed along the arc of a circle 15.0 ft in radius? See Fig. 8.50.

Fig. 8.50

46. An ammeter needle is deflected 52.00° by a current of 0.2500 A. The needle is 3.750 in. long, and a circular scale is used. How long is the scale for a maximum current of 1.500 A?

47. A drill bit $\frac{3}{8}$ in. in diameter rotates at 1200 r/min. What is the linear velocity of a point on its circumference?

48. A helicopter blade is 2.75 m long and is rotating at 420 r/min. What is the linear velocity of the tip of the blade?

49. A waterwheel used to generate electricity has paddles 3.75 m long. The speed of the end of a paddle is one-fourth that of the water. If the water is flowing at the rate of 6.50 m/s, what is the angular velocity of the waterwheel?

50. A jet is traveling westward with the sun directly overhead (the jet is on a line between the sun and the center of the earth). How fast must the jet fly in order to keep the sun directly overhead? (Assume that the earth's radius is 3960 mi, the altitude of the jet is low, and the earth rotates about its axis once in 24.0 h.)

51. A 1500-kW wind turbine (windmill) rotates at 40.0 r/min. What is the linear velocity of a point on the end of a blade, if the blade is 35 ft long (from the center of rotation)?

52. What is the linear velocity of a point in Charleston, South Carolina, which is at a latitude of 32°46′ N? The radius of the earth is 3960 mi.

53. Through what total angle does the drive shaft of a car rotate in 1 s when the tachometer reads 2400 r/min?

54. A baseball field is designed such that the outfield fence is along the arc of a circle with its center at second base. If the radius of the circle is 280 ft, what is the playing area of the field? See Fig. 8.51.

Fig. 8.51

55. The turbine fan blade of a turbojet engine is 1.2 m in diameter and rotates at 250 r/s. How fast is the tip of a blade moving?

56. A patio is in the shape of a circular sector with a central angle of 160.0°. It is enclosed by a railing of which the circular part is 11.6 m long. What is the area of the patio?

57. An oil storage tank 4.25 m long has a flat bottom as shown in Fig. 8.52. The radius of the circular part is 1.10 m. What volume of oil does the tank hold?

Fig. 8.52   1.48 m

58. Two equal beams of light illuminate the area shown in Fig. 8.53. What area is lit by both beams?

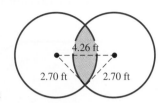

Fig. 8.53

In Exercises 57–60, another use of radians is illustrated.

(W) 59. Use a calculator (in radian mode) to evaluate the ratios $(\sin\theta)/\theta$ and $(\tan\theta)/\theta$ for $\theta = 0.1$, 0.01, 0.001, and 0.0001. From these values, explain why it is possible to say that

$$\sin\theta = \tan\theta = \theta \qquad (8.14)$$

approximately for very small angles.

60. Using Eq. (8.14), evaluate $\tan 0.001°$. Compare with a calculator value.

61. An astronomer observes that a star 12.5 light-years away moves through an angle of 0.2″ in 1 year. Assuming it moved in a straight line perpendicular to the initial line of observation, how many miles did the star move? (1 light-year $= 5.88 \times 10^{12}$ mi.) Use Eq. (8.14).

62. In calculating a back line of a lot, a surveyor discovers an error of 0.05° in an angle measurement. If the lot is 136.0 m deep, by how much is the back-line calculation in error? See Fig. 8.54. Use Eq. (8.14).

Fig. 8.54

Answers to Practice Exercises

1. 81.9°   **2.** 334 in.^{2} **3.** 14,300 ft

CHAPTER 8   **EQUATIONS**

$$\sin\theta = \frac{y}{r} \qquad \cos\theta = \frac{x}{r} \qquad \tan\theta = \frac{y}{x}$$

$$\cot\theta = \frac{x}{y} \qquad \sec\theta = \frac{r}{x} \qquad \csc\theta = \frac{r}{y} \qquad (8.1)$$

$$F(\theta_2) = \pm F(180° - \theta_2) = \pm F(\alpha) \qquad (\alpha \text{ is reference angle}) \qquad (8.3)$$

$$F(\theta_3) = \pm F(\theta_3 - 180°) = \pm F(\alpha) \qquad (8.4)$$

$$F(\theta_4) = \pm F(360° - \theta_4) = \pm F(\alpha) \qquad (8.5)$$

α is reference angle	$\theta = \alpha$ (first quadrant)	
	$\theta = 180° - \alpha$ (second quadrant)	
	$\theta = 180° + \alpha$ (third quadrant)	(8.6)
	$\theta = 360° - \alpha$ (fourth quadrant)	

Negative angles

$$\sin(-\theta) = -\sin\theta \qquad \cos(-\theta) = \cos\theta \qquad \tan(-\theta) = -\tan\theta$$
$$\csc(-\theta) = -\csc\theta \qquad \sec(-\theta) = \sec\theta \qquad \cot(-\theta) = -\cot\theta \tag{8.7}$$

Radian-degree conversions

$$\pi \text{ rad} = 180° \tag{8.8}$$

$$1° = \frac{\pi}{180}\text{ rad} = 0.01745 \text{ rad} \tag{8.9}$$

$$1 \text{ rad} = \frac{180°}{\pi} = 57.30° \tag{8.10}$$

Circular arc length

$$s = \theta r \qquad (\theta \text{ in radians}) \tag{8.11}$$

Circular sector area

$$A = \frac{1}{2}\theta r^2 \qquad (\theta \text{ in radians}) \tag{8.12}$$

Linear and angular velocity

$$v = \omega r \tag{8.13}$$

CHAPTER 8   REVIEW EXERCISES

In Exercises 1–4, find the trigonometric functions of θ. The terminal side of θ passes through the given point.

1. $(6, 8)$

2. $(-12, 5)$

3. $(42, -12)$

4. $(-2, -3)$

In Exercises 5–8, express the given trigonometric functions in terms of the same function of a positive acute angle.

5. $\cos 132°, \tan 194°$

6. $\sin 243°, \cot 318°$

7. $\sin 289°, \sec(-15°)$

8. $\cos 463°, \csc(-100°)$

In Exercises 9–12, express the given angle measurements in terms of π.

9. $40°, 153°$

10. $22.5°, 324°$

11. $408°, 202.5°$

12. $12°, -162°$

In Exercises 13–20, the given numbers represent angle measure. Express the measure of each angle in degrees.

13. $\dfrac{7\pi}{5}, \dfrac{13\pi}{18}$

14. $\dfrac{3\pi}{8}, \dfrac{7\pi}{20}$

15. $\dfrac{\pi}{15}, \dfrac{11\pi}{6}$

16. $\dfrac{27\pi}{10}, \dfrac{5\pi}{4}$

17. 0.560

18. -1.354

19. 36.07

20. 14.5

In Exercises 21–28, express the given angles in radians (not in terms of π).

21. $102°$

22. $305°$

23. $20.25°$

24. $148.38°$

25. $262.05°$

26. $-18.72°$

27. $-636.2°$

28. $385.4°$

In Exercises 29–48, determine the values of the given trigonometric functions directly on a calculator. The angles are approximate. Express answers to Exercises 41–44 to four significant digits.

29. $\cos 245.5°$

30. $\sin 141.3°$

31. $\cot 295°$

32. $\tan 184°$

33. $\csc 247.82°$

34. $\sec 96.17°$

35. $\sin 565.24°$

36. $\cos 326.72°$

37. $\tan 301.4°$

38. $\sin 703.9°$

39. $\tan 436.42°$

40. $\cos(-162.32°)$

41. $\sin \dfrac{9\pi}{5}$

42. $\sec \dfrac{5\pi}{8}$

43. $\cos\dfrac{7\pi}{6}$

44. $\tan\dfrac{23\pi}{12}$

45. $\sin 0.5906$

46. $\tan 0.8035$

47. $\csc 2.153$

48. $\cos(-7.190)$

In Exercises 49–52, find θ in degrees for $0° \le \theta < 360°$.

49. $\tan\theta = 0.1817$

50. $\sin\theta = -0.9323$

51. $\cos\theta = -0.4730$

52. $\cot\theta = 1.196$

In Exercises 53–56, find θ in radians for $0 \le \theta < 2\pi$.

53. $\cos\theta = 0.8387$

54. $\sin\theta = 0.1045$

55. $\sin\theta = -0.8650$

56. $\tan\theta = 8.480$

In Exercises 57–60, find θ in degrees for $0° \le \theta < 360°$.

57. $\cos\theta = -0.7222$, $\sin\theta < 0$

58. $\tan\theta = -1.683$, $\cos\theta < 0$

59. $\cot\theta = 0.4291$, $\cos\theta < 0$

60. $\sin\theta = 0.2626$, $\tan\theta < 0$

In Exercises 61–68, for an arc of length s, area of sector A, and central angle θ of circle of radius r, find the indicated quantity for the given values.

61. $s = 20.3$ in., $\theta = 107.5°$, $r = ?$

62. $s = 5840$ ft, $r = 1060$ ft, $\theta = ?$

63. $A = 265$ mm^2, $r = 12.8$ mm, $\theta = ?$

64. $A = 0.908$ km^2, $\theta = 234.5°$, $r = ?$

65. $r = 4.62$ m, $A = 32.8$ m^2, $s = ?$

66. $\theta = 98.5°$, $A = 0.493$ ft^2, $s = ?$

67. $\theta = 0.85°$, $s = 7.94$ in., $A = ?$

68. $r = 254$ cm, $s = 76.1$ cm, $A = ?$

In Exercises 69–92, solve the given problems.

69. Show that the area A of a minor segment of a circle of radius r intercepted by a central angle θ (in radians) (see Fig. 8.55) is
$$A = \frac{1}{2}r^2(\theta - \sin\theta).$$

Fig. 8.55

70. The cross section of a tunnel is the major segment of a circle of radius 12.0 ft wide. The base of the tunnel is 20.0 ft wide. What is the area of the cross section? See Exercise 69.

71. Find the area of the parcel of land shown in Fig. 8.56. It is a right triangle attached to a circle sector.

Fig. 8.56

20.0°

30.0 m

40.0 m

72. The speedometer of a car is designed to be accurate with tires that are 14.0 in. in radius. If the tires are changed to 15.0 in. in radius, and the speedometer shows 55 mi/h, how fast is the car actually going?

73. The instantaneous power p (in W) input to a resistor in an alternating-current circuit is $p = p_m\sin^2 377t$, where p_m is the maximum power input and t is the time (in s). Find p for $p_m = 0.120$ W and $t = 2.00$ ms. ($\sin^2\theta = (\sin\theta)^2$.)

74. The horizontal distance x through which a pendulum moves is given by $x = a(\theta + \sin\theta)$, where a is a constant and θ is the angle between the vertical and the pendulum. Find x for $a = 45.0$ cm and $\theta = 0.175$.

75. A sector gear with a pitch radius of 8.25 in. and a 6.60-in. arc of contact is shown in Fig. 8.57. What is the sector angle θ?

6.60-in. arc

8.25 in.

Fig. 8.57

76. Two pulleys have radii of 10.0 in. and 6.00 in., and their centers are 40.0 in. apart. If the pulley belt is uncrossed, what must be the length of the belt?

77. A special vehicle for traveling on glacial ice has tires that are 4.8 ft in diameter. If the vehicle travels at 3.5 mi/h, what is the angular velocity (in r/min) of the tire?

78. A rotating circular restaurant at the top of a hotel has a diameter of 32.5 m. If it completes one revolution in 24.0 min, what is the velocity of the outer surface?

79. Find the velocity (in mi/h) of the moon as it revolves about the earth. Assume it takes 28 days for one revolution at a distance of 240,000 mi from the earth.

80. The stopboard of a shot-put circle is a circular arc 1.22 m in length. The radius of the circle is 1.06 m. What is the central angle?

(W) 81. The longitude of Anchorage, Alaska, is 150° W, and the longitude of St. Petersburg, Russia, is 30° E. Both cities are at a latitude of 60° N. (a) Find the great circle distance (see page 249) from Anchorage to St. Petersburg over the north pole. (b) Find the distance between them along the 60° N latitude arc. The radius of the earth is 3960 mi. What do the results show?

82. A piece of circular filter paper 15.0 cm in diameter is folded such that its effective filtering area is the same as that of a sector with central angle of 220°. What is the filtering area?

83. To produce an electric current, a circular loop of wire of diameter 25.0 cm is rotating about its diameter at 60.0 r/s in a magnetic field. What is the greatest linear velocity of any point on the loop?

84. Find the area of the decorative glass panel shown in Fig. 8.58. The panel is made of two equal circular sectors and an isosceles triangle.

2.00 ft

Fig. 8.58   3.75 ft

85. A circular hood is to be used over a piece of machinery. It is to be made from a circular piece of sheet metal 3.25 ft in radius. A hole 0.75 ft in radius and a sector of central angle 80.0° are to be removed to make the hood. What is the area of the top of the hood?

86. The chain on a chain saw is driven by a sprocket 7.50 cm in diameter. If the chain is 108 cm long and makes one revolution in 0.250 s, what is the angular velocity (in r/s) of the sprocket?

87. An *ultracentrifuge*, used to observe the sedimentation of particles such as proteins, may rotate as fast as 80,000 r/min. If it rotates at this rate and is 7.20 cm in diameter, what is the linear velocity of a particle at the outer edge?

88. A computer is programmed to shade in a sector of a pie chart 2.44 cm in radius. If the perimeter of the shaded sector is 7.32 cm, what is the central angle (in degrees) of the sector? See Fig. 8.59.

2.44 cm

Fig. 8.59

89. A Gothic arch, commonly used in medieval European structures, is formed by two circular arcs. In one type, each arc is one-sixth of a circle, with the center of each at the base on the end of the other arc. See Fig. 8.60. Therefore, the width of the arch equals the radius of each arc. For such an arch, find the area of the opening if the width is 15.0 m.

Fig. 8.60   15.0 m

90. The Trans-Alaska Pipeline was assembled in sections 40.0 ft long and 4.00 ft in diameter. If the depth of the oil in one horizontal section is 1.00 ft, what is the volume of oil in this section?

91. A laser beam is transmitted with a "width" of 0.0008° and makes a circular spot of radius 2.50 km on a distant object. How far is the object from the source of the laser beam? Use Eq. (8.14).

92. The planet Venus subtends an angle of 15″ to an observer on earth. If the distance between Venus and earth is 1.04×10^8 mi, what is the diameter of Venus? Use Eq. (8.14).

Writing Exercise

93. Write a paragraph explaining how you determine the units for the result of the following problem: An astronaut in a spacecraft circles the moon once each 1.95 h. If the altitude of the spacecraft is constant at 70.0 mi, what is its velocity? The radius of the moon is 1080 mi. (What is the answer?)

CHAPTER 8   PRACTICE TEST

1. Change 150° to radians in terms of π.
2. Determine the sign of (a) sec 285°, (b) sin 245°, (c) cot 265°.
3. Express sin 205° in terms of the sine of a positive acute angle. Do not evaluate.
4. Find $\sin \theta$ and $\sec \theta$ if θ is in standard position and the terminal side passes through $(-9, 12)$.
5. An airplane propeller blade is 2.80 ft long and rotates at 2200 r/min. What is the linear velocity of a point on the tip of the blade?
6. Given that 3.572 is the measure of an angle, express the angle in degrees.
7. If $\tan \theta = 0.2396$, find θ, in degrees, for $0° \le \theta < 360°$.
8. If $\cos \theta = -0.8244$ and $\csc \theta < 0$, find θ in radians for $0 \le \theta < 2\pi$.
9. The floor of a sunroom is in the shape of a circular sector of arc length 32.0 ft and radius 8.50 ft. What is the area of the floor?
10. A circular sector has an area of 38.5 cm² and a diameter of 12.2 cm. What is the arc length of the sector?

9 Vectors and Oblique Triangles

In many applications, we often deal with such things as forces and velocities. To study them, both their *magnitudes* and *directions* must be known. In general, a quantity for which we must specify *both magnitude and direction* is called a *vector.*

In basic applications, a vector is usually represented by an arrow showing its magnitude and direction, although this was not common before the 1800s. In 1743, the French mathematician d'Alembert published a paper on dynamics in which he used some diagrams, but most of the text was algebraic. In 1788, the French mathematician Lagrange wrote a classic work on *Analytical Mechanics,* but the text was all algebraic and included no diagrams.

In the 1800s, calculus was used to greatly advance the use of vectors, and in turn, these advancements became very important in further developments in scientific fields such as electromagnetic theory. Also in the 1800s, mathematicians defined a vector more generally and opened up study in new areas of advanced mathematics. In this text, we deal only with the basic meaning of a vector. However, a vector is an excellent example of a math concept that came from a basic physics concept.

After studying vectors, we then develop methods of solving *oblique* triangles (triangles that are not right triangles). Although obviously not known in their modern forms, one of these methods, the *law of cosines,* was known to the Greek astronomer Ptolemy (90–168), and the other method, the *law of sines,* was known to Islamic mathematicians of the 1100s. In solving oblique triangles, we often use the trig functions of obtuse angles.

Vectors are of great importance in many fields of science and technology, including physics, engineering, structural design, and navigation. One modern application is illustrated in the picture below. As with right triangles, the applications of oblique triangles are found in many fields of science and technology.

The wind must be considered to find the proper heading for an aircraft. In Section 9.5, we use vectors and oblique triangles to show how this may be done.

9.1 Introduction to Vectors

Scalars and Vectors • Addition of Vectors • Scalar Multiple of a Vector • Subtraction of Vectors

We deal with many quantities that may be described by a number that shows only the magnitude. These include lengths, areas, time intervals, monetary amounts, and temperatures. *Quantities such as these, described only by the* **magnitude,** *are known as* **scalars.**

As we said on the previous page, *many other quantities, called* **vectors,** *are fully described only when both the* **magnitude** *and* **direction** *are specified.* The following example shows the difference between scalars and vectors.

EXAMPLE 1 Scalars and vectors – application

A jet is traveling at 600 mi/h. From this statement alone, we know only the *speed* of the jet. *Speed is a scalar quantity,* and it tells us only the *magnitude* of the rate. Knowing only the speed of the jet, we know the rate at which it is moving, but we do not know where it is headed.

If the phrase "in a direction 10° south of west" is added to the sentence about the jet, we specify the direction of travel as well as the speed. We then know the *velocity* of the jet; that is the *direction* of travel as well as the rate at which it is moving. *Velocity is a* **vector** *quantity.* ∎

For an example of the action of two vectors, consider a boat moving in a river. We will assume that the boat is driven by a motor that can move it at 8 mi/h in still water and that the river's current is going 6 mi/h downstream, as shown in Fig. 9.1. We quickly see that the movement of the boat depends on the direction in which it is headed. If it heads downstream, it moves at 14 mi/h, for the water is moving at 6 mi/h and the boat moves at 8 mi/h with respect to the water. If it heads upstream, however, it moves only at 2 mi/h, since the river is acting directly against the motor. If the boat heads directly across the river, the point it reaches on the other side is not directly opposite the point from which it started. This is so because the river is moving the boat downstream *at the same time* the boat moves across the river.

Checking this last case further, assume that the river is 0.4 mi wide where the boat is crossing. It takes 0.05 h (0.4 mi ÷ 8 mi/h = 0.05 h) to cross. In 0.05 h, the river will carry the boat 0.3 mi (0.05 h × 6 mi/h = 0.3 mi) downstream. This means the boat went 0.3 mi downstream as it went 0.4 mi across the river. From the Pythagorean theorem, we see that it went 0.5 mi from its starting point to its finishing point:

$$d^2 = 0.4^2 + 0.3^2 = 0.25$$
$$d = 0.5 \text{ mi}$$

Since the 0.5 mi was traveled in 0.05 h, the magnitude of the velocity (the *speed*) of the boat was actually

$$v = \frac{d}{t} = \frac{0.5 \text{ mi}}{0.05 \text{ h}} = 10 \text{ mi/h}$$

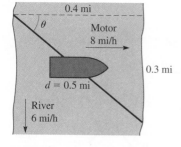

Fig. 9.1

Also, note that the direction of this velocity can be represented along a line that makes an angle θ with the line directed directly across the river, as shown in Fig. 9.1. We can find this angle by noting that

$$\tan \theta = \frac{0.3 \text{ mi}}{0.4 \text{ mi}} = 0.75$$
$$\theta = \tan^{-1} 0.75 = 37°$$

Therefore, when headed directly across the river, the boat's velocity is 10 mi/h directed at an angle of 37° downstream from a line directly across the river.

R = A + B

Fig. 9.2

ADDITION OF VECTORS

We have just seen two velocity vectors being *added*. Note that these vectors are not added the way numbers are added. We must take into account their directions as well as their magnitudes. Reasoning along these lines, let us now define the sum of two vectors.

We will represent a vector quantity by a letter printed in **boldface** type. The same letter in *italic* (lightface) type represents the magnitude only. Thus, **A** is a vector of magnitude *A*. In handwriting, one usually places an arrow over the letter to represent a vector, such as $\vec{A}$.

Let **A** and **B** represent vectors directed from *O* to *P* and *P* to *Q*, respectively (see Fig. 9.2). *The vector sum* **A** + **B** *is the vector* **R,** *from the* **initial point** *O to the* **terminal point** *Q. Here, vector* **R** *is called the* **resultant.** *In general, a resultant is a single vector that is the vector sum of any number of other vectors.*

There are two common methods of adding vectors by means of a diagram. The first is illustrated in Fig. 9.3. To add **B** to **A**, shift **B** parallel to itself until its tail touches the head of **A**. *The vector sum* **A** + **B** *is the resultant vector* **R,** *which is drawn from the tail of* **A** *to the head of* **B.** In using this method, we can move a vector for addition as long as *its magnitude and direction remain unchanged.* (Since the magnitude and direction specify a vector, two vectors in different *locations* are considered the same if they have the same magnitude and direction.) When using a diagram to add vectors, it must be drawn with reasonable accuracy.

CAUTION ▶

Polygon Method

Fig. 9.3
R = A + B

Three or more vectors are added in the same general manner. We place the initial point of the second vector at the terminal point of the first vector, the initial point of the third vector at the terminal point of the second vector, and so on. The resultant is the vector from the initial point of the first vector to the terminal point of the last vector. The order in which they are added does not matter.

EXAMPLE 2 Adding vectors – polygon method

The addition of vectors **A, B,** and **C** is shown in Fig. 9.4.

Practice Exercise

1. For the vectors in Example 2, show that **R = B + C + A.**

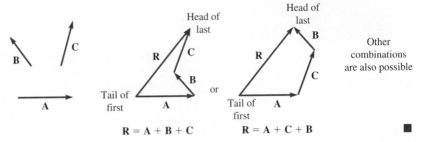

Other combinations are also possible

Fig. 9.4
R = A + B + C **R = A + C + B**

Parallelogram Method   Another method that is convenient when two vectors are being added is to *let the two vectors being added be the sides of a parallelogram. The resultant is then the diagonal of the parallelogram.* The initial point of the resultant is the *common initial point of the two vectors being added.* In using this method, the vectors are first placed tail to tail. This method is illustrated in the following example.

EXAMPLE 3   Adding vectors – parallelogram method

The addition of vectors **A** and **B** is shown in Fig. 9.5.

Fig. 9.5   (1)   (2)   **R = A + B**   (3)   (4)   ■

Scalar Multiple of Vector   If vector **C** is in the same direction as vector **A** and **C** has a magnitude *n* times that of **A**, then **C** = *n***A**, where *the vector n***A** *is called the* **scalar multiple** *of vector* **A.** This means that 2**A** is a vector that is twice as long as **A** but is *in the same direction.* Note carefully that only the magnitudes of **A** and 2**A** are different, and their directions are the same. The addition of scalar multiples of vectors is illustrated in the following example.

EXAMPLE 4   Scalar multiple of a vector

For vectors **A** and **B** in Fig. 9.6, find vector 3**A** + 2**B**.

Fig. 9.6

Subtraction of Vectors   Vector **B** is subtracted from vector **A** by reversing the direction of **B** and proceeding

NOTE ▶   as in vector addition. Thus, **A** − **B** = **A** + (−**B**), where *the minus sign indicates that vector* −**B** *has the opposite direction of vector* **B.** Vector subtraction is illustrated in the following example.

EXAMPLE 5   Subtracting vectors

For vectors **A** and **B** in Fig. 9.7, find vector 2**A** − **B**.

Fig. 9.7   **R = 2A − B**   ■

Practice Exercise

2. For the vectors in Example 5, find vector **B** − 3**A**.

Among the most important applications of vectors is that of the forces acting on a structure or on an object. The next example shows the addition of forces by using the parallelogram method.

(a)

(b)

Fig. 9.8

EXAMPLE 6 Adding vectors – application

Two persons pull horizontally on ropes attached to a car mired in mud. One person pulls with a force of 500 N directly to the right, while the other person pulls with a force of 350 N at an angle of 40° from the first force, as shown in Fig. 9.8(a). Find the resultant force on the car.

We make a scale drawing of the forces as shown in Fig. 9.8(b), measuring the magnitudes of the forces with a ruler and the angles with a protractor. (The scale drawing of the forces is made larger and with a different scale than that in Fig. 9.8(a) in order to get better accuracy.) We then complete the parallelogram and draw in the diagonal that represents the resultant force. Finally, we find that the resultant force is about 800 N and that it acts at an angle of about 16° from the first force. ∎

Two other important vector quantities are *velocity* and *displacement*. Velocity as a vector is illustrated in Example 1. *The **displacement** of an object is the change in its position. Displacement is given by the distance from a reference point and the angle from a reference direction.* The following example illustrates the difference between *distance* and *displacement*.

EXAMPLE 7 Adding vectors – displacement

Fig. 9.9

To avoid a storm, a jet travels at 60° north of east from Detroit for 190 mi and then turns to a direction of 10° north of east for 280 mi to Ottawa. Find the displacement of Ottawa from Detroit.

We make a scale drawing in Fig. 9.9 to show the route taken by the jet. Measuring distances with a ruler and angles with a protractor, we find that Ottawa is about 420 mi from Detroit, at an angle of about 30° north of east. By giving both the magnitude and the **direction**, we have given the displacement.

If the jet returned directly from Ottawa to Detroit, its *displacement* from Detroit would be *zero*, although it traveled a *distance* of about 890 mi. ∎

EXERCISES 9.1

In Exercises 1–4, find the resultant vectors if the given changes are made in the indicated examples of this section.

1. In Example 2, what is the resultant of the three vectors if the direction of vector **A** is reversed?
2. In Example 4, for vectors **A** and **B**, what is vector 2**A** + 3**B**?
3. In Example 5, for vectors **A** and **B**, what is vector 2**B** − **A**?
4. In Example 6, if 20° replaces 40°, what is the resultant force?

(W) *In Exercises 5–8, determine whether a scalar or a vector is described in (a) and (b). Explain your answers.*

5. (a) A soccer player runs 15 m from the center of the field.
   (b) A soccer player runs 15 m from the center of the field toward the opponents' goal.
6. (a) A small-craft warning reports 25 mi/h winds.
   (b) A small-craft warning reports 25 mi/h winds from the north.
7. (a) An arm of an industrial robot pushes with a 10-lb force downward on a part.
   (b) A part is being pushed with a 10-lb force by an arm of an industrial robot.

8. (a) A ballistics test shows that a bullet hit a wall at a speed of 400 ft/s.
   (b) A ballistics test shows that a bullet hit a wall at a speed of 400 ft/s perpendicular to the wall.

In Exercises 9–14, add the given vectors by drawing the appropriate resultant.

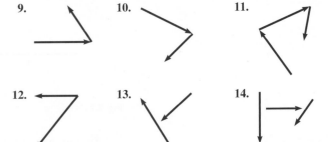

In Exercises 15–18, draw the given vectors and find their sum graphically. The magnitude is shown first, followed by the direction as an angle in standard position.

15. 3.6 cm, 0°; 4.3 cm, 90°   **16.** 2.3 cm, 45°; 5.2 cm, 120°

17. 6.0 cm, 150°; 1.8 cm, 315°   **18.** 7.5 cm, 240°; 2.3 cm, 30°

In Exercises 19–40, find the indicated vector sums and differences with the given vectors by means of diagrams. (You might find graph paper to be helpful.)

A   B   C   D   E

19. A + B   **20.** B + C   **21.** C + D   **22.** D + E
23. A + C + E   **24.** B + D + A   **25.** 2A + D + E
26. B + E + A   **27.** B + 3E   **28.** A + 2C
29. 3C + E   **30.** 2D + A   **31.** A − B
32. C − D   **33.** E − B   **34.** C − 2A
35. 3B + $\frac{1}{2}$A   **36.** 2B − $\frac{3}{2}$E   **37.** B + 2C − E
38. A + 2D − 3B **39.** C − B − $\frac{3}{4}$A   **40.** D − 2C − $\frac{1}{2}$E

In Exercises 41–48, solve the given problems. Use a ruler and protractor as in Examples 6 and 7.

41. Two forces that act on an airplane wing are called the *lift* and the *drag*. Find the resultant of these forces acting on the airplane wing in Fig. 9.10.

Lift = 840 lb

Drag = 320 lb

Fig. 9.10

42. Two electric charges create an electric field intensity, a vector quantity, at a given point. The field intensity is 30 kN/C to the right and 60 kN/C at an angle of 45° above the horizontal to the right. Find the resultant electric field intensity at this point.

43. A ski tow is moving skiers vertically upward at 24 m/min and horizontally at 44 m/min. What is the velocity of the tow?

44. A small plane travels at 120 mi/h in still air. It is headed due south in a wind of 30 mi/h from the northeast. What is the resultant velocity of the plane?

45. A driver takes the wrong road at an intersection and travels 4 mi north, then 6 mi east, and finally 10 mi to the southeast to reach the home of a friend. What is the displacement of the friend's home from the intersection?

46. A ship travels 20 km in a direction of 30° south of east and then turns due south for another 40 km. What is the ship's displacement from its initial position?

47. Three ropes hold a helium-filled balloon in place, but two of the ropes break. The remaining rope holds the balloon with a tension of 510 N at an angle of 80° with the ground due to a wind. The weight (a vertical force) of the balloon and contents is 400 N, and the upward buoyant force is 900 N. The wind creates a horizontal force of 90 N on the balloon. What is the resultant force on the balloon?

48. While unloading a crate weighing 610 N, the chain from a crane supports it with a force of 650 N at an angle of 20° from the vertical. What force must a horizontal rope exert on the crate so that the total force (including its weight) on the crate is zero?

Answers to Practice Exercises

1. Same as **R** in Fig. 9.4.   **2.** (Half scale)

9.2 Components of Vectors

Resolving a Vector into Components • Finding *x*- and *y*-Components • Meaning of Sign of Component

Using diagrams is useful in developing an understanding of vectors. However, unless the diagrams are drawn with great care, the results we get are not too accurate. Therefore, other methods are needed in order to get more accurate results.

In this section, we show how a given vector can be made to be the sum of two other vectors, with any required degree of accuracy. In the next section, we show how this lets us add vectors to get their sum with the required accuracy in the result.

Two vectors that, when added together, have a resultant equal to the original vector, are called **components** *of the original vector.* In the illustration of the boat in Section 9.1, the velocities of 8 mi/h across the river and 6 mi/h downstream are components of the 10 mi/h vector directed at the angle θ.

Resolving a Vector into Components

Certain components of a vector are of particular importance. If the initial point of a vector is placed at the origin of a rectangular coordinate system and its direction is given by an angle in standard position, we may find its *x*- and *y*-**components.** *These components are vectors directed along the axes that, when added together, equal the given vector.* The initial points of these components are at the origin, and the terminal points are at the points where perpendicular lines from the terminal point of the given vector cross the axes. *Finding these component vectors is called* **resolving** *the vector into its components.*

EXAMPLE 1 Components of first-quadrant vector

Find the x- and y-components of the vector **A** shown in Fig. 9.11. The magnitude of **A** is 7.25.

From the figure, we see that A_x, the magnitude of the x-component of $\mathbf{A}_x$, is related to **A** by

$$\frac{A_x}{A} = \cos 62.0°$$

$$A_x = A \cos 62.0°$$

In the same way, A_y, the magnitude of the y-component of $\mathbf{A}_y$, is related to **A** ($\mathbf{A}_y$ could be placed along the vertical dashed line) by

$$\frac{A_y}{A} = \sin 62.0°$$

$$A_y = A \sin 62.0°$$

From these relations, knowing that $A = 7.25$, we have

$$A_x = 7.25 \cos 62.0° = 3.40$$
$$A_y = 7.25 \sin 62.0° = 6.40$$

This means that the x-component is directed along the x-axis to the right and has a magnitude of 3.40. Also, the y-component is directed along the y-axis upward and its magnitude is 6.40. These two component vectors can replace vector **A**, since the effect they have is the same as **A**. ∎

Fig. 9.11

Practice Exercise

1. For the vector in Example 1, change the angle to 25.0° and find the x- and y-components.

EXAMPLE 2 Components of second-quadrant vector

Resolve a vector 14.4 units long and directed at an angle of 126.0° into its x- and y-components. See Fig. 9.12.

Placing the initial point of the vector at the origin and putting the angle in standard position, note that the vector directed along the x-axis, $\mathbf{V}_x$, is related to the vector **V** of magnitude V by

$$V_x = V \cos 126.0°$$

Since the vector directed along the y-axis, $\mathbf{V}_y$, could also be placed along the vertical dashed line, it is related to the vector **V** by

$$V_y = V \sin 126.0°$$

Thus, the vectors $\mathbf{V}_x$ and $\mathbf{V}_y$ have the magnitudes

$$V_x = 14.4 \cos 126.0° = -8.46 \qquad V_y = 14.4 \sin 126.0° = 11.6$$

Therefore, we have resolved the given vector into two components: one, directed along the negative x-axis, of magnitude 8.46, and the other, directed along the positive y-axis, of magnitude 11.6.

It is also possible to use the reference angle, as long as *the proper sign is attached to each component.* In this case, the reference angle is 54.0°, and therefore

$$V_x = -14.4 \cos 54.0° = -8.46 \qquad V_y = +14.4 \sin 54.0° = 11.6$$

directed along negative x-axis   directed along positive y-axis

The minus sign shows that the x-component is directed to the left. ∎

Fig. 9.12

Practice Exercise

2. For the vector in Example 2, change the angle to 306.0° and find the x- and y-components.

From Examples 1 and 2, we can see that the steps used in finding the *x*- and *y*-components of a vector are as follows:

> **STEPS USED IN FINDING THE *x*- AND *y*-COMPONENTS OF A VECTOR**
> 1. *Place vector* **A** *such that θ is in standard position.*
> 2. *Calculate A_x and A_y from $A_x = A \cos\theta$ and $A_y = A \sin\theta$.* We may use the reference angle if we note the direction of the component.
> 3. *Check the components* to see if each is in the correct direction and has a magnitude that is proper for the reference angle.

EXAMPLE 3 Components of third-quadrant vector

Resolve vector **A,** of magnitude 375.4 and direction $\theta = 205.32°$, into its *x*- and *y*-components. See Fig. 9.13.

By placing **A** such that θ is in standard position, we see that

$$A_x = A \cos 205.32° = 375.4 \cos 205.32° = -339.3$$

and

$$A_y = A \sin 205.32° = 375.4 \sin 205.32° = -160.5$$

directed along negative axis

The angle 205.32° places the vector in the third quadrant, and each of the components is directed along the negative axis. This must be the case for a third-quadrant angle. Also, the reference angle is 25.32°, and we see that the magnitude of A_x is greater than the magnitude of A_y, which must be true for a reference angle that is less than 45°. ∎

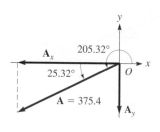

Fig. 9.13

EXAMPLE 4 Vector components – application

The tension **T** in a cable supporting the sign shown in Fig. 9.14(a) is 85.0 lb. If the cable makes an angle of 53.5° with the horizontal, find the horizontal and vertical components of the tension.

The tension is the force the cable exerts on the sign. Showing the tension in Fig. 9.14(b), we see that

$$T_y = T \sin 53.5° = 85.0 \sin 53.5°$$
$$= 68.3 \text{ lb}$$
$$T_x = T \cos 53.5° = 85.0 \cos 53.5°$$
$$= 50.6 \text{ lb}$$

Note that $T_y > T_x$, which should be the case for an acute angle greater than 45°. ∎

Fig. 9.14   (a)   (b)

EXERCISES 9.2

In Exercises 1–4, find the component vectors if the given changes are made in the indicated examples of this section.

1. In Example 2, change 126.0° to 216.0°.

2. In Example 3, change 205.32° to 295.32°.

3. In Example 3, change 205.32° to 270.00°.

4. In Example 4, change 53.5° to 45.0°.

In Exercises 5–10, find the horizontal and vertical components of the vectors shown in the given figures. In each, the magnitude of the vector is 750.

5.   **6.**   **7.**

8.   **9.**   **10.**

In Exercises 11–20, find the x- and y-components of the given vectors by use of the trigonometric functions. The magnitude is shown first, followed by the direction as an angle in standard position.

11. 8.60 lb, $\theta = 68.0°$

12. 9750 N, $\theta = 243.0°$

13. 76.8 m/s, $\theta = 145.0°$

14. 0.0998 ft/s, $\theta = 296.0°$

15. 9040 ft/s², $\theta = 270.0°$

16. 16.4 cm/s², $\theta = 156.5°$

17. 2.65 mN, $\theta = 197.3°$

18. 678 lb, $\theta = 22.5°$

19. 0.8734 ft, $\theta = 157.83°$

20. 509.4 m, $\theta = 360.0°$

In Exercises 21–34, find the required horizontal and vertical components of the given vectors.

21. A nuclear submarine approaches the surface of the ocean at 25.0 km/h and at an angle of 17.3° with the surface. What are the components of its velocity? See Fig. 9.15.

Fig. 9.15

22. Water is flowing downhill at 18.0 ft/s through a pipe that is at an angle of 66.4° with the horizontal. What are the components of its velocity?

23. A car is being unloaded from a ship. It is supported by a cable from a crane and guided into position by a horizontal rope. If the tension in the cable is 2790 lb and the cable makes an angle of 3.5° with the vertical, what are the weight W of the car and the tension T in the rope? (The weight of the cable is negligible to that of the car.) See Fig. 9.16.

Fig. 9.16

24. A car traveling at 25 mi/h hits a stone wall at an angle of 65° with the wall. What is the component of the car's velocity that is perpendicular to the wall?

25. With the sun directly overhead, a plane is taking off at 125 km/h at an angle of 22.0° above the horizontal. How fast is the plane's shadow moving along the runway?

26. A jet is 145 km at a position 37.5° north of east of Tahiti. What are the components of the jet's displacement from Tahiti?

27. The end of a robot arm is 3.50 ft on a line 78.6° above the horizontal from the point where it does a weld. What are the components of the displacement from the end of the robot arm to the welding point?

28. The tension in a rope attached to a boat is 55.0 lb. The rope is attached to the boat 12.0 ft below the level at which it is being drawn in. At the point where there are 36.0 ft of rope out, what force tends to bring the boat toward the wharf, and what force tends to raise the boat? See Fig. 9.17.

Fig. 9.17

29. A person applies a force of 210 N perpendicular to a jack handle that is at an angle of 25° above the horizontal. What are the horizontal and vertical components of the force?

30. A water skier is pulled up the ramp shown in Fig. 9.18 at 28 ft/s. How fast is the skier rising when leaving the ramp?

Fig. 9.18

31. A wheelbarrow is being pushed with a force of 80 lb directed along the handles, which are at an angle of 20° above the horizontal. What is the effective force for moving the wheelbarrow forward?

32. Two upward forces are acting on a bolt. One force of 60.5 lb acts at an angle of 82.4° above the horizontal, and the other force of 37.2 lb acts at an angle of 50.5° below the first force. What is the total upward force on the bolt? See Fig. 9.19.

Fig. 9.19

33. Vertical wind sheer in the lowest 100 m above the ground is of great importance to aircraft when taking off or landing. It is defined as the rate at which the wind velocity changes per meter above ground. If the vertical wind sheer at 50 m above the ground is 0.75 (km/h)/m directed at angle of 40° above the ground, what are its vertical and horizontal components?

34. At one point, the *Pioneer* space probe was entering the gravitational field of Jupiter at an angle of 2.55° below the horizontal with a velocity of 18,550 mi/h. What were the components of its velocity?

Answers to Practice Exercises

1. $A_x = 6.57$, $A_y = 3.06$ **2.** $V_x = 8.46$, $V_y = -11.6$

9.3 Vector Addition by Components

Adding Vectors at Right Angles •
Adding *x*-Components and
Adding *y*-Components •
Standard-position Angle of Resultant

Now that we have developed the meaning of the components of a vector, we are able to add vectors to any degree of required accuracy. To do this, we use the components of the vector, the Pythagorean theorem, and the tangent of the standard-position angle of the resultant. In the following example, two vectors at right angles are added.

EXAMPLE 1 Adding vectors at right angles

Add vectors **A** and **B**, with $A = 14.5$ and $B = 9.10$. The vectors are at right angles, as shown in Fig. 9.20.

We can find the magnitude R of the resultant vector **R** by use of the Pythagorean theorem. This leads to

$$R = \sqrt{A^2 + B^2} = \sqrt{(14.5)^2 + (9.10)^2}$$
$$= 17.1$$

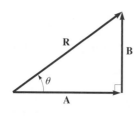

Fig. 9.20

We now determine the direction of the resultant vector **R** by specifying its direction as the angle θ in Fig. 9.20, that is, the angle that **R** makes with vector **A.** Therefore,

$$\tan \theta = \frac{B}{A} = \frac{9.10}{14.5}$$
$$\theta = 32.1°$$

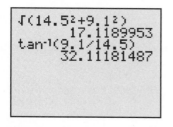

Fig. 9.21

Calculation of the values of R and θ are shown in the calculator window in Fig. 9.21. As we showed in Chapter 4, it is not necessary to record or store the value of the quotient 9.10/14.5, as we can calculate the value of θ directly on the calculator as $\tan^{-1}(9.10/14.5)$.

Therefore, we see that **R** is a vector of magnitude $R = 17.1$ and in a direction 32.1° from vector **A.**

Note that Fig. 9.20 shows vectors **A** and **B** as horizontal and vertical, respectively. This means they are the horizontal and vertical components of the resultant vector **R**. However, we would find the resultant of any two vectors *at right angles* in the same way. ∎

NOTE ▶ If the vectors being added are not at right angles, first *place each vector with its tail at the origin.* Next, *resolve each vector into its x- and y-components.* Then *add the x-components* and *add the y-components to find the x- and y-components of the resultant.* Then, by using the Pythagorean theorem, *find the magnitude of the resultant,* and by use of the tangent, *find the angle that gives the direction of the resultant.*

CAUTION ▶ *Remember, a vector is not completely specified unless both its magnitude and its direction are specified.* A common error is to determine the magnitude, but not to find the angle θ that is used to define its direction.

The following examples illustrate the addition of vectors by first finding the components of the given vectors.

EXAMPLE 2   Adding by first combining components

Find the resultant of two vectors **A** and **B** such that $A = 1200$, $\theta_A = 270.0°$, $B = 1750$, and $\theta_B = 115.0°$.

First, place the vectors on a coordinate system with the tail of each at the origin as shown in Fig. 9.22(a). Then resolve each vector into its x- and y-components, as shown in Fig. 9.22(b) and as calculated below. (Note that **A** is vertical and has no horizontal component.) Next, the components are combined, as in Fig. 9.22(c) and as calculated. Finally, the magnitude of the resultant and the angle θ (to determine the direction), as shown in Fig. 9.22(d), are calculated.

Fig. 9.22

$$A_x = A \cos 270.0° = 1200 \cos 270.0° = 0$$
$$B_x = B \cos 115.0° = 1750 \cos 115.0° = -739.6$$
$$A_y = A \sin 270.0° = 1200 \sin 270.0° = -1200$$
$$B_y = B \sin 115.0° = 1750 \sin 115.0° = 1586$$

— Fig. 9.22(b)

$$R_x = A_x + B_x = 0 - 739.6 = -739.6$$
$$R_y = A_y + B_y = -1200 + 1586 = 386$$

— Fig. 9.22(c)

$$R = \sqrt{R_x^2 + R_y^2} = \sqrt{(-739.6)^2 + 386^2} = 834$$
$$\tan \theta = \frac{R_y}{R_x} = \frac{386}{-739.6} \qquad \theta = 152.4° \longleftarrow 180° - 27.6°$$

— Fig. 9.22(d)

Practice Exercise

1. In Example 2, change 270° to 90° and then add the vectors.

Thus, the resultant has a magnitude of 834 and is directed at a standard-position angle of 152.4°. In finding θ from a calculator,

CAUTION ▶   *the calculator display shows an angle of* $-27.6°$. *However,* θ *is a second-quadrant angle, since* R_x *is negative and* R_y *is positive.*

Therefore, we must use 27.6° as a reference angle. For this reason, it is usually advisable to *find the reference angle first* by disregarding the signs of R_x and R_y when finding θ. Thus,

$$\tan \theta_{\text{ref}} = \left| \frac{R_y}{R_x} \right| = \frac{386}{739.6} \qquad \theta_{\text{ref}} = 27.6°$$

The values shown in this example have been rounded off. In using the calculator, R_x and R_y are each calculated in one step and stored for the calculation of R and θ, and we will show these steps in the next example. However, here we wished to show the individual steps and results to more clearly show the method.

When we found R_y, we saw that we could do a vector addition as a scalar addition. Also, since the magnitude of the components and the resultant are much smaller than either of the original vectors, this vector addition would have been difficult to do accurately by means of a diagram.   ■

EXAMPLE 3 Adding by first combining components

Find the resultant **R** of the two vectors shown in Fig. 9.23(a), **A** of magnitude 8.075 and standard-position angle of 57.26° and **B** of magnitude 5.437 and standard-position angle of 322.15°.

In Fig. 9.23(b), we show the components of vectors **A** and **B,** and then in Fig. 9.23(c), we show the resultant and its components.

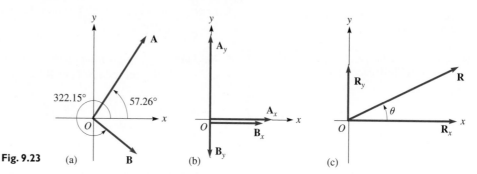

Fig. 9.23   (a)   (b)   (c)

Vector	Magnitude	Angle	x-Component		y-Component	
A	8.075	57.26°	$A_x = 8.075 \cos 57.26°$	$= 4.367$	$A_y = 8.075 \sin 57.26°$	$= 6.792$
B	5.437	322.15°	$B_x = 5.437 \cos 322.15°$	$= 4.293$	$B_y = 5.437 \sin 322.15°$	$= -3.336$
R			$R_x = A_x + B_x$	$= 8.660$	$R_y = A_y + B_y$	$= 3.456$

$$R = \sqrt{R_x^2 + R_y^2} = \sqrt{(8.660)^2 + (3.456)^2} = 9.324$$

$$\theta = \tan^{-1}\frac{R_y}{R_x} = \tan^{-1}\left(\frac{3.456}{8.660}\right) = 21.76° \longleftarrow \text{don't forget the direction}$$

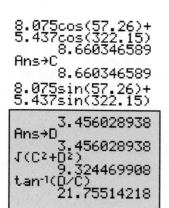

Fig. 9.24

The resultant vector is 9.324 units long and is directed at a standard-position angle of 21.76°, as shown in Fig. 9.23(c). We know that the resultant is in the first quadrant since both R_x and R_y are positive.

In the table above, we have shown rounded-off values for each result. However, when using a calculator, it is necessary only to calculate R_x and R_y in one step each, store these values, and use them to calculate R and θ. This is shown in the calculator window in Fig. 9.24. The lines shown above the window are those that are displaced in proceeding with the solution.

In the calculator solution, $C = R_x$ and $D = R_y$. In finding this solution, we see that both R_x and R_y are positive, which means that θ is a first-quadrant angle. ∎

Some general formulas can be derived from the previous examples. For a given vector **A**, directed at an angle θ, of magnitude A, and with components A_x and A_y, we have the following relations:

$$A_x = A \cos \theta \quad A_y = A \sin \theta \tag{9.1}$$

$$A = \sqrt{A_x^2 + A_y^2} \tag{9.2}$$

$$\theta_{\text{ref}} = \tan^{-1}\frac{|A_y|}{|A_x|} \tag{9.3}$$

The value of θ is found by using the reference angle from Eq. (9.3) and the quadrant in which the resultant lies.

From the previous examples, we see that the following procedure is used for adding vectors.

PROCEDURE FOR ADDING VECTORS BY COMPONENTS

1. *Add the x-components of the given vectors to obtain R_x.*
2. *Add the y-components of the given vectors to obtain R_y.*
3. *Find the magnitude of the resultant* **R.** *Use Eq. (9.2) in the form*

$$R = \sqrt{R_x^2 + R_y^2}$$

4. *Find the standard-position angle θ for the resultant* **R.** *First, find the reference angle θ_{ref} for the resultant* **R** *by using Eq. (9.3) in the form*

$$\theta_{ref} = \tan^{-1}\frac{|R_y|}{|R_x|}$$

■ Some calculators have a specific feature for adding vectors.

■ See Appendix C for a graphing calculator program ADDVCTR. It can be used to add vectors.

EXAMPLE 4 Standard-position angle of resultant

Find the resultant of the three given vectors in Fig. 9.25. The magnitudes of these vectors are $T = 422$, $U = 405$, and $V = 210$.

Fig. 9.25   (a)   (b)   (c)

Practice Exercise

2. In Example 4, find the resultant of vectors **U** and **V**. (Do not include vector **T**.)

We could change the given angles to standard-position angles. However, we will use the given angles, *being careful to give the proper sign to each component*. In the following table, we show the x- and y-components of the given vectors, and the sums of these components give us the components of **R**.

Fig. 9.26

Vector	Magnitude	Ref. Angle	x-Component	y-Component
T	422	0°	$422 \cos 0° =$ 422.0	$422 \sin 0° =$ 0.0
U	405	55.0°	$-405 \cos 55.0° = -232.3$	$-405 \sin 55.0° = -331.8$
V	210	70.0°	$-210 \cos 70.0° = -71.8$	$+210 \sin 70.0° = 197.3$
R			117.9	note the variation −134.5 due to rounding off

In the calculator solution shown in Fig. 9.26, $A = R_x$, $B = R_y$, and abs means absolute value. From this display, we have

$$R = 179 \quad \text{and} \quad \theta_{ref} = 48.8°$$

Since R_x is positive and R_y is negative, we know that θ is **a fourth-quadrant angle.** Therefore, to find θ, we subtract θ_{ref} from 360°. This means

$$\theta = 360° - 48.8° = 311.2° \qquad\qquad ■$$

EXERCISES 9.3

In Exercises 1 and 2, find the resultant vectors if the given changes are made in the indicated examples of this section.

1. In Example 2, find the resultant if θ_A is changed to 0°.

2. In Example 3, find the resultant if θ_B is changed to 232.15°.

*In Exercises 3–6, vectors **A** and **B** are at right angles. Find the magnitude and direction (the angle from vector **A**) of the resultant.*

3. $A = 14.7$
   $B = 19.2$

4. $A = 592$
   $B = 195$

5. $A = 3.086$
   $B = 7.143$

6. $A = 1734$
   $B = 3297$

In Exercises 7–14, with the given sets of components, find R and θ.

7. $R_x = 5.18, R_y = 8.56$

8. $R_x = 89.6, R_y = -52.0$

9. $R_x = -0.982, R_y = 2.56$

10. $R_x = -729, R_y = -209$

11. $R_x = -646, R_y = 2030$

12. $R_x = -31.2, R_y = -41.2$

13. $R_x = 6941, R_v = -1246$

14. $R_x = 7.627, R_y = -6.353$

In Exercises 15–30, add the given vectors by using the trigonometric functions and the Pythagorean theorem.

15. $A = 18.0, \theta_A = 0.0°$
    $B = 12.0, \theta_B = 27.0°$

16. $F = 154, \theta_F = 90.0°$
    $T = 128, \theta_T = 43.0°$

17. $C = 5650, \theta_C = 76.0°$
    $D = 1280, \theta_D = 160.0°$

18. $A = 6.89, \theta_A = 123.0°$
    $B = 29.0, \theta_B = 260.0°$

19. $A = 9.821, \theta_A = 34.27°$
    $B = 17.45, \theta_B = 752.50°$

20. $E = 1653, \theta_E = 36.37°$
    $F = 9807, \theta_F = 253.06°$

21. $A = 21.9, \theta_A = 236.2°$
    $B = 96.7, \theta_B = 11.5°$
    $C = 62.9, \theta_C = 143.4°$

22. $R = 630, \theta_R = 189.6°$
    $F = 176, \theta_F = 320.1°$
    $T = 324, \theta_T = 75.4°$

23. $U = 0.364, \theta_U = 175.7°$
    $V = 0.596, \theta_V = 319.5°$
    $W = 0.129, \theta_W = 100.6°$

24. $A = 6.4, \theta_A = 126°$
    $B = 5.9, \theta_B = 238°$
    $C = 3.2, \theta_C = 72°$

25. The vectors in Fig. 9.27.

26. The vectors in Fig. 9.28.

Fig. 9.27

Fig. 9.28

Answers to Practice Exercises

1. $R = 2880, \theta = 104.9°$ 2. $R = 332, \theta = 203.8°$

27. The vectors in Fig. 9.29.

28. The vectors in Fig. 9.30.

Fig. 9.29

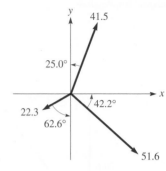

Fig. 9.30

29. In order to move an ocean liner into the channel, three tugboats exert the forces shown in Fig. 9.31. What is the resultant of these forces?

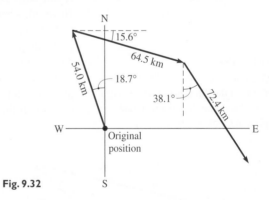

Fig. 9.31

30. A naval cruiser on maneuvers travels 54.0 km at 18.7° west of north, then turns and travels 64.5 km at 15.6° south of east, and finally turns to travel 72.4 km at 38.1° east of south. Find its displacement from its original position. See Fig. 9.32.

Fig. 9.32

In Exercises 31–34, solve the given problems.

31. The angle between two forces of 17 N and 25 N is 37°. What is the magnitude of the resultant force?

32. What is the magnitude of a velocity vector that makes an angle of 18.0° with its horizontal component of 420 mi/h?

33. The angle between two equal-momentum vectors of 15.0 kg·m/s in magnitude is 72.0°. What is the magnitude of the resultant?

34. The resultant **R** of displacements **A** and **B** is perpendicular to **A**. If $A = 25$ m, $B = 35$ m, and $R = 25$ m, find the angle between **A** and **B**.

9.4 Applications of Vectors

Forces • Displacement • Velocity • Other Applications

In Section 9.1, we introduced the important vector quantities of force, velocity, and displacement, and we found vector sums by use of diagrams. Now, we can use the method of Section 9.3 to find sums of these kinds of vectors and others and to use them in various types of applications.

EXAMPLE 1 Forces at right angles

(Top view)

F

8.00 N

θ

Figurine   6.00 N

Fig. 9.33

In centering a figurine on a table, two persons apply forces on it. These forces are at right angles and have magnitudes of 6.00 N and 8.00 N. The angle between their lines of action is 90.0°. What is the resultant of these forces on the figurine?

By means of an appropriate diagram (Fig. 9.33), we may better visualize the actual situation. Note that a good choice of axes (unless specified, it is often convenient to choose the x- and y-axes to fit the problem) is to have the x-axis in the direction of the 6.00-N force and the y-axis in the direction of the 8.00-N force. (This is possible since the angle between them is 90°.) With this choice, note that the two given forces will be the x- and y-components of the resultant. Therefore, we arrive at the following results:

$$F_x = 6.00\,\text{N}, \ F_y = 8.00\,\text{N}$$
$$F = \sqrt{(6.00)^2 + (8.00)^2} = 10.0\,\text{N}$$
$$\theta = \tan^{-1}\frac{F_y}{F_x} = \tan^{-1}\frac{8.00}{6.00}$$
$$= 53.1°$$

The resultant has a magnitude of 10.0 N and acts at an angle of 53.1° from the 6.00-N force. ∎

■ The metric unit of force, the newton (N), is named for the great English mathematician and physicist Sir Isaac Newton (1642–1727). His name will appear on other pages of this text, as some of his many accomplishments are noted.

EXAMPLE 2 Ship displacement

A ship sails 32.50 mi due east and then turns 41.25° north of east. After sailing another 16.18 mi, where is it with reference to the starting point?

In this problem, we are to find the resultant displacement of the ship from the two given displacements. The problem is diagrammed in Fig. 9.34, where the first displacement is labeled vector **A** and the second as vector **B**.

Since east corresponds to the positive x-direction, note that the x-component of the resultant is $A + B_x$ and the y-component of the resultant is B_y. Therefore, we have the following results:

$$R_x = A + B_x = 32.50 + 16.18\cos 41.25°$$
$$= 32.50 + 12.16$$
$$= 44.66\,\text{mi}$$
$$R_y = 16.18\sin 41.25° = 10.67\,\text{mi}$$
$$R = \sqrt{(44.66)^2 + (10.67)^2} = 45.92\,\text{mi}$$
$$\theta = \tan^{-1}\frac{10.67}{44.66}$$
$$= 13.44°$$

Therefore, the ship is 45.92 mi from the starting point, in a direction 13.44° north of east. ∎

Practice Exercise

1. In Example 2, change "41.25° north of east" to "in the direction of 41.25° north of west," and then find the resultant displacement of the ship.

y (N)

41.25°   R   B

16.18 mi   41.25°

θ

O 32.50 mi   A x (E)

Fig. 9.34

EXAMPLE 3 Airplane velocity

An airplane headed due east is in a wind blowing from the southeast. What is the resultant velocity of the plane with respect to the surface of the earth if the velocity of the plane with respect to the air is 600 km/h and that of the wind is 100 km/h? See Fig. 9.35.

Let $\mathbf{v}_{px}$ be the velocity of the plane in the x-direction (east), $\mathbf{v}_{py}$ the velocity of the plane in the y-direction, $\mathbf{v}_{wx}$ the x-component of the velocity of the wind, $\mathbf{v}_{wy}$ the y-component of the velocity of the wind, and $\mathbf{v}_{pa}$ the velocity of the plane with respect to the air. Therefore,

■ An aircraft's *heading* is the direction in which it is pointed. Its *air speed* is the speed at which it travels through the air surrounding it. Due to the wind, the heading and air speed, and its actual direction and speed relative to the ground, will differ.

$$v_{px} = v_{pa} - v_{wx} = 600 - 100(\cos 45.0°) = 529 \text{ km/h}$$
$$v_{py} = v_{wy} = 100(\sin 45.0°) = 70.7 \text{ km/h}$$
$$v = \sqrt{(529)^2 + (70.7)^2} = 534 \text{ km/h}$$
$$\theta = \tan^{-1}\frac{v_{py}}{v_{px}} = \tan^{-1}\frac{70.7}{529}$$
$$= 7.6°$$

Fig. 9.35

The plane is traveling 534 km/h and is flying in a direction 7.6° north of east. From this, we observe that a plane does not necessarily head in the direction of its destination. ■

Equilibrium of Forces

NOTE ▶

As we have seen, an important vector quantity is the force acting on an object. One of the most important applications of vectors involves forces that are in **equilibrium.** *For an object to be in equilibrium,* the net force acting on it in any direction must be zero. This condition is satisfied if the sum of the x-components of the force is zero and the sum of the y-components of the force is also zero. The following two examples illustrate forces in equilibrium.

EXAMPLE 4 Forces on block on inclined plane

A cement block is resting on a straight inclined plank that makes an angle of 30.0° with the horizontal. If the block weighs 80.0 lb, what is the force of friction between the block and the plank?

The weight of the cement block is the force exerted on the block due to gravity. Therefore, the weight is directed vertically downward. The frictional force tends to oppose the motion of the block and is directed upward along the plank. The frictional force must be sufficient to counterbalance that component of the weight of the block that is directed down the plank for the block to be at rest (not moving). The plank itself "holds up" that component of the weight that is perpendicular to the plank. A convenient set of coordinates (see Fig. 9.36) is one with the origin at the center of the block and with the x-axis directed up the plank and the y-axis perpendicular to the plank. The magnitude of the frictional force $\mathbf{F}_f$ is given by

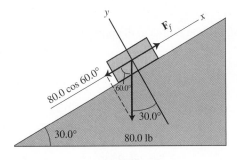

Fig. 9.36

$$F_f = 80.0 \cos 60.0° \qquad \begin{array}{l}\text{component of weight down plank}\\ \text{equals frictional force}\end{array}$$

$$= 40.0 \text{ lb}$$

We have used the 60.0° angle since it is the reference angle. We could have expressed the frictional force as $F_f = 80.0 \sin 30.0°$.

■ Newton's *third law of motion* states that when an object exerts a force on another object, the second object exerts on the first object a force of the same magnitude but in the opposite direction. The force exerted by the plank on the block is an illustration of this law. (Sir Isaac Newton, again. See page 268.)

Here, it is assumed that the block is small enough that we may calculate all forces as though they act at the center of the block (although we know that the frictional force acts at the surface between the block and the plank). ■

Fig. 9.37

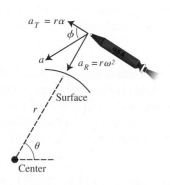

Fig. 9.38

```
6.6E6(1.1E-3)²→A
                 7.986
6.6E6*.42E-6→B
                 2.772
√(A²+B²)
            8.453412329
tan⁻¹(A/B)
            70.85773181
```

Fig. 9.39

EXAMPLE 5 Forces for equilibrium

A 165-lb mountain climber suspended by a rope pushes on the side of a cliff with a horizontal force of 25.0 lb. What is the tension **T** in the rope if the climber is in equilibrium? See Fig. 9.37.

For equilibrium, the tension in the rope must be equal and opposite to the resultant of the climber's weight and force against the cliff. This means the magnitude of the x-component of the tension is 25.0 lb (the reaction force of the cliff—another illustration of Newton's third law—see Example 4) and the magnitude of the y-component is 165 lb (see Fig. 9.37).

$$T = \sqrt{25.0^2 + 165^2}$$
$$= 167 \text{ lb}$$

$$\theta = \tan^{-1}\frac{165}{25.0} = 81.4°$$

As with any vector, we must find the direction in which the tension acts (in this case, along the rope). ∎

EXAMPLE 6 Acceleration of spacecraft

For a spacecraft moving in a circular path around the earth, the tangential component $\mathbf{a}_T$ and the centripetal component $\mathbf{a}_R$ of its acceleration are given by the expressions shown in Fig. 9.38. The radius of the circle through which it is moving (from the center of the earth to the spacecraft) is r, its angular velocity is ω, and its angular acceleration is α (the rate at which ω is changing).

While going into orbit, at one point a spacecraft is moving in a circular path 230 km above the surface of the earth. At this point, $r = 6.60 \times 10^6 \text{ m}$, $\omega = 1.10 \times 10^{-3} \text{ rad/s}$, and $\alpha = 0.420 \times 10^{-6} \text{ rad/s}^2$. Calculate the magnitude of the resultant acceleration and the angle it makes with the tangential component.

$$a_R = r\omega^2 = (6.60 \times 10^6)(1.10 \times 10^{-3})^2$$
$$= 7.99 \text{ m/s}^2 \qquad \text{See Fig. 9.39}$$
$$a_T = r\alpha = (6.60 \times 10^6)(0.420 \times 10^{-6})$$
$$= 2.77 \text{ m/s}^2 \qquad \text{See Fig. 9.39}$$

Since a tangent line to a circle is perpendicular to the radius at the point of tangency, $\mathbf{a}_T$ is perpendicular to $\mathbf{a}_R$. Thus,

$$a = \sqrt{a_T^2 + a_R^2} = \sqrt{7.99^2 + 2.77^2}$$
$$= 8.45 \text{ m/s}^2 \qquad \text{See Fig. 9.39}$$

$$\phi = \tan^{-1}\frac{a_R}{a_T} = \tan^{-1}\frac{7.99}{2.77}$$
$$= 70.9° \qquad \text{See Fig. 9.39} \qquad ∎$$

EXERCISES 9.4

In Exercises 1 and 2, find the necessary quantities if the given changes are made in the indicated examples of this section.

1. In Example 2, find the resultant location if the 41.25° angle is changed to 31.25°.

2. In Example 3, find the resultant velocity if the wind is from the southwest, rather than the southeast.

In Exercises 3–36, solve the given problems.

3. Two hockey players strike the puck at the same time, hitting it with horizontal forces of 5.75 lb and 3.25 lb that are perpendicular to each other. Find the resultant of these forces.

4. To straighten a small tree, two horizontal ropes perpendicular to each other are attached to the tree. If the tensions in the ropes are 18.5 lb and 23.0 lb, what is the resultant force on the tree?

5. In lifting a heavy piece of equipment from the mud, a cable from a crane exerts a vertical force of 6500 N, and a cable from a truck exerts a force of 8300 N at 10.0° above the horizontal. Find the resultant of these forces.

6. At a point in the plane, two electric charges create an electric field (a vector quantity) of 25.9 kN/C at 10.8° above the horizontal to the right and 12.6 kN/C at 83.4 below the horizontal to the right. Find the resultant electric field.

7. A motorboat leaves a dock and travels 1580 ft due west, then turns 35.0° to the south and travels another 1640 ft to a second dock. What is the displacement of the second dock from the first dock?

8. Toronto is 650 km at 19.0° north of east from Chicago. Cincinnati is 390 km at 48.0° south of east from Chicago. What is the displacement of Cincinnati from Toronto?

9. From a fixed point, a surveyor locates a pole at 215.6 ft due east and a building corner at 358.2 ft at 37.72° north of east. What is the displacement of the building from the pole?

10. A rocket is launched with a vertical component of velocity of 2840 km/h and a horizontal component of velocity of 1520 km/h. What is its resultant velocity?

11. In testing the behavior of a tire on ice, a force of 520 lb is exerted to the side, and a force of 780 lb is exerted to the front. What is the resultant force on the tire?

12. To raise a crate, two ropes are attached to its top. If the force in one rope is 240 lb at 25° from the vertical, what must be the force in the second rope at 35° from the vertical in order to lift the crate straight upward?

13. A storm front is moving east at 22.0 km/h and south at 12.5 km/h. Find the resultant velocity of the front.

14. To move forward, a helicopter pilot tilts the helicopter forward. If the rotor generates a force of 3200 lb, with a horizontal component (thrust) of 420 lb, what is the vertical component (lift)?

15. The acceleration (a vector quantity) of gravity on a sky diver is 9.8 m/s². If the force of the wind also causes an acceleration of 1.2 m/s² at an angle of 15° above the horizontal, what is the resultant acceleration of the sky diver?

16. In an accident, a truck with momentum (a vector quantity) of 22,100 kg · m/s strikes a car with momentum of 17,800 kg · m/s from the rear. The angle between their directions of motion is 25.0°. What is the resultant momentum?

17. In an automobile safety test, a shoulder and seat belt exerts a force of 95.0 lb directly backward and a force of 83.0 lb backward at an angle of 20.0° below the horizontal on a dummy. If the belt holds the dummy from moving farther forward, what force did the dummy exert on the belt? See Fig. 9.40.

95.0 lb
20.0°
83.0 lb
Fig. 9.40

18. Two perpendicular forces act on a ring at the end of a chain that passes over a pulley and holds an automobile engine. If the forces have the values shown in Fig. 9.41, what is the weight of the engine?

490 lb
120 lb
Fig. 9.41

19. A plane flies at 550 km/h into a head wind of 60 km/h at 78° with the direction of the plane. Find the resultant velocity of the plane with respect to the ground. See Fig. 9.42.

Wind
60 km/h
78°
550 km/h
Fig. 9.42

20. A ship's navigator determines that the ship is moving through the water at 17.5 mi/h with a heading of 26.3° north of east, but that the ship is actually moving at 19.3 mi/h in a direction of 33.7° north of east. What is the velocity of the current?

21. A space shuttle is moving in orbit at 18,250 mi/h. A satellite is launched to the rear at 120 mi/h at an angle of 5.20° from the direction of the shuttle. Find the velocity of the satellite.

22. A block of ice slides down a (frictionless) ramp with an acceleration of 5.3 m/s². If the ramp makes an angle of 32.7° with the horizontal, find g, the acceleration due to gravity. See Fig. 9.43.

5.3 m/s²
g
32.7°
Fig. 9.43

23. In navigation, one method of expressing *bearing* is to give a single angle measured clockwise from due north. A radar station notes the bearing of two planes, one 15.5 km away on a bearing of 22.0°, and the other 12.0 km away on a bearing of 180.0°. Find the displacement of the second plane from the first. See Fig. 9.44.

N
22.0°
15.5 km
Radar
180.0°
12.0 km
Fig. 9.44

(W) **24.** As a satellite travels around earth at an altitude of 22,320 mi, the magnitude of its linear velocity (see Example 5, page 247) remains unchanged, but its *direction* changes continually, which means there is a change in velocity.

Using a diagram based on Fig. 9.45, draw the difference in velocities (subtract the first velocity from the second) for the satellite as it travels (a) one-sixth, and (b) one-quarter of its orbit. Draw the vectors from the point at which their lines of action cross. What conclusion can be drawn about the direction of the change in velocity?

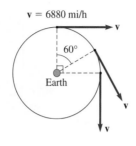

Fig. 9.45

25. A passenger on a cruise ship traveling due east at a speed of 32 km/h notes that the smoke from the ship's funnels makes an angle of 15° with the ship's wake. If the wind is from the southwest, find the speed of the wind.

26. A mine shaft goes due west 75 m from the opening at an angle of 25° below the horizontal surface. It then becomes horizontal and turns 30° north of west and continues for another 45 m. What is the displacement of the end of the tunnel from the opening?

27. A crowbar 1.5 m long is supported underneath 1.2 m from the end by a rock, with the other end under a boulder. If the crowbar is at an angle of 18° with the horizontal and a person pushes down, perpendicular to the crowbar, with a force of 240 N, what vertical force is applied to the boulder? See Fig. 9.46. (The forces are related by $1.2F_1 = 0.3F_2$. See Exercise 50 on page 152.)

Fig. 9.46

28. A scuba diver's body is directed downstream at 75° to the bank of a river. If the diver swims at 25 m/min, and the water is moving at 5.0 m/min, what is the diver's velocity?

29. While starting up, a circular saw blade 8.20 in. in diameter is rotating at 212 rad/min and has an angular acceleration of 318 rad/min². What is the acceleration of the tip of one of the teeth? (See Example 6.)

30. A boat travels across a river, reaching the opposite bank at a point directly opposite that from which it left. If the boat travels 6.00 km/h in still water and the current of the river flows at 3.00 km/h, what was the velocity of the boat in the water?

31. In searching for a boat lost at sea, a Coast Guard cutter leaves a port and travels 75.0 mi due east. It then turns 65° north of east and travels another 75.0 mi, and finally turns another 65.0° toward the west and travels another 75.0 mi. What is its displacement from the port?

32. A car is held stationary on a ramp by two forces. One is the force of 480 lb by the brakes, which hold it from rolling down the ramp. The other is a reaction force by the ramp of 2250 lb, perpendicular to the ramp. This force keeps the car from going through the ramp. See Fig. 9.47. What is the weight of the car, and at what angle with the horizontal is the ramp inclined?

Fig. 9.47

33. A plane is moving at 75.0 m/s, and a package with weather instruments is ejected horizontally from the plane at 15.0 m/s, perpendicular to the direction of the plane. If the vertical velocity v_v (in m/s), as a function of time t (in s) of fall, is given by $v_v = 9.80t$, what is the velocity of the package after 2.00 s (before its parachute opens)?

34. A flat rectangular barge, 48.0 m long and 20.0 m wide, is headed directly across a stream at 4.5 km/h. The stream flows at 3.8 km/h. What is the velocity, relative to the riverbed, of a person walking diagonally across the barge at 5.0 km/h while facing the opposite upstream bank?

35. In Fig. 9.48, a long, straight conductor perpendicular to the plane of the paper carries an electric current i. A bar magnet having poles of strength m lies in the plane of the paper. The vectors $\mathbf{H}_i$, $\mathbf{H}_N$, and $\mathbf{H}_S$ represent the components of the magnetic intensity $\mathbf{H}$ due to the current and to the N and S poles of the magnet, respectively. The magnitudes of the components of $\mathbf{H}$ are given by

$$H_i = \frac{1}{2\pi}\frac{i}{a} \qquad H_N = \frac{1}{4\pi}\frac{m}{b^2} \qquad H_S = \frac{1}{4\pi}\frac{m}{c^2}$$

Given that $a = 0.300$ m, $b = 0.400$ m, $c = 0.300$ m, the length of the magnet is 0.500 m, $i = 4.00$ A, and $m = 2.00$ A $\cdot$ m, calculate the resultant magnetic intensity $\mathbf{H}$. The component $\mathbf{H}_i$ is parallel to the magnet.

Fig. 9.48

36. Solve the problem of Exercise 35 if $\mathbf{H}_i$ is directed away from the magnet, making an angle of 10.0° with the direction of the magnet.

Answers to Practice Exercise

1. $R = 22.96$ mi, $\theta = 27.68°$

9.5 Oblique Triangles, the Law of Sines

Oblique Triangles • Law of Sines •
Case 1: Two Angles and One Side •
Case 2: Two Sides and the
Angle Opposite One of Them •
Ambiguous Case

To this point, we have limited our study of triangle solution to right triangles. However, *many triangles that require solution do not contain a right angle. Such triangles are called* **oblique triangles.** We now discuss solutions of oblique triangles.

In Section 4.4, we stated that we need to know three parts, at least one of them a side, to solve a triangle. There are four possible such combinations of parts, and these combinations are as follows:

Case 1. *Two angles and one side*

Case 2. *Two sides and the angle opposite one of them*

Case 3. *Two sides and the included angle*

Case 4. *Three sides*

There are several ways in which oblique triangles may be solved, but we restrict our attention to the two most useful methods, the **law of sines** and the **law of cosines.** In this section, we discuss the law of sines and show that it may be used to solve Case 1 and Case 2.

Let ABC be an oblique triangle with sides a, b, and c opposite angles A, B, and C, respectively. By drawing a perpendicular h from B to side b, as shown in Fig. 9.49(a), note that $h/c = \sin A$ and $h/a = \sin C$, and therefore

$$h = c \sin A \quad \text{or} \quad h = a \sin C \tag{9.4}$$

This relationship also holds if one of the angles is obtuse. By drawing a perpendicular h to the extension of side b, as shown in Fig. 9.49(b), we see that $h/c = \sin A$.

Also, in Fig. 9.49(b), $h/a = \sin(180° - C)$, where $180° - C$ is also the reference angle for the obtuse angle C. For a second-quadrant reference angle $180° - C$, we have $\sin[180° - (180° - C)] = \sin C$. Putting this all together, $\sin(180° - C) = \sin C$. This means that

$$h = c \sin A \quad \text{or} \quad h = a \sin(180° - C) = a \sin C \tag{9.5}$$

The results are precisely the same in Eqs. (9.4) and (9.5). Setting the results for h equal to each other, we have

$$c \sin A = a \sin C$$

Dividing each side by $\sin A \, \sin C$ and reversing sides gives us

$$\frac{a}{\sin A} = \frac{c}{\sin C} \tag{9.6}$$

By dropping a perpendicular from A to a, we also derive the result

$$c \sin B = b \sin C$$

which, when each side is divided by $\sin B \, \sin C$, and reversing sides, give us

$$\frac{b}{\sin B} = \frac{c}{\sin C} \tag{9.7}$$

If we had dropped a perpendicular to either side a or side c, equations similar to Eqs. (9.6) and (9.7) would have resulted.

(a)

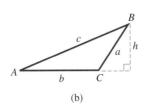

(b)

Fig. 9.49

Combining Eqs. (9.6) and (9.7), *for any triangle with sides a, b, and c, opposite angles A, B, and C, respectively,* such as the one shown in Fig. 9.50, *we have the* **law of sines:**

Law of Sines

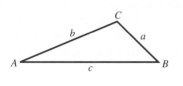

Fig. 9.50

$$\frac{a}{\sin A} = \frac{b}{\sin B} = \frac{c}{\sin C} \qquad (9.8)$$

NOTE ▶

Another form of the law of sines can be obtained by equating the reciprocals of each of the fractions in Eq. (9.8). The law of sines is a statement of proportionality between the sides of a triangle and the sines of the angles opposite them. Note that *there are actually three equations combined in Eq. (9.8).* Of these, we use the one with three known parts of the triangle and we find the fourth part. In finding the complete solution of a triangle, it may be necessary to use two of the three equations.

CASE 1: TWO ANGLES AND ONE SIDE

Now, we see how the law of sines is used in the solution of a triangle in which two angles and one side are known. If two angles are known, the third may be found from the fact that the sum of the angles in a triangle is 180°. At this point, we must be able to find the ratio between the given side and the sine of the angle opposite it. Then, by use of the law of sines, we may find the other two sides.

EXAMPLE 1 Case 1: Two angles and one side

Given $c = 6.00$, $A = 60.0°$, and $B = 40.0°$, find a, b, and C.
First, we can see that

$$C = 180.0° - (60.0° + 40.0°) = 80.0°$$

We now know side c and angle C, which allows us to use Eq. (9.8). Therefore, using the equation relating a, A, c, and C, we have

Fig. 9.51

$$\frac{a}{\sin 60.0°} = \frac{6.00}{\sin 80.0°} \quad \text{or} \quad a = \frac{6.00 \sin 60.0°}{\sin 80.0°} = 5.28$$

Now, using the equation relating b, B, c, and C, we have

$$\frac{b}{\sin 40.0°} = \frac{6.00}{\sin 80.0°} \quad \text{or} \quad b = \frac{6.00 \sin 40.0°}{\sin 80.0°} = 3.92$$

Thus, $a = 5.28$, $b = 3.92$, and $C = 80.0°$. See Fig. 9.51. We could also have used the form of Eq. (9.8) relating a, A, b, and B in order to find b, but any error in calculating a would make b in error as well. Of course, any error in calculating C would make both a and b in error.
The calculator solution is shown in Fig. 9.52. ∎

Fig. 9.52

Practice Exercise

1. In Example 1, change the value of B to 65.0°, and then find a.

NOTE ▶

The solution of a triangle can be checked approximately by noting that *the smallest angle is opposite the shortest side, and the largest angle is opposite the longest side.* Note that this is the case in Example 1, where b (shortest side) is opposite B (smallest angle), and c (longest side) is opposite C (largest angle).

Fig. 9.53

EXAMPLE 2 Case 1: Two angles and one side

Solve the triangle with the following given parts: $a = 63.71$, $A = 56.29°$, and $B = 97.06°$. See Fig. 9.53.

From the figure, we see that we are to find angle C and sides b and c. We first determine angle C:

$$C = 180° - (A + B) = 180° - (56.29° + 97.06°)$$
$$= 26.65°$$

Noting the three angles, we know that c is the shortest side (C is the smallest angle) and b is the longest side (B is the largest angle). This means that the length of a is between c and b, or $c < 63.71$ and $b > 63.71$. Now using the ratio $a/\sin A$ of Eq. (9.8) (the law of sines) to find sides b and c, we have

$$\frac{b}{\sin 97.06°} = \frac{63.71}{\sin 56.29°} \quad \text{or} \quad b = \frac{63.71 \sin 97.06°}{\sin 56.29°} = 76.01$$

$$\frac{c}{\sin 26.65°} = \frac{63.71}{\sin 56.29°} \quad \text{or} \quad c = \frac{63.71 \sin 26.65°}{\sin 56.29°} = 34.35$$

```
180-(56.29+97.06
)
            26.65
63.71sin(97.06)/
sin(56.29)
       76.00696912
63.71sin(26.65)/
sin(56.29)
       34.35256783
```

Fig. 9.54

Thus, $b = 76.01$, $c = 34.35$, and $C = 26.65°$. Note that $c < a$ and $b > a$, as expected. The calculator solution is shown in Fig. 9.54. ∎

If the given information is appropriate, the law of sines may be used to solve applied problems. The following example illustrates the use of the law of sines in such a problem.

EXAMPLE 3 Case 1: Application

Two observers A and B sight a helicopter due east. The observers are 3540 ft apart, and the angles of elevation they each measure to the helicopter are 32.0° and 44.0°, respectively. How far is observer A from the helicopter? See Fig. 9.55.

■ The first successful helicopter was made in the United States by Igor Sikorsky in 1939.

Letting H represent the position of the helicopter, we see that angle B within the triangle ABH is $180° - 44.0° = 136.0°$. This means that the angle at H within the triangle is

$$H = 180° - (32.0° + 136.0°) = 12.0°$$

Now, using the law of sines to find required side b, we have

required side → / opposite known angle $\dfrac{b}{\sin 136.0°} = \dfrac{3540}{\sin 12.0°}$ ← known side / opposite known angle

or

$$b = \frac{3540 \sin 136.0°}{\sin 12.0°} = 11{,}800 \text{ ft}$$

Fig. 9.55

Thus, observer A is about 11,800 ft from the helicopter. ∎

CASE 2: TWO SIDES AND THE ANGLE OPPOSITE ONE OF THEM

For a triangle in which we know two sides and the angle opposite one of the given sides, the solution will be either *one triangle*, or *two triangles*, or even possibly *no triangle*. The following examples illustrate how each of these results is possible.

EXAMPLE 4 Case 2: Two sides and angle opposite

Solve the triangle with the following given parts: $a = 60.0$, $b = 40.0$, and $B = 30.0°$.

First, make a good scale drawing (Fig. 9.56(a)) by drawing angle B and measuring off 60 for a. This will more clearly show that side $b = 40.0$ will intersect side c at either position A or A'. This means there are two triangles that satisfy the given values. Using the law of sines, we solve the case for which A is an acute angle:

$$\frac{60.0}{\sin A} = \frac{40.0}{\sin 30.0°} \quad \text{or} \quad \sin A = \frac{60.0 \sin 30.0°}{40.0}$$

$$A = \sin^{-1}\left(\frac{60.0 \sin 30.0°}{40.0}\right) = 48.6°$$

$$C = 180° - (30.0° + 48.6°) = 101.4°$$

Therefore, $A = 48.6°$ and $C = 101.4°$. Using the law of sines again to find c, we have

$$\frac{c}{\sin 101.4°} = \frac{40.0}{\sin 30.0°}$$

$$c = \frac{40.0 \sin 101.4°}{\sin 30.0°}$$

$$= 78.4$$

Thus, $A = 48.6°$, $C = 101.4°$, and $c = 78.4$. See Fig. 9.56(b).

The other solution is the case in which A', opposite side a, is an obtuse angle. Therefore,

$$A' = 180° - A = 180° - 48.6°$$

$$= 131.4°$$

$$C' = 180° - (30.0° + 131.4°)$$

$$= 18.6°$$

Using the law of sines to find c', we have

$$\frac{c'}{\sin 18.6°} = \frac{40.0}{\sin 30.0°}$$

$$c' = \frac{40.0 \sin 18.6°}{\sin 30.0°}$$

$$= 25.5$$

This means that the second solution is $A' = 131.4°$, $C' = 18.6°$, and $c' = 25.5$. See Fig. 9.56(c).

The complete sequence for the calculator solution is shown in Fig. 9.57. The upper window shows the completion of the solution for A, C, and c. The lower window shows the solution for A', C', and c'. ∎

Fig. 9.56

Fig. 9.57

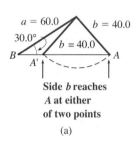

EXAMPLE 5 Case 2: Possible solutions

In Example 4, if $b > 60.0$, only one solution would result. In this case, side b would intercept side c at A. It also intercepts the extension of side c, but this would require that angle B not be included in the triangle (see Fig. 9.58). Thus, only one solution may result if $b > a$.

In Example 4, there would be *no solution* if side b were not at least 30.0. If this were the case, side b would not be long enough to even touch side c. It can be seen that b must at least equal $a \sin B$. If it is just equal to $a \sin B$, there is *one solution,* a right triangle. See Figure 9.59.

Fig. 9.58

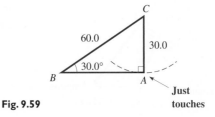

Fig. 9.59   Just touches ∎

Ambiguous Case

Summarizing the results for Case 2 as illustrated in Examples 4 and 5, we make the following conclusions. Given sides a and b and angle A (assuming here that a and A ($A < 90°$) are corresponding parts), we have the following summary of solutions for Case 2.

Practice Exercise

2. Determine which of the four possible solution types occurs if $c = 28$, $b = 48$, and $C = 30°$.

CAUTION ▶

> **SUMMARY OF SOLUTIONS:**
> **TWO SIDES AND THE ANGLE OPPOSITE ONE OF THEM**
> 1. **No solution** *if $a < b \sin A$.* See Fig. 9.60(a).
> 2. **A right triangle solution** *if $a = b \sin A$.* See Fig. 9.60(b).
> 3. **Two solutions** *if $b \sin A < a < b$.* See Fig. 9.60(c).
> 4. **One solution** *if $a > b$.* See Fig. 9.60(d).

(a)

(b)

(c)

(d)

Fig. 9.60

NOTE ▶

Note that *in order to have two solutions, we must know two sides and the angle opposite one of the sides, and the shorter side must be opposite the known angle.*

If there is *no solution,* the calculator will indicate an *error.* If the solution is a *right triangle,* the calculator will show an angle of *exactly* 90° (no extra decimal digits will be displayed.)

For the reason that two solutions may result from it, Case 2 is called the **ambiguous case.** The following example illustrates Case 2 in an applied problem.

EXAMPLE 6 Case 2: Application

Kingston, Jamaica, is 43.2° south of east of Havana, Cuba. What should be the heading of a plane from Havana to Kingston if the wind is from the west at 40.0 km/h and the plane's speed with respect to the air is 300 km/h?

The heading should be set so that the resultant of the plane's velocity with respect to the air $\mathbf{v}_{pa}$ and the velocity of the wind $\mathbf{v}_w$ will be in the direction from Havana to Kingston. This means that the resultant velocity $\mathbf{v}_{pg}$ of the plane with respect to the ground must be at an angle of 43.2° south of east from Havana.

Using the given information, we draw the vector triangle shown in Fig. 9.61. In the triangle, we know that the angle at Kingston is 43.2° by noting the alternate-interior angles (see page 51). By finding θ, the required heading can be found. There can be only one solution, since $v_{pa} > v_w$. Using the law of sines, we have

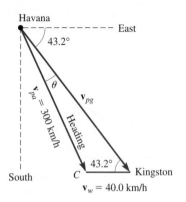

■ See the chapter introduction.

Fig. 9.61

$$\begin{array}{l}\text{known side} \\ \text{opposite} \\ \text{required angle}\end{array} \rightarrow \dfrac{40.0}{\sin\theta} = \dfrac{300}{\sin 43.2°} \leftarrow \begin{array}{l}\text{known side} \\ \text{opposite} \\ \text{known angle}\end{array}$$

$$\sin\theta = \frac{40.0 \sin 43.2°}{300}, \qquad \theta = 5.2°$$

Therefore, the heading should be 43.2° + 5.2° = 48.4° south of east. Compare this example with Example 3 on page 269. ■

If we try to use the law of sines for Case 3 or Case 4, we find that we do not have enough information to complete any of the ratios. These cases can, however, be solved by the law of cosines as shown in the next section.

EXAMPLE 7 Cases 3&4 not solvable by law of sines

Given (case 3) two sides and the included angle of a triangle $a = 2, b = 3, C = 45°$, and (case 4) the three sides (case 4) $a = 5, b = 6, c = 7$, we set up the ratios

$$\text{(case 3)}\ \frac{2}{\sin A} = \frac{3}{\sin B} = \frac{c}{\sin 45°}, \quad \text{and}\ \text{(case 4)}\ \frac{5}{\sin A} = \frac{6}{\sin B} = \frac{7}{\sin C}$$

The solution cannot be found since each of the three possible equations in either case 3 or case 4 contains two unknowns. ■

EXERCISES 9.5

In Exercises 1 and 2, solve the resulting triangles if the given changes are made in the indicated examples of this section.

1. In Example 2, solve the triangle if the value of B is changed to 82.94°.

2. In Example 4, solve the triangle if the value of a is changed to 70.0.

In Exercises 3–22, solve the triangles with the given parts.

3. $a = 45.7, A = 65.0°, B = 49.0°$

4. $b = 3.07, A = 26.0°, C = 120.0°$

5. $c = 4380, A = 37.4°, B = 34.6°$

6. $a = 93.2, B = 0.9°, C = 82.6°$

7. $a = 4.601, b = 3.107, A = 18.23°$

8. $b = 362.2, c = 294.6, B = 69.37°$

9. $b = 7751, c = 3642, B = 20.73°$

10. $a = 150.4, c = 250.9, C = 76.43°$

11. $b = 0.0742, B = 51.0°, C = 3.4°$

12. $c = 729, B = 121.0°, C = 44.2°$

13. $a = 63.8, B = 58.4°, C = 22.2°$

14. $a = 0.130, A = 55.2°, B = 117.5°$

15. $b = 4384, B = 47.43°, C = 64.56°$

16. $b = 283.2, B = 13.79°, C = 76.38°$

17. $a = 5.240, b = 4.446, B = 48.13°$

18. $a = 89.45, c = 37.36, C = 15.62°$

19. $b = 2880, c = 3650, B = 31.4°$

20. $a = 0.841, b = 0.965, A = 57.1°$

21. $a = 450, b = 1260, A = 64.8°$

22. $a = 20, c = 10, C = 30°$

In Exercises 23–40, use the law of sines to solve the given problems.

23. Two angles of a triangle measure 29.0° and 57.0°. The longest side is 52.0 cm long. What is the length of the shortest side?

24. Two angles of a triangle measure 45.0° and 55.0°. The side opposite the 55.0° angle is 8.75 in. long. What is the length of the longest side?

25. The loading ramp at a package delivery service is 9.5 ft long and makes a 22.5° angle with the level ground. If this ramp is replaced with one that is 22.5 ft long, what angle does the new ramp make with the ground?

26. The longest side of a triangular parcel of land is 172 m long, and the shortest side is 105 m long. If the largest angle is 82.0°, what is the length of the third side?

27. The Pentagon (headquarters of the U.S. Department of Defense) is the largest office building in the world. It is a regular pentagon (five sides), 921 ft on a side. Find the greatest straight-line distance from one point on the outside of the building to another outside point (the length of a diagonal).

28. Two ropes hold a 175-lb crate as shown in Fig. 9.62. Find the tensions T_1 and T_2 in the ropes. (*Hint:* Move vectors so that they are tail to head to form a triangle. The vector sum $T_1 + T_2$ must equal 175 lb for equilibrium. See page 269.)

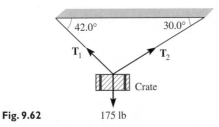

Fig. 9.62

29. Find the tension T in the left guy wire attached to the top of the tower shown in Fig. 9.63. (*Hint:* The horizontal components of the tensions must be equal and opposite for equilibrium. See page 269. Thus, move the tension vectors tail to head to form a triangle with a vertical resultant. This resultant equals the upward force at the top of the tower for equilibrium. This last force is not shown and does not have to be calculated.)

Fig. 9.63

30. Find the distance from Atlanta to Raleigh, North Carolina, from Fig. 9.64.

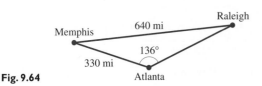

Fig. 9.64

31. Find the distance between Gravois Ave. and Jefferson Ave. along Arsenal St. in St. Louis, from Fig. 9.65.

Fig. 9.65

32. When an airplane is landing at an 8250-ft runway, the angles of depression to the ends of the runway are 10.0° and 13.5°. How far is the plane from the near end of the runway?

33. Find the total length of the path of the laser beam that is shown in Fig. 9.66.

Fig. 9.66

34. In widening a highway, it is necessary for a construction crew to cut into the bank along the highway. The present angle of elevation of the straight slope of the bank is 23.0°, and the new angle is to be 38.5°, leaving the top of the slope at its present position. If the slope of the present bank is 220 ft long, how far horizontally into the bank at its base must they dig?

35. A communications satellite is directly above the extension of a line between receiving towers A and B. It is determined from radio signals that the angle of elevation of the satellite from tower A is 89.2°, and the angle of elevation from tower B is 86.5°. See Fig. 9.67. If A and B are 1290 km apart, how far is the satellite from A? (Neglect the curvature of the earth.)

Fig. 9.67

36. An astronaut on the moon drives a lunar rover 16 km in the direction 60.0° north of east from the base. (a) Through what angle must the rover then be turned so that by driving 12 km farther the astronaut can turn again to return to base along a north-south line? (b) How long is the last leg of the trip? (c) Can the astronaut make it back to base if the maximum range of the rover is 40 km?

37. A boat owner wishes to cross a river 2.60 km wide and go directly to a point on the opposite side 1.75 km downstream. The boat goes 8.00 km/h in still water, and the stream flows at 3.50 km/h. What should the boat's heading be?

38. A motorist traveling along a level highway at 75 km/h directly toward a mountain notes that the angle of elevation of the mountain top changes from about 20° to about 30° in a 20-min period. How much closer on a direct line did the mountain top become?

39. A hillside is inclined at 23° with the horizontal. From a given point on the slope, it has been found that a vein of gold is 55 m directly below. At what angle below the hillside slope from another point downhill must a straight 65-m shaft be dug to reach the vein?

40. Point *P* on the mechanism shown in Fig. 9.68 is driven back and forth horizontally. If the minimum value of angle θ is 32.0°, what is the distance between extreme positions of *P*? What is the maximum possible value of angle θ?

Fig. 9.68

Answers to Practice Exercises

1. $a = 6.34$ **2.** Two solutions

9.6   The Law of Cosines

Law of Cosines •
Case 3: Two Sides & Included Angle •
Case 4: Three Sides •
Summary of Solving Oblique Triangles

As noted in the last section, the law of sines cannot be used for Case 3 (two sides and the included angle) and Case 4 (three sides). In this section, we develop the *law of cosines,* which can be used for Cases 3 and 4. After finding another part of the triangle using the law of cosines, we will often find it easier to complete the solution using the law of sines.

Consider any oblique triangle—for example, either triangle shown in Fig. 9.69. For each triangle, $h/b = \sin A$, or $h = b \sin A$. Also, using the Pythagorean theorem, we obtain $a^2 = h^2 + x^2$ for each triangle. Therefore (with $(\sin A)^2 = \sin^2 A$),

$$a^2 = b^2 \sin^2 A + x^2 \tag{9.9}$$

(a)

In Fig. 9.69(a), note that $(c - x)/b = \cos A$, or $c - x = b \cos A$. Solving for *x*, we have $x = c - b \cos A$. In Fig. 9.69(b), $c + x = b \cos A$, and solving for *x*, we have $x = b \cos A - c$. Substituting these relations into Eq. (9.9), we obtain

$$a^2 = b^2 \sin^2 A + (c - b \cos A)^2$$

and

$$a^2 = b^2 \sin^2 A + (b \cos A - c)^2 \tag{9.10}$$

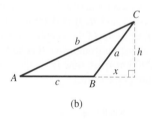

(b)

Fig. 9.69

respectively. When expanded, these both give

$$a^2 = b^2 \sin^2 A + b^2 \cos^2 A + c^2 - 2bc \cos A$$
$$= b^2(\sin^2 A + \cos^2 A) + c^2 - 2bc \cos A \tag{9.11}$$

Recalling the definitions of the trigonometric functions, we know that $\sin \theta = y/r$ and $\cos \theta = x/r$. Thus, $\sin^2 \theta + \cos^2 \theta = (y^2 + x^2)/r^2$. However, $x^2 + y^2 = r^2$, which means

$$\sin^2 \theta + \cos^2 \theta = 1 \tag{9.12}$$

This equation is valid for any angle θ, since we have made no assumptions as to the properties of θ. Thus, by substituting Eq. (9.12) into Eq. (9.11), we arrive at the **law of cosines:**

Law of Cosines

$$a^2 = b^2 + c^2 - 2bc \cos A \tag{9.13}$$

Using the method above, we may also show that

$$b^2 = a^2 + c^2 - 2ac \cos B$$
$$c^2 = a^2 + b^2 - 2ab \cos C$$

CASE 3: TWO SIDES AND THE INCLUDED ANGLE

If two sides and the included angle of a triangle are known, the forms of the law of cosines show that we may directly solve for the side opposite the given angle. Then, as noted earlier, the solution may be completed using the law of sines.

EXAMPLE 1 Case 3: Two sides and included angle

Solve the triangle with $a = 45.0$, $b = 67.0$, and $C = 35.0°$. See Fig. 9.70.

Since angle C is known, first solve for side c, using the law of cosines in the form $c^2 = a^2 + b^2 - 2ab \cos C$. Substituting, we have

Fig. 9.70

Fig. 9.71

$$\underbrace{c^2}_{\text{unknown side opposite known angle}} = \underbrace{45.0^2 + 67.0^2}_{\text{known sides}} - 2(45.0)(67.0) \cos 35.0°$$

$$c = \sqrt{45.0^2 + 67.0^2 - 2(45.0)(67.0) \cos 35.0°} = 39.7$$

From the law of sines, we now have

$$\frac{45.0}{\sin A} = \frac{67.0}{\sin B} = \frac{39.7}{\sin 35.0°} \leftarrow \text{sides opposite} \atop \leftarrow \text{angles}$$

$$\sin A = \frac{45.0 \sin 35.0°}{39.7}, \qquad A = 40.6°$$

The calculator solution is shown in Fig. 9.71. As shown, we found that $B = 104.4°$ using the fact that the sum of the angles is $180°$, rather than using the relation from the law of sines. Therefore, $c = 39.7$, $A = 40.6°$, and $B = 104.4°$.

After finding side c, solving for angle B rather than angle A, the calculator would show $B = 75.5°$. Then, when subtracting the sum of angles B and C from $180°$, we would get $A = 69.5°$. Although this appears to be correct, it is not. Since the largest angle of a triangle might be greater than $90°$, *first solve for the smaller unknown angle.* This smaller angle (opposite the shorter known side) cannot be greater than $90°$, and the larger unknown angle can be found by subtraction as is done in this example. ∎

CAUTION ▶

Practice Exercise

1. In Example 1, change the value of a to 95.0, and find A and B.

EXAMPLE 2 Case 3: Application

Two forces are acting on a bolt. One is a 78.0-N force acting horizontally to the right, and the other is a force of 45.0 N acting upward to the right, $15.0°$ from the vertical. Find the resultant force **F**. See Fig. 9.72.

Moving the 45.0-N vector to the right and using the lower triangle with the $105.0°$ angle, the magnitude of **F** is

$$F = \sqrt{78.0^2 + 45.0^2 - 2(78.0)(45.0) \cos 105.0°}$$
$$= 99.6 \, \text{N}$$

To find θ, use the law of sines:

$$\frac{45.0}{\sin \theta} = \frac{99.6}{\sin 105.0°}, \qquad \sin \theta = \frac{45.0 \sin 105.0°}{99.6}$$

This gives us $\theta = 25.9°$. See Fig. 9.73.

We can also solve this problem using vector components. ∎

Fig. 9.72

Fig. 9.73

CASE 4: THREE SIDES

CAUTION ▸

Given the three sides of a triangle, we may solve for the angle opposite any side using the law of cosines. ***The best procedure is to find the largest angle first.*** This avoids the ambiguous case if we switch to the law of sines and there is an obtuse angle. *The largest angle is opposite the longest side.* Another procedure is to use the law of cosines to find two angles.

EXAMPLE 3   Case 4: Three sides

Solve the triangle with sides $a = 49.33$, $b = 21.61$, and $c = 42.57$. See Fig. 9.74.

Since the longest side is $a = 49.33$, first solve for angle A:

$$a^2 = b^2 + c^2 - 2bc \cos A$$

$$\cos A = \frac{b^2 + c^2 - a^2}{2bc} = \frac{21.61^2 + 42.57^2 - 49.33^2}{2(21.61)(42.57)}$$

$$A = 94.81°$$

From the law of sines, we now have

$$\frac{49.33}{\sin A} = \frac{21.61}{\sin B} = \frac{42.57}{\sin C}$$

The calculator solution is shown in Fig. 9.75. Therefore, $B = 25.88°$ and $C = 59.31°$. ∎

```
cos⁻¹((21.61²+42.
57²-49.33²)/(2*2
1.61*42.57))
          94.80922198
Ans→A
          94.80922198
sin⁻¹(21.61sin(A)
/49.33)
          25.88255393
180-(A+Ans)
          59.30822409
```

Fig. 9.75

Fig. 9.74

EXAMPLE 4   Case 4: Application

A vertical radio antenna is to be built on a hillside with a constant slope. A guy wire is to be attached at a point 145 ft up the antenna, and at a point 110 ft from the base of the antenna up the hillside. If the guy wire is 170 ft long, what angle does the antenna make with the hillside?

From Fig. 9.76, we can set up the equation necessary for the solution.

$$170^2 = 110^2 + 145^2 - 2(110)(145)\cos \theta$$

$$\theta = \cos^{-1} \frac{110^2 + 145^2 - 170^2}{2(110)(145)} = 82.4°$$ ∎

Fig. 9.76

■ Summary of solving an oblique triangle using the law of sines or the law of cosines

---

Practice Exercise

2. Using Fig. 9.76, and only the data given in Example 4, find the angle between the guy wire and the hillside.

SOLVING OBLIQUE TRIANGLES

Case 1: Two Angles and One Side
Find the unknown angle by subtracting the sum of the known angles from 180°. Use the **law of sines** to find the unknown sides.

Case 2: Two Sides and the Angle Opposite One of Them
Use the known side and the known angle opposite it to find the angle opposite the other known side. Find the third angle from the fact that the sum of the angles is 180°. Use the **law of sines** to find the third side. **CAUTION:** There may be two solutions. See page 277 for a summary of Case 2 and the *ambiguous case*.

Case 3: Two Sides and the Included Angle
Find the third side by using the **law of cosines**. Find the *smaller* unknown angle (opposite the shorter side) by using the **law of sines**. Complete the solution using the fact that the sum of the angles is 180°.

Case 4: Three Sides
Find the *largest angle* (opposite the longest side) by using the **law of cosines**. Find a second angle by using the **law of sines**. Complete the solution by using the fact that the sum of the angles is 180°.

Other variations in finding the solutions can be used. For example, after finding the third side in Case 3 or finding the largest angle in Case 4, the solution can be completed by using the law of sines. All angles in Case 4 can be found by using the law of cosines. The methods shown above are those normally used.

EXERCISES 9.6

In Exercises 1 and 2, solve the resulting triangles if the given changes are made in the indicated examples of this section.

1. In Example 1, solve the triangle if the value of C is changed to 145°.

2. In Example 3, solve the triangle if the value of a is changed to 29.33.

In Exercises 3–22, solve the triangles with the given parts.

3. $a = 6.00, b = 7.56, C = 54.0°$

4. $b = 87.3, c = 34.0, A = 130.0°$

5. $a = 4530, b = 924, C = 98.0°$

6. $a = 0.0845, c = 0.116, B = 85.0°$

7. $a = 39.53, b = 45.22, c = 67.15$

8. $a = 2.331, b = 2.726, c = 2.917$

9. $a = 385.4, b = 467.7, c = 800.9$

10. $a = 0.2433, b = 0.2635, c = 0.1538$

11. $a = 320, b = 847, C = 158.0°$

12. $b = 18.3, c = 27.1, A = 8.7°$

13. $a = 2140, c = 428, B = 86.3°$

14. $a = 1.13, b = 0.510, C = 77.6°$

15. $b = 103.7, c = 159.1, C = 104.67°$

16. $a = 49.32, b = 54.55, B = 114.36°$

17. $a = 0.4937, b = 0.5956, c = 0.6398$

18. $a = 69.72, b = 49.30, c = 22.29$

19. $a = 723, b = 598, c = 158$

20. $a = 1.78, b = 6.04, c = 4.80$

21. $a = 1500, A = 15°, B = 140°$

(W) 22. $a = 17, b = 24, c = 42$. Explain your answer.

In Exercises 23–40, use the law of cosines to solve the given problems.

23. For a triangle with sides a, b, and c opposite angles A, B, and C, respectively, show that $1 + \cos A = \dfrac{(b + c + a)(b + c - a)}{2bc}$.

(W) 24. Set up equations (do not solve) to solve the triangle in Fig. 9.77 by the law of cosines. Why is the law of sines easier to use?

Fig. 9.77

(W) 25. On page 277, we saw that it was possible to be given two sides and the angle opposite one of them, and find that there is no triangle that can have those parts. Is it possible in Case 3 or in Case 4 that there is no solution for a set of given parts?

26. Three circles of radii 24 in., 32 in., and 42 in. are externally tangent to each other (each is tangent to the other two). Find the largest angle of the triangle formed by joining their centers.

27. A nuclear submarine leaves its base and travels at 23.5 mi/h. For 2.00 h, it travels along a course of 32.1° north of west. It then turns an additional 21.5° north of west and travels for another 1.00 h. How far from its base is it?

28. The robot arm shown in Fig. 9.78 places packages on a conveyor belt. What is the distance x?

Fig. 9.78

29. Find the angle between the front legs and the back legs of the folding chair shown in Fig. 9.79.

Fig. 9.79

30. In a baseball field, the four bases are at the vertices of a square 90.0 ft on a side. The pitching rubber is 60.5 ft from home plate. See Fig. 9.80. How far is it from the pitching rubber to first base?

Fig. 9.80

31. A plane leaves an airport and travels 624 km due east. It then turns toward the north and travels another 326 km. It then turns again less than 180° and travels another 846 km directly back to the airport. Through what angles did it turn?

32. The apparent depth of an object submerged in water is less than its actual depth. A coin is actually 5.00 in. from an observer's eye just above the surface, but it appears to be only 4.25 in. The real light ray from the coin makes an angle with the surface that is 8.1° greater than the angle the apparent ray makes. How much deeper is the coin than it appears to be? See Fig. 9.81.

Fig. 9.81

33. A nut is in the shape of a regular hexagon (six sides). If each side is 9.53 mm, what opening on a wrench is necessary to tighten the nut? See Fig. 9.82.

9.53 mm

Fig. 9.82

34. Two ropes support a 78.3-lb crate from above. The tensions in the ropes are 50.6 lb and 37.5 lb. What is the angle between the ropes? (See Exercise 28 of Section 9.5.)

35. A ferryboat travels at 11.5 km/h with respect to the water. Because of the river current, it is traveling at 12.7 km/h with respect to the land in the direction of its destination. If the ferryboat's heading is 23.6° from the direction of its destination, what is the velocity of the current?

36. The airline distance from Denver to Dallas is 660 mi. It is 800 mi from Denver to St. Louis and 550 mi from Dallas to St. Louis. Find the angle between the routes from Dallas.

37. An air traffic controller sights two planes that are due east from the control tower and headed toward each other. One is 15.8 mi from the tower at an angle of elevation of 26.4°, and the other is 32.7 mi from the tower at an angle of elevation of 12.4°. How far apart are the planes?

38. A ship's captain notes that a second ship is 14.5 km away at a bearing (see page 271, Exercise 23) of 46.3°, and that a third ship was at a distance of 21.7 km at a bearing of 201.0°. How far apart are the second and third ships?

39. A park is in the shape of a parallelogram with sides of 1.25 km and 1.90 km that meet in a 78.0° angle. The park has two diagonal paths. What is the length of each path?

W **40.** A triangular machine part has sides of 5 cm and 8 cm. Explain why the law of sines, or the law of cosines, is used to start the solution of the triangle if the third known part is (a) the third side, (b) the angle opposite the 5-cm side, or (c) the angle between the 5-cm and 8-cm sides.

Answers to Practice Exercises

1. $B = 43.8°$, $A = 101.2°$ **2.** $57.7°$

CHAPTER 9 EQUATIONS

Vector components	$A_x = A \cos \theta \qquad A_y = A \sin \theta$	(9.1)
	$A = \sqrt{A_x^2 + A_y^2}$	(9.2)
	$\theta_{ref} = \tan^{-1} \dfrac{|A_y|}{|A_x|}$	(9.3)
Law of sines	$\dfrac{a}{\sin A} = \dfrac{b}{\sin B} = \dfrac{c}{\sin C}$	(9.8)
Law of cosines	$a^2 = b^2 + c^2 - 2bc \cos A$	
	$b^2 = a^2 + c^2 - 2ac \cos B$	(9.13)
	$c^2 = a^2 + b^2 - 2ab \cos C$	

CHAPTER 9 REVIEW EXERCISES

In Exercises 1–4, find the x- and y-components of the given vectors by use of the trigonometric functions.

1. $A = 65.0, \theta_A = 28.0°$

2. $A = 8050, \theta_A = 149.0°$

3. $A = 0.9204, \theta_A = 215.59°$

4. $A = 657.1, \theta_A = 343.74°$

*In Exercises 5–8, vectors **A** and **B** are at right angles. Find the magnitude and direction of the resultant.*

5. $A = 327$
$B = 505$

6. $A = 6.8$
$B = 2.9$

7. $A = 4964$
$B = 3298$

8. $A = 26.52$
$B = 89.86$

In Exercises 9–16, add the given vectors by use of the trigonometric functions and the Pythagorean theorem.

9. $A = 780, \theta_A = 28.0°$
$B = 346, \theta_B = 320.0°$

10. $J = 0.0120, \theta_J = 370.5°$
$K = 0.00781, \theta_K = 260.0°$

11. $A = 22.51, \theta_A = 130.16°$
$B = 7.604, \theta_B = 200.09°$

12. $A = 18,760, \theta_A = 110.43°$
$B = 4835, \theta_B = 350.20°$

13. $Y = 51.33, \theta_Y = 12.25°$
$Z = 42.61, \theta_Z = 291.77°$

14. $A = 703.1, \theta_A = 122.54°$
$B = 302.9, \theta_B = 214.82°$

15. $A = 0.750, \theta_A = 15.0°$
$B = 0.265, \theta_B = 192.4°$
$C = 0.548, \theta_C = 344.7°$

16. $S = 8120, \theta_S = 141.9°$
$T = 1540, \theta_T = 165.2°$
$U = 3470, \theta_U = 296.0°$

In Exercises 17–36, solve the triangles with the given parts.

17. $A = 48.0°, B = 68.0°, a = 145$

18. $A = 132.0°, b = 7.50, C = 32.0°$

19. $a = 22.8, B = 33.5°, C = 125.3°$

20. $A = 71.0°, B = 48.5°, c = 8.42$

21. $A = 17.85°, B = 154.16°, c = 7863$

22. $a = 1.985, b = 4.189, c = 3.652$

23. $b = 7607, c = 4053, B = 110.09°$

24. $A = 77.06°, a = 12.07, c = 5.104$

25. $b = 14.5, c = 13.0, C = 56.6°$

26. $B = 40.6°, b = 7.00, c = 18.0$

27. $a = 186, B = 130.0°, c = 106$

28. $b = 750, c = 1100, A = 56°$

29. $a = 7.86, b = 2.45, C = 2.5°$

30. $a = 0.208, c = 0.697, B = 165.4°$

31. $A = 67.16°, B = 96.84°, c = 532.9$

32. $A = 43.12°, a = 7.893, b = 4.113$

33. $a = 17, b = 12, c = 25$

34. $a = 9064, b = 9953, c = 1106$

35. $a = 0.530, b = 0.875, c = 1.25$

36. $a = 47.4, b = 40.0, c = 45.5$

In Exercises 37–70, solve the given problems.

37. For any triangle ABC show that
$$\frac{a^2 + b^2 + c^2}{2abc} = \frac{\cos A}{a} + \frac{\cos B}{b} + \frac{\cos C}{c}$$

(W) 38. In solving a triangle for Case 3 (two sides and the included angle), explain what type of solution is obtained if the included angle is a right angle.

39. Two angles of a triangle measure 22° and 112°. The shortest side is 54 cm. What is the longest side?

40. Two of the angles of a triangle measure 42.0° and 59.5°. The longest side is 5.00 cm longer than the shortest side. What is the perimeter of the triangle?

41. An architect determines the two acute angles and one of the legs of a right triangular wall panel. Show that the area A_t is
$$A_t = \frac{a^2 \sin B}{2 \sin A}$$

42. A surveyor determines the three angles and one side of a triangular tract of land. (a) Show that the area A_t can be found from
$$A_t = \frac{a^2 \sin B \sin C}{2 \sin A}.$$ (b) For a right triangle, show that this agrees with the formula in Exercise 41.

43. Find the horizontal and vertical components of the force shown in Fig. 9.83.

Fig. 9.83   **Fig. 9.84**

44. Find the horizontal and vertical components of the velocity shown in Fig. 9.84.

45. In a ballistics test, a bullet was fired into a block of wood with a velocity of 2200 ft/s and at an angle of 71.3° with the surface of the block. What was the component of the velocity perpendicular to the surface?

46. A storm cloud is moving at 15 mi/h from the northwest. A television tower is 60° south of east of the cloud. What is the component of the cloud's velocity toward the tower?

47. During a 3.00-min period after taking off, the supersonic jet Concorde traveled at 480 km/h at an angle of 24.0° above the horizontal. What was its gain in altitude during this period?

48. A rocket is launched at an angle of 42.0° with the horizontal and with a speed of 2500 ft/s. What are its horizontal and vertical components of velocity?

49. Three forces of 3200 lb, 1300 lb, and 2100 lb act on a bolt as shown in Fig. 9.85. Find the resultant force.

Fig. 9.85

50. In Fig. 9.86, force **F** represents the total surface tension force around the circumference on the liquid in the capillary tube. The vertical component of **F** holds up the liquid in the tube above the liquid surface outside the tube. What is the vertical component of **F**?

Fig. 9.86

51. A helium-filled balloon rises vertically at 3.5 m/s as the wind carries it horizontally at 5.0 m/s. What is the resultant velocity of the balloon?

52. A shearing pin is designed to break and disengage gears before damage is done to a machine. In a test, a vertically upward force of 8250 lb and a 7520 lb at 35.0° below the horizontal are applied to a shearing pin. What is the resultant force?

53. A magnetic force of 0.15 N is applied at an angle of 22.5° above the horizontal on an iron bar. A second magnetic force of 0.20 N is applied from the opposite side at an angle of 15.0° above the horizontal. What is the upward force on the bar?

54. The three smallest countries in Europe are Vatican City (0.4 km²), Monaco (1.9 km²), and San Marino (61 km²). San Marino is 220 km due north of Vatican City, and Monaco is 460 km at 29° north of west of Vatican City. How far is Monaco from San Marino?

55. A water molecule (H_2O) consists of two hydrogen atoms and one oxygen atom. The distance from the nucleus of each hydrogen atom to the nucleus of the oxygen atom is 0.96 pm (see Appendix B.2) and the bond angle (see Fig. 9.87) is 105°. How far is one hydrogen nucleus from the other?

Fig. 9.87

56. A crater on the moon is 150 mi in diameter. If the distance to the moon (to each side of the crater) from the earth is 240,000 mi, what angle is subtended by the crater at an observer's position on the earth?

57. In Fig. 9.88, a damper mechanism in an air-conditioning system is shown. If $\theta = 27.5°$ when the spring is at its shortest and longest lengths, what are these lengths?

Fig. 9.88

58. A bullet is fired from the ground of a level field at an angle of 39.0° above the horizontal. It travels in a straight line at 2200 ft/s for 0.20 s when it strikes a target. The sound of the strike is recorded 0.32 s later on the ground. If sound travels at 1130 ft/s, where is the recording device located?

59. In order to get around an obstruction, an oil pipeline is constructed in two straight sections, one 3.756 km long and the other 4.675 km long, with an angle of 168.85° between the sections where they are joined. How much more pipeline was necessary due to the obstruction?

60. Three pipes of radii 2.50 cm, 3.25 cm, and 4.25 cm are welded together lengthwise. See Fig. 9.89. Find the angles between the center-to-center lines.

Fig. 9.89   End view

61. Two satellites are being observed at the same observing station. One is 22,500 mi from the station, and the other is 18,700 mi away. The angle between their lines of observation is 105.4°. How far apart are the satellites?

62. Find the side x in the truss in Fig. 9.90.

Fig. 9.90

63. The angle of depression of a fire noticed west of a fire tower is 6.2°. The angle of depression of a pond, also west of the tower, is 13.5°. If the fire and pond are at the same altitude, and the tower is 2.25 km from the pond on a direct line, how far is the fire from the pond?

64. A surveyor wishes to find the distance between two points between which there is a security-restricted area. The surveyor measures the distance from each of these points to a third point and finds them to be 226.73 m and 185.12 m. If the angle between the lines of sight from the third point to the other points is 126.724°, how far apart are the two points?

65. Atlanta is 290 mi and 51.0° south of east from Nashville. The pilot of an airplane due north of Atlanta radios Nashville and finds the plane is on a line 10.5° south of east from Nashville. How far is the plane from Nashville?

66. In going around a storm, a plane flies 125 mi south, then 140 mi at 30.0° south of west, and finally 225 mi at 15.0° north of west. What is the displacement of the plane from its original position?

67. One end of a 1450-ft bridge is sighted from a distance of 3250 ft. The angle between the lines of sight of the ends of the bridge is 25.2°. From these data, how far is the observer from the other end of the bridge?

68. A plane is traveling horizontally at 1250 ft/s. A missile is fired horizontally from it 30.0° from the direction in which the plane is traveling. If the missile leaves the plane at 2040 ft/s, what is its velocity 10.0 s later if the vertical component is given by $v_V = -32.0t$ (in ft/s)?

69. A sailboat is headed due north, and its sail is set perpendicular to the wind, which is from the south of west. The component of the force of the wind in the direction of the heading is 480 N, and the component perpendicular to the heading (the *drift* component) is 650 N. What is the force exerted by the wind, and what is the direction of the wind? See Fig. 9.91.

Fig. 9.91

70. Boston is 650 km and 21.0° south of west from Halifax, Nova Scotia. Radio signals locate a ship 10.5° east of south from Halifax and 5.6° north of east from Boston. How far is the ship from each city?

Writing Exercise

71. The resultant of three horizontal forces, 45 N, 35 N, and 25 N, that act on a bolt is zero. Write a paragraph or two explaining how to find the angles between the forces.

CHAPTER 9   PRACTICE TEST

In all triangle solutions, sides a, b, c, are opposite angles A, B, C, respectively.

1. By use of a diagram, find the vector sum $2\mathbf{A} + \mathbf{B}$ for the given vectors.

2. For the triangle in which $a = 22.5$, $B = 78.6°$, and $c = 30.9$, find b.

3. A surveyor locates a tree 36.50 m to the northeast of a set position. The tree is 21.38 m north of a utility pole. What is the displacement of the utility pole from the set position?

4. For the triangle in which $A = 18.9°$, $B = 104.2°$, and $a = 426$, find c.

5. Solve the triangle in which $a = 9.84$, $b = 3.29$, and $c = 8.44$.

6. For vector $\mathbf{R}$, find R and standard position θ if $R_x = -235$ and $R_y = 152$.

7. Find the horizontal and vertical components of a vector of magnitude 871 that is directed at a standard-position angle of 284.3°.

8. A ship leaves a port and travels due west. At a certain point it turns 31.5° north of west and travels an additional 42.0 mi to a point 63.0 mi on a direct line from the port. How far from the port is the point where the ship turned?

9. Find the sum of the vectors for which $A = 449$, $\theta_A = 74.2°$, $B = 285$, and $\theta_B = 208.9°$. Use the trigonometric functions and the Pythagorean theorem.

10. Solve the triangle for which $a = 22.3$, $b = 29.6$, and $A = 36.5°$.

10

Graphs of the Trigonometric Functions

The electronics era is thought by many to have started in the 1880s with the discovery of the vacuum tube by the American inventor Thomas Edison and the discovery of radio waves by the German physicist Heinrich Hertz. Then in the 1890s, the cathode-ray tube was developed and, as an oscilloscope, has been used since that time to analyze various types of wave forms, such as sound waves and radio waves. Since the mid-1900s, devices similar to a cathode-ray tube have been used in TV picture tubes and computer displays.

What is seen on the screen of an oscilloscope are electric signals that are represented by graphs of trigonometric functions. As noted earlier, the basic method of graphing was developed in the mid-1600s, and using trigonometric functions of numbers has been common since the mid-1700s. Therefore, the graphs of the trigonometric functions were well known in the late 1800s and became very useful in the development of electronics.

The graphs of the trigonometric functions are useful in many areas of application, particularly those that involve wave motion and periodic values. Filtering electronic signals in communications, mixing musical sounds on a tape in a recording studio, studying the seasonal temperatures of an area, and analyzing ocean waves and tides illustrate some of the many applications of periodic motion.

As well as their use in applications, the graphs of the trigonometric functions give us one of the clearest ways of showing the properties of the various functions. Therefore, in this chapter, we show graphs of these functions, with emphasis on the sine and cosine functions.

In Section 10.6, we show the resulting curve when an oscilloscope is used to combine and display electric signals.

10.1 Graphs of $y = a \sin x$ and $y = a \cos x$

Graphs of $y = \sin x$ and $y = \cos x$ •
Amplitude • Graphs of
$y = a \sin x$ and $y = a \cos x$

When plotting and sketching the graphs of the trigonometric functions, it is normal to *express the angle in radians.* By using radians, both the independent variable and the dependent variable are real numbers. Therefore,

NOTE ▶ *it is necessary to be able to readily use angles expressed in radians.*

If necessary, review Section 8.3 on radian measure of angles.

In this section, the graphs of the sine and cosine functions are shown. We begin by making a table of values of x and y for the function $y = \sin x$, where x and y are used in the standard way as the *independent variable* and *dependent variable.* We plot the points to obtain the graph in Fig. 10.1.

Fig. 10.1

x	0	$\frac{\pi}{6}$	$\frac{\pi}{3}$	$\frac{\pi}{2}$	$\frac{2\pi}{3}$	$\frac{5\pi}{6}$	π
y	0	0.5	0.87	1	0.87	0.5	0

x	$\frac{7\pi}{6}$	$\frac{4\pi}{3}$	$\frac{3\pi}{2}$	$\frac{5\pi}{3}$	$\frac{11\pi}{6}$	2π
y	-0.5	-0.87	-1	-0.87	-0.5	0

The graph of $y = \cos x$ may be drawn in the same way. The next table gives values for plotting the graph of $y = \cos x$, and the graph is shown in Fig. 10.2.

Fig. 10.2

x	0	$\frac{\pi}{6}$	$\frac{\pi}{3}$	$\frac{\pi}{2}$	$\frac{2\pi}{3}$	$\frac{5\pi}{6}$	π
y	1	0.87	0.5	0	-0.5	-0.87	-1

x	$\frac{7\pi}{6}$	$\frac{4\pi}{3}$	$\frac{3\pi}{2}$	$\frac{5\pi}{3}$	$\frac{11\pi}{6}$	2π
y	-0.87	-0.5	0	0.5	0.87	1

The graphs are continued beyond the values shown in the tables to indicate that *they continue indefinitely in each direction.* To show this more clearly, in Figs. 10.3 and 10.4, note the graphs of $y = \sin x$ and $y = \cos x$ from $x = -10$ to $x = 10$.

Fig. 10.3

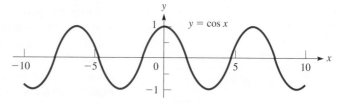

Fig. 10.4

From these tables and graphs, it can be seen that *the graphs of $y = \sin x$ and $y = \cos x$ are of exactly the same shape (called* **sinusoidal**), *with the cosine curve displacement $\pi/2$ units to the left of the sine curve.* The shape of these curves should be recognized readily, with special note as to the points at which they cross the axes. This information will be especially valuable in *sketching* similar curves, since the basic sinusoidal shape remains the same. It will not be necessary to plot numerous points every time we wish to sketch such a curve.

To obtain the graph of $y = a \sin x$, note that all the y-values obtained for the graph of $y = \sin x$ are to be multiplied by the number a. In this case, the greatest value of the sine function is $|a|$. *The number $|a|$ is called the* **amplitude** *of the curve and represents the greatest y-value of the curve. Also, the curve will have no value less than* $-|a|$. This is true for $y = a \cos x$ as well as $y = a \sin x$.

Amplitude

EXAMPLE 1 Plot graph of $y = a \sin x$

Plot the graph of $y = 2 \sin x$.

Since $a = 2$, the amplitude of this curve is $|2| = 2$. This means that the maximum value of y is 2 and the minimum value is $y = -2$. The table of values follows, and the curve is shown in Fig. 10.5.

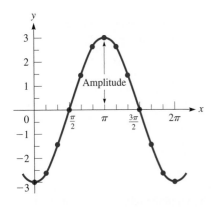

Fig. 10.5

x	0	$\frac{\pi}{6}$	$\frac{\pi}{3}$	$\frac{\pi}{2}$	$\frac{2\pi}{3}$	$\frac{5\pi}{6}$	π
y	0	1	1.73	2	1.73	1	0

x	$\frac{7\pi}{6}$	$\frac{4\pi}{3}$	$\frac{3\pi}{2}$	$\frac{5\pi}{3}$	$\frac{11\pi}{6}$	2π
y	-1	-1.73	-2	-1.73	-1	0

EXAMPLE 2 Plot graph of $y = a \cos x$

Plot the graph of $y = -3 \cos x$.

In this case, $a = -3$, and this means that the amplitude is $|-3| = 3$. Therefore, the maximum value of y is 3, and the minimum value of y is -3. The table of values follows, and the curve is shown in Fig. 10.6.

Fig. 10.6

x	0	$\frac{\pi}{6}$	$\frac{\pi}{3}$	$\frac{\pi}{2}$	$\frac{2\pi}{3}$	$\frac{5\pi}{6}$	π
y	-3	-2.6	-1.5	0	1.5	2.6	3

x	$\frac{7\pi}{6}$	$\frac{4\pi}{3}$	$\frac{3\pi}{2}$	$\frac{5\pi}{3}$	$\frac{11\pi}{6}$	2π
y	2.6	1.5	0	-1.5	-2.6	-3

Note from Example 2 that *the effect of the negative sign with the number a is to* **invert** *the curve about the x-axis.* The effect of the number a can also be seen readily from these examples.

From the previous examples, note that the function $y = a \sin x$ has zeros for $x = 0, \pi, 2\pi$ and that it has its maximum or minimum values for $x = \pi/2, 3\pi/2$. The function $y = a \cos x$ has its zeros for $x = \pi/2, 3\pi/2$ and its maximum or minimum values for $x = 0, \pi, 2\pi$. This is summarized in Table 10.1. Therefore, by knowing the general shape of the sine curve, where it has its zeros, and what its amplitude is, *we can rapidly* **sketch** *curves of the form* $y = a \sin x$ *and* $y = a \cos x$.

Since the graphs of $y = a \sin x$ and $y = a \cos x$ can extend indefinitely to the right and to the left, we see that the domain of each is all real numbers. We should note that the key values of $x = 0, \pi/2, \pi, 3\pi/2$, and 2π are those only for x from 0 to 2π. Corresponding values ($x = 5\pi/2, 3\pi$, and their negatives) could also be used. Also, from the graphs, we can readily see that the range of these functions is $-|a| \le f(x) \le |a|$.

Table 10.1

	$x = 0, \pi, 2\pi$	$\frac{\pi}{2}, \frac{3\pi}{2}$
$y = a \sin x$	zeros	max. or min.
$y = a \cos x$	max. or min.	zeros

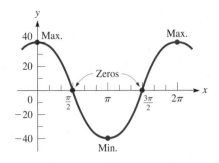

Fig. 10.7

Practice Exercise

1. For the graph of $y = -6 \sin x$, set up a table of key values for $0 \le x \le 2\pi$.

EXAMPLE 3 Using key values to sketch graph

Sketch the graph of $y = 40 \cos x$.

First, we set up a table of values for the points where the curve has its zeros, maximum points, and minimum points:

x	0	$\frac{\pi}{2}$	π	$\frac{3\pi}{2}$	2π
y	40 max.	0	−40 min.	0	40 max.

Now, we plot these points and join them, knowing the basic sinusoidal shape of the curve. See Fig. 10.7. ∎

The graphs of $y = a \sin x$ and $y = a \cos x$ can be displayed easily on a graphing calculator. In the next example, we see that the calculator displays the features of the curve we should expect from our previous discussion.

EXAMPLE 4 Displaying graphs on calculator

Display the graph of $y = -2 \sin x$ on a graphing calculator.

Using radian mode on the calculator, Fig. 10.8(a) shows the calculator view for the key values in the following table:

x	0	$\frac{\pi}{2}$	π	$\frac{3\pi}{2}$	2π
y	0	−2 min.	0	2 max.	0

We see the amplitude of 2 and the effect of the negative sign in inverting the curve, as expected. Fig. 10.8(b) shows the calculator graph for $x = -10$ to $x = 10$. ∎

(a)

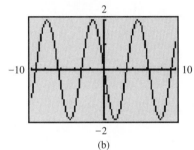
(b)

Fig. 10.8

EXERCISES 10.1

In Exercises 1 and 2, graph the function if the given changes are made in the indicated examples of this section.

1. In Example 2, if the sign of the coefficient of $\cos x$ is changed, plot the graph of the resulting function.

2. In Example 4, if the sign of the coefficient of $\sin x$ is changed, display the graph of the resulting function.

In Exercises 3–6, complete the following table for the given functions and then plot the resulting graphs.

x	$-\pi$	$-\frac{3\pi}{4}$	$-\frac{\pi}{2}$	$-\frac{\pi}{4}$	0	$\frac{\pi}{4}$	$\frac{\pi}{2}$	$\frac{3\pi}{4}$	π
y									

x	$\frac{5\pi}{4}$	$\frac{3\pi}{2}$	$\frac{7\pi}{4}$	2π	$\frac{9\pi}{4}$	$\frac{5\pi}{2}$	$\frac{11\pi}{4}$	3π
y								

3. $y = \sin x$
4. $y = \cos x$
5. $y = 3 \cos x$
6. $y = -4 \sin x$

In Exercises 7–22, sketch the graphs of the given functions. Check each using a graphing calculator.

7. $y = 3 \sin x$
8. $y = 5 \sin x$
9. $y = \frac{5}{2} \sin x$
10. $y = 35 \sin x$
11. $y = 200 \cos x$
12. $y = 0.25 \cos x$
13. $y = 0.8 \cos x$
14. $y = \frac{3}{2} \cos x$
15. $y = -\sin x$
16. $y = -300 \sin x$
17. $y = -1500 \sin x$
18. $y = -0.2 \sin x$
19. $y = -\cos x$
20. $y = -8 \cos x$
21. $y = -50 \cos x$
22. $y = -0.4 \cos x$

Although units of π are convenient, we must remember that π is only a number. Numbers that are not multiples of π may be used. In Exercises 23–26, plot the indicated graphs by finding the values of y that correspond to values of x of 0, 1, 2, 3, 4, 5, 6, and 7 on a calculator. (Remember, the numbers 0, 1, 2, and so on represent radian measure.)

23. $y = \sin x$
24. $y = -30 \sin x$
25. $y = 12 \cos x$
26. $y = 2 \cos x$

In Exercises 27–32, solve the given problems.

27. Find the function and graph it for a function of the form $y = a \sin x$ that passes through $(\pi/2, -2)$.

28. Find the function and graph it for a function of the form $y = a \sin x$ that passes through $(3\pi/2, -2)$.

29. Find the function and graph it for a function of the form $y = a \cos x$ that passes through $(\pi, 2)$.

30. Find the function and graph it for a function of the form $y = a \cos x$ that passes through $(2\pi, -2)$.

31. The graph displayed on an oscilloscope can be represented by $y = -0.05 \sin x$. Display this curve on a graphing calculator.

32. The displacement y (in cm) of the end of a robot arm for welding is $y = 4.75 \cos t$, where t is the time (in s). Display this curve on a graphing calculator.

In Exercises 33–36, the graph of a function of the form $y = a \sin x$ or $y = a \cos x$ is shown. Determine the specific function of each.

33.

34.

35.

36.

In Exercises 37–40, use the trace feature of a graphing calculator to find the value of a for either $y = a \sin x$ or $y = a \cos x$, whichever is correct, such that the given point is on the graph. The amplitude of each function is 2.50. Thereby determine the function. (All points are located such that the x value is between $-\pi$ and π.)

37. $(0.67, -1.55)$

38. $(-1.20, 0.90)$

39. $(2.07, 1.20)$

40. $(-2.47, -1.55)$

Answers to Practice Exercise

1.

x	0	$\frac{\pi}{2}$	π	$\frac{3\pi}{2}$	2π
y	0	-6	0	6	0
		min.		max.	

10.2 Graphs of $y = a \sin bx$ and $y = a \cos bx$

Period of a Function •
Graphs of $y = a \sin bx$ and $y = a \cos bx$ •
Important Values for Sketching

Fig. 10.9

In graphing the function $y = \sin x$, we see that the values of y repeat every 2π units of x. This is because $\sin x = \sin(x + 2\pi) = \sin(x + 4\pi)$, and so forth. For any function F, we say that it has a *period P* if $F(x) = F(x + P)$. For functions that are periodic, such as the sine and cosine, *the **period** is the x-distance between a point and the next corresponding point for which the value of y repeats.*

Let us now plot the curve $y = \sin 2x$. This means that we choose a value of x, multiply this value by 2, and find the sine of the result. This leads to the following table of values for this function:

x	0	$\frac{\pi}{8}$	$\frac{\pi}{4}$	$\frac{3\pi}{8}$	$\frac{\pi}{2}$	$\frac{5\pi}{8}$	$\frac{3\pi}{4}$	$\frac{7\pi}{8}$	π	$\frac{9\pi}{8}$	$\frac{5\pi}{4}$
$2x$	0	$\frac{\pi}{4}$	$\frac{\pi}{2}$	$\frac{3\pi}{4}$	π	$\frac{5\pi}{4}$	$\frac{3\pi}{2}$	$\frac{7\pi}{4}$	2π	$\frac{9\pi}{4}$	$\frac{5\pi}{2}$
y	0	0.7	1	0.7	0	-0.7	-1	-0.7	0	0.7	1

Plotting these points, we have the curve shown in Fig. 10.9.

From the table and Fig. 10.9, note that $y = \sin 2x$ repeats after π units of x. The effect of the 2 is that the period of $y = \sin 2x$ is half the period of the curve of $y = \sin x$. We then conclude that if the period of a function $F(x)$ is P, then the period of $F(bx)$ is P/b. Since each of the functions $\sin x$ and $\cos x$ has a period of 2π, *each of the functions $\sin bx$ and $\cos bx$ has a period of $2\pi/b$.*

NOTE ▶

EXAMPLE 1 Finding period of a function

(a) The period of $\cos 4x$ is $\frac{2\pi}{4} = \frac{\pi}{2}$.   **(b)** The period of $\sin 3\pi x$ is $\frac{2\pi}{3\pi} = \frac{2}{3}$.

(c) The period of $\sin \frac{1}{2}x$ is $\dfrac{2\pi}{\frac{1}{2}} = 4\pi$.   **(d)** The period of $\cos \frac{\pi}{4}x$ is $\dfrac{2\pi}{\frac{\pi}{4}} = 8$.

Practice Exercises

Find the period of each function.

1. $y = \sin \pi x$ **2.** $y = \cos \frac{1}{3}x$

In (a), the period tells us that the curve of $y = \cos 4x$ will repeat every $\pi/2$ (approximately 1.57) units of x. In (b), we see that the curve of $y = \sin 3\pi x$ will repeat every $2/3$ of a unit. In (c) and (d), the periods are longer than those of $y = \sin x$ and $y = \cos x$. ∎

Combining the value of the period with the value of the amplitude from Section 10.1, we conclude that *the functions $y = a \sin bx$ and $y = a \cos bx$ each has an amplitude of $|a|$ and a period of $2\pi/b$.* These properties are very useful in sketching these functions.

EXAMPLE 2 Sketching graph of $y = a \sin bx$

Sketch the graph of $y = 3 \sin 4x$ for $0 \le x \le 2\pi$.

Since $a = 3$, the amplitude is 3. The $4x$ tells us that the period is $2\pi/4 = \pi/2$. This means that $y = 0$ for $x = 0$ and for $y = \pi/2$. Since this sine function is zero halfway between $x = 0$ and $x = \pi/2$, we find that $y = 0$ for $x = \pi/4$. Also, the fact that the graph of the sine function reaches its maximum and minimum values halfway between zeros means that $y = 3$ for $x = \pi/8$, and $y = -3$ for $x = 3\pi/8$. Note that the values of x in the following table are those for which $4x = 0$, $\pi/2$, π, $3\pi/2$, 2π, and so on.

Fig. 10.10

x	0	$\frac{\pi}{8}$	$\frac{\pi}{4}$	$\frac{3\pi}{8}$	$\frac{\pi}{2}$	$\frac{5\pi}{8}$	$\frac{3\pi}{4}$	$\frac{7\pi}{8}$	π
y	0	3	0	-3	0	3	0	-3	0

Using the values from the table and the fact that the curve is sinusoidal in form, we sketch the graph of this function in Fig. 10.10. We see again that knowing the key values and the basic shape of the curve allows us to *sketch* the graph of the curve quickly and easily. ∎

Note from Example 2 that *an important distance in sketching a sine curve or a cosine curve is one-fourth of the period.* For $y = a \sin bx$, it is one-fourth of the period from the origin to the first value of x where y is at its maximum (or minimum) value. Then we proceed another one-fourth period to a zero, another one-fourth period to the next minimum (or maximum) value, another to the next zero (this is where the period is completed), and so on.

Similarly, one-fourth of the period is useful in sketching the graph of $y = \cos bx$. For this function, the maximum (or minimum) value occurs for $x = 0$. At the following one-fourth-period values, there is a zero, a minimum (or maximum), a zero, and a maximum (or minimum) at the start of the next period. Therefore,

NOTE ▶   *by finding one-fourth of the period, we can easily find the important values for sketching the curve*

We now summarize the important values for sketching the graphs of $y = a \sin bx$ and $y = a \cos bx$.

IMPORTANT VALUES FOR SKETCHING $y = a \sin bx$ AND $y = a \cos bx$

1. *The amplitude:* $|a|$

2. *The period:* $2\pi/b$

3. *Values of the function for each one-fourth period*

EXAMPLE 3 Using important values to sketch graph

Sketch the graph of $y = -2 \cos 3x$ for $0 \le x \le 2\pi$.

Note that the amplitude is 2 and the period is $\frac{2\pi}{3}$. This means that one-fourth of the period is $\frac{1}{4} \times \frac{2\pi}{3} = \frac{\pi}{6}$. Since the cosine curve is at a maximum or minimum for $x = 0$, we find that $y = -2$ for $x = 0$ (the negative value is due to the minus sign before the function), which means it is a minimum point. The curve then has a zero at $x = \frac{\pi}{6}$, a maximum value of 2 at $x = 2\left(\frac{\pi}{6}\right) = \frac{\pi}{3}$, a zero at $x = 3\left(\frac{\pi}{6}\right) = \frac{\pi}{2}$, and its next value of -2 at $x = 4\left(\frac{\pi}{6}\right) = \frac{2\pi}{3}$, and so on. Therefore, we have the following table:

x	0	$\frac{\pi}{6}$	$\frac{\pi}{3}$	$\frac{\pi}{2}$	$\frac{2\pi}{3}$	$\frac{5\pi}{6}$	π	$\frac{7\pi}{6}$	$\frac{4\pi}{3}$	$\frac{3\pi}{2}$	$\frac{5\pi}{3}$	$\frac{11\pi}{6}$	2π
y	-2	0	2	0	-2	0	2	0	-2	0	2	0	-2

Using this table and the sinusoidal shape of the cosine curve, we sketch the function in Fig. 10.11. ∎

Fig. 10.11

EXAMPLE 4 Graph on calculator – application

A generator produces a voltage $V = 200 \cos 50\pi t$, where t is the time in seconds (50π has units of rad/s; thus, $50\pi t$ is an angle in radians). Use a graphing calculator to display the graph of V as a function of t for $0 \le t \le 0.06$ s.

The amplitude is 200 V and the period is $2\pi/(50\pi) = 0.04$ s. Since the period is not in terms of π, it is more convenient to use decimal units for t rather than to use units in terms of π as in the previous graphs. Thus, we have the following table of values:

$t(s)$	0	0.01	0.02	0.03	0.04	0.05	0.06
$V(V)$	200	0	-200	0	200	0	-200

For the graphing calculator, use x for t and y for V. This means we graph the function $y_1 = 200 \cos 50\pi x$, as shown in Fig. 10.12. From the amplitude of 200 V and the above table, we choose the *window* values shown in Fig. 10.12. We do not consider negative values of t, for they have no real meaning in this problem. ∎

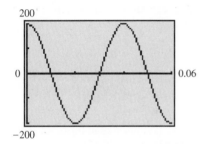

Fig. 10.12

EXERCISES 10.2

In Exercises 1 and 2, graph the function if the given changes are made in the indicated examples of this section.

1. In Example 2, if the coefficient of x is changed from 4 to 6, sketch the graph of the resulting function.

2. In Example 3, if the coefficient of x is changed from 3 to 4, sketch the graph of the resulting function.

In Exercises 3–22, find the period of each function.

3. $y = 2 \sin 6x$

4. $y = 4 \sin 2x$

5. $y = 3 \cos 8x$

6. $y = 28 \cos 10x$

7. $y = -2 \sin 12x$

8. $y = -\frac{1}{5} \sin 5x$

9. $y = -\cos 16x$

10. $y = -4 \cos 2x$

11. $y = 520 \sin 2\pi x$

12. $y = 2 \sin 3\pi x$

13. $y = 3 \cos 4\pi x$

14. $y = 4 \cos 10\pi x$

15. $y = 15 \sin \frac{1}{3} x$

16. $y = -25 \sin \frac{2}{5} x$

17. $y = -\frac{1}{2} \cos \frac{2}{3} x$

18. $y = \frac{1}{3} \cos \frac{1}{4} x$

19. $y = 0.4 \sin \dfrac{2\pi x}{3}$

20. $y = 1.5 \cos \dfrac{\pi x}{10}$

21. $y = 3.3 \cos \pi^2 x$

22. $y = -12.5 \sin \frac{2x}{\pi}$

In Exercises 23–42, sketch the graphs of the given functions. Check each using a graphing calculator. (These are the same functions as in Exercises 3–22.)

23. $y = 2 \sin 6x$

24. $y = 4 \sin 2x$

25. $y = 3 \cos 8x$

26. $y = 28 \cos 10x$

27. $y = -2 \sin 12x$

28. $y = -\frac{1}{5} \sin 5x$

29. $y = -\cos 16x$

30. $y = -4 \cos 2x$

31. $y = 520 \sin 2\pi x$

32. $y = 2 \sin 3\pi x$

33. $y = 3 \cos 4\pi x$

34. $y = 4 \cos 10\pi x$

35. $y = 15 \sin \frac{1}{3}x$

36. $y = -25 \sin \frac{2}{5}x$

37. $y = -\frac{1}{2}\cos \frac{2}{3}x$

38. $y = \frac{1}{3}\cos \frac{1}{4}x$

39. $y = 0.4 \sin \dfrac{2\pi x}{3}$

40. $y = 1.5 \cos \dfrac{\pi x}{10}$

41. $y = 3.3 \cos \pi^2 x$

42. $y = -12.5 \sin \frac{2x}{\pi}$

In Exercises 43–46, the period is given for a function of the form $y = \sin bx$. Write the function corresponding to the given period.

43. $\dfrac{\pi}{3}$ **44.** $\dfrac{2\pi}{5}$ **45.** $\dfrac{1}{3}$ **46.** 6

In Exercises 47–50, graph the given functions. In Exercises 47 and 48, first rewrite the function with a positive angle, and then graph the resulting function.

47. $y = 3 \sin(-2x)$

48. $y = -5 \cos(-4\pi x)$

49. $y = 8\left|\cos\left(\frac{\pi}{2}x\right)\right|$

50. $y = 0.4 \left|\sin 6x\right|$

In Exercises 51–60, solve the given problems.

51. By noting the periods of $\sin 2x$ and $\sin 3x$, find the minimum period of the function $y = \sin 2x + \sin 3x$.

52. By noting the period of $\cos \frac{1}{2}x$ and $\cos \frac{1}{3}x$, find the minimum period of the function $y = \cos \frac{1}{2}x + \cos \frac{1}{3}x$.

53. Find the function and graph it for a function of the form $y = -2 \sin bx$ that passes through $(\pi/4, -2)$ and for which b has the smallest possible positive value.

54. Find the function and graph it for a function of the form $y = 2 \sin bx$ that passes through $(\pi/6, 2)$ and for which b has the smallest possible positive value.

55. Find the function and graph it for a function of the form $y = 2 \cos bx$ that passes through $(\pi, 0)$ and for which b has the smallest possible positive value.

56. Find the function and graph it for a function of the form $y = -2 \cos bx$ that passes through $(\pi/2, 2)$ and for which b has the smallest possible positive value.

57. The standard electric voltage in a 60-Hz alternating-current circuit is given by $V = 170 \sin 120\pi t$, where t is the time in seconds. Sketch the graph of V as a function of t for $0 \le t \le 0.05$ s.

58. To tune the instruments of an orchestra before a concert, an A note is struck on a piano. The piano wire vibrates with a displacement y (in mm) given by $y = 3.20 \cos 880\pi t$, where t is in seconds. Sketch the graph of y vs. t for $0 \le t \le 0.01$ s.

59. The velocity v (in in./s) of a piston is $v = 450 \cos 3600t$, where t is in seconds. Sketch the graph of v vs. t for $0 \le t \le 0.006$ s.

60. On a Florida beach, the tides have water levels about 4 m between low and high tides. The period is about 12.5 h. Find a cosine function that describes these tides if high tide is at midnight of a given day. Sketch the graph.

In Exercises 61–64, the graph of a function of the form $y = a \sin bx$ or $y = a \cos bx$ is shown. Determine the specific function of each.

61.

62.

63.

64.

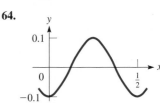

Answers to Practice Exercises

1. 2   **2.** 6π

10.3 Graphs of $y = a \sin(bx + c)$ and $y = a \cos(bx + c)$

Phase Angle • Displacement • Graphs of $y = a \sin(bx + c)$ and $y = a \cos(bx + c)$ • Cycle

In the function $y = a \sin(bx + c)$, c represents the **phase angle**. It is another very important quantity in graphing the sine and cosine functions. Its meaning is illustrated in the following example.

EXAMPLE 1 Sketch function with phase angle

Sketch the graph of $y = \sin\left(2x + \frac{\pi}{4}\right)$.

Here, $c = \pi/4$. Therefore, in order to obtain values for the table, we assume a value for x, multiply it by 2, add $\pi/4$ to this value, and then find the sine of the result. The values shown are those for which $2x + \pi/4 = 0, \pi/4, \pi, 2, 3\pi/4, \pi$, and so on, which are the important values for $y = \sin 2x$.

x	$-\frac{\pi}{8}$	0	$\frac{\pi}{8}$	$\frac{\pi}{4}$	$\frac{3\pi}{8}$	$\frac{\pi}{2}$	$\frac{5\pi}{8}$	$\frac{3\pi}{4}$	$\frac{7\pi}{8}$	π
y	0	0.7	1	0.7	0	-0.7	-1	-0.7	0	0.7

Solving $2x + \pi/4 = 0$, we get $x = -\pi/8$, which gives $y = \sin 0 = 0$. The other values for y are found in the same way. See Fig. 10.13.   ∎

Fig. 10.13

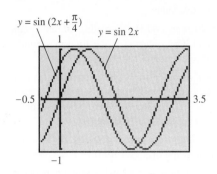

Fig. 10.14

■ See Appendix C for a graphing calculator program SINECURV. It displays the graphs of $y = \sin x$, $y = 2 \sin x$, $y = \sin 2x$, and $y = 2 \sin(2x - \pi/3)$.

■ Carefully note the difference between $y = \sin(bx + c)$ and $y = \sin bx + c$. Writing $\sin(bx + c)$ means to find the sine of the quantity of $bx + c$, whereas $\sin bx + c$ means to find the sine of bx and then add the value c.

Note from Example 1 that *the graph of $y = \sin\left(2x + \frac{\pi}{4}\right)$ is precisely the same as the graph of $y = \sin 2x$, except that it is **shifted** $\pi/8$ units to the left*. In Fig. 10.14, a graphing calculator view shows the graphs of $y = \sin 2x$ and $y = \sin\left(2x + \frac{\pi}{4}\right)$. We see that the shapes are the same and that the graph of $y = \sin\left(2x + \frac{\pi}{4}\right)$ is about 0.4 unit ($\pi/8 \approx 0.39$) to the left of the graph of $y = \sin 2x$.

In general, the effect of c in the equation $y = a \sin(bx + c)$ is to shift the curve of $y = a \sin bx$ to the left if $c > 0$, or shift the curve to the right if $c < 0$. The amount of this shift is given by $-c/b$. Due to its importance in sketching curves, *the quantity $-c/b$ is called the **displacement** (or **phase shift**)*.

We can see the reason that the displacement is $-c/b$ by noting corresponding points on the graphs of $y = \sin bx$ and $y = \sin(bx + c)$. For $y = \sin bx$, when $x = 0$, then $y = 0$. For $y = \sin(bx + c)$, when $x = -c/b$, then $y = 0$. The point $(-c/b, 0)$ on the graph of $y = \sin(bx + c)$ is $-c/b$ units to the left of the point $(0,0)$ on the graph of $y = \sin x$. In Fig. 10.14, $-c/b = -\pi/8$.

Therefore, we use the displacement combined with the amplitude and the period along with the other information from the previous sections to sketch curves of the functions $y = a \sin(bx + c)$ and $y = a \cos(bx + c)$, where $b > 0$.

IMPORTANT QUANTITIES TO DETERMINE FOR SKETCHING GRAPHS OF
$y = a \sin(bx + c)$ **AND** $y = a \cos(bx + c)$

Amplitude $= |a|$

Period $= \dfrac{2\pi}{b}$

Displacement $= -\dfrac{c}{b}$

(10.1)

By use of these quantities and the one-fourth period distance, the graphs of the sine and cosine functions can be readily sketched.

A general illustration of the graph of $y = a \sin(bx + c)$ is shown in Fig. 10.15. Note carefully that

CAUTION ▶

*the displacement is **negative** (to the left) for $c > 0$*   (Fig. 10.15(a))
*the displacement is **positive** (to the right) for $c < 0$*   (Fig. 10.15(b))

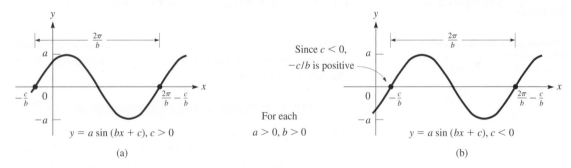

Fig. 10.15

NOTE ▶

Note also that we can find the displacement for the graphs of $y = a \sin(bx + c)$ by solving $bx + c = 0$ for x. We see that $x = -c/b$. This is the same as the horizontal shift for graphs of functions that is shown on page 99.

EXAMPLE 2 Sketching graph of $y = a \sin (bx + c)$

Sketch the graph of $y = 2 \sin(3x - \pi)$.

First, note that $a = 2$, $b = 3$, and $c = -\pi$. Therefore, the amplitude is 2, the period is $2\pi/3$, and the displacement is $-(-\pi/3) = \pi/3$. (We can also get the displacement from $3x - \pi = 0, x = \pi/3$.)

Note that the curve "starts" at $x = \pi/3$ and starts repeating $2\pi/3$ units to the right of this point. Be sure to grasp this point well. *The period tells us the number of units along the x-axis between such corresponding points.* One-fourth of the period is $\frac{1}{4}\left(\frac{2\pi}{3}\right) = \frac{\pi}{6}$.

Important values are at $\frac{\pi}{3}, \frac{\pi}{3} + \frac{\pi}{6} = \frac{\pi}{2}, \frac{\pi}{3} + 2\left(\frac{\pi}{6}\right) = \frac{2\pi}{3}$, and so on. We now make the table of important values and sketch the graph shown in Fig. 10.16.

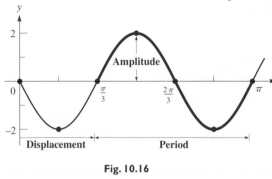

Fig. 10.16

x	0	$\frac{\pi}{6}$	$\frac{\pi}{3}$	$\frac{\pi}{2}$	$\frac{2\pi}{3}$	$\frac{5\pi}{6}$	π	$\leftarrow \frac{1}{4}$(period) $= \frac{\pi}{6}$
y	0	-2	0	2	0	-2	0	

Note that since the period is $2\pi/3$, the curve passes through the origin. ∎

EXAMPLE 3 Sketching graph of $y = a \cos (bx + c)$

Sketch the graph of the function $y = -\cos\left(2x + \frac{\pi}{6}\right)$.

First, we determine that

1. the amplitude is 1
2. the period is $\frac{2\pi}{2} = \pi$
3. the displacement is $-\frac{\pi}{6} \div 2 = -\frac{\pi}{12}$

We now make a table of important values, noting that the curve starts repeating π units to the right of $-\frac{\pi}{12}$.

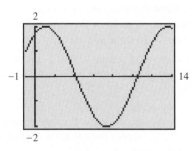

Fig. 10.17

x	$-\frac{\pi}{12}$	$\frac{\pi}{6}$	$\frac{5\pi}{12}$	$\frac{2\pi}{3}$	$\frac{11\pi}{12}$	$\leftarrow \frac{1}{4}$(period) $= \frac{\pi}{4}$
y	-1	0	1	0	-1	

From this table, we sketch the graph in Fig. 10.17. ∎

Each of the heavy portions of the graphs in Figs. 10.16 and 10.17 is called a *cycle* of the curve. A **cycle** *is any section of the graph that includes exactly one period.*

Practice Exercise

1. For the graph of $y = 8 \sin(2x - \pi/3)$, determine amplitude, period, and displacement.

EXAMPLE 4 Graph on calculator

View the graph of $y = 2 \cos\left(\frac{1}{2}x - \frac{\pi}{6}\right)$ on a graphing calculator.

From the values $a = 2$, $b = 1/2$, and $c = -\pi/6$, we determine that

1. the amplitude is 2
2. the period is $2\pi \div \frac{1}{2} = 4\pi$
3. the displacement is $-\left(-\frac{\pi}{6}\right) \div \frac{1}{2} = \frac{\pi}{3}$

We now make a table of important values:

x	$\frac{\pi}{3}$	$\frac{4\pi}{3}$	$\frac{7\pi}{3}$	$\frac{10\pi}{3}$	$\frac{13\pi}{3}$	$\leftarrow \frac{1}{4}$(period) $= \pi$
y	2	0	-2	0	2	

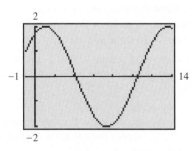

This table helps us choose the values for the *window* settings in Fig. 10.18. We choose Xmin $= -1$ in order to start to the left of the y-axis and Xmax $= 14$ since $13\pi/3 \approx 13.6$. Also, we choose Ymin $= -2$ and Ymax $= 2$ since the amplitude is 2. We see that the graph in Fig. 10.18 is a little more than one cycle. ∎

Fig. 10.18

EXAMPLE 5   Graph on calculator – application

The cross section of a certain water wave is $y = 0.7 \sin\left(\frac{\pi}{2}x + \frac{\pi}{4}\right)$, where x and y are measured in feet. Display two cycles of y vs. x on a graphing calculator.

From the values $a = 0.7$ ft, $b = \pi/2$ ft^{-1} (this means per foot), and $c = \pi/4$, we find the amplitude $= 0.7$ ft, period $= \frac{2\pi}{\pi/2} = 4$ ft, and displacement $= -\frac{\pi/4}{\pi/2} = -0.5$ ft. Using these values, we choose the following values for the *window* settings.

1. Xmin $= -0.5$ (the displacement is -0.5 ft)
2. Xmax $= 7.5$ (the period is 4 ft, and we want two periods, starting at $x = -0.5$)
3. Ymin $= -0.7$, Ymax $= 0.7$ (the amplitude is 0.7 ft)

The graphing calculator view is shown in Fig. 10.19. The negative values of x have the significance of giving points on the wave to the left of the origin. (When *time* is used, no actual physical meaning is generally given to negative values of t.) ∎

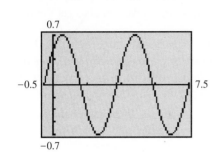

Fig. 10.19

EXERCISES 10.3

In Exercises 1 and 2, graph the function if the given changes are made in the indicated examples of this section.

1. In Example 3, if the sign before $\pi/6$ is changed, sketch the graph of the resulting function.

2. In Example 4, if the sign before $\pi/6$ is changed, sketch the graph of the resulting function.

In Exercises 3–26, determine the amplitude, period, and displacement for each function. Then sketch the graphs of the functions. Check each using a graphing calculator.

3. $y = \sin\left(x - \dfrac{\pi}{6}\right)$

4. $y = 3\sin\left(x + \dfrac{\pi}{4}\right)$

5. $y = \cos\left(x + \dfrac{\pi}{6}\right)$

6. $y = 2\cos\left(x - \dfrac{\pi}{8}\right)$

7. $y = 0.2\sin\left(2x + \dfrac{\pi}{2}\right)$

8. $y = -\sin\left(3x - \dfrac{\pi}{2}\right)$

9. $y = -\cos(2x - \pi)$

10. $y = 0.4\sin\left(3x + \dfrac{\pi}{3}\right)$

11. $y = \dfrac{1}{2}\sin\left(\dfrac{1}{2}x - \dfrac{\pi}{4}\right)$

12. $y = 2\sin\left(\dfrac{1}{4}x + \dfrac{\pi}{2}\right)$

13. $y = 30\cos\left(\dfrac{1}{3}x + \dfrac{\pi}{3}\right)$

14. $y = \dfrac{1}{3}\cos\left(\dfrac{1}{2}x - \dfrac{\pi}{8}\right)$

15. $y = \sin\left(\pi x + \dfrac{\pi}{8}\right)$

16. $y = -2\sin(2\pi x - \pi)$

17. $y = 0.08\cos\left(4\pi x - \dfrac{\pi}{5}\right)$

18. $y = 25\cos\left(3\pi x + \dfrac{\pi}{2}\right)$

19. $y = -0.6\sin(2\pi x - 1)$

20. $y = 1.8\sin\left(\pi x + \dfrac{1}{3}\right)$

21. $y = 40\cos(3\pi x + 2)$

22. $y = 360\cos(6\pi x - 1)$

23. $y = \sin(\pi^2 x - \pi)$

24. $y = -\dfrac{1}{2}\sin\left(2x - \dfrac{1}{\pi}\right)$

25. $y = -\dfrac{3}{2}\cos\left(\pi x + \dfrac{\pi^2}{6}\right)$

26. $y = \pi\cos\left(\dfrac{1}{\pi}x + \dfrac{1}{3}\right)$

In Exercises 27–30, write the equation for the given function with the given amplitude, period, and displacement, respectively.

27. sine, 4, 3π, $-\pi/4$

28. cosine, 8, $2\pi/3$, $\pi/3$

29. cosine, 12, $1/2$, $1/8$

30. sine, 18, 4, -1

In Exercises 31–34, show that the given equations are correct. The method using a graphing calculator is indicated in Exercise 31.

31. By viewing the graphs of $y_1 = \sin x$ and $y_2 = \cos(x - \pi/2)$, show that $\cos(x - \pi/2) = \sin x$.

32. Show that $\cos(2x - 3\pi/8) = \cos(3\pi/8 - 2x)$.

33. Show that $\sin(x/2 - 3\pi/4) = -\sin(3\pi/4 - x/2)$.

34. Show that $2|\sin(x - \pi/3)| = \sin(\pi/3 - x) - \sin(x - \pi/3)$ for $-2\pi/3 < x < \pi/3$.

In Exercises 35–40, solve the given problems. In Exercises 39 and 40, use a graphing calculator to view the indicated curves.

35. Find the function and graph it for a function of the form $y = 2\sin(2x + c)$ that passes through $(-\pi/8, 0)$ and for which c has the smallest possible positive value.

36. Find the function and graph it for a function of the form $y = 2\cos(2x - c)$ that passes through $(\pi/6, 2)$ and for which c has the smallest possible positive value.

37. A wave traveling in a string may be represented by the equation $y = A\sin 2\pi\left(\dfrac{t}{T} - \dfrac{x}{\lambda}\right)$. Here, A is the amplitude, t is the time the wave has traveled, x is the distance from the origin, T is the time required for the wave to travel one *wavelength* λ (the Greek letter lambda). Sketch three cycles of the wave for which $A = 2.00$ cm, $T = 0.100$ s, $\lambda = 20.0$ cm, and $x = 5.00$ cm.

38. The electric current i (in μA) in a certain circuit is given by $i = 3.8\cos 2\pi(t + 0.20)$, where t is the time in seconds. Sketch three cycles of this function.

39. A certain satellite circles the earth such that its distance y, in miles north or south (altitude is not considered) from the equator, is $y = 4500\cos(0.025t - 0.25)$, where t is the time (in min) after launch. View two cycles of the graph.

40. In performing a test on a patient, a medical technician used an ultrasonic signal given by the equation $I = A \sin(\omega t + \theta)$. View two cycles of the graph of I vs. t if $A = 5 \, \text{nW/m}^2$, $\omega = 2 \times 10^5 \, \text{rad/s}$, and $\theta = 0.4$.

(W) *In Exercises 41–44, give the specific form of the equation by evaluating a, b, and c through an inspection of the given curve. Explain how a, b, and c are found.*

41. $y = a \sin(bx + c)$
Fig. 10.20

42. $y = a \cos(bx + c)$
Fig. 10.20

43. $y = a \cos(bx + c)$
Fig. 10.21

44. $y = a \sin(bx + c)$
Fig. 10.21

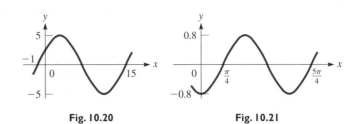

Fig. 10.20   **Fig. 10.21**

Answers to Practice Exercise

1. amp. $= 8$, per. $= \pi$, disp. $= \pi/6$

10.4 Graphs of $y = \tan x$, $y = \cot x$, $y = \sec x$, $y = \csc x$

Graph of $y = \tan x$ • Reciprocal Functions • Asymptotes • Graphs of $y = \cot x, y = \sec x, y = \csc x$

Using a graphing calculator, we now display the graph of $y = \tan x$ in Fig. 10.22. Then, knowing that $\csc x = 1/\sin x$, $\sec x = 1/\cos x$, and $\cot x = 1/\tan x$ (see Section 4.3), we are able to find the values of $y = \csc x$, $y = \sec x$, and $y = \cot x$ from the indicated corresponding reciprocal functions. Using these values, we can graph these functions, and in Figs. 10.24–10.26 we show the resulting graphs. A similar and more extensive graph of $y = \tan x$ is also shown in Fig. 10.23.

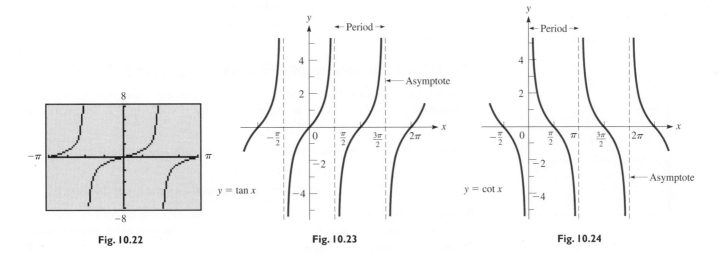

Fig. 10.22   **Fig. 10.23**   **Fig. 10.24**

■ Repeating from above the reciprocal relationships that are the basis for graphing $y = \cot x$, $y = \sec x$, $y = \csc x$, we have

$$\csc x = \frac{1}{\sin x} \qquad \sec x = \frac{1}{\cos x}$$

$$\cot x = \frac{1}{\tan x} \qquad \textbf{(10.2)}$$

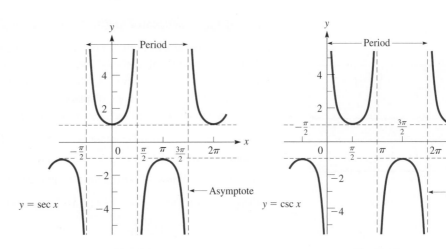

Fig. 10.25   **Fig. 10.26**

From these graphs, note that the period for $y = \tan x$ and $y = \cot x$ is π, and that the period of $y = \sec x$ and $y = \csc x$ is 2π. *The vertical dashed lines are* **asymptotes** (see Sections 3.4 and 21.6). The curves *approach* these lines, but never actually reach them.

The functions are not defined for the values of x for which the curve has asymptotes. This means that the domains do not include these values of x. Thus, we see that the domains of $y = \tan x$ and $y = \sec x$ include all real numbers, except the values $x = -\pi/2, \pi/2, 3\pi/2$, and so on. The domain of $y = \cot x$ and $y = \csc x$ include all real numbers except $x = -\pi, 0, \pi, 2\pi$, and so on.

From the graphs, we see that the ranges of $y = \tan x$ and $y = \cot x$ are all real numbers, but that the ranges of $y = \sec x$ and $y = \csc x$ do not include the real numbers between -1 and 1.

To sketch functions such as $y = a \sec x$, first sketch $y = \sec x$ and then multiply the y-values by a. *Here, a is not an amplitude,* since the ranges of these functions are not limited in the same way they are for the sine and cosine functions.

EXAMPLE 1   Sketching graph of $y = \sec bx$

Sketch the graph of $y = 2 \sec x$.

First, we sketch in $y = \sec x$, shown as the light curve in Fig. 10.27. Then we multiply the y-values of this secant function by 2. Although we can only estimate these values and do this approximately, a reasonable graph can be sketched this way. The desired curve is shown in color in Fig. 10.27. ∎

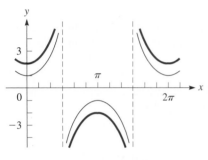

Fig. 10.27

Using a graphing calculator, we can display the graphs of these functions more easily and more accurately than by sketching them. By knowing the general shape and period of the function, the values for the *window* settings can be determined without having to reset them too often.

EXAMPLE 2   Calculator graph of $y = a \cot bx$

View at least two cycles of the graph of $y = 0.5 \cot 2x$ on a graphing calculator.

Since the period of $y = \cot x$ is π, the period of $y = \cot 2x$ is $\pi/2$. Therefore, we choose the *window* settings as follows:

Xmin = 0   ($x = 0$ is one asymptote of the curve)

Xmax = 3.2   ($\pi \approx 3.14$; the period is $\pi/2$; two periods is π)

Ymin = -5, Ymax = 5   (the range is all x; this shows enough of the curve)

Fig. 10.28

We must remember to enter the function as $y_1 = 0.5(\tan 2x)^{-1}$, since $\cot x = (\tan x)^{-1}$. The graphing calculator view is shown in Fig. 10.28. We can view many more cycles of the curve with appropriate *window* settings. ∎

EXAMPLE 3   Calculator graph of $y = a \csc (bx + c)$

View at least two periods of the graph of $y = 2 \csc(2x + \pi/4)$ on a graphing calculator.

Since the period of $\csc x$ is 2π, the period of $\csc(2x + \pi/4)$ is $2\pi/2 = \pi$. Recalling that $\csc x = (\sin x)^{-1}$, the curve will have the same displacement as $y = \sin(2x + \pi/4)$. Therefore, displacement is $-\frac{\pi/4}{2} = -\frac{\pi}{8}$. There is some flexibility in choosing the *window* settings, and as an example, we choose the following settings.

Xmin = -0.5 (the displacement is $-\pi/8 = -0.4$)

Xmax = 6 (displacement = $-\pi/8$; period = π, $-\pi/8 + 2\pi = 15\pi/8 = 5.9$)

Ymin = -6, Ymax = 6 (there is no curve between $y = -2$ and $y = 2$)

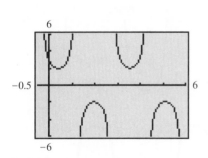

Fig. 10.29

With $y_1 = 2(\sin(2x + \pi/4))^{-1}$, Fig. 10.29 shows the calculator view. ∎

EXERCISES 10.4

In Exercises 1 and 2, view the graphs on a graphing calculator if the given changes are made in the indicated examples of this section.

1. In Example 2, change 0.5 to 5.

2. In Example 3, change the sign before $\pi/4$.

In Exercises 3–6, fill in the following table for each function and plot the graph from these points.

x	$-\frac{\pi}{2}$	$-\frac{\pi}{3}$	$-\frac{\pi}{4}$	$-\frac{\pi}{6}$	0	$\frac{\pi}{6}$	$\frac{\pi}{4}$	$\frac{\pi}{3}$	$\frac{\pi}{2}$	$\frac{2\pi}{3}$	$\frac{3\pi}{4}$	$\frac{5\pi}{6}$	π
y													

3. $y = \tan x$

4. $y = \cot x$

5. $y = \sec x$

6. $y = \csc x$

In Exercises 7–14, sketch the graphs of the given functions by use of the basic curve forms (Figs. 10.23, 10.24, 10.25, and 10.26). See Example 1.

7. $y = 2 \tan x$

8. $y = 3 \cot x$

9. $y = \frac{1}{2} \sec x$

10. $y = \frac{3}{2} \csc x$

11. $y = -8 \cot x$

12. $y = -0.1 \tan x$

13. $y = -3 \csc x$

14. $y = -60 \sec x$

In Exercises 15–24, view at least two cycles of the graphs of the given functions on a graphing calculator.

15. $y = \tan 2x$

16. $y = 2 \cot 3x$

17. $y = \frac{1}{2} \sec 3x$

18. $y = 0.4 \csc 2x$

19. $y = 2 \cot\left(2x + \frac{\pi}{6}\right)$

20. $y = \tan\left(3x - \frac{\pi}{2}\right)$

21. $y = 18 \csc\left(3x - \frac{\pi}{3}\right)$

22. $y = 12 \sec\left(2x + \frac{\pi}{4}\right)$

23. $y = 75 \tan\left(0.5x - \frac{\pi}{8}\right)$

24. $y = 0.5 \sec\left(0.2x + \frac{\pi}{25}\right)$

In Exercises 25 and 26, solve the given problems. In Exercises 27–30, sketch the appropriate graphs, and check each on a calculator.

25. Write the equation of a secant function with zero displacement, a period of 4π, and that passes through $(0, -3)$.

26. Use a graphing calculator to show that $\sin x < \tan x$ for $0 < x < \pi/2$, although $\sin x$ and $\tan x$ are nearly equal for the values near zero.

27. Near Antarctica, an iceberg with a vertical face 200 m high is seen from a small boat. At a distance x from the iceberg, the angle of elevation θ of the top of the iceberg can be found from the equation $x = 200 \cot \theta$. Sketch x as a function of θ.

28. In a laser experiment, two mirrors move horizontally in equal and opposite distances from point A. The laser path from and to point B is shown in Fig. 10.30. From the figure, we see that $x = a \tan \theta$. Sketch the graph of $x = f(\theta)$ for $a = 5.00$ cm.

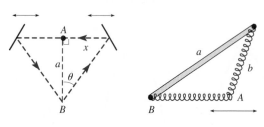

Fig. 10.30   Fig. 10.31

29. A mechanism with two springs is shown in Fig. 10.31, where point A is restricted to move horizontally. From the law of sines, we see that $b = (a \sin B) \csc A$. Sketch the graph of b as a function of A for $a = 4.00$ cm and $B = \pi/4$.

30. A cantilever column of length L will buckle if too large a downward force P is applied d units off center. The horizontal deflection x (see Fig. 10.32) is $x = d(\sec(kL) - 1)$, where k is a constant depending on P, and $0 < kL < \pi/2$. For a constant d, sketch the graph of x as a function of kL.

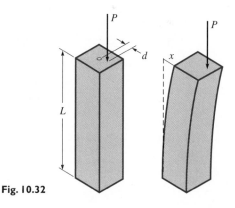

Fig. 10.32

10.5 Applications of the Trigonometric Graphs

**Simple Harmonic Motion •
Alternating Electric Current •
Frequency**

When an object moves in a circular path with constant velocity (see Section 8.4), its *projection* on a diameter moves with **simple harmonic motion.** For example, the shadow of a ball at the end of a string and moving at a constant rate, moves with simple harmonic motion. We now consider this physical concept and some of its applications.

Fig. 10.33

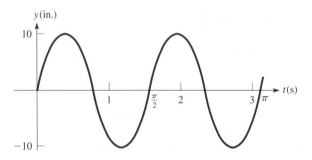

Fig. 10.34

EXAMPLE 1 Simple harmonic motion

In Fig. 10.33, assume that a particle starts at the end of the radius at $(R, 0)$ and moves counterclockwise around the circle with constant angular velocity ω. *The displacement of the projection on the y-axis is d and is given by* $d = R \sin \theta$. The displacement is shown for a few different positions of the end of the radius.

Since $\theta/t = \omega$, or $\theta = \omega t$, we have

$$d = R \sin \omega t \qquad (10.3)$$

as the equation for the displacement of this projection, with time t as the independent variable.

For the case where $R = 10.0$ in. and $\omega = 4.00$ rad/s, we have

$$d = 10.0 \sin 4.00t$$

By sketching or viewing the graph of this function, we can find the displacement d of the projection for a given time t. The graph is shown in Fig. 10.34. ∎

In Example 1, note that *time is the independent variable.* This is motion for which the object (the end of the projection) remains at the same horizontal position $(x = 0)$ and moves only vertically according to a sinusoidal function. In the previous sections, we dealt with functions in which y is a sinusoidal function of the horizontal displacement x. Think of a water wave. At *one point* of the wave, the motion is only vertical and sinusoidal with time. At *one given time,* a picture would indicate a sinusoidal movement from one horizontal position to the next.

EXAMPLE 2 Simple harmonic motion – application

A windmill is used to pump water. The radius of the blade is 2.5 m, and it is moving with constant angular velocity. If the vertical displacement of the end of the blade is timed from the point it is at an angle of 45° ($\pi/4$ rad) from the horizontal (see Fig. 10.35(a)), the displacement d is given by

$$d = 2.5 \sin\left(\omega t + \frac{\pi}{4}\right)$$

If the blade makes an angle of 90° ($\pi/2$ rad) when $t = 0$ (see Fig. 10.35(b)), the displacement d is given by

$$d = 2.5 \sin\left(\omega t + \frac{\pi}{2}\right) \quad \text{or} \quad d = 2.5 \cos \omega t$$

If timing started at the first maximum for the displacement, the resulting curve for the displacement would be that of the cosine function. ∎

(a)   (b)

Fig. 10.35

Practice Exercise

1. If the windmill blade in Example 2 starts at an angle of 135°, what equation gives its displacement in terms of the cosine function?

Other examples of simple harmonic motion are (1) the movement of a pendulum bob through its arc (a very close approximation to simple harmonic motion), (2) the motion of an object "bobbing" in water, (3) the movement of the end of a vibrating rod (which we hear as sound), and (4) the displacement of a weight moving up and down on a spring. Other phenomena that give rise to equations like those for simple harmonic motion are found in the fields of optics, sound, and electricity. The equations for such phenomena have the same mathematical form because they result from vibratory movement or motion in a circle.

EXAMPLE 3 Alternating current

A very important use of the trigonometric curves arises in the study of alternating current, which is caused by the motion of a wire passing through a magnetic field. If the wire is moving in a circular path, with angular velocity ω, the current i in the wire at time t is given by an equation of the form

$$i = I_m \sin(\omega t + \alpha)$$

where I_m is the maximum current attainable and α is the phase angle.

The current may be represented by a sinusoidal wave. Given that $I_m = 6.00$ A, $\omega = 120\pi$ rad/s, and $\alpha = \pi/6$, we have the equation

$$i = 6.00 \sin\left(120\pi t + \frac{\pi}{6}\right)$$

From this equation, note that the amplitude is 6.00 A, the period is $\frac{1}{60}$ s, and the displacement is $-\frac{1}{720}$ s. From these values, we draw the graph as shown in Fig. 10.36. Since the current takes on both positive and negative values, we conclude that it moves alternately in one direction and then the other. ∎

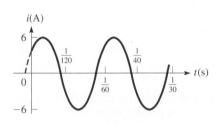

Fig. 10.36

■ Named for the German physicist Heinrich Hertz (1857–1894).

It is a common practice to express the rate of rotation in terms of *the* **frequency** *f, the number of cycles per second,* rather than directly in terms of the angular velocity ω, the number of radians per second. *The unit for frequency is the* **hertz** (Hz), *and* $1\,\text{Hz} = 1$ cycle/s. Since there are 2π rad in one cycle, we have

$$\omega = 2\pi f \tag{10.4}$$

It is the frequency f that is referred to in electric current, on radio stations, for musical tones, and so on.

EXAMPLE 4 Frequency – hertz

For the electric current in Example 3, $\omega = 120\pi$ rad/s. The corresponding frequency f is

$$f = \frac{120\pi}{2\pi} = 60\,\text{Hz}$$

This means that 120π rad/s corresponds to 60 cycles/s. This is the standard frequency used for alternating current. ∎

EXERCISES 10.5

In Exercises 1 and 2, answer the given questions about the indicated examples of this section.

1. In Example 1, what is the equation relating d and t if the end of the radius starts at $(0, R)$?

2. In Example 2, if the blade starts at an angle of $-45°$, what is the equation relating d and t as (a) a sine function? (b) A cosine function?

A graphing calculator may be used in the following exercises.

In Exercises 3 and 4, sketch two cycles of the curve of the projection of Example 1 as a function of time for the given values.

3. $R = 2.40$ cm, $\omega = 2.00$ rad/s   4. $R = 18.5$ ft, $f = 0.250$ Hz

In Exercises 5 and 6, a point on a cam is 8.30 cm from the center of rotation. The cam is rotating with a constant angular velocity, and the vertical displacement $d = 8.30$ cm for $t = 0$ s. See Fig. 10.37. Sketch two cycles of d as a function of t for the given values.

5. $f = 3.20$ Hz

6. $\omega = 3.20$ rad/s

Fig. 10.37

In Exercises 7 and 8, a satellite is orbiting the earth such that its displacement D north of the equator (or south if D < 0) is given by D = A sin(ωt + α). Sketch two cycles of D as a function of t for the given values.

7. $A = 500\,\text{mi}, \omega = 3.60\,\text{rad/h}, \alpha = 0$

8. $A = 850\,\text{km}, f = 1.6 \times 10^{-4}\,\text{Hz}, \alpha = \pi/3$

In Exercises 9 and 10, for an alternating-current circuit in which the voltage e is given by e = E cos(ωt + α), sketch two cycles of the voltage as a function of time for the given values.

9. $E = 170\,\text{V}, f = 60.0\,\text{Hz}, \alpha = -\pi/3$

10. $E = 80\,\text{V}, \omega = 377\,\text{rad/s}, \alpha = \pi/2$

In Exercises 11 and 12, refer to the wave in the string described in Exercise 37 of Section 10.3. For a point on the string, the displacement y is given by $y = A \sin 2\pi\left(\dfrac{t}{T} - \dfrac{x}{\lambda}\right)$. We see that each point on the string moves with simple harmonic motion. Sketch two cycles of y as a function of t for the given values.

11. $A = 3.20\,\text{cm}, T = 0.050\,\text{s}, \lambda = 40.0\,\text{cm}, x = 5.00\,\text{cm}$

12. $A = 0.750\,\text{in.}, T = 0.250\,\text{s}, \lambda = 24.0\,\text{in.}, x = 20.0\,\text{in.}$

In Exercises 13 and 14, the air pressure within a plastic container changes above and below the external atmospheric pressure by $p = p_0 \sin 2\pi f t$. Sketch two cycles of p = f(t) for the given values.

13. $p_0 = 2.80\,\text{lb/in.}^2, f = 2.30\,\text{Hz}$

14. $p_0 = 45.0\,\text{kPa}, f = 0.450\,\text{Hz}$

Answer to Practice Exercise

1. $d = 2.5 \cos(\omega t + \pi/4)$

In Exercises 15–22, sketch the required curves.

15. The angular displacement θ of a certain pendulum bob in terms of its initial displacement θ_0 is $\theta = \theta_0 \cos \omega t$. If $\omega = 2.00\,\text{rad/s}$, and $\theta_0 = \pi/30$ rad, draw two cycles for the resulting equation.

16. A study found that, when breathing normally, the increase in volume V (in L) of air in a person's lungs as a function of the time t (in s) is $V = 0.30 \sin 0.50\pi t$. Sketch two cycles.

17. Sketch two cycles of the radio signal $e = 0.014 \cos(2\pi f t + \pi/4)$ (e in volts, f in hertz, and t in seconds) for a station broadcasting with $f = 950\,\text{kHz}$ ("95" on the AM radio dial).

18. Sketch two cycles of the acoustical intensity I of the sound wave for which $I = A \cos(2\pi f t - \alpha)$, given that t is in seconds, $A = 0.027\,\text{W/cm}^2, f = 240\,\text{Hz}$, and $\alpha = 0.80$.

(W)19. The rotating beacon of a parked police car is 12 m from a straight wall. (a) Sketch the graph of the length L of the light beam, where $L = 12 \sec \pi t$, for $0 \le t \le 2.0$ s. (b) Which part(s) of the graph show meaningful values? Explain.

20. The motion of a piston of a car engine approximates simple harmonic motion. Given that the stroke (twice the amplitude) is 0.100 m, the engine runs at 2800 r/min, and the piston starts at the middle of its stroke, find the equation for the displacement d as a function of t. Sketch two cycles.

21. A riverboat's paddle has a 12-ft radius and rotates at 18 r/min. Find the equation of motion of the vertical displacement (from the center of the wheel) y of the end of a paddle as a function of the time t if the paddle is initially horizontal. Sketch two cycles.

22. The sinusoidal electromagnetic wave emitted by an antenna in a cellular phone system has a frequency of $7.5 \times 10^9\,\text{Hz}$ and an amplitude of $0.045\,\text{V/m}$. Find the equation representing the wave if it starts at the origin. Sketch two cycles.

10.6 Composite Trigonometric Curves

**Additional of Ordinates •
Parametric Equations •
Lissajous Figures**

Many applications involve functions that in themselves are a combination of two or more simpler functions. In this section, we discuss methods by which the curve of such a function can be found by combining values from the simpler functions.

EXAMPLE 1 Function that is sum of simpler functions

Sketch the graph of $y = 2 + \sin 2x$.

This function is the sum of the simpler functions $y_1 = 2$ and $y_2 = \sin 2x$. We may find values for y by adding 2 to each important value of $y_2 = \sin 2x$.

For $y_2 = \sin 2x$, the amplitude is 1, and the period is $2\pi/2 = \pi$. Therefore, we obtain the values in the following table and sketch the graph in Fig. 10.38.

Fig. 10.38

x	0	$\dfrac{\pi}{4}$	$\dfrac{\pi}{2}$	$\dfrac{3\pi}{4}$	π
$\sin 2x$	0	1	0	-1	0
$2 + \sin 2x$	2	3	2	1	2

Note that this is a vertical shift of 2 units of the graph of $y = \sin 2x$, in the same way as discussed on page 99. ∎

ADDITION OF ORDINATES

Another way to sketch the resulting graph is to *first sketch the two simpler curves and then add the y-values graphically. This method is called* **addition of ordinates** *and is* illustrated in the following example.

EXAMPLE 2 Addition of ordinates

Sketch the graph of $y = 2\cos x + \sin 2x$.

On the same set of coordinate axes, we sketch the curves $y = 2\cos x$ and $y = \sin 2x$. These are shown as dashed and solid light curves in Fig. 10.39. For various values of x, we determine the distance above or below the x-axis of each curve and add these distances, noting that those above the axis are positive and those below the axis are negative. We thereby *graphically* **add** *the y-values* of these curves to get points on the resulting curve, shown in color in Fig. 10.39.

At A, add the two lengths (shown side-by-side for clarity) to get the length for y. At B, both lengths are negative, and the value for y is the sum of these negative values. At C, one is positive and the other negative, and we must subtract the lower length from the upper one to get the length for y.

We combine these lengths for enough x-values to get a good curve. Some points are easily found. Where one curve crosses the x-axis, its value is zero, and the resulting curve has its point on the other curve. Here, where $\sin 2x$ is zero, the points for the resulting curve lie on the curve of $2\cos x$.

We should also add values where each curve is at its maximum or its minimum. *Extra care should be taken for those values of x for which one curve is* **positive** *and the other is* **negative.** ∎

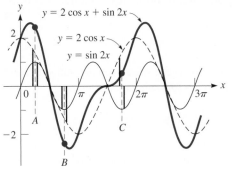

Fig. 10.39

We have seen how a fairly complex curve can be sketched graphically. It is expected that a graphing calculator (or computer *grapher*) will generally be used to view most graphs, particularly ones that are difficult to sketch. A graphing calculator can display such curves much more easily, and with much greater accuracy. Information about the amplitude, period, and displacement is useful in choosing *window* settings. For these graphs, it is important that the calculator is in *radian mode*.

EXAMPLE 3 Graphing calculator addition of ordinates

Use a graphing calculator to display the graph of $y = \frac{x}{2} - \cos x$.

Here, we note that the curve is a combination of the straight line $y = x/2$ and the trigonometric curve $y = \cos x$. There are several good choices for the *window* settings, depending on how much of the curve is to be viewed. To see a little more than one period of $\cos x$, we can make the following choices:

Xmin $= -1$ (to start to left of y-axis)

Xmax $= 7$ (period of $\cos x$ is $2\pi = 6.3$)

Ymin $= -2$ (line passes through $(0, 0)$; amplitude of $y = \cos x$ is 1)

Ymax $= 4$ (slope of line is 1/2)

See Fig. 10.40(a). The graphs of $y = x/2 - \cos x$, $y = x/2$, and $y = -\cos x$ are shown in Fig. 10.40(b).

NOTE ▶ We show $y = -\cos x$ rather than $y = \cos x$ because, if *addition of ordinates* were being used, *it is easier to add graphic values than to subtract them.* Here we can **add** $y_1 = x/2$ and $y_2 = -\cos x$ in order to get $y = x/2 - \cos x$. ∎

(a)

(b)

Fig. 10.40

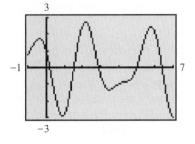

Fig. 10.41

EXAMPLE 4 Graphing calculator addition of ordinates

View the graph of $y = \cos \pi x - 2 \sin 2x$ on a graphing calculator.

The combination of $y = \cos \pi x$ and $y = 2 \sin 2x$ leads to the following choices for the *window* settings:

Xmin $= -1$ (to start to the left of the y-axis)

Xmax $= 7$ (the periods are 2 and π; this shows at least two periods of each)

Ymin $= -3$, Ymax $= 3$ (the sum of the amplitudes is 3)

There are many possible choices for Xmin and Xmax to get a good view of the graph on a calculator. However, since the sum of the amplitudes is 3, note that the curve cannot be below $y = -3$ or above $y = 3$.

The graphing calculator view is shown in Fig. 10.41.

This graph can be constructed by using addition of ordinates, although it is difficult to do very accurately. ■

LISSAJOUS FIGURES

An important application of trigonometric curves is made when they are added at *right angles*. The methods for doing this are shown in the following examples.

EXAMPLE 5 Graphing parametric equations

Plot the graph for which the values of x and y are given by the equations $y = \sin 2\pi t$ and $x = 2 \cos \pi t$. *Equations given in this form, x and y in terms of a third variable, are called* **parametric equations.**

Since both x and y are in terms of t, by assuming values of t, we find corresponding values of x and y and use these values to plot the graph. Since the periods of $\sin 2\pi t$ and $2 \cos \pi t$ are $t = 1$ and $t = 2$, respectively, we will use values of $t = 0, 1/4, 1/2, 3/4, 1$, and so on. These give us convenient values of $0, \pi/4, \pi/2, 3\pi/4, \pi$, and so on to use in the table. We plot the points in Fig. 10.42.

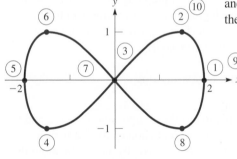

Fig. 10.42

■ Named for the French physicist Jules Lissajous (1822–1880).

t	0	$\frac{1}{4}$	$\frac{1}{2}$	$\frac{3}{4}$	1	$\frac{5}{4}$	$\frac{3}{2}$	$\frac{7}{4}$	2	$\frac{9}{4}$
x	2	1.4	0	-1.4	-2	-1.4	0	1.4	2	1.4
y	0	1	0	-1	0	1	0	-1	0	1
Point number	1	2	3	4	5	6	7	8	9	10

■

Since x and y are trigonometric functions of a third variable t, and since the x-axis is at right angles to the y-axis, values of x and y obtained in this way result in a combination of two trigonometric curves at right angles. *Figures obtained in this way are called* **Lissajous figures.** Note that the Lissajous figure in Fig. 10.42 *is not a function* since there are *two* values of y for each value of x (except $x = -2, 0, 2$) in the domain.

In practice, Lissajous figures can be displayed on an *oscilloscope* by applying an electric signal between a pair of horizontal plates and another signal between a pair of vertical plates. These signals are then seen on the screen of the oscilloscope. This type of screen (a cathode-ray tube) is similar to that used on most older TV sets.

There is additional coverage of parametric equations in Chapter 21. They are also used in some topics in the study of calculus.

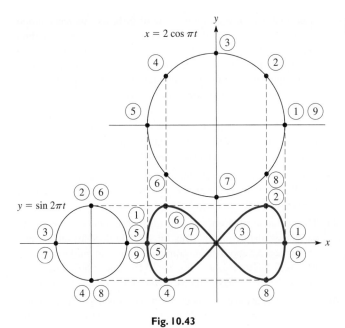

$x = 2 \cos \pi t$

$y = \sin 2\pi t$

Fig. 10.43

EXAMPLE 6 Graphing Lissajous figures

If a circle is placed on the x-axis and another on the y-axis, we may represent the coordinates (x, y) for the curve of Example 5 by the lengths of the projections (see Example 1 of Section 10.5) of a point moving around each circle. A careful study of Fig. 10.43 will clarify this. We note that the radius of the circle giving the x-values is 2 and that the radius of the circle giving the y-values is 1. This is due to the way in which x and y are defined. Also, due to these definitions, the point revolves around the y-circle twice as fast as the corresponding point around the x-circle.

■ See the chapter introduction.

■ On an oscilloscope, the curve would result when two electric signals are used. The first would have twice the amplitude and one-half the frequency of the other.

■

Most graphing calculators can be used to display a curve defined by parametric equations. It is necessary to use the *mode* feature and select *parametric equations*. Use the manual for the calculator, as there are some differences as to how this is done on various calculators.

EXAMPLE 7 Graphing calculator Lissajous figures

Use a graphing calculator to display the graph defined by the parametric equations $x = 2 \cos \pi t$ and $y = \sin 2\pi t$. These are the same equations as those used in Examples 5 and 6.

First, select the parametric equation option from the *mode* feature and enter the parametric equations $x_{1T} = 2 \cos \pi t$ and $y_{1T} = \sin 2\pi t$. Then make the following *window* settings:

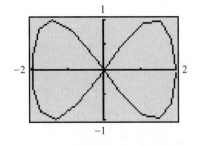

Tmin = 0   (standard default settings, and the usual choice)

Tmax = 2   (the periods are 2 and 1; the longer period is 2)

Tstep = .1047   (standard default setting; curve is smoother with 0.01)

Xmin = -2, Xmax = 2   (smallest and largest possible values of x), Xscl = 1

Ymin = -1, Ymax = 1   (smallest and largest possible values of y), Yscl = 0.5

Fig. 10.44

The calculator graph is shown in Fig. 10.44.   ■

EXERCISES 10.6

In Exercises 1–8, sketch the curves of the given functions by addition of ordinates.

1. $y = 1 + \sin x$

2. $y = 3 - 2 \cos x$

3. $y = \frac{1}{3}x + \sin 2x$

4. $y = x - \sin x$

5. $y = \frac{1}{10}x^2 - \sin \pi x$

6. $y = \frac{1}{4}x^2 + \cos 3x$

7. $y = \sin x + \cos x$

8. $y = \sin x + \sin 2x$

In Exercises 9–20, display the graphs of the given functions on a graphing calculator.

9. $y = x^3 + 10 \sin 2x$

10. $y = \dfrac{1}{x^2 + 1} - \cos \pi x$

11. $y = \sin x - 1.5 \sin 2x$

12. $y = \cos 3x - 3 \sin x$

13. $y = 20 \cos 2x + 30 \sin x$

14. $y = \frac{1}{2} \sin 4x + \cos 2x$

15. $y = 2 \sin x - \cos 1.5x$

16. $y = 8 \sin 0.5x - 12 \sin x$

17. $y = \sin \pi x - \cos 2x$

18. $y = 2 \cos 4x - \cos\left(x - \dfrac{\pi}{4}\right)$

19. $y = 2 \sin\left(2x - \dfrac{\pi}{6}\right) + \cos\left(2x + \dfrac{\pi}{3}\right)$

20. $y = 3 \cos 2\pi x + \sin \frac{\pi}{2}x$

In Exercises 21–24, plot the Lissajous figures.

21. $y = 3 \sin t, x = 2 \sin t$

22. $x = 2 \cos t, y = \cos(t + 4)$

23. $y = 2 \cos 2\pi t, x = \sin \pi t$

24. $x = \cos\left(t + \dfrac{\pi}{4}\right), y = \sin 2t$

In Exercises 25–32, use a graphing calculator to display the Lissajous figures.

25. $x = \cos \pi\left(t + \dfrac{1}{6}\right), y = 2 \sin \pi t$

26. $x = \sin^2 \pi t, y = \cos \pi t$

27. $x = 4 \cos 3t, y = \cos 2t$

28. $x = 2 \sin \pi t, y = 3 \sin 3\pi t$

29. $y = \sin(t + 1), x = \sin 5t$

30. $x = 5 \cos t, y = 3 \sin 5t$

31. $x = 2 \cos \pi t, y = 3 \sin\left(2\pi t - \dfrac{\pi}{4}\right)$

32. $x = 1.5 \cos 3\pi t, y = 0.5 \cos 5\pi t$

In Exercises 33–44, sketch the appropriate curves. A graphing calculator may be used.

33. An object oscillating on a spring has a displacement (in ft) given by $y = 0.4 \sin 4t + 0.3 \cos 4t$, where t is the time (in s). Sketch the graph.

34. The voltage e in a certain electric circuit is given by $e = 50 \sin 50\pi t + 80 \sin 60\pi t$, where t is the time (in s). Sketch the graph.

35. An analysis of the temperature records of Louisville, Kentucky, indicates that the average daily temperature T (in °F) during the year is approximately $T = 56 - 22 \cos\left[\frac{\pi}{6}(x - 0.5)\right]$, where x is measured in months ($x = 0.5$ is Jan. 15, etc.). Sketch the graph of T vs. x for one year.

36. An analysis of data shows that the mean density d (in mg/cm³) of a calcium compound in the bones of women is given by $d = 139.3 + 48.6 \sin(0.0674x - 0.210)$, where x represents the ages of women ($20 \le x \le 80$ years). (A woman is considered to be osteoporotic if $d < 115$ mg/cm³.) Sketch the graph.

37. A normal person with a pulse rate of 60 beats/min has a blood pressure of "120 over 80." This means the pressure is oscillating between a high (systolic) of 120 mm of mercury (shown as mmHg) and a low (diastolic) of 80 mmHg. Assuming a sinusoidal type of function, find the pressure p as a function of the time t if the initial pressure is 120 mmHg. Sketch the graph for the first 5 s.

38. The strain e (dimensionless) on a cable caused by vibration is $e = 0.0080 - 0.0020 \sin 30t + 0.0040 \cos 10t$, where t is measured in seconds. Sketch two cycles of e as a function of t.

39. The electric current i (in mA) in a certain circuit is given by $i = 0.32 + 0.50 \sin t - 0.20 \cos 2t$, where t is in milliseconds. Sketch two cycles of i as a function of t.

40. The available solar energy depends on the amount of sunlight, and the available time in a day for sunlight depends on the time of the year. An approximate correction factor (in min) to standard time is $C = 10 \sin \frac{1}{29}(n - 80) - 7.5 \cos \frac{1}{58}(n - 80)$, where n is the number of the day of the year. Sketch C as a function of n.

41. Two signals are seen on an oscilloscope as being at right angles. The equations for the displacements of these signals are $x = 4 \cos \pi t$ and $y = 2 \sin 3\pi t$. Sketch the figure that appears on the oscilloscope.

42. In the study of optics, light is said to be *elliptically polarized* if certain optic vibrations are out of phase. These may be represented by Lissajous figures. Determine the Lissajous figure for two light waves given by $w_1 = \sin \omega t$ and $w_2 = \sin\left(\omega t + \frac{\pi}{4}\right)$.

43. In checking electric circuit elements, a *square wave* such as that shown in Fig. 10.45 may be displayed on an oscilloscope. Display the graph of $y = 1 + \dfrac{4}{\pi} \sin\left(\dfrac{\pi x}{4}\right) + \dfrac{4}{3\pi} \sin\left(\dfrac{3\pi x}{4}\right)$ on a graphing calculator, and compare it with Fig. 10.45. This equation gives the first three terms of a *Fourier series*. As more terms of the series are added, the approximation to a square wave is better.

Fig. 10.45

44. Another type of display on an oscilloscope may be a *sawtooth wave* such as that shown in Fig. 10.46. Display the graph of $y = 1 - \dfrac{8}{\pi^2}\left(\cos \dfrac{\pi x}{2} + \dfrac{1}{9} \cos \dfrac{3\pi x}{2}\right)$ on a graphing calculator and compare it with Fig. 10.46. See Exercise 43.

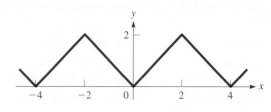

Fig. 10.46

CHAPTER 10   EQUATIONS

For the graphs of $y = a \sin(bx + c)$
and $y = a \cos(bx + c)$

$$\text{Amplitude} = |a| \quad \text{Period} = \frac{2\pi}{b} \quad \text{Displacement} = -\frac{c}{b} \quad \text{(10.1)}$$

$y = a \sin(bx + c), c > 0$

(a)

For each
$a > 0, b > 0$

$y = a \sin(bx + c), c < 0$

(b)

Reciprocal relationships

$$\csc x = \frac{1}{\sin x} \quad \sec x = \frac{1}{\cos x} \quad \cot x = \frac{1}{\tan x} \quad \text{(10.2)}$$

Simple harmonic motion

$$d = R \sin \omega t \quad \text{(10.3)}$$

Angular velocity and frequency

$$\omega = 2\pi f \quad \text{(10.4)}$$

CHAPTER 10   REVIEW EXERCISES

In Exercises 1–28, sketch the curves of the given trigonometric functions. Check each using a graphing calculator.

1. $y = \frac{2}{3} \sin x$
2. $y = -4 \sin x$
3. $y = -2 \cos x$
4. $y = 2.3 \cos(-x)$
5. $y = 2 \sin 3x$
6. $y = 4.5 \sin 12x$
7. $y = 0.4 \cos 4x$
8. $y = 24 \cos 6x$
9. $y = 3 \cos \frac{1}{3}x$
10. $y = 3 \sin(-0.5x)$
11. $y = \sin \pi x$
12. $y = 36 \sin 4\pi x$
13. $y = 5 \cos\left(\frac{\pi x}{2}\right)$
14. $y = -\cos 6\pi x$
15. $y = -0.5 \sin\left(-\frac{\pi x}{6}\right)$
16. $y = 8 \sin \frac{\pi}{4}x$
17. $y = 2 \sin\left(3x - \frac{\pi}{2}\right)$
18. $y = 3 \sin\left(\frac{x}{2} + \frac{\pi}{2}\right)$
19. $y = -2 \cos(4x + \pi)$
20. $y = 0.8 \cos\left(\frac{x}{6} - \frac{\pi}{2}\right)$
21. $y = -\sin\left(\pi x + \frac{\pi}{6}\right)$
22. $y = 250 \sin(3\pi x - \pi)$
23. $y = 8 \cos\left(4\pi x - \frac{\pi}{2}\right)$
24. $y = 3 \cos(2\pi x + \pi)$
25. $y = 0.3 \tan 0.5x$
26. $y = \frac{1}{4} \sec x$
27. $y = -\frac{1}{3} \csc x$
28. $y = -5 \cot \pi x$

In Exercises 29–32, sketch the curves of the given functions by addition of ordinates.

29. $y = 2 + \frac{1}{2} \sin 2x$
30. $y = \frac{1}{2}x - \cos \frac{1}{3}x$
31. $y = \sin 2x + 3 \cos x$
32. $y = \sin 3x + 2 \cos 2x$

In Exercises 33–40, display the curves of the given functions on a graphing calculator.

33. $y = 2 \sin x - \cos 2x$
34. $y = 10 \sin 3x - 20 \cos x$
35. $y = \cos\left(x + \frac{\pi}{4}\right) - 0.4 \sin 2x$
36. $y = 2 \cos \pi x + \cos(2\pi x - \pi)$
37. $y = \dfrac{\sin x}{x}$
38. $y = \sqrt{x} \sin 0.5x$

(W) 39. $y = \sin^2 x + \cos^2 x \quad (\sin^2 x = (\sin x)^2)$
What conclusion can be drawn from the graph?

(W) 40. $y = \sin\left(x + \frac{\pi}{4}\right) - \cos\left(x - \frac{\pi}{4}\right) + 1$
What conclusion can be drawn from the graph?

In Exercises 41–44, give the specific form of the indicated equation by evaluating a, b, and c through an inspection of the given curve.

41. $y = a\sin(bx + c)$
(Figure 10.47)

42. $y = a\cos(bx + c)$
(Figure 10.47)

43. $y = a\cos(bx + c)$
(Figure 10.48)

44. $y = a\sin(bx + c)$
(Figure 10.48)

Fig. 10.47

Fig. 10.48

In Exercises 45–48, display the Lissajous figures on a graphing calculator.

45. $x = -\cos 2\pi t, y = 2\sin \pi t$

46. $x = \sin\left(t + \dfrac{\pi}{6}\right), y = \sin t$

47. $x = 2\cos\left(2\pi t + \dfrac{\pi}{4}\right), y = \cos \pi t$

48. $x = \cos\left(t - \dfrac{\pi}{6}\right), y = \cos\left(2t + \dfrac{\pi}{3}\right)$

In Exercises 49–60, solve the given problems.

49. Display the function $y = 2|\sin 0.2\pi x| - |\cos 0.4\pi x|$ on a graphing calculator.

50. Display the function $y = 0.2|\tan 2x|$ on a graphing calculator.

51. Show that $\cos\left(x + \dfrac{\pi}{4}\right) = \sin\left(\dfrac{\pi}{4} - x\right)$ on a graphing calculator.

52. Show that $\tan\left(x - \dfrac{\pi}{3}\right) = -\tan\left(\dfrac{\pi}{3} - x\right)$ on a graphing calculator.

53. What is the period of the function $y = 2\cos 0.5x + \sin 3x$?

54. What is the period of the function $y = \sin \pi x + 3\sin 0.25\pi x$?

55. Find the function and graph it if it is of the form $y = a\sin x$ and passes through $(5\pi/2, 3)$.

56. Find the function and graph it if it is of the form $y = a\cos x$ and passes through $(4\pi, -3)$.

57. Find the function and graph it if it is of the form $y = 3\cos bx$ and passes through $(\pi/3, -3)$ and b has the smallest possible positive value.

58. Find the function and graph it if it is of the form $y = 3\sin bx$ and passes through $(\pi/3, 0)$ and b has the smallest possible positive value.

59. Find the function and graph it for a function of the form $y = 3\sin(\pi x + c)$ that passes through $(-0.25, 0)$ and for which c has the smallest possible positive value.

60. Write the equation of the cosecant function with zero displacement, a period of 2, and that passes through $(0.5, 4)$.

In Exercises 61–82, sketch the appropriate curves. A graphing calculator may be used.

61. The range R of a rocket is given by $R = \dfrac{v_0^2 \sin 2\theta}{g}$. Sketch R as a function of θ for $v_0 = 1000$ m/s and $g = 9.8$ m/s². See Fig. 10.49.

Fig. 10.49

62. The blade of a saber saw moves vertically up and down at 18 strokes per second. The vertical displacement y (in cm) is given by $y = 1.2\sin 36\pi t$, where t is in seconds. Sketch at least two cycles of the graph of y vs. t.

63. The velocity v (in cm/s) of a piston in a certain engine is given by $v = \omega D\cos \omega t$, where ω is the angular velocity of the crankshaft in radians per second and t is the time in seconds. Sketch the graph of v vs. t if the engine is at 3000 r/min and $D = 3.6$ cm.

64. A light wave for the color yellow can be represented by the equation $y = A\sin 3.4 \times 10^{15} t$. With A as a constant, sketch two cycles of y as a function of t (in s).

65. The electric current i (in A) in a circuit in which there is a *full-wave rectifier* is $i = 10|\sin 120\pi t|$. Sketch the graph of $i = f(t)$ for $0 \le t \le 0.05$ s. What is the period of the current?

66. A circular disk suspended by a thin wire attached to the center of one of its flat faces is twisted through an angle θ. Torsion in the wire tends to turn the disk back in the opposite direction (thus, the name *torsion pendulum* is given to this device). The angular displacement θ (in rad) as a function of time t (in s) is $\theta = \theta_0 \cos(\omega t + \alpha)$, where θ_0 is the maximum angular displacement, ω is a constant that depends on the properties of the disk and wire, and α is the phase angle. Sketch the graph of θ vs. t if $\theta_0 = 0.100$ rad, $\omega = 2.50$ rad/s, and $\alpha = \pi/4$. See Fig. 10.50.

Fig. 10.50

67. The vertical displacement y of a point at the end of a propeller blade of a small boat is $y = 14.0\sin 40.0\pi t$. Sketch two cycles of y (in cm) as a function of t (in s).

68. In optics, two waves are said to interfere destructively if, when they both pass through a medium, the amplitude of the resulting wave is zero. Sketch the graph of $y = \sin x + \cos(x + \pi/2)$ and find whether or not it would represent destructive interference of two waves.

69. The vertical displacement y (in ft) of a buoy floating in water is given by $y = 3.0\cos 0.2t + 1.0\sin 0.4t$, where t is in seconds. Sketch the graph of y as a function of t for the first 40 s.

70. The vertical motion of a rubber raft on a lake approximates simple harmonic motion due to the waves. If the amplitude of the motion is 0.250 m and the period is 3.00 s, find an equation for the vertical displacement y as a function of the time t. Sketch two cycles.

71. A drafting student draws a circle through the three vertices of a right triangle. The hypotenuse of the triangle is the diameter d of the circle, and from Fig. 10.51, we see that $d = a \sec \theta$. Sketch the graph of d as a function of θ for $a = 3.00$ in.

Fig. 10.51

72. The height h (in m) of a certain rocket ascending vertically is given by $h = 800 \tan \theta$, where θ is the angle of elevation from an observer 800 m from the launch pad. Sketch h as a function of θ.

73. At 40° N latitude the number of hours h of daylight each day during the year is approximately $h = 12.2 + 2.8 \sin\left[\frac{\pi}{6}(x - 2.7)\right]$, where x is measured in months ($x = 0.5$ is Jan. 15, etc.). Sketch the graph of h vs. x for one year. (Some of the cities near 40° N are Philadelphia, Madrid, Naples, Ankara, and Beijing.)

(W) 74. The equation in Exercise 73 can be used for the number of hours of daylight at 40° S latitude with the appropriate change. Explain what change is necessary and determine the proper equation. Sketch the graph. (This would be appropriate for southern Argentina and Wellington, New Zealand.)

75. If the upper end of a spring is not fixed and is being moved with a sinusoidal motion, the motion of the bob at the end of the spring is affected. Sketch the curve if the motion of the upper end of a spring is being moved by an external force and the bob moves according to the equation $y = 4 \sin 2t - 2 \cos 2t$.

76. The loudness L (in decibels) of a fire siren as a function of the time t (in s) is approximately $L = 40 - 35 \cos 2t + 60 \sin t$. Sketch this function for $0 \leq t \leq 10$ s.

77. The path of a roller mechanism used in an assembly-line process is given by $x = \theta - \sin \theta$ and $y = 1 - \cos \theta$. Sketch the path for $0 \leq \theta \leq 2\pi$.

78. The equations for two voltage signals that give a resulting curve on an oscilloscope are $x = 6 \sin \pi t$ and $y = 4 \cos 4\pi t$. Sketch the graph of the curve displayed on the oscilloscope.

79. The impedance Z (in Ω) and resistance R (in Ω) for an alternating-current circuit are related by $Z = R \sec \theta$, where θ is called the phase angle. Sketch the graph for Z as a function of θ for $-\pi/2 < \theta < \pi/2$.

80. For an object sliding down an inclined plane at constant speed, the coefficient of friction μ between the object and the plane is given by $\mu = \tan \theta$, where θ is the angle between the plane and the horizontal. Sketch the graph of μ vs. θ.

81. The charge q (in C) on a certain capacitor as a function of the time t (in s) is given by $q = 0.0003(3 - 2 \sin 100t \cos 100t)$. Sketch two cycles of q vs. t.

82. The instantaneous power p (in W) in an electric circuit is defined as the product of the instantaneous voltage e and the instantaneous current i (in A). If we have $e = 100 \cos 200t$ and $i = 2 \cos\left(200t + \frac{\pi}{4}\right)$, plot the graph e vs. t and the graph of i vs. t on the same coordinate system. Then sketch the graph of p vs. t by multiplying appropriate values of e and i.

Writing Exercise

83. A wave passing through a string can be described at any instant by the equation $y = a \sin(bx + c)$. Write one or two paragraphs explaining the change in the wave (a) if a is doubled, (b) if b is doubled, and (c) if c is doubled.

CHAPTER 10   PRACTICE TEST

1. Determine the amplitude, period, and displacement of the function $y = -3 \sin(4\pi x - \pi/3)$.

In Problems 2–5, sketch the graphs of the given functions.

2. $y = 0.5 \cos \frac{\pi}{2} x$ 3. $y = 2 + 3 \sin x$

4. $y = 3 \sec x$ 5. $y = 2 \sin\left(2x - \frac{\pi}{3}\right)$

6. A wave is traveling in a string. The displacement y (in in.) as a function of the time t (in s) from its equilibrium position is given by $y = A \cos(2\pi/T)t$. T is the period (in s) of the motion. If $A = 0.200$ in. and $T = 0.100$ s, sketch two cycles of y vs t.

7. Sketch the graph of $y = 2 \sin x + \cos 2x$ by addition of ordinates.

8. Use a graphing calculator to display the Lissajous figure for which $x = \sin \pi t$ and $y = 2 \cos 2\pi t$.

9. Sketch two cycles of the curve of a projection of the end of a radius on the y-axis. The radius is of length R and it is rotating counterclockwise about the origin at 2.00 rad/s. It starts at an angle of $\pi/6$ with the positive x-axis.

10. Find the function of the form $y = 2 \sin bx$ if its graph passes through $(\pi/3, 2)$ and b is the smallest possible positive value. Then graph the function.

11

Exponents and Radicals

For use in later chapters, we now further develop the use of exponents and radicals. In previous chapters, we have used only exponents that are integers, and by introducing exponents that are fractions we will show the relationship between exponents and radicals (for example, we will show that $\sqrt{x} = x^{1/2}$). In more advanced math and in applications, it is more convenient to use fractional exponents rather than radicals.

As we noted in Chapter 6, the use of symbols led to advances in mathematics and science. As letters became commonly used in the 1600s, it was common to write, for example, x^3 as xxx. For larger powers, this is obviously inconvenient, and the modern use of exponents came into use. The first to use exponents consistently was the French mathematician René Descartes in the 1630s.

The meaning of negative and fractional exponents was first found by the English mathematician Wallis in the 1650s, although he did not write them as we do today. In the 1670s, it was the great English mathematician and physicist Isaac Newton who first used all exponents (positive, negative, and fractional) with modern notation. This improvement in notation made the development of many areas of mathematics, particularly calculus, easier. In this way, it also helped to make many advances in the applications of mathematics.

As we develop the various operations with exponents and radicals, we will show their uses in some technical areas of application. They are used in a number of formulas in areas such as electronics, hydrodynamics, optics, solar energy, and machine design.

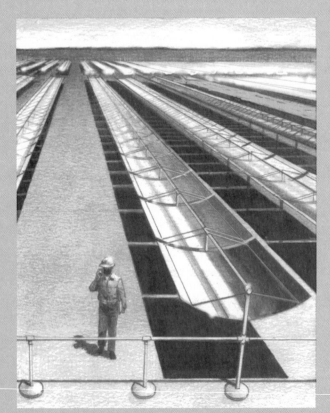

In finding the rate at which solar radiation changes at a solar-energy collector, the following expression is found:

$$\frac{(t^4 + 100)^{1/2} - 2t^3(t + 6)(t^4 + 100)^{-1/2}}{[(t^4 + 100)^{1/2}]^2}$$

In Section 11.2, we show that this can be written in a much simpler form.

11.1 Simplifying Expressions with Integral Exponents

Laws of Exponents • Simplifying Expressions • Care with Zero and Negative Exponents

The laws of exponents were given in Section 1.4. For reference, they are

$$a^m \times a^n = a^{m+n} \tag{11.1}$$

$$\frac{a^m}{a^n} = a^{m-n} \quad \text{or} \quad \frac{a^m}{a^n} = \frac{1}{a^{n-m}} \quad (a \neq 0) \tag{11.2}$$

$$(a^m)^n = a^{mn} \tag{11.3}$$

$$(ab)^n = a^n b^n, \quad \left(\frac{a}{b}\right)^n = \frac{a^n}{b^n} \quad (b \neq 0) \tag{11.4}$$

$$a^0 = 1 \quad (a \neq 0) \tag{11.5}$$

$$a^{-n} = \frac{1}{a^n} \quad (a \neq 0) \tag{11.6}$$

Although Eqs. (11.1) to (11.4) were originally defined for positive integers as exponents, we showed in Section 1.4 that, with the definitions in Eqs. (11.5) and (11.6), they are valid for all integral exponents. Later in the chapter, we will show how fractions are used as exponents. These equations are very important to the developments of this chapter and should be reviewed and learned thoroughly.

EXAMPLE 1 Simplifying basic expressions

(a) Applying Eq. (11.1) and then Eq. (11.6), we have

$$a^3 \times a^{-5} = a^{3-5} = a^{-2} = \frac{1}{a^2}$$

Negative exponents are generally not used in the expression of the final result, unless specified otherwise. However, they are often used in intermediate steps.

(b) Applying Eq. (11.1), then (11.6), and then (11.4), we have

$$(10^3 \times 10^{-4})^2 = (10^{3-4})^2 = (10^{-1})^2 = \left(\frac{1}{10}\right)^2 = \frac{1}{10^2} = \frac{1}{100}$$

The result is in proper form as either $1/10^2$ or $1/100$. If the exponent is large, it is common to leave the exponent in the answer. If the power of ten is being used *as part of scientific notation,* then *the form with the negative exponent* 10^{-2} *is used.* ∎

EXAMPLE 2 Simplifying basic expressions

(a) Applying Eqs. (11.2) and (11.5), we have

$$\frac{a^2 b^3 c^0}{ab^7} = \frac{a^{2-1}(1)}{b^{7-3}} = \frac{a}{b^4}$$

(b) Applying Eqs. (11.4) and (11.3), and then (11.6), we have

$$(x^{-2}y)^3 = (x^{-2})^3(y^3) = x^{-6}y^3 = \frac{y^3}{x^6} \qquad \blacksquare$$

Often, *several different combinations of Eqs. (11.1) to (11.6) can be used to simplify an expression.* This is illustrated in the next example.

EXAMPLE 3 Simplification done in different ways

(a) $(x^2y)^2\left(\dfrac{2}{x}\right)^{-2} = \dfrac{(x^4y^2)}{\left(\dfrac{2}{x}\right)^2} = \dfrac{x^4y^2}{\dfrac{4}{x^2}} = \dfrac{x^4y^2}{1} \times \dfrac{x^2}{4} = \dfrac{x^6y^2}{4}$

(b) $(x^2y)^2\left(\dfrac{2}{x}\right)^{-2} = (x^4y^2)\left(\dfrac{2^{-2}}{x^{-2}}\right) = (x^4y^2)\left(\dfrac{x^2}{2^2}\right) = \dfrac{x^6y^2}{4}$

In (a), we first used Eq. (11.6) and then (11.4). The simplification was completed by changing the division of a fraction to multiplication and using Eq. (11.1). In (b), we first used Eq. (11.3), then (11.6), and finally (11.1). ∎

Practice Exercise

1. Simplify: $x\left(\dfrac{3}{x}\right)^{-3}$

EXAMPLE 4 Exponents and units of measurement

When writing a denominate number, if units of measurement appear in the denominator, they can be written using negative exponents. For example, the metric unit for pressure is the *pascal*, where 1 pa = $1\,N/m^2$. This can be written as

$$1\,Pa = 1\,N/m^2 = 1\,N \cdot m^{-2}$$

where $1/m^2 = m^{-2}$.

The metric unit for energy is the *joule*, where $1\,J = 1\,kg \cdot (m \cdot s^{-1})^2$, or

$$1\,J = 1\,kg \cdot m^2 \cdot s^{-2} = 1\,kg \cdot m^2/s^2$$ ∎

■ Named for the French mathematician and scientist Blaise Pascal (1623–1662).

■ Named for the English physicist James Prescott Joule (1818–1899).

Care must be taken to apply the laws of exponents properly. Certain common problems are pointed out in the following examples.

EXAMPLE 5 Be careful with zero exponents

The expression $(-5x)^0$ equals 1, whereas the expression $-5x^0$ equals –5. For $(-5x)^0$, the parentheses show that the expression $-5x$ is raised to the zero power, whereas for $-5x^0$, only x is raised to the zero power and we have

$$-5x^0 = -5(1) = -5$$

CAUTION ▶ Also, $(-5)^0 = 1$ but $-5^0 = -1$. Again, for $(-5)^0$, parentheses show -5 raised to the zero power, whereas for -5^0, only 5 is raised to the zero power.

Similarly, $(-2)^2 = 4$ and $-2^2 = -4$.

Practice Exercise

2. Simplify: $3/x^0$

CAUTION ▶ For the same reasons, $2x^{-1} = \dfrac{2}{x}$, whereas $(2x)^{-1} = \dfrac{1}{2x}$. ∎

EXAMPLE 6 Simplification done in different ways

(a) $(2a + b^{-1})^{-2} = \dfrac{1}{(2a + b^{-1})^2} = \dfrac{1}{\left(2a + \dfrac{1}{b}\right)^2} = \dfrac{1}{\left(\dfrac{2ab + 1}{b}\right)^2}$

$= \dfrac{1}{\dfrac{(2ab + 1)^2}{b^2}} = \dfrac{b^2}{(2ab + 1)^2}$ not necessary to expand the denominator

(b) $(2a + b^{-1})^{-2} = \left(2a + \dfrac{1}{b}\right)^{-2} = \left(\dfrac{2ab + 1}{b}\right)^{-2}$

$= \dfrac{(2ab + 1)^{-2}}{b^{-2}} = \dfrac{b^2}{(2ab + 1)^2}$ positive exponents used in the final result ∎

EXAMPLE 7 Be careful not to make common error

There is an error that is commonly made in simplifying the type of expression in Example 6. We must be careful to see that

CAUTION ▶ $\qquad (2a + b^{-1})^{-2}$ is *not* equal to $(2a)^{-2} + (b^{-1})^{-2}$, or $\dfrac{1}{4a^2} + b^2$

Remember: As noted in Section 6.1, when raising a binomial (or any multinomial) to a power, we cannot simply raise each term to the power to obtain the result.

■ For reference, Eq. (11.4) is $(ab)^n = a^n b^n$.

However, when raising a product of factors to a power, we use Eq. (11.4). Thus,

$$(2ab^{-1})^{-2} = (2a)^{-2}(b^{-1})^{-2} = \frac{b^2}{(2a)^2} = \frac{b^2}{4a^2}$$

We see that we must be careful to distinguish between the power of a sum of terms and the power of a product of factors.   ■

CAUTION ▶ From the preceding examples, we see that *when a factor is moved from the denominator to the numerator of a fraction, or conversely, the* **sign** *of the exponent is changed.* We should carefully note the word *factor;* this rule does not apply to moving *terms* in the numerator or the denominator.

EXAMPLE 8 Factor moved numerator to denominator

(a) $3L^{-1} - (2L)^{-2} = \dfrac{3}{L} - \dfrac{1}{(2L)^2} = \dfrac{3}{L} - \dfrac{1}{4L^2}$

$$= \frac{12L - 1}{4L^2}$$

Practice Exercise

3. Simplify: $(3a)^{-1} - 3a^{-2}$

(b) $3^{-1}\left(\dfrac{4^{-2}}{3 - 3^{-1}}\right) = \dfrac{1}{3}\left(\dfrac{1}{4^2}\right)\left(\dfrac{1}{3 - \dfrac{1}{3}}\right) = \dfrac{1}{3 \times 4^2}\left(\dfrac{1}{\dfrac{9 - 1}{3}}\right)$

$$= \frac{1}{3 \times 4^2}\left(\frac{3}{8}\right) = \frac{1}{128} \qquad\qquad ■$$

EXAMPLE 9 Simplifying complex expression

terms

$$\frac{1}{x^{-1}}\left(\frac{x^{-1} - y^{-1}}{x^2 - y^2}\right) = \frac{x}{1}\left(\frac{\dfrac{1}{x} - \dfrac{1}{y}}{x^2 - y^2}\right) = x\left(\frac{\dfrac{y - x}{xy}}{x^2 - y^2}\right)$$

$$= \frac{\dfrac{x(y - x)}{xy}}{(x - y)(x + y)}$$

$$= \frac{x(y - x)}{xy} \times \frac{1}{(x - y)(x + y)}$$

$$= \frac{x(y - x)}{xy(x - y)(x + y)} = \frac{-(x - y)}{y(x - y)(x + y)}$$

$$= -\frac{1}{y(x + y)}$$

CAUTION ▶ Note that in this example, *the* x^{-1} *and* y^{-1} *in the numerator could not be moved directly to the denominator with positive exponents* because they are only terms of the original numerator.   ■

EXAMPLE 10 Simplifying expression from calculus

$$3(x + 4)^2(x - 3)^{-2} - 2(x - 3)^{-3}(x + 4)^3$$

$$= \frac{3(x + 4)^2}{(x - 3)^2} - \frac{2(x + 4)^3}{(x - 3)^3} = \frac{3(x - 3)(x + 4)^2 - 2(x + 4)^3}{(x - 3)^3}$$

$$= \frac{(x + 4)^2[3(x - 3) - 2(x + 4)]}{(x - 3)^3}$$

$$= \frac{(x + 4)^2(x - 17)}{(x - 3)^3}$$

Expressions like the one in this example are often found in problems in calculus. ∎

EXERCISES 11.1

In Exercises 1–4, solve the resulting problems if the given changes are made in the indicated examples of this section.

1. In Example 3, change the factor x^2 to x^{-2} and then find the result.

2. In Example 6, change the term $2a$ to $2a^{-1}$ and then find the result.

3. In Example 8(b), change the 3^{-1} in the denominator to 3^{-2} and then find the result.

4. In Example 9, change the sign in the numerator from $-$ to $+$ and then find the result.

In Exercises 5–52, express each of the given expressions in simplest form with only positive exponents.

5. $x^7 x^{-4}$

6. $y^9 y^{-2}$

7. $2a^2 a^{-6}$

8. $5ss^{-5}$

9. $5^0 \times 5^{-3}$

10. $(3^2 \times 4^{-3})^3$

11. $(2\pi x^{-1})^2$

12. $(3xy^{-2})^3$

13. $2(5an^{-2})^{-1}$

14. $4(6s^2 t^{-1})^{-2}$

15. $(-4)^0$

16. -4^0

17. $-7x^0$

18. $(-7x)^0$

19. $3x^{-2}$

20. $(3x)^{-2}$

21. $(7a^{-1}x)^{-3}$

22. $7a^{-1}x^{-3}$

23. $\left(\dfrac{2}{n^3}\right)^{-3}$

24. $\left(\dfrac{3}{x^3}\right)^{-2}$

25. $3\left(\dfrac{a}{b^{-2}}\right)^{-3}$

26. $5\left(\dfrac{2n^{-2}}{D^{-1}}\right)^{-2}$

27. $(a + b)^{-1}$

28. $a^{-1} + b^{-1}$

29. $3x^{-2} + 2y^{-2}$

30. $(3x + 2y)^{-2}$

31. $(2a^{-n})^2\left(\dfrac{3}{2a^n}\right)^{-1}$

32. $(7 \times 3^{-a})\left(\dfrac{3^a}{7}\right)^2$

33. $\left(\dfrac{3a^2}{4b}\right)^{-3}\left(\dfrac{4}{a}\right)^{-5}$

34. $(2np^{-2})^{-2}(4^{-1}p^2)^{-1}$

35. $\left(\dfrac{V^{-1}}{2t}\right)^{-2}\left(\dfrac{t^2}{V^{-2}}\right)^{-3}$

36. $ab\left(\dfrac{a^{-2}}{b^2}\right)^{-3}\left(\dfrac{a^{-3}}{b^5}\right)^2$

37. $2a^{-2} + (2a^{-2})^4$

38. $3(a^{-1}z^2)^{-3} + c^{-2}z^{-1}$

39. $2 \times 3^{-1} + 4 \times 3^{-2}$

40. $5 \times 2^{-2} - 3^{-1} \times 2^3$

41. $(R_1^{-1} + R_2^{-1})^{-1}$

42. $2(2a - b^{-2})^{-1}$

43. $(n^{-2} - 2n^{-1})^2$

44. $2(2^{-3} - 4^{-1})^{-2}$

45. $\dfrac{6^{-1}}{4^{-2} + 2}$

46. $\dfrac{x - y^{-1}}{x^{-1} - y}$

47. $\dfrac{x^{-2} - y^{-2}}{x^{-1} - y^{-1}}$

48. $\dfrac{ax^{-2} + a^{-2}x}{a^{-1} + x^{-1}}$

49. $2t^{-2} + t^{-1}(t + 1)$

50. $3x^{-1} - x^{-3}(y + 2)$

51. $(D - 1)^{-1} + (D + 1)^{-1}$

52. $4(2x - 1)(x + 2)^{-1} - (2x - 1)^2(x + 2)^{-2}$

In Exercises 53–70, solve the given problems.

53. If $x < 0$, is it ever true that $x^{-2} < x^{-1}$?

54. Is it true that $(a + b)^0 = 1$ for all values of a and b?

55. Express $4^2 \times 64$ (a) as a power of 4 and (b) as a power of 2.

56. Express $1/81$ (a) as a power of 9 and (b) as a power of 3.

57. (a) By use of Eqs. (11.4) and (11.6), show that
$$\left(\frac{a}{b}\right)^{-n} = \left(\frac{b}{a}\right)^n$$

(b) Verify the equation in part (a) by evaluating each side with $a = 3.576$, $b = 8.091$, and $n = 7$.

(W) 58. For what integral values of n is $(-3)^{-n} = -3^{-n}$? Explain.

59. For what integral value(s) of n is $n^\pi > \pi^n$?

(W) 60. Evaluate $(8^{19})^{12}/(8^{16})^{14}$. What happens when you try to evaluate this on a calculator?

61. Solve for x: $2^{5x} = 2^7(2^{2x})^2$.

62. In analyzing the tuning of an electronic circuit, the expression $[\omega\omega_0^{-1} - \omega_0\omega^{-1}]^2$ is used. Expand and simplify this expression.

63. The metric unit of energy, the *joule* (J), can be expressed as $kg \cdot s^{-2} \cdot m^2$. Simplify these units and include *newtons* (see Appendix B) and only positive exponents in the final result.

64. The units for the electric quantity called *permittivity* are $C^2 \cdot N^{-1} \cdot m^{-2}$. Given that $1\,F = 1\,C^2 \cdot J^{-1}$, show that the units of permittivity are F/m. See Appendix B.

65. When studying a solar energy system, the units encountered are $kg \cdot s^{-1}(m \cdot s^{-2})^2$. Simplify these units and include *joules* (see Example 4) and only positive exponents in the final result.

66. The metric units for the velocity v of an object are $m \cdot s^{-1}$, and the units for the acceleration a of the object are $m \cdot s^{-2}$. What are the units for v/a?

67. Given that $v = a^p t^r$, where v is the velocity of an object, a is its acceleration, and t is the time, use the metric units given in Exercise 66 to show that $p = r = 1$.

68. An idealized model of the thermodynamic process in a gasoline engine is the *Otto cycle*. The efficiency e of the process is

$$e = \dfrac{\dfrac{T_1 r^\gamma}{r} - \dfrac{T_2 r^\gamma}{r} - T_1 + T_2}{\dfrac{T_1 r^\gamma}{r} - \dfrac{T_2 r^\gamma}{r}}.$$ Show that $e = 1 - \dfrac{1}{r^{\gamma-1}}$.

69. An expression encountered in finance is

$$\dfrac{p(1 + i)^{-1}[(1 + i)^{-n} - 1]}{(1 + i)^{-1} - 1}$$

where n is an integer. Simplify this expression.

70. In optics, the combined focal length F of two lenses is given by $F = [f_1^{-1} + f_2^{-1} + d(f_1 f_2)^{-1}]^{-1}$, where f_1 and f_2 are the focal lengths of the lenses and d is the distance between them. Simplify the right side of this equation.

Answers to Practice Exercises

1. $\dfrac{x^4}{27}$ **2.** 3   **3.** $\dfrac{a - 9}{3a^2}$

11.2 Fractional Exponents

Meaning and Interpretation of Fractional Exponents • Evaluations • Graphs • Simplifying Expressions

In Section 11.1, we reviewed the use of integral exponents, including exponents that are negative integers and zero. We now show how rational numbers may be used as exponents. With the appropriate definitions, all the laws of exponents are valid for all rational numbers as exponents.

Equation (11.3) states that $(a^m)^n = a^{mn}$. If we were to let $m = \frac{1}{2}$ and $n = 2$, we would have $(a^{1/2})^2 = a^1$. However, we already have a way of writing a quantity that, when squared, equals a. This is written as $\sqrt{a}$. To be consistent with previous definitions and to allow the laws of exponents to hold, we define

■ The radical in Eq. (11.7) is the symbol for the nth root of a number. Do not confuse it with $\sqrt{a}$, the square root of a.

$$a^{1/n} = \sqrt[n]{a} \tag{11.7}$$

In order that Eqs. (11.3) and (11.7) may hold at the same time, we define

$$a^{m/n} = \sqrt[n]{a^m} = \left(\sqrt[n]{a}\right)^m \tag{11.8}$$

These definitions are valid for all the laws of exponents. We must note that Eqs. (11.7) and (11.8) are valid as long as $\sqrt[n]{a}$ does not involve the even root of a negative number. Such numbers are imaginary and are considered in Chapter 12.

EXAMPLE 1 Meaning of fractional exponent

We now verify that Eq. (11.1) holds for the above definitions:

$$a^{1/4} a^{1/4} a^{1/4} a^{1/4} = a^{(1/4)+(1/4)+(1/4)+(1/4)} = a^1$$

■ For reference, Eq. (11.1) is $a^m \times a^n = a^{m+n}$.

■ Eq. (11.3) is $(a^m)^n = a^{mn}$.

Now, $a^{1/4} = \sqrt[4]{a}$ by definition. Also, by definition, $\sqrt[4]{a}\,\sqrt[4]{a}\,\sqrt[4]{a}\,\sqrt[4]{a} = a$. Equation (11.1) is thereby verified for $n = 4$ in Eq. (11.7).

Equation (11.3) is verified by the following:

$$(a^{1/4})(a^{1/4})(a^{1/4})(a^{1/4}) = (a^{1/4})^4 = a^1 = \left(\sqrt[4]{a}\right)^4 \qquad ■$$

We may interpret $a^{m/n}$ in Eq. (11.8) as the mth power of the nth root of a, as well as the nth root of the mth power of a. This is illustrated in the following example.

EXAMPLE 2 Interpretation of fractional exponent

$$8^{2/3} = \left(\sqrt[3]{8}\right)^2 = (2)^2 = 4 \quad \text{or} \quad 8^{2/3} = \sqrt[3]{8^2} = \sqrt[3]{64} = 4 \qquad ■$$

Although both interpretations of Eq. (11.8) are possible, as indicated in Example 2, in evaluating numerical expressions involving fractional exponents without a calculator, it is almost always best to *find the root first, as indicated by the denominator* of the fractional exponent. This allows us to find the root of the smaller number, which is normally easier to find.

EXAMPLE 3 Evaluating

To evaluate $(64)^{5/2}$, we should proceed as follows:

$$(64)^{5/2} = [(64)^{1/2}]^5 = 8^5 = 32,768$$

If we raised 64 to the fifth power first, we would have

$$(64)^{5/2} = (64^5)^{1/2} = (1,073,741,824)^{1/2}$$

We would now have to evaluate the indicated square root. This demonstrates why it is preferable to find the indicated root first.   ■

EXAMPLE 4 Evaluating

(a) $(16)^{3/4} = (16^{1/4})^3 = 2^3 = 8$

(b) $4^{-1/2} = \dfrac{1}{4^{1/2}} = \dfrac{1}{2}$ **(c)** $9^{3/2} = (9^{1/2})^3 = 3^3 = 27$

Practice Exercises

Evaluate: **1.** $81^{5/4}$ **2.** $27^{-4/3}$

We note in (b) that Eq. (11.6) must also hold for negative rational exponents. In writing $4^{-1/2}$ as $1/4^{1/2}$, the *sign* of the exponent is changed.   ■

Fractional exponents allow us to find roots of numbers on a calculator. By use of the appropriate key $\big($on most calculators $\boxed{x^y}$ or $\boxed{\wedge}\big)$, we may raise any positive number to any power. For roots, we use the equivalent fractional exponent. Powers that are fractions or decimal in form are entered directly.

■ Named for the Scottish physicist Lord Kelvin (1824–1907).

EXAMPLE 5 Fractional exponents and calculator

The thermodynamic temperature T (in kelvins (K)) is related to the pressure P (in kPa) of a gas by the equation $T = 80.5P^{2/7}$. Find the value of T for $P = 750$ kPa.

Substituting, we have

$$T = 80.5(750)^{2/7}$$

The calculator display for this calculation is shown in Fig. 11.1. Therefore, $T = 534$ K.   ■

```
80.5*750^(2/7)
       533.6321116
```

Fig. 11.1

When finding powers of negative numbers, some calculators will show an error. If this is the case, enter the positive value of the number and then enter a negative sign for the result when appropriate. From Section 1.6, we recall that *an even root of a negative number is imaginary and an odd root of a negative number is negative.* If it is an integral power, the basic laws of signs are used.

EXAMPLE 6 Fractional exponents and graphs

Plot the graph of the function $y = 2x^{1/3}$.

In obtaining points for the graph, we use $(1/3)$ as the power when using the calculator. If your calculator does not evaluate powers of negative numbers, enter positive values for the negative values of x and then make the results negative. Since $x^{1/3} = \sqrt[3]{x}$, we know that $x^{1/3}$ is negative for negative values of x because we have an odd root of a negative number. We get the following table of values:

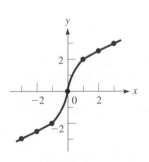

Fig. 11.2

x	-3	-2	-1	0	1	2	3
y	-2.9	-2.5	-2.0	0	2.0	2.5	2.9

The graph is shown in Fig. 11.2. Of course, this curve can easily be shown on a graphing calculator.   ■

Fractional exponents are often easier to use in more complex expressions involving roots. This is true in algebra and in topics from more advanced mathematics. Any expression with radicals can also be expressed with fractional exponents and then simplified.

EXAMPLE 7 Simplifying expressions

(a) $(8a^2b^4)^{1/3} = [(8^{1/3})(a^2)^{1/3}(b^4)^{1/3}]$ using Eq. (11.4)

$\qquad\qquad\quad = 2a^{2/3}b^{4/3}$ using Eqs. (11.7) and (11.3)

(b) $a^{3/4}a^{4/5} = a^{3/4+4/5} = a^{31/20}$ using Eq. (11.1)

(c) $\left(\dfrac{4^{-3/2}x^{2/3}y^{-7/4}}{2^{3/2}x^{-1/3}y^{3/4}}\right)^{2/3} = \left(\dfrac{x^{2/3}x^{1/3}}{2^{3/2}4^{3/2}y^{3/4}y^{7/4}}\right)^{2/3}$ using Eq. (11.6)

$\qquad\qquad\qquad\qquad\qquad = \left(\dfrac{x^{2/3+1/3}}{2^{3/2}4^{3/2}y^{3/4+7/4}}\right)^{2/3}$ using Eq. (11.1)

$\qquad\qquad\qquad\qquad\qquad = \dfrac{x^{(1)(2/3)}}{2^{(3/2)(2/3)}4^{(3/2)(2/3)}y^{(10/4)(2/3)}}$ using Eq. (11.4)

$\qquad\qquad\qquad\qquad\qquad = \dfrac{x^{2/3}}{8y^{5/3}}$

(d) $(4x^4)^{-1/2} - 3x^{-3} = \dfrac{1}{(4x^4)^{1/2}} - \dfrac{3}{x^3}$ using Eq. (11.6)

$\qquad\qquad\qquad\qquad = \dfrac{1}{2x^2} - \dfrac{3}{x^3} = \dfrac{x-6}{2x^3}$ using Eq. (11.7); common denominator ∎

Practice Exercises

Simplify: **3.** $(64a^2c^3)^{2/3}$

          4. $2x^{-4} - (8x^3)^{-1/3}$

EXAMPLE 8 Simplifying expression – application

During a day, the rate R at which the radiation changes at a solar-radiation collector is

$$R = \frac{(t^4 + 100)^{1/2} - 2t^3(t + 6)(t^4 + 100)^{-1/2}}{[(t^4 + 100)^{1/2}]^2}$$

Here, R is measured in kW/(m$^2 \cdot$h), t is the number of hours from noon, and $-6\,\text{h} \le t \le 8\,\text{h}$. Express the right side of this equation in simpler form and find R for $t = 0$ (noon) and for $t = 4\,\text{h}$ (4 P.M.)

Performing the simplification, we have the following steps:

■ See the chapter introduction.

■ Solar collectors supply the power for most space satellites.

■ For reference, basic forms of Eq. (11.1) to (11.7) are as follows:

$a^m \times a^n = a^{m+n}$ (11.1)

$\dfrac{a^m}{a^n} = a^{m-n}$ (11.2)

$(a^m)^n = a^{mn}$ (11.3)

$(ab)^n = a^n b^n$ (11.4)

$a^0 = 1$ (11.5)

$a^{-n} = \dfrac{1}{a^n}$ (11.6)

$a^{1/n} = \sqrt[n]{a}$ (11.7)

$$R = \frac{(t^4 + 100)^{1/2} - \dfrac{2t^3(t+6)}{(t^4 + 100)^{1/2}}}{(t^4 + 100)}$$ ← using Eq. (11.6)

                                     ← using Eq. (11.3)

$$= \frac{\dfrac{(t^4 + 100)^{1/2}(t^4 + 100)^{1/2} - 2t^3(t+6)}{(t^4 + 100)^{1/2}}}{(t^4 + 100)}$$ ← common denominator

$$= \frac{(t^4 + 100) - 2t^3(t+6)}{(t^4 + 100)^{1/2}} \times \frac{1}{t^4 + 100}$$ ← invert divisor and multiply

$$= \frac{100 - 12t^3 - t^4}{(t^4 + 100)^{1/2}(t^4 + 100)} = \frac{100 - 12t^3 - t^4}{(t^4 + 100)^{3/2}}$$ ← using Eq. (11.1)

For $t = 0$: $R = \dfrac{100 - 12(0^3) - 0^4}{(0^4 + 100)^{3/2}} = 0.10\ \text{kW}/(\text{m}^2 \cdot \text{h})$

For $t = 4\,\text{h}$: $R = \dfrac{100 - 12(4^3) - 4^4}{(4^4 + 100)^{3/2}} = -0.14\ \text{kW}/(\text{m}^2 \cdot \text{h})$

This shows that the radiation is increasing at noon, and decreasing at 4 P.M. ∎

EXERCISES 11.2

In Exercises 1–4, solve the resulting problems if the given changes are made in the indicated examples of this section.

1. In Example 2, change the exponent to 4/3.
2. In Example 4(b), change the exponent to −3/2.
3. In Example 6, change the exponent to −1/3.
4. In Example 7(d), change the exponent −1/2 to −3/2.

In Exercises 5–28, evaluate the given expressions.

5. $25^{1/2}$ 6. $27^{1/3}$ 7. $81^{1/4}$ 8. $125^{2/3}$
9. $100^{25/2}$ 10. $-16^{5/4}$ 11. $8^{-1/3}$ 12. $16^{-1/4}$
13. $64^{-2/3}$ 14. $-32^{-4/5}$ 15. $5^{1/2}5^{3/2}$ 16. $(4^4)^{3/2}$
17. $(3^6)^{2/3}$ 18. $\dfrac{121^{-1/2}}{100^{1/2}}$ 19. $\dfrac{1000^{1/3}}{-400^{-1/2}}$ 20. $\dfrac{-7^{-1/2}}{6^{-1}7^{1/2}}$
21. $\dfrac{15^{2/3}}{5^2 15^{-1/3}}$ 22. $\dfrac{(-27)^{1/3}}{6}$ 23. $\dfrac{(-8)^{2/3}}{-2}$ 24. $\dfrac{-4^{-1/2}}{(-64)^{-2/3}}$
25. $125^{-2/3} - 100^{-3/2}$ 26. $32^{0.4} + 25^{-0.5}$
27. $\dfrac{16^{-0.25}}{5} + \dfrac{2^{-0.6}}{2^{0.4}}$ 28. $\dfrac{4^{-1}}{36^{-1/2}} - \dfrac{5^{-1/2}}{5^{1/2}}$

In Exercises 29–32, use a calculator to evaluate each expression.

29. $17.98^{1/4}$ 30. $(-750.81)^{2/3}$
31. $4.0187^{-4/9}$ 32. $0.1863^{-1/6}$

In Exercises 33–56, simplify the given expressions. Express all answers with positive exponents.

33. $B^{2/3}B^{1/2}$ 34. $x^{5/6}x^{-1/3}$ 35. $\dfrac{y^{-1/2}}{-y^{2/5}}$
36. $\dfrac{s^{1/4}s^{2/3}}{s^{-1}}$ 37. $\dfrac{x^{3/10}}{x^{-1/5}x^2}$ 38. $\dfrac{R^{-2/5}R^2}{R^{-3/10}}$
39. $(8a^3b^6)^{1/3}$ 40. $(8b^{-4}c^2)^{2/3}$ 41. $(16a^4b^3)^{-3/4}$
42. $(32C^5D^4)^{-2/5}$ 43. $\left(\dfrac{a^{5/7}}{a^{2/3}}\right)^{7/4}$ 44. $\left(\dfrac{4a^{5/6}b^{-1/5}}{a^{2/3}b^2}\right)^{-1/2}$
45. $\dfrac{1}{2}(4x^2 + 1)^{-1/2}(8x)$ 46. $\dfrac{(x + x^{1/2})(x - x^{1/2})}{x}$
47. $\dfrac{y^{3/8}(y^{5/8} - y^{13/8})}{y^{1/2}(y^{1/2} - y^{-1/2})}$ 48. $\dfrac{3^{-1}a^{1/2}}{4^{-1/2}b} \div \dfrac{9^{1/2}a^{-1/3}}{2b^{-1/4}}$
49. $(T^{-1} + 2T^{-2})^{-1/2}$ 50. $(a^{-2} - a^{-4})^{-1/4}$
51. $(a^3)^{-4/3} + a^{-2}$ 52. $(4N^6)^{-1/2} - 2N^{-1}$
53. $[(a^{1/2} - a^{-1/2})^2 + 4]^{1/2}$ 54. $4x^{1/2} + \dfrac{1}{2}x^{-1/2}(4x + 1)$
55. $x^2(2x - 1)^{-1/2} + 2x(2x - 1)^{1/2}$
56. $(3n - 1)^{-2/3}(1 - n) - (3n - 1)^{1/3}$

In Exercises 57–60, graph the given functions.

57. $f(x) = 3x^{1/2}$ 58. $f(x) = 2x^{2/3}$
59. $f(t) = t^{-4/5}$ 60. $f(V) = 4V^{3/2}$

In Exercises 61–68, perform the indicated operations.

61. Simplify $(x^{n-1} \div x^{n-3})^{1/3}$ and express the result as a radical.

W 62. (a) Simplify $(x^2 - 4x + 4)^{1/2}$. (b) For what values of x is your answer in part (a) valid? Explain.

W 63. A factor used in determining the performance of a solar-energy storage system is $(A/S)^{-1/4}$, where A is the actual storage capacity and S is a standard storage capacity. If this factor is 0.5, explain how to find the ratio A/S.

64. A factor used in measuring the loudness sensed by the human ear is $(I/I_0)^{0.3}$, where I is the intensity of the sound and I_0 is a reference intensity. Evaluate this factor for $I = 3.2 \times 10^{-6}$ W/m^2 (ordinary conversation) and $I_0 = 10^{-12}$ W/m^2.

65. The period T of a satellite circling earth is given by
$$T^2 = kR^3\left(1 + \frac{d}{R}\right)^3,$$
where R is the radius of earth, d is the distance of the satellite above earth, and k is a constant. Solve for R, using fractional exponents in the result.

66. The withdrawal resistance R of a nail of diameter d indicates its holding power. One formula for R is $R = ks^{5/2}dh$, where k is a constant, s is the specific gravity of the wood, and h is the depth of the nail in the wood. Solve for s using fractional exponents in the result.

67. The electric current i (in A) in a circuit with a battery of voltage E, a resistance R, and an inductance L, is $i = \dfrac{E}{R}(1 - e^{-Rt/L})$, where t is the time after the circuit is closed. See Fig. 11.3. Find i for $E = 6.20$ V, $R = 1.20\ \Omega$, $L = 3.24$ H, and $t = 0.00100$ s. (The number e is irrational and can be found from the calculator.)

Fig. 11.3

Fig. 11.4

68. For a heat-seeking rocket in pursuit of an aircraft, the distance d (in km) from the rocket to the aircraft is $d = \dfrac{500(\sin\theta)^{1/2}}{(1 - \cos\theta)^{3/2}}$, where θ is shown in Fig. 11.4. Find d for $\theta = 125.0°$.

Answers to Practice Exercises

1. 243   2. $\dfrac{1}{81}$ 3. $16a^{4/3}c^2$ 4. $\dfrac{4 - x^3}{2x^4}$

11.3 Simplest Radical Form

Operations with Radicals • Removing Perfect *n*th-power Factors • Reducing the Order of a Radical • Rationalizing the Denominator.

As we have said, any expression with radicals can also be expressed with fractional exponents. For adding or subtracting radicals, there is little advantage to changing form, but with multiplication or division of radicals, fractional exponents have some advantages. Therefore, we now define operations with radicals so that they are consistent with the laws of exponents. This will let us use the form that is more convenient for the operation being performed.

$$\sqrt[n]{a^n} = \left(\sqrt[n]{a}\right)^n = a \tag{11.9}$$

$$\sqrt[n]{a}\sqrt[n]{b} = \sqrt[n]{ab} \tag{11.10}$$

$$\sqrt[m]{\sqrt[n]{a}} = \sqrt[mn]{a} \tag{11.11}$$

$$\frac{\sqrt[n]{a}}{\sqrt[n]{b}} = \sqrt[n]{\frac{a}{b}} \quad (b \neq 0) \tag{11.12}$$

NOTE ▶ *The number under the radical is called the **radicand**, and the number indicating the root being taken is called the **order** (or **index**) of the radical.* To avoid difficulties with imaginary numbers (which are considered in the next chapter), *we will assume that all letters represent positive numbers.*

EXAMPLE 1 Operation with radicals

Following are illustrations using Eqs. (11.9) to (11.12).

(a) $\sqrt[5]{4^5} = \left(\sqrt[5]{4}\right)^5 = 4$ using Eq. (11.9)

(b) $\sqrt[3]{2}\sqrt[3]{3} = \sqrt[3]{2 \times 3} = \sqrt[3]{6}$ using Eq. (11.10)

(c) $\sqrt[3]{\sqrt{5}} = \sqrt[3\times2]{5} = \sqrt[6]{5}$ using Eq. (11.11)

(d) $\dfrac{\sqrt{7}}{\sqrt{3}} = \sqrt{\dfrac{7}{3}}$ using Eq. (11.12)   ∎

EXAMPLE 2 Root of sum and root of product

In Example 5 of Section 1.6, we saw that

$$\sqrt{16 + 9} \quad \text{is } \textbf{\textit{not}} \text{ equal to} \quad \sqrt{16} + \sqrt{9}$$

However, using Eq. (11.10),

$$\sqrt{16 \times 9} = \sqrt{16} \times \sqrt{9}$$
$$= 4 \times 3 = 12$$

NOTE ▶ Therefore, we must *be careful to distinguish between the root of a sum of terms and the root of a product of factors.* This is the same as with powers of sums and powers of products, as shown in Example 7 of Section 11.1. It should be the same, as a root can be interpreted as a fractional exponent.   ∎

EXAMPLE 3 Removing perfect square factors

To simplify $\sqrt{75}$, we know that $75 = (25)(3)$ and that $\sqrt{25} = 5$. As in Section 1.6 and now using Eq. (11.10), we write

$$\sqrt{75} = \sqrt{(25)(3)} = \sqrt{25}\sqrt{3} = 5\sqrt{3}$$

        └─ perfect square

This illustrates one step that should always be carried out in simplifying radicals: *Always*

NOTE ▶ *remove all perfect nth-power factors from the radicand of a radical of order n.*   ∎

EXAMPLE 4   Removing perfect nth-power factors

(a) $\sqrt{72} = \sqrt{(36)(2)} = \sqrt{36}\sqrt{2} = 6\sqrt{2}$

$\qquad\qquad$ perfect square

(b) $\sqrt{a^3b^2} = \sqrt{(a^2)(a)(b^2)} = \sqrt{a^2}\sqrt{a}\sqrt{b^2} = ab\sqrt{a}$

$\qquad\qquad\qquad$ perfect squares

(c) cube root $\rightarrow$ $\sqrt[3]{40} = \sqrt[3]{(8)(5)} = \sqrt[3]{8}\sqrt[3]{5} = 2\sqrt[3]{5}$

$\qquad\qquad$ perfect cube

(d) fifth root $\rightarrow$ $\sqrt[5]{64x^8y^{12}} = \sqrt[5]{(32)(2)(x^5)(x^3)(y^{10})(y^2)}$

$\qquad\qquad\qquad\qquad$ perfect fifth powers

$\qquad = \sqrt[5]{(32)(x^5)(y^{10})}\sqrt[5]{2x^3y^2}$

$\qquad = 2xy^2\sqrt[5]{2x^3y^2}$ ∎

The next example illustrates another procedure used to simplify radicals. It is to **NOTE ▶** *reduce the order of the radical,* when it is possible to do so.

EXAMPLE 5   Reducing order of radical

(a) $\sqrt[6]{8} = \sqrt[6]{2^3} = 2^{3/6} = 2^{1/2} = \sqrt{2}$

Here, we started with a sixth root and ended with a square root, thereby reducing the order of the radical. Fractional exponents are often helpful for this.

(b) $\sqrt[8]{16} = \sqrt[8]{2^4} = 2^{4/8} = 2^{1/2} = \sqrt{2}$

(c) $\dfrac{\sqrt[4]{9}}{\sqrt{3}} = \dfrac{\sqrt[4]{3^2}}{\sqrt{3}} = \dfrac{3^{2/4}}{3^{1/2}} = 1$

(d) $\dfrac{\sqrt[6]{8}}{\sqrt{7}} = \dfrac{\sqrt[6]{2^3}}{\sqrt{7}} = \dfrac{2^{1/2}}{7^{1/2}} = \sqrt{\dfrac{2}{7}}$

(e) $\sqrt[9]{27x^6y^{12}} = \sqrt[9]{3^3x^6y^9y^3} = 3^{3/9}x^{6/9}y^{9/9}y^{3/9} = 3^{1/3}x^{2/3}yy^{1/3}$

$\qquad = y\sqrt[3]{3x^2y}$ ∎

If a radical is to be written in its *simplest form,* the two operations illustrated in the last four examples must be performed. Therefore, we have the following:

STEPS TO REDUCE A RADICAL TO SIMPLEST FORM
1. *Remove all perfect nth-power factors from a radical of order n.*
2. *If possible, reduce the order of the radical.*

When working with fractions, it has traditionally been the practice to write a fraction with radicals in a form in which the denominator contains no radicals. Such a fraction was not considered to be in simplest form unless this was done. This step of simplification was performed primarily for ease of calculation, but with a calculator it does not matter to any extent that there is a radical in the denominator. However, the procedure of writing a radical in this form, called **rationalizing the denominator,** is at times useful for other purposes. Therefore, the following examples show how the process of rationalizing the denominator is carried out.

EXAMPLE 6 Rationalizing denominator

To write $\sqrt{\frac{2}{5}}$ in an equivalent form in which the denominator is not included under the radical sign, we *create a perfect square in the denominator* by multiplying the numerator and the denominator under the radical by 5. This gives us $\sqrt{\frac{10}{25}}$, which may be written as $\frac{1}{5}\sqrt{10}$ or $\frac{\sqrt{10}}{5}$. These steps are written as follows:

$$\sqrt{\frac{2}{5}} = \sqrt{\frac{2\times5}{5\times5}} = \sqrt{\frac{10}{25}} = \frac{\sqrt{10}}{\sqrt{25}} = \frac{\sqrt{10}}{5}$$

perfect square

EXAMPLE 7 Rationalizing denominator

(a) $\dfrac{5}{\sqrt{18}} = \dfrac{5}{3\sqrt{2}} = \dfrac{5(\sqrt{2})}{3\sqrt{2}(\sqrt{2})} = \dfrac{5\sqrt{2}}{3\sqrt{4}} = \dfrac{5\sqrt{2}}{6}$

perfect square

(b) cube root → $\sqrt[3]{\dfrac{2}{3}} = \sqrt[3]{\dfrac{2\times9}{3\times9}} = \sqrt[3]{\dfrac{18}{27}} = \dfrac{\sqrt[3]{18}}{\sqrt[3]{27}} = \dfrac{\sqrt[3]{18}}{3}$

perfect cube

In (a), a perfect square was made by multiplying by $\sqrt{2}$. We can indicate this as we have shown or by multiplying the numerator by $\sqrt{2}$ and multiplying the 2 under the radical in the denominator by 2, in which case the denominator would be $3\sqrt{2\times2}$. In (b), we want a perfect cube, since a cube root is being found.

Practice Exercise

Rationalize the denominator: **3.** $\sqrt{\dfrac{a}{3b}}$

EXAMPLE 8 Rationalizing expression – application

The period T (in s) for one cycle of a simple pendulum is given by $T = 2\pi\sqrt{L/g}$, where L is the length of the pendulum and g is the acceleration due to gravity. Rationalize the denominator on the right side of this equation if $L = 3.0$ ft and $g = 32$ ft/s^2.

Substituting, and then rationalizing, we have

$$T = 2\pi\sqrt{\frac{3.0}{32}} = 2\pi\sqrt{\frac{3.0\times2}{32\times2}}$$
$$= 2\pi\sqrt{\frac{6.0}{64}} = \frac{2\pi}{8}\sqrt{6.0}$$
$$= \frac{\pi}{4}\sqrt{6.0} = 1.9\text{ s}$$

■ See Appendix C for a graphing calculator program TBLROOTS. It displays a table of square roots and cube roots.

EXAMPLE 9 Simplifying expression

Simplify $\sqrt{\dfrac{1}{2a^2} + 2b^{-2}}$ and rationalize the denominator.

We will write the expression using only positive exponents, and then perform the required operations.

first combine fractions over lowest common denominator →
$$\sqrt{\frac{1}{2a^2} + \frac{2}{b^2}} = \sqrt{\frac{b^2+4a^2}{2a^2b^2}} = \frac{\sqrt{b^2+4a^2}}{ab\sqrt{2}}$$ ← sum of squares—radical cannot be simplified
$$= \frac{\sqrt{b^2+4a^2}\sqrt{2}}{ab\sqrt{2\times2}}$$
$$= \frac{\sqrt{2(b^2+4a^2)}}{2ab}$$

EXERCISES 11.3

In Exercises 1–4, simplify the resulting expressions if the given changes are made in the indicated examples of this section.

1. In Example 4(b), change the exponent of b to 4 and then find the resulting expression.

2. In Example 5(d), change the 8 to 27 and then find the resulting expression.

3. In Example 7(b), change the root to a fourth root and then find the resulting expression.

4. In Example 9, replace b^{-2} with b^{-4} and then find the resulting expression.

In Exercises 5–64, write each expression in simplest radical form. If a radical appears in the denominator, rationalize the denominator.

5. $\sqrt{24}$
6. $\sqrt{150}$
7. $\sqrt{108}$

8. $\sqrt{98}$
9. $\sqrt{x^2 y^5}$
10. $\sqrt{pq^2 r^7}$

11. $\sqrt{x^2 y^4 z^3}$
12. $\sqrt{12ab^2}$
13. $\sqrt{18R^5 TV^4}$

14. $\sqrt{132M^2 N^3}$
15. $\sqrt[3]{16}$
16. $\sqrt[4]{48}$

17. $\sqrt[5]{96}$
18. $\sqrt[3]{-512}$
19. $\sqrt[3]{8a^2}$

20. $\sqrt[3]{5a^4 b^2}$
21. $\sqrt[4]{64r^3 s^4 t^5}$
22. $\sqrt[5]{16x^5 y^3 z^{11}}$

23. $\sqrt[3]{8}\sqrt[5]{4}$
24. $\sqrt[4]{4}\sqrt[4]{64}$
25. $\sqrt[3]{P}\sqrt[3]{P^2 V}$

26. $\sqrt[6]{3m^5 n^8}\sqrt[6]{9mn}$
27. $\sqrt{\dfrac{3}{2}}$
28. $\sqrt{\dfrac{11}{12}}$

29. $\sqrt[3]{\dfrac{3}{4}}$
30. $\sqrt[4]{\dfrac{2}{25}}$
31. $\sqrt[5]{\dfrac{1}{9}}$

32. $\sqrt[6]{\dfrac{5}{4}}$
33. $\sqrt[4]{400}$
34. $\sqrt[8]{81}$

35. $\sqrt[6]{64}$
36. $\sqrt[9]{27}$
37. $\sqrt{4 \times 10^4}$

38. $\sqrt{4 \times 10^5}$
39. $\sqrt{4 \times 10^6}$
40. $\sqrt[3]{16 \times 10^5}$

41. $\sqrt[4]{4a^2}$
42. $\sqrt[6]{b^2 c^4}$
43. $\sqrt[4]{\dfrac{1}{4}}$

44. $\dfrac{\sqrt[4]{320}}{\sqrt[4]{5}}$
45. $\sqrt[4]{\sqrt[3]{16}}$
46. $\sqrt[5]{\sqrt[4]{9}}$

47. $\sqrt{\sqrt{\sqrt{n}}}$
48. $\sqrt{b^4 \sqrt{a}}$
49. $\sqrt{28u^3 v^{-5}}$

50. $\sqrt{98x^6 y^{-7}}$
51. $\sqrt{64 + 144}$
52. $\sqrt{9 + 81}$

53. $\sqrt{\dfrac{2x}{3c^4}}$
54. $\sqrt{\dfrac{n}{m^3}}$
55. $\sqrt{\dfrac{5}{4} - \dfrac{1}{8}}$

56. $\sqrt{\dfrac{1}{a^2} + \dfrac{1}{b}}$
57. $\sqrt{xy^{-1} + x^{-1}y}$
58. $\sqrt{x^2 + 4^{-1}}$

59. $\sqrt{\dfrac{C - 2}{C + 2}}$
60. $\sqrt{a^2 + 2ab + b^2}$
61. $\sqrt{a^2 + b^2}$

62. $\sqrt{4x^2 - 1}$
63. $\sqrt{9x^2 - 6x + 1}$
64. $\sqrt{\dfrac{1}{2} + 2r + 2r^2}$

In Exercises 65–72, perform the required operation.

65. Change to radicals of the same order: $\sqrt{a}$; $\sqrt[3]{b}$; $\sqrt[6]{c}$.

66. Change to radicals of the same order: $3x^{2/3}$; $2y^{1/2}$; $(5z)^{1/4}$

67. Display the graphs of $y_1 = \sqrt{x + 2}$ and $y_2 = \sqrt{x} + \sqrt{2}$ on a calculator to show that $\sqrt{x + 2}$ is not equal to $\sqrt{x} + \sqrt{2}$.

(W) 68. An approximate equation for the efficiency E (in percent) of an engine is $E = 100\left(1 - 1/\sqrt[5]{R^2}\right)$, where R is the compression ratio. Explain how this equation can be written with fractional exponents and then find E for $R = 7.35$.

69. When analyzing the velocity of an object falling through a great distance, the expression $a\sqrt{2g/a}$ arises. Show by rationalizing the denominator that this expression takes on a simpler form.

70. According to the Doppler effect, the frequency f_0 observed of a sound of frequency f, when the observer is moving away from the source with velocity u, is given by $f_0 = \dfrac{fv}{v - u}\sqrt{1 - \left(\dfrac{u}{v}\right)^2}$, where v is the velocity of sound. Simplify the right side.

71. In analyzing an electronic filter circuit, the expression $\dfrac{8A}{\pi^2 \sqrt{1 + (f_0/f)^2}}$ is used. Rationalize the denominator, expressing the answer without the fraction f_0/f.

72. The expression $\dfrac{1}{2L}\sqrt{R^2 - \dfrac{4L}{C}}$ occurs in the study of electric circuits. Simplify this expression by combining terms under the radical and rationalizing the denominator.

Answers to Practice Exercises

1. $2x^2\sqrt{2x}$ 2. $2a^2\sqrt[3]{2ab^2}$ 3. $\dfrac{\sqrt{3ab}}{3b}$

11.4 Addition and Subtraction of Radicals

Similar Radicals • Adding and Subtracting Radicals • Calculator Verification of Result

When we add or subtract algebraic expressions, we combine similar terms, those that differ only in numerical coefficients. This is true when adding or subtracting radicals. *The radicals must be similar to perform the addition,* rather than simply indicate the addition. *Radicals are* **similar** *if they differ only in their numerical coefficients.* This means they must have the same order and have the same radicand.

In order to add radicals, we first express each radical in its simplest form, rationalize any denominators, and then combine those that are similar. For those that are not similar, we can only indicate the addition.

EXAMPLE 1   Adding similar radicals

(a) $2\sqrt{7} - 5\sqrt{7} + \sqrt{7} = -2\sqrt{7}$ all similar radicals

This result follows the distributive law, as it should. We can write

$$2\sqrt{7} - 5\sqrt{7} + \sqrt{7} = (2 - 5 + 1)\sqrt{7} = -2\sqrt{7}$$

We can also see that the terms combine just as

$$2x - 5x + x = -2x$$

(b) $\sqrt[5]{6} + 4\sqrt[5]{6} - 2\sqrt[5]{6} = 3\sqrt[5]{6}$ all similar radicals
(c) $\sqrt{5} + 2\sqrt{3} - 5\sqrt{5} = 2\sqrt{3} - 4\sqrt{5}$ answer contains two terms

similar radicals

We note in (c) that we are able only to indicate the final subtraction since the radicals are not similar. ∎

EXAMPLE 2   Adding radicals

(a) $\sqrt{2} + \sqrt{8} = \sqrt{2} + \sqrt{4 \times 2} = \sqrt{2} + \sqrt{4}\sqrt{2}$
$= \sqrt{2} + 2\sqrt{2} = 3\sqrt{2}$
(b) $\sqrt[3]{24} + \sqrt[3]{81} = \sqrt[3]{8 \times 3} + \sqrt[3]{27 \times 3} = \sqrt[3]{8}\sqrt[3]{3} + \sqrt[3]{27}\sqrt[3]{3}$
$= 2\sqrt[3]{3} + 3\sqrt[3]{3} = 5\sqrt[3]{3}$

Notice that $\sqrt{8}$, $\sqrt[3]{24}$, and $\sqrt[3]{81}$ were simplified before performing the additions.
CAUTION ▶ We also note that $\sqrt{2} + \sqrt{8}$ **is not** *equal to* $\sqrt{2 + 8}$. ∎

We note in the illustrations of Example 2 that the radicals do not initially appear to be similar. However, after each is simplified, we are able to recognize the similar radicals.

EXAMPLE 3   Simplify each – combine similar radicals

(a) $6\sqrt{7} - \sqrt{28} + 3\sqrt{63} = 6\sqrt{7} - \sqrt{4 \times 7} + 3\sqrt{9 \times 7}$
$= 6\sqrt{7} - 2\sqrt{7} + 3(3\sqrt{7})$
$= 6\sqrt{7} - 2\sqrt{7} + 9\sqrt{7}$
$= 13\sqrt{7}$ all similar radicals

Practice Exercise
Combine: **1.** $5\sqrt{44} - \sqrt{99}$

(b) $3\sqrt{125} - \sqrt{20} + \sqrt{27} = 3\sqrt{25 \times 5} - \sqrt{4 \times 5} + \sqrt{9 \times 3}$
$= 3(5\sqrt{5}) - 2\sqrt{5} + 3\sqrt{3}$
$= 13\sqrt{5} + 3\sqrt{3}$ not similar to others ∎

EXAMPLE 4   Simplify each – combine similar radicals

$$\sqrt{24} + \sqrt{\frac{3}{2}} = \sqrt{4 \times 6} + \sqrt{\frac{3 \times 2}{2 \times 2}} = \sqrt{4}\sqrt{6} + \sqrt{\frac{6}{4}}$$
$$= 2\sqrt{6} + \frac{\sqrt{6}}{2} = \frac{4\sqrt{6} + \sqrt{6}}{2} = \frac{5}{2}\sqrt{6}$$

One radical was simplified by removing the perfect square factor, and in the other, we rationalized the denominator. Note that we would not be able to combine the radicals if we did not rationalize the denominator of the second radical. ∎

Our main purpose in this section is to add radicals in radical form. However, a decimal value can be obtained by use of a calculator, and in the following example, we use decimal values to verify the result of the addition.

EXAMPLE 5 Calculator verification of result

Use a calculator to verify the result of the following addition:

$$\sqrt{32} + 7\sqrt{18} - 2\sqrt{200} = \sqrt{16 \times 2} + 7\sqrt{9 \times 2} - 2\sqrt{100 \times 2}$$
$$= 4\sqrt{2} + 7\left(3\sqrt{2}\right) - 2\left(10\sqrt{2}\right)$$
$$= 4\sqrt{2} + 21\sqrt{2} - 20\sqrt{2}$$
$$= 5\sqrt{2}$$

The calculator display that verifies this addition is shown in Fig. 11.5. ∎

Next is an example of adding radical expressions that contain literal numbers.

```
√(32)+7√(18)-2√(
200)
        7.071067812
5√(2)
        7.071067812
```

Fig. 11.5

EXAMPLE 6 Radicals with literal numbers

$$\sqrt{\frac{2}{3a}} - 2\sqrt{\frac{3}{2a}} = \sqrt{\frac{2(3a)}{3a(3a)}} - 2\sqrt{\frac{3(2a)}{2a(2a)}} = \sqrt{\frac{6a}{9a^2}} - 2\sqrt{\frac{6a}{4a^2}}$$
$$= \frac{1}{3a}\sqrt{6a} - \frac{2}{2a}\sqrt{6a} = \frac{1}{3a}\sqrt{6a} - \frac{1}{a}\sqrt{6a}$$
$$= \frac{\sqrt{6a} - 3\sqrt{6a}}{3a} = \frac{-2\sqrt{6a}}{3a} = -\frac{2}{3a}\sqrt{6a}$$ ∎

Practice Exercise

Combine: **2.** $\sqrt{\dfrac{8x}{5y}} - \sqrt{10xy}$

EXERCISES 11.4

In Exercises 1 and 2, simplify the resulting expressions if the given changes are made in the indicated examples of this section.

1. In Example 3(b), change $\sqrt{27}$ to $\sqrt{45}$ and then find the resulting simplified expression.

2. In Example 5, change $\sqrt{32}$ to $\sqrt{48}$ and then find the resulting simplified expression.

In Exercises 3–38, express each radical in simplest form, rationalize denominators, and perform the indicated operations.

3. $2\sqrt{3} + 5\sqrt{3}$ **4.** $8\sqrt{11} - 3\sqrt{11}$

5. $\sqrt{28} + \sqrt{5} - 3\sqrt{7}$ **6.** $8\sqrt{6} - \sqrt{12} - 5\sqrt{6}$

7. $\sqrt{5} + \sqrt{16 + 4}$ **8.** $\sqrt{7} + \sqrt{36 + 27}$

9. $2\sqrt{3t^2} - 3\sqrt{12t^2}$ **10.** $4\sqrt{2n^2} - \sqrt{50n^2}$

11. $\sqrt{8a} - \sqrt{32a}$ **12.** $\sqrt{27x} + 2\sqrt{18x}$

13. $2\sqrt{28} + 3\sqrt{175}$ **14.** $\sqrt{100 + 25} - 7\sqrt{80}$

15. $2\sqrt{200} - \sqrt{1250} - \sqrt{450}$ **16.** $2\sqrt{44} - \sqrt{99} + \sqrt{2}\sqrt{88}$

17. $3\sqrt{75R} + 2\sqrt{48R} - 2\sqrt{18R}$ **18.** $2\sqrt{28} - \sqrt{108} - 2\sqrt{175}$

19. $\sqrt{60} + \sqrt{\frac{5}{3}}$ **20.** $3\sqrt{84} - \sqrt{\frac{3}{7}}$

21. $\sqrt{\frac{1}{2}} + \sqrt{\frac{25}{2}} - 4\sqrt{18}$ **22.** $\sqrt{6} - \sqrt{\frac{2}{3}} - \sqrt{18}$

23. $\sqrt[3]{81} + \sqrt[3]{3000}$ **24.** $\sqrt[3]{-16} + \sqrt[3]{54}$

25. $\sqrt[4]{32} - \sqrt[8]{4}$ **26.** $\sqrt[6]{\sqrt{2}} - \sqrt[12]{2^{13}}$

27. $5\sqrt{a^3b} - \sqrt{4ab^5}$ **28.** $\sqrt{2R^2I} + \sqrt{8}\sqrt{I^3}$

29. $\sqrt{6}\sqrt{5}\sqrt{3} - \sqrt{40a^2}$ **30.** $3\sqrt{60b^2n} - b\sqrt{135n}$

31. $\sqrt[3]{24a^2b^4} - \sqrt[3]{3a^5b}$ **32.** $\sqrt[5]{32a^6b^4} + 3a\sqrt[5]{243ab^9}$

33. $\sqrt{\dfrac{a}{c^5}} - \sqrt{\dfrac{c}{a^3}}$ **34.** $\sqrt{\dfrac{2x}{3y}} + \sqrt{\dfrac{27y}{8x}}$

35. $\sqrt[3]{ab^{-1}} - \sqrt[3]{8a^{-2}b^2}$ **36.** $\sqrt{\dfrac{2xy^{-1}}{3}} + \sqrt{\dfrac{27x^{-1}y}{8}}$

37. $\sqrt{\dfrac{T-V}{T+V}} - \sqrt{\dfrac{T+V}{T-V}}$ **38.** $\sqrt{\dfrac{16}{x} + 8 + x} - \sqrt{1 - \dfrac{1}{x}}$

In Exercises 39–44, express each radical in simplest form, rationalize denominators, and perform the indicated operations. Then use a calculator to verify the result.

39. $3\sqrt{45} + 3\sqrt{75} - 2\sqrt{500}$ **40.** $2\sqrt{40} + 3\sqrt{90} - 5\sqrt{250}$

41. $2\sqrt{\frac{2}{3}} + \sqrt{24} - 5\sqrt{\frac{3}{2}}$ **42.** $\sqrt{\frac{2}{7}} - 2\sqrt{\frac{7}{2}} + 5\sqrt{56}$

43. $2\sqrt[3]{16} - \sqrt[3]{\frac{1}{4}}$ **44.** $5\sqrt[4]{810} - \sqrt[4]{\frac{5}{8}}$

In Exercises 45–52, solve the given problems.

45. Find the exact sum of the positive roots of $x^2 - 2x - 2 = 0$ and $x^2 + 2x - 11 = 0$.

W **46.** For the quadratic equation $ax^2 + bx + c = 0$, if a, b, and c are integers, the sum of the roots is a rational number. Explain.

(W) 47. Without calculating the actual value, determine whether $10\sqrt{11} - \sqrt{1000}$ is positive or negative. Explain.

48. The adjacent sides of a parallelogram are $\sqrt{12}$ and $\sqrt{27}$ units long. What is the perimeter of the parallelogram?

49. The two legs of a right triangle are $2\sqrt{2}$ and $2\sqrt{6}$ units long. What is the perimeter of the triangle?

50. The current I (in A) passing through a resistor R (in Ω) in which P watts of power are dissipated is $I = \sqrt{P/R}$. If the power dissipated in the resistors shown in Fig. 11.6 is W watts, what is the sum of the currents in radical form?

Fig. 11.6

51. A rectangular piece of plywood 4 ft by 8 ft has corners cut from it, as shown in Fig. 11.7. Find the perimeter of the remaining piece in exact form and in decimal form.

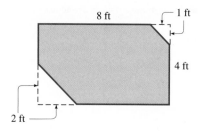

Fig. 11.7

52. Three squares with areas of 150 cm², 54 cm², and 24 cm² are displayed on a computer monitor. What is the sum (in radical form) of the perimeters of these squares?

Answers to Practice Exercises

1. $7\sqrt{11}$ 2. $\dfrac{(2 - 5y)\sqrt{10xy}}{5y}$

11.5 Multiplication and Division of Radicals

Multiplication of Radicals • Power of a Radical • Division of Radicals • Rationalizing the Denominator

When multiplying expressions containing radicals, we use Eq. (11.10), along with the normal procedures of algebraic multiplication. *Note that the orders of the radicals being multiplied in Eq. (11.10) are the same.* The following examples illustrate the method.

EXAMPLE 1 Multiplying monomial radicals

(a) $\sqrt{5}\sqrt{2} = \sqrt{5 \times 2} = \sqrt{10}$

(b) $\sqrt{33}\sqrt{3} = \sqrt{33 \times 3} = \sqrt{99} = \sqrt{9 \times 11} = \sqrt{9}\sqrt{11}$
$\qquad = 3\sqrt{11}$ ⎿ perfect square

or $\quad \sqrt{33}\sqrt{3} = \sqrt{33 \times 3} = \sqrt{11 \times 3 \times 3}$
$\qquad = 3\sqrt{11}$

Note that we express the resulting radical in simplest form.

■ For reference, Eq. (11.10) is $\sqrt[n]{a}\ \sqrt[n]{b} = \sqrt[n]{ab}$.

(c) $\sqrt[3]{6}\sqrt[3]{4} = \sqrt[3]{6(4)} = \sqrt[3]{24} = \sqrt[3]{8}\sqrt[3]{3}$
$\qquad = 2\sqrt[3]{3}$ ⎿ perfect cube

(d) $\sqrt[5]{8a^3b^4}\sqrt[5]{8a^2b^3} = \sqrt[5]{(8a^3b^4)(8a^2b^3)} = \sqrt[5]{64a^5b^7} = \sqrt[5]{32a^5b^5}\sqrt[5]{2b^2}$
$\qquad = 2ab\sqrt[5]{2b^2}$ ⟵ perfect fifth power ■

EXAMPLE 2 Multiplying binomial radicals

(a) $\sqrt{2}(3\sqrt{5} - 4\sqrt{2}) = 3\sqrt{2}\sqrt{5} - 4\sqrt{2}\sqrt{2} = 3\sqrt{10} - 4\sqrt{4}$
$\qquad = 3\sqrt{10} - 4(2) = 3\sqrt{10} - 8$

(b) $(5\sqrt{7} - 2\sqrt{3})(4\sqrt{7} + 3\sqrt{3}) = (5\sqrt{7})(4\sqrt{7}) + (5\sqrt{7})(3\sqrt{3}) - (2\sqrt{3})(4\sqrt{7}) - (2\sqrt{3})(3\sqrt{3})$
$\qquad = (5)(4)\sqrt{7}\sqrt{7} + (5)(3)\sqrt{7}\sqrt{3} - (2)(4)\sqrt{3}\sqrt{7} - (2)(3)\sqrt{3}\sqrt{3}$
$\qquad = 20(7) + 15\sqrt{21} - 8\sqrt{21} - 6(3)$
$\qquad = 140 + 7\sqrt{21} - 18$
$\qquad = 122 + 7\sqrt{21}$

```
(5√(7)-2√(3))(4√
(7)+3√(3))
       154.0780299
122+7√(21)
       154.0780299
```

The calculator display that verifies this multiplication is shown in Fig. 11.8. ■

Fig. 11.8

■ For reference, Eqs. (6.3) and (6.4) are
$$(x + y)^2 = x^2 + 2xy + y^2$$
$$(x - y)^2 = x^2 - 2xy + y^2$$

When raising a single-term radical expression to a power, we use the basic meaning of the power. When raising a binomial to a power, we proceed as with any binomial. We use Eqs. (6.3) and (6.4) along with Eq. (11.10) if the binomial is squared. These are illustrated in the next two examples.

EXAMPLE 3 Power of radical

(a) $(2\sqrt{7})^2 = 2^2(\sqrt{7})^2 = 4(7) = 28$

(b) $(2\sqrt{7})^3 = 2^3(\sqrt{7})^3 = 8(\sqrt{7})^2(\sqrt{7}) = 8(7)\sqrt{7}$
$$= 56\sqrt{7}$$

(c) $(3 + \sqrt{5})^2 = 3^2 + 2(3)\sqrt{5} + (\sqrt{5})^2 = 9 + 6\sqrt{5} + 5$
$$= 14 + 6\sqrt{5}$$

(d) $(\sqrt{a} - \sqrt{b})^2 = (\sqrt{a})^2 - 2\sqrt{a}\sqrt{b} + (\sqrt{b})^2$
$$= a + b - 2\sqrt{ab}$$ ■

Practice Exercises

Multiply: **1.** $2\sqrt{5}(\sqrt{15} - 3\sqrt{3})$
2. $(2\sqrt{6} - \sqrt{3})(3\sqrt{6} + 2\sqrt{3})$

CAUTION ▶ Again, we note that *to multiply radicals and combine them under one radical sign, it is necessary that the order of the radicals be the same.* If necessary, we can make the order of each radical the same by appropriate operations on each radical separately. Fractional exponents are frequently useful for this purpose.

EXAMPLE 4 Use of fractional exponents

(a) $\sqrt[3]{2}\sqrt{5} = 2^{1/3}5^{1/2} = 2^{2/6}5^{3/6} = (2^2 5^3)^{1/6} = \sqrt[6]{500}$

(b) $\sqrt[3]{4a^2b}\sqrt[4]{8a^3b^2} = (2^2a^2b)^{1/3}(2^3a^3b^2)^{1/4} = (2^2a^2b)^{4/12}(2^3a^3b^2)^{3/12}$
$$= (2^8a^8b^4)^{1/12}(2^9a^9b^6)^{1/12} = (2^{17}a^{17}b^{10})^{1/12}$$
$$= 2a(2^5a^5b^{10})^{1/12}$$
$$= 2a\sqrt[12]{32a^5b^{10}}$$ ■

DIVISION OF RADICALS

If a fraction involving a radical is to be changed in form, *rationalizing the denominator* or *rationalizing the numerator* is the principal step. Although calculators have made the rationalization of denominators unnecessary for calculation, this process often makes the form of the fraction simpler. Also, rationalizing numerators is useful at times in more advanced mathematics. We now consider rationalizing when the denominator (or numerator) to be rationalized has more than one term.

NOTE ▶ *If the denominator (or numerator) is the sum (or difference) of two terms, at least one of which is a radical, the fraction is rationalized by multiplying both the numerator and the denominator by the difference (or sum) of the same two terms, if the radicals are square roots.*

EXAMPLE 5 Rationalizing denominator

The fraction $\dfrac{1}{\sqrt{3} - \sqrt{2}}$ can be rationalized by multiplying the numerator and the denominator by $\sqrt{3} + \sqrt{2}$. In this way, the radicals will be removed from the denominator.

$$\frac{1}{\sqrt{3} - \sqrt{2}} \times \frac{\sqrt{3} + \sqrt{2}}{\sqrt{3} + \sqrt{2}} = \frac{\sqrt{3} + \sqrt{2}}{(\sqrt{3})^2 - (\sqrt{2})^2} = \frac{\sqrt{3} + \sqrt{2}}{3 - 2} = \sqrt{3} + \sqrt{2}$$

change sign ■

The reason this technique works is that an expression of the form $a^2 - b^2$ is created in the denominator, where a or b (or both) is a radical. We see that the result is a denominator free of radicals.

EXAMPLE 6 Rationalizing denominator

Rationalize the denominator of $\dfrac{3\sqrt{6} - \sqrt{2}}{\sqrt{6} + \sqrt{2}}$ and simplify the result.

$$\frac{3\sqrt{6} - \sqrt{2}}{\sqrt{6} + \sqrt{2}}\left(\frac{\sqrt{6} - \sqrt{2}}{\sqrt{6} - \sqrt{2}}\right) = \frac{18 - 4\sqrt{12} + 2}{4}$$

change sign

$$= \frac{20 - 4(2\sqrt{3})}{4} = 5 - 2\sqrt{3}$$

We note that, after rationalizing the denominator, the result has a much simpler form than the original expression. ∎

As we noted earlier, in certain types of algebraic operations, it may be necessary to rationalize the numerator of an expression. This procedure is illustrated in the following example.

EXAMPLE 7 Rationalizing – application

In studying the properties of a semiconductor, the expression

$$\frac{C_1 + C_2\sqrt{1 + 2V}}{\sqrt{1 + 2V}}$$

is used. Here, C_1 and C_2 are constants and V is the voltage across a junction of the semiconductor. Rationalize the numerator of this expression.

Multiplying numerator and denominator by $C_1 - C_2\sqrt{1 + 2V}$, we have

$$\frac{C_1 + C_2\sqrt{1 + 2V}}{\sqrt{1 + 2V}} = \frac{(C_1 + C_2\sqrt{1 + 2V})(C_1 - C_2\sqrt{1 + 2V})}{\sqrt{1 + 2V}(C_1 - C_2\sqrt{1 + 2V})}$$

$$= \frac{C_1^2 - C_2^2(\sqrt{1 + 2V})^2}{C_1\sqrt{1 + 2V} - C_2\sqrt{1 + 2V}\sqrt{1 + 2V}}$$

$$= \frac{C_1^2 - C_2^2(1 + 2V)}{C_1\sqrt{1 + 2V} - C_2(1 + 2V)}$$ ∎

Practice Exercise

Rationalize the denominator and simplify:

3. $\dfrac{3 - 2\sqrt{3}}{2 + 7\sqrt{3}}$

EXERCISES 11.5

In Exercises 1–4, perform the indicated operations on the resulting expressions if the given changes are made in the indicated examples of this section.

1. In Example 2(a), change $4\sqrt{2}$ to $4\sqrt{8}$ and then perform the multiplication.

2. In Example 4(a), change $\sqrt{5}$ to $\sqrt[6]{5}$ and then perform the multiplication.

3. In Example 5, change the sign in the denominator from $-$ to $+$ and then rationalize the denominator.

4. In Example 6, rationalize the numerator of the given expression.

In Exercises 5–46, perform the indicated operations, expressing answers in simplest form with rationalized denominators.

5. $\sqrt{3}\sqrt{10}$ **6.** $\sqrt{2}\sqrt{51}$ **7.** $\sqrt{6}\sqrt{2}$ **8.** $\sqrt{7}\sqrt{14}$

9. $\sqrt[3]{4}\sqrt[3]{2}$ **10.** $\sqrt[5]{4}\sqrt[5]{16}$ **11.** $(5\sqrt{2})^2$ **12.** $2(3\sqrt{5})^3$

13. $\sqrt{8}\sqrt{\frac{5}{2}}$ **14.** $\sqrt[6]{\frac{6}{7}}\sqrt[6]{\frac{2}{3}}$

15. $\sqrt{3}(\sqrt{2} - \sqrt{5})$ **16.** $3\sqrt{5}(\sqrt{15} - 2\sqrt{5})$

17. $(2 - \sqrt{5})(2 + \sqrt{5})$ **18.** $4(2 - \sqrt{7})^2$

19. $(3\sqrt{30} - 2\sqrt{3})(6\sqrt{30} + 7\sqrt{3})$

20. $(3\sqrt{7a} - \sqrt{8})(\sqrt{7a} + \sqrt{2})$

21. $(3\sqrt{11} - \sqrt{x})(2\sqrt{11} + 5\sqrt{x})$

22. $(2\sqrt{10} + 3\sqrt{15})(\sqrt{10} - 7\sqrt{15})$

23. $\sqrt{a}(\sqrt{ab} + \sqrt{c^3})$

24. $\sqrt{3x}(\sqrt[3]{3x} - \sqrt{xy})$

25. $\dfrac{\sqrt{6} - 3}{\sqrt{6}}$

26. $\dfrac{5 - \sqrt{10}}{\sqrt[4]{10}}$

27. $(\sqrt{2a} - \sqrt{b})(\sqrt{2a} + 3\sqrt{b})$

28. $(2\sqrt{mn} + 3\sqrt{n})^2$

29. $\sqrt{2}\sqrt[3]{3}$

30. $\sqrt[5]{16}\sqrt[3]{8}$

31. $\dfrac{1}{\sqrt{7} + \sqrt{3}}$

32. $\dfrac{6\sqrt[4]{25}}{5 - 2\sqrt{5}}$

33. $\dfrac{\sqrt{2} - 1}{\sqrt{7} - 3\sqrt{2}}$

34. $\dfrac{2\sqrt{15} - 3}{\sqrt{15} + 4}$

35. $\dfrac{2\sqrt{3} - 5\sqrt{5}}{\sqrt{3} + 2\sqrt{5}}$

36. $\dfrac{\sqrt{15} - 3\sqrt{5}}{2\sqrt{15} - \sqrt{5}}$

37. $\dfrac{2\sqrt{x}}{\sqrt{x} - \sqrt{5}}$

38. $\dfrac{\sqrt{2c} + 3d}{\sqrt{2c} - d}$

39. $\left(\sqrt[5]{\sqrt{6}} - \sqrt{5}\right)\left(\sqrt[5]{\sqrt{6}} + \sqrt{5}\right)$

40. $\sqrt[3]{5 - \sqrt{17}}\sqrt[3]{5 + \sqrt{17}}$

41. $\left(\sqrt{\dfrac{2}{R}} + \sqrt{\dfrac{R}{2}}\right)\left(\sqrt{\dfrac{2}{R}} - 2\sqrt{\dfrac{R}{2}}\right)$

42. $(3 + \sqrt{6 - 2a})(2 - \sqrt{6 - 2a})$

43. $\dfrac{\sqrt{x + y}}{\sqrt{x - y} - \sqrt{x}}$

44. $\dfrac{\sqrt{1 + a}}{a - \sqrt{1 - a}}$

45. $\dfrac{\sqrt{a} + \sqrt{a - 2}}{\sqrt{a} - \sqrt{a - 2}}$

46. $\dfrac{\sqrt{T^4 - V^4}}{\sqrt{V^{-2} - T^{-2}}}$

In Exercises 47–50, perform the indicated operations, expressing answers in simplest form with rationalized denominators. Then verify the result with a calculator.

47. $(\sqrt{11} + \sqrt{6})(\sqrt{11} - 2\sqrt{6})$

48. $(2\sqrt{5} - \sqrt{7})(3\sqrt{5} + \sqrt{7})$

49. $\dfrac{2\sqrt{6} - \sqrt{5}}{3\sqrt{6} - 4\sqrt{5}}$

50. $\dfrac{\sqrt{7} - 4\sqrt{2}}{5\sqrt{7} - 4\sqrt{2}}$

In Exercises 51–54, combine the terms into a single fraction, but do not rationalize the denominators.

51. $2\sqrt{x} + \dfrac{1}{\sqrt{x}}$

52. $\dfrac{3}{2\sqrt{3x - 4}} - \sqrt{3x - 4}$

53. $\dfrac{x^2}{\sqrt{2x + 1}} + 2x\sqrt{2x + 1}$

54. $4\sqrt{x^2 + 1} - \dfrac{4x}{\sqrt{x^2 + 1}}$

In Exercises 55–58, rationalize the numerator of each fraction.

55. $\dfrac{\sqrt{5} + 2\sqrt{2}}{3\sqrt{10}}$

56. $\dfrac{\sqrt{19} - 3}{5}$

57. $\dfrac{\sqrt{x + h} - \sqrt{x}}{h}$

58. $\dfrac{\sqrt{3x + 4} + \sqrt{3x}}{8}$

In Exercises 59–72, solve the given problems.

59. By substitution, show that $x = 1 - \sqrt{2}$ is a solution of the equation $x^2 - 2x - 1 = 0$.

W 60. For the quadratic equation $ax^2 + bx + c = 0$, if a, b, and c are integers, the product of the roots is a rational number. Explain.

61. Determine the relationship between a and c in $ax^2 + bx + c = 0$ if the roots of the equation are reciprocals.

62. Evaluate $r^2 - s^2$ if $r = -b + \sqrt{b^2 - 4ac}$ and $s = -b - \sqrt{b^2 - 4ac}$.

63. Rationalize the denominator of $\dfrac{1}{\sqrt[3]{x^2} + \sqrt[3]{x} + 1}$.

(*Hint:* See Eq. 6.10).

64. One leg of a right triangle is $2\sqrt{7}$ and the hypotenuse is 6. What is the area of the triangle?

65. For an object oscillating at the end of a spring and on which there is a force that retards the motion, the equation $m^2 + bm + k^2 = 0$ must be solved. Here, b is a constant related to the retarding force, and k is the spring constant. By substitution, show that $m = \frac{1}{2}\left(\sqrt{b^2 - 4k^2} - b\right)$ is a solution.

66. Among the products of a specialty furniture company are tables with tops in the shape of a regular octagon (eight sides). Express the area A of a table top as a function of the side s of the octagon.

67. An expression used in determining the characteristics of a spur gear is $\dfrac{50}{50 + \sqrt{V}}$. Rationalize the denominator.

68. When studying the orbits of earth satellites, the expression $\left(\dfrac{GM}{4\pi^2}\right)^{1/3} T^{2/3}$ arises. Express it in simplest rationalized radical form.

69. In analyzing a tuned amplifier circuit, the expression $\dfrac{2Q}{\sqrt{\sqrt{2} - 1}}$ is used. Rationalize the denominator.

70. When analyzing the ratio of resultant forces when forces **F** and **T** act on a structure, the expression $\sqrt{F^2 + T^2}/\sqrt{F^{-2} + T^{-2}}$ arises. Simplify this expression.

71. The resonant frequency ω of a capacitance C in parallel with a resistance R and inductance L (see Fig. 11.9) is

$$\omega = \dfrac{1}{\sqrt{LC}}\sqrt{1 - \dfrac{R^2 C}{L}}.$$

Combine terms under the radical, rationalize the denominator, and simplify.

Fig. 11.9

72. In fluid dynamics, the expression $\dfrac{a}{1 + m/\sqrt{r}}$ arises. Combine terms in the denominator, rationalize the denominator, and simplify.

Answers to Practice Exercises

1. $10\sqrt{3} - 6\sqrt{15}$ **2.** $30 + 3\sqrt{2}$ **3.** $\dfrac{-48 + 25\sqrt{3}}{143}$

CHAPTER 11   EQUATIONS

Exponents $a^m \times a^n = a^{m+n}$ (11.1)

$$\frac{a^m}{a^n} = a^{m-n} \quad \text{or} \quad \frac{a^m}{a^n} = \frac{1}{a^{n-m}} \quad (a \neq 0)$$ (11.2)

$$(a^m)^n = a^{mn}$$ (11.3)

$$(ab)^n = a^n b^n, \quad \left(\frac{a}{b}\right)^n = \frac{a^n}{b^n} \quad (b \neq 0)$$ (11.4)

$$a^0 = 1 \quad (a \neq 0)$$ (11.5)

$$a^{-n} = \frac{1}{a^n} \quad (a \neq 0)$$ (11.6)

Fractional exponents $a^{1/n} = \sqrt[n]{a}$ (11.7)

$$a^{m/n} = \sqrt[n]{a^m} = \left(\sqrt[n]{a}\right)^m$$ (11.8)

Radicals $\sqrt[n]{a^n} = \left(\sqrt[n]{a}\right)^n = a$ (11.9)

$$\sqrt[n]{a}\sqrt[n]{b} = \sqrt[n]{ab}$$ (11.10)

$$\sqrt[m]{\sqrt[n]{a}} = \sqrt[mn]{a}$$ (11.11)

$$\frac{\sqrt[n]{a}}{\sqrt[n]{b}} = \sqrt[n]{\frac{a}{b}} \quad (b \neq 0)$$ (11.12)

CHAPTER 11   REVIEW EXERCISES

In Exercises 1–28, express each expression in simplest form with only positive exponents.

1. $2a^{-2}b^0$ 2. $(2c)^{-1}z^{-2}$ 3. $\dfrac{9c^{-1}}{d^{-3}}$ 4. $\dfrac{-5x^0}{3y^{-1}}$

5. $3(25)^{3/2}$ 6. $32^{2/5}$ 7. $400^{-3/2}$ 8. $7(1000)^{-2/3}$

9. $\left(\dfrac{3}{t^2}\right)^{-2}$ 10. $\left(\dfrac{2x^3}{3}\right)^{-3}$ 11. $\dfrac{-8^{2/3}}{49^{-1/2}}$ 12. $\dfrac{4^{-45}}{2^{-90}}$

13. $(2a^{1/3}b^{5/6})^6$ 14. $(ax^{-1/2}y^{1/4})^8$ 15. $(-32m^{15}n^{10})^{3/5}$

16. $(27x^{-6}y^9)^{2/3}$ 17. $2L^{-2} - 4C^{-1}$ 18. $C^4C^{-1} + (C^4)^{-1}$

19. $\dfrac{2x^{-1}}{2x^{-1} + y^{-1}}$ 20. $\dfrac{9a}{(2a)^{-1} - a}$ 21. $(a - 3b^{-1})^{-1}$

22. $(2s^{-2} + t)^{-2}$ 23. $(x^3 - y^{-3})^{1/3}$ 24. $(8a^3)^{2/3}(a^{-2} + 1)^{1/2}$

25. $(W^2 + 2WH + H^2)^{-1/2}$ 26. $\left[\dfrac{(9a)^0(4x^2)^{1/3}(3b^{1/2})}{(2b^0)^2}\right]^{-6}$

27. $2x(x - 1)^{-2} - 2(x^2 + 1)(x - 1)^{-3}$

28. $4(1 - T^2)^{1/2} - (1 - T^2)^{-1/2}$

In Exercises 29–78, perform the indicated operations and express the result in simplest radical form with rationalized denominators.

29. $\sqrt{68}$ 30. $4\sqrt{96}$ 31. $\sqrt{ab^5c^2}$ 32. $\sqrt{x^3y^4z^6}$

33. $\sqrt{9a^3b^4}$ 34. $\sqrt{8x^5y^2}$ 35. $\sqrt{84st^3u^{-2}}$ 36. $\sqrt{52L^2C^{-5}}$

37. $\dfrac{5}{\sqrt{2s}}$ 38. $\dfrac{3a}{\sqrt{5x}}$ 39. $\sqrt{\dfrac{11}{27}}$ 40. $\sqrt{\dfrac{7}{8V}}$

41. $\sqrt[4]{8m^6n^9}$ 42. $\sqrt[3]{9a^7b^{-3}}$ 43. $\sqrt[4]{\sqrt[3]{64}}$

44. $\sqrt{a^{-3}\sqrt[5]{b^{12}}}$ 45. $\sqrt{36 + 4} - 2\sqrt{10}$ 46. $2\sqrt{68x} - \sqrt{153x}$

47. $\sqrt{63} - 2\sqrt{112} - \sqrt{28}$ 48. $7\sqrt{20} - \sqrt{80} - 2\sqrt{125}$

49. $a\sqrt{2x^3} + \sqrt{8a^2x^3}$ 50. $2\sqrt{m^2n^3} - \sqrt{n^5}$

51. $\sqrt[3]{8a^4} + b\sqrt[3]{a}$ 52. $\sqrt[4]{2xy^5} - \sqrt[4]{32xy}$

53. $5\sqrt{5}(6\sqrt{5} - \sqrt{35})$ 54. $2\sqrt{8}(5\sqrt{2} - \sqrt{6})$

55. $2\sqrt{2}(\sqrt{6} - \sqrt{10})$ 56. $3\sqrt{5}(\sqrt{15m^2} + 2\sqrt{35m^2})$

57. $(2 - 3\sqrt{17B})(3 + \sqrt{17B})$ 58. $(5\sqrt{6} - 4)(3\sqrt{6} + 5)$

59. $(2\sqrt{7} - 3\sqrt{a})(3\sqrt{7} + \sqrt{a})$

60. $(3\sqrt{2} - \sqrt{13})(5\sqrt{2} + 3\sqrt{13})$

61. $\dfrac{\sqrt{3x}}{2\sqrt{3x} - \sqrt{y}}$ 62. $\dfrac{5\sqrt{a}}{2\sqrt{a} - c}$ 63. $\dfrac{\sqrt{2}}{\sqrt{3} - 4\sqrt{2}}$

64. $\dfrac{4}{3 - 2\sqrt{7}}$ 65. $\dfrac{\sqrt{7} - \sqrt{5}}{\sqrt{5} + 3\sqrt{7}}$ 66. $\dfrac{7 - 2\sqrt{6}}{3 + 2\sqrt{6}}$

67. $\dfrac{2\sqrt{x} - a}{3\sqrt{x} + 5a}$ 68. $\dfrac{2\sqrt{3y} - \sqrt{2}}{9\sqrt{3y} - 5\sqrt{2}}$ 69. $\sqrt{4b^2 + 1}$

70. $x^{3/8}(x^{5/8} - 8x^{13/8})$ 71. $(3x^{1/2} - 2x)(3x^{1/2} + 2x)$

72. $(x^{3n-1}/x^{n-1})^{1/n}$

73. $(1 + 6^{1/2})(3^{1/2} + 2^{1/2})(3^{1/2} - 2^{1/2})$

74. $\sqrt{a^{-2} + \dfrac{1}{b^2}}$

75. $\left(\dfrac{2 - \sqrt{15}}{2}\right)^2 - \left(\dfrac{2 - \sqrt{15}}{2}\right)$

76. $\sqrt{2 + a^{-1}b + ab^{-1}} + \sqrt{a^4b^2 + 2a^3b^2 + a^2b^2}$

77. $\sqrt{3 + n}(\sqrt{3 + n} - \sqrt{n})^{-1}$

78. $\sqrt{1 + \sqrt{2}}(\sqrt{1 + \sqrt{2}} + \sqrt{2})^{-1}$

In Exercises 79 and 80, perform the indicated operations and express the result in simplified radical form with rationalized denominators. In Exercises 81 and 82, without a calculator, show that the given equations are true. Then verify each result with a calculator.

79. $(\sqrt{7} - 2\sqrt{15})(3\sqrt{7} - \sqrt{15})$ 80. $\dfrac{2\sqrt{3} - 7\sqrt{14}}{3\sqrt{3} + 2\sqrt{14}}$

81. $\sqrt{\sqrt{2} - 1}(\sqrt{2} + 1) = \sqrt{\sqrt{2} + 1}$

82. $\sqrt{\sqrt{3} + 1}(\sqrt{3} - 1) = \sqrt{2}(\sqrt{3} - 1)$

In Exercises 83–100, perform the indicated operations.

83. The legs of a right triangle are $\sqrt{18}$ and $\sqrt{32}$. Find the perimeter of the triangle.

84. One of the legs of a right triangle is $3\sqrt{5}$ and the hypotenuse is $5\sqrt{3}$. Find the area of the triangle.

85. Evaluate $3x^2 - 2x + 5$ for $x = \frac{1}{2}(2 - \sqrt{3})$.

86. Evaluate x if $x^{-1/3} = 0.2$.

87. The average annual increase i (in %) of the cost of living over n years is given by $i = 100[(C_2/C_1)^{1/n} - 1]$, where C_1 is the cost of living index for the first year and C_2 is the cost of living index for the last year of the period. Evaluate i if $C_1 = 130.7$ and $C_2 = 172.0$ are the values for 1990 and 2000, respectively.

88. Kepler's third law of planetary motion may be given as $T = kr^{3/2}$, where T is the time for one revolution of a planet around the sun, r is its mean radius from the sun, and $k = 1.115 \times 10^{-12}$ year/mi$^{3/2}$. Find the time for one revolution of Venus about the sun if $r = 6.73 \times 10^7$ mi.

89. The speed v of a ship of weight W whose engines produce power P is $v = k\sqrt[3]{P/W}$. Express this equation (a) with a fractional exponent and (b) as a radical with the denominator rationalized.

90. In analyzing the orbit of an earth satellite, the expression

$$\sqrt{1 + \frac{2E}{m}\left(\frac{h}{GM}\right)^2}$$

is used. Combine terms under the radical and express the result with the denominator rationalized.

91. A square plastic sheet of side x is stretched by an amount equal to $\sqrt{x}$ horizontally and vertically. Find the expression for the percent increase in the area of the sheet.

92. The value P of an object originally worth P_0 and that appreciates at an annual rate of r for t years is given by $P = P_0(1 + r)^t$. Solve for r.

93. The expression $(1 - v^2/c^2)^{-1/2}$ arises in the theory of relativity. Simplify, using only positive exponents in the result.

94. The compression ratio r of a certain gasoline engine is related to the efficiency e (in %) of the engine by $r = \left(\dfrac{100 - e}{100}\right)^{-2.5}$. What compression ratio is necessary to have an efficiency of 55%?

95. A square is decreasing in size on a computer screen. To find how fast the side of the square changes, we must rationalize the numerator of $\dfrac{\sqrt{A + h} - \sqrt{A}}{h}$. Perform this operation.

96. A plane takes off 200 mi east of its destination, and a 50 mi/h wind is from the south. If the plane's velocity is 250 mi/h and its heading is always toward its destination, its path can be described as $y = 100[(0.005x)^{0.8} - (0.005x)^{1.2}]$, $0 \le x \le 200$ mi. Sketch the path and check using a graphing calculator.

97. In an experiment, a laser beam follows the path shown in Fig. 11.10. Express the length of the path in simplest radical form.

Fig. 11.10

98. The flow rate V (in ft/s) through a storm drain pipe is found from $V = 110(0.55)^{2/3}(0.0018)^{1/2}$. Find the value of V.

99. The frequency of a certain electric circuit is given by $\dfrac{1}{2\pi\sqrt{\dfrac{LC_1C_2}{C_1 + C_2}}}$. Express this in simplest rationalized radical form.

100. A computer analysis of an experiment showed that the fraction f of viruses surviving X-ray dosages was given by $f = \dfrac{20}{d + \sqrt{3d + 400}}$, where d is the dosage. Express this with the denominator rationalized.

Writing Exercise

101. In calculating the forces on a tower by the wind, it is necessary to evaluate $0.018^{0.13}$. Write a paragraph explaining why this form is preferable to the equivalent radical form.

CHAPTER 11 PRACTICE TEST

In Problems 1–14, simplify the given expressions. For those with exponents, express each result with only positive exponents. For the radicals, rationalize the denominator where applicable.

1. $-5y^0$

2. $(3\pi x^{-4})^{-2}$

3. $\dfrac{100^{3/2}}{8^{-2/3}}$

4. $(as^{-1/3}t^{3/4})^{12}$

5. $2\sqrt{20} - \sqrt{125}$

6. $(2x^{-1} + y^{-2})^{-1}$

7. $(\sqrt{2x} - 3\sqrt{y})^2$

8. $\sqrt[3]{\sqrt[4]{4}}$

9. $\dfrac{3 - 2\sqrt{2}}{2\sqrt{x}}$

10. $\sqrt{27a^4b^3}$

11. $2\sqrt{2}(3\sqrt{10} - \sqrt{6})$

12. $(2x + 3)^{1/2} + (x + 1)(2x + 3)^{-1/2}$

13. $\left(\dfrac{4a^{-1/2}b^{3/4}}{b^{-2}}\right)\left(\dfrac{b^{-1}}{2a}\right)$

14. $\dfrac{2\sqrt{15} + \sqrt{3}}{\sqrt{15} - 2\sqrt{3}}$

15. Express $\dfrac{3^{-1/2}}{2}$ in simplest radical form with a rationalized denominator.

16. In the study of fluid flow in pipes, the expression $0.220N^{-1/6}$ is found. Evaluate this expression for $N = 64 \times 10^6$.

Complex Numbers

12

Among the important areas of research in the mid-nineteenth century was the study of light, which was determined to be a form of electromagnetic radiation. Today, extensive use of electromagnetic waves is made in a great variety of situations. These include the transmission of signals for radio, television, and cell phones. Also, radar, microwave ovens, and X-ray machines are among the many other applications of electromagnetic waves.

In the 1860s, James Clerk Maxwell, a Scottish physicist, used the research of others—and his own—to develop a set of very important equations for electromagnetic radiation. In doing so, he actually predicted *mathematically* the existence of electromagnetic waves, such as radio waves.

It was not until 1887 that Heinrich Hertz, a German physicist, produced and observed radio waves in the laboratory. We see that mathematics was used in predicting the existence of one of the most important devices in use today, over 20 years before it was actually observed.

Important in the study of electromagnetic waves and many areas of electricity and electronics are *complex numbers,* which we study in this chapter. Complex numbers include imaginary numbers as well as real numbers. Other than briefly noting imaginary numbers in Chapters 1 and 7, we have purposely avoided any extended discussion of them until now.

Despite their names, complex numbers and imaginary numbers have very real and useful applications, as we have noted. Besides electricity and electronics, they are used in the studies of mechanical vibrations, optics, and acoustics.

In Section 12.7, we see that tuning in a radio station involves a basic application of complex numbers to electricity.

12.1 Basic Definitions

Imaginary Unit • Rectangular Form of a
Complex Number • Equality of Complex
Numbers • Conjugate

Since the square of a positive number or a negative number is positive, it is not possible to square any real number and have a negative result. Therefore, we must define a number system to include square roots of negative numbers. We will find that such numbers can be used to great advantage in certain applications.

If the radicand in a square root is negative, we can express the indicated root as the product of $\sqrt{-1}$ and the square root of a positive number. *The symbol $\sqrt{-1}$ is defined as the* **imaginary unit** *and is denoted by the symbol j.* Therefore, we have

$$j = \sqrt{-1} \quad \text{and} \quad j^2 = -1 \tag{12.1}$$

Generally, mathematicians use the symbol i for $\sqrt{-1}$, and therefore most nontechnical textbooks use i. However, one of the major technical applications of complex numbers is in electronics, where i represents electric current. Therefore, we will use j for $\sqrt{-1}$, which is also the standard symbol in electronics textbooks.

EXAMPLE 1 Square roots in terms of *j*

Express the following square roots in terms of j.

(a) $\sqrt{-9} = \sqrt{(9)(-1)} = \sqrt{9}\sqrt{-1} = 3j$

(b) $\sqrt{-0.25} = \sqrt{0.25}\sqrt{-1} = 0.5j$

(c) $\sqrt{-5} = \sqrt{5}\sqrt{-1} = \sqrt{5}j = j\sqrt{5}$ the form $j\sqrt{5}$ is better than $\sqrt{5}j$

NOTE ▸ When a radical appears, we write the result in a form with the j before the radical as in the last illustration. This clearly shows that the j is not *under* the radical. ∎

EXAMPLE 2 Be careful in simplifying

$(\sqrt{-4})^2 = (\sqrt{4}\sqrt{-1})^2 = (j\sqrt{4})^2 = 4j^2 = -4$

The simplification of this expression does not follow Eq. (11.10), which states that $\sqrt{ab} = \sqrt{a}\sqrt{b}$ for square roots. This is the reason it was noted as being valid only if a and b are not negative. Therefore, we see that *Eq. (11.10) does not always hold for negative values of a and b.*

In simplifying $(\sqrt{-4})^2$, we can write it as $(\sqrt{-4})(\sqrt{-4})$. However, we *cannot now write this product as* $\sqrt{(-4)(-4)}$, for this leads to an incorrect result of $+4$. In fact, *we cannot use Eq. (11.10) if both a and b are negative.* It cannot be overemphasized that

CAUTION ▸
$$(\sqrt{-4})^2 \text{ is not equal to } \sqrt{(-4)(-4)}$$

As shown at the beginning of this example, for $(\sqrt{-4})^2$, we must first write $\sqrt{-4}$ as $j\sqrt{4}$ before continuing with the simplification. ∎

EXAMPLE 3 First, express number in terms of *j*

To further illustrate the method of handling square roots of negative numbers, consider the difference between $\sqrt{-3}\sqrt{-12}$ and $\sqrt{(-3)(-12)}$. For these expressions, we have

$$\sqrt{-3}\sqrt{-12} = (j\sqrt{3})(j\sqrt{12}) = (\sqrt{3}\sqrt{12})j^2 = (\sqrt{36})j^2$$
$$= 6(-1) = -6$$
$$\sqrt{(-3)(-12)} = \sqrt{36} = 6$$

For $\sqrt{-3}\sqrt{-12}$, we have the product of square roots of negative numbers, whereas for $\sqrt{(-3)(-12)}$, we have the product of negative numbers under the radical. We must be careful to note the difference. ∎

NOTE ▶ From Examples 2 and 3, we see that *when we are dealing with square roots of negative numbers, **each should be expressed in terms of j before proceeding.*** To do this, for any positive real number a we write

$$\sqrt{-a} = j\sqrt{a} \quad (a > 0) \tag{12.2}$$

EXAMPLE 4 Expressing numbers in terms of *j*

Practice Exercises
Simplify: **1.** $\sqrt{-5}\sqrt{-20}$ **2.** $-\sqrt{-48}$

(a) $\sqrt{-6} = \sqrt{(6)(-1)} = \sqrt{6}\sqrt{-1} = j\sqrt{6}$

this step is correct if only
one is negative, as in this case

(b) $-\sqrt{-75} = -\sqrt{(25)(3)(-1)} = -\sqrt{(25)(3)}\sqrt{-1} = -5j\sqrt{3}$

CAUTION ▶ We note that $-\sqrt{-75}$ *is not equal to* $\sqrt{75}$. ∎

At times, we need to raise imaginary numbers to some power. Using the definitions of exponents and of j, we have the following results:

$$
\begin{aligned}
j &= j & j^5 &= j^4 j = j \\
j^2 &= -1 & j^6 &= j^4 j^2 = (1)(-1) = -1 \\
j^3 &= j^2 j = -j & j^7 &= j^4 j^3 = (1)(-j) = -j \\
j^4 &= j^2 j^2 = (-1)(-1) = 1 & j^8 &= j^4 j^4 = (1)(1) = 1
\end{aligned}
$$

The powers of j go through the cycle, $j, -1, -j, 1, j, -1, -j, 1$, and so forth. Noting this—and the fact that j raised to a power that is a multiple of 4 equals 1—allows us to raise j to any integral power almost on sight.

EXAMPLE 5 Powers of *j*

Practice Exercise
Simplify: **3.** $-j^{25}$

(a) $j^{10} = j^8 j^2 = (1)(-1) = -1$

(b) $j^{45} = j^{44} j = (1)(j) = j$

(c) $j^{531} = j^{528} j^3 = (1)(-j) = -j$

└─── exponents 8, 44, 528 are multiples of 4 ∎

RECTANGULAR FORM OF A COMPLEX NUMBER

■ Even negative numbers were not widely accepted by mathematicians until late in the sixteenth century.

Using real numbers and the imaginary unit j, we define a new kind of number. *A* **complex number** *is any number that can be written in the form* $a + bj$, *where a and b are real numbers. If* $a = 0$ *and* $b \neq 0$, *we have a number of the form bj, which is a* **pure imaginary number.** *If* $b = 0$, *then* $a + bj$ *is a real number. The form* $a + bj$ *is known as the* **rectangular form** *of a complex number, where a is known as the* **real part** *and b is known as the* **imaginary part.** We see that complex numbers include all real numbers and all pure imaginary numbers.

■ Imaginary numbers were so named because the French mathematician René Descartes (1596–1650) referred to them as "imaginaries." Most of the mathematical development of them occurred in the eighteenth century.

A comment here about the words *imaginary* and *complex* is in order. The choice of the names of these numbers is historical in nature, and unfortunately it leads to some misconceptions about the numbers. The use of *imaginary* does not imply that the numbers do not exist. Imaginary numbers do in fact exist, as they are defined above. In the same way, the use of *complex* does not imply that the numbers are complicated and therefore difficult to understand. With the proper definitions and operations, we can work with complex numbers, just as with any other type of number.

■ Complex numbers were named by the German mathematician Karl Friedrich Gauss (1777–1855).

A complex number being the sum of a real number and an imaginary number, it is not positive or negative in the usual sense, and it is not equal in the usual way to another complex number. However, the real part is positive or negative (or zero), and the same is true of the imaginary part. Thus, *two complex numbers are equal if and only if their real parts are equal and their imaginary parts are equal.* That is,

NOTE ▶   *two complex numbers, a + bj and x + yj, are equal if a = x and b = y.*

EXAMPLE 6 Equality of complex numbers

(a) $a + bj = 3 + 4j$ if $a = 3$ and $b = 4$

    real part ⬏   ⬑ imaginary part

(b) $x + 3j = yj - 5$

    $x - yj = -5 - 3j$ if $x = -5$ and $y = 3$ ∎

EXAMPLE 7 Equality of complex numbers

Practice Exercise

Evaluate x and y:
4. $4 - 6j - x = j + jy$

What values of x and y satisfy the equation $x + 3(xj + y) = 5 - j - jy$?

Rewriting each side in the form $a + bj$, and equating real parts and equating imaginary parts, we have

$$(x + 3y) + 3xj = 5 + (-1 - y)j$$

For equality, $x + 3y = 5$ and $3x = -1 - y$. The solution of this system of equations is $x = -1$ and $y = 2$.   ∎

Conjugate   *The* **conjugate** *of the complex number $a + bj$ is the complex number $a - bj$.* We see that the sign of the imaginary part is changed to obtain the conjugate.

EXAMPLE 8 Conjugates

(a) $3 - 2j$ is the conjugate of $3 + 2j$. We may also say that $3 + 2j$ is the conjugate of $3 - 2j$. Thus, each is the conjugate of the other.

(b) $-2 - 5j$ and $-2 + 5j$ are conjugates.

(c) $6j$ and $-6j$ are conjugates.

(d) 3 is the conjugate of 3 (imaginary part is zero).   ∎

EXERCISES 12.1

In Exercises 1–4, perform the indicated operations on the resulting expressions if the given changes are made in the indicated examples of this section.

1. In Example 3, put a $-$ sign before the first radical of the first illustration and then simplify.

2. In Example 4(a), put a j in front of the radical and then simplify.

3. In Example 5(a), add 40 to the exponent and then evaluate.

4. In Example 7, change the 5 on the right side of the equation to -5 and then solve.

In Exercises 5–16, express each number in terms of j.

5. $\sqrt{-81}$ **6.** $\sqrt{-121}$ **7.** $-\sqrt{-4}$ **8.** $-\sqrt{-49}$

9. $\sqrt{-0.36}$ **10.** $-\sqrt{-0.01}$ **11.** $4\sqrt{-8}$ **12.** $3\sqrt{-48}$

13. $\sqrt{-\frac{7}{4}}$ **14.** $-\sqrt{-\frac{5}{9}}$ **15.** $-\sqrt{-4e^2}$ **16.** $\sqrt{-\pi^4}$

In Exercises 17–32, simplify each of the given expressions.

17. (a) $(\sqrt{-7})^2$ (b) $\sqrt{(-7)^2}$

18. (a) $\sqrt{(-15)^2}$ (b) $(\sqrt{-15})^2$

19. (a) $\sqrt{(-2)(-8)}$ (b) $\sqrt{-2}\sqrt{-8}$

20. (a) $\sqrt{-9}\sqrt{-16}$ (b) $\sqrt{(-9)(-16)}$

21. $\sqrt{-\frac{1}{15}}\sqrt{-\frac{27}{5}}$ **22.** $-\sqrt{\left(-\frac{4}{7}\right)\left(-\frac{49}{16}\right)}$

23. $-\sqrt{-5}\sqrt{-2}\sqrt{-10}$ **24.** $-\sqrt{(-3)(-7)}\sqrt{-21}$

25. (a) $-j^6$ (b) $(-j)^6$ **26.** (a) $-j^{21}$ (b) $(-j)^{21}$

27. $j^2 - j^6$ **28.** $2j^5 - 1/j^{-2}$

29. $j^{15} - j^{13}$ **30.** $3j^{48} + j^{200}$ **31.** $-\sqrt{(-j)^2}$ **32.** $-\sqrt{-j^2}$

In Exercises 33–44, perform the indicated operations and simplify each complex number to its rectangular form.

33. $2 + \sqrt{-9}$ **34.** $-26 + \sqrt{-64}$ **35.** $3j - \sqrt{-100}$

36. $-\sqrt{1} - \sqrt{-400}$ **37.** $\sqrt{-4j^2} + \sqrt{-4}$ **38.** $5 - 2\sqrt{25j^2}$

39. $2j^2 + 3j$ **40.** $j^3 - 6$ **41.** $\sqrt{18} - \sqrt{-8}$

42. $\sqrt{-27} + \sqrt{12}$ **43.** $(\sqrt{-2})^2 + j^4$ **44.** $(2\sqrt{2})^2 - j^6$

In Exercises 45–48, find the conjugate of each complex number.

45. (a) $6 - 7j$ (b) $8 + j$ **46.** (a) $-3 + 2j$ (b) $-9 - j$

47. (a) $2j$ (b) -4 **48.** (a) 6 (b) $-5j$

In Exercises 49–54, find the values of x and y that satisfy the given equations.

49. $7x - 2yj = 14 + 4j$ **50.** $2x + 3jy = -6 + 12j$

51. $6j - 7 = 3 - x - yj$ **52.** $9 - j = xj + 1 - y$

53. $x - 2j^2 + 7j = yj + 2xj^3$

54. $2x - 6xj^3 - 3j^2 = yj - y + 7j^5$

In Exercises 55–64, answer the given questions.

55. How is a number changed if it is multiplied by (a) j^4; (b) j^2?

56. Evaluate: (a) j^{-8}; (b) j^{-6}

57. Are $8j$ and $-8j$ the solutions to the equation $x^2 + 64 = 0$?

58. Are $2j$ and $-2j$ solutions to the equation $x^4 + 16 = 0$?

59. Evaluate $j + j^2 + j^3 + j^4 + j^5 + j^6 + j^7 + j^8$.

60. What is the smallest positive value of n for which $j^{-1} = j^n$?

(W) 61. Is it possible that a given complex number and its conjugate are equal? Explain.

(W) 62. Is it possible that a given complex number and the negative of its conjugate are equal? Explain.

63. Show that the real part of $x + yj$ equals the imaginary part of $j(x + yj)$.

(W) 64. Explain why a real number is a complex number, but a complex number may not be a real number.

Answers to Practice Exercises

1. -10 **2.** $-4j\sqrt{3}$ **3.** $-j$ **4.** $x = 4, y = -7$

12.2 Basic Operations with Complex Numbers

Addition, Subtraction, Multiplication, and Division of Complex Numbers • Complex Numbers on Calculator

The basic operations of addition, subtraction, multiplication, and division of complex numbers are based on the operations for binomials with real coefficients (see Chapters 1 and 6). In performing these operations, we treat j as we would any other literal number, although we must properly handle any powers of j that might occur. However, *we must be careful to express all complex numbers in terms of j before performing these operations.* Therefore, we now have the definitions for these operations.

NOTE ▶

■ Compare Eq. (12.5) with Eq. (6.6).

> **BASIC OPERATIONS ON COMPLEX NUMBERS**
>
> Addition:
> $$(a + bj) + (c + dj) = (a + c) + (b + d)j \qquad (12.3)$$
>
> Subtraction:
> $$(a + bj) - (c + dj) = (a - c) + (b - d)j \qquad (12.4)$$
>
> Multiplication:
> $$(a + bj)(c + dj) = (ac - bd) + (ad + bc)j \qquad (12.5)$$
>
> Division:
> $$\frac{a + bj}{c + dj} = \frac{(a + bj)(c - dj)}{(c + dj)(c - dj)} = \frac{(ac + bd) + (bc - ad)j}{c^2 + d^2} \qquad (12.6)$$

Eqs. (12.3) and (12.4) show that we *add and subtract complex numbers by combining the real parts and combining the imaginary parts.*

EXAMPLE 1 Adding, subtracting complex numbers

(a) $(3 - 2j) + (-5 + 7j) = (3 - 5) + (-2 + 7)j$
$$= -2 + 5j$$

(b) $(7 + 9j) - (6 - 4j) = (7 - 6) + (9 - (-4))j$
$$= 1 + 13j$$

■

When complex numbers are multiplied, Eq. (12.5) indicates that *we express numbers in terms of j, proceed as with any algebraic multiplication, and note that $j^2 = -1$.*

EXAMPLE 2 Multiplying complex numbers

(a) $(6 - \sqrt{-4})(\sqrt{-9}) = (6 - 2j)(3j)$ write in terms of j

$$= 18j - 6j^2 = 18j - 6(-1)$$

$$= 6 + 18j$$

(b) $(-9.4 - 6.2j)(2.5 + 1.5j) = (-9.4)(2.5) + (-9.4)(1.5j)$

$$+ (-6.2j)(2.5) + (-6.2j)(1.5j)$$

$$= -23.5 - 14.1j - 15.5j - 9.3j^2$$

$$= -23.5 - 29.6j - 9.3(-1)$$

$$= -14.2 - 29.6j \qquad \blacksquare$$

Practice Exercise

Multiply: **1.** $(3 - 7j)(9 + 2j)$

The procedure shown in Eq. (12.6) for dividing by a complex number is the same as that used for rationalizing the denominator of a fraction. The result is in the proper form of a complex number. Therefore, *to divide by a complex number, multiply the numerator and the denominator by the conjugate of the denominator.*

EXAMPLE 3 Dividing complex numbers

■ Most calculators are programmed for complex numbers. The solutions for this example are shown in Fig. 12.1. Note the use of i for j.

(a) $\dfrac{7 - 2j}{3 + 4j} = \dfrac{(7 - 2j)(3 - 4j)}{(3 + 4j)(3 - 4j)}$ multiply by conjugate of denominator

$$= \dfrac{21 - 28j - 6j + 8j^2}{9 - 16j^2} = \dfrac{21 - 34j + 8(-1)}{9 - 16(-1)}$$

$$= \dfrac{13 - 34j}{25}$$

```
(7-2i)/(3+4i)
            .52-1.36i
1/i+2/(3+i)
            .6-1.2i
```

Fig. 12.1

This could be written in the form $a + bj$ as $\frac{13}{25} - \frac{34}{25}j$, but this type of result is generally left as a single fraction. In decimal form, the result would be expressed as $0.52 - 1.36j$.

(b) $\dfrac{1}{j} + \dfrac{2}{3 + j} = \dfrac{(3 + j) + 2j}{j(3 + j)} = \dfrac{3 + 3j}{3j - 1} = \dfrac{3 + 3j}{-1 + 3j} \cdot \dfrac{-1 - 3j}{-1 - 3j}$

$$= \dfrac{-3 - 12j - 9j^2}{1 - 9j^2} = \dfrac{6 - 12j}{10}$$

$$= \dfrac{3 - 6j}{5} \qquad \blacksquare$$

Practice Exercise

Divide: **2.** $\dfrac{5 - 3j}{2 + 7j}$

EXAMPLE 4 Dividing complex numbers – application

In an alternating-current circuit, the voltage E is given by $E = IZ$, where I is the current (in A) and Z is the impedance (in Ω). Each of these can be represented by complex numbers. Find the complex number representation for I if $E = 4.20 - 3.00$ volts and $Z = 5.30 + 2.65$ ohms. (This type of circuit is discussed in Section 12.7.)

Since $I = E/Z$, we have

```
5.3²+2.65²
            35.1125
Ans→D
            35.1125
(4.2*5.3-3*2.65)
/D
            .4075471698
(4.2*-2.65-3*5.3
)/D
            -.7698113208
```

Fig. 12.2

$$I = \dfrac{4.20 - 3.00j}{5.30 + 2.65j} = \dfrac{(4.20 - 3.00j)(5.30 - 2.65j)}{(5.30 + 2.65j)(5.30 - 2.65j)}$$ multiply by conjugate of denominator

$$= \dfrac{22.26 - 11.13j - 15.90j + 7.95j^2}{5.30^2 - 2.65^2 j^2} = \dfrac{22.26 - 7.95 - 27.03j}{5.30^2 + 2.65^2}$$

$$= \dfrac{14.31 - 27.03j}{35.11} = 0.408 - 0.770j \text{ amperes}$$

Not using complex numbers, the calculator solution is shown in Fig. 12.2. Note that we first stored the value of the denominator, treating j^2 as -1.   ■

EXERCISES 12.2

In Exercises 1–4, perform the indicated operations on the resulting expressions if the given changes are made in the indicated examples of this section.

1. In Example 1(b), change the sign in the first parentheses from + to − and then perform the addition.

2. In Example 2(b), change the sign before $6.2j$ from − to +, and then perform the multiplication.

3. In Example 3(a), change the sign in the denominator from + to − and then simplify.

4. In Example 3(b), change the sign in the second denominator from + to − and then simplify.

In Exercises 5–38, perform the indicated operations, expressing all answers in the form $a + bj$.

5. $(3 − 7j) + (2 − j)$
6. $(−4 − j) + (−7 − 4j)$
7. $(7j − 6) − (19 − 3j)$
8. $(5.4 − 3.4j) − (2.9j + 5.5)$
9. $0.23 − (0.46 − 0.19j) + 0.67j$
10. $(7 − j) − (4 − 4j) + (6 − j)$
11. $(12j − 21) − (15 − 18j) − 9j$
12. $(0.062j − 0.073) − 0.030j − (0.121 − 0.051j)$
13. $(7 − j)(7j)$
14. $(−2.2j)(1.5j − 4.0)$
15. $(4 − j)(5 + 2j)$
16. $(8j − 5)(7 + 4j)$
17. $(\sqrt{−18}\sqrt{−4})(3j)$
18. $\sqrt{−6}\sqrt{−12}\sqrt{30}$
19. $7j^3 − 7\sqrt{−9}$
20. $6j − 5j^2\sqrt{−63}$
21. $j\sqrt{−7} − j^6\sqrt{112} + 3j$
22. $j^2\sqrt{−7} − \sqrt{−28} + 8$
23. $(3 − 7j)^2$
24. $(8j + 20)^2$
25. $(1 − j)^3$
26. $(1 + j)(1 − j)^2$
27. $\dfrac{6j}{2 − 5j}$
28. $\dfrac{0.25}{3 − \sqrt{−1}}$
29. $\dfrac{1 − j}{3j}$
30. $\dfrac{6 + 5j}{3 − 4j}$
31. $\dfrac{j\sqrt{2} − 5}{j\sqrt{2} + 3}$
32. $\dfrac{j^5 − j^3}{3 + j}$
33. $\dfrac{j^2 − j}{2j − j^8}$
34. $\dfrac{3}{2j} + \dfrac{5}{6 − j}$
35. $\dfrac{4j}{1 − j} − \dfrac{8 + j}{2 + 3j}$
36. $\dfrac{(6j + 5)(2 − 4j)}{(5 − j)(4j + 1)}$
37. $(4j^5 − 5j^4 + 2j^3 − 3j^2)^2$
38. $(2j^2 − 3j^3 + 2j^4 − 2j^5)^6$

In Exercises 39–42, evaluate each expression on a calculator. Express answers in the form $a + bj$.

39. $(3j^9 − 5j^3)(4j^6 − 6j^8)$
40. $(5j − 4j^2 + 3j^7)(2j^{12} − j^{13})$
41. $\dfrac{(2 − j^3)^4}{(j^8 − j^6)^3} + j$
42. $(1 + j)^{−3}(2 − j)^{−2}$

In Exercises 43–54, answer the given problems.

43. Show that $−1 − j$ is a solution to the equation $x^2 + 2x + 2 = 0$.
44. Show that $1 − j\sqrt{3}$ is a solution to the equation $x^2 + 4 = 2x$.
45. Multiply $−3 + j$ by its conjugate.
46. Divide $2 − 3j$ by its conjugate.
47. Write the reciprocal of $3 − j$ in rectangular form.
48. Write the reciprocal of $2 + 5j$ in rectangular form.
49. Write $j^{−2} + j^{−3}$ in rectangular form.
50. Solve for x: $(x + 2j)^2 = 5 + 12j$
51. Solve for x: $(x + 3j)^2 = 7 − 24j$
52. For $\dfrac{3}{5} + \dfrac{4}{5}j$, find: (a) the conjugate; (b) the reciprocal.
53. If $f(x) = x + \dfrac{1}{x}$, find $f(1 + 3j)$.
54. When finding the current in a certain electric circuit, the expression $(s + 1 + 4j)(s + 1 − 4j)$ occurs. Simplify this expression.

In Exercises 55–58, solve the given problems. Refer to Example 4.

55. If $I = 0.835 − 0.427j$ amperes and $Z = 250 + 170j$ ohms, find the complex-number representation for E.
56. If $E = 5.70 − 3.65j$ volts and $I = 0.360 − 0.525j$ amperes, find the complex-number representation for Z.
57. If $E = 85 + 74j$ volts and $Z = 2500 − 1200j$ ohms, find the complex-number representation for I.
58. In an alternating-current circuit, two impedances Z_1 and Z_2 have a total impedance Z_T of $Z_T = \dfrac{Z_1 Z_2}{Z_1 + Z_2}$. Find Z_T for $Z_1 = 3200 + 4800j$ ohms and $Z_2 = 4800 − 6400j$ ohms.

In Exercises 59–62, answer or explain as indicated.

59. What type of number is the result of (a) adding a complex number to its conjugate and (b) subtracting a complex number from its conjugate?
(W) 60. If the reciprocal of $a + bj$ equals $a − bj$, what condition must a and b satisfy?
(W) 61. Explain why the product of a complex number and its conjugate is real and nonnegative.
(W) 62. Explain how to show that the reciprocal of the imaginary unit is the negative of the imaginary unit.

Answers to Practice Exercises

1. $41 − 57j$ **2.** $\dfrac{−11 − 41j}{53}$

12.3 Graphical Representation of Complex Numbers

Complex Plane • Complex Number as a Point • Adding and Subtracting Complex Numbers Graphically

Graphically, we represent a complex number as a ***point,*** designated as $a + bj$, in the rectangular coordinate system. *The **real part** is the **x-value** of the point, and the **imaginary part** is the **y-value** of the point.* Used in this way, *the coordinate system is called the* **complex plane,** *the horizontal axis is the* **real axis,** *and the vertical axis is the* **imaginary axis.**

EXAMPLE 1 Complex number in complex plane

In Fig. 12.3, A represents the complex number $3 - 2j$, B represents $-1 + j$, and C represents $-2 - 3j$. These complex numbers are represented by the points $(3, -2)$, $(-1, 1)$, and $(-2, -3)$, respectively, of the standard rectangular coordinate system.

Keep in mind that the meaning given to the points representing complex numbers in the complex plane is different from the meaning given to the points in the standard rectangular coordinate system. *A point in the complex plane represents a **single complex number,*** whereas *a point in the rectangular coordinate system represents a **pair of real numbers.*** ∎

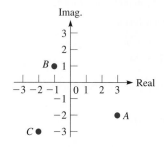

Fig. 12.3

In the complex plane, consider two complex numbers—for example, $1 + 2j$ and $3 + j$—and their sum $4 + 3j$. Drawing lines from the origin to these points (see Fig. 12.4), note that if we think of the complex numbers as being vectors, their sum is the vector sum. Because complex numbers can be used to represent vectors, the numbers are particularly important. *Any complex number can be thought of as representing a vector from the origin to its point in the complex plane.* This leads to the method used to add complex numbers graphically.

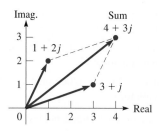

Fig. 12.4

> **STEPS TO ADD COMPLEX NUMBERS GRAPHICALLY**
>
> **1.** *Find the point corresponding to one of the numbers and draw a line from the origin to this point.*
> **2.** *Repeat step 1 for the second number.*
> **3.** *Complete a parallelogram with the lines drawn as adjacent sides. The resulting fourth vertex is the point representing the sum.*

Note that *this is equivalent to adding vectors by graphical means.*

EXAMPLE 2 Adding complex numbers graphically

(a) Add the complex numbers $5 - 2j$ and $-2 - j$ graphically.

The solution is shown in Fig. 12.5. We see that the fourth vertex of the parallelogram is at $3 - 3j$. This is, of course, the algebraic sum of these two complex numbers.

Fig. 12.5

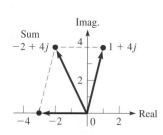

Fig. 12.6

(b) Add the complex numbers -3 and $1 + 4j$.

First, note that $-3 = -3 + 0j$, which means that the point representing -3 is on the negative real axis. In Fig. 12.6, we show the numbers -3 and $1 + 4j$ on the graph and complete the parallelogram. From the graph, we see that the sum is $-2 + 4j$. ∎

EXAMPLE 3 Subtracting complex numbers graphically

Subtract $4 - 2j$ from $2 - 3j$ graphically.

Subtracting $4 - 2j$ is equivalent to adding $-4 + 2j$. Therefore, we complete the solution by adding $-4 + 2j$ and $2 - 3j$, as shown in Fig. 12.7. The result is $-2 - j$. ∎

Fig. 12.7

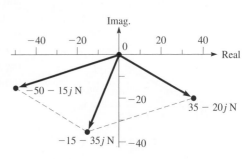

Fig. 12.8

EXAMPLE 4 Adding complex numbers – application

Two forces acting on an overhead bolt can be represented by $35 - 20j$ N and $-50 - 15j$ N. Find the resultant force graphically.

The forces are shown in Fig. 12.8. From the graph, we can see that the sum of the forces, which is the resultant force, is $-15 - 35j$ N. ∎

EXERCISES 12.3

In Exercises 1 and 2, perform the indicated operations for the resulting complex numbers if the given changes are made in the indicated examples of this section.

1. In Example 2(a), change the sign of the imaginary part of the second complex number and then add the numbers graphically.

2. In Example 3, change the sign of the imaginary part of the second complex number and do the subtraction graphically.

In Exercises 3–8, locate the given numbers in the complex plane.

3. $2 + 6j$ 4. $-5 + j$ 5. $-4 - 3j$

6. 10 7. $-3j$ 8. $3 - 4j$

In Exercises 9–28, perform the indicated operations graphically. Check them algebraically.

9. $2 + (3 + 4j)$ 10. $2j + (-2 + 3j)$

11. $(5 - j) + (3 + 2j)$ 12. $(3 - 2j) + (-1 - j)$

13. $5j - (1 - 4j)$ 14. $(0.2 - 0.1j) - 0.1$

15. $(2 - 4j) + (-2 + j)$ 16. $(-1 - 2j) + (6 - j)$

17. $(3 - 2j) - (4 - 6j)$ 18. $(-25 - 40j) - (20 - 55j)$

19. $(80 + 300j) - (260 + 150j)$

20. $(-j - 2) - (-1 - 3j)$

21. $(1.5 - 0.5j) + (3.0 + 2.5j)$ 22. $(3.5 + 2.0j) - (1.5j - 4.0)$

23. $(3 - 6j) - (-1 - 8j)$ 24. $(-6 - 3j) + (2 - 7j)$

25. $(2j + 1) - 3j - (j + 1)$ 26. $(6 - j) - 9 - (2j - 3)$

27. $(j - 6) - j + (j - 7)$ 28. $j - (1 - j) + (3 + 2j)$

In Exercises 29–32, show the given number, its negative, and its conjugate on the same coordinate system.

29. $3 + 2j$ 30. $-2 + 4j$ 31. $-3 - 5j$ 32. $5 - j$

In Exercises 33 and 34, show the numbers $a + bj$, $3(a + bj)$, and $-3(a + bj)$ on the same coordinate system. The multiplication of a complex number by a real number is called **scalar multiplication** *of the complex number.*

33. $3 - j$ 34. $-10 - 30j$

(W) *In Exercises 35–38, perform the indicated vector operations graphically on the complex number $2 + 4j$.*

35. Graph the complex number and its conjugate. Describe the relative positions.

36. Add the number and its conjugate. Describe the result.

37. Subtract the conjugate from the number. Describe the result.

38. Graph the number, the number multiplied by j, the number multiplied by j^2, and the number multiplied by j^3 on the same graph. Describe the result of multiplying a complex number by j.

In Exercises 39 and 40, perform the indicated vector additions graphically. Check them algebraically.

39. Two ropes hold a boat at a dock. The tensions in the ropes can be represented by $40 + 10j$ lb and $50 - 25j$ lb. Find the resultant force.

40. Relative to the air, a plane heads north of west with a velocity that can be represented by $-480 + 210j$ km/h. The wind is blowing from south of west with a velocity that can be represented by $60 + 210j$ km/h. Find the resultant velocity of the plane.

12.4 Polar Form of a Complex Number

**Polar Form $r(\cos\theta + j\sin\theta)$ •
Expressing Numbers in Polar Form**

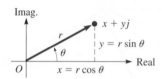

Fig. 12.9

In this section, we use the fact that a complex number can be represented by a vector to write complex numbers in another form. This form has advantages when multiplying and dividing complex numbers, and we will discuss these operations later in this chapter.

In the complex plane, by drawing a vector from the origin to the point that represents the number $x + yj$, an angle θ in standard position is formed. The point $x + yj$ is r units from the origin. In fact, *we can find any point in the complex plane by knowing the angle θ and the value of r*. The equations relating x, y, r, and θ, are similar to those developed for vectors (Eqs. (9.1) to (9.3)). Referring to Fig. 12.9, we see that

$$x = r\cos\theta \qquad y = r\sin\theta \tag{12.7}$$

$$r^2 = x^2 + y^2 \qquad \tan\theta = \frac{y}{x} \tag{12.8}$$

Substituting Eq. (12.7) into the rectangular form $x + yj$ of a complex number, we have $x + yj = r\cos\theta + j(r\sin\theta)$, or

$$x + yj = r(\cos\theta + j\sin\theta) \tag{12.9}$$

*The right side of Eq. (12.9) is called the **polar form** (sometimes the **trigonometric form**). The length r is the **absolute value**, or the **modulus, and** θ is the **argument** of the complex number.* Eq. (12.9), along with Eqs. (12.8), defines the polar form of a complex number.

EXAMPLE 1 Representing number in polar form

Represent the complex number $3 + 4j$ graphically and give its polar form.

From the rectangular form $3 + 4j$, we see that $x = 3$ and $y = 4$. Using Eqs. (12.8), we have

$$r = \sqrt{3^2 + 4^2} = 5 \qquad \theta = \tan^{-1}\frac{4}{3} = 53.1°$$

Thus, the polar form is $5(\cos 53.1° + j\sin 53.1°)$. See Fig. 12.10. ∎

Fig. 12.10

In Example 1, in expressing the polar form as $5(\cos 53.1° + j\sin 53.1°)$, we rounded the angle to the nearest 0.1°, as it is not possible to express the result *exactly* in degrees. In dealing with nonexact numbers, we will express angle to the nearest 0.1°. Other results that cannot be written exactly will be expressed to three significant digits, unless a different accuracy is given in the problem. Of course, in applied situations, most numbers are approximate, as they are derived through measurement.

Another convenient and widely used notation for the polar form is $r\,\underline{/\theta}$. We must remember in using this form that it represents a complex number and is simply a shorthand way of writing $r(\cos\theta + j\sin\theta)$. Therefore,

$$r\,\underline{/\theta} = r(\cos\theta + j\sin\theta) \tag{12.10}$$

Practice Exercise

Write in polar form: **1.** $15 - 8j$

EXAMPLE 2 Polar form $r\,\underline{/\theta}$

(a) $3(\cos 40° + j\sin 40°) = 3\,\underline{/40°}$

(b) $6.26(\cos 217.3° + j\sin 217.3°) = 6.26\,\underline{/217.3°}$

(c) $5\,\underline{/120°} = 5(\cos 120° + j\sin 120°)$

(d) $14.5\,\underline{/306.2°} = 14.5(\cos 306.2° + j\sin 306.2°)$ ∎

Fig. 12.11

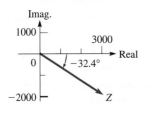

Fig. 12.12

Practice Exercise

Write in rectangular form: **2.** $2.50 \underline{/120°}$

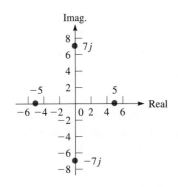

Fig. 12.13

EXAMPLE 3 Rectangular form to polar form

Represent the complex number $-1.04 - 1.56j$ graphically and give its polar form.

The graphical representation is shown in Fig. 12.11. From Eqs. (12.8), we have

$$r = \sqrt{(-1.04)^2 + (-1.56)^2} = 1.87$$

$$\theta_{\text{ref}} = \tan^{-1}\frac{1.56}{1.04} = 56.3° \qquad \theta = 180° + 56.3° = 236.3°$$

Since both the real and imaginary parts are negative, θ is a third-quadrant angle. Therefore, we found the reference angle before finding θ. This means the polar forms are

$$1.87(\cos 236.3° + j \sin 236.3°) = 1.87 \underline{/236.3°} \qquad \blacksquare$$

EXAMPLE 4 Polar form to rectangular form

The impedance Z (in Ω) in an alternating-current circuit is given by $Z = 3560 \underline{/-32.4°}$. Express this in rectangular form.

From the polar form, we have $r = 3560 \ \Omega$ and $\theta = -32.4°$ (it is common to use negative angles in this type of application). This means that we can also write

$$Z = 3560(\cos(-32.4°) + j \sin(-32.4°))$$

See Fig. 12.12. This means that

$$x = 3560 \cos(-32.4°) = 3010$$
$$y = 3560 \sin(-32.4°) = -1910$$

Therefore, the rectangular form is $Z = 3010 - 1910j \ \Omega$. $\qquad \blacksquare$

Complex numbers can be converted between rectangular and polar forms on most calculators. This will be illustrated in the next section.

EXAMPLE 5 Polar form – real and imaginary numbers

Represent the numbers 5, -5, $7j$, and $-7j$ in polar form.

Since a positive real number is on the positive real axis, its polar form is

$$a = a(\cos 0° + j \sin 0°) = a \underline{/0°}$$

Negative real numbers, being on the negative real axis, are written as

$$a = |a|(\cos 180° + j \sin 180°) = |a| \underline{/180°}$$

Thus,

$$5 = 5(\cos 0° + j \sin 0°) = 5 \underline{/0°}$$
$$-5 = 5(\cos 180° + j \sin 180°) = 5 \underline{/180°}$$

Positive pure imaginary numbers lie on the positive imaginary axis and are expressed in polar form by

$$bj = b(\cos 90° + j \sin 90°) = b \underline{/90°}$$

Similarly, negative pure imaginary numbers, being on the negative imaginary axis, are written as

$$bj = |b|(\cos 270° + j \sin 270°) = |b| \underline{/270°}$$

Thus,

$$7j = 7(\cos 90° + j \sin 90°) = 7 \underline{/90°}$$
$$-7j = 7(\cos 270° + j \sin 270°) = 7 \underline{/270°}$$

The *complex numbers* 5, -5, $7j$, and $-7j$ are shown graphically in Fig. 12.13. $\qquad \blacksquare$

EXERCISES 12.4

In Exercises 1 and 2, change the sign of the real part of the complex number in the indicated example of this section and then perform the indicated operations for the resulting complex number.

1. Example 1   **2.** Example 3

In Exercises 3–18, represent each complex number graphically and give the polar form of each.

3. $8 + 6j$ **4.** $-8 - 15j$ **5.** $30 - 40j$

6. $-5 + 12j$ **7.** $-2.00 + 3.00j$ **8.** $7.00 - 5.00j$

9. $-0.55 - 0.24j$ **10.** $460 - 460j$ **11.** $1 + j\sqrt{3}$

12. $\sqrt{2} - j\sqrt{2}$ **13.** $3.514 - 7.256j$ **14.** $62.31 + 95.27j$

15. -3 **16.** 60 **17.** $9j$ **18.** $-2j$

In Exercises 19–36, represent each complex number graphically and give the rectangular form of each.

19. $5.00(\cos 54.0° + j\sin 54.0°)$ **20.** $6(\cos 180° + j\sin 180°)$

21. $160(\cos 150.0° + j\sin 150.0°)$

22. $2.50(\cos 315.0° + j\sin 315.0°)$

23. $3.00(\cos 232.0° + j\sin 232.0°)$

24. $220.8(\cos 155.13° + j\sin 155.13°)$

25. $0.08(\cos 360° + j\sin 360°)$ **26.** $15(\cos 0° + j\sin 0°)$

27. $120(\cos 270° + j\sin 270°)$ **28.** $\cos 600.0° + j\sin 600.0°$

29. $4.75\underline{/172.8°}$ **30.** $1.50\underline{/62.3°}$

31. $0.9326\underline{/229.54°}$ **32.** $277.8\underline{/-342.63°}$

33. $7.32\underline{/-270°}$ **34.** $18.3\underline{/540.0°}$

35. $86.42\underline{/94.62°}$ **36.** $4629\underline{/182.44°}$

In Exercises 37–44, solve the given problems.

37. What is the argument for any negative real number?

38. For $x + yj$, what is the argument if $x = y < 0$?

39. Show that the conjugate of $r\underline{/\theta}$ is $r\underline{/-\theta}$.

40. Find r and $\tan(\theta_{ref})$ for the complex number $a^b(a + bj)$. $(a > 0)$

41. The voltage of a certain generator is represented by $2.84 - 1.06j$ kV. Write this voltage in polar form.

42. Find the magnitude and direction of a force on a bolt that is represented by $40.5 + 24.5j$ newtons.

43. The electric field intensity of a light wave can be described by $12.4\underline{/78.3°}$ V/m. Write this in rectangular form.

44. The current in a certain microprocessor circuit is represented by $3.75\underline{/15.0°}$ μA. Write this in rectangular form.

Answers to Practice Exercises

1. $17(\cos 331.9° + j\sin 331.9°)$ **2.** $-1.25 + 2.17j$

12.5 Exponential Form of a Complex Number

**Exponential Form $re^{j\theta}$ •
Expressing Numbers in Exponential
Form • Summary of Important Forms**

■ The number e is named for the Swiss mathematician Leonhard Euler (1707–1783). His works in mathematics and other fields filled about 70 volumes.

Another important form of a complex number is the *exponential form*. It is commonly used in electronics, engineering, and physics applications. As we will see in the next section, it is also convenient for multiplication and division of complex numbers, as the rectangular form is for addition and subtraction.

The **exponential form** *of a complex number is written as* $re^{j\theta}$, where r and θ have the same meanings as given in the previous section, although θ *is expressed in radians.* *The number e is a special irrational number and has an approximate value*

$$e = 2.7182818284590452$$

This number e is very important in mathematics, and we will see it again in the next chapter. For now, it is necessary to accept the value for e, although in calculus its meaning is shown along with the reason it has the above value. We can find its value on a calculator by using the $\boxed{e^x}$ key, with $x = 1$, or by using the $\boxed{e}$ key.

In advanced mathematics, it is shown that

$$re^{j\theta} = r(\cos\theta + j\sin\theta) \tag{12.11}$$

CAUTION ▶

By expressing θ in radians, the expression $j\theta$ is an exponent, and it can be shown to obey all the laws of exponents as discussed in Chapter 11. Therefore, we will always *express θ in radians when using exponential form.*

EXAMPLE 1 Polar form to exponential form

Express the number $8.50\underline{/136.3°}$ in exponential form.

Since this complex number is in polar form, we note that $r = 8.50$ and that we must express $136.3°$ in radians. Changing $136.3°$ to radians, we have

$$\frac{136.3\pi}{180} = 2.38 \text{ rad}$$

Therefore, the required exponential form is $8.50e^{2.38j}$. This means that

$$8.50\underline{/136.3°} = 8.50e^{2.38j} \qquad \begin{array}{l} \text{degrees to radians} \\[1em] \text{value of } r \end{array}$$

We see that the principal step in changing from polar form to exponential form is to change θ from degrees to radians. ■

EXAMPLE 2 Rectangular form to exponential form

Express the number $3.07 - 7.43j$ in exponential form.

From the rectangular form, we have $x = 3.07$ and $y = -7.43$. Therefore,

$$r = \sqrt{(3.07)^2 + (-7.43)^2} = 8.04$$

$$\theta_{\text{ref}} = \tan^{-1}\frac{7.43}{3.07} = 67.6° \qquad \theta = 360° - 67.6° = 292.4°$$

Since $292.4° = 5.10$ rad, the exponential form is $8.04e^{5.10j}$. (Using radian mode, we could show that $2\pi - \tan^{-1}(7.43/3.07) = 5.10$). This means that

$$3.07 - 7.43j = 8.04e^{5.10j}$$ ■

EXAMPLE 3 Exponential form to other forms

Express the complex number $2.00e^{4.80j}$ in polar and rectangular forms.

We first express 4.80 rad as $275.0°$. From the exponential form, we know that $r = 2.00$. Thus, the polar form is

$$2.00(\cos 275.0° + j\sin 275.0°)$$

Using the distributive law, we rewrite the polar form and then evaluate. Thus,

$$2.00e^{4.80j} = 2.00(\cos 275.0° + j\sin 275.0°)$$
$$= 2.00\cos 275.0° + (2.00\sin 275.0°)j$$
$$= 0.174 - 1.99j$$ ■

EXAMPLE 4 Power of number in exponential form

Express $(1.846e^{1.229j})^2$ in polar and rectangular forms.

First, we note that $(1.846e^{1.229j})^2 = 1.846^2e^{2.458j}$. Since 2.458 rad $= 140.83°$, the polar form is $1.846^2\underline{/140.83°} = 3.408\underline{/140.83°}$.

For the rectangular form, using radian mode, we have

$$(1.846e^{1.229j})^2 = 1.846^2e^{2.458j} = 1.846^2(\cos 2.458 + j\sin 2.458)$$
$$= -2.642 + 2.152j$$ ■

As we noted earlier, an important application of the use of complex numbers is in alternating-current analysis. When an alternating current flows through a given circuit, usually the current and voltage have different phases. That is, they do not reach their peak values at the same time. Therefore, one way of accounting for the magnitude as well as the phase of an electric current or voltage is to write it as a complex number. Here, the modulus is the actual magnitude of the current or voltage, and the argument θ is a measure of the phase.

Practice Exercise

1. Express $25.0(\cos 127.0° + j\sin 127.0°)$ in exponential form.

Practice Exercise

2. Express $-20.5 - 16.8j$ in exponential form.

■ For reference, Eq. (8.8) is π rad $= 180°$.

■ Calculators that operate with complex numbers treat the exponential and polar forms as being the same, since each uses r and θ. Figures 12.14 and 12.15 show the conversions of Examples 4 and 5 on such a calculator.

Fig. 12.14

EXAMPLE 5 Exponential form – application

A current of $2.00 - 4.00j$ amperes flows in a given circuit. Express this current in exponential form and find the magnitude of the current.

From the rectangular form, we have $x = 2.00$ and $y = -4.00$. Therefore,

$$r = \sqrt{(2.00)^2 + (-4.00)^2} = 4.47 \text{ A}$$

$$\theta = \tan^{-1}\frac{-4.00}{2.00} = -63.4°$$

It is normal to express the phase in terms of negative angles when the imaginary part is negative. Therefore, $63.4° = 1.11$ rad, which means the exponential form is $4.47e^{-1.11j}$. The modulus of 4.47 means the magnitude of the current is 4.47 A. Note in the calculator display in Fig. 12.15 that the angle is shown in radians, and is negative. ∎

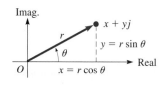

```
abs(2-4i)
         4.472135955
angle(2-4i)
        -1.107148718
```

Fig. 12.15

We now summarize the three important forms of a complex number. See Fig. 12.16.

Rectangular:	$x + yj$
Polar:	$r(\cos\theta + j\sin\theta) = r\underline{/\theta}$
Exponential:	$re^{j\theta}$

It follows that

$$x + yj = r(\cos\theta + j\sin\theta) = r\underline{/\theta} = re^{j\theta} \qquad (12.12)$$

where

$$r^2 = x^2 + y^2 \qquad \tan\theta = \frac{y}{x} \qquad (12.8)$$

Imag.

Fig. 12.16

In Eq. (12.12), the argument θ is the same for exponential and polar forms. It is usually expressed in radians in exponential form and in degrees in polar form.

EXERCISES 12.5

In Exercises 1–4, perform the indicated operations for the resulting complex numbers if the given changes are made in the indicated examples of this section.

1. In Example 1, change 136.3° to 226.3° and then find the exponential form.

2. In Example 2, change the sign before 7.43j from − to + and then find the exponential form.

3. In Example 3, change the exponent to 3.80j and then find the polar and rectangular forms.

4. In Example 4, change the exponent 2 to 3 and then find the polar and rectangular forms.

In Exercises 5–24, express the given numbers in exponential form.

5. $3.00(\cos 60.0° + j\sin 60.0°)$ **6.** $575(\cos 135.0° + j\sin 135°)$

7. $0.450(\cos 282.3° + j\sin 282.3°)$

8. $2.10(\cos 588.7° + j\sin 588.7°)$

9. $375.5[\cos(-95.46°) + j\sin(-95.46°)]$

10. $16.72[\cos(-7.14°) + j\sin(-7.14°)]$

11. $0.515\underline{/198.3°}$ **12.** $4650\underline{/326.5°}$ **13.** $4.06\underline{/-61.4°}$

14. $0.0192\underline{/76.7°}$ **15.** $9245\underline{/296.32°}$ **16.** $82.76\underline{/470.09°}$

17. $3 - 4j$ **18.** $-1 - 5j$ **19.** $-30 + 20j$

20. $600 + 100j$ **21.** $5.90 + 2.40j$ **22.** $47.3 - 10.9j$

23. $-634.6 - 528.2j$ **24.** $-8573 + 5477j$

In Exercises 25–32, express the given complex numbers in polar and rectangular forms.

25. $3.00e^{0.500j}$ **26.** $20.0e^{1.00j}$ **27.** $464e^{1.85j}$

28. $2.50e^{3.84j}$ **29.** $3.20e^{-5.41j}$ **30.** $0.800e^{3.00j}$

31. $0.1724e^{2.391j}$ **32.** $820.7e^{-3.492j}$

In Exercises 33–36, perform the indicated operations and express results in rectangular and polar forms.

33. $(4.55e^{1.32j})^2$ **34.** $(0.926e^{0.253j})^3$

35. $(625e^{3.46j})(4.40e^{1.22j})$ **36.** $(18.0e^{5.13j})(25.5e^{0.77j})$

In Exercises 37–42, perform the indicated operations.

37. Using a calculator, express $3.73 + 5.24j$ in exponential form. See Fig. 12.15.

38. Using a calculator, express $75.6e^{1.25j}$ in rectangular form. See Fig. 12.14.

39. The impedance in an antenna circuit is $375 + 110j$ ohms. Write this in exponential form and find the magnitude of the impedance.

40. The intensity of the signal from a radar microwave signal is $37.0[\cos(-65.3°) + j\sin(-65.3°)]$V/m. Write this in exponential form.

41. In an electric circuit, the *admittance* is the reciprocal of the impedance. If the impedance is $2800 - 1450j$ ohms in a certain circuit, find the exponential form of the admittance.

42. In dealing with the superposition of waves, the expression $E_1E_2(e^{j(\alpha-\beta)} + e^{-j(\alpha-\beta)})$ occurs. Show that it can be written as $2E_1E_2\cos(\alpha - \beta)$.

Answers to Practice Exercises

1. $25.0e^{2.22j}$ **2.** $26.5e^{3.83j}$

12.6 Products, Quotients, Powers, and Roots of Complex Numbers

Multiplication, Division, and Powers in Polar and Exponential Forms • DeMoivre's Theorem • Roots

The operations of multiplication and division can be performed with complex numbers in polar and exponential forms, as well as rectangular form. Doing them in these forms is convenient, but also leads to finding powers and roots of complex numbers.

Using exponential form and laws of exponents, we multiply two complex numbers as

$$[r_1e^{j\theta_1}] \times [r_2e^{j\theta_2}] = r_1r_2e^{j\theta_1+j\theta_2} = r_1r_2e^{j(\theta_1+\theta_2)}$$

We use this equation to express the product of two complex numbers in polar form:

$$[r_1e^{j\theta_1}] \times [r_2e^{j\theta_2}] = [r_1(\cos\theta_1 + j\sin\theta_1)] \times [r_2(\cos\theta_2 + j\sin\theta_2)]$$

and

$$r_1r_2e^{j(\theta_1+\theta_2)} = r_1r_2[\cos(\theta_1 + \theta_2) + j\sin(\theta_1 + \theta_2)]$$

The polar expressions are equal, which means *the product of two complex numbers is*

$$\begin{aligned} r_1(\cos\theta_1 &+ j\sin\theta_1)r_2(\cos\theta_2 + j\sin\theta_2) \\ &= r_1r_2[\cos(\theta_1 + \theta_2) + j\sin(\theta_1 + \theta_2)] \\ (r_1\underline{/\theta_1})(r_2\underline{/\theta_2}) &= r_1r_2\underline{/\theta_1 + \theta_2} \end{aligned}$$

(12.13)

NOTE ▶   *The magnitudes are multiplied,* and *the angles are added.*

EXAMPLE 1   Multiplication in polar form

Multiply the complex numbers $2 + 3j$ and $1 - j$ by using the polar form of each.

$$\text{For } 2 + 3j: \quad r_1 = \sqrt{2^2 + 3^2} = 3.61 \qquad \tan\theta_1 = \frac{3}{2} \qquad \theta_1 = 56.3°$$

$$\text{For } 1 - j: \quad r_2 = \sqrt{1^2 + (-1)^2} = 1.41 \quad \tan\theta_2 = \frac{-1}{1} \qquad \theta_2 = 315.0°$$

$$\underbrace{(3.61)(\cos 56.3° + j\sin 56.3°)(1.41)(\cos 315.0° + j\sin 315.0°)}_{\text{sum}}$$

$$= (3.61)(1.41)[\cos(56.3° + 315.0°) + j\sin(56.3° + 315.0°)]$$
$$= 5.09(\cos 371.3° + j\sin 371.3°) = 5.09(\cos 11.3° + j\sin 11.3°)$$

Note that in the final result, the angle is expressed as $11.3°$. The angle is usually expressed between $0°$ and $360°$, unless specified otherwise. As we have seen, in some applications, it is common to use negative angles. ∎

EXAMPLE 2 Multiplication in $r \underline{/\theta}$ form

When we use the $r \underline{/\theta}$ polar form to multiply the two complex numbers in Example 1, we have

$$r_1 = 3.61 \qquad \theta_1 = 56.3°$$
$$r_2 = 1.41 \qquad \theta_2 = 315.0°$$

$$(3.61 \underline{/56.3°})(1.41 \underline{/315.0°}) = (3.61)(1.41) \underline{/56.3° + 315.0°}$$

$$= 5.09 \underline{/371.3°}$$
$$= 5.09 \underline{/11.3°}$$

Again, note that the angle in the final result is between 0° and 360°. ∎

If we wish to *divide* one complex number in exponential form by another, we arrive at the following result:

$$r_1 e^{j\theta_1} \div r_2 e^{j\theta_2} = \frac{r_1}{r_2} e^{j(\theta_1 - \theta_2)} \tag{12.14}$$

Therefore, *the result of dividing one complex number in polar form by another is given by*

$$\frac{r_1(\cos \theta_1 + j \sin \theta_1)}{r_2(\cos \theta_2 + j \sin \theta_2)} = \frac{r_1}{r_2}[\cos(\theta_1 - \theta_2) + j \sin(\theta_1 - \theta_2)]$$
$$\frac{r_1 \underline{/\theta_1}}{r_2 \underline{/\theta_2}} = \frac{r_1}{r_2} \underline{/\theta_1 - \theta_2} \tag{12.15}$$

NOTE ▶ The *magnitudes are divided,* and the *angles are subtracted.*

EXAMPLE 3 Division in polar form

Divide the first complex number of Example 1 by the second.
  Using the polar forms, we have

$$\frac{3.61(\cos 56.3° + j \sin 56.3°)}{1.41(\cos 315.0° + j \sin 315.0°)} = \frac{3.61}{1.41}[\cos(56.3° - 315.0°) + j\sin(56.3° - 315.0°)]$$

$$= 2.56[\cos(-258.7°) + j \sin(-258.7°)]$$
$$= 2.56(\cos 101.3° + j \sin 101.3°) \quad ∎$$

EXAMPLE 4 Division in $r \underline{/\theta}$ form

Repeating Example 3 using $r \underline{/\theta}$ polar form, we have

$$\frac{3.61 \underline{/56.3°}}{1.41 \underline{/315.0°}} = \frac{3.61}{1.41} \underline{/56.3° - 315.0°} = 2.56 \underline{/-258.7°}$$
$$= 2.56 \underline{/101.3°} \quad ∎$$

CAUTION ▶ We have just seen that multiplying and dividing numbers in polar form can be easily performed. However, if we are to add or subtract numbers in polar form, *we must do the addition or subtraction by using rectangular form.*

Practice Exercise

1. Find the polar form product:
   $(3 \underline{/50°})(5 \underline{/65°})$

Practice Exercise

2. Find the polar form quotient:
   $$\frac{3(\cos 50° + j \sin 50°)}{5(\cos 65° + j \sin 65°)}$$

EXAMPLE 5 Addition in polar form

Perform the addition $1.563\underline{/37.56°} + 3.827\underline{/146.23°}$.

In order to do this addition, we must change each number to rectangular form:

$$1.563\underline{/37.56°} + 3.827\underline{/146.23°}$$
$$= 1.563(\cos 37.56° + j\sin 37.56°) + 3.827(\cos 146.23° + j\sin 146.23°)$$
$$= 1.2390 + 0.9528j - 3.1813 + 2.1273j$$
$$= -1.9423 + 3.0801j$$

Now, we change this to polar form:

$$r = \sqrt{(-1.9423)^2 + (3.0801)^2} = 3.641$$
$$\tan\theta = \frac{3.0801}{-1.9423} \qquad \theta = 122.24°$$

Therefore,

$$1.563\underline{/37.56°} + 3.827\underline{/146.23°} = 3.641\underline{/122.24°} \qquad\blacksquare$$

DEMOIVRE'S THEOREM

To raise a complex number to a power, we may use the exponential form of the number in Eq. (11.3). This leads to

■ For reference, Eq. (11.3) is $(a^m)^n = a^{mn}$.

$$(re^{j\theta})^n = r^n e^{jn\theta} \qquad (12.16)$$

Extending this to polar form, we have

$$\boxed{\begin{array}{c} [r(\cos\theta + j\sin\theta)]^n = r^n(\cos n\theta + j\sin n\theta) \\ (r\underline{/\theta})^n = r^n\underline{/n\theta} \end{array}} \qquad (12.17)$$

■ Named for the mathematician Abraham DeMoivre (1667–1754).

Equation (12.17) is known as **DeMoivre's theorem** *and is valid for all real values of n. It is also used for finding roots of complex numbers if n is a fractional exponent. We note that* the magnitude is raised to the power, *and* the angle is multiplied by the power.

EXAMPLE 6 Power by DeMoivre's theorem

Using DeMoivre's theorem, find $(2 + 3j)^3$.

From Example 1, we know $r = 3.61$ and $\theta = 56.3°$. Therefore,

$$[3.61(\cos 56.3° + j\sin 56.3°)]^3 = (3.61)^3[\cos(3 \times 56.3°) + j\sin(3 \times 56.3°)]$$
$$= 47.0(\cos 168.9° + j\sin 168.9°)$$
$$= 47.0\underline{/168.9°}$$

Expressing θ in radians, we have $\theta = 56.3° = 0.983$ rad. Therefore,

$$(3.61e^{0.983j})^3 = (3.61)^3 e^{3\times0.983j} = 47.0e^{2.95j}$$
$$(2 + 3j)^3 = 47.0(\cos 168.9° + j\sin 168.9°)$$
$$= 47.0\underline{/168.9°}$$
$$= 47.0e^{2.95j} = -46 + 9j \qquad\blacksquare$$

Practice Exercise

3. Find the polar form power: $(3\cos 50°)^8$

EXAMPLE 7 Cube roots by DeMoivre's theorem

Find the cube root of -1.

Since we know that -1 is a real number, we can find its cube root by means of the definition. That is, $(-1)^3 = -1$. We check this by DeMoivre's theorem. Writing -1 in polar form, we have

$$-1 = 1(\cos 180° + j \sin 180°)$$

Applying DeMoivre's theorem, with $n = \frac{1}{3}$, we obtain

$$(-1)^{1/3} = 1^{1/3}(\cos \tfrac{1}{3} 180° + j \sin \tfrac{1}{3} 180°) = \cos 60° + j \sin 60°$$

$$= \frac{1}{2} + j\frac{\sqrt{3}}{2} \qquad \text{exact answer}$$

$$= 0.5000 + 0.8660j \qquad \text{decimal approximation}$$

Observe that we did not obtain -1 as the answer. If we check the answer, in the form $\frac{1}{2} + j\frac{\sqrt{3}}{2}$, by actually cubing it, we obtain -1! Therefore, it is a correct answer.

We should note that it is possible to take $\frac{1}{3}$ of any angle up to 1080° and still have an angle less than 360°. Since 180° and 540° have the same terminal side, let us try writing -1 as $1(\cos 540° + j \sin 540°)$. Using DeMoivre's theorem, we have

$$(-1)^{1/3} = 1^{1/3}(\cos \tfrac{1}{3}540° + j \sin \tfrac{1}{3}540°) = \cos 180° + j \sin 180° = -1$$

We have found the answer we originally anticipated.

Angles of 180° and 900° also have the same terminal side, so we try

$$(-1)^{1/3} = 1^{1/3}\left(\cos \tfrac{1}{3}900° + j \sin \tfrac{1}{3}900°\right) = \cos 300° + j \sin 300°$$

$$= \frac{1}{2} - j\frac{\sqrt{3}}{2} \qquad \text{exact answer}$$

$$= 0.5000 - 0.8660j \qquad \text{decimal approximation}$$

Checking this, we find that it is also a correct root. We may try 1260°, but $\frac{1}{3}(1260°) = 420°$, which has the same functional values as 60°, and would give us the answer $0.5000 + 0.8660j$ again.

We have found, therefore, *three cube roots* of -1. They are

$$-1, \quad \frac{1}{2} + j\frac{\sqrt{3}}{2}, \quad \frac{1}{2} - j\frac{\sqrt{3}}{2}$$

These roots are graphed in Fig. 12.17. Note that they are equally spaced on the circumference of a circle of radius 1. ∎

When the results of Example 7 are generalized, it can be proved that *there are n nth roots of a complex number*. When graphed, these roots are on a circle of radius $r^{1/n}$ and are equally spaced $360°/n$ apart. Following is the method for finding these n roots.

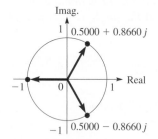

Fig. 12.17

USING DEMOIVRE'S THEOREM TO FIND THE n nth ROOTS OF A COMPLEX NUMBER

1. *Express the number in polar form.*
2. *Express the root as a fractional exponent.*
3. *Use Eq. (12.17) with θ to find one root.*
4. *Use Eq. (12.17) and add 360° to θ, n − 1 times, to find the other roots.*

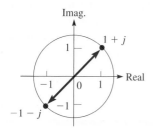

Fig. 12.18

EXAMPLE 8 Square roots by DeMoivre's theorem

Find the two square roots of $2j$.

First, we write $2j$ in polar form as $2j = 2(\cos 90° + j \sin 90°)$. To find square roots, we use the exponent $1/2$. The first square root is

$$(2j)^{1/2} = 2^{1/2}\left(\cos \frac{90°}{2} + j \sin \frac{90°}{2}\right) = \sqrt{2}(\cos 45° + j \sin 45°) = 1 + j$$

To find the other square root, we add $360°$ to $90°$. This gives us

$$(2j)^{1/2} = 2^{1/2}\left(\cos \frac{450°}{2} + j \sin \frac{450°}{2}\right) = \sqrt{2}(\cos 225° + j \sin 225°) = -1 - j$$

Therefore, the two square roots of $2j$ are $1 + j$ and $-1 - j$. We see in Fig. 12.18 that they are on a circle of radius $\sqrt{2}$ and $180°$ apart. ∎

EXAMPLE 9 Sixth roots by DeMoivre's theorem

Find all the roots of the equation $x^6 - 64 = 0$.

Solving for x, we have $x^6 = 64$, or $x = \sqrt[6]{64}$. Therefore, we have to find the six sixth roots of 64. Writing 64 in polar form, we have $64 = 64(\cos 0° + j \sin 0°)$. Using the exponent $1/6$ for the sixth root, we have the following solutions:

First root: $64^{1/6} = 64^{1/6}\left(\cos \dfrac{0°}{6} + j \sin \dfrac{0°}{6}\right) = 2(\cos 0° + j \sin 0°) = 2$

add $360°$

Second root: $64^{1/6} = 64^{1/6}\left(\cos \dfrac{0° + 360°}{6} + j \sin \dfrac{0° + 360°}{6}\right)$

$= 2(\cos 60° + j \sin 60°) = 1 + j\sqrt{3}$

add $2 \times 360°$

Third root: $64^{1/6} = 64^{1/6}\left(\cos \dfrac{0° + 720°}{6} + j \sin \dfrac{0° + 720°}{6}\right)$

$= 2(\cos 120° + j \sin 120°) = -1 + j\sqrt{3}$

add $3 \times 360°$

Fourth root: $64^{1/6} = 64^{1/6}\left(\cos \dfrac{0° + 1080°}{6} + j \sin \dfrac{0° + 1080°}{6}\right)$

$= 2(\cos 180° + j \sin 180°) = -2$

add $4 \times 360°$

Fifth root: $64^{1/6} = 64^{1/6}\left(\cos \dfrac{0° + 1440°}{6} + j \sin \dfrac{0° + 1440°}{6}\right)$

$= 2(\cos 240° + j \sin 240°) = -1 - j\sqrt{3}$

add $5 \times 360°$

Sixth root: $64^{1/6} = 64^{1/6}\left(\cos \dfrac{0° + 1800°}{6} + j \sin \dfrac{0° + 1800°}{6}\right)$

$= 2(\cos 300° + j \sin 300°) = 1 - j\sqrt{3}$

These roots are graphed in Fig. 12.19. Note that they are equally spaced $60°$ apart on the circumference of a circle of radius 2. ∎

■ See Appendix C for a graphing calculator program DEMOIVRE. It can be used to find the roots of a complex number.

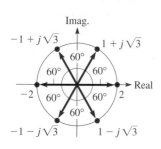

Fig. 12.19

From the text and examples of this and previous sections, we see the advantages for the various forms of writing complex numbers. Rectangular form can be used for all of the basic operations but lends itself best to addition and subtraction. Polar form is used for multiplication, division, raising to powers, and finding roots. Exponential form can be used for multiplication, division, and powers, and also for theoretical purposes (for example, deriving DeMoivre's theorem).

EXERCISES 12.6

In Exercises 1–4, perform the indicated operations for the resulting complex numbers if the given changes are made in the indicated examples of this section.

1. In Example 1, change the sign of the complex part of the second complex number and then perform the multiplication.

2. In Example 2, multiply the same complex numbers as in Exercise 1.

3. In Example 6, change the exponent to 5 and then find the result.

4. In Example 8, replace $2j$ with $-2j$ and then find the roots.

In Exercises 5–20, perform the indicated operations. Leave the result in polar form.

5. $[4(\cos 60° + j \sin 60°)][2(\cos 20° + j \sin 20°)]$

6. $[3(\cos 120° + j \sin 120°)][5(\cos 45° + j \sin 45°)]$

7. $(0.5\underline{/140°})(6\underline{/110°})$

8. $(0.4\underline{/320°})(5.5\underline{/-150°})$

9. $\dfrac{8(\cos 100° + j \sin 100°)}{4(\cos 65° + j \sin 65°)}$

10. $\dfrac{9(\cos 230° + j \sin 230°)}{45(\cos 80° + j \sin 80°)}$

11. $\dfrac{12\underline{/320°}}{5\underline{/-210°}}$

12. $\dfrac{2\underline{/90°}}{4\underline{/75°}}$

13. $[0.2(\cos 35° + j \sin 35°)]^3$

14. $[3(\cos 120° + j \sin 120°)]^4$

15. $(2\underline{/135°})^8$

16. $(1\underline{/142°})^{10}$

17. $\dfrac{(50\underline{/236°})(2\underline{/84°})}{125\underline{/47°}}$

18. $\dfrac{36\underline{/274°}}{(2\underline{/141°})(6\underline{/195°})}$

19. $\dfrac{(4\underline{/24°})(10\underline{/326°})}{(1\underline{/186°})(8\underline{/77°})}$

20. $\dfrac{(25\underline{/194°})(6\underline{/239°})}{(30\underline{/17°})(10\underline{/29°})}$

In Exercises 21–24, perform the indicated operations. Express results in polar form. See Example 5.

21. $2.78\underline{/56.8°} + 1.37\underline{/207.3°}$

22. $15.9\underline{/142.6°} - 18.5\underline{/71.4°}$

23. $7085\underline{/115.62°} - 4667\underline{/296.34°}$

24. $307.5\underline{/326.54°} + 726.3\underline{/96.41°}$

In Exercises 25–36, change each number to polar form and then perform the indicated operations. Express the result in rectangular and polar forms. Check by performing the same operation in rectangular form.

25. $(3 + 4j)(5 - 12j)$

26. $(-2 + 5j)(-1 - j)$

27. $(7 - 3j)(8 + j)$

28. $(1 + 5j)(4 + 2j)$

29. $\dfrac{7}{1 - 3j}$

30. $\dfrac{40j}{7 + 2j}$

31. $\dfrac{30 + 40j}{5 - 12j}$

32. $\dfrac{-2 + 5j}{-1 - j}$

33. $(3 + 4j)^4$

34. $(-1 - j)^8$

35. $(2 + 3j)^5$

36. $(1 - 2j)^6$

In Exercises 37–42, use DeMoivre's theorem to find all the indicated roots. Be sure to find all roots.

37. The two square roots of $4(\cos 60° + j \sin 60°)$

38. The three cube roots of $27(\cos 120° + j \sin 120°)$

39. The three cube roots of $3 - 4j$

40. The two square roots of $-5 + 12j$

41. The square roots of $1 + j$

42. The cube roots of $\sqrt{3} + j$

In Exercises 43–48, find all of the roots of the given equations.

43. $x^4 - 1 = 0$

44. $x^3 - 8 = 0$

45. $x^3 + 27j = 0$

46. $x^4 - j = 0$

47. $x^5 + 32 = 0$

48. $x^6 + 8 = 0$

In Exercises 49–58, perform the indicated operations.

49. Using the results of Example 7, find the cube roots of -125.

50. Using the results of Example 8, find the square roots of $32j$.

51. In Example 7, we showed that one cube root of -1 is $\frac{1}{2} - \frac{1}{2}j\sqrt{3}$. Cube this number in rectangular form and show that the result is -1.

(W) 52. Explain why the two square roots of a complex number are negatives of each other.

53. The cube roots of -1 can be found by solving the equation $x^3 + 1 = 0$. Find these roots by factoring $x^3 + 1$ as the sum of cubes and compare with Example 7.

54. The cube roots of 8 can be found by solving the equation $x^3 - 8 = 0$. Find these roots by factoring $x^3 - 8$ as the difference of cubes and compare with Exercise 44.

55. The electric power p (in W) supplied to an element in a circuit is the product of the voltage e and the current i (in A). Find the expression for the power supplied if $e = 6.80 \underline{/56.3°}$ volts and $i = 0.0705 \underline{/-15.8°}$ amperes.

56. The displacement d (in in.) of a weight suspended on a system of two springs is $d = 6.03 \underline{/22.5°} + 3.26 \underline{/76.0°}$ in. Perform the addition and express the answer in polar form.

57. The voltage across a certain inductor is $V = (8.66 \underline{/90.0°})(50.0 \underline{/135.0°})/(10.0 \underline{/60.0°})$ volts. Simplify this expression and find the magnitude of the voltage.

58. In a microprocessor circuit, the current is $I = 3.75 \underline{/15.0°}$ μA and the impedance is $Z = 2500 \underline{/-35.0°}$ ohms. Find the voltage E in rectangular form. (See Example 4 on page 338.)

Answers to Practice Exercises

1. $15 \underline{/115°}$ **2.** $0.6 \underline{/345°}$ **3.** $6561 \underline{/40°}$

12.7 An Application to Alternating-current (ac) Circuits

Basic Circuit with Resistance, Inductance, and Capacitance • Impedance • Phase Angle • Phasor • Resonance

R C L

Fig. 12.20

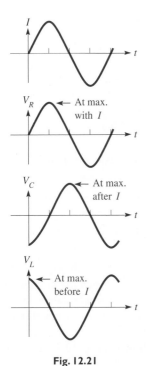

I

V_R ← At max. with I

V_C ← At max. after I

V_L ← At max. before I

Fig. 12.21

■ Eqs. (12.18) are based on Ohm's law, which states that the current is proportional to the voltage for a constant resistance. It is named for the German physicist Georg Ohm (1787–1854). The ohm (Ω) is named for him.

We now show an application of complex numbers in a basic type of alternating-current circuit. We show how the voltage is measured between any two points in a circuit containing a resistance, a capacitance, and an inductance. This circuit is similar to one noted in earlier examples and exercises in this chapter.

A *resistance* is any part of a circuit that tends to obstruct the flow of electric current through the circuit. It is denoted by R (units in ohms, Ω) and in diagrams by ─◊◊◊◊─, as shown in Fig. 12.20. A *capacitance* is two nonconnected plates in a circuit; no current actually flows across the gap between them. In an ac circuit, an electric charge is continually going to and from each plate and, therefore, the current in the circuit is not effectively stopped. It is denoted by C (units in farads, F) and in diagrams by ─┤├─ (see Fig. 12.20). An *inductance* is basically a coil of wire in which current is induced because the current is continually changing in the circuit. It is denoted by L (units in henrys, H) and in diagrams by ◊◊◊◊ (see Fig. 12.20). All these elements affect the voltage in an alternating-current circuit. We state here the relation each has to the voltage and current in the circuit.

In Chapter 10, we noted that the current and voltage in an ac circuit could be represented by a sine or a cosine curve. Therefore, each reaches peak values periodically. *If they reach their respective peak values at the same time, they are **in phase**. If the voltage reaches its peak before the current, the voltage **leads** the current. If the voltage reaches its peak after the current, the voltage **lags** the current.*

In the study of electricity, it is shown that the voltage across a resistance is in phase with the current. The voltage across a capacitor lags the current by 90°, and the voltage across an inductance leads the current by 90°. This is shown in Fig. 12.21, where, in a given circuit, I represents the current, V_R is the voltage across a resistor, V_C is the voltage across a capacitor, V_L is the voltage across an inductor, and t represents time.

Each element in an ac circuit tends to offer a type of resistance to the flow of current. *The effective resistance of any part of the circuit is called the **reactance**,* and it is denoted by X. The voltage across any part of the circuit whose reactance is X is given by $V = IX$, where I is the current (in amperes) and V is the voltage (in volts). Therefore,

the voltage V_R across a resistor with resistance R,

the voltage V_C across a capacitor with reactance X_C, and

the voltage V_L across an inductor with reactance X_L

are, respectively,

$$V_R = IR \qquad V_C = IX_C \qquad V_L = IX_L \tag{12.18}$$

Fig. 12.22

To determine the voltage across a combination of these elements of a circuit, we must account for the reactance, as well as the phase of the voltage across the individual elements. Since the voltage across a resistor is in phase with the current, we represent V_R along the positive real axis as a real number. Since the voltage across an inductance leads the current by 90°, we represent this voltage as a positive, pure imaginary number. In the same way, by representing the voltage across a capacitor as a negative, pure imaginary number, we show that the voltage *lags* the current by 90°. These representations are meaningful since the positive imaginary axis is +90° from the positive real axis, and the negative imaginary axis is −90° from the positive real axis. See Fig. 12.22.

The circuit elements shown in Fig. 12.20 are in *series,* and all circuits we consider (except Exercises 22 and 23) are series circuits. The total voltage across a series of all three elements is given by $V_R + V_L + V_C$, which we represent by V_{RLC}. Therefore,

$$V_{RLC} = IR + IX_Lj - IX_Cj = I[R + j(X_L - X_C)]$$

This expression is also written as

$$V_{RLC} = IZ \tag{12.19}$$

where the symbol Z is called the **impedance** *of the circuit. It is the total effective resistance to the flow of current by a combination of the elements in the circuit,* taking into account the phase of the voltage in each element. From its definition, we see that Z is a complex number.

$$Z = R + j(X_L - X_C) \tag{12.20}$$

with a magnitude

$$|Z| = \sqrt{R^2 + (X_L - X_C)^2} \tag{12.21}$$

Also, as a complex number, it makes an angle θ with the x-axis, given by

$$\theta = \tan^{-1}\frac{X_L - X_C}{R} \tag{12.22}$$

(a)

(b)

Fig. 12.23

All these equations are based on phase relations of voltages with respect to the current. Therefore, *the angle θ represents the phase angle between the current and the voltage.* The standard way of expressing θ is to *use a positive angle if the voltage leads the current* and *use a negative angle if the voltage lags the current.* Using Eq. (12.22), a calculator will give the correct angle even when $\tan \theta < 0$.

If the voltage leads the current, then $X_L > X_C$ as shown in Fig. 12.23(a). If the voltage lags the current, then $X_L < X_C$ as shown in Fig. 12.23(b).

In the examples and exercises of this section, the commonly used units and symbols for them are used. For a summary of these units and symbols, including prefixes, see Appendix B.

EXAMPLE 1 Finding impedance and voltage

In the series circuit shown in Fig. 12.24(a), $R = 12.0\ \Omega$ and $X_L = 5.00\ \Omega$. A current of 2.00 A is in the circuit. Find the voltage across each element, the impedance, the voltage across the combination, and the phase angle between the current and the voltage.

The voltage across the resistor (between points a and b) is the product of the current and the resistance $(V = IR)$. This means $V_R = (2.00)(12.0) = 24.0\ \text{V}$. The voltage across the inductor (between points b and c) is the product of the current and the reactance, or $V_L = (2.00)(5.00) = 10.0\ \text{V}$.

To find the voltage across the combination, between points a and c, we must first find the magnitude of the impedance. Note that *the voltage is not the arithmetic sum of V_R and V_L,* as we must account for the phase.

By Eq. (12.20), the impedance is (there is no capacitor)

$$Z = 12.0 + 5.00j$$

with magnitude

$$|Z| = \sqrt{R^2 + X_L^2} = \sqrt{(12.0)^2 + (5.00)^2} = 13.0\ \Omega$$

Thus, the magnitude of the voltage across the combination of the resistor and the inductance is

$$|V_{RL}| = (2.00)(13.0) = 26.0\ \text{V}$$

The phase angle between the voltage and the current is found by Eq. (12.22). This gives

$$\theta = \tan^{-1}\frac{5.00}{12.0} = 22.6°$$

The voltage *leads* the current by 22.6°, and this is shown in Fig. 12.24(b). ∎

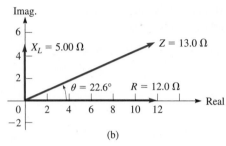

Fig. 12.24

■ See Appendix C for a graphing calculator program IMPEDANC. It can be used to calculate the impedance in a circuit.

EXAMPLE 2 Finding impedance and phase angle

For a circuit in which $R = 8.00\ \Omega$, $X_L = 7.00\ \Omega$, and $X_C = 13.0\ \Omega$, find the impedance and the phase angle between the current and the voltage.

By the definition of impedance, Eq. (12.20), we have

$$Z = 8.00 + (7.00 - 13.0)j = 8.00 - 6.00j$$

where the magnitude of the impedance is

$$|Z| = \sqrt{(8.00)^2 + (-6.00)^2} = 10.0\ \Omega$$

The phase angle is found by

$$\theta = \tan^{-1}\frac{-6.00}{8.00} = -36.9°$$

The angle $\theta = -36.9°$ is given directly by the calculator, and it is the angle we want. As we noted after Eq. (12.22), we express θ as a negative angle if the voltage lags the current, as it does in this example. See Fig. 12.25.

From the values above, we write the impedance in polar form as $Z = 10.0\underline{/-36.9°}$ ohms. ∎

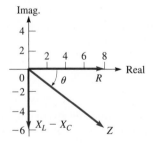

Fig. 12.25

Note that the resistance is represented in the same way as a vector along the positive *x*-axis. Actually, resistance is not a vector quantity but is represented in this manner in order to assign an angle as the phase of the current. The important concept in this analysis is that *the phase **difference** between the current and voltage is constant,* and therefore any direction may be chosen arbitrarily for one of them. Once this choice is made, other phase angles are measured with respect to this direction. A common choice, as above, is to make the phase angle of the current zero. If an arbitrary angle is chosen, it is necessary to treat the current, voltage, and impedance as complex numbers.

EXAMPLE 3   Finding voltage

In a particular circuit, the current is $2.00 - 3.00j$ A and the impedance is $6.00 + 2.00j$ ohms. The voltage across this part of the circuit is

$$V = (2.00 - 3.00j)(6.00 + 2.00j) = 12.0 - 14.0j - 6.00j^2$$
$$= 12.0 - 14.0j + 6.00$$
$$= 18.0 - 14.0j \text{ volts}$$

The magnitude of the voltage is

$$|V| = \sqrt{(18.0)^2 + (-14.0)^2} = 22.8 \text{ V} \qquad \blacksquare$$

Since the voltage across a resistor is in phase with the current, this voltage can be represented as having a phase difference of zero with respect to the current. Therefore, the resistance is indicated as an arrow in the positive real direction, denoting the fact that the current and the voltage are in phase. *Such a representation is called a **phasor.*** The arrow denoted by *R*, as in Fig. 12.24, is actually the phasor representing the voltage across the resistor. Remember, the positive real axis is arbitrarily chosen as the direction of the phase of the current.

To show properly that the voltage across an inductance leads the current by 90°, its reactance (effective resistance) is multiplied by *j*. We know that there is a positive 90° angle between a positive real number and a positive imaginary number. In the same way, by multiplying the capacitive reactance by $-j$, we show the 90° difference in phase between the voltage and the current in a capacitor, with the current leading. Therefore, jX_L represents the phasor for the voltage across an inductor and $-jX_C$ is the phasor for the voltage across the capacitor. The phasor for the voltage across the combination of the resistance, inductance, and capacitance is *Z*, where the phase difference between the voltage and the current for the combination is the angle θ.

From this, we see that *multiplying a phasor by j means to perform the operation of rotating it through* 90°. For this reason, *j* is also called the *j-operator*.

EXAMPLE 4   Multiplication by *j*

Multiplying a positive real number *A* by *j*, we have $A \times j = Aj$, which is a positive imaginary number. In the complex plane, Aj is 90° from *A*, which means that by multiplying *A* by *j* we rotated *A* by 90°. Similarly, we see that $Aj \times j = Aj^2 = -A$, which is a negative real number, rotated 90° from Aj. Therefore, successive multiplications of *A* by *j* give us

$A \times j = Aj$	positive imaginary number
$Aj \times j = Aj^2 = -A$	negative real number
$-A \times j = -Aj$	negative imaginary number
$-Aj \times j = -Aj^2 = A$	positive real number

See Fig. 12.26. (See Exercise 38 on page 341.) $\blacksquare$

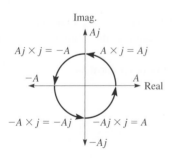

Fig. 12.26

An alternating current is produced by a coil of wire rotating through a magnetic field. If the angular velocity of the wire is ω, the capacitive and inductive reactances are given by

$$X_C = \frac{1}{\omega C} \quad \text{and} \quad X_L = \omega L \tag{12.23}$$

Therefore, if ω, C, and L are known, the reactance of the circuit can be found.

EXAMPLE 5 Finding voltage-current phase difference

If $R = 12.0\,\Omega$, $L = 0.300$ H, $C = 250\,\mu$F, and $\omega = 80.0$ rad/s, find the impedance and the phase difference between the current and the voltage.

$$X_C = \frac{1}{(80.0)(250 \times 10^{-6})} = 50.0\,\Omega$$
$$X_L = (0.300)(80.0) = 24.0\,\Omega$$
$$Z = 12.0 + (24.0 - 50.0)j = 12.0 - 26.0j$$
$$|Z| = \sqrt{(12.0)^2 + (-26.0)^2} = 28.6\,\Omega$$
$$\theta = \tan^{-1}\frac{-26.0}{12.0} = -65.2°$$
$$Z = 28.6\,\underline{/-65.2°}\,\Omega$$

The voltage *lags* the current (see Fig. 12.27). ∎

Fig. 12.27

Recall from Section 10.5 that the angular velocity ω is related to the frequency f by the relation $\omega = 2\pi f$. It is very common to use frequency when discussing alternating current.

An important concept in the application of this theory is that of **resonance.** *For resonance, the impedance of any circuit is a minimum, or the total impedance is R.* Thus, $X_L - X_C = 0$. Also, it can be seen that the current and the voltage are in phase under these conditions. Resonance is required for the tuning of radio and television receivers.

EXAMPLE 6 Resonance

In the antenna circuit of a radio, the inductance is 4.20 mH, and the capacitance is variable. What range of values of capacitance is necessary for the radio to receive the AM band of radio stations, with frequencies from 530 kHz to 1600 kHz?

For proper tuning, the circuit should be in resonance, or $X_L = X_C$. This means that

$$2\pi f L = \frac{1}{2\pi f C} \quad \text{or} \quad C = \frac{1}{(2\pi f)^2 L}$$

For $f_1 = 530$ kHz $= 5.30 \times 10^5$ Hz and $L = 4.20$ mH $= 4.20 \times 10^{-3}$ H,

$$C_1 = \frac{1}{(2\pi)^2(5.30 \times 10^5)^2(4.20 \times 10^{-3})} = 2.15 \times 10^{-11}\,\text{F} = 21.5\,\text{pF}$$

and for $f_2 = 1600$ kHz $= 1.60 \times 10^6$ Hz and $L = 4.20 \times 10^{-3}$ H, we have

$$C_2 = \frac{1}{(2\pi)^2(1.60 \times 10^6)^2(4.20 \times 10^{-3})} = 2.36 \times 10^{-12}\,\text{F} = 2.36\,\text{pF}$$

The capacitance should be capable of varying from 2.36 pF to 21.6 pF. ∎

■ See the chapter introduction.

■ From Appendix B, the following prefixes are defined as follows:

Prefix	Factor	Symbol
pico	10^{-12}	p
milli	10^{-3}	m
kilo	10^3	k

EXERCISES 12.7

In Exercises 1 and 2, perform the indicated operations if the given changes are made in the indicated examples of this section.

1. In Example 1, change the value of X_L to 16.0 Ω and then solve the given problem.

2. In Example 5, double the values of L and C and then solve the given problem.

In Exercises 3–6, use the circuit shown in Fig. 12.28. The current in the circuit is 5.75 mA. Determine the indicated quantities.

Fig. 12.28

3. The voltage across the resistor (between points a and b).

4. The voltage across the inductor (between points b and c).

5. (a) The magnitude of the impedance across the resistor and the inductor (between points a and c).
   (b) The phase angle between the current and the voltage for this combination.
   (c) The voltage across this combination.

6. (a) The magnitude of the impedance across the resistor, inductor, and capacitor (between points a and d).
   (b) The phase angle between the current and the voltage for this combination.
   (c) The voltage across this combination.

In Exercises 7–10, an ac circuit contains the given combination of circuit elements from among a resistor ($R = 45.0\,\Omega$), a capacitor ($C = 86.2\,\mu F$), and an inductor ($L = 42.9$ mH). If the frequency in the circuit is $f = 60.0$ Hz, find (a) the magnitude of the impedance and (b) the phase angle between the current and the voltage.

7. The circuit has the inductor and the capacitor (an *LC* circuit).

8. The circuit has the resistor and the capacitor (an *RC* circuit).

9. The circuit has the resistor and the inductor (an *RL* circuit).

10. The circuit has the resistor, the inductor, and the capacitor (an *RLC* circuit).

In Exercises 11–24, solve the given problems.

11. Given that the current in a given circuit is $3.90 - 6.04j$ mA and the impedance is $5.16 + 1.14j$ kΩ, find the magnitude of the voltage.

12. Given that the voltage in a given circuit is $8.375 - 3.140j$ V and the impedance is $2.146 - 1.114j$ Ω, find the magnitude of the current.

13. A resistance ($R = 25.3\,\Omega$) and a capacitance ($C = 2.75$ nF) are in an AM radio circuit. If $f = 1200$ kHz, find the impedance across the resistor and the capacitor.

14. A resistance ($R = 64.5\,\Omega$) and an inductance ($L = 1.08$ mH) are in a telephone circuit. If $f = 8.53$ kHz, find the impedance across the resistor and inductor.

15. The reactance of an inductor is 1200 Ω for $f = 280$ Hz. What is the inductance?

16. A resistor, an inductor, and a capacitor are connected in series across an ac voltage source. A voltmeter measures 12.0 V, 15.5 V, and 10.5 V, respectively, when placed across each element separately. What is the voltage of the source?

17. An inductance of 12.5 μH and a capacitance of 47.0 nF are in series in an amplifier circuit. Find the frequency for resonance.

18. A capacitance ($C = 95.2$ nF) and an inductance are in series in the circuit of a receiver for navigation signals. Find the inductance if the frequency for resonance is 50.0 kHz.

19. In Example 6, what should be the capacitance in order to receive a 680-kHz radio signal?

20. A 220-V source with $f = 60.0$ Hz is connected in series to an inductance ($L = 2.05$ H) and a resistance R in an electric-motor circuit. Find R if the current is 0.250 A.

21. The power P (in W) supplied to a series combination of elements in an ac circuit is $P = VI \cos\theta$, where V is the effective voltage, I is the effective current, and θ is the phase angle between the current and voltage. If $V = 225$ mV across the resistor, capacitor, and inductor combination in Exercise 10, determine the power supplied to these elements.

22. For two impedances Z_1 and Z_2 in parallel, the reciprocal of the combined impedance Z_C is the sum of the reciprocals of Z_1 and Z_2. Find the combined impedance for the parallel circuit elements in Fig. 12.29 if the current in the circuit has a frequency of 60.0 Hz.

Fig. 12.29   **Fig. 12.30**

23. Find the combined impedance of the circuit elements in Fig. 12.30. The frequency of the current in the circuit is 60.0 Hz. See Exercise 22.

24. (a) If the complex number j, in polar form, is multiplied by itself, what is the resulting number in polar and rectangular forms?
    (b) In the complex plane, where is the resulting complex number in relation to j?

CHAPTER 12   EQUATIONS

Chapter Equations for Complex Numbers

Imaginary unit

$$j = \sqrt{-1} \quad \text{and} \quad j^2 = -1 \tag{12.1}$$

$$\sqrt{-a} = j\sqrt{a} \quad (a > 0) \tag{12.2}$$

Basic operations

$$(a + bj) + (c + dj) = (a + c) + (b + d)j \tag{12.3}$$

$$(a + bj) - (c + dj) = (a - c) + (b - d)j \tag{12.4}$$

$$(a + bj)(c + dj) = (ac - bd) + (ad + bc)j \tag{12.5}$$

$$\frac{a + bj}{c + dj} = \frac{(a + bj)(c - dj)}{(c + dj)(c - dj)} = \frac{(ac + bd) + (bc - ad)j}{c^2 + d^2} \tag{12.6}$$

Complex number forms

Rectangular: $x + yj$
Polar: $r(\cos\theta + j\sin\theta) = r\underline{/\theta}$
Exponential: $re^{j\theta}$

$$x = r\cos\theta \quad y = r\sin\theta \tag{12.7}$$

$$r^2 = x^2 + y^2 \quad \tan\theta = \frac{y}{x} \tag{12.8}$$

$$x + yj = r(\cos\theta + j\sin\theta) = r\underline{/\theta} = re^{j\theta} \tag{12.12}$$

Product in polar form

$$r_1(\cos\theta_1 + j\sin\theta_1)r_2(\cos\theta_2 + j\sin\theta_2) = r_1 r_2[\cos(\theta_1 + \theta_2) + j\sin(\theta_1 + \theta_2)]$$

$$(r_1\underline{/\theta_1})(r_2\underline{/\theta_2}) = r_1 r_2\underline{/\theta_1 + \theta_2} \tag{12.13}$$

Quotient in polar form

$$\frac{r_1(\cos\theta_1 + j\sin\theta_1)}{r_2(\cos\theta_2 + j\sin\theta_2)} = \frac{r_1}{r_2}[\cos(\theta_1 - \theta_2) + j\sin(\theta_1 - \theta_2)] \tag{12.15}$$

$$\frac{r_1\underline{/\theta_1}}{r_2\underline{/\theta_2}} = \frac{r_1}{r_2}\underline{/\theta_1 - \theta_2}$$

DeMoivre's theorem

$$[r(\cos\theta + j\sin\theta)]^n = r^n(\cos n\theta + j\sin n\theta) \tag{12.17}$$

$$(r\underline{/\theta})^n = r^n\underline{/n\theta}$$

Chapter Equations for Alternating-current Circuits

Voltage, current, reactance

$$V_R = IR \quad V_C = IX_C \quad V_L = IX_L \tag{12.18}$$

Impedance

$$V_{RLC} = IZ \tag{12.19}$$

$$Z = R + j(X_L - X_C) \tag{12.20}$$

$$|Z| = \sqrt{R^2 + (X_L - X_C)^2} \tag{12.21}$$

Phase angle

$$\theta = \tan^{-1}\frac{X_L - X_C}{R} \tag{12.22}$$

Capacitive reactance and inductive reactance

$$X_C = \frac{1}{\omega C} \quad \text{and} \quad X_L = \omega L \tag{12.23}$$

CHAPTER 12 REVIEW EXERCISES

In Exercises 1–16, perform the indicated operations, expressing all answers in simplest rectangular form.

1. $(6 - 2j) + (4 + j)$

2. $(12 + 7j) + (-8 + 6j)$

3. $(18 - 3j) - (12 - 5j)$

4. $(-4 - 2j) - \sqrt{-49}$

5. $(2 + j)(4 - j)$

6. $(-5 + 3j)(8 - 4j)$

7. $(2j^5)(6 - 3j)(4 + 3j)$

8. $j(3 - 2j) - (j^3)(5 + j)$

9. $\dfrac{3}{7 - 6j}$

10. $\dfrac{24j}{2 + 9j}$

11. $\dfrac{6 - \sqrt{-16}}{\sqrt{-4}}$

12. $\dfrac{3 + \sqrt{-4}}{4 - j}$

13. $\dfrac{5j - (3 - j)}{4 - 2j}$

14. $\dfrac{2 + (j - 6)}{1 - 2j}$

15. $\dfrac{j(7 - 3j)}{2 - j^7}$

16. $\dfrac{(2 - j)(3 + 2j)}{4 + 3j^3}$

In Exercises 17–20, find the values of x and y for which the equations are valid.

17. $3x - 2j = yj - 9$

18. $2xj - 2y = (y + 3)j - 3$

19. $(3 - 2j)(x + jy) = 4 + j^9$

20. $(x + jy)(7j - 4) = j(x - 5)$

In Exercises 21–24, perform the indicated operations graphically. Check them algebraically.

21. $(-1 + 5j) + (4 + 6j)$

22. $(7 - 2j) + (-5 + 4j)$

23. $(9 + 2j) - (5 - 6j)$

24. $(8 + 4j) - (11 - 3j)$

In Exercises 25–32, give the polar and exponential forms of each of the complex numbers.

25. $1 - j$

26. $4 + 3j$

27. $-22 - 77j$

28. $60 - 20j$

29. $1.07 + 4.55j$

30. $-327 + 158j$

31. 5000

32. $-4j^5$

In Exercises 33–44, give the rectangular form of each number:

33. $2(\cos 225° + j \sin 225°)$

34. $48(\cos 60° + j \sin 60°)$

35. $5.011(\cos 123.82° + j \sin 123.82°)$

36. $2.417(\cos 656.26° + j \sin 656.26°)$

37. $0.62 \underline{/-72°}$

38. $20 \underline{/160°}$

39. $27.08 \underline{/346.27°}$

40. $1.689 \underline{/194.36°}$

41. $2.00e^{0.25j}$

42. $e^{-3.62j}$

43. $(35.37e^{1.096j})^2$

44. $(13.6e^{2.158j})(3.27e^{3.888j})$

In Exercises 45–60, perform the indicated operations. Leave the result in polar form.

45. $[3(\cos 32° + j \sin 32°)][5(\cos 52° + j \sin 52°)]$

46. $[2.5(\cos 162° + j \sin 162°)][8(\cos 115° + j \sin 115°)]$

47. $(40 \underline{/18°})(0.5 \underline{/245°})$

48. $(0.1254 \underline{/172.38°})(27.17 \underline{/204.34°})$

49. $\dfrac{24(\cos 165° + j \sin 165°)}{3(\cos 106° + j \sin 106°)}$

50. $\dfrac{18(\cos 403° + j \sin 403°)}{4(\cos 192° + j \sin 192°)}$

51. $\dfrac{245.6 \underline{/326.44°}}{17.19 \underline{/192.83°}}$

52. $\dfrac{4 \underline{/206°}}{100 \underline{/-320°}}$

53. $0.983 \underline{/47.2°} + 0.366 \underline{/95.1°}$

54. $17.8 \underline{/110.4°} - 14.9 \underline{/226.3°}$

55. $7644 \underline{/294.36°} - 6871 \underline{/17.86°}$

56. $4.944 \underline{/327.49°} + 8.009 \underline{/7.37°}$

57. $[2(\cos 16° + j \sin 16°)]^{10}$

58. $[3(\cos 36° + j \sin 36°)]^6$

59. $(7 \underline{/110.5°})^3$

60. $(536 \underline{/220.3°})^4$

In Exercises 61–64, change each number to polar form and then perform the indicated operations. Express the final result in rectangular and polar forms. Check by performing the same operation in rectangular form.

61. $(1 - j)^{10}$

62. $(\sqrt{3} + j)^8(1 + j)^5$

63. $\dfrac{(5 + 5j)^4}{(-1 - j)^6}$

64. $(\sqrt{3} - j)^{-8}$

In Exercises 65–68, find all the roots of the given equations.

65. $x^3 + 8 = 0$

66. $x^3 - 1 = 0$

67. $x^4 + j = 0$

68. $x^5 - 32j = 0$

In Exercises 69–72, determine the rectangular form and the polar form of the complex number for which the graphical representation is shown in the given figure.

69.

70.

71.

72.

In Exercises 73–86, solve the given problems.

73. Evaluate $x^2 - 2x + 4$ for $x = 5 - 2j$.

74. Evaluate $2x^2 + 5x - 7$ for $x = -8 + 7j$.

75. Using the quadratic formula, solve for x: $x^2 + 3jx - 2 = 0$.

76. Using the quadratic formula, solve for x: $jx^2 - 2x - 3j = 0$.

77. Find a quadratic equation with roots $2 + j$ and $2 - j$.

78. Find a quadratic equation with roots $-3 + 4j$ and $-3 - 4j$.

79. Are $1 - j$ and $-1 - j$ solutions to the equation $x^2 - 2x + 2 = 0$?

80. Show that $\frac{1}{2}(1 + j\sqrt{3})$ is the reciprocal of its conjugate.

81. Solve for x: $(1 + jx)^2 = 1 + j - x^2$

82. What is the argument for any negative imaginary number?

83. If $f(x) = 2x - (x - 1)^{-1}$, find $f(1 + 2j)$.

84. If $f(x) = x^{-2} + 3x^{-1}$, find $f(4 + j)$.

85. Using a calculator, express $5 - 3j$ in polar form.

86. Using a calculator, express $25.0e^{2.25j}$ in rectangular form.

In Exercises 87–98, find the required quantities.

87. A 60-V ac voltage source is connected in series across a resistor, an inductor, and a capacitor. The voltage across the inductor is 60 V, and the voltage across the capacitor is 60 V. What is the voltage across the resistor?

88. In a series ac circuit with a resistor, an inductor, and a capacitor, $R = 6.50\ \Omega$, $X_C = 3.74\ \Omega$, and $Z = 7.50\ \Omega$. Find X_L.

89. In a series ac circuit with a resistor, an inductor, and a capacitor, $R = 6250\ \Omega$, $Z = 6720\ \Omega$, and $X_L = 1320\ \Omega$. Find the phase angle θ.

90. A coil of wire rotates at 120.0 r/s. If the coil generates a current in a circuit containing a resistance of 12.07 Ω, an inductance of 0.1405 H, and an impedance of 22.35 Ω, what must be the value of a capacitor (in F) in the circuit?

91. What is the frequency f for resonance in a circuit for which $L = 2.65$ H and $C = 18.3\ \mu\text{F}$?

92. The displacement of an electromagnetic wave is given by $d = A(\cos \omega t + j \sin \omega t) + B(\cos \omega t - j \sin \omega t)$. Find the expressions for the magnitude and phase angle of d.

93. Two cables lift a crate. The tensions in the cables can be represented by $2100 - 1200j$ N and $1200 + 5600j$ N. Express the resultant tension in polar form.

94. A boat is headed across a river with a velocity (relative to the water) that can be represented as $6.5 + 1.7j$ mi/h. The velocity of the river current can be represented as $-1.1 - 4.3j$ mi/h. Express the resultant velocity of the boat in polar form.

95. In the study of shearing effects in the spinal column, the expression $\dfrac{1}{\mu + j\omega n}$ is found. Express this in rectangular form.

96. In the theory of light reflection on metals, the expression
$$\frac{\mu(1 - kj) - 1}{\mu(1 - kj) + 1}$$
is encountered. Simplify this expression.

97. Show that $e^{j\pi} = -1$.

98. Show that $(e^{j\pi})^{1/2} = j$.

Writing Exercises

99. A computer programmer is writing a program to determine the n nth roots of a real number. Part of the program is to show the number of real roots and the number of pure imaginary roots. Write one or two paragraphs explaining how these numbers of roots can be determined without actually finding the roots.

CHAPTER 12   PRACTICE TEST

1. Add, expressing the resultant in rectangular form: $(3 - \sqrt{-4}) + (5\sqrt{-9} - 1)$.

2. Multiply, expressing the resultant in polar form: $(2\underline{/130°})(3\underline{/45°})$.

3. Express $2 - 7j$ in polar form.

4. Express in terms of j: (a) $-\sqrt{-64}$ (b) $-j^{15}$.

5. Add graphically: $(4 - 3j) + (-1 + 4j)$.

6. Simplify, expressing the result in rectangular form: $\dfrac{2 - 4j}{5 + 3j}$.

7. Express $2.56(\cos 125.2° + j \sin 125.2°)$ in exponential form.

8. For an ac circuit in which $R = 3.50\ \Omega$, $X_L = 6.20\ \Omega$, and $X_C = 7.35\ \Omega$, find the impedance and the phase angle between the current and the voltage.

9. Express $3.47 - 2.81j$ in exponential form.

10. Find the values of x and y: $x + 2j - y = yj - 3xj$.

11. What is the capacitance of the circuit in a radio that has an inductance of 8.75 mH if it is to receive a station with frequency 600 kHz?

12. Find the cube roots of j.

13

Exponential and Logarithmic Functions

By the early 1600s, astronomy had progressed to the point of finding accurate information about the motion of the heavenly bodies. Also, navigation had led to a more systematic exploration of earth. In making the accurate measurements needed in astronomy and navigation, many lengthy calculations had to be performed, and all such calculations had to be done by hand.

Noting that astronomers' calculations usually involved sines of angles, John Napier (1550–1617), a Scottish mathematician, constructed a table of values that allowed multiplication of these sines by addition of values from the table. These tables of *logarithms* first appeared in 1614. Therefore, logarithms were essentially invented to make multiplications by means of addition, thereby making them easier.

Napier's logarithms were not in base 10, and the English mathematician Henry Briggs (1561–1631) realized that logarithms in base 10 would make the calculations even easier. He spent many years laboriously developing a table of base 10 logarithms, which was not completed until after his death.

Logarithms were enthusiastically received by mathematicians and scientists as a long-needed tool for lengthy calculations. The great French mathematician Pierre Laplace (1749–1827) stated "by shortening the labors doubled the life of the astronomer." Logarithms were commonly used for calculations until the 1970s when the scientific calculator came into use.

In this chapter, we study the *logarithmic function* and the *exponential function.* Although logarithms are no longer used directly for calculations, they are of great importance in many scientific and technical applications and in advanced mathematics. The units used to measure the intensity of sound, the intensity of earthquakes the distinction of a base from an acid, the power gains and losses in electrical transmission lines are all measured in logarithmic units. Exponential functions are used in electronics, mechanical systems, thermodynamics, nuclear physics, biology in studying population growth, and in business to calculate compound interest.

The growth of population can often be measured using an exponential function. This is illustrated in Section 13.6.

13.1 Exponential Functions

The Exponential Function •
Graphing Exponential Functions •
Features of Exponential Functions

In Chapter 11, we showed that any rational number can be used as an exponent. Now letting *the exponent be a variable,* we define the **exponential function** *as*

$$y = b^x \tag{13.1}$$

where $b > 0$, $b \neq 1$, and x is any real number. *The number b is called the* **base.**

For an exponential function, we use only real numbers. Therefore, $b > 0$, because if b were negative and x were a fractional exponent with an even-number denominator, y would be imaginary. Also, $b \neq 1$, since 1 to any real power is 1 (y would be constant).

EXAMPLE 1   Exponential functions

From the definition, $y = 3^x$ is an exponential function, but $y = (-3)^x$ is not because the base is negative. However, $y = -3^x$ is an exponential function, because it is -1 times 3^x, and any real-number multiple of an exponential function is also an exponential function.

Also, $y = (\sqrt{3})^x$ is an exponential function since it can be written as $y = 3^{x/2}$. As long as x is a real number, so is $x/2$. Therefore, the exponent of 3 is real.

The function $y = 3^{-x}$ is an exponential function. If x is real, so is $-x$.

Other exponential functions are: $y = -2(8^{-0.55x})$ and $y = 35(1.0001)^x$. ∎

EXAMPLE 2   Evaluating an exponential function

Evaluate the function $y = -2(4^x)$ for the given values of x.

(a) If $x = 2$, $y = -2(4^2) = -2(16) = -32$.

(b) If $x = -2$, $y = -2(4^{-2}) = -2/16 = -1/8$.

(c) If $x = 3/2$, $y = -2(4^{3/2}) = -2(8) = -16$.

(d) If $x = \sqrt{2}$, $y = -2(4^{\sqrt{2}}) = -14.206$ (calculator evaluation—see Fig. 13.1). ∎

```
-2*4^√(2)
        -14.2059866
```

Fig. 13.1

GRAPHING EXPONENTIAL FUNCTIONS

We now show some representative graphs of the exponential function.

EXAMPLE 3   Graphing an exponential function

Plot the graph of $y = 2^x$.

For this function, we have the values in the following table:

x	-3	-2	-1	0	1	2	3
y	$\frac{1}{8}$	$\frac{1}{4}$	$\frac{1}{2}$	1	2	4	8

$2^{-3} = \frac{1}{8}$ $2^0 = 1$ $2^3 = 8$

The curve is shown in Fig. 13.2(a). In Chapter 1, we used only integral exponents, and the enlarged points are for these values. In Chapter 11, we introduced rational exponents, and using them would fill in many points between those for integers, but all the points for irrational numbers would be missing and the curve would be dotted (see Fig. 13.2(b)). Using all real numbers, including the irrational numbers, for exponents we have all points on the curve shown in Fig. 13.2(a).

We see that the x-axis is an *asymptote* of the curve. As noted on page 93, an asymptote is a line that the curve gets closer and closer to as values of x increase (or decrease) without bound, although the curve never actually touches the asymptote. ∎

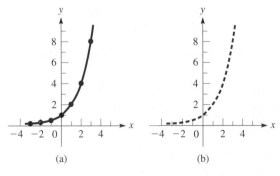

Fig. 13.2

Practice Exercises

Evaluate $y = 16^x$ for:

1. $x = 3/2$ **2.** $x = -0.5$

The bases b of the exponential function of greatest importance in applications are greater than 1. However, in order to understand how the exponential function differs somewhat if $b < 1$, we now use a graphing calculator to display such a graph.

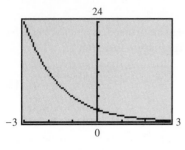

Fig. 13.3

EXAMPLE 4 Exponential function on a calculator

Display the graph of $y = 3\left(\frac{1}{2}\right)^x$ on a graphing calculator.

For this function, $(1/2)^x = 1/2^x = 2^{-x}$. Since x may be any real number, *as x becomes more negative, y increases more rapidly*. Therefore, on a calculator, we let $y_1 = 3(1/2)^x$ (or $y_1 = .5^x$) and have the display in Fig. 13.3 with the *window* settings shown. ∎

Any exponential curve where $b > 1$ will be similar in shape to that shown in Fig. 13.2, and if $b < 1$, it will be similar to the curve in Fig. 13.3. From these examples, we can see that exponential functions have the following basic features.

> **BASIC FEATURES OF EXPONENTIAL FUNCTIONS**
>
> 1. *The domain is all values of x; the range is $y > 0$.*
> 2. *The x-axis is an asymptote of the graph of $y = b^x$.*
> 3. *As x increases, b^x increases if $b > 1$, and b^x decreases if $b < 1$.*

As we have noted, exponential functions are important in many applications. We now illustrate an application in the next example, and others are shown in the exercises.

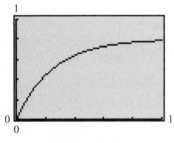

Fig. 13.4

EXAMPLE 5 Exponential function – application

In an electric circuit in which there is a battery, an inductor, and a resistor, the current i (in A) as a function of the time t (in s) is $i = 0.8(1 - e^{-4t})$. Display the graph of this function on a graphing calculator. Here, e is the same number we introduced on page 344 and is equal to approximately 2.718.

Here, $e^{-4t} = 1$ for $t = 0$, and this means $i = 0$ for $t = 0$. Also, e^{-4t} becomes very small in a very short time, and this means i cannot be greater than 0.8 A, and that $i = 0.8$ A is an asymptote. With these considerations, we have the settings and curve shown in Fig. 13.4. Here, we used y for i, x for t, and $y_1 = 0.8(1 - e^{-4x})$. ∎

EXERCISES 13.1

In Exercises 1 and 2, perform the indicated operations if the given changes are made in the indicated examples of this section.

1. In Example 2(c), change the sign of x and then evaluate.

2. In Example 3, change the sign of the exponent and then plot the graph.

In Exercises 3–6, determine if the given functions are exponential functions.

3. (a) $y = 5^x$ (b) $y = 5^{-x}$ **4.** (a) $y = -7^x$ (b) $y = (-7)^{-x}$

5. (a) $y = -7(-5)^{-x}$ (b) $y = -7(5^{-x})$

6. (a) $y = \left(\sqrt{5}\right)^{-x}$ (b) $y = -\left(\sqrt{-5}\right)^x$

In Exercises 7–12, evaluate the exponential function $y = 9^x$ for the given values of x.

7. $x = 0.5$ **8.** $x = 4$ **9.** $x = -2$

10. $x = -0.5$ **11.** $x = -3/2$ **12.** $x = 5/2$

In Exercises 13–18, plot the graphs of the given functions.

13. $y = 4^x$ **14.** $y = 0.25^x$ **15.** $y = 0.2(10^{-x})$

16. $y = -5(1.6^{-x})$ **17.** $y = 0.5\pi^x$ **18.** $y = 2e^x$

In Exercises 19–24, display the graphs of the given functions on a graphing calculator.

19. $y = 0.3(2.55)^x$ **20.** $y = -1.5(4.15)^x$

21. $y = 0.1(0.25^{2x})$ **22.** $y = 0.4(0.95)^x$

23. $i = 1.2(2 + 6^{-t})$ **24.** $y = 0.5e^{-x}$

In Exercises 25–38, solve the given problems.

25. Find the base b of the function $y = b^x$ if its graph passes through the point $(3, 64)$.

26. Find the base b of the function $y = b^x$ if its graph passes through the point $(-2, 64)$.

27. For the function $f(x) = b^x$, show that $f(a + b) = f(a) \cdot f(b)$.

28. For the function $f(x) = b^x$, show that $f(a - b) = f(a)/f(b)$.

29. Use a graphing calculator to graph the function $y = 2^{|x|}$.

30. To show the *damping effect* of an exponential function, use a graphing calculator to display the graph of $y = (x^3)(2^{-x})$. Be sure to use appropriate window settings.

31. Use a graphing calculator to find the value(s) for which $x^2 = 2^x$.

32. Use a graphing calculator to find the integral values of x for which $x^3 > 3^x$.

33. The value V of a bank account in which \$250 is invested at 5.00% interest, compounded annually, is $V = 250(1.0500)^t$, where t is the time in years. Find the value of the account after 4 years.

34. The intensity I of an earthquake is given by $I = I_0(10)^R$, where I_0 is a minimum intensity for comparison and R is the Richter scale magnitude of the earthquake. Evaluate I in terms if I_0 if $R = 5.5$.

35. The electric current i (in mA) in the circuit shown in Fig. 13.5 is $i = 2.5(1 - e^{-0.10t})$, where t is the time (in s). Evaluate i for $t = 5.0$ ms.

36. A projection of the annual growth rate p (in %) of the number of users of the Internet is $p = 8.5(1.2^{-t} + 1)$, where t is the number of years after 2005. Display the graph of this function on a graphing calculator from 2005 to 2015.

37. The flash unit on a camera operates by releasing the stored charge on a capacitor. For a particular unit, the charge q (in μC) as a function of the time t (in s) is $q = 100e^{-10t}$. Display the graph on a graphing calculator.

38. The height y (in m) of the Gateway Arch in St. Louis (see Fig. 13.6) is given by $y = 230.9 - 19.5(e^{x/38.9} + e^{-x/38.9})$, where x is the distance (in m) from the point on the ground level directly below the top. Display the graph on a graphing calculator.

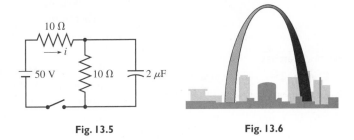

Fig. 13.5   Fig. 13.6

Answers to Practice Exercises

1. 64   **2.** 1/4

13.2 Logarithmic Functions

Exponential Form • Logarithmic Form • Logarithmic Function • Graphing Logarithmic Functions • Features of Logarithmic Functions • Inverse Functions

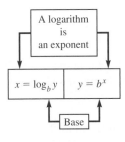

Fig. 13.7

For many uses in mathematics and for many applications, it is necessary to express the exponent x in the exponential function $y = b^x$ in terms of y and the base b. This is done by defining a *logarithm*. Therefore, *if $y = b^x$, the exponent x is the* **logarithm** *of the number y to the base b.* We write this as

$$\text{If } y = b^x, \text{ then } x = \log_b y. \qquad (13.2)$$

This means that x is the power to which the base b must be raised in order to equal the number y. That is, x is a logarithm, and *a logarithm is an exponent.* As with the exponential function, for the equation $x = \log_b y$, x may be any real number, b is a positive number other than 1, and y is a positive real number. In Eq. (13.2),

$y = b^x$ *is the* **exponential form,** and $x = \log_b y$ *is the* **logarithmic form.**

See Fig. 13.7.

EXAMPLE 1 Exponential form and logarithmic form

The equation $y = 2^x$ is written as $x = \log_2 y$ when written in logarithmic form. When we choose values of y to find corresponding values of x from this equation, we ask ourselves "2 raised to what power x gives y?"

This means that if $y = 8$, we ask "what power of 2 gives us 8?" Then knowing that $2^3 = 8$, we know that $x = 3$. Therefore, $3 = \log_2 8$. ∎

EXAMPLE 2 Changing to logarithmic form

(a) $3^2 = 9$ in logarithmic form is $2 = \log_3 9$.

(b) $4^{-1} = 1/4$ in logarithmic form is $-1 = \log_4(1/4)$.

CAUTION ▶ Remember, *the exponent may be negative.* The *base* must be positive. ∎

EXAMPLE 3 Changing between forms

(a) $(64)^{1/3} = 4$ in logarithmic form is $\frac{1}{3} = \log_{64} 4$.

(b) $(32)^{3/5} = 8$ in logarithmic form is $\frac{3}{5} = \log_{32} 8$.

(c) $\log_2 32 = 5$ in exponential form is $32 = 2^5$.

(d) $\log_6 \left(\frac{1}{36}\right) = -2$ in exponential form is $\frac{1}{36} = 6^{-2}$. ■

EXAMPLE 4 Evaluating by changing form

(a) Find b, given that $-4 = \log_b \left(\frac{1}{81}\right)$.

Writing this in exponential form, we have $\frac{1}{81} = b^{-4}$. Thus, $\frac{1}{81} = \frac{1}{b^4}$ or $\frac{1}{3^4} = \frac{1}{b^4}$. Therefore, $b = 3$.

(b) Given $\log_4 y = 1/2$, in exponential form it becomes $y = 4^{1/2}$, or $y = 2$. ■

Practice Exercises

1. Change $125^{2/3} = 25$ to logarithmic form.
2. Change $\log_3(1/3) = -1$ to exponential form.

We see that exponential form is very useful for determining values written in logarithmic form. For this reason, it is important that you learn to transform readily from one form to the other.

In order to change a function of the form $y = ab^x$ into logarithmic form, we must first write it as $y/a = b^x$. **The coefficient of b^x must be equal to 1,** which is the form of Eq. (13.1). In the same way, the coefficient of $\log_b y$ must be 1 in order to change it into exponential form.

CAUTION ▶

EXAMPLE 5 Evaluating – application

The power supply P (in W) of a certain satellite is given by $P = 75e^{-0.005t}$, where t is the time (in days) after launch. By writing this equation in logarithmic form, solve for t.

In order to have the equation in the exponential form of Eq. (13.1), we must have only $e^{-0.005t}$ on the right. Therefore, by dividing by 75, we have

$$\frac{P}{75} = e^{-0.005t}$$

Writing this in logarithmic form, we have $\log_e(P/75) = -0.005t$, or

$$t = \frac{\log_e\left(\dfrac{P}{75}\right)}{-0.005} = -200 \log_e\left(\frac{P}{75}\right) \qquad \frac{1}{-0.005} = -200$$

We recall from Section 12.5 that e is the special irrational number equal to about 2.718. It is important as a base of logarithms, and this will be discussed in Section 13.5. ■

The Logarithmic Function

When we are working with functions, we must keep in mind that a function is defined by the operation being performed on the independent variable, and not by the letter chosen to represent it. However, for consistency, it is standard practice to let y represent the dependent variable and x represent the independent variable. Therefore, *the* **logarithmic function** *is*

$$\boxed{y = \log_b x} \tag{13.3}$$

As with the exponential function, $b > 0$ and $b \neq 1$.

NOTE ▶

Equations (13.2) and (13.3) do not represent different *functions,* due to the difference in location of the variables, since they represent the *same operation* on the independent variable that appears in each. However, Eq. (13.3) expresses the function with the standard dependent and independent variables.

Practice Exercise

3. For the function in Example 6, evaluate y for $x = 8$.

EXAMPLE 6 Logarithmic function

For the logarithmic function $y = \log_2 x$, we have the standard independent variable x and the standard dependent variable y.

If $x = 16$, $y = \log_2 16$, which means that $y = 4$, since $2^4 = 16$.

If $x = \frac{1}{16}$, $y = \log_2 \left(\frac{1}{16}\right)$, which means that $y = -4$, since $2^{-4} = \frac{1}{16}$. ∎

GRAPHING LOGARITHMIC FUNCTIONS

We now show some representative graphs for the logarithmic function. From these graphs, we can see the basic properties of the logarithmic function.

EXAMPLE 7 Graphing logarithmic function

Plot the graph of $y = \log_2 x$.

We can find the points for this graph more easily if we first put the equation in exponential form: $x = 2^y$. By assuming values for y, we can find the corresponding values for x.

x	$\frac{1}{4}$	$\frac{1}{2}$	1	2	4	8
y	-2	-1	0	1	2	3

Using these values, we construct the graph seen in Fig. 13.8. ∎

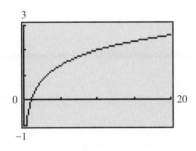

Fig. 13.8

The log keys on the calculator are ⟨log⟩ (for $\log_{10} x$) and ⟨ln⟩ (for $\log_e x$). Therefore, for now, we restrict graphing logarithmic functions to bases 10 and e on the graphing calculator. In Section 13.5, we will see how to use the calculator to graph a logarithmic function with any positive real-number base.

EXAMPLE 8 Logarithmic graph on calculator

Display the graph of $y = 2\log_{10} x$ on a graphing calculator.

On a graphing calculator, we set $y_1 = 2 \log x$, and the curve is displayed as shown in Fig. 13.9. The settings for Xmin and Xmax were selected since the domain is $x > 0$ and $\log_{10} x$ increases very slowly. The settings for Ymin and Ymax were selected since $\log_{10} 0.1 = -1$, and y does not reach 4 until $x = 100$. ∎

Fig. 13.9

From these graphs, we can see that logarithmic functions have the following features. We consider only the features for which $b > 1$, for these are the bases of importance.

BASIC FEATURES OF LOGARITHMIC FUNCTIONS ($b > 1$)

1. *The domain is $x > 0$; the range is all values of y.*

2. *The negative y-axis is an asymptote of the graph of $y = \log_b x$.*

3. *If $0 < x < 1$, $\log_b x < 0$; if $x = 1$, $\log_b x = 0$; if $x > 1$, $\log_b x > 0$.*

4. *If $x > 1$, x increases more rapidly than $\log_b x$.*

Table 1

x	1	4	16	64
$\log_2 x$	0	2	4	6
2^x	2	16	65,536	1.8×10^{19}

We just noted that if $b > 1$ and $x > 1$, x increases more rapidly than $\log_b x$. It is also true that b^x increases more rapidly than x. Actually, as x becomes larger, $\log_b x$ increases very slowly, and b^x increases very rapidly. Using $\log_2 x$ and 2^x and a calculator, we have the table of values shown to the left. This shows that we must choose values of x carefully when graphing these functions.

INVERSE FUNCTIONS

For the exponential function $y = b^x$ and the logarithmic function $y = \log_b x$, if we solve for the independent variable in one of the functions by changing the form, then interchange the variables, we obtain the other function. *Such functions are called* **inverse functions.**

This means that the x- and y-coordinates of inverse functions are interchanged. As a result, the graphs of inverse functions are mirror images of each other across the line $y = x$. This is illustrated in the following example.

EXAMPLE 9 Inverse functions

The functions $y = 2^x$ and $y = \log_2 x$ are inverse functions. We show this by solving $y = 2^x$ for x and then interchanging x and y.

Writing $y = 2^x$ in logarithmic form gives us $x = \log_2 y$. Then interchanging x and y, we have $y = \log_2 x$, which is the inverse function.

Making a table of values for each function, we have

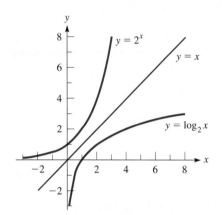

Fig. 13.10

$y = 2^x$:

x	-3	-2	-1	0	1	2	3
y	$\frac{1}{8}$	$\frac{1}{4}$	$\frac{1}{2}$	1	2	4	8

$y = \log_2 x$:

x	$\frac{1}{8}$	$\frac{1}{4}$	$\frac{1}{2}$	1	2	4	8
y	-3	-2	-1	0	1	2	3

We see that the coordinates are interchanged. In Fig. 13.10, note that the graphs of these two functions reflect each other across the line $y = x$. ∎

For a function, there is exactly one value of y in the range for each value of x in the domain. This must also hold for the inverse function. Thus, for a function to have an inverse function, there must be only one x for each y. This is true for $y = b^x$ and $y = \log_b x$, as we have seen earlier in this section.

EXERCISES 13.2

In Exercises 1–4, perform the indicated operations if the given changes are made in the indicated examples of this section.

1. In Example 3(b), change the exponent to 4/5 and then make any other necessary changes.

2. In Example 4(b), change the 1/2 to 5/2 and then make any other necessary changes.

3. In Example 6, change the logarithm base to 4 and then make any other necessary changes.

4. In Example 7, change the logarithm base to 4 and then plot the graph.

In Exercises 5–16, express the given equations in logarithmic form.

5. $3^3 = 27$

6. $5^2 = 25$

7. $4^4 = 256$

8. $2^7 = 128$

9. $7^{-2} = \dfrac{1}{49}$

10. $3^{-2} = \frac{1}{9}$

11. $2^{-6} = \frac{1}{64}$

12. $(12)^0 = 1$

13. $8^{1/3} = 2$

14. $(81)^{3/4} = 27$

15. $\left(\frac{1}{4}\right)^2 = \frac{1}{16}$

16. $\left(\frac{1}{2}\right)^{-2} = 4$

In Exercises 17–28, express the given equations in exponential form.

17. $\log_3 81 = 4$

18. $\log_{11} 121 = 2$

19. $\log_9 9 = 1$

20. $\log_{15} 1 = 0$

21. $\log_{25} 5 = \frac{1}{2}$

22. $\log_8 16 = \frac{4}{3}$

23. $\log_{243} 3 = 0.2$

24. $\log_{32}\left(\frac{1}{8}\right) = -0.6$

25. $\log_{10} 0.1 = -1$

26. $\log_7\left(\frac{1}{49}\right) = -2$

27. $\log_{0.5} 16 = -4$

28. $\log_{1/3} 3 = -1$

In Exercises 29–44, determine the value of the unknown.

29. $\log_4 16 = x$

30. $\log_5 125 = x$

31. $\log_{10} 0.01 = x$ **32.** $\log_{16}\left(\frac{1}{4}\right) = x$

33. $\log_7 y = 3$ **34.** $\log_8(N + 1) = 3$

35. $\log_8(A - 2) = -\frac{2}{3}$ **36.** $\log_7 y = -2$

37. $\log_b 5 = 2$ **38.** $\log_b 625 = 4$

39. $\log_b 4 = -\frac{1}{3}$ **40.** $\log_b 4 = \frac{2}{3}$

41. $\log_{10} 10^{0.2} = x$ **42.** $\log_5 5^{2.3} = R + 1$

43. $\log_3 27^{-1} = x + 1$ **44.** $\log_b\left(\frac{1}{4}\right) = -0.5$

In Exercises 45–50, plot the graphs of the given functions.

45. $y = \log_3 x$ **46.** $y = \log_4 x$

47. $y = \log_{0.5} x$ **48.** $y = 3\log_2 x$

49. $N = 0.2\log_4 v$ **50.** $A = 2.4\log_{10}(2r)$

In Exercises 51–54, display the graphs of the given functions on a graphing calculator.

51. $y = 3\log_e x$ **52.** $y = 5\log_{10}|x|$

53. $y = -\log_{10}(-x)$ **54.** $y = -\log_e(-2x)$

In Exercises 55–72, perform the indicated operations.

55. Evaluate $\log_4 x$ for (a) $x = 1/64$ and (b) $x = -1/2$.

56. Evaluate: (a) $\log_b b$ (b) $\log_b 1$

57. If $f(x) = \log_5 x$, find: (a) $f(\sqrt{5})$ (b) $f(0)$

58. If $f(x) = \log_b x$ and $f(3) = 2$, find $f(9)$.

59. On a graphing calculator, display the graphs of: (a) $y_1 = \log_{10} x$ (b) $y_2 = \log_{10}(x + 2)$ (c) $y_3 = \log_{10}(x - 2)$ (See page 99.)

(W) 60. On a graphing calculator, display the graphs of $y_1 = 2\log_{10} x$ and $y_2 = \log_{10} x^2$. Describe any similarities or differences.

61. Using a graphing calculator, solve the equation $\log_{10} x = x - 2$.

(W) 62. Use a graphing calculator to display the graphs (on the same screen) of $y = \log_e(x^2 + c)$, with $c = -4, 0, 4$. Describe the results.

63. The magnitudes (visual brightness), m_1 and m_2, of two stars are related to their (actual) brightnesses, b_1 and b_2, by the equation $m_1 - m_2 = 2.5\log_{10}(b_2/b_1)$. Solve for b_2.

64. The velocity v of a rocket when its fuel is completely burned is given by $v = u\log_e(w_0/w)$, where u is the exhaust velocity, w_0 is the liftoff weight, and w is the burnout weight. Solve for w.

65. An equation relating the number N of atoms of radium at any time t in terms of the number N_0 of atoms at $t = 0$ is $\log_e(N/N_0) = -kt$, where k is a constant. Solve for N.

66. The capacitance C of a cylindrical capacitor is given by $C = k/\log_e(R_2/R_1)$, where R_1 and R_2 are the inner and outer radii. Solve for R_1.

67. The work W done by a sample of nitrogen gas during an isothermal (constant temperature) change from volume V_1 to volume V_2 is given by $W = k\log_e(V_2/V_1)$. Solve for V_1.

68. An equation used in measuring the flow of water in a channel is $C = -a\log_{10}(b/R)$. Solve for R.

69. The time t (in ps) required for N calculations by a certain computer design is $t = N + \log_2 N$. Sketch the graph of this function.

70. When a tractor-trailer turns a right angle corner, its rear wheels follow a curve called a *tractrix*, the equation for which is $y = \log_e\left(\dfrac{1 + \sqrt{1 - x^2}}{x}\right) - \sqrt{1 - x^2}$. Display the curve on a graphing calculator.

(W) 71. Display the graphs of $y = (1/3)^x$ and $y = 3^{-x}$ and then compare them. Explain what your comparison shows.

(W) 72. In Exercise 47, the graph of $y = \log_{0.5} x$ is plotted. By inspecting the graph and noting the properties of $\log_{0.5} x$, describe some of the differences of logarithms to a base less than 1 from those to a base greater than 1.

In Exercises 73–76, show that the given functions are inverse functions of each other. Then display the graphs of each function and the line $y = x$ on a graphing calculator and note that each is the mirror image of the other across $y = x$.

73. $y = 10^{x/2}$ and $y = 2\log_{10} x$

74. $y = e^x$ and $y = \log_e x$

75. $y = 3x$ and $y = x/3$

76. $y = 2x + 4$ and $y = 0.5x - 2$

Answers to Practice Exercises

1. $2/3 = \log_{125} 25$ **2.** $1/3 = 3^{-1}$ **3.** 3

13.3 Properties of Logarithms

Sum of Logarithms for Product • Difference of Logarithms for Quotient • Multiple of Logarithm for Power • Logarithms of 1 and b

Since a logarithm is an exponent, it must follow the laws of exponents. The laws used in this section to derive the very useful properties of logarithms are listed here for reference.

■ A logarithm is an exponent. However, a historical curiosity is that logarithms were developed before exponents were used.

$$b^u b^v = b^{u+v} \tag{13.4}$$

$$\frac{b^u}{b^v} = b^{u-v} \tag{13.5}$$

$$(b^u)^n = b^{nu} \tag{13.6}$$

The next example shows the reasoning used in deriving the properties of logarithms.

EXAMPLE 1 Sum of logarithms for product

We know that $8 \times 16 = 128$. Writing these numbers as powers of 2, we have

$$8 = 2^3 \qquad 16 = 2^4 \qquad 128 = 2^7 = 2^{3+4}$$

The logarithmic forms can be written as

$$3 = \log_2 8 \qquad 4 = \log_2 16 \qquad 3 + 4 = \log_2 128$$

This means that

$$\log_2 8 + \log_2 16 = \log_2 128$$

where

$$8 \times 16 = 128$$

The *sum of the logarithms* of 8 and 16 equals the logarithm of 128, where the *product* of 8 and 16 equals 128. ∎

■ For reference, Eqs. (13.4) to (13.6) are

$$b^u b^v = b^{u+v}$$
$$\frac{b^u}{b^v} = b^{u-v}$$
$$(b^u)^n = b^{nu}$$

Following Example 1, if we let $u = \log_b x$ and $v = \log_b y$ and write these equations in exponential form, we have $x = b^u$ and $y = b^v$. Therefore, forming the product of x and y, we obtain

$$xy = b^u b^v = b^{u+v} \quad \text{or} \quad xy = b^{u+v}$$

Writing this last equation in logarithmic form yields

$$u + v = \log_b xy$$

or

Logarithm of a Product

$$\boxed{\log_b xy = \log_b x + \log_b y} \qquad (13.7)$$

Equation (13.7) states the property that *the logarithm of the product of two numbers is equal to the sum of the logarithms of the numbers.*

Using the same definitions of u and v to form the quotient of x and y, we then have

$$\frac{x}{y} = \frac{b^u}{b^v} = b^{u-v} \quad \text{or} \quad \frac{x}{y} = b^{u-v}$$

Writing this last equation in logarithmic form, we have

$$u - v = \log_b\left(\frac{x}{y}\right)$$

or

Logarithm of a Quotient

$$\boxed{\log_b\left(\frac{x}{y}\right) = \log_b x - \log_b y} \qquad (13.8)$$

Equation (13.8) states the property that *the logarithm of the quotient of two numbers is equal to the logarithm of the numerator minus the logarithm of the denominator.*

If we again let $u = \log_b x$ and write this in exponential form, we have $x = b^u$. To find the nth power of x, we write

$$x^n = (b^u)^n = b^{nu}$$

Expressing this equation in logarithmic form yields

$$nu = \log_b(x^n)$$

or

Logarithm of a Power

$$\log_b(x^n) = n \log_b x \qquad (13.9)$$

Equation (13.9) states that *the logarithm of the nth power of a number is equal to n times the logarithm of the number.* The exponent n may be any real number, which, of course, includes all rational and irrational numbers.

■ In advanced mathematics, the logarithms of negative and imaginary numbers are defined.

In Section 13.2, we showed that the base b of logarithms must be a positive number. Since $x = b^u$ and $y = b^v$, this means that x and y are also positive numbers. Therefore, the *properties of logarithms that have just been derived are valid only for positive values of x and y.*

EXAMPLE 2 Logarithms of product, quotient, power

(a) Using Eq. (13.7), we may express $\log_4 15$ as a sum of logarithms:

$$\log_4 15 = \log_4(3 \times 5) = \log_4 3 + \log_4 5 \qquad \text{logarithm of product} \atop \text{sum of logarithms}$$

(b) Using Eq. (13.8), we may express $\log_4\left(\frac{5}{3}\right)$ as the difference of logarithms:

$$\log_4\left(\frac{5}{3}\right) = \log_4 5 - \log_4 3 \qquad \text{logarithm of quotient} \atop \text{difference of logarithms}$$

(c) Using Eq. (13.9), we may express $\log_4(t^2)$ as twice $\log_4 t$:

$$\log_4(t^2) = 2\log_4 t \qquad \text{logarithm of power} \atop \text{multiple of logarithm}$$

(d) Using Eq. (13.8) and then Eq. (13.7), we have

$$\log_4\left(\frac{xy}{z}\right) = \log_4(xy) - \log_4 z$$
$$= \log_4 x + \log_4 y - \log_4 z \qquad ■$$

Practice Exercises

Express as sum of difference of logarithms.

1. $\log_3 10$ **2.** $\log_3(2a/5)$

EXAMPLE 3 Sum of logarithms as single quantity

We may also express a sum or difference of logarithms as the logarithm of a single quantity.

(a) $\log_4 3 + \log_4 x = \log_4(3 \times x) = \log_4 3x \qquad$ using Eq. (13.7)

(b) $\log_4 3 - \log_4 x = \log_4\left(\frac{3}{x}\right) \qquad$ using Eq. (13.8)

(c) $\log_4 3 + 2\log_4 x = \log_4 3 + \log_4(x^2) = \log_4 3x^2 \qquad$ using Eqs. (13.7) and (13.9)

(d) $\log_4 3 + 2\log_4 x - \log_4 y = \log_4\left(\frac{3x^2}{y}\right) \qquad$ using Eqs. (13.7), (13.8), and (13.9)

■

In Section 13.2, we noted that $\log_b 1 = 0$. Also, since $b = b^1$ in logarithmic form is $\log_b b = 1$, we have $\log_b(b^n) = n \log_b b = n(1) = n$.

Summarizing these properties, we have

$$\boxed{\log_b 1 = 0 \qquad \log_b b = 1} \tag{13.10}$$

$$\boxed{\log_b(b^n) = n} \tag{13.11}$$

These equations may be used to find exact values of certain logarithms.

EXAMPLE 4 Exact values for certain logarithms

(a) We may evaluate $\log_3 9$ using Eq. (13.11):

$$\log_3 9 = \log_3(3^2) = 2$$

We can establish the exact value since the base of logarithms and the number being raised to the power are the same. Of course, this could have been evaluated directly from the definition of a logarithm.

(b) Using Eq. (13.11), we can write $\log_3(3^{0.4}) = 0.4$. Although we did not evaluate $3^{0.4}$, we can evaluate $\log_3(3^{0.4})$. ∎

Practice Exercise

3. Find the exact value of $2 \log_2 8$.

EXAMPLE 5 Using the properties of logarithms

(a) $\log_2 6 = \log_2(2 \times 3) = \log_2 2 + \log_2 3 = 1 + \log_2 3$
(b) $\log_5 \frac{1}{5} = \log_5 1 - \log_5 5 = 0 - 1 = -1$
(c) $\log_7 \sqrt{7} = \log_7(7^{1/2}) = \frac{1}{2}\log_7 7 = \frac{1}{2}$ ∎

EXAMPLE 6 Evaluation in two ways

The following illustration shows the evaluation of a logarithm in two different ways. Either method is appropriate.

(a) $\log_5\left(\frac{1}{25}\right) = \log_5 1 - \log_5 25 = 0 - \log_5(5^2) = -2$
(b) $\log_5\left(\frac{1}{25}\right) = \log_5(5^{-2}) = -2$ ∎

EXAMPLE 7 Solving equation with logarithms

Use the basic properties of logarithms to solve the following equation for y in terms of x: $\log_b y = 2 \log_b x + \log_b a$.

Using Eq. (13.9) and then Eq. (13.7), we have

$$\log_b y = \log_b(x^2) + \log_b a = \log_b(ax^2)$$

Since we have the logarithm to the base b of different expressions on each side of the resulting equation, the expressions must be equal. Therefore,

$$y = ax^2$$ ∎

EXAMPLE 8 Solving equation – application

An equation for the current i and the time t in an electric circuit containing a resistance R and a capacitance C is $\log_e i - \log_e I = -t/RC$ (where R and C are both factors in the denominator). Here, I is the current for $t = 0$. Solve for i as a function of t.

Using Eq. (13.8), we rewrite the left side of this equation, obtaining

$$\log_e\left(\frac{i}{I}\right) = -\frac{t}{RC}$$

Rewriting this in exponential form, we have

$$\frac{i}{I} = e^{-t/RC} \quad \text{or} \quad i = Ie^{-t/RC}$$ ∎

EXERCISES 13.3

In Exercises 1–8, perform the indicated operations on the resulting expressions if the given changes are made in original expressions of the indicated examples of this section.

1. In Example 2(a), change the 15 to 21.
2. In Example 2(d), change the x to 2 and the z to 3.
3. In Example 3(b), change the 3 to 5.
4. In Example 3(c), change the 2 to 3.
5. In Example 4(a), change the 9 to 27.
6. In Example 5(a), change the 6 to 10.
7. In Example 5(c), change the 7's to 5's.
8. In Example 7, change the 2 to 3.

In Exercises 9–20, express each as a sum, difference, or multiple of logarithms. See Example 2.

9. $\log_5 33$
10. $\log_3 14$
11. $\log_7\left(\frac{5}{3}\right)$
12. $\log_3\left(\frac{2}{11}\right)$
13. $\log_2(a^3)$
14. $2\log_8(n^5)$
15. $\log_6 abc$
16. $\log_2\left(\frac{xy}{z^2}\right)$
17. $8\log_5\sqrt[4]{y}$
18. $\log_4\sqrt[7]{x}$
19. $\log_2\left(\frac{\sqrt{x}}{a^2}\right)$
20. $\log_3\left(\frac{\sqrt[3]{y}}{7}\right)$

In Exercises 21–28, express each as the logarithm of a single quantity. See Example 3.

21. $\log_b a + \log_b c$
22. $\log_2 3 + \log_2 x$
23. $\log_5 9 - \log_5 3$
24. $-\log_8 R + \log_8 V$
25. $-\log_b\sqrt{x} + \log_b x^2$
26. $\log_4 3^3 + \log_4 9$
27. $2\log_e 2 + 3\log_e \pi$
28. $\frac{1}{2}\log_b a - 2\log_b 5$

In Exercises 29–36, determine the exact value of each of the given logarithms.

29. $\log_2\left(\frac{1}{32}\right)$
30. $\log_3\left(\frac{1}{81}\right)$
31. $\log_2(2^{2.5})$
32. $\log_5(5^{0.1})$
33. $6\log_7\sqrt{7}$
34. $\pi\log_6\sqrt[3]{6}$
35. $\log_3\sqrt[4]{27}$
36. $\log_5\sqrt[3]{25}$

In Exercises 37–44, express each as a sum, difference, or multiple of logarithms. In each case, part of the logarithm may be determined exactly.

37. $\log_3 18$
38. $\log_5 75$
39. $\log_2\left(\frac{1}{6}\right)$
40. $\log_{10}(0.05)$
41. $\log_3\sqrt{6}$
42. $\log_2\sqrt[3]{24}$
43. $\log_{10} 3000$
44. $\log_{10}(40^2)$

In Exercises 45–56, solve for y in terms of x.

45. $\log_b y = \log_b 2 + \log_b x$
46. $\log_b y = \log_b 6 - \log_b x$
47. $\log_4 y = \log_4 x - \log_4 5 + \log_4 3$
48. $\log_3 y = -2\log_3(x + 1) + \log_3 7$
49. $\log_{10} y = 2\log_{10} 7 - 3\log_{10} x$
50. $\log_b y = 3\log_b\sqrt{x} + 2\log_b 10$
51. $5\log_2 y - \log_2 x = 3\log_2 4 + \log_2 a$
52. $4\log_2 x - 3\log_2 y = \log_2 27$
53. $\log_2 x + \log_2 y = 1$
54. $\pi\log_4 x + \log_4 y = 1$
55. $\frac{2\log_5 x}{\log_5 3} - \log_5 y = 2$
56. $\log_8 x = 2\log_8 y + 4$

In Exercises 57–68, solve the given problems.

(W) 57. Explain why $\log_{10}(x + 3)$ is not equal to $\log_{10} x + \log_{10} 3$.

(W) 58. Evaluate $2\log_2(2x) - \log_2 x^2$. For what values of x is the value of this expression valid? Explain.

(W) 59. Display the graphs of $y = \log_e(e^2 x)$ and $y = 2 + \log_e x$ on a graphing calculator and explain why they are the same.

60. If $x = \log_b 2$ and $y = \log_b 3$, express $\log_b 12$ in terms of x and y.

61. If $\log_b x = 2$ and $\log_b y = 3$, find $\log_b\sqrt{x^2 y^4}$.

62. Is it true that $\log_b(ab)^x = x\log_b a + x$?

(W) 63. On the same screen of a graphing calculator, display the graphs of $y_1 = \log_{10} x - \log_{10}(x^2 + 1)$ and $y_2 = \log_{10}\dfrac{x}{x^2 + 1}$. What conclusion can be drawn from the display?

64. The use of the insecticide DDT was banned in the United States in 1972. A computer analysis shows that an expression relating the amount A still present in an area, the original amount A_0, and the time t (in years) since 1972 is $\log_{10} A = \log_{10} A_0 + 0.1t\log_{10} 0.8$. Solve for A as a function of t.

65. A study of urban density shows that the population density D (in persons/mi^2) is related to the distance r (in mi) from the city center by $\log_e D = \log_e a - br + cr^2$, where a, b, and c are positive constants. Solve for D as a function of r.

66. When a person ingests a medication capsule, it is found that the rate R (in mg/min) that it enters the bloodstream in time t (in min) is given by $\log_{10} R - \log_{10} 5 = t\log_{10} 0.95$. Solve for R as a function of t.

67. A container of water is heated to 90°C and then placed in a room at 0°C. The temperature T of the water is related to the time t (in min) by $\log_e T = \log_e 90.0 - 0.23\,t$. Find T as a function of t.

68. In analyzing the power gain in an electric circuit, the equation $N = 10(2\log_{10} I_1 - 2\log_{10} I_2 + \log_{10} R_1 - \log_{10} R_2)$ is used. Express this with a single logarithm on the right side.

Answers to Practice Exercises

1. $\log_3 2 + \log_3 5$ 2. $\log_3 2 + \log_3 a - \log_3 5$ 3. 6

13.4 Logarithms to the Base 10

Common Logarithms • Antilogarithm • Calculations Using Base 10 Logarithms

In Section 13.2, we stated that a base of logarithms must be a positive number, not equal to 1. In the examples and exercises of the previous sections, we used a number of different bases. There are, however, only two bases that are generally used. They are 10 and e, where e is the irrational number approximately equal to 2.718 that we introduced in Section 12.5 and have used in the previous sections of this chapter.

Base 10 logarithms were developed for calculational purposes and were used a great deal for making calculations until the 1970s, when the modern scientific calculator became widely available. Base 10 logarithms are still used in several scientific measurements, and therefore a need still exists for them. Base e logarithms are used extensively in technical and scientific work; we consider them in detail in the next section.

Logarithms to the base 10 are called **common logarithms.** They may be found directly by use of a calculator, and the $\boxed{\log}$ key is used for this purpose. This, of course, is the same key we have used with logarithmic functions to the base 10 on the graphing calculator. This calculator key indicates the common notation. *When no base is shown, it is assumed to be the base 10.*

NOTE ▶

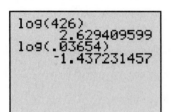

```
log(426)
       2.629409599
log(.03654)
      -1.437231457
```

Fig. 13.11

EXAMPLE 1 Base 10 logarithms on calculator

Using a graphing calculator to find log 426, as shown in the first two lines of the display in Fig. 13.11, we find that

$$\log 426 = 2.629$$

↑—— no base shown means base is 10

when the result is rounded off. The decimal part of a logarithm is normally expressed to the same accuracy as that of the number of which it is the logarithm, although showing one additional digit in the logarithm is generally acceptable.

Since $10^2 = 100$ and $10^3 = 1000$, and in this case

$$10^{2.629} = 426$$

we see that the 2.629 power of 10 gives a number between 100 and 1000. ∎

EXAMPLE 2 Negative base 10 logarithm

Finding log 0.03654, as shown in the third and fourth lines of the calculator display in Fig. 13.11, we see that

$$\log 0.03654 = -1.4372$$

We note that the logarithm here is negative. This should be the case when we recall the meaning of a logarithm. Raising 10 to a negative power gives us a number between 0 and 1, and here we have

$$10^{-1.4372} = 0.03654$$ ∎

We may also use a calculator to find a number N if we know log N. *In this case, we refer to N as the* **antilogarithm** *of log N.* On the calculator, we use the $\boxed{10^x}$ key. We note that it shows the basic definition of a logarithm. (On many scientific calculators, the key sequence $\boxed{\text{inv}}$ $\boxed{\log}$ is used. Note that this sequence shows that the exponential and logarithmic functions are inverse functions.)

```
10^(1.1854)
        15.32498295
10log(15.8/.625)
        14.0277707
```

Fig. 13.12

EXAMPLE 3 Antilogarithm – inverse logarithm

Given $\log N = 1.1854$, we find N as shown in the first two lines of the calculator display in Fig. 13.12. Therefore,

$$N = 15.32$$

where the result has been rounded off. Since $10^1 = 10$ and $10^2 = 100$, we see that $N = 10^{1.1854}$ and is a number between 10 and 100. ∎

The following example illustrates an application in which a measurement requires the direct use of the value of a logarithm.

EXAMPLE 4 Application – decibels

The power gain G (in decibels (dB)) of an electronic device is given by $G = 10\log(P_0/P_i)$, where P_0 is the output power (in W) and P_i is the input power. Determine the power gain for an amplifier for which $P_0 = 15.8$ W and $P_i = 0.625$ W.

Substituting the given values, we have

$$G = 10\log\frac{15.8}{0.625} = 14.0\,\text{dB}$$

where lines 3 and 4 of the display in Fig. 13.12 show the evaluation on a calculator. ∎

■ The unit of sound intensity level (used for power gain), the bel (B), is named for the U.S. inventor Alexander Graham Bell (1847–1922). The decibel is the commonly used unit.

As noted in the chapter introduction, logarithms were developed for calculational purposes. They were first used in the seventeenth century for making tedious and complicated calculations that arose in astronomy and navigation. These complicated calculations were greatly simplified, since logarithms allowed them to be performed by means of basic additions, subtractions, multiplications, and divisions. Performing calculations in this way provides an opportunity to understand better the meaning and properties of logarithms. Also, certain calculations cannot be done directly on a calculator but can be done by logarithms.

Practice Exercises

Evaluate x using a calculator.

1. $x = \log 0.5392$ **2.** $\log x = 2.2901$

EXAMPLE 5 Calculation using logarithms

A certain computer design has 64 different sequences of ten binary digits so that the total number of possible states is $(2^{10})^{64} = 1024^{64}$. Evaluate 1024^{64} using logarithms. (It is very possible that your calculator cannot do this calculation directly.)

Since $\log x^n = n\log x$, we know that $\log 1024^{64} = 64\log 1024$. Although most calculators will not directly evaluate 1024^{64}, we can use one to find the value of $64\log 1024$. Since 1024^{64} is *exact*, we will show ten calculator digits until we round off the result. We therefore evaluate 1024^{64} as follows:

```
64log(1024)
        192.6591972
10^(.6591972)
        4.562440356
```

Fig. 13.13

Let $N = 1024^{64}$

$\log N = \log 1024^{64} = 64\log 1024$ using Eq. (13.9): $\log_b x^n = n\log_b x$

$\quad = 192.6591972$

$N = 10^{192.6591972}$ meaning of logarithm

$\quad = 10^{192} \times 10^{0.6591972}$ using Eq. (13.4): $b^u b^v = b^{u+v}$

$\quad = (10^{192}) \times (4.5624)$ antilogarithm of 0.6591972 is 4.5624 (rounded off)

$\quad = 4.5624 \times 10^{192}$

By using Eq. (13.4), $10^{0.6591972}$ represents a number between 1 and 10 ($10^0 = 1$ and $10^1 = 10$), and we can write the result immediately in scientific notation.

Although we used a calculator to find 192.6591972 and 4.5624 as shown in Fig. 13.13, the calculation was done essentially by logarithms. ∎

Example 5 shows that calculations using logarithms are based on Eqs. (13.7), (13.8), and (13.9). Multiplication is performed by addition of logarithms, division is performed by the subtraction of logarithms, and a power is found by a multiple of a logarithm. A root of a number is found by using the fractional exponent form of the power.

EXERCISES 13.4

In Exercises 1 and 2, find the indicated values if the given changes are made in the indicated examples of this section.

1. In Example 2, change 0.03654 to 0.3654 and then find the required value.

2. In Example 3, change 1.1854 to 2.1854 and then find the required value.

In Exercises 3–12, find the common logarithm of each of the given numbers by using a calculator.

3. 567

4. 0.0640

5. 9.24×10^6

6. 3.19^3

7. 1.174^{-4}

8. 8.043×10^{-8}

9. $\cos 12.5°$

10. $\tan 12.6$

11. $\sqrt{274}$

12. $\log_2 16$

In Exercises 13–20, find the antilogarithm of each of the given logarithms by using a calculator.

13. 4.437

14. 0.929

15. −1.3045

16. −6.9788

17. 3.30112

18. 8.82436

19. −2.23746

20. −10.336

In Exercises 21–24, use logarithms to evaluate the given expressions.

21. $(5.98)(14.3)$

22. $\dfrac{895}{73.4^{86}}$

23. $\left(\sqrt[10]{7.32}\right)(2470)^{30}$

24. $\dfrac{126{,}000^{20}}{2.63^{2.5}}$

In Exercises 25–28, use a calculator to verify the given values.

25. $\log 14 + \log 0.5 = \log 7$

26. $\log 500 - \log 20 = \log 25$

27. $\log 81 = 4 \log 3$

28. $\log 6 = 0.5 \log 36$

In Exercises 29–32, find the logarithms of the given numbers.

29. The signal used by some cell phones has a frequency of 9.00×10^8 Hz.

30. A large sunspot may be 3.5×10^4 km in diameter.

31. About 1.3×10^{-14}% of carbon atoms are carbon-14, the radioactive isotope used in determining the age of samples from ancient sites.

32. In an air sample taken in an urban area, $5/10^6$ of the air was carbon monoxide.

In Exercises 33–36, find the indicated values.

33. Find T (in K) if $\log T = 8$, where T is the temperature sufficient for nuclear fission.

34. Find v (in m/s) if $\log v = 7.423$, where v is the speed of an electron in a TV picture tube.

35. Find e if $\log e = -0.35$, where e is the efficiency of a certain gasoline engine.

36. Find E (in J) if $\log E = -18.49$, where E is the energy of a photon of visible light.

In Exercises 37–42, solve the given problems by evaluating the appropriate logarithms.

37. Simplify: $\dfrac{\log_b x^2}{\log 100}$.

38. Evaluate: $\log(\log 10^{100})$ (See Exercise 47 on page 23.)

39. Evaluate: $2(10^{\log 0.1}) + 3(10^{\log 0.01})$.

40. A stereo amplifier has an input power of 0.750 W and an output power of 25.0 W. What is the power gain? (See Example 4.)

41. Measured on the Richter scale, the magnitude of an earthquake of intensity I is defined as $R = \log(I/I_0)$, where I_0 is a minimum level for comparison. What is the Richter scale reading for the 2007 Peruvian earthquake for which $I = 79{,}000{,}000 I_0$?

42. How many times greater was the intensity of the 2001 Peruvian earthquake $(R = 8.1)$ than the 1969 San Francisco Bay earthquake $(R = 7.1)$? (See Exercise 41.)

In Exercises 43 and 44, use logarithms to perform the indicated calculations.

43. A certain type of optical switch in a fiber-optic system allows a light signal to continue in either of two fibers. How many possible paths could a light signal follow if it passes through 400 such switches?

44. The peak current I_m (in A) in an alternating-current circuit is given by $I_m = \sqrt{\dfrac{2P}{Z \cos \theta}}$, where P is the power developed, Z is the magnitude of the impedance, and θ is the phase angle between the current and voltage. Evaluate I_m for $P = 5.25$ W, $Z = 320\ \Omega$, and $\theta = 35.4°$.

Answers to Practice Exercises

1. −0.2683   **2.** 195.0

13.5 Natural Logarithms

Natural Logarithm (ln x) •
Change of Base •
Calculator Values and Graphs

As we have noted, another number important as a base of logarithms is the number e. *Logarithms to the base e are called* **natural logarithms.** Since e is an irrational number equal to about 2.718, it may appear to be a very unnatural choice as a base of logarithms. However, in calculus, the reason for its choice and the fact that it is a very natural number for a base of logarithms are shown.

Just as log x refers to logarithms to the base 10, the notation ln x is used to denote logarithms to the base e. We briefly noted this in Section 13.2 in discussing the graphing calculator. Due to the extensive use of natural logarithms, the notation ln x is more convenient than **$\log_e x$,** although they mean the same thing.

Since more than one base is important, at times it is useful to change a logarithm from one base to another. If $u = \log_b x$, then $b^u = x$. Taking logarithms of both sides of this last expression to the base a, we have

$$\log_a b^u = \log_a x$$
$$u \log_a b = \log_a x$$
$$u = \frac{\log_a x}{\log_a b}$$

However, $u = \log_b x$, which means that

$$\log_b x = \frac{\log_a x}{\log_a b} \tag{13.12}$$

Equation (13.12) allows us to change a logarithm in one base to a logarithm in another base. The following examples illustrate the method of performing this operation.

EXAMPLE 1 Change of base to find natural log

Change log 20 to a logarithm with base e; that is, find ln 20.

Using Eq. (13.12) with $a = 10$, $b = e$, and $x = 20$, we have

$$\log_e 20 = \frac{\log_{10} 20}{\log_{10} e}$$

or $$\ln 20 = \frac{\log 20}{\log e} = 2.996 \qquad \text{see lines 1 and 2 of calculator display in Fig. 13.14}$$

This means that $e^{2.996} = 20$.   ∎

EXAMPLE 2 Change of base to find log to base 5

Find $\log_5 560$.

In Eq. (13.12), if we let $a = 10$ and $b = 5$, we have

$$\log_5 x = \frac{\log x}{\log 5}$$

In this example, $x = 560$. Therefore, we have

$$\log_5 560 = \frac{\log 560}{\log 5} = 3.932 \qquad \text{see lines 3 and 4 of calculator display in Fig. 13.14}$$

From the definition of a logarithm, this means that

$$5^{3.932} = 560$$ ∎

Practice Exercise

1. Find $\log_3 23$.

```
log(20)/log(e^(1
))
         2.995732274
log(560)/log(5)
         3.931768187
```

Fig. 13.14

Since natural logarithms are used extensively, it is often convenient to have Eq. (13.12) written specifically for use with logarithms to the base 10 and natural logarithms. First, using $a = 10$ and $b = e$, we have

$$\ln x = \frac{\log x}{\log e} \qquad \text{(13.13)}$$

Then with $a = e$ and $b = 10$, we have

$$\log x = \frac{\ln x}{\ln 10} \qquad \text{(13.14)}$$

Note that we really found ln 20 in Example 1 by using Eq. (13.13).

Values of natural logarithms can be found directly on a calculator. The $\boxed{\ln}$ key is used for this purpose. In order to find the antilogarithm of a natural logarithm, we use the $\boxed{e^x}$ key. The following example illustrates finding a natural logarithm and an antilogarithm on a graphing calculator.

EXAMPLE 3 Natural log on calculator

(a) From the first two lines of the calculator display in Fig. 13.15, we find that

$$\ln 236.5 = 5.4659$$

which means that $e^{5.4659} = 236.5$.

(b) Given that $\ln N = -0.8729$, we determine N by finding $e^{-0.8729}$ on the calculator. This gives us (see lines 3 and 4 of the display in Fig. 13.15)

$$N = 0.4177 \qquad \blacksquare$$

```
ln(236.5)
       5.465948208
e^(-.8729)
       .4177383497
```

Fig. 13.15

Using Eq. (13.12), we can display the graph of a logarithmic function with any base on a graphing calculator, as we show in the next example.

EXAMPLE 4 Calculator graph of log to base 2

Display the graph of $y = \log_2 x$ on a graphing calculator.

To display this graph, we use the fact that

$$\log_2 x = \frac{\log x}{\log 2} \quad \text{or} \quad \ln_2 x = \frac{\ln x}{\ln 2}$$

Therefore, we enter the function

$$y = \frac{3 \log x}{\log 2} \qquad \left(\text{or } y = \frac{3 \ln x}{\ln 2} \right)$$

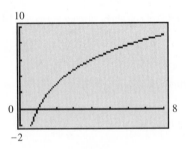

Fig. 13.16

in the calculator. We can select the *window* settings by considering the domain and range of the logarithmic function and by the fact that $3/\log 2 \approx 10$. This tells us that the range setting should be about ten times that of the curve $y = \log x$. The curve is shown in the calculator display in Fig. 13.16. $\blacksquare$

Applications of natural logarithms are found in many fields of technology. One such application is shown in the next example, and others are found in the exercises.

Practice Exercise

2. Use a calculator to evaluate ln 57.2.

EXAMPLE 5   Natural log – application

Fig. 13.17

The electric current i in a circuit containing a resistance and an inductance (see Fig. 13.17) is given by $\ln(i/I) = -Rt/L$, where I is the current at $t = 0$, R is the resistance, t is the time, and L is the inductance. Calculate how long (in s) it takes i to reach 0.430 A, if $I = 0.750$ A, $R = 7.50\ \Omega$, and $L = 1.25$ H.

Solving for t and then evaluating, we have

$$t = -\frac{L\ln(i/I)}{R} = -\frac{L(\ln i - \ln I)}{R} \qquad \text{either form can be used}$$

$$= -\frac{1.25(\ln 0.430 - \ln 0.750)}{7.50} = 0.0927\text{ s} \qquad \text{evaluating}$$

Therefore, the current changes from 0.750 A to 0.430 A in 0.0927 s. ∎

EXERCISES 13.5

In Exercises 1 and 2, find the indicated values if the given changes are made in the indicated examples of this section.

1. In Example 1, change 20 to 200 and then evaluate.

2. In Example 2, change the base to 4 and then evaluate.

In Exercises 3–8, use logarithms to the base 10 to find the natural logarithms of the given numbers.

3. 26.0   **4.** 631   **5.** 1.562

6. 0.5017   **7.** 0.0073267   **8.** 0.00044348

In Exercises 9–14, use logarithms to the base 10 to find the indicated logarithms.

9. $\log_7 42$ **10.** $\log_2 86$ **11.** $\log_\pi 245$

12. $\log_{12} 122$ **13.** $\log_{40} 750$ **14.** $\log_{100} 3720$

In Exercises 15–22, find the natural logarithms of the given numbers.

15. 51.4   **16.** 293   **17.** 1.394

18. 6552   **19.** 0.9917   **20.** 0.002086

21. $(0.012937)^4$ **22.** $\sqrt{0.000060808}$

In Exercises 23–26, use Eq. (13.14) to find the common logarithms of the given numbers.

23. 45.17   **24.** 8765

25. 0.68528   **26.** 0.0014298

In Exercises 27–34, find the natural antilogarithms of the given logarithms.

27. 2.190   **28.** 5.420   **29.** 0.0084210

30. 0.632   **31.** −0.7429   **32.** −2.94218

33. −23.504   **34.** −0.00804

In Exercises 35–38, use a graphing calculator to display the indicated graphs.

35. The graph of $y = \log_5 x$ **36.** The graph of $y = 2\log_8 x$

37. Graphically show that $y = 2^x$ and $y = \log_2 x$ are inverse functions. (See Example 9 of Section 13.2.)

W 38. Explain what is meant by the expression $\ln \ln x$. Display the graph of $y = \ln \ln x$ on a calculator.

In Exercises 39–42, use a calculator to verify the given values.

39. $\ln 5 + \ln 8 = \ln 40$ **40.** $2\ln 6 - \ln 3 = \ln 12$

41. $4\ln 3 = \ln 81$ **42.** $\ln 5 - 0.5\ln 25 = \ln 1$

In Exercises 43–56, solve the given problems.

43. Evaluate: $\sqrt{\ln e^9}$.

44. Solve for y in terms of x: $\ln y + 2\ln x = 1 + \ln 5$.

45. Solve for x: $\ln(\log x) = 0$.

46. If $\ln x = 3$ and $\ln y = 4$, find $\sqrt{x^2 y}$.

47. If $x = \ln 4$ and $y = \ln 5$, express $\ln 80$ in terms of x and y.

48. On a graphing calculator, compare the graphs of $y_1 = \ln(x^2 + 1)$ and $y_2 = \ln(x^2 + 1)^{-1}$.

49. Find f (in Hz) if $\ln f = 21.619$, where f is the frequency of the microwaves in a microwave oven.

50. Find k (in 1/Pa) if $\ln k = -21.504$, where k is the compressibility of water.

51. If interest is compounded continuously (daily compounded interest closely approximates this), with an interest rate i, a bank account will double in t years according to $i = (\ln 2)/t$. Find i if the account is to double in 8.5 years.

52. One approximate formula for world population growth is $T = 50.0\ln 2$, where T is the number of years for the population to double. According to this formula, how long does it take for the population to double?

53. For the electric circuit of Example 5, find how long it takes the current to reach 0.1 of the initial value of 0.750 A.

54. The velocity v (in m/s) of a rocket increases as fuel is consumed and ejected. Considering fuel as part of the mass m of a rocket in flight and m_0 as its original mass, the velocity is given by $v = 2500(\ln m_0 - \ln m)$. If 0.75 of the original mass is fuel, find v when all the fuel is used.

55. The distance x traveled by a motorboat in t seconds after the engine is cut off is given by $x = k^{-1} \ln(k v_0 t + 1)$, where v_0 is the velocity of the boat at the time the engine is cut and k is a constant. Find how long it takes a boat to go 150 m if $v_0 = 12.0$ m/s and $k = 6.80 \times 10^{-3}$/m.

56. The electric current i (in A) in a circuit that has a 1-H inductor, a 10-Ω resistor, and a 6-V battery, and the time t (in s) are related by the equation $10t = -\ln(1 - i/0.6)$. Solve for i.

Answers to Practice Exercises

1. 2.854   **2.** 4.05

13.6 Exponential and Logarithmic Equations

Exponential Equations •
Logarithmic Equations

EXPONENTIAL EQUATIONS

An equation in which the variable occurs in an exponent is called an **exponential equation.** Although some may be solved by changing to logarithmic form, they are generally solved by *taking the logarithm of each side* and then using the properties of logarithms.

EXAMPLE 1 Exponential equation – solved two ways

(a) We can solve the exponential equation $2^x = 8$ by writing it in logarithmic form. This gives

$$x = \log_2 8 = 3 \qquad 2^3 = 8$$

This method is good if we can directly evaluate the resulting logarithm.

■ For reference, Eqs. (13.7), (13.8), and (13.9) are

$\log_b xy = \log_b x + \log_b y$

$\log_b\left(\dfrac{x}{y}\right) = \log_b x - \log_b y$

$\log_b(x^n) = n \log_b x$

(b) Since 2^x and 8 are equal, the logarithms of 2^x and 8 are also equal. Therefore, we can also solve $2^x = 8$ in a more general way by taking logarithms (to any proper base) of both sides and equating these logarithms. This gives us

$$\log 2^x = \log 8 \qquad \text{or} \qquad \ln 2^x = \ln 8$$
$$x \log 2 = \log 8 \qquad\qquad x \ln 2 = \ln 8 \qquad \text{using Eq. (13.9)}$$
$$x = \frac{\log 8}{\log 2} = 3 \qquad\qquad x = \frac{\ln 8}{\ln 2} = 3 \qquad \text{using a calculator}$$

■

EXAMPLE 2 Solving an exponential equation

Solve the equation $3^{x-2} = 5$.

Taking logarithms of each side and equating them, we have

$$\log 3^{x-2} = \log 5$$
$$(x - 2)\log 3 = \log 5 \qquad \text{using Eq. (13.9)}$$
$$x = 2 + \frac{\log 5}{\log 3} = 3.465$$

Practice Exercise

1. Solve for x: $2^{x+1} = 7$

This solution means that

$$3^{3.465-2} = 3^{1.465} = 5$$

which can be checked by a calculator.

■

EXAMPLE 3 Solving an exponential equation

Solve the equation $2(4^{x-1}) = 17^x$.

By taking logarithms of each side, we have the following:

$$\log 2 + (x - 1)\log 4 = x \log 17 \qquad \text{using Eqs. (13.7) and (13.9)}$$
$$x \log 4 - x \log 17 = \log 4 - \log 2$$
$$x(\log 4 - \log 17) = \log 4 - \log 2$$
$$x = \frac{\log 4 - \log 2}{\log 4 - \log 17} = \frac{\log(4/2)}{\log 4 - \log 17} \qquad \text{using Eq. (13.8)}$$
$$= \frac{\log 2}{\log 4 - \log 17} = -0.479$$

■

EXAMPLE 4 Exponential equation – application

At constant temperature, the atmospheric pressure p (in Pa) at an altitude h (in m) is given by $p = p_0 e^{kh}$, where p_0 is the pressure where $h = 0$ (usually taken as sea level). Given that $p_0 = 101.3$ kPa (atmospheric pressure at sea level) and $p = 68.9$ kPa for $h = 3050$ m, find the value of k.

Since the equation is defined in terms of e, we can solve it most easily by taking natural logarithms of *each side*. By doing this, we have the following solution:

$$\ln p = \ln(p_0 e^{kh}) = \ln p_0 + \ln e^{kh} \qquad \text{using Eq. (13.7)}$$

$$= \ln p_0 + kh \ln e = \ln p_0 + kh \qquad \text{using Eq. (13.9), } \ln e = 1$$

$$\ln p - \ln p_0 = kh$$

$$k = \frac{\ln p - \ln p_0}{h}$$

Substituting the given values, we have

$$k = \frac{\ln(68.9 \times 10^3) - \ln(101.3 \times 10^3)}{3050} = -0.000126/\text{m} \qquad \blacksquare$$

LOGARITHMIC EQUATIONS

Some of the important measurements in scientific and technical work are defined in terms of logarithms. Using these formulas can lead to solving a **logarithmic equation,** *which is an equation with the logarithm of an expression involving the variable.* In solving logarithmic equations, we use the basic properties of logarithms to help change them into a usable form. There is, however, no general algebraic method for solving such equations, and we consider only some special cases.

EXAMPLE 5 Logarithmic equation – application

The human ear responds to sound on a scale that is approximately proportional to the logarithm of the intensity of the sound. Therefore, the loudness of sound (measured in dB) is defined by the equation $b = 10 \log(I/I_0)$, where I is the intensity of the sound and I_0 is the minimum intensity detectable.

A busy street has a loudness of 70 dB, and riveting has a loudness of 100 dB. To find how many times greater the intensity I_r of the sound of riveting is than the intensity I_c of the sound of the city street, we substitute in the above equation. This gives

$$70 = 10 \log\left(\frac{I_c}{I_0}\right) \quad \text{and} \quad 100 = 10 \log\left(\frac{I_r}{I_0}\right)$$

To solve for I_c and I_r, we divide each side by 10 and then use exponential form:

$$7.0 = \log\left(\frac{I_c}{I_0}\right) \quad \text{and} \quad 10 = \log\left(\frac{I_r}{I_0}\right)$$

$$\frac{I_c}{I_0} = 10^{7.0} \qquad\qquad \frac{I_r}{I_0} = 10^{10}$$

$$I_c = I_0(10^{7.0}) \qquad I_r = I_0(10^{10})$$

■ This demonstrates that sound intensity levels change much more than loudness levels. (See Exercise 50.)

Since we want the number of times I_r is greater than I_c, we divide I_r by I_c:

$$\frac{I_r}{I_c} = \frac{I_0(10^{10})}{I_0(10^{7.0})} = \frac{10^{10}}{10^{7.0}} = 10^{3.0} \quad \text{or} \quad I_r = 10^{3.0}I_c = 1000I_c$$

Thus, the sound of riveting is 1000 times as intense as the sound of the city street. ■

■ See the chapter introduction.
For reference,

$$\log_b xy = \log_b x + \log_b y \qquad (13.7)$$

$$\log_b\left(\frac{x}{y}\right) = \log_b x - \log_b y \qquad (13.8)$$

$$\log_b(x^n) = n\log_b x \qquad (13.9)$$

EXAMPLE 6 Logarithmic equation – application

An analysis of the population of Canada from 1980 to 2000 led to the equation $\log_2 P = \log_2 23.9 + 0.450$, where P is the projected population (in millions) in 2010. Determine this population.

The solution is as follows:

$$\log_2 P - \log_2 23.9 = 0.450$$
$$\log_2(P/23.9) = 0.450 \qquad \text{using Eq. (13.8)}$$
$$P/23.9 = 2^{0.450} \qquad \text{exponential form}$$
$$P = 23.9(2^{0.450}) = 32.6 \text{ million} \qquad \text{projected 2010 population} \quad ■$$

EXAMPLE 7 Solving logarithmic equation

Solve the logarithmic equation $2\ln 2 + \ln x = \ln 3$.

Using the properties of logarithms, we have the following solution:

$$2\ln 2 + \ln x = \ln 3$$
$$\ln 2^2 + \ln x - \ln 3 = 0 \qquad \text{using Eq. (13.9)}$$
$$\ln \frac{4x}{3} = 0 \qquad \text{using Eqs. (13.7) and (13.8)}$$
$$\frac{4x}{3} = e^0 = 1 \qquad \text{exponential form}$$
$$4x = 3, \qquad x = 3/4$$

Fig. 13.18

Since $\ln(3/4) = \ln 3 - \ln 4$, this solution checks in the *original equation*.

The calculator solution of this equation, using the *Solver* feature, is displayed in Fig. 13.18. It can also be solved graphically by finding the *zero* of the function $y_1 = 2\ln 2 + \ln x - \ln 3$, or the *intersection* of $y_1 = 2\ln 2 + \ln x$ and $y_2 = \ln 3$. ■

EXAMPLE 8 Solving logarithmic equation

Solve the logarithmic equation $2\log x - 1 = \log(1 - 2x)$.

$$\log x^2 - \log(1 - 2x) = 1$$
$$\log \frac{x^2}{1 - 2x} = 1 \qquad \text{using Eq. (13.8)}$$
$$\frac{x^2}{1 - 2x} = 10^1 \qquad \text{exponential form}$$
$$x^2 = 10 - 20x$$
$$x^2 + 20x - 10 = 0$$
$$x = \frac{-20 \pm \sqrt{400 + 40}}{2} = -10 \pm \sqrt{110}$$

Practice Exercise

2. Solve for x: $3\log 2 - \log(x + 1) = 1$

Since logarithms of negative numbers are not defined and $-10 - \sqrt{110}$ is negative and cannot be used in the first term of the original equation, we have

$$x = -10 + \sqrt{110} = 0.488 \qquad \text{use a calculator to check this result} \quad ■$$

EXERCISES 13.6

In Exercises 1 and 2, find the indicated values if the given changes are made in the indicated examples of this section.

1. In Example 2, change the sign in the exponent from $-$ to $+$ and then solve the equation.

2. In Example 7, change ln 3 to ln 6 and then solve the equation.

In Exercises 3–30, solve the given equations.

3. $2^x = 16$ **4.** $3^x = 1/81$ **5.** $5^x = 0.3$

6. $\pi^x = 15$ **7.** $3^{-x} = 0.525$ **8.** $e^{-x} = 17.54$

9. $6^{x+1} = 78$ **10.** $5^{x-1} = 0.07$ **11.** $3(14^x) = 400$

12. $0.8^x = 0.4$ **13.** $0.6^x = 2^{x^2}$ **14.** $15.6^{x+2} = 23^x$

15. $3\log_8 x = -2$ **16.** $5\log_{32} x = -3$

17. $\log x^2 = (\log x)^2$ **18.** $x^{\log x} = 1000x^2$

19. $\log_2 x + \log_2 7 = \log_2 21$ **20.** $2\log_2 3 - \log_2 x = \log_2 45$

21. $2\log(3 - x) = 1$ **22.** $3\log(2x - 1) = 1$

23. $\log 12x^2 - \log 3x = 3$ **24.** $\ln x - \ln(1/3) = 1$

25. $3\ln 2 + \ln(x - 1) = \ln 24$ **26.** $\log_2 x + \log_2(x + 2) = 3$

27. $\frac{1}{2}\log(x + 2) + \log 5 = 1$ **28.** $2\log_x 2 + \log_2 x = 3$

29. $\log(2x - 1) + \log(x + 4) = 1$

30. $\ln(2x - 1) - 2\ln 4 = 3\ln 2$

In Exercises 31–38, use a graphing calculator to solve the given equations.

31. $15^{-x} = 1.326$ **32.** $e^{2x} = 3.625$

33. $4(3^x) = 5$ **34.** $5^{x+2} = e^{2x}$

35. $3\ln 2x = 2$ **36.** $\log 4x + \log x = 2$

37. $2\ln 2 - \ln x = -1$ **38.** $\log(x - 3) + \log x = \log 4$

In Exercises 39–56, find the indicated quantities.

39. Solve for x: $e^x + e^{-x} = 3$. (*Hint:* Multiply each term by e^x and then it can be treated as a quadratic equation in e^x.)

40. Solve for x: $3^x + 3^{-x} = 4$. See Exercise 39.

41. If $y = 1.5e^{-0.90x}$, find y when $x = 7.1$.

42. What values of x cannot be solutions of the equation $y = \log(2x - 5) + \log(x^2 + 1)$?

43. Use logarithms to find the x-intercept of the graph of $y = 3 - 4^{x+2}$.

44. Use a graphing calculator to find the point of intersection of the curves of $3x + 5y = 6$ and $y = 1.5\ln(x + 3)$.

45. In computer design, the number N of bits of memory is often expressed as a power of 2, or $N = 2^x$. Find x if $N = 2.68 \times 10^8$ bits.

46. Referring to Exercise 64 on page 373, in what year will the amount of DDT be 25% of the original amount?

47. Forensic scientists determine the temperature T (in °C) of a body t hours after death from the equation $T = T_0 + (37 - T_0)0.97^t$, where T_0 is the air temperature. If a body is discovered at midnight with a body temperature of 27°C in a room at 22°C, at what time did death occur?

48. When a camera flash goes off, the batteries recharge the flash's capacitor to a charge Q according to $Q = Q_0(1 - e^{-kt})$, where Q_0 is the maximum charge. How long does it take to recharge the capacitor to 90% of capacity if $k = 0.5$?

49. In chemistry, the pH value of a solution is a measure of its acidity. The pH value is defined by $pH = -\log(H^+)$, where H^+ is the hydrogen-ion concentration. If the pH of a sample of rainwater is 4.764, find the hydrogen-ion concentration. (If pH < 7, the solution is acid. If pH > 7, the solution is basic.) Acid rain has a pH between 4 and 5, and normal rain is slightly acidic with a pH of about 5.6.

50. Referring to Example 5, show that if the difference in loudness of two sounds is d decibels, the louder sound is $10^{d/10}$ more intense than the quieter sound.

51. Measured on the Richter scale, the magnitude of an earthquake of intensity I is defined as $R = \log(I/I_0)$, where I_0 is a minimum level for comparison. How many times I_0 was the 1906 San Francisco earthquake whose magnitude was 8.3 on the Richter scale?

52. How many times more intense was the 1906 San Francisco earthquake than the 1994 Northridge (southern California) earthquake, $R = 6.8$? (See Exercise 51.)

53. Studies have shown that the concentration c (in mg/cm^3 of blood) of aspirin in a typical person is related to the time t (in h) after the aspirin reaches maximum concentration by the equation $\ln c = \ln 15 - 0.20t$. Solve for c as a function of t.

54. In an electric circuit containing a resistor and a capacitor with an initial charge q_0, the charge q on the capacitor at any time t after closing the switch can be found by solving the equation $\ln q = -\dfrac{t}{RC} + \ln q_0$. Here, R is the resistance, and C is the capacitance. Solve for q as a function of t.

55. An earth satellite loses 0.1% of its remaining power each week. An equation relating the power P, the initial power P_0, and the time t (in weeks) is $\ln P = t\ln 0.999 + \ln P_0$. Solve for P as a function of t.

56. In finding the path of a certain plane, the equation $\ln r = \ln a - \ln\cos\theta - (v/w)\ln(\sec\theta + \tan\theta)$ is used. Solve for r. (The conditions of the plane's flight are similar to those described in Exercise 96, p. 332.)

Many exponential and logarithmic equations cannot be solved algebraically as we did in this section. However, they can be solved graphically. For example, such equations can be solved by either of the methods noted at the end of Example 7. In Exercises 57–60, solve the given equations by use of a graphing calculator.

57. $2^x + 3^x = 50$ **58.** $4^x + x^2 = 25$

59. The curve in which a uniform wire or rope hangs under its own weight is called a *catenary*. An example of a catenary that we see every day is a wire strung between utility poles, as shown in Fig. 13.19. For a particular wire, the equation of the catenary it forms is $y = 2(e^{x/4} + e^{-x/4})$, where (x, y) is a point on the curve. Find x for $y = 5.8$ m.

Catenary

Fig. 13.19

60. In finding the current i in a certain electric circuit, the equation $i = 2\ln(t + 2) - t$ relates the current and the time t (in s). Find t for $i = 0.05$ A.

Answers to Practice Exercises

1. 1.807   **2.** $-1/5$

13.7 Graphs on Logarithmic and Semilogarithmic Paper

Logarithmic Scale •
Semilogarithmic (Semilog) Paper •
Logarithmic (Log-Log) Paper

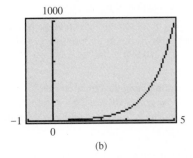

Fig. 13.20

When constructing the graphs of some functions, one of the variables changes much more rapidly than the other. We saw this in graphing the exponential and logarithmic functions in Sections 13.1 and 13.2. The following example illustrates this point.

EXAMPLE 1 Graph of exponential function

Plot the graph of $y = 4(3^x)$.

Constructing the following table of values,

x	-1	0	1	2	3	4	5
y	1.3	4	12	36	108	324	972

we then plot these values as shown in Fig. 13.20(a).

We see that as x changes from -1 to 5, y changes much more rapidly, from about 1 to nearly 1000. Also, because of the scale that must be used, we see that it is not possible to show accurately the differences in the y-values on the graph.

Even a graphing calculator cannot show the graph accurately for the values near $x = 0$, if we wish to view all of this part of the curve. In fact, the graphing calculator view shows the curve as being on the axis for these values, as we see in Fig. 13.20(b). Note in this figure that we have shown the same values of the range of the function as we did in Fig. 13.20(a). ■

It is possible to graph a function with a large change in values, for one or both variables, more accurately than can be done on the standard rectangular coordinate system. This is done by using a scale marked off in distances proportional to the logarithms of the values being represented. *Such a scale is called a* **logarithmic scale.** For example, $\log 1 = 0$, $\log 2 = 0.301$, and $\log 10 = 1$. Thus, on a logarithmic scale, the 2 is placed 0.301 unit of distance from the 1 to the 10. Figure 13.21 shows a logarithmic scale with the numbers represented and the distance used for each.

On a logarithmic scale, the distances between the integers are not equal, but this scale does allow for a much greater range of values and much greater accuracy for many of the values. There is another advantage to using logarithmic scales. Many equations that would have more complex curves when graphed on the standard rectangular coordinate system will have simpler curves, often straight lines, when graphed using logarithmic scales. In many cases, this makes the analysis of the curve much easier.

Zero and negative numbers do not appear on the logarithmic scale. In fact, all numbers used on the logarithmic scale must be positive, since the domain of the logarithmic function includes only positive real numbers. Thus, the logarithmic scale must start at some number greater than zero. This number is a power of 10 and can be very small, say, $10^{-6} = 0.000001$, but it is positive.

If we wish to use a large range of values for only one of the variables, we use what is known as **semilogarithmic,** or **semilog,** graph paper. On this graph paper, only one axis (usually the y-axis) uses a logarithmic scale. If we wish to use a large range of values for both variables, we use **logarithmic,** or **log-log,** graph paper. Both axes are marked with logarithmic scales.

The following examples illustrate the use of semilog and log-log graph paper.

Fig. 13.21

Fig. 13.22

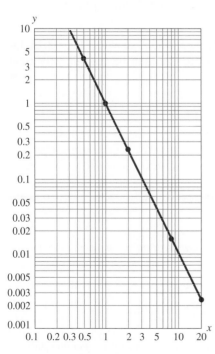

Fig. 13.23

EXAMPLE 2 Graph on semilogarithmic paper

Construct the graph of $y = 4(3^x)$ on semilogarithmic graph paper.

This is the same function as in Example 1, and we repeat the table of values:

x	-1	0	1	2	3	4	5
y	1.3	4	12	36	108	324	972

Again, we see that the range of y-values is large. When we plotted this curve on the rectangular coordinate system in Example 1, we had to use large units along the y-axis. This made the values of 1.3, 4, 12, and 36 appear at practically the same level. However, when we use semilog graph paper, we can label each axis such that all y-values are accurately plotted as well as the x-values.

The logarithmic scale is shown in **cycles,** and we must label the base line of the first cycle as 1 times a power of 10 (0.01, 0.1, 1, 10, 100, and so on) with the following cycle labeled with the next power of 10. The lines between are labeled with 2, 3, 4, and so on, times the proper power of 10. See the vertical scale in Fig. 13.22. We now plot the points in the table on the graph. The resulting graph is a straight line, as we see in Fig. 13.22. Taking logarithms of each side of the equation, we have

$$\log y = \log[4(3^x)] = \log 4 + \log 3^x \qquad \text{using Eq. (13.7)}$$
$$= \log 4 + x \log 3 \qquad \text{using Eq. (13.9)}$$

However, since $\log y$ was plotted automatically (because we used semilogarithmic paper), the graph really represents

$$u = \log 4 + x \log 3$$

where $u = \log y$; $\log 3$ and $\log 4$ are constants, and therefore this equation is of the form $u = mx + b$, which is a straight line (see Section 5.2).

The logarithmic scale in Fig. 13.22 has *three cycles,* since all values of three powers of 10 are represented. ∎

EXAMPLE 3 Graph on logarithmic paper

Construct the graph of $x^4 y^2 = 1$ on logarithmic paper.

First, we solve for y and make a table of values. Considering positive values of x and y, we have

$$y = \sqrt{\frac{1}{x^4}} = \frac{1}{x^2}$$

x	0.5	1	2	8	20
y	4	1	0.25	0.0156	0.0025

We plot these values on log-log paper on which both scales are logarithmic, as shown in Fig. 13.23. We again see that we have a straight line. Taking logarithms of both sides of the equation, we have

$$\log(x^4 y^2) = \log 1$$
$$\log x^4 + \log y^2 = 0 \qquad \text{using Eq. (13.7)}$$
$$4 \log x + 2 \log y = 0 \qquad \text{using Eq. (13.9)}$$

If we let $u = \log y$ and $v = \log x$, we then have

$$4v + 2u = 0 \quad \text{or} \quad u = -2v$$

which is the equation of a straight line, as shown in Fig. 13.23. Note, however, that not all graphs on logarithmic paper are straight lines. ∎

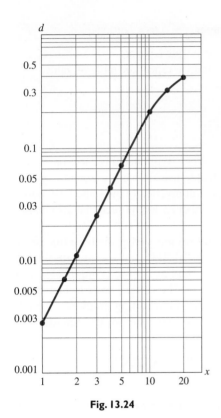

Fig. 13.24

EXAMPLE 4 Graph of application on log-log paper

The deflection (in ft) of a certain cantilever beam as a function of the distance x (in ft) from one end is

$$d = 0.0001(30x^2 - x^3)$$

If the beam is 20.0 ft long, plot a graph of d as a function of x on log-log paper.
    Constructing a table of values, we have

x (ft)	1.00	1.50	2.00	3.00	4.00
d (ft)	0.00290	0.00641	0.0112	0.0243	0.0416

x (ft)	5.00	10.0	15.0	20.0
d (ft)	0.0625	0.200	0.338	0.400

Since the beam is 20.0 ft long, there is no meaning to values of x greater than 20.0 ft. The graph is shown in Fig. 13.24. ■

Logarithmic and semilogarithmic paper may be useful for plotting data derived from experimentation. Often, the data cover too large a range of values to be plotted on ordinary graph paper. The next example illustrates the use of semilogarithmic paper to plot data.

EXAMPLE 5 Graph of application on semilog paper

The vapor pressure of water depends on the temperature. The following table gives the vapor pressure (in kPa) for the corresponding values of temperature (in °C):

T (°C)	10	20	40	60	80	100	120	140	160
P (kPa)	1.19	2.33	7.34	19.9	47.3	101	199	361	617

These data are then plotted on semilogarithmic paper, as shown in Fig. 13.25. Intermediate values of temperature and pressure can then be read directly from the graph. ■

Fig. 13.25

EXERCISES 13.7

In Exercises 1 and 2, make the given changes in the indicated examples, and then draw the graphs.

1. In Example 2, change the 4 to 2 and then make the graph.

2. In Example 3, change the 1 to 4 and then make the graph.

In Exercises 3–10, plot the graphs of the given functions on semilogarithmic paper.

3. $y = 2^x$ **4.** $y = 5^x$ **5.** $y = 5(4^x)$

6. $y = 6^{-x}$ **7.** $y = x^3$ **8.** $y = 2x^4$

9. $y = 2x^3 + 6x$ **10.** $y = 4x^3 + 2x^2$

In Exercises 11–18, plot the graphs of the given functions on log-log paper.

11. $y = 0.01x^4$ **12.** $y = \sqrt{x}$ **13.** $y = x^{2/3}$

14. $y = 8x^{0.25}$ **15.** $xy = 40$ **16.** $x^2y^3 = 16$

17. $x^2y^2 = 25$ **18.** $x^3y = 8$

In Exercises 19–26, determine the type of graph paper on which the graph of the given function is a straight line. Using the appropriate paper, sketch the graph.

19. $y = 3^{-x}$ **20.** $y = 0.2x^3$ **21.** $y = 3x^6$

22. $y = 5(10^{-x})$ **23.** $y = 4^{x/2}$ **24.** $xy^3 = 10$

25. $x\sqrt{y} = 4$ **26.** $y(2^x) = 3$

In Exercises 27–38, plot the indicated graphs.

27. On the moon, the distance s (in ft) a rock will fall due to gravity is $s = 2.66t^2$, where t is the time (in s) of fall. Plot the graph of s as a function of t for $0 \leq t \leq 10$ s on (a) a regular rectangular coordinate system and (b) a semilogarithmic coordinate system.

28. By pumping, the air pressure in a tank is reduced by 18% each second. Thus, the pressure p (in kPa) in the tank is given by $p = 101(0.82)^t$, where t is the time (in s). Plot the graph of p as a function of t for $0 \leq t \leq 30$ s on (a) a regular rectangular coordinate system and (b) a semilogarithmic coordinate system.

29. Strontium-90 decays according to the equation $N = N_0 e^{-0.028t}$, where N is the amount present after t years and N_0 is the original amount. Plot N as a function of t on semilog paper if $N_0 = 1000$ g.

30. The electric power P (in W) in a certain battery as a function of the resistance R (in Ω) in the circuit is given by $P = \dfrac{100R}{(0.50 + R)^2}$. Plot P as a function of R on semilog paper, using the logarithmic scale for R and values of R from $0.01\,\Omega$ to $10\,\Omega$. Compare the graph with that in Fig. 3.20 on page 93.

31. The acceleration g (in m/s^2) produced by the gravitational force of the earth on a spacecraft is given by $g = 3.99 \times 10^{14}/r^2$, where r is the distance from the center of the earth to the spacecraft. On log-log paper, graph g as a function of r from $r = 6.37 \times 10^6$ m (the earth's surface) to $r = 3.91 \times 10^8$ m (the distance to the moon).

32. In undergoing an adiabatic (no *heat* gained or lost) expansion of a gas, the relation between the pressure p (in kPa) and the volume v (in m^3) is $p^2 v^3 = 850$. On log-log paper, graph p as a function of v from $v = 0.10$ m^3 to $v = 10$ m^3.

33. The number of cell phone subscribers in the United States from 1991 to 2006 is shown in the following table. Plot N as a function of the year on semilog paper.

Year	1991	1994	1997	2000	2003	2006
$N\,(\times 10^6)$	7.56	24.1	55.3	109	159	233

34. The period T (in years) and mean distance d (given as a ratio of that of earth) from the sun to the planets (Mercury, Venus, Earth, Mars, Jupiter, Saturn, Uranus, Neptune, Pluto) are given below. Plot T as a function of d on log-log paper.

Planet	M	V	E	M	J	S	U	N	P
d	0.39	0.72	1.00	1.52	5.20	9.54	19.2	30.1	39.5
T	0.24	0.62	1.00	1.88	11.9	29.5	84.0	165	249

35. The intensity level B (in dB) and the frequency f (in Hz) for a sound of constant loudness were measured as shown in the table that follows. Plot the data for B as a function of f on semilog paper, using the log scale for f.

f (Hz)	100	200	500	1000	2000	5000	10,000
B (dB)	40	30	22	20	18	24	30

36. The atmospheric pressure p (in kPa) at a given altitude h (in km) is given in the following table. On semilog paper, plot p as a function of h.

h (km)	0	10	20	30	40
p (kPa)	101	25	6.3	2.0	0.53

37. One end of a very hot steel bar is sprayed with a stream of cool water. The rate of cooling R (in °F/s) as a function of the distance d (in in.) from one end of the bar is then measured, with the results shown in the following table. On log-log paper, plot R as a function of d. Such experiments are made to determine the hardness of steel.

d (in.)	0.063	0.13	0.19	0.25
R (°F/s)	600	190	100	72

d (in.)	0.38	0.50	0.75	1.0	1.5
R (°F/s)	46	29	17	10	6.0

38. The magnetic intensity H (in A/m) and flux density B (in teslas) of annealed iron are given in the following table. Plot H as a function of B on log-log paper.

B (T)	0.0042	0.043	0.67	1.01
H (A/m)	10	50	100	150

B (T)	1.18	1.44	1.58	1.72
H (A/m)	200	500	1000	10,000

In Exercises 39 and 40, plot the indicated semilogarithmic graphs for the following application.

In a particular electric circuit, called a low-pass filter, the input voltage V_i is across a resistor and a capacitor, and the output voltage V_0 is across the capacitor (see Fig. 13.26). The voltage gain G (in dB) is given by

Fig. 13.26

$$G = 20 \log \frac{1}{\sqrt{1 + (\omega T)^2}}$$

where $\tan \phi = -\omega T$.

Here, ϕ is the phase angle of V_0/V_i. For values of ωT of 0.01, 0.1, 0.3, 1.0, 3.0, 10.0, 30.0, and 100, plot the indicated graphs. These graphs are called a Bode diagram for the circuit.

39. Calculate values of G for the given values of ωT and plot a semilogarithmic graph of G vs. ωT.

40. Calculate values of ϕ (as negative angles) for the given values of ωT and plot a semilogarithmic graph of ϕ vs. ωT.

CHAPTER 13 EQUATIONS

Exponential function	$y = b^x$	(13.1)
Logarithmic form	$x = \log_b y$	(13.2)
Logarithmic function	$y = \log_b x$	(13.3)
Laws of exponents	$b^u b^v = b^{u+v}$	(13.4)
	$\dfrac{b^u}{b^v} = b^{u-v}$	(13.5)
	$(b^u)^n = b^{nu}$	(13.6)
Properties of logarithms	$\log_b xy = \log_b x + \log_b y$	(13.7)
	$\log_b\left(\dfrac{x}{y}\right) = \log_b x - \log_b y$	(13.8)
	$\log_b(x^n) = n\log_b x$	(13.9)
	$\log_b 1 = 0 \qquad \log_b b = 1$	(13.10)
	$\log_b(b^n) = n$	(13.11)
Changing base of logarithms	$\log_b x = \dfrac{\log_a x}{\log_a b}$	(13.12)
	$\ln x = \dfrac{\log x}{\log e}$	(13.13)
	$\log x = \dfrac{\ln x}{\ln 10}$	(13.14)

CHAPTER 13 REVIEW EXERCISES

In Exercises 1–12, determine the value of x.

1. $\log_{10} x = 4$
2. $\log_9 x = 3$
3. $\log_5 x = -1$
4. $\log_4(\sin x) = -0.5$
5. $2\log_2 8 = x$
6. $\log_{12} 144 = x - 3$
7. $\log_8 32 = x$
8. $\log_9 27 = x$
9. $\log_x 36 = 2$
10. $\log_x(1/243) = 5$
11. $\log_x 10 = \frac{1}{2}$
12. $\log_x 8 = 0$

In Exercises 13–24, express each as a sum, difference, or multiple of logarithms. Wherever possible, evaluate logarithms of the result.

13. $\log_3 2x$
14. $\log_5\left(\dfrac{7}{a}\right)$
15. $\log_3(t^2)$
16. $\log_6 \sqrt{5}$
17. $\log_2 28$
18. $\log_7 98$
19. $\log_4 \sqrt{48}$
20. $\log \sqrt[V]{32y}$
21. $\log_3\left(\dfrac{9}{x}\right)$
22. $\log_6\left(\dfrac{5}{36}\right)$
23. $\log_{10}(1000x^4)$
24. $\log_3(9^2 \times 6^3)$

In Exercises 25–36, solve for y in terms of x.

25. $\log_6 y = \log_6 4 - \log_6 x$
26. $2\ln y = \ln e^2 - 3\ln x$
27. $3\ln y = 2 + 3\ln x$
28. $2(\log_9 y + 2\log_9 x) = 1$
29. $\log_3 y = \frac{1}{2}\log_3 7 + \frac{1}{2}\log_3 x$
30. $(\log_2 3)(\log_2 y) - \log_2 x = 3$
31. $\log_5 x + \log_5 y = \log_5 3 + 1$
32. $\log_7 y = 2\log_7 5 + \log_7 x + 2$
33. $2(\log_4 y - 3\log_4 x) = 3$
34. $\dfrac{\log_7 x}{\log_7 4} - \log_7 y = 1$
35. $2^y = e^x$
36. $10^y = 3^{x+1}$

In Exercises 37–44, display the graphs of the given functions on a graphing calculator.

37. $y = 0.5(5^x)$

38. $y = 3(2^{-x})$

39. $R = 0.2 \log_4 r$

40. $y = 10 \log_{16} x$

41. $y = \log_{3.15} x$

42. $y = 0.1 \log_{4.05} x$

43. $y = 1 - e^{-|x|}$

44. $s = 2(1 - e^{-0.2t})$

In Exercises 45–48, use logarithms to the base 10 to find the natural logarithms of the given numbers.

45. 8.86

46. 0.303

47. sin 2.07

48. $\sqrt{0.542}$

In Exercises 49–52, use natural logarithms to find logarithms to the base 10 of the given numbers.

49. 65.89

50. 0.0781

51. 0.1197^3

52. log 1000

In Exercises 53–60, solve the given equations.

53. $e^{2x} = 5$

54. $2(5^x) = 15$

55. $3^{x+2} = 5^x$

56. $2^x/3^{1-x} = 12^{x+1}$

57. $\log_4 z + \log_4 6 = \log_4 12$

58. $\log_8(x + 2) = 2 - \log_8 2$

59. $2 \log_3 2 - \log_3(x + 1) = \log_3 5$

60. $\log(n + 2) + \log n = 0.4771$

In Exercises 61 and 62, plot the graphs of the given functions on semi-logarithmic paper. In Exercises 63 and 64, plot the graphs of the given functions on log-log paper.

61. $y = 8^x$

62. $y = 5x^3$

63. $y = \sqrt[3]{x}$

64. $xy^4 = 16$

If x is eliminated between Eqs. (13.1) and (13.2), we have

$$y = b^{\log_b y} \qquad (13.15)$$

In Exercises 65–68, evaluate the given expressions using Eq. (13.15).

65. $10^{\log 4}$

66. $2e^{\ln 7.5}$

67. $3e^{2\ln 2}$

68. $5(10^{2\log 3})$

In Exercises 69–104, solve the given problems.

69. Use a calculator to verify that $2 \log 3 - \log 6 = \log 1.5$.

70. Use a calculator to verify that $3 \ln 2 + 0.5 \ln 64 = 3 \ln 4$.

71. Evaluate $\sqrt[3]{\ln e^8} - \sqrt{\log 10^4}$.

72. Solve for x: $2^x + 32(2^{-x}) = 12$.

73. If $f(x) = 2 \log_b x$ and $f(8) = 3$, find $f(4)$.

74. Evaluate: $\log(2 \log 1000)$.

75. Evaluate: $7(10^{\log 0.1}) + 6000(100^{\log 0.001})$.

76. If $x = \log_b 7$ and $y = \log_b 2$, express $\log_b 42$ in terms of x and y.

77. Use a graphing calculator to solve the equation $\log_5 x = 2x - 7$.

78. Use logarithms to find the x-intercept of the graph of $y = 2 - 5^{2x-3}$.

79. If an amount of P dollars is invested at an annual interest rate r (expressed as a decimal), the value V of the investment after t years is $V = P(1 + r/n)^{nt}$, if interest is compounded n times a year. If $1000 is invested at an annual interest rate of 6%, compounded semiannually, express V as a function of t and solve for t.

80. The current i (in A) in a certain electric circuit is given by $i = 16(1 - e^{-250t})$, where t is the time (in s). Solve for t.

81. The formula $\ln(I/I_0) = -\beta h$ is used in estimating the thickness of the ozone layer. Here, I_0 is the intensity of a wavelength of sunlight before reaching the earth's atmosphere, I is the intensity of the light after passing through h cm of the ozone layer, and β is a constant. Solve for I.

82. The amount A of cesium-137 (a dangerous radioactive element) remaining after t years is given by $A = A_0(0.5^{t/30.3})$, where A_0 is the initial amount. In what year will the cesium-137 be 10% of that released at the Chernobyl disaster in 1986?

83. The bending moment M (in N·m) of a particular concrete column is given by $\log M = 6.663$. What is the value of M?

84. A state lottery pays $500 for a $1 ticket if a person picks the correct three-digit number determined by the random draw of three numbered balls. The probability of a 50% chance of winning in x drawings is $1 - 0.999^x = 0.5$. For how many drawings does a person have to buy a ticket to have a 50% chance of winning?

85. An approximate formula for the population (in millions) of the state of Kansas since 1970 is $P = 2.25e^{0.00486t}$, where t is the number of years since 1970. Sketch the graph of P vs. t for 1970 to 2000.

86. A computer analysis of the luminous efficiency E (in lumens/W) of a tungsten lamp as a function of its input power P (in W) is given by $E = 22.0(1 - 0.65e^{-0.008P})$. Sketch the graph of E as a function of P for $0 \le P \le 1000$ W.

87. An original amount of 100 mg of radium radioactively decomposes such that N mg remain after t years. The function relating t and N is $t = 2350(\ln 100 - \ln N)$. Sketch the graph.

88. The time t (in s) to chemically change 5 kg of a certain substance into another is given by $t = -5\log\left(\dfrac{5 - x}{5}\right)$, where x is the number of kilograms that have been changed at any time. Sketch the graph.

89. An equation that may be used for the angular velocity ω of the slider mechanism in Fig. 13.27 is $2 \ln \omega = \ln 3g + \ln \sin \theta - \ln l$, where g is the acceleration due to gravity. Solve for $\sin \theta$.

Fig. 13.27

90. Taking into account the weight loss of fuel, the maximum velocity v_m of a rocket is $v_m = u(\ln m_0 - \ln m_s) - gt_f$, where m_0 is the initial mass of the rocket and fuel, m_s is the mass of the rocket shell, t_f is the time during which fuel is expended, u is the velocity of the expelled fuel, and g is the acceleration due to gravity. Solve for m_0.

91. An equation used to calculate the capacity C (in bits/s) of a telephone channel is $C = B \log_2(1 + R)$. Solve for R.

92. An equation used in studying the action of a protein molecule is $\ln A = \ln \theta - \ln(1 - \theta)$. Solve for θ.

93. The magnitudes (visual brightnesses), m_1 and m_2, of two stars are related to their (actual) brightnesses, b_1 and b_2, by the equation $m_1 - m_2 = 2.5 \log(b_2/b_1)$. As a result of this definition, magnitudes may be negative, and *magnitudes decrease as brightnesses increase.* The magnitude of the brightest star, Sirius, is -1.4, and the magnitudes of the faintest stars observable with the naked eye are about 6.0. How much brighter is Sirius than these faintest stars?

94. The power gain of an electronic device such as an amplifier is defined as $n = 10 \log(P_0/P_i)$, where n is measured in decibels, P_0 (in W) is the power output, and P_i (in W) is the power input. If $P_0 = 10.0$ W and $P_i = 0.125$ W, calculate the power gain. (See Example 4 on page 375.)

95. In studying the frictional effects on a flywheel, the revolutions per minute R that it makes as a function of the time t (in min) is given by $R = 4520(0.750)^{2.50t}$. Find t for $R = 1950$ r/min.

96. The efficiency e of a gasoline engine as a function of its compression ratio r is given by $e = 1 - r^{1-\gamma}$, where γ is a constant. Find γ for $e = 0.55$ and $r = 7.5$.

97. The intensity I of light decreases from its value I_0 as it passes a distance x through a medium. Given that $x = k(\ln I_0 - \ln I)$, where k is a constant depending on the medium, find x for $I = 0.850 I_0$ and $k = 5.00$ cm.

98. Under certain conditions, the temperature T and pressure p are related by the following equation, where T_0 is the temperature at pressure p_0: $\dfrac{T}{T_0} = \left(\dfrac{p}{p_0}\right)^{\frac{k-1}{k}}$. Solve for k.

99. The temperature T of an object with an initial temperature T_1 in water at temperature T_0 as a function of the time t is given by $T = T_1 + (T_0 - T_1)e^{-kt}$. Solve for t.

100. Assume the U.S. national debt D increases at an annual rate of 5% and that the debt in 2000 was \$5.7 trillion. This means $D = 5.7(1.05^t)$, where t is the number of years after 2000. In what year will the debt have doubled?

101. Pure water is running into a brine solution, and the same amount of solution is running out. The number n of kilograms of salt in the solution after t min is found by solving the equation $\ln n = -0.04t + \ln 20$. Solve for n as a function of t.

102. For the circuit in Fig. 13.28, the current i (in mA) is given by $i = 1.6 \, e^{-100t}$. Plot the graph of i as a function of t for the first 0.05 s on semilog paper.

Fig. 13.28

103. For a particular solar-energy system, the collector area A required to supply a fraction F of the total energy is given by $A = 480 \, F^{2.2}$. Plot A (in m^2) as a function of F, from $F = 0.1$ to $F = 0.9$, on semilog paper.

104. The current I (in μA) and resistance R (in Ω) were measured as follows in a certain microcomputer circuit:

R (Ω)	100	200	500	1000	2000	5000	10,000
I (μA)	81	41	16	8.2	4.0	1.6	0.8

Plot I as a function of R on log-log paper.

Writing Exercise

105. While checking logarithmic curves on a calculator, a machine-design student noted that a certain robotic arm was shaped like part of the graph of $y = \ln(2x^{2/3})$. As a check, the student rewrote the equation as $y = (2 \ln x)/3 + \ln 4 - \ln(\ln e^2)$. Write a paragraph explaining (a) whether the second equation is equivalent to the first, and (b) if the graphs of the two equations are identical.

CHAPTER 13 PRACTICE TEST

In Problems 1–4, determine the value of x.

1. $\log_9 x = -\frac{1}{2}$

2. $\log_3 x - \log_3 2 = 2$

3. $\log_x 64 = 3$

4. $3^{3x+1} = 8$

5. Graph the function $y = 2 \log_4 x$.

6. Graph the function $y = 2(3^x)$ on semilog paper.

7. Express $\log_5\left(\dfrac{4a^3}{7}\right)$ as a combination of a sum, difference, and multiple of logarithms, including $\log_5 2$.

8. Solve for y in terms of x: $3 \log_7 x - \log_7 y = 2$.

9. An equation used for a certain electric circuit is $\ln i - \ln I = -t/RC$. Solve for i.

10. Evaluate: $\dfrac{2 \ln 0.9523}{\log 6066}$.

11. Evaluate: $\log_5 732$.

12. If A_0 dollars are invested at 8%, compounded continuously for t years, the value A of the investment is given by $A = A_0 e^{0.08t}$. Determine how long it takes for the investment to double in value.

Additional Types of Equations and Systems of Equations

14

In this chapter, we discuss graphical and algebraic solutions of systems of equations of types different from those of earlier chapters. We also consider solutions of two special types of equations.

One of the methods involves the use of graphs for the solutions, similar to that we used in Chapter 3. The intersection of curves was very much part of the basic method of solution of equations used in the mid-1600s by Rene Descartes, who developed the coordinate system. Since that time, graphical solutions of equations have been very common and very useful in science and technology.

In the study of optics in the 1800s, it was found that light traveled much faster in free space than in other mediums, such as glass and water. Scientists then defined n, the *index of refraction,* to be the ratio of the speed of

light in free space to the speed of light in a particular substance. In 1836, the French mathematician Cauchy developed the equation $n = A + B\lambda^{-2} + C\lambda^{-4}$ that related the index of refraction with the wavelength of light λ in a medium. Once the constants A, B, and C were found for a particular medium (by solving three simultaneous equations as in Chapter 5), this equation can be solved for λ for a particular value of n using a method known at the time that is used in this chapter (see Exercise 36 on page 402). Again, we see that an earlier mathematical method was useful in dealing with a new scientific discovery.

Applications of the types of equations and systems of equations of this chapter are found in many fields of science and technology. These include physics, electricity, business, and structural design.

A three-dimensional holographic image is formed without the use of a lens. In Section 14.4, the dimensions of a holograph are determined.

Graphical Solution of Systems of Equations

Conic Sections •
Parabola, Circle, Ellipse, Hyperbola •
Solving Systems of Equations

In this section, we first discuss the graphs of the *circle, parabola, ellipse,* and *hyperbola,* which are known as the **conic sections.** We saw the parabola earlier when discussing quadratic functions in Chapter 7. Then we will find graphical solutions of systems of equations involving these and other nonlinear equations.

EXAMPLE 1 Calculator graph – parabola

Graph the equation $y = 3x^2 - 6x$.

We graphed equations of this form in Section 7.4. Since the general quadratic function is $y = ax^2 + bx + c$, for $y = 3x^2 - 6x$, we have $a = 3$, $b = -6$, and $c = 0$. Therefore, $-b/(2a) = 1$, which means the x-coordinate of the vertex is $-(-6)/6 = 1$. Since $y = -3$ for $x = 1$, the vertex is $(1, -3)$. It is a minimum point since $a > 0$.

Knowing the vertex and the fact that the graph goes through the origin ($c = 0$), we choose appropriate *window* settings and have the display shown in Fig. 14.1.

As we showed in Section 7.4, the curve is a **parabola,** and a parabola always results if the equation is of the form of the quadratic function $y = ax^2 + bx + c$. ∎

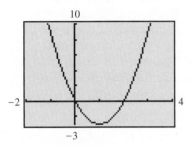

Fig. 14.1

EXAMPLE 2 Plotted graph – circle

Plot the graph of the equation $x^2 + y^2 = 25$.

We first solve this equation for y, and we obtain $y = \sqrt{25 - x^2}$, or $y = -\sqrt{25 - x^2}$, which we write as $y = \pm\sqrt{25 - x^2}$. We now assume values for x and find the corresponding values for y.

x	0	± 1	± 2	± 3	± 4	± 5
y	± 5	± 4.9	± 4.6	± 4	± 3	0

If $x > 5$, values of y are imaginary. We cannot plot these because x and y must both be real. When we show $y = \pm 3$ for $x = \pm 4$, this is a short way of representing four points. These points are $(4, 3)$, $(4, -3)$, $(-4, 3)$, and $(-4, -3)$.

In Fig. 14.2, *the resulting curve is a* **circle.** A circle with its center at the origin results from an equation of the form $x^2 + y^2 = r^2$, where r is the radius. ∎

Fig. 14.2

■ See the *vertical-line test* on page 94.

From the graph of the circle in Fig. 14.2, we see that *the equation of a circle does not represent a function.* There are *two* values of y for most of the values of x in the domain. We must take this into account when displaying the graph of such an equation on a graphing calculator. This is illustrated in the next example.

EXAMPLE 3 Calculator graph – ellipse

Display the graph of the equation $2x^2 + 5y^2 = 10$ on a graphing calculator.

First, solving for y, we get $y = \pm\sqrt{\dfrac{10 - 2x^2}{5}}$. To display the graph of this equation on a calculator, *we must enter both functions,* one as $y_1 = \sqrt{(10 - 2x^2)/5}$, and the other as $y_2 = -\sqrt{(10 - 2x^2)/5}$.

Trying some *window* settings (or noting that the domain is from $-\sqrt{5}$ to $\sqrt{5}$ and the range is from $-\sqrt{2}$ to $\sqrt{2}$), we get the graphing calculator display that is shown in Fig. 14.3.

The curve is an **ellipse.** An ellipse will be the resulting curve if the equation is of the form $ax^2 + by^2 = c$, where the constants a, b, and c have the same sign, and for which $a \neq b$. ∎

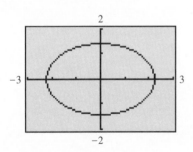

Fig. 14.3

■ A more complete discussion of the conic sections is found in Chapter 21.

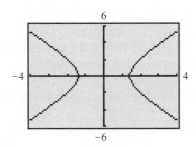

Fig. 14.4

When these types of curves are displayed on the graphing calculator, it is possible that there are small gaps in the curve. The appearance of such gaps depends on the pixel width and the functional values used by the calculator.

EXAMPLE 4 Calculator graph – hyperbola

Display the graph of $2x^2 - y^2 = 4$ on a graphing calculator.

Solving for y, we get

$$y = \pm\sqrt{2x^2 - 4}$$

As in Example 3, we enter two functions in the calculator, one with the plus sign and the other with the minus sign, and we have the display shown in Fig. 14.4. The *window* settings are chosen by noting that the values $-\sqrt{2} < x < \sqrt{2}$ are not in the domain of either $y_1 = \sqrt{2x^2 - 4}$ or $y_2 = -\sqrt{2x^2 - 4}$. These values of x would lead to imaginary values of y.

The curve is a **hyperbola,** which results when we have an equation of the form $ax^2 + by^2 = c$, if a and b have *different* signs. ■

SOLVING SYSTEMS OF EQUATIONS

As in solving systems of linear equations, we solve any system by finding the values of x and y that satisfy both equations at the same time. To solve a system graphically, we graph the equations and find the coordinates of all points of intersection. If the curves do not intersect, the system has no real solutions.

EXAMPLE 5 Calculator – solving a system

For proper ventilation, the vent for a hot-air heating system is to have a rectangular cross-sectional area of 4.5 ft² and is to be made from sheet metal 10.0 ft wide. Find the dimensions of this cross-sectional area of the vent.

In Fig. 14.5, we have let l = the length and w = the width of the area. Since the area is 4.5 ft², we have $lw = 4.5$. Also, since the sheet metal is 10.0 ft wide, this is the perimeter of the area. This gives us $2l + 2w = 10.0$, or $l + w = 5.0$. This means that the system of equations to be solved is

$$lw = 4.5$$
$$l + w = 5.0$$

Solving each equation for l, we have

$$l = 4.5/w \quad \text{and} \quad l = 5.0 - w$$

We now display the graphs of these two equations on a graphing calculator, using x for w and y for l. Since negative values of l and w have no meaning to the solution and since the straight line $l = 5.0 - w$ has intercepts of $(5.0, 0)$ and $(0, 5.0)$, we choose the *window* settings as shown in Fig. 14.6.

Using the *intersect* feature (or the *trace* and *zoom* features) with the graph in Fig. 14.6, we find that the solutions are approximately $(1.2, 3.8)$ and $(3.8, 1.2)$. Using the length as the longer dimension, we have the solution of

$$l = 3.8 \text{ ft} \quad \text{and} \quad w = 1.2 \text{ ft}$$

We see that this checks with the statement of the problem. ■

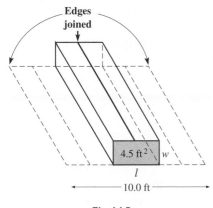

Edges joined

4.5 ft² w

l

10.0 ft

Fig. 14.5

■ See the chapter introduction.

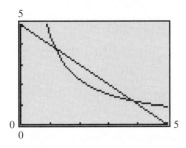

Fig. 14.6

In Example 5, we graphed the equation $xy = 4.5$ (having used x for w and y for l). The graph of this equation is also a *hyperbola,* another form of which is $xy = c$. We now show the solutions of two more systems of equations.

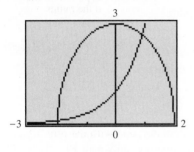

Fig. 14.7

EXAMPLE 6 Calculator – solving a system

Graphically solve the system of equations:

$$9x^2 + 4y^2 = 36$$
$$y = 3^x$$

Solving the first equation for y, we have $y = \frac{1}{2}\sqrt{36 - 9x^2}$. Since this is an ellipse, as shown in Example 3, its domain extends from $x = -2$ to $x = 2$ ($9x^2$ cannot be greater than 36). The exponential curve cannot be negative and increases rapidly, as discussed in Chapter 13. This means we need only graph the upper part of the ellipse (using the $+$ sign). Therefore, we choose the *window* settings as shown in the calculator display in Fig. 14.7.

Using the *intersection* feature, we find the approximate solutions of $x = -2.00$, $y = 0.11$ and $x = 0.90$, $y = 2.68$. ∎

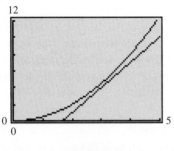

Fig. 14.8

EXAMPLE 7 System of equations – no solution

Graphically solve the system of equations:

$$x^2 = 2y$$
$$3x - y = 5$$

Solving the first equation for y, we get $y = \frac{1}{2}x^2$. This parabola has its vertex at the origin, and it opens upward since $a > 0$. This means there cannot be a point of intersection for $y < 0$.

The straight line has intercepts of $(0, -5)$ and $(5/3, 0)$. This means that any possible point of intersection must be in the first quadrant. Therefore, we have the *window* settings as shown in the calculator display in Fig. 14.8.

From the figure, we see that the two curves do not intersect. This means that there are *no real solutions* to this system of equations. ∎

EXERCISES 14.1

In Exercises 1–4, make the given changes in the indicated examples of this section, and then perform the indicated operations.

1. In Example 1, change the $-$ sign before $6x$ to $+$.

2. In Example 2, change the $+$ sign before y^2 to $-$.

3. In Example 6, change the coefficient of x^2 to 25.

4. In Example 7, change the coefficient of y in the first equation to 3.

In Exercises 5–30, solve the given systems of equations graphically by using a graphing calculator. Find all values to at least the nearest 0.1.

5. $y = 2x$
$x^2 + y^2 = 16$

6. $3x - y = 4$
$y = 6 - 2x^2$

7. $x^2 + 2y^2 = 8$
$x - 2y = 4$

8. $y = 3x - 6$
$xy = 6$

9. $y = x^2 - 2$
$4y = 12x - 17$

10. $4x^2 + 25y^2 = 21$
$10y = 31 - 9x$

11. $8y = 11x^2$
$xy = 30$

12. $y = -2x^2$
$y = x^2 - 6$

13. $y = -x^2 + 4$
$x^2 + y^2 = 9$

14. $y = x^3$
$x^2 + 2y^2 = 16$

15. $x^2 - 4y^2 = 16$
$x^2 + y^2 = 1$

16. $y = 2x^2 - 4x$
$x^2 y = -4$

17. $2x^2 + 3y^2 = 19$
$x^2 + y^2 = 9$

18. $x^2 - y^2 = 4$
$2x^2 + y^2 = 16$

19. $x^2 + y^2 = 7$
$y(x + 2) = 3$

20. $x^2 + y^2 = 0.25$
$x^2 - y^2 = 0.07$

21. $y = x^2$
$y = \sin x$

22. $y = 3 + 2x - x^2$
$y = 2\cos 2x$

23. $y = e^{-x}$
$x + y = 2$

24. $y = 2^{x+1}$
$x^2 + y^2 = 4$

25. $x^2 - y^2 = 7$
$y = 4\log_2 x$

26. $x^2 + 4y^2 = 16$
$y = 2\ln x$

27. $y = \ln(x - 1)$
$y = \sin \frac{1}{2} x$

28. $y = \cos x$
$y = \log_3 x$

29. $10^{x+y} = 150$
$y = x^2$

30. $e^{x^2+y^2} = 20$
$xy = 4$

In Exercise 31, draw the appropriate figures. In Exercises 32–38, set up systems of equations and solve them graphically.

31. By drawing rough sketches, show that a parabola and an ellipse can have 0, 1, 2, 3, or 4 possible points of intersection.

32. A rectangular security area is to be enclosed by fencing and divided in two equal parts of 1600 m² each by a fence parallel to the shorter sides. Find the dimensions of the security area if the total amount of fencing is 280 m.

33. A helicopter is located 5.2 mi north of east of a radio tower such that it is three times as far north as it is east from the tower. Find the northern and eastern components of the displacement from the tower.

34. A 4.60-m insulating strip is placed completely around a rectangular solar panel with an area of 1.20 m². What are the dimensions of the panel?

35. The power developed in an electric resistor is i^2R, where i is the current. If a first current passes through a 2.0-Ω resistor and a second current passes through a 3.0-Ω resistor, the total power produced is 12 W. If the resistors are reversed, the total power produced is 16 W. Find the currents (in A) if $i > 0$.

W **36.** A circular hot tub is located on the square deck of a home. The side of the deck is 24 ft more than the radius of the hot tub, and there are 780 ft² of deck around the tub. Find the radius of the hot tub and the length of the side of the deck. Explain your answer.

37. Assume earth is a sphere, with $x^2 + y^2 = 41$ as the equation of a circumference (distance in thousands of km). If a meteorite approaching earth has a path described as $y^2 = 20x + 140$, will the meteorite strike earth? If so, where?

38. Two people meet at the intersection of two perpendicular roads. Each then leaves along a different road with one walking 1.0 km/h faster than the other. If they are 7.0 km apart (on a direct line) after 1.0 h, how fast is each walking?

14.2 Algebraic Solution of Systems of Equations

**Solution by Substitution •
Solution by Addition or Subtraction**

Often, the graphical method is the easiest way to solve a system of equations. With a graphing calculator, it is possible to find the result with good accuracy. However, the graphical method does not usually give the *exact* answer. Using algebraic methods to find exact solutions for some systems of equations is either not possible or quite involved. There are systems, however, for which there are relatively simple algebraic solutions. In this section, we consider two useful methods, both of which we discussed before when we were studying systems of linear equations.

SOLUTION BY SUBSTITUTION

The first method is *substitution*. If we can solve one of the equations for one of its variables, we can substitute this solution into the other equation. We then have only one unknown in the resulting equation, and we can then solve this equation by methods discussed in earlier chapters.

EXAMPLE 1 Solution by substitution

By substitution, solve the system of equations

$$2x - y = 4$$
$$x^2 - y^2 = 4$$

We solve the first equation for y, obtaining $y = 2x - 4$. We now substitute $2x - 4$ for y in the second equation, getting

$$x^2 - (2x - 4)^2 = 4 \qquad \text{in second equation, } y \text{ replaced by } 2x - 4$$

When simplified, this gives a quadratic equation.

$$x^2 - (4x^2 - 16x + 16) = 4$$
$$-3x^2 + 16x - 20 = 0$$
$$x = \frac{-16 \pm \sqrt{256 - 4(-3)(-20)}}{-6} = \frac{-16 \pm \sqrt{16}}{-6} = \frac{-16 \pm 4}{-6} = \frac{10}{3}, 2$$

We now find the corresponding values of y by substituting into $y = 2x - 4$. Thus, we have the solutions $x = \frac{10}{3}$, $y = \frac{8}{3}$, and $x = 2$, $y = 0$. As a check, we find that these values also satisfy the equation $x^2 - y^2 = 4$. Compare these solutions with those that would be obtained from Fig. 14.9. ∎

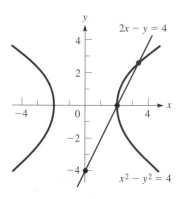

Fig. 14.9

EXAMPLE 2 Solution by substitution

By substitution, solve the system of equations

$$xy = -2$$
$$2x + y = 2$$

From the first equation, we have $y = -2/x$. Substituting this into the second equation, we have

in second equation, y replaced by $-\dfrac{2}{x}$

$$2x + \left(-\frac{2}{x}\right) = 2$$
$$2x^2 - 2 = 2x$$
$$x^2 - x - 1 = 0$$
$$x = \frac{1 \pm \sqrt{1 + 4}}{2} = \frac{1 \pm \sqrt{5}}{2}$$

By substituting these values for x into either of the original equations, we find the corresponding values of y, and we have the solutions

$$x = \frac{1 + \sqrt{5}}{2}, y = 1 - \sqrt{5} \quad \text{and} \quad x = \frac{1 - \sqrt{5}}{2}, y = 1 + \sqrt{5}$$

These can be checked by substituting in the original equations. In decimal form, they are

$$x \approx 1.618, y \approx -1.236 \quad \text{and} \quad x \approx -0.618, y \approx 3.236$$

The graphical solutions are shown in Fig. 14.10.

The solutions can also be found by first solving the second equation for y, or either equation for x, and then substituting in the other equation. ∎

SOLUTION BY ADDITION OR SUBTRACTION

The other algebraic method is that of elimination by *addition or subtraction*. This method is most useful if both equations have only squared terms and constants.

EXAMPLE 3 Solution by addition

By addition or subtraction, solve the system of equations

$$2x^2 + y^2 = 9$$
$$x^2 - y^2 = 3$$

We note that if we add the corresponding sides of each equation, y^2 is eliminated. This leads to the solution.

$$
\begin{aligned}
2x^2 + y^2 &= 9 \\
\underline{x^2 - y^2 = 3} & \\
3x^2 &= 12 \qquad \text{add} \\
x^2 &= 4 \\
x &= \pm 2
\end{aligned}
$$

For $x = 2$, we have two corresponding y-values, $y = \pm 1$. Also, for $x = -2$, we have two corresponding y-values, $y = \pm 1$. Thus, we have four solutions:

$$x = 2, y = 1 \qquad x = 2, y = -1 \qquad x = -2, y = 1 \qquad x = -2, y = -1$$

Each solution checks in the original equations. The graphical solutions are shown in Fig. 14.11. ∎

Practice Exercise

1. Solve by substitution: $2x^2 + y^2 = 3$
$\qquad\qquad\qquad\quad x - y = 2$

Fig. 14.10

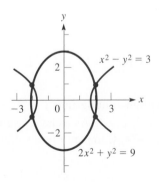

Fig. 14.11

Practice Exercise

2. Solve by addition or subtraction:
$$x^2 + y^2 = 6$$
$$2x^2 - y^2 = 6$$

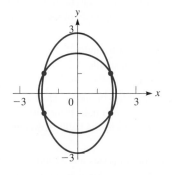

Fig. 14.12

EXAMPLE 4 Solution by subtraction

By addition or subtraction, solve the system of equations

$$5x^2 + 2y^2 = 17$$
$$x^2 + y^2 = 4$$

Multiplying the second equation by 2, and subtracting the resulting equations, we have

$$
\begin{aligned}
5x^2 + 2y^2 &= 17 \\
\underline{2x^2 + 2y^2} &= 8 \qquad \text{each term of second equation multiplied by 2} \\
3x^2 \quad\;\; &= 9 \qquad \text{subtract} \\
x^2 = 3, \quad x &= \pm\sqrt{3}
\end{aligned}
$$

The corresponding values of y for each value of x are ± 1. Again, we have four solutions:

$$x = \sqrt{3}, y = 1 \qquad x = \sqrt{3}, y = -1$$
$$x = -\sqrt{3}, y = 1 \qquad x = -\sqrt{3}, y = -1$$

Each solution checks when substituted in the original equations. The graphical solutions are shown in Fig. 14.12. ∎

EXAMPLE 5 Solution by substitution – application

A certain number of machine parts cost \$1000. If they cost \$5 less per part, ten additional parts could be purchased for the same amount of money. What is the cost of each part?

Let c = the cost per part, and n = the number of parts. From the first statement, we see that $cn = 1000$. From the second statement, $(c - 5)(n + 10) = 1000$. Rewriting these equations, we have

$$n = 1000/c$$
$$cn + 10c - 5n - 50 = 1000$$
$$c\left(\frac{1000}{c}\right) + 10c - 5\left(\frac{1000}{c}\right) - 50 = 1000 \qquad \text{substituting first equation in second equation}$$
$$1000 + 10c - \frac{5000}{c} - 50 = 1000$$
$$10c - \frac{5000}{c} - 50 = 0$$
$$c^2 - 5c - 500 = 0$$
$$(c + 20)(c - 25) = 0$$
$$c = -20, 25$$

Since a negative answer has no significance in this particular situation, we see that the solution is $c = \$25$ per part. Checking with the original statement of the problem, we see that this is correct. ∎

EXERCISES 14.2

In Exercises 1–4, make the given changes in the indicated examples of this section and then solve the resulting systems of equations.

1. In Example 1, change the sign before y in the first equation from $-$ to $+$ and then solve the system.

2. In Example 2, change the right side of the second equation from 2 to 3 and then solve the system.

3. In Example 3, change the coefficient of x^2 in the first equation from 2 to 1 and then solve the system.

4. In Example 4, change the left side of the first equation from $5x^2 + 2y^2$ to $2x^2 + 5y^2$ and then solve the system.

In Exercises 5–28, solve the given systems of equations algebraically.

5. $y = x + 1$
$y = x^2 + 1$

6. $y = 2x - 1$
$y = 2x^2 + 2x - 3$

7. $x + 2y = 3$
$x^2 + y^2 = 26$

8. $p^2 + 4h^2 = 4$
$h = p + 1$

9. $x + y = 1$
$x^2 - y^2 = 1$

10. $x + y = 2$
$2x^2 - y^2 = 1$

11. $2x - y = 2$
$2x^2 + 3y^2 = 4$

12. $6y - x = 6$
$x^2 + 3y^2 = 36$

13. $wh = 1$
$w + h = 2$

14. $xy = 100$
$x + y = 20$

15. $xy = 4$
$4x - 3y = 2$

16. $xy = -4$
$2x + y = -2$

17. $y = x^2$
$y = 3x^2 - 50$

18. $M = L^2 - 1$
$2L^2 - M^2 = 2$

19. $x^2 - y = -1$
$x^2 + y^2 = 5$

20. $s^2 + t^2 = 8$
$3s - t = 2$

21. $D^2 - 1 = R$
$D^2 - 2R^2 = 1$

22. $2y^2 - 4x = 7$
$y^2 + 2x^2 = 3$

23. $x^2 + y^2 = 25$
$x^2 - 2y^2 = 7$

24. $3x^2 - y^2 = 4$
$x^2 + 4y^2 = 10$

25. $x^2 + 3y^2 = 37$
$2x^2 - 9y^2 = 14$

26. $5x^2 - 4y^2 = 15$
$3y^2 + 4x^2 = 12$

27. $x^2 + y^2 + 4x = 1$
$x^2 + y^2 - 2y = 9$

28. $x^2 + y^2 - 4x - 2y + 4 = 0$
$x^2 + y^2 - 2x - 4y + 4 = 0$

(Hint for Exercises 27 and 28: First, subtract one equation from the other to get an equation relating x and y. Then substitute this equation in either given equation.)

In Exercises 29–44, solve the indicated systems of equations algebraically. In Exercises 33–44, it is necessary to set up the systems of equations properly.

29. Solve for x and y: $x^2 - y^2 = a^2 - b^2$; $x - y = a - b$.

30. For what value of b are the two solutions of the system $x^2 - 2y = 5$; $y = 3x + b$ equal to each other? For this value of b, what is true of the graphs of the two functions?

31. A rocket is fired from behind a ship and follows the path given by $h = 3x - 0.05x^2$, where h is its altitude (in mi) and x is the horizontal distance traveled (in mi). A missile fired from the ship follows the path given by $h = 0.8x - 15$. For $h > 0$ and $x > 0$, find where the paths of the rocket and missile cross.

32. A 2-kg block collides with an 8-kg block. Using the physical laws of conservation of energy and conservation of momentum, along with given conditions, the following equations involving the velocities are established:

$v_1^2 + 4v_2^2 = 41$
$2v_1 + 8v_2 = 12$

Find these velocities (in m/s) if $v_2 > 0$.

33. One face of a washer has an area of 37.7 cm². The inner radius is 2.00 cm less than the outer radius. What are the radii?

34. A right triangular sail has a perimeter of 28.0 ft and a hypotenuse of 12.5 ft. Find the lengths of the sides of the sail.

35. A rectangular computer chip has a surface area of 2.1 cm² and a perimeter of 5.8 cm. Find the length and the width of the chip.

36. The impedance Z in an alternating-current circuit is 2.00 Ω. If the resistance R is numerically equal to the square of the reactance X, find R and X. See Section 12.7.

37. A roof truss is in the shape of a right triangle. If there are 4.60 m of lumber in the truss and the longest side is 2.20 m long, what are the lengths of the other two sides of the truss?

38. In a certain roller mechanism, the radius of one steel ball is 2.00 cm greater than the radius of a second steel ball. If the difference in their masses is 7100 g, find the radii of the balls. The density of steel is 7.70 g/cm³.

39. A set of equal electrical resistors in series has a total resistance (the sum of the resistances) of 78.0 Ω. Another set of two fewer equal resistors in series also has a total resistance of 78.0 Ω. If each resistor in the second set is 1.3 Ω greater than each of the first set, how many are in each set?

40. Security fencing encloses a rectangular storage area of 1600 m² that is divided into two sections by additional fencing parallel to the shorter sides. Find the dimensions of the storage area if 220 m of fencing are used.

41. An open liner for a carton is to be made from a rectangular sheet of cardboard of area 216 in.² by cutting equal 2.00-in. squares from each corner and bending up the sides. If the volume within the liner is 224 in.³, what are dimensions of the cardboard sheet?

42. Two guy wires, one 140 ft long and the other 120 ft long, are attached at the same point of a TV tower with the longer one secured in the (level) ground 30 ft farther from the base of the tower than the shorter one. How high up on the tower are they attached?

43. A jet travels at 610 mi/h relative to the air. It takes the jet 1.6 h longer to travel the 3660 mi from London to Washington, D.C., against the wind than it takes from Washington to London with the wind. Find the velocity of the wind.

44. In a marketing survey, a company found that the total gross income for selling t tables at a price of p dollars each was $35,000. It then increased the price of each table by $100 and found that the total income was only $27,000 because 40 fewer tables were sold. Find p and t.

Answers to Practice Exercises

1. $x = 1/3, y = -5/3$; $x = 1, y = -1$

2. $x = 2, y = \sqrt{2}$; $x = 2, y = -\sqrt{2}$;
$x = -2, y = \sqrt{2}$; $x = -2, y = -\sqrt{2}$

14.3 Equations in Quadratic Form

Substituting to Fit Quadratic Form •
Solving Equations in Quadratic Form •
Extraneous Roots

NOTE ▶

Often, we encounter equations that can be solved by methods applicable to quadratic equations, even though these equations are not actually quadratic. They do have the property, however, that *with a proper substitution **they may be written in the form of a quadratic equation.*** All that is necessary is that the equation have terms including some variable quantity, its square, and perhaps a constant term. The following example illustrates these types of equations.

EXAMPLE 1 Identifying quadratic form

(a) The equation $x - 2\sqrt{x} - 5 = 0$ is an equation in quadratic form, because if we let $y = \sqrt{x}$, we have $x = (\sqrt{x})^2 = y^2$, and the resulting equation is $y^2 - 2y - 5 = 0$.

(b) $t^{-4} - 5t^{-2} + 3 = 0$

$$\overset{\overset{\displaystyle (t^{-2})^2}{\big\downarrow}}{}$$

By letting $y = t^{-2}$, we have $y^2 - 5y + 3 = 0$.

(c) $t^3 - 3t^{3/2} - 7 = 0$

$$\overset{\overset{\displaystyle (t^{3/2})^2}{\big\downarrow}}{}$$

By letting $y = t^{3/2}$, we have $y^2 - 3y - 7 = 0$.

(d) $(x + 1)^4 - (x + 1)^2 - 1 = 0$

$$\overset{\overset{\displaystyle [(x+1)^2]^2}{\big\downarrow}}{}$$

By letting $y = (x + 1)^2$, we have $y^2 - y - 1 = 0$.

(e) $x^{10} - 2x^5 + 1 = 0$

$$\overset{\overset{\displaystyle (x^5)^2}{\big\downarrow}}{}$$

By letting $y = x^5$, we have $y^2 - 2y + 1 = 0$.   ∎

The following examples illustrate the method of solving equations in quadratic form.

EXAMPLE 2 Solving equation in quadratic form

Solve the equation $2x^4 + 7x^2 = 4$.

We first let $y = x^2$ to write the equation in quadratic form. We will then solve the resulting quadratic equation for y. However, solutions for x are required, so we again let $y = x^2$ to solve for x.

$$2y^2 + 7y - 4 = 0 \qquad \text{let } y = x^2$$
$$(2y - 1)(y + 4) = 0 \qquad \text{factor and solve for } y$$
$$y = \tfrac{1}{2} \qquad \text{or} \quad y = -4$$
$$x^2 = \tfrac{1}{2} \qquad \text{or} \quad x^2 = -4 \qquad y = x^2 \ \text{(to solve for } x)$$
$$x = \pm\frac{1}{\sqrt{2}} \qquad \text{or} \quad x = \pm 2j$$

Fig. 14.13

We can let $y_1 = 2x^4 + 7x^2$ and $y_2 = 4$ to solve the system on a graphing calculator. The display is shown in Fig. 14.13, and we see that there are two points of intersection, one for $x = 0.7071$ and the other for $x = -0.7071$. Since $1/\sqrt{2} = 0.7071$, this verifies the real solutions. The imaginary solutions cannot be found graphically. Substitution of each value *in the original equation* shows each value to be a solution.   ∎

Two of the solutions in Example 2 are complex numbers. We were able to find these solutions directly from the definition of the square root of a negative number. In some cases (see Exercise 22 of this section), it is necessary to use the method of Section 12.6 to find such complex-number solutions.

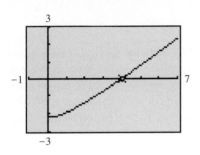

Fig. 14.14

EXAMPLE 3 Solving equation with square root

Solve the equation $x - \sqrt{x} - 2 = 0$.

By letting $y = \sqrt{x}$, we have

$$y^2 - y - 2 = 0$$
$$(y - 2)(y + 1) = 0$$
$$y = 2 \quad \text{or} \quad y = -1$$

Since $y = \sqrt{x}$, we note that y cannot be negative, and this means $y = -1$ cannot lead to a solution. For $y = 2$, we have $x = 4$. Checking, we find that $x = 4$ satisfies the original equation. Therefore, the only solution is $x = 4$.

The graph of $y_1 = x - \sqrt{x} - 2$ is shown in the graphing calculator display in Fig. 14.14. Note that $x = 4$ is the only solution shown. ∎

Extraneous Roots

CAUTION ▶

Example 3 illustrates a very important point. *Whenever an operation involving the unknown is performed on an equation, this operation may introduce roots into a subsequent equation that are not roots of the original equation. Therefore, we must **check all answers in the original equation.*** Only operations involving constants—that is, adding, subtracting, multiplying by, or dividing by constants—are certain not to introduce the **extraneous roots.** We first encountered the concept of an extraneous root in Section 6.8, when we discussed equations involving fractions.

EXAMPLE 4 Solving equation with negative exponents

Solve the equation $x^{-2} + 3x^{-1} + 1 = 0$.

By substituting $y = x^{-1}$, we have $y^2 + 3y + 1 = 0$. To solve this equation, we may use the quadratic formula:

$$y = \frac{-3 \pm \sqrt{9 - 4}}{2} = \frac{-3 \pm \sqrt{5}}{2}$$

Since $x = 1/y$, we have

$$x = \frac{2}{-3 + \sqrt{5}} \quad \text{or} \quad x = \frac{2}{-3 - \sqrt{5}}$$

These answers in decimal form are

$$x \approx -2.618 \quad \text{or} \quad x \approx -0.382$$

These results check when substituted in the original equation. In checking these decimal answers, it is more accurate to use the calculator values, before rounding them off. This can be done by storing the calculator values in memory. ∎

Practice Exercise

1. Solve for x: $x^{-4} - 8x^{-2} + 16 = 0$

EXAMPLE 5 Solving equation with grouped terms

Solve the equation $(x^2 - x)^2 - 8(x^2 - x) + 12 = 0$.

By substituting $y = x^2 - x$, we have

$$y^2 - 8y + 12 = 0$$
$$(y - 2)(y - 6) = 0$$
$$y = 2 \quad \text{or} \quad y = 6$$
$$x^2 - x = 2 \quad \text{or} \quad x^2 - x = 6 \qquad y = x^2 - x$$

Solving each of these equations, we have

$$x^2 - x - 2 = 0 \qquad\qquad x^2 - x - 6 = 0$$
$$(x - 2)(x + 1) = 0 \qquad (x - 3)(x + 2) = 0$$
$$x = 2 \quad \text{or} \quad x = -1 \qquad x = 3 \quad \text{or} \quad x = -2$$

Each value checks when substituted in the original equation. ∎

EXAMPLE 6 Quadratic form – application

Fig. 14.15

A rectangular photograph has an area of 120 cm². The diagonal of the photograph is 17 cm. Find its length and width. See Fig. 14.15.

Since the required quantities are the length and width, let l = the length of the photograph and let w = its width. Since the area is 120 cm², $lw = 120$. Also, using the Pythagorean theorem and the fact that the diagonal is 17 cm, we have the equation $l^2 + w^2 = 17^2 = 289$. Therefore, we are to solve the system of equations

$$lw = 120 \qquad l^2 + w^2 = 289$$

Solving the first equation for l, we have $l = 120/w$. Substituting this into the second equation, we have

$$\left(\frac{120}{w}\right)^2 + w^2 = 289$$

$$\frac{14{,}400}{w^2} + w^2 = 289$$

$$14{,}400 + w^4 = 289w^2$$

Let $x = w^2$.

$$x^2 - 289x + 14{,}400 = 0$$

$$x = \frac{-(-289) \pm \sqrt{(-289)^2 - 4(1)(14{,}400)}}{2(1)}$$

$$x = 225 \quad \text{or} \quad x = 64$$

Therefore, $w^2 = 225$ or $w^2 = 64$.

Solving for w, we get $w = \pm 15$ or $w = \pm 8.0$. Only the positive values are meaningful in this problem, which means if $w = 8.0$ cm, then $l = 15$ cm. Normally, we designate the longer dimension as the length (although by letting $w = 15$ cm, we get $l = 8.0$ cm). Checking with the statement of the problem, we see that these dimensions for the photograph give an area of 120 cm² and a diagonal of 17 cm. ■

■ The first permanent photograph was taken in 1816 by the French inventor Joseph Niepce (1765–1833).

EXERCISES 14.3

In Exercises 1 and 2, make the given changes in the indicated examples of this section and then solve the resulting equations.

1. In Example 2, change the + before the $7x^2$ to − and then solve the equation.

2. In Example 3, change the 2 to 6 and then solve the equation.

In Exercises 3–26, solve the given equations algebraically. In Exercise 10, explain your method.

3. $x^4 - 13x^2 + 36 = 0$

4. $4R^4 + 15R^2 = 4$

5. $x^{-2} - 2x^{-1} - 8 = 0$

6. $10x^{-2} + 3x^{-1} - 1 = 0$

7. $x^{-4} + 2x^{-2} = 24$

8. $x^{-1} - x^{-1/2} = 2$

9. $2x - 7\sqrt{x} + 5 = 0$

(W)10. $4x + 3\sqrt{x} = 1$

11. $3\sqrt[3]{x} - 5\sqrt[6]{x} + 2 = 0$

12. $\sqrt{x} + 3\sqrt[4]{x} = 28$

13. $x^{2/3} - 2x^{1/3} - 15 = 0$

14. $x^3 + 2x^{3/2} - 80 = 0$

15. $2n^{1/2} - 5n^{1/4} = 3$

16. $4x^{4/3} + 9 = 13x^{2/3}$

17. $(x - 1) - \sqrt{x - 1} = 20$

18. $(C + 1)^{-2/3} + 5(C + 1)^{-1/3} - 6 = 0$

19. $(x^2 - 2x)^2 - 11(x^2 - 2x) + 24 = 0$

20. $(x^2 - 1)^2 + (x^2 - 1)^{-2} = 2$

21. $x - 3\sqrt{x - 2} = 6$ (Let $y = \sqrt{x - 2}$.)

22. $x^6 + 7x^3 - 8 = 0$

23. $\dfrac{1}{s^2 + 1} + \dfrac{2}{s^2 + 3} = 1$

24. $\left(x + \frac{2}{x}\right)^2 - 6x - \frac{12}{x} = -9$

25. $e^{2x} - e^x = 0$

26. $10^{2x} - 2(10^x) = 0$

In Exercises 27–32, solve the given equations algebraically and check the solutions with a graphing calculator.

27. $x^4 - 20x^2 + 64 = 0$

28. $x^{-2} - x^{-1} - 42 = 0$

29. $x + 2 = 3\sqrt{x}$

30. $x^{2/3} - 4x^{1/3} = 12$

31. $(\log x)^2 - 3 \log x + 2 = 0$

32. $2^x + 32(2^{-x}) = 12$

In Exercises 33–40, solve the given problems algebraically.

33. Solve for x: $\log(x^4 + 4) - \log(5x^2) = 0$.

34. If $f(x) = x^2 + 1$, find x if $f(x^2 - 3) = 5$.

35. The equivalent resistance R_T of two resistors R_1 and R_2 in parallel is given by $R_T^{-1} = R_1^{-1} + R_2^{-1}$. If $R_T = 1.00 \ \Omega$ and $R_2 = \sqrt{R_1}$, find R_1 and R_2.

36. An equation used in the study of the dispersion of light is $\mu = A + B\lambda^{-2} + C\lambda^{-4}$. Solve for λ. (See the chapter introduction.)

37. In the theory dealing with optical interferometers, the equation $\sqrt{F} = 2\sqrt{p}/(1 - p)$ is used. Solve for p if $F = 16$.

38. A special washer is made from a circular disc 3.50 cm in radius by removing a rectangular area of 12.0 cm² from the center. If each corner of the rectangular area is 0.50 cm from the outer edge of the washer, what are the dimensions of the area that is removed?

39. A rectangular TV screen has an area of 347 in.² and a diagonal of 27.0 in. Find the dimensions of the screen.

40. A roof truss in the shape of a right triangle has a perimeter of 90 ft. If the hypotenuse is 1 ft longer than one of the other sides, what are the sides of the truss?

Answer to Practice Exercise

1. $1/2, 1/2, -1/2, -1/2$

14.4 Equations with Radicals

Solve by Squaring Both Sides •
Solve by Isolating a Radical •
Solve by Using Quadratic Form

Equations with radicals in them are normally solved by squaring both sides of the equation if the radical represents a square root or by a similar operation for the other roots. However, when we do this, we often introduce *extraneous roots*. Thus, it is very important that *all solutions be checked in the original equation.*

EXAMPLE 1 Solve by squaring both sides

Solve the equation $\sqrt{x - 4} = 2$.

By squaring both sides of the equation, we have

$$(\sqrt{x - 4})^2 = 2^2$$
$$x - 4 = 4$$
$$x = 8$$

This solution checks when put into the original equation. ∎

EXAMPLE 2 Solve by squaring both sides

Solve the equation $2\sqrt{3x - 1} = 3x$.

Squaring both sides of the equation gives us

$$(2\sqrt{3x - 1})^2 = (3x)^2 \leftarrow \text{don't forget to square the 2}$$
$$4(3x - 1) = 9x^2$$
$$12x - 4 = 9x^2$$
$$9x^2 - 12x + 4 = 0$$
$$(3x - 2)^2 = 0$$
$$x = \frac{2}{3} \qquad \text{(double root)}$$

Checking this solution in the original equation, we have

$$2\sqrt{3\left(\tfrac{2}{3}\right) - 1} \overset{?}{=} 3\left(\tfrac{2}{3}\right), \qquad 2\sqrt{2 - 1} \overset{?}{=} 2, \qquad 2 = 2$$

Therefore, the solution $x = \frac{2}{3}$ checks.

We can check this solution graphically by letting $y_1 = 2\sqrt{3x - 1}$ and $y_2 = 3x$. The calculator display is shown in Fig. 14.16. The *intersection* feature shows that the only x-value that the curves have in common is $x = 0.6667$, which agrees with the solution of $x = 2/3$. This also means the line y_1 is tangent to the curve of y_2. ∎

Practice Exercise

1. Solve for x: $\sqrt{2x + 3} = x$

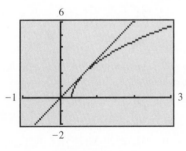

Fig. 14.16

EXAMPLE 3 Solve by cubing both sides

Solve the equation $\sqrt[3]{x - 8} = 2$.

Cubing both sides of the equation, we have

$$x - 8 = 8$$
$$x = 16$$

Checking this solution in the original equation, we get

$$\sqrt[3]{16 - 8} \overset{?}{=} 2, \qquad 2 = 2$$

Therefore, the solution checks.   ■

If a radical and other terms are on one side of the equation, we *first isolate the radical*. That is, we rewrite the equation with the radical on one side and all other terms on the other side.

EXAMPLE 4 Solve by isolating radical

Solve the equation $\sqrt{x - 1} + 3 = x$.

We first isolate the radical by subtracting 3 from each side. This gives us

$$\sqrt{x - 1} = x - 3$$

We now square both sides and proceed with the solution:

CAUTION ◗

$$(\sqrt{x - 1})^2 = (x - 3)^2 \quad \rceil \text{ square the expression on each side,}$$
$$x - 1 = x^2 - 6x + 9 \ \leftarrow \rfloor \text{ not just the terms separately}$$
$$x^2 - 7x + 10 = 0$$
$$(x - 5)(x - 2) = 0$$
$$x = 5 \quad \text{or} \quad x = 2$$

Practice Exercise

2. Solve for x: $\sqrt{x + 4} + 2 = x$

The solution $x = 5$ checks, but the solution $x = 2$ gives $4 = 2$. Thus, the solution is $x = 5$. The value $x = 2$ is an extraneous root.   ■

EXAMPLE 5 Solve by isolating radical

Solve the equation $\sqrt{x + 1} + \sqrt{x - 4} = 5$.

This is most easily solved by first isolating one of the radicals by placing the other radical on the right side of the equation. We then square both sides of the resulting equation.

CAUTION ◗

$$\sqrt{x + 1} = 5 - \sqrt{x - 4} \ \leftarrow \text{ two terms}$$
$$(\sqrt{x + 1})^2 = (5 - \sqrt{x - 4})^2$$
$$x + 1 = 25 - 10\sqrt{x - 4} + (\sqrt{x - 4})^2 \leftarrow \text{ be careful!}$$
$$= 25 - 10\sqrt{x - 4} + x - 4$$

Now, isolating the radical on one side of the equation and squaring again, we have

$$10\sqrt{x - 4} = 20$$
$$\sqrt{x - 4} = 2 \qquad \text{divide by 10}$$
$$x - 4 = 4 \qquad \text{square both sides}$$
$$x = 8$$

```
√(X+1)+√(X-4)…=0
∎X=8
 bound={-1E99,1…
∎left-rt=0
```

Fig. 14.17

This solution checks. The calculator solution of this equation, using the *Solver* feature, is displayed in Fig. 14.17.

We note again that in squaring $5 - \sqrt{x - 4}$, we do not simply square 5 and $\sqrt{x - 4}$. This is similar to $(5a - 2)^2$ in Example 2 of Section 6.1.   ■

EXAMPLE 6 Solve by using quadratic form

Solve the equation $\sqrt{x} - \sqrt[4]{x} = 2$.

We can solve this most easily by handling it as an equation in quadratic form. By letting $y = \sqrt[4]{x}$, we have

$$y^2 - y - 2 = 0$$
$$(y - 2)(y + 1) = 0$$
$$y = 2 \quad \text{or} \quad y = -1$$

Since $y = \sqrt[4]{x}$, we know that a negative value of y does not lead to a solution of the original equation. Therefore, we see that $y = -1$ cannot give us a solution. For the other value, $y = 2$, we have

$$\sqrt[4]{x} = 2, \qquad x = 16$$

This checks, because $\sqrt{16} - \sqrt[4]{16} = 4 - 2 = 2$. Therefore, the only solution of the original equation is $x = 16$. ∎

■ See the chapter introduction.

EXAMPLE 7 Squaring both sides – application

Each cross section of a holographic image is in the shape of a right triangle. The perimeter of the cross section is 60 cm, and its area is 120 cm². Find the length of each of the three sides.

If we let the two legs of the triangle be x and y, as shown in Fig. 14.18, from the formulas for the perimeter p and the area A of a triangle, we have

$$p = x + y + \sqrt{x^2 + y^2} \quad \text{and} \quad A = \tfrac{1}{2}xy$$

where the hypotenuse was found by use of the Pythagorean theorem. Using the information given in the statement of the problem, we arrive at the equations

$$x + y + \sqrt{x^2 + y^2} = 60 \quad \text{and} \quad xy = 240$$

Isolating the radical in the first equation and then squaring both sides, we have

$$\sqrt{x^2 + y^2} = 60 - x - y$$
$$x^2 + y^2 = 3600 - 120x - 120y + x^2 + 2xy + y^2$$
$$0 = 3600 - 120x - 120y + 2xy$$

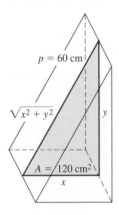

Fig. 14.18

Solving the second of the original equations for y, we have $y = 240/x$. Substituting, we have

$$0 = 3600 - 120x - 120\left(\frac{240}{x}\right) + 2x\left(\frac{240}{x}\right)$$

$$0 = 3600x - 120x^2 - 120(240) + 480x \qquad \text{multiply by } x$$
$$0 = 30x - x^2 - 240 + 4x \qquad \text{divide by 120}$$
$$x^2 - 34x + 240 = 0 \qquad \text{collect terms on left}$$
$$(x - 10)(x - 24) = 0$$
$$x = 10 \text{ cm} \quad \text{or} \quad x = 24 \text{ cm}$$

■ Holography is a method of producing a three-dimensional image without the use of a lens. The theory of holography was developed in the late 1940s by the British engineer Dennis Gabor (1900–1979). After the invention of lasers, the first holographs were produced in the early 1960s.

If $x = 10$ cm, then $y = 24$ cm, or if $x = 24$ cm, then $y = 10$ cm. Therefore, the legs of the holographic cross section are 10 cm and 24 cm, and the hypotenuse is 26 cm. For these sides, $p = 60$ cm and $A = 120$ cm². We see that these values check with the statement of the problem. ∎

EXERCISES 14.4

In Exercises 1–4, make the given changes in the indicated examples of this section, and then solve the resulting equations.

1. In Example 2, change the $3x$ on the right to 3.

2. In Example 3, change the 8 under the radical to 19.

3. In Example 4, change the 3 on the left to 7.

4. In Example 5, change the 4 under the second radical to 14.

In Exercises 5–34, solve the given equations. In Exercises 19 and 22, explain how the extraneous root is introduced.

5. $\sqrt{x-8} = 2$

6. $\sqrt{x+4} = 3$

7. $\sqrt{8-2x} = x$

8. $2\sqrt{2P+5} = P$

9. $\sqrt{3x+2} = 3x$

10. $\sqrt{5x-1} + 3 = x$

11. $2\sqrt{3-x} - x = 5$

12. $x - 3\sqrt{2x+1} = -5$

13. $\sqrt[3]{y-5} = 3$

14. $\sqrt[4]{5-x} = 2$

15. $5\sqrt{s} - 6 = s$

16. $3\sqrt{x} + 2 = 2x$

17. $\sqrt{x^2-9} = 4$

18. $t^2 = 3 - \sqrt{2t^2-3}$

(W) 19. $\sqrt{x+4} + 8 = x$

20. $\sqrt{x+15} + 5 = x$

21. $\sqrt{5+\sqrt{x}} = \sqrt{x} - 1$

(W) 22. $\sqrt{13+\sqrt{x}} = \sqrt{x} + 1$

23. $3\sqrt{1-2t} + 1 = 2t$

24. $1 - 2\sqrt{y+4} = y$

25. $2\sqrt{x+2} - \sqrt{3x+4} = 1$

26. $\sqrt{x-1} + \sqrt{x+2} = 3$

27. $\sqrt{5x+1} - 1 = 3\sqrt{x}$

28. $\sqrt{x-7} = \sqrt{x} - 7$

29. $\sqrt{6x-5} - \sqrt{x+4} = 2$

30. $\sqrt{5x-4} - \sqrt{x} = 2$

31. $\sqrt{x-9} = \dfrac{36}{\sqrt{x-9}} - \sqrt{x}$

32. $\sqrt[4]{x+10} = \sqrt{x-2}$

33. $\sqrt{x-2} = \sqrt[4]{x-2} + 12$

34. $\sqrt{3x + \sqrt{3x+4}} = 4$

In Exercises 35–38, solve the given equations algebraically and check the solutions with a graphing calculator.

35. $\sqrt{3x+4} = x$

36. $\sqrt{x-2} + 3 = x$

37. $\sqrt{2x+1} + 3\sqrt{x} = 9$

38. $\sqrt{2x+1} - \sqrt{x+4} = 1$

In Exercises 39–52, solve the given problems.

39. If $f(x) = \sqrt{x+3}$, find x if $f(x+6) = 5$.

40. Solve $\sqrt{x - \sqrt{2x}} = 2$ algebraically, and check the solution from the graph on a calculator.

41. Solve $\sqrt{x-1} + x = 3$ algebraically. Then compare the solution with that of Example 4. Noting that the algebraic steps after isolating the radical are identical, why is the solution different?

42. The resonant frequency f in an electric circuit with an inductance L and a capacitance C is given by $f = \dfrac{1}{2\pi\sqrt{LC}}$. Solve for L.

43. A formula used in calculating the range R for radio communication is $R = \sqrt{2rh + h^2}$. Solve for h.

44. An equation used in analyzing a certain type of concrete beam is $k = \sqrt{2np + (np)^2} - np$. Solve for p.

45. In the study of spur gears in contact, the equation $kC = \sqrt{R_1^2 - R_2^2} + \sqrt{r_1^2 - r_2^2} - A$ is used. Solve for r_1^2.

46. The focal length f of a lens, in terms of the image distance q and the object distance p, is given by $\frac{1}{f} = \frac{1}{p} + \frac{1}{q}$. Find p and q if $f = 4$ cm and $p = \sqrt{q}$.

47. One section of a roof is in the shape of an isosceles triangle. Its perimeter is three times the height to its base of 4.0 m. What is the area of this section of the roof?

48. A smaller of two cubical boxes is centered on the larger box, and they are taped together with a wide adhesive that just goes around both boxes (see Fig. 14.19). If the edge of the larger box is 1.00 in. greater than that of the smaller box, what are the lengths of the edges of the boxes if 100.0 in. of tape is used?

Fig. 14.19

49. A freighter is 5.2 km farther from a Coast Guard station on a straight coast than from the closest point A on the coast. If the station is 8.3 km from A, how far is it from the freighter?

50. The velocity v of an object that falls through a distance h is given by $v = \sqrt{2gh}$, where g is the acceleration due to gravity. Two objects are dropped from heights that differ by 10.0 m such that the sum of their velocities when they strike the ground is 20.0 m/s. Find the heights from which they are dropped if $g = 9.80$ m/s^2.

51. A point D on Denmark's largest island is 2.4 mi from the nearest point S on the coast of Sweden (assume the coast is straight, which is nearly the case). A person in a motorboat travels straight from D to a point on the beach x km from S and then travels x km farther along the beach away from S. Find x if the person traveled a total of 4.5 mi. See Fig. 14.20.

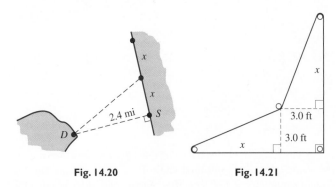

Fig. 14.20   **Fig. 14.21**

52. The length of the roller belt in Fig. 14.21 is 28.0 ft. Find x.

Answers to Practice Exercises

1. 3   **2.** 5

CHAPTER 14   REVIEW EXERCISES

In Exercises 1–10, solve the given systems of equations by use of a graphing calculator.

1. $x + 2y = 6$
$y = 4x^2$

2. $x + y = 3$
$x^2 + y^2 = 25$

3. $3x + 2y = 6$
$x^2 + 4y^2 = 4$

4. $x^2 - 2y = 0$
$y = 3x - 5$

5. $y = x^2 + 1$
$4x^2 + 16y^2 = 29$

6. $\dfrac{x^2}{4} + y^2 = 1$
$x^2 - y^2 = 1$

7. $y = 11 - x^2$
$y = 2x^2 - 1$

8. $x^2y = 63$
$y = 25 - 2x^2$

9. $y = x^2 - 2x$
$y = 1 - e^{-x}$

10. $y = \ln x$
$y = \sin x$

In Exercises 11–20, solve each of the given systems of equations algebraically.

11. $y = 4x^2$
$y = 8x$

12. $x + y = 12$
$xy = 20$

13. $2R = L^2$
$R^2 + L^2 = 3$

14. $y = x^2$
$2x^2 - y^2 = 1$

15. $4u^2 + v = 3$
$2u + 3v = 1$

16. $x^2 + 7y^2 = 56$
$2x^2 - 8y^2 = 90$

17. $4x^2 - 7y^2 = 21$
$x^2 + 2y^2 = 99$

18. $s - t = 6$
$\sqrt{s} - \sqrt{t} = 1$

19. $4x^2 + 3xy = 4$
$x + 3y = 4$

20. $\dfrac{6}{x} + \dfrac{3}{y} = 4$
$\dfrac{36}{x^2} + \dfrac{36}{y^2} = 13$

In Exercises 21–38, solve the given equations.

21. $x^4 - 20x^2 + 64 = 0$

22. $t^6 - 26t^3 - 27 = 0$

23. $x^{3/2} - 9x^{3/4} + 8 = 0$

24. $x^{1/2} + 3x^{1/4} - 28 = 0$

25. $D^{-2} + 4D^{-1} - 21 = 0$

26. $2x - 3\sqrt{x} - 5 = 0$

27. $4(\ln x)^2 - \ln x^2 = 0$

28. $e^x + e^{-x} = 2$

29. $\dfrac{4}{r^2 + 1} + \dfrac{7}{2r^2 + 1} = 2$

30. $(x^2 + 5x)^2 - 5(x^2 + 5x) = 6$

31. $3\sqrt{2Z + 4} = 2Z$

32. $\sqrt[3]{x - 2} = 3$

33. $\sqrt{5x + 9} + 1 = x$

34. $2\sqrt{5x - 3} - 1 = 2x$

35. $\sqrt{x + 1} + \sqrt{x} = 2$

36. $\sqrt{3x^2 - 2} - \sqrt{x^2 + 7} = 1$

37. $\sqrt{n + 4} + 2\sqrt{n + 2} = 3$

38. $\sqrt{3x - 2} - \sqrt{x + 7} = 1$

(W) *In Exercises 39 and 40, find the value of x. (In each, the expression on the right is called a* **continued radical.** *Also, ... means that the pattern continues indefinitely.) (Hint: Square both sides.) Noting the result, complete the solution and explain your method.*

39. $x = \sqrt{2 + \sqrt{2 + \sqrt{2 + \ldots}}}$

40. $x = \sqrt{6 - \sqrt{6 - \sqrt{6 - \ldots}}}$

In Exercises 41–46, solve the given equations algebraically and check the solutions with a graphing calculator.

41. $x^3 - 2x^{3/2} - 48 = 0$

42. $(x + 1)^4 - 54 = 3(x + 1)^2$

43. $2\sqrt{3x + 1} - \sqrt{x - 1} = 6$

44. $3\sqrt{x} + \sqrt{x - 9} = 11$

45. $\sqrt[3]{x^3 - 7} = x - 1$

46. $\sqrt{x^2 + 7} + \sqrt[4]{x^2 + 7} = 6$

In Exercises 47–58, solve the given problems.

47. Solve for x and y: $x^2 - y^2 = 2a + 1$; $x - y = 1$

48. Solve for x: $\log(\sqrt{x} + 38) - \log x = 1$

49. Solve $\sqrt{\sqrt{x} - 1} = 2$ for x. Check using the graph on a calculator.

50. If $f(x) = \sqrt{8 - 2x}$ and $f(x + 1) = 2$, find x.

51. Use a graphing calculator to solve the following system of three equations: $x^2 + y^2 = 13$, $y = x - 1$, $xy = 6$.

(W) **52.** Algebraically solve the following system of three equations: $y = -x^2$, $y = x - 1$, $xy = 1$. Explain the results.

53. In the study of atomic structure, the equation $L = \dfrac{h}{2\pi}\sqrt{l(l + 1)}$ is used. Solve for l ($l > 0$).

54. The frequency ω of a certain *RLC* circuit is given by $\omega = \dfrac{\sqrt{R^2 + 4(L/C)} + R}{2L}$. Solve for C.

55. In the theory dealing with a suspended cable, the equation $y = \sqrt{s^2 - m^2} - m$ is used. Solve for m.

56. The equation $V = e^2cr^{-2} - e^2Zr^{-1}$ is used in spectroscopy. Solve for r.

57. In an experiment, an object is allowed to fall, stops, and then falls for twice the initial time. The total distance the object falls is 45 ft. The equations relating the times t_1 and t_2 (in s) of fall are $16t_1^2 + 16t_2^2 = 45$ and $t_2 = 2t_1$. Find the times of fall.

58. If two objects collide and the kinetic energy remains constant, the collision is termed perfectly elastic. Under these conditions, if an object of mass m_1 and initial velocity u_1 strikes a second object (initially at rest) of mass m_2, such that the velocities after collision are v_1 and v_2, the following equations are found:

$$m_1u_1 = m_1v_1 + m_2v_2$$
$$\tfrac{1}{2}m_1u_1^2 = \tfrac{1}{2}m_1v_1^2 + \tfrac{1}{2}m_2v_2^2$$

Solve these equations for m_2 in terms of u_1, v_1, and m_1.

In Exercises 59–70, set up the appropriate equations and solve them.

59. A wrench is dropped by a worker at a construction site. Four seconds later the worker hears it hit the ground below. How high is the worker above the ground? (The velocity of sound is 1100 ft/s, and the distance the wrench falls as a function of time is $s = 16t^2$.)

60. The rectangular flat screen for a computer monitor has a perimeter of 122.6 cm, and a diagonal of 43.6 cm. Find the dimensions of the screen.

61. The perimeter of a banner in the shape of an isosceles triangle is 72 in., and its area is 240 in.^{2}. Using a calculator, graphically find the lengths of the sides of the banner.

62. A rectangular field is enclosed by fencing and a wall along one side and half of an adjacent side. See Fig. 14.22. If the area of the field is 9000 ft^2 and 240 ft of fencing are used, what are the dimensions of the field?

Wall

$A = 9000$ ft^2

240 ft of fencing

Fencing

Fig. 14.22

63. A circuit on a computer chip is designed to be within the area shown in Fig. 14.23. If this part of the chip has an area of 9.0 mm^2 and a perimeter of 16 mm, find x and y.

Fig. 14.23

64. For the plywood piece shown in Fig. 14.24, find x and y.

Fig. 14.24   7.50 ft

65. The viewing window on a graphing calculator has an area of 1770 mm^2 and a diagonal of 62 mm. What are the length and width of the rectangle?

66. A trough is made from a piece of sheet metal 12.0 in. wide. The cross section of the trough is shown in Fig. 14.25. Find x.

Fig. 14.25   ← 12.0 in. →

67. The circular solar cell and square solar cell shown in Fig. 14.26 have a combined surface area of 40.0 cm^2. Find the radius of the circular cell and the side of the square cell.

Fig. 14.26   ← 7.00 cm →

68. A plastic band 19.0 cm long is bent into the shape of a triangle with sides $\sqrt{x-1}$, $\sqrt{5x-1}$, and 9. Find x.

69. A Coast Guard ship travels from Houston to Mobile, and later it returns to Houston at a speed that is 6.0 mi/h faster. If Houston is 510 mi from Mobile and the total travel time is 35 h, find the speed of the ship in each direction.

70. Two trains are approaching the same crossing on tracks that are at right angles to each other. Each is traveling at 60.0 km/h. If one is 6.00 km from the crossing when the other is 3.00 km from it, how much later will they be 4.00 km apart (on a direct line)?

Writing Exercise

71. Using a computer, an engineer designs a triangular support structure with sides (in m) of x, $\sqrt{x-1}$, and 4.00 m. If the perimeter is to be 9.00 m, the equation to be solved is $x + \sqrt{x-1} + 4.00 = 9.00$. This equation can be solved by either of two methods used in this chapter. Write one or two paragraphs identifying the methods and explaining how they are used to solve the equation.

CHAPTER 14   PRACTICE TEST

1. Solve for x: $x^{1/2} - 2x^{1/4} = 3$.

2. Solve for x: $3\sqrt{x-2} - \sqrt{x+1} = 1$.

3. Solve for x: $x^4 - 17x^2 + 16 = 0$.

4. Solve for x and y algebraically:
$$x^2 - 2y = 5$$
$$2x + 6y = 1$$

5. Solve for x: $\sqrt[3]{2x+5} = 5$.

6. The velocity v of an object falling under the influence of gravity in terms of its initial velocity v_0, the acceleration due to gravity g, and the height h fallen is given by $v = \sqrt{v_0^2 + 2gh}$. Solve for h.

7. Solve for x and y graphically: $x^2 - y^2 = 4$
$$xy = 2$$

8. A rectangular desktop has a perimeter of 14.0 ft and an area of 10.0 ft^2. Find the length and the width of the desktop.

15 Equations of Higher Degree

The errors in the hand-compiled mathematical tables used by the British mathematician Charles Babbage (1792–1871) led him to design a machine for solving polynomial equations. A model of his *difference engine* was well received, but even after many years with government financing, a full-scale model was never successfully produced. Despite this setback, he then designed an *analytical engine,* which he hoped would perform many kinds of calculations. Although never built, it did have the important features of a modern computer: input, storage, control unit, and output. Because of this design, Babbage is often considered the father of the computer. (In 1992, his design was used successfully to make the device.)

We see that polynomials played an important role in the development of computers. Today, among many things, computers are used to solve equations, including polynomial equations, which include linear equations (first degree) and quadratic equations (second degree).

Various methods of solving higher-degree polynomial equations were developed in the 1500s and 1600s. Although known prior to 1650, the *fundamental theorem of algebra* (which we state later in the chapter) was proven by Karl Friedrich Gauss in 1799, and this was a big advance in the solution of these equations. Today, Gauss is considered by many as the greatest mathematician of all time.

In this chapter, we study some of the methods that have been developed for solving higher-degree polynomial equations. The solutions we find include all possible roots, including complex-number roots. Some calculators are programmed to find all such roots, but graphical methods cannot be used to find the complex roots.

Applications of higher-degree equations arise in a number of technical areas. Included in these are the resistance in an electric circuit, finding the dimensions of a container or structure, robotics, and calculating various business production costs.

In Section 15.3, we see how the design of a box to hold a product involves the solution of a higher-degree equation.

15.1 The Remainder and Factor Theorems; Synthetic Division

**Polynomial Function •
Remainder Theorem • Factor Theorem •
Synthetic Division**

In solving higher-degree polynomial equations, the quadratic formula can be used for second-degree equations, and methods have been found for certain third- and fourth-degree equations. It can also be proven that polynomial equations of degree higher than 4 cannot in general be solved algebraically.

In this section, we present two theorems and a simplified method for algebraic division. These will help us in solving polynomial equations later in the chapter.

Any function of the form

$$f(x) = a_0 x^n + a_1 x^{n-1} + \cdots + a_n \tag{15.1}$$

where $a_0 \neq 0$ and n is a positive integer or zero is called a **polynomial function.** We will be considering only polynomials in which the coefficients $a_0, a_1, \ldots, a_n$ are real numbers.

If we divide a polynomial by $x - r$, we find a result of the form

$$f(x) = (x - r)q(x) + R \tag{15.2}$$

where $q(x)$ is the quotient and R is the remainder.

EXAMPLE 1 Division with remainder

Divide $f(x) = 3x^2 + 5x - 8$ by $x - 2$.

The division is shown at the left, and it shows that

$$3x^2 + 5x - 8 = (x - 2)(3x + 11) + 14$$

where, for this function $f(x)$ with $r = 2$, we identify $q(x)$ and R as

$$q(x) = 3x + 11 \qquad R = 14$$

$$\begin{array}{r} 3x + 11 \\ x - 2 \overline{)\,3x^2 + 5x - 8} \\ \underline{3x^2 - 6x} \\ 11x - 8 \\ \underline{11x - 22} \\ 14 \end{array}$$

THE REMAINDER THEOREM

If we now set $x = r$ in Eq. (15.2), we have $f(r) = q(r)(r - r) + R = f(r)(0) + R$, or

$$f(r) = R \tag{15.3}$$

This leads us to the **remainder theorem,** which states that *if a polynomial $f(x)$ is divided by $x - r$ until a constant remainder R is obtained, then $f(r) = R$.* As we can see, this means the remainder equals the value of the function of x at $x = r$.

EXAMPLE 2 Verifying remainder theorem

In Example 1, $f(x) = 3x^2 + 5x - 8$, $R = 14$, and $r = 2$.

We find that

$$f(2) = 3(2^2) + 5(2) - 8$$
$$= 12 + 10 - 8$$
$$= 14$$

Therefore, $f(2) = 14$ verifies that $f(r) = R$ for this example.

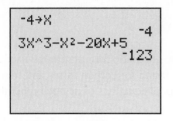

Fig. 15.1

EXAMPLE 3 Using remainder theorem

By using the remainder theorem, determine the remainder when $3x^3 - x^2 - 20x + 5$ is divided by $x + 4$.

In using the remainder theorem, we determine the remainder when the function is divided by $x - r$ by evaluating the function for $x = r$. To have $x + 4$ in the proper form to identify r, we write it as $x - (-4)$. *This means that $r = -4$, and we therefore evaluate the function $f(x) = 3x^3 - x^2 - 20x + 5$ for $x = -4$, or find $f(-4)$:

$$f(-4) = 3(-4)^3 - (-4)^2 - 20(-4) + 5 = -192 - 16 + 80 + 5$$
$$= -123$$

The remainder is -123 when $3x^3 - x^2 - 20x + 5$ is divided by $x + 4$.

NOTE ▶ To evaluate the function with a calculator, *first store in memory the value to be substituted.* Figure 15.1 shows a calculator display for this example. ■

THE FACTOR THEOREM

The remainder theorem leads to another important theorem known as the **factor theorem.** It states that *if $f(r) = R = 0$, then $x - r$ is a factor of $f(x)$.* We see in Eq. (15.2) that if the remainder $R = 0$, then $f(x) = (x - r)q(x)$, and this shows that $x - r$ is a factor of $f(x)$. Therefore, we have the following meanings for $f(r) = 0$.

ZERO, FACTOR, AND ROOT OF THE FUNCTION $f(x)$

*If $f(r) = 0$, then $x = r$ is a **zero** of $f(x)$,*

$x - r$ *is a **factor** of $f(x)$, and*

$x = r$ *is a **root** of the equation $f(x) = 0$*

EXAMPLE 4 Using factor theorem

(a) We determine that $t + 1$ is a factor of $f(t) = t^3 + 2t^2 - 5t - 6$ because $f(-1) = 0$, as we now show:

$$f(-1) = (-1)^3 + 2(-1)^2 - 5(-1) - 6 = -1 + 2 + 5 - 6 = 0$$

(b) However, $t + 2$ is not a factor of $f(t)$ because $f(-2)$ is not zero, as we now show:

$$f(-2) = (-2)^3 + 2(-2)^2 - 5(-2) - 6 = -8 + 8 + 10 - 6 = 4$$ ■

EXAMPLE 5 Checking zero with remainder theorem

Determine if $\frac{2}{3}$ is a zero of $f(x) = 3x^3 + 4x^2 - 16x + 8$.

Evaluating $f\left(\frac{2}{3}\right)$, we have

$$f\left(\frac{2}{3}\right) = 3\left(\frac{2}{3}\right)^3 + 4\left(\frac{2}{3}\right)^2 - 16\left(\frac{2}{3}\right) + 8$$

$$= \frac{8}{9} + \frac{16}{9} - \frac{32}{3} + 8 = \frac{8 + 16 - 96 + 72}{9} = 0$$

Since $f\left(\frac{2}{3}\right) = 0$, $\frac{2}{3}$ is a zero of the function. ■

Practice Exercise

1. Use the remainder theorem to determine whether 1/2 is a zero of $f(x) = 2x^3 + 9x^2 - 11x + 3$.

SYNTHETIC DIVISION

In the sections that follow, we will find that division of a polynomial by the factor $x - r$ is also useful in solving polynomial equations. Therefore, we now develop a simplified form of long division, known as **synthetic division.** It allows us to easily find the coefficients of the quotient and the remainder. If the degree of the equation is high, it is easier to use synthetic division than to calculate $f(r)$. The method for synthetic division is developed in the following example.

EXAMPLE 6 Developing synthetic division

Divide $x^4 + 4x^3 - x^2 - 16x - 14$ by $x - 2$.

We first perform this division in the usual manner:

$$
\begin{array}{r}
x^3 + 6x^2 + 11x + 6 \\
x - 2\, \overline{)\, x^4 + 4x^3 - x^2 - 16x - 14} \\
\underline{x^4 - 2x^3} \\
6x^3 - x^2 \\
\underline{6x^3 - 12x^2} \\
11x^2 - 16x \\
\underline{11x^2 - 22x} \\
6x - 14 \\
\underline{6x - 12} \\
-2
\end{array}
$$

In doing the division, notice that we repeat many terms and that the only important numbers are the coefficients. This means there is no need to write in the powers of x. To the left of the division, we write it without x's and without identical terms.

All numbers below the dividend may be written in two lines. Then all coefficients of the quotient, except the first, appear in the bottom line. Therefore, the line above the dividend is omitted, and we have the form at the left.

Now, write the first coefficient (in this case, 1) in the bottom line. Also, change the -2 to 2, which is the actual value of r. Then in the form at the left, write the 2 on the right. *In this form, the 1, 6, 11, and 6 are the coefficients of the x^3, x^2, x, and constant term of the quotient. The -2 is the remainder.*

Finally, it is easier to use addition rather than subtraction in the process, so we change the signs of the numbers in the middle row. Remember that originally the bottom line was found by subtraction. Therefore, we have the last form on the left.

In the last form at the left, we have 1 (of the bottom row) $\times 2\,(= r) = 2$, the first number of the middle row. In the second column, $4 + 2 = 6$, the second number in the bottom row. Then, $6 \times 2\,(= r) = 12$, the second number of the second row; $-1 + 12 = 11; 11 \times 2 = 22; 22 + (-16) = 6; 6 \times 2 = 12;$ and $12 + (-14) = -2$.

We read the bottom line of the last form, the one we use in *synthetic division,* as

$$1x^3 + 6x^2 + 11x + 6 \text{ with a remainder of } -2$$

The method of synthetic division shown in the last form is outlined below. ■

PROCEDURE FOR SYNTHETIC DIVISION

1. *Write the coefficients of $f(x)$. Be certain that the powers are in descending order and that zeros are inserted for missing powers.*

2. *Carry down the left coefficient, then multiply it by r, and place this product under the second coefficient of the top line.*

3. *Add the two numbers in the second column and place the result below. Multiply this sum by r and place the product under the third coefficient of the top line.*

4. *Continue this process until the bottom row has as many numbers as the top row.*

■ See Appendix C for a graphing calculator program SYNTHDIV. It can be used to find the coefficients and remainder if a polynomial is divided by $x - R$.

Using synthetic division, the last number in the bottom row is the remainder, and the other numbers are the respective coefficients of the quotient. The first term of the quotient is of degree one less than the first term of the dividend.

EXAMPLE 7 Using synthetic division

Divide $x^5 + 2x^4 - 4x^2 + 3x - 4$ by $x + 3$ using synthetic division.

Since the powers of x are in descending order, write down the coefficients of $f(x)$. In doing so, we must be certain to include a zero for the missing x^3 term. Next, note that the divisor is $x + 3$, which means that $r = -3$. The -3 is placed to the right. This gives us a top line of

$$\text{coefficients} \rightarrow 1 \quad 2 \quad 0 \quad -4 \quad 3 \quad -4 \quad \underline{|-3} \leftarrow r$$

Next, we carry the left coefficient, 1, to the bottom line and multiply it by r, -3, placing the product, -3, in the middle line under the second coefficient, 2. We then add the 2 and the -3 and place the result, -1, below. This gives

Now, multiply the -1 by -3 ($= r$) and place the result, 3, in the middle line under the zero. Now, add and continue the process, obtaining the following result:

$$
\begin{array}{rrrrrr|r}
1 & 2 & 0 & -4 & 3 & -4 & -3 \\
  & -3 & 3 & -9 & 39 & -126 & \\
\hline
1 & -1 & 3 & -13 & 42 & -130 &
\end{array}
$$

↑ ↑ ↑ ↑ ↑   ↑— remainder
coefficients and constant
of quotient

NOTE ▶

Since the degree of the dividend is 5, the degree of the quotient is 4. This means that the quotient is $x^4 - x^3 + 3x^2 - 13x + 42$ and the remainder is -130. In turn, this means that for $f(x) = x^5 + 2x^4 - 4x^2 + 3x - 4$, we have $f(-3) = -130$.   ∎

EXAMPLE 8 Using synthetic division

By synthetic division, divide $3x^4 - 5x + 6$ by $x - 4$.

$$
\begin{array}{rrrrr|r}
3 & 0 & 0 & -5 & 6 & 4 \\
  & 12 & 48 & 192 & 748 & \\
\hline
3 & 12 & 48 & 187 & 754 &
\end{array}
$$

The quotient is $3x^3 + 12x^2 + 48x + 187$, and the remainder is 754.   ∎

EXAMPLE 9 Checking factor with synthetic division

By synthetic division, determine whether or not $t + 4$ is a factor of $t^4 + 2t^3 - 15t^2 - 32t - 16$.

$$
\begin{array}{rrrrr|r}
1 & 2 & -15 & -32 & -16 & -4 \\
  & -4 & 8 & 28 & 16 & \\
\hline
1 & -2 & -7 & -4 & 0 &
\end{array}
$$

Since the remainder is zero, $t + 4$ is a factor. We may also conclude that

$$f(t) = (t + 4)(t^3 - 2t^2 - 7t - 4)$$ ∎

EXAMPLE 10 Checking rational factor

By using synthetic division, determine whether $2x - 3$ is a factor of $2x^3 - 3x^2 + 8x - 12$.

CAUTION ▶ We first note that the coefficient of x in the possible factor is not 1. Thus, *we cannot use r = 3, since the factor is not of the form x − r.* However, $2x - 3 = 2\left(x - \frac{3}{2}\right)$, which means that if $2\left(x - \frac{3}{2}\right)$ is a factor of the function, $2x - 3$ is a factor. If we use $r = \frac{3}{2}$ and find that the remainder is zero, then $x - \frac{3}{2}$ is a factor.

$$
\begin{array}{rrrr|l}
2 & -3 & 8 & -12 & \frac{3}{2} \\
  & 3 & 0 & 12 & \\
\hline
2 & 0 & 8 & 0 &
\end{array}
$$

Since the remainder is zero, $x - \frac{3}{2}$ is a factor. Also, the quotient is $2x^2 + 8$, which may be factored into $2(x^2 + 4)$. Thus, 2 is also a factor of the function. This means that $2\left(x - \frac{3}{2}\right)$ is a factor of the function, and this in turn means that $2x - 3$ is a factor. This tells us that

$$2x^3 - 3x^2 + 8x - 12 = (2x - 3)(x^2 + 4)$$ ∎

EXAMPLE 11 Checking a zero

Determine whether or not -12.5 is a zero of the function $f(x) = 6x^3 + 61x^2 - 171x + 100$.

If -12.5 is a zero of $f(x)$, then $x - (-12.5)$, or $x + 12.5$, is a factor of $f(x)$, and $f(-12.5) = 0$. We can find the remainder by direct use of the remainder theorem or by synthetic division.

Using synthetic division and a calculator to make the calculations, we have the following setup and calculator sequence:

$$
\begin{array}{rrrr|l}
6 & 61 & -171 & 100 & -12.5 \\
  & -75 & 175 & -50 & \\
\hline
6 & -14 & 4 & 50 &
\end{array}
$$

Since the remainder is 50, and not zero, -12.5 is not a zero of $f(x)$.   ∎

Practice Exercise

2. Use synthetic division to determine whether $x + 2$ is a factor of $2x^3 + x^2 - 12x - 8$.

EXERCISES 15.1

In Exercises 1–4, make the given changes in the indicated examples of this section, and then perform the indicated operations.

1. In Example 3, change the $x + 4$ to $x + 3$ and then find the remainder.

2. In Example 4(a), change the $t + 1$ to $t - 1$ and then determine if $t - 1$ is a factor.

3. In Example 7, change the $x + 3$ to $x + 2$ and then perform the synthetic division.

4. In Example 10, change the $2x - 3$ to $2x + 3$ and then determine whether $2x + 3$ is a factor.

In Exercises 5–10, find the remainder by long division.

5. $(x^3 + 2x + 3) \div (x + 1)$

6. $(x^4 - 4x^3 - x^2 + x - 100) \div (x + 3)$

7. $(2x^5 - x^2 + 8x + 44) \div (x + 2)$

8. $(4s^3 - 9s^2 - 24s - 17) \div (s - 5)$

9. $(2x^4 - 3x^3 - 2x^2 - 15x - 16) \div (2x - 3)$

10. $(2x^4 - 10x^2 + 30x - 60) \div (x + 4)$

In Exercises 11–16, find the remainder using the remainder theorem. Do not use synthetic division.

11. $(R^4 + R^3 - 9R^2 + 3) \div (R + 4)$

12. $(4x^4 - x^2 + 5x - 7) \div (x - 3)$

13. $(2x^4 - 7x^3 - x^2 + 8) \div (x - 3)$

14. $(3n^4 - 13n^2 + 10n - 10) \div (n + 4)$

15. $(x^5 - 3x^3 + 5x^2 - 10x + 6) \div (x - 2)$

16. $(3x^4 - 12x^3 - 60x + 4) \div (x - 0.5)$

In Exercises 17–22, use the factor theorem to determine whether or not the second expression is a factor of the first expression. Do not use synthetic division.

17. $4x^3 + x^2 - 16x - 4$, $x - 2$

18. $3x^3 + 14x^2 + 7x - 4$, $x + 4$

19. $3V^4 - 7V^3 + V + 8, V - 2$

20. $x^5 - 2x^4 + 3x^3 - 6x^2 - 4x + 8, x - 2$

21. $x^{61} - 1, x + 1$

22. $x^7 - 128^{-1}, x + 2^{-1}$

In Exercises 23–32, perform the indicated divisions by synthetic division.

23. $(x^3 + 2x^2 - x - 2) \div (x - 1)$

24. $(x^3 - 3x^2 - x + 2) \div (x - 2)$

25. $(x^3 + 2x^2 - 3x + 4) \div (x + 1)$

26. $(2x^3 - 4x^2 + x - 1) \div (x + 2)$

27. $(p^6 - 6p^3 - 2p^2 - 6) \div (p - 2)$

28. $(x^5 + 4x^4 - 8) \div (x + 1)$

29. $(x^7 - 128) \div (x - 2)$

30. $(20x^4 + 11x^3 - 89x^2 + 60x - 77) \div (x + 2.75)$

31. $(2x^4 + x^3 + 3x^2 - 1) \div (2x - 1)$

32. $(6t^4 + 5t^3 - 10t + 4) \div (3t - 2)$

In Exercises 33–40, use the factor theorem and synthetic division to determine whether or not the second expression is a factor of the first.

33. $2x^5 - x^3 + 3x^2 - 4$; $x + 1$

34. $t^5 - 3t^4 - t^2 - 6$; $t - 3$

35. $4x^3 - 6x^2 + 2x - 2$; $x - \frac{1}{2}$

36. $3x^3 - 5x^2 + x + 1$; $x + \frac{1}{3}$

37. $2Z^4 - Z^3 - 4Z^2 + 1$; $2Z - 1$

38. $6x^4 + 5x^3 - x^2 + 6x - 2$; $3x - 1$

39. $4x^4 + 2x^3 - 8x^2 + 3x + 12$; $2x + 3$

40. $3x^4 - 2x^3 + x^2 + 15x + 4$; $3x + 4$

In Exercises 41–44, use synthetic division to determine whether or not the given numbers are zeros of the given functions.

41. $x^4 - 5x^3 - 15x^2 + 5x + 14$; 7

42. $r^4 + 5r^3 - 18r - 8$; -4

43. $85x^3 + 348x^2 - 263x + 120$; -4.8

44. $2x^3 + 13x^2 + 10x - 4$; $\frac{1}{2}$

In Exercises 45–58, solve the given problems.

45. If $f(x) = 2x^3 + 3x^2 - 19x - 4$, and $f(x) = (x + 4)g(x)$, find $g(x)$.

46. Using synthetic division, divide $ax^2 + bx + c$ by $x + 1$.

(W) 47. By division, show that $2x - 1$ is a factor of $f(x) = 4x^3 + 8x^2 - x - 2$. May we therefore conclude that $f(1) = 0$? Explain.

(W) 48. By division, show that $x^2 + 2$ is a factor of $f(x) = 3x^3 - x^2 + 6x - 2$. May we therefore conclude that $f(-2) = 0$? Explain.

49. For what value of k is $x - 2$ a factor of $f(x) = 2x^3 + kx^2 - x + 14$?

50. For what value of k is $x + 1$ a factor of $f(x) = 3x^4 + 3x^3 + 2x^2 + kx - 4$?

51. Use synthetic division: $(x^3 - 3x^2 + x - 3) \div (x + j)$.

52. Use synthetic division: $(2x^3 - 7x^2 + 10x - 6) \div [x - (1 + j)]$.

(W) 53. If $f(x) = -g(x)$, do the functions have the same zeros? Explain.

(W) 54. Do the functions $f(x)$ and $f(-x)$ have the same zeros? Explain.

55. In finding the electric current in a certain circuit, it is necessary to factor the denominator of $\dfrac{2s}{s^3 + 5s^2 + 4s + 20}$. Is (a) $(s - 2)$ or (b) $(s + 5)$ a factor?

56. In the theory of the motion of a sphere moving through a fluid, the function $f(r) = 4r^3 - 3ar^2 - a^3$ is used. Is (a) $r = a$ or (b) $r = 2a$ a zero of $f(r)$?

57. In finding the volume V (in cm^3) of a certain gas in equilibrium with a liquid, it is necessary to solve the equation $V^3 - 6V^2 + 12V = 8$. Use synthetic division to determine if $V = 2$ cm^3.

58. An architect is designing a window in the shape of a segment of a circle. An approximate formula for the area is $A = \dfrac{h^3}{2w} + \dfrac{2wh}{3}$, where A is the area, w is the width, and h is the height of the segment. If the width is 1.500 m and the area is 0.5417 m^2, use synthetic division to show that $h = 0.500$ m.

Answers to Practice Exercises

1. Yes ($R = 0$) **2.** No ($R = 4$)

15.2 The Roots of an Equation

**The Fundamental Theorem of Algebra •
Linear Factors • Complex Roots •
Remaining Quadratic Factor**

In this section, we present certain theorems that are useful in determining the number of roots in the equation $f(x) = 0$ and the nature of some of these roots. In dealing with polynomial equations of higher degree, it is helpful to have as much of this kind of information as we can find before actually solving for all of the roots, including any possible complex roots. A graph does not show the complex roots an equation may have, but we can verify the real roots, as we show by use of a graphing calculator in some of the examples that follow.

■ This theorem was first proved in 1799 by the German mathematician Karl Gauss (1777–1855) for his doctoral thesis. See the chapter introduction on page 408.

The first of these theorems is so important that it is called the **fundamental theorem of algebra.** It states that

every polynomial equation has at least one (real or complex) root.

The proof of this theorem is of an advanced nature, and therefore we accept its validity at this time. However, using the fundamental theorem, we can show the validity of other theorems that are useful in solving equations.

Let us now assume that we have a polynomial equation $f(x) = 0$ and that we are looking for its roots. By the fundamental theorem, we know that it has at least one root. Assuming that we can find this root by some means (the factor theorem, for example), we call this root r_1. Thus,

$$f(x) = (x - r_1)f_1(x)$$

where $f_1(x)$ is the polynomial quotient found by dividing $f(x)$ by $(x - r_1)$. However, since the fundamental theorem states that any polynomial equation has at least one root, this must apply to $f_1(x) = 0$ as well. Let us assume that $f_1(x) = 0$ has the root r_2. Therefore, this means that $f(x) = (x - r_1)(x - r_2)f_2(x)$. Continuing this process until one of the quotients is a constant a, we have

$$f(x) = a(x - r_1)(x - r_2) \cdots (x - r_n)$$

Note that one linear factor appears each time a root is found and that the degree of the quotient is one less each time. Thus, there are n linear factors, if the degree of $f(x)$ is n.

Therefore, based on the fundamental theorem of algebra, we have the following two related theorems.

1. *A polynomial of the nth degree can be factored into n linear factors.*

2. *A polynomial equation of degree n has exactly n roots.*

These theorems are illustrated in the following example.

EXAMPLE 1   Illustrating fundamental theorem

For the equation $f(x) = 2x^4 - 3x^3 - 12x^2 + 7x + 6 = 0$, we are given the factors that we show. In the next section, we will see how to find these factors.

For the function $f(x)$, we have

$$2x^4 - 3x^3 - 12x^2 + 7x + 6 = (x - 3)(2x^3 + 3x^2 - 3x - 2)$$
$$2x^3 + 3x^2 - 3x - 2 = (x + 2)(2x^2 - x - 1)$$
$$2x^2 - x - 1 = (x - 1)(2x + 1)$$
$$2x + 1 = 2(x + \tfrac{1}{2})$$

Therefore,

$$2x^4 - 3x^3 - 12x^2 + 7x + 6 = 2(x - 3)(x + 2)(x - 1)(x + \tfrac{1}{2}) = 0$$

The degree of $f(x)$ is 4. There are four linear factors: $(x - 3)$, $(x + 2)$, $(x - 1)$, and $\left(x + \tfrac{1}{2}\right)$. There are four roots of the equation: 3, -2, 1, and $-\tfrac{1}{2}$. Thus, we have verified each of the theorems above for this example.   ■

It is not necessary for each root of an equation to be different from the other roots. For example, the equation $(x - 1)^2 = 0$ has two roots, both of which are 1. Such roots are referred to as *multiple* (or *repeated*) *roots*.

When we solve the equation $x^2 + 1 = 0$, the roots are j and $-j$. In fact, for any equation (with real coefficients) that has a root of the form $a + bj$ ($b \neq 0$), there is also a root of the form $a - bj$. This is so because we can find the solutions of an equation of the form $ax^2 + bx + c = 0$ from the quadratic formula as

$$\frac{-b + \sqrt{b^2 - 4ac}}{2a} \quad \text{and} \quad \frac{-b - \sqrt{b^2 - 4ac}}{2a}$$

and the only difference between these roots is the sign before the radical. Thus,

NOTE ◆ *if the coefficients of the equation $f(x) = 0$ are real and $a + bj$ ($b \neq 0$) is a complex root, then its conjugate, $a - bj$, is also a root.*

EXAMPLE 2 Illustrating complex roots

Consider the equation $f(x) = (x - 1)^3(x^2 + x + 1) = 0$.

The factor $(x - 1)^3$ shows that there is a triple root of 1, and there is a total of five roots, since the highest-power term would be x^5 if we were to multiply out the function. To find the other two roots, we use the quadratic formula on the *factor* $(x^2 + x + 1)$. This is permissible, since we are finding the values of x for

$$x^2 + x + 1 = 0$$

For this, we have

$$x = \frac{-1 \pm \sqrt{1 - 4}}{2}$$

Thus,

$$x = \frac{-1 + j\sqrt{3}}{2} \quad \text{and} \quad x = \frac{-1 - j\sqrt{3}}{2}$$

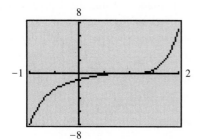

Fig. 15.2

Therefore, the roots of $f(x) = 0$ are $1, 1, 1, \dfrac{-1 + j\sqrt{3}}{2}$, and $\dfrac{-1 - j\sqrt{3}}{2}$. In the graph of $f(x)$ shown in Fig. 15.2, the *zero* feature gives a value of $x = 1$, but the calculator display does not show that it is a triple root, or that there are complex roots. ■

NOTE ◆ From Example 2, we can see that *whenever enough roots are known so that the remaining factor is quadratic, it is possible to find the remaining roots from the quadratic formula.* This is true for finding real or complex roots.

EXAMPLE 3 Solving equation – given one root

Solve the equation $3x^3 + 10x^2 - 16x - 32 = 0$; $-\frac{4}{3}$ is a root.

Using synthetic division and the given root, we have the setup shown at the left. From this, we see that

$$3x^3 + 10x^2 - 16x - 32 = \left(x + \tfrac{4}{3}\right)(3x^2 + 6x - 24)$$

We know that $x + \frac{4}{3}$ is a factor from the given root and that $3x^2 + 6x - 24$ is a factor found from synthetic division. This second factor can be factored as

$$3x^2 + 6x - 24 = 3(x^2 + 2x - 8) = 3(x + 4)(x - 2)$$

Therefore, we have

$$3x^3 + 10x^2 - 16x - 32 = 3\left(x + \tfrac{4}{3}\right)(x + 4)(x - 2)$$

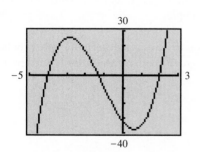

Fig. 15.3

This means the roots are $-\frac{4}{3}$, -4, and 2. Since the three roots are real, they can be found from a calculator display, as shown in Fig. 15.3. ■

$\blacksquare$ $\begin{array}{rrrrr} 1 & 3 & -4 & -10 & -4 \end{array}$ $\underline{|-1}$
$\begin{array}{rrrr} -1 & -2 & 6 & 4 \end{array}$
$\begin{array}{rrrrr} \hline 1 & 2 & -6 & -4 & 0 \end{array}$

$\blacksquare$ $\begin{array}{rrrr} 1 & 2 & -6 & -4 \end{array}$ $\underline{|2}$
$\begin{array}{rrr} 2 & 8 & 4 \end{array}$
$\begin{array}{rrrr} \hline 1 & 4 & 2 & 0 \end{array}$

Practice Exercise

1. Solve $x^4 - x^3 - 2x^2 - 4x - 24 = 0$, given that -2 and 3 are roots.

EXAMPLE 4 Solving equation – given two roots

Solve $x^4 + 3x^3 - 4x^2 - 10x - 4 = 0$; -1 and 2 are roots.

Using synthetic division and the root -1, the first setup at the left shows that

$$x^4 + 3x^3 - 4x^2 - 10x - 4 = (x + 1)(x^3 + 2x^2 - 6x - 4)$$

We now know that $x - 2$ must be a factor of $x^3 + 2x^2 - 6x - 4$, since it is a factor of the original function. Again, using synthetic division and this time the root 2, we have the second setup at the left. Thus,

$$x^4 + 3x^3 - 4x^2 - 10x - 4 = (x + 1)(x - 2)(x^2 + 4x + 2)$$

Since the original equation can now be written as

$$(x + 1)(x - 2)(x^2 + 4x + 2) = 0$$

the remaining two roots are found by solving

$$x^2 + 4x + 2 = 0$$

by the quadratic formula. This gives us

$$x = \frac{-4 \pm \sqrt{16 - 8}}{2} = \frac{-4 \pm 2\sqrt{2}}{2} = -2 \pm \sqrt{2}$$

Therefore, the roots are -1, 2, $-2 + \sqrt{2}$, and $-2 - \sqrt{2}$. $\blacksquare$

EXAMPLE 5 Solving equation – given double root

Solve the equation $3x^4 - 26x^3 + 63x^2 - 36x - 20 = 0$, given that 2 is a double root.

Using synthetic division, we have the first setup at the left. It tells us that

$$3x^4 - 26x^3 + 63x^2 - 36x - 20 = (x - 2)(3x^3 - 20x^2 + 23x + 10)$$

Also, since 2 is a double root, it must be a root of $3x^3 - 20x^2 + 23x + 10 = 0$. Using synthetic division again, we have the second setup at the left. This second quotient $3x^2 - 14x - 5$ factors into $(3x + 1)(x - 5)$. The roots are 2, 2, $-\frac{1}{3}$, and 5.

Since the quotient of the first division is the dividend for the second division, both divisions can be done without rewriting the first quotient as follows:

$\blacksquare$ $\begin{array}{rrrrr} 3 & -26 & 63 & -36 & -20 \end{array}$ $\underline{|2}$
$\begin{array}{rrrr} 6 & -40 & 46 & 20 \end{array}$
$\begin{array}{rrrrr} \hline 3 & -20 & 23 & 10 & 0 \end{array}$

NOTE $\blacktriangleright$

$\begin{array}{rrrr} 3 & -20 & 23 & 10 \end{array}$ $\underline{|2}$
$\begin{array}{rrr} 6 & -28 & -10 \end{array}$
$\begin{array}{rrrr} \hline 3 & -14 & -5 & 0 \end{array}$

first division $\longrightarrow$ $\begin{array}{rrrrr} 3 & -26 & 63 & -36 & -20 \end{array}$ $\underline{|2}$
$\begin{array}{rrrr} 6 & -40 & 46 & 20 \end{array}$
$\begin{array}{rrrrr} \hline 3 & -20 & 23 & 10 & 0 \end{array}$ $\underline{|2}$
second division $\longrightarrow$ $\begin{array}{rrr} 6 & -28 & -10 \end{array}$
$\begin{array}{rrr} \hline 3 & -14 & -5 & 0 \end{array}$ $\blacksquare$

EXAMPLE 6 Solving equation – given complex root

Solve the equation $2x^4 - 5x^3 + 11x^2 - 3x - 5 = 0$, given that $1 + 2j$ is a root.

Since $1 + 2j$ is a root, we know that $1 - 2j$ is also a root. Using synthetic division twice, we can then reduce the remaining factor to a quadratic function.

$$\begin{array}{rrrrr} 2 & -5 & 11 & -3 & -5 \end{array} \quad \underline{|1 + 2j}$$
$$\begin{array}{rrrr} 2 + 4j & -11 - 2j & 4 - 2j & 5 \end{array}$$
$$\begin{array}{rrrrr} \hline 2 & -3 + 4j & -2j & 1 - 2j & 0 \end{array} \quad \underline{|1 - 2j}$$
$$\begin{array}{rrr} 2 - 4j & -1 + 2j & -1 + 2j \end{array}$$
$$\begin{array}{rrrr} \hline 2 & -1 & -1 & 0 \end{array}$$

The quadratic factor $2x^2 - x - 1$ factors into $(2x + 1)(x - 1)$. Therefore, the roots of the equation are $1 + 2j$, $1 - 2j$, 1, and $-\frac{1}{2}$. As we can see, the calculator display in Fig. 15.4 shows only the real roots. $\blacksquare$

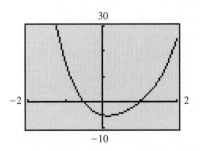

Fig. 15.4

Graphical Features Indicating Number and Types of Roots

If $f(x)$ is of degree n and the roots of $f(x) = 0$ are real and different, the graph of $f(x)$ crosses the x-axis n times and $f(x)$ changes signs as it crosses. See Fig. 15.5, where $n = 4$, the equation has four changes of sign, and the graph crosses four times.

In Fig. 15.5, we see that for two points on the graph, *the graph must cross the x-axis an odd number of times between points on the different sides of the x-axis and an even number of times between points on the same side of the x-axis, if at all.*

For each pair of complex roots, the number of times the graph crosses the x-axis is reduced by two (Fig. 15.6). For multiple roots, the graph crosses the x-axis once if the multiple is odd, or is tangent to the x-axis if the multiple is even (Fig. 15.7).

The graph must cross the x-axis at least once if the degree of $f(x)$ is odd, because the range includes all real numbers (Figs. 15.6 and 15.7). The curve may not cross the x-axis if the degree of $f(x)$ is even, because the range is bounded at a minimum point or a maximum point (Fig. 15.8), as we have seen for the quadratic function.

$$y = f(x)$$
$$= x^4 - 5x^3 + 5x^2 + 5x - 6$$
$$= (x + 1)(x - 1)(x - 2)(x - 3)$$

Fig. 15.5

$$y = f(x)$$
$$= x^3 - 2x^2 + 3$$
$$= (x + 1)(x^2 - 3x + 3)$$

Fig. 15.6

$$y = f(x)$$
$$= x^5 + x^4 - 2x^3 - 2x^2 + x + 1$$
$$= (x + 1)^3 (x - 1)^2$$

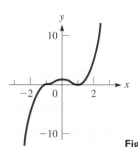

Fig. 15.7

$$y = f(x)$$
$$= x^4 + 4$$
$$= (x^2 - 2x + 2)(x^2 + 2x + 2)$$

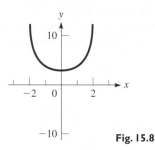

Fig. 15.8

EXERCISES 15.2

In Exercises 1 and 2, make the given changes in the indicated examples of this section and then solve the resulting equation.

1. In Example 2, change the middle term of the second factor to $2x$.

2. In Example 6, change the middle three terms to the left of the $=$ sign to $-3x^3 + 7x^2 + 7x$, given the same root.

In Exercises 3–6, find the roots of the given equations by inspection.

3. $(x - 2)(x^2 - 9) = 0$

4. $x(2x + 5)^2(x^2 - 64) = 0$

5. $(x^2 + 6x + 9)(x^2 + 4) = 0$

6. $(4x^2 + 9)(4x^2 + 4x + 1) = 0$

In Exercises 7–26, solve the given equations using synthetic division, given the roots indicated.

7. $x^3 + x^2 - 8x - 12 = 0$ $(r_1 = -2)$

8. $R^3 + 1 = 0$ $(r_1 = -1)$

9. $2x^3 + 11x^2 + 20x + 12 = 0$ $\left(r_1 = -\frac{3}{2}\right)$

10. $4x^3 - 20x^2 - x + 5 = 0$ $\left(r_1 = \frac{1}{2}\right)$

11. $3x^3 + 2x^2 + 3x + 2 = 0$ $(r_1 = j)$

12. $x^3 + 5x^2 + 9x + 5 = 0$ $(r_1 = -2 + j)$

13. $t^4 + t^3 - 2t^2 + 4t - 24 = 0$ $(r_1 = 2, r_2 = -3)$

14. $x^4 - 2x^3 - 20x^2 - 8x - 96 = 0$ $(r_1 = 6, r_2 = -4)$

15. $2x^4 - 19x^3 + 39x^2 + 35x - 25 = 0$ (5 is a double root)

16. $4n^4 + 28n^3 + 61n^2 + 42n + 9 = 0$ (-3 is a double root)

17. $6x^4 + 5x^3 - 15x^2 + 4 = 0$ $\left(r_1 = -\frac{1}{2}, r_2 = \frac{2}{3}\right)$

18. $6x^4 - 5x^3 - 14x^2 + 14x - 3 = 0$ $\left(r_1 = \frac{1}{3}, r_2 = \frac{3}{2}\right)$

19. $2x^4 - x^3 - 4x^2 + 10x - 4 = 0$ $(r_1 = 1 + j)$

20. $s^4 - 8s^3 - 72s - 81 = 0$ $(r_1 = 3j)$

21. $x^5 - 3x^4 + 4x^3 - 4x^2 + 3x - 1 = 0$ (1 is a triple root)

22. $12x^5 - 7x^4 + 41x^3 - 26x^2 - 28x + 8 = 0$
$\left(r_1 = 1, r_2 = \frac{1}{4}, r_3 = -\frac{2}{3}\right)$

23. $P^5 - 3P^4 - P + 3 = 0$ $(r_1 = 3, r_2 = j)$

24. $4x^5 + x^3 - 4x^2 - 1 = 0$ $\left(r_1 = 1, r_2 = \frac{1}{2}j\right)$

25. $(x^6 + 2x^5 - 4x^4 - 10x^3 - 41x^2 - 72x - 36 = 0$
(-1 is a double root; $2j$ is a root)

26. $x^6 - x^5 - 2x^3 - 3x^2 - x - 2 = 0$ (j is a double root)

In Exercises 27 and 28, answer the given questions.

W 27. Why cannot a third-degree polynomial function with real coefficients have zeros of 1, 2, and j?

W 28. How can the graph of a fourth-degree polynomial equation have its only x-intercepts as 0, 1, and 2?

In Exercises 29 and 30, form the indicated equations.

29. Form a polynomial equation of degree 3 and with integral coefficients, having a root of $1 + j$, and for which $f(2) = 4$.

30. Form a polynomial equation of the smallest possible degree and with integral coefficients, having a double root of 3, and a root of j.

Answer to Practice Exercise

1. $-2,\ 3,\ 2j,\ -2j$

15.3 Rational and Irrational Roots

**Possible Rational Roots •
Descartes' Rule of Signs •
Roots of a Polynomial Equation**

The product $(x + 2)(x - 4)(x + 3)$ equals $x^3 + x^2 - 14x - 24$. Here, we find that the constant 24 is determined only by the 2, 4, and 3. These numbers represent the roots of the equation if the given function is set equal to zero. In fact, if we find all the integral roots of an equation with integral coefficients and represent the equation in the form

$$f(x) = (x - r_1)(x - r_2)\cdots(x - r_k)f_{k+1}(x) = 0$$

where all the roots indicated are integers, the constant term of $f(x)$ must have factors of $r_1, r_2, \ldots, r_k$. This leads us to the theorem which states that

in a polynomial equation $f(x) = 0$, if the coefficient of the highest power is 1, then any integral roots are factors of the constant term of $f(x)$.

EXAMPLE 1 Possible integral roots if $a_0 = 1$

The equation $x^5 - 4x^4 - 7x^3 + 14x^2 - 44x + 120 = 0$ can be written as

$$(x - 5)(x + 3)(x - 2)(x^2 + 4) = 0$$

We now note that $5(3)(2)(4) = 120$. Thus, the roots 5, -3, and 2 are numerical factors of $|120|$. The theorem states nothing about the signs involved. ∎

If the coefficient a_0 of the highest-power term of $f(x)$ is an integer not equal to 1, the polynomial equation $f(x) = 0$ may have rational roots that are not integers. We can factor a_0 from every term of $f(x)$. Thus, any polynomial equation $f(x) = a_0x^n + a_1x^{n-1} + \cdots + a_n = 0$ with integral coefficients can be written in the form

$$f(x) = a_0\left(x^n + \frac{a_1}{a_0}x^{n-1} + \cdots + \frac{a_n}{a_0}\right) = 0$$

Since a_n and a_0 are integers, a_n/a_0 is a rational number. Using the same reasoning as with integral roots applied to the polynomial within the parentheses, we see that any rational roots are factors of a_n/a_0. This leads to the following theorem:

Any rational root r_r of a polynomial equation (with integral coefficients)

$$f(x) = a_0x^n + a_1x^{n-1} + \cdots + a_n = 0$$

is an integral factor of a_n divided by an integral factor of a_0. Therefore,

$$r_r = \frac{\text{integral factor of } a_n}{\text{integral factor of } a_0} \tag{15.4}$$

EXAMPLE 2 Possible rational roots

If $f(x) = 4x^3 - 3x^2 - 25x - 6$, any possible rational roots must be integral factors of 6 divided by integral factors of 4. These factors of 6 are 1, 2, 3, and 6, and these factors of 4 are 1, 2, and 4. Forming all possible positive and negative quotients, any possible rational roots that exist will be found in the following list: ±1, $\pm\frac{1}{2}$, $\pm\frac{1}{4}$, ±2, ±3, $\pm\frac{3}{2}$, $\pm\frac{3}{4}$, ±6.

The roots of this equation are -2, 3, and $-\frac{1}{4}$. ∎

■ Named for the French mathematician René Descartes (1596–1650). (See page 89).

There are 16 different possible rational roots in Example 2, but we cannot tell which of these are the actual roots. Therefore we now present a rule, known as *Descartes' rule of signs,* which will help us to find these roots.

DESCARTES' RULE OF SIGNS

1. *The number of positive roots of a polynomial equation* $f(x) = 0$, *cannot exceed the number of changes in sign in* $f(x)$ *in going from one term to the next in* $f(x)$.

2. *The number of negative roots cannot exceed the number of sign changes in* $f(-x)$.

We can reason this way: If $f(x)$ has all positive terms, then any positive number substituted in $f(x)$ must give a positive value for $f(x)$. This indicates that the number substituted in the function is not a root. Thus, there must be at least one negative and one positive term in the function for any positive number to be a root. This is not a proof, but it does indicate the type of reasoning used in developing the theorem.

EXAMPLE 3 Using Descartes' rule of signs

By Descartes' rule of signs, determine the maximum possible number of positive and negative roots of $3x^3 - x^2 - x + 4$.

Here, $f(x) = 3x^3 - x^2 - x + 4$. The first term is positive, and the second is negative, which indicates a change of sign. The third term is also negative; there is no change of sign from the second to the third term. The fourth term is positive, thus giving us a second change of sign, from the third to the fourth term. Hence, there are two changes in sign, which we can show as follows:

$$f(x) = 3x^3 - x^2 - x + 4$$

$$\underbrace{\qquad}_{1} \qquad \underbrace{\qquad}_{2} \longleftarrow \text{two sign changes}$$

Practice Exercise

1. Determine the maximum possible number of positive and negative roots of $9x^4 - 12x^3 - 5x^2 + 12x - 4 = 0$.

Since there are *two* changes of sign in $f(x)$, there are *no more than two* positive roots of $f(x) = 0$.

To find the maximum possible number of negative roots, we must find the number of sign changes in $f(-x)$. Thus,

$$f(-x) = 3(-x)^3 - (-x)^2 - (-x) + 4$$
$$= -3x^3 - x^2 + x + 4$$

$$\underbrace{\qquad}_{} \longleftarrow \text{one sign change}$$

NOTE ▶ There is only one change of sign in $f(-x)$; therefore, there is one negative root. *When there is just one change of sign in* $f(x)$, *there is a positive root, and when there is just one change of sign in* $f(-x)$, *there is a negative root.* ■

EXAMPLE 4 Using Descartes' rule of signs

For the equation $4x^5 - x^4 - 4x^3 + x^2 - 5x - 6 = 0$, we write

$$f(x) = 4x^5 - x^4 - 4x^3 + x^2 - 5x - 6 \qquad \longleftarrow \text{three sign changes}$$

$$f(-x) = -4x^5 - x^4 + 4x^3 + x^2 + 5x - 6 \qquad \longleftarrow \text{two sign changes}$$

Thus, there are no more than three positive and two negative roots. ■

At this point, let us summarize the information we can determine about the roots of a polynomial equation $f(x) = 0$ of degree n and with real coefficients.

ROOTS OF A POLYNOMIAL EQUATION OF DEGREE n

1. *There are n roots.*

2. *Complex roots appear in conjugate pairs.*

3. *Any rational roots must be factors of the constant term divided by factors of the coefficient of the highest-power term.*

4. *The maximum number of positive roots is the number of sign changes in $f(x)$, and the maximum number of negative roots is the number of sign changes in $f(-x)$.*

5. *Once we find $n - 2$ of the roots, we can find the remaining roots by the quadratic formula.*

■ Some calculators have a specific feature for solving higher-degree equations.

Since synthetic division is easy to perform, it is usually used to try possible roots. When a root is found, the quotient is of one degree less than the degree of the dividend. Each root found makes the ensuing work easier. The following examples show the complete method, as well as two other helpful rules.

EXAMPLE 5 Finding roots of equation

Find the roots of the equation $2x^3 + x^2 + 5x - 3 = 0$.

Since $n = 3$, there are three roots. If we can find one of these roots, we can use the quadratic formula to find the other two. We have

$$f(x) = 2x^3 + x^2 + 5x - 3 \quad \text{and} \quad f(-x) = -2x^3 + x^2 - 5x - 3$$

which shows there is one positive root and no more than two negative roots, which may or may not be rational. The *possible* rational roots are ± 1, $\pm\frac{1}{2}$, $\pm\frac{3}{2}$, ± 3.

First, trying the root 1 (always a possibility if there are positive roots), we have the synthetic division shown at the left. The remainder of 5 tells us that 1 is not a root, but we have gained some additional information, if we observe closely. If we try any positive number larger than 1, the results in the last row will be larger positive numbers than we now have. The products will be larger, and therefore the sums will also be larger. Thus, there is no positive root larger than 1. This leads to the following rule: *When we are trying a positive root, if the bottom row contains all positive numbers, then there are no roots larger than the value tried.* This rule tells us that there is no reason to try $+\frac{3}{2}$ and $+3$ as roots.

Now, let us try $+\frac{1}{2}$, as shown at the left. The zero remainder tells us that $+\frac{1}{2}$ is a root, and the remaining factor is $2x^2 + 2x + 6$, which itself factors to $2(x^2 + x + 3)$. By the quadratic formula, we find the remaining roots by solving the equation $x^2 + x + 3 = 0$. This gives us

$$x = \frac{-1 \pm \sqrt{1 - 12}}{2} = \frac{-1 \pm j\sqrt{11}}{2}$$

The three roots are $+\frac{1}{2}$, $\frac{-1 + j\sqrt{11}}{2}$, and $\frac{-1 - j\sqrt{11}}{2}$. There are no negative roots since the nonpositive roots are complex. Proceeding this way, we did not have to try any negative roots. The calculator display in Fig. 15.9 verifies the one real root at $x = 1/2$. ■

Practice Exercise

2. Find the roots of the equation
$9x^4 - 12x^3 - 5x^2 + 12x - 4 = 0$.

■
```
2   1   5  -3 ⌊1
        2   3   8
    2   3   8   5
```

■
```
2   1   5  -3 ⌊½
        1   1   3
    2   2   6   0
```

Fig. 15.9

Fig. 15.10

EXAMPLE 6 Solving equation – application

During a cycle of the movement of the weight on the double spring in Fig. 15.10, the force F (in N) on the weight by the spring is

$$F = x^4 - 7x^3 + 12x^2 + 4x$$

where x is the displacement (in cm) of the top of the double spring. For what values of x is $F = 16$ N?

Substituting 16 for F, we see that we are to solve the equation

$$x^4 - 7x^3 + 12x^2 + 4x - 16 = 0$$

To solve this equation, we write

$$f(x) = x^4 - 7x^3 + 12x^2 + 4x - 16$$
$$f(-x) = x^4 + 7x^3 + 12x^2 - 4x - 16$$

We see that there are four roots; there are no more than three positive roots, and there is one negative root. Since the coefficient of x^4 is 1, any possible rational roots must be integers. These possible rational roots are ±1, ±2, ±4, ±8, and ±16. Since there is only one negative root, we look for this one first. Trying -2, we have

$$
\begin{array}{rrrrr|r}
1 & -7 & +12 & +4 & -16 & \underline{-2} \\
  & -2 & +18 & -60 & +112 & \\
\hline
1 & -9 & +30 & -56 & +96 &
\end{array}
$$

NOTE ▶ If we were to try any negative roots less than -2 (remember, -3 is less than -2), we would find that the numbers would still alternate from term to term in the quotient. Thus, we have this rule: *When we are trying a negative root, if the signs alternate in the bottom row, then there are no roots less than the value tried.* In this case, we know that -4, -8, and -16 cannot be roots.

$$
\begin{array}{rrrrr|r}
\blacksquare\ 1 & -7 & +12 & +4 & -16 & \underline{-1} \\
  & -1 & 8 & -20 & 16 & \\
\hline
1 & -8 & 20 & -16 & 0 &
\end{array}
$$

Next, we try -1, as shown at the left. The remainder of zero tells us that -1 is the negative root.

Now that we have found the one negative root, we look for the positive roots.

$$
\begin{array}{rrrr|r}
\blacksquare\ 1 & -8 & 20 & -16 & \underline{1} \\
  & 1 & -7 & -13 & \\
\hline
1 & -7 & 13 & -3 &
\end{array}
$$

Trying 1, we have the setup at the left. The remainder of -3 tells us that 1 is not a root. Next, we try 2, as shown at the left. We see that 2 is a root.

$$
\begin{array}{rrrr|r}
\blacksquare\ 1 & -8 & 20 & -16 & \underline{2} \\
  & 2 & -12 & 16 & \\
\hline
1 & -6 & 8 & 0 &
\end{array}
$$

It is not necessary to find any more roots by trial and error. We may now use the quadratic formula or factoring on the equation $x^2 - 6x + 8 = 0$. The remaining roots are 2 and 4. Thus, the roots are $-1, 2, 2$, and 4. (Note that 2 is a double root.)

These roots now indicate that $F = 16$ N for displacements of -1 cm, 2 cm, and 4 cm. ∎

By the methods we have presented, we can look for *all roots* of a polynomial equation. These include any possible *complex roots* and *exact values of the rational and irrational roots,* if they exist. These methods allow us to solve a great many polynomial equations for these roots, but there are numerous other equations for which these methods are not sufficient.

When a polynomial equation has more than two irrational roots, we cannot generally find these roots by the methods we have developed. Approximate values can be found on a calculator either graphically or by evaluating the function. These methods are illustrated in the following two examples.

EXAMPLE 7 Finding roots with calculator

Find the roots of the equation $x^3 - 2x^2 - 3x + 2 = 0$, using a graphing calculator.

Since the degree of the equation is 3, we know there are three roots, and since the degree is odd, there is at least one real root. Using Descartes' rule of signs, we have

$$f(x) = x^3 - 2x^2 - 3x + 2 \qquad \text{two changes of sign}$$
$$f(-x) = -x^3 - 2x^2 + 3x + 2 \qquad \text{one change of sign}$$

This means there is one negative root and no more than two positive roots, if any. Therefore, setting

$$y = x^3 - 2x^2 - 3x + 2$$

and using the *window* settings (after two or three trials) shown in the calculator display in Fig. 15.11, the view shows one negative and two positive roots. Using the *zero* feature (or the *trace* and *zoom* features), these roots are

$$-1.34, 0.53, 2.81 \qquad \text{to the nearest 0.01}$$

Of course, we can use a graphing calculator to help locate and determine the number of real roots of any equation. ■

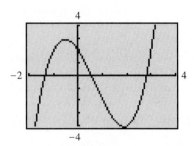

Fig. 15.11

EXAMPLE 8 Solving applied problem with calculator

The bottom part of a box to hold a jigsaw puzzle is to be made from a rectangular piece of cardboard 37.0 cm by 31.0 cm by cutting out equal squares from the corners, bending up the sides, and taping the corners. See Fig. 15.12. If the volume of the bottom part is to be 2770 cm³, find the side of the square that is to be cut out.

Let $x = $ the side of the square to be cut out. This means

$$2770 = x(37.0 - 2x)(31.0 - 2x) \qquad \text{volume} = 2770 \text{ cm}^3$$
$$= 4x^3 - 136x^2 + 1147x$$
$$4x^3 - 136x^2 + 1147x - 2770 = 0 \qquad \text{simplify with terms on the left}$$

We evaluate $f(x) = 4x^3 - 136x^2 + 1147x - 2770$ for integral values of x and find that $f(4) = -102$ and $f(5) = 65$. This means that $f(x) = 0$ between $x = 4$ and $x = 5$. By entering $f(x)$ as Y_1 in a graphing calculator, we can evaluate the function as shown in the calculator display in Fig. 15.13. This shows the following values:

$$f(4) = -102$$
$$f(5) = 65 \qquad \text{different signs means at least one root between 4 and 5}$$
$$f(4.4) = -15.42 \qquad \text{one root between 4.4 and 4.5}$$
$$f(4.5) = 2$$
$$f(4.48) = -1.353 \qquad \text{root between 4.48 and 4.49, closer to 4.49}$$
$$f(4.49) = 0.3318$$

From these values, $x = 4.49$ cm, to three significant digits. This value gives a volume

$$V = 4.49(37.0 - 2(4.49))(31.0 - 2(4.49)) = 2770 \text{ cm}^3$$

which means that the solution checks.

Checking values greater than $x = 5$, we find that there is another root between 6 and 7. Using the same procedure as above, we find that $x = 6.79$ cm is also a solution. (A third root $x = 22.7$ cm is obviously too large.) Therefore, there are two possible squares that can be cut from each corner to get a bottom part with a volume of 2770 cm³. ■

■ See the chapter introduction.

37.0 cm

Fig. 15.12

Fig. 15.13

■ Methods of evaluating functions on a graphing calculator are shown on page 82.

EXERCISES 15.3

In Exercises 1 and 2, make the given changes in the indicated examples of this section and then perform the indicated operation.

1. In Example 4, change all signs between terms.
2. In Example 5, change the + sign before the $5x$ to −.

In Exercises 3–20, solve the given equations without using a graphing calculator.

3. $x^3 + 2x^2 - x - 2 = 0$
4. $x^3 + x^2 - 5x + 3 = 0$
5. $x^3 + 2x^2 - 5x - 6 = 0$
6. $t^3 - 12t - 16 = 0$
7. $3x^3 - x - 2 = 0$
8. $3t^3 + 8t^2 - 1 = 0$
9. $3x^3 + 11x^2 + 5x - 3 = 0$
10. $4x^3 - 5x^2 - 23x + 6 = 0$
11. $x^4 - 11x^2 - 12x + 4 = 0$
12. $8x^4 - 32x^3 - x + 4 = 0$
13. $5n^4 - 2n^3 + 40n - 16 = 0$
14. $4n^4 - 17n^2 + 14n - 3 = 0$
15. $12x^4 + 44x^3 + 21x^2 - 11x = 6$
16. $9x^4 - 3x^3 + 34x^2 - 12x = 8$
17. $D^5 + D^4 - 9D^3 - 5D^2 + 16D + 12 = 0$
18. $x^6 - x^4 - 14x^2 + 24 = 0$
19. $4x^5 - 24x^4 + 49x^3 - 38x^2 + 12x - 8 = 0$
20. $2x^5 + 5x^4 - 4x^3 - 19x^2 - 16x = 4$

In Exercises 21–24, use a graphing calculator to solve the given equations to the nearest 0.01.

21. $x^3 - 2x^2 - 5x + 4 = 0$
22. $2x^4 - 15x^2 - 7x + 3 = 0$
23. $8x^4 + 36x^3 + 35x^2 - 4x - 4 = 0$
24. $2x^5 - 3x^4 + 8x^3 - 4x^2 - 4x + 2 = 0$

In Exercises 25–28, use a calculator to find the irrational root (to the nearest 0.01) that lies between the given values. (See Example 8.)

25. $x^3 - 6x^2 + 10x - 4 = 0$ (0 and 1)
26. $r^4 - r^3 - 3r^2 - r - 4 = 0$ (2 and 3)
27. $3x^3 + 13x^2 + 3x - 4 = 0$ (−1 and 0)
28. $3x^4 - 3x^3 - 11x^2 - x - 4 = 0$ (−2 and −1)

In Exercises 29–46, solve the given problems. Use a graphing calculator in Exercises 38, 39, and 42.

29. Solve the following system algebraically:
$y = x^4 - 11x^2; y = 12x - 4$

30. Find rational values of a such that $(x - a)$ will divide into $x^3 + x^2 - 4x - 4$ with a remainder of zero.

31. Where does the graph of the function $f(x) = 4x^3 + 3x^2 - 20x - 15$ cross the x-axis?

32. Where does the graph of the function $f(s) = 2s^4 - s^3 - 5s^2 + 7s - 6$ cross the s-axis?

33. The angular acceleration α (in rad/s^2) of the wheel of a car is given by $\alpha = -0.2t^3 + t^2$, where t is the time (in s). For what values of t is $\alpha = 2.0$ rad/s^2?

34. In finding one of the dimensions d (in in.) of the support columns of a building, the equation $3d^3 + 5d^2 - 400d - 18,000 = 0$ is found. What is this dimension?

35. The deflection y of a beam at a horizontal distance x from one end is given by $y = k(x^4 - 2Lx^3 - L^3x)$, where L is the length of the beam and k is a constant. For what values of x is the deflection zero?

36. The specific gravity s of a sphere of radius r that sinks to a depth h in water is given by $s = \dfrac{3rh^2 - h^3}{4r^3}$. Find the depth to which a spherical buoy of radius 4.0 cm sinks if $s = 0.50$.

37. A variable electric voltage in a circuit is given by $V = 0.1t^4 - 1.0t^3 + 3.5t^2 - 5.0t + 2.3$, where t is the time (in s). If the voltage is on for 5.0 s, when is $V = 0$?

38. The pressure difference p (in kPa) at a distance x (in km) from one end of an oil pipeline is given by $p = x^5 - 3x^4 - x^2 + 7x$. If the pipeline is 4 km long, where is $p = 0$?

39. A rectangular tray is made from a square piece of sheet metal 10.0 cm on a side by cutting equal squares from each corner, bending up the sides, and then welding them together. How long is the side of the square that must be cut out if the volume of the tray is 70.0 cm^3?

40. The angle θ of a robot arm with the horizontal as a function of time t (in s) is given by $\theta = 15 + 20t^2 - 4t^3$ for $0 \le t \le 5$ s. Find t for $\theta = 40°$.

41. The radii of four different-sized ball bearings differ by 1.00 mm in radius from one size to the next. If the volume of the largest equals the volumes of the other three combined, find the radii.

42. A rectangular safe is to be made of steel of uniform thickness, including the door. The inside dimensions are 1.20 m, 1.20 m, and 2.00 m. If the volume of steel is 1.25 m^3, find its thickness.

43. For electrical resistors connected in parallel, the reciprocal of the combined resistance equals the sum of the reciprocals of the individual resistances. If three resistors are connected in parallel such that the second resistance is 1400 Ω more than the first and the third is 5600 Ω more than the first, find the resistances for a combined resistance of 1400 Ω.

44. Each of three revolving doors has a perimeter of 6.60 m and revolves through a volume of 9.50 m^3 in one revolution about their common vertical side. What are the door's dimensions?

45. If a, b, and c are positive integers, find the combinations of the possible positive, negative, and nonreal complex roots if $f(x) = ax^3 - bx^2 + c = 0$.

(W) 46. An equation $f(x) = 0$ involves only odd powers of x with positive coefficients. Explain why this equation has no real root except $x = 0$.

Answers to Practice Exercises

1. 3 positive, 1 negative   2. −1, 2/3, 2/3, 1

CHAPTER 15   EQUATIONS

Polynomial function	$f(x) = a_0 x^n + a_1 x^{n-1} + \cdots + a_n$	(15.1)
Remainder theorem	$f(x) = (x - r)q(x) + R$	(15.2)
	$f(r) = R$	(15.3)
Rational roots	$r_r = \dfrac{\text{integral factor of } a_n}{\text{integral factor of } a_0}$	(15.4)

CHAPTER 15   REVIEW EXERCISES

In Exercises 1–4, find the remainder of the indicated division by the remainder theorem.

1. $(2x^3 - 4x^2 - x + 4) \div (x - 1)$

2. $(x^3 - 2x^2 + 9) \div (x + 2)$

3. $(4n^3 + n + 4) \div (n + 3)$

4. $(x^4 - 5x^3 + 8x^2 + 15x - 2) \div (x - 3)$

In Exercises 5–8, use the factor theorem to determine whether or not the second expression is a factor of the first.

5. $x^4 + x^3 + x^2 - 2x - 3; \quad x + 1$

6. $2s^3 - 6s - 4; \quad s - 2$

7. $2t^4 - 10t^3 - t^2 - 3t + 10; \quad t + 5$

8. $9v^3 + 6v^2 + 4v + 2; \quad 3v + 1$

In Exercises 9–16, use synthetic division to perform the indicated divisions.

9. $(x^3 + 3x^2 + 6x + 1) \div (x - 1)$

10. $(3x^3 - 2x^2 + 7) \div (x - 3)$

11. $(2x^3 - 3x^2 - 4x + 3) \div (x + 2)$

12. $(3D^3 + 8D^2 - 16) \div (D + 4)$

13. $(x^4 + 3x^3 - 20x^2 - 2x + 56) \div (x + 6)$

14. $(x^4 - 6x^3 + x - 8) \div (x - 3)$

15. $(2m^5 - 46m^3 + m^2 - 9) \div (m - 5)$

16. $(x^6 + 63x^3 + 5x^2 - 9x - 8) \div (x + 4)$

In Exercises 17–20, use synthetic division to determine whether or not the given numbers are zeros of the given functions.

17. $y^3 + 5y^2 - 6 = 0; \quad -3$

18. $8y^4 - 32y^3 - y + 4; \quad 4$

19. $2x^4 - x^3 + 2x^2 + x - 1; \quad \frac{1}{2}$

20. $6W^4 - 7W^3 + 2W^2 - 9W - 6; \quad -\frac{2}{3}$

In Exercises 21–32, find all the roots of the given equations, using synthetic division and the given roots.

21. $x^3 - 4x^2 - 7x + 10 = 0 \quad (r_1 = 5)$

22. $3B^3 - B^2 - 24B + 28 = 0 \quad (r_1 = 2)$

23. $x^4 - 10x^3 + 35x^2 - 50x + 24 = 0 \quad (r_1 = 1, r_2 = 2)$

24. $x^4 - x^3 - 5x^2 - x - 6 = 0 \quad (r_1 = 3, r_2 = -2)$

25. $4p^4 - p^2 - 18p + 9 = 0 \quad \left(r_1 = \frac{1}{2}, r_2 = \frac{3}{2}\right)$

26. $15x^4 + 4x^3 + 56x^2 + 16x - 16 = 0$ $(r_1 = 2/5, r_2 = -2/3)$

27. $4x^4 + 4x^3 + x^2 + 4x - 3 = 0 \quad (r_1 = j)$

28. $x^4 + 2x^3 - 4x - 4 = 0 \quad (r_1 = -1 + j)$

29. $s^5 + 3s^4 - s^3 - 11s^2 - 12s - 4 = 0 \quad (-1 \text{ is a triple root})$

30. $24x^5 + 10x^4 + 7x^2 - 6x + 1 = 0$ $\left(r_1 = -1, r_2 = \frac{1}{4}, r_3 = \frac{1}{3}\right)$

31. $V^5 + 4V^4 + 5V^3 - V^2 - 4V - 5 = 0$ $(r_1 = 1, r_2 = -2 + j)$

32. $2x^5 - x^4 + 8x - 4 = 0 \quad \left(r_1 = \frac{1}{2}, r_2 = 1 + j\right)$

In Exercises 33–40, solve the given equations.

33. $x^3 + x^2 - 10x + 8 = 0$

34. $x^3 - 8x^2 + 20x = 16$

35. $2r^3 - 3r^2 + 1 = 0$

36. $2x^3 - 3x^2 - 11x + 6 = 0$

37. $6x^3 - x^2 - 12x = 5$

38. $6y^3 + 19y^2 + 2y = 3$

39. $4t^4 - 17t^2 + 14t - 3 = 0$

40. $2x^4 + 5x^3 - 14x^2 - 23x + 30 = 0$

In Exercises 41–62, solve the given problems. Where appropriate, set up the required equations.

41. What are the possible numbers of real zeros (double roots count as two, etc.) for a polynomial with real coefficients and of degree 5?

42. What are the possible combinations of real and nonreal complex zeros (double roots count as two, etc.) of a fourth-degree polynomial?

43. If a calculator shows a real root, how many nonreal complex roots are possible for a sixth-degree polynomial equation $f(x) = 0$.

44. Find rational values of a such that $(x - a)$ will divide into $x^3 - 13x + 3$ with a remainder of -9.

45. Explain how to find k if $x + 2$ is a factor of $f(x) = 3x^3 + kx^2 - 8x - 8$. What is k?

46. Explain how to find k if $x - 3$ is a factor of $f(x) = kx^4 - 15x^2 - 5x - 12$. What is k?

47. Form a polynomial equation of degree 3 with integral coefficients and having roots of j and 5.

48. If $f(x) = 3x^4 - 18x^3 - 2x^2 + 13x - 6$, and $f(x) = g(x)(x - 6)$, find $g(x)$.

49. Solve the following system algebraically: $x^2 = y + 3$; $xy = 2$

50. Where does the graph of the function $f(x) = 2x^4 - 7x^3 + 11x^2 - 28x + 12$ cross the x-axis?

51. Find the irrational root of the equation $3x^3 - x^2 - 8x - 2 = 0$ that lies between 1 and 2.

52. The edge of a cube is 10 cm greater than the radius of a sphere. If the volumes of the figures are equal, what is this volume?

53. A computer analysis of the number of crimes committed each month in a certain city for the first 10 months of a year showed that $n = x^3 - 9x^2 + 15x + 600$. Here, n is the number of monthly crimes and x is the number of the month (as of the last day). In what month were 580 crimes committed?

54. A company determined that the number s (in thousands) of computer chips that it could supply at a price p of less than \$5 is given by $s = 4p^2 - 25$, whereas the demand d (in thousands) for the chips is given by $d = p^3 - 22p + 50$. For what price is the supply equal to the demand?

55. In order to find the diameter d (in cm) of a helical spring subject to given forces, it is necessary to solve the equation $64d^3 - 144d^2 + 108d - 27 = 0$. Solve for d.

56. A cubical tablet for purifying water is wrapped in a sheet of foil 0.500 mm thick. The total volume of tablet and foil is 33.1% greater than the volume of the tablet alone. Find the length of the edge of the tablet.

57. For the mirror shown in Fig. 15.14, the reciprocal of the focal distance f equals the sum of the reciprocals of the object distance p and image distance q (in in.). Find p, if $q = p + 4$ and $f = (p + 1)/p$.

Fig. 15.14   Image

58. Three electric capacitors are connected in series. The capacitance of the second is 1 μF more than that of the first, and the third is 2 μF more than the second. The capacitance of the combination is 1.33 μF. The equation used to determine C, the capacitance of the first capacitor, is

$$\frac{1}{C} + \frac{1}{C + 1} + \frac{1}{C + 3} = \frac{3}{4}$$

Find the values of the capacitances.

59. The height of a cylindrical oil tank is 3.2 m more than the radius. If the volume of the tank is 680 m³, what are the radius and the height of the tank?

60. A grain storage bin has a square base, each side of which is 5.5 m longer than the height of the bin. If the bin holds 160 m³ of grain, find its dimensions.

61. A rectangular door has a diagonal brace that is 0.900 ft longer than the height of the door. If the area of the door is 24.3 ft², find its dimensions.

62. The radius of one ball bearing is 1.0 mm greater than the radius of a second ball bearing. If the sum of their volumes is 100 mm³, find the radius of each.

Writing Exercise

63. A computer science student is to write a computer program that will print out the values of n for which $x + r$ is a factor of $x^n + r^n$. Write a paragraph that states which are the values of n and explains how they are found.

CHAPTER 15   PRACTICE TEST

1. Is -3 a zero for the function $2x^3 + 3x^2 + 7x - 6$? Explain.

2. Find the remaining roots of the equation $x^4 - 2x^3 - 7x^2 + 20x - 12 = 0$; 2 is a double root.

3. Use synthetic division to perform the division $(x^3 - 5x^2 + 4x - 9) \div (x - 3)$.

4. Use the factor theorem and synthetic division to determine whether or not $2x + 1$ is a factor of $2x^4 + 15x^3 + 23x^2 - 16$.

5. Use the remainder theorem to find the remainder of the division $(x^3 + 4x^2 + 7x - 9) \div (x + 4)$.

6. Solve for x: $2x^4 - x^3 + 5x^2 - 4x - 12 = 0$.

7. The ends of a 10-ft beam are supported at different levels. The deflection y of the beam is given by $y = kx^2(x^3 + 436x - 4000)$, where x is the horizontal distance from one end and k is a constant. Find the values of x for which the deflection is zero.

8. A cubical metal block is heated such that its edge increases by 1.0 mm and its volume is doubled. Find the edge of the cube to tenths.

Matrices; Systems of Linear Equations

While working with systems of linear equations in the 1850s and 1860s, the English mathematician Arthur Cayley developed the use of a *matrix* (as we will show, a matrix is simply a rectangular array of numbers). His interest was in the mathematical methods involved, and for many years matrices were used by mathematicians with little or no reference to possible applications. Of course, it was known they could be used when simultaneous linear equations were involved.

It was not until the 1920s that one of the first major applications of matrices was made when physicists used them in developing theories about the elementary particles within the atom. Their work is still very important in atomic and nuclear physics.

Since about 1950, matrices have become a very important and useful tool in many other areas of application, such as social science and economics. They are now used extensively in business and industry in making appropriate decisions in research, development, and production.

Here, in the case of matrices, we have a mathematical method that was developed only for its usefulness in mathematics, but has been found later to be very useful in many areas of application.

Prior to the development of matrices, other methods of solving systems of linear equations were used. As noted in Chapter 5, Cramer's rule was known in the mid-1700s, but its use for systems that did not have integers for coefficients could involve a great deal of calculation. This led the German mathematician Karl Friedrich Gauss in the early 1800s to develop a systematic method of solving systems of equations, now known as *Gaussian elimination*. This led to a matrix method that is now used in programming computers to solve systems of equations.

In the final two sections of this chapter, we develop the method of Gaussian elimination, and show the methods that were used to evaluate determinants before the extensive use of calculators and computers.

In Section 16.2, we see how to determine the amount of material and worker time required for the production of automobile parts.

Matrices: Definitions and Basic Operations

Elements of Matrix • Square Matrix • Equality of Matrices • Matrix Addition and Subtraction • Scalar Multiplication

In this section, we introduce the definitions and some basic operations with matrices. In the three sections that follow, we develop additional operations and show how they are used in solving systems of linear equations. This is only an introduction to the use of matrices, and additional uses and operations with matrices are shown in other sources.

A **matrix** *is an ordered rectangular array of numbers.* To distinguish such an array from a determinant, we enclose it within brackets. As with a determinant, *the individual members are called* **elements** *of the matrix.*

■ We first introduced the term *matrix* on page 164 when we showed the evaluation of a determinant on a calculator.

EXAMPLE 1 Illustrations of matrices

Some examples of matrices are shown here.

$$\begin{bmatrix} 2 & 8 \\ 1 & 0 \end{bmatrix} \quad \begin{bmatrix} 2 & -4 & 6 \\ -1 & 0 & 5 \end{bmatrix} \quad \begin{bmatrix} 4 & 6 \\ 0 & -1 \\ -2 & 5 \\ 3 & 0 \end{bmatrix}$$

$$\begin{bmatrix} -1 & 8 & 6 & 7 & 9 \\ 2 & 6 & 0 & 4 & 3 \\ 5 & -1 & 8 & 10 & 2 \end{bmatrix} \quad \begin{bmatrix} -1 & 2 & 0 & 9 \end{bmatrix}$$ ■

As we can see, it is not necessary for the number of columns and number of rows to be the same, although such is the case for a determinant. However, *if the number of rows does equal the number of columns, the matrix is called a* **square matrix.** We will find that square matrices are of special importance. *If all the elements of a matrix are zero, the matrix is called a* **zero matrix.** It is convenient to designate a given matrix by a capital letter.

CAUTION ▶ We must be careful to distinguish between a matrix and a determinant. *A matrix is simply **any** rectangular array **of numbers**, whereas a determinant is a specific value associated with a **square** matrix.*

EXAMPLE 2 Illustrating rows and columns

Consider the following matrices:

$$A = \begin{bmatrix} 5 & 0 & -1 \\ 1 & 2 & 6 \\ 0 & -4 & -5 \end{bmatrix} \quad B = \begin{bmatrix} 9 \\ 8 \\ 1 \\ 5 \end{bmatrix} \quad C = \begin{bmatrix} -1 & 6 & 8 & 9 \end{bmatrix} \quad O = \begin{bmatrix} 0 & 0 \\ 0 & 0 \end{bmatrix}$$

Matrix A is an example of a square matrix, matrix B is an example of a matrix with four rows and one column, matrix C is an example of a matrix with one row and four columns, and matrix O is an example of a zero matrix. ■

To be able to refer to specific elements of a matrix and to give a general representation, a double-subscript notation is usually employed. That is,

$$A = \begin{bmatrix} a_{11} & a_{12} & a_{13} \\ a_{21} & a_{22} & a_{23} \\ a_{31} & a_{32} & a_{33} \end{bmatrix}$$

$$\text{row} \nearrow \nwarrow \text{column}$$

We see that the first subscript refers to the row in which the element lies and the second subscript refers to the column in which the element lies.

NOTE ▶   *Two matrices are said to be* **equal** *if and only if they are identical.* That is, they must have the same number of columns, the same number of rows, and the elements must respectively be equal.

EXAMPLE 3 Equality of matrices

(a)
$$\begin{bmatrix} a_{11} & a_{12} & a_{13} \\ a_{21} & a_{22} & a_{23} \end{bmatrix} = \begin{bmatrix} 1 & -5 & 0 \\ 4 & 6 & -3 \end{bmatrix}$$

if and only if $a_{11} = 1$, $a_{12} = -5$, $a_{13} = 0$, $a_{21} = 4$, $a_{22} = 6$, and $a_{23} = -3$.

(b) The matrices

$$\begin{bmatrix} 1 & 2 & 3 \\ -1 & -2 & -5 \end{bmatrix} \text{ and } \begin{bmatrix} 1 & 2 & -5 \\ -1 & -2 & 3 \end{bmatrix}$$

are not equal, since the elements in the third column are reversed.

(c) The matrices

$$\begin{bmatrix} 2 & 3 \\ -1 & 5 \end{bmatrix} \text{ and } \begin{bmatrix} 2 & 3 & 0 \\ -1 & 5 & 0 \end{bmatrix}$$

are not equal, since the number of columns is different. This is true despite the fact that both elements of the third column are zeros.   ■

EXAMPLE 4 Matrix equation

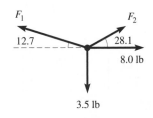

Fig. 16.1

The forces acting on a bolt are in equilibrium, as shown in Fig. 16.1. Analyzing the horizontal and vertical components as in Section 9.4, we find the following matrix equation. Find forces F_1 and F_2.

$$\begin{bmatrix} 0.98F_1 - 0.88F_2 \\ 0.22F_1 + 0.47F_2 \end{bmatrix} = \begin{bmatrix} 8.0 \\ 3.5 \end{bmatrix}$$

From the equality of matrices, we know that $0.98F_1 - 0.88F_2 = 8.0$ and $0.22F_1 + 0.47F_2 = 3.5$. Therefore, to find the forces F_1 and F_2, we must solve the system of equations

$$0.98F_1 - 0.88F_2 = 8.0$$
$$0.22F_1 + 0.47F_2 = 3.5$$

Using determinants, we have

$$F_1 = \frac{\begin{vmatrix} 8.0 & -0.88 \\ 3.5 & 0.47 \end{vmatrix}}{\begin{vmatrix} 0.98 & -0.88 \\ 0.22 & 0.47 \end{vmatrix}} = \frac{8.0(0.47) - 3.5(-0.88)}{0.98(0.47) - 0.22(-0.88)} = 10.5 \text{ lb}$$

Using determinants again, or by substituting this value into either equation, we find that $F_2 = 2.6$ lb. These values check when substituted into the original matrix equation.   ■

MATRIX ADDITION AND SUBTRACTION

If two matrices have the same number of rows and the same number of columns, their **sum** *is defined as the matrix consisting of the sums of the corresponding elements.* If the number of rows or the number of columns of the two matrices is not equal, they cannot be added.

EXAMPLE 5 Adding matrices

(a)

$$\begin{bmatrix} 8 & 1 & -5 & 9 \\ 0 & -2 & 3 & 7 \end{bmatrix} + \begin{bmatrix} -3 & 4 & 6 & 0 \\ 6 & -2 & 6 & 5 \end{bmatrix} = \begin{bmatrix} 8+(-3) & 1+4 & -5+6 & 9+0 \\ 0+6 & -2+(-2) & 3+6 & 7+5 \end{bmatrix}$$

$$= \begin{bmatrix} 5 & 5 & 1 & 9 \\ 6 & -4 & 9 & 12 \end{bmatrix}$$

(b) The matrices

$$\begin{bmatrix} 3 & -5 & 8 \\ 2 & 9 & 0 \\ 4 & -2 & 3 \end{bmatrix} \text{ and } \begin{bmatrix} 3 & -5 & 8 & 0 \\ 2 & 9 & 0 & 0 \\ 4 & -2 & 3 & 0 \end{bmatrix}$$

cannot be added since the second matrix has one more column than the first matrix. This is true even though the extra column contains only zeros. ∎

*The product of a number and a matrix (known as **scalar multiplication** of a matrix) is defined as the matrix whose elements are obtained by multiplying each element of the given matrix by the given number.* Thus, we obtain matrix kA by multiplying the elements of matrix A by k. In this way, $A + A$ and $2A$ are the same matrix.

EXAMPLE 6 Scalar multiplication

For the matrix A, where

$$A = \begin{bmatrix} -5 & 7 \\ 3 & 0 \end{bmatrix} \text{ we have } 2A = \begin{bmatrix} 2(-5) & 2(7) \\ 2(3) & 2(0) \end{bmatrix} = \begin{bmatrix} -10 & 14 \\ 6 & 0 \end{bmatrix}$$

Figure 16.2 shows a calculator display for A and $2A$. ∎

By combining the definitions for the addition of matrices and for the scalar multiplication of a matrix, we can define the subtraction of matrices. That is, the **difference** of matrices A and B is given by $A - B = A + (-B)$. Therefore, we change the sign of each element of B, and proceed as in addition.

EXAMPLE 7 Subtracting matrices

$$\begin{bmatrix} 7 & -4 \\ -9 & 3 \end{bmatrix} - \begin{bmatrix} -2 & 6 \\ -8 & 5 \end{bmatrix} = \begin{bmatrix} 7 & -4 \\ -9 & 3 \end{bmatrix} + \begin{bmatrix} 2 & -6 \\ 8 & -5 \end{bmatrix} = \begin{bmatrix} 9 & -10 \\ -1 & -2 \end{bmatrix}$$ ∎

The operations of addition, subtraction, and multiplication of a matrix by a number are like those for real numbers. For these operations, the algebra of matrices is like the algebra of real numbers. We see that the following laws hold for matrices.

$A + B = B + A$	(commutative law)	**(16.1)**
$A + (B + C) = (A + B) + C$	(associative law)	**(16.2)**
$k(A + B) = kA + kB$		**(16.3)**
$A + O = A$		**(16.4)**

Here, we have let O represent the zero matrix. We will find in the next section that not all laws for matrix operations are like those for real numbers.

Practice Exercise

1. For matrices A and B, find $A + B$.
$$A = \begin{bmatrix} 2 & -4 \\ -7 & 5 \end{bmatrix} \quad B = \begin{bmatrix} -9 & -6 \\ 7 & 3 \end{bmatrix}$$

```
[A]
          [[-5 7]
           [3  0]]
2[A]
          [[-10 14]
           [6   0]]
```

Fig. 16.2

Practice Exercise

2. For matrices A and B in Practice Exercise 1, find $A - 2B$.

EXERCISES 16.1

In Exercises 1 and 2, make the given changes in the indicated examples of this section and then perform the indicated operations.

1. In Example 5(a), interchange the second and third columns of the second matrix and then add the matrices.

2. In Example 6, find the matrix $-2A$.

In Exercises 3–10, determine the value of the literal numbers in each of the given matrix equalities.

3. $\begin{bmatrix} a & b \\ c & d \end{bmatrix} = \begin{bmatrix} 1 & -3 \\ 4 & 7 \end{bmatrix}$ 4. $\begin{bmatrix} x \\ x+y \end{bmatrix} = \begin{bmatrix} 2 \\ 5 \end{bmatrix}$

5. $\begin{bmatrix} x & 2y & z \\ r/4 & -s & -5t \end{bmatrix} = \begin{bmatrix} -2 & 10 & -9 \\ 12 & -4 & 5 \end{bmatrix}$

6. $[a+bj \quad 2c-dj \quad 3e+fj] = [5j \quad a+6 \quad 3b+c]$
   $(j = \sqrt{-1})$

7. $\begin{bmatrix} C+D \\ 2C-D \\ D-2E \end{bmatrix} = \begin{bmatrix} 5 \\ 4 \\ 6 \end{bmatrix}$ 8. $\begin{bmatrix} 2x-3y \\ x+4y \end{bmatrix} = \begin{bmatrix} 13 \\ 1 \end{bmatrix}$

9. $\begin{bmatrix} x-3 & x+y \\ x-z & y+z \\ x+t & y-t \end{bmatrix} = \begin{bmatrix} 5 & 3 \\ 4 & -1 \end{bmatrix}$ 10. $\begin{bmatrix} x \\ x+y \end{bmatrix} = \begin{bmatrix} 2 & 0 \\ 4 & -3 \end{bmatrix}$

In Exercises 11–14, find the indicated sums of matrices.

11. $\begin{bmatrix} 2 & 3 \\ -5 & 4 \end{bmatrix} + \begin{bmatrix} -1 & 7 \\ 5 & -2 \end{bmatrix}$

12. $\begin{bmatrix} 1 & 0 & 9 \\ 3 & -5 & -2 \end{bmatrix} + \begin{bmatrix} 4 & -1 & 7 \\ 2 & 0 & -3 \end{bmatrix}$

13. $\begin{bmatrix} 50 & -82 \\ -34 & 57 \\ -15 & 62 \end{bmatrix} + \begin{bmatrix} -55 & 82 \\ 45 & 14 \\ 26 & -67 \end{bmatrix}$

14. $\begin{bmatrix} 4.7 & 2.1 & -9.6 \\ -6.8 & 4.8 & 7.4 \\ -1.9 & 0.7 & 5.9 \end{bmatrix} + \begin{bmatrix} -4.9 & -9.6 & -2.1 \\ 3.4 & 0.7 & 0.0 \\ 5.6 & 10.1 & -1.6 \end{bmatrix}$

In Exercises 15–26, use matrices A, B, and C to find the indicated matrices.

$A = \begin{bmatrix} -1 & 4 & -7 \\ 2 & -6 & 11 \end{bmatrix}$ $B = \begin{bmatrix} 7 & 9 & -6 \\ 4 & -1 & -8 \end{bmatrix}$ $C = \begin{bmatrix} 3 & 19 \\ 7 & -5 \end{bmatrix}$

15. $A + B$ 16. $A - B$ 17. $A + C$ 18. $B + C$
19. $2A + B$ 20. $2B + A$ 21. $A - 2B$ 22. $3A - B$
23. $-4A$ 24. $-3B$ 25. $-C - A$ 26. $-\frac{1}{2}A + B$

In Exercises 27–34, use the matrices for Exercises 15–26 and find the indicated matrices using a graphing calculator.

27. $3A$ 28. $-2C$ 29. $A + 3B$ 30. $B - 0.5A$
31. $A - C$ 32. $-4A - 3B$ 33. $-6B + A$ 34. $A - B + 2C$

In Exercise 35–38, use matrices A and B to show that the indicated laws hold for these matrices. In Exercise 35, explain the meaning of the result.

$A = \begin{bmatrix} -1 & 2 & 3 & 7 \\ 0 & -3 & -1 & 4 \\ 9 & -1 & 0 & -2 \end{bmatrix}$ $B = \begin{bmatrix} 4 & -1 & -3 & 0 \\ 5 & 0 & -1 & 1 \\ 1 & 11 & 8 & 2 \end{bmatrix}$

(W) 35. $A + B = B + A$ 36. $A + O = A$
37. $-(A - B) = B - A$ 38. $3(A + B) = 3A + 3B$

In Exercises 39 and 40, find the unknown quantities in the given matrix equations.

39. An airplane is flying in a direction 21.0° north of east at 235 km/h but is headed 14.5° north of east. The wind is from the southeast. Find the speed of the wind v_w and the speed of the plane v_p relative to the wind from the given matrix equation. See Fig. 16.3.

$$\begin{bmatrix} v_p \cos 14.5° - v_w \cos 45.0° \\ v_p \sin 14.5° + v_w \sin 45.0° \end{bmatrix} = \begin{bmatrix} 235 \cos 21.0° \\ 235 \sin 21.0° \end{bmatrix}$$

Fig. 16.3

40. Find the electric currents shown in Fig. 16.4 by solving the following matrix equation:

$$\begin{bmatrix} I_1 + I_2 + I_3 \\ -2I_1 + 3I_2 \\ -3I_2 + 6I_3 \end{bmatrix} = \begin{bmatrix} 0 \\ 24 \\ 0 \end{bmatrix}$$

Fig. 16.4

In Exercises 41–44, perform the indicated matrix operations.

41. The contractor of a housing development constructs four different types of houses, with either a carport, a one-car garage, or a two-car garage. The following matrix shows the number of houses of each type and the type of garage.

	Type A	Type B	Type C	Type D
Carport	96	75	0	0
1-car garage	62	44	24	0
2-car garage	0	35	68	78

If the contractor builds two additional identical developments, find the matrix showing the total number of each house-garage type built in the three developments.

42. The inventory of a drug supply company shows that the following numbers of cases of bottles of vitamins C and B_3 (niacin) are in stock: Vitamin C—25 cases of 100-mg bottles, 10 cases of 250-mg bottles, and 32 cases of 500-mg bottles; vitamin B_3—30 cases of 100-mg bottles, 18 cases of 250-mg bottles, and 40 cases of 500-mg bottles. This is represented by matrix A below. After two shipments are sent out, each of which can be represented by matrix B below, find the matrix that represents the remaining inventory.

$$A = \begin{bmatrix} 25 & 10 & 32 \\ 30 & 18 & 40 \end{bmatrix} \qquad B = \begin{bmatrix} 10 & 5 & 6 \\ 12 & 4 & 8 \end{bmatrix}$$

43. One serving of brand K of breakfast cereal provides the given percentages of the given vitamins and minerals: vitamin A, 15%; vitamin C, 25%; calcium, 10%; iron, 25%. One serving of brand G

provides: vitamin A, 10%; vitamin C, 10%; calcium, 10%; iron, 45%. One serving of tomato juice provides: vitamin A, 15%; vitamin C, 30%; calcium, 3%; iron, 3%. One serving of orange-pineapple juice provides vitamin A, 0%; vitamin C, 100%; calcium, 2%; iron, 2%. Set up a two-row, four-column matrix B to represent the data for the cereals and a similar matrix J for the juices.

(W) 44. Referring to Exercise 43, find the matrix $B + J$ and explain the meaning of its elements.

Answers to Practice Exercises

1. $\begin{bmatrix} -7 & -10 \\ 0 & 8 \end{bmatrix}$ **2.** $\begin{bmatrix} 20 & 8 \\ -21 & -1 \end{bmatrix}$

16.2 Multiplication of Matrices

Multiplication of Matrices • Identity Matrix • Inverse of a Matrix

The definition for the multiplication of matrices does not have an intuitive basis. However, through the solution of a system of linear equations we can, at least in part, show why multiplication is defined as it is. Consider Example 1.

EXAMPLE 1 Reasoning for definition of multiplication

If we solve the system of equations

$$2x + y = 1$$
$$7x + 3y = 5$$

we get $x = 2$, $y = -3$. Checking this solution in each of the equations, we get

$$2(2) + 1(-3) = 1$$
$$7(2) + 3(-3) = 5$$

Let us represent the coefficients of the equations by the matrix $\begin{bmatrix} 2 & 1 \\ 7 & 3 \end{bmatrix}$ and the solutions by the matrix $\begin{bmatrix} 2 \\ -3 \end{bmatrix}$. If we now indicate the multiplications of these matrices and perform it as shown

$$\begin{bmatrix} 2 & 1 \\ 7 & 3 \end{bmatrix}\begin{bmatrix} 2 \\ -3 \end{bmatrix} = \begin{bmatrix} 2(2) + 1(-3) \\ 7(2) + 3(-3) \end{bmatrix} = \begin{bmatrix} 1 \\ 5 \end{bmatrix}$$

we note that we obtain a matrix that properly represents the right-side values of the equations. (Note the products and sums in the resulting matrix.) ∎

Following reasons along the lines indicated in Example 1, we now define the **multiplication of matrices.** If the number of columns in a first matrix equals the number of rows in a second matrix, the product of these matrices is formed as follows: *The element in a specified row and a specified column of the product matrix is the sum of the products formed by multiplying each element in the specified row of the first matrix by the corresponding element in the specific column of the second matrix.* The product matrix will have the same number of rows as the first matrix and the same number of columns as the second matrix. Consider the following examples.

EXAMPLE 2 Multiplying matrices

Find the product AB, where

$$A = \begin{bmatrix} 2 & 1 \\ -3 & 0 \\ 1 & 2 \end{bmatrix} \qquad B = \begin{bmatrix} -1 & 6 & 5 & -2 \\ 3 & 0 & 1 & -4 \end{bmatrix}$$

Practice Exercise

1. Find the product AB.

$$A = \begin{bmatrix} -1 & 4 \\ 8 & -2 \\ 0 & 12 \end{bmatrix} \quad B = \begin{bmatrix} 5 & -3 \\ -7 & 10 \end{bmatrix}$$

With two columns in matrix A and two rows in matrix B, the product can be formed. The element in the first row and first column of the product is the sum of the products of the corresponding elements of the first row of A and first column of B. The elements in the first row and second column of the product is the sum of the products of corresponding elements of the first row of A and second column of B. We continue until we have three rows (the number in A) and four columns (the number in B).

$$\begin{bmatrix} 2 & 1 \\ -3 & 0 \\ 1 & 2 \end{bmatrix}\begin{bmatrix} -1 & 6 & 5 & -2 \\ 3 & 0 & 1 & -4 \end{bmatrix} = \begin{bmatrix} 2(-1)+1(3) & 2(6)+1(0) & 2(5)+1(1) & 2(-2)+1(-4) \\ -3(-1)+0(3) & -3(6)+0(0) & -3(5)+0(1) & -3(-2)+0(-4) \\ 1(-1)+2(3) & 1(6)+2(0) & 1(5)+2(1) & 1(-2)+2(-4) \end{bmatrix}$$

$$= \begin{bmatrix} 1 & 12 & 11 & -8 \\ 3 & -18 & -15 & 6 \\ 5 & 6 & 7 & -10 \end{bmatrix}$$

The elements used to form the element in the first row and first column and the element in the third row and second column of the product are outlined in color. Figure 16.5 shows the calculator display for matrices A and B and their product AB.

In trying to form the product BA, we see that B has four columns and A has three rows. Since these numbers are not the same, the product cannot be formed. Therefore, $AB \neq BA$, which means *matrix multiplication is not commutative* (*except in special cases*), and therefore differs from multiplication of real numbers. ■

Fig. 16.5

EXAMPLE 3

The product of two matrices below may be formed because the first matrix has four columns and the second matrix has four rows. The matrix is formed as shown.

$$\begin{bmatrix} -1 & 9 & 3 & -2 \\ 2 & 0 & -7 & 1 \end{bmatrix}\begin{bmatrix} 6 & -2 \\ 1 & 0 \\ 3 & -5 \\ 3 & 9 \end{bmatrix} = \begin{bmatrix} -1(6)+9(1)+3(3)+(-2)(3) & -1(-2)+9(0)+3(-5)+(-2)(9) \\ 2(6)+0(1)+(-7)(3)+1(3) & 2(-2)+0(0)+(-7)(-5)+1(9) \end{bmatrix}$$

$$= \begin{bmatrix} -6+9+9-6 & 2+0-15-18 \\ 12+0-21+3 & -4+0+35+9 \end{bmatrix}$$

$$= \begin{bmatrix} 6 & -31 \\ -6 & 40 \end{bmatrix}$$ ■

■ Note that, if the second matrix is to the left of the first, then the product has four rows and four columns.

IDENTITY MATRIX

There are two special matrices of particular importance in the multiplication of matrices. The first of these is the **identity matrix I,** *which is a square matrix with 1's for elements of the principal diagonal with all other elements zero.* (The principal diagonal starts with the element a_{11}.) It has the property that if it is multiplied by another square matrix with the same number of rows and columns, then the second matrix equals the product matrix.

EXAMPLE 4 Identity matrix

Show that $AI = IA = A$ for the matrix

$$A = \begin{bmatrix} 2 & -3 \\ 4 & 1 \end{bmatrix}$$

Since A has two rows and two columns, we choose I with two rows and two columns. Therefore, for this case

$$I = \begin{bmatrix} 1 & 0 \\ 0 & 1 \end{bmatrix} \leftarrow \text{elements of principal diagonal are 1's}$$

Forming the indicated products, we have results as follows:

$$AI = \begin{bmatrix} 2 & -3 \\ 4 & 1 \end{bmatrix}\begin{bmatrix} 1 & 0 \\ 0 & 1 \end{bmatrix}$$

$$= \begin{bmatrix} 2(1) + (-3)(0) & 2(0) + (-3)(1) \\ 4(1) + 1(0) & 4(0) + 1(1) \end{bmatrix} = \begin{bmatrix} 2 & -3 \\ 4 & 1 \end{bmatrix}$$

$$IA = \begin{bmatrix} 1 & 0 \\ 0 & 1 \end{bmatrix}\begin{bmatrix} 2 & -3 \\ 4 & 1 \end{bmatrix}$$

$$= \begin{bmatrix} 1(2) + 0(4) & 1(-3) + 0(1) \\ 0(2) + 1(4) & 0(-3) + 1(1) \end{bmatrix} = \begin{bmatrix} 2 & -3 \\ 4 & 1 \end{bmatrix}$$

Therefore, we see that $AI = IA = A$. ∎

INVERSE OF A MATRIX

For a given square matrix A, its **inverse** A^{-1} is the other important special matrix. *The matrix A and its inverse A^{-1} have the property that*

$$AA^{-1} = A^{-1}A = I \tag{16.5}$$

If the product of two square matrices equals the identity matrix, the matrices are called inverses of each other. Under certain conditions, the inverse of a given square matrix may not exist, although for most square matrices, the inverse does exist. In the next section, we develop the procedure for finding the inverse of a square matrix, and the section that follows shows how the inverse is used in the solution of systems of equations. At this point, we simply show that the product of certain matrices equals the identity matrix and that therefore these matrices are inverses of each other.

EXAMPLE 5 Inverse matrix

For the given matrices A and B, show that $AB = BA = I$, and therefore that $B = A^{-1}$.

$$A = \begin{bmatrix} 1 & -3 \\ -2 & 7 \end{bmatrix} \quad B = \begin{bmatrix} 7 & 3 \\ 2 & 1 \end{bmatrix}$$

Practice Exercise

2. Show that $AB = BA = I$.

$$A = \begin{bmatrix} 5 & -7 \\ -2 & 3 \end{bmatrix} \quad B = \begin{bmatrix} 3 & 7 \\ 2 & 5 \end{bmatrix}$$

Forming the products AB and BA, we have the following:

$$AB = \begin{bmatrix} 1 & -3 \\ -2 & 7 \end{bmatrix}\begin{bmatrix} 7 & 3 \\ 2 & 1 \end{bmatrix} = \begin{bmatrix} 7 - 6 & 3 - 3 \\ -14 + 14 & -6 + 7 \end{bmatrix} = \begin{bmatrix} 1 & 0 \\ 0 & 1 \end{bmatrix}$$

$$BA = \begin{bmatrix} 7 & 3 \\ 2 & 1 \end{bmatrix}\begin{bmatrix} 1 & -3 \\ -2 & 7 \end{bmatrix} = \begin{bmatrix} 7 - 6 & -21 + 21 \\ 2 - 2 & -6 + 7 \end{bmatrix} = \begin{bmatrix} 1 & 0 \\ 0 & 1 \end{bmatrix}$$

Since $AB = I$ and $BA = I$, $B = A^{-1}$ and $A = B^{-1}$. ∎

EXAMPLE 6 Matrix multiplication – application

A company makes three types of automobile parts. In one day, it produces 40 of type X, 50 of type Y, and 80 of type Z. Required for production are 4 units of material and 1 worker-hour for type X, 5 units of material and 2 worker-hours for type Y, and 3 units of material and 2 worker-hours for type Z. By representing the number of each type produced as matrix A and the material and time requirements as matrix B, we have

■ See the chapter introduction.

$$A = \begin{bmatrix} 40 & 50 & 80 \end{bmatrix} \qquad B = \begin{bmatrix} 4 & 1 \\ 5 & 2 \\ 3 & 2 \end{bmatrix}$$

type X, type Y, type Z — number of each type produced

units of material, worker-hours — ← type X, ← type Y, ← type Z — material and time required for each

The product AB gives the total number of units of material and the total number of worker-hours needed for the day's production in a one-row, two-column matrix:

$$AB = \begin{bmatrix} 40 & 50 & 80 \end{bmatrix} \begin{bmatrix} 4 & 1 \\ 5 & 2 \\ 3 & 2 \end{bmatrix}$$

total units of material, total worker-hours

$$= \begin{bmatrix} 160 + 250 + 240 & 40 + 100 + 160 \end{bmatrix} = \begin{bmatrix} 650 & 300 \end{bmatrix}$$

Therefore, 650 units of material and 300 worker-hours are required. ■

Some Properties of Matrix Algebra We have seen that *matrix multiplication is not commutative* (see Example 2); that is, $AB \neq BA$ in general. This differs from the multiplication of real numbers. Another difference is that it is possible that $AB = O$, even though neither A nor B is O. There are also some similarities, for example, $AI = A$, where I and the number 1 are equivalent. Also, *the distributive property $A(B + C) = AB + AC$ holds for matrix multiplication.*

EXERCISES 16.2

In Exercises 1 and 2, make the given changes in the indicated examples of this section and then perform the indicated multiplications.

1. In Example 2, interchange columns 1 and 2 in matrix A and then do the multiplication.

2. In Example 5, in A change -2 to 2 and -3 to 3, in B change 2 to -2 and 3 to -3, and then do the multiplications.

In Exercises 3–10, perform the indicated multiplications.

3. $\begin{bmatrix} 4 & -2 \end{bmatrix} \begin{bmatrix} -1 & 0 \\ 2 & 6 \end{bmatrix}$

4. $\begin{bmatrix} -\frac{1}{2} & 6 & -\frac{2}{3} \end{bmatrix} \begin{bmatrix} 4 & 8 \\ \frac{1}{3} & -\frac{1}{2} \\ 0 & 9 \end{bmatrix}$

5. $\begin{bmatrix} 2 & -3 & 1 \\ 0 & 7 & -3 \end{bmatrix} \begin{bmatrix} 90 \\ -25 \\ 50 \end{bmatrix}$

6. $\begin{bmatrix} 0 & -1 & 2 \\ 4 & 11 & 2 \end{bmatrix} \begin{bmatrix} 3 & -1 \\ 1 & 2 \\ 6 & 1 \end{bmatrix}$

7. $\begin{bmatrix} -8 & \frac{3}{4} \\ \frac{7}{2} & -8 \\ -6 & \frac{4}{5} \end{bmatrix} \begin{bmatrix} \frac{1}{4} & -3 \\ 2 & 5 \end{bmatrix}$

8. $\begin{bmatrix} 12 & -47 \\ 43 & -18 \\ 36 & -22 \end{bmatrix} \begin{bmatrix} 25 \\ 66 \end{bmatrix}$

9. $\begin{bmatrix} -1 & 7 \\ 3 & 5 \\ 10 & -1 \\ -5 & 12 \end{bmatrix} \begin{bmatrix} 2 & 1 \\ 5 & -3 \end{bmatrix}$

10. $\begin{bmatrix} 5 & 4 \end{bmatrix} \begin{bmatrix} 4 & -4 \\ -5 & 5 \end{bmatrix}$

In Exercises 11–14, use a graphing calculator to perform the indicated multiplications.

11. $\begin{bmatrix} 2 & -3 \\ 5 & -1 \end{bmatrix} \begin{bmatrix} 3 & 0 & -1 \\ 7 & -5 & 8 \end{bmatrix}$

12. $\begin{bmatrix} -7 & 8 \\ 5 & 0 \end{bmatrix} \begin{bmatrix} -90 & 100 \\ 10 & 40 \end{bmatrix}$

13. $\begin{bmatrix} -9.2 & 2.3 & 0.5 \\ -3.8 & -2.4 & 9.2 \end{bmatrix} \begin{bmatrix} 6.5 & -5.2 \\ 4.9 & 1.7 \\ -1.8 & 6.9 \end{bmatrix}$

14. $\begin{bmatrix} 1 & 2 & -6 & 6 & 1 \\ -2 & 4 & 0 & 1 & 2 \end{bmatrix} \begin{bmatrix} 1 \\ -1 \\ 0 \\ 5 \\ 2 \end{bmatrix}$

In Exercises 15–18, find, if possible, AB and BA. If it is not possible, explain why.

15. $A = \begin{bmatrix} 1 & -3 & 8 \end{bmatrix}$ $B = \begin{bmatrix} -1 \\ 5 \\ 7 \end{bmatrix}$

16. $A = \begin{bmatrix} -3 & 2 & 0 \\ 1 & -4 & 5 \end{bmatrix}$ $B = \begin{bmatrix} -2 & 0 \\ 4 & -6 \\ 5 & 1 \end{bmatrix}$

17. $A = \begin{bmatrix} -10 & 25 & 40 \\ 42 & -5 & 0 \end{bmatrix}$ $B = \begin{bmatrix} 6 \\ -15 \\ 12 \end{bmatrix}$

18. $A = \begin{bmatrix} -2 & 1 & 7 \\ 3 & -1 & 0 \\ 0 & 2 & -1 \end{bmatrix}$ $B = \begin{bmatrix} 4 & -1 & 5 \end{bmatrix}$

In Exercises 19–22, show that $AI = IA = A$.

19. $A = \begin{bmatrix} 1 & 8 \\ -2 & 2 \end{bmatrix}$

20. $A = \begin{bmatrix} -15 & 28 \\ -5 & 64 \end{bmatrix}$

21. $A = \begin{bmatrix} 1 & 3 & -5 \\ 2 & 0 & 1 \\ 1 & -2 & 4 \end{bmatrix}$

22. $A = \begin{bmatrix} -1 & 2 & 0 \\ 4 & -3 & 1 \\ 2 & 1 & 3 \end{bmatrix}$

In Exercises 23–26, determine whether or not $B = A^{-1}$.

23. $A = \begin{bmatrix} 5 & -2 \\ -2 & 1 \end{bmatrix}$ $B = \begin{bmatrix} 1 & 2 \\ 2 & 5 \end{bmatrix}$

24. $A = \begin{bmatrix} 3 & -4 \\ 5 & -7 \end{bmatrix}$ $B = \begin{bmatrix} 7 & -4 \\ 5 & -2 \end{bmatrix}$

25. $A = \begin{bmatrix} 1 & -2 & 3 \\ 2 & -5 & 7 \\ -1 & 3 & -5 \end{bmatrix}$ $B = \begin{bmatrix} 4 & -1 & 1 \\ 3 & -2 & -1 \\ 1 & -1 & -1 \end{bmatrix}$

26. $A = \begin{bmatrix} 1 & -1 & 3 \\ 3 & -4 & 8 \\ -2 & 3 & -4 \end{bmatrix}$ $B = \begin{bmatrix} 8 & -5 & -4 \\ 4 & -2 & -1 \\ -1 & 1 & 1 \end{bmatrix}$

In Exercises 27–30, determine by matrix multiplication whether or not A is the proper matrix of solution values.

27. $3x - 2y = -1$ $A = \begin{bmatrix} 1 \\ 2 \end{bmatrix}$
$4x + y = 6$

28. $4x + y = -5$ $A = \begin{bmatrix} -2 \\ 3 \end{bmatrix}$
$3x + 4y = 6$

29. $3x + y + 2z = 1$
$x - 3y + 4z = -3$ $A = \begin{bmatrix} -1 \\ 2 \\ 1 \end{bmatrix}$
$2x + 2y + z = 1$

30. $2x - y + z = 7$
$x - 3y + 2z = 6$ $A = \begin{bmatrix} 3 \\ -2 \\ -1 \end{bmatrix}$
$3x + y - z = 8$

In Exercises 31–34, perform the indicated matrix multiplications on a graphing calculator, using the following matrices. For matrix A, $A^2 = A \times A$.

$$A = \begin{bmatrix} 2 & -3 & -5 \\ -1 & 4 & 5 \\ 1 & -3 & -4 \end{bmatrix} B = \begin{bmatrix} 1 & -2 & -6 \\ -3 & 2 & 9 \\ 2 & 0 & -3 \end{bmatrix} C = \begin{bmatrix} 1 & -3 & -4 \\ -1 & 3 & 4 \\ 1 & -3 & -4 \end{bmatrix}$$

31. Show that $A^2 = A$.

32. Show that $C^2 = O$.

33. Show that $B^3 = B$

34. Show that $A^3B^3 = AB$

In Exercises 35–46, perform the indicated matrix multiplications.

35. For matrices $A = \begin{bmatrix} a & b \\ b & a \end{bmatrix}$ and $B = \begin{bmatrix} c & d \\ d & c \end{bmatrix}$, show that
$AB = BA$.

36. For matrix $A = \begin{bmatrix} 1 & 2 & 2 \\ 2 & 1 & 2 \\ 2 & 2 & 1 \end{bmatrix}$, show that $A^2 - 4A - 5I = O$.

37. Using two rows and columns, show that $(-I)^2 = I$.

(W) **38.** For $J = \begin{bmatrix} j & 0 \\ 0 & j \end{bmatrix}$, where $j = \sqrt{-1}$, show that $J^2 = -I$, $J^3 = -J$, and $J^4 = I$. Explain the similarity with j^2, j^3, and j^4.

39. Show that $A^2 - I = (A + I)(A - I)$ for $A = \begin{bmatrix} 2 & 4 \\ 3 & 5 \end{bmatrix}$.

40. In the study of polarized light, the matrix product
$$\begin{bmatrix} 1 & 0 \\ 0 & -j \end{bmatrix} \begin{bmatrix} 1 & 0 \\ 1 & -j \end{bmatrix} \begin{bmatrix} 1 \\ 1 \end{bmatrix}$$ occurs $(j = \sqrt{-1})$. Find this product.

41. In studying the motion of electrons, one of the Pauli spin matrices used is $s_y = \begin{bmatrix} 0 & -j \\ j & 0 \end{bmatrix}$, where $j = \sqrt{-1}$. Show that $s_y^2 = I$.

42. In analyzing the motion of a robotic mechanism, the following matrix multiplication is used. Perform the multiplication and evaluate each element of the result.
$$\begin{bmatrix} \cos 60° & -\sin 60° & 0 \\ \sin 60° & \cos 60° & 0 \\ 0 & 0 & 1 \end{bmatrix} \begin{bmatrix} 2 \\ 4 \\ 0 \end{bmatrix}$$

43. In an *ammeter*, nearly all the electric current flows through a *shunt*, and the remaining known fraction of current is measured by the meter. See Fig. 16.6. From the given matrix equation, find voltage v_2 and current i_2 in terms of v_1, i_1, and resistance R, whichever may be applicable.

$$\begin{bmatrix} v_2 \\ i_2 \end{bmatrix} = \begin{bmatrix} 1 & 0 \\ -\dfrac{1}{R} & 1 \end{bmatrix} \begin{bmatrix} v_1 \\ i_1 \end{bmatrix}$$

Fig. 16.6

44. In the theory related to the reproduction of color photography, the equation

$$\begin{bmatrix} X \\ Y \\ Z \end{bmatrix} = \begin{bmatrix} 1.0 & 0.1 & 0 \\ 0.5 & 1.0 & 0.1 \\ 0.3 & 0.4 & 1.0 \end{bmatrix} \begin{bmatrix} x \\ y \\ z \end{bmatrix}$$

is found. The X, Y, and Z represent the red, green, and blue densities of the reproductions, respectively, and the x, y, and z represent the red, green, and blue densities, respectively, of the subject. Give the equations relating X, Y, and Z and x, y, and z.

45. The path of an earth satellite can be written as

$$[x \quad y] \begin{bmatrix} 2.76 & -1 \\ 1 & 2.81 \end{bmatrix} \begin{bmatrix} x \\ y \end{bmatrix} = [7.76 \times 10^7]$$

where distances are in miles. What type of curve is represented? (See Section 14.1.)

46. Using Kirchhoff's laws on the circuit shown in Fig. 16.7, the following matrix equation is found. By matrix multiplication, find the resulting system of equations.

$$\begin{bmatrix} R_1 + R_2 & -R_2 & 0 \\ -R_2 & R_2 + R_3 + R_4 & -R_4 \\ 0 & -R_4 & R_4 + R_5 \end{bmatrix} \begin{bmatrix} I_1 \\ I_2 \\ I_3 \end{bmatrix} = \begin{bmatrix} V_1 \\ 0 \\ -V_2 \end{bmatrix}$$

Fig. 16.7

<hr>

Answers to Practice Exercises

1. $\begin{bmatrix} -33 & 43 \\ 54 & -44 \\ -84 & 120 \end{bmatrix}$ 2. $AB = BA = \begin{bmatrix} 1 & 0 \\ 0 & 1 \end{bmatrix}$

16.3 Finding the Inverse of a Matrix

Inverse of a 2 × 2 Matrix •
Gauss–Jordan Method •
Inverse on a Calculator

We now show how to find the inverse of a matrix, and in the next section, we show how it is used in solving a system of linear equations. First, we show two methods of finding the inverse of a two-row, two-column (2×2) matrix. The first method is as follows:

Inverse of a 2 × 2 Matrix

1. *Interchange the elements on the principal diagonal.*

2. *Change the signs of the off-diagonal elements.*

3. *Divide each resulting element by the determinant of the given matrix.*

This method, which can be used with second-order square matrices *but not with higher-order matrices,* is illustrated in the following example.

EXAMPLE 1 Inverse of 2 × 2 matrix – method 1

Find the inverse of the matrix $A = \begin{bmatrix} 2 & -3 \\ 4 & -7 \end{bmatrix}$.

First, we interchange the elements on the principal diagonal and change the signs of the off-diagonal elements. This gives us the matrix

$$\begin{bmatrix} -7 & 3 \\ -4 & 2 \end{bmatrix} \quad \begin{array}{l} \leftarrow \text{ signs changed} \\ \leftarrow \text{ elements interchanged} \end{array}$$

■ If the value of the determinant is zero, the inverse matrix does not exist.

Now, we find the determinant of the original matrix, which means we evaluate

$$\begin{vmatrix} 2 & -3 \\ 4 & -7 \end{vmatrix} = -14 - (-12) = -2$$

We now divide each element of the second matrix by -2. This gives

Practice Exercise

1. Find the inverse:

$$A = \begin{bmatrix} 3 & -8 \\ 1 & -2 \end{bmatrix}$$

$$A^{-1} = \frac{1}{-2} \begin{bmatrix} -7 & 3 \\ -4 & 2 \end{bmatrix} = \begin{bmatrix} \dfrac{-7}{-2} & \dfrac{3}{-2} \\ \dfrac{-4}{-2} & \dfrac{2}{-2} \end{bmatrix} = \begin{bmatrix} \dfrac{7}{2} & -\dfrac{3}{2} \\ 2 & -1 \end{bmatrix} \quad \leftarrow \text{ inverse}$$

Checking by multiplication gives

■ See Exercise 37.

$$AA^{-1} = \begin{bmatrix} 2 & -3 \\ 4 & -7 \end{bmatrix} \begin{bmatrix} \dfrac{7}{2} & -\dfrac{3}{2} \\ 2 & -1 \end{bmatrix} = \begin{bmatrix} 7 - 6 & -3 + 3 \\ 14 - 14 & -6 + 7 \end{bmatrix} = \begin{bmatrix} 1 & 0 \\ 0 & 1 \end{bmatrix} = I$$

Since $AA^{-1} = I$, the matrix A^{-1} is the proper inverse matrix. ■

Gauss–Jordan Method

■ Named for the German mathematician Karl Gauss (1777–1855) and the German geodesist Wilhelm Jordan (1842–1899).

The second method, called the *Gauss–Jordan method,* involves *transforming the given matrix into the identity matrix while* **transforming the identity matrix into the inverse.** There are three types of steps allowable in making these transformations:

1. *Any two rows may be interchanged.*
2. *Every element in any row may be multiplied by any number other than zero.*
3. *Any row may be replaced by a row whose elements are the sum of a nonzero multiple of itself and a nonzero multiple of another row.*

NOTE ▶ Note that these are **row operations,** not column operations, and are the operations used in solving a system of equations by addition and subtraction.

EXAMPLE 2 2 × 2 Inverse – Gauss-Jordan method

Find the inverse of the matrix

$$A = \begin{bmatrix} 2 & -3 \\ 4 & -7 \end{bmatrix} \qquad \text{this is the same matrix as in Example 1}$$

Practice Exercise

2. Find the inverse using the Gauss–Jordan method: $A = \begin{bmatrix} 3 & -8 \\ 1 & -2 \end{bmatrix}$

First, we set up the given matrix with the identity matrix as follows:

$$\begin{bmatrix} 2 & -3 & | & 1 & 0 \\ 4 & -7 & | & 0 & 1 \end{bmatrix}$$

The vertical line simply shows the separation of the two matrices.

We wish to transform the left matrix into the identity matrix. Therefore, the first requirement is a 1 for element a_{11}. Therefore, we divide all elements of the first row by 2. This gives the following setup:

$$\begin{bmatrix} 1 & -\frac{3}{2} & | & \frac{1}{2} & 0 \\ 4 & -7 & | & 0 & 1 \end{bmatrix}$$

Next, we want to have a zero for element a_{21}. Therefore, we subtract 4 times each element of row 1 from the corresponding element in row 2, replacing the elements of row 2. This gives us the following setup:

■ As in this example, (1) always work one column at a time, from left to right and (2) never undo the work in a previously completed column.

$$\begin{bmatrix} 1 & -\frac{3}{2} & | & \frac{1}{2} & 0 \\ 4 - 4(1) & -7 - 4(-\frac{3}{2}) & | & 0 - 4(\frac{1}{2}) & 1 - 4(0) \end{bmatrix} \quad \text{or} \quad \begin{bmatrix} 1 & -\frac{3}{2} & | & \frac{1}{2} & 0 \\ 0 & -1 & | & -2 & 1 \end{bmatrix}$$

Next, we want to have 1, not −1, for element a_{22}. Therefore, we multiply each element of row 2 by −1. This gives

$$\begin{bmatrix} 1 & -\frac{3}{2} & | & \frac{1}{2} & 0 \\ 0 & 1 & | & 2 & -1 \end{bmatrix}$$

Finally, we want zero for element a_{12}. Therefore, we add $\frac{3}{2}$ times each element of row 2 to the corresponding elements of row 1, replacing row 1. This gives

$$\begin{bmatrix} 1 + \frac{3}{2}(0) & -\frac{3}{2} + \frac{3}{2}(1) & | & \frac{1}{2} + \frac{3}{2}(2) & 0 + \frac{3}{2}(-1) \\ 0 & 1 & | & 2 & -1 \end{bmatrix} \quad \text{or} \quad \begin{bmatrix} 1 & 0 & | & \frac{7}{2} & -\frac{3}{2} \\ 0 & 1 & | & 2 & -1 \end{bmatrix}$$

At this point, we have transformed the given matrix into the identity matrix, and the identity matrix into the inverse. Therefore, the matrix to the right of the vertical bar in the last setup is the required inverse. Thus,

$$A^{-1} = \begin{bmatrix} \frac{7}{2} & -\frac{3}{2} \\ 2 & -1 \end{bmatrix}$$

See Fig. 16.8 for a calculator window showing matrix A and its inverse A^{-1}. ■

```
[A]
        [[2  -3]
         [4  -7]]
[A]⁻¹
        [[3.5 -1.5]
         [2   -1  ]]
```

Fig. 16.8

In transforming a matrix into the identity matrix, we work on one column at a time, transforming the columns in order from left to right. It is generally best to make the element on the principal diagonal for the column 1 first and then make all other elements in the column 0. This was done in Example 2, and we now illustrate it with another 2 × 2 matrix, and then we find the inverse of a 3 × 3 matrix. The method is applicable for any square matrix.

EXAMPLE 3   2 × 2 Inverse – Gauss-Jordan method

Find the inverse of the matrix $\begin{bmatrix} -3 & 6 \\ 4 & 5 \end{bmatrix}$.

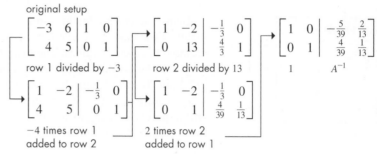

Therefore, $A^{-1} = \begin{bmatrix} -\frac{5}{39} & \frac{2}{13} \\ \frac{4}{39} & \frac{1}{13} \end{bmatrix}$, which can be checked by multiplication.   ∎

EXAMPLE 4   3 × 3 Inverse – Gauss-Jordan method

Find the inverse of the matrix $\begin{bmatrix} 1 & 2 & -1 \\ 3 & 5 & -1 \\ -2 & -1 & -2 \end{bmatrix}$.

■ On many calculators, the third column of A^{-1} (shown to the right of the window) is found by *scrolling* the screen.

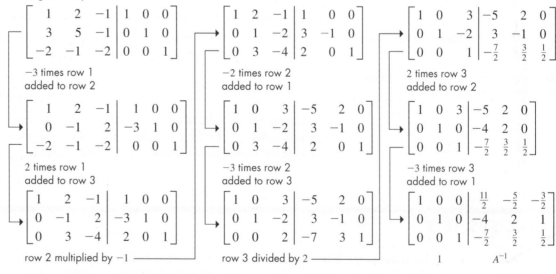

Therefore, the required inverse matrix is

$$\begin{bmatrix} \frac{11}{2} & -\frac{5}{2} & -\frac{3}{2} \\ -4 & 2 & 1 \\ -\frac{7}{2} & \frac{3}{2} & \frac{1}{2} \end{bmatrix}$$

which may be checked by multiplication. See Fig. 16.9 for a calculator window showing A and its inverse A^{-1}.   ∎

```
[A]
    [[1   2   -1]
     [3   5   -1]
     [-2  -1  -2]]
[A]⁻¹
    [[5.5  -2.5  -1.…   -1.5]
     [-4    2     1  …   1   ]
     [-3.5  1.5   .5 …   .5  ]]
```

Fig. 16.9

In the next example, we review entering a matrix, displaying it, and finding its inverse on a typical calculator. The manual for any particular model should be used to see how the various operations are done on it.

EXAMPLE 5 4 × 4 inverse on a calculator

By using a graphing calculator, find the inverse of the matrix

$$A = \begin{bmatrix} 2 & -2 & 3 & 2 \\ 3 & 1 & 5 & 2 \\ -2 & 5 & 2 & -3 \\ 4 & -5 & -1 & 4 \end{bmatrix}$$

Using the *matrix* feature, in Fig. 16.10(a) we show the window when the elements of the matrix are entered. Scrolling by use of the arrow key will probably be required to enter all values. The matrix is then displayed as in Fig. 16.10(b).

To get the display for the inverse in Fig. 16.10(c), we enter [A] and then use the $\boxed{x^{-1}}$ key. To avoid long decimal approximations in the display, we have used the *mode* feature to round off values to three decimal places. On most calculators, this still requires scrolling to see all values, but they are easier to read. We have shown those that need scrolling to the right of the window.

(a)

(b)

Fig. 16.10

(c)

From these calculator windows, we have found the inverse matrix to be

$$A^{-1} = \begin{bmatrix} 0.435 & -0.652 & 1.652 & 1.348 \\ -0.739 & 0.609 & -0.609 & -0.391 \\ 0.391 & -0.087 & 0.087 & -0.087 \\ -1.261 & 1.391 & -2.391 & -1.609 \end{bmatrix}$$

The elements of A^{-1} can be checked directly on the calculator by showing that the product $AA^{-1} = I$ (if the rounded values are used, the values of I will be approximate but should be sufficiently close to the necessary 1's and 0's). ∎

■ On a calculator, the *fraction* feature can be used to change decimals to fractions.

■ Note that calculator entries such as 1E⁻10 mean that it is zero.

EXERCISES 16.3

In Exercises 1 and 2, make the given changes in the indicated examples of this section and then find the matrix inverses.

1. In Example 1, change the element −7 to −5 and then find the inverse using the same method.

2. In Example 3, change the element −3 to −2 and then find the inverse using the same method.

In Exercises 3–10, find the inverse of each of the given matrices by the method of Example 1 of this section.

3. $\begin{bmatrix} 2 & -5 \\ -2 & 4 \end{bmatrix}$

4. $\begin{bmatrix} -6 & 3 \\ 3 & -2 \end{bmatrix}$

5. $\begin{bmatrix} -1 & 5 \\ 4 & 10 \end{bmatrix}$

6. $\begin{bmatrix} 8 & -1 \\ -4 & -5 \end{bmatrix}$

7. $\begin{bmatrix} 0 & -4 \\ 2 & 6 \end{bmatrix}$

8. $\begin{bmatrix} 7 & -3 \\ -6 & 4 \end{bmatrix}$

9. $\begin{bmatrix} -50 & -45 \\ 26 & 80 \end{bmatrix}$

10. $\begin{bmatrix} 7.2 & -3.6 \\ -1.3 & -5.7 \end{bmatrix}$

In Exercises 11–24, find the inverse of each of the given matrices by transforming the identity matrix, as in Examples 2–4.

11. $\begin{bmatrix} 1 & 2 \\ 2 & 3 \end{bmatrix}$

12. $\begin{bmatrix} 1 & 5 \\ -1 & -4 \end{bmatrix}$

13. $\begin{bmatrix} 2 & 4 \\ -1 & -1 \end{bmatrix}$

14. $\begin{bmatrix} -2 & 6 \\ 3 & -4 \end{bmatrix}$

15. $\begin{bmatrix} 2 & 5 \\ -1 & 2 \end{bmatrix}$

16. $\begin{bmatrix} -2 & 3 \\ -3 & 5 \end{bmatrix}$

17. $\begin{bmatrix} 25 & 30 \\ -10 & -14 \end{bmatrix}$

18. $\begin{bmatrix} 1 & -3 \\ 7 & -5 \end{bmatrix}$

19. $\begin{bmatrix} 1 & -3 & -2 \\ -2 & 7 & 3 \\ 1 & -1 & -3 \end{bmatrix}$

20. $\begin{bmatrix} 1 & 2 & -1 \\ 3 & 7 & -5 \\ -1 & -2 & 0 \end{bmatrix}$

21. $\begin{bmatrix} 1 & 3 & 2 \\ -2 & -5 & -1 \\ 2 & 4 & 0 \end{bmatrix}$

22. $\begin{bmatrix} 1 & 3 & 4 \\ -1 & -4 & -2 \\ 4 & 9 & 20 \end{bmatrix}$

23. $\begin{bmatrix} 2 & 4 & 0 \\ 3 & 4 & -2 \\ -1 & 1 & 2 \end{bmatrix}$

24. $\begin{bmatrix} -2 & 6 & 1 \\ 0 & 3 & -3 \\ 4 & -7 & 3 \end{bmatrix}$

In Exercises 25–34, find the inverse of each of the given matrices by using a graphing calculator, as in Example 5. The matrices in Exercises 27–29 are the same as those in Exercises 21–23.

25. $\begin{bmatrix} 2 & 8 \\ -1 & 6 \end{bmatrix}$

26. $\begin{bmatrix} 20 & -45 \\ -12 & 24 \end{bmatrix}$

27. $\begin{bmatrix} 1 & 3 & 2 \\ -2 & -5 & -1 \\ 2 & 4 & 0 \end{bmatrix}$

28. $\begin{bmatrix} 1 & 3 & 4 \\ -1 & -4 & -2 \\ 4 & 9 & 20 \end{bmatrix}$

29. $\begin{bmatrix} 2 & 4 & 0 \\ 3 & 4 & -2 \\ -1 & 1 & 2 \end{bmatrix}$

30. $\begin{bmatrix} 10 & -5 & 30 \\ -2 & 4 & -5 \\ 20 & 5 & 5 \end{bmatrix}$

31. $\begin{bmatrix} 1 & -2 & 1 & 0 \\ 1 & -2 & 2 & -3 \\ 0 & 1 & -1 & 1 \\ -2 & 3 & -2 & 3 \end{bmatrix}$

32. $\begin{bmatrix} 3 & -2 & -1 & 4 \\ 2 & 0 & 5 & 1 \\ -1 & 2 & 1 & -2 \\ 4 & 1 & 3 & 5 \end{bmatrix}$

33. $\begin{bmatrix} 0.2 & 1.2 & -0.8 & -0.5 \\ -0.4 & 3.0 & -1.6 & 0.4 \\ 1.0 & -2.4 & 3.2 & 1.5 \\ -0.1 & 0.4 & 0.0 & 3.0 \end{bmatrix}$

34. $\begin{bmatrix} 12.5 & -2.6 & 1.2 & 7.6 \\ -4.6 & 10.0 & -4.7 & -6.8 \\ 5.7 & -3.7 & 7.3 & 11.0 \\ 8.8 & 6.8 & 14.0 & 4.7 \end{bmatrix}$

In Exercises 35–42, solve the given problems.

35. Show that the matrix $\begin{bmatrix} 1 & 1 \\ 1 & 1 \end{bmatrix}$ has no inverse.

(W) 36. Find the determinant of the matrix $\begin{bmatrix} 1 & -2 & 0 \\ -2 & 4 & 8 \\ 3 & -6 & 6 \end{bmatrix}$. Explain what this tells us about its inverse.

37. For the matrix $A = \begin{bmatrix} a & b \\ c & d \end{bmatrix}$, show that

$$\frac{1}{ad - bc} \begin{bmatrix} a & b \\ c & d \end{bmatrix} \begin{bmatrix} d & -b \\ -c & a \end{bmatrix} = \begin{bmatrix} 1 & 0 \\ 0 & 1 \end{bmatrix}$$

This verifies the method of Example 1.

(W) 38. Describe the relationship between the elements of the matrix $\begin{bmatrix} a & 0 & 0 \\ 0 & b & 0 \\ 0 & 0 & c \end{bmatrix}$ and the elements of its inverse.

39. For matrices $A = \begin{bmatrix} 2 & -4 \\ -1 & 3 \end{bmatrix}$, $B = (A + I)^{-1}$, and $C = A - I$, show that $BC = CB$.

40. The matrix $A = \begin{bmatrix} 2 & -1 & 1 \\ -1 & 4 & -3 \\ 1 & -3 & 2 \end{bmatrix}$ is *symmetric* (note the elements on opposite sides of the main diagonal are equal). Show that A^{-1} is also symmetric.

41. For the *four-terminal network* shown in Fig. 16.11, it can be shown that the voltage matrix V is related to the coefficient matrix A and the current matrix I by $V = A^{-1}I$, where

$$V = \begin{bmatrix} v_1 \\ v_2 \end{bmatrix} \qquad A = \begin{bmatrix} a_{11} & a_{12} \\ a_{21} & a_{22} \end{bmatrix} \qquad I = \begin{bmatrix} i_1 \\ i_2 \end{bmatrix}$$

Fig. 16.11

Find the individual equations for v_1 and v_2 that give each in terms of i_1 and i_2.

42. The rotations of a robot arm such as that shown in Fig. 16.12 are often represented by matrices. The values represent trigonometric functions of the angles of rotation. For the following rotation matrix R, find R^{-1}.

$$R = \begin{bmatrix} 0.8 & 0.0 & -0.6 \\ 0.0 & 1.0 & 0.0 \\ 0.6 & 0.0 & 0.8 \end{bmatrix}$$

Fig. 16.12

Answers to Practice Exercises

1. and **2.** $\begin{bmatrix} -1 & 4 \\ -\frac{1}{2} & \frac{3}{2} \end{bmatrix}$

16.4 Matrices and Linear Equations

Multiply Matrix of Constants by Inverse of Matrix of Coefficients • Solving Systems of Equations • Matrix Solution on a Calculator

As we stated at the beginning of Section 16.1, matrices can be used to solve systems of equations, and in this section, we show one method by which this is done. As we develop this method, it will be apparent that there is a great deal of numerical work involved. However, methods such as this one are easily programmed for use on a computer, which can do the arithmetic work very rapidly. Also, most graphing calculators can perform these operations, and we will show an example at the end of the section in which a calculator is used to solve a system of four equations more readily than with earlier methods. It is the *method* of solving the system of equations that is of primary importance here.

Let us consider the system of equations

$$a_1 x + b_1 y = c_1$$
$$a_2 x + b_2 y = c_2$$

Recalling the definition of equality of matrices, we can write this system as

$$\begin{bmatrix} a_1 x + b_1 y \\ a_2 x + b_2 y \end{bmatrix} = \begin{bmatrix} c_1 \\ c_2 \end{bmatrix} \qquad (16.6)$$

If we let

$$A = \begin{bmatrix} a_1 & b_1 \\ a_2 & b_2 \end{bmatrix} \qquad X = \begin{bmatrix} x \\ y \end{bmatrix} \qquad C = \begin{bmatrix} c_1 \\ c_2 \end{bmatrix} \qquad (16.7)$$

the left side of Eq. (16.6) can be written as the product of matrices A and X.

$$AX = C \qquad (16.8)$$

If we now multiply (on the left) each side of this matrix equation by A^{-1}, we have

$$A^{-1}AX = A^{-1}C$$

Since $A^{-1}A = I$, we have

$$IX = A^{-1}C$$

However, $IX = X$. Therefore,

$$X = A^{-1}C \qquad (16.9)$$

NOTE ▶ Equation (16.9) states that *we can solve a system of linear equations by multiplying the one-column matrix of the constants on the right by the inverse of the matrix of the coefficients.* The result is a one-column matrix whose elements are the required values for the solution. Also, note that

CAUTION ▶
$$X = A^{-1}C \quad \text{and } \textit{not} \quad CA^{-1}$$

as the order of matrix multiplication must be carefully followed.

EXAMPLE 1 Matrix solution – two equations

Use matrices to solve the system of equations

$$2x - y = 7$$
$$5x - 3y = 18$$

We set up the matrix of coefficients and the matrix of constants as

$$A = \begin{bmatrix} 2 & -1 \\ 5 & -3 \end{bmatrix} \quad \text{and} \quad C = \begin{bmatrix} 7 \\ 18 \end{bmatrix}$$

By either of the methods of the previous section, we can determine the inverse of matrix A to be

$$A^{-1} = \begin{bmatrix} 3 & -1 \\ 5 & -2 \end{bmatrix}$$

We now form the matrix product $A^{-1}C$.

$$A^{-1}C = \begin{bmatrix} 3 & -1 \\ 5 & -2 \end{bmatrix} \begin{bmatrix} 7 \\ 18 \end{bmatrix} = \begin{bmatrix} 21 - 18 \\ 35 - 36 \end{bmatrix} = \begin{bmatrix} 3 \\ -1 \end{bmatrix}$$

Since $X = A^{-1}C$, this means that

$$\begin{bmatrix} x \\ y \end{bmatrix} = \begin{bmatrix} 3 \\ -1 \end{bmatrix}$$

Therefore, the required solution is $x = 3$ and $y = -1$, which checks when these values are substituted into the original equations. ∎

EXAMPLE 2 Matrix solution two equations – application

For the electric circuit shown in Fig. 16.13, the equations used to find the currents (in amperes) i_1 and i_2 are

$$\begin{matrix} 2.30i_1 + 6.45(i_1 + i_2) = 15.0 \\ 1.25i_2 + 6.45(i_1 + i_2) = 12.5 \end{matrix} \quad \text{or} \quad \begin{matrix} 8.75i_1 + 6.45i_2 = 15.0 \\ 6.45i_1 + 7.70i_2 = 12.5 \end{matrix}$$

Using matrices to solve this system of equations, we set up the matrix A of coefficients, the matrix C of constants, and the matrix X of currents as

$$A = \begin{bmatrix} 8.75 & 6.45 \\ 6.45 & 7.70 \end{bmatrix} \quad C = \begin{bmatrix} 15.0 \\ 12.5 \end{bmatrix} \quad X = \begin{bmatrix} i_1 \\ i_2 \end{bmatrix}$$

We now find the inverse of A as

$$A^{-1} = \frac{1}{8.75(7.70) - 6.45(6.45)} \begin{bmatrix} 7.70 & -6.45 \\ -6.45 & 8.75 \end{bmatrix} = \begin{bmatrix} 0.2988 & -0.2503 \\ -0.2503 & 0.3395 \end{bmatrix}$$

Therefore,

$$X = A^{-1}C = \begin{bmatrix} 0.2988 & -0.2503 \\ -0.2503 & 0.3395 \end{bmatrix} \begin{bmatrix} 15.0 \\ 12.5 \end{bmatrix}$$

$$= \begin{bmatrix} 0.2988(15.0) - 0.2503(12.5) \\ -0.2503(15.0) + 0.3395(12.5) \end{bmatrix} = \begin{bmatrix} 1.35 \\ 0.49 \end{bmatrix}$$

Therefore, the required currents are $i_1 = 1.35$ A and $i_2 = 0.49$ A. These values check when substituted into the original equations. ∎

Practice Exercise

1. Use matrices to solve the system of equations
$$x - 3y = 6$$
$$2x + y = 5$$

15.0 V
2.30 Ω
6.45 Ω
1.25 Ω
12.5 V
i_1
i_2

Fig. 16.13

EXAMPLE 3 Matrix solution – three equations

Use matrices to solve the system of equations

$$x + 4y - z = 4$$
$$x + 3y + z = 8$$
$$2x + 6y + z = 13$$

Setting up matrices A, C, and X, we have

$$A = \begin{bmatrix} 1 & 4 & -1 \\ 1 & 3 & 1 \\ 2 & 6 & 1 \end{bmatrix} \qquad C = \begin{bmatrix} 4 \\ 8 \\ 13 \end{bmatrix} \qquad X = \begin{bmatrix} x \\ y \\ z \end{bmatrix}$$

To give another example of finding the inverse of a 3×3 matrix, we briefly show the steps for finding A^{-1}:

$$\left[\begin{array}{ccc|ccc} 1 & 4 & -1 & 1 & 0 & 0 \\ 1 & 3 & 1 & 0 & 1 & 0 \\ 2 & 6 & 1 & 0 & 0 & 1 \end{array}\right] \rightarrow \left[\begin{array}{ccc|ccc} 1 & 4 & -1 & 1 & 0 & 0 \\ 0 & 1 & -2 & 1 & -1 & 0 \\ 0 & -2 & 3 & -2 & 0 & 1 \end{array}\right] \rightarrow \left[\begin{array}{ccc|ccc} 1 & 0 & 7 & -3 & 4 & 0 \\ 0 & 1 & -2 & 1 & -1 & 0 \\ 0 & 0 & 1 & 0 & 2 & -1 \end{array}\right]$$

$$\left[\begin{array}{ccc|ccc} 1 & 4 & -1 & 1 & 0 & 0 \\ 0 & -1 & 2 & -1 & 1 & 0 \\ 2 & 6 & 1 & 0 & 0 & 1 \end{array}\right] \rightarrow \left[\begin{array}{ccc|ccc} 1 & 0 & 7 & -3 & 4 & 0 \\ 0 & 1 & -2 & 1 & -1 & 0 \\ 0 & -2 & 3 & -2 & 0 & 1 \end{array}\right] \rightarrow \left[\begin{array}{ccc|ccc} 1 & 0 & 7 & -3 & 4 & 0 \\ 0 & 1 & 0 & 1 & 3 & -2 \\ 0 & 0 & 1 & 0 & 2 & -1 \end{array}\right]$$

$$\left[\begin{array}{ccc|ccc} 1 & 4 & -1 & 1 & 0 & 0 \\ 0 & -1 & 2 & -1 & 1 & 0 \\ 0 & -2 & 3 & -2 & 0 & 1 \end{array}\right] \rightarrow \left[\begin{array}{ccc|ccc} 1 & 0 & 7 & -3 & 4 & 0 \\ 0 & 1 & -2 & 1 & -1 & 0 \\ 0 & 0 & -1 & 0 & -2 & 1 \end{array}\right] \rightarrow \left[\begin{array}{ccc|ccc} 1 & 0 & 0 & -3 & -10 & 7 \\ 0 & 1 & 0 & 1 & 3 & -2 \\ 0 & 0 & 1 & 0 & 2 & -1 \end{array}\right]$$

Thus, $A^{-1} = \begin{bmatrix} -3 & -10 & 7 \\ 1 & 3 & -2 \\ 0 & 2 & -1 \end{bmatrix}$ and

$$X = A^{-1}C = \begin{bmatrix} -3 & -10 & 7 \\ 1 & 3 & -2 \\ 0 & 2 & -1 \end{bmatrix} \begin{bmatrix} 4 \\ 8 \\ 13 \end{bmatrix} = \begin{bmatrix} -12 - 80 + 91 \\ 4 + 24 - 26 \\ 0 \quad 16 - 13 \end{bmatrix} = \begin{bmatrix} -1 \\ 2 \\ 3 \end{bmatrix}$$

This means that $x = -1$, $y = 2$, and $z = 3$.   ∎

EXAMPLE 4 Matrix solution – three equations

Use matrices to solve the system of equations

$$x + 2y - z = -4$$
$$3x + 5y - z = -5$$
$$-2x - y - 2z = -5$$

Setting up matrices A, C, and X, we have

$$A = \begin{bmatrix} 1 & 2 & -1 \\ 3 & 5 & -1 \\ -2 & -1 & -2 \end{bmatrix} \qquad C = \begin{bmatrix} -4 \\ -5 \\ -5 \end{bmatrix} \qquad X = \begin{bmatrix} x \\ y \\ z \end{bmatrix}$$

Finding A^{-1} (see Example 4 of Section 16.3) and solving for X, we have

$$A^{-1} = \begin{bmatrix} \frac{11}{2} & -\frac{5}{2} & -\frac{3}{2} \\ -4 & 2 & 1 \\ -\frac{7}{2} & \frac{3}{2} & \frac{1}{2} \end{bmatrix}$$

$$X = A^{-1}C = \begin{bmatrix} \frac{11}{2} & -\frac{5}{2} & -\frac{3}{2} \\ -4 & 2 & 1 \\ -\frac{7}{2} & \frac{3}{2} & \frac{1}{2} \end{bmatrix} \begin{bmatrix} -4 \\ -5 \\ -5 \end{bmatrix} = \begin{bmatrix} -2 \\ 1 \\ 4 \end{bmatrix}$$

This means that $x = -2$, $y = 1$, and $z = 4$.   ∎

Fig. 16.14

EXAMPLE 5 Matrix solution on calculator – 4 equations

Use a calculator to perform the necessary matrix operations in solving the following system of equations:

$$2r + 4s - t + u = 5$$
$$r - 2s + 3t - u = -4$$
$$3r + s + 2t - 4u = 8$$
$$4r + 5s - t + 3u = -1$$

First, we set up matrices A, X, and C:

$$A = \begin{bmatrix} 2 & 4 & -1 & 1 \\ 1 & -2 & 3 & -1 \\ 3 & 1 & 2 & -4 \\ 4 & 5 & -1 & 3 \end{bmatrix} \quad X = \begin{bmatrix} r \\ s \\ t \\ u \end{bmatrix} \quad C = \begin{bmatrix} 5 \\ -4 \\ 8 \\ -1 \end{bmatrix}$$

It is now necessary only to enter matrices A and C in the calculator and find the matrix product $A^{-1}C$, as shown in the upper window in Fig. 16.14. (There is no need to record or display A^{-1}.) This shows that the solution is

$$r = -2 \quad s = 3 \quad t = 0.5 \quad u = -2.5$$

This solution can be checked on the calculator by storing the resulting matrix X as matrix B and finding the matrix product AB, which should equal matrix C as shown in Fig. 16.14. The product AB is equivalent to substituting each value into the original equations, as shown in Eq. (16.8). ∎

EXERCISES 16.4

In Exercises 1 and 2, make the given changes in the indicated examples of this section and then solve the systems of equations.

1. In Example 1, change the 18 to 19 and then solve the system of equations.

2. In Example 3, change the 8 to 7 and the 13 to 12 and then solve the system of equations.

In Exercises 3–10, solve the given systems of equations by using the inverse of the coefficient matrix. The numbers in parentheses refer to exercises from Section 16.3, where the inverses may be checked.

3. $2x - 5y = -14$ (3)
$-2x + 4y = 11$

4. $-x + 5y = 4$ (5)
$4x + 10y = -4$

5. $x + 2y = 7$ (11)
$2x + 3y = 11$

6. $2x + 4y = -9$ (13)
$-x - y = 2$

7. $2x + 5y = -6$ (15)
$-x + 2y = -6$

8. $x - 3y - 2z = -8$ (19)
$-2x + 7y + 3z = 19$
$x - y - 3z = -3$

9. $x + 3y + 2z = 5$ (21)
$-2x - 5y - z = -1$
$2x + 4y = -2$

10. $2x + 4y = -2$ (23)
$3x + 4y - 2z = -6$
$-x + y + 2z = 5$

In Exercises 11–18, solve the given systems of equations by using the inverse of the coefficient matrix.

11. $2x - 3y = 3$
$4x - 5y = 4$

12. $x + 2y = 3$
$3x + 4y = 11$

13. $2.5x + 2.8y = -3.0$
$3.5x - 1.6y = 9.6$

14. $12x - 5y = -400$
$31x + 25y = 180$

15. $x + 2y + 2z = -4$
$4x + 9y + 10z = -18$
$-x + 3y + 7z = -7$

16. $x - 4y - 2z = -7$
$-x + 5y + 5z = 18$
$3x - 7y + 10z = 38$

17. $2x + 4y + z = 5$
$-2x - 2y - z = -6$
$-x + 2y + z = 0$

18. $4x + y = 2$
$-2x - y + 3z = -18$
$2x + y - z = 8$

In Exercises 19–26, solve the given systems of equations by using the inverse of the coefficient matrix. Use a calculator to perform the necessary matrix operations and display the results and the check. See Example 5.

19. $2x - y - z = 7$
$4x - 3y + 2z = 4$
$3x + 5y + z = -10$

20. $6x + 2y + 9z = 13$
$7x + 6y - 6z = 6$
$5x - 4y + 3z = 15$

21. $u - 3v - 2w = 9$
$3u + 2v + 6w = 20$
$4u - v + 3w = 25$

22. $2x + y - z = 1$
$3x - 2y - 8z = -3$
$x + 3y + z = 10$

23. $x - 5y + 2z - t = -18$
$3x + y - 3z + 2t = 17$
$4x - 2y + z - t = -1$
$-2x + 3y - z + 4t = 11$

24. $2p + q + 5r + s = 5$
$p + q - 3r - 4s = -1$
$3p + 6q - 2r + s = 8$
$2p + 2q + 2r - 3s = 2$

25. $2v + 3w + x - y - 2z = 6$
$6v - 2w - x + 3y - z = 21$
$v + 3w - 4x + 2y + 3z = -9$
$3v - w - x + 7y + 4z = 5$
$v + 6w + 6x - 4y - z = -4$

26. $4x - y + 2z - 2t + u = -15$
$8x + y - z + 4t - 2u = 26$
$2x - 6y - 2z + t - u = 10$
$2x + 5y + z - 3t + 8u = -22$
$4x - 3y + 2z + 4t + 2u = -4$

In Exercises 27–38, solve the indicated systems of equations using the inverse of the coefficient matrix. In Exercises 33–38, it is necessary to set up the appropriate equations.

27. For the following system of equations, solve for x^2 and y using the matrix methods of this section, and then solve for x and y.

$x^2 + y = 2$
$2x^2 - y = 10$

28. For the following system of equations, solve for x^2 and y^2 using the matrix methods of this section, and then solve for x and y.

$x^2 - y^2 = 8$
$x^2 + y^2 = 10$

(W) 29. Solve any pair of the system of equations $2x - y = 4$, $3x + y = 1$, and $x - 2y = 5$. Show that the solution is valid for any pair chosen. What conclusions can be drawn about the graphs of the three equations?

(W) 30. In solving the system of equations $3x - 4y = 5$, $8y - 6x = 7$, what conclusion can be drawn?

31. Forces **A** and **B** hold up a beam that weighs 2540 N, as shown in Fig. 16.15. The equations used to find the forces are

$A \sin 47.2° + B \sin 64.4° = 2540$
$A \cos 47.2° - B \cos 64.4° = 0$

Find the magnitude of each force.

Fig. 16.15

32. In applying Kirchhoff's laws (see Exercise 36 on page 168) to the circuit shown in Fig. 16.16, the following equations are found. Determine the indicated currents (in A).

$I_A + I_B + I_C = 0$
$2I_A - 5I_B = 6$
$5I_B - I_C = -3$

Fig. 16.16

33. Two batteries in an electric circuit have a combined voltage of 18 V, and one battery produces 6 V less than twice the other. What is the voltage of each?

34. An investment of $5000 is invested partly at 3.00% and the remainder at 3.50%. If the total annual income is $167, how much is invested at each rate?

35. What volume of each of a 20% acid solution and a 50% acid solution should be combined to form 48 mL of a 25% solution?

36. Two computer programs together require 51.8 megabytes of memory. If one program requires 2.0 megabytes more than twice the other, what are the memory requirements of each?

37. A research chemist wants to make 10.0 L of gasoline containing 2.0% of a new experimental additive. Gasoline without additive and two mixtures of gasoline with additive, one with 5.0% and the other with 6.0%, are to be used. If four times as much gasoline without additive as the 5.0% mixture is to be used, how much of each is needed?

38. A river tour boat takes 5.0 h to cruise downstream and 7.0 h for the return upstream. If the river flows at 4.0 mi/h, how fast does the boat travel in still water, and how far downstream does the boat go before starting the return trip?

Answer to Practice Exercise

1. $x = 3$, $y = -1$

16.5 Gaussian Elimination

Gaussian Elimination Form •
Solving Systems of Equations •
Number of Possible Solutions •
Consistent and Inconsistent Systems

We now show a general method that can be used to solve a system of linear equations. The procedure used in this method is similar to that used in finding the inverse of a matrix in Section 16.3. It is known as **Gaussian elimination,** and as noted in the chapter introduction, it was developed in the early 1800s by Karl Gauss. Today, it is commonly used in computer programs for the solutions of systems of linear equations.

When using this method, we first rewrite the system of equations in a different form. In doing so, there are very specific operations that may be used on the equations. The following example illustrates the form that is to be obtained, and we then state the operations that are used to obtain this form.

EXAMPLE 1 Gaussian elimination form

■ $x - 3y - z = 1$
$\quad\quad y + 2z = 5$
$\quad\quad\quad\quad z = 2$

In solving the system of linear equations to the left, we see that the third equation directly gives us the value $z = 2$. Since the second equation contains only y and z, we can now substitute $z = 2$ into the second equation to get $y = 1$. Then we can find x by substituting $y = 1$ and $z = 2$ into the first equation, and get $x = 6$. Therefore, after writing the system of equations in the form to the left, the solution is completed by substituting back, starting with the last equation. ∎

The method of Gaussian elimination is based on first rewriting the given system of equations in a form like that shown in Example 1, so that the last equation shows the value of one of the unknowns, and the others are found by substituting back, as in Example 1. The method to obtain the proper form is based on the use of any of the following three operations on the equations of the original system of equations.

Operations Used to Obtain Proper Form

1. *Any two equations may be interchanged.*
2. *Both sides of an equation may be multiplied by a nonzero constant.*
3. *A nonzero multiple of one equation may be added to another equation.*

Using three linear equations in three unknowns as an example, by using the above operations, we can change the system

$$a_1x + b_1y + c_1z = d_1$$
$$a_2x + b_2y + c_2z = d_2$$
$$a_3x + b_3y + c_3z = d_3$$

(16.10)

into the equivalent system

$$x + b_4y + c_4z = d_4$$
$$y + c_5z = d_5$$
$$z = d_6$$

(16.11)

The solution is completed by substituting the value of z into the second equation, and then substituting the values of y and z into the first equation, as in Example 1.

EXAMPLE 2 Solving a system of two equations

Solve the given system of equations by Gaussian elimination.

Given system of equations (equations and solution at left)

■ This is in general form of Eq. (16.10).
$2x + y = 4$
$3x - 2y = 3$

$x + \frac{1}{2}y = 2$
$3x - 2y = 3$

$x + \frac{1}{2}y = 2$
$\quad - \frac{7}{2}y = -3$

This is in general form of Eq. (16.11).
$x + \frac{1}{2}y = 2$
$\quad\quad y = \frac{6}{7}$

$x + \frac{1}{2}\left(\frac{6}{7}\right) = 2$
$\quad\quad x = \frac{11}{7}$

We want the coefficient of x in the first equation to be 1. Therefore, divide the first equation by 2, the coefficient of x.

We next eliminate x in the second equation by subtracting 3 times the first equation from the second equation.

Now, we solve the second equation for y by dividing by $-\frac{7}{2}$.

To find the value of x, we substitute the value of $y = \frac{6}{7}$ into the first equation.

The solution is $x = \frac{11}{7}$, $y = \frac{6}{7}$, which checks when substituted into the original equations. ∎

EXAMPLE 3 Solving system of three equations

Solve the given system of equations by Gaussian elimination.

Given system of equations (equations and solution at left)

■ This is in the general form of Eq. (16.10)

$$\begin{aligned} x + 3y - 2z &= -5 \\ 2x - y + 4z &= 7 \\ -3x + 2y - 3z &= -1 \end{aligned}$$

Since the coefficient of x in the first equation is 1, proceed to the next step. Subtract 2 times the first equation from the second equation and add 3 times the first equation to the third equation to eliminate x from the second and third equations.
To get the coefficient of y equal to 1 in the second equation, divide it by -7.

$$\begin{aligned} x + 3y - 2z &= -5 \\ - 7y + 8z &= 17 \\ 11y - 9z &= -16 \end{aligned}$$

$$\begin{aligned} x + 3y - 2z &= -5 \\ y - \tfrac{8}{7}z &= -\tfrac{17}{7} \\ 11y - 9z &= -16 \end{aligned}$$

To eliminate y from the third equation, subtract 11 times the second equation from the third equation.

$$\begin{aligned} x + 3y - 2z &= -5 \\ y - \tfrac{8}{7}z &= -\tfrac{17}{7} \\ \tfrac{25}{7}z &= \tfrac{75}{7} \end{aligned}$$

Solve for z by multiplying the third equation by $\tfrac{7}{25}$ (or dividing by $\tfrac{25}{7}$).

■ This is in the form of Eq. (16.11)

$$\begin{aligned} x + 3y - 2z &= -5 \\ y - \tfrac{8}{7}z &= -\tfrac{17}{7} \\ z &= 3 \end{aligned}$$

The value of y is found by substituting $z = 3$ into the second equation.

$$\begin{aligned} y - \tfrac{8}{7}(3) &= -\tfrac{17}{7} \\ y &= 1 \end{aligned}$$

The value of x is found by substituting $y = 1$ and $z = 3$ back into the first equation.
Thus, $x = -2$, $y = 1$, $z = 3$, and this solution checks in the original equations. ■

$$\begin{aligned} x + 3(1) - 2(3) &= -5 \\ x &= -2 \end{aligned}$$

EXAMPLE 4 Solving system with $a_0 = 0$

Solve the given system of equations by Gaussian elimination.

Given system of equations (equations and solution at left)

■ This is in general form of Eq. (16.10).

$$\begin{aligned} 4y + z &= 2 \\ 2x + 6y - 2z &= 3 \\ 4x + 8y - 5z &= 4 \end{aligned}$$

Since the first equation does not contain x, which means that $a_1 = 0$, we cannot divide by a_1. Therefore, interchange the first and second equations and then divide the new first equation by 2.

$$\begin{aligned} x + 3y - z &= \tfrac{3}{2} \\ 4y + z &= 2 \\ 4x + 8y - 5z &= 4 \end{aligned}$$

Eliminate x in the third equation by subtracting 4 times the first equation from the third equation.

$$\begin{aligned} x + 3y - z &= \tfrac{3}{2} \\ 4y + z &= 2 \\ - 4y - z &= -2 \end{aligned}$$

Make the coefficient of y in the second equation equal to 1 by dividing the second equation by 4.

$$\begin{aligned} x + 3y - z &= \tfrac{3}{2} \\ y + \tfrac{1}{4}z &= \tfrac{1}{2} \\ - 4y - z &= -2 \end{aligned}$$

Eliminate y from the third equation by adding 4 times the second equation to the third equation.

■ This is the form that Eq. (16.11) takes in this case.

$$\begin{aligned} x + 3y - z &= \tfrac{3}{2} \\ y + \tfrac{1}{4}z &= \tfrac{1}{2} \\ 0 &= 0 \end{aligned}$$

Since the third equation, $0 = 0$, is correct, we can continue. Although there is no specific value for z, it is possible to express both x and y in terms of z.

$$\begin{aligned} x + 3y - z &= \tfrac{3}{2} \\ y &= \tfrac{1}{2} - \tfrac{1}{4}z \end{aligned}$$

Solve the second equation for y. In this case, it is expressed in terms of z. We no longer need to include the third equation.

$$x + 3\left(\tfrac{1}{2} - \tfrac{1}{4}z\right) - z = \tfrac{3}{2}$$

$$x = \tfrac{7}{4}z$$

Substitute the solution for y in the first equation and solve for x in terms of z.

CAUTION ▶

We have expressed the solution as $x = \tfrac{7}{4}z$ and $y = \tfrac{1}{2} - \tfrac{1}{4}z$. Since both x and y are expressed in terms of z and there is no specific value of z, the value of z can be chosen arbitrarily. This means *there is an unlimited number of solutions.* For example, if $z = 4$, then $x = 7$ and $y = -\tfrac{1}{2}$. If $z = -2$, then $x = -\tfrac{7}{2}$ and $y = 1$. ■

Possible Number of Solutions

When we solved systems of linear equations in Chapter 5, we found that not all systems have unique solutions, as they do in Examples 2 and 3. In Example 4, we illustrated the use of Gaussian elimination on a system of equations for which the solution is not unique.

In Example 4, one of the equations became $0 = 0$, and there was an unlimited number of solutions. If any of the equations of a system becomes $0 = a$, $a \neq 0$, then the system is inconsistent and there is no solution. Example 5 illustrates such a system.

If a system has more unknowns than equations, or it can be written this way, as in Example 4, it usually has an unlimited number of solutions. It is possible, however, that such a system is inconsistent.

If a system has more equations than unknowns, it is inconsistent unless enough equations become $0 = 0$ such that at least one solution is found. The following example illustrates two systems in which there are more equations than unknowns.

Practice Exercise

1. Use Gaussian elimination to solve the system of equations
$$2x - y = 7$$
$$4x + 3y = -1$$

EXAMPLE 5 Consistent and inconsistent systems

Solve the following systems of equations by Gaussian elimination.

The solutions are shown at the left. We note that each system has three equations and two unknowns.

In the solution of the first system, the third equation becomes $0 = 0$, and only two equations are needed to find the solution $x = 1$, $y = 2$.

In the solution of the second system, the third equation becomes $0 = -4$, which means the system is inconsistent and there is no solution.

Graphing the systems, note that in Fig. 16.17, each of the lines passes through the point $(1, 2)$, whereas in Fig. 16.18, there is no point common to the three lines.

First system
$$
\begin{aligned}
x + 2y &= 5 \\
3x - y &= 1 \\
4x + y &= 6
\end{aligned}
$$
$$
\begin{aligned}
x + 2y &= 5 \\
-7y &= -14 \\
-7y &= -14
\end{aligned}
$$
$$
\begin{aligned}
x + 2y &= 5 \\
y &= 2 \\
-7y &= -14
\end{aligned}
$$
$$
\begin{aligned}
x + 2y &= 5 \\
y &= 2 \\
0 &= 0 \\
x &= 1
\end{aligned}
$$

Second system
$$
\begin{aligned}
x + 2y &= 5 \\
3x - y &= 1 \\
4x + y &= 2
\end{aligned}
$$
$$
\begin{aligned}
x + 2y &= 5 \\
-7y &= -14 \\
-7y &= -18
\end{aligned}
$$
$$
\begin{aligned}
x + 2y &= 5 \\
y &= 2 \\
-7y &= -18
\end{aligned}
$$
$$
\begin{aligned}
x + 2y &= 5 \\
y &= 2 \\
0 &= -4
\end{aligned}
$$

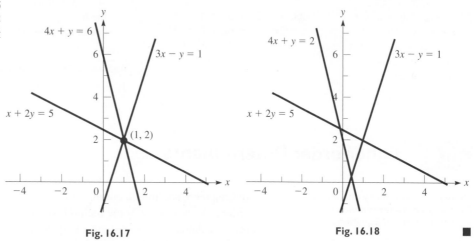

Fig. 16.17   Fig. 16.18

EXERCISES 16.5

In Exercises 1 and 2, make the given changes in the indicated examples of this section, and then solve the resulting systems using Gaussian elimination.

1. In Example 2, interchange the two equations. (After the interchange, do not use Procedure 1 on page 447.)

2. In Example 5, change the third equation of the second system to $5x + 3y = 11$.

In Exercises 3–28, solve the given systems of equations by Gaussian elimination. If there is an unlimited number of solutions, find two of them.

3. $x + 2y = 4$
   $3x - y = 5$

4. $2x + y = 1$
   $5x + 2y = 1$

5. $5x - 3y = 2$
   $-2x + 4y = 3$

6. $-3x + 2y = 4$
   $4x + y = -5$

7. $2x - 3y + z = 4$
   $6y - 4x - 2z = 9$

8. $3s + 4t - u = -5$
   $2u - 6s - 8t = 10$

9. $x + 3y + 3z = -3$
   $2x + 2y + z = -5$
   $-2x - y + 4z = 6$

10. $3x - y + 2z = 3$
    $4x - 2y + z = 3$
    $6x + 6y + 3z = 4$

11. $w + 2x - y + 3z = 12$
$2w - 2y - z = 3$
$3x - y - z = -1$
$-w + 2x + y + 2z = 3$

12. $3x - 2y = -11$
$5x + y = -1$
$2x + 3y = 10$
$x - 5y = -21$

13. $x - 4y + z = 2$
$3x - y + 4z = -4$

14. $4x + z = 6$
$2x - y - 2z = -2$

15. $2x - y + z = 5$
$3x + 2y - 2z = 4$
$5x + 8y - 8z = 5$

16. $3u + 6v + 2w = -2$
$u + 3v - 4w = 2$
$2u - 3v - 2w = -2$

17. $x + 3y + z = 4$
$2x - 6y - 3z = 10$
$4x - 9y + 3z = 4$

18. $30x + 20y - 10z = 30$
$4x - 2y - 6z = 4$
$-5x + 20y - 25z = -5$

19. $2x - 4y = 7$
$3x + 5y = -6$
$9x - 7y = 15$

20. $4x - y = 5$
$2x + 2y = 3$
$6x - 4y = 7$
$2x + y = 4$

21. $3x + 5y = -2$
$24x - 18y = 13$
$15x - 33y = 19$
$6x + 68y = -33$

22. $x + 3y - z = 1$
$3x - y + 4z = 4$
$-2x + 2y + 3z = 17$
$3x + 7y + 5z = 23$

23. $4x - 8y - 8z = 12$
$10x + 5y + 15z = 20$
$-6x - 3y - 3z = 15$
$3x + 3y - 2z = 2$

24. $2x - y - 2z - t = 4$
$4x + 2y + 3z + 2t = 3$
$-2x - y + 4z = -2$

25. $s + 2t - 3u = 2$
$6t + 3s - 9u = 6$
$7s + 14t - 21u = 13$

26. $x + 2y - 3z + 2t = 3$
$4y - 6z + 4t = 1$

27. $r - s - 3t - u = 1$
$2r + 4s - 2u = 2$
$r + 5s + 3t - u = 1$
$3r + 4s - 2t = 0$
$r + 2t - 3u = 3$

28. $x + 2y - 3z = 4$
$2x - y - 6z + 2t = 2$
$x + 3y + 3z - t = 1$

In Exercises 29 and 30, solve the given problems using Gaussian elimination.

29. Solve the system $a_1x + b_1y = c_1$, $a_2x + b_2y = c_2$ and show that the result is the same as that obtained using determinants as in Section 5.5. See page 154.

30. Solve the system $x + 2y = 6$, $2x + ay = 4$ and show that the solution depends on the value of a. What value of a does the solution show may not be used?

In Exercises 31–34, set up systems of equations and solve by Gaussian elimination.

31. One computer can perform x calculations per second, and a second computer can perform y calculations per second. If each operates for 0.200 s, 25.0 million calculations are performed. If the first operates for 0.400 s and the second for 0.300 s, 43.2 million calculations are performed. Find x and y.

32. The voltage across an electric resistor equals the current (in A) times the resistance (in Ω). If a current of 3.00 A passes through each of two resistors, the sum of the voltages is 10.5 V. If 2.00 A passes through the first resistor and 4.00 A passes through the second resistor, the sum of the voltages is 13.0 V. Find the resistances.

33. Three machines together produce 650 parts each hour. Twice the production of the second machine is 10 parts/h more than the sum of the production of the other two machines. If the first operates for 3.00 h and the others operate for 2.00 h, 1550 parts are produced. Find the production rate of each machine.

34. A total of $12,000 is invested, part at 6.5%, part at 6.0%, and part at 5.5%, yielding a total annual interest of $726. The income from the 6.5% part yields $128 less than that for the other two parts combined. How much is invested at each rate?

Answer to Practice Exercise

1. $x = 2$, $y = -3$

16.6 Higher-order Determinants

Minors • Expansion by Minors •
Properties of Determinants •
Solving Systems of Equations

In Chapter 5, we limited our discussion of determinants to those of the second and third orders. We now show some methods of evaluating higher-order determinants, and use these methods to solve systems of equations.

From Section 5.7, we recall the definition of a third-order determinant. Then rearranging terms and factoring a_1, $-a_2$, and a_3, we have

$$\begin{vmatrix} a_1 & b_1 & c_1 \\ a_2 & b_2 & c_2 \\ a_3 & b_3 & c_3 \end{vmatrix} = a_1b_2c_3 + a_3b_1c_2 + a_2b_3c_1 - a_3b_2c_1 - a_1b_3c_2 - a_2b_1c_3$$

$$= a_1(b_2c_3 - b_3c_2) - a_2(b_1c_3 - b_3c_1) + a_3(b_1c_2 - b_2c_1)$$

$$= a_1\begin{vmatrix} b_2 & c_2 \\ b_3 & c_3 \end{vmatrix} - a_2\begin{vmatrix} b_1 & c_1 \\ b_3 & c_3 \end{vmatrix} + a_3\begin{vmatrix} b_1 & c_1 \\ b_2 & c_2 \end{vmatrix} \qquad (16.12)$$

In Eq. (16.12), the third-order determinant is expanded as products of the elements of the first column and second-order determinants, known as *minors*. In general, *the* **minor** *of an element of a determinant is the determinant that results by deleting the row and column in which the element lies.*

EXAMPLE 1 Minors

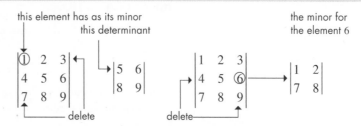

Equation (16.12) is only one way of expressing the expansion of the third-order determinant, but it does lead to a general theorem for the expansion of a determinant of any order. It is not a proof of the following theorem, but does provide a basis for it.

EXPANSION OF A DETERMINANT BY MINORS

The value of a determinant of order n may be found by forming the n products of the elements of any column (or row) and their minors. A product is given a plus sign if the sum of the number of the column and the number of the row in which the element lies is even, and a minus sign if this sum is odd. The algebraic sum of these terms is the value of the determinant.

EXAMPLE 2 Expansion by minors

In evaluating the following determinant, note that the third column has two zeros. This means that expanding by the third column will require less numerical work. Therefore,

$$\begin{vmatrix} 3 & -2 & 0 & 2 \\ 1 & 0 & -1 & 4 \\ -3 & 1 & 2 & -2 \\ 2 & -1 & 0 & -1 \end{vmatrix} = +(0)\begin{vmatrix} 1 & 0 & 4 \\ -3 & 1 & -2 \\ 2 & -1 & -1 \end{vmatrix} - (-1)\begin{vmatrix} 3 & -2 & 2 \\ -3 & 1 & -2 \\ 2 & -1 & -1 \end{vmatrix} + (2)\begin{vmatrix} 3 & -2 & 2 \\ 1 & 0 & 4 \\ 2 & -1 & -1 \end{vmatrix} - (0)\begin{vmatrix} 3 & -2 & 2 \\ 1 & 0 & 4 \\ -3 & 1 & -2 \end{vmatrix}$$

$$= \begin{vmatrix} 3 & -2 & 2 \\ -3 & 1 & -2 \\ 2 & -1 & -1 \end{vmatrix} + 2\begin{vmatrix} 3 & -2 & 2 \\ 1 & 0 & 4 \\ 2 & -1 & -1 \end{vmatrix}$$

$$= \left[3\begin{vmatrix} 1 & -2 \\ -1 & -1 \end{vmatrix} - (-3)\begin{vmatrix} -2 & 2 \\ -1 & -1 \end{vmatrix} + 2\begin{vmatrix} -2 & 2 \\ 1 & -2 \end{vmatrix} \right] + 2\left[-(-2)\begin{vmatrix} 1 & 4 \\ 2 & -1 \end{vmatrix} + 0\begin{vmatrix} 3 & 2 \\ 2 & -1 \end{vmatrix} - (-1)\begin{vmatrix} 3 & 2 \\ 1 & 4 \end{vmatrix} \right]$$

expanding first determinant by first column   expanding second determinant by second column

$$= [3(-1 - 2) + 3(2 + 2) + 2(4 - 2)] + 2[2(-1 - 8) + (12 - 2)]$$

$$= [-9 + 12 + 4] + 2[-18 + 10] = 7 + 2(-8) = -9$$

PROPERTIES OF DETERMINANTS

We can expand any determinant by minors, but even a fourth-order determinant may require many calculations. We now state some properties of determinants with which they can be evaluated, often with less work than with minors, and not requiring a calculator or computer.

1. *If each element above or each element below the principal diagonal of a determinant is zero, then the product of the elements of the principal diagonal is the value of the determinant.*

2. *If **all** corresponding rows and columns of a determinant are interchanged, the value of the determinant is unchanged.*

EXAMPLE 3 Properties of determinants 1 and 2

(a) $\begin{vmatrix} 2 & 1 & 8 \\ 0 & -5 & 9 \\ 0 & 0 & -6 \end{vmatrix} = 2(-5)(-6) = 60$ property 1

(b) $\begin{vmatrix} 1 & 3 & -1 \\ 2 & 0 & 4 \\ -2 & 5 & -6 \end{vmatrix}$ we obtain the determinant $\begin{vmatrix} 1 & 2 & -2 \\ 3 & 0 & 5 \\ -1 & 4 & -6 \end{vmatrix}$ property 2
the value of each determinant is −18 ∎

3. *If two columns (or rows) of a determinant are identical, the value of the determinant is zero.*

4. *If two columns (or rows) of a determinant are interchanged, the value of the determinant is changed in sign.*

EXAMPLE 4 Properties of determinants 3 and 4

(a)   identical $\begin{vmatrix} 3 & 5 & 2 \\ -4 & 6 & 9 \\ -4 & 6 & 9 \end{vmatrix} = 0$ property 3

(b) $\begin{vmatrix} 3 & 0 & 2 \\ 1 & 1 & 5 \\ 2 & 1 & 3 \end{vmatrix}$ and $\begin{vmatrix} 2 & 0 & 3 \\ 5 & 1 & 1 \\ 3 & 1 & 2 \end{vmatrix}$ property 4
the value of the first is −8, and the value of the second is 8 ∎

5. *If all elements of a column (or row) are multiplied by the same number k, the value of the determinant is multiplied by k.*

6. *If all the elements of any column (or row) are multiplied by the same number k, and the resulting numbers are added to the corresponding elements of another column (or row), the value of the determinant is unchanged.*

EXAMPLE 5 Properties of determinants 5 and 6

(a) $\begin{vmatrix} -1 & 0 & 6 \\ 6 & 3 & -6 \\ 0 & 5 & 3 \end{vmatrix} = 3 \begin{vmatrix} -1 & 0 & 6 \\ 2 & 1 & -2 \\ 0 & 5 & 3 \end{vmatrix}$ property 5
the value of the first is 141, and the value of the second is 47
141 = 3(47)

(b) The value of the following determinant is unchanged if we multiply each element of the first row by 2 and add these numbers to the corresponding elements of the second row. The value of each determinant is −37. The great value in using property 6 is that we can purposely place zeros in the resulting determinant.

$4 \times 2 = 8$ $-1 \times 2 = -2$ $3 \times 2 = 6$

$\begin{vmatrix} 4 & -1 & 3 \\ 2 & 2 & 1 \\ 1 & 0 & -3 \end{vmatrix} = \begin{vmatrix} 4 & -1 & 3 \\ 2+8 & 2+(-2) & 1+6 \\ 1 & 0 & -3 \end{vmatrix} = \begin{vmatrix} 4 & -1 & 3 \\ 10 & 0 & 7 \\ 1 & 0 & -3 \end{vmatrix}$ or $\begin{vmatrix} 4 & -1 & 3 \\ 10 & 0 & 7 \\ 1 & 0 & -3 \end{vmatrix} = \begin{vmatrix} 4 & -1 & 3 \\ 2 & 2 & 1 \\ 1 & 0 & -3 \end{vmatrix}$ ∎

With the use of these six properties, determinants of higher order can be evaluated much more easily. The technique is to

obtain zeros in a given column (or row) in all positions except one.

We can then expand by this column (or row), thereby reducing the order of the determinant. Property 6 is probably the most valuable for obtaining the zeros.

EXAMPLE 6 Evaluating determinant using properties

Using these properties, we evaluate the following determinant.

$$\begin{vmatrix} 3 & 2 & -1 & 1 \\ -1 & 1 & 2 & 3 \\ 2 & 2 & 1 & 4 \\ 0 & -1 & -2 & 2 \end{vmatrix} = \begin{vmatrix} 0 & 5 & 5 & 10 \\ -1 & 1 & 2 & 3 \\ 2 & 2 & 1 & 4 \\ 0 & -1 & -2 & 2 \end{vmatrix}$$

Each element of the second row is multiplied by 3, and the resulting numbers are added to the corresponding elements of the first row. Here, we have used property 6. In this way, a zero has been placed in column 1, row 1.

$$= \begin{vmatrix} 0 & 5 & 5 & 10 \\ -1 & 1 & 2 & 3 \\ 0 & 4 & 5 & 10 \\ 0 & -1 & -2 & 2 \end{vmatrix}$$

Each element of the second row is multiplied by 2, and the resulting numbers are added to the corresponding elements of the third row. Again, we have used property 6. Also, a zero has been placed in the first column, third row. We now have three zeros in the first column.

$$= -(-1)\begin{vmatrix} 5 & 5 & 10 \\ 4 & 5 & 10 \\ -1 & -2 & 2 \end{vmatrix}$$

Expand the determinant by the first column. We have now reduced the determinant to a third-order determinant.

$$= 5\begin{vmatrix} 1 & 1 & 2 \\ 4 & 5 & 10 \\ -1 & -2 & 2 \end{vmatrix}$$

Factor 5 from each element of the first row. Here, we are using property 5.

$$= 5(2)\begin{vmatrix} 1 & 1 & 1 \\ 4 & 5 & 5 \\ -1 & -2 & 1 \end{vmatrix}$$

Factor 2 from each element of the third column. Again, we are using property 5. Also, by doing this, we have reduced the size of the numbers, and the resulting numbers are somewhat easier to work with.

$$= 10\begin{vmatrix} 1 & 1 & 1 \\ 0 & 1 & 1 \\ -1 & -2 & 1 \end{vmatrix}$$

Each element of the first row is multiplied by -4, and the resulting numbers are added to the corresponding elements of the second row. Here, we are using property 6. We have placed a zero in the first column, second row.

$$= 10\begin{vmatrix} 1 & 1 & 1 \\ 0 & 1 & 1 \\ 0 & -1 & 2 \end{vmatrix}$$

Each element of the first row is added to the corresponding element of the third row. Again, we have used property 6. A zero has been placed in the first column, third row. We now have two zeros in the first column.

$$= 10(1)\begin{vmatrix} 1 & 1 \\ -1 & 2 \end{vmatrix}$$

Expand the determinant by the first column.

$$= 10(2+1) = 30$$

Expand the second-order determinant.

A somewhat more systematic method is to place zeros below the principal diagonal and then use property 1. ∎

■ To display and evaluate a determinant on a calculator, see page 164.

Fig. 16.19

The methods of this section and that of expansion by minors illustrate the ways in which determinants were evaluated before the extensive use of computers and calculators. As we noted in Chapter 5, most graphing calculators can be used to quickly evaluate determinants. In fact, most can evaluate determinants up to the sixth- (or possibly higher) order determinants. Figure 16.19 shows the calculator evaluation of the fourth-order determinant of Example 6.

SOLVING SYSTEMS OF LINEAR EQUATIONS BY DETERMINANTS

We can use the expansion of determinants by minors to solve systems of linear equations. **Cramer's rule** *for solving systems of linear equations, as stated in Section 5.7, is valid for any system of n equations in n unknowns.*

EXAMPLE 7　Solving system, of four equations

Solve the following system of equations:

$$
\begin{aligned}
x + 2y + z &= 5 \\
2x \qquad + z + 2t &= 1 \\
x - y + 3z + 4t &= -6 \\
4x - y \qquad - 2t &= 0
\end{aligned}
$$

constants ↓ 　　　　　　　　　　　　　　　　　expanding by fourth row

$$
x = \frac{\begin{vmatrix} 5 & 2 & 1 & 0 \\ 1 & 0 & 1 & 2 \\ -6 & -1 & 3 & 4 \\ 0 & -1 & 0 & -2 \end{vmatrix}}{\begin{vmatrix} 1 & 2 & 1 & 0 \\ 2 & 0 & 1 & 2 \\ 1 & -1 & 3 & 4 \\ 4 & -1 & 0 & -2 \end{vmatrix}} = \frac{-(0)\begin{vmatrix} 2 & 1 & 0 \\ 0 & 1 & 2 \\ -1 & 3 & 4 \end{vmatrix} + (-1)\begin{vmatrix} 5 & 1 & 0 \\ 1 & 1 & 2 \\ -6 & 3 & 4 \end{vmatrix} - (0)\begin{vmatrix} 5 & 2 & 0 \\ 1 & 0 & 2 \\ -6 & -1 & 4 \end{vmatrix} + (-2)\begin{vmatrix} 5 & 2 & 1 \\ 1 & 0 & 1 \\ -6 & -1 & 3 \end{vmatrix}}{(1)\begin{vmatrix} 0 & 1 & 2 \\ -1 & 3 & 4 \\ -1 & 0 & -2 \end{vmatrix} - 2\begin{vmatrix} 2 & 1 & 2 \\ 1 & 3 & 4 \\ 4 & 0 & -2 \end{vmatrix} + (1)\begin{vmatrix} 2 & 0 & 2 \\ 1 & -1 & 4 \\ 4 & -1 & -2 \end{vmatrix} - (0)\begin{vmatrix} 2 & 0 & 1 \\ 1 & -1 & 3 \\ 4 & -1 & 0 \end{vmatrix}}
$$

expanding by first row

$$
= \frac{-(-26) - 2(-14)}{1(0) - 2(-18) + 1(18)} = \frac{26 + 28}{36 + 18} = \frac{54}{54} = 1
$$

We could use minors or the properties of determinants for the evaluations. Noting the two zeros in the fourth row of the numerator, we expanded by minors. For the denominator, which we need to evaluate only once, we used minors of the first row, although we could have used the properties to create zeros in the first column.

In solving for y, we again note the two zeros in the fourth row:

expanding by fourth row

$$
y = \frac{\begin{vmatrix} 1 & 5 & 1 & 0 \\ 2 & 1 & 1 & 2 \\ 1 & -6 & 3 & 4 \\ 4 & 0 & 0 & -2 \end{vmatrix}}{54} = \frac{-4\begin{vmatrix} 5 & 1 & 0 \\ 1 & 1 & 2 \\ -6 & 3 & 4 \end{vmatrix} + (-2)\begin{vmatrix} 1 & 5 & 1 \\ 2 & 1 & 1 \\ 1 & -6 & 3 \end{vmatrix}}{54}
$$

$$
= \frac{-4(-26) - 2(-29)}{54} = \frac{104 + 58}{54} = \frac{162}{54} = 3
$$

Substituting $x = 1$ and $y = 3$ in the first equation gives us $z = -2$. Then substituting $x = 1$ and $y = 3$ in the fourth equation gives us $t = 1/2$. Therefore, the required solution is $x = 1$, $y = 3$, $z = -2$, $t = 1/2$. We can check the solution by substituting in the second or third equation (we used the first and fourth to *find* values of z and t.)　■

EXERCISES 16.6

In Exercises 1 and 2, make the given changes in the indicated examples of this section. Then evaluate the resulting determinants.

1. In Example 4(b), change the third column of the first determinant to　0 0 3.

2. In Example 5(a), change the third column of the first determinant to　2 −2 1.

In Exercises 3–6, evaluate each determinant by inspection. Observation will allow evaluation by using the properties of this section.

3. $\begin{vmatrix} 4 & -5 & 8 \\ 0 & 3 & -8 \\ 0 & 0 & -5 \end{vmatrix}$

4. $\begin{vmatrix} 3 & 0 & 0 \\ 0 & 10 & 0 \\ -9 & -1 & -5 \end{vmatrix}$

5.
$$\begin{vmatrix} 3 & -2 & 4 & 2 \\ 5 & -1 & 2 & -1 \\ 3 & -2 & 4 & 2 \\ 0 & 3 & -6 & 0 \end{vmatrix}$$

6.
$$\begin{vmatrix} -12 & -24 & -24 & 15 \\ 12 & 32 & 32 & -35 \\ -22 & 18 & 18 & 18 \\ 44 & 0 & 0 & -26 \end{vmatrix}$$

In Exercises 7–10, use the given value of the determinant at the right and the properties of this section to evaluate the following determinants.

$$\begin{vmatrix} 2 & -3 & 1 \\ -4 & 1 & 3 \\ 1 & -3 & -2 \end{vmatrix} = 40$$

7.
$$\begin{vmatrix} 2 & 1 & -3 \\ -4 & 3 & 1 \\ 1 & -2 & -3 \end{vmatrix}$$

8.
$$\begin{vmatrix} 2 & -3 & 1 \\ -4 & 1 & 3 \\ 2 & -6 & -4 \end{vmatrix}$$

9.
$$\begin{vmatrix} 2 & -3 & -1 \\ -4 & 1 & -3 \\ 1 & -3 & 2 \end{vmatrix}$$

10.
$$\begin{vmatrix} 2 & -4 & 1 \\ -3 & 1 & -3 \\ 1 & 3 & -2 \end{vmatrix}$$

In Exercises 11–20, evaluate the given determinants by expansion by minors.

11.
$$\begin{vmatrix} 3 & 0 & 0 \\ -2 & 1 & 4 \\ 4 & -2 & 5 \end{vmatrix}$$

12.
$$\begin{vmatrix} 10 & 0 & -3 \\ -2 & -4 & 1 \\ 3 & 0 & 2 \end{vmatrix}$$

13.
$$\begin{vmatrix} 3 & 1 & 0 \\ -2 & 3 & -1 \\ 4 & 2 & 5 \end{vmatrix}$$

14.
$$\begin{vmatrix} -40 & 30 & -20 \\ -8 & 8 & 16 \\ -15 & 75 & -45 \end{vmatrix}$$

15.
$$\begin{vmatrix} 4 & 3 & 6 & 0 \\ 3 & 0 & 0 & 4 \\ 5 & 0 & 1 & 2 \\ 2 & 1 & 1 & 7 \end{vmatrix}$$

16.
$$\begin{vmatrix} 6 & -3 & -6 & 3 \\ -2 & 1 & 2 & -1 \\ 18 & 7 & -1 & 5 \\ 0 & -1 & 10 & 10 \end{vmatrix}$$

17.
$$\begin{vmatrix} 1 & 3 & -3 & 5 \\ 4 & 2 & 1 & 2 \\ 3 & 2 & -2 & 2 \\ 0 & 1 & 2 & -1 \end{vmatrix}$$

18.
$$\begin{vmatrix} -2 & 2 & 1 & 3 \\ 1 & 4 & 3 & 1 \\ 4 & 3 & -2 & -2 \\ 3 & -2 & 1 & 5 \end{vmatrix}$$

19.
$$\begin{vmatrix} 1 & 2 & 0 & 1 & 0 \\ 0 & 2 & 1 & 0 & 1 \\ 1 & 0 & -1 & 1 & -1 \\ -2 & 0 & -1 & 2 & 1 \\ 1 & 0 & 2 & -1 & -2 \end{vmatrix}$$

20.
$$\begin{vmatrix} -1 & 3 & 5 & 0 & -5 \\ 0 & 1 & 7 & 3 & -2 \\ 5 & -2 & -1 & 0 & 3 \\ -3 & 0 & 2 & -1 & 3 \\ 6 & 2 & 1 & -4 & 2 \end{vmatrix}$$

In Exercises 21–30, use the determinants for Exercises 11–20 and evaluate each using the properties of determinants. Do not evaluate directly more than one second-order determinant.

In Exercises 31–34, solve the given systems of equations by determinants. Evaluate by expansion by minors.

31. $x + t = 0$
$3x + y + z = -1$
$2y - z + 3t = 1$
$2z - 3t = 1$

32. $2x + y + z = 4$
$2y - 2z - t = 3$
$3y - 3z + 2t = 1$
$6x - y + t = 0$

33. $x + 2y - z = 6$
$y - 2z - 3t = -5$
$3x - 2y + t = 2$
$2x + y + z - t = 0$

34. $2p + 3r + s = 4$
$p - 2r - 3s + 4t = -1$
$3p + r + s - 5t = 3$
$-p + 2r + s + 3t = 2$

In Exercises 35–38, solve the given systems of equations by determinants. Evaluate by using the properties of determinants.

35. $2x + y + z = 2$
$3y - z + 2t = 4$
$y + 2z + t = 0$
$3x + 2z = 4$

36. $2x + y + z = 0$
$x - y + 2t = 2$
$2y + z + 4t = 2$
$5x + 2z + 2t = 4$

37. $D + E + 2F = 1$
$2D - E + G = -2$
$D - E - F - 2G = 4$
$2D - E + 2F - G = 0$

38. $3x + y + t = 0$
$3z + 2t = 8$
$6x + 2y + 2z + t = 3$
$3x - y - z - t = 0$

In Exercises 39–42, make the indicated changes in the determinant at the right, and then solve the indicated problem. Assume the elements are nonzero, unless otherwise specified

$$\begin{vmatrix} a & b & c \\ d & e & f \\ g & h & i \end{vmatrix}$$

39. Evaluate the determinant if $a = c$, $d = f$, and $g = i$.

40. Evaluate the determinant if $b = c = f = 0$.

41. By what factor is the value of the determinant changed if all elements are doubled?

42. How is the value changed if a is added to g, b added to h, and c added to i?

In Exercises 43–48, solve the given problems by using determinants.

43. In applying Kirchhoff's laws (see Exercise 42 on page 158) to the circuit shown in Fig. 16.20, the following equations are found. Determine the indicated currents (in A).

$I_A + I_B + I_C + I_D + I_E = 0$
$-2I_A + 3I_B = 0$
$3I_B - 3I_C = 6$
$-3I_C + I_D = 0$
$-I_D + 2I_E = 0$

Fig. 16.20

44. In analyzing the forces A, B, C, and D shown on the beam in Fig. 16.21, the following equations are used. Find these forces.

$A + B = 850$
$A + B + 400 = 0.8C + 0.6D$
$0.6C = 0.8D$
$5A - 5B + 4C - 3D = 0$

Fig. 16.21

45. In analyzing the motion of four equally spaced particles along a string, the equation at the right is found. Here, C depends on the string and the mass of each object. Solve for C $(C > 0)$.

$$\begin{vmatrix} C & -1 & 0 & 0 \\ -1 & C & -1 & 0 \\ 0 & -1 & C & -1 \\ 0 & 0 & -1 & C \end{vmatrix} = 0$$

46. An alloy is to be made from four other alloys containing copper (Cu), nickel (Ni), zinc (Zn), and Iron (Fe). The first is 80% Cu and 20% Ni. The second is 60% Cu, 20% Ni, and 20% Zn. The third is 30% Cu, 60% Ni, and 10% Fe. The fourth is 20% Ni, 40% Zn, and 40% Fe. How much of each is needed so that the final alloy has 56 g Cu, 28 g Ni, 10 g Zn, and 6 g Fe?

47. In testing for air pollution, a given air sample contained 6.0 parts per million (ppm) of four pollutants, sulfur dioxide (SO_2), nitric oxide (NO), nitrogen dioxide (NO_2), and carbon monoxide (CO). The ppm of CO was 10 times that of SO_2, which in turn equaled those of NO and NO_2. There was a total of 0.8 ppm of SO_2 and NO. How many ppm of each were present in the air sample?

48. Three computer programs, A, B, and C, use 15% of a computer's 6.0 GB (gigabyte) hard-drive memory. If two additional programs are added, one requiring the same memory as A and the other half the memory of C, 22% of the memory will be used. However, if two other programs are added to A, B, and C, one requiring half the memory of B and the other the same memory as C, 25% of the memory will be used. How many megabytes of memory are required for each of A, B, and C?

CHAPTER 16 EQUATIONS

Basic laws for matrices

$$A + B = B + A \quad \text{(commutative law)} \tag{16.1}$$
$$A + (B + C) = (A + B) + C \quad \text{(associative law)} \tag{16.2}$$
$$k(A + B) = kA + kB \tag{16.3}$$
$$A + O = A \tag{16.4}$$

Inverse matrix

$$AA^{-1} = A^{-1}A = I \tag{16.5}$$

Solving systems of equations by matrices

$$\begin{bmatrix} a_1x + b_1y \\ a_2x + b_2y \end{bmatrix} = \begin{bmatrix} c_1 \\ c_2 \end{bmatrix} \tag{16.6}$$

$$A = \begin{bmatrix} a_1 & b_1 \\ a_2 & b_2 \end{bmatrix} \quad X = \begin{bmatrix} x \\ y \end{bmatrix} \quad C = \begin{bmatrix} c_1 \\ c_2 \end{bmatrix} \tag{16.7}$$

$$AX = C \tag{16.8}$$
$$X = A^{-1}C \tag{16.9}$$

Gaussian elimination

$$\begin{aligned} a_1x + b_1y + c_1z &= d_1 & x + b_4y + c_4z &= d_4 \\ a_2x + b_2y + c_2z &= d_2 & y + c_5z &= d_5 \\ a_3x + b_3y + c_3z &= d_3 & z &= d_6 \end{aligned} \tag{16.10}\ (16.11)$$

Higher-order determinants

$$\begin{vmatrix} a_1 & b_1 & c_1 \\ a_2 & b_2 & c_2 \\ a_3 & b_3 & c_3 \end{vmatrix} = a_1\begin{vmatrix} b_2 & c_2 \\ b_3 & c_3 \end{vmatrix} - a_2\begin{vmatrix} b_1 & c_1 \\ b_3 & c_3 \end{vmatrix} + a_3\begin{vmatrix} b_1 & c_1 \\ b_2 & c_2 \end{vmatrix} \tag{16.12}$$

CHAPTER 16 REVIEW EXERCISES

In Exercises 1–6, determine the values of the literal numbers.

1. $\begin{bmatrix} 2a \\ a - b \end{bmatrix} = \begin{bmatrix} 8 \\ 5 \end{bmatrix}$

2. $\begin{bmatrix} x - y \\ 2x + 2z \\ 4y + z \end{bmatrix} = \begin{bmatrix} 1 \\ 3 \\ -1 \end{bmatrix}$

3. $\begin{bmatrix} 2x & 3y & 2z \\ x + y & 2y + z & z - x \end{bmatrix} = \begin{bmatrix} 4 & -9 & 5 \\ a & b & c \end{bmatrix}$

4. $\begin{bmatrix} a + bj & b \\ aj & b - aj \end{bmatrix} = \begin{bmatrix} 6j & 2d \\ 2cj & ej^2 \end{bmatrix}$ $(j = \sqrt{-1})$

5. $\begin{bmatrix} \cos \pi & \sin \frac{\pi}{6} \\ x + y & x - y \end{bmatrix} = \begin{bmatrix} x & y \\ a & b \end{bmatrix}$

6. $\begin{bmatrix} \ln e & \log 100 \\ a^2 & b^2 \end{bmatrix} = \begin{bmatrix} a + b & a - b \\ x & y \end{bmatrix}$

In Exercises 7–12, use the given matrices and perform the indicated operations.

$$A = \begin{bmatrix} 2 & -3 \\ 4 & 1 \\ -5 & 0 \\ 2 & -3 \end{bmatrix} \quad B = \begin{bmatrix} -1 & 0 \\ 4 & -6 \\ -3 & -2 \\ 1 & -7 \end{bmatrix} \quad C = \begin{bmatrix} 5 & -6 \\ 2 & 8 \\ 0 & -2 \end{bmatrix}$$

7. $A + B$ **8.** $2C$

9. $B - A$ **10.** $2C - B$

11. $2A - 3B$ **12.** $2(A - B)$

In Exercises 13–16, perform the indicated matrix multiplications.

13. $\begin{bmatrix} 2 & -1 \\ -2 & 1 \end{bmatrix} \begin{bmatrix} 1 & -1 \\ 2 & -2 \end{bmatrix}$

14. $\begin{bmatrix} 6 & -4 & 1 & 0 \\ 2 & 0 & -4 & 3 \end{bmatrix} \begin{bmatrix} 7 & -1 & 6 \\ 4 & 0 & 1 \\ 3 & -2 & 5 \\ 9 & 1 & 0 \end{bmatrix}$

15. $\begin{bmatrix} -0.1 & 0.7 \\ 0.2 & 0.0 \\ 0.4 & -0.1 \end{bmatrix} \begin{bmatrix} 0.1 & -0.4 & 0.5 \\ 0.5 & 0.1 & 0.0 \end{bmatrix}$

16. $\begin{bmatrix} 0 & -1 & 6 \\ 8 & 1 & 4 \\ 7 & -2 & -1 \end{bmatrix} \begin{bmatrix} 5 & -1 & 7 & 1 & 5 \\ 0 & 1 & 0 & 4 & 1 \\ 1 & -2 & 3 & 0 & 1 \end{bmatrix}$

In Exercises 17–24, find the inverses of the given matrices. Check each by using a calculator.

17. $\begin{bmatrix} 2 & -5 \\ 2 & -4 \end{bmatrix}$ **18.** $\begin{bmatrix} -1 & -6 \\ 2 & 10 \end{bmatrix}$

19. $\begin{bmatrix} 0.07 & -0.01 \\ 0.04 & 0.08 \end{bmatrix}$ **20.** $\begin{bmatrix} 50 & -12 \\ 42 & -80 \end{bmatrix}$

21. $\begin{bmatrix} 1 & 1 & -2 \\ -1 & -2 & 1 \\ 0 & 3 & 4 \end{bmatrix}$ **22.** $\begin{bmatrix} -1 & -1 & 2 \\ 2 & 3 & 0 \\ 1 & 4 & 1 \end{bmatrix}$

23. $\begin{bmatrix} 2 & -4 & 3 \\ 4 & -6 & 5 \\ -2 & 1 & -1 \end{bmatrix}$ **24.** $\begin{bmatrix} 3 & 1 & -4 \\ -3 & 1 & -2 \\ -6 & 0 & 3 \end{bmatrix}$

In Exercises 25–32, solve the given systems of equations using the inverse of the coefficient matrix.

25. $2x - 3y = -9$
$4x - y = -13$

26. $5A - 7B = 62$
$6A + 5B = -6$

27. $33x + 52y = -450$
$45x - 62y = 1380$

28. $0.24x - 0.26y = -3.1$
$0.40x + 0.34y = -1.3$

29. $2u - 3v + 2w = 7$
$3u + v - 3w = -6$
$u + 4v + w = -13$

30. $2x + 2y - z = 8$
$x + 4y + 2z = 5$
$3x - 2y + z = 17$

31. $x + 2y + 3z = 1$
$3x - 4y - 3z = 2$
$7x - 6y + 6z = 2$

32. $3x + 2y + z = 2$
$2x + 3y - 6z = 3$
$x + 3y + 3z = 1$

In Exercises 33–40, solve the given systems of equations by Gaussian elimination. For Exercises 33–38, use those that are indicated from Exercises 25–32.

33. Exercise 25 **34.** Exercise 26 **35.** Exercise 29

36. Exercise 30 **37.** Exercise 31 **38.** Exercise 32

39. $2x + 3y - z = 10$
$x - 2y + 6z = -6$
$5x + 4y + 4z = 14$

40. $x - 3y + 4z - 2t = 6$
$2x + y - 2z + 3t = 7$
$3x - 9y + 12z - 6t = 12$

In Exercises 41–44, solve the systems of equations by determinants, using the properties of determinants. As indicated, use the systems from Exercises 29–32.

41. Exercise 29 **42.** Exercise 30

43. Exercise 31 **44.** Exercise 32

In Exercises 45–52, solve the given systems of equations by using the coefficient matrix. Use a calculator to perform the necessary matrix operations and display the results and the check.

45. $3x - 2y + z = 6$
$2x + 3z = 3$
$4x - y + 5z = 6$

46. $7n + p + 2r = 3$
$4n - 2p + 4r = -2$
$2n + 3p - 6r = 3$

47. $2x - 3y + z - t = -8$
$4x + 3z + 2t = -3$
$2y - 3z - t = 12$
$x - y - z + t = 3$

48. $3x + 2y - 2z - 2t = 0$
$5y + 3z + 4t = 3$
$6y - 3z + 4t = 9$
$6x - y + 2z - 2t = -3$

49. $3x - y + 6z - 2t = 8$
$2x + 5y + z + 2t = 7$
$4x - 3y + 8z + 3t = -17$
$3x + 5y - 3z + t = 8$

50. $A + B + 2C - 3D = 15$
$3A + 3B - 8C - 2D = 9$
$6A - 4B + 6C + D = -6$
$2A + 2B - 4C - 2D = 8$

51. $4r - s + 8t - 2u + 4v = -1$
$3r + 2s - 4t + 3u - v = 4$
$3r + 3s + 2t + 5u + 6v = 13$
$6r - s + 2t - 2u + v = 0$
$r - 2s + 4t - 3u + 3v = 1$

52. $2v + 3w + 2x - 2y + 5z = -1$
$7v + 8w + 3x + y - 4z = 3$
$v - 2w - 4x - 4y - 8z = -9$
$3v - w + 7x + 5y - 3z = -18$
$4v + 5w + x + 3y - 6z = 7$

In Exercises 53–56, use matrices A and B.

$$A = \begin{bmatrix} 1 & 0 \\ 3 & 4 \end{bmatrix} \quad B = \begin{bmatrix} 0 & 1 & 0 \\ 0 & 0 & 1 \\ 1 & 0 & 0 \end{bmatrix}$$

53. Find A^2, A^3, and A^4. **54.** Show that $(A^2)^2 = A^4$.

55. Show that $B^3 = I$. **56.** Show that $B^4 = B$.

In Exercises 57–60, evaluate the given determinants by expansion by minors.

57. $\begin{vmatrix} 4 & 2 & 3 \\ 1 & -5 & -2 \\ -3 & 4 & -3 \end{vmatrix}$ **58.** $\begin{vmatrix} 4 & -5 & -1 \\ 6 & 1 & 6 \\ -2 & 4 & -3 \end{vmatrix}$

59. $\begin{vmatrix} 1 & 4 & 0 & -3 \\ 3 & 1 & 2 & 5 \\ -2 & -2 & -4 & 1 \\ -1 & 6 & 3 & -4 \end{vmatrix}$ **60.** $\begin{vmatrix} -2 & 6 & 6 & -1 \\ 1 & -2 & -5 & 2 \\ 5 & -4 & 4 & 3 \\ -3 & 1 & -2 & -3 \end{vmatrix}$

In Exercises 61–64, use the determinants for Exercises 57–60 and evaluate each using properties of determinants. Do not evaluate directly more than one second-order determinant.

In Exercises 65 and 66, use the matrix N.

$$N = \begin{bmatrix} 0 & -1 \\ 1 & 0 \end{bmatrix}$$

65. Show that $N^{-1} = -N$. **66.** Show that $N^2 = -I$.

In Exercises 67 and 68, solve the given problems.

67. For any real number n, show that $\begin{bmatrix} n & 1+n \\ 1-n & -n \end{bmatrix}^2 = I$.

(W) **68.** For the matrix $N = \begin{bmatrix} 1 & 1 \\ 1 & 1 \end{bmatrix}$, find (a) N^2, (b) N^3, (c) N^4. What is N^{20}? Explain.

In Exercises 69–72, use matrices A and B.

$$A = \begin{bmatrix} 1 & -2 \\ 0 & 3 \end{bmatrix} \quad B = \begin{bmatrix} -3 & 1 \\ 2 & -1 \end{bmatrix}$$

69. Show that $(A + B)(A - B) \neq A^2 - B^2$.

70. Show that $(A + B)^2 \neq A^2 + 2AB + B^2$.

71. Show that the inverse of $2A$ is $A^{-1}/2$.

72. Show that the inverse of $B/2$ is $2B^{-1}$.

In Exercises 73–76, solve the given systems of equations by use of matrices as in Section 16.4.

73. Two electric resistors, R_1 and R_2, are tested with currents and voltages such that the following equations are found:

$2R_1 + 3R_2 = 26$
$3R_1 + 2R_2 = 24$

Find the resistances R_1 and R_2 (in Ω).

74. A company produces two products, each of which is processed in two departments. Considering the worker time available, the numbers x and y of each product produced each week can be found by solving the system of equations

$4.0x + 2.5y = 1200$
$3.2x + 4.0y = 1200$

Find x and y.

75. A beam is supported as shown in Fig. 16.22. Find the force F and the tension T by solving the following system of equations:

$0.500F = 0.866T$
$0.866F + 0.500T = 350$

Fig. 16.22

76. To find the electric currents (in A) indicated in Fig. 16.23, it is necessary to solve the following equations.

$$\begin{aligned} I_A + I_B + I_C &= 0 \\ 5I_A - 2I_B &= -4 \\ 2I_B - I_C &= 0 \end{aligned}$$

Find I_A, I_B, and I_C.

Fig. 16.23   4 V

In Exercises 77–80, solve the system of equations in Exercises 73–76 by Gaussian elimination.

In Exercises 81–84, solve the given problems by setting up the necessary equations and solving them by any appropriate method of this chapter.

81. A crime suspect passes an intersection in a car traveling at 110 mi/h. The police pass the intersection 3.0 min later in a car traveling at 135 mi/h. How long is it before the police overtake the suspect?

82. A contractor needs a backhoe and a generator for two different jobs. Renting the backhoe for 5.0 h and the generator for 6.0 h costs $425 for one job. On the other job, renting the backhoe for 2.0 h and the generator for 8.0 h costs $310. What are the hourly charges for the backhoe and the generator?

83. By mass, three alloys have the following percentages of lead, zinc, and copper.

	Lead	Zinc	Copper
Alloy A	60%	30%	10%
Alloy B	40%	30%	30%
Alloy C	30%	70%	

How many grams of each of alloys A, B, and C must be mixed to get 100 g of an alloy that is 44% lead, 38% zinc, and 18% copper?

84. On a 750-mi trip from Salt Lake City to San Francisco that took a total of 5.5 h, a person took a limousine to the airport, then a plane, and finally a car to reach the final destination. The limousine took as long as the final car trip and the time for connections. The limousine averaged 55 mi/h, the plane averaged 400 mi/h, and the car averaged 40 mi/h. The plane traveled four times as far as the limousine and car combined. How long did each part of the trip and the connections take?

In Exercises 85–88, perform the indicated matrix operations.

85. An automobile maker has two assembly plants at which cars with either 4, 6, or 8 cylinders and with either standard or automatic transmission are assembled. The annual production at the first plant of cars with the number of cylinders–transmission type (standard, automatic) is as follows:
4: 12,000, 15,000; 6: 24,000, 8000; 8: 4000, 30,000
At the second plant the annual production is
4: 15,000, 20,000; 6: 12,000, 3000; 8: 2000, 22,000
Set up matrices for this production and by matrix addition, find the matrix for the total production by the number of cylinders and type of transmission.

86. Set up a matrix representing the information given in Exercise 83. A given shipment contains 500 g of alloy A, 800 g of alloy B, and 700 g of alloy C. Set up a matrix for this information. By multiplying these matrices, obtain a matrix that gives the total weight of lead, zinc, and copper in the shipment.

87. The matrix equation

$$\left[\begin{bmatrix} R_1 & -R_2 \\ -R_2 & R_1 \end{bmatrix} + R_2 \begin{bmatrix} 1 & 0 \\ 0 & 1 \end{bmatrix} \right] \begin{bmatrix} i_1 \\ i_2 \end{bmatrix} = \begin{bmatrix} 6 \\ 0 \end{bmatrix}$$

may be used to represent the system of equations relating the currents and resistances of the circuit in Fig. 16.24. Find this system of equations by performing the indicated matrix operations.

Fig. 16.24

88. A person prepared a meal of the following items, each having the given number of grams of protein, carbohydrates, and fat, respectively. Beef stew: 25, 21, 22; coleslaw: 3, 10, 10; (light) ice cream: 7, 25, 6. If the calorie count of each gram of protein, carbohydrate, and fat is 4.1 Cal/g, 3.9 Cal/g, and 8.9 Cal/g, respectively, find the total calorie count of each item by matrix multiplication. (1 Cal = 1 kcal.)

Writing Exercise

89. A hardware company has 60 different retail stores in which 1500 different products are sold. Write a paragraph or two explaining why matrices provide an efficient method of inventory control, and what matrix operations in this chapter would be of use.

CHAPTER 16   PRACTICE TEST

1. For matrices A and B, find $A - 2B$.
$$A = \begin{bmatrix} 3 & -1 & 4 \\ 2 & 0 & -2 \end{bmatrix} \quad B = \begin{bmatrix} 1 & 4 & 5 \\ -1 & -2 & 3 \end{bmatrix}$$

2. Evaluate the literal symbols.
$$\begin{bmatrix} 2x & x - y & z \\ x + z & 2y & y + z \end{bmatrix} = \begin{bmatrix} 6 & -2 & 4 \\ a & b & c \end{bmatrix}$$

3. For matrices C and D, find CD and DC.
$$C = \begin{bmatrix} 1 & 0 & 4 \\ 2 & -2 & 1 \\ -1 & 3 & 2 \end{bmatrix} \quad D = \begin{bmatrix} 2 & -2 \\ 4 & -5 \\ 6 & 1 \end{bmatrix}$$

4. Determine whether or not $B = A^{-1}$.
$$A = \begin{bmatrix} 2 & -5 \\ 1 & -2 \end{bmatrix} \quad B = \begin{bmatrix} -2 & 5 \\ -1 & 2 \end{bmatrix}$$

5. For matrix C of Problem 3, find C^{-1}.

6. Solve by using the inverse of the coefficient matrix.
$$2x - 3y = 11$$
$$x + 2y = 2$$

7. Solve the system of equations in Problem 6 by Gaussian elimination.

8. Evaluate the following determinant by expansion by minors.
$$\begin{vmatrix} 2 & -4 & -3 \\ -3 & 6 & 2 \\ 5 & -1 & 5 \end{vmatrix}$$

9. Evaluate the determinant in Problem 8 by using the properties of determinants.

10. Solve the following system of equations by using the inverse of the coefficient matrix. Use a calculator to perform the necessary matrix operations and display the result and check. Write down the display shown in the window with the solutions.
$$7x - 2y + z = 6$$
$$2x + 3y - 4z = 6$$
$$4x - 5y + 2z = 10$$

11. Fifty shares of stock A and 30 shares of stock B cost $2600. Thirty shares of stock A and 40 shares of stock B cost $2000. What is the price per share of each stock? Solve by setting up the appropriate equations and then using the inverse of the coefficient matrix.

17

Inequalities

Having devoted a great deal of time to the solution of equations and systems of equations, we now turn our attention to solving inequalities and systems of inequalities. In doing so, we will find it necessary to find *all values* of the variable or variables that satisfy the inequality or system of inequalities.

There are numerous technical applications of inequalities. For example, in electricity it might be necessary to find the values of a current that are *greater than* a specified value. In designing a link in a robotic mechanism, it might be necessary to find the forces that are *less than* a specified value. Computers can be programmed to switch from one part of a program to another, based upon a result that is greater than (or less than) some given value.

Systems of linear *equations* have been studied for better than 2000 years, but almost no attention was given to systems of linear *inequalities* until World War II in the 1940s. Problems of deploying personnel and aircraft effectively and allocating supplies efficiently led the U.S. Air Force to have a number of scientists, economists, and mathematicians look for solutions. From this, a procedure for analyzing such problems was devised in 1947 by George Danzig and his colleagues. Their system involved using systems of linear inequalities and is today called *linear programming*. We introduce the basic method of linear programming in the final section of this chapter.

Here, we see that a mathematical method was developed as a result of a military need. Today, linear programming is widely used in business and industry in order to set production levels for maximizing profits and minimizing costs.

In Section 17.6, we use inequalities to show how an airline can minimize operating costs when determining how many planes it needs.

17.1 Properties of Inequalities

**Solution of an Inequality •
Conditional Inequality •
Absolute Inequality •
Properties of Inequalities**

In Chapter 1, we first introduced the signs of inequality. To this point, only a basic understanding of their meanings has been necessary to show certain intervals associated with a variable. In this section, we review the meanings and develop certain basic properties of inequalities. We also show the meaning of the solution of an inequality and how it is shown on the number line.

The expression $a < b$ is read as "a is less than b," and the expression $a > b$ is read as "a is greater than b." *These signs define what is known as the* **sense** *(indicated by the direction of the sign) of the inequality.* Two inequalities are said to have the same sense if the signs of inequality point in the same direction. They are said to have the opposite sense if the signs of inequality point in opposite directions. *The two sides of the inequality are called* **members** *of the inequality.*

EXAMPLE 1 Sense of inequality

The inequalities $x + 3 > 2$ and $x + 1 > 0$ have the same sense, as do the inequalities $3x - 1 < 4$ and $x^2 - 1 < 3$.

The inequalities $x - 4 < 0$ and $x > -4$ have the opposite sense, as do the inequalities $2x + 4 > 1$ and $3x^2 - 7 < 1$. ∎

The **solution** *of an inequality consists of all values of the variable that make the inequality a true statement.* Most inequalities with which we deal are **conditional inequalities,** *which are true for some, but not all, values of the variable.* Also, *some inequalities are true for all values of the variable, and they are called* **absolute inequalities.** A solution of an inequality consists of only real numbers, as the terms *greater than* or *less than* have not been defined for complex numbers.

EXAMPLE 2 Conditional and absolute inequalities

NOTE ▶

The inequality $x + 1 > 0$ is true for all values of x greater than -1. Therefore, the values of x that satisfy this inequality are written as $x > -1$. This illustrates the difference between the solution of an equation and the solution of an inequality. The solution of an equation normally consists of a few specific numbers, whereas *the solution to an inequality normally consists of an interval of values of the variable.* Any and all values within this interval are termed solutions of the inequality. Since the inequality $x + 1 > 0$ is satisfied only by the values of x in the interval $x > -1$, it is a *conditional inequality.*

The inequality $x^2 + 1 > 0$ is true for all real values of x, since x^2 is never negative. It is an *absolute inequality.* ∎

There are occasions when it is convenient to combine an inequality with an equality. For such purposes, the symbols $\leq$, meaning *less than or equal to,* and $\geq$, meaning *greater than or equal to,* are used.

EXAMPLE 3 Greater (or less) than or equal to

If we wish to state that x is positive, we can write $x > 0$. However, the value zero is not included in the solution. If we wish to state that x is not negative, we write $x \geq 0$. Here, zero is part of the solution.

In order to state that x is less than or equal to -5, we write $x \leq -5$. ∎

In the sections that follow, we will solve inequalities. It is often useful to show the solution on the number line. The next example shows how this is done.

Fig. 17.1

(a) To graph $x > 2$, we draw a small open circle at 2 on the number line (which is equivalent to the x-axis). Then we draw a solid line to the right of the point and with an arrowhead pointing to the right, indicating all values greater than 2. See Fig. 17.1(a). The *open circle* shows that the point is not part of the indicated solution.

(b) To graph $x \leq 1$, we follow the same basic procedure as in part (a), except that we use a solid circle and the arrowhead points to the left. See Fig. 17.1(b). The *solid circle* shows that the point is part of the indicated solution. ∎

PROPERTIES OF INEQUALITIES

We now show the basic operations performed on inequalities. These are the same operations as those performed on equations, but in certain cases the results take on a different form. *The following are the* **properties of inequalities.**

> **1.** *The sense of an inequality is not changed when the same number is added to—or subtracted from—both members of the inequality.* Symbolically, this may be stated as "if $a > b$, then $a + c > b + c$ and $a - c > b - c$."

EXAMPLE 5 Illustrations of property 1

Using property 1 on the inequality $9 > 6$, we have the following results:

$$9 > 6 \qquad\qquad 9 > 6$$

add 4 to each member subtract 12 from each member

$$9 + 4 > 6 + 4 \qquad 9 - 12 > 6 - 12$$

$$13 > 10 \qquad\qquad -3 > -6$$

In Fig. 17.2, we see that 9 is to the right of 6, 13 is to the right of 10, and -3 is to the right of -6. ∎

Fig. 17.2

> **2.** *The sense of an inequality is not changed if both members are multiplied or divided by the same positive number.* Symbolically, this is stated as "if $a > b$, then $ac > bc$, and $a/c > b/c$, provided that $c > 0$."

EXAMPLE 6 Illustrations of property 2

Using property 2 on the inequality $8 < 15$, we have the following results:

$$8 < 15 \qquad\qquad 8 < 15$$

multiply both members by 2 divide both members by 2

$$2(8) < 2(15) \qquad \frac{8}{2} < \frac{15}{2}$$

$$16 < 30 \qquad\qquad 4 < \frac{15}{2}$$ ∎

CAUTION ▸

> **3.** *The sense of an inequality is reversed if both members are multiplied or divided by the same negative number.* Symbolically, this is stated as "if $a > b$, then $ac < bc$, and $a/c < b/c$, provided that $c < 0$."

NOTE ▸

In using this property of inequalities, be very careful to note that *the inequality sign remains the same if both members are multiplied or divided by a positive number,* but that *the **inequality sign changes if both members are multiplied or divided by a negative number**.* Most of the errors made in dealing with inequalities occur when using this property.

EXAMPLE 7 Be careful in using property 3

Using property 3 on the inequality $4 > -2$, we have the following results:

In Fig. 17.3, we see that 4 is to the right of -2, but that -12 *is to the left of* 6 and that -2 *is to the left of* 1. This is consistent with reversing the sense of the inequality when it is multiplied by -3 and when it is divided by -2. ∎

Fig. 17.3

Practice Exercises

For the inequality $-6 < 3$, state the inequality that results.

1. Multiply both members by 4.
2. Divide both members by -3

---

4. *If both members of an inequality are positive numbers and n is a positive integer, then the inequality formed by taking the nth power of each member, or the nth root of each member, is in the same sense as the given inequality.* Symbolically, this is stated as "if $a > b$, then $a^n > b^n$, and $\sqrt[n]{a} > \sqrt[n]{b}$, provided that $n > 0$, $a > 0$, $b > 0$."

EXAMPLE 8 Illustrations of property 4

Using property 4 on the inequality $16 > 9$, we have

$$16 > 9 \qquad\qquad 16 > 9$$

$$\text{square both members} \qquad \text{take square root of both members}$$

$$16^2 > 9^2 \qquad\qquad \sqrt{16} > \sqrt{9}$$

$$256 > 81 \qquad\qquad 4 > 3$$ ∎

Many inequalities have more than two members. In fact, inequalities with three members are common, and care must be used in stating these inequalities.

EXAMPLE 9 Inequalities with three members

(a) To state that 5 is less than 6, and also greater than 2, we may write $2 < 5 < 6$, or $6 > 5 > 2$. (Generally, the *less than* form is preferred.)

(b) To state that a number x may be greater than -1 *and* also less than or equal to 3, we write $-1 < x \le 3$. (It can also be written as $x > -1$ *and* $x \le 3$.) This is shown in Fig. 17.4(a). Note the use of the open circle and the solid circle.

(c) By writing $x \le -4$ *or* $x > 2$, we state that x is less than or equal to -4, or greater than 2. **It may not be shown as $2 < x \le -4$,** for this shows x as being less than -4, and also greater than 2, and *no such numbers exist.* See Fig. 17.4(b). ∎

CAUTION ▸

NOTE ▸ Note carefully that **and** *is used when the solution consists of values that make* **both** *statements true. The word* **or** *is used when the solution consists of values that make* **either** *statement true.* (In everyday speech, *or* can sometimes mean that one statement is true or another statement is true, but *not* that both are true.)

EXAMPLE 10 Meaning of *and* – meaning of *or*

The inequality $x^2 - 3x + 2 > 0$ is satisfied if x is either greater than 2 *or* less than 1. This is written as $x > 2$ *or* $x < 1$, but it is incorrect to state it as $1 > x > 2$. (If we wrote it this way, we would be saying that the same value of x is less than 1 *and* at the same time greater than 2. Of course, as noted for this type of situation in Example 9, no such number exists.) However, we could say that the inequality is not satisfied for $1 \le x \le 2$, which means those values of x greater than *or* equal to 1 *and* less than or equal to 2 (between or equal to 1 and 2). ∎

Practice Exercise

3. Graph the inequality $-1 < x \le 3$ on the number line.

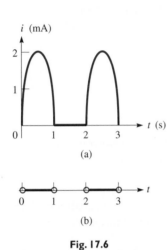

Fig. 17.5

Fig. 17.6

EXAMPLE 11 Setting up inequality – application

The design of a rectangular solar panel shows that the length l is between 80 cm and 90 cm and the width w is between 40 cm and 80 cm. See Fig. 17.5. Find the values of area the panel may have.

Since l is to be less than 90 cm and w less than 80 cm, the area must be less than $(90\,\text{cm})(80\,\text{cm}) = 7200\,\text{cm}^2$. Also, since l is to be greater than 80 cm and w greater than 40 cm, the area must be greater than $(80\,\text{cm})(40\,\text{cm}) = 3200\,\text{cm}^2$. Therefore, the area A may be represented as

$$3200\,\text{cm}^2 < A < 7200\,\text{cm}^2$$

This means the area is greater than $3200\,\text{cm}^2$ *and* less than $7200\,\text{cm}^2$. ∎

EXAMPLE 12 Setting up inequality – application

A semiconductor *diode* has the property that an electric current flows through it in only one direction. If it is an alternating-current circuit, the current in the circuit flows only during the half-cycle when the diode allows it to flow. If a source of current given by $i = 2\sin \pi t$ (i in mA, t in seconds) is connected in series with a diode, write the inequalities for the current and the time. Assume the source is on for 3.0 s and a positive current passes through the diode.

We are to find the values of t that correspond to $i > 0$. From the properties of the sine function, we know that $2\sin \pi t$ has a period of $2\pi/\pi = 2.0$ s. Therefore, the current is zero for $t = 0$, 1.0 s, 2.0 s, and 3.0 s.

The source current is positive for $0 < t < 1.0$ s and for $2.0\,\text{s} < t < 3.0$ s.
The source current is negative for $1.0\,\text{s} < t < 2.0$ s.
Therefore, in the circuit

$$i > 0 \quad \text{for} \quad 0 < t < 1.0\,\text{s} \quad \text{and} \quad 2.0\,\text{s} < t < 3.0\,\text{s}$$
$$i = 0 \quad \text{for} \quad t = 0, 1.0\,\text{s} \le t \le 2.0\,\text{s}$$

A graph of the current in the circuit as a function of time is shown in Fig. 17.6(a). In Fig. 17.6(b), the values of t for which $i > 0$ are shown. ∎

EXERCISES 17.1

In Exercises 1–4, make the given changes in the indicated examples of this section and then perform the indicated operations.

 1. In Example 2, in the first paragraph, change the $>$ to $<$ and then complete the meaning of the resulting inequality as in the first sentence. Rewrite the meaning as in the second line.

2. In Example 4(b), change $\le$ to $>$ and then graph the resulting inequality.

3. In Example 7, change the inequality to $-2 > -4$ and then perform the two operations shown in color.

4. In Example 9(b), change the -1 to -3 and the 3 to 1 and then write the two forms in which an inequality represents the statement.

In Exercises 5–12, for the inequality $4 < 9$, state the inequality that results when the given operations are performed on both members.

5. Add 3.

6. Subtract 16.

7. Multiply by 5.

8. Multiply by -2.

9. Divide by -1.

10. Divide by 0.5.

11. Square both.

12. Take square roots.

In Exercises 13–24, give the inequalities equivalent to the following statements about the number x.

13. Greater than -2

14. Less than 0.7

15. Less than or equal to 45

16. Greater than or equal to -6

17. Greater than 1 and less than 7

18. Greater than or equal to -200 and less than 650.

19. Less than -9, or greater than or equal to -4

20. Less than or equal to 8, or greater than or equal to 12

21. Less than 1, or greater than 3 and less than or equal to 5

22. Greater than or equal to 0 and less than or equal to 2, or greater than 5

23. Greater than -2 and less than 2, or greater than or equal to 3 and less than 4

24. Less than -4, or greater than or equal to 0 and less than or equal to 1, or greater than or equal to 5

(W) *In Exercises 25–28, give verbal statements equivalent to the given inequalities involving the number x.*

25. $0 < x \le 2$ **26.** $x < 5$ or $x > 7$

27. $x < -10$ or $10 \le x < 20$

28. $-1 \le x < 3$ or $5 < x < 7$

In Exercises 29–44, graph the given inequalities on the number line.

29. $x < 3$ **30.** $x \ge -1$

31. $x \le 1$ or $x > 3$ **32.** $x < -300$ or $x \ge 0$

33. $0 \le x < 5$ **34.** $-4 < y < -2$

35. $x \ge -3$ and $x < 5$ **36.** $x > 4$ and $x < 3$

37. $x < -1$ or $1 \le x < 4$

38. $-3 < x < 0$ or $x > 3$

39. $-3 < x < -1$ or $1 < x \le 3$

40. $1 < x \le 2$ or $3 \le x < 4$

41. $t < -0.3$ or $t > -0.3$

42. $x < 1$ or $1 < x \le 4$

43. $(x \le 5$ or $x \ge 8)$ and $(3 < x < 10)$

44. $(x < 7$ and $x > 2)$ or $(x > 10$ or $x < 1)$

In Exercises 45–48, answer the given questions about the inequality $0 < a < b$.

45. Is $a^2 < b^2$ a conditional inequality or an absolute inequality?

46. Is $|a - b| < b - a$?

47. If each member of the inequality $2 > 1$ is multiplied by $a - b$, is the result $2(a - b) > (a - b)$?

(W) **48.** What is wrong with the following sequence of steps?
$$a < b, \; ab < b^2, \; ab - b^2 < 0, \; b(a - b) < 0, \; b < 0$$

In Exercises 49–52, solve the given problems.

49. Write the relationship between $(|x| + |y|)$ and $|x + y|$ if $x > 0$ and $y < 0$.

50. Write the relationship between $|xy|$ and $|x||y|$ if $x > 0$ and $y < 0$.

(W) **51.** Explain the error in the following "proof" that $3 < 2$:
(1) $1/8 < 1/4$ (2) $0.5^3 < 0.5^2$ (3) $\log 0.5^3 < \log 0.5^2$
(4) $3 \log 0.5 < 2 \log 0.5$ (5) $3 < 2$

52. If $x \ne y$, show that $x^2 + y^2 > 2xy$.

In Exercises 53–60, some applications of inequalities are shown.

53. An electron microscope can magnify an object from 2000 times to 1,000,000 times. Assuming these values are exact, express these magnifications M as an inequality and graph them.

54. A busy person glances at a digital clock that shows 9:36. Another glance a short time later shows the clock at 9:44. Express the amount of time t (in min) that could have elapsed between glances by use of inequalities. Graph these values of t.

55. An earth satellite put into orbit near the earth's surface will have an elliptic orbit if its velocity v is between 18,000 mi/h and 25,000 mi/h. Write this as an inequality and graph these values of v.

56. Fossils found in Jurassic rocks indicate that dinosaurs flourished during the Jurassic geological period, 140 MY (million years ago) to 200 MY. Write this as an inequality, with t representing past time. Graph the values of t.

57. In executing a program, a computer must perform a set of calculations. Any one of the calculations takes no more than 2565 steps. Express the number n of steps required for a given calculation by an inequality. (Note that n is a positive *integer*.)

58. The velocity v of an ultrasound wave in soft human tissue may be represented as 1550 ± 60 m/s, where the ± 60 m/s gives the possible variation in the velocity. Express the possible velocities by an inequality.

59. The electric intensity E within a charged spherical conductor is zero. The intensity on the surface and outside the sphere equals a constant k divided by the square of the distance r from the center of the sphere. State these relations for a sphere of radius a by using inequalities, and graph E as a function of r.

60. If the current from the source in Example 12 is $i = 5 \cos 4\pi t$ and the diode allows only negative current to flow, write the inequalities and draw the graph for the current in the circuit as a function of time for $0 \le t \le 1$ s.

Answers to Practice Exercises

1. $-24 < 12$ **2.** $2 > -1$ **3.**

17.2 Solving Linear Inequalities

**Linear Inequalities •
Inequalities with Three Members •
Solving an Inequality with a Calculator**

Using the properties and definitions discussed in Section 17.1, we can now proceed to solve inequalities. In this section, we solve linear inequalities in one variable. Similar to linear functions as defined in Chapter 5, *a* **linear inequality** *is one in which each term contains only one variable and the exponent of each variable is* 1. We will consider linear inequalities in two variables in Section 17.5.

The procedure for solving a linear inequality in one variable is like that we used in solving basic equations in Chapter 1. We solve the inequality by isolating the variable, and to do this we perform the same operations on each member of the inequality. The operations are based on the properties given in Section 17.1.

EXAMPLE 1 Solutions using basic operations

In each of the following inequalities, by performing the indicated operation, we isolate x and thereby solve the inequality.

$x + 2 < 4$	$\dfrac{x}{2} > 4$	$2x \leq 4$
Subtract 2 from each member.	Multiply each member by 2.	Divide each member by 2.
$x < 2$	$x > 8$	$x \leq 2$

Each solution can be checked by substituting any number in the indicated interval into the original inequality. For example, any value less than 2 will satisfy the first inequality, whereas 2 or any number less than 2 will satisfy the third inequality. ∎

EXAMPLE 2 Solving linear inequality

Solve the following inequality: $3 - 2x \geq 15$.

We have the following solution:

$$3 - 2x \geq 15 \qquad \text{original inequality}$$
$$-2x \geq 12 \qquad \text{subtract 3 from each member}$$

inequality reversed ⟶

$$x \leq -6 \qquad \text{divide each member by } -2$$

CAUTION ▶   Again, carefully note that *the sign of inequality was reversed when each number was divided by* **−2.** We check the solution by substituting -7 in the original inequality, obtaining $17 \geq 15$. ∎

EXAMPLE 3 Solving linear inequality

Solve the inequality $2x \leq 3 - x$.

The solution proceeds as follows:

$$2x \leq 3 - x \qquad \text{original inequality}$$
$$3x \leq 3 \qquad \text{add } x \text{ to each member}$$
$$x \leq 1 \qquad \text{divide each member by 3}$$

This solution checks and is represented in Fig. 17.7, as we showed in Section 17.1.

This inequality could have been solved by combining x-terms on the right. In doing so, we would obtain $1 \geq x$. Since this might be misread, it is best to combine the variable terms on the left, as we did above. ∎

Part of solution

Fig. 17.7

EXAMPLE 4 Solving linear inequality

Solve the inequality $\frac{3}{2}(1 - x) > \frac{1}{4} - x$.

$$\frac{3}{2}(1 - x) > \frac{1}{4} - x \qquad \text{original inequality}$$
$$6(1 - x) > 1 - 4x \qquad \text{multiply each member by 4}$$
$$6 - 6x > 1 - 4x \qquad \text{remove parentheses}$$
$$-6x > -5 - 4x \qquad \text{subtract 6 from each member}$$
$$-2x > -5 \qquad \text{add } 4x \text{ to each member}$$
$$x < \frac{5}{2} \qquad \text{divide each member by } -2$$

Note that the sense of the inequality was reversed when we divided by -2. This solution is shown in Fig. 17.8. Any value of $x < 5/2$ checks when substituted into the original inequality. ∎

Practice Exercise

1. Solve the inequality $2(3 - x) > 5 + 4x$

Not part of solution

Fig. 17.8

The following example illustrates an application that involves the solution of an inequality.

EXAMPLE 5 Linear inequality – application

The velocity v (in ft/s) of a missile in terms of the time t (in s) is given by $v = 960 - 32t$. For how long is the velocity positive? (Since velocity is a vector, this can also be interpreted as asking "how long is the missile moving upward?")

In terms of inequalities, we are asked to find the values of t for which $v > 0$. This means that we must solve the inequality $960 - 32t > 0$. The solution is as follows:

$$960 - 32t > 0 \qquad \text{original inequality}$$
$$-32t > -960 \qquad \text{subtract 960 from each member}$$
$$t < 30 \text{ s} \qquad \text{divide each member by } -32$$

Negative values of t have no meaning in this problem. Checking $t = 0$, we find that $v = 960 \text{ ft/s}$. Therefore, the complete solution is $0 \le t < 30 \text{ s}$.

In Fig. 17.9(a), we show the graph of $v = 960 - 32t$, and in Fig. 17.9(b), we show the solution $0 \le t < 30 \text{ s}$ on the number line (which is really the t-axis in this case). Note that the values of v are above the t-axis for those values of t that are part of the solution. This shows the relationship of the graph of v as a function of t, and the solution as graphed on the number line (the t-axis). ∎

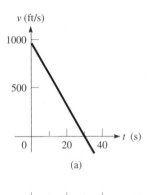

(a)

(b)

Fig. 17.9

INEQUALITIES WITH THREE MEMBERS

EXAMPLE 6 Solving inequality with three members

Solve: $-1 < 2x + 3 < 6$.

We have the following solution.

$$-1 < 2x + 3 < 6 \qquad \text{original inequality}$$
$$-4 < 2x < 3 \qquad \text{subtract 3 from each member}$$
$$-2 < x < \frac{3}{2} \qquad \text{divide each member by 2}$$

The solution is shown in Fig. 17.10. ∎

Fig. 17.10

EXAMPLE 7 Solving inequality with three members

Solve the inequality $2x < x - 4 \le 3x + 8$.

Since we cannot isolate x in the middle member (or in any member), we rewrite the inequality as

$$2x < x - 4 \quad \text{and} \quad x - 4 \le 3x + 8$$

We then solve each of the inequalities, keeping in mind that the solution must satisfy both of them. Therefore, we have

$$2x < x - 4 \quad \text{and} \quad x - 4 \le 3x + 8$$
$$-2x \le 12$$
$$x < -4 \qquad\qquad x \ge -6$$

We see that the solution is $x < -4$ *and* $x \ge -6$, which can also be written as $-6 \le x < -4$. This second form is generally preferred since it is more concise and more easily interpreted. The solution checks and is shown in Fig. 17.11. ∎

Practice Exercise

2. Solve the inequality: $-2 \le 4x - 3 < 5$

Fig. 17.11

EXAMPLE 8   Inequality – three members – application

In emptying a wastewater tank, one pump can remove no more than 40 L/min. If it operates for 8.0 min and a second pump operates for 5.0 min, what must be the pumping rate of the second pump if 480 L are to be removed?

Let x = the pumping rate of the first pump and y = the pumping rate of the second pump. Since the first operates for 8.0 min and the second for 5.0 min to remove 480 L, we have

first   second
pump   pump   total ◄— amounts pumped

$$8.0x + 5.0y = 480$$

Since we know that the first pump can remove no more than 40 L/min, which means that $0 \le x \le 40$ L/min, we solve for x, then substitute in this inequality:

$x = 60 - 0.625y$	solve for x
$0 \le 60 - 0.625y \le 40$	substitute in inequality
$-60 \le -0.625y \le -20$	subtract 60 from each member
$96 \ge y \ge 32$	divide each member by -0.625
$32 \le y \le 96$ L/min	use $\le$ symbol (optional step)

Fig. 17.12

This means that the second pump must be able to pump at least 32 L/min and no more than 96 L/min. See Fig. 17.12.

Although this was a three-member inequality combined with equalities, the solution was done in the same way as with a two-member inequality. ■

SOLVING AN INEQUALITY WITH A CALCULATOR

A graphing calculator can be used to show the solution of an inequality. On most calculators, a value of 1 is shown if the inequality is satisfied, and a value of 0 is shown if the inequality is not satisfied. This means that if, for example, we enter $8 > 3$ (see the manual to see how $>$ is entered), the calculator displays 1, and if we enter $3 > 8$, it shows 0, as in the display in Fig. 17.13. We can also use an inequality feature to show graphically the solution of an inequality.

Fig. 17.13

EXAMPLE 9   Inequality solution on calculator

Display the solution of the inequality $2x \le 3 - x$ (see Example 3) on a graphing calculator.

We set y_1 *equal to the inequality,* as shown in Fig. 17.14(a) and then graph y_1. From Fig. 17.14(b), we see that $y_1 = 1$ up to $x = 1$. Thus, the solution that is shown is

NOTE ► $x < 1$. We should note that *this method does not distinguish between $<$ and $\le$ (or $>$ and $\ge$),* so the same solution would be displayed for $2x < 3 - x$. Substituting $x = 1$ into the inequality, we find $2 = 2$, and therefore the complete solution is $x \le 1$. In Fig. 17.14(b) we used a heavier line setting for the solution so that the solution of zero from $x = 1$ to $x = 4$ is easily seen, Compare Fig. 17.14(b) with Fig. 17.7.

■ Depending on the calculator model, and the settings, the solution may show a vertical connection where the solution changes from 0 to 1, or 1 to 0.

Fig. 17.14   (a)   (b)   ■

EXAMPLE 10   Inequality solution on calculator

Display the solution of the inequality $-1 < 2x + 3 < 6$ (see Example 6) on a graphing calculator.

(a)

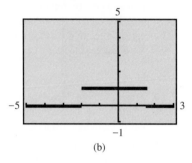

(b)

Fig. 17.15

In order to display the solution, we must write the inequality as

$$-1 < 2x + 3 \quad \text{and} \quad 2x + 3 < 6 \qquad \text{see Example 7}$$

Then enter $y_1 = -1 < 2x + 3$ and $2x + 3 < 6$ (consult the manual to determine how "and" is entered) in the calculator, as shown in Fig. 17.15(a). From Fig. 17.15(b), we see that the solution is $-2 < x < 1.5$. The *trace* and *zoom* features can be used to get accurate values. Compare Fig. 17.15(b) with Fig. 17.10.   ■

EXERCISES 17.2

In Exercises 1–4, make the given changes in the indicated examples of this section and then perform the indicated operations.

1. In Example 2, change the 3 to 21 and then solve the resulting inequality.

2. In Example 4, change the 1/4 to 7/4 and then solve and display the resulting inequality.

3. In Example 6, change the + in the middle member to − and then solve the resulting inequality. Graph the solution.

4. In Example 10, change the + in the middle member to − and then display the solution on a graphing calculator.

In Exercises 5–28, solve the given inequalities. Graph each solution.

5. $x - 3 > -4$

6. $x + 2 \le 6$

7. $\dfrac{1}{2}x < 32$

8. $-4t > 12$

9. $3x - 5 \le -11$

10. $\dfrac{1}{3}x + 2 \ge 1$

11. $6 - y > 8$

12. $32 - 5x < -8$

13. $\dfrac{4x - 5}{2} \le x$

14. $1.50 - 5.24x > 3.75 + 2.25x$

15. $180 - 6(T + 12) > 14T + 285$

16. $-2[x - (3 - 2x)] > \dfrac{1 - 5x}{3} + 2$

17. $2.50(1.50 - 3.40x) < 3.84 - 8.45x$

18. $(2x - 7)(x + 1) \le 4 - x(1 - 2x)$

19. $\dfrac{1}{3} - \dfrac{L}{2} < L + \dfrac{3}{2}$

20. $\dfrac{x}{5} - 2 > \dfrac{2}{3}(x + 3)$

21. $-1 < 2x + 1 < 3$

22. $2 < 3R + 1 \le 8$

23. $-4 \le 1 - x < -1$

24. $0 \le 3 - 2x \le 6$

25. $2x < x - 1 \le 3x + 5$

26. $x + 19 \le 25 - x < 2x$

27. $2s - 3 < s - 5 < 3s - 3$

28. $0 < 1 - x \le 3 \quad \text{or} \quad -1 < 2x - 3 < 5$

In Exercises 29–36, solve the inequalities by displaying the solutions on a graphing calculator. See Examples 9 and 10.

29. $3x - 2 < 8 - x$

30. $40(x - 2) > x + 60$

31. $\dfrac{1}{2}(x + 15) \ge 5 - 2x$

32. $\dfrac{1}{3}x - 2 \le \dfrac{1}{2}x + 1$

33. $0.1 < 0.5 - 0.2t < 0.9$

34. $-3 < 2 - \dfrac{s}{3} \le -1$

35. $x - 3 < 2x + 5 < 6x + 7$

36. $n - 3 < 2n + 4 \le 1 - n$

In Exercises 37–52, solve the given problems by setting up and solving appropriate inequalities. Graph each solution.

37. Determine the values of x that are in the domain of the function $f(x) = \sqrt{2x - 10}$.

38. Determine the values of x that are in the domain of the function $f(x) = 1/\sqrt{3 - 0.5x}$.

39. For what values of k are the roots of the equation $x^2 - kx + 9 = 0$ imaginary?

40. For what values of k are the roots of the equation $2x^2 + 3x + k = 0$ real and unequal?

41. A contractor is considering two similar jobs, each of which is estimated to take n hours to complete. One pays \$350 plus \$15 per hour, and the other pays \$25 per hour. For what values of n will the contractor make more at the second position?

42. In designing plastic pipe, if the inner radius r is increased by 5.00 cm, and the inner cross-sectional area is increased by between 125 cm^2, and 175 cm^2, what are the possible inner radii of the pipe?

43. The relation between the temperature in degrees Fahrenheit F and degrees Celsius C is $9C = 5(F - 32)$. What temperatures F correspond to temperatures between $10°$ C and $20°$ C?

44. The voltage drop V across a resistor is the product of the current i (in A) and the resistance R (in Ω). Find the possible voltage drops across a variable resistor R, if the minimum and maximum resistances are 1.6 kΩ and 3.6 kΩ, respectively, and the current is constant at 2.5 mA.

45. A rectangular PV (photovoltaic) solar panel is designed to be 1.42 m long and supply 130 W/m^2 of power. What must the width of the panel be in order to supply between 100 W and 150 W?

46. A beam is supported at each end, as shown in Fig. 17.16. Analyzing the forces leads to the equation $F_1 = 13 - 3d$. For what values of d is F_1 more than 6 N?

Fig. 17.16

47. The mass m (in g) of silver plate on a dish is increased by electroplating. The mass of silver on the plate is given by $m = 125 + 15.0t$, where t is the time (in h) of electroplating. For what values of t is m between 131 g and 164 g?

48. For a ground temperature of T_0 (in °C), the temperature T (in °C) at a height h (in m) above the ground is given approximately by $T = T_0 - 0.010h$. If the ground temperature is 25°C, for what heights is the temperature above 10°C?

49. During a given rush hour, the numbers of vehicles shown in Fig. 17.17 go in the indicated directions in a one-way-street section of a city. By finding the possible values of x and the equation relating x and y, find the possible values of y.

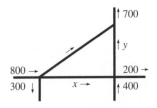

Fig. 17.17

50. The minimum legal speed on a certain interstate highway is 45 mi/h, and the maximum legal speed is 65 mi/h. What legal distances can a motorist travel in 4 h on this highway without stopping?

51. The route of a rapid transit train is 40 km long, and the train makes five stops of equal length. If the train is actually moving for 1 h and each stop must be at least 2 min, what are the lengths of the stops if the train maintains an average speed of at least 30 km/h, including stop times?

52. An oil company plans to install eight storage tanks, each with a capacity of x liters, and five additional tanks, each with a capacity of y liters, such that the total capacity of all tanks is 440,000 L. If capacity y will be at least 40,000 L, what are the possible values of capacity x?

Answers to Practice Exercises

1. $x < 1/6$ **2.** $1/4 \le x < 2$

17.3 Solving Nonlinear Inequalities

Critical Values • Using Linear Factors • Polynomial and Rational Inequalities • Solving Inequalities with Calculator • Solving Inequalities Graphically

Fig. 17.18

In this section, we develop methods of solving inequalities with polynomials, rational expressions (expressions involving fractions), and nonalgebraic expressions. To develop the basic method for solving these types of inequalities, we now take another look at a linear inequality.

In Fig. 17.18, we see that all values of the linear function $f(x) = ax + b$ $(a \neq 0)$ are positive on one side of the point at which $f(x) = 0$, and all values of $f(x)$ are negative on the opposite side of the same point. The means that *we can solve a linear inequality by expressing it with **zero** on the right and then finding the **sign** of the resulting function on either side of zero.*

EXAMPLE 1 Sign of function with zero on right

Solve the inequality $2x - 5 > 1$.

Finding the equivalent inequality with zero on the right, we have $2x - 6 > 0$. Setting the left member equal to zero, we have

$$2x - 6 = 0 \quad \text{for} \quad x = 3$$

which means $f(x) = 2x - 6$ has one sign for $x < 3$, and the other sign for $x > 3$. Testing values in these intervals, we find, for example, that

$$f(x) = -2 \quad \text{for} \quad x = 2 \quad \text{and} \quad f(x) = +2 \quad \text{for} \quad x = 4$$

Therefore, the solution to the original inequality is $x > 3$. The solution in Fig. 17.19(b) corresponds to the positive values of $f(x)$ in Fig. 17.19(a).

We could have solved this inequality by methods of the previous section, but the important idea here is to use the *sign of the function,* with zero on the right. ∎

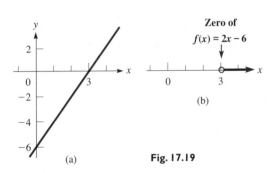

Fig. 17.19

We can extend this method to solving inequalities with polynomials of higher degree. We first find the equivalent inequality with zero on the right and then *factor the function on the left into linear factors and any quadratic factors that lead to complex roots.* As all values of x are considered, each linear factor can change sign at the value for which it is zero. The quadratic factors do not change sign.

This method is especially useful in solving inequalities involving fractions. If a linear factor occurs in the numerator, the function is zero at the value of x for which the linear factor is zero. If such a factor appears in the denominator, the function is undefined where the factor is zero. *The values of x for which a function is zero or undefined are called the* **critical values** *of the function.* As all negative and positive values of x are considered (starting with negative numbers of large absolute value, proceeding through zero, and ending with large positive numbers), a function can change sign only at a critical value.

Critical Values

We now outline the method of solving an inequality by using the critical values of the function. Several examples of the method follow.

USING CRITICAL VALUES TO SOLVE AN INEQUALITY

1. *Determine the equivalent inequality with zero on the right.*
2. *Find all linear factors of the function.*
3. *To find the critical values, set each linear factor equal to zero and solve for x.*
4. *Determine the sign of the function to the left of the leftmost critical value, between critical values, and to the right of the rightmost critical value.*
5. *Those intervals in which the function has the proper sign satisfy the inequality.*

EXAMPLE 2 Solving quadratic inequality

Solve the inequality $x^2 - 3 > 2x$.

We first find the equivalent inequality with zero on the right. Therefore, we have $x^2 - 2x - 3 > 0$. We then factor the left member and have

$$(x + 1)(x - 3) > 0$$

Setting each factor equal to zero, we find the left critical value is -1 and the right critical value is 3. All values of x to the left of -1 give the same sign for the function. All values between -1 and 3 give the function the same sign. All values of x to the right of 3 give the same sign to the function. Therefore, *we must determine the sign of $f(x)$ for each of the intervals $x < -1$, $-1 < x < 3$, and $x > 3$.*

For $x < -1$, both factors are negative, which means their product is positive, or $(x + 1)(x - 3) > 0$. For $-1 < x < 3$, the left factor is positive, and the right factor is negative, which means their product is negative, or $(x + 1)(x - 3) < 0$. For $x > 3$, both factors are positive, which means their product is positive, or $(x + 1)(x - 3) > 0$.

Summarizing these results for $f(x) = (x + 1)(x - 3)$, we have

$$\text{If } x < -1, f(x) > 0 \quad \text{if } -1 < x < 3, f(x) < 0 \quad \text{if } x > 3, f(x) > 0$$

Therefore, the solution to the inequality is $x < -1$ or $x > 3$.

The solution that is shown in Fig. 17.20(b) corresponds to the positive values of the function $f(x) = x^2 - 2x - 3$, shown in Fig. 17.20(a). ∎

(a)

$f(x) > 0 \qquad f(x) < 0 \qquad f(x) > 0$

Critical values

(b)

Fig. 17.20

(a)

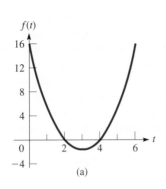

(b)

Fig. 17.21

Practice Exercise

1. Solve the inequality $x^2 - x - 42 < 0$

(a)

(b)

Fig. 17.22

EXAMPLE 3 Solving cubic inequality

Solve the inequality $x^3 - 4x^2 + x + 6 < 0$.

By methods developed in Chapter 15, we factor the function on the left and obtain $(x + 1)(x - 2)(x - 3) < 0$. The critical values are $-1, 2, 3$. We wish to determine the sign of the left member for the intervals $x < -1$, $-1 < x < 2$, $2 < x < 3$, and $x > 3$. The following table shows each interval, the sign of each factor in each interval, and the resulting sign of the function.

$$f(x) = (x + 1)(x - 2)(x - 3)$$

Interval	$(x + 1)(x - 2)(x - 3)$			Sign of $f(x)$
$x < -1$	$-$	$-$	$-$	$-$
$-1 < x < 2$	$+$	$-$	$-$	$+$
$2 < x < 3$	$+$	$+$	$-$	$-$
$x > 3$	$+$	$+$	$+$	$+$

Since we want values for $f(x) < 0$, the solution is $x < -1$ or $2 < x < 3$. The solution shown in Fig. 17.21(b) corresponds to the values of $f(x) < 0$ in Fig. 17.21(a). ∎

The following example illustrates an applied situation that involves the solution of an inequality.

EXAMPLE 4 Solving quadratic inequality – application

The force F (in N) acting on a cam varies according to the time t (in s), and it is given by the function $F = 2t^2 - 12t + 20$. For what values of t, $0 \le t \le 6$ s, is the force at least 4 N?

For a force of at least 4 N, we know that $F \ge 4$ N, or $2t^2 - 12t + 20 \ge 4$. This means we are to solve the inequality $2t^2 - 12t + 16 \ge 0$, and the solution is as follows:

$$2t^2 - 12t + 16 \ge 0$$
$$t^2 - 6t + 8 \ge 0$$
$$(t - 2)(t - 4) \ge 0$$

The critical values are $t = 2$ and $t = 4$, which lead to the following table:

Interval	$(t - 2)(t - 4)$		Sign of $(t - 2)(t - 4)$
$0 \le t < 2$	$-$	$-$	$+$
$2 < t < 4$	$+$	$-$	$-$
$4 < t \le 6$	$+$	$+$	$+$

We see that the values of t that satisfy the *greater than* part of the problem are $0 \le t < 2$ and $4 < t \le 6$. Since we know that $(t - 2)(t - 4) = 0$ for $t = 2$ and $t = 4$, the solution is

$$0 \le t \le 2 \text{ s} \quad \text{or} \quad 4 \text{ s} \le t \le 6 \text{ s}$$

The graph of $f(t) = 2t^2 - 12t + 16$ is shown in Fig. 17.22(a). The solution, shown in Fig. 17.22(b), corresponds to the values of $f(t)$ that are zero or positive or for which $F \ge 4$ N.

We note here that if the cam rotates in 6-s intervals, the force on the cam is periodic, varying from 2 N to 20 N. ∎

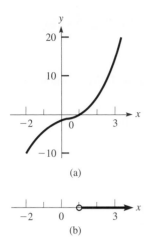

(a)

(b)

Fig. 17.23

EXAMPLE 5 Solving cubic inequality

Solve the inequality $x^3 - x^2 + x - 1 > 0$.

In order to factor this function, we can use the methods of Chapter 15, or we might note that it is factorable by grouping (see Section 6.2). Either method leads to the inequality

$$(x^2 + 1)(x - 1) > 0$$

In this case, we have a quadratic factor $x^2 + 1$ that leads to imaginary roots (j and $-j$) and is never negative. This means we have only one linear factor and therefore only one critical value.

Setting $x - 1 = 0$, we get the critical value $x = 1$. Since $x - 1$ is positive for $x > 1$ and negative for $x < 1$, the solution to the inequality is $x > 1$.

The solution is shown in Fig. 17.23(b), and we see that this solution corresponds to the positive values of $f(x) = x^3 - x^2 + x - 1$ shown in Fig. 17.23(a). ∎

EXAMPLE 6 Values representing real numbers

Find the values of x for which $\sqrt{\dfrac{x - 3}{x + 4}}$ represents a real number.

For the expression to represent a real number, the fraction under the radical must be greater than or equal to zero. This means we must solve the inequality.

$$\frac{x - 3}{x + 4} \geq 0$$

The critical values are found from the factors that are in the numerator or in the denominator. Thus, the critical values are -4 and 3. Considering now the *greater than* part of the $\geq$ sign, we set up the following table:

Interval	$\dfrac{x - 3}{x + 4}$	Sign of $\dfrac{x - 3}{x + 4}$
$x < -4$	$\dfrac{-}{-}$	$+$
$-4 < x < 3$	$\dfrac{-}{+}$	$-$
$x > 3$	$\dfrac{+}{+}$	$+$

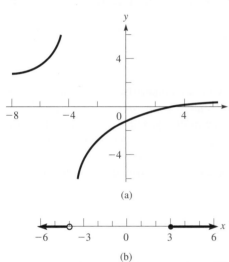

(a)

(b)

Fig. 17.24

Thus, the values that satisfy the *greater than* part of the problem are those for which $x < -4$ or for which $x > 3$. Now, considering the equality part of the $\geq$ sign, we note that $x = 3$ is valid, for the fraction is zero. However,

if $x = -4$, we have division by zero, and therefore x may not equal -4.

Therefore, the solution of the inequality is $x < -4$ or $x \geq 3$, and this means these are the values for which the original expression represents a real number. The graph of $f(x) = \dfrac{x - 3}{x + 4}$ is shown in Fig. 17.24(a), and the graph of the solution is shown in Fig. 17.24(b). ∎

Practice Exercise

2. Solve the inequality: $\dfrac{x + 3}{2x - 5} \geq 0$

EXAMPLE 7 Solving rational inequality

Solve the inequality $\dfrac{(x-2)^2(x+3)}{4-x} < 0$.

The critical values are $-3, 2,$ and 4. Thus, we have the following table:

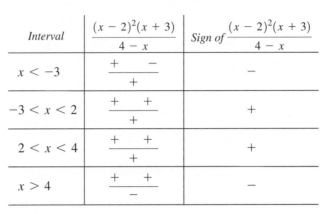

Interval	$\dfrac{(x-2)^2(x+3)}{4-x}$	Sign of $\dfrac{(x-2)^2(x+3)}{4-x}$
$x < -3$	$\dfrac{+\quad-}{+}$	$-$
$-3 < x < 2$	$\dfrac{+\quad+}{+}$	$+$
$2 < x < 4$	$\dfrac{+\quad+}{+}$	$+$
$x > 4$	$\dfrac{+\quad+}{-}$	$-$

Thus, the solution is $x < -3$ or $x > 4$. The graph of the function is shown in Fig. 17.25(a), and the graph of the solution is shown in Fig. 17.25(b). ∎

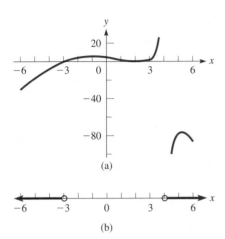

Fig. 17.25

EXAMPLE 8 Solving rational inequality

Solve the inequality $\dfrac{x+3}{x-1} > 2$.

CAUTION ▶ It seems that all we have to do is multiply both members by $x-1$ and solve the resulting linear inequality. However, *this will lead to an incorrect result.* The reason is that we assume $x-1$ is positive when we multiply, but $x-1$ is negative for $x < 1$. To avoid this problem, *first subtract 2 from each member* and combine terms on the left.

NOTE ▶ *This procedure should be used with any inequality involving fractions with the variable in the denominator.* Therefore, the solution is

$$\frac{x+3}{x-1} - 2 > 0 \qquad \text{subtract 2 from both members}$$

$$\frac{x+3-2(x-1)}{x-1} > 0 \qquad \text{combine over the common denominator}$$

$$\frac{5-x}{x-1} > 0 \qquad \text{this is the form to use}$$

The critical values are 1 and 5, and we have the following table of signs:

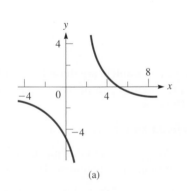

Interval	$(5-x)/(x-1)$	Sign of $(5-x)/(x-1)$
$x < 1$	$+\ /\ -$	$-$
$1 < x < 5$	$+\ /\ +$	$+$
$x > 5$	$-\ /\ +$	$-$

Thus, the solution is $1 < x < 5$. The graph of $f(x) = (5-x)/(x-1)$ is shown in Fig. 17.26(a), and the graph of the solution is shown in Fig. 17.26(b). ∎

Fig. 17.26

Calculator Solution   We can use the graphing calculator method of displaying the solution that was developed in Examples 9 and 10 on pages 468 and 469. This method can be used for any of the inequalities of this section, and we illustrate it again in the next example.

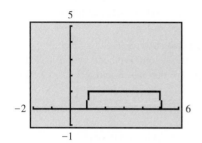

Fig. 17.27

EXAMPLE 9 Solving rational inequality on calculator

Solve the inequality of Example 8, $\dfrac{x+3}{x-1} > 2$.

On the calculator, we set $y_1 = (x+3)/(x-1) > 2$, and we then have the display shown in Fig. 17.27. This confirms the solution of $1 < x < 5$. ∎

SOLVING INEQUALITIES GRAPHICALLY

We can get an approximate solution of an inequality from the graph of a function, including functions that are not factorable or not algebraic. The method follows several of the earlier examples. First, write the equivalent inequality with zero on the right and then graph this function. Those values of x corresponding to the proper values of y (either above or below the x-axis) are those that satisfy the inequality.

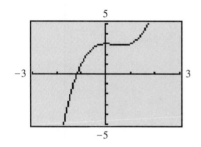

Fig. 17.28

EXAMPLE 10 Solving inequality graphically

Use a graphing calculator to solve the inequality $x^3 > x^2 - 3$.

Finding the equivalent inequality with zero on the right, we have $x^3 - x^2 + 3 > 0$. On the calculator, we then let $y_1 = x^3 - x^2 + 3$ and graph this function. This gives us the calculator display shown in Fig. 17.28.

We can see that the curve crosses the x-axis only once, between $x = -2$ and $x = -1$. Using the *zero* feature, we find that this value is approximately $x = -1.17$. Since we want positive values of x that correspond to positive values of y, we see that the solution is $x > -1.17$. ∎

EXERCISES 17.3

In Exercises 1–4, make the given changes in the indicated examples of this section and then solve the resulting inequalities.

1. In Example 2, change the − sign before the 3 to + and change $2x$ to $4x$, then solve the resulting inequality, and graph the solution.

2. In Example 5, change both − signs to +, then solve the resulting inequality, and graph the solution.

3. In Example 7, change the exponent on $(x-2)$ from 2 to 3, then solve the resulting inequality, and graph the solution.

4. In Example 9, on the right, change the 2 to 3, then solve the resulting inequality, and graph the solution.

In Exercises 5–24, solve the given inequalities. Graph each solution. It is suggested that you also graph the function on a graphing calculator as a check.

5. $x^2 - 16 < 0$

6. $x^2 + 3x \geq 0$

7. $2x^2 \leq 4x$

8. $x^2 - 4x > 21$

9. $2x^2 - 12 \leq -5x$

10. $9t^2 + 6t > -1$

11. $x^2 + 4x \leq -4$

12. $6x^2 + 1 < 5x$

13. $R^2 + 4 > 0$

14. $x^4 + 2 < 1$

15. $x^3 + x^2 - 2x > 0$

16. $x^3 - 2x^2 + x \geq 0$

17. $s^3 + 2s^2 - s \geq 2$

18. $n^4 - 2n^3 + 8n + 12 \leq 7n^2$

19. $\dfrac{2x-3}{x+6} \leq 0$

20. $\dfrac{x+15}{x-9} > 0$

21. $\dfrac{x^2 - 6x - 7}{x+5} > 0$

22. $\dfrac{(x-2)^2(5-x)}{(4-x)^3} \leq 0$

23. $\dfrac{x}{x+1} > 1$

24. $\dfrac{2p}{p-1} > 3$

In Exercises 25–32, solve the inequalities by displaying the solutions on a graphing calculator. See Example 9.

25. $3x^2 + 5x \geq 2$

26. $12x^2 + x > 1$

27. $\dfrac{T-8}{3-T} \leq 0$

28. $\dfrac{3x+1}{x+3} \geq 0$

29. $\dfrac{6-x}{3-x-4x^2} \geq 0$

30. $\dfrac{4-x}{3+2x-x^2} > 0$

31. $\dfrac{x^4(9-x)(x-5)(2-x)}{(4-x)^5} > 0$

32. $\dfrac{2}{x-3} < 4$

In Exercises 33–36, determine the values for x for which the radicals represent real numbers.

33. $\sqrt{(x-1)(x+2)}$ **34.** $\sqrt{x^2-3x}$

35. $\sqrt{-x-x^2}$ **36.** $\sqrt{\dfrac{x^3+6x^2+8x}{3-x}}$

In Exercises 37–44, solve the given inequalities graphically by using a graphing calculator. See Example 10.

37. $x^3-x>2$ **38.** $0.5x^3<3-2x^2$

39. $x^4<x^2-2x-1$ **40.** $3x^4+x+1>5x^2$

41. $2^x>x+2$ **42.** $\log x<1-2x^2$

43. $\sin x<0.1x^2-1$ **44.** $4\cos 2x>2x-3$

In Exercises 45–52, use inequalities to solve the given problems.

45. Is $x^2>x$ for all x? Explain.

46. Is $x>1/x$ for all x? Explain.

47. Find an inequality of the form $ax^2+bx+c<0$ with $a>0$ for which the solution is $-1<x<4$.

48. Find an inequality of the form $ax^3+bx<0$ with $a>0$ for which the solution is $x<-1$ or $0<x<1$.

49. Algebraically find the values of x for which $2^{x+2}>3^{2x-3}$.

50. Graphically find the values of x for which $2\log_2 x<\log_3(x+1)$.

51. For what values of real numbers a and b does the inequality $(x-a)(x-b)<0$ have real solutions?

52. Algebraically find the intervals for which $f(x)=2x^4-5x^3+3x^2$ is positive and those for which it is negative. Using only this information, draw a rough sketch of the graph of the function.

In Exercises 53–64, answer the given questions by solving the appropriate inequalities.

53. The electric power p (in W) delivered to part of a circuit is given by $p=6i-4i^2$, where i is the current (in A). For what positive values of i is the power greater than 2 W?

54. The weight w (in tons) of fuel in a rocket after launch is $w=2000-t^2-140t$, where t is the time (in min). During what period of time is the weight of fuel greater than 500 tons?

55. In programming a computer, the formula $63n=2^x-1$ may be used. Here, n is the number of items to be added and x is the number of bits needed to represent the sum. Find x if $n<100$.

56. The object distance p (in cm) and image distance q (in cm) for a camera of focal length 3.00 cm is given by $p=3.00q/(q-3.00)$. For what values of q is $p>12.0$ cm?

57. The total capacitance C of capacitors C_1 and C_2 in series is $C^{-1}=C_1^{-1}+C_2^{-1}$. If $C_2=4.00\,\mu$F, find C_1 if $C>1.00\,\mu$F.

58. A rectangular field is to be enclosed by a fence and divided down the middle by another fence. The middle fence costs \$4/ft and the other fence cost \$8/ft. If the area of the field is to be 8000 ft², and the cost of the fence cannot exceed \$4000, what are the possible dimensions of the field?

59. The weight w (in N) of an object h meters above the surface of earth is $w=r^2w_0/(r+h)^2$, where r is the radius of earth and w_0 is the weight of the object at sea level. Given that $r=6380$ km, if an object weighs 200 N at sea level, for what altitudes is its weight less than 100 N?

60. One type of machine part costs \$10 each, and a second type costs \$20 each. How many of the second type can be purchased if at least 30 of the first type are purchased and a total of \$1000 is spent?

61. The length of a rectangular microprocessor chip is 2.0 mm more than its width. If its area is less than 35 mm², what values are possible for the width if it must be at least 3.0 mm?

62. A laser source is 2.0 in. from the nearest point P on a flat mirror, and the laser beam is directed at a point Q that is on the mirror and is x in. from P. The beam is then reflected to the receiver, which is x in. from Q. What is x if the total length of the beam is greater than 6.5 in.? See Fig. 17.29.

Fig. 17.29 Source

63. A plane takes off from Winnipeg and flies due east at 620 km/h. At the same time, a second plane takes off from the surface of Lake Winnipeg 310 km due north of Winnipeg and flies due north at 560 km/h. For how many hours are the planes less than 1000 km apart?

64. An open box (no top) is formed from a piece of cardboard 8.00 in. square by cutting equal squares from the corners, turning up the resulting sides, and taping the edges together. Find the edges of the squares that are cut out in order that the volume of the box is greater than 32.0 in³. See Fig. 17.30.

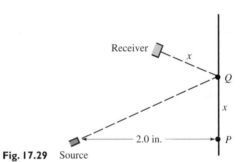

Fig. 17.30 8.00 in.

Answers to Practice Exercises

1. $-6<x<7$ **2.** $x\le-3$ or $x>5/2$

17.4 Inequalities Involving Absolute Values

Absolute Value Greater Than Given Value • Absolute Value Less Than Given Value

Inequalities involving absolute values are often useful in later topics in mathematics such as calculus and in applications such as the accuracy of measurements. In this section, we show the meaning of such inequalities and how they are solved.

If we wish to write the inequality $|x| > 1$ without absolute-value signs, we must note that we are considering values of x that are *numerically* larger than 1. Thus, we may write this inequality in the equivalent form $x < -1$ or $x > 1$. We now note that *the original inequality, with an absolute-value sign, can be written in terms of two equivalent inequalities, neither involving absolute values.* If we are asked to write the inequality $|x| < 1$ without absolute-value signs, we write $-1 < x < 1$, since we are considering values of x numerically less than 1.

Following reasoning similar to this, whenever absolute values are involved in inequalities, the following two relations allow us to write equivalent inequalities without absolute values. For $n > 0$,

$$\text{If } |f(x)| > n, \text{ then } f(x) < -n \text{ or } f(x) > n. \tag{17.1}$$

$$\text{If } |f(x)| < n, \text{ then } -n < f(x) < n. \tag{17.2}$$

EXAMPLE 1 Absolute value less than

Solve the inequality $|x - 3| < 2$.

Here, we want values of x such that $x - 3$ is numerically smaller than 2, or the values of x within 2 units of $x = 3$. These are given by the inequality $1 < x < 5$. Now, using Eq. (17.2), we have

$$-2 < x - 3 < 2$$

By adding 3 to all three members of this inequality, we have

$$1 < x < 5$$

which is the proper interval. See Fig. 17.31. ∎

$|x - 3| < 2$

2 units from 3

Fig. 17.31

EXAMPLE 2 Absolute value greater than

Solve the inequality $|2x - 1| > 5$.

By using Eq. (17.1), we have

$$2x - 1 < -5 \quad \text{or} \quad 2x - 1 > 5$$

Completing the solution, we have

$$2x < -4 \quad \text{or} \quad 2x > 6 \qquad \text{add 1 to each member}$$
$$x < -2 \qquad \quad x > 3 \qquad \text{divide each member by 2}$$

This means that the given inequality is satisfied for $x < -2$ or for $x > 3$. We must be very careful to remember that *we cannot write this as* $3 < x < -2$. The solution is shown in Fig. 17.32

The meaning of this inequality is that the numerical value of $2x - 1$ is greater than 5. By considering values in these intervals, we can see that this is true for values of x less than -2 or greater than 3. ∎

Fig. 17.32

EXAMPLE 3 Absolute value greater than or equal to

Solve the inequality $2\left|\dfrac{2x}{3} + 1\right| \geq 4$.

The solution is as follows:

$$2\left|\frac{2x}{3} + 1\right| \geq 4 \qquad \text{original inequality}$$

$$\left|\frac{2x}{3} + 1\right| \geq 2 \qquad \text{divide each member by 2}$$

$$\frac{2x}{3} + 1 \leq -2 \quad \text{or} \quad \frac{2x}{3} + 1 \geq 2 \qquad \text{using Eq. (17.1)}$$

$$2x + 3 \leq -6 \qquad\qquad 2x + 3 \geq 6$$

$$2x \leq -9 \qquad\qquad 2x \geq 3$$

$$x \leq -\frac{9}{2} \qquad\qquad x \geq \frac{3}{2} \qquad \text{solution}$$

This solution is shown in Fig. 17.33. Note that the sign of equality does not change the method of solution. It simply indicates that $-\frac{9}{2}$ and $\frac{3}{2}$ are included in the solution. ∎

Practice Exercise

1. Solve the inequality $|2x - 9| > 3$.

Fig. 17.33

EXAMPLE 4 Absolute value less than

Solve the inequality $|3 - 2x| < 3$.

We have the following solution.

$$|3 - 2x| < 3 \qquad \text{original inequality}$$

$$-3 < 3 - 2x < 3 \qquad \text{using Eq. (17.2)}$$

$$-6 < -2x < 0$$

$$3 > x > 0 \qquad \text{divide by } -2 \text{ and reverse signs of inequality}$$

$$0 < x < 3 \qquad \text{solution}$$

The meaning of the inequality is that the numerical value of $3 - 2x$ is less than 3. This is true for values of x between 0 and 3. The solution is shown in Fig. 17.34. ∎

Fig. 17.34

Practice Exercise

2. Solve the inequality $|4 - x| \leq 2$.

EXAMPLE 5 Absolute value – application

A technician measures an electric current and reports that it is 0.036 A with a possible error of ± 0.002 A. Write this result for the current i, using an inequality with absolute values.

The statement of the problem tells us that the current is no less than 0.034 A and no more than 0.038 A. Another way of stating this is that the numerical difference between the true value of i (unknown exactly) and the measured value, 0.036 A, is less than or equal to 0.002 A. Using an absolute-value inequality, this is written as

$$|i - 0.036| \leq 0.002$$

where values are in amperes.

We can see that this inequality is correct by using (Eq. 17.2):

$$-0.002 \leq i - 0.036 \leq 0.002$$

$$0.034 \leq i \leq 0.038 \qquad \text{add 0.036 to each member}$$

This verifies that i should not be less than 0.034 A or more than 0.038 A. The solution is shown in Fig. 17.35. ∎

Fig. 17.35

Calculator Solution

Most graphing calculators can be used to display the solution of an inequality that involves absolute values as they can display the solutions to other inequalities (see Example 9 on page 468, Example 10 on page 469, and Example 9 on page 475). Since calculators differ in their operation, check the manual of your calculator to see how an absolute value is entered and displayed. The following example shows the display of the solution of an inequality on a graphing calculator.

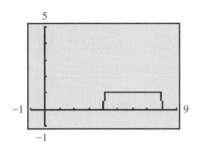

Fig. 17.36

EXAMPLE 6

Display the solution to the inequality $\left|\dfrac{x}{2} - 3\right| < 1$ on a graphing calculator.

On the calculator, set $y_1 = \text{abs}(x/2 - 3) < 1$ and obtain the display shown in Fig. 17.36. From this display, we see that the solution is $4 < x < 8$. It is possible that it will take two or three *window* settings to get an accurate solution.   ∎

EXERCISES 17.4

In Exercises 1 and 2, make the given changes in the indicated examples of this section and then solve the resulting inequalities.

1. In Example 2, change the $>$ to $<$, solve the resulting inequality, and graph the solution.

2. In Example 4, change the $<$ to $>$, solve the resulting inequality, and graph the solution.

In Exercises 3–24, solve the given inequalities. Graph each solution.

3. $|x - 4| < 1$

4. $|x + 1| < 3$

5. $|5x + 4| > 6$

6. $\left|\dfrac{1}{2}N - 1\right| > 1$

7. $1 + |6x - 5| \le 5$

8. $|30 - 42x| \le 0$

9. $|3 - 4x| > 3$

10. $3 + |3x + 1| \ge 5$

11. $\left|\dfrac{t + 1}{5}\right| < 5$

12. $\left|\dfrac{2x - 9}{4}\right| < 1$

13. $|20x + 85| \le 43$

14. $|2.6x - 9.1| > 10.4$

15. $2|x - 24| > 84$

16. $3|4 - 3x| \le 10$

17. $8 + 3|3 - 2x| < 11$

18. $5 - 4|1 - 7x| > 13$

19. $4|2 - 5x| \ge 6$

20. $2.5|7.1 - 2.0x| \le 6.5$

21. $\left|\dfrac{3R}{5} + 1\right| < 8$

22. $\left|\dfrac{4x}{3} - 5\right| \ge 7$

23. $\left|6.5 - \dfrac{x}{2}\right| \ge 2.3$

24. $\left|\dfrac{2w + 5}{w + 1}\right| \ge 1$

In Exercises 25–28, solve the given inequalities by displaying the solutions on a graphing calculator. See Example 6.

25. $|2x - 5| < 3$

26. $2|6 - T| > 5$

27. $\left|4 - \dfrac{x}{2}\right| \ge 1$

28. $\left|\dfrac{y}{3} + 12\right| \le 25$

In Exercises 29–32, solve the given quadratic inequalities. Check each by displaying the solution on a graphing calculator.

29. $|x^2 + x - 4| > 2$
(After using Eq. (17.1), you will have two inequalities. The solution includes the values of x that satisfy *either* of the inequalities.)

30. $|x^2 + 3x - 1| > 3$ (See Exercise 29.)

31. $|x^2 + x - 4| < 2$
(Use Eq. (17.2), then treat the resulting inequality as two inequalities of the form $f(x) > -n$ and $f(x) < n$. The solution includes the values of x that satisfy *both* of the inequalities.)

32. $|x^2 + 3x - 1| < 3$ (See Exercise 31.)

In Exercises 33–38, solve the given problems.

(W) 33. Solve for x if $|x| < a$ and $a \le 0$. Explain.

34. Solve for x if $|x - 1| < 4$ and $x \ge 0$.

35. Solve for x: $a - |bx| < c$ given that $a - c > 0$.

36. Solve for x: $1 < |x - 2| < 3$.

37. The thickness t (in km) of earth's crust varies and can be described as $|t - 27| \le 23$. What are the minimum and maximum values of the thickness of earth's crust?

38. The temperature T (in °C) at which a certain integrated-circuit computer chip is designed to operate is given by the inequality $|T - 110| \le 10$. What are the minimum and maximum temperatures of this range?

In Exercises 39–44, use inequalities involving absolute values to solve the given problems.

(W) 39. The production p (in barrels) of oil at a refinery is estimated at $2{,}000{,}000 \pm 200{,}000$. Express p using an inequality with absolute values and describe the production in a verbal statement.

40. The *Mach number M* of a moving object is the ratio of its velocity v to the velocity of sound v_s, and v_s varies with temperature. A jet traveling at 1650 km/h changes its altitude from 500 m to 5500 m. At 500 m (with the temperature at 27°C), $v_s = 1250$ km/h, and at 5500 m ($-3°$C), $v_s = 1180$ km/h. Express the range of M, using an inequality with absolute values.

41. The diameter d of a certain type of tubing is 3.675 cm with a tolerance of 0.002 cm. Express this as an inequality with absolute values.

42. The moon is at a maximum distance of 4.07×10^5 km from earth and at a minimum distance of 3.57×10^5 km. Express this as an inequality with absolute values.

43. The voltage V in a certain circuit is given by $V = 6.0 - 200i$, where i is the current (in A). For what values of the current is the absolute value of the voltage less than 2.0 V?

44. A rocket is fired from a plane flying horizontally at 9000 ft. The height h (in ft) of the rocket above the plane is given by $h = 560t - 16t^2$, where t is the time (in s) of flight of the rocket. When is the rocket more than 4000 ft above or below the plane? See Fig. 17.37.

Fig. 17.37

Answers to Practice Exercises

1. $x < 3$ or $x > 6$ **2.** $2 \leq x \leq 6$

17.5 Graphical Solution of Inequalities with Two Variables

Points Above and Below Curve •
Use of Dashed Curve or Solid Curve •
Solution Using Calculator

To this point, we have considered inequalities with one variable and certain methods of solving them. We may also graphically solve inequalities involving two variables, such as x and y. In this section, we consider the solution of such inequalities.

Let us consider the function $y = f(x)$. We know that the coordinates of points on the graph satisfy the equation $y = f(x)$. However, for points above the graph of the function, we have $y > f(x)$, and for points below the graph of the function, we have $y < f(x)$. Consider the following example.

EXAMPLE 1 Checking points above and below line

Consider the linear function $y = 2x - 1$, the graph of which is shown in Fig. 17.38. This equation is satisfied for points on the line. For example, the point $(2, 3)$ is on the line, and we have $3 = 2(2) - 1 = 3$. Therefore, for points on the line, we have $y = 2x - 1$, or $y - 2x + 1 = 0$.

The point $(2, 4)$ is above the line, since we have $4 > 2(2) - 1$, or $4 > 3$. Therefore, for points above the line, we have $y > 2x - 1$, or $y - 2x + 1 > 0$. In the same way, for points below the line, $y < 2x - 1$ or $y - 2x + 1 < 0$. We note this is true for the point $(2, 1)$, since $1 < 2(2) - 1$, or $1 < 3$.

The line for which $y = 2x - 1$ and the regions for which $y > 2x - 1$ and for which $y < 2x - 1$ are shown in Fig. 17.38.

Summarizing,

$$y > 2x - 1 \quad \text{for points } \textit{above} \text{ the line}$$
$$y = 2x - 1 \quad \text{for points } \textit{on} \text{ the line}$$
$$y < 2x - 1 \quad \text{for points } \textit{below} \text{ the line}$$ ∎

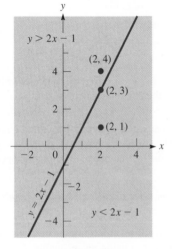

Fig. 17.38

The illustration in Example 1 leads us to the graphical method of indicating the points that satisfy an inequality with two variables. First, we solve the inequality for y and then determine the graph of the function $y = f(x)$. *If we wish to solve the inequality $y > f(x)$, we indicate the appropriate points by shading in the region above the curve. For the inequality $y < f(x)$, we indicate the appropriate points by shading in the region below the curve.* We note that the complete solution to the inequality consists of all points in an entire region of the plane.

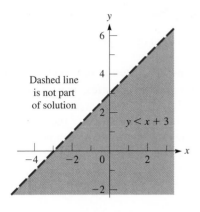

Dashed line is not part of solution

$y < x + 3$

Fig. 17.39

EXAMPLE 2 Sketching $y < f(x)$ inequality

Draw a sketch of the graph of the inequality $y < x + 3$.

First, we draw the function $y = x + 3$, as shown by the dashed line in Fig. 17.39. Since we wish to find all the points that satisfy the inequality $y < x + 3$, we show these points by shading in the region below the line. The line is shown as a *dashed line* to indicate that points on it do not satisfy the inequality.

Most graphing calculators can be used to display the solution of an inequality involving two variables by shading in an area above a curve, below a curve, or between curves. The manner in which this is done varies according to the model of the calculator. Therefore, the manual should be used to determine how this is done on any particular model of calculator. In Fig. 17.40, such a graphing calculator display is shown for this inequality.

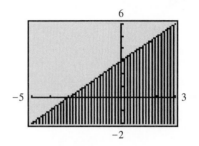

Fig. 17.40

EXAMPLE 3 Sketching $y \le f(x)$ inequality – application

After a snowstorm, it is estimated that it will take 30 min to plow each mile of Route 15 and 45 min to plow each mile of Route 80. If no more than 60 plowing-hours are available, what combinations of Route 15 and Route 80 can be plowed?

Let x = miles of Route 15 that can be plowed and y = miles of Route 80 that can be plowed. The time to plow along each route is the product of the time for each mile and the number of miles to be plowed. This gives us

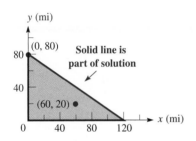

Solid line is part of solution

Fig. 17.41

$$\underset{\substack{\uparrow \\ 30\,\text{min}}}{\underbrace{(0.50\,\text{h/mi})(x\,\text{mi})}_{\text{time to plow Rt. 15}}} + \underset{\substack{\uparrow \\ 45\,\text{min}}}{\underbrace{(0.75\,\text{h/mi})(y\,\text{mi})}_{\text{time to plow Rt. 80}}} \le \underset{\text{max. available time}}{60\,\text{h}}$$

$$0.50x + 0.75y \le 60$$
$$y \le 80 - 0.67x$$

Noting that negative values of x and y do not have meaning, we have the graph in Fig. 17.41, shading in the region below the line since we have $y < 80 - 0.67x$ for that region. Any point in the shaded region, or on the axes or the line around the shaded region, gives a solution. The *solid line* indicates that points on it are part of the solution.

The point $(0, 80)$, for example, is a solution and tells us that 80 mi of Route 80 can be plowed if none of Route 15 is plowed. In this case, all 60 h of plowing time are used for Route 80. Another possibility is shown by the point $(60, 20)$, which indicates that 60 mi of Route 15 and 20 mi of Route 80 can be plowed. In this case, not all of the 60 plowing hours are used. ∎

EXAMPLE 4 Draw a sketch of the graph of the inequality $y > x^2 - 4$.

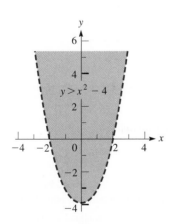

$y > x^2 - 4$

Fig. 17.42

Although the graph of $y = x^2 - 4$ is not a straight line, the method of solution is the same. We graph the function $y = x^2 - 4$ as a dashed curve, since it is not part of the solution, as shown in Fig. 17.42. We then shade in the region above the curve to indicate the points that satisfy the inequality. ∎

Fig. 17.43

■ Some calculators can show solid lines and dashed lines when graphing inequalities.

EXAMPLE 5 Solution of system of inequalities

Draw a sketch of the region that is defined by the system of inequalities $y \geq -x - 2$ and $y + x^2 < 0$.

Similar to the solution of a system of equations, *the solution of a system of inequalities is any pair of values (x, y) that satisfies both inequalities.* This means we want the region common to both inequalities. In Fig. 17.43, we first shade in the region above the line $y = -x - 2$ and then shade in the region below the parabola $y = -x^2$. The region defined by this system is the darkly shaded region below the parabola that is also above and on the line. The calculator display for this region is shown in Fig. 17.44.

If we are asked to find the region defined by $y \geq -x - 2$ *or* $y + x^2 < 0$, it consists of both shaded regions and all points on the line.

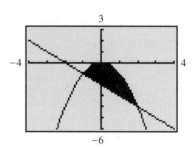

Fig. 17.44

■

EXERCISES 17.5

In Exercises 1 and 2, make the given changes in the indicated examples of this section and then draw the graph of the resulting inequality.

1. In Example 2, change $x + 3$ to $3 - x$ and then draw the graph of the resulting inequality.

2. In Example 4, change $x^2 - 4$ to $4 - x^2$ and then draw the graph of the resulting inequality.

In Exercises 3–22, draw a sketch of the graph of the given inequality.

3. $y > x - 1$
4. $y < 3x - 2$
5. $y \geq 2x + 5$
6. $y \leq 15 - 3x$
7. $3x + 2y + 6 > 0$
8. $x + 4y - 8 < 0$
9. $y < x^2$
10. $y \leq 2x^2 - 3$
11. $2x^2 - 4x - y > 0$
12. $y \leq x^3 - 1$
13. $y < 32x - x^4$
14. $y \leq \sqrt{2x + 5}$
15. $y > \dfrac{10}{x^2 + 1}$
16. $y < \ln x$
17. $y > 1 + \sin 2x$
18. $y > |x| - 3$
19. $-4 < y \leq -2$
20. $y > |x + 3|$
21. $|x| < |y|$
22. $|x - 3y| > 2$

In Exercises 23–32, draw a sketch of the graph of the region in which the points satisfy the given system of inequalities.

23. $y > x$
    $y > 1 - x$
24. $y \leq 2x$
    $y \geq x - 1$
25. $y \leq 2x^2$
    $y > x - 2$
26. $y > x^2$
    $y < x + 4$
27. $y > \frac{1}{2}x^2$
    $y \leq 4x - x^2$
28. $y < 4 - x$
    $y < \sqrt{16 - x^2}$

29. $y \geq 0$
    $y \leq \sin x$
    $0 \leq x \leq 3\pi$
30. $y > 0$
    $y > 1 - x$
    $y < e^x$
31. $|y + 2| < 5$
    $|x - 3| \leq 2$
32. $16x + 3y - 12 > 0$
    $y > x^2 - 2x - 3$
    $|2x - 3| < 3$

In Exercises 33–42, use a graphing calculator to display the solution of the given inequality or system of inequalities.

33. $2x + y < 5$
34. $4x - y > 1$
35. $y \geq 1 - x^2$
36. $y < |4 - 2x|$
37. $y > 2x - 1$
    $y < x^4 - 8$
38. $y < 3 - x$
    $y > 3x - x^3$
39. $y > x^2 + 2x - 8$
    $y < \dfrac{1}{x} - 2$
40. $y > -2x^2$
    $y < 1 - e^{-x}$
41. $y \leq |2x - 3|$
    $y > 1 - 2x^2$
42. $y \geq |4 - x^2|$
    $y < 2 \ln |x|$

In Exercises 43–48, solve the given problems.

43. By an inequality, define the region below the line $4x - 2y + 5 = 0$.

44. By an inequality, define the region that is bounded by or includes the parabola $x^2 - 2y = 0$, and that contains the point $(1, 0.4)$.

45. For $Ax + By > C$, if $B < 0$, would you shade above or below the line?

46. Find a system of inequalities that would describe the region within the triangle with vertices (0, 0) (0, 4), and (2, 0).

47. Draw a graph of the solution of the system $y \geq 2x^2 - 6$ and $y = x - 3$.

48. Draw a graph of the solution of the system $y < |x + 2|$ and $y = x^2$.

In Exercises 49–54, set up the necessary inequalities and sketch the graph of the region in which the points satisfy the indicated system of inequalities.

49. A telephone company is installing two types of fiber-optic cable in an area. It is estimated that no more than 300 m of type A cable, and at least 200 m but no more than 400 m of type B cable, are needed. Graph the possible lengths of cable that are needed.

50. A refinery can produce gasoline and diesel fuel, in amounts of any combination, except that equipment restricts total production to 2500 gal/day. Graph the different possible production combinations of the two fuels.

51. The elements of an electric circuit dissipate p watts of power. The power p_R dissipated by a resistor in the circuit is given by $p_R = Ri^2$, where R is the resistance (in Ω) and i is the current (in A). Graph the possible values of p and i for $p > p_R$ and $R = 0.5 \, \Omega$.

52. The cross-sectional area A (in m²) of a certain trapezoid culvert in terms of its depth d (in m) is $A = 2d + d^2$. Graph the possible values of d and A if A is between 1 m² and 2 m².

53. One pump can remove wastewater at the rate of 75 gal/min, and a second pump works at the rate of 45 gal/min. Graph the possible values of the time (in min) that each of these pumps operates such that together they pump more than 4500 gal.

54. A rectangular computer chip is being designed such that its perimeter is no more than 30 mm, its width at least 3 mm and its length at least 8 mm. Graph the possible values of the width w and the length l.

17.6 Linear Programming

Constraints • Objective Function • Feasible Point • Vertices of Region of Feasible Points

■ This section is an introduction to linear programming. Other methods are developed in a more complete coverage.

An important area in which graphs of inequalities with two or more variables are used is in the branch of mathematics known as **linear programming** (in this context, "programming" does not mean computer programming). This subject, which we mentioned in the chapter introduction, is widely applied in industry, business, economics, and technology. The analysis of many social problems can also be made by use of linear programming.

Linear programming is used to analyze problems such as those related to maximizing profits, minimizing costs, or the use of materials with certain constraints of production. First, we look at a similar type of mathematical problem.

EXAMPLE 1 Maximum value of F

Find the maximum value of F, where $F = 2x + 3y$ and x and y are subject to the conditions that

$$x \geq 0, y \geq 0$$
$$x + y \leq 6$$
$$x + 2y \leq 8$$

These four inequalities that define the conditions on x and y are known as the **constraints** of the problem, and F is known as the **objective function.**

We now graph this set of inequalities, as shown in Fig. 17.45. Each point in the shaded region (including the line segments on the edges) satisfies all the constraints and is known as a **feasible point.**

The maximum value of F must be found at one of the feasible points. Testing for values at the vertices of the region, we have the values in the following table:

Point	(0, 0)	(6, 0)	(4, 2)	(0, 4)
Value of F	0	12	14	12

If we evaluate F at any other feasible point, we will find that $F < 14$. Therefore, the maximum value of F under the given constraints is 14. ■

Fig. 17.45

Practice Exercise

1. In Example 1, what is the maximum value of F if the third constraint is changed to $x + y \leq 5$?

In Example 1, we found the maximum value of a *linear* objective function, subject to *linear* constraints (thus the name *linear programming*). We found the maximum value of F, subject to the given constraints, to be at one of the vertices of the region of feasible points. In fact, in the theory of linear programming it is established that

the maximum and minimum values of the objective function will occur at a vertex of the region of feasible points

(or at all points along a line segment connecting two vertices).

EXAMPLE 2 Maximum and minimum values of *F*

Find the maximum and minimum values of the objective function $F = 3x + y$, subject to the constraints

$$x \geq 2, y \geq 0$$
$$x + y \leq 8$$
$$2y - x \leq 1$$

The constraints are graphed as shown in Fig. 17.46. We then locate the vertices, and evaluate F at these vertices as follows.

Vertex	(2, 0)	(8, 0)	(5, 3)	(2, 1.5)
Value of F	6	24	18	7.5

Therefore, we see that the maximum value of F is 24, and the minimum value of F is 6. If we check the value of F at any other feasible point, we will find a value between 6 and 24. ■

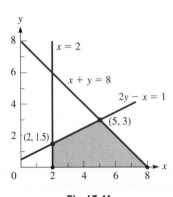

Fig. 17.46

The following two examples show the use of linear programming in finding a maximum value and a minimum value in applied situations.

EXAMPLE 3 Linear programming – application

A company makes two types of stereo speaker systems, their good-quality system and their highest-quality system. The production of these systems requires assembly of the speaker system itself and the production of the cabinets in which they are installed. The good-quality system requires 3 worker-hours for speaker assembly and 2 worker-hours for cabinet production for each complete system. The highest-quality system requires 4 worker-hours for speaker assembly and 6 worker-hours for cabinet production for each complete system. Available skilled labor allows for a maximum of 480 worker-hours per week for speaker assembly and a maximum of 540 worker-hours per week for cabinet production. It is anticipated that all systems will be sold and that the profit will be $30 for each good-quality system and $75 for each highest-quality system. How many of each system should be produced to provide the greatest profit?

First, let x = the number of good-quality systems and y = the number of highest-quality systems made in one week. Thus, the profit P is given by

$$P = 30x + 75y$$

We know that negative numbers are not valid for either x or y, and therefore we have $x \geq 0$ and $y \geq 0$.

The number of available worker-hours per week for each part of the production also restricts the number of systems that can be made. In the speaker-assembly shop, $3x$ worker-hours are needed for each good-quality system and $4y$ worker-hours are needed for each highest-quality system. The constraint that only 480 hours are available in the speaker-assembly shop means that $3x + 4y \leq 480$.

■ The loudspeaker was developed in the early 1920s by the U.S. inventor Kellogg Rice.

In the cabinet shop, it takes $2x$ worker-hours for a good-quality system and $6y$ worker-hours for a highest-quality system. The constraint that only 540 hours are available in the cabinet shop means that $2x + 6y \leq 540$.

Therefore, we want to maximize the profit $P = 30x + 75y$, which is the objective function, under the constraints

$x \geq 0, y \geq 0$	number of systems produced cannot be negative
$3x + 4y \leq 480$	worker-hours for speaker assembly
$2x + 6y \leq 540$	worker-hours for cabinet production

The constraints are graphed as shown in Fig. 17.47. We locate the vertices and evaluate the profit P at these points as follows:

Vertex	(0, 0)	(160, 0)	(72, 66)	(0, 90)
Profit ($)	0	4800	7110	6750

Therefore, we see that the greatest profit of $7110 is made by producing 72 good-quality systems and 66 highest-quality systems.

A way of showing the number of each system to be made for the greatest profit is to assume values of the profit P and graph these lines. For example, for $P = \$3000$ or $P = \$6000$, we have the lines shown in Fig. 17.47. Both amounts of profit are possible with various combinations of speaker systems being produced. However, we note that the line for $P = \$6000$ passes through feasible points farther from the origin. It is also clear that the lines for the profits are parallel and that *the greatest profit attainable is given by the line passing through A,* where $3x + 4y = 40$ and $2x + 6y = 540$ intersect. This also illustrates why the greatest profit is found at one of the vertices of the region of feasible points.

Fig. 17.47   ■

The problems in linear programming that arise in business and industry involve many more variables than in the simplified examples in this text. They are solved by computers using matrices, but the basic idea is the same as that shown in this section.

■ See the chapter introduction.

EXAMPLE 4 Linear programming – application

An airline plans to open new routes and use two types of planes, A and B, on these routes. It is expected there would be at least 400 first-class passengers and 2000 economy-class passengers on these routes each day. Plane A costs $18,000/day to operate and has seats for 40 first class and 80 economy-class passengers. Plane B costs $16,000/day to operate and has seats for 20 first-class and 160 economy-class passengers. How many of each type of plane should be used for these routes to minimize operating costs? (Assume that the planes will be used only on these routes.)

We first let x = the number of A planes and y = the number of B planes to be used. Then the operating cost $C = 18{,}000x + 16{,}000y$ is the objective function. The constraints are

$$x \geq 0, y \geq 0 \qquad \text{number of planes cannot be negative}$$
$$40x + 20y \geq 400 \qquad \text{at least 400 first-class passengers}$$
$$80x + 160y \geq 2000 \qquad \text{at least 2000 economy-class passengers}$$

Fig. 17.48

The constraints are graphed as shown in Fig. 17.48. We see that the region of feasible points is unlimited (there could be more than 2400 passengers), but we still want to evaluate the operating cost C at the vertices. Evaluating C at these points, we have

Vertex	(25, 0)	(5, 10)	(0, 20)
Cost ($)	450,000	250,000	320,000

Therefore, five type-A planes and ten type-B planes should be used on these routes to keep the operating costs at a minimum of $250,000. Obviously, if the number of passengers does not meet the numbers expected, these numbers of planes should be changed accordingly.   ■

EXERCISES 17.6

In Exercises 1 and 2, make the given changes in the indicated examples of this section and then find the indicated values.

1. In Example 1, change the last constraint to $2x + y \leq 8$. Then graph the feasible points and find the maximum value of F.

2. In Example 2, change the last constraint to $3y - x \leq 4$. Then graph the feasible points and find the maximum and minimum values of F.

In Exercises 3–16, find the indicated maximum and minimum values by the linear programming method of this section. For Exercises 5–16, the constraints are shown below the objective function.

3. Graphing the constraints of a linear programming problem shows the consecutive vertices of the region of feasible points to be $(0,0)$, $(12,0)$, $(10,7)$, $(0,5)$, and $(0,0)$. What are the maximum and minimum values of the objective function $F = 3x + 4y$ in this region?

4. Graphing the constraints of a linear programming problem shows the consecutive vertices of the region of feasible points to be $(1,3)$, $(8,0)$, $(9,7)$, $(5,8)$, $(0,6)$, and $(1,3)$. What are the maximum and minimum values of the objective function $F = 2x + 5y$ in this region?

5. Maximum P:
$P = 3x + 5y$
$x \geq 0, y \geq 0$
$2x + y \leq 6$

6. Maximum P:
$P = 2x + 7y$
$x \geq 1, y \geq 0$
$x + 4y \leq 8$

7. Maximum P:
$P = 5x + 9y$
$x \geq 0, y \geq 0$
$x + 2y \leq 6$

8. Minimum C:
$C = 10x + 20y$
$x \geq 0, y \geq 0$
$3x + 5y \geq 30$

9. Minimum C:
$C = 4x + 6y$
$x \geq 0, y \geq 0$
$x + y \geq 5$
$x + 2y \geq 7$

10. Minimum C:
$C = 6x + 4y$
$x \geq 0, y \geq 0$
$2x + y \geq 6$
$x + y \geq 5$

11. Maximum and minimum F:
$F = x + 3y$
$x \geq 1, y \geq 2$
$y - x \leq 3$
$y + 2x \leq 8$

12. Maximum and minimum F:
$F = 3x - y$
$x \geq 1, y \geq 0$
$x + 4y \leq 8$
$4x + y \leq 8$

13. Maximum P:
$P = 9x + 2y$
$x \geq 0, y \geq 0$
$2x + 5y \leq 10$
$4x + 3y \leq 12$

14. Minimum C:
$C = 3x + 8y$
$x \geq 0, y \geq 0$
$6x + y \geq 6$
$x + 4y \geq 4$

15. Minimum C:
$C = 6x + 4y$
$y \geq 2$
$x + y \leq 12$
$x + 2y \geq 12$
$2x + y \geq 12$

16. Maximum P:
$P = 3x + 4y$
$2x + y \geq 2$
$x + 2y \geq 2$
$x + y \leq 2$

In Exercises 17–22, solve the given linear programming problems.

17. A person wants to invest no more than $9000, part at 6% and part at 5%, with no more than twice as much being invested at 6% as at 5%. How much should be invested at each rate to maximize the income from the investments?

18. An oil refinery refines types A and B of crude oil and can refine as much as 4000 barrels each week. Type A crude has 2 kg of impurities per barrel, type B has 3 kg of impurities per barrel, and the refinery can handle no more than 9000 kg of these impurities each week. How much of each type should be refined in order to maximize profits, if the profit is $4/barrel for type A and $5/barrel for type B?

19. A manufacturer produces a business calculator and a graphing calculator. Each calculator is assembled in two sets of operations, where each operation is in production 8 h during each day. The average time required for a business calculator in the first operation is 3 min, and 6 min is required in the second operation. The graphing calculator averages 6 min in the first operation and 4 min in the second operation. All calculators can be sold; the profit for a business calculator is $8, and the profit for a graphing calculator is $10. How many of each type of calculator should be made each day in order to maximize profit?

(W) 20. Using the information given in Example 3, with the one change that the profit on each good-quality speaker system is $60 (instead of $30), how many of each system should be made? Explain why this one change in data makes such a change in the solution.

21. Brands A and B of breakfast cereal are both enriched with vitamins P and Q. The necessary information about these cereals is as follows:

	Cereal A	Cereal B	RDA
Vitamin P	1 unit/oz	2 units/oz	10 units
Vitamin Q	5 units/oz	3 units/oz	30 units
Cost	12¢/oz	18¢/oz	

(RDA is the *Recommended Daily Allowance*.) Find the amount of each cereal that together satisfies the RDA of vitamins P and Q at the lowest cost.

22. A computer company makes parts A and B in each of two different plants. It costs $4000 per day to operate the first plant and $5000 per day to operate the second plant. Each day the first plant produces 100 of part A and 200 of part B, while at the second plant 250 of part A and 100 of part B are produced. How many days should each plant operate to produce 2000 of each part and keep operating costs at a minimum?

Answers to Practice Exercise

1. 13 (at (2, 3))

If $|f(x)| > n$, then $f(x) < -n$ or $f(x) > n$.   **(17.1)**

If $|f(x)| < n$, then $-n < f(x) < n$.   **(17.2)**

CHAPTER 17   **REVIEW EXERCISES**

In Exercises 1–16, solve each of the given inequalities algebraically. Graph each solution.

1. $2x - 12 > 0$

2. $2.4(T - 4.0) \geq 5.5 - 2.4T$

3. $4 < 2x - 1 < 11$

4. $2x < x + 1 < 4x + 7$

5. $5x^2 + 9x < 2$

6. $x^2 + 2x > 63$

7. $6n^2 - n > 35$

8. $2x^3 + 4 \leq x^2 + 8x$

9. $\dfrac{(2x - 1)(3 - x)}{x + 4} > 0$

10. $x^4 + x^2 \leq 0$

11. $\dfrac{8}{x} < 2$

12. $\dfrac{1}{x - 2} < \dfrac{1}{4}$

13. $|3x + 2| \leq 4$

14. $|4 - 3x| \geq 7$

15. $|3 - 5x| > 7$

16. $\left| 2 - \dfrac{y}{2} \right| \leq 0$

In Exercises 17–24, solve the given inequalities on a graphing calculator such that the display is the graph of the solution.

17. $5 - 3x < 0$

18. $6 \leq 4x - 2 < 9$

19. $2 \leq \dfrac{4n - 2}{3} < 3$

20. $3x < 2x + 1 < x - 5$

21. $\dfrac{8 - R}{2R + 1} \leq 0$

22. $\dfrac{(3 - x)^2}{2x + 7} \leq 0$

23. $|x - 30| > 48$

24. $2|2x - 9| < 8$

In Exercises 25–28, use a graphing calculator to solve the given inequalities. Graph the appropriate function and from the graph determine the solution.

25. $x^3 + x + 1 < 0$

26. $\dfrac{2}{R+2} > 3$

27. $e^{-t} > 0.5$

28. $\sin 2x < 0.8 \quad (0 < x < 4)$

In Exercises 29–40, draw a sketch of the region in which the points satisfy the given inequality or system of inequalities.

29. $y > 12 - 3x$

30. $y < \dfrac{1}{2}x + 2$

31. $2y - 3x - 4 \le 0$

32. $3y - x + 6 \ge 0$

33. $y > 6 - 2x^2$

34. $y \le \dfrac{6}{x^2 - 49}$

35. $y - |x + 1| < 0$

36. $2y + 2x^3 + 6x > 3$

37. $y > x + 1$
    $y < 4 - x^2$

38. $y > 2x - x^2$
    $y \ge -2$

39. $y \le \dfrac{4}{x^2 + 1}$
    $y < x - 3$

40. $y < \cos\dfrac{1}{2}x$
    $y > \dfrac{1}{2}e^x$
    $-\pi < x < \pi$

In Exercises 41–48, use a graphing calculator to display the region in which the points satisfy the given inequality or system of inequalities.

41. $y < 3x + 5$

42. $y > 2 - \dfrac{1}{4}x$

43. $y > 8 + 7x - x^2$

44. $y < x^3 + 4x^2 - x - 4$

45. $y < 32x - x^4$

46. $y > 2x - 1$
    $y < 6 - 3x^2$

47. $y > 1 - x\sin 2x$
    $y < 5 - x^2$

48. $y > |x - 1|$
    $y < 4 + \ln x$

In Exercises 49–52, determine the values of x for which the given radicals represent real numbers.

49. $\sqrt{3 - x}$

50. $\sqrt{x + 5}$

51. $\sqrt{x^2 + 4x}$

52. $\sqrt{\dfrac{x - 1}{x + 2}}$

In Exercises 53–56, find the indicated maximum and minimum values by the method of linear programming. The constraints are shown below the objective function.

53. Maximum P:
    $P = 2x + 9y$
    $x \ge 0, y \ge 0$
    $x + 4y \le 13$
    $3y - x \le 8$

54. Maximum P:
    $P = x + 2y$
    $x \ge 1, y \ge 0$
    $3x + y \le 6$
    $2x + 3y \le 8$

55. Minimum C:
    $C = 3x + 4y$
    $x \ge 0, y \ge 1$
    $2x + 3y \ge 6$
    $4x + 2y \ge 5$

56. Minimum C:
    $C = 2x + 4y$
    $x \ge 0, y \ge 0$
    $x + 3y \ge 6$
    $4x + 7y \ge 18$

In Exercises 57–82, solve the given problems using inequalities. (All data are accurate to at least two significant digits.)

(W) 57. Under what conditions is $|a + b| < |a| + |b|$?

(W) 58. Is $|a - b| < |a| + |b|$ always true? Explain.

59. If $a > 0$, $a \ne 1$, show that $a + a^{-1} > 2$.

60. Solve for x: $a + |bx| < c$ given that $a - c < 0$.

61. Form an inequality of the form $ax^2 + bx + c < 0$ with $a > 0$ for which the solution is $-2 < x < 5$.

62. By means of an inequality, define the region above the line $x - 3y - 6 = 0$.

63. Draw a graph of the system $y < 1 - x^2$ and $y = x^2$.

64. Find a system of inequalities that would describe the region within the quadrilateral with vertices $(0, 0)$, $(4, 4)$, $(0, -3)$ and $(4, -3)$.

65. Find the values for which $f(x) = (x - 2)(x - 3)$ is positive, zero, and negative. Use this information along with $f(0)$ and $f(5)$ to make a rough sketch of the graph of $f(x)$.

66. Follow the same instructions as in Exercise 65 for the function $f(x) = (x - 2)/(x - 3)$.

67. If two adjacent sides of a square design on a sheet of metal are extended 6.0 cm and 10.0 cm, respectively, how long is each side of the square if the perimeter of the rectangle is at least twice as long as the perimeter of the square? See Fig. 17.49.

Fig. 17.49

68. The value V (in \$) of each building lot in a development is estimated as $V = 65{,}000 + 5000t$, where t is the time in years from now. For how long is the value of each lot no more than \$90,000?

69. The cost C of producing two of one type of calculator and five of a second type is \$50. If the cost of producing each of the second type is between \$5 and \$8, what are the possible costs of producing each of the first type?

70. City A is 600 km from city B. One car starts from A for B 1 h before a second car. The first car averages 60 km/h, and the second car averages 80 km/h for the trip. For what times after the first car starts is the second car ahead of the first car?

71. The pressure p (in kPa) at a depth d (in m) in the ocean is given by $p = 101 + 10.1d$. For what values of d is $p > 500$ kPa?

72. After conducting tests, it was determined that the stopping distance x (in ft) of a car traveling 60 mi/h was $|x - 290| \le 35$. Express this inequality without absolute values and find the interval of stopping distances that were found in the tests.

73. A heating unit with 80% efficiency and a second unit with 90% efficiency deliver 360,000 Btu of heat to an office complex. If the first unit consumes an amount of fuel that contains no more than 261,000 Btu, what is the Btu content of the fuel consumed by the second unit?

74. A rectangular parking lot is to have a perimeter of 100 m and an area no greater than 600 m^2. What are the possible dimensions of the lot?

75. The electric power p (in W) dissipated in a resistor is given by $p = Ri^2$, where R is the resistance (in Ω) and i is the current (in A). For a given resistor, $R = 12.0\ \Omega$, and the power varies between 2.50 W and 8.00 W. Find the values of the current.

76. The reciprocal of the total resistance of two electric resistances in parallel equals the sum of the reciprocals of the resistances. If a 2.0-Ω resistance is in parallel with a resistance R, with a total resistance greater than 0.5 Ω, find R.

77. The efficiency e (in %) of a certain gasoline engine is given by $e = 100(1 - r^{-0.4})$, where r is the *compression ratio* for the engine. For what values of r is $e > 50\%$?

78. A rocket is fired such that its height h (in mi) is given by $h = 41t - t^2$. For what values of t (in min) is the height greater than 400 mi?

79. In developing a new product, a company estimates that it will take no more than 1200 min of computer time for research and no more than 1000 min of computer time for development. Graph the possible combinations of the computer times that are needed.

80. A natural-gas supplier has a maximum of 120 worker-hours per week for delivery and for customer service. Graph the possible combinations of times available for these two services.

81. A company produces two types of cell phones, the regular model and the deluxe model. For each regular model produced, there is a profit of $8, and for each deluxe model the profit is $15. The same amount of materials is used to make each model, but the supply is sufficient only for 450 cell phones per day. The deluxe model requires twice the time to produce as the regular model. If only regular models were made, there would be time enough to produce 600 per day. Assuming all cell phones will be sold, how many of each model should be produced if the profit is to be a maximum?

82. A company that manufactures DVD/CD players gets two different parts, A and B, from two different suppliers. Each package of parts from the first supplier costs $2.00 and contains 6 of each type of part. Each package of parts from the second supplier costs $1.50 and contains 4 of A and 8 of B. How many packages should be bought from each supplier to keep the total cost to a minimum, if production requirements are 600 of A and 900 of B?

Writing Exercises

83. In planning a new city development, an engineer uses a rectangular coordinate system to locate points within the development. A park in the shape of a quadrilateral has corners at $(0, 0)$, $(0, 20)$, $(40, 20)$, and $(20, 40)$ (measurements in meters). Write two or three paragraphs explaining how to describe the park region with inequalities and find these inequalities.

CHAPTER 17   **PRACTICE TEST**

1. State conditions on x and y in terms of inequalities if the point (x, y) is in the second quadrant.

In Problems 2–7, solve the given inequalities algebraically and graph each solution.

2. $\dfrac{-x}{2} \geq 3$

3. $3x + 1 < -5$

4. $-1 < 1 - 2x < 5$

5. $\dfrac{x^2 + x}{x - 2} \leq 0$

6. $|2x + 1| \geq 3$

7. $|2 - 3x| < 8$

8. Sketch the region in which the points satisfy the following system of inequalities:
$$y < x^2$$
$$y \geq x + 1$$

9. Determine the values of x for which $\sqrt{x^2 - x - 6}$ represents a real number.

10. The length of a rectangular lot is 20 m more than its width. If the area is to be at least 4800 m^2, what values may the width be?

11. Type A wire costs $0.10 per foot, and type B wire costs $0.20 per foot. Show the possible combinations of lengths of wire that can be purchased for less than $5.00.

12. The range of the visible spectrum in terms of the wavelength λ of light ranges from about $\lambda = 400$ nm (violet) to about $\lambda = 700$ nm (red). Express these values using an inequality with absolute values.

13. Solve the inequality $x^2 > 12 - x$ on a graphing calculator such that the display is the graph of the solution.

14. By using linear programming, find the maximum value of the objective function $P = 5x + 3y$ subject to the following constraints: $x \geq 0$, $y \geq 0$, $2x + 3y \leq 12$, $4x + y \leq 8$.

18

Variation

As millions watched on television, men walked on the moon for the first time in 1969. For hundreds of years before that, men on the moon were subjects of dreams and science fiction.

Numerous discoveries in science, engineering, and mathematics helped make it possible for a spacecraft to travel to the moon. These included, for example, discovery and development of rocket flight and numerous developments in electronics in the mid-1900s. However, it can be argued that the beginning came with the formulation of the *universal law of gravitation* by Newton in the 1680s, which was based on earlier discoveries regarding the motion of the planets and the moon. This, along with developments in advanced mathematics, made it possible to determine what conditions must be met in order to have successful travel in space.

The universal law of gravitation is stated using the terminology of *variation,* the principal topic of this chapter. Using the language of variation, we state how one variable changes as other related variables change. Among the examples we show is that of the universal law of gravitation in Section 18.2.

We begin this chapter by reviewing the meanings of *ratio* and *proportion,* which were first introduced in Chapter 1. Then we see how ratio and proportion lead to variation and setting up numerous relationships. Many of these relationships are based on experimentation and observation, just as in the case of gravitation.

Applications of variation are found in all areas of technology. It is used in acoustics, biology, chemistry, computer technology, economics, electronics, environmental technology, hydrodynamics, mechanics, navigation, optics, physics, space technology, thermodynamics, and other fields.

In Section 18.2, we discuss a space-age application of Newton's universal law of gravitation.

18.1 Ratio and Proportion

Ratio • Proportion

In order to develop the meaning of *variation,* we now review and expand our discussion of *ratio* and *proportion.* First, from Chapter 1, recall that *the quotient a/b is the* **ratio** *of a to b.* Therefore, a fraction is a ratio.

A measurement is the ratio of the measured magnitude to an accepted unit of measurement. For example, measuring the length of an object as 5 cm means it is five times as long as the accepted unit of length, the centimeter. Other examples of ratios are density (weight/volume), relative density (density of object/density of water), and pressure (force/area). Thus, ratios compare quantities of the same kind (for example, the trigonometric ratios) or express the division of magnitudes of different quantities (such a ratio is also called a **rate**).

EXAMPLE 1   Ratio – applications – units of measurement

The approximate airline distance from Chicago to Dallas is 800 mi, and the approximate airline distance from Chicago to Cleveland is 300 mi. The ratio of these distances is

$$\frac{800 \text{ mi}}{300 \text{ mi}} = \frac{8}{3}$$

Since both units are in miles, the resulting ratio is a dimensionless number.

If a jet travels from Chicago to Dallas in 2 h, its average speed is

$$\frac{800 \text{ mi}}{2 \text{ h}} = 400 \text{ mi/h}$$

■ The first jet-propelled airplane was flown in Germany in 1928.

In this case, we must attach the proper units to the resulting ratio.   ■

As we noted in Example 1, we must be careful to attach the proper units to the resulting ratio. Generally, the ratio of measurements of the same kind should be expressed as a dimensionless number. Consider the following example.

EXAMPLE 2   Ratio of measurements of same kind

The length of a certain room is 24 ft, and the width of the room is 18 ft. Therefore, the ratio of the length to the width is $\frac{24}{18}$, or $\frac{4}{3}$.

If the width of the room is expressed as 6 yd, we have the ratio 24 ft/6 yd = 4 ft/1 yd. However, this does not clearly show the ratio. It is better and more meaningful first to change the units of one of the measurements to the units of the other measurement. Changing the length from 6 yd to 18 ft, we express the ratio as $\frac{4}{3}$, as we saw above. From this ratio, we can easily see that the length is $\frac{4}{3}$ as long as the width.   ■

Dimensionless ratios are often used in definitions in mathematics and in technology. For example, the irrational number π is the dimensionless ratio of the circumference of a circle to its diameter. The specific gravity of a substance is the ratio of its density to the density of water. Other illustrations are found in the exercises for this section.

From Chapter 1, also recall that *an equation stating that two ratios are equal is called a* **proportion.** By this definition, a proportion is

$$\frac{a}{b} = \frac{c}{d} \qquad\qquad \textbf{(18.1)}$$

Consider the following example.

EXAMPLE 3 Proportion – application – scale of map

On a certain map, 1 in. represents 10 mi. Therefore, on this map we have a ratio of distances of 1 in./10 mi. To find the distance represented by 3.5 in. on the map, we can set up the proportion

$$\frac{3.5 \text{ in.}}{x} = \frac{1 \text{ in.}}{10 \text{ mi.}} \quad \longleftarrow \text{ map distances} \atop \longleftarrow \text{ land distances}$$

$$(10x)\left(\frac{3.5}{x}\right) = 10x\left(\frac{1}{10}\right) \qquad \text{multiply each side by LCD} = 10x$$

$$35 = x \quad \text{or} \quad x = 35 \text{ mi}$$

The ratio 1 in./10 mi is the *scale* of the map and has a special meaning, relating map distances in inches to land distances in miles. In a case like this, we should not change either unit to the other, even though they are both units of length. ∎

EXAMPLE 4 Proportion – change units of measurement

Given that 1 in. = 2.54 cm, what is the length in centimeters of the diagonal of a flat computer screen that is 17.0 in. long? See Fig. 18.1.

If we equate the ratio of known lengths to the ratio of the given length to the required length, we can find the required length by solving the resulting proportion (which is an equation). This gives us

$$\frac{1 \text{ in.}}{2.54 \text{ cm}} = \frac{17.0 \text{ in.}}{x \text{ cm}}$$

$$x = (17.0)(2.54)$$

$$= 43.2 \text{ cm}$$

Therefore, the diagonal of the flat computer screen is 17.0 in., or 43.2 cm. ∎

In Appendix B, there are additional illustrations of changing units. Also, a listing of all the various units used in the text is included.

EXAMPLE 5 Proportion – application

The magnitude of an electric field E is the ratio of the force F on a charge q to the magnitude of q. We can write this as $E = F/q$. If we know the force exerted on a particular charge at some point in the field, we can determine the force that would be exerted on another charge placed at the same point. For example, if we know that a force of 10 nN is exerted on a charge of 4.0 nC, we can then determine the force that would be exerted on a charge of 6.0 nC by the proportion

$$\frac{10 \times 10^{-9}}{4.0 \times 10^{-9}} = \frac{F}{6.0 \times 10^{-9}} \quad \longleftarrow \text{ forces at point} \atop \longleftarrow \text{ charges at point}$$

$$F = \frac{(6.0 \times 10^{-9})(10 \times 10^{-9})}{4.0 \times 10^{-9}}$$

$$= 15 \times 10^{-9} = 15 \text{ nN} \quad ∎$$

Fig. 18.1

Practice Exercise

1. In a certain electric field a force of 21 nN is exerted on a charge of 6.0 nC. At the same point, what is the force on a charge of 16 nC?

Practice Exercise

2. An alloy is 7 parts zinc and 5 parts lead. How much lead is there in 48 g of the alloy?

EXAMPLE 6 Proportion – application

An alloy is 5 parts tin and 3 parts lead. How many grams of each are in 40 g of the alloy?

First, let $x =$ the number of grams of tin in the 40 g of the alloy. Next, we note that there are 8 total parts of alloy, of which 5 are tin. Thus, 5 is to 8 as x is to 40. Therefore,

$$\text{parts tin} \longrightarrow \frac{5}{8} = \frac{x}{40} \longleftarrow \text{grams of tin} \atop \text{total grams}$$
$$\text{total parts} \longrightarrow$$

$$x = 40\left(\frac{5}{8}\right) = 25 \text{ g}$$

There are 25 g of tin and 15 g of lead. The ratio 25/15 is the same as 5/3. ■

EXERCISES 18.1

In Exercises 1 and 2, make the given changes in the indicated examples of this section and then solve the indicated problem.

1. In Example 3, change 10 mi to 16 mi.

2. In Example 6, change 5 parts tin to 7 parts tin.

In Exercises 3–10, express the ratios in the simplest form.

3. 18 V to 3 V   4. 63 ft² to 18 ft²   5. 96 h to 3 days

6. 120 s to 4 min   7. 20 qt to 2.5 gal   8. 25 cm² to 75 mm²

9. 0.14 kg to 3500 mg   10. 2000 μm to 6 mm

In Exercises 11–24, find the required ratios.

11. The *efficiency* of a power amplifier is defined as the ratio of the power output to the power input. Find the efficiency of an amplifier for which the power output is 2.6 W and the power input is 9.6 W.

12. A virus 3.0×10^{-5} cm long appears to be 1.2 cm long through a microscope. What is the *magnification* (ratio of image length to object length) of the microscope?

13. The *coefficient of friction* for two contacting surfaces is the ratio of the frictional force between them to the perpendicular force that presses them together. If it takes 450 N to overcome friction to move a 1.10-kN crate along the floor, what is the coefficient of friction between the crate and the floor? See Fig. 18.2.

Fig. 18.2

14. The *Mach number* of a moving object is the ratio of its velocity to the velocity of sound (1200 km/h). Find the Mach number of a jet traveling at 2000 km/h.

15. The *capacitance C* of a capacitor is defined as the ratio of its charge q (in C) to the voltage V. Find C (in F) for which $q = 5.00 \mu$C and $V = 200$ V. (1 F = 1 C/1 V.)

16. The *atomic mass* of an atom of carbon is defined to be 12 u. The ratio of the atomic mass of an atom of oxygen to that of an atom of carbon is $\frac{4}{3}$. What is the atomic mass of an atom of oxygen? (The symbol u represents the *unified atomic mass unit*, where $1 u = 1.66 \times 10^{-27}$ kg.)

17. An important design feature of an aircraft wing is its *aspect ratio*. It is defined as the ratio of the square of the span of the wing (wingtip to wingtip) to the total area of the wing. If the span of the wing for a certain aircraft is 32.0 ft and the area is 195 ft², find the aspect ratio.

18. For an automobile engine, the ratio of the cylinder volume to compressed volume is the *compression ratio*. If the cylinder volume of 820 cm³ is compressed to 110 cm³, find the compression ratio.

19. The *specific gravity* of a substance is the ratio of its density to the density of water. If the density of steel is 487 lb/ft³ and that of water is 62.4 lb/ft³, what is the specific gravity of steel?

20. The *percent grade* of a road is the ratio of vertical rise to the horizontal change in distance (expressed in percent). If a highway rises 75 m for each 1.2 km along the horizontal, what is the percent grade?

21. The *percent error* in a measurement is the ratio of the error in the measurement to the measurement itself, expressed as a percent. When writing a computer program, the memory remaining is determined as 2450 bytes and then it is correctly found to be 2540 bytes. What is the percent error in the first reading?

22. The electric *current* in a given circuit is the ratio of the voltage to the resistance. What is the current $(1 \text{ V}/1 \Omega = 1 \text{ A})$ for a circuit where the voltage is 24.0 mV and the resistance is 10.0 Ω?

23. The *mass* of an object is the ratio of its weight to the acceleration g due to gravity. If a space probe weighs 8.46 kN on earth, where $g = 9.80$ m/s², find its mass. (See Appendix B.)

24. *Power* is defined as the ratio of work done to the time required to do the work. If an engine performs 3.65 kJ of work in 15.0 s, find the power developed by the engine. (See Appendix B.)

In Exercises 25–28, find the required quantities from the given proportions.

25. According to Boyle's law, the relation $p_1/p_2 = V_1/V_2$ holds for pressures p_1 and p_2 and volumes V_1 and V_2 of a gas at constant temperature. Find V_1 if $p_1 = 36.6$ kPa, $p_2 = 84.4$ kPa, and $V_2 = 0.0447$ m³.

26. For two connected gears, the relation

$$\frac{d_1}{d_2} = \frac{N_1}{N_2}$$

holds, where d is the diameter of the gear and N is the number of teeth. Find N_1 if $d_1 = 2.60$ in., $d_2 = 11.7$ in., and $N_2 = 45$. The ratio N_2/N_1 is called the *gear ratio*. See Fig. 18.3.

Fig. 18.3

27. In an electric instrument called a "Wheatstone bridge," electric resistances are related by

$$\frac{R_1}{R_2} = \frac{R_3}{R_4}$$

Find R_2 if $R_1 = 6.00\ \Omega$, $R_3 = 62.5\ \Omega$, and $R_4 = 15.0\ \Omega$. See Fig. 18.4.

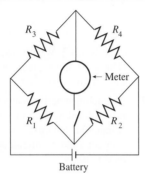

Fig. 18.4

28. In a transformer, an electric current in one coil of wire induces a current in a second coil. For a transformer,

$$\frac{i_1}{i_2} = \frac{t_2}{t_1}$$

where i is the current and t is the number of windings in each coil. In a neon sign amplifier, $i_1 = 1.2$ A and the *turns ratio* $t_2/t_1 = 160$. Find i_2. See Fig. 18.5.

Fig. 18.5

In Exercises 29–48, answer the given questions by setting up and solving the appropriate proportions.

29. Given that $1.00\ \text{in.}^2 = 6.45\ \text{cm}^2$, what area in square inches is $36.3\ \text{cm}^2$?

30. Given that $1.000\ \text{kg} = 2.205\ \text{lb}$, what mass in kilograms is equivalent to $175.5\ \text{lb}$?

31. Given that $1.00\ \text{hp} = 746\ \text{W}$, what power in horsepower is $250\ \text{W}$?

32. Given that $1.50\ \text{L} = 1.59\ \text{qt}$, what capacity in quarts is $2.75\ \text{L}$?

33. Given that $2.00\ \text{km} = 1.24\ \text{mi}$, what distance in kilometers is $750\ \text{mi}$?

34. Given that $10^4\ \text{cm}^2 = 10^6\ \text{mm}^2$, what area in square centimeters is $2.50 \times 10^5\ \text{mm}^2$?

35. How many meters per second are equivalent to $45.0\ \text{km/h}$?

36. How many gallons per hour are equivalent to $540\ \text{mL/min}$?

37. A particular type of automobile engine produces $62,500\ \text{cm}^3$ of carbon monoxide in $2.00\ \text{min}$. How much carbon monoxide is produced in $45.0\ \text{s}$?

38. An airplane consumes $36.0\ \text{gal}$ of gasoline in flying $425\ \text{mi}$. Under similar conditions, how far can it fly on $52.5\ \text{gal}$?

39. Two separate sections of a roof have the same slope. If the rise and run on one section are, respectively, $3.0\ \text{m}$ and $6.3\ \text{m}$, what is the run on the other section if its rise is $4.2\ \text{m}$?

40. When a bullet is fired from a loosely held rifle, the ratio of the mass of the bullet to that of the rifle equals the negative of the reciprocal of the ratio of the velocity of the bullet to that of the rifle. If a $3.0\ \text{kg}$ rifle fires a $5.0\ \text{g}$ bullet and the velocity of the bullet is $300\ \text{m/s}$, what is the recoil velocity of the rifle?

41. A board $15.0\ \text{ft}$ long is cut into two pieces, the lengths of which are in the ratio $2/3$. Find the lengths of the pieces.

42. If c/d is in *inverse ratio* to a/b, then $a/b = d/c$ (see Exercises 27 and 28). The current i (in A) in an electric circuit is in inverse ratio to the resistance R (in Ω). If $i = 0.25\ \text{mA}$ when $R = 2.8\ \Omega$, what is i when $R = 7.2\ \Omega$?

43. By weight, the ratio of chlorine to sodium in table salt is 35.46 to 23.00. How much sodium is contained in $50.00\ \text{kg}$ of salt?

44. An industrial cleaner is diluted 2 parts of cleaner to 5 parts of water. What volume (in mL) of cleaner should be used to get $350\ \text{mL}$ of diluted solution?

45. In testing for quality control, it was found that 17 of every 500 computer chips produced by a company in a day were defective. If a total of 595 defective parts were found, what was the total number of chips produced during that day?

46. An electric current of $0.772\ \text{mA}$ passes into two wires in which it is divided into currents in the ratio of 2.83 to 1.09. What are the currents in the two wires?

47. The ratio of the width to the height of an HDTV screen is $16/9$. What is the length of the diagonal of an HDTV screen if it is $38.6\ \text{in}$. wide?

48. Of the earth's water area, the Pacific Ocean covers 46.0%, and the Atlantic Ocean covers 23.9%. Together they cover a total of $2.53 \times 10^8\ \text{km}^2$. What is the area of each?

Answers to Practice Exercises

1. 56 nN   **2.** 20 g

18.2 Variation

Direct Variation • Inverse Variation •
Joint Variation • Calculating Constant
of Proportionality

■ Named for the French physicist, Jacques
Charles (1746–1823).

Scientific laws are often stated in terms of ratios and proportions. For example, Charles' law can be stated as "for a perfect gas under constant pressure, the ratio of any two volumes this gas may occupy equals the ratio of the absolute temperatures." Symbolically, this could be stated as $V_1/V_2 = T_1/T_2$. Thus, if the ratio of the volumes and one of the values of the temperature are known, we can easily find the other temperature.

By multiplying both sides of the proportion of Charles' law by V_2/T_1, we can change the form of the proportion to $V_1/T_1 = V_2/T_2$. This statement says that the ratio of the volume to the temperature (for constant pressure) is constant. Thus, if any pair of values of volume and temperature is known, this ratio of V_1/T_1 can be calculated. This ratio of V_1/T_1 can be called a constant k, which means that Charles' law can be written as $V/T = k$. We now have the statement that the ratio of the volume to temperature is always constant; or, as it is normally stated, "The volume is proportional to the temperature." Therefore, we write $V = kT$, the clearest and most informative statement of Charles' law.

Thus, *for any two quantities always in the same proportion, we say that one is* **proportional to** (*or* **varies directly as**) *the second. To show that y is proportional to x (or varies directly as x), we write*

Direct Variation

$$y = kx \qquad (18.2)$$

where k is the **constant of proportionality.** This type of relationship is known as **direct variation.**

EXAMPLE 1 Direct variation – applications

(a) The circumference c of a circle is proportional to (varies directly as) the radius r. We write this as $c = 2\pi r$. Since we know that $c = 2\pi r$ for a circle, we know that in this case the constant of proportionality is $k = 2\pi$.

(b) The fact that the resistance R of a wire varies directly as (is proportional to) its length l is written as $R = kl$. As the length of the wire increases (or decreases), this equation tells us that the resistance increases (or decreases) proportionally. In this case, the constant of proportionality is different for different wires, although it remains constant for any given wire.   ■

It is very common that, when two quantities are related, the product of the two quantities remains constant. In such a case, $yx = k$, or

Inverse Variation

$$y = \frac{k}{x} \qquad (18.3)$$

This is read as "*y* **varies inversely as** *x*" *or* "*y* **is inversely proportional to** *x*." This type of relationship is known as **inverse variation.**

EXAMPLE 2 Inverse variation – application

Boyle's law states that "at a given temperature, the pressure p of an ideal gas varies inversely as the volume V." We write this as $p = k/V$. In this case, as the volume of the gas increases, the pressure decreases, or as the volume decreases, the pressure increases.   ■

■ Named for the English physicist, Robert
Boyle (1627–1691).

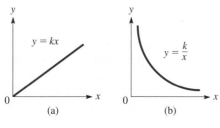

Fig. 18.6

In Fig. 18.6(a), the graph of the equation for direct variation $y = kx$, where $(x \geq 0)$, is shown. It is a straight line of slope k $(k > 0)$ and y-intercept of 0. We see that y increases as x increases, or y decreases as x decreases. In Fig. 18.6(b), the graph of the equation for inverse variation $y = k/x$ $(k > 0, x > 0)$ is shown. It is a *hyperbola* (a different form of the equation from that of Example 4 of Section 14.1). As x increases, y decreases, or as x decreases, y increases.

For many relationships, one quantity varies as a specific power of another quantity. The terms *varies directly* and *varies inversely as* are used in the following example with a specific power of the independent variable.

EXAMPLE 3   Vary as specific power – applications

(a) The statement that the volume V of a sphere varies directly as the cube of its radius is written as $V = kr^3$. In this case, we know that $k = 4\pi/3$. We see that as the radius increases, the volume increases much more rapidly. For example, if $r = 2.00$ cm, $V = 33.5$ cm^3, and if $r = 3.00$ cm, $V = 113$ cm^3.

(b) A company finds that the number n of units of a product that are sold is inversely proportional to the square of the price p of the product. This is written as $n = k/p^2$. As the price is raised, the number of units that are sold decreases much more rapidly. ∎

One quantity may vary as the product of two or more other quantities. Such variation is called **joint variation.** We write

Joint Variation

$$\boxed{y = kxz} \tag{18.4}$$

to show that y varies jointly as x and z.

EXAMPLE 4   Joint variation – application

The cost C of a piece of sheet metal varies jointly as the area A of the piece and the cost c per unit area. This is written as $C = kAc$. Here, C increases as the product Ac increases, and decreases as the product Ac decreases. ∎

Direct, inverse, and joint variations may be combined. A given relationship may be a combination of two or all three of these types of variation.

EXAMPLE 5   Combined variation – application

Newton's *universal law of gravitation* can be stated as follows: "The force F of gravitation between two objects varies jointly as the masses m_1 and m_2 of the objects, and inversely as the square of the distance r between their centers." We write this as

■ Formulated by the great English mathematician and physicist, Isaac Newton (1642–1727).

$$F = \frac{Gm_1m_2}{r^2} \quad \begin{array}{l}\leftarrow \text{force varies jointly as masses} \\ \quad \text{and} \\ \leftarrow \text{inversely as the square of the distance}\end{array}$$

where G is the constant of proportionality.

Note the use of the word *and* in this example. It is used to indicate that F varies in more than one way, but *it is **not** interpreted as addition.* ∎

CAUTION ▶

Practice Exercise

1. Express the relationship that y varies directly as the square of x and inversely as z.

CALCULATING THE CONSTANT OF PROPORTIONALITY

Once we have used the given statement to set up a general equation in terms of the variables and the constant of proportionality, we may calculate the value of the constant of proportionality if *one complete set of values* of the variables is known. *This value can then be substituted into the general equation to find the specific equation* relating the variables. We can then find the value of any one of the variables for any set of the others.

EXAMPLE 6 Find one value – given others

If y varies inversely as x, and $x = 15$ when $y = 4$, find the value of y when $x = 12$.
  The solution is as follows:

$$y = \frac{k}{x} \qquad \text{general equation showing inverse variation}$$

$$4 = \frac{k}{15}, \quad k = 60 \qquad \text{evaluate constant of proportionality}$$

$$y = \frac{60}{x} \qquad \text{specific equation relating } y \text{ and } x$$

$$y = \frac{60}{12} = 5 \qquad \text{evaluating } y \text{ for } x = 12 \qquad \blacksquare$$

Practice Exercise

2. If y varies directly as x, and $y = 5$ when $x = 20$, find the value of y when $x = 12$.

EXAMPLE 7 Vary as square root – application

The frequency f of vibration of a wire varies directly as the square root of the tension T of the wire. If $f = 420$ Hz when $T = 1.14$ N, find f when $T = 3.40$ N.
  The steps in making this evaluation are outlined below:

$$f = k\sqrt{T} \qquad \text{set up general equation: } f \text{ varies directly as } \sqrt{T}$$

$$420\,\text{Hz} = k\sqrt{1.14\,\text{N}} \qquad \text{substitute given set of values and evaluate } k$$

$$k = 393\,\text{Hz}/\text{N}^{1/2}$$

$$f = 393\sqrt{T} \qquad \text{substitute value of } k \text{ to get specific equation}$$

$$f = 393\sqrt{3.40} \qquad \text{evaluate } f \text{ for } T = 3.40\ N$$

$$= 725\,\text{Hz}$$

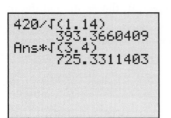

```
420/√(1.14)
        393.3660409
Ans*√(3.4)
        725.3311403
```

Fig. 18.7

We note that k has a set of units associated with it, and this usually will be the case in applied situations. As long as we do not change the units that are used for any of the variables, the units for the final variable that is evaluated will remain the same. The calculator screen for this calculation is shown in Fig. 18.7. $\blacksquare$

EXAMPLE 8 Vary as square – application

The heat H developed in an electric resistor varies jointly as the time t and the square of the current i in the resistor. If H_0 joules of heat are developed in t_0 seconds with i_0 amperes passing through the resistor, how much heat is developed if both the time and the current are doubled?

$$H = kti^2 \qquad \text{set up general equation}$$

$$H_0\,\text{J} = k(t_0\,\text{s})(i_0\,\text{A})^2 \qquad \text{substitute given values and}$$

$$k = \frac{H_0}{t_0 i_0^2}\text{J}/(\text{s} \cdot \text{A}^2) \qquad \text{evaluate } k$$

$$H = \frac{H_0 t i^2}{t_0 i_0^2} \qquad \text{substitute for } k \text{ to get specific equation}$$

We are asked to determine H when both the time and the current are doubled. This means we are to substitute $t = 2t_0$ and $i = 2i_0$. Making this substitution,

$$H = \frac{H_0(2t_0)(2i_0)^2}{t_0 i_0^2} = \frac{8H_0 t_0 i_0^2}{t_0 i_0^2} = 8H_0$$

The heat developed is eight times that for the original values of i and t. $\blacksquare$

■ See the chapter introduction.

■ The first landing on the moon was by the crew of the U. S. spacecraft *Apollo 11* in July 1969.

EXAMPLE 9 Space age application

In Example 5, we stated Newton's universal law of gravitation. This law was formulated in the late seventeenth century, but it has numerous modern space-age applications. Use this law to solve the following problem.

A spacecraft is traveling from the earth to the moon, which are 240,000 mi apart. The mass of the moon is 0.0123 that of the earth. How far from the earth is the gravitational force of the earth on the spacecraft equal to the gravitational force of the moon on the spacecraft?

From Example 5, we have the gravitational force between two objects as

$$F = \frac{Gm_1m_2}{r^2}$$

where the constant of proportionality G is the same for any two objects. Since we want the force between the earth and the spacecraft to equal the force between the moon and the spacecraft, we have

$$\frac{Gm_sm_e}{r^2} = \frac{Gm_sm_m}{(240,000 - r)^2}$$

where m_s, m_e, and m_m are the masses of the spacecraft, the earth, and the moon, respectively; r is the distance from the earth to the spacecraft; and $240,000 - r$ is the distance from the moon to the spacecraft. Since $m_m = 0.0123m_e$, we have

$$\frac{Gm_sm_e}{r^2} = \frac{Gm_s(0.0123m_e)}{(240,000 - r)^2}$$

$$\frac{1}{r^2} = \frac{0.0123}{(240,000 - r)^2} \qquad \text{divide each side by } Gm_sm_e$$

$$(240,000 - r)^2 = 0.0123r^2 \qquad \text{multiply each side by LCD}$$

$$240,000 - r = 0.111r \qquad \text{take square roots}$$

$$1.111r = 240,000$$

$$r = 216,000 \text{ mi}$$

Therefore, the spacecraft is 216,000 mi from the earth and 24,000 mi from the moon when the gravitational forces are equal. See Fig. 18.8. ■

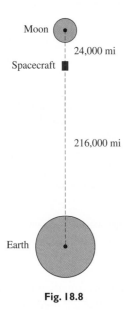

Moon
24,000 mi
Spacecraft
216,000 mi
Earth

Fig. 18.8

EXERCISES 18.2

In Exercises 1–4, make the given changes in the indicated examples of this section and solve the indicated problems.

1. In Example 1(a), change radius r to diameter d, write the appropriate equation, and find the value of the constant of proportionality.

2. In Example 6, change "inversely as x" to "inversely as the square of x" and then solve the resulting problem.

3. In Example 7, change 1.14 N to 1.35 N and then solve the resulting problem.

4. In Example 8, change "both the time and current are doubled" to "the time is halved and the current is doubled" and then solve the resulting problem.

In Exercises 5–12, set up the general equations from the given statements.

5. The speed v at which a galaxy is moving away from earth varies directly as its distance r from earth.

6. The demand D for a product varies inversely as its price P.

7. The electric resistance R of a wire varies inversely as the square of its diameter d.

8. The volume V of silt carried by a river is proportional to the sixth power of the velocity v of the river.

9. In a tornado, the pressure P that a roof will withstand is inversely proportional to the square root of the area A of the roof.

10. During an adiabatic (no heat loss or gain) expansion of a gas, the pressure p is inversely proportional to the 3/2 power of the volume V.

11. The stiffness S of a beam varies jointly as its width w and the cube of its depth d.

12. The average electric power P entering a load varies jointly as the resistance R of the load and the square of the effective voltage V, and inversely as the square of the impedance Z.

(W) *In Exercises 13–16, express the meaning of the given equation in a verbal statement, using the language of variation. (k and π are constants.)*

13. $A = \pi r^2$

14. $s = \dfrac{k}{t^{1.2}}$

15. $f = \dfrac{kL}{\sqrt{m}}$

16. $V = \pi r^2 h$

In Exercises 17–20, give the specific equation relating the variables after evaluating the constant of proportionality for the given set of values.

17. V varies directly as the square of H, and $V = 2$ when $H = 64$.

18. n is inversely proportional to the square of p, and $n = \frac{1}{27}$ when $p = 3$.

19. p is proportional to q and inversely proportional to the cube of r, and $p = 6$ when $q = 3$ and $r = 2$.

20. v is proportional to t and the square root of s, and $v = 80$ when $s = 4$ and $t = 5$.

In Exercises 21–28, find the required value by setting up the general equation and then evaluating.

21. Find y when $x = 10$ if y varies directly as x, and $y = 20$ when $x = 8$.

22. Find y when $x = 5$ if y varies directly as the square of x, and $y = 6$ when $x = 8$.

23. Find s when $t = 720$ if s is inversely proportional to t, and $s = 800$ when $t = 0.030$.

24. Find p for $q = 0.8$ if p is inversely proportional to the square of q, and $p = 18$ when $q = 0.2$.

25. Find y for $x = 6$ and $z = 5$ if y varies directly as x and inversely as z, and $y = 60$ when $x = 4$ and $z = 10$.

26. Find r when $n = 16$ if r varies directly as the square root of n and $r = 4$ when $n = 25$.

27. Find f when $p = 2$ and $c = 4$ if f varies jointly as p and the cube of c, and $f = 8$ when $p = 4$ and $c = 0.1$.

28. Find v when $r = 2$, $s = 3$, and $t = 4$ if v varies jointly as r and s and inversely as the square of t, and $v = 8$ when $r = 2$, $s = 6$, and $t = 6$.

In Exercises 29 and 30, A varies directly as x, and B varies directly as x, although not in the same proportion as A. All numbers are positive.

29. Show that $A + B$ varies directly as x.

30. Show that $\sqrt{AB}$ varies directly as x.

In Exercises 31–60, solve the given applied problems involving variation.

31. The time t it takes to download a program onto a computer hard drive is proportional to the memory M required for the program. If it takes 45.0 min to download a program with 27.5 MB (megabytes) of memory required, what is the rate of downloading in kB (kilobytes) per second?

32. The volume V of carbon dioxide (CO_2) that is exhausted from a room in a given time varies directly as the initial volume V_0 that is present. If 75 ft^3 of CO_2 are removed in 1 h from a room with an initial volume of 160 ft^3, how much is removed in 1 h if the initial volume is 130 ft^3?

33. The amount of heat H required to melt ice is proportional to the mass m of ice that is melted. If it takes 2.93×10^5 J to melt 875 g of ice, how much heat is required to melt 625 g?

34. In electroplating, the mass m of the material deposited varies directly as the time t during which the electric current is on. Set up the equation for this relationship if 2.50 g are deposited in 5.25 h.

35. Hooke's law states that the force needed to stretch a spring is proportional to the amount the spring is stretched. If 10.0 lb stretches a certain spring 4.00 in, how much will the spring be stretched by a force of 6.00 lb?

36. The rate H of heat removal by an air conditioner is proportional to the electric power input P. The constant of proportionality is the *performance coefficient*. Find the performance coefficient of an air conditioner for which $H = 1.8$ kW and $P = 720$ W.

37. The energy E available daily from a solar collector varies directly as the percent p that the sun shines during the day. If a collector provides 1200 kJ for 75% sunshine, how much does it provide for a day during which there is 35% sunshine?

38. The distance d that can be seen from horizon to horizon from an airplane varies directly as the square root of the altitude h of the airplane. If $d = 133$ mi for $h = 12{,}000$ ft, find d for $h = 16{,}000$ ft.

39. The time t required to empty a wastewater-holding tank is inversely proportional to the cross-sectional area A of the drainage pipe. If it takes 2.0 h to empty a tank with a drainage pipe for which $A = 48$ in.^{2}, how long will it take to empty the tank if $A = 68$ in.^{2}?

(W) **40.** The time t required to make a particular trip is inversely proportional to the average speed v. If a jet takes 2.75 h at an average speed of 520 km/h, how long will it take at an average speed of 620 km/h? Explain the meaning of the constant of proportionality.

(W) **41.** In a physics experiment, a given force was applied to three objects. The mass m and the resulting acceleration a were recorded as follows:

m (g)	2.0	3.0	4.0
a (cm/s^2)	30	20	15

(a) Is the relationship $a = f(m)$ one of direct or inverse variation? Explain. (b) Find $a = f(m)$.

W **42.** The lift L of each of three model airplane wings of width w was measured and recorded as follows:

w (cm)	20	40	60
L (N)	10	40	90

(a) Is the relationship $L = f(w)$ one of direct or inverse variation? Explain. (b) Find $L = f(w)$.

43. The power P required to propel a ship varies directly as the cube of the speed s of the ship. If 5200 hp will propel a ship at 12.0 mi/h, what power is required to propel it at 15.0 mi/h?

44. The f-number lens setting of a camera varies directly as the square root of the time t that the film is exposed. If the f-number is 8 (written as $f/8$) for $t = 0.0200$ s, find the f-number for $t = 0.0098$ s.

45. The force F on the blade of a wind generator varies jointly as the blade area A and the square of the wind velocity v. Find the equation relating F, A, and v if $F = 19.2$ lb when $A = 3.72$ ft^2 and $v = 31.4$ ft/s.

46. The escape velocity v a spacecraft needs to leave the gravitational field of a planet varies directly as the square root of the product of the planet's radius R and its acceleration due to gravity g. For Mars and earth, $R_M = 0.533R_e$ and $g_M = 0.400g_e$. Find v_M for Mars if $v_e = 11.2$ km/s.

47. The force F between two parallel wires carrying electric currents is inversely proportional to the distance d between the wires. If a force of 0.750 N exists between wires that are 1.25 cm apart, what is the force between them if they are separated by 1.75 cm?

48. The velocity v of a pulse traveling in a string varies directly as the square root of the tension T in the string. If the velocity of a pulse in a string is 450 ft/s when the tension is 20.0 lb, find the velocity when the tension is 30.0 lb.

49. The average speed s of oxygen molecules in the air is directly proportional to the square root of the absolute temperature T. If the speed of the molecules is 460 m/s at 273 K, what is the speed at 300 K?

50. The time t required to test a computer memory unit varies directly as the square of the number n of memory cells in the unit. If a unit with 4800 memory cells can be tested in 15.0 s, how long does it take to test a unit with 8400 memory cells?

51. The electric resistance R of a wire varies directly as its length l and inversely as its cross-sectional area A. Find the relation between resistance, length, and area for a wire that has a resistance of 0.200 Ω for a length of 225 ft and cross-sectional area of 0.0500 in.^{2}.

52. The general gas law states that the pressure P of an ideal gas varies directly as the thermodynamic temperature T and inversely as the volume V. If $P = 610$ kPa for $V = 10.0$ cm^3 and $T = 290$ K, find V for $P = 400$ kPa and $T = 400$ K.

53. The power P in an electric circuit varies jointly as the resistance R and the square of the current I. If the power is 10.0 W when the current is 0.500 A and the resistance is 40.0 Ω, find the power if the current is 2.00 A and the resistance is 20.0 Ω.

54. The difference $m_1 - m_2$ in magnitudes (visual brightnesses) of two stars varies directly as the base 10 logarithm of the ratio b_2/b_1 of their actual brightnesses. For two particular stars, if $b_2 = 100b_1$ for $m_1 = 7$ and $m_2 = 2$, find the equation relating m_1, m_2, b_1, and b_2.

55. The power gain G by a parabolic microwave dish varies directly as the square of the diameter d of the opening and inversely as the square of the wavelength λ of the wave carrier. Find the equation relating G, d, and λ if $G = 5.5 \times 10^4$ for $d = 2.9$ m and $\lambda = 3.0$ cm.

56. The intensity I of sound varies directly as the power P of the source and inversely as the square of the distance r from the source. Two sound sources are separated by a distance d, and one has twice the power output of the other. Where should an observer be located on a line between them such that the intensity of each sound is the same?

57. The x-component of the acceleration of an object moving around a circle with constant angular velocity ω varies jointly as $\cos \omega t$ and the square of ω. If the x-component of the acceleration is -11.4 ft/s^2 when $t = 1.00$ s for $\omega = 0.524$ rad/s, find the x-component of the acceleration when $t = 2.00$ s.

58. The tangent of the proper banking angle θ of the road for a car making a turn is directly proportional to the square of the car's velocity v and inversely proportional to the radius r of the turn. If 7.75° is the proper banking angle for a car traveling at 20.0 m/s around a turn of radius 300 m, what is the proper banking angle for a car traveling at 30.0 m/s around a turn of radius 250 m? See Fig. 18.9.

Fig. 18.9

59. The acoustical intensity I of a sound wave is proportional to the square of the pressure amplitude P and inversely proportional to the velocity v of the wave. If $I = 0.474$ W/m^2 for $P = 20.0$ Pa and $v = 346$ m/s, find I if $P = 15.0$ Pa and $v = 320$ m/s.

60. To cook a certain vegetable mix in a microwave oven, the instructions are to cook 4.0 oz for 2.5 min or 8.0 oz for 3.5 min. Assuming the cooking time t is proportional to some power (not necessarily integral) of the weight w, use logarithms to find t as a function of w.

Answers to Practice Exercises

1. $y = kx^2/z$ **2.** $y = 3$

CHAPTER 18   EQUATIONS

Proportion	$\dfrac{a}{b} = \dfrac{c}{d}$	**(18.1)**
Direct variation	$y = kx$	**(18.2)**
Inverse variation	$y = \dfrac{k}{x}$	**(18.3)**
Joint variation	$y = kxz$	**(18.4)**

CHAPTER 18   REVIEW EXERCISES

In Exercises 1–14, find the indicated ratios.

1. 4 Mg to 20 kg

2. 300 nm to 6 μm

3. 375 mL to 25 cL

4. 12 ks to 2 h

5. The number π equals the ratio of the circumference c of a circle to its diameter d. To check the value of π, a technician used computer simulation to measure the circumference and diameter of a metal cylinder and found the values to be $c = 4.2736$ cm and $d = 1.3603$ cm. What value of π did the technician get?

6. The ratio of the diagonal d of a square to the side s of the square is $\sqrt{2}$. The diagonal and side of the face of a glass cube are found to be $d = 35.375$ mm and $s = 25.014$ mm. What is the value of $\sqrt{2}$ found from these measurements?

7. The mechanical advantage of a lever is the ratio of the output force F_0 to the input force F_i. Find the mechanical advantage if $F_0 = 28$ kN and $F_i = 5000$ N.

8. For an automobile, the ratio of the number n_1 of teeth on the ring gear to the number n_2 of teeth on the pinion gear is the *rear axle ratio* of the car. Find this ratio if $n_1 = 64$ and $n_2 = 20$.

9. The pressure p exerted on a surface is the ratio of the force F on the surface to its area A. Find the pressure on a square patch, 2.25 in. on a side, on a tank if the force on the patch is 37.4 lb.

10. The electric resistance R of a resistor is the ratio of the voltage V across the resistor to the current i in the resistor. Find R if $V = 0.632$ V and $i = 2.03$ mA.

11. The *heat of vaporization* of a substance is the amount of heat needed to change a unit amount from liquid to vapor. Experimentation shows 7910 J are needed to change 3.50 g of water to steam. What is the heat of vaporization of water?

12. The electric *current gain G* is the ratio of a constant α to $1 - \alpha$. What is the current gain for $\alpha = 0.95$?

13. The ratio of the commission for selling a home to the selling price of the home is the commission rate. What is this rate if the commission of $20,900 is charged for selling a home priced at $380,000?

14. A total cholesterol level of 200 mg/dL is considered too high, but many doctors consider the ratio of the total cholesterol to HDL (high density lipids – the "good" cholesterol) a more important measure of risk to coronary heart disease. What is this ratio if the total cholesterol is 179 mg/dL and the HDL is 39 mg/dL? (Average for women is about 4.5, and for men is about 5.0.)

In Exercises 15 and 16, given that $(a/b = c/d,$ b and d not zero), show that indicated proportions are correct.

15. $\dfrac{a + b}{b} = \dfrac{c + d}{d}$

16. $\dfrac{a - b}{b} = \dfrac{c - d}{d}$

In Exercises 17–32, answer the given questions by setting up and solving the appropriate proportions.

17. On a map of Hawaii, 1.3 in. represents 20.0 mi. If the distance on the map between Honolulu and Kahului on Maui is 6.0 in., how far is Kahului from Honolulu?

18. Given that 1.000 lb = 453.6 g, what is the weight in pounds of a 14.0-g computer disk?

19. Given that 1.00 Btu = 1060 J, how much heat in joules is produced by a heating element that produces 2660 Btu?

20. Given that 1.00 L = 61.0 in.^{3}, what capacity in liters has a cubical box that is 3.23 in. along an edge?

21. A computer printer can print 50 pages in 2 min 5 s. How many pages can it print in 10.0 min?

22. A solar heater with a collector area of 58.0 m^2 is required to heat 2560 kg of water. Under the same conditions, how much water can be heated by a rectangular solar collector 9.50 m by 8.75 m?

23. The dosage of a certain medicine is 25 mL for each 10 lb of the patient's weight. What is the dosage for a person weighing 56 kg?

24. A woman invests $50,000 and a man invests $20,000 in a partnership. If profits are to be shared in the ratio that each invested in the partnership, how much does each receive from $10,500 in profits?

25. On a certain blueprint, a measurement of 25.0 ft is represented by 2.00 in. What is the actual distance between two points if they are 5.75 in. apart on the blueprint?

26. The chlorine concentration in a water supply is 0.12 part per million. How much chlorine is there in a cylindrical holding tank 4.22 m in radius and 5.82 m high filled from the water supply?

27. One fiber-optic cable carries 60.0% as many messages as another fiber-optic cable. Together they carry 12,000 messages. How many does each carry?

28. Two types of roadbed material, one 50% rock and the other 100% rock, are used in the ratio of 4 to 1 to form a roadbed. If a total of 150 tons are used, how much rock is in the roadbed?

29. To neutralize 80.0 kg of sodium hydroxide, 98.0 kg of sulfuric acid are needed. How much sodium hydroxide can be neutralized with 37.0 kg of sulfuric acid?

30. For analysis, 105 mg of DNA sample is divided into two vials in the ratio of 2 to 5. How many milligrams are in each vial?

31. A total of 322 bolts are in two containers. The ratio of the number of bolts in the first container to the number in the second container is 5/9. How many are in each container?

32. A gasoline company sells octane-87 gas and octane-91 gas in the ratio of 9 to 2. How many of each octane are sold of a total of 16.5 million gallons?

In Exercises 33–36, give the specific equation relating the variables after evaluating the constant of proportionality for the given set of values.

33. y varies directly as the square of x, and $y = 27$ when $x = 3$.

34. f varies inversely as l, and $f = 5$ when $l = 8$.

35. v is directly proportional to x and inversely proportional to the cube of y, and $v = 10$ when $x = 5$ and $y = 4$.

36. r varies jointly as u, v, and the square of w, and $r = 8$ when $u = 2$, $v = 4$, and $w = 3$.

In Exercises 37–76, solve the given applied problems.

37. For the lever balanced at the fulcrum, the relation

$$\frac{F_1}{F_2} = \frac{L_2}{L_1}$$

holds, where F_1 and F_2 are forces on opposite sides of the fulcrum at distances L_1 and L_2 (see Fig. 18.10). Find L_2 if $F_1 = 4.50$ lb, $F_2 = 6.75$ lb, and $L_1 = 17.5$ in.

Fig. 18.10

38. A company finds that the volume V of sales of a certain item and the price P of the item are related by

$$\frac{P_1}{P_2} = \frac{V_2}{V_1}$$

Find V_2 if $P_1 = \$8.00$, $P_2 = \$6.00$, and $V_1 = 3000$ per week.

39. The image height h and object height H for the lens shown in Fig. 18.11 are related to the image distance q and object distance p by

$$\frac{h}{H} = \frac{q}{p}$$

Find q if $h = 24$ cm, $H = 84$ cm, and $p = 36$ cm.

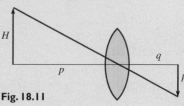

Fig. 18.11

40. For two pulleys connected by a belt, the relation

$$\frac{d_1}{d_2} = \frac{n_2}{n_1}$$

holds, where d is the diameter of the pulley and n is the number of revolutions per unit time it makes. Find n_2 if $d_1 = 4.60$ in., $d_2 = 8.30$ in., and $n_1 = 18.0$ r/min.

41. An apartment owner charges rent R proportional to the floor area A of the apartment. Find the equation relating R and A if an apartment of 900 ft^2 rents for \$850/month.

42. The number r of aluminum cans that can be made by recycling n used cans is proportional to n. How many cans can be made from 50,000 used cans if $r = 115$ cans for $n = 125$ cans?

43. Under certain conditions, the rate of increase v of bacteria is proportional to the number N of bacteria present. Find v for $N = 7500$ bacteria, if $v = 800$ bacteria/h for $N = 4000$ bacteria.

44. The index of refraction n of a medium varies inversely as the velocity of light v within it. For quartz, $n = 1.46$ and $v = 2.05 \times 10^8$ m/s. What is n for a diamond, in which $v = 1.24 \times 10^8$ m/s?

45. The force F needed to tighten a bolt varies as the length L of the wrench handle. See Fig. 18.12. If $F = 250$ N for $L = 22$ cm, and $F = 550$ N for $L = 10$ cm, determine the equation relating F and L.

Fig. 18.12

46. At a certain jeweler, the cost C of a diamond varies as the weight w of the diamond. If a 0.5-carat diamond costs \$320, and a 2.0-carat diamond costs \$5120, what is the equation relating C and w?

47. The period T of a pendulum varies directly as the square root of its length L. If $T = \pi/2$ s for $L = 2.00$ ft, find T for $L = 4.00$ ft.

48. The component of velocity v_x of an object moving in a circle with constant angular velocity ω varies jointly with ω and $\sin \omega t$. If $\omega = \pi/6$ rad/s, and $v_x = -4\pi$ ft/s when $t = 1.00$ s, find v_x when $t = 9.00$ s.

49. The charge C on a capacitor varies directly as the voltage V across it. If the charge is $6.3 \, \mu C$ with a voltage of 220 V across a capacitor, what is the charge on it with a voltage of 150 V across it?

50. The amount of natural gas burned is proportional to the amount of oxygen consumed. If 24.0 lb of oxygen is consumed in burning 15.0 lb of natural gas, how much air, which is 23.2% oxygen by weight, is consumed to burn 50.0 lb of natural gas?

51. The power P of a gas engine is proportional to the area A of the piston. If an engine with a piston area of 8.00 in.^{2} can develop 30.0 hp, what power is developed by an engine with a piston area of 6.0 in.^{2}?

52. The decrease in temperature above a region is directly proportional to the altitude above the region. If the temperature T at the base of the Rock of Gibraltar is 22.0°C and a plane 3.50 km above notes that the temperature is 1.0°C, what is the temperature at the top of Gibraltar, the altitude of which is 430 m? (Assume there are no other temperature effects.)

53. The distance d an object falls under the influence of gravity varies directly as the square of the time t of fall. If an object falls 64.0 ft in 2.00 s, how far will it fall in 3.00 s?

54. The kinetic energy E of a moving object varies jointly as the mass m of the object and the square of its velocity v. If a 5.00-kg object, traveling at 10.0 m/s, has a kinetic energy of 250 J, find the kinetic energy of an 8.00-kg object moving at 50.0 m/s.

55. In a particular computer design, N numbers can be sorted in a time proportional to the square of $\log N$. How many times longer does it take to sort 8000 numbers than to sort 2000 numbers?

56. The velocity v of a jet of fluid flowing from an opening in the side of a container is proportional to the square root of the depth d of the opening. If the velocity of the jet from an opening at a depth of 1.22 m is 4.88 m/s, what is the velocity of a jet from an opening at a depth of 7.62 m? See Fig. 18.13.

Fig. 18.13

57. In an electric circuit containing an inductance L and a capacitance C, the resonant frequency f is inversely proportional to the square root of the capacitance. If the resonant frequency in a circuit is 25.0 Hz and the capacitance is $95.0 \, \mu F$, what is the resonant frequency of this circuit if the capacitance is $25.0 \, \mu F$?

58. The rate of emission R of radiant energy from the surface of a body is proportional to the fourth power of the theromodynamic temperature T. Given that a 25.0-W (the rate of emission) lamp has an operating temperature of 2500 K, what is the operating temperature of a similar 40.0-W lamp?

59. The frequency f of a radio wave is inversely proportional to its wavelength λ. The constant of proportionality is the velocity of the wave, which equals the speed of light. Find this velocity if an FM radio wave has a frequency of 90.9 MHz and a wavelength of 3.29 m.

60. The acceleration of gravity g on a satellite in orbit around the earth varies inversely as the square of its distance r from the center of the earth. If $g = 8.7 \, m/s^2$ for a satellite at an altitude of 400 km above the surface of the earth, find g if it is 1000 km above the surface. The radius of the earth is $6.4 \times 10^6 \, m$.

61. If the weight w of an airplane varies directly as the cube of its length L, and $w = 3500$ lb for $L = 42$ ft, what would be the weight of a plane that is 15 m long?

62. The thrust T of a propeller varies jointly as the square of the number n of revolutions per second and the fourth power of its diameter d. What is the effect on T if n is doubled and d is halved?

63. Using *holography* (a method of producing an image without using a lens), an image of concentric circles is formed. The radius r of each circle varies directly as the square root of the wavelength λ of the light used. If $r = 3.56$ cm for $\lambda = 575$ nm, find r if $\lambda = 483$ nm.

64. A metal circular ring has a circular cross section of radius r. If R is the radius of the ring (measured to the middle of the cross section), the volume V of metal in the ring varies directly as R and the square of r. If $V = 2550 \, mm^3$ for $r = 2.32$ mm and $R = 24.0$ mm, find V for $r = 3.50$ mm and $R = 32.0$ mm. See Fig. 18.14.

Fig. 18.14   **Fig. 18.15**

65. The range R of a projectile varies jointly as the square of its initial velocity v_0 and the sine of twice the angle θ from the horizontal at which it is fired. See Fig. 18.15. A bullet for which $v_0 = 850$ m/s and $\theta = 22.0°$ has a range of 5.12×10^4 m. Find the range if $v_0 = 750$ m/s and $\theta = 43.2°$.

66. Kepler's third law of planetary motion states that the square of the period of any planet is proportional to the cube of the mean radius (about the sun) of that planet, with the constant of proportionality being the same for all planets. Using the fact that the period of the earth is 1 year and its mean radius is 93.0 million miles, calculate the mean radius for Venus, given that its period is 7.38 months.

67. The stopping distance d of a car varies directly as the square of the velocity v of the car when the brakes are applied. A car moving at 32 mi/h can stop in 52 ft. What is the stopping distance for the car if it is moving at 55 mi/h?

68. The load L that a helical spring can support varies directly as the cube of its wire diameter d and inversely as its coil diameter D. A spring for which $d = 0.120$ in. and $D = 0.953$ in. can support 45.0 lb. What is the coil diameter of a similar spring that supports 78.5 lb and for which $d = 0.156$ in.?

69. The volume rate of flow R of blood through an artery varies directly as the fourth power of the radius r of the artery and inversely as the distance d along the artery. If an operation is successful in effectively increasing the radius of an artery by 25% and decreasing its length by 2%, by how much is the volume rate of flow increased?

70. The safe, uniformly distributed load L on a horizontal beam, supported at both ends, varies jointly as the width w and the square of the depth d and inversely as the distance D between supports. Given that one beam has double the dimensions of another, how many times heavier is the safe load it can support than the first can support?

71. A bank statement exactly 30 years old is discovered. It states, "This 10-year-old account is now worth \$185.03 and pays 4% interest compounded annually." An investment with annual compound interest varies directly as $1 + r$ to the power n, where r is the interest rate expressed as a decimal and n is the number of years of compounding. What was the value of the original investment, and what is it worth now?

72. The distance s that an object falls due to gravity varies jointly as the acceleration g due to gravity and the square of the time t of fall. The acceleration due to gravity on the moon is 0.172 of that on earth. If a rock falls for t_0 seconds on earth, how many times farther would the rock fall on the moon in $3t_0$ seconds?

73. The heat loss L through fiberglass insulation varies directly as the time t and inversely as the thickness d of the fiberglass. If the loss through 8.0 in. of fiberglass is 1200 Btu in 30 min, what is the loss through 6.0 in. in 1 h 30 min?

74. A quantity important in analyzing the rotation of an object is its *moment of inertia I.* For a ball bearing, the moment of inertia varies directly as its mass m and the square of its radius r. Find the general expression for I if $I = 39.9\,\text{g} \cdot \text{cm}^2$ for $m = 63.8\,\text{g}$ and $r = 1.25\,\text{cm}$.

75. In the study of polarized light, the intensity I is proportional to the square of the cosine of the angle θ of transmission. If $I = 0.025\,\text{W/m}^2$ for $\theta = 12.0°$, find I for $\theta = 20.0°$.

76. The force F that acts on a pendulum bob is proportional to the mass m of the bob and the sine of the angle θ the pendulum makes with the vertical. If $F = 0.120\,\text{N}$ for $m = 0.350\,\text{kg}$ and $\theta = 2.00°$, find F for $m = 0.750\,\text{kg}$ and $\theta = 3.50°$.

Writing Exercise

77. A fruit-packing company plans to reduce the size of its fruit juice can (a right circular cylinder) by 10% and keep the price of each can the same (effectively raising the price). The radius and the height of the new can are to be equally proportional to those of the old can. Write one or two paragraphs explaining how to determine the percent decrease in the radius and the height of the old can that is required to make the new can.

CHAPTER 18 **PRACTICE TEST**

1. Express the ratio of 180 s to 4 min in simplest form.

2. A person 1.8 m tall is photographed, and the film image is 20.0 mm high. Under the same conditions, how tall is a person whose film image is 14.5 mm high?

3. The change L in length of a copper rod varies directly as the change T in temperature of the rod. Set up an equation for this relationship if $L = 2.7$ cm for $T = 150°$C.

4. Given that 1.00 in = 2.54 cm, what length in inches is 7.24 cm?

5. The perimeter of a rectangular solar panel is 210.0 in. The ratio of the length to the width is 7 to 3. What are the dimensions of the panel?

6. The difference p in pressure in a fluid between that at the surface and that at a point below varies jointly as the density d of the fluid and the depth h of the point. The density of water is $1000\,\text{kg/m}^3$, and the density of alcohol is $800\,\text{kg/m}^3$. This difference in pressure at a point 0.200 m below the surface of water is 1.96 kPa. What is the difference in pressure at a point 0.300 m below the surface of alcohol? (All data are accurate to three significant digits.)

7. The crushing load L of a pillar varies directly as the fourth power of its radius r and inversely as the square of its length l. If one pillar has twice the radius and three times the length of a second pillar, what is the ratio of the crushing load of the first pillar to that of the second pillar?

Sequences and the Binomial Theorem

19

If a person is saving for the future and invests $1000 at 5%, compounded annually, the value of the investment 40 years later would be about $7040. Even better, if the interest is compounded daily, which is a common method today, it would be worth about $7390 in 40 years. A person saving for retirement would do much better by putting aside $1000 each year and letting the interest accumulate. If $1000 is invested each year at 5%, compounded annually, the total investment would be worth about $126,840 after 40 years. If the interest is compounded daily, it would be worth about $131,000 in 40 years. Obviously, if more is invested, or the interest rate is higher, the value of these investments would be higher.

Each of these values can be found quickly since the values of the annual investments

form what is called a *geometric sequence,* and formulas can be formed for such sums. Such formulas involving compound interest are widely used in calculating values such as monthly car payments, home mortgages, and annuities.

A *sequence* is a set of numbers arranged in some specific way and usually follows a pattern. Sequences have been of interest to people for centuries. There are records that date back to at least 1700 B.C.E. showing calculations involving sequences. Euclid (see page 49) in his *Elements* dealt with sequences in about 300 B.C.E. More advanced forms of sequences were used extensively in the study of advanced mathematics in the 1700s and 1800s. These advances in mathematics have been very important in many areas of science and technology.

Of the many types of sequences, we study certain basic ones in this chapter. Included are those used in the expansion of a binomial to a power. We show applications in areas such as physics and chemistry in studying radioactivity, biology in studying population growth, and, of course, in business when calculating interest.

Sequences are basic to many calculations in business, including compound interest. In Section 19.2, we show such a calculation.

505

19.1 Arithmetic Sequences

Arithmetic Sequence •
Common Difference •
Finite and Infinite Sequences •
Sum of *n* Terms

A *sequence* of numbers may consist of numbers chosen in any way we may wish to select. This could include a random selection of numbers, although the sequences that are useful follow a pattern. We consider only those sequences that include real numbers or literal numbers that represent real numbers.

An **arithmetic sequence** (*or* **arithmetic progression**) *is a set of numbers in which each number after the first can be obtained from the preceding one by adding to it a fixed number called the* **common difference.** This definition can be expressed in terms of the *recursion formula*

$$a_n = a_{n-1} + d \tag{19.1}$$

where a_n is any term, a_{n-1} is the preceding term, and d is the common difference.

EXAMPLE 1 Illustrations of arithmetic sequences

(a) The sequence $2, 5, 8, 11, 14, \ldots$, is an arithmetic sequence with a common difference $d = 3$. We can obtain any term by adding 3 to the previous term. We see that the fifth term is $a_5 = a_4 + d$, or $14 = 11 + 3$.

(b) The sequence $7, 2, -3, -8, \ldots$, is an arithmetic sequence with $d = -5$. We can get any term after the first by adding -5 to the previous term.

The three dots after the 14 in part (a) and after -8 in part (b) mean that the sequences continue. ∎

If we know the first term of an arithmetic sequence, we can find any other term by adding the common difference enough times to get the desired term. This, however, is very inefficient, and there is a general way of finding a particular term.

If a_1 is the first term and d is the common difference, the second term is $a_1 + d$, the third term is $a_1 + 2d$, and so on. For the *n*th term, we need to add d to the first term $n - 1$ times. Therefore, *the nth term, a_n, of the arithmetic sequence is given by*

_n_th term

$$a_n = a_1 + (n - 1)d \tag{19.2}$$

Equation (19.2) can be used to find any given term in any arithmetic sequence. We can refer to a_n as the *last term* of an arithmetic sequence if no terms beyond it are included in the sequence. Such a sequence is called a *finite* sequence. If the terms in a sequence continue without end, the sequence is called an *infinite* sequence.

EXAMPLE 2 Finding specified term

Find the tenth term of the arithmetic sequence $2, 5, 8, \ldots$.

By subtracting any term from the following term, we find the common difference $d = 3$, and see that the first term $a_1 = 2$. Therefore, the tenth term, a_{10}, is

$$
\begin{array}{ccc}
a_1 & n & d \\
\downarrow & \downarrow & \downarrow
\end{array}
$$

$$a_{10} = 2 + (10 - 1)3 = 2 + (9)(3)$$
$$= 29$$

The three dots after the 8 show that the sequence continues. With no additional information given, this indicates that it is an infinite arithmetic sequence. ∎

EXAMPLE 3 Finding common difference

Find the common difference between successive terms of the arithmetic sequence for which the third term is 5 and the 34th term is -119.

In order to use Eq. (19.2), we could calculate a_1, but we can also treat the third term as a_1 and the 34th term as a_{32}. This gives us the same sequence from 5 to -119. Therefore, using the values $a_1 = 5$, $a_{32} = -119$, and $n = 32$, we can find the value of d. Substituting these values in Eq. (19.2) gives

$$-119 = 5 + (32 - 1)d$$
$$31d = -124$$
$$d = -4$$

There is no information as to whether this is a finite or an infinite sequence. The solution is the same in either case. ■

Practice Exercise

1. Find d if $a_1 = 4$ and $a_{13} = 34$.

EXAMPLE 4 Finding number of terms

How many integers between 10 and 1000 are divisible by 6?

We must first find the smallest and the largest numbers in this range that are divisible by 6. These numbers are 12 and 996. Obviously, the common difference between one multiple of 6 and the next is 6. Thus, we can solve this as an arithmetic sequence with $a_1 = 12$, $a_n = 996$, and $d = 6$. Substituting these values in Eq. (19.2), we have

$$996 = 12 + (n - 1)6$$
$$6n = 990$$
$$n = 165$$

Thus, 165 numbers between 10 and 1000 are divisible by 6.

All the positive multiples of 6 are included in the infinite arithmetic sequence $6, 12, 18, \ldots$, whereas those between 10 and 1000 are included in the finite arithmetic sequence $12, 18, 24, \ldots, 996$. ■

EXAMPLE 5 Arithmetic sequence – application

A package delivery company uses a metal (very low friction) ramp to slide packages from the sorting area down to the loading area below. If a given package is pushed such that it starts down the ramp at 25 cm/s, and the package accelerates as it slides down the ramp such that it gains 35 cm/s during each second, after how many seconds is its velocity 305 cm/s? See Fig. 19.1.

Here, we see that the velocity (in cm/s) of the package after each second is

$$60, 95, 130, \ldots, 305, \ldots$$

Therefore, $a_1 = 60$ (the 25 cm/s was at the beginning, that is, after 0 s), $d = 35$, $a_n = 305$, and we are to find n, which in this case, represents the number of seconds during which the velocity increases. Therefore,

$$305 = 60 + (n - 1)(35)$$
$$245 = 35n - 35$$
$$35n = 280$$
$$n = 8.0$$

This means that the velocity of a package sliding down the ramp is 305 cm/s after 8.0 s. ■

Fig. 19.1

SUM OF n TERMS

Another important quantity related to an arithmetic sequence is the sum of the first n terms. We can indicate this sum by starting the sum either with the first term or with the last term, as shown by these two equations:

$$S_n = a_1 + (a_1 + d) + (a_1 + 2d) + \cdots + (a_n - d) + a_n$$

or

$$S_n = a_n + (a_n - d) + (a_n - 2d) + \cdots + (a_1 + d) + a_1$$

If we now add the corresponding members of these two equations, we obtain the result

$$2S_n = (a_1 + a_n) + (a_1 + a_n) + (a_1 + a_n) + \cdots + (a_1 + a_n) + (a_1 + a_n)$$

Each term on the right in parentheses has the same expression $(a_1 + a_n)$, and there are n such terms. This tells us that *the sum of the first n terms is given by*

$$\boxed{S_n = \frac{n}{2}(a_1 + a_n)}$$ (19.3)

The use of Eq. (19.3) is illustrated in the following examples.

EXAMPLE 6 Finding sum of terms

Find the sum of the first 1000 positive integers.

The first 1000 positive integers form a finite arithmetic sequence for which $a_1 = 1, a_{1000} = 1000, n = 1000$, and $d = 1$. Substituting these values in Eq. (19.3) (in which we do not use the value of d), we have

$$S_{1000} = \frac{1000}{2}(1 + 1000) = 500(1001)$$
$$= 500{,}500 \qquad \blacksquare$$

■ Karl Friedrich Gauss (1777–1856) is considered as one of the greatest mathematicians of all time. A popular story about him is that at the age of 10 his schoolmaster asked his class to add all the numbers from 1 to 100, assuming it would take some time to do. He had barely finished stating the problem when Gauss handed in the correct answer of 5050. He had done it essentially as we did the solution in Example 6.

EXAMPLE 7 Finding sum of terms

Find the sum of the first ten terms of the arithmetic sequence in which the first term is 4 and the common difference is -5.

We are to find S_n, given that $n = 10, a_1 = 4$, and $d = -5$. Since Eq. (19.3) uses the value of a_n but not the value of d, we first find a_{10} by using Eq. (19.2). This gives us

$$a_{10} = 4 + (10 - 1)(-5) = 4 - 45$$
$$= -41$$

Now, we can solve for S_{10} by using Eq. (19.3):

$$S_{10} = \frac{10}{2}(4 - 41) = 5(-37)$$
$$= -185 \qquad \blacksquare$$

Practice Exercise

2. Find S_9 if $d = -2$ and $a_9 = 1$.

If any three of the five values a_1, a_n, n, d, and S_n are given for a particular arithmetic sequence, the other two may be found from Eqs. (19.2) and (19.3). Consider the following example.

EXAMPLE 8 Finding n and a_n

For an arithmetic sequence, given that $a_1 = 2, d = 1.5$, and $S_n = 72$, find n and a_n.

First, we substitute the given values in Eqs. (19.2) and (19.3) in order to identify what is known, and how to proceed. Making these substitutions, we have

$$a_n = 2 + (n - 1)(1.5) \qquad \text{substituting in Eq. (19.2)}$$

$$72 = \frac{n}{2}(2 + a_n) \qquad \text{substituting in Eq. (19.3)}$$

$$72 = \frac{n}{2}[2 + 2 + (n - 1)(1.5)] \qquad \text{substituting first equation into second equation}$$

$$= 2n + 0.75n(n - 1)$$

$$288 = 8n + 3n^2 - 3n \qquad \text{multiplying each term by 4 to clear decimals}$$

$$3n^2 + 5n - 288 = 0$$

$$n = \frac{-5 \pm \sqrt{25 - 4(3)(-288)}}{6} = \frac{-5 \pm \sqrt{3481}}{6} = \frac{-5 \pm 59}{6}$$

Since n must be a positive integer, we find that $n = (-5 + 59)/6 = 9$. Therefore, since $n = 9$, $a_9 = 2 + (9 - 1)(1.5) = 14$. The values $n = 9$ and $a_9 = 14$ check with the given values. ∎

EXAMPLE 9 Sum of terms – application

The voltage across a resistor increases such that during each second the increase is 0.002 mV less than during the previous second. Given that the increase during the first second is 0.350 mV, what is the total voltage increase during the first 10.0 s?

We are asked to find the sum of the voltage increases 0.350 mV, 0.348 mV, 0.346 mV, ..., so as to include ten increases. This means we want the sum of an arithmetic sequence for which $a_1 = 0.350, d = -0.002$, and $n = 10$. Since we need a_n to use Eq. (19.3), we first calculate it, using Eq. (19.2):

$$a_{10} = 0.350 + (10 - 1)(-0.002) = 0.332 \text{ mV}$$

Now, we use Eq. (19.3) to find the sum with $a_1 = 0.350, a_{10} = 0.332$, and $n = 10$:

$$S_{10} = \frac{10}{2}(0.350 + 0.332) = 3.410 \text{ mV}$$

Thus, the total voltage increase is 3.410 mV. ∎

EXERCISES 19.1

In Exercises 1–4, make the given changes in the indicated examples of this section and then solve the resulting problems.

1. In Example 3, change −119 to −88 and then find the common difference.

2. In Example 4, change 6 to 5 and then find out how many numbers there are.

3. In Example 6, change 1000 to 500 and then find the sum.

4. In Example 7, change −5 to 5 and then find the sum.

In Exercises 5–8, write the first five terms of the arithmetic sequence with the given values.

5. $a_1 = 4, d = 2$

6. $a_1 = 6, d = -\frac{1}{2}$

7. $a_1 = 2.5, a_5 = -1.5$

8. $a_2 = -2, a_5 = 43$

In Exercises 9–16, find the nth term of the arithmetic sequence with the given values.

9. $1, 4, 7, \ldots; n = 8$

10. $-6, -4, -2, \ldots; n = 10$

11. $\frac{7\pi}{4}, \frac{3\pi}{2}, \frac{5\pi}{4}, \ldots; n = 17$

12. $2, 0.5, -1, \ldots; n = 25$

13. $a_1 = -0.7, d = 0.4, n = 80$

14. $a_1 = \frac{3}{2}, d = \frac{1}{6}, n = 601$

15. $a_1 = b, d = 2b, n = 25$

16. $a_1 = -c, d = 3c, n = 30$

In Exercises 17–20, find the sum of the n terms of the indicated arithmetic sequence.

17. $n = 20, a_1 = 4, a_{20} = 40$

18. $n = 8, a_1 = -12, d = -2$

19. $-2, -\frac{5}{2}, -3, \ldots; n = 10$

20. $3k, \frac{10}{3}k, \frac{11}{3}k, \ldots; n = 40$

In Exercises 21–32, find any of the values of a_1, d, a_n, n, or S_n that are missing for an arithmetic sequence.

21. $a_1 = 5$, $d = 8$, $a_n = 45$

22. $a_1 = -2$, $n = 60$, $a_n = 28$

23. $a_1 = \frac{5}{3}$, $n = 20$, $S_{20} = \frac{40}{3}$

24. $a_1 = 0.1$, $a_n = -5.9$, $S_n = -8.7$

25. $d = 3$, $n = 30$, $S_{30} = 1875$

26. $d = 9$, $a_n = 86$, $S_n = 455$

27. $a_1 = 7.4$, $d = -0.5$, $a_n = 23.1$

28. $a_1 = -\frac{9}{7}$, $n = 19$, $a_{19} = -\frac{36}{7}$

29. $a_1 = 5k$, $d = 0.5k$, $S_n = 104k$

30. $d = -2c$, $n = 50$, $S_{50} = 0$

31. $a_1 = -c$, $a_n = \dfrac{b}{2}$, $S_n = 2b - 4c$

32. $a_1 = 3b$, $n = 7$, $d = \dfrac{b}{3}$

In Exercises 33–60, find the indicated quantities for the appropriate arithmetic sequence.

33. $a_6 = 560$, $a_{10} = 720$ (find a_1, d, S_n for $n = 10$)

34. $a_{17} = -91$, $a_2 = -73$ (find a_1, d, S_n for $n = 40$)

(W) **35.** Is $\ln 3$, $\ln 6$, $\ln 12, \ldots$ an arithmetic sequence? Explain. If it is, what is the fifth term?

(W) **36.** Is $\sin 2°$, $\sin 4°$, $\sin 6°, \ldots$ an arithmetic sequence? Explain. If it is, what is the fifth term?

37. Find a formula with variable n for the nth term of the arithmetic sequence with $a_1 = 3$ and $a_{n+1} = a_n + 2$ for $n = 1, 2, 3, \ldots$.

38. If a, b, and c are the first three terms of an arithmetic sequence, find their sum in terms of b only.

39. The terms a, m, b form an arithmetic sequence. Express a formula for the common difference d in terms of a and b. (The term m is the *arithmetic mean* of the terms a and b.)

40. The terms a, m, n, b form an arithmetic sequence. Express m and n in terms of a and b. (The terms m and n are *arithmetic means* of the terms a and b.)

41. Write the first five terms of the arithmetic sequence for which the second term is b and the third term is c.

42. If the sum of the first two terms of an arithmetic sequence equals the sum of the first three terms, find the sum of the first five terms.

43. Find the sum of the first 100 positive integers. (See the margin note on page 508.)

44. Find the number of multiples of 8 between 99 and 999.

45. Find x if $3 - x$, $-x$, and $\sqrt{9 - 2x}$ are the first three terms of an arithmetic sequence.

46. Is x, $x + 2y$, $2x + 3y, \ldots$ an arithmetic sequence? If it is, find the sum of the first 100 terms.

47. A beach now has an area of 9500 m^2 but is eroding such that it loses 100 m^2 more of its area each year than during the previous year. If it lost 400 m^2 during the last year, what will be its area 8 years from now?

48. During a period of heavy rains, on a given day 110,000 ft^3/s of water was being released from a dam. In order to minimize downstream flooding, engineers then reduced the releases by 10,000 ft^3/s each day thereafter. How much water was released during the first week of these releases?

49. At a logging camp, 15 layers of logs are so piled that there are 20 logs in the bottom layer, and each layer has 1 less log than the layer below it. How many logs are in the pile?

50. In order to prevent an electric current surge in a circuit, the resistance R in the circuit is stepped down by 4.0 Ω after each 0.1 s. If the voltage V is constant at 120 V, do the resulting currents I (in A) form an arithmetic sequence if $V = IR$?

51. There are 12 seats in the first row around a semicircular stage. Each row behind the first has 4 more seats than the row in front of it. How many rows of seats are there if there is a total of 300 seats?

52. A bank loan of $8000 is repaid in annual payments of $1000 plus 10% interest on the unpaid balance. What is the total amount of interest paid?

53. A car depreciates $1800 during the first year after it is bought. Each year thereafter it depreciates $150 less than the year before. How many years after it was bought will it be considered to have no value, and what was the original cost?

54. The sequence of ships' bells is as follows: 12:30 A.M. one bell is rung, and each half hour later one more bell is rung than the previous time until eight bells are rung. The sequence is then repeated starting at 4:30 A.M., again until eight bells are rung. This pattern is followed throughout the day. How many bells are rung in one day?

55. If a stone released from a spacecraft near the surface of Mars falls 1.85 m during the first second, 5.55 m during the second second, 9.25 m during third second, 12.95 m during fourth second, and so on, how far from the surface is the spacecraft if it takes the stone 10.0 s to reach the surface?

56. In preparing a bid for constructing a new building, a contractor determines that the foundation and basement will cost $605,000 and the first floor will cost $360,000. Each floor above the first will cost $15,000 more than the one below it. How much will the building cost if it is to be 18 floors high?

57. Derive a formula for S_n in terms of n, a_1, and d.

(W) **58.** A *harmonic sequence* is a sequence of numbers whose reciprocals form an arithmetic sequence. Is a harmonic sequence also an arithmetic sequence? Explain.

59. Show that the sum of the first n positive integers is $\frac{1}{2}n(n + 1)$.

60. Show that the sum of the first n positive odd integers is n^2.

Answers to Practice Exercises

1. $d = 2.5$ **2.** $S_9 = 81$

19.2 Geometric Sequences   511

19.2 Geometric Sequences

**Geometric Sequence •
Common Ratio •
Sum of n Terms**

A second type of important sequence of numbers is the **geometric sequence** (or **geometric progression**). *In a geometric sequence, each number after the first can be obtained from the preceding one by multiplying it by a fixed number, called the* **common ratio.** We can express this definition in terms of the *recursion formula*

$$a_n = ra_{n-1}$$ (19.4)

where a_n is any term, a_{n-1} is the preceding term, and r is the common ratio. One important application of geometric sequences is in computing compound interest on savings accounts. Other applications are found in areas such as biology and physics.

EXAMPLE 1 Illustrations of geometric sequences

(a) The sequence 2, 4, 8, 16, ..., is a geometric sequence with a common ratio of 2. Any term after the first can be obtained by multiplying the previous term by 2. We see that the fourth term $a_4 = ra_3$, or $16 = 2(8)$.

(b) The sequence 9, -3, 1, $-1/3$, ..., is a geometric sequence with a common ratio of $-1/3$. We can obtain any term after the first by multiplying the previous term by $-1/3$. ∎

If we know the first term, we can find any other term by multiplying by the common ratio a sufficient number of times. When we do this for a general geometric sequence, we can find the nth term in terms of the first term a_1, the common ratio r, and n. Thus, the second term is $a_1 r$, the third term is $a_1 r^2$, and so forth. In general, the expression for the nth term is

nth term

$$a_n = a_1 r^{n-1}$$ (19.5)

EXAMPLE 2 Finding specified term

Find the eighth term of the geometric sequence 8, 4, 2, ...

By dividing any term by the previous term, we find the common ratio to be $\frac{1}{2}$. From the terms given, we see that $a_1 = 8$. From the statement of the problem, we know that $n = 8$. Thus, we substitute into Eq. (19.5) to find a_8:

$$a_8 = 8\left(\frac{1}{2}\right)^{8-1} = \frac{8}{2^7} = \frac{1}{16}$$

where a_1, r, n are indicated. ∎

EXAMPLE 3 Finding common ratio

Fig. 19.2

Calculator display:
```
((-3125*625)/(64
*8))^9-1
          -2.5
```

Find the common ratio r if $a_1 = \frac{8}{625}$ and $a_{10} = -\frac{3125}{64}$.

Using Eq. (19.5), we have

$$-\frac{3125}{64} = \frac{8}{625}r^{10-1} = \frac{8}{625}r^9 \qquad \text{substituting in Eq. (19.5)}$$

$$r^9 = -\frac{(3125)(625)}{(64)(8)} \qquad \text{solving for } r$$

$$r = \left[-\frac{(3125)(625)}{(64)(8)}\right]^{1/9} = -2.5 \qquad \text{see calculator display in Fig. 19.2}$$
∎

EXAMPLE 4   Finding specified term

Find the seventh term of the geometric sequence for which the second term is 3, the fourth term is 9, and $r > 0$.

We can find r if we let $a_1 = 3, a_3 = 9$, and $n = 3$. (At this point, we are considering a sequence made up of 3, the next number, and 9. These are the second, third, and fourth terms of the original sequence.) Thus,

$$9 = 3r^2, \qquad r = \sqrt{3} \qquad (\text{since } r > 0)$$

We can now find a_1 of the original sequence by considering just the first two terms of the sequence, a_1 and $a_2 = 3$:

$$3 = a_1\left(\sqrt{3}\right)^{2-1}, \qquad a_1 = \sqrt{3}$$

We can now find the seventh term, using $a_1 = \sqrt{3}, r = \sqrt{3}$, and $n = 7$:

$$a_7 = \sqrt{3}\left(\sqrt{3}\right)^{7-1} = \sqrt{3}(3^3) = 27\sqrt{3}$$

We could have shortened this procedure one step by letting the second term be the first term of a new sequence of six terms. If the first term is of no importance in itself, this is acceptable.   ∎

Practice Exercise

1. Find the sixth term of the geometric sequence for which $a_1 = 81$ and $r = -1/3$.

EXAMPLE 5   Geometric sequence – application

In an experiment, 22.0% of a substance changes chemically each 10.0 min. If there is originally 120 g of the substance, how much will remain after 45.0 min?

Let P = the portion of the substance remaining after each minute. From the statement of the problem, we know that $r = 0.780$, since 78.0% remains after each 10.0-min period. We also know that $a_1 = 120$ g, and we let n represent the number of minutes of elapsed time. This means that $P = 120(0.780)^{n/10.0}$. It is necessary to divide by 10.0 because the ratio is given for a 10.0-min period. In order to find P when $n = 45.0$ min, we write

$$P = 120(0.780)^{45.0/10.0} = 120(0.780)^{4.50}$$
$$= 39.2 \text{ g}$$

This means that 39.2 g remain after 45.0 min. Note that the power 4.50 represents 4.50 ten-minute periods.   ∎

SUM OF n TERMS

A general expression for the sum S_n of the first n terms of a geometric sequence may be found by directly forming the sum and multiplying this equation by r:

$$S_n = a_1 + a_1 r + a_1 r^2 + \cdots + a_1 r^{n-1}$$
$$rS_n = a_1 r + a_1 r^2 + a_1 r^3 + \cdots + a_1 r^n$$

If we now subtract the second of these equations from the first, we get $S_n - rS_n = a_1 - a_1 r^n$. All other terms cancel by subtraction. Now, factoring S_n from the terms on the left and a_1 from the terms on the right, we solve for S_n. Thus, *the sum S_n of the first n terms of a geometric sequence is*

$$S_n = \frac{a_1(1 - r^n)}{1 - r} \qquad (r \neq 1)$$

(19.6)

EXAMPLE 6   Finding sum of terms

Find the sum of the first seven terms of the geometric sequence for which the first term is 2 and the common ratio is 1/2.

We are to find S_n, given that $a_1 = 2, r = \frac{1}{2}$, and $n = 7$. Using Eq. (19.6), we have

$$S_7 = \frac{2\left(1 - \left(\frac{1}{2}\right)^7\right)}{1 - \frac{1}{2}} = \frac{2\left(1 - \frac{1}{128}\right)}{\frac{1}{2}} = 4\left(\frac{127}{128}\right) = \frac{127}{32} \qquad \blacksquare$$

Practice Exercise

2. Find the sum of the six terms of the geometric sequence in Practice Exercise 1.

EXAMPLE 7   Finding sum of terms – application

If $100 is invested each year at 5% interest compounded annually, what would be the total amount of the investment after 10 years (before the 11th deposit is made)?

After 1 year, the amount invested will have added to it the interest for the year. Therefore, for the last (10th) $100 invested, its value will become

$$\$100(1 + 0.05) = \$100(1.05) = \$105$$

■ See the chapter introduction.

The next to last $100 will have interest added twice. After 1 year, its value becomes $100(1.05)$, and after 2 years it is $100(1.05)(1.05) = \$100(1.05)^2$. In the same way, the value of the first $100 becomes $100(1.05)^{10}$, since it will have interest added 10 times. This means that we are to find the sum of the sequence

■ It is reported that Albert Einstein was once asked what was the greatest discovery ever made, and his reply was "compound interest."

$$100(1.05) + 100(1.05)^2 + 100(1.05)^3 + \cdots + 100(1.05)^{10}$$

| 1 year | 2 years | 3 years | 10 years |
| in account | in account | in account | in account |

or

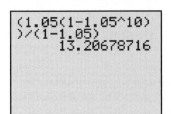

Fig. 19.3

$$100\left[1.05 + (1.05)^2 + (1.05)^3 + \cdots + (1.05)^{10}\right]$$

For the sequence in the brackets, we have $a_1 = 1.05, r = 1.05$, and $n = 10$. Thus,

$$S_{10} = \frac{1.05\left[1 - (1.05)^{10}\right]}{1 - 1.05} = 13.2068$$

The total value of the $100 investments is $100(13.2068) = \$1320.68$. We see that $320.68 in interest has been earned. See Fig. 19.3. $\qquad \blacksquare$

EXERCISES 19.2

In Exercises 1 and 2, make the given changes in the indicated examples of this section and then solve the given problems.

1. In Example 4, change "seventh" to "tenth."

2. In Example 6, change $\frac{1}{2}$ to $\frac{1}{3}$ and then find the sum.

In Exercises 3–6, write down the first five terms of the geometric sequence with the given values.

3. $a_1 = 6400, \ r = 0.25$

4. $a_1 = 0.09, \ r = -\frac{2}{3}$

5. $a_1 = \frac{1}{6}, \ r = 3$

6. $a_1 = -3, \ r = 2$

In Exercises 7–14, find the nth term of the geometric sequence with the given values.

7. $\frac{1}{2}, 1, 2, \ldots; n = 9$

8. $10, 1, 0.1, \ldots; n = 8$

9. $125, -25, 5, \ldots; n = 7$

10. $0.1, 0.3, 0.9, \ldots; n = 5$

11. $a = -2700, \ r = -\frac{1}{3}, \ n = 10$

12. $a_1 = 48, \ r = \frac{1}{2}, \ n = 12$

13. $10^{100}, -10^{98}, 10^{96}, \ldots; n = 51$

14. $-2, 4k, -8k^2, \ldots; n = 6$

In Exercises 15–20, find the sum of the first n terms of the indicated geometric sequence with the given values.

15. $a_1 = \frac{1}{8}, \ r = 4, \ n = 5$

16. $162, -54, 18, \ldots; n = 6$

17. $384, 192, 96, \ldots, n = 7$

18. $a_1 = 9, \ a_n = -243, \ n = 4$

19. $a_1 = 96, \ r = -\frac{k}{2}, \ n = 10$

20. $\log 2, \log 4, \log 16, \ldots; n = 6$

In Exercises 21–28, find any of the values of a_1, r, a_n, n, or S_n that are missing.

21. $a_1 = \frac{1}{16}, \ r = 4, \ n = 6$

22. $r = 0.2, \ a_n = 0.00032, \ n = 7$

23. $r = \frac{3}{2}$, $n = 5$, $S_5 = 211$ **24.** $r = -\frac{1}{2}$, $a_n = \frac{1}{8}$, $n = 7$

25. $a_n = 27$, $n = 4$, $S_4 = 40$ **26.** $a_1 = 3$, $n = 7$, $a_7 = 192$

27. $a_1 = 75$, $r = \frac{1}{5}$, $a_n = \frac{3}{625}$ **28.** $r = -2$, $n = 6$, $S_6 = 42$

In Exercises 29–58, find the indicated quantities.

(W) **29.** Is $3, 3^{x+1}, 3^{2x+1}, \dots$ a geometric sequence? Explain. If it is, find a_{20}.

30. The first two terms of a geometric sequence are added together to form the first term of a second sequence. The third and fourth terms of the first sequence are added to form the second term of the second sequence, and so on. If n is even for the first sequence, is the second sequence geometric?

31. Show that a, $\sqrt{ab}$, b are three successive terms of a geometric sequence ($a > 0, b > 0$).

32. Find x if $\sqrt{x}, 4, \sqrt{15x + 4}$ are the first three terms of a geometric sequence.

33. Find x if $2, 6, 2x + 8$ are successive terms of a geometric sequence.

34. Write the first five terms of a geometric sequence of which the first two terms are a and b.

35. Using Eq. (19.6), find the sum of the terms of a geometric sequence if $r = -1$, and n is (a) odd, (b) even.

36. If a positive number c is added to each term of a geometric sequence, is the resulting sequence geometric?

37. The sum of the first three terms of a geometric sequence equals seven times the first term. Find the common ratio.

38. The sum of the first three terms of an arithmetic sequence is 3. What are the numbers if their squares form a geometric sequence?

39. Each stroke of a pump removes 8.2% of the remaining air from a container. What percent of the air remains after 50 strokes?

40. From 1990 to 2000, Raleigh, North Carolina, had a 2.34% average growth rate of population, compounded annually. If the population in 1990 was 219,000, what was the 2000 population?

41. An electric current decreases by 12.5% each $1.00\ \mu s$. If the initial current is 3.27 mA, what is the current after $8.20\ \mu s$?

42. A copying machine is set to reduce the dimensions of material copied by 10%. A drawing 12.0 cm wide is reduced, and then the copies are in turn reduced. What is the width of the drawing on the sixth reduction?

43. How much is an investment of $250 worth after 8 years if it earns annual interest of 7.2% compounded monthly? (7.2% annual interest compounded monthly means that 0.6% (7.2%/12) interest is added each month.)

44. A chemical spill pollutes a stream. A monitoring device finds 620 ppm (parts per million) of the chemical 1.0 mi below the spill, and the readings decrease by 12.5% for each mile farther downstream. How far downstream is the reading 100 ppm?

45. Measurements show that the temperature of a distant star is presently 9800°C and is decreasing by 10% every 800 years. What will its temperature be in 4000 years?

46. The chlorine in a swimming pool was measured (in ppm, parts per million) to be 1.80, 1.53, 1.30, and 1.10 on four successive days. Noting that these values approximate a geometric sequence, what would be the reading three days after the last reading?

47. The strength of a signal in a fiber-optic cable decreases 12% for every 15 km along the cable. What percent of the signal remains after 100 km?

48. A series of deposits, each of value A and made at equal time intervals, earns an interest rate of i for the time interval. The deposits have a total value of

$$A(1 + i) + A(1 + i)^2 + A(1 + i)^3 + \cdots + A(1 + i)^n$$

after n time intervals (just before the next deposit). Find a formula for this sum.

49. A thermometer is removed from hot water at 100.0°C into a room at 20.0°C. The temperature difference D between the thermometer and the air decreases by 35.0% each minute. What is the temperature reading on the thermometer 10.0 min later?

50. The power on a space satellite is supplied by a radioactive isotope. On a given satellite, the power decreases by 0.2% each day. What percent of the initial power remains after 1 year?

51. If you decided to save money by putting away 1¢ on a given day, 2¢ one week later, 4¢ a week later, and so on, how much would you have to put away 6 months (26 weeks) after putting away the 1¢?

52. How many direct ancestors (parents, grandparents, and so on) does a person have in the ten generations that preceded him or her (assuming that no ancestor appears in more than one line of descent)?

53. Derive a formula for S_n in terms of a_1, r, and a_n.

(W) **54.** Write down several terms of a general geometric sequence. Then take the logarithm of each term. Explain why the resulting sequence is an arithmetic sequence.

(W) **55.** Do the squares of the terms of a geometric sequence also form a geometric sequence? Explain.

(W) **56.** If $a_1, a_2, a_3, \dots$ is an arithmetic sequence, explain why $2^{a_1}, 2^{a_2}, 2^{a_3}, \dots$ is a geometric sequence.

57. The numbers $8, x, y$ form an arithmetic sequence, and the numbers $x, y, 36$ form a geometric sequence. Find all of the possible sequences.

58. For $f(x) = ab^x$, is $f(1), f(3), f(5)$ an arithmetic sequence or a geometric sequence? What is the common difference (or common ratio)?

Answers to Practice Exercises

1. $a_6 = -1/3$ **2.** $S_6 = 182/3$

19.3 Infinite Geometric Series

Series • Infinity • Limit • Repeating Decimal

In the previous sections, we developed formulas for the sum of the first n terms of an arithmetic sequence and of a geometric sequence. *The indicated sum of the terms of a sequence is called a* **series.**

EXAMPLE 1 Illustrations of series

(a) The indicated sum of the terms of the arithmetic sequence $2, 5, 8, 11, 14, \ldots$ is the series $2 + 5 + 8 + 11 + 14 + \cdots$.

(b) The indicated sum of terms of the geometric sequence $1, \frac{1}{2}, \frac{1}{4}, \frac{1}{8}, \ldots$ is the series $1 + \frac{1}{2} + \frac{1}{4} + \frac{1}{8} + \cdots$. ∎

The series associated with a finite sequence will sum up to a real number. The series associated with an infinite arithmetic sequence will not sum up to a real number, as the terms being added become larger and larger numerically. The sum is unbounded, as we can see in Example 1(a). The series associated with an infinite geometric sequence may or may not sum up to a real number, as we now show.

Let us now consider the sum of the first n terms of the infinite geometric sequence $1, \frac{1}{2}, \frac{1}{4}, \ldots$. This is the sum of the n terms of the associated geometric series

$$1 + \frac{1}{2} + \frac{1}{4} + \cdots + \frac{1}{2^{n-1}}$$

Here, $a_1 = 1$ and $r = \frac{1}{2}$, and we find that we get the values of S_n for the given values of n in the following table:

n	2	3	4	5	6	7	8	9	10
S_n	$\frac{3}{2}$	$\frac{7}{4}$	$\frac{15}{8}$	$\frac{31}{16}$	$\frac{63}{32}$	$\frac{127}{64}$	$\frac{255}{128}$	$\frac{511}{256}$	$\frac{1023}{512}$

$1 + \frac{1}{2} + \frac{1}{4} + \frac{1}{8}$ ↑   ↑ $1 + \frac{1}{2} + \frac{1}{4} + \frac{1}{8} + \frac{1}{16} + \frac{1}{32} + \frac{1}{64}$

The series for $n = 4$ and $n = 7$ are shown. We see that as n gets larger, the numerator of each fraction becomes more nearly twice the denominator. In fact, we find that if we continue to compute S_n as n becomes larger, S_n can be found as close to the value 2 as desired, although it will never actually reach the value 2. For example, if $n = 100$, $S_{100} = 2 - 1.6 \times 10^{-30}$, which could be written as

$$1.9999999999999999999999999999984$$

to 32 significant digits. For the sum of the first n terms of a geometric sequence

$$S_n = a_1 \frac{1 - r^n}{1 - r}$$

the term r^n becomes exceedingly small if $|r| < 1$, and if n is sufficiently large, this term is effectively zero. *If this term were exactly zero,* the sum would be

$$S_n = 1 \frac{1 - 0}{1 - \frac{1}{2}} = 2$$

The only problem is that we cannot find any number large enough for n to make $\left(\frac{1}{2}\right)^n$ zero. There is, however, an accepted notation for this. This notation is

$$\lim_{n \to \infty} r^n = 0 \quad (\text{if } |r| < 1)$$

and it is read as "the limit, as n *approaches* infinity, of r to the nth power is zero."

■ The symbol ∞ for infinity was first used by the English mathematician, John Wallis (1616–1703).

CAUTION ▶ The symbol ∞ is read as **infinity,** *but it must not be thought of as a number.* It is simply a symbol that stands for a *process* of considering numbers that become large without bound. The number called the **limit** of the sums is simply the number the sums get closer and closer to, as n is considered to approach infinity. This notation and terminology are of particular importance in the calculus.

If we consider values of r such that $|r| < 1$ and let the values of n become unbounded, we find that $\lim_{n \to \infty} r^n = 0$. The formula for *the sum of the terms of an infinite geometric series then becomes*

$$S = \frac{a_1}{1 - r} \quad (|r| < 1) \tag{19.7}$$

where a_1 is the first term and r is the common ratio. If $|r| \geq 1$, S is unbounded in value.

EXAMPLE 2 Sum of infinite geometric series

Find the sum of the infinite geometric series

$$4 - \frac{1}{2} + \frac{1}{16} - \frac{1}{128} + \cdots$$

Here, we see that $a_1 = 4$. We find r by dividing any term by the previous term, and we find that $r = -\frac{1}{8}$. We then find the sum by substituting in Eq. (19.7). This gives us

$$S = \frac{4}{1 - \left(-\frac{1}{8}\right)} = \frac{4}{1 + \frac{1}{8}}$$

$$= \frac{4}{1} \times \frac{8}{9} = \frac{32}{9} \qquad \blacksquare$$

Practice Exercise

1. Find the sum of the infinite geometric series $9 + 3 + 1 + 1/3 + \ldots$.

EXAMPLE 3 Decimal form to fraction

Find the fraction that has its decimal form $0.121212\ldots$.

This decimal form can be considered as being

$$0.12 + 0.0012 + 0.000012 + \cdots$$

which means that we have an infinite geometric series in which $a_1 = 0.12$ and $r = 0.01$. Thus,

$$S = \frac{0.12}{1 - 0.01} = \frac{0.12}{0.99}$$

$$= \frac{4}{33}$$

Practice Exercise

2. Find the fraction that has as its decimal form $0.272727\ldots$.

Therefore, the decimal $0.121212\ldots$ and the fraction $\frac{4}{33}$ represent the same number. ∎

The decimal in Example 3 is called a **repeating decimal,** because *a particular sequence of digits in the decimal form repeats endlessly.* This example verifies the theorem that any repeating decimal represents a rational number. However, not all repeating decimals start repeating immediately. If the numbers never do repeat, the decimal represents an irrational number. For example, there are no repeating decimals that represent π, $\sqrt{2}$, or e. The decimal form of the number e does repeat at one point, but the repetition stops. As we noted in Section 12.5, the decimal form of e to 16 decimal places is 2.7 1828 1828 4590 452. We see that the sequence of digits 1828 repeats only once.

■ A calculator can be used to change a repeating decimal to a fraction. Fig. 19.4 shows this for the decimals in Examples 3 and 4.

```
.1212121212121212▶
Frac
                4/33
.50345345345345▶
Frac
            3353/6660
```

Fig. 19.4

EXAMPLE 4 Repeating decimal

Find the fraction that has its decimal form the repeating decimal 0.50345345345....

We first separate the decimal into the beginning, nonrepeating part, and the infinite repeating decimal, which follows. Thus, we have

$$0.50345345345\ldots = 0.50 + 0.00345345345\ldots$$

This means that we are to add $\frac{50}{100}$ to the fraction that represents the sum of the terms of the infinite geometric series $0.00345 + 0.00000345 + \cdots$. For this series, $a_1 = 0.00345$ and $r = 0.001$. We find this sum to be

$$S = \frac{0.00345}{1 - 0.001} = \frac{0.00345}{0.999} = \frac{115}{33,300} = \frac{23}{6660}$$

Therefore,

$$0.50345345\ldots = \frac{5}{10} + \frac{23}{6660} = \frac{5(666) + 23}{6660} = \frac{3353}{6660}$$ ■

EXAMPLE 5 Infinite geometric series – application

Each swing of a certain pendulum bob is 95% as long as the previous swing. How far does the bob travel in coming to rest if the first swing is 40.0 in. long?

We are to find the sum of the terms of an infinite geometric series for which $a_1 = 40.0$ and $r = 95\% = \frac{19}{20}$. Substituting these values into Eq. (19.7), we obtain

$$S = \frac{40.0}{1 - \frac{19}{20}} = \frac{40.0}{\frac{1}{20}} = (40.0)(20) = 800 \text{ in.}$$

The pendulum bob travels 800 in. (about 67 ft) in coming to rest. ■

EXERCISES 19.3

In Exercises 1 and 2, make the given changes in the indicated examples of this section, and then solve the resulting problems.

1. In Example 2, change all − signs to + in the series

2. In Example 3, change the decimal form to 0.012012012....

In Exercises 3–6, find the indicated quantity for an infinite geometric series.

3. $a_1 = 4$, $r = \frac{1}{2}$, $S = ?$ 4. $a_1 = 68$, $r = -\frac{1}{3}$, $S = ?$

5. $a_1 = 0.5$, $S = 0.625$, $r = ?$

6. $S = 4 + 2\sqrt{2}$, $r = \dfrac{1}{\sqrt{2}}$, $a_1 = ?$

In Exercises 7–14, find the sums of the given infinite geometric series.

7. $20 - 1 + 0.05 - \cdots$ 8. $9 + 8.1 + 7.29 + \cdots$

9. $1 + \frac{7}{8} + \frac{49}{64} + \cdots$ 10. $6 - 4 + \frac{8}{3} - \cdots$

11. $1 + 10^{-4} + 10^{-8} + \cdots$ 12. $1000 - 300 + 90 - \cdots$

13. $\left(2 + \sqrt{3}\right) + 1 + \left(2 - \sqrt{3}\right) + \cdots$

14. $\left(1 + \sqrt{2}\right) - 1 + \left(\sqrt{2} - 1\right) - \cdots$

In Exercises 15–28, find the fractions equal to the given decimals.

15. 0.33333...   16. 0.55555...

17. 0.499999...   18. 0.999999...

19. 0.404040...   20. 0.070707...

21. 0.181818...   22. 0.336336336...

23. 0.0273273273...   24. 0.82222...

25. 0.366666...   26. 0.66424242...

27. 0.100841841841...   28. 0.184561845618456...

In Exercises 29–38, solve the given problems by use of the sum of an infinite geometric series.

29. There are two infinite geometric series with $a_1 = 50$ and $a_3 = 2$. Find the sum of each.

(W) 30. Explain why there is no infinite geometric series with $a_1 = 5$ and $S = 2$.

31. Liquid is continuously collected in a wastewater-holding tank such that during a given hour only 92.0% as much liquid is collected as in the previous hour. If 28.0 gal are collected in the first hour, what must be the minimum capacity of the tank?

32. If 75% of all aluminum cans are recycled, what is the total number of recycled cans that can be made from 400,000 cans that are recycled over and over until all the aluminum from these cans is used up? (Assume no aluminum is lost in the recycling process.)

33. The amounts of plutonium-237 that decay each day because of radioactivity form a geometric sequence. Given that the amounts that decay during each of the first 4 days are 5.882 g, 5.782 g, 5.684 g, and 5.587 g, respectively, what total amount will decay?

34. A helium-filled balloon rose 120 ft in 1.0 min. Each minute after that, it rose 75% as much as in the previous minute. What was its maximum height?

35. A bicyclist traveling at 10 m/s then coasts to a stop as the bicycle travels 0.90 as far each second as in the previous second. How far does the bicycle travel in coasting to a stop?

(W) 36. A square has sides of 20 cm. Another square is inscribed in the first square by joining the midpoints of the sides. Assuming that such inscribed squares can be formed endlessly, find the sum of the areas of all the squares and explain how the sum is found. See Fig. 19.5.

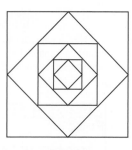

Fig. 19.5

37. Find x if the sum of the terms of the infinite geometric series $1 + 2x + 4x^2 + \dots$ is 2/3.

38. Find the sum of the terms of the infinite series $1 + 2x + 3x^2 + 4x^3 + \dots$ for $|x| < 1$. (*Hint:* Use $S - xS$.)

Answers to Practice Exercises

1. $S = 27/2$ **2.** 3/11

19.4 The Binomial Theorem

Properties of the Binomial $(a + b)^n$ • Factorial Notation • Binomial Theorem • Pascal's Triangle • Binomial Series

To expand $(x + 2)^5$ would require a number of multiplications that would be a tedious operation. We now develop *the* **binomial theorem,** *by which it is possible to expand binomials to any given power without direct multiplication.* Using this type of expansion, it is also possible to expand some expressions for which direct multiplication is not possible. The binomial theorem is used to develop expressions needed in certain mathematics topics and in technical applications.

By direct multiplication, we may obtain the following expansions of the binomial $a + b$:

$$(a + b)^0 = 1$$
$$(a + b)^1 = a + b$$
$$(a + b)^2 = a^2 + 2ab + b^2$$
$$(a + b)^3 = a^3 + 3a^2b + 3ab^2 + b^3$$
$$(a + b)^4 = a^4 + 4a^3b + 6a^2b^2 + 4ab^3 + b^4$$
$$(a + b)^5 = a^5 + 5a^4b + 10a^3b^2 + 10a^2b^3 + 5ab^4 + b^5$$

Inspection shows these expansions have certain properties, and we assume that these properties are valid for the expansion of $(a + b)^n$, where n is any positive integer.

PROPERTIES OF THE BINOMIAL $(a + b)^n$

1. *There are $n + 1$ terms.*

2. *The first term is a^n, and the final term is b^n.*

3. *Progressing from the first term to the last, the exponent of a decreases by 1 from term to term, the exponent of b increases by 1 from term to term, and the sum of the exponents of a and b in each term is n.*

4. *If the coefficient of any term is multiplied by the exponent of a in that term and this product is divided by the number of that term, we obtain the coefficient of the next term.*

5. *The coefficients of terms equidistant from the ends are equal.*

EXAMPLE 1 Using basic binomial properties

Using the basic properties, develop the expansion for $(a + b)^5$.

Since the exponent of the binomial is 5, we have $n = 5$.

From Property 1, we know that there are six terms.

From Property 2, we know that the first term is a^5 and the final term is b^5.

From Property 3, we know that the factors of a and b in terms 2, 3, 4, and 5 are a^4b, a^3b^2, a^2b^3, and ab^4, respectively.

From Property 4, we obtain the coefficients of terms 2, 3, 4, and 5. In the first term, a^5, the coefficient is 1. Multiplying by 5, the power of a, and dividing by 1, the number of the term, we obtain 5, which is the coefficient of the second term. Thus, the second term is $5a^4b$. Again using Property 4, we obtain the coefficient of the third term. The coefficient of the second term is 5. Multiplying by 4, and dividing by 2, we obtain 10. This means that the third term is $10a^3b^2$.

From Property 5, we know that the coefficient of the fifth term is the same as the second and that the coefficient of the fourth term is the same as the third. These properties are illustrated in the following diagram:

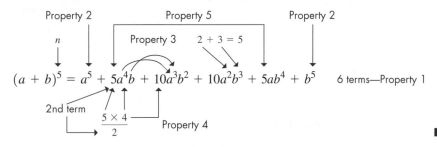

It is not necessary to use the above properties directly to expand a given binomial. If they are applied to $(a + b)^n$, a general formula for the expansion of a binomial may be obtained. In developing and stating the general formula, it is convenient to use the **factorial notation $n!$,** where

$$n! = n(n - 1)(n - 2) \cdots (2)(1) \qquad (19.8)$$

We see that $n!$, read "n factorial," represents the product of the first n positive integers.

EXAMPLE 2 Evaluating factorials

(a) $3! = (3)(2)(1) = 6$

(b) $5! = (5)(4)(3)(2)(1) = 120$

(c) $8! = (8)(7)(6)(5)(4)(3)(2)(1) = 40,320$

(d) $3! + 5! = 6 + 120 = 126$

(e) $\dfrac{4!}{2!} = \dfrac{(4)(3)(2)(1)}{(2)(1)} = 12$

In evaluating factorials, we must remember that they represent products of numbers. In parts (d) and (e), we see that

CAUTION ▶ $3! + 5!$ **is not** $(3 + 5)!$ and $4!/2!$ **is not** $2!$.

The evaluation of the factorials from parts (c) and (d) is shown in the calculator display in Fig. 19.6.

8!
                40320
3!+5!
                126

Fig. 19.6

Practice Exercise

1. Evaluate $9!/7!$.

THE BINOMIAL FORMULA

■ See Appendix C for a graphing calculator program BINEXPAN. It gives the first K coefficients of $(AX + B)^N$.

Based on the binomial properties, *the* **binomial theorem** *states that the following* **binomial formula** *is valid for all positive integer values of* n (the binomial theorem is proven through advanced methods).

$$(a + b)^n = a^n + na^{n-1}b + \frac{n(n-1)}{2!}a^{n-2}b^2 + \frac{n(n-1)(n-2)}{3!}a^{n-3}b^3 + \cdots + b^n \qquad (19.9)$$

EXAMPLE 3 Using binomial formula

Using the binomial formula, expand $(2x + 3)^6$.

In using the binomial formula for $(2x + 3)^6$, we use $2x$ for a, 3 for b, and 6 for n. Thus,

$$(2x + 3)^6 = (2x)^6 + 6(2x)^5(3) + \frac{(6)(5)}{2}(2x)^4(3^2) + \frac{(6)(5)(4)}{(2)(3)}(2x)^3(3^3) + \frac{(6)(5)(4)(3)}{(2)(3)(4)}(2x)^2(3^4) + \frac{(6)(5)(4)(3)(2)}{(2)(3)(4)(5)}(2x)(3^5) + 3^6$$

$$= 64x^6 + 576x^5 + 2160x^4 + 4320x^3 + 4860x^2 + 2916x + 729 \qquad ■$$

Pascal's Triangle

■ Named for the French scientist and mathematician Blaise Pascal (1623–1662).

For the first few integral powers of a binomial $a + b$, the coefficients can be obtained by setting them up in the following pattern, known as **Pascal's triangle.**

$n = 0$						1					
$n = 1$					1		1				
$n = 2$				1		2		1			
$n = 3$			1		3		3		1		
$n = 4$		1		4		6		4		1	
$n = 5$	1		5		10		10		5		1
$n = 6$	1	6	15	20	15	6	1				

see expansions on page 518

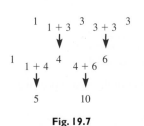

Fig. 19.7

We note that the first and last coefficients shown in each row are 1, and the second and next-to-last coefficients are equal to n. Other coefficients are obtained by adding the two nearest coefficients in the row above, as illustrated in Fig. 19.7 for the indicated section of Pascal's triangle. This pattern may be continued indefinitely, although use of Pascal's triangle is cumbersome for high values of n.

EXAMPLE 4 Using Pascal's triangle

Using Pascal's triangle, expand $(5s - 2t)^4$.

Here, we note that $n = 4$. Thus, the coefficients of the five terms are 1, 4, 6, 4, and 1, respectively. Also, here we use $5s$ for a and $-2t$ for b. We are expanding this expression as $[(5s) + (-2t)]^4$. Therefore,

from Pascal's triangle for $n = 4$

$$(5s - 2t)^4 = (5s)^4 + 4(5s)^3(-2t) + 6(5s)^2(-2t)^2 + 4(5s)(-2t)^3 + (-2t)^4$$

$$= 625s^4 - 1000s^3t + 600s^2t^2 - 160st^3 + 16t^4 \qquad ■$$

In certain uses of a binomial expansion, it is not necessary to obtain all terms. Only the first few terms are required. The following example illustrates finding the first four terms of an expansion.

EXAMPLE 5 Using binomial formula

Find the first four terms of the expansion of $(x + 7)^{12}$.

Here, we use x for a, 7 for b, and 12 for n. Thus, from the binomial formula, we have

$$(x + 7)^{12} = x^{12} + 12x^{11}(7) + \frac{(12)(11)}{2}x^{10}(7^2) + \frac{(12)(11)(10)}{(2)(3)}x^9(7^3) + \cdots$$

$$= x^{12} + 84x^{11} + 3234x^{10} + 75{,}460x^9 + \cdots$$ ∎

Practice Exercise

2. Find the first three terms of the expansion of $(x - 4)^9$.

If we let $a = 1$ and $b = x$ in the binomial formula, we obtain the **binomial series**

Binomial Series

$$(1 + x)^n = 1 + nx + \frac{n(n - 1)}{2!}x^2 + \frac{n(n - 1)(n - 2)}{3!}x^3 + \cdots \qquad (19.10)$$

which, through advanced methods, can be shown to be valid for any real number n if $|x| < 1$. When n is either negative or a fraction, we obtain an infinite series. In such a case, we calculate as many terms as may be needed although such a series is not obtainable through direct multiplication. The binomial series may be used to develop important expressions that are used in applications and more advanced mathematics topics.

EXAMPLE 6 Binomial series – application

In the analysis of forces on beams, the expression $1/(1 + m^2)^{3/2}$ is used. Use the binomial series to find the first four terms of the expansion.

Using negative exponents, we have

$$1/(1 + m^2)^{3/2} = (1 + m^2)^{-3/2}$$

Now, in using Eq. (19.10), we have $n = -3/2$ and $x = m^2$:

$$(1 + m^2)^{-3/2} = 1 + \left(-\frac{3}{2}\right)(m^2) + \frac{\left(-\frac{3}{2}\right)\left(-\frac{3}{2} - 1\right)}{2!}(m^2)^2 + \frac{\left(-\frac{3}{2}\right)\left(-\frac{3}{2} - 1\right)\left(-\frac{3}{2} - 2\right)}{3!}(m^2)^3 + \cdots$$

Therefore,

$$\frac{1}{(1 + m^2)^{3/2}} = 1 - \frac{3}{2}m^2 + \frac{15}{8}m^4 - \frac{35}{16}m^6 + \cdots$$ ∎

EXAMPLE 7 Evaluation using binomial series

Approximate the value of 0.97^7 by use of the binomial series.

We note that $0.97 = 1 - 0.03$, which means $0.97^7 = [1 + (-0.03)]^7$. Using four terms of the binomial series, we have

$$0.97^7 = [1 + (-0.03)]^7$$

$$= 1 + 7(-0.03) + \frac{7(6)}{2!}(-0.03)^2 + \frac{7(6)(5)}{3!}(-0.03)^3$$

$$= 1 - 0.21 + 0.0189 - 0.000945 = 0.807955$$

From a calculator, we find that $0.97^7 = 0.807983$ (to six decimal places), which means these values agree to four significant digits with a value 0.8080. Greater accuracy can be found using the binomial series if more terms are used. ∎

EXERCISES 19.4

In Exercises 1 and 2, make the given changes in the indicated examples of this section and then solve the given problems.

1. In Example 3, change the exponent from 6 to 5.

2. In Example 7, change 0.97 to 0.98.

In Exercises 3–12, expand and simplify the given expressions by use of the binomial formula.

3. $(t + 4)^3$

4. $(x - 2)^3$

5. $(2x - 1)^4$

6. $(x^2 + 5)^4$

7. $(6 + 0.1)^5$

8. $(xy - z)^5$

9. $(n + 2\pi)^5$

10. $(1 - j)^6 \ (j = \sqrt{-1})$

11. $(2a - b^2)^6$

12. $\left(\dfrac{a}{x} + x\right)^6$

In Exercises 13–16, expand and simplify the given expressions by use of Pascal's triangle.

13. $(5x - 3)^4$ **14.** $(x - 4)^5$ **15.** $(2a + 1)^6$ **16.** $(x - 3)^7$

In Exercises 17–24, find the first four terms of the indicated expansions.

17. $(x + 2)^{10}$

18. $(x - 4)^8$

19. $(2a - 1)^7$

20. $(3b + 2)^9$

21. $(x^{1/2} - 4y)^{12}$

22. $(2a - x^{-1})^{11}$

23. $\left(b^2 + \dfrac{1}{2b}\right)^{20}$

24. $\left(2x^2 + \dfrac{y}{3}\right)^{15}$

In Exercises 25–28, approximate the value of the given expression to three decimal places by using three terms of the appropriate binomial series. Check using a calculator.

25. 1.05^6 **26.** $\sqrt{0.927}$ **27.** $\sqrt[3]{1.045}$ **28.** 0.98^{-7}

In Exercises 29–36, find the first four terms of the indicated expansions by use of the binomial series.

29. $(1 + x)^8$

30. $(1 + x)^{-1/3}$

31. $(1 - 3x)^{-2}$

32. $(1 - 2\sqrt{x})^9$

33. $\sqrt{1 + x}$

34. $\dfrac{1}{\sqrt{1 - x}}$

35. $\dfrac{1}{\sqrt{9 - 9x}}$

36. $\sqrt{4 + x^2}$

In Exercises 37–40, solve the given problems involving factorials.

37. Using a calculator, evaluate (a) $17! + 4!$, (b) $21!$, (c) $17! \times 4!$, and (d) $68!$.

38. Using a calculator, evaluate (a) $8! - 7!$, (b) $8!/7!$, (c) $8! \times 7!$ and (d) $56!$.

(W) 39. Show that $n! = n \times (n - 1)!$ for $n \geq 2$. To use this equation for $n = 1$, explain why it is necessary to define $0! = 1$ (this is a standard definition of $0!$).

40. Show that $\dfrac{(n + 1)!}{(n - 2)!} = n^3 - n$ for $n \geq 2$. See Exercise 39.

In Exercises 41–44, find the indicated terms by use of the following information. The $r + 1$ term of the expansion of $(a + b)^n$ is given by

$$\frac{n(n - 1)(n - 2)\cdots(n - r + 1)}{r!}a^{n-r}b^r$$

41. The term involving b^5 in $(a + b)^8$

42. The term involving y^6 in $(x + y)^{10}$

43. The fifth term of $(2x - 3b)^{12}$

44. The sixth term of $\left(\sqrt{a} - \sqrt{b}\right)^{14}$

In Exercises 45–56, solve the given problems.

(W) 45. Explain why $n!$ ends in a zero if $n > 4$.

46. Expand $[(a + b) + c]^3$ using the binomial theorem. Group terms as indicated.

47. In the expansion of $(a - x)^n$, where n is a positive integer, show that the sum of the coefficients is zero.

48. If x is very small, show that $(1 + x)^{-n}$ is approximately $(1 - nx)$.

49. Approximate $\sqrt{6}$ to hundredths by noting that $\sqrt{6} = \sqrt{4(1.5)} = 2\sqrt{1 + 0.5}$ and using four terms of the appropriate binomial series.

50. Approximate $\sqrt[3]{10}$ by using the method of Exercise 49.

51. A company purchases a piece of equipment for A dollars, and the equipment depreciates at a rate of r each year. Its value V after n years is $V = A(1 - r)^n$. Expand this expression for $n = 5$.

52. In finding the rate of change of emission of energy from the surface of a body at temperature T, the expression $(T + h)^4$ is used. Expand this expression.

53. In the theory associated with the magnetic field due to an electric current, the expression $1 - \dfrac{x}{\sqrt{a^2 + x^2}}$ is found. By expanding $(a^2 + x^2)^{-1/2}$, find the first three nonzero terms that could be used to approximate the given expression.

54. In the theory related to the dispersion of light, the expression $1 + \dfrac{A}{1 - \lambda_0^2/\lambda^2}$ arises. (a) Let $x = \lambda_0^2/\lambda^2$ and find the first four terms of the expansion of $(1 - x)^{-1}$. (b) Find the same expansion by using long division. (c) Write the original expression in expanded form, using the results of (a) and (b).

55. In finding the rate at which the resistance of a wire changes with its radius, the expression $\dfrac{k}{(r + h)^2} - \dfrac{k}{r^2}$ is used. Use three terms of the binomial expansion of $k(r + h)^{-2}$ to approximate this expression.

56. Find the first four terms of the expansion of $(1 + x)^{-1}$ and then divide $1 + x$ into 1. Compare the results.

Answers to Practice Exercises

1. 72   **2.** $x^9 - 36x^8 + 576x^7 - \cdots$

CHAPTER 19   EQUATIONS

Arithmetic sequences	Recursion formula	$a_n = a_{n-1} + d$	**(19.1)**
	nth term	$a_n = a_1 + (n-1)d$	**(19.2)**
	Sum of n terms	$S_n = \dfrac{n}{2}(a_1 + a_n)$	**(19.3)**

Geometric sequences	Recursion formula	$a_n = ra_{n-1}$	**(19.4)**
	nth term	$a_n = a_1 r^{n-1}$	**(19.5)**
	Sum of n terms	$S_n = \dfrac{a_1(1 - r^n)}{1 - r} \quad (r \neq 1)$	**(19.6)**

Sum of geometric series $S = \dfrac{a_1}{1 - r} \quad (|r| < 1)$ **(19.7)**

Factorial notation $n! = n(n-1)(n-2)\cdots(2)(1)$ **(19.8)**

Binomial formula $(a+b)^n = a^n + na^{n-1}b + \dfrac{n(n-1)}{2!}a^{n-2}b^2 + \dfrac{n(n-1)(n-2)}{3!}a^{n-3}b^3 + \cdots + b^n$ **(19.9)**

Binomial series $(1+x)^n = 1 + nx + \dfrac{n(n-1)}{2!}x^2 + \dfrac{n(n-1)(n-2)}{3!}x^3 + \cdots$ **(19.10)**

CHAPTER 19   REVIEW EXERCISES

In Exercises 1–8, find the indicated term of each sequence.

1. $1, 6, 11, \ldots$ (17th)

2. $1, -3, -7, \ldots$ (21st)

3. $500, 100, 20, \ldots$ (9th)

4. $0.025, 0.01, 0.004, \ldots$ (7th)

5. $-1, 3.5, 8, \ldots$ (16th)

6. $-1, -\frac{5}{3}, -\frac{7}{3}, \ldots$ (25th)

7. $\frac{3}{4}, \frac{1}{2}, \frac{1}{3}, \ldots$ (7th)

8. $5^{-2}, 5^0, 5^2, \ldots$ (7th)

In Exercises 9–12, find the sum of each sequence with the indicated values.

9. $a_1 = -4$, $n = 15$, $a_{15} = 17$ (arith.)

10. $a_1 = 300$, $d = -\frac{20}{3}$, $n = 10$

11. $a_1 = 16$, $r = -\frac{1}{2}$, $n = 14$

12. $a_1 = 64$, $a_n = 729$, $n = 7$ (geom., $r > 0$)

In Exercises 13–24, find the indicated quantities for the appropriate sequences.

13. $a_1 = 17$, $d = -2$, $n = 9$, $S_9 = ?$

14. $d = \frac{4}{3}$, $a_1 = -3$, $a_n = 17$, $n = ?$

15. $a_1 = 4$, $r = \sqrt{2}$, $n = 7$, $a_7 = ?$

16. $a_n = \frac{49}{8}$, $r = -\frac{2}{7}$, $S_n = \frac{1911}{32}$, $a_1 = ?$

17. $a_1 = 80$, $a_n = -25$, $S_n = 220$, $d = ?$

18. $a_1 = 2$, $d = 0.2$, $n = 11$, $S_{11} = ?$

19. $n = 6$, $r = -0.25$, $S_6 = 204.75$, $a_6 = ?$

20. $a_1 = 10$, $r = 0.1$, $S_n = 11.111$, $n = ?$

21. $a_1 = -1$, $a_n = 32$, $n = 12$, $S_{12} = ?$ (arith.)

22. $a_1 = 100$, $a_n = 6400$, $S_n = 32,500$, $n = ?$ (arith.)

23. $a_1 = 1$, $n = 7$, $a_7 = 64$, $S_7 = ?$

24. $a_1 = \frac{1}{4}$, $n = 6$, $a_6 = 8$, $S_6 = ?$

In Exercises 25–28, find the sums of the given infinite geometric series.

25. $0.9 + 0.6 + 0.4 + \cdots$

26. $1280 - 320 + 80 - \cdots$

27. $1 + 1.02^{-1} + 1.02^{-2} + \cdots$

28. $3 - \sqrt{3} + 1 - \cdots$

In Exercises 29–32, find the fractions equal to the given decimals.

29. $0.030303\ldots$

30. $0.363363\ldots$

31. $0.0727272\ldots$

32. $0.25399399399\ldots$

In Exercises 33–36, expand and simplify the given expression. In Exercises 37–40, find the first four terms of the appropriate expansion.

33. $(x - 2)^4$

34. $(3 + 0.1)^4$

35. $(x^2 + 4)^5$

36. $(3n^{1/2} - a)^6$

37. $(a + 2e)^{10}$

38. $\left(\dfrac{x}{4} - y\right)^{12}$

39. $\left(p^2 - \dfrac{q}{6}\right)^9$

40. $\left(2s^2 - \tfrac{3}{2}t^{-1}\right)^{14}$

In Exercises 41–48, find the first four terms of the indicated expansions by use of the binomial series.

41. $(1 + x)^{12}$

42. $(1 - 2x)^{10}$

43. $\sqrt{1 + x^2}$

44. $\left(4 - 4\sqrt{x}\right)^{-1}$

45. $\sqrt{1 - a^2}$

46. $\sqrt{1 + 2b^4}$

47. $(2 - 4x)^{-3}$

48. $(1 + 4x)^{-1/4}$

In Exercises 49–92, solve the given problems by use of an appropriate sequence or expansion. All numbers are accurate to at least two significant digits.

49. Find the sum of the first 1000 positive even integers.

50. How many integers divisible by 4 lie between 23 and 121?

51. What is the fifth term of the arithmetic sequence in which the first term is a and the second term is b?

52. Find three consecutive numbers in an arithmetic sequence such that their sum is 15 and the sum of their squares is 77.

53. For a geometric sequence, is it possible that $a_3 = 6$, $a_5 = 9$, and $a_7 = 12$?

54. Find three consecutive numbers in a geometric sequence such that their product is 64 and the sum of their squares is 84.

55. Find a formula with variable n of the arithmetic sequence with $a_1 = -5$, $a_{n+1} = a_n - 3$, for $n = 1, 2, 3, \ldots$.

56. Find the ninth term of the sequence $(a + 2b), b, -a, \ldots$.

57. Approximate the value of $(1.06)^{-6}$ by using three terms of the appropriate binomial series. Check using a calculator.

58. Approximate the value of $\sqrt[4]{0.94}$ by using three terms of the appropriate binomial series. Check using a calculator.

59. Approximate the value of $\sqrt{30}$ by noting that $\sqrt{30} = \sqrt{25(1.2)} = 5\sqrt{1 + 0.2}$ and using three terms of the appropriate binomial series.

60. Approximate $\sqrt[3]{29.7}$ by using the *method* of Exercise 59.

61. Each stroke of a pile driver moves a post 2 in. less than the previous stroke. If the first stroke moves the post 24 in., which stroke moves the post 4 in.?

62. During each hour, an exhaust fan removes 15.0% of the carbon dioxide present in the air in a room at the beginning of the hour. What percent of the carbon dioxide remains after 10.0 h?

63. Each 1.0 mm of a filter through which light passes reduces the intensity of the light by 12%. How thick should the filter be to reduce the intensity of the light to 20%?

64. A pile of dirt and ten holes are in a straight line. It is 20 ft from the dirt pile to the nearest hole, and the holes are 8 ft apart. If a backhoe takes two trips to fill each hole, how far must it travel in filling all the holes if it starts and ends at the dirt pile?

65. A roof support with equally spaced vertical pieces is shown in Fig. 19.8. Find the total length of the vertical pieces if the shortest one is 10.0 in. long.

84.8°

10.0 in.

Fig. 19.8   ⟵ 224.0 in. ⟶

66. During each microsecond, the current in an electric circuit decreases by 9.3%. If the initial current is 2.45 mA, how long does it take to reach 0.50 mA?

67. A machine that costs $8600 depreciates 1.0% in value each month. What is its value 5 years after it was purchased?

68. The level of chemical pollution in a lake is 4.50 ppb (parts per billion). If the level increases by 0.20 ppb in the following month and by 5.0% less each month thereafter, what will be the maximum level?

69. A piece of paper 0.0040 in. thick is cut in half. These two pieces are then placed one on the other and cut in half. If this is repeated such that the paper is cut in half 40 times, how high will the pile be?

70. After the power is turned off, an object on a nearly frictionless surface slows down such that it travels 99.9% as far during 1 s as during the previous second. If it travels 100 cm during the first second after the power is turned off, how far does it travel while stopping?

71. Under gravity, an object falls 16 ft during the first second, 48 ft during the second second, 80 ft during the third second, and so on. How far will it fall during the 20th second?

72. For the object in Exercise 71, what is the total distance fallen during the first 20 s?

73. An object suspended on a spring is oscillating up and down. If the first oscillation is 10.0 cm and each oscillation thereafter is 9/10 of the preceding one, find the total distance the object travels in coming to rest.

74. During each oscillation, a pendulum swings through 85% of the distance of the previous oscillation. If the pendulum swings through 80.8 cm in the first oscillation, through what total distance does it move in 12 oscillations?

75. A person invests $1000 each year at the beginning of the year. What is the total value of these investments after 20 years if they earn 7.5% annual interest, compounded semiannually?

76. A well driller charges $10.00 for drilling the first meter of a well and for each meter thereafter charges 0.20% more than for the preceding meter. How much is charged for drilling a 150-m well?

77. When new, an article cost $250. It was then sold at successive yard sales at 40% of the previous price. What was the price at the fourth yard sale?

78. Each side of an equilateral triangle is 2 cm in length. The midpoints are joined to form a second equilateral triangle. The midpoints of the second triangle are joined to form a third equilateral triangle. Find the sum of all of the perimeters of the triangles if this process is continued indefinitely. See Fig. 19.9.

Fig. 19.9

79. In testing a type of insulation, the temperature in a room was made to fall to 2/3 of the initial temperature after 1.0 h, to 2/5 of the initial temperature after 2.0 h, to 2/7 of the initial temperature after 3.0 h, and so on. If the initial temperature was 50.0°C, what was the temperature after 12.0 h? (This is an illustration of a *harmonic sequence*.)

80. Two competing businesses make the same item and sell it initially for $100. One increases the price by $8 each year for 5 years, and the other increases the price by 8% each year for 5 years. What is the difference in price after 5 years?

81. In hydrodynamics, while studying compressible fluid flow, the expression $\left(1 + \dfrac{a-1}{2}m^2\right)^{a/(a-1)}$ arises. Find the first three terms of the expansion of this expression.

82. In finding the partial pressure P_F of the fluorine gas under certain conditions, the equation
$$P_F = \frac{(1 + 2 \times 10^{-10}) - \sqrt{1 + 4 \times 10^{-10}}}{2} \text{ atm}$$
is found. By using three terms of the expansion for $\sqrt{1 + x}$, approximate the value of this expression.

83. During 1 year a beach eroded 1.2 m to a line 48.3 m from the wall of a building. If the erosion is 0.1 m more each year than the previous year, when will the waterline reach the wall?

84. On a highway with a steep incline, the runaway truck ramp is constructed so that a vehicle that has lost its brakes can stop. The ramp is designed to slow a truck in succeeding 20-m distances by 10 km/h, 12 km/h, 14 km/h, ... If the ramp is 160 m long, will it stop a truck moving at 120 km/h when it reaches the ramp?

85. Each application of an insecticide destroys 75% of a certain insect. How many applications are needed to destroy at least 99.9% of the insects?

86. A house valued at $375,000 increases in value 5.00%/year for the next six years and then decreases 3.00%/year for the following four years. What is its value at the end of the ten years?

87. Oil pumped from a certain oil field decreases 10% each year. How long will it take for production to be 10% of the first year's production?

88. A wire hung between two poles is parabolic in shape. To find the length of wire between two points on the wire, the expression $\sqrt{1 + 0.08x^2}$ is used. Find the first three terms of the binomial expansion of this expression.

89. Show that the middle term of a finite arithmetic sequence equals S/n, if n is odd.

90. The sum of three terms of a geometric sequence is -3, and the second is 6 more than the first. Find these terms.

(W) 91. Do the reciprocals of the terms of a geometric sequence form a geometric sequence? Explain.

92. The terms a, $a + 12$, $a + 24$ form an arithmetic sequence, and the terms a, $a + 24$, $a + 12$ form a geometric sequence. Find these sequences.

Writing Exercise

93. Derive a formula for the value V after 1 year of an amount A invested at $r\%$ (as a decimal) annual interest, compounded n times during the year. If $A = \$1000$ and $r = 0.10$ (10%), write two or three paragraphs explaining why the amount of interest increases as n increases and stating your approach to finding the maximum possible amount of interest.

CHAPTER 19   PRACTICE TEST

1. Write the first five terms of the sequence for which (a) $a_1 = 8$ and $d = -1/2$;   (b) $a_1 = 8$ and $r = -1/2$.

2. Find the sum of the first seven terms of the sequence $6, -2, \frac{2}{3}, \ldots$.

3. For a given sequence $a_1 = 6$, $d = 4$, and $S_n = 126$. Find n.

4. Find the fraction equal to the decimal $0.454545\ldots$.

5. Find the first three terms of the expansion of $\sqrt{1 - 4x}$.

6. Expand and simply the expression $(2x - y)^5$.

7. What is the value after 20 years of an investment of $2500 if it draws 5% annual interest compounded annually?

8. Find the sum of the first 100 even integers.

9. A ball is dropped from a height of 8.00 ft, and on each rebound, it rises to 1/2 of the height it last fell. If it bounces indefinitely, through what total distance will it move?

20 Additional Topics in Trigonometry

As the use of electricity became widespread in the 1880s, there was a serious debate over the best way to distribute electric power. The American inventor Thomas Edison favored the use of direct current because it was safer and did not vary with time. Another American inventor and engineer, George Westinghouse, favored alternating current because the voltage could be stepped up and down with transformers during transmission.

Also favoring the use of alternating current was Nikola Tesla, an American (born in Croatia) electrical engineer, and he had a strong influence on the fact that alternating current came to be used for transmission. Tesla developed many electrical devices, among them the *polyphase generator* that allowed alternating current to be transmitted with constant instantaneous power. Using this type of generator, power losses are greatly reduced in transmission lines, which allows for smaller conductors, and the power can be generated far from where it is used.

Three-phase systems are used in most commercial electric generators, and in Section 20.2 we show how a relationship involving trigonometric functions can be used to show a basic property of the current produced by a three-phase generator. Many such relationships among the trigonometric functions can be found from the definitions and other known relationships.

The trigonometric relationships that we develop in this chapter are important for a number of reasons. In fact, we already made use of some of them in Section 10.4 when we graphed certain trigonometric functions, and in Chapter 9 in deriving the law of cosines. In calculus, certain problems use trigonometric relationships, even including some in which these functions do not appear in the initial problem or final answer. Also, they are useful in a number of technical applications in areas such as electronics, optics, solar energy, and robotics.

Later in the chapter, we see that various trigonometric relationships are used in solving equations with trigonometric functions. Also, we develop the concept of the inverse trigonometric functions that were introduced in Chapter 4.

In Section 20.2, we show how trigonometric relationships are used in analyzing the voltage produced by a three-phase generator, the most widely used type of polyphase generator of alternating current.

20.1 Fundamental Trigonometric Identities

Trigonometric Identities •
Basic Identities •
Proving Trigonometric Identities

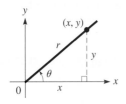

Fig. 20.1

From Chapters 4 and 8, from the definitions, recall that $\sin\theta = y/r$ and $\csc\theta = r/y$ (see Fig. 20.1). Since $y/r = 1/(r/y)$, we see that $\sin\theta = 1/\csc\theta$. These definitions hold for *any* angle, which means that $\sin\theta = 1/\csc\theta$ is true for *any* angle. *This type of relation, which is true for any value of the variable, is called an* **identity.** Of course, values where division by zero would be indicated are excluded.

In this section, we develop several important identities among the trigonometric functions. We also show how the basic identities are used to verify other identities.

From the definitions, we have

$$\sin\theta\,\csc\theta = \frac{y}{r}\times\frac{r}{y} = 1 \quad \text{or} \quad \sin\theta = \frac{1}{\csc\theta} \quad \text{or} \quad \csc\theta = \frac{1}{\sin\theta}$$

$$\cos\theta\,\sec\theta = \frac{x}{r}\times\frac{r}{x} = 1 \quad \text{or} \quad \cos\theta = \frac{1}{\sec\theta} \quad \text{or} \quad \sec\theta = \frac{1}{\cos\theta}$$

$$\tan\theta\,\cot\theta = \frac{y}{x}\times\frac{x}{y} = 1 \quad \text{or} \quad \tan\theta = \frac{1}{\cot\theta} \quad \text{or} \quad \cot\theta = \frac{1}{\tan\theta}$$

$$\frac{\sin\theta}{\cos\theta} = \frac{y/r}{x/r} = \frac{y}{x} = \tan\theta \qquad \frac{\cos\theta}{\sin\theta} = \frac{x/r}{y/r} = \frac{x}{y} = \cot\theta$$

Also, from the definitions and the Pythagorean theorem in the form of $x^2 + y^2 = r^2$, we arrive at the following identities.

By dividing the Pythagorean relation through by r^2, we have

$$\left(\frac{x}{r}\right)^2 + \left(\frac{y}{r}\right)^2 = 1 \quad \text{which leads us to} \quad \cos^2\theta + \sin^2\theta = 1$$

By dividing the Pythagorean relation by x^2, we have

$$1 + \left(\frac{y}{x}\right)^2 = \left(\frac{r}{x}\right)^2 \quad \text{which leads us to} \quad 1 + \tan^2\theta = \sec^2\theta$$

By dividing the Pythagorean relation by y^2, we have

$$\left(\frac{x}{y}\right)^2 + 1 = \left(\frac{r}{y}\right)^2 \quad \text{which leads us to} \quad \cot^2\theta + 1 = \csc^2\theta$$

Basic Identities   The term $\cos^2\theta$ is the common way of writing $(\cos\theta)^2$, and it means to square the value of the cosine of the angle. Obviously, the same holds true for the other functions. Summarizing these results, we have the following important identities.

$$\sin\theta = \frac{1}{\csc\theta} \quad \textbf{(20.1)} \qquad \tan\theta = \frac{\sin\theta}{\cos\theta} \quad \textbf{(20.4)} \qquad \sin^2\theta + \cos^2\theta = 1 \quad \textbf{(20.6)}$$

$$\cos\theta = \frac{1}{\sec\theta} \quad \textbf{(20.2)} \qquad \cot\theta = \frac{\cos\theta}{\sin\theta} \quad \textbf{(20.5)} \qquad 1 + \tan^2\theta = \sec^2\theta \quad \textbf{(20.7)}$$

$$\tan\theta = \frac{1}{\cot\theta} \quad \textbf{(20.3)} \qquad\qquad\qquad\qquad\qquad\qquad 1 + \cot^2\theta = \csc^2\theta \quad \textbf{(20.8)}$$

In using the basic identities, θ may stand for any angle or number or expression representing an angle or a number.

EXAMPLE 1 Using basic identities

(a) $\sin ax = \dfrac{1}{\csc ax}$ using Eq. (20.1)

(b) $\tan 157° = \dfrac{\sin 157°}{\cos 157°}$ using Eq. (20.4)

(c) $\sin^2 \frac{\pi}{4} + \cos^2 \frac{\pi}{4} = 1$ using Eq. (20.6) ∎

EXAMPLE 2 Checking identities with numerical values

Let us check the last two illustrations of Example 1 for the given values of θ.

(a) Using a calculator, we find that

$$\frac{\sin 157°}{\cos 157°} = \frac{0.3907311285}{-0.9205048535} = -0.4244748162 \text{ and } \tan 157° = -0.4244748162$$

We see that $\sin 157°/\cos 157° = \tan 157°$.

(b) Checking Example 1(c), refer to Example 2 on page 116, or to Fig. 20.2.

$$\sin \frac{\pi}{4} = \sin 45° = \frac{1}{\sqrt{2}} = \frac{\sqrt{2}}{2} \text{ and } \cos \frac{\pi}{4} = \cos 45° = \frac{\sqrt{2}}{2}$$

$$\sin^2 \frac{\pi}{4} + \cos^2 \frac{\pi}{4} = \left(\frac{\sqrt{2}}{2}\right)^2 + \left(\frac{\sqrt{2}}{2}\right)^2 = \frac{1}{2} + \frac{1}{2} = 1$$

We see that this checks with Eq. (20.6) for these values. ∎

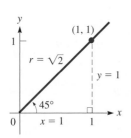

Fig. 20.2

EXAMPLE 3 Simplifying basic expressions

(a) Multiply and simplify the expression $\sin \theta \tan \theta (\csc \theta + \cot \theta)$.

$$\sin \theta \tan \theta (\csc \theta + \cot \theta) = \sin \theta \tan \theta \csc \theta + \sin \theta \tan \theta \cot \theta$$

$$= \left(\frac{1}{\csc \theta}\right) \tan \theta \csc \theta + \sin \theta \left(\frac{1}{\cot \theta}\right) \cot \theta$$

$$= \tan \theta + \sin \theta$$

using Eqs. (20.1) and (20.4)

Practice Exercise

1. Multiply and simplify:
$\cos x(\sec x + \tan x)$

(b) Factor and simplify the expression $\tan^3 \alpha + \tan \alpha$.

$$\tan^3 \alpha + \tan \alpha = \tan \alpha(\tan^2 \alpha + 1)$$

$$= \tan \alpha \sec^2 \alpha$$

using Eq. (20.7)

We see that when algebraic operations are performed on trigonometric terms, they are handled in just the same way as with algebraic terms. This is true for the basic operations of addition, subtraction, multiplication and division, as well as other operations used in simplifying expressions, such as factoring or taking roots. ∎

PROVING TRIGONOMETRIC IDENTITIES

A great many identities exist among the trigonometric functions. We are going to use the basic identities that have been developed in Eqs. (20.1) through (20.8), along with a few additional ones developed in later sections to prove the validity of still other identities.

CAUTION ▶

The ability to prove trigonometric identities depends to a large extent on being very familiar with the basic identities so that you can *recognize them in somewhat different forms.*

If you do not learn these basic identities and learn them well, you will have difficulty in following the examples and doing the exercises. The more readily you recognize these forms, the more easily you will be able to prove such identities.

In proving identities, look for combinations that appear in, or are very similar to, those in the basic identities. This is illustrated in the following examples.

EXAMPLE 4 Proving trig identity

In proving the identity

$$\sin x = \frac{\cos x}{\cot x}$$

we know that $\cot x = \frac{\cos x}{\sin x}$. Since $\sin x$ appears on the left, substituting for $\cot x$ on the right will eliminate $\cot x$ and introduce $\sin x$. This should help us proceed in proving the identity. Thus,

$$\sin x = \frac{\cos x}{\cot x} = \frac{\cos x}{\dfrac{\cos x}{\sin x}} = \frac{\cos x}{1} \times \frac{\sin x}{\cos x}$$

Eq. (20.5)   invert

$$= \sin x \quad \text{cancel } \cos x \text{ factors}$$

By showing that the right side may be changed exactly to $\sin x$, the expression on the left side, we have proved the identity. ∎

Practice Exercise

2. Prove the identity $\dfrac{\cos x \csc x}{\cot^2 x} = \tan x$.

■ As we point out here, performing the *algebraic operations* carefully and correctly is very important when working with trigonometric expressions. Operations such as substituting, factoring, and simplifying fractions are frequently used.

Some important points should be made in relation to the proof of the identity of Example 4. We must recognize what basic identities may be useful. The proof of an identity requires the use of the basic algebraic operations, and these must be done carefully and correctly. Although in Example 4, we changed the right side to the form on the left, we could have changed the left to the form on the right. From this and the fact that various substitutions are possible, we see that a variety of procedures can be used to prove any given identity.

EXAMPLE 5 Trig identity – two different solutions

Prove that $\tan \theta \csc \theta = \sec \theta$.

In proving this identity, we know that $\tan \theta = \dfrac{\sin \theta}{\cos \theta}$ and also that $\dfrac{1}{\cos \theta} = \sec \theta$. Thus, by substituting for $\tan \theta$, we introduce $\cos \theta$ in the denominator, which is equivalent to introducing $\sec \theta$ in the numerator. Therefore, changing only the left side, we have

$$\tan \theta \csc \theta = \frac{\sin \theta}{\cos \theta} \csc \theta = \frac{\sin \theta}{\cos \theta} \frac{1}{\sin \theta}$$

Eq. (20.4)   Eq. (20.1)

$$= \frac{1}{\cos \theta} \quad \text{cancel } \sin \theta \text{ factors}$$

$$= \sec \theta \quad \text{using Eq. (20.2)}$$

Having changed the left side into the form on the right side, we have proven the identity.

NOTE ▸ Many variations of the preceding steps are possible. Also, we could have changed the right side to obtain the form on the left. For example,

$$\tan \theta \csc \theta = \sec \theta = \frac{1}{\cos \theta} \quad \text{using Eq. (20.2)}$$

$$= \frac{\sin \theta}{\cos \theta \sin \theta} = \frac{\sin \theta}{\cos \theta} \frac{1}{\sin \theta} \quad \text{multiply numerator and denominator by } \sin \theta \text{ and rewrite}$$

$$= \tan \theta \csc \theta \quad \text{using Eqs. (20.4) and (20.1)} \quad ∎$$

In proving the identities in Examples 4 and 5, we showed that the expression on one side of the equal sign can be shanged into the expressiuon on the other side. Although not entirely necessary, *we will restrict the method of proof to changing only one side into the same form as the other side.* In this way, we know the form we are to obtain, and by looking ahead, we are better able to make the proper changes.

CAUTION ▶ There is no set procedure for working with identities. The most important factors are to (1) *recognize the proper forms,* (2) *try to identify what effect a change may have* before performing it, and (3) *perform all steps correctly.* Normally, *it is easier to change the form of the more complicated side* to the same form as the less complicated side. If the forms are about the same, a first close look often suggests possible steps to use.

EXAMPLE 6 Proving trig identity

Prove the identity $\dfrac{\sec^2 y}{\cot y} - \tan^3 y = \tan y$.

Here, we simplify the left side. We can remove $\cot y$ from the denominator, since $\cot y = 1/\tan y$. Also, the presence of $\sec^2 y$ suggests the use of Eq. (20.7). Therefore, we have

$$\frac{\sec^2 y}{\cot y} - \tan^3 y = \frac{\sec^2 y}{\dfrac{1}{\tan y}} - \tan^3 y = \sec^2 y \tan y - \tan^3 y$$

$$= \tan y (\sec^2 y - \tan^2 y) = \tan y(1)$$

$$= \tan y$$

Here, we have used Eq. (20.7) in the form $\sec^2 y - \tan^2 y = 1$. ∎

EXAMPLE 7 Proving trig identity

Prove the identity $\dfrac{1 - \sin x}{\sin x \cot x} = \dfrac{\cos x}{1 + \sin x}$.

The combination $1 - \sin x$ also suggests $1 - \sin^2 x$, since multiplying $(1 - \sin x)$ by $(1 + \sin x)$ gives $1 - \sin^2 x$, which can then be replaced by $\cos^2 x$. Thus, changing only the left side, we have

$$\frac{1 - \sin x}{\sin x \cot x} = \frac{(1 - \sin x)(1 + \sin x)}{\sin x \cot x (1 + \sin x)} \qquad \text{multiply numerator and denominator by } 1 + \sin x$$

$$= \frac{1 - \sin^2 x}{\sin x \left(\dfrac{\cos x}{\sin x}\right)(1 + \sin x)} = \frac{\cos^2 x}{\cos x (1 + \sin x)} \longleftarrow \text{Eq. (20.6)}$$

cancel $\sin x$

$$= \frac{\cos x}{1 + \sin x} \qquad \text{cancel } \cos x$$

∎

EXAMPLE 8 Trig identity – application

In finding the radiation rate of an accelerated electric charge, it is necessary to show that $\sin^3 \theta = \sin \theta - \sin \theta \cos^2 \theta$. Show this by changing the left side.

Since each term on the right has a factor of $\sin \theta$, we see that we can proceed by writing $\sin^3 \theta$ as $\sin \theta (\sin^2 \theta)$. Then the factor $\sin^2 \theta$ and the $\cos^2 \theta$ on the right suggest the use of Eq. (20.6). Thus, we have

$$\sin^3 \theta = \sin \theta (\sin^2 \theta) = \sin \theta (1 - \cos^2 \theta)$$

$$= \sin \theta - \sin \theta \cos^2 \theta \qquad \text{multiplying}$$

Since we substituted for $\sin^2 \theta$, we used Eq. (20.6) in the form $\sin^2 \theta = 1 - \cos^2 \theta$. ∎

EXAMPLE 9   Simplifying trig expression

Simplify the expression $\dfrac{\csc x}{\tan x + \cot x}$.

We proceed with a simplification in a manner similar to proving an identity, although we do not know what the result should be. Following is one procedure for this simplification:

$$\frac{\csc x}{\tan x + \cot x} = \frac{\csc x}{\tan x + \dfrac{1}{\tan x}} = \frac{\csc x}{\dfrac{\tan^2 x + 1}{\tan x}}$$

$$= \frac{\csc x \tan x}{\tan^2 x + 1} = \frac{\csc x \tan x}{\sec^2 x} = \frac{\dfrac{1}{\sin x}\dfrac{\sin x}{\cos x}}{\dfrac{1}{\cos^2 x}}$$

$$= \frac{1}{\sin x}\frac{\sin x}{\cos x}\frac{\cos^2 x}{1} = \cos x \quad\blacksquare$$

A graphing calculator can be used to check an identity or a simplification. This is done by graphing the function on each side of an identity, or the initial expression and the final expression for simplification. If the two graphs are the same, the identity or simplification is probably shown to be correct, although this is not strictly a proof.

EXAMPLE 10   Trig identity – verifying by calculator

Use a graphing calculator to verify the identity of Example 7.

Noting this identity as $\dfrac{1 - \sin x}{\sin x \cot x} = \dfrac{\cos x}{1 + \sin x}$, on a graphing calculator, we let

$$y_1 = (1 - \sin x)/(\sin x/\tan x) \qquad \text{noting that } \cot x = 1/\tan x$$
$$y_2 = \cos x/(1 + \sin x)$$

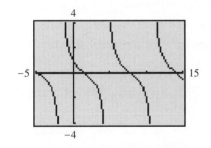

Fig. 20.3

We then graph these two functions as shown in Fig. 20.3. We choose a domain that obviously includes more than one period of each function. Also, after the first curve is plotted, we must watch the screen carefully to see if any new points are plotted for the second curve (use a heavier curve if possible). Since these curves are the same, the identity appears to be verified (although it has not been *proven*).   ∎

EXERCISES 20.1

In Exercises 1 and 2, make the given changes in the indicated examples of this section and then prove the resulting identities.

1. In Example 4, change the right side to $\tan x/\sec x$.

2. In Example 6, change the first term on the left to $\sin y/\cos^3 y$.

In Exercises 3–6, use a calculator to check the indicated basic identities for the given angles.

3. Eq. (20.3) for $\theta = 56°$

4. Eq. (20.5) for $\theta = 280°$

5. Eq. (20.6) for $\theta = 4\pi/3$

6. Eq. (20.7) for $\theta = 5\pi/6$

In Exercises 7–12, multiply and simplify. In Exercises 13–18, factor and simplify.

7. $\cos x(\tan x - \sec x)$

8. $\tan y(\cot y + 3\cos y)$

9. $\cos\theta\cot\theta(\sec\theta - 2\tan\theta)$

10. $(\csc x - 1)(\csc x + 1)$

11. $\tan u(\cot u + \tan u)$

12. $\cos^2 t(1 + \tan^2 t)$

13. $\sin x + \sin x \tan^2 x$

14. $\sec\theta + \sec^3\theta$

15. $\sin^3 t \cos t + \sin t \cos^3 t$

16. $\tan^2 u \sec^2 u - \tan^4 u$

17. $\csc^4 y - 1$

18. $\sin x + \sin x \cot^2 x$

In Exercises 19–42, prove the given identities.

19. $\dfrac{\sin x}{\tan x} = \cos x$

20. $\dfrac{\csc\theta}{\sec\theta} = \cot\theta$

21. $\sin x \sec x = \tan x$

22. $\cot\theta \sec\theta = \csc\theta$

23. $\csc^2 x(1 - \cos^2 x) = 1$

24. $\sec\theta(1 - \sin^2\theta) = \cos\theta$

25. $\sin x(1 + \cot^2 x) = \csc x$

26. $\csc x(\csc x - \sin x) = \cot^2 x$

27. $\cos \theta \cot \theta + \sin \theta = \csc \theta$

28. $\csc x \sec x - \tan x = \cot x$

29. $\cot \theta \sec^2 \theta - \cot \theta = \tan \theta$

30. $\sin y + \sin y \cot^2 y = \csc y$

31. $\tan x + \cot x = \sec x \csc x$

32. $\tan x + \cot x = \tan x \csc^2 x$

33. $\cos^2 x - \sin^2 x = 1 - 2\sin^2 x$

34. $\dfrac{1 + \cos x}{\sin x} = \dfrac{\sin x}{1 - \cos x}$

35. $\dfrac{\sin \theta}{\csc \theta} + \dfrac{\cos \theta}{\sec \theta} = 1$

36. $\dfrac{\sec \theta}{\cos \theta} - \dfrac{\tan \theta}{\cot \theta} = 1$

37. $2\sin^4 x - 3\sin^2 x + 1 = \cos^2 x(1 - 2\sin^2 x)$

38. $\dfrac{\sin^2 \theta + 2\cos \theta - 1}{\sin^2 \theta + 3\cos \theta - 3} = \dfrac{1}{1 - \sec \theta}$

39. $\dfrac{1}{2}\sin \pi t \left(\dfrac{\sin \pi t}{1 - \cos \pi t} + \dfrac{1 - \cos \pi t}{\sin \pi t} \right) = 1$

40. $\dfrac{\cot \omega t}{\sec \omega t - \tan \omega t} - \dfrac{\cos \omega t}{\sec \omega t + \tan \omega t} = \sin \omega t + \csc \omega t$

41. $1 + \sin^2 x + \sin^4 x + \cdots = \sec^2 x$

42. $1 - \tan^2 x + \tan^4 x - \cdots = \cos^2 x \quad \left(-\dfrac{\pi}{4} < x < \dfrac{\pi}{4} \right)$

In Exercises 43–50, simplify the given expressions. The result will be one of $\sin x$, $\cos x$, $\tan x$, $\cot x$, $\sec x$, or $\csc x$.

43. $\dfrac{\tan x \csc^2 x}{1 + \tan^2 x}$

44. $\dfrac{\cos x - \cos^3 x}{\sin x - \sin^3 x}$

45. $\cot x(\sec x - \cos x)$

46. $\sin x(\tan x + \cot x)$

47. $\dfrac{\tan x + \cot x}{\csc x}$

48. $\dfrac{1 + \tan x}{\sin x} - \sec x$

49. $\dfrac{\cos x + \sin x}{1 + \tan x}$

50. $\dfrac{\sec x - \cos x}{\tan x}$

In Exercises 51–54, use a graphing calculator to verify the given identities by comparing the graphs of each side.

51. $\sin x(\csc x - \sin x) = \cos^2 x$

52. $\cos y(\sec y - \cos y) = \sin^2 y$

53. $\dfrac{\sec x + \csc x}{1 + \tan x} = \csc x$

54. $\dfrac{\cot x + 1}{\cot x} = 1 + \tan x$

In Exercises 55–58, use a graphing calculator to determine whether the given equations are identities.

55. $\sec \theta \tan \theta \csc \theta = \tan^2 \theta + 1$

56. $\sin x \cos x \tan x = \cos^2 x - 1$

57. $\dfrac{2\cos^2 x - 1}{\sin x \cos x} = \tan x - \cot x$

58. $\cos^3 x \csc^3 x \tan^3 x = \csc^2 x - \cot^2 x$

In Exercises 59–62, solve the given problems involving trigonometric identities.

59. When designing a solar-energy collector, it is necessary to account for the latitude and longitude of the location, the angle of the sun, and the angle of the collector. In doing this, the equation $\cos \theta = \cos A \cos B \cos C + \sin A \sin B$ is used. If $\theta = 90°$, show that $\cos C = -\tan A \tan B$.

60. The path of a point on the circumference of a circle, such as a point on the rim of a bicycle wheel as it rolls along, traces out a curve called a *cycloid*. To find the distance through which the point moves, it is necessary to simplify the expression $(1 - \cos \theta)^2 + \sin^2 \theta$. Perform this simplification.

61. Show that the length l of the straight brace shown in Fig. 20.4 can be found from the equation
$$l = \dfrac{a(1 + \tan \theta)}{\sin \theta}$$

Fig. 20.4

62. In determining the path of least time between two points under certain conditions, it is necessary to show that
$$\sqrt{\dfrac{1 + \cos \theta}{1 - \cos \theta}}\, \sin \theta = 1 + \cos \theta$$
Show this by transforming the left-hand side.

In Exercises 63–68, solve the given problems.

(W) 63. Explain how to transform $\sin \theta \tan \theta + \cos \theta$ into $\sec \theta$.

(W) 64. Explain how to transform $\tan^2 \theta \cos^2 \theta + \cot^2 \theta \sin^2 \theta$ into 1.

65. Show that $\sin^2 x(1 - \sec^2 x) + \cos^2 x(1 + \sec^4 x)$ has a constant value.

66. Show that $\cot y \csc y \sec y - \csc y \cos y \cot y$ has a constant value.

67. Prove that $\sec^2 \theta + \csc^2 \theta = \sec^2 \theta \csc^2 \theta$ by expressing each function in terms of its x, y, and r definition.

68. Prove that $\dfrac{\csc \theta}{\tan \theta + \cot \theta} = \cos \theta$ by expressing each function in terms of its x, y, and r definition.

In Exercises 69–72, use the given substitutions to show that the given equations are valid. In each, $0 < \theta < \pi/2$.

69. If $x = \cos \theta$, show that $\sqrt{1 - x^2} = \sin \theta$.

70. If $x = 3\sin \theta$, show that $\sqrt{9 - x^2} = 3\cos \theta$.

71. If $x = 2\tan \theta$, show that $\sqrt{4 + x^2} = 2\sec \theta$.

72. If $x = 4\sec \theta$, show that $\sqrt{x^2 - 16} = 4\tan \theta$.

Answers to Practice Exercises

1. $1 + \sin x$ 2. $\tan x$

20.2 The Sum and Difference Formulas

Formulas for $\sin(\alpha + \beta)$ and $\cos(\alpha + \beta)$ •
Formulas for $\sin(\alpha - \beta)$ and $\cos(\alpha - \beta)$ •
Formulas for $\tan(\alpha + \beta)$ and $\tan(\alpha - \beta)$

There are other important relations among the trigonometric functions. The most important and useful relations are those that involve twice an angle and half an angle. To obtain these relations, in this section, we derive the expressions for the sine and cosine of the sum and difference of two angles. These expressions will lead directly to the desired relations of double and half angles that we will derive in the following sections.

Equation (12.13) gives the polar (or trigonometric) form of the product of two complex numbers. We can use this formula to derive the expressions for the sine and cosine of the sum and difference of two angles.

Using Eq. (12.13) to find the product of the complex numbers $\cos \alpha + j \sin \alpha$ and $\cos \beta + j \sin \beta$, which are represented in Fig. 20.5, we have

$$(\cos \alpha + j \sin \alpha)(\cos \beta + j \sin \beta) = \cos(\alpha + \beta) + j \sin(\alpha + \beta)$$

Expanding the left side, and then switching sides, we have

$$\cos(\alpha + \beta) + j \sin(\alpha + \beta) = (\cos \alpha \cos \beta - \sin \alpha \sin \beta) + j(\sin \alpha \cos \beta + \cos \alpha \sin \beta)$$

Since two complex numbers are equal if their real parts are equal and their imaginary parts are equal, we have the following formulas:

$$\sin(\alpha + \beta) = \sin \alpha \cos \beta + \cos \alpha \sin \beta \qquad \textbf{(20.9)}$$
$$\cos(\alpha + \beta) = \cos \alpha \cos \beta - \sin \alpha \sin \beta \qquad \textbf{(20.10)}$$

■ For reference, Eq. (12.13) is
$$r_1(\cos \theta_1 + j \sin \theta_1)r_2(\cos \theta_2 + j \sin \theta_2)$$
$$= r_1 r_2[\cos(\theta_1 + \theta_2) + j \sin(\theta_1 + \theta_2)]$$

Imag.

$\cos \alpha + j \sin \alpha$ $\cos \beta + j \sin \beta$

Fig. 20.5

EXAMPLE 1 Verifying $\sin(\alpha + \beta)$ formula

Verify that $\sin 90° = 1$, by finding $\sin(60° + 30°)$.

$$\sin 90° = \sin(60° + 30°) = \sin 60° \cos 30° + \cos 60° \sin 30° \qquad \text{using Eq. (20.9)}$$
$$= \frac{\sqrt{3}}{2} \times \frac{\sqrt{3}}{2} + \frac{1}{2} \times \frac{1}{2} \qquad \text{for values, see Section 4.3}$$
$$= \frac{3}{4} + \frac{1}{4} = 1$$

CAUTION ▶   It should be obvious from this example that $\sin(\alpha + \beta)$ *is* not *equal to* $\sin \alpha + \sin \beta$, which is something that many students simply assume before they become familiar with the formulas and ideas of this section. If we used such a formula, we would get $\sin 90° = \frac{1}{2}\sqrt{3} + \frac{1}{2} = 1.366$ for the combination $(60° + 30°)$. This is not possible, since the values of the sine never exceed 1 in value. Also, if we used the combination $(45° + 45°)$, we would get 1.414, a different value for the same number, $\sin 90°$. ■

EXAMPLE 2 Using $\cos(\alpha + \beta)$ with numerical values

Given that $\sin \alpha = \frac{5}{13}$ (α in the first quadrant) and $\sin \beta = -\frac{3}{5}$ (for β in the third quadrant), find $\cos(\alpha + \beta)$.

Since $\sin \alpha = \frac{5}{13}$ for α in the first quadrant, from Fig. 20.6, we have $\cos \alpha = \frac{12}{13}$. Also, since $\sin \beta = -\frac{3}{5}$ for β in the third quadrant, from Fig. 20.6, we also have $\cos \beta = -\frac{4}{5}$.

Then, by using Eq. (20.10), we have

$$\cos(\alpha + \beta) = \cos \alpha \cos \beta - \sin \alpha \sin \beta$$
$$= \frac{12}{13}\left(-\frac{4}{5}\right) - \frac{5}{13}\left(-\frac{3}{5}\right)$$
$$= -\frac{48}{65} + \frac{15}{65} = -\frac{33}{65}$$

■

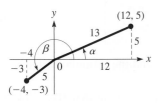

Fig. 20.6

From Eqs. (20.9) and (20.10), we can easily find expressions for $\sin(\alpha - \beta)$ and $\cos(\alpha - \beta)$. This is done by finding $\sin[\alpha + (-\beta)]$ and $\cos[\alpha + (-\beta)]$. Thus, we have

$$\sin(\alpha - \beta) = \sin[\alpha + (-\beta)] = \sin\alpha\cos(-\beta) + \cos\alpha\sin(-\beta)$$

Since $\cos(-\beta) = \cos\beta$ and $\sin(-\beta) = -\sin\beta$ (see Eq. (8.7) on page 239), we have

$$\boxed{\sin(\alpha - \beta) = \sin\alpha\cos\beta - \cos\alpha\sin\beta} \qquad \textbf{(20.11)}$$

In the same manner, we find that

$$\boxed{\cos(\alpha - \beta) = \cos\alpha\cos\beta + \sin\alpha\sin\beta} \qquad \textbf{(20.12)}$$

EXAMPLE 3 Using $\cos(\alpha - \beta)$ formula

Find $\cos 15°$ from $\cos(45° - 30°)$.

$$\cos 15° = \cos(45° - 30°) = \cos 45°\cos 30° + \sin 45°\sin 30° \qquad \text{using Eq. (20.12)}$$

$$= \frac{\sqrt{2}}{2} \times \frac{\sqrt{3}}{2} + \frac{\sqrt{2}}{2} \times \frac{1}{2} = \frac{\sqrt{6} + \sqrt{2}}{4} \qquad \text{(exact)}$$

$$= 0.9659 \qquad \blacksquare$$

EXAMPLE 4 Sin$(\alpha - \beta)$ formula – application

In analyzing the motion of an object oscillating up and down at the end of a spring, the expression $\sin(\omega t + \alpha)\cos\alpha - \cos(\omega t + \alpha)\sin\alpha$ occurs. Simplify this expression.

If we let $x = \omega t + \alpha$, the expression becomes $\sin x\cos\alpha - \cos x\sin\alpha$, which is the form for $\sin(x - \alpha)$. Therefore,

$$\sin(\omega t + \alpha)\cos\alpha - \cos(\omega t + \alpha)\sin\alpha = \sin x\cos\alpha - \cos x\sin\alpha = \sin(x - \alpha)$$

$$= \sin(\omega t + \alpha - \alpha) = \sin\omega t \qquad \blacksquare$$

EXAMPLE 5 Using $\cos(\alpha + \beta)$ formula

Evaluate $\cos 23°\cos 67° - \sin 23°\sin 67°$.

We note that this expression fits the form of the right side of Eq. (20.10), so

$$\cos 23°\cos 67° - \sin 23°\sin 67° = \cos(23° + 67°)$$

$$= \cos 90°$$

$$= 0$$

Practice Exercise

1. Evaluate:
   $\sin 115°\cos 25° - \cos 115°\sin 25°$

NOTE ▶ Again, we are able to evaluate this expression by *recognizing the form* of the given expression. Evaluation by a calculator will verify the result. $\qquad \blacksquare$

By dividing the right side of Eq. (20.9) by that of Eq. (20.10), we can determine expressions for $\tan(\alpha + \beta)$, and by dividing the right side of Eq. (20.11) by that of Eq. (20.12), we can determine an expression for $\tan(\alpha - \beta)$. The derivation of these formulas is Exercise 33 of this section. These formulas can be written together, as

$$\boxed{\tan(\alpha \pm \beta) = \frac{\tan\alpha \pm \tan\beta}{1 \mp \tan\alpha\tan\beta}} \qquad \textbf{(20.13)}$$

The formula for $\tan(\alpha + \beta)$ uses the upper signs, and the formula for $\tan(\alpha - \beta)$ uses the lower signs.

Certain trigonometric identities can be proven by the formulas derived in this section. The following examples illustrate this use of these formulas.

EXAMPLE 6 Proving trig identity with $\tan(\alpha \pm \beta)$

Show that $\tan(\alpha + \beta) \tan(\alpha - \beta) = \dfrac{\tan^2 \alpha - \tan^2 \beta}{1 - \tan^2 \alpha \tan^2 \beta}$.

Using both of Eqs. (20.13), we have

$$\tan(\alpha + \beta) \tan(\alpha - \beta) = \left(\frac{\tan \alpha + \tan \beta}{1 - \tan \alpha \tan \beta}\right)\left(\frac{\tan \alpha - \tan \beta}{1 + \tan \alpha \tan \beta}\right)$$

$$= \frac{\tan^2 \alpha - \tan^2 \beta}{1 - \tan^2 \alpha \tan^2 \beta} \qquad \blacksquare$$

EXAMPLE 7 $\sin(\alpha + \beta)$ – calculator verification

Prove that $\sin(180° + x) = -\sin x$.

$$\sin(180° + x) = \sin 180° \cos x + \cos 180° \sin x \qquad \text{using Eq. (20.9)}$$

$$= (0) \cos x + (-1) \sin x \qquad \sin 180° = 0, \cos 180° = -1$$

$$= -\sin x$$

Although x may or may not be an acute angle, this agrees with the results for the sine of a third-quadrant angle as discussed in Section 8.2. See Fig. 20.7. The calculator check of this identity, with $y_1 = \sin(\pi + x)$ and $y_2 = -\sin x$ is shown in Fig. 20.8. $\blacksquare$

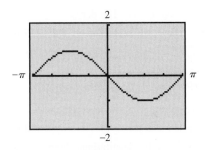

Fig. 20.7

Fig. 20.8

Practice Exercise

2. Simplify: $\tan(180° + x)$

EXAMPLE 8 Trig identity using $\sin(\alpha - \beta)$

Simplify the expression $\dfrac{\sin(\alpha - \beta)}{\sin \alpha \sin \beta}$.

$$\frac{\sin(\alpha - \beta)}{\sin \alpha \sin \beta} = \frac{\sin \alpha \cos \beta - \cos \alpha \sin \beta}{\sin \alpha \sin \beta} \qquad \text{using Eq. (20.11)}$$

$$= \frac{\sin \alpha \cos \beta}{\sin \alpha \sin \beta} - \frac{\cos \alpha \sin \beta}{\sin \alpha \sin \beta}$$

$$= \frac{\cos \beta}{\sin \beta} - \frac{\cos \alpha}{\sin \alpha}$$

$$= \cot \beta - \cot \alpha \qquad \text{using Eq. (20.5)} \qquad \blacksquare$$

EXAMPLE 9 Application using $\sin(\alpha - \beta)$

Alternating electric current is produced essentially by a coil of wire rotating in a magnetic field, and this is the basis for designing generators of alternating current. A three-phase generator uses three coils of wire and thereby produces three electric currents at the same time. This is the most widely used type of *polyphase generator* as mentioned in the chapter introduction on page 526.

■ See the chapter introduction.

The voltages induced in a three-phase generator can be represented as

$$E_1 = E_0 \sin \omega t \qquad E_2 = E_0 \sin\left(\omega t - \tfrac{2\pi}{3}\right) \qquad E_3 = E_0 \sin\left(\omega t - \tfrac{4\pi}{3}\right)$$

where E_0 is the maximum voltage and ω is the angular velocity of rotation. Show that the sum of these voltages at any time t is zero.

Setting up the sum $E_1 + E_2 + E_3$ and using Eq. (20.11), we have

$$E_1 + E_2 + E_3 = E_0\left[\sin \omega t + \sin\left(\omega t - \tfrac{2\pi}{3}\right) + \sin\left(\omega t - \tfrac{4\pi}{3}\right)\right]$$

$$= E_0\left(\sin \omega t + \sin \omega t \cos \tfrac{2\pi}{3} - \cos \omega t \sin \tfrac{2\pi}{3} + \sin \omega t \cos \tfrac{4\pi}{3} - \cos \omega t \sin \tfrac{4\pi}{3}\right)$$

$$= E_0\left[\sin \omega t + (\sin \omega t)\left(-\tfrac{1}{2}\right) - (\cos \omega t)\left(\tfrac{1}{2}\sqrt{3}\right) + (\sin \omega t)\left(-\tfrac{1}{2}\right) - (\cos \omega t)\left(-\tfrac{1}{2}\sqrt{3}\right)\right]$$

$$= E_0\left[(\sin \omega t)\left(1 - \tfrac{1}{2} - \tfrac{1}{2}\right) + (\cos \omega t)\left(\tfrac{1}{2}\sqrt{3} - \tfrac{1}{2}\sqrt{3}\right)\right] = 0 \qquad \blacksquare$$

EXERCISES 20.2

In Exercises 1 and 2, make the given changes in the indicated examples of this section and then solve the given problems.

1. In Example 2, change $\frac{5}{13}$ to $\frac{12}{13}$ and then find the value of $\cos(\alpha + \beta)$.

2. In Example 7, change $180° + x$ to $180° - x$ and then determine what other changes result.

In Exercises 3–6, determine the values of the given functions as indicated.

3. Find $\sin 105°$ by using $105° = 60° + 45°$.

4. Find $\tan 75°$ by using $75° = 30° + 45°$.

5. Find $\cos 15°$ by using $15° = 60° - 45°$.

6. Find $\sin 15°$ by using $15° = 45° - 30°$.

In Exercises 7–10, evaluate the given functions with the following information: $\sin \alpha = 4/5$ (α in first quadrant) and $\cos \beta = -12/13$ (β in second quadrant).

7. $\sin(\alpha + \beta)$ **8.** $\tan(\beta - \alpha)$

9. $\cos(\alpha + \beta)$ **10.** $\sin(\alpha - \beta)$

In Exercises 11–20, simplify the given expressions.

11. $\sin x \cos 2x + \sin 2x \cos x$ **12.** $\sin 3x \cos x - \sin x \cos 3x$

13. $\cos \pi \cos x + \sin \pi \sin x$

14. $\dfrac{\tan(x - y) + \tan x}{1 - \tan(x - y) \tan y}$

15. $\sin(180° - x)$ **16.** $\cos(2\pi - x)$

17. $\tan(x - \pi)$ **18.** $\sin(x + \pi/2)$

19. $\sin 3x \cos(3x - \pi) - \cos 3x \sin(3x - \pi)$

20. $\cos(x + \pi) \cos(x - \pi) + \sin(x + \pi) \sin(x - \pi)$

In Exercises 21–24, evaluate each expression by first changing the form. Verify each by use of a calculator.

21. $\sin 122° \cos 32° - \cos 122° \sin 32°$

22. $\cos 250° \cos 70° + \sin 250° \sin 70°$

23. $\cos \frac{\pi}{5} \cos \frac{3\pi}{10} - \sin \frac{\pi}{5} \sin \frac{3\pi}{10}$ **24.** $\dfrac{\tan 18° + \tan 27°}{1 - \tan 18° \tan 27°}$

In Exercises 25–28, prove the given identities.

25. $\sin(x + y)\sin(x - y) = \sin^2 x - \sin^2 y$

26. $\cos(x + y)\cos(x - y) = \cos^2 x - \sin^2 y$

27. $\cos(\alpha + \beta) + \cos(\alpha - \beta) = 2 \cos \alpha \cos \beta$

(W) 28. $\tan(90° + x) = -\cot x$ (Explain why Eq. (20.13) cannot be used for this, but Eqs. (20.9) and (20.10) can be used.)

In Exercises 29–32, verify each identity by comparing the graph of the left side with the graph of the right side on a graphing calculator.

29. $\cos(30° + x) = \dfrac{\sqrt{3} \cos x - \sin x}{2}$

30. $\sin(120° - x) = \dfrac{\sqrt{3} \cos x + \sin x}{2}$

31. $\tan\left(\frac{\pi}{4} + x\right) = \dfrac{1 + \tan x}{1 - \tan x}$ **32.** $\cos\left(\frac{\pi}{2} - x\right) = \sin x$

In Exercises 33–36, derive the given equations as indicated. Equations (20.14)–(20.16) are known as the product formulas.

33. By dividing the right side of Eq. (20.9) by that of Eq. (20.10), and dividing the right side of Eq. (20.11) by that of Eq. (20.12), derive Eq. (20.13).

$$\tan(\alpha \pm \beta) = \dfrac{\tan \alpha \pm \tan \beta}{1 \mp \tan \alpha \tan \beta} \qquad (20.13)$$

(*Hint:* Divide numerator and denominator by $\cos \alpha \cos \beta$.)

34. By adding Eqs. (20.9) and (20.11), derive the equation

$$\sin \alpha \cos \beta = \tfrac{1}{2}[\sin(\alpha + \beta) + \sin(\alpha - \beta)] \qquad (20.14)$$

35. By adding Eqs. (20.10) and (20.12), derive the equation

$$\cos \alpha \cos \beta = \tfrac{1}{2}[\cos(\alpha + \beta) + \cos(\alpha - \beta)] \qquad (20.15)$$

36. By subtracting Eq. (20.10) from Eq. (20.12), derive

$$\sin \alpha \sin \beta = \tfrac{1}{2}[\cos(\alpha - \beta) - \cos(\alpha + \beta)] \qquad (20.16)$$

In Exercises 37–40, derive the given equations by letting $\alpha + \beta = x$ and $\alpha - \beta = y$, which leads to $\alpha = \frac{1}{2}(x + y)$ and $\beta = \frac{1}{2}(x - y)$. The resulting equations are known as the factor formulas.

37. Use Eq. (20.14) and the substitutions above to derive the equation

$$\sin x + \sin y = 2 \sin \tfrac{1}{2}(x + y)\cos \tfrac{1}{2}(x - y) \qquad (20.17)$$

38. Use Eqs. (20.9) and (20.11) and the substitutions above to derive the equation

$$\sin x - \sin y = 2 \sin \tfrac{1}{2}(x - y)\cos \tfrac{1}{2}(x + y) \qquad (20.18)$$

39. Use Eq. (20.15) and the substitutions above to derive the equation

$$\cos x + \cos y = 2 \cos \tfrac{1}{2}(x + y)\cos \tfrac{1}{2}(x - y) \qquad (20.19)$$

40. Use Eq. (20.16) and the substitutions above to derive the equation

$$\cos x - \cos y = -2 \sin \tfrac{1}{2}(x + y)\sin \tfrac{1}{2}(x - y) \qquad (20.20)$$

In Exercises 41–52, solve the given problems.

41. Show that $\frac{\sin 2x}{\sin x} = 2 \cos x$. (*Hint:* $\sin 2x = \sin(x + x)$.)

42. Using graphs displayed on a graphing calculator, verify the identity in Exercise 41.

(W) 43. Explain how the exact value of $\sin 75°$ can be found using either Eq. (20.9) or Eq. (20.11).

44. Express $\cos(A + B + C)$ in terms of $\sin A$, $\sin B$, $\sin C$, $\cos A$, $\cos B$, and $\cos C$.

45. The design of a certain three-phase alternating-current generator uses the fact that the sum of the currents $I \cos(\theta + 30°)$, $I \cos(\theta + 150°)$, and $I \cos(\theta + 270°)$ is zero. Verify this.

46. Use a calculator to verify the result for the three-phase generator shown in Example 9. Let $y_1 = \sin x$, $y_2 = \sin(x - 2\pi/3)$, $y_3 = \sin(x - 4\pi/3)$, and $y_4 = y_1 + y_2 + y_3$. Use a heavier curve for y_4 so as to show the result clearly.

47. For voltages $V_1 = 20 \sin 120\pi t$ and $V_2 = 20 \cos 120\pi t$, show that $V = V_1 + V_2 = 20\sqrt{2} \sin(120\pi t + \pi/4)$. Use a calculator to verify this result.

48. The displacements y_1 and y_2 of two waves traveling through the same medium are given by $y_1 = A \sin 2\pi(t/T - x/\lambda)$ and $y_2 = A \sin 2\pi(t/T + x/\lambda)$. Find an expression for the displacement $y_1 + y_2$ of the combination of the waves.

49. An alternating electric current i is given by the equation $i = i_0 \sin(\omega t + \alpha)$. Show that this can be written as $i = i_1 \sin \omega t + i_2 \cos \omega t$, where $i_1 = i_0 \cos \alpha$ and $i_2 = i_0 \sin \alpha$.

50. A weight **w** is held in equilibrium by forces **F** and **T** as shown in Fig. 20.9. Equations relating w, F, and T are

$$F \cos \theta = T \sin \alpha$$
$$w + F \sin \theta = T \cos \alpha$$

Show that $w = \dfrac{T \cos(\theta + a)}{\cos \theta}$.

Fig. 20.9

51. For the two bevel gears shown in Fig. 20.10, the equation

$$\tan \alpha = \frac{\sin \beta}{R + \cos \beta} \text{ is used.}$$

Here, R is the ratio of gear 1 to gear 2. Show that

$$R = \frac{\sin(\beta - \alpha)}{\sin \alpha}.$$

Fig. 20.10

52. In the analysis of the angles of incidence i and reflection r of a light ray subject to certain conditions, the following expression is found:

$$E_2\left(\frac{\tan r}{\tan i} + 1\right) = E_1\left(\frac{\tan r}{\tan i} - 1\right)$$

Show that $E_2 = E_1 \dfrac{\sin(r - i)}{\sin(r + i)}$.

Answers to Practice Exercises

1. 1   **2.** $\tan x$

20.3 Double-angle Formulas

Formula for $\sin 2\alpha$ •
Formulas for $\cos 2\alpha$ •
Formula for $\tan 2\alpha$

If we let $\beta = \alpha$ in Eqs. (20.9) and (20.10), we can derive the important double-angle formulas. By making this substitution in Eq. (20.9), we have

$$\sin(\alpha + \alpha) = \sin(2\alpha) = \sin \alpha \cos \alpha + \cos \alpha \sin \alpha = 2 \sin \alpha \cos \alpha$$

Using the same substitution in Eq. (20.10), we have

$$\cos(\alpha + \alpha) = \cos \alpha \cos \alpha - \sin \alpha \sin \alpha = \cos^2 \alpha - \sin^2 \alpha$$

Again, using this substitution in the $\tan(\alpha + \beta)$ form of Eq. (20.13), we have

$$\tan(\alpha + \alpha) = \frac{\tan \alpha + \tan \alpha}{1 - \tan \alpha \tan \alpha} = \frac{2 \tan \alpha}{1 - \tan^2 \alpha}$$

Then using the basic identity Eq. (20.6), other forms of the equation for $\cos 2\alpha$ may be derived. Summarizing these forms, we have

$$\sin 2\alpha = 2 \sin \alpha \cos \alpha \qquad (20.21)$$
$$\cos 2\alpha = \cos^2 \alpha - \sin^2 \alpha \qquad (20.22)$$
$$= 2 \cos^2 \alpha - 1 \qquad (20.23)$$
$$= 1 - 2 \sin^2 \alpha \qquad (20.24)$$
$$\tan 2\alpha = \frac{2 \tan \alpha}{1 - \tan^2 \alpha} \qquad (20.25)$$

These double-angle formulas are widely used in applications of trigonometry, especially in calculus. They should be recognized quickly in any of the above forms.

EXAMPLE 1 Using double-angle formulas

(a) If $\alpha = 30°$, we have

$$\cos 60° = \cos 2(30°) = \cos^2 30° - \sin^2 30° = \left(\frac{\sqrt{3}}{2}\right)^2 - \left(\frac{1}{2}\right)^2 = \frac{1}{2} \quad \text{using Eq. (20.22)}$$

(b) If $\alpha = 3x$, we have

$$\sin 6x = \sin 2(3x) = 2 \sin 3x \cos 3x \quad \text{using Eq. (20.21)}$$

(c) If $2\alpha = x$, we may write $\alpha = x/2$, which means that

$$\sin x = \sin 2\left(\frac{x}{2}\right) = 2 \sin \frac{x}{2} \cos \frac{x}{2} \quad \text{using Eq. (20.21)}$$

(d) If $\alpha = \frac{\pi}{6}$, we have

$$\tan \frac{\pi}{3} = \tan 2\left(\frac{\pi}{6}\right) = \frac{2 \tan \frac{\pi}{6}}{1 - \tan^2\left(\frac{\pi}{6}\right)} = \frac{2(\sqrt{3}/3)}{1 - (\sqrt{3}/3)^2} = \sqrt{3} \quad \text{using Eq. (20.25)} \quad \blacksquare$$

EXAMPLE 2 Simplification using $\cos 2\alpha$ formula

Simplify the expression $\cos^2 2x - \sin^2 2x$.

Since this is the difference of the square of the cosine of an angle and the square of the sine of the same angle, it fits the right side of Eq. (20.22). Therefore, letting $\alpha = 2x$, we have

$$\cos^2 2x - \sin^2 2x = \cos 2(2x) = \cos 4x \quad \blacksquare$$

EXAMPLE 3 Using $\sin 2\alpha$ formula – application

To find the area A of a right triangular tract of land, a surveyor may use the formula $A = \frac{1}{4}c^2 \sin 2\theta$, where c is the hypotenuse and θ is *either* of the acute angles. Derive this formula.

In Fig. 20.11, we see that $\sin \theta = a/c$ and $\cos \theta = b/c$, which gives us

$$a = c \sin \theta \quad \text{and} \quad b = c \cos \theta$$

The area is given by $A = \frac{1}{2}ab$, which leads to the solution

$$A = \frac{1}{2}ab = \frac{1}{2}(c \sin \theta)(c \cos \theta)$$

$$= \frac{1}{2}c^2 \sin \theta \cos \theta = \frac{1}{2}c^2\left(\frac{1}{2}\sin 2\theta\right) \quad \text{using Eq. (20.21)}$$

$$= \frac{1}{4}c^2 \sin 2\theta$$

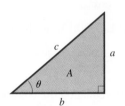

Fig. 20.11

In using Eq. (20.21), we divided both sides by 2 to get $\sin \theta \cos \theta = \frac{1}{2}\sin 2\theta$.

If we had labeled the upper acute angle in Fig. 20.11 as θ, we would have $a = c \cos \theta$ and $b = c \sin \theta$. Using these values in the formula for the area gives the same solution. $\blacksquare$

EXAMPLE 4 Verifying values

(a) Verifying the values of $\sin 90°$, using the functions of $45°$, we have

$$\sin 90° = \sin 2(45°) = 2 \sin 45° \cos 45° = 2\left(\frac{\sqrt{2}}{2}\right)\left(\frac{\sqrt{2}}{2}\right) = 1 \quad \text{using Eq. (20.21)}$$

(b) Using Eq. (20.25), $\tan 142° = \dfrac{2 \tan 71°}{1 - \tan^2 71°}$. Using a calculator to verify this, we have

$$\tan 142° = -0.7812856265 \quad \text{and} \quad \frac{2 \tan 71°}{1 - \tan^2 71°} = -0.7812856265 \quad \blacksquare$$

(a)

(b)

Fig. 20.12

Practice Exercise

1. Simplify: $\dfrac{\sin 2x}{\cos^2 x - \sin^2 x}$

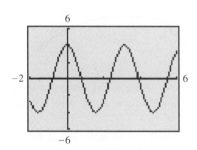

Fig. 20.13

EXAMPLE 5 Evaluation using $\sin 2\alpha$ formula

Knowing that $\cos \alpha = 3/5$ for an angle in the fourth quadrant, we see from Fig. 20.12(a) that $\sin \alpha = -4/5$. Therefore, we have

$$\sin 2\alpha = 2 \sin \alpha \cos \alpha \qquad \text{Eq. (20.21)}$$

$$= 2\left(-\frac{4}{5}\right)\left(\frac{3}{5}\right) = -\frac{24}{25}$$

In Fig. 20.12(b), angle 2α is shown. It is a third-quadrant angle, which verifies the sign of the result. (Since $\cos \alpha = 3/5$, $\alpha = 307°$ and $2\alpha = 614°$, which is a third-quadrant angle.) ∎

EXAMPLE 6 Simplification using $\cos 2\alpha$

Simplify the expression $\dfrac{2}{1 + \cos 2x}$.

$$\frac{2}{1 + \cos 2x} = \frac{2}{1 + (2\cos^2 x - 1)} \qquad \text{using Eq. (20.23)}$$

$$= \frac{2}{2\cos^2 x} = \sec^2 x \qquad \text{using Eq. (20.2)} \quad ∎$$

EXAMPLE 7 Trig identity – calculator verification

Prove the identity $\dfrac{\sin 3x}{\sin x} + \dfrac{\cos 3x}{\cos x} = 4 \cos 2x$.

Since the left side is the more complex side, we change it to the form on the right:

$$\frac{\sin 3x}{\sin x} + \frac{\cos 3x}{\cos x} = \frac{\sin 3x \cos x + \cos 3x \sin x}{\sin x \cos x} \qquad \text{combining fractions}$$

$$= \frac{\sin(3x + x)}{\frac{1}{2}\sin 2x} \qquad \begin{matrix} \longleftarrow \text{using Eq. (20.9)} \\ \longleftarrow \text{using Eq. (20.21)} \end{matrix}$$

$$= \frac{2 \sin 4x}{\sin 2x} = \frac{2(2 \sin 2x \cos 2x)}{\sin 2x} \qquad \text{using Eq.(20.21)}$$

$$= 4 \cos 2x$$

Checking this identity by comparing the graphs of

$$y_1 = (\sin 3x)/(\sin x) + (\cos 3x)/(\cos x) \quad \text{and} \quad y_2 = 4 \cos 2x$$

on a graphing calculator, both curves are shown by the graph in Fig. 20.13. ∎

EXERCISES 20.3

In Exercises 1–4, make the given changes in the indicated examples of this section and then solve the resulting problems.

1. In Example 1(d), change $\frac{\pi}{6}$ to $\frac{\pi}{3}$ and then evaluate $\tan \frac{2\pi}{3}$.

2. In Example 2, change $2x$ to $3x$ and then simplify.

3. In Example 5, change $3/5$ to $4/5$ and then evaluate $\sin 2\alpha$.

4. In Example 6, change the $+$ in the denominator to $-$ and then simplify the expression on the left.

In Exercises 5–8, determine the values of the indicated functions in the given manner.

5. Find $\sin 60°$ by using the functions of $30°$.

6. Find $\sin 120°$ by using the functions of $60°$.

7. Find $\tan 120°$ by using the functions of $60°$.

8. Find $\cos 60°$ by using the functions of $30°$.

In Exercises 9–14, use a calculator to verify the values found by using the double-angle formulas.

9. Find $\sin 258°$ directly and by using functions of $129°$.

10. Find $\tan 84°$ directly and by using functions of $42°$.

11. Find $\cos 96°$ directly and by using functions of $48°$.

12. Find $\cos 276°$ directly and by using functions of $138°$.

13. Find $\tan \frac{2\pi}{5}$ directly and by using functions of $\frac{\pi}{5}$.

14. Find $\sin(0.2\pi)$ directly and by using functions of 0.1π.

In Exercises 15–18, evaluate the indicated functions with the given information.

15. Find $\sin 2x$ if $\cos x = \frac{4}{5}$ (in first quadrant).

16. Find $\cos 2x$ if $\sin x = -\frac{12}{13}$ (in third quadrant).

17. Find $\tan 2x$ if $\sin x = 0.5$ (in second quadrant).

18. Find $\sin 4x$ if $\tan x = 0.6$ (in first quadrant).

In Exercises 19–30, simplify the given expressions.

19. $4 \sin 4x \cos 4x$

20. $4 \sin^2 x \cos^2 x$

21. $1 - 2 \sin^2 4x$

22. $\dfrac{4 \tan 4\theta}{1 - \tan^2 4\theta}$

23. $2 \cos^2 \frac{1}{2}x - 1$

24. $2 \sin \frac{1}{2}x \cos \frac{1}{2}x$

25. $4 \sin^2 2x - 2$

26. $\cos 3x \sin 3x$

27. $\dfrac{\sin 4\theta}{\sin 2\theta}$

28. $\cos^4 u - \sin^4 u$

29. $\dfrac{\sin 3x}{\sin x} - \dfrac{\cos 3x}{\cos x}$

30. $\dfrac{\cos 3x}{\sin x} + \dfrac{\sin 3x}{\cos x}$

In Exercises 31–40, prove the given identities.

31. $\cos^2 \alpha - \sin^2 \alpha = 2 \cos^2 \alpha - 1$

32. $\cos^2 \alpha - \sin^2 \alpha = 1 - 2 \sin^2 \alpha$

33. $\dfrac{\cos x - \tan x \sin x}{\sec x} = \cos 2x$ **34.** $2 + \dfrac{\cos 2\theta}{\sin^2 \theta} = \csc^2 \theta$

35. $\dfrac{\sin 2\theta}{1 + \cos 2\theta} = \tan \theta$ **36.** $\dfrac{2 \tan \alpha}{1 + \tan^2 \alpha} = \sin 2\alpha$

37. $1 - \cos 2\theta = \dfrac{2}{1 + \cot^2 \theta}$

38. $\dfrac{\cos^3 \theta + \sin^3 \theta}{\cos \theta + \sin \theta} = 1 - \dfrac{1}{2} \sin 2\theta$

39. $\ln(1 - \cos 2x) - \ln(1 + \cos 2x) = 2 \ln \tan x$

40. $\log(20 \sin^2 \theta + 10 \cos 2\theta) = 1$

In Exercises 41–44, verify each identity by comparing the graph of the left side with the graph of the right side on a graphing calculator.

41. $\tan 2\theta = \dfrac{2}{\cot \theta - \tan \theta}$ **42.** $\dfrac{1 - \tan^2 x}{\sec^2 x} = \cos 2x$

43. $(\sin x + \cos x)^2 = 1 + \sin 2x$

44. $2 \csc 2x \tan x = \sec^2 x$

In Exercises 45–60, solve the given problems.

45. Express $\sin 3x$ in terms of $\sin x$ only.

46. Express $\cos 3x$ in terms of $\cos x$ only.

47. Express $\cos 4x$ in terms of $\cos x$ only.

48. Express $\sin 4x$ in terms of $\sin x$ and $\cos x$.

49. Show that $\cos 2x + \sin 2x \tan x$ reduces to 1.

50. For an acute angle θ, show that $2 \sin \theta > \sin 2\theta$.

51. Without graphing, determine the amplitude and period of the function $y = 4 \sin x \cos x$. Explain.

52. Without graphing, determine the amplitude and period of the function $y = \cos^2 x - \sin^2 x$.

53. The path of a bouncing ball is given by $y = \sqrt{(\sin x + \cos x)^2}$. Show that this path can also be shown as $y = \sqrt{1 + \sin 2x}$. Use a calculator to show that this can also be shown as $y = |\sin x + \cos x|$.

54. The equation for the trajectory of a missile fired into the air at an angle α with velocity v_0 is $y = x \tan \alpha - \dfrac{g}{2v_0^2 \cos^2 \alpha} x^2$. Here, g is the acceleration due to gravity. On the right of the equal sign, combine terms and simplify.

55. The CN Tower in Toronto is 553 m high, and it has an observation deck at the 335-m level. How far from the top of the CN Tower must a 553-m-high helicopter be in order that the angle subtended at the helicopter by the part of the tower above the deck equals the angle subtended by the part of the tower below the deck?

56. The cross section of a radio-wave reflector is defined by $x = \cos 2\theta$, $y = \sin \theta$. Find the relation between x and y by eliminating θ.

57. To find the horizontal range R of a projectile, the equation $R = vt \cos \alpha$ is used, where α is the angle between the line of fire and the horizontal, v is the initial velocity of the projectile, and t is the time of flight. It can be shown that $t = (2v \sin \alpha)/g$, where g is the acceleration due to gravity. Show that $R = (v^2 \sin 2\alpha)/g$. See Fig. 20.14.

Fig. 20.14 R

58. In analyzing light reflection from a cylinder onto a flat surface, the expression $3 \cos \theta - \cos 3\theta$ arises. Show that this equals $2 \cos \theta \cos 2\theta + 4 \sin \theta \sin 2\theta$.

59. The instantaneous electric power p in an inductor is given by the equation $p = vi \sin \omega t \sin(\omega t - \pi/2)$. Show that this equation can be written as $p = -\frac{1}{2} vi \sin 2\omega t$.

60. In the study of the stress at a point in a bar, the equation $s = a \cos^2 \theta + b \sin^2 \theta - 2t \sin \theta \cos \theta$ arises. Show that this equation can be written as $s = \frac{1}{2}(a + b) + \frac{1}{2}(a - b) \cos 2\theta - t \sin 2\theta$.

Answer to Practice Exercise

1. $\tan 2x$

20.4 Half-Angle Formulas

Formula for $\sin(\alpha/2)$ • Formula for $\cos(\alpha/2)$

If we let $\theta = \alpha/2$ in the identity $\cos 2\theta = 1 - 2\sin^2\theta$ and then solve for $\sin(\alpha/2)$,

$$\sin\frac{\alpha}{2} = \pm\sqrt{\frac{1 - \cos\alpha}{2}} \qquad (20.26)$$

Also, with the same substitution in the identity $\cos 2\theta = 2\cos^2\theta - 1$, which is then solved for $\cos(\alpha/2)$, we have

$$\cos\frac{\alpha}{2} = \pm\sqrt{\frac{1 + \cos\alpha}{2}} \qquad (20.27)$$

CAUTION ▶ In each of Eqs. (20.26) and (20.27), *the sign chosen depends on the quadrant in which $\frac{\alpha}{2}$ lies.*

EXAMPLE 1 Evaluation using $\cos(\alpha/2)$ formula

We can find $\cos 165°$ by using the relation

$$\cos 165° = -\sqrt{\frac{1 + \cos 330°}{2}} \qquad \text{using Eq. (20.27)}$$

$$= -\sqrt{\frac{1 + 0.8660}{2}} = -0.9659$$

Here, the minus sign is used, since $165°$ is in the second quadrant, and the cosine of a second-quadrant angle is negative. ■

EXAMPLE 2 Evaluation using $\sin(\alpha/2)$ formula

Simplify $\sqrt{\dfrac{1 - \cos 114°}{2}}$ by expressing the result in terms of one-half the given angle. Then, using a calculator, show that the values are equal.

We note that the given expression fits the form of the right side of Eq. (20.26), which means that

$$\sqrt{\frac{1 - \cos 114°}{2}} = \sin\tfrac{1}{2}(114°) = \sin 57°$$

Using a calculator shows that

$$\sqrt{\frac{1 - \cos 114°}{2}} = 0.8386705679 \quad \text{and} \quad \sin 57° = 0.8386705679$$

which verifies the equation for these values. ■

EXAMPLE 3 Simplification using $\cos(\alpha/2)$ formula

Simplify the expression $\sqrt{\dfrac{9 + 9\cos 6x}{2}}$.

$$\sqrt{\frac{9 + 9\cos 6x}{2}} = \sqrt{\frac{9(1 + \cos 6x)}{2}} = 3\sqrt{\frac{1 + \cos 6x}{2}}$$

$$= 3\cos\tfrac{1}{2}(6x) \qquad \text{using Eq. (20.27) with } \alpha = 6x$$

$$= 3\cos 3x$$

Practice Exercise

1. Simplify: $\sqrt{\dfrac{25 - 25\cos 4x}{2}}$

Noting the original expression, we see that $\cos 3x$ cannot be negative. ■

EXAMPLE 4   Trig identity with $\sin(\alpha/2)$ – application

In the kinetic theory of gases, the expression
$\sqrt{(1 - \cos\alpha)^2 + \sin^2\alpha}$ is found. Show that this expression equals $2\sin\frac{1}{2}\alpha$:

$$\sqrt{(1 - \cos\alpha)^2 + \sin^2\alpha} = \sqrt{1 - 2\cos\alpha + \cos^2\alpha + \sin^2\alpha} \qquad \text{expanding}$$
$$= \sqrt{1 - 2\cos\alpha + 1} \qquad \text{using Eq. (20.26)}$$
$$= \sqrt{2 - 2\cos\alpha}$$
$$= \sqrt{2(1 - \cos\alpha)} \qquad \text{factoring}$$

This last expression is very similar to that for $\sin\frac{1}{2}\alpha$, except that no 2 appears in the denominator. Therefore, multiplying the numerator and the denominator under the radical by 2 leads to the solution:

$$\sqrt{2(1 - \cos\alpha)} = \sqrt{\frac{4(1 - \cos\alpha)}{2}} = 2\sqrt{\frac{1 - \cos\alpha}{2}}$$
$$= 2\sin\frac{1}{2}\alpha \qquad \text{using Eq. (20.26)}$$

Noting the original expression, we see that $\sin\frac{1}{2}\alpha$ cannot be negative. ∎

EXAMPLE 5   Evaluation using $\cos(\alpha/2)$ formula

Given that $\tan\alpha = \frac{8}{15}$ $(180° < \alpha < 270°)$, find $\cos\left(\frac{\alpha}{2}\right)$.

Knowing that $\tan\alpha = \frac{8}{15}$ for a third-quadrant angle, we determine from Fig. 20.15 that $\cos\alpha = -\frac{15}{17}$. This means

$$\cos\frac{\alpha}{2} = -\sqrt{\frac{1 + (-15/17)}{2}} = -\sqrt{\frac{2}{34}} \qquad \text{using Eq. (20.27)}$$
$$= -\frac{1}{17}\sqrt{17} = -0.2425$$

Since $180° < \alpha < 270°$, we know that $90° < \frac{\alpha}{2} < 135°$ and therefore $\frac{\alpha}{2}$ is in the second quadrant. Since the cosine is negative for second-quadrant angles, we use the negative value of the radical. ∎

Fig. 20.15

EXAMPLE 6   Simplification – calculator verification

Show that $2\cos^2\frac{x}{2} - \cos x = 1$.

The first step is to substitute for $\cos\frac{x}{2}$, which will result in each term on the left being in terms of x and no $\frac{x}{2}$ terms will exist. This might allow us to combine terms. So we perform this operation, and we have for the left side

$$2\cos^2\frac{x}{2} - \cos x = 2\left(\frac{1 + \cos x}{2}\right) - \cos x \qquad \substack{\text{using Eq. (20.27) with} \\ \text{both sides squared}}$$
$$= 1 + \cos x - \cos x = 1$$

From Fig. 20.16, we verify that the graph of $y_1 = 2[\cos(x/2)]^2 - \cos x$ is the same as the graph of $y_2 = 1$. ∎

Fig. 20.16

Practice Exercise

2. Find the formula for $\csc(x/2)$.

EXAMPLE 7   Formulas for other functions of $\alpha/2$.

We can find relations for other functions of $\frac{\alpha}{2}$ by expressing these functions in terms of $\sin\left(\frac{\alpha}{2}\right)$ and $\cos\left(\frac{\alpha}{2}\right)$. For example,

$$\sec\frac{\alpha}{2} = \frac{1}{\cos\frac{\alpha}{2}} = \pm\frac{1}{\sqrt{\frac{1 + \cos\alpha}{2}}} \qquad \text{using Eq. (20.27)}$$
$$= \pm\sqrt{\frac{2}{1 + \cos\alpha}}$$

EXERCISES 20.4

In Exercises 1 and 2, make the given changes in the indicated examples of this section and then solve the given problem.

1. In Example 2, change the $-$ sign in the numerator to $+$.

2. In Example 5, change $\frac{8}{15}$ $(180° < \alpha < 270°)$ to $-\frac{8}{15}$ $(270° < \alpha < 360°)$.

In Exercises 3–8, use the half-angle formulas to evaluate the given functions.

3. $\cos 15°$
4. $\sin 22.5°$
5. $\sin 105°$
6. $\cos 112.5°$
7. $\cos \frac{7\pi}{8}$
8. $\sin \frac{11\pi}{12}$

In Exercises 9–12, simplify the given expressions by giving the results in terms of one-half the given angle. Then use a calculator to verify the result.

9. $\sqrt{\dfrac{1 - \cos 236°}{2}}$
10. $\sqrt{\dfrac{1 + \cos 98°}{2}}$
11. $\sqrt{1 + \cos 164°}$
12. $\sqrt{2 - 2\cos 328°}$

In Exercises 13–20, simplify the given expressions.

13. $\sqrt{\dfrac{1 - \cos 6x}{2}}$
14. $\sqrt{\dfrac{4 + 4\cos 8\beta}{2}}$
15. $\sqrt{8 + 8\cos 4x}$
16. $\sqrt{2 - 2\cos 16x}$
17. $\sqrt{4 - 4\cos 10\theta}$
18. $\sqrt{18 + 18\cos 1.4x}$
19. $2\sin^2 \dfrac{x}{2} + \cos x$
20. $2\cos^2 \dfrac{\theta}{2} \sec \theta$

In Exercises 21–24, evaluate the indicated functions.

21. Find the value of $\sin\left(\dfrac{\alpha}{2}\right)$ if $\cos \alpha = \frac{12}{13}$ $(0° < \alpha < 90°)$.
22. Find the value of $\cos\left(\dfrac{\alpha}{2}\right)$ if $\sin \alpha = -\frac{4}{5}$ $(180° < \alpha < 270°)$.
23. Find the value of $\cos\left(\dfrac{\alpha}{2}\right)$ if $\tan \alpha = -0.2917$ $(90° < \alpha < 180°)$.
24. Find the value of $\sin\left(\dfrac{\alpha}{2}\right)$ if $\cos \alpha = 0.4706$ $(270° < \alpha < 360°)$.

In Exercises 25–28, derive the required expressions.

25. Derive an expression for $\csc\left(\dfrac{\alpha}{2}\right)$ in terms of $\sec \alpha$
26. Derive an expression for $\sec\left(\dfrac{\alpha}{2}\right)$ in terms of $\sec \alpha$.
27. Derive an expression for $\tan\left(\dfrac{\alpha}{2}\right)$ in terms of $\sin \alpha$ and $\cos \alpha$.
28. Derive an expression for $\cot\left(\dfrac{\alpha}{2}\right)$ in terms of $\sin \alpha$ and $\cos \alpha$.

In Exercises 29–32, prove the given identities.

29. $\sin \dfrac{\alpha}{2} = \dfrac{1 - \cos \alpha}{2\sin \frac{\alpha}{2}}$
30. $\cos \dfrac{\theta}{2} = \dfrac{\sin \theta}{2\sin \frac{\theta}{2}}$
31. $2\cos \dfrac{x}{2} = (1 + \cos x)\sec \dfrac{x}{2}$
32. $\cos^2 \dfrac{x}{2}\left[1 + \left(\dfrac{\sin x}{1 + \cos x}\right)^2\right] = 1$

In Exercises 33–36, verify each identity by comparing the graph of the left side with the graph of the right side on a graphing calculator.

33. $2\sin^2 \dfrac{\alpha}{2} - \cos^2 \dfrac{\alpha}{2} = \dfrac{1 - 3\cos \alpha}{2}$
34. $\cos^2 \dfrac{A}{2} - \sin^2 \dfrac{A}{2} = \dfrac{\sin 2A}{2\sin A}$
35. $2\sin^2 \dfrac{\theta}{2} = \dfrac{\sin^2 \theta}{1 + \cos \theta}$
36. $\tan \dfrac{\alpha}{2} = \dfrac{\sin \alpha}{1 + \cos \alpha}$

In Exercises 37–44, use the half-angle formulas to solve the given problems.

37. Find $\tan \theta$ if $\sin(\theta/2) = 3/5$.

38. In a right triangle with sides and angles as shown in Fig. 20.17, show that $\sin^2 \dfrac{A}{2} = \dfrac{c - b}{2c}$.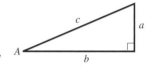

Fig. 20.17

39. In finding the path of a sliding particle, the expression $\sqrt{8 - 8\cos \theta}$ is used. Simplify this expression.

40. In designing track for a railway system, the equation $d = 4r\sin^2 \dfrac{A}{2}$ is used. Solve for d in terms of $\cos A$.

41. In electronics, in order to find the *root-mean-square current* in a circuit, it is necessary to express $\sin^2 \omega t$ in terms of $\cos 2\omega t$. Show how this is done.

42. In studying interference patterns of radio signals, the expression $2E^2 - 2E^2 \cos(\pi - \theta)$ arises. Show that this can be written as $4E^2 \cos^2(\theta/2)$.

43. The index of refraction n, the angle A of a prism, and the minimum angle of deflection ϕ are related by $n = \dfrac{\sin \frac{1}{2}(A + \phi)}{\sin \frac{1}{2}A}$. See Fig. 20.18. Show that an equivalent expression is
$$n = \sqrt{\dfrac{1 - \cos A\cos \phi + \sin A\sin \phi}{1 - \cos A}}$$

Fig. 20.18

Fig. 20.19

44. For the structure shown in Fig. 20.19, show that $x = 2l\sin^2 \frac{1}{2}\theta$.

Answers to Practice Exercises

1. $5\sin 2x$
2. $\pm\sqrt{\dfrac{2}{1 - \cos x}}$

20.5 Solving Trigonometric Equations

Solve for Function - Then Angle • Use Algebraic Methods and Trigonometric Identities • Use of Calculator

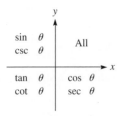

Positive functions

Fig. 20.20

One of the most important uses of the trigonometric identities is in the solution of equations involving trigonometric functions. The solution of this type of equation consists of the angles that satisfy the equation. When solving for the angle, we generally first solve for a value of a function of the angle and then find the angle from this value of the function.

When equations are written in terms of more than one function, the identities provide a way of changing many of them to equations or factors involving only one function of the same angle. Thus, *the solution is found by using algebraic methods and trigonometric identities and values.* From Chapter 8, recall that we must be careful regarding the sign of the value of a trigonometric function in finding the angle. Figure 20.20 shows again the quadrants in which the functions are positive. Functions not listed are negative.

EXAMPLE 1 First solve for $\cos\theta$

Solve the equation $2\cos\theta - 1 = 0$ for all values of θ such that $0 \le \theta \le 2\pi$.

Solving the equation for $\cos\theta$, we obtain $\cos\theta = \frac{1}{2}$. The problem asks for all values of θ from 0 to 2π that satisfy the equation. We know that the cosines of angles in the first and fourth quadrants are positive. Also, we know that $\cos\frac{\pi}{3} = \frac{1}{2}$, which means that $\frac{\pi}{3}$ is the reference angle. Therefore, the solution proceeds as follows:

$$2\cos\theta - 1 = 0$$
$$2\cos\theta = 1 \qquad \text{solve for } \cos\theta$$
$$\cos\theta = \frac{1}{2}$$
$$\theta = \frac{\pi}{3}, \frac{5\pi}{3} \qquad \theta \text{ in quadrants I and IV} \qquad \blacksquare$$

EXAMPLE 2 Use identity – factor – solve for $\sin x$

Solve the equation $2\cos^2 x - \sin x - 1 = 0 \quad (0 \le x < 2\pi)$.

By use of the identity $\sin^2 x + \cos^2 x = 1$, this equation may be put in terms of $\sin x$ only. Thus, we have

$$2(1 - \sin^2 x) - \sin x - 1 = 0 \qquad \text{use identity}$$
$$-2\sin^2 x - \sin x + 1 = 0 \qquad \text{solve for } \sin x$$
$$2\sin^2 x + \sin x - 1 = 0$$
$$(2\sin x - 1)(\sin x + 1) = 0 \qquad \text{factor}$$

Setting each factor equal to zero, we find $\sin x = 1/2$, or $\sin x = -1$. For the domain 0 to 2π, $\sin x = 1/2$ gives $x = \pi/6, 5\pi/6$, and $\sin x = -1$ gives $x = 3\pi/2$. Therefore,

$$x = \frac{\pi}{6}, \frac{5\pi}{6}, \frac{3\pi}{2}$$

These values check when substituted in the original equation. $\qquad \blacksquare$

Graphical Solutions   As with algebraic equations, graphical solutions of trigonometric equations are approximate, whereas algebraic solutions often give exact solutions. As before, *we collect all terms on the left of the equal sign, with zero on the right.* We then *graph the function on the left to find its zeros by finding the values of x where the graph crosses (or is tangent to) the x-axis.*

Practice Exercise

1. Solve for x ($0 \le x < 2\pi$):
$2\sin x + 1 = 0$

Fig. 20.21

EXAMPLE 3 Solution using calculator

Graphically solve the equation $2\cos^2 x - \sin x - 1 = 0$ $(0 \le x < 2\pi)$ by using a graphing calculator. (This is the same equation as in Example 2.)

Since all the terms of the equation are on the left, with zero on the right, we now set $y = 2\cos^2 x - \sin x - 1$. We enter this function in the graphing calculator as Y_1, and the graph is displayed in Fig. 20.21. Since angles are expressed in radians and $2\pi = 6.3$, Xmax was chosen as 6.4. Using the *trace* and *zoom* features (or the *zero* feature, or by letting $y_1 = 2\cos^2 x - \sin x - 1$, $y_2 = 0$ and using the *intersect* feature) of a graphing calculator, we find that $y = 0$ for

$$x = 0.52, 2.62, 4.71$$

These values are the same as in Example 2. We note that for $x = 4.71$, the curve *touches* the x-axis but does not cross it. This means it is *tangent* to the x-axis. ∎

EXAMPLE 4 Solve – check for extraneous solutions

Solve the equation $\cos(x/2) = 1 + \cos x$ $(0 \le x < 2\pi)$.

By using the half-angle formula for $\cos(x/2)$ and then squaring both sides of the resulting equation, this equation can be solved:

$$\pm\sqrt{\frac{1+\cos x}{2}} = 1 + \cos x \qquad \text{using identity}$$

$$\frac{1+\cos x}{2} = 1 + 2\cos x + \cos^2 x \qquad \text{squaring both sides}$$

$$2\cos^2 x + 3\cos x + 1 = 0 \qquad \text{simplifying}$$

$$(2\cos x + 1)(\cos x + 1) = 0 \qquad \text{factoring}$$

$$\cos x = -\frac{1}{2}, -1$$

$$x = \frac{2\pi}{3}, \frac{4\pi}{3}, \pi$$

Fig. 20.22

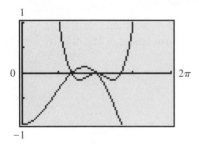

Fig. 20.23

In finding this solution, we squared both sides of the original equation. In doing this, we may have introduced extraneous solutions (see Section 14.3). Thus, we must check each solution in the original equation to see if it is valid. Hence,

$$\cos\frac{\pi}{3} \overset{?}{=} 1 + \cos\frac{2\pi}{3} \quad \text{or} \quad \frac{1}{2} \overset{?}{=} 1 + \left(-\frac{1}{2}\right) \quad \text{or} \quad \frac{1}{2} = \frac{1}{2}$$

$$\cos\frac{2\pi}{3} \overset{?}{=} 1 + \cos\frac{4\pi}{3} \quad \text{or} \quad -\frac{1}{2} \overset{?}{=} 1 + \left(-\frac{1}{2}\right) \quad \text{or} \quad -\frac{1}{2} \neq \frac{1}{2}$$

$$\cos\frac{\pi}{2} \overset{?}{=} 1 + \cos\pi \quad \text{or} \quad 0 \overset{?}{=} 1 - 1 \quad \text{or} \quad 0 = 0$$

Thus, the apparent solution $x = \frac{4\pi}{3}$ is not a solution of the original equation. The correct solutions are $x = \frac{2\pi}{3}$ and $x = \pi$.

We can see that these values agree with the values for x for which the graph of $y_1 = \cos(x/2) - 1 - \cos x$ crosses the x-axis in Fig. 20.22. In Fig. 20.23 the graph of this function and the graph of $y_2 = 2\cos^2 x + 3\cos x + 1$ are compared. We see $x = 0$ for both curves at $x = 2\pi/3$ and $x = \pi$, but that the curve for y_2 is quite different from that of y_1, and $y_2 = 0$ for $x = 4\pi/3$. This also shows us that, *when checking the solutions, we must use the original equation,* and not an equation that has been derived after any algebraic step has been performed that could basically change the function. In this case, as we mentioned, this operation was squaring both sides of the original equation. ∎

Fig. 20.24

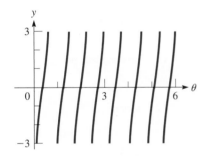

Fig. 20.25

Practice Exercise

2. Solve for $x (0 \le x < 2\pi)$:
$\sec^2 x + 2 \tan x = 0$

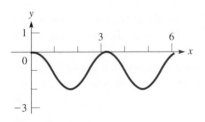

Fig. 20.26

EXAMPLE 5 Trig equation – application

The vertical displacement y of an object at the end of a spring, which itself is being moved up and down, is given by $y = 3.50 \sin t + 1.20 \sin 2t$. Find the first two values of t (in seconds) for which $y = 0$.

Using the double-angle formula for $\sin 2t$ leads to the solution.

$$3.50 \sin t + 1.20 \sin 2t = 0 \qquad \text{setting } y = 0$$
$$3.50 \sin t + 2.40 \sin t \cos t = 0 \qquad \text{using identities}$$
$$\sin t (3.50 + 2.40 \cos t) = 0 \qquad \text{factoring}$$
$$\sin t = 0 \qquad \text{or} \quad \cos t = -1.46$$
$$t = 0.00, 3.14, \dots$$

Since $\cos t$ cannot be numerically larger than 1, there are no values of t for which $\cos t = -1.46$. Thus, the required times are $t = 0.00\,\text{s}, 3.14\,\text{s}$.

We can see that these values agree with the values of t for which the graph of $y = 3.50 \sin t + 1.20 \sin 2t$ crosses the t-axis in Fig. 20.24. (In using a graphing calculator, use x for t.) ■

EXAMPLE 6 Solve for 2θ – then θ

Solve the equation $\tan 2\theta - \cot 2\theta = 0$ $(0 \le \theta < 2\pi)$.

$$\tan 2\theta - \frac{1}{\tan 2\theta} = 0 \qquad \text{using } \cot 2\theta = \frac{1}{\tan 2\theta}$$

$$\tan^2 2\theta = 1 \qquad \text{multiplying by } \tan 2\theta \text{ and adding 1 to each side}$$

$$\tan 2\theta = \pm 1 \qquad \text{taking square roots}$$

For $0 \le \theta < 2\pi$, we must have values of 2θ such that $0 \le 2\theta < 4\pi$. Therefore,

$$2\theta = \frac{\pi}{4}, \frac{3\pi}{4}, \frac{5\pi}{4}, \frac{7\pi}{4}, \frac{9\pi}{4}, \frac{11\pi}{4}, \frac{13\pi}{4}, \frac{15\pi}{4}$$

This means that the solutions are

$$\theta = \frac{\pi}{8}, \frac{3\pi}{8}, \frac{5\pi}{8}, \frac{7\pi}{8}, \frac{9\pi}{8}, \frac{11\pi}{8}, \frac{13\pi}{8}, \frac{15\pi}{8}$$

These values satisfy the original equation. Since we multiplied through by $\tan 2\theta$ in the solution, any value of θ that leads to $\tan 2\theta = 0$ would not be valid, since this would indicate division by zero in the original equation.

We see that these solutions agree with the values of θ for which the graph of $y = \cos 2\theta - \cot 2\theta$ crosses the θ-axis in Fig. 20.25. (In using a graphing calculator, use x for θ.) ■

EXAMPLE 7 Recognize trigonometric form – solve

Solve the equation $\cos 3x \cos x + \sin 3x \sin x = 1$ $(0 \le x < 2\pi)$.

The left side of this equation is of the general form $\cos(A - x)$, where $A = 3x$. Therefore,

$$\cos 3x \cos x + \sin 3x \sin x = \cos(3x - x) = \cos 2x$$

The original equation becomes

$$\cos 2x = 1$$

This equation is satisfied if $2x = 0$ or $2x = 2\pi$. The solutions are $x = 0$ and $x = \pi$. Only through recognition of the proper trigonometric form can we readily solve this equation.

We see that these solutions agree with the two values of x for which the graph of $y = \cos 3x \cos x + \sin 3x \sin x - 1$ touches the x-axis in Fig. 20.26. ■

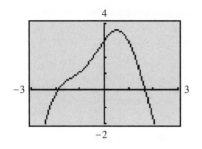

Fig. 20.27

EXAMPLE 8 Solve using calculator

Solve the equation $\sin 2x + 3 = x^2$.

Although we can substitute for $\sin 2x$, we cannot express x^2 in terms of a trigonometric function. However, we can graphically find an approximate solution.

Collecting terms on the left, we have $\sin 2x + 3 - x^2 = 0$. Then let $y = 2x + 3 - x^2$. Entering this function into the calculator as Y_1, we find from its graph where y is zero. See Fig. 20.27. Using the *zero* feature, we find that $x = -1.90$ and $x = 1.67$ are the only values for which $y = 0$.

The solution can also be found by letting $y_1 = \sin 2x + 3$, $y_2 = x^2$, and using the *intersect* feature of a graphing calculator. ∎

EXERCISES 20.5

In Exercises 1–4, make the given changes in the indicated examples of this section and then solve the resulting problems.

1. In Example 1, change $2 \cos \theta$ to $\tan \theta$.

2. In Example 2, change $2 \cos^2 x$ to $2 \sin^2 x$.

3. In Example 4, change $\cos \frac{1}{2}x$ to $\sin \frac{1}{2}x$ and on the right of the equal sign change the $+$ to $-$.

4. In Example 7, change the $+$ to $-$.

In Exercises 5–20, solve the given trigonometric equations analytically (using identities when necessary for exact values when possible) for values of x for $0 \le x < 2\pi$.

5. $\sin x - 1 = 0$

6. $2 \cos x + 1 = 0$

7. $\sin(x - \frac{\pi}{4}) = \cos(x - \frac{\pi}{4})$

8. $4 \tan x + 2 = 3(1 + \tan x)$

9. $4 \cos^2 x - 1 = 0$

10. $3 \tan^2 x - 1 = 0$

11. $2 \sin^2 x - \sin x = 0$

12. $\sin 4x - \sin 2x = 0$

13. $\sin 2x \sin x + \cos x = 0$

14. $\sin x - \sin \frac{x}{2} = 0$

15. $2 \cos^2 x - 2 \cos 2x - 1 = 0$

16. $\tan^2 x + 6 = 5 \tan x$

17. $4 \tan x - \sec^2 x = 0$

18. $\sin x \sin \frac{1}{2}x = 1 - \cos x$

19. $\sin 2x \cos x - \cos 2x \sin x = 0$

20. $\cos 3x \cos x - \sin 3x \sin x = 0$

In Exercises 21–38, solve the given trigonometric equations analytically and by use of a graphing calculator. Compare results. Use values of x for $0 \le x < 2\pi$.

21. $\tan x + 1 = 0$

22. $2 \sin x + 1 = 0$

23. $3 - 4 \cos x = 7 - (2 - \cos x)$

24. $7 \sin x - 2 = 3(2 - \sin x)$

25. $4 \sin^2 x - 3 = 0$

26. $|\sin x| = \frac{1}{2}$

27. $\sin 4x - \cos 2x = 0$

28. $3 \cos x - 4 \cos^2 x = 0$

29. $2 \sin x = \tan x$

30. $\cos 2x + \sin^2 x = 0$

31. $\sin^2 x - 2 \sin x = 1$

32. $2 \cos^2 2x + 1 = 3 \cos 2x$

33. $\tan x + 3 \cot x = 4$

34. $\tan^2 x + 4 = 2 \sec^2 x$

35. $\sin 2x + \cos 2x = 0$

36. $2 \sin 4x + \csc 4x = 3$

37. $2 \sin 2x - \cos x \sin^3 x = 0$

38. $\tan^4 x = 1$

In Exercises 39–52, solve the indicated equations analytically.

39. $\sin 3x + \sin x = 0$ (*Hint*: See Eq. (20.17).)

40. $\cos 3x - \cos x = 0$ (*Hint*: See Eq. (20.20).)

41. Is there any positive acute angle θ for which $\sin \theta + \cos \theta + \tan \theta + \cot \theta + \sec \theta + \csc \theta = 1$? Explain.

42. Use a graphing calculator to determine the minimum value of the function to the left of the equal sign in Exercise 41 (for a positive acute angle).

43. Solve the system of equations $r = \sin \theta$, $r = \sin 2\theta$, for $0 \le \theta < 2\pi$.

44. Solve the system of equations $r = \sin \theta$, $r = \cos 2\theta$, for $0 \le \theta < 2\pi$.

45. The acceleration due to gravity g (in m/s²) varies with latitude, approximately given by $g = 9.7805(1 + 0.0053 \sin^2 \theta)$, where θ is the latitude in degrees. Find θ for $g = 9.8000$.

46. Under certain conditions, the electric current i (in A) in the circuit shown in Fig. 20.28 is given below. For what value of t (in s) is the current first equal to zero?

Fig. 20.28

$i = -e^{-100t}(32.0 \sin 624.5t + 0.200 \cos 624.5t)$.

47. The vertical displacement y (in m) of the end of a robot arm is given by $y = 2.30 \cos 0.1t - 1.35 \sin 0.2t$. Find the first four values of t (in s) for which $y = 0$.

48. In finding the maximum illuminance from a point source of light, it is necessary to solve the equation $\cos\theta \sin 2\theta - \sin^3\theta = 0$. Find θ if $0 < \theta < 90°$.

49. To find the angle θ subtended by a certain object on a camera film, it is necessary to solve the equation $\dfrac{p^2 \tan\theta}{0.0063 + p\tan\theta} = 1.6$, where p is the distance from the camera to the object. Find θ if $p = 4.8$ m.

50. The velocity of a certain piston is maximum when the acute crank angle θ satisfies the equation $8\cos\theta + \cos 2\theta = 0$. Find this angle.

51. Resolve a force of 500.0 N into two components, perpendicular to each other, for which the sum of their magnitudes is 700.0 N, by using the angle between a component and the resultant.

52. Looking for a lost ship in the North Atlantic Ocean, a plane flew from Reykjavik, Iceland, 160 km west. It then turned and flew due north and then made a final turn to fly directly back to Reykjavik.

If the total distance flown was 480 km, how long were the final two legs of the flight? Solve by setting up and solving an appropriate trigonometric equation. (The Pythagorean theorem may be used only as a check.)

In Exercises 53–60, solve the given equations graphically.

53. $3\sin x - x = 0$

54. $4\cos x + 3x = 0$

55. $2\sin 2x = x^2 + 1$

56. $\sqrt{x} - \sin 3x = 1$

57. $2\ln x = 1 - \cos 2x$

58. $e^x = 1 + \sin x$

59. In finding the frequencies of vibration of a vibrating wire, the equation $x\tan x = 2.00$ occurs. Find x if $0 < x < \pi/2$.

60. An equation used in astronomy is $\theta - e\sin\theta = M$. Solve for θ for $e = 0.25$ and $M = 0.75$.

Answers to Practice Exercises

1. $x = 7\pi/6,\ 11\pi/6$ **2.** $x = 3\pi/4,\ 7\pi/4$

20.6 The Inverse Trigonometric Functions

Inverse Functions •
Inverse Trigonometric Functions •
Ranges of the Inverse Trig Functions

When we studied the exponential and logarithmic functions, we often changed an expression from one form to the other. The exponential function $y = b^x$, written in logarithmic form with x as a function of y, is $x = \log_b y$. Then we wrote the logarithmic function as $y = \log_b x$, since it is standard practice to use y as the dependent variable and x as the independent variable.

These two functions, the exponential function $y = b^x$ and the logarithmic function $y = \log_b x$, are called **inverse functions.** *This means that if we solve for the independent variable in terms of the dependent variable in one function, we will arrive at the functional relationship expressed by the other.* It also means that, for every value of x, there is only one corresponding value of y.

Just as we are able to solve $y = b^x$ for the exponent by writing it in logarithmic form, at times it is necessary to solve for the independent variable (the angle) in trigonometric functions. Therefore, we define the **inverse sine function**

$$y = \sin^{-1} x \qquad \left(-\frac{\pi}{2} \le y \le \frac{\pi}{2}\right) \qquad \textbf{(20.28)}$$

where y is the angle whose sine is x. This means that x is the value of the sine of the angle y, or $x = \sin y$. (It is necessary to show the range as $-\pi/2 \le y \le \pi/2$, as we will see shortly.)

CAUTION ▶

*In Eq. (20.28), the -1 is **not** an exponent.*

The -1 in $\sin^{-1} x$ is the notation showing the inverse function. We introduced this notation in Chapter 4 when finding the angle with a known value of one of the functions.

The notations Arcsin x, arcsin x, Sin^{-1} x are also used to designate the inverse sine. Some calculators use the [inv] key and then the [sin] key to find values of the inverse sine. However, since most calculators use the notation $\sin^{-1}$ as a second function on the [sin] key, we will continue to use $\sin^{-1} x$ for the inverse sine.

Similar definitions are used for the other inverse trigonometric functions. They also have meanings similar to that of Eq. (20.28).

EXAMPLE 1 Meaning of inverse trig functions

(a) $y = \cos^{-1} x$ is read as "y is the angle whose cosine is x." In this case, $x = \cos y$.

(b) $y = \tan^{-1} 2x$ is read as "y is the angle whose tangent is $2x$." In this case, $2x = \tan y$.

(c) $y = \csc^{-1}(1 - x)$ is read as "y is the angle whose cosecant is $1 - x$." In this case, $1 - x = \csc y$, or $x = 1 - \csc y$. ∎

We have seen that $y = \sin^{-1} x$ means that $x = \sin y$. From our previous work with the trigonometric functions, we know that there is an unlimited number of possible values of y for a given value of x in $x = \sin y$. Consider the following example.

EXAMPLE 2 Values of trig functions

(a) For $x = \sin y$, we know that

$$\sin \frac{\pi}{6} = \frac{1}{2} \quad \text{and} \quad \sin \frac{5\pi}{6} = \frac{1}{2}$$

In fact, $x = \frac{1}{2}$ also for values of y of $-\frac{7\pi}{6}, \frac{13\pi}{6}, \frac{17\pi}{6}$, and so on.

(b) For $x = \cos y$, we know that

$$\cos 0 = 1 \quad \text{and} \quad \cos 2\pi = 1$$

In fact, $\cos y = 1$ for y equal to any even multiple of π. ∎

From Chapter 3, we know that *to have a properly defined **function**, there must be only one value of the dependent variable for a given value of the independent variable.* (A *relation*, on the other hand, may have more than one such value.) Therefore, as in Eq. (20.28), in order to have only one value of y for each value of x in the domain of the inverse trigonometric functions, it is not possible to include all values of y in the range. For this reason, *the range of each of the* **inverse trigonometric functions** *is defined as follows:*

$$-\frac{\pi}{2} \le \sin^{-1} x \le \frac{\pi}{2} \quad 0 \le \cos^{-1} x \le \pi \quad -\frac{\pi}{2} < \tan^{-1} x < \frac{\pi}{2} \quad 0 < \cot^{-1} x < \pi$$

$$0 \le \sec^{-1} x \le \pi \quad \left(\sec^{-1} x \ne \frac{\pi}{2} \right) \quad -\frac{\pi}{2} \le \csc^{-1} x \le \frac{\pi}{2} \quad (\csc^{-1} x \ne 0) \tag{20.29}$$

We must choose a value of y in the range as defined in Eqs. (20.29) that corresponds to a given value of x in the domain. We will discuss the domains and the reasons for these definitions following the next two examples.

EXAMPLE 3 Range of inverse sine and cosine

(a) $\sin^{-1}\left(\frac{1}{2}\right) = \frac{\pi}{6}$ first-quadrant angle

This is the only value of the function that lies within the defined range. The value $\frac{5\pi}{6}$ is not correct, even though $\sin\left(\frac{5\pi}{6}\right) = \frac{1}{2}$, since $\frac{5\pi}{6}$ lies outside the defined range.

(b) $\cos^{-1}\left(-\frac{1}{2}\right) = \frac{2\pi}{3}$ second-quadrant angle

Other values such as $\frac{4\pi}{3}$ and $-\frac{2\pi}{3}$ are not correct, since they are not within the defined range for the function $\cos^{-1} x$. ∎

EXAMPLE 4 Range of inverse tangent

$$\tan^{-1}(-1) = -\frac{\pi}{4} \qquad \text{fourth-quadrant angle}$$

CAUTION ▶ This is the only value within the defined range for the function $\tan^{-1} x$. We must re-member that ***when x is negative for*** $\sin^{-1} x$ ***and*** $\tan^{-1} x$, ***the value of y is a fourth-quadrant angle, expressed as a negative angle.*** This is a direct result of the definition. (The single exception is $\sin^{-1}(-1) = -\pi/2$, which is a quadrantal angle and is not in the fourth quadrant.) ∎

In choosing these values to be the ranges of the inverse trigonometric functions, we first note that the *domain* of $y = \sin^{-1} x$ and $y = \cos^{-1} x$ are each $-1 \le x \le 1$, since the sine and cosine functions take on only these values. Therefore, for each value in this domain, we use only one value of y in the range of the function. Although the domain of $y = \tan^{-1} x$ is all real numbers, we still use only one value of y in the range.

The ranges of the inverse trigonometric functions are chosen so that if x is positive, the resulting value is an angle in the first quadrant. However, care must be taken in choosing the range for negative values of x.

Since the sine of a second-quadrant angle is positive, we cannot choose these angles for $\sin^{-1} x$ for negative values of x. Therefore, we chose fourth-quadrant angles in the form of negative angles in order to have a continuous range of values for $\sin^{-1} x$. The range for $\tan^{-1} x$ is chosen in the same way for similar reasons. However, since the co-sine of a fourth-quadrant angle is positive, the range for $\cos^{-1} x$ cannot be the same. To keep a continuous range of values for $\cos^{-1} x$, second-quadrant angles are used.

Values for the other functions are chosen such that the result is also an angle in the first quadrant if x is positive. As for negative values of x, it rarely makes any differ-ence, since either positive values of x arise, or we can use one of the other functions. Our definitions, however, are those that are generally used.

The graphs of the inverse trigonometric functions can be used to show the domains and ranges. We can obtain the graph of the inverse sine function by first sketching the sine curve $x = \sin y$ *along the y-axis.* We then mark the specific part of this curve for which $-\frac{\pi}{2} \le y \le \frac{\pi}{2}$ as the graph of the inverse sine function. The graphs of the other inverse trigonometric functions are found in the same manner. In Figs. 20.29, 20.30, and 20.31, the graphs $x = \sin y$, $x = \cos y$, and $x = \tan y$, respectively, are shown. The heavier, colored portions indicate the graphs of the respective inverse trigonometric functions.

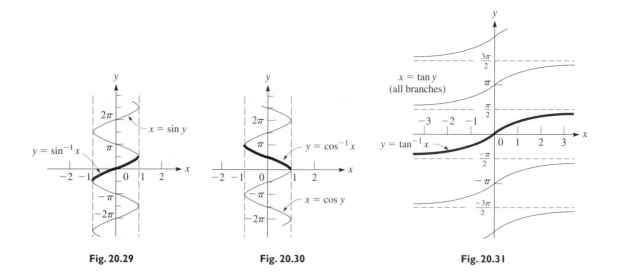

Fig. 20.29   Fig. 20.30   Fig. 20.31

The following examples further illustrate the values and meanings of the inverse trigonometric functions.

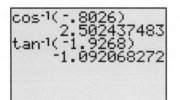

Fig. 20.32

EXAMPLE 5 Values of inverse trig functions

(a) $\sin^{-1}\left(-\sqrt{3}/2\right) = -\pi/3$ **(b)** $\cos^{-1}(-1) = \pi$

(c) $\tan^{-1} 0 = 0$ **(d)** $\tan^{-1}\left(\sqrt{3}\right) = \pi/3$

Using a calculator in radian mode, we find the following values:

(e) $\sin^{-1} 0.6294 = 0.6808$ **(f)** $\sin^{-1}(-0.1568) = -0.1574$

(g) $\cos^{-1}(-0.8026) = 2.5024$ **(h)** $\tan^{-1}(-1.9268) = -1.0921$

In Fig. 20.32, we show the evaluations for parts (g) and (h) on a calculator. We note that in each case the calculator gives the value of the inverse function in the defined range for the given function. ∎

EXAMPLE 6 Given inverse function – solve for x

Given that $y = \pi - \sec^{-1} 2x$, solve for x.

We first find the expression for $\sec^{-1} 2x$ and then use the meaning of the inverse secant. The solution follows:

$$y = \pi - \sec^{-1} 2x$$
$$\sec^{-1} 2x = \pi - y \qquad \text{solve for } \sec^{-1} 2x$$
$$2x = \sec(\pi - y) \qquad \text{use meaning of inverse secant}$$
$$x = -\frac{1}{2}\sec y \qquad \sec(\pi - y) = -\sec y$$

As $\sec 2x$ and $2 \sec x$ are different functions, $\sec^{-1} 2x$ and $2 \sec^{-1} x$ are also different functions. Since the values of $\sec^{-1} 2x$ are restricted, so are the resulting values of y. ∎

EXAMPLE 7 Solve for angle – application

The instantaneous power p in an electric inductor is given by the equation $p = vi \sin \omega t \cos \omega t$. Solve for t.

Noting the product $\sin \omega t \cos \omega t$ suggests using $\sin 2\alpha = 2 \sin \alpha \cos \alpha$. Then, using the meaning of the inverse sine, we can complete the solution:

$$p = vi \sin \omega t \cos \omega t$$
$$= \frac{1}{2} vi \sin 2\omega t \qquad \text{using double-angle formula}$$
$$\sin 2\omega t = \frac{2p}{vi}$$
$$2\omega t = \sin^{-1}\left(\frac{2p}{vi}\right) \qquad \text{using meaning of inverse sine}$$
$$t = \frac{1}{2\omega} \sin^{-1}\left(\frac{2p}{vi}\right) \qquad\qquad ∎$$

If we know the value of one of the inverse functions, we can find the trigonometric functions of the angle. If general relations are desired, a representative triangle is very useful. The following examples illustrate these methods.

Practice Exercise

1. Evaluate: $\tan^{-1}(-\sqrt{3}/3)$

EXAMPLE 8 Angle in terms of inverse functions

(a) Find $\cos(\sin^{-1} 0.5)$.

Knowing that the values of inverse trigonometric functions are *angles,* we see that $\sin^{-1} 0.5$ is a first-quadrant angle. Thus, we find $\sin^{-1} 0.5 = \pi/6$. The problem is now to find $\cos(\pi/6)$. This is, of course, $\sqrt{3}/2$, or 0.8660. Thus,

$$\cos(\sin^{-1} 0.5) = \cos(\pi/6) = 0.8660. \qquad \blacksquare$$

Practice Exercise
2. Evaluate: $\sin[\cos^{-1}(-0.5)]$

(b) $\sin(\cot^{-1} 1) = \sin(\pi/4)$ first-quadrant angle

$$= \frac{\sqrt{2}}{2} = 0.7071$$

(c) $\tan[\cos^{-1}(-1)] = \tan \pi$ quadrantal angle

$$= 0$$

(d) $\cos[\sin^{-1}(-0.2395)] = 0.9709$ using a calculator $\qquad \blacksquare$

EXAMPLE 9 Trig function of inverse trig function

Find $\sin(\tan^{-1} x)$.

We know that $\tan^{-1} x$ is another say of stating "the angle whose tangent is x." Thus, let us draw a right triangle (as in Fig. 20.33) and label one of the acute angles as θ, the side opposite θ as x, and the side adjacent to θ as 1. In this way, we see that, by definition, $\tan \theta = \frac{x}{1}$, or $\theta = \tan^{-1} x$, which means θ is the desired angle. By the Pythagorean theorem, the hypotenuse of this triangle is $\sqrt{x^2 + 1}$. Now, we find that $\sin \theta$, which is the same as $\sin(\tan^{-1} x)$, is $x/\sqrt{x^2 + 1}$. Thus,

$$\sin(\tan^{-1} x) = \frac{x}{\sqrt{x^2 + 1}} \qquad \blacksquare$$

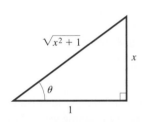

Fig. 20.33

EXAMPLE 10 Trig function of inverse trig function

Find $\cos(2 \sin^{-1} x)$.

From Fig. 20.34, we see that $\theta = \sin^{-1} x$. From the double-angle formulas, we have

$$\cos 2\theta = 1 - 2 \sin^2 \theta$$

Thus, since $\sin \theta = x$, we have

$$\cos(2 \sin^{-1} x) = 1 - 2x^2 \qquad \blacksquare$$

Fig. 20.34

EXAMPLE 11 Inverse trig function – application

A triangular brace of sides a, b, and c supports a shelf, as shown in Fig. 20.35. Find the expression for the angle between sides b and c.

The law of cosines leads to the solution:

$$a^2 = b^2 + c^2 - 2bc \cos A \qquad \text{law of cosines}$$
$$2bc \cos A = b^2 + c^2 - a^2 \qquad \text{solving for cos } A$$
$$\cos A = \frac{b^2 + c^2 - a^2}{2bc}$$
$$A = \cos^{-1}\left(\frac{b^2 + c^2 - a^2}{2bc}\right) \qquad \text{using meaning of inverse cosine} \qquad \blacksquare$$

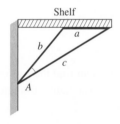

Fig. 20.35

EXERCISES 20.6

In Exercises 1–4, make the given changes in the indicated examples of this section and then solve the resulting problems.

  1. In Example 1(b), change $2x$ to $3A$.

2. In Example 3(a), change $1/2$ to -1.

3. In Example 8(a), change 0.5 to 1.

4. In Example 9, change sin to cos.

In Exercises 5–10, write down the meaning of each of the given equations. See Example 1.

5. $y = \cot^{-1} 3x$ **6.** $y = \csc^{-1} 4x$

7. $y = 2\sin^{-1} x$ **8.** $y = 3\tan^{-1} x$

9. $y = 5\cos^{-1}(2x - 1)$ **10.** $y = 4\sin^{-1}(3x + 2)$

In Exercises 11–28, evaluate exactly the given expressions.

11. $\cos^{-1} 0.5$ **12.** $\sin^{-1} 1$ **13.** $\tan^{-1} 1$

14. $\cos^{-1} 2$ **15.** $\tan^{-1}\left(-\sqrt{3}\right)$ **16.** $\sin^{-1}(-0.5)$

17. $\sec^{-1} 0.5$ **18.** $\cot^{-1} \sqrt{3}$ **19.** $\sin^{-1}\left(-\sqrt{2}/2\right)$

20. $\cos^{-1}\left(-\sqrt{3}/2\right)$ **21.** $\sin\left(\tan^{-1} \sqrt{3}\right)$ **22.** $\tan\left[\sin^{-1}(2/3)\right]$

23. $\cos^{-1}\left[\cos(-\pi/4)\right]$ **24.** $\tan^{-1}\left[\tan(2\pi/3)\right]$

25. $\cos\left[\tan^{-1}(-5)\right]$ **26.** $\sec\left[\cos^{-1}(-0.5)\right]$

27. $\cos(2\sin^{-1} 1)$ **28.** $\sin(2\tan^{-1} 2)$

In Exercises 29–32, find the exact value of x.

29. $\tan^{-1} x = \sin^{-1}\frac{2}{5}$ **30.** $\cot^{-1} x = \cos^{-1}\frac{1}{3}$

31. $\sec^{-1} x = -\sin^{-1}\left(-\frac{1}{2}\right)$ **32.** $\sin^{-1} x = -\tan^{-1}(-1)$

In Exercises 33–40, use a calculator to evaluate the given expressions.

33. $\tan^{-1}(-3.7321)$ **34.** $\cos^{-1}(-0.6561)$

35. $\sin^{-1} 0.0219$ **36.** $\tan^{-1} 0.2846$

37. $\tan\left[\cos^{-1}(-0.6281)\right]$ **38.** $\cos\left[\tan(-7.2256)\right]$

39. $\sin\left[\tan^{-1}(-0.2297)\right]$ **40.** $\tan\left[\sin^{-1}(-0.3019)\right]$

In Exercises 41–46, solve the given equations for x.

41. $y = \sin 3x$ **42.** $y = \cos(2x - \pi)$

43. $y = \tan^{-1}(x/4)$ **44.** $y = 2\sin^{-1}(x/6)$

45. $1 - y = \cos^{-1}(1 - x)$ **46.** $2y = \cot^{-1} 3x - 5$

In Exercises 47–54, find an algebraic expression for each of the given expressions.

47. $\tan(\sin^{-1} x)$ **48.** $\sin(\cos^{-1} x)$

49. $\sin(\sin^{-1} x + \cos^{-1} y)$ **50.** $\cos(\sin^{-1} x - \cos^{-1} y)$

51. $\sec(\csc^{-1} 3x)$ **52.** $\tan(\sin^{-1} 2x)$

53. $\sin(2\sin^{-1} x)$ **54.** $\cos(2\tan^{-1} x)$

In Exercises 55–60, solve the given problems with the use of the inverse trigonometric functions.

55. Is $\sin^{-1}(\sin x) = x$ for all x? Explain.

56. Show that the area A of a segment of a circle of radius r, bounded by a chord at a distance d from the center, is given by $A = r^2\cos^{-1}(d/r) - d\sqrt{r^2 - d^2}$.

57. In the analysis of ocean tides, the equation $y = A\cos 2(\omega t + \phi)$ is used. Solve for t.

58. For an object of weight w on an inclined plane that is at an angle θ to the horizontal, the equation relating w and θ is $\mu w \cos \theta = w \sin \theta$, where μ is the coefficient of friction between the surfaces in contact. Solve for θ.

59. The electric current in a certain circuit is given by $i = I_m[\sin(\omega t + \alpha)\cos \phi + \cos(\omega t + \alpha)\sin \phi]$. Solve for t.

60. The time t as a function of the displacement d of a piston is given by $t = \dfrac{1}{2\pi f}\cos^{-1}\dfrac{d}{A}$. Solve for d.

In Exercises 61 and 62, prove that the given expressions are equal. Use the relation for $\sin(\alpha + \beta)$ and show that the sine of the sum of the angles on the left equals the sine of the angle on the right.

61. $\sin^{-1}\dfrac{3}{5} + \sin^{-1}\dfrac{5}{13} = \sin^{-1}\dfrac{56}{65}$

62. $\tan^{-1}\dfrac{1}{3} + \tan^{-1}\dfrac{1}{2} = \dfrac{\pi}{4}$

In Exercises 63–66, evaluate the given expressions.

63. $\sin^{-1} 0.5 + \cos^{-1} 0.5$ **64.** $\tan^{-1} \sqrt{3} + \cot^{-1} \sqrt{3}$

65. $\sin^{-1} x + \sin^{-1}(-x)$ **66.** $\sin^{-1} x + \cos^{-1} x$

In Exercises 67–70, solve for the angle A for the given triangles in the given figures in terms of the given sides and angles. For Exercises 67 and 68, explain your method.

67.

Fig. 20.36

68.

Fig. 20.37

69.

Fig. 20.38

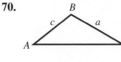

70.

Fig. 20.39

In Exercises 71–74, solve the given problems.

71. The height of the Statue of Liberty is 151 ft. See Fig. 20.40. From the deck of a boat at a horizontal distance d from the statue, the angles of elevation of the top of the statue and the top of its pedestal are α and β, respectively. Show that

$$\alpha = \tan^{-1}\left(\frac{151}{d} + \tan \beta\right).$$

151 ft

Fig. 20.40

(W) 72. Explain why $\sin^{-1} 2x$ is not equal to $2 \sin^{-1} x$.

73. If a TV camera is x m from a launch pad of a 50-m rocket that is y m above the ground, find an expression for θ, the angle subtended by the rocket at the camera lens.

74. Show that the length L of the pulley belt shown in Fig. 20.41 is

$$L = 24 + 11\pi + 10 \sin^{-1}\frac{5}{13}.$$

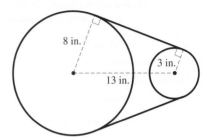

8 in.

3 in.

13 in.

Fig. 20.41

Answers to Practice Exercises

1. $-\pi/6$ **2.** $\sqrt{3}/2$

CHAPTER 20   EQUATIONS

Basic trigonometric identities

$$\sin \theta = \frac{1}{\csc \theta} \quad \text{(20.1)}$$

$$\cos \theta = \frac{1}{\sec \theta} \quad \text{(20.2)}$$

$$\tan \theta = \frac{1}{\cot \theta} \quad \text{(20.3)}$$

$$\tan \theta = \frac{\sin \theta}{\cos \theta} \quad \text{(20.4)}$$

$$\cot \theta = \frac{\cos \theta}{\sin \theta} \quad \text{(20.5)}$$

$$\sin^2 \theta + \cos^2 \theta = 1 \quad \text{(20.6)}$$

$$1 + \tan^2 \theta = \sec^2 \theta \quad \text{(20.7)}$$

$$1 + \cot^2 \theta = \csc^2 \theta \quad \text{(20.8)}$$

Sum and difference identities

$$\sin(\alpha + \beta) = \sin \alpha \cos \beta + \cos \alpha \sin \beta \quad \text{(20.9)}$$

$$\cos(\alpha + \beta) = \cos \alpha \cos \beta - \sin \alpha \sin \beta \quad \text{(20.10)}$$

$$\sin(\alpha - \beta) = \sin \alpha \cos \beta - \cos \alpha \sin \beta \quad \text{(20.11)}$$

$$\cos(\alpha - \beta) = \cos \alpha \cos \beta + \sin \alpha \sin \beta \quad \text{(20.12)}$$

$$\tan(\alpha \pm \beta) = \frac{\tan \alpha \pm \tan \beta}{1 \mp \tan \alpha \tan \beta} \quad \text{(20.13)}$$

Double-angle formulas

$$\sin 2\alpha = 2 \sin \alpha \cos \alpha \quad \text{(20.21)}$$

$$\cos 2\alpha = \cos^2 \alpha - \sin^2 \alpha \quad \text{(20.22)}$$

$$= 2 \cos^2 \alpha - 1 \quad \text{(20.23)}$$

$$= 1 - 2 \sin^2 \alpha \quad \text{(20.24)}$$

$$\tan 2\alpha = \frac{2 \tan \alpha}{1 - \tan^2 \alpha} \quad \text{(20.25)}$$

Half-angle formulas
$$\sin\frac{\alpha}{2} = \pm\sqrt{\frac{1-\cos\alpha}{2}} \qquad (20.26)$$

$$\cos\frac{\alpha}{2} = \pm\sqrt{\frac{1+\cos\alpha}{2}} \qquad (20.27)$$

Inverse trigonometric functions
$$y = \sin^{-1}x \quad \left(-\frac{\pi}{2} \le y \le \frac{\pi}{2}\right) \qquad (20.28)$$

$$-\frac{\pi}{2} \le \sin^{-1}x \le \frac{\pi}{2} \qquad 0 \le \cos^{-1}x \le \pi \qquad -\frac{\pi}{2} < \tan^{-1}x < \frac{\pi}{2} \qquad 0 < \cot^{-1}x < \pi$$

$$0 \le \sec^{-1}x \le \pi \quad \left(\sec^{-1}x \ne \frac{\pi}{2}\right) \qquad -\frac{\pi}{2} \le \csc^{-1}x \le \frac{\pi}{2} \quad (\csc^{-1}x \ne 0) \qquad (20.29)$$

CHAPTER 20   REVIEW EXERCISES

In Exercises 1–8, determine the values of the indicated functions in the given manner.

1. Find $\sin 120°$ by using $120° = 90° + 30°$.
2. Find $\cos 30°$ by using $30° = 90° - 60°$.
3. Find $\sin 315°$ by using $315° = 360° - 45°$.
4. Find $\tan\frac{5\pi}{4}$ by using $\frac{5\pi}{4} = \pi + \frac{\pi}{4}$.
5. Find $\cos \pi$ by using $\pi = 2\left(\frac{\pi}{2}\right)$.
6. Find $\sin 180°$ by using $180° = 2(90°)$.
7. Find $\tan 60°$ by using $60° = 2(30°)$.
8. Find $\cos 45°$ by using $45° = \frac{1}{2}(90°)$.

In Exercises 9–16, simplify the given expressions by using one of the basic formulas of the chapter. Then use a calculator to verify the result by finding the value of the original expression and the value of the simplified expression.

9. $\sin 14° \cos 38° + \cos 14° \sin 38°$
10. $\cos^2 148° - \sin^2 148°$
11. $2\sin\frac{\pi}{12}\cos\frac{\pi}{12}$
12. $1 - 2\sin^2\frac{\pi}{8}$
13. $\cos 73° \cos(-142°) + \sin 73° \sin(-142°)$
14. $\cos 3° \cos 215° - \sin 3° \sin 215°$
15. $\dfrac{4\tan 12°}{1 - \tan^2 12°}$
16. $\sqrt{\dfrac{1 - \cos 166°}{2}}$

In Exercises 17–24, simplify each of the given expressions. Expansion of any term is not necessary; recognition of the proper form leads to the proper result.

17. $\sin 2x \cos 3x + \cos 2x \sin 3x$
18. $\cos 7x \cos 3x + \sin 7x \sin 3x$
19. $8\sin 6x \cos 6x$
20. $\dfrac{\tan x + \tan 2x}{1 - \tan x \tan 2x}$
21. $2 - 4\sin^2 6x$
22. $\cos^2 2x - \sin^2 2x$
23. $\sqrt{2 + 2\cos 2x}$
24. $\sqrt{32 - 32\cos 4x}$

In Exercises 25–32, evaluate the given expressions.

25. $\sin^{-1}(-1)$
26. $\sec^{-1}\sqrt{2}$
27. $\cos^{-1} 0.9659$
28. $\tan^{-1}(-6.249)$
29. $\tan[\sin^{-1}(-0.5)]$
30. $\cos\left[\tan^{-1}\left(-\sqrt{3}\right)\right]$
31. $\sin^{-1}[\sin(7\pi/6)]$
32. $\cos^{-1}[\tan(-\pi/4)]$

In Exercises 33–36, simplify the given expressions.

33. $\dfrac{\sec y}{\cos y} - \dfrac{\tan y}{\cot y}$
34. $\dfrac{\sin 2\theta}{2\csc\theta} - \cos^3\theta$
35. $\sin x(\csc x - \sin x)$
36. $\cos y(\sec y - \cos y)$

In Exercises 37–44, prove the given identities.

37. $\dfrac{\sec^4 x - 1}{\tan^2 x} = 2 + \tan^2 x$
38. $\cos^2 y - \sin^2 y = \dfrac{1 - \tan^2 y}{1 + \tan^2 y}$
39. $2\csc 2x \cot x = 1 + \cot^2 x$
40. $\sin x \cot^2 x = \csc x - \sin x$
41. $\dfrac{1 - \sin^2\theta}{1 - \cos^2\theta} = \cot^2\theta$
42. $\dfrac{\cos 2\theta}{\cos^2\theta} = 1 - \tan^2\theta$
43. $\sin\dfrac{\theta}{2}\cos\dfrac{\theta}{2} = \dfrac{\sin\theta}{2}$
44. $\cos(x - y)\cos y - \sin(x - y)\sin y = \cos x$

In Exercises 45–52, simplify the given expressions. The result will be one of $\sin x$, $\cos x$, $\tan x$, $\cot x$, $\sec x$, or $\csc x$.

45. $\dfrac{\sec x}{\sin x} - \sec x \sin x$
46. $\cos x \cot x + \sin x$
47. $\sin x \tan x + \cos x$
48. $\dfrac{\tan x \csc x}{\sin x} - \cot x$
49. $\dfrac{\sin x \cot x + \cos x}{2\cot x}$
50. $\dfrac{1 + \cos 2x}{2\cos x}$

51. $\dfrac{\sin 2x \sec x}{2}$

52. $(\sec x + \tan x)(1 - \sin x)$

In Exercises 53–60, verify each identity by comparing the graph of the left side with the graph of the right side on a graphing calculator.

53. $\dfrac{\cos \theta - \sin \theta}{\cos \theta + \sin \theta} = \dfrac{\cot \theta - 1}{\cot \theta + 1}$

54. $\sin 3y \cos 2y - \cos 3y \sin 2y = \sin y$

55. $\sin 4x (\cos^2 2x - \sin^2 2x) = \dfrac{\sin 8x}{2}$

56. $\csc 2x + \cot 2x = \cot x$

57. $\dfrac{\sin x}{\csc x - \cot x} = 1 + \cos x$

58. $\cos x - \sin \dfrac{x}{2} = \left(1 - 2 \sin \dfrac{x}{2}\right)\left(1 + \sin \dfrac{x}{2}\right)$

59. $\tan \dfrac{\alpha}{2} = \csc \alpha - \cot \alpha$

60. $\sec \dfrac{x}{2} + \csc \dfrac{x}{2} = \dfrac{2\left(\sin \frac{x}{2} + \cos \frac{x}{2}\right)}{\sin x}$

In Exercises 61–64, solve for x.

61. $y = 2 \cos 2x$

62. $y - 2 = 2 \tan\left(x - \dfrac{\pi}{2}\right)$

63. $y = \dfrac{\pi}{4} - 3 \sin^{-1} 5x$

64. $2y = \sec^{-1} 4x - 2$

In Exercises 65–76, solve the given equations for x $(0 \le x < 2\pi)$.

65. $3(\tan x - 2) = 1 + \tan x$

66. $5 \sin x = 3 - (\sin x + 2)$

67. $2(1 - 2 \sin^2 x) = 1$

68. $\sec x = 2 \tan^2 x$

69. $2 \sin^2 \theta + 3 \cos \theta - 3 = 0$

70. $2 \sin 2x + 1 = 0$

71. $\sin x = \sin \frac{x}{2}$

72. $\cos 2x = \sin(-x)$

73. $\sin 2x = \cos 3x$

74. $\cos 3x \cos x + \sin 3x \sin x = 0$

75. $\sin^2\left(\dfrac{x}{2}\right) - \cos x + 1 = 0$

76. $\sin x + \cos x = 1$

In Exercises 77–80, determine whether the equality is an identity or an equation. If it is an identity, prove it. If it is an equation, solve it for $0 \le x < 2\pi$.

77. $\tan x + \cot x = \csc x \sec x$

78. $\tan x - \sin^2 x = \cos^2 x - \sec^2 x$

79. $\sin x \cos x - 1 = \cos x - \sin x$

80. $2 \tan x = \sin 2x \sec^2 x$

In Exercises 81–84, solve the given equations graphically.

81. $x + \ln x - 3 \cos^2 x = 2$

82. $e^{\sin x} - 2 = x \cos^2 x$

83. $2 \tan^{-1} x + x^2 = 3$

84. $3 \sin^{-1} x = 6 \sin x + 1$

In Exercises 85–90, find an algebraic expression for each of the given expressions.

85. $\tan(\cot^{-1} x)$

86. $\cos(\csc^{-1} x)$

87. $\sin(2 \cos^{-1} x)$

88. $\cos(\pi - \tan^{-1} x)$

89. $\cos(\sin^{-1} x + \tan^{-1} y)$

90. $\sin(\cos^{-1} x - \tan^{-1} y)$

In Exercises 91–94, use the given substitutions to show that the equations are valid for $0 \le \theta < \pi/2$.

91. If $x = 2 \cos \theta$, show that $\sqrt{4 - x^2} = 2 \sin \theta$.

92. If $x = 2 \sec \theta$, show that $\sqrt{x^2 - 4} = 2 \tan \theta$.

93. If $x = \tan \theta$, show that $\dfrac{x}{\sqrt{1 + x^2}} = \sin \theta$.

94. If $x = \cos \theta$, show that $\dfrac{\sqrt{1 - x^2}}{x} = \tan \theta$.

In Exercises 95–114, use the methods and formulas of this chapter to solve the given problems.

95. Prove: $(\cos \theta + j \sin \theta)^2 = \cos 2\theta + j \sin 2\theta$ $(j = \sqrt{-1})$

96. Prove: $(\cos \theta + j \sin \theta)^3 = \cos 3\theta + j \sin 3\theta$ $(j = \sqrt{-1})$

97. Solve the inequality $\sin 2x > 2 \cos x$ for $0 \le x < 2\pi$.

98. Show that $(\cos 2\alpha + \sin^2 \alpha) \sec^2 \alpha$ has a constant value.

99. Show that $y = A \sin 2t + B \cos 2t$ may be written as $y = C \sin(2t + \alpha)$, where $C = \sqrt{A^2 + B^2}$ and $\tan \alpha = B/A$. (*Hint:* Let $A/C = \cos \alpha$ and $B/C = \sin \alpha$.)

100. In a right triangle with sides a and b and hypotenuse c, show that $\sin(A/2) = \sqrt{(c - b)/2c}$, where angle A is opposite a.

101. Forces $\mathbf{A}$ and $\mathbf{B}$ act on a bolt such that $\mathbf{A}$ makes an angle θ with the x-axis and $\mathbf{B}$ makes an angle θ with the y-axis as shown in Fig. 20.42. The resultant $\mathbf{R}$ has components $R_x = A \cos \theta - B \sin \theta$ and $R_y = A \sin \theta + B \cos \theta$. Using these components, show that $R = \sqrt{A^2 + B^2}$.

Fig. 20.42

102. For a certain alternating-current generator, the expression $I \cos \theta + I \cos(\theta + 2\pi/3) + I \cos(\theta + 4\pi/3)$ arises. Simplify this expression.

103. Some comets follow a parabolic path that can be described by the equation $r = (k/2) \csc^2(\theta/2)$, where r is the distance to the sun and k is a constant. Show that this equation can be written as $r = k/(1 - \cos \theta)$.

104. In studying the interference of light waves, the identity
$$\frac{\sin\frac{3}{2}x}{\sin\frac{1}{2}x}\sin x = \sin x + \sin 2x$$ is used. Prove this identity.
$\left(Hint: \sin\frac{3}{2}x = \sin\left(x + \frac{1}{2}x\right).\right)$

105. In the study of chemical spectroscopy, the equation
$$\omega t = \sin^{-1}\frac{\theta - \alpha}{R}$$ arises. Solve for θ.

106. The power p in a certain electric circuit is given by $p = 2.5[\cos\alpha\sin(\omega t + \phi) - \sin\alpha\cos(\omega t + \phi)]$. Solve for t.

107. In surveying, when determining an azimuth (a measure used for reference purposes), it might be necessary to simplify the expression $(2\cos\alpha\cos\beta)^{-1} - \tan\alpha\tan\beta$. Perform this operation by expressing it in the simplest possible form when $\alpha = \beta$.

108. In analyzing the motion of an automobile universal joint, the equation $\sec^2 A - \sin^2 B\tan^2 A = \sec^2 C$ is used. Show that this equation is true if $\tan A\cos B = \tan C$.

109. The instantaneous power p in a certain electric circuit is given by $p = VI\cos\phi\cos^2\omega t - VI\sin\phi\cos\omega t\sin\omega t$. Simplify this expression.

110. In studying waveforms, a sawtooth wave may often be approximated by $y = \sin\pi x + \frac{1}{2}\sin 2\pi x + \frac{1}{3}\sin 3\pi x$. For what values of x, $0 \le x < 2$, is the sum of the first two terms equal to zero?

111. A roof truss is in the shape of an isosceles triangle of height 6.4 ft. If the total length of the three members is 51.2 ft, what is the length of a rafter? Solve by setting up and solving an appropriate trigonometric equation. (The Pythagorean theorem may be used only as a check.)

112. The angle of elevation of the top of the Washington Monument from a point on level ground 250 ft from the center of its base is twice the angle of elevation of the top from a point 610 ft farther away. Find the height of the monument.

113. If a plane surface inclined at angle θ moves horizontally, the angle for which the lifting force of the air is a maximum is found by solving the equation $2\sin\theta\cos^2\theta - \sin^3\theta = 0$, where $0 < \theta < 90°$. Solve for θ.

114. To determine the angle between two sections of a certain robot arm, the equation $1.20\cos\theta + 0.135\cos 2\theta = 0$ is to be solved. Find the required angle θ if $0° < \theta < 180°$.

Writing Exercise

115. In checking the angles of a section of a bridge support, an engineer finds the expression $\cos(2\sin^{-1}0.40)$. Write a paragraph explaining how the value of this expression can be found without the use of a calculator.

CHAPTER 20   **PRACTICE TEST**

1. Prove that $\sec\theta - \dfrac{\tan\theta}{\csc\theta} = \cos\theta$.

2. Solve for x $(0 \le x < 2\pi)$ analytically, using trigonometric relations where necessary: $\sin 2x + \sin x = 0$.

3. Find an algebraic expression for $\cos(\sin^{-1}x)$.

4. The electric current as a function of the time for a particular circuit is given by $i = 8.00e^{-20t}(1.73\cos 10.0t - \sin 10.0t)$. Find the time (in s) when the current is first zero.

5. Prove that $\dfrac{\tan\alpha + \tan\beta}{\tan\alpha - \tan\beta} = \dfrac{\sin(\alpha + \beta)}{\sin(\alpha - \beta)}$.

6. Prove that $\cot^2 x - \cos^2 x = \cot^2 x\cos^2 x$.

7. Find $\cos\frac{1}{2}x$ if $\sin x = -\frac{3}{5}$ and $270° < x < 360°$.

8. The intensity of a certain type of polarized light is given by $I = I_0\sin 2\theta\cos 2\theta$. Solve for θ.

9. Solve graphically: $x - 2\cos x = 5$.

10. Find the exact value of x: $\cos^{-1}x = -\tan^{-1}(-1)$.

21 Plane Analytic Geometry

The development of geometry made little progress from the time of the ancient Greeks until the 1600s. Then in the 1630s, two French mathematicians, Rene Descartes and Pierre de Fermat, independently introduced algebra into the study of geometry. Each used algebraic notation and a coordinate system, and this allowed them to analyze the properties of curves. Fermat briefly indicated his methods in a letter he wrote in 1636. However, in 1637 Descartes included a much more complete work as an appendix to his *Discourse on Method,* and he is therefore now considered as the founder of *analytic geometry.*

At the time, it was known, for example, that a projectile follows a parabolic path and that the orbits of the planets are ellipses, and there was a renewed interest in curves of various kinds because of their applications. However, Descartes was primarily interested in studying the relationship of algebra to geometry. In doing so, he started the development of analytic geometry, which has many applications and also was very important in the invention of calculus that came soon thereafter. In turn, through these advances in mathematics, many more areas of science and technology were able to be developed.

The underlying principle of analytic geometry is the relationship of an algebraic equation and the geometric properties of the curve that represents the equation. In this chapter, we develop equations for a number of important curves and find their properties through an analysis of their equations. Most important among these curves are the *conic sections,* which we briefly introduced in Chapter 14.

As we have noted, analytic geometry has applications in the study of projectile motion and planetary orbits. Other important applications range from the design of gears, airplane wings, and automobile headlights to the construction of bridges and nuclear towers.

In Section 21.4, we show an important application of analytic geometry in the design of a TV satellite dish.

21.1 Basic Definitions

The Distance Formula •
The Slope of a Line • Inclination •
Parallel Lines • Perpendicular Lines

As we have noted, analytic geometry deals with the relationship between an algebraic equation and the geometric curve it represents. In this section, we develop certain basic concepts that will be needed for future use in establishing the proper relationships between an equation and a curve.

THE DISTANCE FORMULA

The first of these concepts involves the distance between any two points in the coordinate plane. If these points lie on a line parallel to the x-axis, the distance from the first point (x_1, y) to the second point (x_2, y) is $|x_2 - x_1|$. The absolute value is used since we are interested only in the magnitude of the distance. Therefore, we could also denote the distance as $|x_1 - x_2|$. Similarly, the distance between two points (x, y_1) and (x, y_2) that lie on a line parallel to the y-axis is $|y_2 - y_1|$ or $|y_1 - y_2|$.

EXAMPLE 1 Distance between points

The line segment joining $A(-1, 5)$ and $B(-4, 5)$ in Fig. 21.1 is parallel to the x-axis. Therefore, the distance between these points is

$$d = |-4 - (-1)| = 3 \quad \text{or} \quad d = |-1 - (-4)| = 3$$

Also, in Fig. 21.1, the line segment joining $C(2, -3)$ and $D(2, 6)$ is parallel to the y-axis. The distance d between these points is

$$d = |6 - (-3)| = 9 \quad \text{or} \quad d = |-3 - 6| = 9 \qquad ■$$

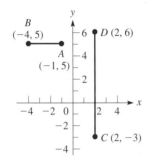

Fig. 21.1

We now wish to find the length of a line segment joining any two points in the plane. If these points are on a line that is not parallel to either axis (see Fig. 21.2), we use the Pythagorean theorem to find the distance between them. By making a right triangle with the line segment joining the points as the hypotenuse and line segments parallel to the axes as legs, we have *the* **distance formula,** *which gives the distance between any two points in the plane.* This formula is

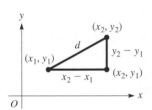

Fig. 21.2

$$\boxed{d = \sqrt{(x_2 - x_1)^2 + (y_2 - y_1)^2}} \qquad (21.1)$$

Here, we choose the positive square root since we are concerned only with the magnitude of the length of the line segment.

EXAMPLE 2 Distance formula – either point as (x_1, y_1)

The distance between $(3, -1)$ and $(-2, -5)$ is given by

$$d = \sqrt{[(-2) - 3]^2 + [(-5) - (-1)]^2}$$
$$= \sqrt{(-5)^2 + (-4)^2} = \sqrt{25 + 16}$$
$$= \sqrt{41} = 6.403$$

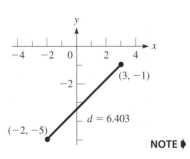

Fig. 21.3

See Fig. 21.3.

NOTE ▶ *It makes no difference which point is chosen as (x_1, y_1) and which is chosen as (x_2, y_2),* since the differences in the x-coordinates and the y-coordinates are squared. We obtain the same value for the distance when we calculate it as

$$d = \sqrt{[3 - (-2)]^2 + [(-1) - (-5)]^2}$$
$$= \sqrt{5^2 + 4^2} = \sqrt{41} = 6.403 \qquad ■$$

Practice Exercise

1. Find the distance between $(-2, 6)$ and $(5, -3)$.

THE SLOPE OF A LINE

Another important quantity for a line is its *slope,* which we defined in Chapter 5. Here, we review its definition and develop its meaning in more detail.

The **slope** *of a line through two points is defined as the difference in the y-coordinates (rise) divided by the difference in the x-coordinates (run).* Therefore, the slope, *m*, which gives a measure of the direction of a line, is defined as

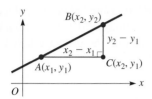

Fig. 21.4

$$m = \frac{y_2 - y_1}{x_2 - x_1} \tag{21.2}$$

See Fig. 21.4. Again, we may interpret either of the points as (x_1, y_1) and the other as (x_2, y_2), although *we must be careful to place the* $y_2 - y_1$ *in the numerator and* $x_2 - x_1$ *in the denominator.* When the line is horizontal, $y_2 - y_1 = 0$ and $m = 0$. When the line is vertical, $x_2 - x_1 = 0$ and the slope is undefined.

EXAMPLE 3 Slope of line

The slope of a line through $(3, -5)$ and $(-2, -6)$ is

$$m = \frac{-6 - (-5)}{-2 - 3} = \frac{-6 + 5}{-5} = \frac{1}{5}$$

See Fig. 21.5. Again, we may interpret either of the points as (x_1, y_1) and the other as (x_2, y_2). We can also obtain the slope of this same line from

$$m = \frac{-5 - (-6)}{3 - (-2)} = \frac{1}{5} \qquad \blacksquare$$

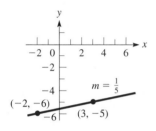

Fig. 21.5

The larger the numerical value of the slope of a line, the more nearly vertical is the line. Also, *a line rising to the right has a positive slope, and a line falling to the right has a negative slope.*

EXAMPLE 4 Magnitude and sign of slope

(a) The line in Example 3 has a positive slope, which is numerically small. From Fig. 21.5, it can be seen that the line rises slightly to the right.

(b) The line joining $(3, 4)$ and $(4, -6)$ has a slope of

$$m = \frac{4 - (-6)}{3 - 4} = -10 \quad \text{or} \quad m = \frac{-6 - 4}{4 - 3} = -10$$

This line falls sharply to the right, as shown in Fig. 21.6. $\blacksquare$

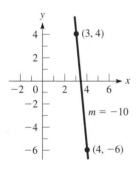

Fig. 21.6

If a given line is extended indefinitely in either direction, it must cross the *x*-axis at some point unless it is parallel to the *x*-axis. *The angle measured from the x-axis in a positive direction to the line is called the* **inclination** *of the line* (see Fig. 21.7). The inclination of a line parallel to the *x*-axis is defined to be zero. *An alternate definition of slope, in terms of the inclination* α, *is*

Fig. 21.7

$$m = \tan \alpha \quad (0° \le \alpha < 180°) \tag{21.3}$$

Since the slope can be defined in terms of any two points on the line, we can choose the *x*-intercept and any other point. Therefore, from the definition of the tangent of an angle, we see that Eq. (21.3) is in agreement with Eq. (21.2).

Fig. 21.8

Fig. 21.9

Fig. 21.10

Fig. 21.11

Practice Exercise

2. What is the slope of a line perpendicular to the line through $(-1, 6)$ and $(-3, -3)$?

EXAMPLE 5 Inclination

(a) The slope of a line with an inclination of 45° is

$$m = \tan 45° = 1.000$$

(b) If a line has a slope of -1.732, we know that $\tan \alpha = -1.732$. Since $\tan \alpha$ is negative, α must be a second-quadrant angle. Therefore, using a calculator to show that $\tan 60° = 1.732$, we find that $\alpha = 180° - 60° = 120°$. See Fig. 21.8.

We see that if the inclination is an acute angle, the slope is positive and the line rises to the right. If the inclination is negative, the slope is negative and the line falls to the right. ∎

Any two parallel lines crossing the x-axis have the same inclination. Therefore, as shown in Fig. 21.9, the *slopes of parallel lines are equal.* This can be stated as

$$\boxed{m_1 = m_2} \qquad \text{(for } \parallel \text{ lines)} \qquad \textbf{(21.4)}$$

If two lines are perpendicular, this means that there must be 90° between their inclinations (Fig. 21.10). The relation between their inclinations is

$$\alpha_2 = \alpha_1 + 90°$$
$$90° - \alpha_2 = -\alpha_1$$

If neither line is vertical (the slope of a vertical line is undefined) and we take the tangent in this last relation, we have

$$\tan(90° - \alpha_2) = \tan(-\alpha_1)$$

or

$$\cot \alpha_2 = -\tan \alpha_1$$

since a function of the complement of an angle equals the cofunction of that angle (see page 120) and since $\tan(-\alpha) = -\tan \alpha$ (see page 239). But $\cot \alpha = 1/\tan \alpha$, which means $1/\tan \alpha_2 = -\tan \alpha_1$. Using the inclination definition of slope, we have *as the relation between slopes of perpendicular lines,*

$$\boxed{m_2 = -\dfrac{1}{m_1} \quad \text{or} \quad m_1 m_2 = -1} \qquad \text{for } \perp \text{ lines} \qquad \textbf{(21.5)}$$

EXAMPLE 6 Perpendicular lines

The line through $(3, -5)$ and $(2, -7)$ has a slope of

$$m_1 = \frac{-5 + 7}{3 - 2} = 2$$

The line through $(4, -6)$ and $(2, -5)$ has a slope of

$$m_2 = \frac{-6 - (-5)}{4 - 2} = -\frac{1}{2}$$

Since the slopes of the two lines are negative reciprocals, we know that the lines are perpendicular. See Fig. 21.11. ∎

■ See Appendix C for a graphing calculator program SLOPEDIS. It calculates the slope of a line through two points and the distance between the points.

Using the formulas for distance and slope, we can show certain basic geometric relationships. The following examples illustrate the use of the formulas and thereby show the use of algebra in solving problems that are basically geometric. This illustrates the methods of analytic geometry.

EXAMPLE 7 Use of algebra in geometric problems

(a) Show that the line segments joining $A(-5, 3)$, $B(6, 0)$, and $C(5, 5)$ form a right triangle. See Fig. 21.12.

If these points are vertices of a right triangle, the slopes of two of the sides must be negative reciprocals. This would show perpendicularity. These slopes are

$$m_{AB} = \frac{3 - 0}{-5 - 6} = -\frac{3}{11} \qquad m_{AC} = \frac{3 - 5}{-5 - 5} = \frac{1}{5} \qquad m_{BC} = \frac{0 - 5}{6 - 5} = -5$$

Fig. 21.12

We see that the slopes of AC and BC are negative reciprocals, which means $AC \perp BC$. From this we conclude that the triangle is a right triangle.

(b) Find the area of the triangle in part (a). See Fig. 21.13.

Since the right angle is at C, the legs of the triangle are AC and BC. The area is one-half the product of the lengths of the legs of a right triangle. The lengths of the legs are

$$d_{AC} = \sqrt{(-5 - 5)^2 + (3 - 5)^2} = \sqrt{104} = 2\sqrt{26}$$
$$d_{BC} = \sqrt{(6 - 5)^2 + (0 - 5)^2} = \sqrt{26}$$

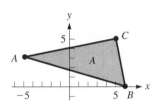

Fig. 21.13

Therefore, the area is $A = \frac{1}{2}(2\sqrt{26})(\sqrt{26}) = 26$. ■

EXERCISES 21.1

In Exercises 1–4, make the given changes in the indicated examples of this section and then solve the resulting problems.

1. In Example 2, change $(-2, -5)$ to $(-2, 5)$.

2. In Example 3, change $(3, -5)$ to $(-3, -5)$.

3. In Example 5(b), change -1.732 to -0.5774.

4. In Example 7(a), change $B(6, 0)$ to $B(-4, -2)$.

In Exercises 5–14, find the distance between the given pairs of points.

5. $(3, 8)$ and $(-1, -2)$

6. $(-1, 3)$ and $(-8, -4)$

7. $(4, -5)$ and $(4, -8)$

8. $(-3, 7)$ and $(2, 10)$

9. $(-12, 20)$ and $(32, -13)$

10. $(23, -9)$ and $(-25, 11)$

11. $(\sqrt{32}, -\sqrt{18})$ and $(-\sqrt{50}, \sqrt{8})$

12. $(e, -\pi)$ and $(-2e, -\pi)$

13. $(1.22, -3.45)$ and $(-1.07, -5.16)$

14. (a, h^2) and $(a + h, (a + h)^2)$

In Exercises 15–24, find the slopes of the lines through the points in Exercises 5–14.

In Exercises 25–28, find the slopes of the lines with the given inclinations.

25. $30°$ **26.** $62.5°$ **27.** $162.7°$ **28.** $93.5°$

In Exercises 29–32, find the inclinations of the lines with the given slopes.

29. 0.364 **30.** 0.824 **31.** -6.691 **32.** -0.721

In Exercises 33–36, determine whether the lines through the two pairs of points are parallel or perpendicular.

33. $(6, -1)$ and $(4, 3)$; $(-5, 2)$ and $(-7, 6)$

34. $(-3, 9)$ and $(4, 4)$; $(9, -1)$ and $(4, -8)$

35. $(-1, -4)$ and $(2, 3)$; $(-5, 2)$ and $(-19, 8)$

36. $(-a, -2b)$ and $(3a, 6b)$; $(2a, -6b)$ and $(5a, 0)$

In Exercises 37–40, determine the value of k.

37. The distance between $(-1, 3)$ and $(11, k)$ is 13.

38. The distance between $(k, 0)$ and $(0, 2k)$ is 10.

39. Points $(6, -1)$, $(3, k)$, and $(-3, -7)$ are on the same line.

40. The points in Exercise 39 are the vertices of a right triangle, with the right angle at $(3, k)$.

In Exercises 41–44, show that the given points are vertices of the given geometric figures.

41. $(2, 3)$, $(4, 9)$, and $(-2, 7)$ are vertices of an isosceles triangle.

42. $(-1, 3)$, $(3, 5)$, and $(5, 1)$ are the vertices of a right triangle.

43. $(-5, -4)$, $(7, 1)$, $(10, 5)$, and $(-2, 0)$ are the vertices of a parallelogram.

44. $(-5, 6)$, $(0, 8)$, $(-3, 1)$, and $(2, 3)$ are the vertices of a square.

In Exercises 45–48, find the indicated areas and perimeters.

45. Find the area of the triangle in Exercise 42.

46. Find the area of the square in Exercise 44.

47. Find the perimeter of the triangle in Exercise 41.

48. Find the perimeter of the parallelogram in Exercise 43.

In Exercises 49–52, use the following definition to find the midpoints between the given points on a straight line.

The *midpoint* between points (x_1, y_1) and (x_2, y_2) on a straight line is the point.

$$\left(\frac{x_1 + x_2}{2}, \frac{y_1 + y_2}{2} \right)$$

49. $(-4, 9)$ and $(6, 1)$ **50.** $(-1, 6)$ and $(-13, -8)$

51. $(-12.4, 25.7)$ and $(6.8, -17.3)$

52. $(2.6, 5.3)$ and $(-4.2, -2.7)$

In Exercises 53–62, solve the given problems.

53. Find the relation between x and y such that (x, y) is always 3 units from the origin.

54. Find the relation between x and y such that (x, y) is always equidistant from the y-axis and $(2, 0)$.

55. Show that the diagonals of a square are perpendicular to each other. (*Hint:* Use $(0, 0)$, $(a, 0)$, and $(0, a)$ as three of the vertices.)

56. The center of a circle is $(2, -3)$, and one end of a diameter is $(-1, 2)$. What are the coordinates of the other end of the diameter?

57. A line segment has a slope of 3 and one endpoint at $(-2, 5)$. If the other endpoint is on the x-axis, what are its coordinates?

58. The points $(-1, 3)$, $(5, x)$, and $(2, 4)$ are collinear (on the same line). Find x.

59. Find the coordinates of the point on the y-axis that is equidistant from $(-3, -5)$ and $(2, 4)$.

60. The points $(-6, 2)$, $(5, -1)$, and (a, a) $(a > 0)$ are vertices of a right triangle, with the right angle at (a, a). Find a.

61. An equilateral triangle has one vertex at the origin, another vertex on the x-axis, and the third vertex in the first quadrant. Find the slopes of the sides of the triangle.

62. A person is working out on a treadmill inclined at 12% (the slope of the treadmill expressed in percent). What is the angle between the treadmill and the horizontal?

Answers to Practice Exercises

1. $d = \sqrt{130} = 11.4$ **2.** $m = -2/9$

21.2 The Straight Line

Point-slope Form of Equation • Vertical Line • Horizontal Line • Slope-intercept Form of Equation • General Form of Equation

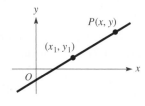

Fig. 21.14

In Chapter 5, we derived the *slope-intercept form* of the equation of the straight line. Here, we extend the development to include other forms of the equation of a straight line. Also, other methods of finding and applying these equations are shown. For completeness, we review some of the material in Chapter 5.

Using the definition of slope, we can derive the general type of equation that represents a straight line. This is another basic method of analytic geometry. That is, equations of a particular form can be shown to represent a particular type of curve. When we recognize the form of the equation, we know the kind of curve it represents. As we have seen, this is of great assistance in sketching the graph.

A straight line can be defined as a *curve* with a constant slope. This means that the value for the slope is the same for any two different points on the line that might be chosen. Thus, considering point (x_1, y_1) on a line to be fixed (Fig. 21.14) and another point $P(x, y)$ that *represents* any other point on the line, we have

$$m = \frac{y - y_1}{x - x_1}$$

which can be written as

$$\boxed{y - y_1 = m(x - x_1)} \tag{21.6}$$

Equation (21.6) is the **point-slope form** *of the equation of a straight line.* It is useful when we know the slope of a line and some point through which the line passes.

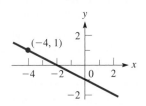

Fig. 21.15

EXAMPLE 1 Point-slope form

Find the equation of the line that passes through $(-4, 1)$ with a slope of $-1/2$. See Fig. 21.15.

Substituting in Eq. (21.6), we find that

$$y - 1 = \left(-\tfrac{1}{2}\right)[x - (-4)]$$

slope

coordinates

Simplifying, we have

$$2y - 2 = -x - 4$$
$$2y + x + 2 = 0$$ ∎

EXAMPLE 2 Equation of line through two points

Find the equation of the line through $(2, -1)$ and $(6, 2)$.

We first find the slope of the line through these points:

$$m = \frac{2 + 1}{6 - 2} = \frac{3}{4}$$

■ See Appendix C for a graphing calculator program GRAPHLIN. It displays the graph of a line through two given points.

Then by using either of the two known points and Eq. (21.6), we can find the equation of the line (see Fig. 21.16):

$$y - (-1) = \frac{3}{4}(x - 2)$$
$$4y + 4 = 3x - 6$$
$$4y - 3x + 10 = 0$$ ∎

Fig. 21.16

Equation (21.6) can be used for any line except for one parallel to the y-axis. Such a line has an undefined slope. However, it does have the property that all points on it have the same x-coordinate, regardless of the y-coordinate. *We represent a line parallel to the y-axis as*

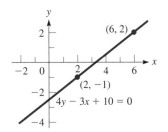

Fig. 21.17   **Fig. 21.18**

$$\boxed{x = a} \qquad \text{see Fig. 21.17} \qquad \textbf{(21.7)}$$

A line parallel to the x-axis has a slope of zero. From Eq. (21.6), we find its equation is $y = y_1$. To keep the same form as Eq. (21.7), we write this as

$$\boxed{y = b} \qquad \text{see Fig. 21.18} \qquad \textbf{(21.8)}$$

EXAMPLE 3 Vertical line – horizontal line

(a) The line $x = 2$ is a line parallel to the y-axis and 2 units to the right of it. This line is shown in Fig. 21.19.

Practice Exercise

1. Find the equation of the straight line that passes through $(-5, 2)$ and $(-1, -6)$.

Fig. 21.19

Fig. 21.20

(b) The line $y = -4$ is a line parallel to the x-axis and 4 units below it. This line is shown in Fig. 21.20. ∎

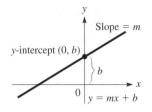

Fig. 21.21

If we choose the special point $(0, b)$, which is the y-intercept of the line, as the point to use in Eq. (21.6), we have $y - b = m(x - 0)$, or

$$y = mx + b \qquad \textbf{(21.9)}$$

Equation (21.9) is the **slope-intercept form** *of the equation of a straight line,* and we first derived it in Chapter 5. Its primary usefulness lies in the fact that once we find the equation of a line and then write it in slope-intercept form, we know that the slope of the line is the coefficient of the x-term and that it crosses the y-axis at the coordinate indicated by the constant term. See Fig. 21.21.

EXAMPLE 4   Slope – intercept form

Find the slope and the y-intercept of the straight line for which the equation is $2y + 4x - 5 = 0$.

We write this equation in slope-intercept form:

$$2y = -4x + 5$$

$$\underset{\text{slope}}{} \qquad \underset{\text{y-coordinate of intercept}}{}$$

$$y = -2x + \frac{5}{2}$$

Fig. 21.22

Since the coefficient of x in this form is -2, the slope is -2. The constant on the right is $5/2$, which means that the y-intercept is $(0, 5/2)$. See Fig. 21.22. ∎

EXAMPLE 5   Straight line – application

The pressure p_0 at the surface of a body of water (due to the atmosphere) is 101 kPa. The pressure p at a depth of 10.0 m is 199 kPa. In general, the pressure difference $p - p_0$ below the surface varies directly as the depth h. Find the equation of p as a function of h, and sketch its graph.

The solution is as follows:

$$
\begin{aligned}
p - p_0 &= kh && \text{direct variation} \\
199 - 101 &= k(10.0) && \text{substitute given values} \\
k &= 9.80 \text{ kPa/m} && \\
p - 101 &= 9.80h && \text{substitute in first equation} \\
p &= 9.80h + 101 &&
\end{aligned}
$$

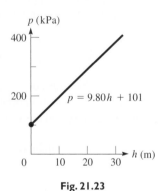

Fig. 21.23

We see that this is the equation of a straight line. The slope is 9.80, and the p-intercept is $(0, 101)$. Negative values do not have any physical meaning. The graph is shown in Fig. 21.23. ∎

From Eqs. (21.6) and (21.9), and from the examples of this section, we see that the equation of the straight line has certain characteristics: We have a term in y, a term in x, and a constant term if we simplify as much as possible. *This form is represented by the equation*

■ In Eq. (21.10), *A* and *B* cannot both be zero.

$$Ax + By + C = 0 \qquad \textbf{(21.10)}$$

which is known as the **general form** *of the equation of the straight line.* We saw this form before in Chapter 5. Now we have shown why it represents a straight line.

Fig. 21.24

Fig. 21.25

Fig. 21.26

Practice Exercise

2. Write the general form of the equation of the straight line that passes through $(0, -1)$ with a slope of 3.

EXAMPLE 6 General form

Find the general form of the equation of the line parallel to the line $3x + 2y - 6 = 0$ and that passes through the point $(-1, 2)$.

Since the line whose equation we want is parallel to the line $3x + 2y - 6 = 0$, it has the same slope. Thus, writing $3x + 2y - 6 = 0$ in slope-intercept form,

$$2y = -3x + 6 \qquad \text{solving for } y$$

$$y = -\frac{3}{2}x + 3$$

Since the slope of $3x + 2y - 6 = 0$ is $-3/2$, the slope of the required line is also $-3/2$. Using $m = -3/2$, the point $(-1, 2)$, and the point-slope form, we have

$$y - 2 = -\frac{3}{2}(x + 1)$$

$$2y - 4 = -3(x + 1)$$

$$3x + 2y - 1 = 0$$

This is the general form of the equation. Both lines are shown in Fig. 21.24.

With $y_1 = -3x/2 + 3$ and $y_2 = -3x/2 + 1/2$, these lines are shown in the calculator display in Fig. 21.25. ∎

In many physical situations, a linear relationship exists between variables. A few examples of this are (1) the distance traveled by an object and the elapsed time, when the velocity is constant, (2) the amount a spring stretches and the force applied, (3) the change in electric resistance and the change in temperature, (4) the force applied to an object and the resulting acceleration, and (5) the pressure at a certain point within a liquid and the depth of the point.

EXAMPLE 7 Straight line – application

For a period of 6.0 s, the velocity v of a car varies linearly with the elapsed time t. If $v = 40$ ft/s when $t = 1.0$ s and $v = 55$ ft/s when $t = 4.0$ s, find the equation relating v and t and graph the function. From the graph, find the initial velocity and the velocity after 6.0 s. What is the meaning of the slope of the line?

With v as the dependent variable and t as the independent variable, the slope is

$$m = \frac{v_2 - v_1}{t_2 - t_1}$$

Using the information given in the statement of the problem, we have

$$m = \frac{55 - 40}{4.0 - 1.0} = 5.0$$

Then using the point-slope form of the equation of a straight line, we have

$$v - 40 = 5.0(t - 1.0)$$

$$v = 5.0t + 35$$

The given values are sufficient to graph the line in Fig. 21.26. There is no need to include negative values of t, since they have no physical meaning. We see that the line crosses the v-axis at 35. This means that the initial velocity (for $t = 0$) is 35 ft/s. Also, when $t = 6.0$ s, we see that $v = 65$ ft/s.

The slope is the ratio of the change in velocity to the change in time. This is the car's *acceleration*. Here, the speed of the car increases 5.0 ft/s each second. We can express this acceleration as $5.0\ (\text{ft/s})/\text{s} = 5.0\ \text{ft/s}^2$. ∎

EXERCISES 21.2

In Exercises 1–4, make the given changes in the indicated examples of this section and then solve the resulting problems.

1. In Example 1, change $(-4, 1)$ to $(4, -1)$.
2. In Example 2, change $(2, -1)$ to $(-2, 1)$.
3. In Example 4, change the $+$ before $4x$ to $-$.
4. In Example 6, change the $+$ before $2y$ to $-$.

In Exercises 5–20, find the equation of each of the lines with the given properties. Sketch the graph of each line.

5. Passes through $(-3, 8)$ with a slope of 4.
6. Passes through $(-2, -1)$ with a slope of -2.
7. Passes through $(2, -5)$ and $(4, 2)$.
8. Has an x-intercept $(4, 0)$ and a y-intercept of $(0, -6)$.
9. Passes through $(-7, 12)$ with an inclination of $45°$.
10. Has a y-intercept $(0, -2)$ and an inclination of $120°$.
11. Passes through $(5.3, -2.7)$ and is parallel to the x-axis.
12. Passes through $(-15, 9)$ and is perpendicular to the x-axis.
13. Is parallel to the y-axis and is 3 units to the left of it.
14. Is parallel to the x-axis and is 4.1 units below it.
15. Perpendicular to line with slope of -3; passes through $(1, -2)$.
16. Parallel to line through $(-1, 7)$ and $(3, 1)$; passes through $(1, 2)$.
17. Has equal intercepts and passes through $(5, 2)$.
18. Is perpendicular to the line $6.0x - 2.4y - 3.9 = 0$ and passes through $(7.5, -4.7)$.
19. Has a slope of -3 and passes through the intersection of the lines $5x - y = 6$ and $x + y = 12$.
20. Passes through the point of intersection of $2x + y - 3 = 0$ and $x - y - 3 = 0$ and through the point $(4, -3)$.

In Exercises 21–28, reduce the equations to slope-intercept form and find the slope and the y-intercept. Sketch each line.

21. $4x - y = 8$
22. $2x - 3y - 6 = 0$
23. $3x + 5y - 10 = 0$
24. $4y = 6x - 9$
25. $3x - 2y - 1 = 0$
26. $4x + 2y - 5 = 0$
27. $11.2x + 1.6 = 3.2y$
28. $11.5x + 4.60y = 5.98$

In Exercises 29–36, determine whether the given lines are parallel, perpendicular, or neither.

29. $3x - 2y + 5 = 0$ and $4y = 6x - 1$
30. $8x - 4y + 1 = 0$ and $4x + 2y - 3 = 0$
31. $6x - 3y - 2 = 0$ and $x + 2y - 4 = 0$
32. $3y - 2x = 4$ and $6x - 9y = 5$
33. $5x + 2y - 3 = 0$ and $10y = 7 - 4x$
34. $48y - 36x = 71$ and $52x = 17 - 39y$

35. $4.5x - 1.8y = 1.7$ and $2.4x + 6.0y = 0.3$
36. $3.5y = 4.3 - 1.5x$ and $3.6x + 8.4y = 1.7$

In Exercises 37–60, solve the given problems. Exercises 45–56 show some applications of straight lines.

37. Find k if the lines $4x - ky = 6$ and $6x + 3y + 2 = 0$ are parallel.
38. Find k if the lines given in Exercise 37 are perpendicular.
(W)39. Find k if the lines $3x - y = 9$ and $kx + 3y = 5$ are perpendicular. Explain how this value is found.
(W)40. Find k such that the line through $(k, 2)$ and $(3, 1 - k)$ is perpendicular to the line $x - 2y = 5$. Explain your method.
41. Find the slope of the line joining points on the graph of $y = x^2$ that have x-coordinates of $-a$ and b $(a > 0, b > 0)$.
42. Show that the *intercept form* $\frac{x}{a} + \frac{y}{b} = 1$ is the equation of a line with x-intercept $(a, 0)$ and y-intercept $(0, b)$.
43. Find the distance from $(4, 1)$ to the line $4x - 3y + 12 = 0$.
44. Find the acute angle between the lines $x + y = 3$ and $2x - 5y = 4$.
45. Show that the following lines intersect to form a parallelogram. $8x + 10y = 3$; $2x - 3y = 5$; $4x - 6y = -3$; $5y + 4x = 1$.
46. For nonzero values of a, b, and c, find the intercepts of the line $ax + by + c = 0$.
47. For nonzero values of a, b, c, and d, show that (a) lines $ax + by + c = 0$ and $ax + by + d = 0$ are parallel, and (b) lines $ax + by + c = 0$ and $bx - ay + d = 0$ are perpendicular.
48. Find the equation of the line with positive intercepts that passes through $(3, 2)$ and forms with the axes a triangle of area 12.
49. In the 1700s, the French physicist Reaumur established a temperature scale on which the freezing point of water was $0°$ and the boiling point was $80°$. Set up an equation for the Fahrenheit temperature F (freezing point $32°$, boiling point $212°$) as a function of the Reaumur temperature R.
50. The voltage V across part of an electric circuit is given by $V = E - iR$, where E is a battery voltage, i is the current, and R is the resistance. If $E = 6.00$ V and $V = 4.35$ V for $i = 9.17$ mA, find V as a function of i. Sketch the graph (i and V may be negative.)
51. The velocity of sound v increases 0.607 m/s for each increase in temperature T of $1.00°$C. If $v = 343$ m/s for $T = 20.0°$C, express v as a function of T.
52. An acid solution is made from x liters of a 20% solution and y liters of a 30% solution. If the final solution contains 20 L of acid, find the equation relating x and y.
(W)53. A wall is 15 cm thick. At the outside, the temperature is $3°$C, and at the inside, it is $23°$C. If the temperature changes at a constant rate through the wall, write an equation of the temperature T in the wall as a function of the distance x from the outside to the inside of the wall. What is the meaning of the slope of the line?

54. An oil-storage tank is emptied at a constant rate. At 10 A.M., 1800 barrels remain, and at 2 P.M., 600 barrels remain. If pumping started at 8 A.M., find the equation relating the number of barrels n at time t (in h) from 8 A.M. When will the tank be empty?

55. The power output P (in W) of a computer chip operating at 120°C is proportional to $120 - T_S$, where T_S is the temperature of the surroundings. If $P = 1.0$ W for $T_S = 80$°C, find the equation relating P and T_S.

(W) 56. The length of a rectangular solar cell is 10 cm more than the width w. Express the perimeter p of the cell as a function of w. What is the meaning of the slope of the line?

57. A light beam is reflected off the edge of an optic fiber at an angle of 0.0032°. The diameter of the fiber is 48 μm. Find the equation of the reflected beam with the x-axis (at the center of the fiber) and the y-axis as shown in Fig. 21.27.

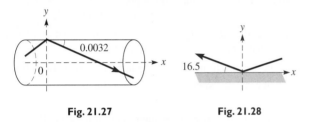

Fig. 21.27   **Fig. 21.28**

58. A police report stated that a bullet caromed upward off a floor at an angle of 16.5° with the floor, as shown in Fig. 21.28. What is the equation of the bullet's path after impact?

(W) 59. A survey of the traffic on a particular highway showed that the number of cars passing a particular point each minute varied linearly from 6:30 A.M. to 8:30 A.M. on workday mornings. The study showed that an average of 45 cars passed the point in 1 min at 7 A.M. and that 115 cars passed in 1 min at 8 A.M. If n is the number of cars passing the point in 1 min, and t is the number of minutes after 6:30 A.M., find the equation relating n and t, and graph the equation. From the graph, determine n at 6:30 A.M. and at 8:30 A.M. What is the meaning of the slope of the line?

60. In a research project on cancer, a tumor was determined to weigh 30 mg when first discovered. While being treated, it grew smaller by 2 mg each month. Find the equation relating the weight w of the tumor as a function of the time t in months. Graph the equation.

In Exercises 61–64, treat the given nonlinear functions as linear functions in order to sketch their graphs. At times, this can be useful in showing certain values of a function. For example, $y = 2 + 3x^2$ can be shown as a straight line by graphing y as a function of x^2. A table of values for this graph is shown along with the corresponding graph in Fig. 21.29.

x	0	1	2	3	4	5
x^2	0	1	4	9	16	25
y	2	5	14	29	50	77

Fig. 21.29

61. The number n of memory cells of a certain computer that can be tested in t seconds is given by $n = 1200\sqrt{t}$. Sketch n as a function of $\sqrt{t}$.

62. The force F (in lb) applied to a lever to balance a certain weight on the opposite side of the fulcrum is given by $F = 40/d$, where d is the distance (in ft) of the force from the fulcrum. Sketch F as a function of $1/d$.

63. A spacecraft is launched such that its altitude h (in km) is given by $h = 300 + 2t^{3/2}$ for $0 \le t < 100$ s. Sketch this as a linear function.

64. The current i (in A) in a certain electric circuit is given by $i = 6(1 - e^{-t})$. Sketch this as a linear function.

In Exercises 65–68, show that the given nonlinear functions are linear when plotted on semilogarithmic or logarithmic paper. In Section 13.7, we noted that graphs on this paper often become straight lines.

65. A function of the form $y = ax^n$ is straight when plotted on logarithmic paper, since $\log y = \log a + n \log x$ is in the form of a straight line. The variables are $\log y$ and $\log x$; the slope can be found from $(\log y - \log a)/\log x = n$, and the intercept is a. (To get the slope from the graph, it is necessary to measure vertical and horizontal distances between two points. The log y-intercept is found where $\log x = 0$, and this occurs when $x = 1$.) Plot $y = 3x^4$ on logarithmic paper to verify this analysis.

66. A function of the form $y = a(b^x)$ is a straight line on semilogarithmic paper, since $\log y = \log a + x \log b$ is in the form of a straight line. The variables are $\log y$ and x, the slope is $\log b$, and the intercept is a. (To get the slope from the graph, we calculate $(\log y - \log a)/x$ for some set of values x and y. The intercept is read directly off the graph where $x = 0$.) Plot $y = 3(2^x)$ on semilogarithmic paper to verify this analysis.

67. If experimental data are plotted on logarithmic paper and the points lie on a straight line, it is possible to determine the function (see Exercise 65). The following data come from an experiment to determine the functional relationship between the pressure p and the volume V of a gas undergoing an adiabatic (no heat loss) change. From the graph on logarithmic paper, determine p as a function of V.

V (m^3)	0.100	0.500	2.00	5.00	10.0
p (kPa)	20.1	2.11	0.303	0.0840	0.0318

68. If experimental data are plotted on semilogarithmic paper, and the points lie on a straight line, it is possible to determine the function (see Exercise 66). The following data come from an experiment designed to determine the relationship between the voltage across an inductor and the time after the switch is opened. Determine v as a function of t.

v (V)	40	15	5.6	2.2	0.8
t (ms)	0.0	20	40	60	80

Answers to Practice Exercises

1. $2x + y + 8 = 0$ **2.** $3x - y - 1 = 0$

21.3 The Circle

Standard Equation • Symmetry • General Equation • Calculator Graph

We have found that we can obtain a general equation that represents a straight line by considering a fixed point on the line and then a general point $P(x, y)$ that can represent any other point on the same line. Mathematically, we can state this as "the line is the **locus** of a point $P(x, y)$ that *moves* from a fixed point with constant slope along the line." That is, the point $P(x, y)$ can be considered as a variable point that moves along the line.

In this way, we can define a number of important curves. *A **circle** is defined as the locus of a point $P(x, y)$ that moves so that it is always equidistant from a fixed point. We call this fixed distance the **radius,** and we call the fixed point the **center** of the circle.* Thus, using this definition, calling the fixed point (h, k) and the radius r, we have

$$\sqrt{(x - h)^2 + (y - k)^2} = r$$

or, by squaring both sides, we have

$$(x - h)^2 + (y - k)^2 = r^2 \qquad \textbf{(21.11)}$$

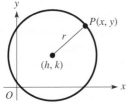

Fig. 21.30

*Equation (21.11) is called the **standard equation** of a circle with center at (h, k) and radius r.* See Fig. 21.30.

EXAMPLE 1 Standard equation

The equation $(x - 1)^2 + (y + 2)^2 = 16$ represents a circle with center at $(1, -2)$ and a radius of 4. We determine these values by considering the equation of the circle to be in the form of Eq. (21.11) as

$$(x \overset{\downarrow}{-} 1)^2 + [y \overset{\downarrow}{-} (-2)]^2 = 4^2$$

form requires − signs
coordinates of center
radius

Note carefully the way in which we found the y-coordinate of the center. *We must have a minus sign before each of the coordinates.* Here, to get the y-coordinate, we had to write $+2$ as $-(-2)$. This circle is shown in Fig. 21.31. ∎

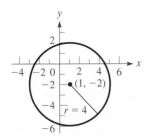

Fig. 21.31

EXAMPLE 2 Find equation of circle

Find the equation of the circle with center at $(2, 1)$ and that passes through $(4, -6)$.

In Eq. (21.11), we can determine the equation of this circle if we can find h, k, and r. From the given information, the center is $(2, 1)$, which means $h = 2$ and $k = 1$. To find r, we use the fact that *all points on the circle must satisfy the equation of the circle.* The point $(4, -6)$ must satisfy Eq. (21.11), with $h = 2$ and $k = 1$. This means

$$(4 - 2)^2 + (-6 - 1)^2 = r^2 \quad \text{or} \quad r^2 = 53$$

Therefore, the equation of the circle is

$$(x - 2)^2 + (y - 1)^2 = 53$$

This circle is shown in Fig. 21.32. ∎

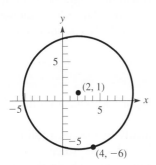

Fig. 21.32

Practice Exercise

1. Find the center and radius of the circle
$(x + 7)^1 + (y - 2)^2 = 1.$

Fig. 21.33

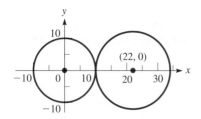

Fig. 21.34

If the center of the circle is at the origin, which means that the coordinates of the center are $(0, 0)$, the equation of the circle (see Fig. 21.33) becomes

$$x^2 + y^2 = r^2 \tag{21.12}$$

The following example illustrates an application using this type of circle and one with its center not at the origin.

EXAMPLE 3 Circle – application

A student is drawing a friction drive in which two circular disks are in contact with each other. They are represented by circles in the drawing. The first has a radius of 10.0 cm, and the second has a radius of 12.0 cm. What is the equation of each circle if the origin is at the center of the first circle and the positive x-axis passes through the center of the second circle? See Fig. 21.34.

Since the center of the smaller circle is at the origin, we can use Eq. (21.12). Given that the radius is 10.0 cm, we have as its equation

$$x^2 + y^2 = 100$$

The fact that the two disks are in contact tells us that they meet at the point $(10.0, 0)$. Knowing that the radius of the larger circle is 12.0 cm tells us that its center is at $(22.0, 0)$. Thus, using Eq. (21.11) with $h = 22.0$, $k = 0$, and $r = 12.0$,

$$(x - 22.0)^2 + (y - 0)^2 = 12.0^2$$

or

$$(x - 22.0)^2 + y^2 = 144$$

as the equation of the larger circle. ∎

Symmetry
A circle with its center at the origin exhibits an important property of the graphs of many equations. *It is* **symmetric** *to the x-axis and also to the y-axis.* Symmetry to the x-axis can be thought of as meaning that the lower half of the curve is a reflection of the upper half, and conversely. It can be shown that *if $-y$ can replace y in an equation without changing the equation, the graph of the equation is* **symmetric to the x-axis.** Symmetry to the y-axis is similar. *If $-x$ can replace x in the equation without changing the equation, the graph is symmetric to the y-axis.*

NOTE ▶

This type of circle is also symmetric to the origin as well as being symmetric to both axes. The meaning of symmetry to the origin is that the origin is the midpoint of any two points (x, y) and $(-x, -y)$ that are on the curve. Thus, *if $-x$ can replace x and $-y$ can replace y at the same time, without changing the equation, the graph of the equation is symmetric to the origin.*

NOTE ▶

EXAMPLE 4 Symmetry

The equation of the circle with its center at the origin and with a radius of 6 is $x^2 + y^2 = 36$.

The symmetry of this circle can be shown analytically by the substitutions mentioned above. Replacing x by $-x$, we obtain $(-x)^2 + y^2 = 36$. Since $(-x)^2 = x^2$, this equation can be rewritten as $x^2 + y^2 = 36$. Since this substitution did not change the equation, the graph is symmetric to the y-axis.

Replacing y by $-y$, we obtain $x^2 + (-y)^2 = 36$, which is the same as $x^2 + y^2 = 36$. This means that the curve is symmetric to the x-axis.

Replacing x by $-x$ and simultaneously replacing y by $-y$, we obtain $(-x)^2 + (-y)^2 = 36$, which is the same as $x^2 + y^2 = 36$. This means that the curve is symmetric to the origin. This circle is shown in Fig. 21.35. ∎

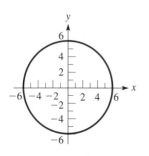

Fig. 21.35

If we multiply out each of the terms in Eq. (21.11), we may combine the resulting terms to obtain

$$x^2 - 2hx + h^2 + y^2 - 2ky + k^2 = r^2$$
$$x^2 + y^2 - 2hx - 2ky + (h^2 + k^2 - r^2) = 0 \qquad \textbf{(21.13)}$$

Since each of h, k, and r is constant for any given circle, the coefficients of x and y and the term within parentheses in Eq. (21.13) are constants. Equation (21.13) can then be written as

$$\boxed{x^2 + y^2 + Dx + Ey + F = 0} \qquad \textbf{(21.14)}$$

Equation (21.14) is called the **general equation** *of the circle.* It tells us that any equation that can be written in that form will represent a circle.

EXAMPLE 5 General equation

Find the center and radius of the circle.

$$x^2 + y^2 - 6x + 8y - 24 = 0$$

CAUTION ▶ We can find this information if we write the given equation in standard form. To do so, *we must complete the square in the x-terms and also in the y-terms.* This is done by first writing the equation in the form

$$(x^2 - 6x \quad) + (y^2 + 8y \quad) = 24$$

To complete the square of the x-terms, we take half of -6, which is -3, square it, and add the result, 9, to each side of the equation. In the same way, we complete the square of the y-terms by adding 16 to each side of the equation, which gives

$$(x^2 - 6x + 9) + (y^2 + 8y + 16) = 24 + 9 + 16$$
$$(x - 3)^2 + (y + 4)^2 = 49$$
$$(x - 3)^2 + (y - (-4))^2 = 7^2$$

Thus, the center is $(3, -4)$, and the radius is 7 (see Fig. 21.36). ∎

EXAMPLE 6 Circle – application

A certain pendulum is found to swing through an arc of the circle $3x^2 + 3y^2 - 9.60y - 2.80 = 0$. What is the length (in m) of the pendulum, and from what point is it swinging?

We see that this equation represents a circle by dividing through by 3. This gives us $x^2 + y^2 - 3.20y - 2.80/3 = 0$. The length of the pendulum is the radius of the circle, and the point from which it swings is the center. These are found as follows:

$$x^2 + (y^2 - 3.20y + 1.60^2) = 1.60^2 + 2.80/3 \quad \text{complete squares in both } x\text{- and } y\text{-terms}$$
$$x^2 + (y - 1.60)^2 = 3.493 \quad \text{standard form}$$

Since $\sqrt{3.493} = 1.87$, the length of the pendulum is 1.87 m. The point from which it is swinging is $(0, 1.60)$. See Fig. 21.37.

Replacing x by $-x$, the equation does not change. Replacing y by $-y$, the equation does change (the $3.20y$ term changes sign). Thus, the circle is symmetric only to the y-axis. ∎

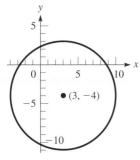

Fig. 21.36

Practice Exercise

2. Find the center and radius of the circle $x^2 + y^2 - 8x + 6y + 21 = 0$.

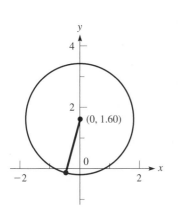

Fig. 21.37

NOTE ▶ In Section 14.1, we noted that the equation of a circle does not represent a *function* since there are two values of y for most values of x in the domain. In fact, *it might be necessary to use the quadratic formula to find the two functions to enter into a graphing calculator* in order to view the curve. This is illustrated in the following example.

EXAMPLE 7 Calculator graph of circle

Display the graph of the circle $3x^2 + 3y^2 + 6y - 20 = 0$ on a graphing calculator.

To fit the form of a quadratic equation in y, we write

$$3y^2 + 6y + (3x^2 - 20) = 0$$

Now, using the quadratic formula to solve for y, we let

$$a = 3 \quad b = 6 \quad c = 3x^2 - 20$$

$$y = \frac{-6 \pm \sqrt{6^2 - 4(3)(3x^2 - 20)}}{2(3)}$$

which means we get the *two functions*

$$y_1 = \frac{-6 + \sqrt{276 - 36x^2}}{6} \quad \text{and} \quad y_2 = \frac{-6 - \sqrt{276 - 36x^2}}{6}$$

which are entered into the calculator to get the view shown in Fig. 21.38.

The *window* values were chosen so that the length along the x-axis is about 1.5 times that along the y-axis so as to have less distortion in the circle. There may be gaps at the left and right sides of the circle. ■

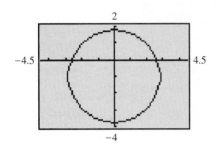

Fig. 21.38

EXERCISES 21.3

In Exercises 1–4, make the given changes in the indicated examples of this section and then solve the resulting problems.

1. In Example 1, change $(y + 2)^2$ to $(y + 1)^2$.
2. In Example 2, change $(2, 1)$ to $(-2, 1)$.
3. In Example 5, change the $+$ before $8y$ to $-$.
4. In Example 7, change the $+$ before $6y$ to $-$.

In Exercises 5–8, determine the center and the radius of each circle.

5. $(x - 2)^2 + (y - 1)^2 = 25$ 6. $(x - 3)^2 + (y + 4)^2 = 49$
7. $4(x + 1)^2 + 4y^2 = 121$ 8. $9x^2 + 9(y - 6)^2 = 64$

In Exercises 9–24, find the equation of each of the circles from the given information.

9. Center at $(0, 0)$, radius 3   10. Center at $(0, 0)$, radius 12
11. Center at $(2, 2)$, radius 4   12. Center at $\left(\frac{3}{2}, -2\right)$, radius $\frac{5}{2}$
13. Center at $(12, -15)$, radius 18
14. Center at $(-3, -5)$, radius $2\sqrt{3}$
15. The origin and $(-6, 8)$ are ends of a diameter
16. The points $(3, 8)$ and $(-3, 0)$ are the ends of a diameter.
17. Concentric with the circle $(x - 2)^2 + (y - 1)^2 = 4$ and passes through $(4, -1)$

18. Concentric with the circle $x^2 + y^2 + 2x - 8y + 8 = 0$ and passes through $(-2, 3)$
19. Center at $(-3, 5)$ and tangent to line $y = 10$
20. Center at $(-7, 1)$, tangent to y-axis
21. Tangent to both axes and the lines $y = 4$ and $x = -4$
22. Tangent to line $y = 2$ and $y = 8$, center on line $y = x$
23. Center at the origin, tangent to the line $x + y = 2$
24. Center at $(5, 12)$, tangent to the line $y = 2x - 3$

In Exercises 25–36, determine the center and radius of each circle. Sketch each circle.

25. $x^2 + (y - 3)^2 = 4$
26. $(x - 2)^2 + (y + 3)^2 = 49$
27. $4(x + 1)^2 + 4(y - 5)^2 = 81$
28. $4(x + 7)^2 + 4(y + 11)^2 = 169$
29. $x^2 + y^2 - 2x - 8 = 0$
30. $x^2 + y^2 - 4x - 6y - 12 = 0$
31. $x^2 + y^2 + 4.20x - 2.60y = 3.51$
32. $x^2 + y^2 + 22x + 14y = 26$
33. $4x^2 + 4y^2 - 16y = 9$
34. $9x^2 + 9y^2 + 18y = 7$

35. $2x^2 + 2y^2 - 4x - 8y - 1 = 0$

36. $3x^2 + 3y^2 - 12x + 4 = 0$

In Exercises 37–40, determine whether the circles with the given equations are symmetric to either axis or to the origin.

37. $x^2 + y^2 = 100$

38. $x^2 + y^2 - 4x - 5 = 0$

39. $3x^2 + 3y^2 + 24y = 8$

40. $5x^2 + 5y^2 - 10x + 20y = 3$

In Exercises 41–62, solve the given problems.

41. Determine whether the circle $x^2 - 6x + y^2 - 7 = 0$ crosses the x-axis.

42. Find the points of intersection of the circle $x^2 + y^2 - x - 3y = 0$ and the line $y = x - 1$.

(W) 43. Find the locus of a point $P(x, y)$ that moves so that its distance from $(2, 4)$ is twice its distance from $(0, 0)$. Describe the locus.

(W) 44. Find the equation of the locus of a point $P(x, y)$ that moves so that the line joining it and $(2, 0)$ is always perpendicular to the line joining it and $(-2, 0)$. Describe the locus.

45. Use a graphing calculator to view the circle $x^2 + y^2 + 5y - 4 = 0$.

46. Use a graphing calculator to view the circle $2x^2 + 2y^2 + 2y - x - 1 = 0$.

(W) 47. What type of graph is represented by the equations
(a) $y = \sqrt{9 - (x - 2)^2}$?   (b) $y = -\sqrt{9 - (x - 2)^2}$?
(c) Are the equations in parts (a) and (b) functions? Explain.

48. What type of graph is represented by the equations
(a) $x^2 + (y - 1)^2 = 0$?   (b) $x^2 + (y - 1)^2 = -1$?

49. Is the point $(0.1, 3.1)$ inside, outside, or on the circle $x^2 + y^2 - 2x - 4y + 3 = 0$?

50. Find the equation of the line along which the diameter of the circle $x^2 + y^2 - 2x - 4y - 4 = 0$ lies, if the diameter is parallel to the line $3x + 5y = 4$.

51. For the equation $(x - h)^2 + (y - k)^2 = p$, how does the value of p indicate whether the graph is a circle, a point, or does not exist?

52. Determine whether the graph of $x^2 + y^2 - 2x + 10y + 29 = 0$ is a circle, a point, or does not exist.

53. The inner and outer circles of the cross section of a pipe are represented by the equations $2.00x^2 + 2.00y^2 = 5.73$ and $2.80x^2 + 2.80y^2 = 8.91$. How thick is the pipe wall?

54. A 12-ft pole leaning against a wall slips to the ground. Find the equation that represents the path of the mid-point of the pole. See Fig. 21.39.

Fig. 21.39

55. In a hoisting device, two of the pulley wheels may be represented by $x^2 + y^2 = 14.5$ and $x^2 + y^2 - 19.6y + 86.0 = 0$. How far apart (in in.) are the wheels?

56. The design of a machine part shows it as a circle represented by the equation $x^2 + y^2 = 42.5$, with a circular hole represented by $x^2 + y^2 + 3.06y - 1.24 = 0$ cut out. What is the least distance (in in.) from the edge of the hole to the edge of the machine part?

57. A wire is rotating in a circular path through a magnetic field to induce an electric current in the wire. The wire is rotating at 60.0 Hz with a constant velocity of 37.7 m/s. Taking the origin at the center of the circle of rotation, find the equation of the path of the wire.

58. A communications satellite remains stationary at an altitude of 22,500 mi over a point on the earth's equator. It therefore rotates once each day about the earth's center. Its velocity is constant, but the horizontal and vertical components, v_H and v_V, of the velocity constantly change. Show that the equation relating v_H and v_V (in mi/h) is that of a circle. The radius of the earth is 3960 mi.

59. Find the equation describing the rim of a circular porthole 2 ft in diameter if the top is 6 ft below the surface of the water. Take the origin at the water surface directly above the center of the porthole.

60. An earthquake occurred 37° north of east of a seismic recording station. If the tremors travel at 4.8 km/s and were recorded 25 s later at the station, find the equation of the circle that represents the tremor recorded at the station. Take the station to be at the center of the coordinate system.

61. In analyzing the strain on a beam, *Mohr's circle* is often used. To form it, normal strain is plotted as the x-coordinate and shear strain is plotted as the y-coordinate. The center of the circle is midway between the minimum and maximum values of normal strain on the x-axis. Find the equation of Mohr's circle if the minimum normal strain is 100×10^{-6} and the maximum normal strain is 900×10^{-6} (strain is unitless). Sketch the graph.

62. An architect designs a Norman window, which has the form of a semicircle surmounted on a rectangle, as in Fig. 21.40. Find the area (in m²) of the window if the circular part is on the circle $x^2 + y^2 - 3.00y + 1.25 = 0$.

Fig. 21.40

Answers to Practice Exercises

1. $C(-7, 2), r = 1$ 2. $C(4, -3), r = 2$

21.4 The Parabola

Directrix • Focus • Axis • Vertex •
Standard Form of Equation •
Axis along x-axis • Axis along y-axis •
Calculator Display

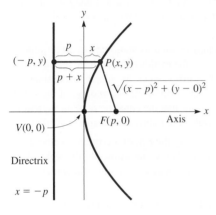

Fig. 21.41

In Chapter 7, we showed that the graph of a quadratic function is a *parabola*. We now define the parabola more generally and find the general form of its equation.

A **parabola** *is defined as the locus of a point $P(x, y)$ that moves so that it is always equidistant from a given line (the **directrix**) and a given point (the **focus**). The line through the focus that is perpendicular to the directrix is the **axis** of the parabola. The point midway between the focus and directrix is the **vertex.**

Using the definition, we now find the equation of the parabola with the focus at $(p, 0)$ and the directrix $x = -p$. With these choices, we find a general equation of a parabola with its vertex at the origin.

From the definition, the distance from $P(x, y)$ on the parabola to the focus $(p, 0)$ must equal the distance from $P(x, y)$ to the directrix $x = -p$. The distance from P to the focus is found from the distance formula. The distance from P to the directrix is the perpendicular distance and is along a line parallel to the x-axis. These distances are shown in Fig. 21.41. Therefore, we have

$$\sqrt{(x - p)^2 + (y - 0)^2} = x + p$$
$$(x - p)^2 + y^2 = (x + p)^2 \qquad \text{squaring both sides}$$
$$x^2 - 2px + p^2 + y^2 = x^2 + 2px + p^2$$

Simplifying, we obtain

$$\boxed{y^2 = 4px} \qquad \qquad \textbf{(21.15)}$$

Equation (21.15) is called the **standard form** *of the equation of a parabola with its axis along the x-axis and the vertex at the origin.* Its symmetry to the x-axis can be proven since $(-y)^2 = 4px$ is the same as $y^2 = 4px$.

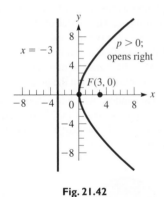

Fig. 21.42

EXAMPLE 1

Find the coordinates of the focus and the equation of the directrix and sketch the graph of the parabola $y^2 = 12x$.

Since the equation of this parabola fits the form of Eq. (21.15), we know that the vertex is at the origin. The coefficient of 12 tells us that

$$4p = 12, \qquad p = 3$$

The focus is the point $(3, 0)$, and the directrix is the line $x = -3$, as shown in Fig. 21.42. ∎

EXAMPLE 2 Find equation given directrix and vertex

If the focus is to the left of the origin, with the directrix an equal distance to the right, the coefficient of the x-term is negative. This tells us that the parabola opens to the left, rather than to the right, as in the case when the focus is to the right of the origin. For example, the parabola $y^2 = -8x$ has its vertex at the origin, its focus at $(-2, 0)$, and the line $x = 2$ as its directrix. We determine this from the equation as follows:

$$y^2 = -8x \qquad 4p = -8, \qquad p = -2$$

Since $p = -2$, we find

the focus is $(-2, 0)$

the directrix is the line $x = -(-2)$, or $x = 2$

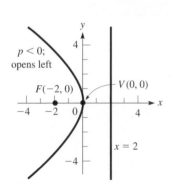

Fig. 21.43

The parabola opens to the left, as shown in Fig. 21.43. ∎

Fig. 21.44

If we chose the focus as the point $(0, p)$ and the directrix as the line $y = -p$ (see Fig. 21.44), we would find that the resulting equation is

$$x^2 = 4py \qquad \text{(21.16)}$$

This is the standard form of the equation of a parabola with the y-axis as its axis and the vertex at the origin. Its symmetry to the y-axis can be proved, since $(-x)^2 = 4py$ is the same as $x^2 = 4py$. We note that **the difference between this equation and Eq. (21.15) is that x is squared and y appears to the first power in Eq. (21.16), rather than the reverse, as in Eq. (21.15).**

EXAMPLE 3 Standard form – axis along y-axis

(a) The parabola $x^2 = 4y$ fits the form of Eq. (21.16). Therefore, its axis is along the y-axis and its vertex is at the origin. From the equation, we find the value of p, which in turn tells us the location of the vertex and the directrix.

$$x^2 = 4y \qquad 4p = 4, \qquad p = 1$$

Focus $(0, p)$ is $(0, 1)$; directrix $y = -p$ is $y = -1$. The parabola is shown in Fig. 21.45, and we see in this case that it opens upward.

(b) The parabola $2x^2 = -9y$ fits the form of Eq. (21.16) if we write it in the form

$$x^2 = -\frac{9}{2}y$$

Here, we see that $4p = -9/2$. Therefore, its axis is along the y-axis, and its vertex is at the origin. Since $4p = -9/2$, we have

$$p = -\frac{9}{8} \qquad \text{focus}\left(0, -\frac{9}{8}\right) \qquad \text{directrix } y = \frac{9}{8}$$

The parabola opens downward, as shown in Fig. 21.46. ∎

Fig. 21.45

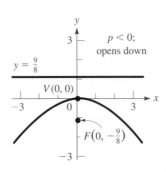

Fig. 21.46

■ See the chapter introduction.

Fig. 21.47   **Fig. 21.48**

EXAMPLE 4 Parabola – application

In calculus, it can be shown that an electromagnetic wave (light wave, television signal, etc.) parallel to the axis of a parabolic reflector will pass through the focus of the parabola. Applications of this property of a parabola are many, including a satellite television dish and a spotlight (the light from the focus is reflected as a beam).

A parabolic satellite television dish is 46.0 cm across and 4.50 cm deep, as shown in Fig. 21.47. Find where the receiver should be located to receive all of the waves reflected off the parabolic surface.

With the vertex at the origin and the axis along the x-axis, we will use the general form $y^2 = 4px$ of the parabola. Since the parabolic opening is 46.0 cm and is 4.50 cm deep, the point (4.50, 23.0) will be on any of the parabolic cross sections. Therefore, we find p by substituting (4.50, 23.0) in the equation. This means

$$23.0^2 = 4p(4.50), \qquad p = 29.4 \text{ cm}$$

and the receiver should be placed 29.4 cm from the vertex as shown in Fig. 21.48. The equation of the parabolic cross section is $y^2 = 118x$. ∎

Practice Exercises

Find the coordinates of the focus and the equation of the directrix for each parabola.

1. $y^2 = -24x$ **2.** $x^2 = 40y$

Equations (21.15) and (21.16) give the general forms of the equation of a parabola with vertex at the origin and focus on one of the coordinate axes. We now use the definition to find the equation of a parabola that has its vertex at a point other than the origin.

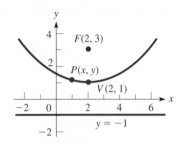

Fig. 21.49

EXAMPLE 5 Find equation from definition

Using the definition of the parabola, find the equation of the parabola with its focus at $(2, 3)$ and its directrix the line $y = -1$. See Fig. 21.49.

Choosing a general point $P(x, y)$ on the parabola and equating the distances from this point to $(2, 3)$ and to the line $y = -1$, we have

$$\sqrt{(x - 2)^2 + (y - 3)^2} = y + 1$$

distance P to F = distance P to $y = -1$

$$(x - 2)^2 + (y - 3)^2 = (y + 1)^2 \qquad \text{squaring both sides}$$

$$x^2 - 4x + 4 + y^2 - 6y + 9 = y^2 + 2y + 1$$

$$8y = 12 - 4x + x^2$$

We note that this type of equation has appeared frequently in earlier chapters. The x-term and the constant (12 in this case) are characteristic of a parabola that does not have its vertex at the origin if the directrix is parallel to the x-axis. ∎

To view a parabola on a graphing calculator, if the axis is along the y-axis, or parallel to it, as in Examples 3 and 5, we solve for y and use this function. If the axis is along the x-axis, or parallel to it, as in Examples 1, 2, and 4, we get *two* functions to graph. This is similar to Example 7 on page 572. These cases are shown in the next example.

EXAMPLE 6 Calculator graph of parabola

To display the graph of the parabola in Example 5 on a graphing calculator, we solve for y and enter this function in the calculator. Therefore, we enter the function

$$y_1 = (12 - 4x + x^2)/8$$

in the calculator, and we get the display in Fig. 21.50.

To display the graph of the parabola in Example 1, where $y^2 = 12x$, we enter

$$y_1 = \sqrt{12x} \quad \text{and} \quad y_2 = -\sqrt{12x}$$

in the calculator, and we get the display in Fig. 21.51. ∎

Fig. 21.50

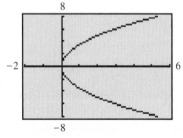

Fig. 21.51

We see that *the equation of a parabola is characterized by the square of either (but not both) x and y and a first power term in the other.* We will consider the parabola further in Sections 21.7, 21.8, and 21.9.

The parabola has numerous technical applications. The reflection property illustrated in Example 4 has other important applications, such as the design of a radar antenna. The path of a projectile is parabolic. The cables of a suspension bridge are parabolic.

EXERCISES 21.4

In Exercises 1–4, make the given changes in the indicated examples of this section and then solve the resulting problems. In each, find the focus and directrix, and sketch the parabola.

1. In Example 1, change $12x$ to $20x$.

2. In Example 2, change $-8x$ to $-20x$.

3. In Example 3(a), change $4y$ to $-6y$.

4. In Example 3(b), change $-9y$ to $7y$.

In Exercises 5–16, determine the coordinates of the focus and the equation of the directrix of the given parabolas. Sketch each curve.

5. $y^2 = 4x$
6. $y^2 = 16x$
7. $y^2 = -4x$
8. $y^2 = -36x$
9. $x^2 = 72y$
10. $x^2 = y$
11. $x^2 = -4y$
12. $x^2 + 12y = 0$
13. $2y^2 - 5x = 0$
14. $3x^2 = 8y$
15. $y = 0.48x^2$
16. $x = 7.6y^2$

In Exercises 17–30, find the equations of the parabolas satisfying the given conditions. The vertex of each is at the origin.

17. Focus $(3, 0)$ **18.** Focus $(0, 0.4)$

19. Focus $(0, -0.5)$ **20.** Focus $(2.5, 0)$

21. Directrix $y = -0.16$ **22.** Directrix $x = 20$

23. Directrix $x = -84$ **24.** Directrix $y = 2.3$

25. Axis $x = 0$, passes through $(-1, 8)$

26. Symmetric to x-axis, passes through $(2, -1)$

27. Passes through $(3, 5)$ and $(3, -5)$

28. Passes through $(6, -1)$ and $(-6, -1)$

29. Passes through $(3, 3)$ and $(12, 6)$

30. Passes through $(-5, -5)$ and $(-10, -20)$

In Exercises 31–58, solve the given problems.

31. Sketch the graph of the inequality $x^2 < 8y$.

32. Sketch the graph of the inequality $9y \le 4x^2$.

33. At what point(s) do the parabolas $y^2 = 2x$ and $x^2 = -16y$ intersect?

34. Using trigonometric identities, show that the parametric equations $x = \sin t$, $y = 2(1 - \cos^2 t)$ are the equations of a parabola.

35. Find the equation of the parabola with focus $(6, 1)$ and directrix $x = 0$, by use of the definition. Sketch the curve.

36. Find the equation of the parabola with focus $(1, 1)$ and directrix $y = 5$, by use of the definition. Sketch the curve.

37. Use a graphing calculator to view the parabola $y^2 + 2x + 8y + 13 = 0$.

38. Use a graphing calculator to view the parabola $y^2 - 2x - 6y + 19 = 0$.

39. The equation of a parabola with vertex (h, k) and axis parallel to the x-axis is $(y - k)^2 = 4p(x - h)$. (This is shown in Section 21.7.) Sketch the parabola for which (h, k) is $(2, -3)$ and $p = 2$.

40. The equation of a parabola with vertex (h, k) and axis parallel to the y-axis is $(x - h)^2 = 4p(y - k)$. (This is shown in Section 21.7.) Sketch the parabola for which (h, k) is $(-1, 2)$ and $p = -3$.

41. The chord of a parabola that passes through the focus and is parallel to the directrix is called the *latus rectum* of the parabola. Find the length of the latus rectum of the parabola $y^2 = 4px$.

42. Find the equation of the circle that has the focus and the vertex of the parabola $x^2 = 8y$ as the ends of a diameter.

43. Find the standard equation of the parabola with vertex $(0, 0)$ and that passes through $(2, 2)$ and $(8, 4)$.

(W) 44. For either standard form of the equation of a parabola, describe what happens to the shape of the parabola as $|p|$ increases.

45. The Golden Gate Bridge at San Francisco Bay is a suspension bridge, and its supporting cables are parabolic. See Fig. 21.52.

With the origin at the low point of the cable, what equation represents the cable if the towers are 4200 ft apart and the maximum sag is 300 ft?

Fig. 21.52

46. The entrance to a building is a parabolic arch 5.6 m high at the center and 7.4 m wide at the base. What equation represents the arch if the vertex is at the top of the arch?

47. The rate of development of heat H (in W) in a resistor of resistance R (in Ω) of an electric circuit is given by $H = Ri^2$, where i is the current (in A) in the resistor. Sketch the graph of H vs. i, if $R = 6.0\ \Omega$.

48. What is the length of the horizontal bar across the parabolically shaped window shown in Fig. 21.53?

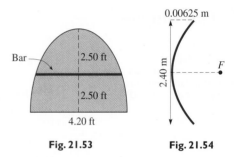

Fig. 21.53   **Fig. 21.54**

49. The primary mirror in the Hubble space telescope has a parabolic cross section, which is shown in Fig. 21.54. What is the focal length (vertex to focus) of the mirror?

50. A rocket is fired horizontally from a plane. Its horizontal distance x and vertical distance y from the point at which it was fired are given by $x = v_0 t$ and $y = \frac{1}{2}gt^2$, where v_0 is the initial velocity of the rocket, t is the time, and g is the acceleration due to gravity. Express y as a function of x and show that it is the equation of a parabola.

51. To launch a spacecraft to the moon, it is first put into orbit around earth and then into a parabolic path towards the moon. Assume the parabolic path is represented by $x^2 = 4py$ and the spacecraft is later observed at $(10, 3)$ (units in thousands of km) after launch. Will this path lead directly to the moon at $(110, 340)$?

52. Under certain load conditions, a beam fixed at both ends is approximately parabolic in shape. If a beam is 4.0 m long and the deflection in the middle is 2.0 cm, find an equation to represent the shape of the beam.

53. A spotlight with a parabolic reflector is 15.0 cm wide and is 6.50 cm deep. See Fig. 21.55 and Example 4. Where should the filament of the bulb be located so as to produce a beam of light?

15.0 cm

6.50 cm

Filament

Fig. 21.55

54. A wire is fastened 36.0 ft up on each of two telephone poles that are 200 ft apart. Halfway between the poles the wire is 30.0 ft above the ground. Assuming the wire is parabolic, find the height of the wire 50.0 ft from either pole.

55. The total annual fraction f of energy supplied by solar energy to a home is given by $f = 0.065\sqrt{A}$, where A is the area of the solar collector. Sketch the graph of f as a function of A $(0 < A \le 200 \text{ m}^2)$.

56. The velocity v (in ft/s) of a jet of water flowing from an opening in the side of a certain container is given by $v = 8\sqrt{h}$, where h is the depth (in ft) of the opening. Sketch a graph of v vs. h.

57. A small island is 4 km from a straight shoreline. A ship channel is equidistant between the island and the shoreline. Write an equation for the channel.

58. Under certain circumstances, the maximum power P (in W) in an electric circuit varies as the square of the voltage of the source E_0 and inversely as the internal resistance R_i (in Ω) of the source. If 10 W is the maximum power for a source of 2.0 V and internal resistance of 0.10 Ω, sketch the graph of P vs. E_0 if R_i remains constant.

Answers to Practice Exercises

1. $F(-6, 0)$; dir. $x = 6$ **2.** $F(0, 10)$; dir. $y = -10$

21.5 The Ellipse

Foci • Vertices • Major Axis • Minor Axis • Standard Equation – Axis along x-axis • Standard Equation – Axis along y-axis • Calculator Display

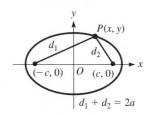

Fig. 21.56

The next important curve is the ellipse. *An* **ellipse** *is defined as the locus of a point $P(x, y)$ that moves so that the sum of its distances from two fixed points is constant. These fixed points are the* **foci** *of the ellipse.* Letting this sum of distances be $2a$ and the foci be the points $(-c, 0)$ and $(c, 0)$, we have

$$\sqrt{(x - c)^2 + y^2} + \sqrt{(x + c)^2 + y^2} = 2a$$

See Fig. 21.56. The ellipse has its center at the origin such that c is the length of the line segment from the center to a focus. We will also see that a has a special meaning. Now, from Section 14.4, we see that we should move one radical to the right and then square each side. This leads to the following steps:

$$\sqrt{(x + c)^2 + y^2} = 2a - \sqrt{(x - c)^2 + y^2}$$
$$(x + c)^2 + y^2 = 4a^2 - 4a\sqrt{(x - c)^2 + y^2} + (\sqrt{(x - c)^2 + y^2})^2$$
$$x^2 + 2cx + c^2 + y^2 = 4a^2 - 4a\sqrt{(x - c)^2 + y^2} + x^2 - 2cx + c^2 + y^2$$
$$4a\sqrt{(x - c)^2 + y^2} = 4a^2 - 4cx$$
$$a\sqrt{(x - c)^2 + y^2} = a^2 - cx$$
$$a^2(x^2 - 2cx + c^2 + y^2) = a^4 - 2a^2cx + c^2x^2$$
$$(a^2 - c^2)x^2 + a^2y^2 = a^2(a^2 - c^2)$$

We now define $a^2 - c^2 = b^2$ (this reason will be shown presently). Therefore,

$$b^2x^2 + a^2y^2 = a^2b^2$$

Dividing through by a^2b^2, we have

$$\boxed{\frac{x^2}{a^2} + \frac{y^2}{b^2} = 1}$$

(21.17)

A graphical analysis of this equation is found on the next page.

The *x*-intercepts are $(-a, 0)$ and $(a, 0)$. This means that $2a$ (the sum of distances used in the derivation) is also the distance between the *x*-intercepts. *The points $(a, 0)$ and $(-a, 0)$ are the* **vertices** *of the ellipse, and the line between them is the* **major axis** [see Fig. 21.57(a)]. Thus, *a is the length of the* **semimajor axis.**

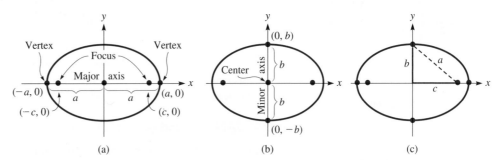

Fig. 21.57   (a)   (b)   (c)

We can now state that *Eq. (21.17) is called the* **standard equation** *of the ellipse with its major axis along the x-axis and its center at the origin.*

The *y*-intercepts of this ellipse are $(0, -b)$ and $(0, b)$. *The line joining these intercepts is called the* **minor axis** *of the ellipse* (Fig. 21.57(b)), *which means b is the length of the* **semiminor** *axis.* The intercept $(0, b)$ is equidistant from $(-c, 0)$ and $(c, 0)$. Since the sum of the distances from these points to $(0, b)$ is $2a$, the distance $(c, 0)$ to $(0, b)$ must be a. Thus, we have a right triangle with line segments of lengths a, b, and c, with a as hypotenuse (Fig. 21.57(c)). Therefore,

$$a^2 = b^2 + c^2 \qquad (21.18)$$

is the relation between distances a, b, and c. This also shows why b was defined as it was in the derivation of Eq. (21.17).

If we choose points on the *y*-axis as the foci, *the standard equation of the ellipse, with its center at the origin and its major axis along the y-axis, is*

$$\frac{y^2}{a^2} + \frac{x^2}{b^2} = 1 \qquad (21.19)$$

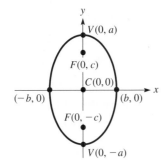

Fig. 21.58

In this case, the vertices are $(0, a)$ and $(0, -a)$, the foci are $(0, c)$ and $(0, -c)$, and the ends of the minor axis are $(b, 0)$ and $(-b, 0)$. See Fig. 21.58.

The ellipses represented by Eqs. (21.17) and (21.19) are both symmetric to both axes and to the origin.

EXAMPLE 1 Standard equation – major axis along *x*-axis

The ellipse $\dfrac{x^2}{25} + \dfrac{y^2}{9} = 1$ seems to fit the form of either Eq. (21.17) or Eq. (21.19). Since $a^2 = b^2 + c^2$, we know that *a is always larger than b.* Since the square of the larger number appears under x^2, we know the equation is in the form of Eq. (21.17). Therefore, $a^2 = 25$ and $b^2 = 9$, or $a = 5$ and $b = 3$. This means that the vertices are $(5, 0)$ and $(-5, 0)$, and the minor axis extends from $(0, -3)$ to $(0, 3)$. See Fig. 21.59.

We find c from the relation $c^2 = a^2 - b^2$. This means that $c^2 = 16$ and the foci are $(4, 0)$ and $(-4, 0)$. ∎

CAUTION ▶

Fig. 21.59

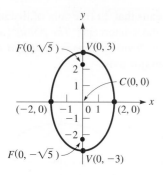

Fig. 21.60

EXAMPLE 2 Standard equation – major axis along y-axis

The equation of the ellipse

$$\frac{x^2}{4} + \frac{y^2}{9} = 1$$

$$b^2 \uparrow \qquad \uparrow a^2$$

has the larger denominator, 9, under the y^2. Therefore, it fits the form of Eq. (21.19) with $a^2 = 9$ and $b^2 = 4$. This means that the vertices are $(0, 3)$ and $(0, -3)$, and the minor axis extends from $(-2, 0)$ to $(2, 0)$. In turn, we know that $c^2 = a^2 - b^2 = 9 - 4 = 5$ and that the foci are $(0, \sqrt{5})$ and $(0, -\sqrt{5})$. This ellipse is shown in Fig. 21.60. ■

EXAMPLE 3 Find vertices, foci, minor axis

Find the coordinates of the vertices, the ends of the minor axis, and the foci of the ellipse $4x^2 + 16y^2 = 64$.

This equation must be put in standard form first, which we do by dividing through by 64. When this is done, we obtain

$$\frac{x^2}{16} + \frac{y^2}{4} = 1 \longleftarrow$$

form requires
+ and 1

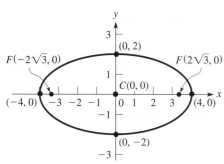

Fig. 21.61

We see that $a^2 = 16$ and $b^2 = 4$, which tells us that $a = 4$ and $b = 2$. Then $c = \sqrt{16 - 4} = \sqrt{12} = 2\sqrt{3}$. Since a^2 appears under x^2, the vertices are $(4, 0)$ and $(-4, 0)$. The ends of the minor axis are $(0, 2)$ and $(0, -2)$, and the foci are $(2\sqrt{3}, 0)$ and $(-2\sqrt{3}, 0)$. See Fig. 21.61. ■

EXAMPLE 4 Ellipse – application

A satellite to study the earth's atmosphere has a minimum altitude of 600 mi and a maximum altitude of 2000 mi. If the path of the satellite about the earth is an ellipse with the center of the earth at one focus, what is the equation of its path? Assume that the radius of the earth is 4000 mi.

We set up the coordinate system such that the center of the ellipse is at the origin and the center of the earth is at the right focus, as shown in Fig. 21.62. We know that the distance between vertices is

$$2a = 2000 + 4000 + 4000 + 600 = 10{,}600 \text{ mi}$$
$$a = 5300 \text{ mi}$$

From the right focus to the right vertex is 4600 mi. This tells us

$$c = a - 4600 = 5300 - 4600 = 700 \text{ mi}$$

We can now calculate b^2 as

$$b^2 = a^2 - c^2 = 5300^2 - 700^2 = 2.76 \times 10^7 \text{ mi}^2$$

Since $a^2 = 5300^2 = 2.81 \times 10^7 \text{ mi}^2$, the equation is

$$\frac{x^2}{2.81 \times 10^7} + \frac{y^2}{2.76 \times 10^7} = 1$$

or

$$2.76x^2 + 2.81y^2 = 7.76 \times 10^7 \qquad ■$$

Fig. 21.62

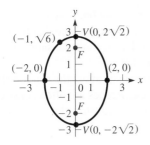

Fig. 21.63

Practice Exercises

Find the vertices and foci of each ellipse.

1. $\dfrac{x^2}{9} + y^2 = 1$ **2.** $25x^2 + 4y^2 = 25$

EXAMPLE 5 Find equation – given points

Find the equation of the ellipse with its center at the origin and an end of its minor axis at $(2, 0)$ and which passes through $(-1, \sqrt{6})$.

Since the center is at the origin and an end of the minor axis is at $(2, 0)$, we know that the ellipse is of the form of Eq. (21.19) and that $b = 2$. Thus, we have

$$\frac{y^2}{a^2} + \frac{x^2}{2^2} = 1$$

In order to find a^2, we use the fact that the ellipse passes through $(-1, \sqrt{6})$. This means that these coordinates satisfy the equation of the ellipse. This gives

$$\frac{(\sqrt{6})^2}{a^2} + \frac{(-1)^2}{4} = 1, \qquad \frac{6}{a^2} = \frac{3}{4}, \qquad a^2 = 8$$

Therefore, the equation of the ellipse, shown in Fig. 21.63, is

$$\frac{y^2}{8} + \frac{x^2}{4} = 1$$ ∎

The following example illustrates the use of the definition of the ellipse to find the equation of an ellipse with its center at a point other than the origin.

EXAMPLE 6 Find equation from definition – calculator

Using the definition, find the equation of the ellipse with foci at $(1, 3)$ and $(9, 3)$, with major axis of 10.

Recalling that the sum of distances in the definition equals the length of the major axis, we now use the same method as in the derivation of Eq. (21.17).

$\sqrt{(x-1)^2 + (y-3)^2} + \sqrt{(x-9)^2 + (y-3)^2} = 10$	use definition of ellipse
$\sqrt{(x-1)^2 + (y-3)^2} = 10 - \sqrt{(x-9)^2 + (y-3)^2}$	isolate a radical
$x^2 - 2x + 1 + y^2 - 6y + 9 = 100 - 20\sqrt{(x-9)^2 + (y-3)^2}$	square both sides
$+ x^2 - 18x + 81 + y^2 - 6y + 9$	and simplify
$20\sqrt{(x-9)^2 + (y-3)^2} = 180 - 16x$	isolate radical
$5\sqrt{(x-9)^2 + (y-3)^2} = 45 - 4x$	divide by 4
$25(x^2 - 18x + 81 + y^2 - 6y + 9) = 2025 - 360x + 16x^2$	square both sides
$9x^2 - 90x + 25y^2 - 150y + 225 = 0$	simplify

The additional x- and y-terms are characteristic of the equation of an ellipse whose center is not at the origin (see Fig. 21.64).

To view this ellipse on a graphing calculator as shown in Fig. 21.65, we solve for y to get the two functions needed. The solutions are $y = \dfrac{15 \pm 3\sqrt{10x - x^2}}{5}$. ∎

Fig. 21.64

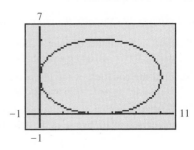

Fig. 21.65

■ The Polish astronomer Nicolaus Copernicus (1473–1543) is credited as being the first to suggest that the earth revolved about the sun, rather than the previously held belief that the earth was the center of the universe.

We see that *the equation of an ellipse is characterized by both an x^2-term and a y^2-term, having different coefficients (in value but not in sign).* We note that the coefficients of the squared terms differ, whereas for the circle, they are the same. We will consider the ellipse further in Sections 21.7, 21.8, and 21.9.

The ellipse has many applications. The orbits of the planets about the sun are elliptical. Gears, cams, and springs are often elliptical in shape. Arches are often constructed in the form of a semiellipse. These and other applications are illustrated in the exercises.

EXERCISES 21.5

In Exercises 1 and 2, make the given changes in the indicated examples of this section and then solve the given problems.

1. In Example 1, change the 9 to 36; find the vertices, ends of the minor axis, and foci; and sketch the ellipse.

2. In Example 2, interchange the 4 and 9; find the vertices, ends of the minor axis, and foci; and sketch the ellipse.

In Exercises 3–16, find the coordinates of the vertices and foci of the given ellipses. Sketch each curve.

3. $\dfrac{x^2}{4} + \dfrac{y^2}{1} = 1$ 4. $\dfrac{x^2}{100} + \dfrac{y^2}{64} = 1$ 5. $\dfrac{x^2}{25} + \dfrac{y^2}{144} = 1$

6. $\dfrac{x^2}{49} + \dfrac{y^2}{81} = 1$ 7. $\dfrac{4x^2}{25} + \dfrac{y^2}{4} = 1$ 8. $x^2 + \dfrac{9y^2}{25} = 1$

9. $4x^2 + 9y^2 = 324$ 10. $x^2 + 36y^2 = 144$

11. $49x^2 + 4y^2 = 196$ 12. $y^2 = 25(1 - x^2)$

13. $y^2 = 8(2 - x^2)$ 14. $2x^2 + 3y^2 = 600$

15. $4x^2 + 25y^2 = 0.25$ 16. $9x^2 + 4y^2 = 0.09$

In Exercises 17–28, find the equations of the ellipses satisfying the given conditions. The center of each is at the origin.

17. Vertex $(15, 0)$, focus $(9, 0)$ 18. Minor axis 8, vertex $(0, -5)$

19. Focus $(0, 8)$, major axis 34   20. Vertex $(0, 13)$, focus $(0, -5)$

21. End of minor axis $(0, 12)$, focus $(8, 0)$

22. Sum of lengths of major and minor axes 18, focus $(3, 0)$

23. Vertex $(8, 0)$, passes through $(2, 3)$

24. Focus $(0, 2)$, passes through $(-1, \sqrt{3})$

25. Passes through $(2, 2)$ and $(1, 4)$

26. Passes through $(-2, 2)$ and $(1, \sqrt{6})$

27. The sum of distances from (x, y) to $(6, 0)$ and $(-6, 0)$ is 20.

28. The sum of distances from (x, y) to $(0, 2)$ and $(0, -2)$ is 5.

In Exercises 29–56, solve the given problems.

29. Find any point(s) of intersection of the graphs of the ellipse $4x^2 + 9y^2 = 40$ and the parabola $y^2 = 4x$.

30. Find the equation of the circle that has the same center as the ellipse $4x^2 + 9y^2 = 36$ and is internally tangent to the ellipse.

31. Find the equation of the ellipse with foci $(-2, 1)$ and $(4, 1)$ and a major axis of 10, by use of the definition. Sketch the curve.

32. Find the equation of the ellipse with foci $(1, 4)$ and $(1, 0)$ that passes through $(4, 4)$, by use of the definition. Sketch the curve.

33. Use a graphing calculator to view the ellipse $4x^2 + 3y^2 + 16x - 18y + 31 = 0$.

34. Use a graphing calculator to view the ellipse $4x^2 + 8y^2 + 4x - 24y + 1 = 0$.

35. The equation of an ellipse with center (h, k) and major axis parallel to the x-axis is $\dfrac{(x - h)^2}{a^2} + \dfrac{(y - k)^2}{b^2} = 1$. (This is shown in Section 21.7.) Sketch the ellipse that has a major axis of 6, a minor axis of 4, and for which (h, k) is $(2, -1)$.

36. The equation of an ellipse with center (h, k) and major axis parallel to the y-axis is $\dfrac{(y - k)^2}{a^2} + \dfrac{(x - h)^2}{b^2} = 1$. (This is shown in Section 21.7.) Sketch the ellipse that has a major axis of 8, a minor axis of 6, and for which (h, k) is $(1, 3)$.

(W) 37. For what values of k does the ellipse $x^2 + ky^2 = 1$ have its vertices on the y-axis? Explain how these values are found.

(W) 38. For what value of k does the ellipse $x^2 + k^2y^2 = 25$ have a focus at $(3, 0)$? Explain how this value is found.

39. Show that the ellipse $2x^2 + 3y^2 - 8x - 4 = 0$ is symmetric to the x-axis.

40. Show that the ellipse $5x^2 + y^2 - 3y - 7 = 0$ is symmetric to the y-axis.

41. Graph the inequality $100x^2 + 49y^2 \leq 4900$.

42. Graph the inequality $5x^2 + 4y^2 > 20$.

43. Show that the parametric equations $x = 2 \sin t$ and $y = 3 \cos t$ define an ellipse.

44. For what values of k does $y^2 = 1 + kx^2$ represent an ellipse with foci on the y-axis?

45. The electric power P (in W) dissipated in a resistance R (in Ω) is given by $P = Ri^2$, where i is the current (in A) in the resistor. Find the equation for the total power of 64 W dissipated in two resistors, with resistances 2.0 Ω and 8.0 Ω, respectively, and with currents i_1 and i_2, respectively. Sketch the graph, assuming that negative values of current are meaningful.

46. The *eccentricity e* of an ellipse is defined as $e = c/a$. A cam in the shape of an ellipse can be described by the equation $x^2 + 9y^2 = 81$. Find the eccentricity of this elliptical cam.

47. A space object (dubbed 2003 UB313), larger and more distant than Pluto, was discovered in 2003. In its elliptical orbit with the sun at one focus, it is 3.5 billion miles from the sun at the closest, and 9.0 billion miles at the farthest. What is the eccentricity of its orbit? See Exercise 46.

W **48.** Halley's Comet has an elliptical orbit with $a = 17.94$ AU (AU is astronomical unit, 1 AU $= 9.3 \times 10^7$ mi) and $b = 4.552$ AU, with the sun at one focus. What is the closest that the comet comes to the sun? Explain your method.

49. A draftsman draws a series of triangles with a base from $(-3, 0)$ to $(3, 0)$ and a perimeter of 14 cm (all measurements in centimeters). Find the equation of the curve on which all of the third vertices of the triangles are located.

50. A lithotripter is used to break up a kidney stone by placing the kidney stone at one focus of an ellipsoid end-section and a source of shock waves at the focus of the other end-section. If the vertices of the end-sections are 30.0 cm apart and a minor axis of the ellipsoid is 6.0 cm, how far apart are the foci? In a lithotripter, the shock waves are reflected as the sound waves noted in Exercise 51.

51. An ellipse has a focal property such that a light ray or sound wave emanating from one focus will be reflected through the other focus. Many buildings, such as Statuary Hall in the U.S. Capitol and the Taj Mahal, are built with elliptical ceilings with the property that a sound from one focus is easily heard at the other focus. If a building has a ceiling whose cross sections are part of an ellipse that can be described by the equation $36x^2 + 225y^2 = 8100$ (measurements in meters), how far apart must two persons stand in order to whisper to each other using this focal property?

52. An airplane wing is designed such that a certain cross section is an ellipse 8.40 ft wide and 1.20 ft thick. Find the equation that can be used to describe the perimeter of this cross section.

53. A road passes through a tunnel with a semielliptical cross section 64 ft wide and 18 ft high at the center. What is the height of the tallest vehicle that can pass through the tunnel at a point 22 ft from the center? See Fig. 21.66.

Fig. 21.66

54. An architect designs a window in the shape of an ellipse 4.50 ft wide and 3.20 ft high. Find the perimeter of the window from the formula $p = \pi(a + b)$. This formula gives a good *approximation* for the perimeter when a and b are nearly equal.

55. The ends of a horizontal tank 20.0 ft long are ellipses, which can be described by the equation $9x^2 + 20y^2 = 180$, where x and y are measured in feet. The area of an ellipse is $A = \pi ab$. Find the volume of the tank.

56. A laser beam 6.80 mm in diameter is incident on a plane surface at an angle of 62.0°, as shown in Fig. 21.67. What is the elliptical area that the laser covers on the surface? (See Exercise 55.)

Fig. 21.67

Answers to Practice Exercises

1. $V(3, 0), V(-3, 0); F(2\sqrt{2}, 0), F(-2\sqrt{2}, 0)$
2. $V(0, 5/2), V(0, -5/2); F(0, \sqrt{21}/2), F(0, -\sqrt{21}/2)$

21.6 The Hyperbola

Foci • Vertices • Asymptotes • Transverse Axis • Conjugate Axis • Standard Equation – Axis along x-axis • Standard Equation – Axis along y-axis • Calculator Display • xy = c Hyperbola

A **hyperbola** *is defined as the locus of a point $P(x, y)$ that moves so that the difference of the distances from two fixed points (the* **foci**) *is constant.* We choose the foci to be $(-c, 0)$ and $(c, 0)$ (see Fig. 21.68), and the constant difference to be $2a$. As with the ellipse, c is the length of the line segment from the center to a focus, and a (as we will see) the length of the line segment from the center to a vertex. Therefore,

$$\sqrt{(x + c)^2 + y^2} - \sqrt{(x - c)^2 + y^2} = 2a$$

Following the same procedure as with the ellipse, the equation of the hyperbola is

$$\frac{x^2}{a^2} - \frac{y^2}{b^2} = 1 \qquad (21.20)$$

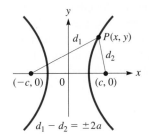

Fig. 21.68

CAUTION ▸ When we derive this equation, *we have a definition of the relation between a, b, and c that is different from that for the ellipse.* This relation is

$$c^2 = a^2 + b^2 \qquad (21.21)$$

In Eq. (21.20), if we let $y = 0$, we find that the x-intercepts are $(-a, 0)$ and $(a, 0)$, just as they are for the ellipse. *These are the* **vertices** *of the hyperbola.* For $x = 0$, we find that we have imaginary solutions for y, which means there are no points on the curve that correspond to a value of $x = 0$.

To find the meaning of b, we solve Eq. (21.20) for y in the special form:

$$\frac{y^2}{b^2} = \frac{x^2}{a^2} - 1$$

$$= \frac{x^2}{a^2} - \frac{a^2 x^2}{a^2 x^2} = \frac{x^2}{a^2}\left(1 - \frac{a^2}{x^2}\right)$$

$$y^2 = \frac{b^2 x^2}{a^2}\left(1 - \frac{a^2}{x^2}\right) \qquad \text{multiply through by } b^2 \text{ and take square root of each side}$$

$$y = \pm\frac{bx}{a}\sqrt{1 - \frac{a^2}{x^2}} \tag{21.22}$$

We note that, if large values of x are assumed in Eq. (21.22), the quantity under the radical becomes approximately 1. In fact, the larger x becomes, the nearer 1 this expression becomes, since the x^2 in the denominator of a^2/x^2 makes this term nearly zero. Thus, for large values of x, Eq. (21.22) is approximately

$$\boxed{y = \pm\frac{bx}{a}} \tag{21.23}$$

NOTE ▶   Equation (21.23) is seen to represent two straight lines, each of which passes through the origin. One has a slope of b/a, and the other has a slope of $-b/a$. *These lines are called the* **asymptotes** *of the hyperbola.* An **asymptote** *is a line that the curve approaches as one of the variables approaches some particular value.* The graph of the tangent function also has asymptotes, as we saw in Fig. 10.23. We can designate this limiting procedure with notation introduced in Chapter 19 by

$$y \to \frac{bx}{a} \quad \text{as} \quad x \to \pm\infty$$

The easiest way to sketch a hyperbola is to draw its asymptotes and then draw the hyperbola out from each vertex so that it comes closer and closer to each asymptote as x becomes numerically larger. To draw the asymptotes, first draw a small rectangle $2a$ by $2b$ with the origin at the center, as shown in Fig. 21.69. Then straight lines, the asymptotes, are drawn through opposite vertices of the rectangle. This shows us that the significance of the value of b is in the slopes of the asymptotes.

Equation (21.20) is called the **standard equation** *of the hyperbola with its center at the origin. It has a* **transverse axis** *of length $2a$ along the x-axis and a* **conjugate axis** *of length $2b$ along the y-axis.* This means that a represents the length of the semitransverse axis and b represents the length of the semiconjugate axis. See Fig. 21.70. From the definition of c, it is the length of the line segment from the center to a focus. Also, c is the length of the semidiagonal of the rectangle, as shown in Fig. 21.70. This shows us the geometric meaning of the relationship among a, b, and c given in Eq. (21.21).

Fig. 21.69

Fig. 21.70

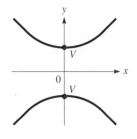

Fig. 21.71

If the transverse axis is along the y-axis and the conjugate axis is along the x-axis, the equation of the hyperbola with its center at the origin (see Fig. 21.71) is

$$\frac{y^2}{a^2} - \frac{x^2}{b^2} = 1 \qquad\qquad \textbf{(21.24)}$$

The hyperbolas represented by Eqs. (21.20) and (21.24) are both symmetric to both axes and to the origin.

EXAMPLE 1 Standard equation – transverse axis on x-axis

The hyperbola $\dfrac{x^2}{16} - \dfrac{y^2}{9} = 1$

$$a^2 \uparrow \qquad \uparrow b^2$$

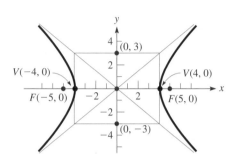

Fig. 21.72

fits the form of Eq. (21.20). We know that it fits Eq. (21.20) and not Eq. (21.24) since the x^2-term is the positive term with 1 on the right. From the equation, we see that $a^2 = 16$ and $b^2 = 9$, or $a = 4$ and $b = 3$. In turn, this means the vertices are $(4, 0)$ and $(-4, 0)$ and the conjugate axis extends from $(0, -3)$ to $(0, 3)$.

Since $c^2 = a^2 + b^2$, we find that $c^2 = 25$, or $c = 5$. The foci are $(-5, 0)$ and $(5, 0)$.

Drawing the rectangle and the asymptotes in Fig. 21.72, we then sketch in the hyperbola from each vertex toward each asymptote. ∎

EXAMPLE 2 Standard equation – transverse axis on y-axis

The hyperbola $\dfrac{y^2}{4} - \dfrac{x^2}{16} = 1$

$$a^2 \uparrow \qquad \uparrow b^2$$

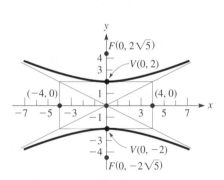

Fig. 21.73

has vertices at $(0, -2)$ and $(0, 2)$. Its conjugate axis extends from $(-4, 0)$ to $(4, 0)$. The foci are $(0, -2\sqrt{5})$ and $(0, 2\sqrt{5})$. We find this directly from the equation since the y^2-term is the positive term with 1 on the right. This means the equation fits the form of Eq. (21.24) with $a^2 = 4$ and $b^2 = 16$. Also, $c^2 = 20$, which means that $c = \sqrt{20} = 2\sqrt{5}$.

Since $2a$ extends along the y-axis, we see that the equations of the asymptotes are $y = \pm(a/b)x$. This is not a contradiction of Eq. (21.23) but the extension of it for a hyperbola with its transverse axis along the y-axis. The ratio a/b gives the slope of the asymptote. The hyperbola is shown in Fig. 21.73. ∎

EXAMPLE 3 Find vertices, foci

Determine the coordinates of the vertices and foci of the hyperbola

$$4x^2 - 9y^2 = 36$$

First, by dividing through by 36, we have

$$\frac{x^2}{9} - \frac{y^2}{4} = 1 \longleftarrow$$

form requires $-$ and 1

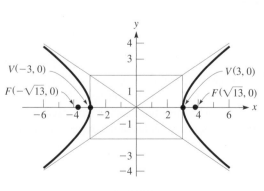

Fig. 21.74

From this form, we see that $a^2 = 9$ and $b^2 = 4$. In turn, this tells us that $a = 3$, $b = 2$, and $c = \sqrt{9 + 4} = \sqrt{13}$. Since a^2 appears under x^2, the equation fits the form of Eq. (21.20). Therefore, the vertices are $(-3, 0)$ and $(3, 0)$ and the foci are $\left(-\sqrt{13}, 0\right)$ and $\left(\sqrt{13}, 0\right)$. The hyperbola is shown in Fig. 21.74. ∎

EXAMPLE 4 Hyperbola – application

In physics, it is shown that where the velocity of a fluid is greatest, the pressure is the least. In designing an experiment to study this effect in the flow of water, a pipe is constructed such that its lengthwise cross section is hyperbolic. The pipe is 1.0 m long, 0.2 m in diameter at the narrowest point in the middle, and 0.4 m in diameter at each end. What is the equation that represents the cross section of the pipe as shown in Fig. 21.75?

As shown, the hyperbola has its transverse axis along the y-axis and its center at the origin. This means the general equation is given by Eq. (21.24). Since the radius at the middle of the pipe is 0.1 m, we know that $a = 0.1$ m. Also, since it is 1.0 m long and the radius at the end is 0.2 m, we know the point $(0.5, 0.2)$ is on the hyperbola. This point must satisfy the equation.

Fig. 21.75

$$\frac{y^2}{a^2} - \frac{x^2}{b^2} = 1 \qquad \text{Eq. (21.24)}$$

point $(0.5, 0.2)$ satisfies equation

$$a = 0.1 \rightarrow \quad \frac{0.2^2}{0.1^2} - \frac{0.5^2}{b^2} = 1$$

$$4 - \frac{0.25}{b^2} = 1, \qquad 3b^2 = 0.25, \qquad b^2 = 0.083$$

$$\frac{y^2}{0.1^2} - \frac{x^2}{0.083} = 1 \qquad \text{substituting } a = 0.1, b^2 = 0.083 \text{ in Eq. (21.24)}$$

$$100y^2 - 12x^2 = 1 \qquad \text{equation of cross section} \qquad \blacksquare$$

Practice Exercises

Find the vertices and foci of each hyperbola.

1. $\frac{x^2}{36} + \frac{y^2}{13} = 1$ 2. $4y^2 - x^2 = 4$

EXAMPLE 5 Hyperbola – calculator display

If we use a graphing calculator to display a hyperbola represented by either Eq. (21.20) or Eq. (21.24), we have two functions when we solve for y. One represents the upper half of the hyperbola, and the other represents the lower half. For the hyperbola in Example 3, the functions are $y_1 = \sqrt{(4x^2 - 36)/9}$ and $y_2 = -\sqrt{(4x^2 - 36)/9}$. This hyperbola is shown in Fig. 21.76.

For the hyperbola in Example 4, the functions are $y_1 = \sqrt{(12x^2 + 1)/100}$ and $y_2 = -\sqrt{(12x^2 + 1)/100}$. See Fig. 21.77. $\blacksquare$

Fig. 21.76

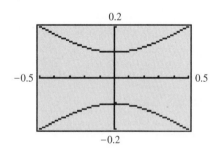

Fig. 21.77

Equations (21.20) and (21.24) give us the standard forms of the equation of the hyperbola with its center at the origin and its foci on one of the coordinate axes. There is another important equation form that represents a hyperbola, and it is

$$\boxed{xy = c} \qquad \qquad \textbf{(21.25)}$$

The asymptotes of this hyperbola are the coordinate axes, and the foci are on the line $y = x$ if c is positive or on the line $y = -x$ if c is negative.

The hyperbola represented by Eq. (21.25) is symmetric to the origin, for if $-x$ replaces x and $-y$ replaces y at the same time, we obtain $(-x)(-y) = c$, or $xy = c$. The equation is unchanged. However, if $-x$ replaces x or $-y$ replaces y, but not both, the sign on the left is changed. This means it is not symmetric to either axis. Here, c represents a constant and is not related to the focus. Two examples of this type of hyperbola are shown on the next page.

EXAMPLE 6 $xy = c$ Hyperbola

Plot the graph of the hyperbola $xy = 4$.

We find the values in the table below and then plot the appropriate points. Here, it is permissible to use a limited number of points, since we know the equation represents a hyperbola. Therefore, using $y = 4/x$, we obtain the values

x	-8	-4	-1	$-\frac{1}{2}$	$\frac{1}{2}$	1	4	8
y	$-\frac{1}{2}$	-1	-4	-8	8	4	1	$\frac{1}{2}$

Note that neither x nor y may equal zero. The hyperbola is shown in Fig. 21.78.

If the constant on the right is negative (for example, if $xy = -4$), then the two branches of the hyperbola are in the second and fourth quadrants.   ∎

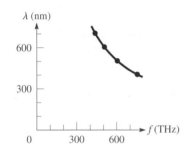

Fig. 21.78

EXAMPLE 7 $xy = c$ Hyperbola – application

For a light wave, the product of its frequency f of vibration and its wavelength λ is a constant, and this constant is the speed of light c. For green light, for which $f = 600$ THz, $\lambda = 500$ nm. Graph λ as a function of f for any light wave.

From the statement above, we know that $f\lambda = c$, and from the given values, we have

$$(600\,\text{THz})(500\,\text{nm}) = (6.0 \times 10^{14}\,\text{Hz})(5.0 \times 10^{-7}\,\text{m}) = 3.0 \times 10^8\,\text{m/s}$$

which means $c = 3.0 \times 10^8$ m/s. We are to sketch $f\lambda = 3.0 \times 10^8$. Solving for λ as $\lambda = 3.0 \times 10^8/f$, **we have the table at the left** (only positive values have meaning). See Fig. 21.79. (Violet light has wavelengths of about 400 nm, orange light has wavelengths of about 600 nm, and red light has wavelengths of about 700 nm.)   ∎

f (THz)	750	600	500	430
λ (nm)	400	500	600	700

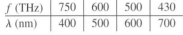

Fig. 21.79

We can conclude that *the equation of a hyperbola is characterized by the presence of both an x^2-term and a y^2-term, having different signs, or by the presence of an xy-term with no squared terms.* We will consider the equation of the hyperbola further in Sections 21.7, 21.8, and 21.9.

The hyperbola has some very useful applications. The LORAN radio navigation system is based on the use of hyperbolic paths. Some reflecting telescopes use hyperbolic mirrors. The paths of comets that never return to pass by the sun are hyperbolic. Some applications are illustrated in the exercises.

■ The first reasonable measurement of the speed of light was made by the Danish astronomer Olaf Roemer (1644–1710). He measured the time required for light to come from the moons of Jupiter across the earth's orbit.

EXERCISES 21.6

In Exercises 1 and 2, make the given changes in the indicated examples of this section and then solve the resulting problems.

1. In Example 2, interchange the denominators of 4 and 16; find the vertices, ends of the conjugate axis, and foci. Sketch the curve.

2. In Example 3, change $-9y^2$ to $-y^2$ and then follow the same instructions as in Exercise 1.

In Exercises 3–16, find the coordinates of the vertices and the foci of the given hyperbolas. Sketch each curve.

3. $\dfrac{x^2}{25} - \dfrac{y^2}{144} = 1$

4. $\dfrac{x^2}{16} - \dfrac{y^2}{4} = 1$

5. $\dfrac{y^2}{9} - \dfrac{x^2}{1} = 1$

6. $\dfrac{y^2}{4} - \dfrac{x^2}{21} = 1$

7. $\dfrac{4x^2}{25} - \dfrac{y^2}{4} = 1$

8. $\dfrac{9y^2}{25} - x^2 = 1$

9. $4x^2 - y^2 = 4$

10. $x^2 - 9y^2 = 1$

11. $2y^2 - 5x^2 = 10$

12. $9y^2 - 16x^2 = 144$

13. $y^2 = 4(x^2 + 1)$

14. $y^2 = 9(x^2 - 1)$

15. $4x^2 - y^2 = 0.64$

16. $9y^2 - x^2 = 0.36$

In Exercises 17–28, find the equations of the hyperbolas satisfying the given conditions. The center of each is at the origin.

17. Vertex $(3, 0)$, focus $(5, 0)$

18. Vertex $(0, 1)$, focus $(0, \sqrt{3})$

19. Conjugate axis $= 48$, vertex $(0, 10)$

20. Sum of lengths of transverse and conjugate axes 28, focus $(10, 0)$

21. Passes through $(2, 3)$, focus $(2, 0)$

22. Passes through $(8, \sqrt{3})$, vertex $(4, 0)$

23. Passes through $(5, 4)$ and $(3, \frac{4}{5}\sqrt{5})$

24. Passes through $(1, 2)$ and $(2, 2\sqrt{2})$

25. Asymptote $y = 2x$, vertex $(1, 0)$

26. Asymptote $y = -4x$, vertex $(0, 4)$

27. The difference of distances to (x, y) from $(10, 0)$ and $(-10, 0)$ is 12.

28. The difference of distances to (x, y) from $(0, 4)$ and $(0, -4)$ is 6.

In Exercises 29–54, solve the given problems.

29. Sketch the graph of the hyperbola $xy = 2$.

30. Sketch the graph of the hyperbola $xy = -4$.

31. Show that the parametric equations $x = \sec t$, $y = \tan t$ define a hyperbola.

32. Show that all hyperbolas $\dfrac{x^2}{\cos^2 \theta} - \dfrac{y^2}{\sin^2 \theta} = 1$ have foci at $(\pm 1, 0)$ for all values of θ.

33. Find any points of intersection of the ellipse $2x^2 + y^2 = 17$ and the hyperbola $y^2 - x^2 = 5$.

34. Find any points of intersection of the hyperbolas $x^2 - 3y^2 = 22$ and $xy = 5$.

35. Find the equation of the hyperbola with foci $(1, 2)$ and $(11, 2)$, and a transverse axis of 8, by use of the definition. Sketch the curve.

36. Find the equation of the hyperbola with vertices $(-2, 4)$ and $(-2, -2)$, and a conjugate axis of 4, by use of the definition. Sketch the curve.

37. Use a graphing calculator to view the hyperbola $x^2 - 4y^2 + 4x + 32y - 64 = 0$.

38. Use a graphing calculator to view the hyperbola $5y^2 - 4x^2 + 8x + 40y + 56 = 0$.

39. The equation of a hyperbola with center (h, k) and transverse axis parallel to the x-axis is $\dfrac{(x - h)^2}{a^2} - \dfrac{(y - k)^2}{b^2} = 1$. (This is shown in Section 21.7). Sketch the hyperbola that has a transverse axis of 4, a conjugate axis of 6, and for which (h, k) is $(-3, 2)$.

40. The equation of a hyperbola with center (h, k) and transverse axis parallel to the y-axis is $\dfrac{(y - k)^2}{a^2} - \dfrac{(x - h)^2}{b^2} = 1$. (This is shown in Section 21.7). Sketch the hyperbola that has a transverse axis of 2, a conjugate axis of 8, and for which (h, k) is $(5, 0)$.

41. Two concentric (same center) hyperbolas are called conjugate hyperbolas if the transverse and conjugate axes of one are, respectively, the conjugate and transverse axes of the other. What is the equation of the hyperbola conjugate to the hyperbola in Exercise 18?

42. As with an ellipse, the *eccentricity e* of a hyperbola is defined as $e = c/a$. Find the eccentricity of the hyperbola $2x^2 - 3y^2 = 24$.

43. Graph the inequality $100x^2 - 49y^2 \le 4900$.

44. Draw a sketch of the graph of the region in which the points satisfy the system of inequalities $xy < 4$, $y > x$.

45. Find the equation of the hyperbola that has the same foci as the ellipse $\frac{x^2}{169} + \frac{y^2}{144} = 1$ and passes through $(4\sqrt{2}, 3)$.

W **46.** Explain how a branch of a hyperbola differs from a parabola.

47. The cross section of the roof of a storage building shown in Fig. 21.80 is hyperbolic with the horizontal beam passing through the focus. Find the equation of the hyperbola such that its center is at the origin.

Fig. 21.80	**Fig. 21.81**

48. A plane is flying at a constant altitude of 2000 m. Show that the equation relating the horizontal distance x and the direct-line distance l from a control tower to the plane is that of a hyperbola. Sketch the graph of l as a function of x. See Fig. 21.81.

49. A jet travels 600 km at a speed of v km/h for t hours. Graph the equation relating v as a function of t.

50. A drain pipe 100 m long has an inside diameter d (in m) and an outside diameter D (in m). If the volume of material of the pipe itself is 0.50 m³, what is the equation relating d and D? Graph D as a function of d.

51. Ohm's law in electricity states that the product of the current i and the resistance R equals the voltage V across the resistance. If a battery of 6.00 V is placed across a variable resistor R, find the equation relating i and R and sketch the graph of i as a function of R.

52. A ray of light directed at one focus of a hyperbolic mirror is reflected toward the other focus. Find the equation that represents the hyperbolic mirror shown in Fig. 21.82.

Fig. 21.82

53. A radio signal is sent simultaneously from stations A and B 600 km apart on the Carolina coast. A ship receives the signal from A 1.20 ms before it receives the signal from B. Given that radio signals travel at 300 km/ms, draw a graph showing the possible locations of the ship. This problem illustrates the basis of LORAN.

54. Maximum intensity for monochromatic (single-color) light from two sources occurs where the difference in distances from the sources is an integral number of wavelengths. Find the equation of the curves of maximum intensity in a thin film between the sources where the difference in paths is two wavelengths and the sources are four wavelengths apart. Let the sources be on the x-axis and the origin midway between them. Use units of one wavelength for both x and y.

Answers to Practice Exercises

1. $V(6, 0), V(-6, 0)$; $F(7, 0), F(-7, 0)$

2. $V(0, 1), V(0, -1)$; $F(0, \sqrt{5}), F(0, -\sqrt{5})$

21.7 Translation of Axes

Centers of Curves Not at Origin •
Axes of Curves Parallel to
Coordinate Axes • Translation of Axes

The equations we have considered for the parabola, the ellipse, and the hyperbola are those for which the center of the ellipse or hyperbola, or vertex of the parabola, is at the origin. In this section, we consider, without specific use of the definition, the equations of these curves for the cases in which the axis of the curve is parallel to one of the coordinate axes. This is done by **translation of axes.**

In Fig. 21.83, we choose a point (h, k) in the xy-coordinate plane as the origin of another coordinate system, the $x'y'$-coordinate system. The x'-axis is parallel to the x-axis and the y'-axis is parallel to the y-axis. Every point now has two sets of coordinates (x, y) and (x', y'). We see that

$$x = x' + h \quad \text{and} \quad y = y' + k \qquad \textbf{(21.26)}$$

Equation (21.26) can also be written in the form

$$x' = x - h \quad \text{and} \quad y' = y - k \qquad \textbf{(21.27)}$$

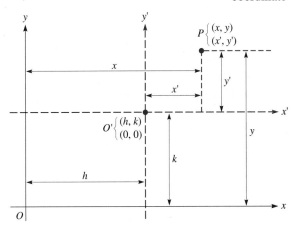

Fig. 21.83

EXAMPLE 1 Find equation – given vertex and focus

Find the equation of the parabola with vertex $(2, 4)$ and focus $(4, 4)$.

If we let the origin of the $x'y'$-coordinate system be the point $(2, 4)$, then the point $(4, 4)$ is the point $(2, 0)$ in the $x'y'$-system. This means $p = 2$ and $4p = 8$. See Fig. 21.84. In the $x'y'$-system, the equation is

$$(y')^2 = 8(x')$$

Using Eqs. (21.27), we have

$$(y - 4)^2 = 8(x - 2)$$

coordinates of vertex $(2, 4)$

as the equation of the parabola in the xy-coordinate system. ∎

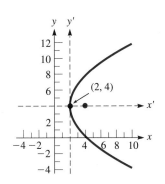

Fig. 21.84

Following the method of Example 1, by writing the equation of the curve in the $x'y'$-system and then using Eqs. (21.27), we have the following more general forms of the equations of the parabola, ellipse, and hyperbola.

Parabola, vertex (h, k):	$(y - k)^2 = 4p(x - h)$	(axis parallel to x-axis)	**(21.28)**
	$(x - h)^2 = 4p(y - k)$	(axis parallel to y-axis)	**(21.29)**
Ellipse, center (h, k):	$\dfrac{(x - h)^2}{a^2} + \dfrac{(y - k)^2}{b^2} = 1$	(major axis parallel to x-axis)	**(21.30)**
	$\dfrac{(y - k)^2}{a^2} + \dfrac{(x - h)^2}{b^2} = 1$	(major axis parallel to y-axis)	**(21.31)**
Hyperbola, center (h, k):	$\dfrac{(x - h)^2}{a^2} - \dfrac{(y - k)^2}{b^2} = 1$	(transverse axis parallel to x-axis)	**(21.32)**
	$\dfrac{(y - k)^2}{a^2} - \dfrac{(x - h)^2}{b^2} = 1$	(transverse axis parallel to y-axis)	**(21.33)**

Fig. 21.85

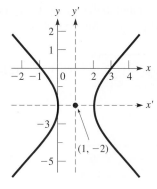

Fig. 21.86

Practice Exercise

1. Find the center and foci of ellipse
$2x^2 + 3y^2 - 8x + 18y + 29 = 0$.

Fig. 21.87

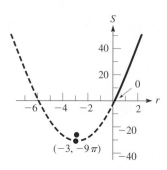

Fig. 21.88

EXAMPLE 2 Describing curve – given equation

Describe the curve of the equation

$$\frac{(x-3)^2}{25} + \frac{(y+2)^2}{9} = 1$$

We see that this equation fits the form of Eq. (21.30) with $h = 3$ and $k = -2$. It is the equation of an ellipse with its center at $(3, -2)$ and its major axis parallel to the x-axis. The semimajor axis is $a = 5$, and the semiminor axis is $b = 3$. The ellipse is shown in Fig. 21.85. ∎

EXAMPLE 3 Finding center of hyperbola

Find the center of the hyperbola $2x^2 - y^2 - 4x - 4y - 4 = 0$.

To analyze this curve, we first complete the square in the x-terms and in the y-terms. This will allow us to recognize properly the choice of h and k.

$$2x^2 - 4x - y^2 - 4y = 4$$
$$2(x^2 - 2x \quad) - (y^2 + 4y \quad) = 4$$
$$2(x^2 - 2x + 1) - (y^2 + 4y + 4) = 4 + 2 - 4$$

We note here that when we added 1 to complete the square of the x-terms within the parentheses, *we were actually adding 2 to the left side.* Thus, we added 2 to the right side. Similarly, when we added 4 to the y-terms within the parentheses, *we were actually subtracting 4 from the left side.* Continuing, we have

$$2(x-1)^2 - (y+2)^2 = 2$$

coordinates of center $(1, -2)$

$$\frac{(x-1)^2}{1} - \frac{(y+2)^2}{2} = 1$$

Therefore, the center of the hyperbola is $(1, -2)$. See Fig. 21.86. ∎

EXAMPLE 4 Translation of axes – application

Cylindrical glass beakers are to be made with a height of 3 in. Express the surface area in terms of the radius of the base and sketch the curve.

The total surface area S of a beaker is the sum of the area of the base and the lateral surface area of the side. In general, S in terms of the radius r of the base and height h of the side is $S = \pi r^2 + 2\pi rh$. Since $h = 3$ in., we have

$$S = \pi r^2 + 6\pi r$$

which is the desired relationship. See Fig. 21.87.

To analyze the equation relating S and r, we complete the square of the r terms:

$$S = \pi(r^2 + 6r)$$
$$S + 9\pi = \pi(r^2 + 6r + 9) \qquad \text{complete the square}$$
$$S + 9\pi = \pi(r + 3)^2$$

vertex $(-3, -9\pi)$

$$(r + 3)^2 = \frac{1}{\pi}(S + 9\pi)$$

This represents a parabola with vertex $(-3, -9\pi)$. Since $4p = 1/\pi$, $p = 1/(4\pi)$, the focus is $\left(-3, \frac{1}{4\pi} - 9\pi\right)$ as shown in Fig. 21.88. The part of the graph for negative r is dashed since only positive values have meaning. ∎

EXERCISES 21.7

In Exercises 1 and 2, make the given changes in the indicated examples of this section and then solve the resulting problem.

1. In Example 2, change $y + 2$ to $y - 2$ and change the sign before the second term from $+$ to $-$. Then describe and sketch the curve.

2. In Example 3, change the fourth and fifth terms from $-4y - 4$ to $+ 6y - 9$ and then find the center.

In Exercises 3–10, describe the curve represented by each equation. Identify the type of curve and its center (or vertex if it is a parabola). Sketch each curve.

3. $(y - 2)^2 = 4(x + 1)$

4. $\dfrac{(x + 4)^2}{4} + \dfrac{(y - 1)^2}{1} = 1$

5. $\dfrac{(x - 1)^2}{4} - \dfrac{(y - 2)^2}{9} = 1$

6. $(y + 5)^2 = -8(x - 2)$

7. $\dfrac{(x + 1)^2}{1} + \dfrac{y^2}{9} = 1$

8. $\dfrac{(y - 4)^2}{16} - \dfrac{(x + 2)^2}{4} = 1$

9. $(x + 3)^2 = -12(y - 1)$

10. $\dfrac{x^2}{0.16} + \dfrac{(y + 1)^2}{0.25} = 1$

In Exercises 11–22, find the equation of each of the curves described by the given information.

11. Parabola: vertex $(-1, 3)$, focus $(3, 3)$

12. Parabola: focus $(2, -5)$, directrix $y = 3$

13. Parabola: axis, directrix are coordinate axes, focus $(12, 0)$

14. Parabola: vertex $(4, 4)$, vertical directrix, passes through $(0, 1)$

15. Ellipse: center $(-2, 2)$, focus $(-5, 2)$, vertex $(-7, 2)$

16. Ellipse: center $(0, 3)$, focus $(12, 3)$, major axis 26 units

17. Ellipse: center $(-2, 1)$, vertex $(-2, 5)$, passes through $(0, 1)$

18. Ellipse: foci $(1, -2)$ and $(1, 10)$, minor axis 5 units

19. Hyperbola: vertex $(-1, 1)$, focus $(-1, 4)$, center $(-1, 2)$

20. Hyperbola: foci $(2, 1)$ and $(8, 1)$, conjugate axis 6 units

21. Hyperbola: vertices $(2, 1)$ and $(-4, 1)$, focus $(-6, 1)$

22. Hyperbola: center $(1, -4)$, focus $(1, 1)$, transverse axis 8 units

In Exercises 23–40, determine the center (or vertex if the curve is a parabola) of the given curve. Sketch each curve.

23. $x^2 + 2x - 4y - 3 = 0$

24. $y^2 - 2x - 2y - 9 = 0$

25. $x^2 + 4y = 24$

26. $x^2 + 4y^2 = 32y$

27. $4x^2 + 9y^2 + 24x = 0$

28. $2x^2 + y^2 + 8x = 8y$

29. $9x^2 - y^2 + 8y = 7$

30. $2x^2 - 4x = 9y - 2$

31. $5x^2 - 4y^2 + 20x + 8y = 4$

32. $0.04x^2 + 0.16y^2 = 0.01y$

33. $4x^2 - y^2 + 32x + 10y + 35 = 0$

34. $2x^2 + 2y^2 - 24x + 16y + 95 = 0$

35. $16x^2 + 25y^2 - 32x + 100y - 284 = 0$

36. $9x^2 - 16y^2 - 18x + 96y - 279 = 0$

37. $5x^2 - 3y^2 + 95 = 40x$

38. $5x^2 + 12y + 18 = 2y^2$

39. $9x^2 + 9y^2 + 14 = 6x + 24y$

40. $4y^2 + 29 = 15x + 12y$

In Exercises 41–52, solve the given problems.

41. Find the equation of the hyperbola with asymptotes $x - y = -1$ and $x + y = -3$ and vertex $(3, -1)$.

42. The circle $x^2 + y^2 + 4x - 5 = 0$ passes through the foci and the ends of the minor axis of an ellipse that has its major axis along the x-axis. Find the equation of the ellipse.

43. The vertex and focus of one parabola are, respectively, the focus and vertex of a second parabola. Find the equation of the first parabola, if $y^2 = 4x$ is the equation of the second.

44. Identify the curve represented by $4y^2 - x^2 - 6x - 2y = 14$ and view it on a graphing calculator.

45. What is the general form of the equation of a *family* of parabolas if each vertex and focus is on the x-axis?

46. What is the general form of the equation of a *family* of ellipses with foci on the y-axis if each passes through the origin?

47. An electric current (in A) is $i = 2 + \sin\left(2\pi t - \frac{\pi}{3}\right)$. What is the equation for the current if the origin of the (t', i') system is taken as $\left(\frac{1}{6}, 2\right)$ of the (t, i) system?

48. The stopping distance d (in ft) of a car traveling at v mi/h is represented by $d = 0.05v^2 + v$. Where is the vertex of the parabola that represents d?

49. The stream from a fire hose follows a parabolic curve and reaches a maximum height of 60 ft at a horizontal distance of 95 ft from the nozzle. Find the equation that represents the stream, with the origin at the nozzle. Sketch the graph.

50. For a constant capacitive reactance and a constant resistance, sketch the graph of the impedance and inductive reactance (as abscissas) for an alternating-current circuit. (See Section 12.7.)

51. Two wheels in a friction drive assembly are equal ellipses, as shown in Fig. 21.89. They are always in contact, with the left wheel fixed in position and the right wheel able to move horizontally. Find the equation that can be used to represent the circumference of each wheel in the position shown.

Fig. 21.89

52. An agricultural test station is to be divided into rectangular sections, each with a perimeter of 480 m. Express the area A of each section in terms of its width w and identify the type of curve represented. Sketch the graph of A as a function of w. For what value of w is A the greatest?

Answer to Practice Exercise

1. $C(2, -3)$, $F(1, -3)$, $F(3, -3)$

21.8 The Second-degree Equation

**Second-degree Equation •
Coefficients A, B, C Determine
Type of Curve • Conic Sections**

The equations of the circle, parabola, ellipse, and hyperbola are all special cases of the same general equation. In this section, we discuss this equation and how to identify the particular form it takes when it represents a specific type of curve.

*Each of these curves can be represented by a **second-degree equation** of the form*

$$Ax^2 + Bxy + Cy^2 + Dx + Ey + F = 0 \qquad (21.34)$$

The coefficients of the second-degree equation terms determine the type of curve that results. Recalling the discussions of the general forms of the equations of the circle, parabola, ellipse, and hyperbola from the previous sections of this chapter, Eq. (21.34) represents the indicated curve for given conditions of A, B, and C, as follows:

1. If $A = C$, $B = 0$, a circle.
2. If $A \neq C$ (but they have the same sign), $B = 0$, an ellipse.
3. If A and C have different signs, $B = 0$, a hyperbola.
4. If $A = 0$, $C = 0$, $B \neq 0$, a hyperbola.
5. If either $A = 0$ or $C = 0$ (both not both), $B = 0$, a parabola. (Special cases, such as a single point or no real locus, can also result.)

Another conclusion about Eq. (21.34) is that, if either $D \neq 0$ or $E \neq 0$ (or both), the center of the curve (or vertex of a parabola) is not at the origin. If $B \neq 0$, the axis of the curve has been rotated. We have considered only one such case (the hyperbola $xy = c$) so far in this chapter. In the following section, *rotation of axes* is taken up, and if $B \neq 0$, we will see that the type of curve depends on the value $B^2 - 4AC$. The following examples illustrate how the type of curve is identified from the equation according to the five criteria listed above.

EXAMPLE 1 Identify circle from equation

The equation $2x^2 = 3 - 2y^2$ represents a circle. This can be seen by putting the equation in the form of Eq. (21.34). This form is

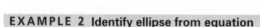

$$2x^2 + 2y^2 - 3 = 0$$

$A = 2 \qquad C = 2$

We see that $A = C$. Also, since there is no xy-term, we know that $B = 0$. This means that the equation represents a circle. If we write it as $x^2 + y^2 = \frac{3}{2}$, we see that it fits the form of Eq. (21.12). The circle is shown in Fig. 21.90. ∎

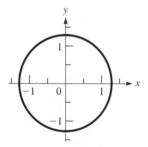

Fig. 21.90

EXAMPLE 2 Identify ellipse from equation

The equation $3y^2 = 6y - x^2 + 3$ represents an ellipse. Before analyzing the equation, we should put it in the form of Eq. (21.34). For this equation, this form is

$$x^2 + 3y^2 - 6y - 3 = 0$$

$A = 1 \qquad C = 3$

Here, we see that $B = 0$, A and C have the same sign, and $A \neq C$. Therefore, it is an ellipse. The $-6y$ term indicates that the center of the ellipse is not at the origin. This ellipse is shown in Fig. 21.91. ∎

Fig. 21.91

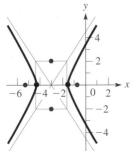

Fig. 21.92

Identify the type of curve represented by each equation.

1. $2y^2 - 4y + 5 = 4x - x^2$

2. $x^2 + y^2 - 8y + 12x = x^2 - y^2$

EXAMPLE 3 Identify hyperbola from equation

Identify the curve represented by $2x^2 + 12x = y^2 - 14$. Determine the appropriate quantities for the curve, and sketch the graph.

Writing this equation in the form of Eq. (21.34), we have

$$2x^2 - y^2 + 12x + 14 = 0$$

$$A = 2 \qquad\qquad C = -1$$

We identify this equation as representing a hyperbola, since A and C have different signs and $B = 0$. We now write it in the standard form of a hyperbola:

$$2x^2 + 12x - y^2 = -14$$

$$2(x^2 + 6x \quad) - y^2 = -14 \qquad \text{complete the square}$$

$$2(x^2 + 6x + 9) - y^2 = -14 + 18$$

$$2(x + 3)^2 - y^2 = 4$$

center is $(-3, 0)$

$$\frac{(x + 3)^2}{2} - \frac{y^2}{4} = 1 \qquad \frac{x'^2}{2} - \frac{y'^2}{4} = 1$$

Thus, we see that the center (h, k) of the hyperbola is the point $(-3, 0)$. Also, $a = \sqrt{2}$ and $b = 2$. This means that the vertices are $(-3 + \sqrt{2}, 0)$ and $(-3 - \sqrt{2}, 0)$, and the conjugate axis extends from $(-3, 2)$ to $(-3, -2)$. Also, $c^2 = 2 + 4 = 6$, which means that $c = \sqrt{6}$. The foci are $(-3 + \sqrt{6}, 0)$ and $(-3 - \sqrt{6}, 0)$. The graph is shown in Fig. 21.92. ∎

EXAMPLE 4 Identify parabola – calculator display

Identify the curve represented by $4y^2 - 23 = 4(4x + 3y)$ and find the appropriate important quantities. Then view it on a graphing calculator.

Writing the equation in the form of Eq. (21.34), we have

$$4y^2 - 16x - 12y - 23 = 0$$

Therefore, we recognize the equation as representing a parabola, since $A = 0$ and $B = 0$. Now, writing the equation in the standard form of a parabola, we have

$$4y^2 - 12y = 16x + 23$$

$$4(y^2 - 3y \quad) = 16x + 23 \qquad \text{complete the square}$$

$$4\left(y^2 - 3y + \frac{9}{4}\right) = 16x + 23 + 9$$

$$4\left(y - \frac{3}{2}\right)^2 = 16(x + 2)$$

vertex $\left(-2, \frac{3}{2}\right)$

$$\left(y - \frac{3}{2}\right)^2 = 4(x + 2) \quad \text{or} \quad y'^2 = 4x'$$

We now note that the vertex is $(-2, 3/2)$ and that $p = 1$. This means that the focus is $(-1, 3/2)$ and the directrix is $x = -3$.

To view the graph of this equation on a graphing calculator, we first solve the equation for y and get $y = (3 \pm 4\sqrt{x + 2})/2$. Entering these two functions in the calculator, we get the view shown in Fig. 21.93. ∎

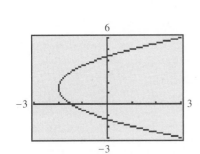

Fig. 21.93

■ See Appendix C for a graphing calculator program GRAPHCON. It displays the graph of the conic $Ax^2 + Cy^2 + Dx + Ey + F = 0$.

In Chapter 14, when these curves were first introduced, they were referred to as **conic sections.** If a plane is passed through a cone, the intersection of the plane and the cone results in one of these curves; the curve formed depends on the angle of the plane with respect to the axis of the cone. This is shown in Fig. 21.94.

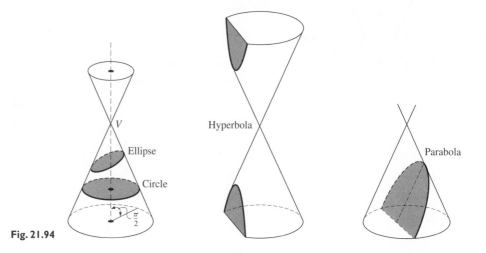

Fig. 21.94

EXERCISES 21.8

In Exercises 1 and 2, make the given changes in the indicated examples of this section and then solve the indicated problem.

1. In Example 1, change the − before the $2y^2$ to + and then determine what type of curve is represented.

2. In Example 3, change $y^2 − 14$ to $14 − y^2$ and then determine what type of curve is represented.

In Exercises 3–24, identify each of the equations as representing either a circle, a parabola, an ellipse, a hyperbola, or none of these.

3. $x^2 + 2y^2 − 2 = 0$

4. $x^2 − y = 0$

5. $2x^2 − y^2 − 1 = 0$

6. $y(y + x^2) = 4$

7. $8x^2 + 2y^2 = 6y(1 − y)$

8. $x(x − 3) = y(1 − 2y^2)$

9. $2.2x^2 − x − y = 1.6$

10. $2x^2 + 4y^2 = y + 2x$

11. $x^2 = (y − 1)(y + 1)$

12. $3.2x^2 = 2.1y(1 − 2y)$

13. $36x^2 = 12y(1 − 3y) + 1$

14. $y = 3(1 − 2x)(1 + 2x)$

15. $y(3 − 2x) = x(5 − 2y)$

16. $x(13 − 5x) = 5y^2$

17. $2xy + x − 3y = 6$

18. $(y + 1)^2 = x^2 + y^2 − 1$

19. $2x(x − y) = y(3 − y − 2x)$

20. $15x^2 = x(x − 12) + 4y(y − 6)$

21. $(x + 1)^2 + (y + 1)^2 = 2(x + y + 1)$

22. $(2x + y)^2 = 4x(y − 2) − 16$

23. $x(y + 3x) = x^2 + xy − y^2 + 1$

24. $4x(x − 1) = 2x^2 − 2y^2 + 3$

In Exercises 25–30, identify the curve represented by each of the given equations. Determine the appropriate important quantities for the curve and sketch the graph.

25. $x^2 = 8(y − x − 2)$

26. $x^2 = 6x − 4y^2 − 1$

27. $y^2 = 2(x^2 − 2x − 2y)$

28. $4x^2 + 4 = 9 − 8x − 4y^2$

29. $y^2 + 42 = 2x(10 − x)$

30. $x^2 − 4y = y^2 + 4(1 − x)$

In Exercises 31–36, identify the type of curve for each equation, and then view it on a graphing calculator.

31. $x^2 + 2y^2 − 4x + 12y + 14 = 0$

32. $4y^2 − x^2 + 40y − 4x + 60 = 0$

33. $4(y^2 − 4x − 2) = 5(4y − 5)$

34. $2(2x^2 − y) = 8 − y^2$

35. $4(y^2 + 6y + 1) = x(x − 4) − 24$

36. $8x + 31 − xy = y(y − 2 − x)$

In Exercises 37–42, use the given values to determine the type of curve represented.

37. For the equation $x^2 + ky^2 = a^2$, what type of curve is represented if (a) $k = 1$, (b) $k < 0$, and (c) $k > 0$ $(k \neq 1)$?

38. For the equation $\dfrac{x^2}{4 − C} − \dfrac{y^2}{C} = 1$, what type of curve is represented if (a) $C < 0$ and (b) $0 < C < 4$? (For $C > 4$, see Exercise 40.)

W **39.** In Eq. (21.34), if $A > C > 0$ and $B = D = E = F = 0$, describe the locus of the equation.

W **40.** For the equation in Exercise 38, describe the locus of the equation if $C > 4$.

W **41.** In Eq. (21.34), if $A = B = C = 0$, $D \neq 0$, $E \neq 0$, and $F \neq 0$, describe the locus of the equation.

W **42.** In Eq. (21.34), if $A = -C \neq 0$, $B = D = E = 0$ and $F = C$, describe the locus of the equation if $C > 0$.

In Exercises 43–48, determine the type of curve from the given information.

43. The diagonal brace in a rectangular metal frame is 3.0 cm longer than the length of one of the sides. Determine the type of curve represented by the equation relating the lengths of the sides of the frame.

44. One circular solar cell has a radius that is 2.0 in. less than the radius r of a second circular solar cell. Determine the type of curve represented by the equation relating the total area A of both cells and r.

45. A flashlight emits a cone of light onto the floor. What type of curve is the perimeter of the lighted area on the floor, if the floor cuts completely through the cone of light?

46. What type of curve is the perimeter of the lighted area on the floor of the cone of light in Exercise 45, if the upper edge of the cone is parallel to the floor?

47. A supersonic jet creates a conical shock wave behind it. What type of curve is outlined on the surface of a lake by the shock wave if the jet is flying horizontally?

48. In Fig. 21.94, if the plane cutting the cones passes through the intersection of the upper and lower cones, what type of curve is the intersection of the plane and cones?

Answers to Practice Exercises

1. Ellipse   **2.** Parabola

21.9 Rotation of Axes

Angle of Rotation of Axes •
Value of $B^2 - 4AC$ Determines Curve
Calculator Display

In this chapter, we have discussed the circle, parabola, ellipse, and hyperbola, and in the last section, we showed how these curves are represented by the second-degree equation

$$Ax^2 + Bxy + Cy^2 + Dx + Ey + F = 0 \qquad \textbf{(21.34)}$$

The discussion of these curves included their equations with the center (vertex of a parabola) at the origin. However, as noted in the last section, except for the special case of the hyperbola $xy = c$, we did not show what happens when the axes are *rotated* about the origin.

If a set of axes is rotated about the origin through an angle θ, as shown in Fig. 21.95, we say that there has been a **rotation of axes***.* In this case, each point P in the plane has two sets of coordinates, (x, y) in the original system and (x', y') in the rotated system.

If we now let r equal the distance from the origin O to point P and let ϕ be the angle between the x'-axis and the line OP, we have

$$x' = r \cos \phi \qquad\qquad y' = r \sin \phi \qquad \textbf{(21.35)}$$
$$x = r \cos (\theta + \phi) \qquad y = r \sin(\theta + \phi) \qquad \textbf{(21.36)}$$

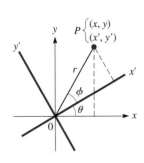

Fig. 21.95

Using the cosine and sine of the sum of two angles, we can write Eqs. (21.36) as

$$x = r \cos \phi \cos \theta - r \sin \phi \sin \theta$$
$$y = r \cos \phi \sin \theta + r \sin \phi \cos \theta \qquad \textbf{(21.37)}$$

Now, using Eqs. (21.35), we have

$$\boxed{\begin{aligned} x &= x' \cos \theta - y' \sin \theta \\ y &= x' \sin \theta + y' \cos \theta \end{aligned}} \qquad \textbf{(21.38)}$$

In our derivation, we have used the special case when θ is acute and P is in the first quadrant of both sets of axes. When simplifying equations of curves using Eqs. (21.38), we find that a rotation through a positive acute angle θ is sufficient. It can be shown, however, that Eqs. (21.38) hold for any θ and position of P.

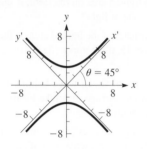

Fig. 21.96

EXAMPLE 1 Rotation through 45°

Transform $x^2 - y^2 + 8 = 0$ by rotating the axes through 45°.

When $\theta = 45°$, the rotation equations (21.38) become

$$x = x' \cos 45° - y' \sin 45° = \frac{x'}{\sqrt{2}} - \frac{y'}{\sqrt{2}}$$

$$y = x' \sin 45° + y' \cos 45° = \frac{x'}{\sqrt{2}} + \frac{y'}{\sqrt{2}}$$

Substituting into the equation $x^2 - y^2 + 8 = 0$ gives

$$\left(\frac{x'}{\sqrt{2}} - \frac{y'}{\sqrt{2}} \right)^2 - \left(\frac{x'}{\sqrt{2}} + \frac{y'}{\sqrt{2}} \right)^2 + 8 = 0$$

$$\frac{1}{2} x'^2 - x'y' + \frac{1}{2} y'^2 - \frac{1}{2} x'^2 - x'y' - \frac{1}{2} y'^2 + 8 = 0$$

$$x'y' = 4$$

The graph and both sets of axes are shown in Fig. 21.96. The original equation represents a hyperbola. In this example, we have shown that the $xy = c$ type of hyperbola is obtained by a 45° rotation of the standard form given by Eq. (21.34). ∎

In Section 21.8, when we showed the type of conic section represented by the second-degree equation Eq. (21.34), the standard forms of the parabola, ellipse, and hyperbola required that there is no xy-term, which means that $B = 0$. Remember, the $xy = c$ type of hyperbola does not fit the standard form of a hyperbola.

Therefore, we see that if we can remove the xy-term from the second-degree equation of the form (21.34), the analysis of the graph is simplified. By a proper rotation of axes, we find that Eq. (21.34) can be transformed into an equation containing no xy-term. We now show how this is done.

By substituting Eqs. (21.38) into Eq. (21.34) and then simplifying, we have

$$(A \cos^2 \theta + B \sin \theta \cos \theta + C \sin^2 \theta)x'^2 + [B \cos 2\theta - (A - C)\sin 2\theta]x'y'$$
$$+ (A \sin^2 \theta - B \sin \theta \cos \theta + C \cos^2 \theta)y'^2 + (D \cos \theta + E \sin \theta)x'$$
$$+ (E \cos \theta - D \sin \theta)y' + F = 0$$

If there is to be no $x'y'$-term, its coefficient must be zero. This means that $B \cos 2\theta - (A - C)\sin 2\theta = 0$, or

Angle of Rotation

$$\tan 2\theta = \frac{B}{A - C} \qquad (A \neq C) \tag{21.39}$$

Eq. (21.39) gives the angle of rotation except when $A = C$. In this case, the coefficient of the $x'y'$-term is $B \cos 2\theta$, which is zero if $2\theta = 90°$. Thus,

$$\theta = 45° \qquad (A = C) \tag{21.40}$$

Consider the following example.

■ For reference: Eq. (20.2) is

$\cos\theta = \dfrac{1}{\sec\theta}$

Eq. (20.7) is $1 + \tan^2\theta = \sec^2\theta$

Eq. (20.26) is $\sin\dfrac{\alpha}{2} = \pm\sqrt{\dfrac{1-\cos\alpha}{2}}$

Eq. (20.27) is $\cos\dfrac{\alpha}{2} = \pm\sqrt{\dfrac{1+\cos\alpha}{2}}$

EXAMPLE 2 Rotating axes – eliminating *xy*-term

By rotation of axes, transform $8x^2 + 4xy + 5y^2 = 9$ into a form without an xy-term. Identify and sketch the curve.

Here, $A = 8$, $B = 4$, and $C = 5$. Therefore, using Eq. (21.39), we have

$$\tan 2\theta = \frac{4}{8-5} = \frac{4}{3}$$

Since $\tan 2\theta$ is positive, we may take 2θ as an acute angle, which means θ is also acute. For the transformation, we need $\sin\theta$ and $\cos\theta$. We find these values by first finding the value of $\cos 2\theta$ and then using the half-angle formulas:

$$\cos 2\theta = \frac{1}{\sec 2\theta} = \frac{1}{\sqrt{1+\tan^2 2\theta}} = \frac{1}{\sqrt{1+\left(\frac{4}{3}\right)^2}} = \frac{3}{5} \qquad \text{using Eqs. (20.2) and (20.7)}$$

Now, using the half-angle formulas, Eqs. (20.26) and (20.27), we have

$$\sin\theta = \sqrt{\frac{1-\cos 2\theta}{2}} = \sqrt{\frac{1-\frac{3}{5}}{2}} = \frac{1}{\sqrt{5}} \qquad \cos\theta = \sqrt{\frac{1+\cos 2\theta}{2}} = \sqrt{\frac{1+\frac{3}{5}}{2}} = \frac{2}{\sqrt{5}}$$

Here, θ is about $26.6°$. Now, substituting these values into Eqs. (21.38), we have

$$x = x'\left(\frac{2}{\sqrt{5}}\right) - y'\left(\frac{1}{\sqrt{5}}\right) = \frac{2x'-y'}{\sqrt{5}} \qquad y = x'\left(\frac{1}{\sqrt{5}}\right) + y'\left(\frac{2}{\sqrt{5}}\right) = \frac{x'+2y'}{\sqrt{5}}$$

Now, substituting into the equation $8x^2 + 4xy + 5y^2 = 9$ gives

$$8\left(\frac{2x'-y'}{\sqrt{5}}\right)^2 + 4\left(\frac{2x'-y'}{\sqrt{5}}\right)\left(\frac{x'+2y'}{\sqrt{5}}\right) + 5\left(\frac{x'+2y'}{\sqrt{5}}\right)^2 = 9$$

$$8(4x'^2 - 4x'y' + y'^2) + 4(2x'^2 + 3x'y' - 2y'^2) + 5(x'^2 + 4x'y' + 4y'^2) = 45$$

$$45x'^2 + 20y'^2 = 45$$

$$\frac{x'^2}{1} + \frac{y'^2}{\frac{9}{4}} = 1$$

This is an ellipse with semimajor axis of $3/2$ and semiminor axis of 1. See Fig. 21.97. ■

Fig. 21.97

In Example 2, $\tan 2\theta$ was positive, and we made 2θ and θ positive. If, when using Eq. (21.39), $\tan 2\theta$ is negative, we then make 2θ obtuse ($90° < 2\theta < 180°$). In this case, $\cos 2\theta$ will be negative, but θ will be acute ($45° < \theta < 90°$).

In Section 21.7, we showed the use of translation of axes in writing an equation in standard form if $B = 0$. In this section, we have seen how rotation of axes is used to eliminate the xy-term. It is possible that both a translation of axes and a rotation of axes are needed to write an equation in standard form.

In Section 21.8, we identified a conic section by inspecting the values of A and C when $B = 0$. If $B \neq 0$, these curves are identified as follows:

1. If $B^2 - 4AC = 0$, a parabola
2. If $B^2 - 4AC < 0$, an ellipse
3. If $B^2 - 4AC > 0$, a hyperbola

Special cases such as a point, parallel or intersecting lines, or no curve may result.

Practice Exercises

Identify the type of curve represented by each equation.

1. $3x^2 - 2xy + 3y^2 = 8$
2. $x^2 - 4xy + y^2 = -5$
3. $4x^2 + 4xy + y^2 - 24x + 38y - 19 = 0$

EXAMPLE 3 Rotating axes – graph – calculator display

For the equation $16x^2 - 24xy + 9y^2 + 20x - 140y - 300 = 0$, identify the curve and simplify it to standard form. Sketch the graph and display it on a graphing calculator.

With $A = 16$, $B = -24$, and $C = 9$, using Eq. (21.39), we have

$$\tan 2\theta = \frac{-24}{16 - 9} = -\frac{24}{7}$$

In this case, $\tan 2\theta$ is negative, and we take 2θ to be an obtuse angle. We then find that $\cos 2\theta = -7/25$. In turn, we find that $\sin\theta = 4/5$ and $\cos\theta = 3/5$. Here, θ is about 53.1°. Using these values in Eqs. (21.38), we find that

$$x = \frac{3x' - 4y'}{5} \qquad y = \frac{4x' + 3y'}{5}$$

Substituting these into the original equation and simplifying, we get

$$y'^2 - 4x' - 4y' - 12 = 0$$

This equation represents a parabola with its axis parallel to the x'-axis. The vertex is found by completing the square:

$$(y' - 2)^2 = 4(x' + 4)$$

The vertex is the point $(-4, 2)$ in the $x'y'$-rotated system. Therefore,

$$y''^2 = 4x''$$

is the equation in the $x''y''$-rotated and then translated system. The graph and the coordinate systems are shown in Fig. 21.98.

To display the curve on a graphing calculator, we solve for y by using the quadratic formula, writing the equation as

$$9y^2 + (-24x - 140)y + (16x^2 + 20x - 300) = 0$$

Therefore, we see that in using the quadratic formula, $a = 9$, $b = -24x - 140$, and $c = 16x^2 + 20x - 300$. Now, solving for y, we have

$$y = \frac{24x + 140 \pm \sqrt{(-24x - 140)^2 - 4(9)(16x^2 + 20x - 300)}}{18}$$

We *enter both functions indicated by the ± sign* to get the display in Fig. 21.99. ∎

Fig. 21.98

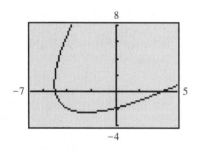

Fig. 21.99

EXERCISES 21.9

In Exercises 1–4, transform the given equations by rotating the axes through the given angle. Identify and sketch each curve.

1. $x^2 - y^2 = 25$, $\theta = 45°$

2. $x^2 + y^2 = 16$, $\theta = 60°$

3. $8x^2 - 4xy + 5y^2 = 36$, $\theta = \tan^{-1} 2$

4. $2x^2 + 24xy - 5y^2 = 8$, $\theta = \tan^{-1}\frac{3}{4}$

In Exercises 5–10, identify the type of curve that each equation represents by evaluating $B^2 - 4AC$.

5. $x^2 + 2xy + x - 7 - 3 = 0$

6. $8x^2 - 4xy + 2y^2 + 7 = 0$

7. $x^2 - 2xy + y^2 + 3y = 0$

8. $4xy + 3y^2 - 8x + 16y + 19 = 0$

9. $13x^2 + 10xy + 13y^2 + 6x - 42y - 27 = 0$

10. $x^2 - 4xy + 4y^2 + 36x + 28y + 24 = 0$

In Exercises 11–16, transform each equation to a form without an xy-term by a rotation of axes. Identify and sketch each curve. Then display each curve on a graphing calculator.

11. $x^2 + 2xy + y^2 - 2x + 2y = 0$

12. $5x^2 - 6xy + 5y^2 = 32$

13. $3x^2 + 4xy = 4$

14. $9x^2 - 24xy + 16y^2 - 320x - 240y = 0$

15. $11x^2 - 6xy + 19y^2 = 20$

16. $x^2 + 4xy - 2y^2 = 6$

In Exercises 17 and 18, transform each equation to a form without an xy-term by a rotation of axes. Then transform the equation to a stan-dard form by a translation of axes. Identify and sketch each curve. Then display each curve on a graphing calculator.

17. $16x^2 - 24xy + 9y^2 - 60x - 80y + 400 = 0$

18. $73x^2 - 72xy + 52y^2 + 100x - 200y + 100 = 0$

In Exercises 19 and 20, solve the given problems.

19. (a) In Equation (21.34), if A and C have opposite signs, what type of curve is represented? (b) If $B \neq 0$, and either A or C is zero, what type of curve is represented?

20. An elliptical cam can be represented by the equation $x^2 - 3xy + 5y^2 - 13 = 0$. Through what angle is the cam rotated from its standard position?

Answers to Practice Exercises

1. Ellipse   2. Hyperbola   3. Parabola

21.10 Polar Coordinates

Pole • Polar Axis • Polar Coordinates • Polar and Rectangular Coordinates

■ The Swiss mathematician Jakob Bernoulli (1654–1705) was among the first to make significant use of polar coordinates.

Fig. 21.100

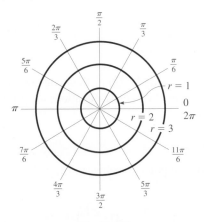

Fig. 21.101

To this point, we have graphed all curves in the rectangular coordinate system. However, for certain types of curves, other coordinate systems are better adapted. We discuss one of these systems here.

Instead of designating a point by its x- and y-coordinates, we can specify its location by its radius vector and the angle the radius vector makes with the x-axis. Thus, the r and θ that are used in the definitions of the trigonometric functions can also be used as the coordinates of points in the plane. The important aspect of choosing coordinates is that, for each set of values, there must be only one point that corresponds to this set. We can see that this condition is satisfied by the use of r and θ as coordinates. *In* **polar coordinates,** *the origin is called the* **pole,** *and the half-line for which the angle is zero (equivalent to the positive x-axis) is called the* **polar axis.** The coordinates of a point are designated as (r, θ). We will use radians when measuring the value of θ. See Fig. 21.100.

When using polar coordinates, we generally label the lines for some of the values of θ; namely, those for $\theta = 0$ (the polar axis), $\theta = \pi/2$ (equivalent to the positive y-axis), $\theta = \pi$ (equivalent to the negative x-axis), $\theta = 3\pi/2$ (equivalent to the negative y-axis), and possibly others. In Fig. 21.101, these lines and those for multiples of $\pi/6$ are shown. Also, the circles for $r = 1$, $r = 2$, and $r = 3$ are shown in this figure.

EXAMPLE 1 Polar coordinates of a point

(a) If $r = 2$ and $\theta = \pi/6$, we have the point shown in Fig. 21.102. The point corresponds to $(\sqrt{3}, 1)$ in rectangular coordinates.

(b) The polar coordinate point $(1, 3\pi/4)$ is also shown. It is equivalent to $(-\sqrt{2}/2, \sqrt{2}/2)$ in rectangular coordinates.

(c) The polar coordinate point $(2, 5)$ is also shown. It is equivalent approximately to $(0.6, -1.9)$ in rectangular coordinates. Remember, the 5 is an angle in radians. ■

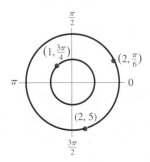

Fig. 21.102

One difference between rectangular coordinates and polar coordinates is that, for each point in the plane, there are limitless possibilities for the polar coordinates of that point. For example, the point $\left(2, \frac{\pi}{6}\right)$ can also be represented by $\left(2, \frac{13\pi}{6}\right)$ since the angles $\frac{\pi}{6}$ and $\frac{13\pi}{6}$ are coterminal. We also remove one restriction on r that we imposed in the definition of the trigonometric functions. That is, r is allowed to take on positive and negative values. If r is negative, θ is located as before, but ***the point is found r units from the pole but on the opposite side*** from that on which it is positive.

NOTE ▶

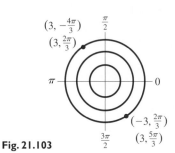

$\left(3, -\frac{4\pi}{3}\right)$
$\left(3, \frac{2\pi}{3}\right)$

$\left(-3, \frac{2\pi}{3}\right)$
$\left(3, \frac{5\pi}{3}\right)$

Fig. 21.103

EXAMPLE 2 Polar coordinates – negative *r*

The polar coordinates $(3, 2\pi/3)$ and $(3, -4\pi/3)$ represent the same point. However, the point $(-3, 2\pi/3)$ is on the opposite side of the pole from $(3, 2\pi/3)$, 3 units from the pole. Another possible set of polar coordinates for the point $(-3, 2\pi/3)$ is $(3, 5\pi/3)$. These points are shown in Fig. 21.103. ■

When locating and plotting a point in polar coordinates, it is generally best to *first locate the terminal side of θ, and then measure r along this terminal side.* This is illustrated in the following example.

EXAMPLE 3 Locate terminal side first

$\theta = \frac{5\pi}{6}$
B
A
$\theta = -2.4$

Fig. 21.104

Plot the points $A(2, 5\pi/6)$ and $(-3.2, -2.4)$ in the polar coordinate system.

To locate A, we determine the terminal side of $\theta = 5\pi/6$ and then determine $r = 2$. See Fig. 21.104.

To locate B, we find the terminal side of $\theta = -2.4$, measuring clockwise from the polar axis (and recalling that $\pi = 3.14 = 180°$). Then we locate $r = -3.2$ on the opposite side of the pole. See Fig. 21.104.

We will find that points with negative values of r occur frequently when plotting curves in polar coordinates. ■

POLAR AND RECTANGULAR COORDINATES

The relationships between the polar coordinates of a point and the rectangular coordinates of the same point come from the definitions of the trigonometric functions. Those most commonly used are (see Fig. 21.105)

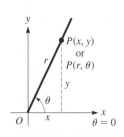

$P(x, y)$
or
$P(r, \theta)$

$\theta = 0$

Fig. 21.105

$$x = r\cos\theta \quad y = r\sin\theta \tag{21.41}$$

$$\tan\theta = \frac{y}{x} \quad r = \sqrt{x^2 + y^2} \tag{21.42}$$

The following examples show the use of Eqs. (21.41) and (21.42) in changing coordinates in one system to coordinates in the other system. Also, these equations are used to transform equations from one system to the other.

EXAMPLE 4 Polar to rectangular coordinates

$(2\sqrt{2}, 2\sqrt{2})$
$r = 4$
$2\sqrt{2}$
$2\sqrt{2}$

Fig. 21.106

Using Eqs. (21.41), and (21.42) we can transform the polar coordinates $(4, \pi/4)$ into the rectangular coordinates $(2\sqrt{2}, 2\sqrt{2})$, since

$$x = 4\cos\frac{\pi}{4} = 4\left(\frac{\sqrt{2}}{2}\right) = 2\sqrt{2} \quad \text{and} \quad y = 4\sin\frac{\pi}{4} = 4\left(\frac{\sqrt{2}}{2}\right) = 2\sqrt{2}$$

See Fig. 21.106. ■

Fig. 21.107   **Fig. 21.108**

■ The cyclotron was invented in 1931 at the University of California. It was the first accelerator to deflect particles into circular paths.

Fig. 21.109

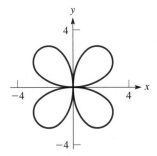

Fig. 21.110

EXAMPLE 5 Rectangular to polar coordinates

Using Eqs. (21.42), we can transform the rectangular coordinates $(3, -5)$ into polar coordinates, as follows.

$$\tan \theta = -\frac{5}{3}, \quad \theta = 5.25 \quad (\text{or } -1.03)$$

$$r = \sqrt{3^2 + (-5)^2} = 5.83$$

We know that θ is a fourth-quadrant angle since x is positive and y is negative. Therefore, the point $(3, -5)$ in rectangular coordinates can be expressed as the point $(5.83, 5.25)$ in polar coordinates (see Fig. 21.107). Other polar coordinates for the point are also possible.

Calculators are programmed to make conversions between rectangular and polar coordinates. The manual for any model should be consulted to determine how any particular model is used for these conversions. For a calculator that uses the *angle* feature, the display for the conversions of this example is shown in Fig. 21.108. ■

EXAMPLE 6 Rectangular to polar equation of curve

If an electrically charged particle enters a magnetic field at right angles to the field, the particle follows a circular path. This fact is used in the design of nuclear particle accelerators.

A proton (positively charged) enters a magnetic field such that its path may be described by the rectangular equation $x^2 + y^2 = 2x$, where measurements are in meters. Find the polar equation of this circle.

We change this equation expressed in the rectangular coordinates x and y into an equation expressed in the polar coordinates r and θ by using the relations $r^2 = x^2 + y^2$ and $x = r \cos \theta$ as follows:

$$
\begin{array}{ll}
x^2 + y^2 = 2x & \text{rectangular equation} \\
r^2 = 2r \cos \theta & \text{substitute} \\
r = 2 \cos \theta & \text{divided by } r
\end{array}
$$

This is the polar equation of the circle, which is shown in Fig. 21.109. ■

EXAMPLE 7 Polar to rectangular equation of curve

Find the rectangular equation of the *rose* $r = 4 \sin 2\theta$.

Using the trigonometric identity $\sin 2\theta = 2 \sin \theta \cos \theta$ and Eqs. (21.41) and (21.42) leads to the solution:

$$
\begin{array}{ll}
r = 4 \sin 2\theta & \text{polar equation} \\
= 4(2 \sin \theta \cos \theta) = 8 \sin \theta \cos \theta & \text{using identity} \\
\sqrt{x^2 + y^2} = 8\left(\dfrac{y}{r}\right)\left(\dfrac{x}{r}\right) = \dfrac{8xy}{r^2} = \dfrac{8xy}{x^2 + y^2} & \text{using Eqs. (21.35) and (21.36)} \\
x^2 + y^2 = \dfrac{64x^2 y^2}{(x^2 + y^2)^2} & \text{squaring both sides} \\
(x^2 + y^2)^3 = 64x^2 y^2 & \text{simplifying}
\end{array}
$$

Plotting the graph of this equation from the rectangular equation would be complicated. However, as we will see in the next section, plotting this graph in polar coordinates is quite simple. The curve is shown in Fig. 21.110. ■

EXERCISES 21.10

In Exercises 1–4, make the given changes in the indicated examples of this section and then solve the indicated problems.

1. In Example 2, change $2\pi/3$ to $\pi/3$ and then find another set of coordinates for each point, similar to those shown for the points in the example.

2. In Example 4, change $\pi/4$ to $\pi/6$

3. In Example 5, change 3 to -3 and -5 to 5.

4. In Example 7, change sin to cos.

In Exercises 5–16, plot the given polar coordinate points on polar coordinate paper.

5. $\left(3, \dfrac{\pi}{6}\right)$ 6. $(2, \pi)$ 7. $\left(\dfrac{5}{2}, -\dfrac{2\pi}{5}\right)$

8. $\left(5, -\dfrac{\pi}{3}\right)$ 9. $\left(-8, \dfrac{7\pi}{6}\right)$ 10. $\left(-5, \dfrac{\pi}{4}\right)$

11. $\left(-3, -\dfrac{5\pi}{4}\right)$ 12. $\left(-4, -\dfrac{5\pi}{3}\right)$ 13. $(2, 2)$

14. $(-6, -6)$ 15. $(0.5, -8.4)$ 16. $(-2.2, 18.8)$

In Exercises 17–20, find a set of polar coordinates for each of the points for which the rectangular coordinates are given.

17. $(\sqrt{3}, 1)$ 18. $(-1, -1)$

19. $\left(-\dfrac{\sqrt{3}}{2}, -\dfrac{1}{2}\right)$ 20. $(-10, 8)$

In Exercises 21–24, find the rectangular coordinates for each of the points for which the polar coordinates are given.

21. $\left(8, \dfrac{4\pi}{3}\right)$ 22. $(-4, -\pi)$

23. $(3.0, -0.40)$ 24. $(-1.0, 1.0)$

In Exercises 25–36, find the polar equation of each of the given rectangular equations.

25. $x = 3$ 26. $y = -x$
27. $x + 2y + 3 = 0$ 28. $x^2 + y^2 = 0.81$
29. $x^2 + (y - 2)^2 = 4$ 30. $x^2 - y^2 = 0.01$
31. $x^2 + 4y^2 = 4$ 32. $y^2 = 4x$
33. $x^2 + y^2 = 6y$ 34. $xy = 9$
35. $x^3 + y^3 - 4xy = 0$ 36. $y = \dfrac{2x}{x^2 + 1}$

In Exercises 37–48, find the rectangular equation of each of the given polar equations. In Exercises 37–44, identify the curve that is represented by the equation.

37. $r = \sin\theta$ 38. $r = 4\cos\theta$
39. $r\cos\theta = 4$ 40. $r\sin\theta = -2$

41. $r = \dfrac{2}{\cos\theta - 3\sin\theta}$ 42. $r = e^{r\cos\theta}\csc\theta$

43. $r = 4\cos\theta + 2\sin\theta$ 44. $r\sin(\theta + \pi/6) = 3$
45. $r = 2(1 + \cos\theta)$ 46. $r = 1 - \sin\theta$
47. $r^2 = \sin 2\theta$ 48. $r^2 = 16\cos 2\theta$

In Exercises 49–60, solve the given problems. All coordinates given are polar coordinates.

49. Is the point $(2, 3\pi/4)$ on the curve $r = 2\sin 2\theta$?

50. Is the point $(1/2, 3\pi/2)$ on the curve $r = \sin(\theta/3)$?

51. Show that the polar coordinate equation $r = a\sin\theta + b\cos\theta$ represents a circle by changing it to a rectangular equation.

52. Find the distance between the points, $(3, \pi/6)$ and $(4, \pi/2)$ by using the law of cosines.

53. The center of a regular hexagon is at the pole with one vertex at $(2, \pi)$. What are the coordinates of the other vertices?

54. Find the distance between the points $(4, \pi/6)$ and $(5, 5\pi/3)$.

55. Under certain conditions, the x- and y-components of a magnetic field B are given by the equations

$$B_x = \dfrac{-ky}{x^2 + y^2} \quad \text{and} \quad B_y = \dfrac{kx}{x^2 + y^2}$$

Write these equations in terms of polar coordinates.

56. In designing a domed roof for a building, an architect uses the equation $x^2 + \dfrac{y^2}{k^2} = 1$, where k is a constant. Write this equation in polar form.

57. The shape of a cam can be described by the polar equation $r = 3 - \sin\theta$. Find the rectangular equation for the shape of the cam.

58. The polar equation of the path of a weather satellite of the earth is $r = \dfrac{4800}{1 + 0.14\cos\theta}$, where r is measured in miles. Find the rectangular equation of the path of this satellite. The path is an ellipse, with the earth at one of the foci.

59. The control tower of an airport is taken to be at the pole, and the polar axis is taken as due east in a polar coordinate graph. How far apart (in km) are planes, at the same altitude, if their positions on the graph are $(6.10, 1.25)$ and $(8.45, 3.74)$?

(W) 60. The perimeter of a certain type of machine part can be described by the equation $r = a\sin\theta + b\cos\theta$ $(a > 0, b > 0)$. Explain why all such machine parts are circular.

Answers to Practice Exercises

1. $(2\sqrt{3}, -2)$ 2. $r = -2\cos\theta$

21.11 Curves in Polar Coordinates

**Curve Sketched from Equations •
Plotting Polar Coordinate Curves •
Calculator Display of Polar Curves**

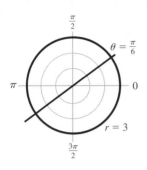

Fig. 21.111

The basic method for finding a curve in polar coordinates is the same as in rectangular coordinates. We assume values of the independent variable—in this case, θ—and then find the corresponding values of the dependent variable r. These points are then plotted and joined, thereby forming the curve that represents the relation in polar coordinates.

Before using the basic method, it is useful to point out that certain basic curves can be sketched directly from the equation. This is done by noting the meaning of each of the polar coordinate variables, r and θ. This is illustrated in the following example.

EXAMPLE 1 Curves sketched from equations

(a) The graph of the polar equation $r = 3$ is a circle of radius 3, with center at the pole. This can be seen to be the case, since $r = 3$ for all possible values of θ. It is not necessary to find specific points for this circle, which is shown in Fig. 21.111.

(b) The graph of $\theta = \pi/6$ is a straight line through the pole. It represents all points for which $\theta = \pi/6$ for all possible values of r. This line is shown in Fig. 21.111. ∎

EXAMPLE 2 Plotting a cardioid

Plot the graph of $r = 1 + \cos \theta$.

We find the following values of r corresponding to the chosen values of θ.

θ	0	$\frac{\pi}{4}$	$\frac{\pi}{2}$	$\frac{3\pi}{4}$	π	$\frac{5\pi}{4}$	$\frac{3\pi}{2}$	$\frac{7\pi}{4}$	2π
r	2	1.7	1	0.3	0	0.3	1	1.7	2
Point Number	1	2	3	4	5	6	7	8	9

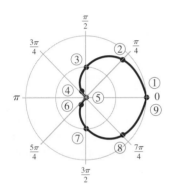

Fig. 21.112

We now see that the points are repeating, and it is unnecessary to find additional points. This curve is called a **cardioid** and is shown in Fig. 21.112. ∎

EXAMPLE 3 Plotting a limaçon

Plot the graph of $r = 1 - 2 \sin \theta$.

Choosing values of θ and then finding the corresponding values of r, we find the following table of values.

θ	0	$\frac{\pi}{4}$	$\frac{\pi}{2}$	$\frac{3\pi}{4}$	π	$\frac{5\pi}{4}$	$\frac{3\pi}{2}$	$\frac{7\pi}{4}$	2π
r	1	−0.4	−1	−0.4	1	2.4	3	2.4	1
Point Number	1	2	3	4	5	6	7	8	9

Fig. 21.113

Particular care should be taken in plotting the points for which r is negative. This curve is known as a **limaçon** and is shown in Fig. 21.113. ∎

Practice Exercise

Determine the type of curve represented by each polar coordinate equation.

1. $\theta = 2\pi/5$ **2.** $r = -4$

EXAMPLE 4 Polar curve – application

A cam is shaped such that the edge of the upper "half" is represented by the equation $r = 2.0 + \cos\theta$ and the lower "half" by the equation $r = \dfrac{3.0}{2.0 - \cos\theta}$, where measurements are in inches. Plot the curve that represents the shape of the cam.

We get the points for the edge of the cam by using values of θ from 0 to π for the upper "half" and from π to 2π for the lower "half". The table of values follows:

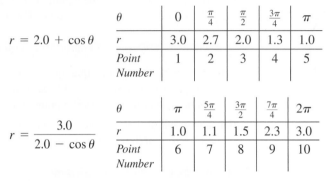

θ	0	$\frac{\pi}{4}$	$\frac{\pi}{2}$	$\frac{3\pi}{4}$	π
$r = 2.0 + \cos\theta$ r	3.0	2.7	2.0	1.3	1.0
Point Number	1	2	3	4	5

θ	π	$\frac{5\pi}{4}$	$\frac{3\pi}{2}$	$\frac{7\pi}{4}$	2π
$r = \dfrac{3.0}{2.0 - \cos\theta}$ r	1.0	1.1	1.5	2.3	3.0
Point Number	6	7	8	9	10

Fig. 21.114

The upper "half" is part of a limaçon and the lower "half" is a semiellipse. The cam is shown in Fig. 21.114. ∎

EXAMPLE 5 Plotting rose

Plot the graph of $r = 2\cos 2\theta$.

In finding values of r, we must be careful first to multiply the values of θ by 2 before finding the cosine of the angle. Also, for this reason, we take values of θ as multiples of $\pi/12$, so as to get enough useful points. The table of values follows:

θ	0	$\frac{\pi}{12}$	$\frac{\pi}{6}$	$\frac{\pi}{4}$	$\frac{\pi}{3}$	$\frac{5\pi}{12}$	$\frac{\pi}{2}$
r	2	1.7	1	0	-1	-1.7	-2

θ	$\frac{7\pi}{12}$	$\frac{2\pi}{3}$	$\frac{3\pi}{4}$	$\frac{5\pi}{6}$	$\frac{11\pi}{12}$	π
r	-1.7	-1	0	1	1.7	2

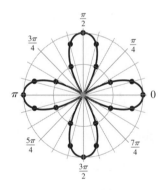

Fig. 21.115

For values of θ starting with π, the values of θ repeat. We have a four-leaf **rose,** as shown in Fig. 21.115. ∎

EXAMPLE 6 Plotting lemniscate

Plot the graph of $r^2 = 9\cos 2\theta$.

Choosing the indicated values of θ, we get the values of r as shown in the following table of values:

θ	0	$\frac{\pi}{8}$	$\frac{\pi}{4}$	$\cdots$	$\frac{3\pi}{4}$	$\frac{7\pi}{8}$	π
r	± 3	± 2.5	0		0	± 2.5	± 3

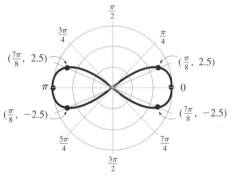

Fig. 21.116

There are no values of r corresponding to values of θ in the range $\pi/4 < \theta < 3\pi/4$, since twice these angles are in the second and third quadrants and the cosine is negative for such angles. The value of r^2 cannot be negative. Also, the values of r repeat for $\theta > \pi$. The figure is called a **lemniscate** and is shown in Fig. 21.116. ∎

EXAMPLE 7 Calculator display of polar curve

View the graph of $r = 1 - 2\cos\theta$ on a graphing calculator.

Using the *mode* feature, a polar curve is displayed using the polar graph option or the parametric graph option, depending on the calculator. (Review the manual for the calculator.) The graph displayed will be the same with either method.

With the polar graph option, the function is entered directly. The values for the viewing window are determined by settings for x, y and the angles θ that will be used. These values are set in a manner similar to those used for parametric equations. (See page 307 for an example of graphing parametric equations.)

With the parametric graph option, to graph $r = f(\theta)$, we note that $x = r\cos\theta$ and $y = r\sin\theta$. This tells us that

$$x = f(\theta)\cos\theta$$
$$y = f(\theta)\sin\theta$$

Thus, for $r = 1 - 2\cos\theta$, by using

$$x = (1 - 2\cos\theta)\cos\theta$$
$$y = (1 - 2\cos\theta)\sin\theta$$

the graph can be displayed, as shown in Fig. 21.117. ∎

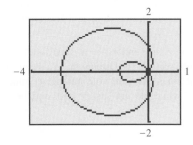

Fig. 21.117

EXERCISES 21.11

In Exercises 1–4, make the given changes in the indicated examples of this section and then make the indicated graphs.

1. In Example 1(b), change $\pi/6$ to $5\pi/6$.

2. In Example 3, change the $-$ before the $2\sin\theta$ to $+$.

3. In Example 5, change cos to sin.

4. In Example 7, change the $-$ before the $2\cos\theta$ to $+$.

In Exercises 5–32, plot the curves of the given polar equations in polar coordinates.

5. $r = 5$ **6.** $r = -2$ **7.** $\theta = 3\pi/4$

8. $\theta = -1.5$ **9.** $r = 4\sec\theta$ **10.** $r = 4\csc\theta$

11. $r = 2\sin\theta$ **12.** $r = 3\cos\theta$

13. $1 - r = \cos\theta$ (cardioid) **14.** $r + 1 = \sin\theta$ (cardioid)

15. $r = 2 - \cos\theta$ (limaçon) **16.** $r = 2 + 3\sin\theta$ (limaçon)

17. $r = 4\sin 2\theta$ (rose) **18.** $r = 2\sin 3\theta$ (rose)

19. $r^2 = 4\sin 2\theta$ (lemniscate) **20.** $r^2 = 2\sin\theta$

21. $r = 2^\theta$ (spiral) **22.** $r = 1.5^{-\theta}$ (spiral)

23. $r = 4|\sin 3\theta|$ **24.** $r = 2\sin\theta\tan\theta$ (cissoid)

25. $r = \dfrac{3}{2 - \cos\theta}$ (ellipse) **26.** $r = \dfrac{2}{1 - \cos\theta}$ (parabola)

27. $r - 2r\cos\theta = 6$ (hyperbola)

28. $3r - 2r\sin\theta = 6$ (ellipse)

29. $r = 4\cos\frac{1}{2}\theta$ **30.** $r = 2 + \cos 3\theta$

31. $r = 2[1 - \sin(\theta - \pi/4)]$ **32.** $r = 4\tan\theta$

In Exercises 33–42, view the curves of the given polar equations on a graphing calculator.

33. $r = \theta$ $(-20 \le \theta \le 20)$ **34.** $r = 0.5^{\sin\theta}$

35. $r = 2\sec\theta + 1$ **36.** $r = 2\cos(\cos 2\theta)$

37. $r = 3\cos 4\theta$ **38.** $r = 3\sin 5\theta$

39. $r = 2\cos\theta + 3\sin\theta$ **40.** $r = 1 + 3\cos\theta - 2\sin\theta$

41. $r + 2 = \cos 2\theta$ **42.** $2r\cos\theta + r\sin\theta = 2$

In Exercises 43–54, solve the given problems: sketch or display the indicated curves.

43. What is the graph of $\tan\theta = 1$?

44. Find the polar equation of the line through the polar points $(1, 0)$ and $(2, \pi/2)$.

45. Using a graphing calculator, show that the curves $r = 2\sin\theta$ and $r = 2\cos\theta$ intersect at right angles. Proper *window* settings are necessary.

46. Using a graphing calculator, determine what type of graph is displayed by $r = 3\sec^2(\theta/2)$.

47. An architect designs a patio shaped such that it can be described as the area within the polar curve $r = 4.0 - \sin\theta$, where measurements are in meters. Sketch the curve that represents the perimeter of the patio.

48. The radiation pattern of a certain television transmitting antenna can be represented by $r = 120(1 + \cos\theta)$, where distances (in km) are measured from the antenna. Sketch the radiation pattern.

49. The joint between two links of a robot arm moves in an elliptical path (in cm), given by $r = \frac{25}{10 + 4\cos\theta}$, Sketch the path.

50. A missile is fired at an airplane and is always directed toward the airplane. The missile is traveling at twice the speed of the airplane. An equation that describes the distance r between the missile and the airplane is $r = \frac{70\sin\theta}{(1 - \cos\theta)^2}$, where θ is the angle between their directions at all times. See Fig. 21.118. This is a *relative pursuit curve*. Sketch the graph of this equation for $\pi/4 \le \theta \le \pi$.

Fig. 21.118 Airplane

51. In studying the photoelectric effect, an equation used for the rate R at which photoelectrons are ejected at various angles θ is $R = \frac{\sin^2\theta}{(1 - 0.5\cos\theta)^2}$. Sketch the graph.

52. In order to display the graph of the rectangular equation
$$4(x^6 + 3x^4y^2 + 3x^2y^4 + y^6 - x^4 - 2x^2y^2 - y^4) + y^2 = 0$$
on a computer screen, it is first transformed into polar coordinates. Transform the equation and then sketch the graph. (*Hint:* the equation can be written as $4(x^2 + y^2)^3 - 4(x^2 + y^2)^2 + y^2 = 0$.)

(W) **53.** Noting the graphs in Exercises 17, 18, 37, and 38, what conclusion do you draw about the value of n and the graph of $y = a\sin n\theta$ or $y = a\cos n\theta$?

(W) **54.** (a) Solve the equations $r = 2\sin\theta$ and $r = 1 + \sin\theta$ simultaneously to find a point of intersection. (b) Display the graphs of the equations on a graphing calculator and note another common point, the coordinates of which do not satisfy both equations. Explain why this occurs.

Answers to Practice Exercises

1. Straight line   **2.** Circle

CHAPTER 21   EQUATIONS

Distance formula	Fig. 21.2	$d = \sqrt{(x_2 - x_1)^2 + (y_2 - y_1)^2}$	**(21.1)**
Slope	Fig. 21.4	$m = \dfrac{y_2 - y_1}{x_2 - x_1}$	**(21.2)**
	Fig. 21.7	$m = \tan\alpha \quad (0° \le \alpha < 180°)$	**(21.3)**
	Fig. 21.9	$m_1 = m_2 \quad$ (for $\parallel$ lines)	**(21.4)**
	Fig. 21.10	$m_2 = -\dfrac{1}{m_1}$ or $m_1 m_2 = -1 \quad$ (for $\perp$ lines)	**(21.5)**
Straight line	Fig. 21.14	$y - y_1 = m(x - x_1)$	**(21.6)**
	Fig. 21.17	$x = a$	**(21.7)**
	Fig. 21.18	$y = b$	**(21.8)**
	Fig. 21.21	$y = mx + b$	**(21.9)**
		$Ax + By + C = 0$	**(21.10)**
Circle	Fig. 21.30	$(x - h)^2 + (y - k)^2 = r^2$	**(21.11)**
	Fig. 21.33	$x^2 + y^2 = r^2$	**(21.12)**
		$x^2 + y^2 + Dx + Ey + F = 0$	**(21.14)**
Parabola	Fig. 21.41	$y^2 = 4px$	**(21.15)**
	Fig. 21.44	$x^2 = 4py$	**(21.16)**

Ellipse	Fig. 21.57	$\dfrac{x^2}{a^2} + \dfrac{y^2}{b^2} = 1$	(21.17)
	Fig. 21.57	$a^2 = b^2 + c^2$	(21.18)
	Fig. 21.58	$\dfrac{y^2}{a^2} + \dfrac{x^2}{b^2} = 1$	(21.19)
Hyperbola	Fig. 21.70	$\dfrac{x^2}{a^2} - \dfrac{y^2}{b^2} = 1$	(21.20)
	Fig. 21.70	$c^2 = a^2 + b^2$	(21.21)
	Fig. 21.69	$y = \pm\dfrac{bx}{a}$ (asymptotes)	(21.23)
	Fig. 21.71	$\dfrac{y^2}{a^2} - \dfrac{x^2}{b^2} = 1$	(21.24)
	Fig. 21.78	$xy = c$	(21.25)
Translation of axes	Fig. 21.83	$x = x' + h$ and $y = y' + k$	(21.26)
		$x' = x - h$ and $y' = y - k$	(21.27)
Parabola, vertex (h, k)		$(y - k)^2 = 4p(x - h)$ (axis parallel to x-axis)	(21.28)
		$(x - h)^2 = 4p(y - k)$ (axis parallel to y-axis)	(21.29)
Ellipse, center (h, k)		$\dfrac{(x - h)^2}{a^2} + \dfrac{(y - k)^2}{b^2} = 1$ (major axis parallel to x-axis)	(21.30)
		$\dfrac{(y - k)^2}{a^2} + \dfrac{(x - h)^2}{b^2} = 1$ (major axis parallel to y-axis)	(21.31)
Hyperbola, center (h, k)		$\dfrac{(x - h)^2}{a^2} - \dfrac{(y - k)^2}{b^2} = 1$ (transverse axis parallel to x-axis)	(21.32)
		$\dfrac{(y - k)^2}{a^2} - \dfrac{(x - h)^2}{b^2} = 1$ (transverse axis parallel to y-axis)	(21.33)
Second-degree equation		$Ax^2 + Bxy + Cy^2 + Dx + Ey + F = 0$	(21.34)
Rotation of axes	Fig. 21.95	$x = x'\cos\theta - y'\sin\theta$	
		$y = x'\sin\theta - y'\cos\theta$	(21.38)
Angle of rotation		$\tan 2\theta = \dfrac{B}{A - C}$ $(A \neq C)$	(21.39)
		$\theta = 45°$ $(A = C)$	(21.40)
Polar coordinates	Fig. 21.105	$x = r\cos\theta$ $y = r\sin\theta$	(21.41)
		$\tan\theta = \dfrac{y}{x}$ $r = \sqrt{x^2 + y^2}$	(21.42)

CHAPTER 21 REVIEW EXERCISES

In Exercises 1–12, find the equation of the indicated curve, subject to the given conditions. Sketch each curve.

1. Straight line: passes through $(1, -7)$ with a slope of 4
2. Straight line: passes through $(-1, 5)$ and $(-2, -3)$
3. Straight line: perpendicular to $3x - 2y + 8 = 0$ and has a y-intercept of $(0, -1)$
4. Straight line: parallel to $2x - 5y + 1 = 0$ and has an x-intercept of $(2, 0)$
5. Circle: concentric with $x^2 + y^2 = 6x$, passes through $(4, -3)$
6. Circle: tangent to lines $x = 3$ and $x = 9$, center on line $y = 2x$
7. Parabola: focus $(3, 0)$, vertex $(0, 0)$
8. Parabola: vertex $(0, 0)$, passes through $(1, 1)$ and $(-2, 4)$
9. Ellipse: vertex $(10, 0)$, focus $(8, 0)$, tangent to $x = -10$
10. Ellipse: center $(0, 0)$, passes through $(0, 3)$ and $(2, 1)$
11. Hyperbola: $V(0, 13)$, $C(0, 0)$, conj. axis of 24
12. Hyperbola: vertex $(0, 8)$, asymptotes $y = 2x$, $y = -2x$

In Exercises 13–26, find the indicated quantities for each of the given equations. Sketch each curve.

13. $x^2 + y^2 + 6x - 7 = 0$, center and radius
14. $2x^2 + 2y^2 + 4x - 8y - 15 = 0$, center and radius
15. $x^2 = -20y$, focus and directrix
16. $y^2 = 24x$, focus and directrix
17. $4x^2 + y^2 = 1$, vertices and foci
18. $2y^2 - 9x^2 = 18$, vertices and foci
19. $2x^2 - 5y^2 = 0.25$, vertices and foci
20. $2x^2 + 25y^2 = 800$, vertices and foci
21. $x^2 - 8x - 4y - 16 = 0$, vertex and focus
22. $y^2 - 4x + 4y + 24 = 0$, vertex and directrix
23. $4x^2 + y^2 - 16x + 2y + 13 = 0$, center
24. $x^2 - 2y^2 + 4x + 4y + 6 = 0$, center
25. $x^2 - 2xy + y^2 + 4x + 4y = 0$, vertex
26. $3x^2 - 3xy + 7y^2 - 5 = 0$, center

In Exercises 27–34, plot the given curves in polar coordinates.

27. $r = 4(1 + \sin \theta)$
28. $r = 1 - 3\cos \theta$
29. $r = 4\cos 3\theta$
30. $r = 3\sin \theta - 4\cos \theta$
31. $r = \dfrac{3}{\sin \theta + 2\cos \theta}$
32. $r = \dfrac{1}{2(\sin \theta - 1)}$
33. $r = 2\sin\left(\dfrac{\theta}{2}\right)$
34. $r = 1 - \cos 2\theta$

In Exercises 35–38, find the polar equation of each of the given rectangular equations.

35. $y = 2x$
36. $2xy = 1$
37. $x^2 + xy + y^2 = 2$
38. $x^2 + (y + 3)^2 = 16$

In Exercises 39–42, find the rectangular equation of each of the given polar equations.

39. $r = 2\sin 2\theta$
40. $r^2 = 9\sin \theta$
41. $r = \dfrac{4}{2 - \cos \theta}$
42. $r = 4\tan \theta \sec \theta$

In Exercises 43–48, determine the number of real solutions of the given systems of equations by sketching the indicated curves. (See Section 14.1.)

43. $x^2 + y^2 = 9$
    $4x^2 + y^2 = 16$
44. $y = e^x$
    $x^2 - y^2 = 1$
45. $x^2 + y^2 - 4y - 5 = 0$
    $y^2 - 4x^2 - 4 = 0$
46. $x^2 - 4y^2 + 2x - 3 = 0$
    $y^2 - 4x - 4 = 0$
47. $y = 2\sin x$
    $y = 2 - x^2$
48. $y = 4\ln x$
    $xy = 6$

In Exercises 49–58, view the curves of the given equations on a graphing calculator.

49. $x^2 + 3y + 2 - (1 + x)^2 = 0$
50. $y^2 = 4x + 6$
51. $2x^2 + 2y^2 + 4y - 3 = 0$
52. $2x^2 + (y - 3)^2 - 5 = 0$
53. $x^2 - 4y^2 + 4x + 24y - 48 = 0$
54. $x^2 + 2xy + y^2 - 3x + 8y = 0$
55. $r = 3\cos(3\theta/2)$
56. $r = 5 - 2\sin 4\theta$
57. $r = 2 - 3\csc \theta$
58. $r = 2\sin(\cos 3\theta)$

In Exercises 59–62, find the equation of the locus of a point $P(x, y)$ that moves as stated.

59. Always 4 units from $(3, -4)$
60. Passes through $(7, -5)$ with a constant slope of -2
61. The sum of its distances from $(1, -3)$ and $(7, -3)$ is 8.
62. The difference of its distances from $(3, -1)$ and $(3, -7)$ is 4.

In Exercises 63–106, solve the given problems.

63. Considering Eq. (21.30) of an ellipse, describe the graph if $a = b$.
64. Show that the ellipse $x^2 + 9y^2 = 9$ has the same foci as the hyperbola $x^2 - y^2 = 4$.
65. The points $(-2, -5)$, $(3, -3)$, and $(13, x)$ are collinear. Find x.
66. For the polar coordinate point $(-5, \pi/4)$, find another set of polar coordinates such that $r < 0$ and $-2\pi < \theta < 0$.
67. Find the distance between the polar coordinate points $(3, \pi/6)$ and $(6, -\pi/3)$.

68. Show that the parametric equations $y = \cot\theta$ and $x = \csc\theta$ define a hyperbola.

69. In two ways, show that the line segments joining $(-3, 11)$, $(2, -1)$, and $(14, 4)$ form a right triangle.

70. Find the equation of the circle that passes through $(3, -2)$, $(-1, -4)$, and $(2, -5)$.

71. Graph the inequality $y > 4(x + 2)^2$.

72. Graph the inequality $4x^2 + 9(y - 2)^2 < 36$.

73. What type of curve is represented by
$(x + jy)^2 + (x - jy)^2 = 2?$ $(j = \sqrt{-1})$

74. For the ellipse in Fig. 21.119, show that the product of the slopes PA and PB is $-b^2/a^2$.

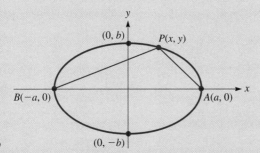

Fig. 21.119

75. Find the area of the square that can be inscribed in the ellipse $7x^2 + 2y^2 = 18$.

76. Using a graphing calculator, determine the number of points of intersection of the polar curves $r = 4|\cos 2\theta|$ and $r = 6\sin[\cos(\cos 3\theta)]$.

77. By means of the definition of a parabola, find the equation of the parabola with focus at $(3, 1)$ and directrix the line $y = -3$. Find the same equation by the method of translation of axes.

78. For what value(s) of k does $x^2 - ky^2 = 1$ represent an ellipse with vertices on the y-axis?

79. The total resistance R_T of two resistances in series in an electric circuit is the sum of the resistances. If a variable resistor R is in series with a 2.5-Ω resistor, express R_T as a function of R and sketch the graph.

80. The acceleration of an object is defined as the change in velocity v divided by the corresponding change in time t. Find the equation relating the velocity v and time t for an object for which the acceleration is $20\,\text{ft/s}^2$ and $v = 5.0\,\text{ft/s}$ when $t = 0\,\text{s}$.

81. The velocity v of a crate sliding down a ramp is given by $v = v_0 + at$, where v_0 is the initial velocity, a is the acceleration, and t is the time. If $v_0 = 5.75\,\text{ft/s}$ and $v = 18.5\,\text{ft/s}$ when $t = 5.50\,\text{s}$, find v as a function of t. Sketch the graph.

82. An airplane touches down when landing at $100\,\text{mi/h}$. Its velocity v while coming to a stop is given by $v = 100 - 20{,}000t$, where t is the time in hours. Sketch the graph of v vs. t.

83. It takes 2.010 kJ of heat to raise the temperature of 1.000 kg of steam by 1.000°C. In a steam generator, a total of y kJ is used to raise the temperature of 50.00 kg of steam from 100°C to T°C. Express y as a function of T and sketch the graph.

84. The temperature in a certain region is 27°C, and at an altitude of 2500 m above the region it is 12°C. If the equation relating the temperature T and the altitude h is linear, find the equation.

85. The radar gun on a police helicopter 490 ft above a multilane highway is directed vertically down onto the highway. If the radar gun signal is cone-shaped with a vertex angle of 14°, what area of the highway is covered by the signal?

86. One of the most famous Ferris wheels is Vienna's (Austria) *Prata*. Find the equation representing its circumference, given that $c = 191$ m. Place the origin of the coordinate system 2.0 m below the bottom of the wheel, and the center of the wheel on the y-axis.

87. The arch of a small bridge across a stream is parabolic. If, at water level, the span of the arch is 80 ft and the maximum height above water level is 20 ft, what is the equation that represents the arch? Choose the most convenient point for the origin of the coordinate system.

88. A laser source is 2.00 cm from a spherical surface of radius 3.00 cm, and the laser beam is tangent to the surface. By placing the center of the sphere at the origin, and the source on the positive x-axis, find the equation of the line along which the beam shown in Fig. 21.120 is directed.

Fig. 21.120

89. The top horizontal cross section of a dam is parabolic. The open area within this cross section is 80 ft across and 50 ft from front to back. Find the equation of the edge of the open area with the vertex at the origin of the coordinate system and the axis along the x-axis.

90. The *quality factor* Q of a series resonant electric circuit with resistance R, inductance L, and capacitance C is given by $Q = \dfrac{1}{R}\sqrt{\dfrac{L}{C}}$. Sketch the graph of Q and L for a circuit in which $R = 1000\,\Omega$ and $C = 4.00\,\mu\text{F}$.

91. At very low temperatures, certain metals have an electric resistance of zero. This phenomenon is called *superconductivity*. A magnetic field also affects the superconductivity. A certain level of magnetic field H_T, the threshold field, is related to the thermodynamic temperature T by $H_T/H_0 = 1 - (T/T_0)^2$, where H_0 and T_0 are specifically defined values of magnetic field and temperature. Sketch the graph of H_T/H_0 vs. T/T_0.

92. A rectangular parking lot is to have a perimeter of 600 m. Express the area A in terms of the width w and sketch the graph.

93. The electric power P (in W) supplied by a battery is given by $P = 12.0i - 0.500i^2$, where i is the current (in A). Sketch the graph of P vs. i.

94. The Colosseum in Rome is in the shape of an ellipse 188 m long and 156 m wide. Find the area of the Colosseum. ($A = \pi ab$ for an ellipse.)

95. A specialty electronics company makes an ultrasonic device to repel animals. It emits a $20 - 25$ kHz sound (above those heard by people), which is unpleasant to animals. The sound covers an elliptical area starting at the device, with the longest dimension extending 120 ft from the device and the focus of the area 15 ft from the device. Find the area covered by the signal. ($A = \pi ab$)

96. A study indicated that the fraction f of cells destroyed by various dosages d of X rays is given by the graph in Fig. 21.121. Assuming that the curve is a quarter-ellipse, find the equation relating f and d for $0 \leq f \leq 1$ and $0 < d \leq 10$ units.

Fig. 21.121

97. A machine-part designer wishes to make a model for an elliptical cam by placing two pins in a design board, putting a loop of string over the pins, and marking off the outline by keeping the string taut. (Note that the definition of the ellipse is being used.) If the cam is to measure 10 cm by 6 cm, how long should the loop of string be and how far apart should the pins be?

98. Soon after reaching the vicinity of the moon, *Apollo 11* (the first spacecraft to land a man on the moon) went into an elliptical lunar orbit. The closest the craft was to the moon in this orbit was 70 mi, and the farthest it was from the moon was 190 mi. What was the equation of the path if the center of the moon was at one of the foci of the ellipse? Assume that the major axis is along the x-axis and that the center of the ellipse is at the origin. The radius of the moon is 1080 mi.

99. The vertical cross section of the cooling tower of a nuclear power plant is hyperbolic, as shown in Fig. 21.122. Find the radius r of the smallest circular horizontal cross section.

Fig. 21.122

100. Tremors from an earthquake are recorded at the California Institute of Technology (Pasadena, California) 36 s before they are recorded at Stanford University (Palo Alto, California). If the seismographs are 510 km apart and the shock waves from the tremors travel at 5.0 km/s, what is the curve on which lies the point where the earthquake occurred?

101. An electronic instrument located at point P records the sound of a rifle shot and the impact of the bullet striking the target at the same instant. Show that P lies on a branch of a hyperbola.

102. A 60-ft rope passes over a pulley 10 ft above the ground, and a crate on the ground is attached at one end. The other end of the rope is held at a level of 4 ft above the ground and is drawn away from the pulley. Express the height of the crate over the ground in terms of the distance the person is from directly below the crate. Sketch the graph of distance and height. See Fig. 21.123. (Neglect the thickness of the crate.)

Fig. 21.123

103. A satellite at an altitude proper to make one revolution per day around the center of the earth will have for an excellent approximation of its projection on the earth of its path the curve $r^2 = R^2 \cos 2(\theta + \frac{\pi}{2})$, where R is the radius of the earth. Sketch the path of the projection.

104. The vertical cross sections of two pipes as drawn on a drawing board are shown in Fig. 21.124. Find the polar equation of each.

Fig. 21.124

105. The path of a certain plane is $r = 200(\sec \theta + \tan \theta)^{-5}/\cos \theta$, $0 < \theta < \pi/2$. Sketch the path and check it on a graphing calculator. (This path is the same as for the plane in Exercise 96 on page 332.)

106. The sound produced by a jet engine was measured at a distance of 100 m in all directions. The loudness of the sound d (in decibels) was found to be $d = 115 + 10 \cos \theta$, where the $0°$ line for the angle θ is directed in front of the engine. Sketch the graph of d vs. θ in polar coordinates (use d as r).

Writing Exercise

107. Under a force that varies inversely as the square of the distance from an attracting object (such as the sun exerts on the earth), it can be shown that the equation of the path an object follows is given in general by

$$\frac{1}{r} = a + b\cos\theta$$

where a and b are constants for a particular path. First, transform this equation into rectangular coordinates. Then write one or two paragraphs explaining why this equation represents one of the conic sections, depending on the values of a and b. It is through this kind of analysis that we know the paths of the planets and comets are conic sections.

CHAPTER 21   PRACTICE TEST

1. (a) Find the distance between $(4, -1)$ and $(6, 3)$. (b) Find the slope of the line perpendicular to the line segment joining the points in part (a).

2. Identify the type of curve represented by the equation $2(x^2 + x) = 1 - y^2$.

3. Sketch the graph of the straight line $4x - 2y + 5 = 0$ by finding its slope and y-intercept.

4. Find the polar equation of the curve whose rectangular equation is $x^2 = 2x - y^2$.

5. Find the vertex and the focus of the parabola $x^2 = -12y$. Sketch the graph.

6. Find the equation of the circle with center at $(-1, 2)$ and that passes through $(2, 3)$.

7. Find the equation of the straight line that passes through $(-4, 1)$ and $(2, -2)$.

8. Where is the focus of a parabolic reflector that is 12.0 cm across and 4.00 cm deep?

9. A hallway 16 ft wide has a ceiling whose cross section is a semi-ellipse. The ceiling is 10 ft high at the walls and 14 ft high at the center. Find the height of the ceiling 4 ft from each wall.

10. Plot the polar curve $r = 3 + \cos\theta$.

11. Find the center and vertices of the conic section $4y^2 - x^2 - 4x - 8y - 4 = 0$. Show completely the sketch of the curve.

12. (a) What type of curve is represented by $8x^2 - 4xy + 5y^2 = 36$? (b) Through what angle must the curve in part (a) be rotated in order that there is no $x'y'$-term?

22

Introduction to Statistics

After the invention of the steam engine in the late 1700s by the Scottish engineer James Watt, the production of machine-made goods became widespread during the 1800s. However, it was not until the 1920s that much attention was paid to the quality control of the goods being produced.

In 1924, Walter Shewhart of Bell Telephone Laboratories used a statistical chart for controlling product variables; in the 1940s, quality control was used in much of wartime production.

Quality control is one of the modern uses of *statistics*, the branch of mathematics in which data are collected, analyzed, and interpreted. Today it is nearly impossible to read a newspaper or watch television news without seeing some type of study, in areas such as medicine or politics, that involves statistics. Other fields in which statistical methods are used include biology, physics, psychology, sociology, economics, business, education, electronics, and others.

The first significant use of statistics was made in the 1660s by John Graunt and in the 1690s by Edmund Halley (Halley's Comet), when each published some conclusions about the population in England based on mortality tables. There was little development of statistics until the 1800s, when statistical measures became more widely used. Examples are the scientist Francis Galton, who used statistics in the study of human heredity, and the nurse Florence Nightingale, who used statistical graphs to show that more soldiers died in the Crimean War (in the 1850s) from unsanitary conditions than from combat wounds.

This chapter is an introduction to some of the basic concepts and uses of statistics. We first consider certain basic statistical measures. Then a section is devoted to very large sets of statistical data, followed by a section on *control charts* that are used in statistical process control in industry. The final sections show how to start with a set of points on a graph and find an equation that best "fits" the data. This equation, which shows a basic relationship between the variables, can be useful in research.

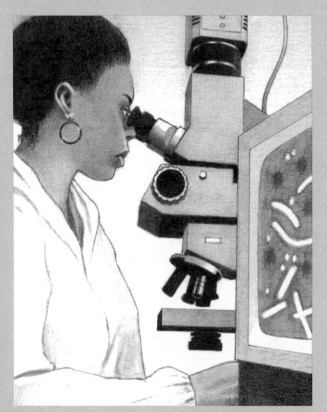

Statistical analysis is used extensively in medical research. In Section 22.6, a method of finding the equation of a line that "fits" given data is shown.

22.1 Frequency Distributions

Array • Class • Frequency •
Frequency Distribution Table •
Relative Frequency • Frequency Polygon •
Histogram • Class Mark • Cumulative
Frequency • Ogive

In statistics, we first collect data and thereby form a set of numbers. These numbers could be measurements of electric current, test scores, lengths of objects, or numerous other possibilities. *Data that have been collected but not yet organized are called* **raw data.** In order to obtain useful information from the data, it is necessary to organize it in some way. Normally, a first step in organizing the data is to arrange the numerical values in ascending (or descending) order, called an **array.** An example of raw data and an array is shown in the following example.

EXAMPLE 1 Illustrating array

A survey of 50 users of home computers asked each user to estimate carefully the number of hours they spent each week on-line on the Internet. Following are the estimates.

12, 20, 15, 14, 7, 10, 12, 25, 18, 5, 10, 24, 16, 3, 12, 14, 28, 8, 13, 18,

15, 8, 11, 15, 14, 22, 14, 19, 6, 10, 18, 4, 16, 24, 18, 5, 13, 20, 12, 12,

25, 11, 8, 12, 20, 5, 10, 15, 13, 8

As we can see, no clear pattern can be seen from this raw data. Arranging these in numerical order to form an array, we can summarize the array by showing the number of persons reporting each estimate as follows:

(hours-persons) 3-1, 4-1, 5-3, 6-1, 7-1, 8-4, 10-4, 11-2, 12-6

13-3, 14-4, 15-4, 16-2, 18-4, 19-1, 20-3, 22-1, 24-2, 25-2, 28-1 ∎

In Example 1, although a pattern is somewhat clearer from the array than from the raw data, a still clearer pattern is found by *grouping the data.* In the process of grouping, the detail of the raw data is lost, but the advantage is that a much clearer overall pattern of the data can be obtained.

The grouping of data is done by first defining what values are to be included in each group and then tabulating the number of values that are in each group. *Each group is called a* **class,** and *the number of values in each class is called the* **frequency.** *The table is called a* **frequency distribution table.** This is illustrated in the following example.

EXAMPLE 2 Frequency distribution table

In Example 1, we note that the estimates vary from 3 h to 28 h. We see that if we form *classes* of 0–4 h, 5–9 h, etc., we will have five possible estimates in each class and that there will be six classes. This gives us the following frequency distribution table of values showing the number of persons (frequency) reporting the indicated estimate of the hours spent on the Internet.

Estimate (hours)	0–4	5–9	10–14	15–19	20–24	25–29
Frequency (persons)	2	9	19	11	6	3

This table shows us the frequency distribution. The 0, 5, 10, and so on are the *lower class limits,* and the 4, 9, 14, and so on are the *upper class limits.* Each class includes five values, which is the *class width.* As with the class limits we have chosen, it is generally preferable to have the same width for each class.

We can see from this frequency distribution table that the pattern of hours on the Internet by the persons responding to the survey is clearer. ∎

At times, it is also helpful to know the **relative frequency** *of the class, which is the frequency of the class divided by the total frequency of all classes.* The relative frequency can be expressed as a fraction, decimal, or percent.

EXAMPLE 3   Relative frequency

The relative frequency of each class for the data in Example 2 can be shown as in the following table:

Estimated Hours on Internet	Frequency	Relative Frequency (%)
0–4	2	4
5–9	9	18
10–14	19	38
15–19	11	22
20–24	6	12
25–29	3	6
Total	50	100

$2/50 = 0.04 = 4\%$

■

Practice Exercise

1. Assuming the data in Example 1 are divided into classes of 0–3 h, 4–7 h, etc., for the 8–11 h class find:
(a) the frequency;
(b) the relative frequency

Just as graphs are useful in representing algebraic functions, so are they a very convenient method of representing frequency distributions. There are several useful types of graphs for such distributions. *Among the most important of these are the* **histogram** *and the* **frequency polygon.** These graphical representations are generally constructed to represent the data in a frequency distribution table.

In order to represent *grouped* data, where the raw data values are generally not all the same within a given class, we find it necessary to use a representative value for each class. For this, we use the **class mark,** *which is found by dividing the sum of the lower and upper class limits by* 2. The following examples illustrate the use of a class mark with a histogram and a frequency polygon.

EXAMPLE 4   Histogram

A histogram represents a particular set of data by displaying each class of the data as a rectangle. Each rectangle is labeled at the center of its base by the *class mark.* The width of each rectangle represents the *class width,* and the height of the rectangle represents the *frequency* of the class.

For the data in Example 2 on estimated hours on the Internet, the class marks are $(0 + 4)/2 = 2$, $(5 + 9)/2 = 7$, and so on. Using these values, a histogram representing these data is shown in Fig. 22.1.   ■

Fig. 22.1

EXAMPLE 5   Frequency polygon

A frequency polygon is used to represent a set of data by plotting the class marks as abscissas (x-values) and the frequencies as ordinates (y-values). The resulting points are joined by straight-line segments.

A frequency polygon representing the data in Example 2 on estimated hours on the Internet is shown in Fig. 22.2.

If the polygon is not completed as shown in Fig. 22.2, and the figure starts at the first class mark, and ends at the last class mark, it is then referred to as a *broken-line graph.*   ■

Complete polygon to next (empty) class mark.

Fig. 22.2

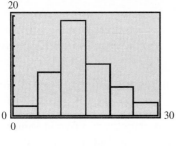

Fig. 22.3

A graphing calculator may be used to display histograms and frequency polygons. The manual should be reviewed to see how this is done on any particular model. On a calculator, it is necessary to enter both sets of numbers, choose proper *window* settings, and select the correct type of statistical graph. A calculator display for the histogram in Fig. 22.1 is shown in Fig. 22.3.

When dividing data into classes, we must decide how many classes to use. A general guideline is that *the number of classes should be no less than* 5 *nor greater than* 15 (this might go to 20 if we have thousands of values to analyze). The principal consideration is that we get a good pattern of the distribution. This is illustrated in the following example.

EXAMPLE 6 Illustration of too many classes

If we divided the data of estimated Internet hours of Example 1 into nine classes, the frequency table would be as follows:

Estimate (hours)	3–5	6–8	9–11	12–14	15–17	18–20	21–23	24–26	27–29
Freq. (respondents)	5	6	6	13	6	8	1	4	1

■ Computer spreadsheets are very useful for this type of analysis.

We can see that this does not give nearly as good an idea of the pattern as the six classes shown in Example 2. ■

Another way of analyzing data is to use *cumulative* totals. The way this is generally done is to change the frequency into a "less than" or a "more than" **cumulative frequency.** To do this, we *add the class frequencies,* starting either at the greatest class boundary or the lowest class boundary. In our discussion, we will restrict our attention to a "less than" cumulative frequency. The graphical display that is generally used for cumulative frequency is called an **ogive** (pronounced oh-jive).

EXAMPLE 7 Cumulative frequency – ogive

For the data on estimated hours on the Internet in Examples 2 and 3, the cumulative frequency is shown in the following table:

Fig. 22.4

Estimated Hours on Internet	*Cumulative Frequency*
Less than 5	2
Less than 10	11
Less than 15	30
Less than 20	41
Less than 25	47
Less than 30	50

The ogive showing the cumulative frequency for the values in this table is shown in Fig. 22.4. The vertical scale shows the *frequency,* and the horizontal scale shows the *class boundaries.*

One important use of an ogive is to determine the number of values above or below a certain value. For example, to *approximate* the number of respondents that use the Internet *less than* 18 hours per week, we draw a line from the horizontal axis to the ogive and then to the vertical axis as shown in Fig. 22.4. From this, we see that *about* 37 respondents use the Internet *less than* 18 hours per week. ■

If the data with which we are dealing has only a limited number of values and we do not divide it into classes, we can use the methods we have developed. We use the specific values rather than *class* values. Consider the following example.

EXAMPLE 8 Using specific values

A test station measured the loudness of the sound of jet aircraft taking off from a certain airport. The decibel (dB) readings measured to the nearest 5 dB for the first 20 jets were as follows:

110, 95, 100, 115, 105, 110, 120, 110, 115, 105, 90, 95, 105, 110, 100, 115, 105, 120, 95, 110

Since there are only seven different values for the 20 readings, the best idea of the pattern is found by using these values, as shown in the following frequency table:

dB Reading	90	95	100	105	110	115	120
Frequency	1	3	2	4	5	3	2

The histogram for this table is shown in Fig. 22.5, and the frequency polygon is shown in Fig. 22.6. ∎

Fig. 22.5

Fig. 22.6

EXERCISES 22.1

In Exercises 1–4, divide the data in Example 1 into five classes of hours (0–5, 6–11, etc.) on the Internet and then do the following:

1. Form a frequency distribution table.

2. Find the relative frequencies.

3. Draw a histogram.   4. Draw an ogive.

In Exercises 5–12, use the following data. An automobile company tested a new engine, and found the following results in twenty tests of the number of miles traveled by a certain model car on each gallon of gasoline.

53, 58, 56, 54, 59, 54, 60, 58, 58, 54,
63, 56, 57, 56, 57, 59, 55, 61, 59, 58

5. Form a frequency distribution table.

6. Find the relative frequencies.

7. Form a frequency distribution table with five classes.

8. Find the relative frequencies for the data in the frequency distribution table in Exercise 7.

9. Draw a histogram for the data of Exercise 7.

10. Draw a frequency polygon for the data of Exercise 7.

11. Form a cumulative frequency table for the data of Exercise 7.

12. Draw an ogive for the data of Exercise 7.

In Exercises 13–32, find the indicated quantities.

13. In testing a computer system, the number of instructions it could perform in 1 ns was measured at different points in a program. The numbers of instructions were recorded as follows:

19, 21, 22, 25, 22, 20, 18, 21, 20, 19, 22, 21, 19, 23, 21

Form a frequency distribution table for these values.

14. For the data of Exercise 13, draw a histogram.

15. For the data of Exercise 13, draw a frequency polygon.

16. For the data of Exercise 13, form a relative frequency distribution table.

17. A strobe light is designed to flash every 2.25 s at a certain setting. Sample bulbs were tested with the following results:

Time (s) between Flashes (class mark)	2.21	2.22	2.23	2.24
Number of Bulbs	2	7	18	41

Time (s)	2.25	2.26	2.27	2.28	2.29
No. Bulbs	56	32	8	3	3

Draw a histogram for these data.

18. For the data of Exercise 17, draw a frequency polygon.

19. For the data of Exercise 17, form a cumulative frequency distribution table.

20. For the data of Exercise 17, draw an ogive.

21. In testing a braking system, the distance required to stop a car from 70 mi/h was measured in 120 trials. The results are shown in the following distribution table:

Stopping Distance (ft)	155–159	160–164	165–169	170–174
Times Car Stopped	2	15	32	36

Stopping Distance (ft)	175–179	180–184	185–189
Times Car Stopped	24	10	1

Form a relative frequency distribution table for these data.

22. For the data in Exercise 21, form a cumulative frequency distribution table.

23. For the data of Exercise 21, draw an ogive.

24. From the ogive in Exercise 23, estimate the number of cars that stopped in less than 176 ft.

25. The dosage, in milliroentgens (mR), given by a particular X-ray machine, was measured 20 times, with the following readings:

4.25, 4.36, 3.96, 4.21, 4.44, 3.83, 4.37, 4.27, 4.33, 4.34, 4.15, 3.90, 4.41, 4.51, 4.18, 4.26, 4.29, 4.09, 4.36, 4.23

Form a histogram with six classes and the lowest class mark of 3.80 mR.

26. For the data used for the histogram in Exercise 25, draw a frequency polygon.

27. The life of a certain type of battery was measured for a sample of batteries with the following results (in number of hours):

34, 30, 32, 35, 31, 28, 29, 30, 32, 25, 31, 30, 28, 36, 33, 34, 30, 33, 31, 34, 29, 30, 32

Draw a frequency polygon using six classes.

28. For the data in Exercise 27, draw a cumulative frequency distribution table using six classes.

Answers to Practice Exercise

1. (a) 10   (b) 20%

29. The diameters of a sample of fiber-optic cables were measured with the following results (diameters are class marks):

Diam. (mm)	0.0055	0.0056	0.0057	0.0058	0.0059	0.0060
No. Cables	4	15	32	36	59	64

Diam. (mm)	0.0061	0.0062	0.0063	0.0064	0.0065	0.0066
No. Cables	22	18	10	12	4	4

Draw a histogram for these data.

30. For the data of Exercise 29, draw a histogram with six classes. Compare the pattern of distribution with that of the histogram in Exercise 29.

W 31. Toss four coins 50 times and tabulate the number of heads that appear for each toss. Draw a frequency polygon showing the number of tosses for which 0, 1, 2, 3, or 4 heads appeared. Describe the distribution (is it about what should be expected?).

W 32. Most calculators can generate random numbers (between 0 and 1). On a calculator, display 50 random numbers and record the first digit. Draw a histogram showing the number of times for which each first digit (0, 1, 2, ..., 9) appeared. Describe the distribution (is it about what should be expected?).

22.2 Measures of Central Tendency

Median • Arithmetic Mean • Mode

Tables and graphical representations give a general description of data. However, it is often useful and convenient to find representative values for the location of the center of the distribution, and other numbers to give a measure of the deviation from this central value. In this way, we can obtain an arithmetical description of the data. We now discuss the values commonly used to measure the location of the center of the distribution. These are referred to as *measures of central tendency.*

Median   The first of these measures of central tendency is the **median.** *The median is the middle number, that number for which there are as many above it as below it in the distribution.* If there is no middle number, the median is that number halfway between the two numbers nearest the middle of the distribution.

EXAMPLE 1 Median – odd or even number of values

Given the numbers 5, 2, 6, 4, 7, 4, 7, 2, 8, 9, 4, 11, 9, 1, 3, we first arrange them in numerical order. This arrangement is

middle number

1, 2, 2, 3, 4, 4, 4, 5, 6, 7, 7, 8, 9, 9, 11

Since there are 15 numbers, the middle number is the eighth. Since the eighth number is 5, the median is 5.

If the number 11 is not included in this set of numbers and there are only 14 numbers in all, the median is that number halfway between the seventh and eighth numbers. Since the seventh is 4 and the eighth is 5, the median is 4.5. ∎

■ The median may be expressed to one more significant digit than that given in the data.

EXAMPLE 2 Median – even number of values

In the distribution of Internet hours in Example 1 of Section 22.1, the median is 13 h. There are 50 values in all, and when listed in order, the 24th to 26th values are each 13 h. The number halfway between the 25th and 26th values is the median. Since both are 13 h, the median is 13 h. ∎

Arithmetic Mean
Another very widely applied measure of central tendency is the **arithmetic mean.** *The mean is calculated by finding the sum of all the values and then dividing by the number of values.* (The arithmetic mean is the number most people call the "average." However, in statistics the word *average* has the more general meaning of a measure of central tendency.)

EXAMPLE 3 Arithmetic mean

The arithmetic mean of the numbers given in Example 1 is determined by finding the sum of all the numbers and dividing by 15. Therefore, by letting $\bar{x}$ (read as "x bar") represent the mean, we have

$$\bar{x} = \frac{5 + 2 + 6 + 4 + 7 + 4 + 7 + 2 + 8 + 9 + 4 + 11 + 9 + 1 + 3}{15}$$

sum of values

$$= \frac{82}{15} = 5.5$$

number of values

Thus, the mean is 5.5. (The mean is usually calculated to one more significant digit than was present in the original data.) ∎

If we wish to find the arithmetic mean of a large number of values, and if some of them appear more than once, the calculation can be simplified. The mean can be calculated by multiplying each value by its frequency, adding these results, and then dividing by the total number of values (the sum of the frequencies). Letting $\bar{x}$ represent the mean of the values $x_1, x_2, \ldots, x_n$, which occur with frequencies $f_1, f_2, \ldots, f_n$, respectively, we have

■ This is called a *weighted mean* since each value is given a weighting based on the number of times it occurs.

$$\bar{x} = \frac{x_1 f_1 + x_2 f_2 + \cdots + x_n f_n}{f_1 + f_2 + \cdots + f_n} \qquad (22.1)$$

EXAMPLE 4 Arithmetic mean using frequencies

Using Eq. (22.1) to find the arithmetic mean of the numbers of Example 1, we first set up a table of values and their respective frequencies, as follows:

Values	1	2	3	4	5	6	7	8	9	11
Frequency	1	2	1	3	1	1	2	1	2	1

We now calculate the arithmetic mean $\bar{x}$ by using Eq. (22.1):

multiply each value by its frequency and add results

$$\bar{x} = \frac{1(1) + 2(2) + 3(1) + 4(3) + 5(1) + 6(1) + 7(2) + 8(1) + 9(2) + 11(1)}{1 + 2 + 1 + 3 + 1 + 1 + 2 + 1 + 2 + 1}$$

sum of frequencies

$$= \frac{82}{15} = 5.5$$

We see that this agrees with the result of Example 3. ∎

Summations such as those in Eq. (22.1) occur frequently in statistics and other branches of mathematics. In order to simplify writing these sums, the symbol Σ is used to indicate the process of summation. (Σ is the Greek capital letter sigma.) Σx means the sum of the x's.

EXAMPLE 5 Summation symbol Σ

We can show the sum of the numbers $x_1, x_2, x_3, \ldots + x_n$ as

$$\Sigma x = x_1 + x_2 + x_3 + \cdots + x_n$$

If these numbers are 3, 7, 2, 6, 8, 4, and 9, we have

$$\Sigma x = 3 + 7 + 2 + 6 + 8 + 4 + 9 = 39 \qquad \blacksquare$$

Practice Exercise

For the following numbers, find the indicated value.
12, 17, 16, 12, 14, 18, 14, 12, 15, 18
1. The median.   **2.** The arithmetic mean

Using the summation symbol Σ, we can write Eq. (22.1) for the arithmetic mean as

$$\bar{x} = \frac{x_1 f_1 + x_2 f_2 + x_3 f_3 + \cdots + x_n f_n}{f_1 + f_2 + f_3 + \cdots + f_n} = \frac{\Sigma x f}{\Sigma f} \qquad (22.1)$$

The summation notation Σx is an abbreviated form of the more general notation $\sum_{i=l}^{n} x_i$. This more general form can be used to indicate the sum of the first n numbers of a sequence or to indicate the sum of a certain set within the sequence. For example, for a set of at least 5 numbers, $\sum_{i=3}^{5} x_i$ indicates the sum of the third through the fifth of these numbers (in Example 5, $\sum_{i=3}^{5} x_i = 16$). We will use the abbreviated form Σx to indicate the sum of all the numbers being considered.

EXAMPLE 6 Arithmetic mean using summation

■ The arithmetic mean is one of a number of statistical measures that can be found on a calculator. The use of a calculator will be shown in the next section.

We find the arithmetic mean of the Internet hours in Example 1 of Section 22.1 (page 613) by

$$\bar{x} = \frac{\Sigma x f}{\Sigma f} = \frac{3(1) + 4(1) + 5(3) + \cdots + 12(6) + \cdots + 28(1)}{50}$$

$$= \frac{687}{50} = 13.7\,\text{h} \qquad \text{(rounded off to tenths)} \qquad \blacksquare$$

Mode

Another measure of central tendency is *the* **mode,** *which is the value that appears most frequently.* If two or more values appear with the same greatest frequency, each is a mode. If no value is repeated, there is no mode.

EXAMPLE 7 Mode

(a) The mode of the numbers in Example 1 is 4, since it appears three times and no other value appears more than twice.

(b) The modes of the numbers

$$1, 2, 2, 4, 5, 5, 6, 7$$

are 2 and 5, since each appears twice and no other number is repeated.

(c) There is no mode for the values

$$1, 2, 5, 6, 7, 9$$

since none of the values is repeated. $\qquad \blacksquare$

EXAMPLE 8 Arithmetic mean – median – mode

To find the frictional force between two specially designed surfaces, the force to move a block with one surface along an inclined plane with the other surface is measured ten times. The results, with forces in newtons, are

$$2.2, 2.4, 2.1, 2.2, 2.5, 2.2, 2.4, 2.7, 2.1, 2.5$$

Find the mean, median, and mode of these forces.

To find the mean, we sum the values of the forces and divide this total by 10. This gives

$$\bar{F} = \frac{\sum F}{10} = \frac{2.2 + 2.4 + 2.1 + 2.2 + 2.5 + 2.2 + 2.4 + 2.7 + 2.1 + 2.5}{10}$$

$$= \frac{23.3}{10} = 2.33 \, \text{N}$$

■ The mean is useful with many statistical methods and is used extensively. The median is also commonly used and is a good choice if there are some extreme values. The mode is occasionally used (if there is only one) for an estimate of data that is approximately symmetric.

The median is found by arranging the values in order and finding the middle value. The values in order are

$$2.1, 2.1, 2.2, 2.2, 2.2, 2.4, 2.4, 2.5, 2.5, 2.7$$

Since there are ten values, we see that the fifth value is 2.2 and the sixth is 2.4. The value midway between these is 2.3, which is the median. Therefore, the median force is 2.3 N.

The mode is 2.2 N, since this value appears three times, which is more than any other value. ■

EXERCISES 22.2

In Exercises 1–4, delete the 5 from the data numbers given for Example 1 and then do the following with the resulting data.

1. Find the median.

2. Find the arithmetic mean using the definition, as in Example 3.

3. Find the arithmetic mean using Eq. (22.1), as in Example 4.

4. Find the mode, as in Example 7.

In Exercises 5–16, use the following sets of numbers.

A: 3, 6, 4, 2, 5, 4, 7, 6, 3, 4, 6, 4, 5, 7, 3
B: 25, 26, 23, 24, 25, 28, 26, 27, 23, 28, 25
C: 0.48, 0.53, 0.49, 0.45, 0.55, 0.49, 0.47, 0.55, 0.48, 0.57, 0.51, 0.46, 0.53, 0.50, 0.49, 0.53
D: 105, 108, 103, 108, 106, 104, 109, 104, 110, 108, 108, 104, 113, 106, 107, 106, 107, 109, 105, 111, 109, 108

In Exercises 5–8, determine the median of the numbers of the given set.

5. Set A

6. Set B

7. Set C

8. Set D

In Exercises 9–12, determine the arithmetic mean of the numbers of the given set.

9. Set A

10. Set B

11. Set C

12. Set D

In Exercises 13–16, determine the mode of the numbers of the given set.

13. Set A

14. Set B

15. Set C

16. Set D

In Exercises 17–34, the required data are those in Exercises 22.1. Find the indicated measures of central tendency.

17. Median of miles per gallon of fuel usage in Exercise 5

18. Mean of miles per gallon of fuel usage in Exercise 5

19. Mode of miles per gallon of fuel usage in Exercise 5

20. Median of computer instructions in Exercise 13

21. Mean of computer instructions in Exercise 13

22. Mode of computer instructions in Exercise 13

23. Median of strobe light times in Exercise 17

24. Mean of strobe light times in Exercise 17

25. Median of stopping distances in Exercise 21. (Use the class mark for each class.)

26. Mean of stopping distances in Exercise 21. (Use the class mark for each class.)

27. Mean of X-ray dosages in Exercise 25

28. Median of X-ray dosages in Exercise 25

29. Mode of X-ray dosages in Exercise 25

30. Mean of battery lives in Exercise 27

31. Median of battery lives in Exercise 27

32. Mode of battery lives in Exercise 27

33. Mean of cable diameters in Exercise 29

34. Median of cable diameters in Exercise 29

In Exercises 35–46, find the indicated measure of central tendency.

35. The weekly salaries (in dollars) for the workers in a small factory are as follows:

600, 750, 625, 575, 525, 700, 550, 750, 625, 800, 700, 575, 600, 700

Find the median and the mode of the salaries.

36. Find the mean salary for the salaries in Exercise 35.

37. In a particular month, the electrical usage, rounded to the nearest 100 kW·h (kilowatt-hours), of 1000 homes in a certain city was summarized as follows:

Usage	500	600	700	800	900	1000	1100	1200
No. Homes	22	80	106	185	380	122	90	15

Find the mean of the electrical usage.

38. Find the median and mode of electrical usage in Exercise 37.

39. A test of air pollution in a city gave the following readings of the concentration of sulfur dioxide (in parts per million) for 18 consecutive days:

0.14, 0.18, 0.27, 0.19, 0.15, 0.22, 0.20, 0.18, 0.15, 0.17, 0.24, 0.23, 0.22, 0.18, 0.32, 0.26, 0.17, 0.23

Find the median and the mode of these readings.

40. Find the mean of the readings in Exercise 39.

41. The *midrange*, another measure of central tendency, is found by finding the sum of the lowest and the highest values and dividing this sum by 2. Find the midrange of the salaries in Exercise 35.

42. Find the midrange of the sulfur dioxide readings in Exercise 39. (See Exercise 41.)

(W)43. Add $100 to each of the salaries in Exercise 35. Then find the median, mean, and mode of the resulting salaries. State any conclusion that might be drawn from the results.

(W)44. Multiply each of the salaries in Exercise 35 by 2. Then find the median, mean, and mode of the resulting salaries. State any conclusion that might be drawn from the results.

(W)45. Change the final salary in Exercise 35 to $4000, with all other salaries being the same. Then find the mean of these salaries. State any conclusion that might be drawn from the result. (The $4000 here is called an *outlier*, which is an extreme value.)

(W)46. Find the median and mode of the salaries indicated in Exercise 45. State any conclusion that might be drawn from the results.

Answers to Practice Exercises

1. 14.5   **2.** 14.8

22.3 Standard Deviation

Population • Sample •
Standard Deviation of Sample •
Population Standard Deviation

In using statistics, we generally collect a sample of data and draw certain conclusions about the complete collection of possible values. In statistics, *the complete collection of values (measurements, people, scores, etc.) is called the* **population,** *and a* **sample** *is a subset of the population.* In the applied examples and exercises of the previous two sections, we were using samples of larger populations. For example, in Example 1 on page 613 we used estimates of the time on the Internet from a *sample* of 50 persons, as it would be impractical to get estimates from the total *population* of *all* users of the Internet.

In the preceding section, we discussed measures of central tendency of the data in the various samples. However, regardless of the measure that may be used, it does not tell us whether the values of the population tend to be grouped closely together or spread out over a large range of values. Therefore, we also need some measure of the deviation, or spread, of the values from the median or the mean. If the spread is small and the numbers are grouped closely together, the measure of central tendency is more reliable and descriptive of the data than in the case in which the spread is greater.

In statistics, there are several measures of spread that may be defined. In this section, we discuss one that is very widely used: the *standard deviation.* It is defined on the next page.

The **standard deviation** *of a set of* **sample** *values is defined by the equation*

$$s = \sqrt{\frac{\sum(x - \bar{x})^2}{n - 1}}$$ (22.2)

The definition of *s* shows that the following steps are used in computing its value.

STEPS FOR CALCULATING STANDARD DEVIATION

1. *Find the arithmetic mean $\bar{x}$ of the numbers of the set.*
2. *Subtract the mean from each number of the set.*
3. *Square these differences.*
4. *Find the sum of these squares.*
5. *Divide this sum by $n - 1$.*
6. *Find the square root of this result.*

NOTE ▶

The standard deviation *s* is a positive number. It is a *deviation from the mean*, regardless of whether the individual numbers are greater than or less than the mean. Numbers close together will have a small standard deviation, whereas numbers further apart have a larger standard deviation. Therefore, *the standard deviation becomes larger as the spread of data increases.*

Following the steps shown above, we use Eq. (22.2) for the calculation of standard deviation in the following examples.

EXAMPLE 1 Standard deviation – using Eq. (22.2)

Find the standard deviation of the following numbers: 1, 5, 4, 2, 6, 2, 1, 1, 5, 3.

A table of the necessary values is shown below, and steps 1–6 are indicated:

	step 2	step 3
x	$x - \bar{x}$	$(x - \bar{x})^2$
1	−2	4
5	2	4
4	1	1
2	−1	1
6	3	9
2	−1	1
1	−2	4
1	−2	4
5	2	4
3	0	0
30		32

step 4

$$\bar{x} = \frac{30}{10} = 3 \qquad \text{step 1}$$

$$\frac{\sum(x - \bar{x})^2}{n - 1} = \frac{32}{10 - 1} = \frac{32}{9} \qquad \text{step 5}$$

$$s = \sqrt{\frac{32}{9}} = 1.9 \qquad \text{step 6}$$

NOTE ▶

In calculating the standard deviation, it is usually rounded off to one more significant digit than was present in the original data. ∎

If some of the values in the data are repeated, we can use the frequency of those values that occur more than once in calculating the standard deviation. This is illustrated in the following example.

EXAMPLE 2 Standard deviation – using Eq. (22.2)

Find the standard deviation of the numbers in Example 1 of Section 22.2.

Since several of the numbers appear more than once, it is helpful to use the frequency of each number in the table, as follows:

x	f	fx	$x - \bar{x}$	$(x - \bar{x})^2$	$f(x - \bar{x})^2$
1	1	1	−4.5	20.25	20.25
2	2	4	−3.5	12.25	24.50
3	1	3	−2.5	6.25	6.25
4	3	12	−1.5	2.25	6.75
5	1	5	−0.5	0.25	0.25
6	1	6	0.5	0.25	0.25
7	2	14	1.5	2.25	4.50
8	1	8	2.5	6.25	6.25
9	2	18	3.5	12.25	24.50
11	1	11	5.5	30.25	30.25
	15	82			123.75

step 2   step 3   (step 1)   step 4

$$\bar{x} = \frac{82}{15} = 5.5$$

$$\frac{\sum f(x - \bar{x})^2}{n - 1} = \frac{123.75}{15 - 1} = \frac{123.75}{14} \quad \text{step 5}$$

$$s = \sqrt{\frac{123.75}{14}} = 3.0 \quad \text{step 6}$$

In some sources, standard deviation is defined such that n, rather than $n - 1$, is used in the denominator of Eq. (22.2). Using n gives us the *population standard deviation*. However, we generally will be using samples of the population, and the use of $n - 1$ gives better estimates of a population standard deviation when using samples. Therefore, in this text we will use Eq. (22.2) and refer to the sample standard deviation simply as the standard deviation.

It is possible to reduce the computational work required to find the standard deviation. Algebraically, it can be shown (although we will not do so here) that the following equation is another form of Eq. (22.2) and therefore gives the same results.

$$s = \sqrt{\frac{n\left(\sum x^2\right) - \left(\sum x\right)^2}{n(n - 1)}} \tag{22.3}$$

Although the form of this equation appears more involved, it does reduce the amount of calculation that is necessary. Consider the following examples.

EXAMPLE 3 Standard deviation – using Eq. (22.3)

Using Eq. (22.3), find s for the numbers in Example 1.

x	x^2
1	1
5	25
4	16
2	4
6	36
2	4
1	1
1	1
5	25
3	9
30	122

$$n = 10$$
$$\sum x^2 = 122$$
$$\left(\sum x\right)^2 = 30^2 = 900$$

$$s = \sqrt{\frac{10(122) - 900}{10(9)}} = 1.9$$

Practice Exercise

1. Find the standard deviation of the first eight numbers in Example 1.

EXAMPLE 4 Standard deviation – application

An ammeter measures the electric current in a circuit. In an ammeter, two resistances are connected in parallel, with most of the current passing through a very low resistance called the *shunt*. The resistance of each shunt in a sample of 100 shunts was measured. The results were grouped, and the class mark and frequency for each class are shown in the following table. Calculate the arithmetic mean and the standard deviation of the resistances of the shunts.

R (ohms)	f	fR	fR^2
0.200	1	0.200	0.0400
0.210	3	0.630	0.1323
0.220	5	1.100	0.2420
0.230	10	2.300	0.5290
0.240	17	4.080	0.9792
0.250	40	10.000	2.5000
0.260	13	3.380	0.8788
0.270	6	1.620	0.4374
0.280	3	0.840	0.2352
0.290	2	0.580	0.1682
	100	24.730	6.1421

$$\overline{R} = \frac{24.730}{100} = 0.247 \ \Omega$$

$$n = 100$$

$$\sum R^2 = 6.1421$$

$$(\sum R)^2 = 24.730^2$$

$$s = \sqrt{\frac{100(6.1421) - 24.730^2}{100(99)}} = 0.016$$

The arithmetic mean of the resistances is $0.247 \ \Omega$, with a standard deviation of $0.016 \ \Omega$. ∎

EXAMPLE 5 Standard deviation – application

Find the standard deviation of the estimated hours on the Internet as grouped in Example 2 of Section 22.1 (page 613). In doing this, we assume that each value in the class is the same as the class mark. The method is not exact, but with a large set of numbers, it provides a good approximation with less arithmetic work.

■ The statistical measures $x, \sum x, \sum x^2, s_x$, σ_x (the population standard deviation), and n (and possibly others) can be displayed on a graphing calculator as shown in Fig. 22.7 for the data of Example 5.

```
1-Var Stats
 x̄=13.9
 Σx=695
 Σx²=11455
 Sx=6.051648451
 σx=5.99082632
↓n=50
```

Fig. 22.7

Interval	x	f	fx	fx^2
0–4	2	2	4	8
5–9	7	9	63	441
10–14	12	19	228	2,736
15–19	17	11	187	3,179
20–24	22	6	132	2,904
25–29	27	3	81	2,187
		50	695	11,455

$$n = 50$$

$$\sum x^2 = 11,455$$

$$(\sum x)^2 = 695^2$$

$$s = \sqrt{\frac{50(11,455) - 695^2}{50(49)}} = 6.1$$

Thus, $s = 6.1$ h. Compare the mean calculated in Example 6 on page 619 (not using class marks) with that shown in the calculator display to the left (using class marks). ∎

In using the statistical measures we have discussed, we must be careful in using and interpreting such measures. Consider the following example.

EXAMPLE 6 Interpreting standard deviation

(a) The numbers 1, 2, 3, 4, 5 have a mean of 3, a median of 3, and a standard deviation of 1.6. These values fairly well describe the center and distribution of the numbers in the set.

(b) The numbers 1, 2, 3, 4, 100 have a mean of 22, a median of 3, and a standard deviation of 44. The large difference between the median and the mean and the very large range of values within one standard deviation of the mean (-22 to 66) indicate that this set of measures does not describe this set of numbers well. In a case like this, the 100 should be checked to see if it is in error. ∎

Example 6 illustrates that the statistical measures can be misleading if the numbers in a set are unevenly distributed. Misleading statistics can also come from the source of the data. Consider the probable results of a survey to find the percent of persons in favor of raising income taxes for the wealthy if the survey is taken at the entrance to a welfare office or if it is taken at the entrance to a stock brokerage firm. There are many other considerations in the proper use and interpretation of statistical measures.

EXERCISES 22.3

In Exercises 1 and 2, in Example 1, change the first 1 to 6 and the first 2 to 7 and then find the standard deviation of the resulting data as directed.

1. Find *s* from the definition, as in Example 1.

2. Find *s* using Eq. 22.3, as in Example 3.

In Exercises 3–14, use the following sets of numbers. They are the same as those used in Exercise 22.2.

A: 3, 6, 4, 2, 5, 4, 7, 6, 3, 4, 6, 4, 5, 7, 3
B: 25, 26, 23, 24, 25, 28, 26, 27, 23, 28, 25
C: 0.48, 0.53, 0.49, 0.45, 0.55, 0.49, 0.47, 0.55, 0.48, 0.57, 0.51, 0.46, 0.53, 0.50, 0.49, 0.53
D: 105, 108, 103, 108, 106, 104, 109, 104, 110, 108, 108, 104, 113, 106, 107, 106, 107, 109, 105, 111, 109, 108

In Exercises 3–6, use Eq. (22.2) to find the standard deviation s for the indicated sets of numbers.

3. Set *A*   4. Set *B*   5. Set *C*   6. Set *D*

In Exercises 7–10, use Eq. (22.3) to find the standard deviation s for the indicated sets of numbers.

7. Set *A*   8. Set *B*   9. Set *C*   10. Set *D*

In Exercises 11–14, use the statistical feature of a calculator to find the arithmetic mean and the standard deviation s for the indicated sets of numbers.

11. Set *A*   12. Set *B*   13. Set *C*   14. Set *D*

In Exercises 15–22, find the standard deviation for the indicated sets of numbers from Exercises 22.1 and 22.2.

15. The miles per gallon of fuel usage in Exercise 5 of Section 22.1, using Eq. (22.2)

16. The miles per gallon of fuel usage in Exercise 5 of Section 22.1, using Eq. (22.3)

17. The computer instructions in Exercise 13 of Section 22.1

18. The X-ray dosages in Exercise 25 of Section 22.1

19. The battery lives in Exercise 27 of Section 22.1

20. The salaries in Exercise 35 of Section 22.2

21. The strobe light times in Exercise 17 of Section 22.1

22. The stopping distances in Exercise 21 of Section 22.1

In Exercises 23–26, find the standard deviation s for the indicated sets of numbers using the statistical feature on a graphing calculator.

23. The air pollution data in Exercise 39 of Section 22.2

24. The following data giving the mean number of days of rain for Vancouver, B.C., for the 12 months of the year.

20, 17, 17, 14, 12, 11, 7, 8, 9, 16, 19, 22

25. The fiber-optic cable diameters in Exercise 29 of Section 22.1

26. The electric power usages in Exercise 37 of Section 22.2

Answer to Practice Exercise

1. 2.0

22.4 Normal Distributions

Normal Distribution of a Population • Standard Normal Distribution • Standard Score (z-Score) • Standard Errors

Fig. 22.8

The distributions in the previous sections have been for a limited number of values. Let us now consider a very large population, such as the useable lifetime of all of the AA batteries sold in the world in a year. It could have a large number of classes with many values within each class.

We would expect a frequency polygon for this very large population to have its maximum frequency very near the mean and taper off to smaller frequencies on either side. It would probably be shaped very close to the curve shown in Fig. 22.8.

Fig. 22.8

The smooth bell-shaped curve in Fig. 22.8 shows the **normal distribution** of a population large enough that the distribution is considered to be *continuous*. (We can think of this as the points on the frequency distribution curve are so close that they can be considered to be touching.) Using advanced methods, its equation is found to be

$$y = \frac{e^{-(x-\mu)^2/2\sigma^2}}{\sigma\sqrt{2\pi}}$$ **(22.4)**

Here, μ is the *population mean* and σ is the *population standard deviation*, and π and e are the familiar numbers first used in Chapters 2 and 12, respectively.

From Eq. (22.4), we can see that any particular normal distribution for a large population depends on the values of μ and σ. The horizontal location of the curve depends on μ, and the shape (how spread out the curve is) depends on σ, but the bell shape remains. This is illustrated in general in the following example.

EXAMPLE 1 Population standard deviation

In Fig. 22.9, for the left curve, $\mu = 10$ and $\sigma = 5$, whereas for the right curve, $\mu = 20$ and $\sigma = 10$.

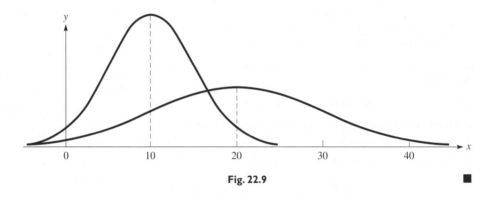

Fig. 22.9   ∎

STANDARD NORMAL DISTRIBUTION

As we have just seen, there are innumerable possible normal distributions. However, there is one of particular interest. *The **standard normal distribution** is the normal distribution for which the mean is* 0 *and the standard deviation is* 1. Making these substitutions in Eq. (22.4), we have

$$y = \frac{1}{\sqrt{2\pi}}e^{-x^2/2}$$ **(22.5)**

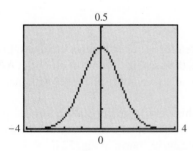

Fig. 22.10

as the equation of the standard normal distribution curve.

From Eq. (22.5), we see that the standard normal distribution curve is symmetric to the y-axis (if we replace x by $-x$, the equation remains unchanged). Since the mean is zero, this means that the curve is symmetric about the mean. We can also see from the exponent of e that y becomes smaller, but is always positive, as x becomes larger numerically, and this in turn means that the x-axis is an asymptote. The graph of the standard normal distribution curve is shown in the graphing calculator display in Fig. 22.10.

Using advanced mathematics, it can be shown that the total area under the standard normal distribution curve (above the x-axis) is exactly one unit. As shown in Fig. 22.11, it is found that the area between $x = 0$ and $x = 1$ is 0.3413, the area between $x = 1$ and $x = 2$ is 0.1359, and the area between $x = 2$ and $x = 3$ is 0.0215. Since the curve is symmetric to the y-axis, areas for negative values of x are equivalent. This means the total area between $x = -3$ and $x = 3$ is about 0.9974 and nearly all the area lies between $x = -3$ and $x = 3$.

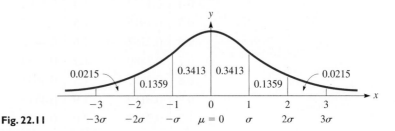

Fig. 22.11

From Fig. 22.11, we can see that *about 68% of the area is within one standard deviation of the mean (in the interval $\mu - \sigma$ to $\mu + \sigma$), and about 95% of the area is within two standard deviations of the mean (in the interval $\mu - 2\sigma$ to $\mu + 2\sigma$).* These percentages are often useful in data analysis.

Since a normal distribution curve gives a measure of the frequency of a particular value within the distribution, *the area under any part of the standard normal curve gives the relative frequency of those values of the distribution.* This is illustrated in the following example.

EXAMPLE 2 Normal distribution

For a normal distribution of values, 0.3413 (34.13%) of the values are between $x = 0$ and $x = 1$, and 68.26% of the values are between $x = -1$ and $x = 1$. For this distribution, 81.85% of the values lie between $x = -2$ and $x = 1$.
$(0.1359 + 0.3413 + 0.3413 = 0.8185)$. ∎

We can find the relative frequency of values for any normal distribution by use of the **standard score** z (or z-score), which is defined as

$$z = \frac{x - \mu}{\sigma}$$ (22.6)

For the normal standard distribution, where $\mu = 0$, if we let $x = \sigma$, then $z = 1$. If we let $x = 2\sigma$, $z = 2$. Therefore, we can see that *a value of z tells us the number of standard deviations the given value of x is above or below the mean.* From the discussion above, we can see that the value of z can tell us the area under the curve between the mean and the value of x corresponding to that value of z. In turn, *this tells us the relative frequency of all values between the mean and the value of x.*

On the next page, Table 22.1 gives the area under the standard normal distribution curve for the given values of z. The table includes values only to $z = 3$ since nearly all of the area is between $z = -3$ and $z = 3$. Since the curve is symmetric to the y-axis, the values shown are also valid for negative values of z.

Table 22.1 Standard Normal (z) Distribution

z	Area	z	Area	z	Area
0.0	0.0000	1.0	0.3413	2.0	0.4772
0.1	0.0398	1.1	0.3643	2.1	0.4821
0.2	0.0793	1.2	0.3849	2.2	0.4861
0.3	0.1179	1.3	0.4032	2.3	0.4893
0.4	0.1554	1.4	0.4192	2.4	0.4918
0.5	0.1915	1.5	0.4332	2.5	0.4938
0.6	0.2257	1.6	0.4452	2.6	0.4953
0.7	0.2580	1.7	0.4554	2.7	0.4965
0.8	0.2881	1.8	0.4641	2.8	0.4974
0.9	0.3159	1.9	0.4713	2.9	0.4981
1.0	0.3413	2.0	0.4772	3.0	0.4987

The following examples illustrate the use of Eq. (22.6) and z-scores.

EXAMPLE 3 Normal score (z-score)

For a normal distribution curve based on values of $\mu = 20$ and $\sigma = 5$, find the area between $x = 24$ and $x = 32$. To find this area, we use Eq. (22.6) to find the corresponding values of z and then find the difference between these z-scores. These z-scores are

$$z = \frac{24 - 20}{5} = 0.8 \quad \text{and} \quad z = \frac{32 - 20}{5} = 2.4$$

For $z = 0.8$, the area is 0.2881, and for $z = 2.4$, the area is 0.4918. Therefore, the area between $x = 24$ and $x = 32$ (see Fig. 22.12) is

$$0.4918 - 0.2881 = 0.2037$$

This means that the relative frequency of the values between $x = 24$ and $x = 32$ is 20.37%. If we have a large set of measured values with $\mu = 20$ and $\sigma = 5$, we should expect that about 20% of them are between $x = 24$ and $x = 32$. ∎

Fig. 22.12

Practice Exercise

1. For values of $\mu = 40$ and $\sigma = 8$, find the area between $x = 36$ and $x = 48$.

EXAMPLE 4 z-score – application

The lifetimes of a certain type of watch battery are normally distributed. The mean lifetime is 400 days, and the standard deviation is 50 days. For a sample of 5000 new batteries, determine how many batteries will last **(a)** between 360 days and 460 days, **(b)** more than 320 days, and **(c)** less than 280 days.

(a) For this distribution, $\mu = 400$ days and $\sigma = 50$ days. Using Eq. (22.6), we find the z-scores for $x = 360$ days and $x = 460$ days. They are

$$z = \frac{360 - 400}{50} = -0.8 \quad \text{and} \quad z = \frac{460 - 400}{50} = 1.2$$

For $z = -0.8$, the area is to the left of the mean, and since the curve is symmetric about the mean, we use the $z = 0.8$ value of the area and add it to the area for $z = 1.2$. Therefore, the area is

$$0.2881 + 0.3849 = 0.6730$$

Fig. 22.13

See Fig. 22.13. This means that 67.30% of the 5000 batteries, or 3365 of the batteries, will last between 360 and 460 days.

Fig. 22.14

Fig. 22.15

◼ A more complete development of statistics, which usually requires at least an entire book, would also include topics in probability. Here, we are pointing out that these standard errors should be considered when dealing with normal distributions.

(b) To determine the number of batteries that will last more than 320 days, we first find the z-score for $x = 320$. It is $z = (320 - 400)/50 = -1.6$. This means we want the total area to the right of $z = -1.6$. In this case, we add the area for $z = 1.6$ to the total area to the right of the mean. Since the total area under the curve is 1.0000, the total area on either side of the mean is 0.5000. Therefore, the area to the right of $z = -1.6$ is $0.4452 + 0.5000 = 0.9452$. See Fig. 22.14. This means that $0.9452 \times 5000 = 4726$ batteries will last more than 320 days.

(c) To find the number of batteries that will last less than 280 days, we first find that $z = (280 - 400)/50 = -2.4$ for $x = 280$. Since we want the total area to the left of $z = -2.4$, we subtract the area for $z = 2.4$ from 0.5000, the total area to the left of the mean. Since the area for $z = 2.4$ is 0.4918, the total area to the left of $z = -2.4$ is $0.5000 - 0.4918 = 0.0082$. See Fig. 22.15. Therefore, $0.0082 \times 5000 = 41$ batteries will last less than 280 days. ◼

STANDARD ERRORS

In Example 4, we assumed that the lifetimes of the batteries were normally distributed. Of course, for any set of 5000 batteries, or any number of batteries for that matter, the lifetimes that actually occur will not follow normal distribution *exactly*. There will be some variation from normal distribution, but for a large sample, this variation should be small. The mean and the standard deviation for any sample would probably vary somewhat from that of the population.

In the study of probability, it is shown that if we select all possible samples of size n from a population with a mean μ and standard deviation σ, the mean of the sample means is also μ. Also, *the standard deviation of the sample means,* denoted by $\sigma_{\bar{x}}$, and called **the standard error of the mean,** is

$$\sigma_{\bar{x}} = \frac{\sigma}{\sqrt{n}} \qquad (22.7)$$

Also, the *standard error of the standard deviation s of the sample,* denoted by σ_s, and called **the standard error of s,** is

$$\sigma_s = \frac{\sigma}{\sqrt{2n}} \qquad (22.8)$$

We can see from these formulas that as the sample size gets larger (*large* is usually considered to be over 30), the less variation there should be in the mean and standard deviation of the sample. This agrees with what we should expect.

EXAMPLE 5 Standard errors

For the sample of 5000 watch batteries in Example 4, we know that $\sigma = 50$ days. Therefore, the standard error in the mean $\bar{x}$ is $50/\sqrt{5000} = 0.7$ day. This means that of all samples of 5000 batteries, about 68% should have a mean lifetime of 400 ± 0.7 day (between 399.3 days and 400.7 days).

The standard error in the standard deviation s is $50/\sqrt{2(5000)} = 0.5$ day. This means that of all samples of 5000 batteries, about 68% should have a standard deviation of 50 ± 0.5 day (between 49.5 days and 50.5 days).

(Considering the significant digits of these values, 68% (within one standard deviation) or even 95% (within two standard deviations) of the sample values of $\bar{x}$ and s should not vary by more than 1 day.) ◼

EXERCISES 22.4

In Exercises 1–4, make the given changes in the indicated examples of this section and then solve the indicated problem.

1. In Example 1, change the second σ from 10 to 5 and then describe the curve that would result in terms of either or both curves shown in Fig. 22.9.

2. In Example 2, change the final $x = 1$ to $x = 2$ and then find the resulting percent of values for that part of the example.

3. In Example 3, change $x = 32$ to $x = 33$ and then find the resulting area.

4. In Example 4(b), change 320 to 360 and then find the resulting number of batteries.

In Exercises 5–8, use a graphing calculator to display the indicated graph of Eq. (22.4).

5. Display the graph of the normal distribution of values for which $\mu = 10$ and $\sigma = 5$. Compare with the graph shown in Fig. 22.9.

6. Display the graph of the normal distribution of values for which $\mu = 20$ and $\sigma = 10$. Compare with the graph shown in Fig. 22.9.

7. Sketch a graph of a normal distribution of values for which $\mu = 100$ and $\sigma = 10$. Then compare with the graph displayed by a graphing calculator.

8. Sketch a graph of a normal distribution of values for which $\mu = 100$ and $\sigma = 30$. Then compare with the graph displayed by a graphing calculator. How does this graph differ from that of Exercise 7?

In Exercises 9–12, use the following data and refer to Fig. 22.11. A sample of 200 bags of cement are weighed as a quality check. Over a long period, it has been found that the mean value and standard deviation for this size bag are known and that the weights are normally distributed. Determine how many bags within this sample should have weights that satisfy the following conditions.

9. Within one standard deviation of the mean

10. Within two standard deviations of the mean

11. Between the mean and two standard deviations above the mean

12. Between one standard deviation below the mean and three standard deviations above the mean

In Exercises 13–16, use the following data. Each AA battery in a sample of 500 batteries is checked for its voltage. It has been previously established for this type of battery (when newly produced) that the voltages are distributed normally with $\mu = 1.50$ V and $\sigma = 0.05$ V.

13. How many batteries have voltages between 1.45 V and 1.55 V?

14. How many batteries have voltages between 1.52 V and 1.58 V?

15. What percent of the batteries have voltages below 1.54 V?

16. What percent of the batteries have voltages above 1.64 V?

In Exercises 17–24, use the following data. The lifetimes of a certain type of automobile tire have been found to be distributed normally with a mean lifetime of 100,000 km and a standard deviation of 10,000 km. Answer the following questions for a sample of 5000 of these tires.

17. How many tires will last between 85,000 km and 100,000 km?

18. How many tires will last between 95,000 km and 115,000 km?

19. How many tires will last more than 118,000 km?

20. If the manufacturer guarantees to replace all tires that do not last 75,000 km, what percent of the tires may have to be replaced under this guarantee?

21. What is the standard error in the mean for all samples of 5000 of these tires? Explain the meaning of this result.

22. What is the standard error in the standard deviation of all samples of 5000 of these tires? Explain the meaning of this result.

23. What percent of the samples of 5000 of these tires should have a mean lifetime of more than 100,282 km?

24. What percent of the samples of 5000 of these tires should have a mean standard deviation of between 9900 km and 10,100 km?

In Exercises 25–32, solve the given problems.

25. With 75.8% of the area under the normal curve to the right of z, find the z-value.

26. With 21% of the area under the normal curve between z_1 and z_2, to the right of $z_1 = 0.8$, find z_2.

27. With 59% of the area under the normal curve between z_1 and z_2, to the left of $z_2 = 1.1$, find z_1.

28. With 5.8% of the area under the normal curve between z_1 and z_2, to the left of $z_2 = 2.0$, find z_1.

29. For the strobe light times in Exercise 17 of Section 22.1, find the percent of times within 1 standard deviation of the mean. From Exercise 23 of Section 22.2, we find that $\bar{x} = 2.248$ s, and from Exercise 21 of Section 22.3, we find that $s = 0.014$ s. Compare the results with that of a normal distribution.

30. Follow the same instructions as in Exercise 29 for the fiber-optic diameters in Exercise 29 of Section 22.1. From Exercise 33 of Section 22.2, $\bar{x} = 0.00595$ mm, and from Exercise 25 of Section 22.3, $s = 0.00022$ mm.

31. Follow the same instructions as in Exercise 29 for the hours estimated on the Internet in Example 1 of Section 22.1. From Example 6 of Section 22.2, we find that $\bar{x} = 13.7$ h, and from Example 5 of Section 22.3, we find that $s = 6.1$ h.

32. Discuss the results found in Exercises 29 and 31, considering the methods used to find the mean and the standard deviation.

Answer to Practice Exercise

1. $z = 0.5328$

22.5 Statistical Process Control

Control Charts • Central Line • Range • Central Line • Upper and Lower Control Limits • Variable • Attribute

■ This is intended only as a brief introduction to this topic. A complete development requires at least a chapter in a statistics book.

One of the most important uses of statistics in industry is Statistical Process Control (SPC), which is used to maintain and improve product quality. Samples are tested during the production at specified intervals to determine whether the production process needs adjustment to meet quality requirements.

A particular industrial process is considered to be *in control* if it is stable and predictable, and sample measurements fall within upper and lower control limits. The process is *out of control* if it has an unpredictable amount of variation and there are sample measurements outside the control limits due to special causes.

EXAMPLE 1 Process – in control – out of control

The manufacturer of 1.5-V batteries states that the voltage of its batteries is no less than 1.45 V or greater than 1.55 V and has designed the manufacturing process to meet these specifications.

If all samples of batteries that are tested have voltages in the proper range with only expected minor variations, the production process is *in control*.

However, if some samples have batteries with voltages out of the proper range, the process is *out of control*. This would indicate some special cause for the problem, such as an improperly operating machine or an impurity getting into the process. The process would probably be halted until the cause is determined. ■

■ Minor variations may be expected, for example, from very small fluctuations in voltage, temperature, or material composition. Special causes resulting in an out-of-control process could include line stoppage, material defect, or an incorrect applied pressure.

CONTROL CHARTS

An important device used in SPC is the *control chart*. It is used to show a trend of a production characteristic over time. Samples are measured at specified intervals of time to see if the measurements are within acceptable limits. The measurements are plotted on a chart to check for trends and abnormalities in the production process.

In making a control chart, we must determine what the mean should be. For a stable process for which previous data are known, it can be based on a production specification or on previous data. For a new or recently modified process, it may be necessary to use present data, although the value may have to be revised for future charts. On a control chart, *this value is used as the population mean, μ.*

It is also necessary to establish the upper and lower control limits. The standard generally used is that 99.7% of the sample measurements should fall within these control limits. This assumes a normal distribution, and we note that this is within three sample standard deviations of the population mean. We will establish these limits by use of a table or a formula that has been made using statistical measures developed in a more complete coverage of quality control. This does follow the normal practice of using a formula or a more complete table in setting up the control limits.

In Fig. 22.16, we show a sample control chart, and on the following pages, we illustrate how control charts are made.

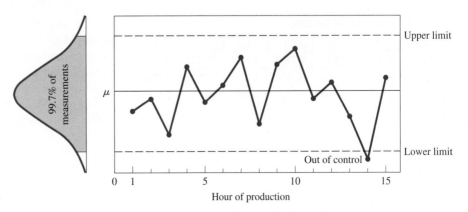

Fig. 22.16

EXAMPLE 2 Making $\bar{x}$ and R control charts

A pharmaceutical company makes a capsule of a prescription drug that contains 500 mg of the drug, according to the label. In a newly modified process of making the capsule, five capsules are tested every 15 min to check the amount of the drug in each capsule. Testing over a 5-h period gave the following results for the 20 subgroups of samples.

Subgroup	Amount of Drug (in mg) of Five Capsules					Mean x	Range R
1	503	501	498	507	502	502.2	9
2	497	499	500	495	502	498.6	7
3	496	500	507	503	502	501.6	11
4	512	503	488	500	497	500.0	24
5	504	505	500	508	502	503.8	8
6	495	495	501	497	497	497.0	6
7	503	500	507	499	498	501.4	9
8	494	498	497	501	496	497.2	7
9	502	504	505	500	502	502.6	5
10	500	502	500	496	497	499.0	6
11	502	498	510	503	497	502.0	13
12	497	498	496	502	500	498.6	6
13	504	500	495	498	501	499.6	9
14	500	499	498	501	494	498.4	7
15	498	496	502	501	505	500.4	9
16	500	503	504	499	505	502.2	6
17	487	496	499	498	494	494.8	12
18	498	497	497	502	497	498.2	5
19	503	501	500	498	504	501.2	6
20	496	494	503	502	501	499.2	9
					Sums	9998.0	174
					Means	499.9	8.7

As we noted from the table, *the range R of each sample is the difference between the highest value and the lowest value of the sample.*

From this table of values, we can make an $\bar{x}$ control chart and an R control chart. The $\bar{x}$ chart maintains a check on the average quality level, whereas the R chart maintains a check on the dispersion of the production process. These two control charts are often plotted together and referred to as the $\bar{x}$–R chart.

In order to define the **central line** of the $\bar{x}$ chart, which ideally is equivalent to the value of the population mean μ, we use the mean of the sample means $\bar{\bar{x}}$. For the central line of the R chart, we use $\bar{R}$. From the table, we see that

$$\bar{\bar{x}} = 499.9 \text{ mg} \quad \text{and} \quad \bar{R} = 8.7 \text{ mg}$$

Table 22.2 Control Chart Factors

n	d_2	A	A_2	D_1	D_2	D_3	D_4
5	2.326	1.342	0.577	0.000	4.918	0.000	2.115
6	2.534	1.225	0.483	0.000	5.078	0.000	2.004
7	2.704	1.134	0.419	0.205	5.203	0.076	1.924

The **upper control limit** (UCL) and the **lower control limit** (LCL) for each chart are defined in terms of the mean range $\bar{R}$ and an appropriate constant taken from a table of control chart factors. These factors, which are related to the sample size n, are determined by statistical considerations found in a more complete coverage of quality control. At the left is a brief table of control chart factors (Table 22.2).

The UCL and LCL for the $\bar{x}$ chart are found as follows:

$$\text{UCL}(\bar{x}) = \bar{\bar{x}} + A_2 \bar{R} = 499.9 + 0.577(8.7) = 504.9 \text{ mg}$$
$$\text{LCL}(\bar{x}) = \bar{\bar{x}} - A_2 \bar{R} = 499.9 - 0.577(8.7) = 494.9 \text{ mg}$$

The UCL and LCL for the R chart are found as follows:

$$\text{LCL}(R) = D_3 \bar{R} = 0.000(8.7) = 0.0 \text{ mg}$$
$$\text{UCL}(R) = D_4 \bar{R} = 2.115(8.7) = 18.4 \text{ mg}$$

Using these central lines and control limit lines, we now plot the $\bar{x}$ control chart in Fig. 22.17 and the R control chart in Fig. 22.18.

Fig. 22.17

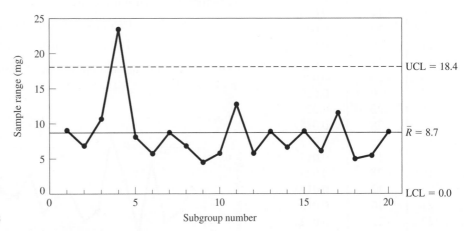

Fig. 22.18

This would be considered a *well-centered process* since $\bar{\bar{x}} = 499.9$ mg, which is very near the target value of 500.0 mg. We do note, however, that subgroup 17 was at a control limit and this might have been due to some special cause, such as the use of a substandard mixture of ingredients. We also note that the process was *out of control* due to some special cause of subgroup 4 since the range was above the upper control limit. We should keep in mind that there are numerous considerations, including human factors, that should be taken into account when making and interpreting control charts and that this is only a very brief introduction to this important industrial use of statistics. ■

In Example 2, the weight (in mg) of a prescription drug was tested. In quality control, weight is a **variable,** *which is a characteristic that can be* **measured.** Other examples of variables are length, voltage, and pressure.

A characteristic that can be **counted** *is an* **attribute,** which is determined to be either acceptable or not acceptable in testing. Examples of attributes are color (acceptable or not), entries on a customer account (correct or incorrect), and defects (not acceptable) in a product. To monitor an attribute in a production process, we get the *proportion* of defective parts by *dividing the number of defective parts in a sample by the total number of parts in the sample,* and then make a *p control chart.* This is illustrated in the following example.

EXAMPLE 3 Making a *p* control chart

The manufacturer of video discs has 1000 DVDs checked each day for defects (surface scratches, for example). The data for this procedure for 25 days are shown in the table at the left.

For the *p* control chart, the central line is the value of $\bar{p}$, which in this case is

$$\bar{p} = \frac{490}{25{,}000} = 0.0196$$

The control limits are each three standard deviations from $\bar{p}$. If *n* is the size of the subgroup, the standard deviation σ_p of a proportion is given by

$$\sigma_p = \sqrt{\frac{\bar{p}(1 - \bar{p})}{n}} = \sqrt{\frac{0.0196(1 - 0.0196)}{1000}} = 0.00438$$

Therefore, the control limits are

$$\text{UCL}(p) = 0.0196 + 3(0.00438) = 0.0327$$
$$\text{LCL}(p) = 0.0196 - 3(0.00438) = 0.0065$$

Using this central line and these control limit lines, we now plot the *p* control chart in Fig. 22.19.

Day	Defective DVDs	Proportion Defective
1	22	0.022
2	16	0.016
3	14	0.014
4	18	0.018
5	12	0.012
6	25	0.025
7	36	0.036
8	16	0.016
9	14	0.014
10	22	0.022
11	20	0.020
12	17	0.017
13	26	0.026
14	20	0.020
15	22	0.022
16	28	0.028
17	17	0.017
18	15	0.015
19	25	0.025
20	12	0.012
21	16	0.016
22	22	0.022
23	19	0.019
24	16	0.016
25	20	0.020
Sum	490	

Fig. 22.19

Practice Exercise

1. In Example 3, change Day 9 datum from 14 to 24 defective parts. Then find UCL(*p*) and LCL(*p*).

According to the proportion mean of 0.0196, the process produces about 2% defective DVDs. We note that the process was out of control on day 7. An adjustment to the production process was probably made to remove the special cause of the additional defective DVDs. ∎

EXERCISES 22.5

In Exercises 1–4, in Example 2, change the first subgroup to 497, 499, 502, 493, and 498 and then proceed as directed.

1. Find UCL ($\bar{x}$) and LCL ($\bar{x}$).

2. Find LCL (R) and UCL (R).

(W) 3. How would the $\bar{x}$ control chart differ from Fig. 22.17?

(W) 4. How would the R control chart differ from Fig. 22.18?

In Exercises 5–8, use the following data.

Five automobile engines are taken from the production line each hour and tested for their torque (in N · m) when rotating at a constant frequency. The measurements of the sample torques for 20 h of testing are as follows:

Hour	Torques (in N · m) of Five Engines				
1	366	352	354	360	362
2	370	374	362	366	356
3	358	357	365	372	361
4	360	368	367	359	363
5	352	356	354	348	350
6	366	361	372	370	363
7	365	366	361	370	362
8	354	363	360	361	364
9	361	358	356	364	364
10	368	366	368	358	360
11	355	360	359	362	353
12	365	364	357	367	370
13	360	364	372	358	365
14	348	360	352	360	354
15	358	364	362	372	361
16	360	361	371	366	346
17	354	359	358	366	366
18	362	366	367	361	357
19	363	373	364	360	358
20	372	362	360	365	367

5. Find the central line, UCL, and LCL for the mean.

6. Find the central line, UCL, and LCL for the range.

7. Plot an $\bar{x}$ chart.   8. Plot an R chart.

In Exercise 9–12, use the following data.

Five AC adaptors that are used to charge batteries of a cellular phone are taken from the production line each 15 minutes and tested for their direct-current output voltage. The output voltages for 24 sample subgroups are as follows:

Subgroup	Output Voltages of Five Adaptors				
1	9.03	9.08	8.85	8.92	8.90
2	9.05	8.98	9.20	9.04	9.12
3	8.93	8.96	9.14	9.06	9.00
4	9.16	9.08	9.04	9.07	8.97
5	9.03	9.08	8.93	8.88	8.95
6	8.92	9.07	8.86	8.96	9.04
7	9.00	9.05	8.90	8.94	8.93
8	8.87	8.99	8.96	9.02	9.03
9	8.89	8.92	9.05	9.10	8.93
10	9.01	9.00	9.09	8.96	8.98
11	8.90	8.97	8.92	8.98	9.03
12	9.04	9.06	8.94	8.93	8.92
13	8.94	8.99	8.93	9.05	9.10
14	9.07	9.01	9.05	8.96	9.02
15	9.01	8.82	8.95	8.99	9.04
16	8.93	8.91	9.04	9.05	8.90
17	9.08	9.03	8.91	8.92	8.96
18	8.94	8.90	9.05	8.93	9.01
19	8.88	8.82	8.89	8.94	8.88
20	9.04	9.00	8.98	8.93	9.05
21	9.00	9.03	8.94	8.92	9.05
22	8.95	8.95	8.91	8.90	9.03
23	9.12	9.04	9.01	8.94	9.02
24	8.94	8.99	8.93	9.05	9.07

9. Find the central line, UCL, and LCL for the mean.

10. Find the central line, UCL, and LCL for the range.

11. Plot an $\bar{x}$ chart.

12. Plot an R chart.

In Exercises 13–16, use the following information.

For a production process for which there is a great deal of data since its last modification, the population mean μ and population standard deviation σ are assumed known. For such a process, we have the following values (using additional statistical analysis):

$\bar{x}$ chart: central line $= \mu$, UCL $= \mu + A\sigma$, LCL $= \mu - A\sigma$

R chart: central line $= d_2\sigma$, UCL $= D_2\sigma$, LCL $= D_1\sigma$

The values of A, d_2, D_2, and D_1 are found in the table of control chart factors in Example 2 (Table 22.2).

13. In the production of robot links and tests for their lenghs, it has been found that $\mu = 2.725$ in. and $\sigma = 0.032$ in. Find the central length, UCL, and LCL for the mean if the sample subgroup size is 5.

14. For the robot link samples of Exercise 13, find the central line, UCL, and LCL for the range.

15. After bottling, the volume of soft drink in six sample bottles is checked each 10 minutes. For this process $\mu = 750.0$ mL and $\sigma = 2.2$ mL. Find the central line, UCL, and LCL for the range.

16. For the bottling process of Exercise 15, find the central line, UCL, and LCL for the mean.

In Exercises 17 and 18, use the following data.

A telephone company rechecks the entries for 1000 of its new customers each week for name, address, and phone number. The data collected regarding the number of new accounts with errors, along with the proportion of these accounts with errors, is given in the following table for a 20-wk period:

Week	Accounts with Errors	Proportion with Errors
1	52	0.052
2	36	0.036
3	27	0.027
4	58	0.058
5	44	0.044
6	21	0.021
7	48	0.048
8	63	0.063
9	32	0.032
10	38	0.038
11	27	0.027
12	43	0.043
13	22	0.022
14	35	0.035
15	41	0.041
16	20	0.020
17	28	0.028
18	37	0.037
19	24	0.024
20	42	0.042
Total	738	

17. For a p chart, find the values for the central line, UCL, and LCL.

18. Plot a p chart.

Answers to Practice Exercise

1. $\text{UCL}(p) = 0.0333, \text{LCL}(p) = 0.0067$

In Exercises 19 and 20, use the following data.

The maker of electric fuses checks 500 fuses each day for defects. The number of defective fuses, along with the proportion of defective fuses for 24 days, is shown in the following table.

Day	Number Defective	Proportion Defective
1	26	0.052
2	32	0.064
3	37	0.074
4	16	0.032
5	28	0.056
6	31	0.062
7	42	0.084
8	22	0.044
9	31	0.062
10	28	0.056
11	24	0.048
12	35	0.070
13	30	0.060
14	34	0.068
15	39	0.078
16	26	0.052
17	23	0.046
18	33	0.066
19	25	0.050
20	25	0.050
21	32	0.064
22	23	0.046
23	34	0.068
24	20	0.040
Total	696	

19. For a p chart, find the values for the central line, UCL, and LCL.

20. Plot a p chart.

22.6 Linear Regression

Regression • Linear Regression •
Method of Least Squares • Deviation •
Least-squares Line

We have considered statistical methods for dealing with one variable. We have discussed methods of tabulating, graphing, and measuring the central tendency and the deviations from this value for one variable and have shown an important industrial application in the area of quality control. We now discuss how to find an equation relating two variables for which a set of points is known.

In this section, we show a method of finding the equation of a straight line that passes through a set of data points, and in this way we *fit* the line to the points. In general, *the fitting of a curve to a set of points is called* **regression.** Fitting a straight line to a set of points is *linear regression,* and fitting some other type of curve is called *nonlinear regression.* We consider nonlinear regression in the next section.

Some of the reasons for using regression to find the equation of a curve that passes through a set of points, and thereby "fit" the curve to the points are (**1**) to express a concise relationship between the variables, (**2**) to use the equation to predict certain fundamental results, (**3**) to determine the reliability of certain sets of data, and (**4**) to use the data for testing certain theoretical concepts.

For a given set of several (at least 5 or 6) points for representing pairs of data values, we cannot reasonably expect that the curve of any given equation will pass through all of the points *exactly*. Therefore, when we fit the curve of an equation to the points, we are finding the curve that best approximates passing through the points. It is possible that the curve that best fits the data will not actually pass directly through any of the points, although it should come reasonably close to most of them. Consider the following example.

EXAMPLE 1 Fitting line to set of points

All the students enrolled in a mathematics course took an entrance test. To study the reliability of this test as an indicator of future success, an instructor tabulated the test scores of ten students (selected at random), along with their course averages at the end of the course, and made a graph of the data. See the table below and Fig. 22.20.

Student	Entrance Test Score, Based on 40	Course Average, Based on 100
A	29	63
B	33	88
C	22	77
D	17	67
E	26	70
F	37	93
G	30	72
H	32	81
I	23	47
J	30	74

Fig. 22.20

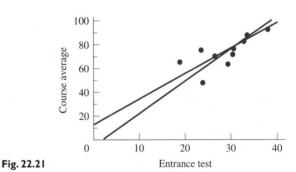

Fig. 22.21

We now ask whether there is a functional relationship between the test scores and the course grades. Certainly, no clear-cut relationship exists, but in general we see that the higher the test score, the higher the course grade. This leads to the possibility that there might be some straight line, from which none of the points would vary too significantly. If such a line could be found, then it could be the basis of predictions as to the possible success a student might have in the course, on the basis of his or her grade on the entrance test. Assuming that such a straight line exists, the problem is to find the equation of this line. Figure 22.21 shows two such possible lines. ■

There are a number of different methods of determining the straight line that best fits the given data points. We employ the method that is most widely used: the **method of least squares.** *The basic principle of this method is that the sum of the squares of the deviations of all data points from the best line (in accordance with this method) has the least value possible. By* **deviation,** *we mean the difference between the y-value of the line and the y-value for the point (of original data) for a particular value of x.*

Fig. 22.22

EXAMPLE 2 Deviation

In Fig. 22.22, the deviations of some of the points of Example 1 are shown. The point $(29, 63)$ (student A of Example 1) has a deviation of 8 from the indicated line in the figure. Thus, we square the value of this deviation to obtain 64. In order to find the equation of the straight line that best fits the given points, the method of least squares requires that the sum of all such squares be a minimum.

Therefore, in applying this method of least squares, it is necessary to use the equation of a straight line and the coordinates of the points of the data. The deviations of all of these data points are determined, and these values are then squared. It is then necessary to determine the constants for the slope m and the y-intercept b in the equation of a straight line $y = mx + b$ for which the sum of the squared values is a minimum. To do this requires certain methods of advanced mathematics.   ∎

Using the methods that are required from advanced mathematics, we show that *the equation of the* **least-squares line**

$$y = mx + b \qquad\qquad (22.9)$$

can be found by calculating the values of the slope m and the y-intercept b by using the formulas

$$m = \frac{n \sum xy - \left(\sum x \right)\left(\sum y \right)}{n \sum x^2 - \left(\sum x \right)^2} \qquad\qquad (22.10)$$

and

$$b = \frac{\left(\sum x^2 \right)\left(\sum y \right) - \left(\sum xy \right)\left(\sum x \right)}{n \sum x^2 - \left(\sum x \right)^2} \qquad\qquad (22.11)$$

In Eqs. (22.10) and (22.11), *the x's and y's are the values of the coordinates of the points in the given data,* and n is the number of points of data. We can reduce the calculational work in finding the values of m and b by noting that the denominators in Eqs. (22.10) and (22.11) are the same. Therefore, in using a calculator, the value of this denominator can be stored in memory.

EXAMPLE 3 Finding equation of least-squares line

Find the equation of the least-squares line for the points indicated in the following table. Graph the line and data points on the same graph.

x	1	2	3	4	5
y	3	6	6	8	12

We see from Eqs. (22.10) and (22.11) that we need the sums of x, y, xy, and x^2 in order to find m and b. Thus, we set up a table for these values, along with the necessary calculations, as follows:

Fig. 22.23

x	y	xy	x^2
1	3	3	1
2	6	12	4
3	6	18	9
4	8	32	16
5	12	60	25
sums → 15	35	125	55
↑ $\sum x$	↑ $\sum y$	↑ $\sum xy$	↑ $\sum x^2$

$n = 5$ (5 points)

$$m = \frac{5(125) - (15)(35)}{5(55) - (15)^2} = \frac{100}{50} = 2$$

$$b = \frac{(55)(35) - (125)(15)}{50} = \frac{50}{50} = 1$$

This means that the equation of the least-squares line is $y = 2x + 1$. This line and the data points are shown in Fig. 22.23. ∎

EXAMPLE 4 Finding equation of least-squares line

Find the least-squares line for the data of Example 1.

Here, the x-values will be the entrance-test scores and the y-values are the course averages.

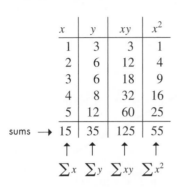

Fig. 22.24

x	y	xy	x^2
29	63	1,827	841
33	88	2,904	1089
22	77	1,694	484
17	67	1,139	289
26	70	1,820	676
37	93	3,441	1369
30	72	2,160	900
32	81	2,592	1024
23	47	1,081	529
30	74	2,220	900
279	732	20,878	8101

$n = 10$

$$m = \frac{10(20,878) - 279(732)}{10(8101) - 279^2} = 1.44$$

$$b = \frac{8101(732) - 20,878(279)}{10(8101) - 279^2} = 33.1$$

Thus, the equation of the least-squares line is $y = 1.44x + 33.1$. The line and data points are shown in Fig. 22.24. This line best fits the data, although the fit is obviously approximate. It can be used to predict the approximate course average that a student might be expected to attain, based on the entrance test. ∎

A graphing calculator can be used to find the slope and the intercept and to display the least-squares line. In the following example, we compare the values using Eqs. (22.9), (22.10), and (22.11) with the corresponding calculator displays.

EXAMPLE 5 Least-squares line – application

In a research project to determine the amount of a drug that remains in the bloodstream after a given dosage, the amounts y (in mg of drug/dL of blood) were recorded after t h, as shown in the following table. Find the least-squares line for these data, expressing y as a function of t. Sketch the graph of the line and data points.

The calculations are as follows:

■ See the chapter introduction.

(a)

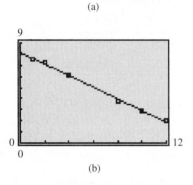

(b)

Fig. 22.26

t (h)	1.0	2.0	4.0	8.0	10.0	12.0
y (mg/dL)	7.6	7.2	6.1	3.8	2.9	2.0

Fig. 22.25

t	y	ty	t^2
1.0	7.6	7.6	1.0
2.0	7.2	14.4	4.0
4.0	6.1	24.4	16.0
8.0	3.8	30.4	64.0
10.0	2.9	29.0	100
12.0	2.0	24.0	144
37.0	29.6	129.8	329

$$n = 6$$

$$m = \frac{6(129.8) - 37.0(29.6)}{6(329) - 37.0^2} = -0.523$$

$$b = \frac{(329)(29.6) - (129.8)(37.0)}{6(329) - 37.0^2} = 8.16$$

The equation of the least-squares line is $y = -0.523t + 8.16$. The line and the data points are shown in Fig. 22.25. This line is useful in determining the effectiveness of the drug. It can also be used to determine when additional medication may be administered.

Using the linear regression feature of a calculator, Fig. 22.26(a) shows the display of the coefficients of the line $y = ax + b$ (note that the slope is a). The value of r is the *coefficient of correlation* (see Exercises 15–18). Figure 22.26(b) shows a calculator display of the points and the line. We see that they agree with Fig. 22.25. ■

EXERCISES 22.6

In Exercises 1–14, find the equation of the least-squares line for the given data. Graph the line and data points on the same graph.

1. In Example 3, replace the y-values with 3, 7, 9, 9, and 12. Then follow the instructions above.

2.

x	1	2	3	4	5	6	7
y	10	17	28	37	49	56	72

3.

x	20	26	30	38	48	60
y	160	145	135	120	100	90

4.

x	1	3	6	5	8	10	4	7	3	8
y	15	12	10	8	9	2	11	9	11	7

5. In Example 5, change the y (mg of drug/dL of blood) values to 8.7, 8.4, 7.7, 7.3, 5.7, 5.2. Then proceed to find y as a function of t, as in Example 5.

6. The velocity v (in m/s) of a falling object was found each second by use of an electronic device, as shown in the following table. Find v as a function of t.

t (s)	1.00	2.00	3.00	4.00	5.00	6.00	7.00
v (m/s)	9.70	19.5	29.5	39.4	49.2	58.9	68.6

7. In an electrical experiment, the following data were found for the values of current and voltage for a particular element of the circuit. Find the voltage V as a function of the current i.

Current (mA)	15.0	10.8	9.30	3.55	4.60
Voltage (V)	3.00	4.10	5.60	8.00	10.50

8. A particular muscle was tested for its speed of shortening as a function of the force applied to it. The results appear below. Find the speed as a function of the force.

Force (N)	60.0	44.2	37.3	24.2	19.5
Speed (m/s)	1.25	1.67	1.96	2.56	3.05

9. The altitude h (in m) of a rocket was measured at several positions at a horizontal distance x (in m) from the launch site, shown in the table. Find the least-squares line for h as a function of x.

x (m)	0	500	1000	1500	2000	2500
h (m)	0	1130	2250	3360	4500	5600

10. In testing an air-conditioning system, the temperature T in a building was measured during the afternoon hours with the results shown in the table. Find the least-squares line for T as a function of the time t from noon.

t (h)	0.0	1.0	2.0	3.0	4.0	5.0
T (°C)	20.5	20.6	20.9	21.3	21.7	22.0

11. The pressure p was measured along an oil pipeline at different distances from a reference point, with results as shown. Find the least-squares line for p as a function of x. Check the values and line with a graphing calculator.

x (ft)	0	100	200	300	400
p (lb/in.^{2})	650	630	605	590	570

12. The heat loss L per hour through various thicknesses of a particular type of insulation was measured as shown in the table. Find the least-squares line for L as a function of t. Check the values and line with a graphing calculator.

t (in.)	3.0	4.0	5.0	6.0	7.0
L (Btu)	5900	4800	3900	3100	2450

13. In an experiment on the photoelectric effect, the frequency of light being used was measured as a function of the stopping potential (the voltage just sufficient to stop the photoelectric effect) with the results given below. Find the least-squares line for V as a function of f. The frequency for $V = 0$ is known as the *threshold frequency*. From the graph determine the threshold frequency. Check the values and curve with a graphing calculator.

f (PHz)	0.550	0.605	0.660	0.735	0.805	0.880
V (V)	0.350	0.600	0.850	1.10	1.45	1.80

14. If gas is cooled under conditions of constant volume, it is noted that the pressure falls nearly proportionally as the temperature. If this were to happen until there was no pressure, the theoretical temperature for this case is referred to as *absolute zero*. In an elementary experiment, the following data were found for pressure and temperature under constant volume.

T (°C)	0.0	20	40	60	80	100
P (kPa)	133	143	153	162	172	183

Find the least-squares line for P as a function of T, and from the graph determine the value of absolute zero found in this experiment. Check the values and curve with a graphing calculator.

The linear coefficient of correlation, a measure of the relatedness of two variables, is defined by $r = m(s_x/s_y)$, where s_x and s_y are the standard deviations of the x-values and y-values, respectively. Due to its definition, the values of r lie in the range $-1 \le r \le 1$. If r is near 1, the correlation is considered good. For the values of r between -0.5 and $+0.5$, the correlation is poor. If r is near -1, the variables are said to be negatively correlated; that is, one increases as the other decreases. In Exercises 15–18, compute r for the given data.

15. Exercise 1   **16.** Exercise 2

17. Exercise 4   **18.** Example 1

22.7   Nonlinear Regression

**Nonlinear Regression •
Least-squares Curve •
Types of Curves on Calculator**

If the experimental points do not appear to be on a straight line, but we recognize them as being approximately on some other type of curve, the method of least squares can be extended to use on these other curves. For example, if the points are apparently on a parabola, we could use the function $y = a + bx^2$. To use the above method, we extend the least-squares line to

$$y = m[f(x)] + b \qquad (22.12)$$

Here, $f(x)$ must be calculated first, and then the problem can be treated as a least-squares line to find the values of m and b. Some of the functions $f(x)$ that may be considered for use are x^2, $1/x$, and 10^x.

Fig. 22.27

EXAMPLE 1   Fitting $y = mx^2 + b$ to set of points

Find the least-squares curve $y = mx^2 + b$ for the following points:

x	0	1	2	3	4	5
y	1	5	12	24	53	76

In using Eq. (22.12), $f(x) = x^2$. Our first step is to calculate values of x^2, and then we use x^2 as we used x in finding the equation of the least-squares line.

x	$f(x) = x^2$	y	x^2y	$(x^2)^2$
0	0	1	0	0
1	1	5	5	1
2	4	12	48	16
3	9	24	216	81
4	16	53	848	256
5	25	76	1900	625
	55	171	3017	979

$n = 6$

$$m = \frac{6(3017) - 55(171)}{6(979) - 55^2} = 3.05$$

$$b = \frac{(979)(171) - (3017)(55)}{6(979) - 55^2} = 0.52$$

Therefore, the required equation is $y = 3.05x^2 + 0.52$. The graph of this equation and the data points are shown in Fig. 22.27.   ■

EXAMPLE 2   Fitting $y = m(1/x) + b$ to set of points

In a physics experiment, the pressure p and volume V of a gas were measured at constant temperature. When the points were plotted, they were seen to approximate the hyperbola $y = c/x$. Find the least-squares approximation to the hyperbola $y = m(1/x) + b$ for the given data. See Fig. 22.28.

P(kPa)

Fig. 22.28

P (kPa)	V (cm^3)	$x\ (= V)$	$f(x) = \frac{1}{x}$	$y\ (= P)$	$\left(\frac{1}{x}\right)y$	$\left(\frac{1}{x}\right)^2$
120.0	21.0	21.0	0.0476190	120.0	5.7142857	0.0022676
99.2	25.0	25.0	0.0400000	99.2	3.9680000	0.0016000
81.3	31.8	31.8	0.0314465	81.3	2.5566038	0.0009889
60.6	41.1	41.1	0.0243309	60.6	1.4744526	0.0005920
42.7	60.1	60.1	0.0166389	42.7	0.7104825	0.0002769
			0.1600353	403.8	14.4238246	0.0057254

(*Calculator note:* The final digits for the values shown may vary depending on the calculator and how the values are used. Here, all individual values are shown with eight digits (rounded off), although more digits were used. The value of $1/x$ was found from the value of x, with the eight digits shown. However, the values of $(1/x)y$ and $(1/x)^2$ were found from the value of $1/x$, using the extra digits. The sums were found using the rounded-off values shown. However, since the data contain only three digits, any variation in the final digits for $1/x$, $(1/x)y$, or $(1/x)^2$, will not matter.)

$$m = \frac{5(14.4238246) - 0.1600353(403.8)}{5(0.0057254) - 0.1600353^2} = 2490$$

$$b = \frac{(0.0057254)(403.8) - (14.4238246)(0.1600353)}{5(0.0057254) - 0.1600353^2} = 1.2$$

The equation of the hyperbola $y = m(1/x) + b$ is

$$y = \frac{2490}{x} + 1.2$$

This hyperbola and data points are shown in Fig. 22.29.   ■

Fig. 22.29

Fig. 22.30

Fig. 22.31

Fig. 22.32

EXAMPLE 3 Finding $y = m(10^x) + b$ to set of points

It has been found experimentally that the tensile strength of brass (a copper-zinc alloy) increases (within certain limits) with the percent of zinc. The following table shows the values that have been found. See Fig. 22.30.

Tensile Strength (10^5 lb/in.^{2})	0.32	0.36	0.40	0.44	0.48
Percent of Zinc	0	5	13	22	34

Fit a curve of the form $y = m(10^x) + b$ to the data. Let x = tensile strength ($\times 10^5$) and y = percent of zinc.

x	$f(x) = 10^x$	y	$(10^x)y$	$(10^x)^2$
0.32	2.0892961	0	0.000000	4.3651583
0.36	2.2908677	5	11.454338	5.2480746
0.40	2.5118864	13	32.654524	6.3095734
0.44	2.7542287	22	60.593031	7.5857758
0.48	3.0199517	34	102.67836	9.1201084
	12.6662306	74	207.38025	32.6286905

(See the note on calculator use in Example 2.)

$$m = \frac{5(207.38025) - 12.6662306(74)}{5(32.6286905) - 12.6662306^2} = 36.8$$

$$b = \frac{32.6286905(74) - 207.38025(12.6662306)}{5(32.6286905) - 12.6662306^2} = -78.3$$

The equation of the curve is $y = 36.8(10^x) - 78.3$. It must be remembered that for practical purposes, y must be positive. The graph of the equation is shown in Fig. 22.31, with the solid portion denoting the meaningful part of the curve. The points of the data are also shown. ∎

As we noted and illustrated in the previous section, a graphing calculator can be used to determine the equation of the regression curve, and to display its graph. A typical calculator can fit a number of different types of equations to data, and the regression equations available on a typical calculator are as follows:

Linear:	$y = ax + b$
Quadratic:	$y = ax^2 + bx + c$
Cubic:	$y = ax^3 + bx^2 + cx + d$
Quartic:	$y = ax^4 + bx^3 + cx^2 + dx + e$
Logarithmic:	$y = a + b \ln x$
Exponential:	$y = ab^x$
Power:	$y = ax^b$
Logistic:	$y = \dfrac{c}{1 - ae^{-bx}}$
Sinusoidal:	$y = a \sin(bx + c) + d$

In comparing an equation obtained using Eq. (22.12) with that obtained using a calculator, we must note the difference in the form of the equation. For example, using the power form $y = ax^b$ on a calculator for Example 2, we get the calculator display shown in Fig. 22.32. The differences are due to (1) a power of -1 is assumed in our use of Eq. (22.12), and (2) there is a b-term in Eq. (22.12) and no b-term in the power form on the calculator.

EXERCISES 22.7

In the following exercises, find the equation of the indicated least-squares curve. Sketch the curve and plot the data points on the same graph.

1. In Example 1, replace the y-values with 2, 3, 10, 25, 44, and 65. Then follow the instructions above.

2. For the points in the following table, find the least-squares curve $y = m\sqrt{x} + b$.

x	0	4	8	12	16
y	1	9	11	14	15

3. In Example 2, change the V (volume of the gas) values to 19.9, 24.5, 29.4, 39.4, 56.0. Then find $y\,(=P)$ as a function of $x\,(=V)$, as in Example 2.

4. In Example 3, change the y (percent of zinc) values to 2, 8, 15, 23, 32. Then find y as a function of x (tensile strength), as in Example 3.

5. The following data were found for the distance y that an object rolled down an inclined plane in time t. Determine the least-squares curve $y = mt^2 + b$. Compare the equation with that using the *quadratic regression* feature on a graphing calculator.

t (s)	1.0	2.0	3.0	4.0	5.0
y (cm)	6.0	23	55	98	148

6. The increase in length y of a certain metallic rod was measured in relation to particular increases x in temperature. Find the least-squares curve $y = mx^2 + b$. Compare the equation with that using the *quadratic regression* feature of a graphing calculator.

x (°C)	50.0	100	150	200	250
y (cm)	1.00	4.40	9.40	16.4	24.0

7. The pressure p at which Freon, a refrigerant, vaporizes for temperature T is given in the following table. Find the least-squares curve $p = mT^2 + b$. Compare the equation with that using the *quadratic regression* feature of a graphing calculator.

T (°F)	0	20	40	60	80
p (lb/in.^{2})	23	35	49	68	88

8. A fraction f of annual hot-water loads at a certain facility are heated by solar energy. The fractions f for certain values of the collector area A are given in the following table. Find the least-squares curve $f = m\sqrt{A} + b$. Compare the equation with that using the *power regression* feature of a graphing calculator.

A (m^2)	0	12	27	56	90
f	0.0	0.2	0.4	0.6	0.8

9. The makers of a special blend of coffee found that the demand for the coffee depended on the price charged. The price P per pound and the monthly sales S are shown in the following table. Find the least-squares curve $P = m(1/S) + b$.

S (thousands)	240	305	420	480	560
P (dollars)	5.60	4.40	3.20	2.80	2.40

10. The resonant frequency f of an electric circuit containing a 4-μF capacitor was measured as a function of an inductance L in the circuit. The following data were found. Find the least-squares curve $f = m(1/\sqrt{L}) + b$.

L (H)	1.0	2.0	4.0	6.0	9.0
f (Hz)	490	360	250	200	170

11. The displacement y of an object at the end of a spring at given times t is shown in the following table. Find the least-squares curve $y = me^{-t} + b$.

t (s)	0.0	0.5	1.0	1.5	2.0	3.0
y (cm)	6.1	3.8	2.3	1.3	0.7	0.3

12. The average daily temperatures T (in °F) for each month in Minneapolis (National Weather Service records) are given in the following table.

t	J	F	M	A	M	J	J	A	S	O	N	D
T(°F)	11	18	29	46	57	68	73	71	61	50	33	19

Find the least-squares curve $T = m\cos[\frac{\pi}{6}(t - 0.5)] + b$. Assume the average temperature is for the 15th of each month. Then the values of t (in months) are $0.5, 1.5, \ldots, 11.5$. (The fit is fairly good.) Compare the equation using the *sinusoidal regression* feature of a graphing calculator. What are the main reasons for the differences in the equations?

CHAPTER 22   EQUATIONS

Arithmetic mean	$\bar{x} = \dfrac{x_1 f_1 + x_2 f_2 + \cdots + x_n f_n}{f_1 + f_2 + \cdots + f_n}$	**(22.1)**
Standard deviation	$s = \sqrt{\dfrac{\sum (x - \bar{x})^2}{n - 1}}$	**(22.2)**
	$s = \sqrt{\dfrac{n \left(\sum x^2 \right) - \left(\sum x \right)^2}{n(n - 1)}}$	**(22.3)**
Normal distribution	$y = \dfrac{e^{-(x - \mu)^2 / 2\sigma^2}}{\sigma \sqrt{2\pi}}$	**(22.4)**
Standard normal distribution	$y = \dfrac{1}{\sqrt{2\pi}} e^{-x^2 / 2}$	**(22.5)**
Standard (z) score	$z = \dfrac{x - \mu}{\sigma}$	**(22.6)**
Standard error of $\bar{x}$	$\sigma_x = \dfrac{\sigma}{\sqrt{n}}$	**(22.7)**
Standard error of s	$\sigma_s = \dfrac{\sigma}{\sqrt{2n}}$	**(22.8)**
Least-squares lines	$y = mx + b$	**(22.9)**
	$m = \dfrac{n \sum xy - \left(\sum x \right) \left(\sum y \right)}{n \sum x^2 - \left(\sum x \right)^2}$	**(22.10)**
	$b = \dfrac{\left(\sum x^2 \right) \left(\sum y \right) - \left(\sum xy \right) \left(\sum x \right)}{n \sum x^2 - \left(\sum x \right)^2}$	**(22.11)**
Nonlinear curves	$y = m[f(x)] + b$	**(22.12)**

CHAPTER 22   REVIEW EXERCISES

In Exercises 1–10, use the following data. An airline's records showed that the percent of on-time flights each day for a 20-day period was as follows:

    72, 75, 76, 70, 77, 73, 80, 75, 82, 85,
    77, 78, 74, 86, 72, 77, 67, 78, 69, 80

1. Determine the median.   2. Determine the mode.

3. Determine the mean.

4. Determine the standard deviation.

5. Construct a frequency distribution table with five classes and a lowest class limit of 67.

6. Draw a frequency polygon for the data in Exercise 5.

7. Draw a histogram for the data in Exercise 5.

8. Construct a relative frequency table for the data in Exercise 5.

9. Construct a cumulative frequency table for the data of Exercise 5.

10. Draw an ogive for the data of Exercise 5.

In Exercises 11–16, use the following data: An important property of oil is its coefficient of viscosity, which gives a measure of how well it flows. In order to determine the viscosity of a certain motor oil, a refinery took samples from 12 different storage tanks and tested them at 50°C. The results (in pascal-seconds) were 0.24, 0.28, 0.29, 0.26, 0.27, 0.26, 0.25, 0.27, 0.28, 0.26, 0.26, 0.25.

11. Find the mean.

12. Find the median.

13. Find the standard deviation.

14. Draw a histogram.

15. Draw a frequency polygon.

16. Determine the mode.

In Exercises 17–24, use the following data: A sample of wind generators was tested for power output when the wind speed was 30 km/h. The following table gives the class marks of the powers produced and the number of generators in each class.

Power (W)	650	660	670	680	690
No. Generators	3	2	7	12	27

Power (W)	700	710	720	730
No. Generators	34	15	16	5

17. Find the median.

18. Find the mean.

19. Find the mode.

20. Draw a histogram.

21. Find the standard deviation.

22. Draw a frequency polygon.

23. Make a cumulative frequency table.

24. Draw an ogive.

In Exercises 25–28, use the following data: A Geiger counter records the presence of high-energy nuclear particles. Even though no apparent radioactive source is present, some particles will be recorded. These are primarily cosmic rays, which are caused by very high-energy particles from outer space. In an experiment to measure the amount of cosmic radiation, the number of counts were recorded during 200 5-s intervals. The following table gives the number of counts and the number of 5-s intervals having this number of counts. Draw a frequency curve for these data.

Counts	0	1	2	3	4	5	6	7	8	9	10
Intervals	3	10	25	45	29	39	26	11	7	2	3

25. Find the median.

26. Find the mean.

27. Draw a histogram.

28. Make a relative frequency table.

In Exercises 29–32, use the following data: Police radar on Interstate 80 recorded the speeds of 110 cars in a 65 mi/h zone. The following table shows the class marks of the speeds recorded and the number of cars in each class.

Speed (mi/h)	40	45	50	55	60	65	70	75	80	85
No. cars	3	4	4	5	8	22	48	10	4	2

29. Find the mean.

30. Find the median.

31. Find the standard deviation.

32. Draw an ogive.

In Exercises 33 and 34, use the following information: A company that makes electric light bulbs tests 500 bulbs each day for defects.

The number of defective bulbs, along with the proportion of defective bulbs for 20 days, is shown in the following table.

Day	Number Defective	Proportion Defective
1	23	0.046
2	31	0.062
3	19	0.038
4	27	0.054
5	29	0.058
6	39	0.078
7	26	0.052
8	17	0.034
9	28	0.056
10	33	0.066
11	22	0.044
12	29	0.058
13	20	0.040
14	35	0.070
15	21	0.042
16	32	0.064
17	25	0.050
18	23	0.046
19	29	0.058
20	32	0.064
Total	540	

33. For a p chart, find the values of the central line, UCL, and LCL.

34. Plot a p chart.

In Exercises 35 and 36, use the following information: Five ball bearings are taken from the production line every 15 min and their diameters are measured. The diameters of the sample ball bearings for 16 successive subgroups are given in the following table.

Subgroup	Diameters (mm) of Five Ball Bearings				
1	4.98	4.92	5.02	4.91	4.93
2	5.03	5.01	4.94	5.06	5.07
3	5.05	5.03	5.00	5.02	4.96
4	5.01	4.92	4.91	4.99	5.03
5	4.92	4.97	5.02	4.95	4.94
6	5.02	4.95	5.01	5.07	5.15
7	4.93	5.03	5.02	4.96	4.99
8	4.85	4.91	4.88	4.92	4.90
9	5.02	4.95	5.06	5.04	5.06
10	4.98	4.98	4.93	5.01	5.00
11	4.90	4.97	4.93	5.05	5.02
12	5.03	5.05	4.92	5.03	4.98
13	4.90	4.96	5.00	5.02	4.97
14	5.09	5.04	5.05	5.02	4.97
15	4.88	5.00	5.02	4.97	4.94
16	5.02	5.09	5.03	4.99	5.03

35. Plot an $\bar{x}$ chart.

36. Plot an R chart.

In Exercises 37–40, use the following data: After analyzing data for a long period of time, it was determined that samples of 500 readings of an organic pollutant for an area are distributed normally. For this pollutant, $\mu = 2.20\ \mu g/m^3$ and $\sigma = 0.50\ \mu g/m^3$.

37. In a sample, how many readings are between $1.50\ \mu g/m^3$ and $2.50\ \mu g/m^3$?

38. In a sample, how many readings are between $2.50\ \mu g/m^3$ and $3.50\ \mu g/m^3$?

39. In a sample, how many readings are above $1.00\ \mu g/m^3$?

40. In a sample, how many readings are below $2.00\ \mu g/m^3$?

In Exercises 41–50, find the indicated least-squares curve. Sketch the curve and data points on the same graph.

41. In a certain experiment, the resistance R of a certain resistor was measured as a function of the temperature T. The data found are shown in the following table. Find the least-squares line, expressing R as a function of T.

T (°C)	0.0	20.0	40.0	60.0	80.0	100
R (Ω)	25.0	26.8	28.9	31.2	32.8	34.7

42. An air-pollution monitoring station took samples of air each hour during the later morning hours and tested each sample for the number n of parts per million (ppm) of carbon monoxide. The results are shown in the table, where t is the number of hours after 6 A.M. Find the least-squares line for n as a function of t.

t (h)	0.0	1.0	2.0	3.0	4.0	5.0	6.0
n (ppm)	8.0	8.2	8.8	9.5	9.7	10.0	10.7

43. The *Mach number* of a moving object is the ratio of its speed to the speed of sound ($740\ mi/h$). The following table shows the speed s of a jet aircraft, in terms of Mach numbers, and the time t after it starts to accelerate. Find the least-squares line of s as a function of t. Compare the equation with that using the *linear regression* feature of a graphing calculator.

t (min)	0.00	0.60	1.20	1.80	2.40	3.00
s (Mach number)	0.88	0.97	1.03	1.11	1.19	1.25

44. In an experiment to determine the relation between the load y on a spring and the length x of the spring, the following data were found. Find the least-squares line that expresses y as a function of x. Compare the equation with that using the *linear regression* feature of a graphing calculator.

Load (lb)	0.0	1.0	2.0	3.0	4.0	5.0
Length (in.)	10.0	11.2	12.3	13.4	14.6	15.9

45. The distance s of a missile above the ground at time t after being released from a plane is given by the following table. Find the least-squares curve of the form $s = mt^2 + b$ for these data. Compare the equation with that using the *quadratic regression* feature of a graphing calculator.

t (s)	0.0	3.0	6.0	9.0	12.0	15.0	18.0
s (m)	3000	2960	2820	2600	2290	1900	1410

46. In an elementary experiment that measured the wavelength L of sound as a function of the frequency f, the following results were obtained.

Frequency (Hz)	240	320	400	480	560
Wavelength (cm)	140	107	81.0	70.0	60.0

Find the least-squares curve of the form $L = m(1/f) + b$ for these data.

47. Measurements were made on the current i (in A) in an electric circuit as a function of the time t (in s). The circuit contained a resistance of $5.00\ \Omega$ and an inductance of $10.0\ H$. The following data were found.

$t(s)$	0.00	2.00	4.00	6.00	8.00
$i(A)$	0.00	2.52	3.45	3.80	3.92

Find the least-squares curve $i = m(e^{-0.500t}) + b$ for these data. From the equation, determine the value of the current as t approaches infinity.

48. The power P (in W) generated by a wind turbine was measured for various wind velocities v (in mi/h), as shown in the following table.

v (mi/h)	10	15	20	25	30	40
P (W)	75	250	600	1200	2100	4800

Find the least-squares curve of the form $P = mv^3 + b$ for these data.

49. The vertical distance y of the cable of a suspension bridge above the surface of the bridge is measured at a horizontal distance x along the bridge from its center. See Fig. 22.33. The results are as follows:

x (m)	0	100	200	300	400	500
y (m)	15	17	23	33	47	65

Plot these points and choose an appropriate function $f(x)$ for $y = m[f(x)] + b$. Then find the equation of the least-squares curve.

Fig. 22.33

50. After being heated, the temperature T of an insulated liquid is measured at times t as follows:

t (h)	0	2	4	6	8	10
T (°C)	100	85	72	63	54	48

Plot these points and choose an appropriate function $f(x)$ for $y = m[f(x)] + b$. Then find the equation of the least-squares curve.

In Exercises 51–54, use the regression feature of a graphing calculator to solve the given problems related to the following data. Using aerial photography, the area A (in km^2) of an oil spill as a function of the time t (in h) after the spill was found to be as follows:

t (h)	1.0	2.0	4.0	6.0	8.0	10.0
A (km^2)	1.4	2.5	4.7	6.8	8.8	10.2

51. Find the linear equation $y = ax + b$ to fit these data.

52. Find the quadratic equation $y = ax^2 + bx + c$ to fit these data.

53. Find the power equation $y = ax^b$ to fit these data.

54. Compare the values of the coefficient of correlation r, to determine whether the linear equation or power equation seems to fit the data better.

In Exercises 55–60, solve the given problems.

55. With 30.5% of the area under the normal curve between $z_1 = 0.5$ and z_2, to the right of z_1, find z_2.

56. With 79.8% of the area under the normal curve between z_1 and $z_2 = 2.1$, find z_1.

57. The nth root of the product of n positive numbers is the *geometric mean* of the numbers. Find the geometric mean of the carbon monoxide readings in Exercise 42.

58. One use of the geometric mean (see Exercise 57) is to find an average ratio. By finding the geometric mean, find the average Mach number for the jet in Exercise 43.

59. Show that Eqs. (22.10) and (22.11) satisfy the equation $\bar{y} = m\bar{x} + b$.

60. Given that $\sum(x - \bar{x})^2 = \sum x^2 - n\bar{x}^2$, derive Eq. (22.3) from Eq. (22.2).

Writing Exercise

61. A research institute is planning a study of the effect of education on the income of workers. Write two or three paragraphs explaining what data should be collected and which of the measures discussed in this chapter would be useful in analyzing the data.

CHAPTER 22 PRACTICE TEST

In Problems 1–3, use the following set of numbers.

5, 6, 1, 4, 9, 5, 7, 3, 8, 10, 5, 8, 4, 9, 6

1. Find the median.

2. Find the mode.

3. Draw a histogram with five classes and the lowest class limit at 1.

In Problems 4–8, use the following data: Two machine parts are considered satisfactorily assembled if their total thickness (to the nearest 0.01 in.) is between or equal to 0.92 in. and 0.94 in. One hundred assemblies are tested, and the class mark of the thicknesses and the number of assemblies in each class are given in the following table.

Total Thickness (in.)	0.90	0.91	0.92	0.93	0.94	0.95	0.96
Number	3	9	31	38	12	5	2

4. Find the mean.

5. Find the standard deviation.

6. Draw a frequency polygon.

7. Make a relative frequency table.

8. Draw an ogive (less than).

9. For a set of values that are normally distributed, what percent of them is below the value (greater than the mean) for which the z-score is 0.2257?

10. The machine-part assemblies in Problems 4–8 were tested in groups of five each hour for 20 hours. Explain, in general, how to use the data from the test subgroups to plot an R chart.

11. Find the equation of the least-squares line for the points indicated in the following table. Graph the line and data points on the same graph.

x	1	3	5	7	9
y	5	11	17	20	27

12. The period of a pendulum as a function of its length was measured, giving the following results:

Length (ft)	1.00	3.00	5.00	7.00	9.00
Period (s)	1.10	1.90	2.50	2.90	3.30

Find the equation of the least-squares curve of the form $y = m\sqrt{x} + b$, which expresses the period as a function of the length.

APPENDIX A
Solving Word Problems

Drill-type problems require a working knowledge of the methods presented, and some algebraic steps to change the algebraic form may be required to complete the solution. *Word problems,* however, require a proper interpretation of the statement of the problem before they can be put in a form for solution.

We have to put word problems in symbolic form in order to solve them, and it is this procedure that most students find difficult. Because such problems require more than going through a certain routine, they demand more analysis and appear to be more difficult. Among the reasons for the student's difficulty at solving word problems are (1) unsuccessful previous attempts at solving word problems, leading the student to believe that all word problems are "impossible," (2) a poorly organized approach to the solution, and (3) failure to read the problem carefully, thereby having an improper and incomplete interpretation of the statement given. These can be overcome with proper attitude and care.

NOTE ▶
A specific procedure for solving word problems is shown on page 41, when word problems are first covered in our study of algebra. There are over 120 completely worked examples of word problems (as well as numerous other examples that show a similar analysis) throughout this text, illustrating proper interpretations and approaches to these problems.

RISERS

The procedure shown on page 41 is similar to that used by most instructors and texts. One of the variations that a number of instructors use is called *RISERS.* This is a word formed from the first letters (an *acronym*) of the words that outline the procedure. These are *Read, Imagine, Sketch, Equate, Relate,* and *Solve.* We now briefly outline this procedure here.

Read the statement of the problem carefully.

Imagine. Take time to get a mental image of the situation described.

Sketch a figure.

Equate, *on the sketch,* the known and unknown quantities.

Relate the known and unknown quantities with an equation.

Solve the equation.

There are problems where a *sketch* may simply be words and numbers placed so that we may properly *equate* the known and unknown quantities. For example, in Example 2 on page 42, the *sketch, equate,* and *relate* steps might look like this:

sketch		34 resistors	56 Ω
	1.5-Ω resistors	2.0-Ω resistors	
equate	x	$34 - x$	
resistance	$1.5x$	$2.0(34 - x)$	56
relate	$1.5x$	$+\ 2.0(34 - x) = 56$	

NOTE ▶
If you follow the method on page 41, or this RISERS variation, or any appropriate step-by-step method, and *write out the solution neatly,* you will find that word problems lend themselves to solution more readily than you have previously found.

A.1

Units of Measurement; The Metric System

B.1 Introduction

Most scientific and technical calculations involve numbers that represent a measurement or count of a specific physical quantity. *Such numbers are called* **denominate numbers,** *and associated with these denominate numbers are* **units of measurement.** For calculations and results to be meaningful, we must know these units. For example, if we measure the length of an object to be 12, we must know whether it is being measured in feet, yards, or some other specified unit of length.

Certain universally accepted **base units** *are used to measure fundamental quantities.* The units for numerous other quantities are expressed in terms of the base units. Fundamental quantities for which base units are defined are (1) length, (2) mass or force, depending on the system of units being used, (3) time, (4) electric current, (5) temperature, (6) amount of substance, and (7) luminous intensity. *Other units, referred to as* **derived units,** *are expressible in terms of the units for these quantities.*

Even though all other quantities can be expressed in terms of the fundamental ones, many have units that are given a specified name. This is done primarily for those quantities that are used very commonly, although it is not done for all such quantities. For example, the volt is defined as a meter^2-kilogram/second^3-ampere, which is in terms of (a unit of length)^2(a unit of mass)/(a unit of time)^3(a unit of electric current). The unit for acceleration has no special name and is left in terms of the base units, for example, feet/second^2. For convenience, special symbols are usually used to designate units. The units for acceleration would be written as ft/s^2.

Two basic systems of units, the **SI metric system** and the **United States Customary system,** traditionally known as the *British system*, are in use today. The SI metric system is used worldwide in all scientific work and in businesses related to international commerce. Most countries in the world use the SI metric system for all measurements used in business, industry, and everyday activities.

In the United States, both systems are used. In the 1970s and 1980s, the U.S. Congress passed legislation defining the SI metric system as the preferred system for trade and commerce. It required federal agencies to use the metric system in their business-related activities. In this period there was a trend to converting units to the metric system, but in the 1990s this trend stopped for the most part. Today, for measurements used in daily life, metric units are used for some, but the U.S. Customary system is used for most of them. In Great Britain there was also a trend in the 1970s and 1980s of conversion to metric units, followed by a period of reverting to traditional units for daily use.

Therefore, both systems are of importance. For that reason, where units are used, some exercises and examples in this text have metric units and others have U.S. Customary units. Technicians and engineers must be able to use both systems.

SI Metric System

■ The abbreviation SI, used throughout the world, comes from the French *Le Système International d'Unités*, the official name of the metric system.

■ The failure to convert units caused the $125,000,000 Mars *Climate Orbiter* to fly too close to Mars and break up in the Martian atmosphere in September 1999. A spacecraft team submitted force data in pounds, but the mission controllers assumed the data were in newtons. The system for checking data did not note the change in units. (1 lb = 4.448 N)

Although more than one system has been developed in which metric units are used, the system now accepted as the metric system is the **International System of Units (SI).** This was established in 1960 and uses some different definitions for base units from the previously developed metric units. However, the measurement of the base units in the SI system is more accessible, and the differences are very slight. *Therefore, when we refer to the metric system, we are using SI units.*

As we have stated, in each system the base units are specified, and all other units are then expressible in terms of these base units. Table B.1 on the next page lists the fundamental quantities, as well as many other commonly used quantities, along with their symbols and the names of units used to represent each quantity shown.

In the U.S. Customary system, the base unit of length is the *foot,* and that of force is the *pound.* In the metric system, the base unit of length is the *meter,* and that of mass is the *kilogram.* Here, we see a difference in the definition of the systems that causes some difficulty when units are converted from one system to the other. That is, a base unit of the U.S. Customary system is a unit of force, and a base unit in the metric system is a unit of mass.

The distinction between mass and force is very significant in physics, and the weight of an object is the force with which it is attracted to the earth. Weight, which is therefore a force, is different from mass, which is a measure of the inertia an object exhibits. Although they are different quantities, mass and weight are, however, very closely related. In fact, the weight of an object equals its mass multiplied by the acceleration due to gravity. Near the surface of the earth, the acceleration due to gravity is nearly constant, although it decreases as the distance from the earth increases. Therefore, near the surface of the earth, the weight of an object is directly proportional to its mass. However, at great distances from the earth, the weight of an object will be zero, whereas its mass does not change.

Since force and mass are different, it is not strictly correct to convert pounds to kilograms. However, since the pound is the base unit in the U.S. Customary system and the kilogram is the base unit in the metric system, at the earth's surface 1 kg corresponds to 2.21 lb, in the sense that the force of gravity on a 1-kg mass is 2.21 lb.

When designating units for weight, we use pounds in the U.S. Customary system. In the metric system, although kilograms are used for weight, it is preferable to specify the mass of an object in kilograms. The force of gravity on an object is designated in newtons, and the use of the term *weight* is avoided unless its meaning is completely clear.

As for the other fundamental quantities, both systems use the *second* as the base unit of time. In the SI system, the *ampere* is defined as the base unit of electric current, and this can also be used in the U.S. Customary system. As for temperature, *degrees Fahrenheit* is used with the U.S. Customary system, and *degrees Celsius* (formerly centigrade) is used with the metric system (actually, the *kelvin* is defined as the base unit, where the temperature in kelvins is the temperature in degrees Celsius plus 273.16). In the SI system, the base unit for the amount of a substance is the *mole,* and the base unit of luminous intensity is the *candela.* These last two are of limited importance to our use in this text.

The metric system was first introduced in France in 1791, soon after the French Revolution. Originally the meter was defined as one ten-millionth of the surface distance from the North Pole to the equator, and has since had changes in the official definition. Today, it is defined as the distance traveled by light (in a vacuum) in 1/299,792,458 s, which effectively *defines* the speed of light to be exactly 299,792,458 m/s. The second is now defined as the time required for 9,192,631,770 cycles of radiation between the two lowest energy states of the cesium-133 atom.

Table B.1 Quantities and Their Associated Units

Quantity	Quantity Symbol	U.S. Customary Name	Symbol	Metric (SI) Name	Symbol	In Terms of Other SI Units
Length	s	foot	ft	**meter**	m	
Mass	m	slug		**kilogram**	kg	
Force	F	pound	lb	newton	N	$m \cdot kg/s^2$
Time	t	second	s	**second**	s	
Area	A		ft^2		m^2	
Volume	V		ft^3		m^3	
Capacity	V	gallon	gal	liter	L	$(1\,L = 1\,dm^3)$
Velocity	v		ft/s		m/s	
Acceleration	a		ft/s^2		m/s^2	
Density	d, ρ		lb/ft^3		kg/m^3	
Pressure	p		lb/ft^2	pascal	Pa	N/m^2
Energy, work	E, W		$ft \cdot lb$	joule	J	$N \cdot m$
Power	P	horsepower	hp	watt	W	J/s
Period	T		s		s	
Frequency	f		1/s	hertz	Hz	1/s
Angle	θ	radian	rad	radian	rad	
Electric current	I, i	ampere	A	**ampere**	A	
Electric charge	q	coulomb	C	coulomb	C	$A \cdot s$
Electric potential	V, E	volt	V	volt	V	$J/(A \cdot s)$
Capacitance	C	farad	F	farad	F	s/Ω
Inductance	L	henry	H	henry	H	$\Omega \cdot s$
Resistance	R	ohm	Ω	ohm	Ω	V/A
Thermodynamic temperature	T			**kelvin**	K	(temp. interval
Temperature	T	degrees Fahrenheit	°F	degrees Celsius	°C	1°C = 1 K)
Quantity of heat	Q	British thermal unit	Btu	joule	J	
Amount of substance	n			**mole**	mol	
Luminous intensity	I	candlepower	cp	**candela**	cd	

Special Notes:

1. The SI base units are shown in boldface type.

2. The unit symbols shown above are those that are used in the text. Many of them were adopted with the adoption of the SI system. This means, for example, that we use s rather than sec for seconds and A rather than amp for amperes. Also, other units, such as volt, are not spelled out, a common practice in the past. When a given unit is used with both systems, we use the SI symbol for the unit.

3. The liter and degree Celsius are not actually SI units. However, they are recognized for use with the SI system due to their practical importance. Also, the symbol for liter has several variations. Presently L is recognized for use in the United States and Canada, I is recognized by the International Committee of Weights and Measures, and ℓ is also recognized for use in several countries.

4. Other units of time, along with their symbols, which are recognized for use with the SI system and are used in this text, are minute, min; hour, h; day, d.

5. Many additional specialized units are used with the SI system. However, most of those that appear in this text are shown in the table. A few of the specialized units are noted when used in the text. One which is frequently used is that for revolution, r.

6. Other common U.S. units used in the text are inch, in.; yard, yd; mile, mi; ounce, oz.; ton; quart, qt; acre.

7. There are a number of units that were used with the metric system prior to the development of the SI system. However, many of these are not to be used with the SI system. Among those that were commonly used are the dyne, erg, and calorie.

Due to greatly varying sizes of certain quantities, the metric system employs certain prefixes to units to denote different orders of magnitude. These prefixes, with their meanings and symbols, are shown in Table B.2.

Table B.2 Metric Prefixes

Prefix	Factor	Symbol	Prefix	Factor	Symbol
exa	10^{18}	E	deci	10^{-1}	d
peta	10^{15}	P	centi	10^{-2}	c
tera	10^{12}	T	milli	10^{-3}	m
giga	10^{9}	G	micro	10^{-6}	μ
mega	10^{6}	M	nano	10^{-9}	n
kilo	10^{3}	k	pico	10^{-12}	p
hecto	10^{2}	h	femto	10^{-15}	f
deca	10^{1}	da	atto	10^{-18}	a

EXAMPLE 1 Metric units and prefixes

Some commonly used units from Table B.2, with appropriate prefixes, along with their symbols and meanings, are shown below.

Unit	Symbol	Meaning	Unit	Symbol	Meaning
megohm	MΩ	10^{6} ohms	milligram	mg	10^{-3} gram
kilometer	km	10^{3} meters	microfarad	μF	10^{-6} farad
centimeter	cm	10^{-2} meter	nanosecond	ns	10^{-9} second

(Mega is shortened to meg when used with "ohm.")   ■

When designating units of area or volume, where square or cubic units are used, we use exponents in the designation. For example, we use m^2 rather than sq m and in.^{3} rather than cu in.

B.2 Reductions and Conversions

Reduction • Conversion • Algebraic Operations with Units • Reduction and Conversion Factors in Fractions Equal to 1

When using denominate numbers, it may be necessary to change from one set of units to another. *A change within a given system is called a* **reduction,** and *a change from one system to another is called a* **conversion.** Table B.3 gives some basic conversion factors. Some calculators are programmed to do conversions.

Table B.3 Conversion Factors

1 in. = 2.54 cm (exact)	1 ft^3 = 28.32 L	1 lb = 453.6 g	1 Btu = 778.0 ft · lb
1 km = 0.6214 mi	1 L = 1.057 qt	1 kg = 2.205 lb	1 hp = 550 ft · lb/s (exact)
		1 lb = 4.448 N	1 hp = 746.0 W

The advantages of the metric system are evident. Reductions within the system are made by using powers of 10, which amounts to moving the decimal point. Reductions in the U.S. Customary system are made by using many different multiples. Comparisons and changes within the metric system are much simpler.

NOTE ▶

To change a given number of one set of units into another set of units, *we perform algebraic operations with units in the same manner as we do with any algebraic symbol.* Consider the following example.

EXAMPLE 1 Algebraic operations with units

If we have a denominate number representing feet per second to be multiplied by another denominate number representing seconds per minute, as far as the units are concerned, we have

$$\frac{\text{ft}}{\text{s}} \times \frac{\text{s}}{\text{min}} = \frac{\text{ft} \times \cancel{\text{s}}}{\cancel{\text{s}} \times \text{min}} = \frac{\text{ft}}{\text{min}}$$

This means that the final result would be in feet per minute. ∎

In changing a number of one set of units to another set of units, we use reduction and conversion factors and the principle illustrated in Example 1. The convenient way to use the values in the tables is in the form of fractions. Since the given values are equal to each other, their quotient is 1. For example, since 1 in. = 2.54 cm,

NOTE ▶

$$\frac{1 \text{ in.}}{2.54 \text{ cm}} = 1 \quad \text{or} \quad \frac{2.54 \text{ cm}}{1 \text{ in.}} = 1$$

since each fraction represents the division of a certain length by itself. Multiplying a quantity by 1 does not change its value. The following examples illustrate the reduction and conversion of units.

EXAMPLE 2 Metric system reduction

Reduce 20 kg to milligrams:

$$20 \text{ kg} = 20 \cancel{\text{kg}} \left(\frac{10^3 \cancel{\text{g}}}{1 \cancel{\text{kg}}}\right)\left(\frac{10^3 \text{ mg}}{1 \cancel{\text{g}}}\right)$$
$$= 20 \times 10^6 \text{ mg} = 2.0 \times 10^7 \text{ mg}$$

We note that this result is found essentially by moving the decimal point three places when changing from kilograms to grams and another three places when changing from grams to milligrams. ∎

EXAMPLE 3 U.S. Customary system reduction

Reduce 30 mi/h to feet per second:

$$30 \frac{\text{mi}}{\text{h}} = \left(30 \frac{\cancel{\text{mi}}}{\cancel{\text{h}}}\right)\left(\frac{5280 \text{ ft}}{1 \cancel{\text{mi}}}\right)\left(\frac{1 \cancel{\text{h}}}{60 \cancel{\text{min}}}\right)\left(\frac{1 \cancel{\text{min}}}{60 \text{ s}}\right) = \frac{(30)(5280) \text{ ft}}{(60)(60) \text{ s}} = 44 \frac{\text{ft}}{\text{s}}$$

The only units remaining after the division are those required. ∎

EXAMPLE 4 Metric system reduction

Reduce 575 g/cm³ to kilograms per cubic meter:

$$575 \frac{\text{g}}{\text{cm}^3} = \left(575 \frac{\text{g}}{\text{cm}^3}\right)\left(\frac{100 \text{ cm}}{1 \text{ m}}\right)^3 \left(\frac{1 \text{ kg}}{1000 \text{ g}}\right)$$
$$= \left(575 \frac{\cancel{\text{g}}}{\cancel{\text{cm}^3}}\right)\left(\frac{10^6 \cancel{\text{cm}^3}}{1 \text{ m}^3}\right)\left(\frac{1 \text{ kg}}{10^3 \cancel{\text{g}}}\right)$$
$$= 575 \times 10^3 \frac{\text{kg}}{\text{m}^3} = 5.75 \times 10^5 \frac{\text{kg}}{\text{m}^3}$$

∎

EXAMPLE 5 U.S. system to metric system conversion

Convert $62.8 \, \text{lb/in.}^2$ to newtons per square meter:

$$62.8 \frac{\text{lb}}{\text{in.}^2} = \left(62.8 \frac{\text{lb}}{\text{in.}^2}\right)\left(\frac{4.448 \, \text{N}}{1 \, \text{lb}}\right)\left(\frac{1 \, \text{in.}}{2.54 \, \text{cm}}\right)^2\left(\frac{100 \, \text{cm}}{1 \, \text{m}}\right)^2$$

$$= \left(62.8 \frac{\cancel{\text{lb}}}{\cancel{\text{in.}^2}}\right)\left(\frac{4.448 \, \text{N}}{1 \, \cancel{\text{lb}}}\right)\left(\frac{1 \, \cancel{\text{in.}^2}}{2.54^2 \, \cancel{\text{cm}^2}}\right)\left(\frac{10^4 \, \cancel{\text{cm}^2}}{1 \, \text{m}^2}\right)$$

$$= \frac{(62.8)(4.448)(10^4)}{(2.54)^2} \frac{\text{N}}{\text{m}^2} = 4.33 \times 10^5 \frac{\text{N}}{\text{m}^2}$$ ∎

EXERCISES B

In Exercises 1–4, give the symbol and the meaning for the given unit.

1. megahertz **2.** kilowatt **3.** millimeter **4.** picosecond

In Exercises 5–8, give the name and the meaning for the units whose symbols are given.

5. kV **6.** GΩ **7.** mA **8.** pF

In Exercises 9–52, make the indicated reductions and conversions.

9. 1 km to centimeters
10. 1 kV to millivolts
11. 1 mi to inches
12. 1 gal to pints
13. 5.25 in. to centimeters
14. 65.0 kg to pound
15. 15.7 qt to liters
16. 185 km to miles
17. $1 \, \text{ft}^2$ to square inches
18. $1 \, \text{yd}^3$ to cubic feet
19. $250 \, \text{mm}^2$ to square meters
20. $0.125 \, \text{M}\Omega$ to milliohms
21. 45.0 lb to grams
22. 0.360 in. to meters
23. $829 \, \text{in.}^3$ to liters
24. 0.0680 kL to cubit feet
25. 22.5 hp to newton centimeters per second
26. 875 Btu to joules
27. 25 kW·h to megajoules
28. 30 ns to minutes
29. 7.50 W to horsepower
30. 326 mL to quarts

31. Determine how many minutes there are in a microcentury (not a commonly used unit). Compare with the length of a standard college class.

32. Determine how many decades there are in 1 Ts.

33. Determine how many meters light travels in one year. (See the last paragraph on page A.3)

34. Determine the speed (in km/h) of the earth moving around the sun. Assume it is a circular path of radius 93,000,000 mi.

35. A weather satellite orbiting the earth weighs 6500 lb. How many tons is this?

36. An airplane is flying at 37,000 ft. What is its altitude in miles?

37. A car's gasoline tank holds 56 L. Convert this capacity to gallons.

38. A hockey puck has a mass of about 0.160 kg. What is its weight in pounds?

39. The velocity of some seismic waves is 6800 m/s. What is this velocity in miles per hour?

40. The memory of a 1985 computer was 64 kB (B is the symbol for byte), and the memory of a 2010 computer is 128 GB. How many times greater is the memory of the 2010 computer?

41. The recorded surface area of a DVD is $17.2 \, \text{in.}^2$. What is this area in square centimeters?

42. A solar panel can generate 0.024 MW·h each day. Convert this to horsepower-hours.

43. The speed of sound is about 1130 ft/s. Change this speed to kilometers per hour.

44. Water flows from a kitchen faucet at the rate of 8.5 gal/min. What is this rate in liters per second?

45. The speedometer of a car is calibrated in kilometers per hour. If the speed of such a car reads 60, how fast in miles per hour is the car moving?

46. The acceleration due to gravity is about $980 \, \text{cm/s}^2$. Convert this to feet per squared second.

47. Fifteen grams of a medication are to be dissolved in 0.060 L of water. Express this concentration in milligrams per deciliter.

48. The earth's surface receives energy from the sun at the rate of $1.35 \, \text{kW/m}^2$. Reduce this to joules per second square centimeter.

49. At sea level, atmospheric pressure is about $14.7 \, \text{lb/in.}^2$. Express this in pascals.

50. The density of water is about $62.4 \, \text{lb/ft}^3$. Convert this to kilograms per cubic meter.

51. A typical electric current density in a wire is $1.2 \times 10^6 \, \text{A/m}^2$. Express this in milliamperes per square centimeter.

52. A certain automobile engine produces a maximum torque of $110 \, \text{N·m}$. Convert this to foot pounds.

In Exercises 53–56, use the following information. A commercial jet with 230 passengers on a 2850-km flight from Vancouver to Chicago averaged 765 km/h and used fuel at the rate of 5650 L/h.

53. How many hours long was the flight?

54. How long in seconds did it take to use 1.0 L of fuel?

55. What was the fuel consumption in km/L?

56. What was the fuel consumption in L/passenger?

The Graphing Calculator

C.1 Introduction

Until the 1970s, the personal calculating device for engineers and scientists was the slide rule. The microprocessor chip became commercially available in 1971, and the scientific calculator became widely used during the 1970s. Then in the 1980s, the graphing calculator was developed and is now used extensively.

The scientific calculator is still used by many when only accurate calculations are required. However, the graphing calculator can perform all the operations of a scientific calculator and numerous other operations. It also has the major advantage that entries can be seen in the viewing *window*, and this allows a visual check of entered data.

For these reasons, the graphing calculator is now used much more than the scientific calculator by engineers and scientists and in the more advanced mathematics courses. Therefore, as we stated on page 11, we restrict our coverage of calculator use in this text to graphing calculators.

This appendix includes a brief discussion of the features of a graphing calculator, followed by a listing of the examples with sample screen displays included throughout the text. Also included is a set of exercises that use the basic calculational and graphing features, and some graphing calculator programs that can be used to perform more extensive operations with fewer steps.

C.2 The Graphing Calculator

**Graphing Calculator Features •
List of Calculator Examples**

Since their introduction in the 1980s, many types and models of graphing calculators have been developed. All models can do all the calculational operations (and more) and display any of the graphs required in this text. Some models can also do symbolic operations. However, regardless of the model you are using, *to determine how the features of a particular model are used, refer to the manual for that model.*

In using a calculator, keep in mind that certain operations do not have defined results. If such an operation is attempted, the calculator will show an error display. Operations that can result in an error display include division by zero, square root of a negative number, logarithm of a negative number, and the inverse trigonometric function of a value outside the defined interval. Also, an error display may result if an improper sequence of keys is used.

On a graphing calculator, negative numbers are entered by using the $\boxed{(-)}$ key, and subtraction is entered by using the $\boxed{-}$ key. There is additional discussion of this on page 11 of Section 1.3. Use of the $\boxed{-}$ key for the $\boxed{(-)}$ key will usually result in an error display.

Most of the keys are used to access two or three different features. These features may also have subroutines that may be used. To activate a second use, the *2nd* (or *shift*) key is used. The *alpha* key is used for entering alphabetical characters when using more than one literal symbol (x is usually available directly) or when entering a calculator program.

GRAPHING CALCULATOR FEATURES

We now list some of the more important features of a graphing calculator with an explanation of their use. (This listing uses the designations on a TI-83 or TI-84 calculator. Your calculator may use a different designation for some features.)

Feature	Use
MODE	Determines how numbers and graphs are displayed. A *menu* is displayed that shows settings that may be used. The *mode* can be set, for example, (1) for the number of decimal places displayed in numbers; (2) to display angles in degrees or in radians; (3) to display the graph of a function or of parametric equations; (4) to display a graph in rectangular or polar coordinates. See the *mode* feature for other possible settings.
Y=	Used to enter functions. These functions are to be graphed or used in some other way.
WINDOW	Used to set the boundaries of the viewed portion of a graph. (On some calculators this is the *range* feature.)
ZOOM	Used to adjust the viewing window, usually to magnify the view of a particular part of a graph.
TRACE	Used to move the cursor from pixel to pixel along a displayed graph.
GRAPH	Used to display the graph of a function. (On some models this feature puts the calculator in graphing mode.)
CALC	For a function Y=, calculates a value (for given value of x) a zero, minimum value, maximum value, point of intersection with another graph, value of derivative at a point, or value of definite integral between values of x.
TABLE	Used with Y= and TBLSET to set up a table of values.
STO▶	Used to store in memory using variable names.
STAT	Used to access various statistical features.
MATH	Used for features such as fraction, nth roots, maximum or minimum values, absolute values, numerical derivatives or integrals, equation solving, and others.
TEST	Used to enter special symbols such as $>$ and $<$ and to enter special logic words such as *and* and *or.*
MATRX	Used to enter matrix values and perform matrix operations such as the evaluation of a determinant.
DRAW	Used to make certain types of drawings, such as shading in an area between curves or a tangent line to a curve.
PRGM	Used to enter and access calculator programs. A set of sample programs is given in the last section of this appendix.

There are many other features on a graphing calculator that are used to perform specific types of calculations and operations. Again, *to determine how the features of a particular model are used, refer to the manual for that model.*

Examples of the use of these features are found in the text examples in the listing that follows on the next page.

■ A *Graphing Calculator Lab Manual* supplement for this text is also available. It provides general directions for using all graphing calculator features and specific key steps for some models.

EXAMPLES OF CALCULATOR USE

■ The first calculational use of a graphing calculator is shown in Section 1.3 on page 11. The first graphical use is shown in Section 3.5 on page 96.

The following list shows the pages on which examples of the use of a TI-83 or TI-84 graphing calculator are given in the text. Calculator *window screens* are shown with each example. The list is divided into calculator uses that are primarily calculational and those that are primarily graphical.

EXERCISES C.2

The following exercises provide an opportunity for practice in performing calculations and a few basic graphs on a graphing calculator. There are many additional exercises throughout the book that require these types of calculations. Many additional exercises that require graphing of various types of curves are found after graphing is introduced in Chapter 3.

In Exercises 1–92, perform the indicated calculations on a graphing calculator.

1. $47.08 + 8.94$

2. $654.1 + 407.7$

3. $4724 - 561.9$

4. $0.9365 - 8.077$

5. 0.0396×471

6. 26.31×0.9393

7. $76.7 \div 194$

8. $52{,}060 \div 75.09$

9. 3.76^2

10. 0.986^2

11. $\sqrt{0.2757}$

12. $\sqrt{60.36}$

13. $\dfrac{1}{0.0749}$

14. $\dfrac{1}{607.9}$

15. $(19.66)^{2.3}$

16. $(8.455)^{1.75}$

17. $\sin 47.3°$

18. $\sin 1.15$

19. $\cos 3.85$

20. $\cos 119.1°$

21. $\tan 306.8°$

22. $\tan 0.537$

23. $\sec 6.11$

24. $\csc 242.0°$

25. $\sin^{-1} 0.6607$ (in degrees)

26. $\cos^{-1}(-0.8311)$ (in radians)

27. $\tan^{-1}(-2.441)$ (in radians)

28. $\sin^{-1} 0.0737$ (in degrees)

29. $\log 3.857$

30. $\log 0.9012$

31. $\ln 808$

32. $\ln 70.5$

33. $10^{0.545}$

34. $10^{-0.0915}$

35. $e^{-5.17}$

36. $e^{1.672}$

37. $(4.38 + 9.07) \div 6.55$

38. $(382 + 964) \div 844$

39. $4.38 + (9.07 \div 6.55)$

40. $382 + (964 \div 844)$

41. $\dfrac{5.73 \times 10^{11}}{20.61 - 7.88}$

42. $\dfrac{7.09 \times 10^{23}}{284 + 839}$

43. $50.38\pi^2$

44. $\dfrac{5\pi}{14.6}$

45. $\sqrt{1.65^2 + 6.44^2}$

46. $\sqrt{0.735^2 + 0.409^2}$

47. $3(3.5)^4 - 4(3.5)^2$

48. $\dfrac{3(-1.86)}{(-1.86)^2 + 1}$

49. $29.4\cos 72.5°$

50. $\dfrac{477}{\sin 58.7°}$

51. $\dfrac{4 + \sqrt{(-4)^2 - 4(3)(-9)}}{2(3)}$

52. $\dfrac{-5 - \sqrt{5^2 - 4(4)(-7)}}{2(4)}$

53. $\dfrac{0.176(180)}{\pi}$

54. $\dfrac{209.6\pi}{180}$

55. $\dfrac{1}{2}\left(\dfrac{51.4\pi}{180}\right)(7.06)^2$

56. $\dfrac{1}{2}\left(\dfrac{148.2\pi}{180}\right)(49.13)^2$

57. $\sin^{-1}\dfrac{27.3\sin 36.5°}{46.8}$

58. $\dfrac{0.684\sin 76.1°}{\sin 39.5°}$

59. $\sqrt{3924^2 + 1762^2 - 2(3924)(1762)\cos 106.2°}$

60. $\cos^{-1}\dfrac{8.09^2 + 4.91^2 - 9.81^2}{2(8.09)(4.91)}$

61. $\sqrt{5.81 \times 10^8} + \sqrt[3]{7.06 + 10^{11}}$

62. $(6.074 \times 10^{-7})^{2/5} - (1.447 \times 10^{-5})^{4/9}$

63. $\dfrac{3}{2\sqrt{7} - \sqrt{6}}$

64. $\dfrac{7\sqrt{5}}{4\sqrt{5} - \sqrt{11}}$

65. $\tan^{-1}\dfrac{7.37}{5.06}$

66. $\tan^{-1}\dfrac{46.3}{-25.5}$

67. $2 + \dfrac{\log 12}{\log 7}$

68. $\dfrac{10^{0.4115}}{\pi}$

69. $\dfrac{26}{2}(-1.450 + 2.075)$

70. $\dfrac{4.55(1 - 1.08^{15})}{1 - 1.08}$

71. $\sin^2\left(\dfrac{\pi}{7}\right) + \cos^2\left(\dfrac{\pi}{7}\right)$

72. $\sec^2\left(\dfrac{2}{9}\pi\right) - \tan^2\left(\dfrac{2}{9}\pi\right)$

73. $\sin 31.6°\cos 58.4° + \sin 58.4°\cos 31.6°$

74. $\cos^2 296.7° - \sin^2 296.7°$

75. $\sqrt{(1.54 - 5.06)^2 + (-4.36 - 8.05)^2}$

76. $\sqrt{(7.03 - 2.94)^2 + (3.51 - 6.44)^2}$

77. $\dfrac{(4.001)^2 - 16}{4.001 - 4}$

78. $\dfrac{(2.001)^2 + 3(2.001) - 10}{2.001 - 2}$

79. $\dfrac{4\pi}{3}(8.01^3 - 8.00^3)$

80. $4\pi(76.3^2 - 76.0^2)$

81. $0.01\left(\dfrac{1}{2}\sqrt{2} + \sqrt{2.01} + \sqrt{2.02} + \dfrac{1}{2}\sqrt{2.03}\right)$

82. $0.2\left[\dfrac{1}{2}(3.5)^2 + 3.7^2 + 3.9^2 + \dfrac{1}{2}(4.1)^2\right]$

83. $\dfrac{e^{0.45} - e^{-0.45}}{e^{0.45} + e^{-0.45}}$

84. $\ln \sin 2e^{-0.055}$

85. $\ln \dfrac{2 - \sqrt{2}}{2 - \sqrt{3}}$

86. $\sqrt{\dfrac{9}{2} + \dfrac{9 \sin 0.2\pi}{8\pi}}$

87. $2 + \dfrac{0.3}{4} - \dfrac{(0.3)^2}{64} + \dfrac{(0.3)^3}{512}$

88. $\dfrac{1}{2} + \dfrac{\pi \sqrt{3}}{360} - \dfrac{1}{4}\left(\dfrac{\pi}{180}\right)^2$

89. $160(1 - e^{-1.50})$

90. $e^{-3.60}(\cos 1.20 + 2 \sin 1.20)$

91. $\dfrac{(10)(9)(8)(7)(6)}{5!}$

92. $\dfrac{20! - 15!}{20! + 15!}$

In Exercises 93–108, display the graphs of the given functions on a graphing calculator. Use window settings to properly display the curve.

93. $y = 2x$ **94.** $y = 6 - x$

95. $y = x^2$ **96.** $y = 4 - 2x^2$

97. $y = 0.5x^3$ **98.** $y = 2x^2 - x^4$

99. $y = \sqrt{4 - x}$ **100.** $y = \sqrt{25 - x^2}$

101. $y = \log x$ **102.** $y = 2^x$

103. $y = 2e^x$ **104.** $y = 3 \ln x$

105. $y = \sin 2x$ **106.** $y = 2 \cos x$

107. $y = \tan^{-1} x$ **108.** $y = \dfrac{6}{x}$

C.3 Graphing Calculator Programs

Programs like those used on a computer can be stored in the memory of a graphing calculator. As mentioned earlier, such programs are used to perform more extensive operations with fewer steps, often significantly fewer steps. Generally, it is necessary only to enter certain data to obtain the required results.

Following are 14 programs that were written for use on a TI-83 or TI-84 graphing calculator. For other calculator models, it may be necessary to adapt the steps indicated for format and designated operations of that model.

The chapter, program title, and page on which the program reference appears are shown. A brief description of each program is also given.

Chapter 2   PYTHAGTH   page 56

This program solves for any side of a right triangle, given the other two sides. The input is chosen from the menu.

```
:Menu("PYTHAGTH", "FINDLEGA",A, "FINDLEGB",B,
"FINDHYPC",C)
:Lbl A:Prompt B,C
:Disp "A=", √ (C²−B²):Stop
:Lbl B:Prompt A,C
:Disp "B=", √ (C²−A²):Stop
:Lbl C:Prompt A,B
:Disp "C=", √ (A²+B²)
```

Chapter 4   SLVRTTRI   page 122

This program solves a right triangle, given both legs.

```
:Prompt A,B
:Disp "C=",√ (A²+B²)
:Disp " θA=", tan⁻¹(A/B)
:Disp "θB=", tan⁻¹(B/A)
```

Chapter 7   QUADFORM   page 221

This program solves the quadratic equation $Ax^2 + Bx + C = 0$ for real roots.

```
:Prompt A,B,C
:B²−4AC → D
:If D<0: Then:Disp "IMAGINARY ROOTS":Stop:End
:Disp "ROOTS ARE",(⁻B+√ (D))/(2A),(⁻B−√ (D))/(2A)
```

Chapter 9   ADDVCTR   page 266

This program adds two vectors, given their magnitudes and angles.

```
:Disp "ENTER MAGNITUDES"
:Input "A=",A:Input"B=",B
:Disp "ENTER ANGLES"
:Input "θA=",P:Input "θB=",Q
:Acos(P)+Bcos(Q) → R
:Asin(P)+Bsin(Q) → S
:tan⁻¹(S/R) → T
:Disp "R=", √ (R²+S²)
:If R<0:Goto A:If S<0:Goto B
:Disp "θR=",T:Stop
:Lbl A:Disp "θR=",T+180:Stop
:Lbl B:Disp "θR=",T+360
```

Chapter 10   SINECURV   page 296

This program displays the graphs of $y_1 = \sin x$, $y_2 = 2 \sin x$, $y_3 = \sin 2x$, and $y_4 = 2 \sin(2x - \pi/3)$. There is no input.

```
:⁻2 → Xmin:7 → Xmax
:⁻2 → Ymin:2 → Ymax
:"sin(X)" → Y₁:"2sin(X)" → Y₂
:"sin(2X)" → Y₃:"2sin(2X−π/3)" → Y₄
:Disp "Y₁=sin(X)"
:Disp "Y₂=2sin(X)"
:Disp "Y₃=sin(2X)"
:Disp "Y₄=2sin(2X − π/3)"
:Pause:DispGraph:Pause:ClrHome
```

Chapter 11   TBLROOTS   page 323

This program displays a table of square roots and cube roots. Input to TblSet should be made before running the program.

```
:"√(X)" → Y₁
:"³√(X)" → Y₂
:DispTable:Pause:ClrHome
```

Chapter 12   DEMOIVRE   page 351

This program finds the nth roots of the complex number $a + bj$. Displayed are the real and imaginary parts of each root.

```
:Prompt A,B,N
:√((A²+B²)^(1/N)) → R
:If A<0:(tan⁻¹(B/A)+180)/N → θ
:If A>0 and B<0:(tan⁻¹(B/A)+360)/N → θ
:If A=0 and B>0:90/N → θ
:If A=0 and B<0:270/N → θ
:If A>0 and B≥0:(tan⁻¹(B/A))/N → θ
:For(I,1,N)
:Disp "ROOT",I
:Disp "REAL PART",Rcos(θ)
:Disp "IMAG PART",Rsin(θ)
:θ+ 360/N → θ:Pause:End
```

Chapter 12   IMPEDANC   page 355

This program calculates the impedance Z and phase angle θ, given a resistance R, inductance L, capacitance C, and frequency F. (Any of R, L, or C can be zero.)

```
:Prompt R,L,F,C
:2πFL → I
:If C=0:Goto A
:(2πFC)⁻¹ → Q
:If R=0:Goto B
:Disp "Z=",√(R²+(I−Q)²)
:Disp "θ=",tan⁻¹((I−Q)/R):Stop
:Lbl A:Disp "Z=",√(R²+I²):Disp "θ=",tan⁻¹(I/R):Stop
:Lbl B:Disp "Z=",abs(I−Q)
:If I>Q:Disp "θ=90"
:If I<Q:Disp "θ=−90"
```

Chapter 15   SYNTHDIV   page 411

This program gives the coefficients and the remainder if a polynomial is divided by $x - R$. The coefficients are entered as a {LIST}.

```
:Disp "LIST COEFFICIENTS"
:Input L₁
:Input "R=",R
:L₁(1) → F:dim(L₁) → D
:Disp "QUOTIENT="
:For(N,1,D−1)
:Pause F:RF+L₁(N+1) → F:End
:Disp "REMAINDER=",F
```

Chapter 17   LINPROG   page 485

This program double shades the area of feasible points for two linear constraints. *Trace* is used to find the point of maximum value or minimum value.

```
:0 → Xmin:0 → Ymin
:Input "Xmax=",Xmax
:Input "Xscl=",Xscl
:Input "Ymax=",Ymax
:Input "Yscl=",Yscl
:Input "Y₁=",Y₁
:Input "Y₂=",Y₂
:Disp "FOR MINIMUM"
:Disp "ENTER A=1"
:Disp "FOR MAXIMUM"
:Disp "ENTER A=2"
:Prompt A
:If A=1: Goto A
:Shade(0,Y₁,0,Xmax,1,3)
:Shade(0,Y₂,0,Xmax,2,3)
:Trace:Pause:ClrHome:Stop
:Lbl A
:Shade(Y₁,Ymax,0,Xmax,1,3)
:Shade(Y₂,Ymax,0,Xmax,2,3)
:Trace:Pause:ClrHome
```

Chapter 19   BINEXPAN   page 520

This program gives the first k coefficients of $(ax + b)^n$.

```
:Prompt K, A,B,N
:A^N → T
:Disp "COEFFICIENTS ="
:For (C,1,K): Disp T
:T(N−C+1)B/AC) → T
:Pause:End
```

Chapter 21 SLOPEDIS page 562

This program calculates the slope m of a line through two points (a, c) and (e, f) and the distance between the points. It also displays the line segment on a split screen. Set proper *window* values before running the program.

:Prompt A,C,E,F
:Horiz
:Output (2,1,"M=")
:Output (2,4,(F−C)/(E −A))
:Output (3,1,"D=")
:Output (3,4,√((E−A)²+(F−C)²))
:Line (A,C,E,F)
:Pause:Full

Chapter 21 GRAPHLIN page 564

This program displays the graph of a line through two points (a, c) and (d, e). The *window* values should be set appropriately before running the program. Default *window* is $(−9, 9)$ by $(−6, 6)$.

:Prompt A,C,D,E
:(E−C)/(D−A) → M:C−MA → B
:"MX+B" → Y₁
:−9 → Xmin:9 → Xmax:1 → Xscl
:−6 → Ymin:6 → Ymax:1 → Yscl
:DispGraph
:Pause:ClrHome

Chapter 21 GRAPHCON page 594

This program displays the graph of a conic of the form $ax^2 + cy^2 + dx + ey + f = 0$. The *window* settings may have to be changed in the program. The default *window* is $(−9, 9)$ by $(−6, 6)$.

:Prompt A,C,D,E,F
:If C=0:Goto A
:(−E+√(E²−4C(AX²+DX+F)))/(2C) → Y₁
:(−E−√(E²−4C(AX²+DX+F)))/(2C) → Y₂
:Lbl B
:−9 → Xmin:9 → Xmax:1 → Xscl
:−6 → Ymin:6 → Ymax:1 → Yscl
:DispGraph
:Pause:ClrHome:Stop
:Lbl A: "(AX²+DX+F)/(−E)" → Y₁
:Goto B

EXERCISES C.3

In each of the following exercises, write and test a graphing calculator program to perform the indicated operations for a topic in the indicated chapter.

1. (Chapter 2) To find the total surface area and volume of a right circular cylinder for given values of the radius r and height h.

2. (Chapter 5) To solve the system of equations $ax + by = c$ and $dx + ey = f$ for given values of the constants.

3. (Chapter 6) To find the coefficients, e, f, and g for the product of binomials $(ax + b)(cx + d) = ex^2 + fx + g$ for given values of the constants.

4. (Chapter 8) To find the arc length and area of a circular sector for given values of the radius r and angle θ.

5. (Chapter 9) To solve a triangle given angles A, B, and included side C.

6. (Chapter 13) To find the logarithm to any given base b.

7. (Chapter 18) To find the constant of proportionality k for $y = kr^a s^b / t^c$, given the values of y, r, a, s, b, t, and c.

8. (Chapter 19) For a geometric sequence, to calculate the nth term, sum of n terms, and the sum of the infinite series for given values of the first term a, ratio r, and n.

ANSWERS TO ODD-NUMBERED EXERCISES

Since statements will vary for writing exercises (W), answers here are in abbreviated form. Answers are not included for end-of-chapter writing exercises.

Exercises 1.1, page 5

1. Change $\frac{5}{1}$ to $\frac{-3}{1}$ and $\frac{-19}{1}$ to $\frac{14}{1}$.　　3. Change $2 > -4$, 2 is to the right of -4, to $-6 < -4$, -6 is to the left of -4. Also the figure must be changed to show -6 to left of -4.

5. 3: integer, rational, real; $\sqrt{-4}$: imag.; $-\dfrac{\pi}{6}$: irrational, real

7. $3, 4, \dfrac{\pi}{2}$　　9. $6 < 8$　　11. $\pi > -3.2$　　13. $-4 < -|-3|$

15. $-\dfrac{1}{3} > -\dfrac{1}{2}$　　17. $\dfrac{1}{3}, -\dfrac{\sqrt{3}}{4}, \dfrac{b}{y}$

19.

$$-\tfrac{12}{5} \qquad \sqrt{3} \quad 2.5$$

$$-3\ -2\ -1\ \ 0\ \ 1\ \ 2\ \ 3$$

21. No, $|0| = 0$

23. The number itself　　25. $\dfrac{3}{22}$ or $\dfrac{3}{23}$

27. $-3.1, -|-3|, -1, \sqrt{5}, \pi, |-8|, 9$

29. (a) Positive integer (b) Negative integer (c) Positive rational number less than 1

31. (a) Yes (b) Yes　　33. (a) To right of origin (b) To left of -4

35. Between 0 and 1　　37. $a =$ any real number; $b = 0$

39. 0.0008 F　　41. $N = 1000an$

43. Yes; -20 is to right of -30

Exercises 1.2, page 10

1. 22　　3. -4　　5. 4　　7. 6　　9. -3　　11. -24

13. 35　　15. 20　　17. 40　　19. -1　　21. -17

23. Undefined　　25. 20　　27. 16　　29. -9　　31. 24

33. -6　　35. 3　　37. Commutative law of multiplication

39. Distributive law　　41. Associative law of addition

43. Associative law of multiplication　　45. d　　47. b

49. (a) Positive (b) Negative

51. Correct. (For $x > 0$, x is positive; for $x < 0$, $-x$ is positive.)

53. (a) Negative reciprocals of each other (b) They may not be equal.

55. $-2.4\,\text{kW}\cdot\text{h}$　　57. $2°\text{C}$　　59. 10 V

61. $100\,\text{m} + 200\,\text{m} = 200\,\text{m} + 100\,\text{m}$; commutative law of addition

63. $4(8\,\text{min} + 6\,\text{min})$; distributive law

Exercises 1.3, page 15

1. 0.390 has 3 sig. digits; the zero is not needed for proper location of the decimal point.

3. 75.7　　5. 8 is exact; 55 is approx.　　7. 24 and 1440 are exact

9. 3, 4　　11. 3, 3　　13. 1, 5　　15. (a) 0.01 (b) 30.8

17. (a) Same (b) 78.0　　19. (a) 0.004 (b) Same

21. (a) 4.94 (b) 4.9　　23. (a) 50,900 (b) 51,000

25. (a) 9550 (b) 9500　　27. (a) 0.945 (b) 0.94

29. (a) 12 (b) 12.20　　31. (a) 0.015 (b) 0.0114

33. (a) 0.1 (b) 0.1356　　35. (a) 6.5 (b) 6.086

37. 15.8788　　39. 204.2　　41. 2.745 MHz, 2.755 MHz

43. Too many sig. digits; time has only 2 sig. digits

45. (a) 19.3 (b) 27　　47. (a) 2 (b) 2 (c) -2 (d) 0 (e) Undefined

49. Example: $(456789 - 987654) \div 9 = -58985$; integer

51. (a) $3.1416 > \pi$; (b) $22/7 > \pi$

53. (a) $\frac{1}{3} = 0.33333333\ldots$ (b) $\frac{5}{11} = 0.4545454545\ldots$
(c) $\frac{2}{5} = 0.400000000$ (0 repeats)

55. 95.3 MJ　　57. 262,144 bytes　　59. 59.14%

Exercises 1.4, page 20

1. x^6　　3. $\dfrac{a^6 x^2}{b^4 t^2}$　　5. x^7　　7. $2b^6$　　9. m^2　　11. $\dfrac{1}{7n^4}$

13. P^8　　15. $8\pi^3$　　17. $a^{30}T^{60}$　　19. $\dfrac{8}{b^3}$　　21. $\dfrac{x^8}{16}$

23. 1　　25. -3　　27. $\dfrac{1}{6}$　　29. R^2　　31. $-t^{14}$

33. $\dfrac{1}{64v^{12}}$　　35. $-L^2$　　37. $\dfrac{1}{8}$　　39. 1　　41. $\dfrac{a}{x^2}$

43. $\dfrac{64s^6}{g^2}$　　45. $\dfrac{x^3}{64a^3}$　　47. $\dfrac{5n^3}{T}$　　49. -53　　51. 253

53. -0.421　　55. 9990　　57. Yes　　59. 625　　61. 1

63. $\dfrac{G^2 k^5 T^5}{h}$　　65. $\dfrac{r}{6}$　　67. $3212.27

Exercises 1.5, page 23

1. 8060　　3. 45,000　　5. 0.00201　　7. 3.23

9. 18.6　　11. 4×10^3　　13. 8.7×10^{-3}　　15. 6.09×10^0

17. 6.3×10^{-2}　　19. 1×10^0　　21. 5.6×10^{13}

23. 2.2×10^8　　25. 3.2×10^{-34}　　27. 1.728×10^{87}

29. 4.85×10^{10}　　31. 1.59×10^7　　33. 9.965×10^{-3}

35. 3.38×10^{16}　　37. $6.5 \times 10^6\,\text{kW}$　　39. $3 \times 10^{-6}\,\text{W}$

41. $2 \times 10^9\,\text{Hz}$　　43. 0.0000000000016 W

45. (a) 2.3×10^3 (b) 230×10^{-3} (c) 23×10^0

47. (a) 1×10^{100} (b) $10^{10^{100}}$　　49. $4.2 \times 10^{-8}\,\text{s}$

51. $2.46 \times 10^{-1}\,\text{s}$　　53. $3.32 \times 10^{-18}\,\text{kg}$　　55. 3.433 Ω

Exercises 1.6, page 25

1. -4　　3. 12　　5. 9　　7. -11　　9. -7　　11. 0.3

13. 5　　15. -6　　17. 5　　19. 47　　21. 53　　23. $20\sqrt{3}$

25. $4\sqrt{21}$　　27. $2\sqrt{5}$　　29. 4　　31. 7　　33. 10

35. $3\sqrt{10}$　　37. 9.24　　39. 0.6877　　41. (a) 60 (b) 84

43. (a) 0.0388 (b) 0.0246　　45. 60 mi/h　　47. 1450 m/s

49. 42.0 in.   **51.** no, not true if $a < 0$
53. (a) 12.9 (b) -0.598 **55.** (a) Imag. (b) Real

Exercises 1.7, page 29

1. $3x - 3y$ **3.** $4ax + 5s$ **5.** $8x$ **7.** $y + 4x$
9. $5F - 3T - 2$ **11.** $-a^2b - a^2b^2$ **13.** $3s - 4$
15. $5x - v - 4$ **17.** $5a - 5$ **19.** $-5a + 2$
21. $-2t + 5u$ **23.** $7r + 8s$ **25.** $-50 + 19j$
27. $-9 + 3n$ **29.** $18 - 2t^2$ **31.** $6a$ **33.** $2aZ + 1$
35. $4c - 6$ **37.** $8p - 5q$ **39.** $-4x^2 + 22$ **41.** $7V^2 - 3$
43. $-6t + 13$ **45.** $4Z - 24R$ **47.** $2D + d$
49. $3B - 2\alpha$ **51.** $40x + 250$
53. (a) $x^2 + 2y + 2a - b$ (b) $3x^2 - 4y + 2a + b$
55. Yes; $|a - b| = |-(b - a)| = |b - a|$

Exercises 1.8, page 31

1. $-8s^8t^{13}$ **3.** $x^2 - 5x + 6$ **5.** a^3x **7.** $-a^2c^3x^3$
9. $-8a^3x^5$ **11.** $i^2R + 2i^2r$ **13.** $-3s^3 + 15st$
15. $5m^3n + 15m^2n$ **17.** $-3M^2 - 3MN + 6M$
19. $acx^4 + acx^3y^3$ **21.** $x^2 + 2x - 15$ **23.** $2x^2 + 9x - 5$
25. $6a^2 - 7ab + 2b^2$ **27.** $6s^2 + 11st - 35t^2$
29. $2x^3 + 5x^2 - 2x - 5$ **31.** $x^2 - 4xy + 4y^2 - 16$
33. $2a^2 - 16a - 18$ **35.** $18T^2 - 15T - 18$
37. $-2L^3 + 6L^2 + 8L$ **39.** $4x^2 - 20x + 25$
41. $x_1^2 + 6x_1x_2 + 9x_2^2$ **43.** $x^2y^2z^2 - 4xyz + 4$
45. $2x^2 + 32x + 128$ **47.** $-x^3 + 2x^2 + 5x - 6$
49. $6T^3 + 9T^2 - 6T$ **51.** (a) $49 \neq 9 + 16$ (b) $1 \neq 9 - 16$
53. $n^2 - 1 = (n - 1)(n + 1)$ for any n
55. $(x + y)^3 = x^3 + 3x^2y + 3xy^2 + y^3$
57. $P + 0.02Pr + 0.0001Pr^2$ **59.** $3R^2 - 4RX$
61. $n^2 + 200n + 10,000$ **63.** $R_1^2 - R_2^2$

Exercises 1.9, page 34

1. $\dfrac{3}{y^3}$ **3.** $3x - 2$ **5.** $-4x^2y$ **7.** $\dfrac{4t^4}{r^2}$ **9.** $4x^2$
11. $-6a$ **13.** $a^2 + 2y$ **15.** $t - 2rt^2$ **17.** $q + 2p - 4q^3$
19. $\dfrac{2L}{R} - R$ **21.** $\dfrac{1}{3a} - \dfrac{2b}{3a} + 1$ **23.** $3y^n - 2ay$
25. $2x + 1$ **27.** $x - 1$ **29.** $4x^2 - x - 1, R = -3$
31. $Z - 2 - \dfrac{1}{4Z + 3}$ **33.** $x^2 + x - 6$ **35.** $2a^2 + 8$
37. $x^2 - 2x + 4$ **39.** $x - y$ **41.** $x - y + z$ **43.** -5
45. $\dfrac{x^4 + 1}{x + 1} = x^3 - x^2 + x - 1 + \dfrac{2}{x + 1}$
47. $A + \dfrac{\mu^2E^2}{2A} - \dfrac{\mu^4E^4}{8A^3}$ **49.** $\dfrac{GMm}{R}$
51. $s^2 + 2s + 6 + \dfrac{16s + 16}{s^2 - 2s - 2}$

Exercises 1.10, page 38

1. $-9, -15, -36, -4$ **3.** $\dfrac{1}{8}$ **5.** 9   **7.** -1 **9.** -10
11. -5 **13.** -3 **15.** 4   **17.** $-\dfrac{7}{2}$ **19.** 8
21. $\dfrac{10}{3}$ **23.** 2   **25.** -2.5 **27.** 0   **29.** 8
31. 9 or -9 **33.** 9.5   **35.** -2.1 **37.** 5.7   **39.** 0.85
41. (a) Identity (b) Conditional equation   **43.** $x = 3.75$
45. 55 mi/h   **47.** 120°C   **49.** 750 gal   **51.** 460 mi

Exercises 1.11, page 40

1. $\dfrac{v - v_0}{t}$ **3.** $\dfrac{V_0 + bTV_0 - V}{bV_0}$ **5.** $\dfrac{E}{I}$ **7.** $g_2 - rL$
9. $\dfrac{Q}{Sd^2}$ **11.** $\dfrac{p - p_a}{dg}$ **13.** $\dfrac{APV}{R}$ **15.** $\dfrac{0.3t - ct^2}{c}$
17. $Tv - c$ **19.** $\dfrac{K_1m_1 - K_2m_1}{K_2}$ **21.** $\dfrac{2mg - 2am}{a}$
23. $\dfrac{C_0^2 - C_1^2}{2C_1^2}$ **25.** $\dfrac{rA - N}{r}$ **27.** $100T_1 - 100T_2$
29. $\dfrac{Q_1 + PQ_1}{P}$ **31.** $\dfrac{N + N_2 - N_2T}{T}$
33. $\dfrac{L - \pi r_2 - 2x_1 - x_2}{\pi}$ **35.** $\dfrac{gJP + V_1^2}{V_1}$
37. $\dfrac{Cd(k_1 + k_2)}{2Ak_1k_2}$ **39.** $\dfrac{CN - NV}{C}$ **41.** 1040 K
43. 32.3°C   **45.** 3.22 Ω   **47.** $\dfrac{d - 4v_2 - 2v_1}{v_1}$

Exercises 1.12, page 44

1. 29 1.5-Ω resistors, 5 2.5-Ω resistors   **3.** 3.000 h
5. $22,000, $27,000
7. 1.9 million the first year, 2.6 million the second year
9. 50 acres at $200/acre, 90 acres at $300/acre
11. 60 ppm/h   **13.** 20 girders
15. $-2.3\,\mu A, -4.6\,\mu A, 6.9\,\mu A$ **17.** 6.9 km, 9.5 km
19. 32 CDs, 22 DVDs   **21.** 900 m
23. 84.2 km/h, 92.2 km/h   **25.** 390 s, first car
27. 64 mi from A   **29.** 4 L   **31.** 79 km/h

Review Exercises for Chapter 1, page 46

1. -10 **3.** -20 **5.** -22 **7.** -25 **9.** -4
11. 5   **13.** $4r^2t^4$ **15.** $-\dfrac{24t}{m^2n}$ **17.** $\dfrac{8T^3}{N}$ **19.** $3\sqrt{5}$
21. (a) 3 (b) 8800   **23.** (a) 4 (b) 9.0   **25.** 18.0
27. 1.3×10^{-4} **29.** $-a - 2ab$ **31.** $7LC - 3$
33. $2x^2 + 9x - 5$ **35.** $x^2 + 16x + 64$ **37.** $hk - 3h^2k^4$
39. $7R - 6r$ **41.** $13xy - 10z$ **43.** $2x^3 - x^2 - 7x - 3$
45. $-3x^2y + 24xy^2 - 48y^3$ **47.** $-9p^2 + 3pq + 18p^2q$
49. $\dfrac{6q}{p} - 2 + \dfrac{3q^4}{p^3}$ **51.** $2x - 5$ **53.** $x^2 - 2x + 3$

55. $4x^3 - 2x^2 + 6x, R = -1$ **57.** $15r - 3s - 3t$

59. $y^2 + 5y - 1, R = 4$ **61.** $-\dfrac{9}{2}$ **63.** $\dfrac{21}{10}$ **65.** $-\dfrac{7}{3}$

67. 3   **69.** $-\dfrac{19}{5}$ **71.** 1   **73.** 6×10^{10} bytes

75. 1.92×10^8 km   **77.** 25,300,000,000,000 mi

79. 0.000000000001 W/m^2 **81.** 0.15 Bq/L   **83.** $\dfrac{R}{n^2}$

85. $\dfrac{PL^2}{\pi^2 I}$ **87.** $\dfrac{Rr - Pp}{Q}$ **89.** $\dfrac{d + A}{A}$

91. $\dfrac{N_1 + TN_3 - N_3}{T}$ **93.** $\dfrac{RH + AT_1}{A}$

95. $\dfrac{d - 3kbx^2 + kx^3}{3kx^2}$ **97.** 8.2×10^5 **99.** 110 m

101. $0.0188\ \Omega$ **103.** $4x + 4a$

105. $4t + 4h - 2t^2 - 4th - 2h^2$ **107.** Yes (18 to 0)

109. Equation is an identity.

111. $x^3 - 3x^2 y + 3xy^2 - y^3 = -y^3 + 3y^2 x - 3yx^2 + x^3$

113. 4×10^{-7} **115.** \$59, \$131

117. 160 cm^3, 80 cm^3, 320 cm^3 **119.** 1900 Ω, 3100 Ω

121. 6.7 h   **123.** 34.6 h after second ship enters Red Sea

125. 400 L, 600 L   **127.** 290 ft^2

Exercises 2.1, page 52

1. 90°   **3.** 4   **5.** $\angle EBD, \angle DBC$ **7.** $\angle ABC$ **9.** 25°

11. BD, BC **13.** 140°   **15.** 40°   **17.** 35°   **19.** 62°

21. 28°   **23.** 134°   **25.** 44°   **27.** 46°   **29.** 4.53 m

31. 3.40 m   **33.** 133°   **35.** 920 m   **37.** 180°

39. Sum of angles is 180°.

Exercises 2.2, page 58

1. 65°   **3.** 7.02 m   **5.** 56°   **7.** 48°   **9.** 8.4 ft^2

11. 32,300 cm^2 **13.** 4.41 ft^2 **15.** 0.390 m^2 **17.** 942 cm

19. 64.5 cm   **21.** 26.6 ft   **23.** 522 cm   **25.** 67°

27. 227.2 cm   **29.** 135°   **31.** Equilateral triangle

33. The smaller triangles are similar.

35. $\angle K = \angle N = 90°;\ \angle LMK = \angle OMN;$
$\angle KLM = \angle NOM; \triangle MKL \sim \triangle MNO$

37. 8   **39.** 65°   **41.** 1150 cm^2 **43.** 60 ft^2 **45.** 18.3 ft

47. 23 ft   **49.** 7.5 m, 9.0 m   **51.** 9.6 ft   **53.** 20 mi

Exercises 2.3, page 62

1. Trapezoid   **3.** 8900 ft^2 **5.** 260 m   **7.** 3.324 in.

9. 12.8 m   **11.** 214.4 ft   **13.** 7.3 mm^2 **15.** 0.683 in.^{2}

17. 9.3 m^2 **19.** 2000 ft^2 **21.** $p = 4a + 2b$

23. $A = bh + a^2$ **25.** It is a rectangle.   **27.** 288 cm^2

29. The diagonal always divides the rhombus into two congruent triangles. All outer sides are always equal.

31. 344 m   **33.** 2.4 ft, 9.6 ft   **35.** 0.84 gal   **37.** 3.04 km^2

39. 360°. A diagonal divides a quadrilateral into two triangles, and the interior angles of each triangle are 180°.

Exercises 2.4, page 65

1. 18°   **3.** $p = 11.6$ in., $A = 8.30$ in.^{2} **5.** (a) AD (b) AF

7. (a) $AF \perp OE$ (b) $\triangle OEC$ **9.** 1730 ft   **11.** 72.6 mm

13. 0.0285 yd^2 **15.** 4.26 m^2 **17.** 25°   **19.** 25°

21. 120°   **23.** 40°   **25.** 0.393 rad   **27.** 2.185 rad

29. $p = \frac{1}{2}\pi r + 2r$ **31.** $A = \frac{1}{4}\pi r^2 - \frac{1}{2}r^2$

33. All are on the same diameter.   **35.** 10.3 in.^{2}

37. 24,900 mi   **39.** $\frac{4}{9}$ **41.** 35.7 in.   **43.** 9500 cm^2

45. 1030 ft^2 **47.** Horizontally and opposite to original direction

Exercises 2.5, page 70

1. Simpson's rule. Arcs give better estimate.

3. Simpson's rule should be more accurate in that it accounts better for the arcs between points on the upper curve.

5. 84 m^2 **7.** 4.9 ft^2 **9.** 9.8 mi^2 **11.** 19,000 km^2

13. 120,000 ft^2 **15.** 8100 ft^2

17. 2.73 in.^{2} The trapezoids are inside the boundary and do not include some of the area.

19. 2.98 in.^{2} The ends of the areas are curved so that they can get closer to the boundary.

Exercises 2.6, page 73

1. The volume is four times as much.   **3.** 771 cm^3 **5.** 366 ft^3

7. 3.99×10^6 m^2 **9.** 2.83 yd^3 **11.** 20,500 cm^2

13. 250,000 in.^{3} **15.** 3.358 m^2 **17.** 0.15 yd^3 **19.** 72.3 cm^2

21. $V = \frac{1}{6}\pi d^3$ **23.** $\frac{6}{1}$ **25.** $\frac{4}{1}$ **27.** 604 in.^{2}

29. 0.00034 mi$^3 = 5.0 \times 10^7$ ft^3 **31.** 3.3×10^6 yd^3

33. 2,350,000 ft^3 **35.** 1560 mm^2 **37.** 447 in.^{3}

39. 1.10 in.^{3}

Review Exercises for Chapter 2, page 76

1. 32°   **3.** 32°   **5.** 41   **7.** 700   **9.** 7.36   **11.** 21.1

13. 25.5 mm   **15.** 3.06 ft^2 **17.** 309 mm   **19.** 3320 in.^{2}

21. 6190 cm^3 **23.** 160,000 ft^3 **25.** 1.62 m^2 **27.** 66.6 in.^{2}

29. 25°   **31.** 65°   **33.** 53°   **35.** 2.4

37. $p = \pi a + b + \sqrt{4a^2 + b^2}$ **39.** $A = ab + \frac{1}{2}\pi a^2$

41. yes   **43.** $n^2; A = \pi(nr)^2 = n^2(\pi r^2)$

45. $a/d = b/c$ **47.** 71°   **49.** 7.9 m   **51.** 12 cm

53. 30 m   **55.** 6.0 m   **57.** 3.73 mi   **59.** 26,200 mi

61. 30 ft^2 **63.** 1.0×10^6 m^2 **65.** 190 m^3 **67.** 10 ft

69. 159 gal   **71.** $w = 93.0$ cm, $h = 52.5$ cm

Exercises 3.1, page 83

1. -13 **3.** $f(T - 10) = 9.1 + 0.08T + 0.001T^2$

5. (a) $A = \pi r^2$ (b) $A = \frac{1}{4}\pi d^2$ **7.** $d = \sqrt[3]{\dfrac{6V}{\pi}}$

9. $A = s^2, s = \sqrt{A}$ **11.** $A = 4r^2 - \pi r^2$ **13.** $3, -1$

15. $5, 5$ **17.** $\dfrac{3 - 2\pi^2}{2\pi}, -\dfrac{1}{2}$ **19.** $\frac{1}{4}a + \frac{1}{2}a^2, 0$

21. $3s^2 + s + 6, 12s^2 - 2s + 6$ **23.** -8 **25.** 62.9, 260

27. -300.0

29. Square the value of the independent variable and add 2.

31. Cube the value of the independent variable and subtract this from 6 times the value of the independent variable.

33. Multiply by 3 the sum of twice r added to 5 and subtract 1.

35. Subtract 3 from twice t and divide this quantity by the sum of t and 2.

37. $A = 5e^2$, $f(e) = 5e^2$

39. $A = 8430 - 140t$, $f(t) = 8430 - 140t$ **41.** 10.4 m

43. 75 ft, $2v + 0.2v^2$, 240 ft, 240 ft   **45.** $d = 55t$

47. (a) $f[f(x)]$ means "function of the function of x". (b) $8x^4$

Exercises 3.2, page 87

1. $-x^2 + 2$ is never greater than 2. The range is all real numbers $f(x) \le 2$.

3. 4 mA, 0 mA

5. Domain: all real numbers; range: all real numbers

7. Domain: all real numbers except 0; range: all real numbers except 0

9. Domain: all real numbers except 0; range: all real numbers $f(s) > 0$

11. Domain: all real numbers $h \ge 0$; range: all real numbers $H(h) \ge 1$; $\sqrt{h}$ cannot be negative

13. Domain: all real numbers; range: all real numbers $y \ge 0$

15. All real numbers $y > 2$

17. All real numbers except -4 and 2   **19.** 2, not defined

21. 2, 0.75   **23.** $d = 80 + 55t$ **25.** $w = 5500 - 2t$

27. $m = 0.5h - 390$ **29.** $C = 5l + 250$

31. (a) $y = 3000 - 0.25x$ (b) 2900 L

33. $y = 110 - \dfrac{3x}{5}$ **35.** $A = \dfrac{1}{16}p^2 + \dfrac{(60 - p)^2}{4\pi}$

37. $d = \sqrt{14,400 + h^2}$; domain: $h \ge 0$; range: $d \ge 120 \ m$

39. $s = \dfrac{300}{t}$; domain: $t > 0$; range: $s > 0$ (upper limits depend on truck)

41. Domain is all values of $C > 0$, with some upper limit depending on the circuit.

43. $m = \begin{cases} 0.5h - 390 \text{ for } h > 1000 \\ 110 \text{ for } 0 \le h \le 1000 \end{cases}$

45. (a) $V = 4w^2 - 24w + 32$ (b) $w \ge 4$ in.   **47.** 1

49. All real numbers $y \ge 2$.

Exercises 3.3, page 90

1. $(-1, 1)$ **3.** $(2, 1), (-1, 2), (-2, -3)$

5.

7. Isosceles triangle   **9.** Rectangle   **11.** $(5, 4)$

13. $(3, -2)$

15. On a line parallel to the y-axis, 1 unit to the right

17. On a line parallel to the x-axis, 3 units above

19. On a line through the origin that bisects the first and third quadrants

21. 0   **23.** To the right of the y-axis

25. To the left of a line that is parallel to the y-axis, 1 unit to its left

27. In first or third quadrant   **29.** On either axis

31. Third quadrant   **33.** (a) 8 (b) 6

Exercises 3.4, page 94

1.

3.

5.

7.

9.

11.

13.

15.

17.

19.

21.

23.

25.

27.

29.

59.

61. They are the same except for (b); there is no point for $x = 2$.

63. Yes   **65.** No

Exercises 3.5, page 100

31.

33.

1. $-2.4, 0.4$

3. All real numbers greater than about 15

35.

37.

39.

41.

5.

7.

9.

11.

43.

45.

47.

49. $A = 100w - w^2$
$30\text{ m} \le w \le 70\text{ m}$

13.

15.

17.

19. $-2.791, 1.791$

51.

53. Only integral values of n have meaning.

21. 2.104

23. 4.321

55. No. $f(1) = 2$, but $f(2)$ is not determined.

57. $y = x$ is the same as $y = |x|$ for $x \ge 0$
$y = |x|$ is the same as $y = -x$ for $x < 0$.

25. 1.4

27. No values

57. Graph for (b)

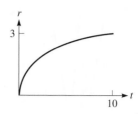

29. All real numbers $y > 0$ or $y \le -1$

59. A curve with negative c is inverted from the curve with positive c.

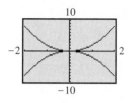

Exercises 3.6, page 104

1.

3.

31. All real numbers $y \ge 0$ or $y \le -4$

5.

7.

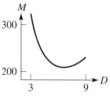

33. All real numbers $Y(y) > 3.46$ (approx.)

9. (a) 132.1°C   (b) 0.7 min   **11.** (a) 1.5 mm (b) 3.2 mm

13. 0.30 H   **15.** 7.2%   **17.** (a) 3.4 ft (b) 17 ft³/s

19. 14 ft³/s   **21.** 0.34

23. 76 m²   **25.** 130.3°C

27. 36 ft³/s

35. All real numbers

37. $y = 3x + 1$

Review Exercises for Chapter 3, page 105

1. $A = 4\pi t^2$ **3.** $y = 50{,}000 - 1.25x$ **5.** 16, −47

7. 3, $\sqrt{1 - 4h}$ **9.** $h^3 + 11h^2 + 36h$ **11.** −3

13. −3.67, 16.7   **15.** 0.16503, −0.21476

17. Domain: all real numbers; range: all real numbers $f(x) \ge 1$

19. Domain: all real numbers $t > -4$; range: all real numbers $g(t) > 0$

21. Domain: all real numbers except 5; range: all real numbers $f(n) > 1$

39. $y = \sqrt{x - 3}$

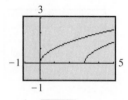

41. $y = -2(x + 2)^2 - 3$

43. $y = \sqrt{2x + 3} + 1$

23.

25.

27.

45. 6.0 V   **47.** 24.4 cm, 29.4 cm   **49.** 18 cm, 30 cm

51. 67 ft   **53.** 3.936 cm, 3.936 cm, 1.936 cm   **55.** 0.25 ft/min

29.

31.

33. 0.4

35. 0.2, 5.8

37. 1.4

39. −0.7, 0.7

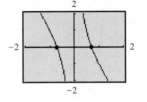

41. All real numbers $y \geq -6.25$

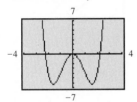

43. All real numbers $y \leq -2.83$ or $y \geq 2.83$

45. Either a or b is positive, the other is negative.

47. $(1, \sqrt{3})$ or $(1, -\sqrt{3})$ **49.** In any quadrant (not on an axis)

51. Many possibilities (two shown)   **53.** $y = \sqrt{x+1} + 1$

55. They are reflections of each other across the y-axis.

57. Yes; no two x-values are the same.

59. 13.4   **61.** 72.0°

63.

65.

67. $L = 2\pi r + 12$

69. $f(309) = 2.63\%$

71.

73.

75.

77. 11.3 ft

79. 3.5 h   **81.** 33°C   **83.** 1.03 ft   **85.** 6.5 h

Exercises 4.1, page 111

1. 865.6°   **3.** −135°

5.

7.

9. 405°, −315°   **11.** 210°, −510°   **13.** 430° 30′, −289° 30′
15. 638.1°, −81.9°   **17.** 15.18°   **19.** 82.91°   **21.** 18.85°
23. −235.49°   **25.** 0.977   **27.** 6.72   **29.** 47° 30′
31. −5° 37′   **33.** 15.20°   **35.** 301.27°

37.

39.

41.

43.

45. I, IV **47.** I, quadrantal **49.** I, II **51.** III, I

53. 21.710° **55.** 86° 16′ 26″

Exercises 4.2, page 115

1. $\sin\theta = \dfrac{3}{5}$, $\cos\theta = \dfrac{4}{5}$, $\tan\theta = \dfrac{3}{4}$, $\cot\theta = \dfrac{4}{3}$,

$\sec\theta = \dfrac{5}{4}$, $\csc\theta = \dfrac{5}{3}$

3. $\sin\theta = \dfrac{4}{5}$, $\cos\theta = \dfrac{3}{5}$, $\tan\theta = \dfrac{4}{3}$, $\cot\theta = \dfrac{3}{4}$,

$\sec\theta = \dfrac{5}{3}$, $\csc\theta = \dfrac{5}{4}$

5. $\sin\theta = \dfrac{8}{17}$, $\cos\theta = \dfrac{15}{17}$, $\tan\theta = \dfrac{8}{15}$, $\cot\theta = \dfrac{15}{8}$,

$\sec\theta = \dfrac{17}{15}$, $\csc\theta = \dfrac{17}{8}$

7. $\sin\theta = \dfrac{40}{41}$, $\cos\theta = \dfrac{9}{41}$, $\tan\theta = \dfrac{40}{9}$, $\cot\theta = \dfrac{9}{40}$,

$\sec\theta = \dfrac{41}{9}$, $\csc\theta = \dfrac{41}{40}$

9. $\sin\theta = \dfrac{\sqrt{15}}{4}$, $\cos\theta = \dfrac{1}{4}$, $\tan\theta = \sqrt{15}$, $\cot\theta = \dfrac{1}{\sqrt{15}}$,

$\sec\theta = 4$, $\csc\theta = \dfrac{4}{\sqrt{15}}$

11. $\sin\theta = \dfrac{1}{\sqrt{2}}$, $\cos\theta = \dfrac{1}{\sqrt{2}}$, $\tan\theta = 1$, $\cot\theta = 1$,

$\sec\theta = \sqrt{2}$, $\csc\theta = \sqrt{2}$

13. $\sin\theta = \dfrac{2}{\sqrt{29}}$, $\cos\theta = \dfrac{5}{\sqrt{29}}$, $\tan\theta = \dfrac{2}{5}$, $\cot\theta = \dfrac{5}{2}$,

$\sec\theta = \dfrac{\sqrt{29}}{5}$, $\csc\theta = \dfrac{\sqrt{29}}{2}$

15. $\sin\theta = 0.808$, $\cos\theta = 0.589$, $\tan\theta = 1.37$, $\cot\theta = 0.729$,

$\sec\theta = 1.70$, $\csc\theta = 1.24$

17. $\dfrac{5}{13}, \dfrac{12}{5}$ **19.** $\dfrac{2}{\sqrt{5}}, \sqrt{5}$ **21.** 0.882, 1.33

23. 0.246, 3.94 **25.** $\sin\theta = \dfrac{4}{5}$, $\tan\theta = \dfrac{4}{3}$

27. $\tan\theta = \dfrac{1}{3}$, $\sec\theta = \dfrac{\sqrt{10}}{3}$ **29.** 1 **31.** $\cos\theta = \sqrt{1 - y^2}$

33. $-\dfrac{7}{3}$ **35.** $\sec\theta$

Exercises 4.3, page 118

1. 20.65° **3.** 71.0°

5. $\sin 40° = 0.64$, $\cos 40° = 0.77$, $\tan 40° = 0.84$, $\cot 40° = 1.19$, $\sec 40° = 1.31$, $\csc 40° = 1.56$

7. $\sin 15° = 0.26$, $\cos 15° = 0.97$, $\tan 15° = 0.27$, $\cot 15° = 3.73$, $\sec 15° = 1.04$, $\csc 15° = 3.86$

9. 0.381 **11.** 1.58 **13.** 0.9626 **15.** 1.0 **17.** 0.4085
19. 1.569 **21.** 116.9 **23.** 0.07063 **25.** 70.97°
27. 65.70° **29.** 11.7° **31.** 49.453° **33.** 53.44°
35. 86.5° **37.** 74.1° **39.** 7.0° **41.** 0.9556 = 0.9556
43. 2.747 = 2.747 **45.** y is always less than r.
47. For a given r, x decreases as θ increases from 0° to 90°.
49. 0.8885 **51.** 0.93614 **53.** 71.6°
55. 87 dB **57.** 48.6°

Exercises 4.4, page 123

1. $\sin A = 0.868$, $\cos A = 0.496$, $\tan A = 1.75$, $\sin B = 0.496$, $\cos B = 0.868$, $\tan B = 0.571$

3. $a = 44.68$, $A = 34.17°$, $B = 55.83°$

5. **7.**

9. $B = 12.2°$, $b = 1450$, $c = 6850$
11. $A = 25.8°$, $B = 64.2°$, $b = 311$
13. $A = 57.9°$, $a = 20.2$, $b = 12.6$
15. $A = 21°$, $a = 32$, $B = 69°$
17. $a = 30.21$, $B = 57.90°$, $b = 48.16$
19. $A = 52.15°$, $B = 37.85°$, $c = 71.85$
21. $A = 52.5°$, $b = 0.661$, $c = 1.09$
23. $A = 15.82°$, $a = 0.5239$, $c = 1.922$
25. $A = 65.886°$, $B = 24.114°$, $c = 648.46$
27. $a = 0.75835$, $c = 14.612$, $B = 87.025°$
29. $a = 0.0959$, $b = 0.0162$, $A = 80.44°$
31. Cannot solve: need one more given part.
33. 4.45 **35.** 40.24° **37.** 43.1° **39.** 788
41. $a = c\sin A$, $b = c\cos A$, $B = 90° - A$
43. $A = 90° - B$, $b = a\tan B$, $c = \dfrac{a}{\cos B}$

Exercises 4.5, page 126

1. 2090 ft **3.** 97 ft **5.** 44.0 ft **7.** 0.4° **9.** 850.1 cm
11. 25.4 ft **13.** 0.21 mi **15.** 26.6°, 63.4°, 90.0°
17. 3.4° **19.** 23.5° **21.** 8.1° **23.** 3.07 cm **25.** 651 ft
27. 30.2° **29.** 2.48 in. **31.** 47.3 m **33.** 2100 ft
35. $d = 2x\tan(0.5\theta)$ **37.** $A = a(b + a\cos\theta)\sin\theta$
39. 35.3°

Review Exercises for Chapter 4, page 129

1. $377.0°, -343.0°$ **3.** $142.5°, -577.5°$ **5.** $31.9°$

7. $-38.1°$ **9.** $17° 30'$ **11.** $749° 42'$

13. $\sin\theta = \dfrac{7}{25}, \cos\theta = \dfrac{24}{25}, \tan\theta = \dfrac{7}{24}, \cot\theta = \dfrac{24}{7},$

$\sec\theta = \dfrac{25}{24}, \csc\theta = \dfrac{25}{7}$

15. $\sin\theta = \dfrac{1}{\sqrt{2}}, \cos\theta = \dfrac{1}{\sqrt{2}}, \tan\theta = 1, \cot\theta = 1,$

$\sec\theta = \sqrt{2}, \csc\theta = \sqrt{2}$

17. $0.923, 2.40$ **19.** $0.447, 1.12$ **21.** 0.952 **23.** 42.12

25. 1.05 **27.** 0 **29.** $18.2°$ **31.** $57.57°$ **33.** $12.25°$

35. $87.7°$ **37.** $8.00°$ **39.** $88.85°$

41. $a = 1.83, B = 73.0°, c = 6.27$

43. $A = 51.5°, B = 38.5°, c = 104$

45. $B = 52.5°, b = 15.6, c = 19.7$

47. $A = 31.61°, a = 4.006, B = 58.39°$

49. $a = 0.6292, B = 40.33°, b = 0.5341$

51. $A = 48.813°, B = 41.187°, b = 10.196$

53. 4.6 **55.** 61.2 **57.** 10.5

59. $x/h = \cot A, y/h = \cot B, c = x + y = h\cot A + h\cot B$

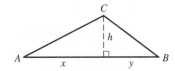

61. $\sqrt{x^2 + 1}/x$ **63.** $16.7°$ **65.** $80.3°$ **67.** $12.0°$

69. (a) $A = \frac{1}{2}bh = \frac{1}{2}b(a\sin C) = \frac{1}{2}ab\sin C$ (b) 679.2 m^2

71. 4.92 km **73.** 9.77 ft^2 **75.** 135 ft^2 **77.** 4.43 m

79. 34 m **81.** 56% **83.** 10.2 in.

85. Middle: $\dfrac{580}{\tan 1.1°} = 30{,}200\text{ m}$; end: $\dfrac{1160}{\tan 2.2°} = 30{,}200\text{ m}$

87. 1.83 km

89. (a) $147{,}000\text{ mm}^2$ (b) $155{,}000\text{ mm}^2$

(c) Curved surfaces covering each pentagon or hexagon would have greater area than plane area.

91. 464 m **93.** 73.3 cm

Exercises 5.1, page 137

1. Yes. The equation is then linear.   **3.** $x = 4, y = 4$

5. Yes; no   **7.** Yes; yes   **9.** $-3, -\dfrac{21}{2}$ **11.** $\dfrac{1}{4}, -0.6$

13. $0, -\dfrac{28}{9}$ **15.** Yes   **17.** No   **19.** No   **21.** Yes

23. Yes   **25.** 6 **27.** Yes   **29.** No

Exercises 5.2, page 141

1. $m = \frac{1}{4}$. The line rises 1 unit for each 4 units in going from left to right.

3. The y-intercept is $(0, 2)$. The checkpoint is $\left(1, \dfrac{4}{3}\right)$.

5. 4 **7.** -5 **9.** $\dfrac{2}{7}$ **11.** $\dfrac{1}{2}$

13.

15.

17.

19.

21. $m = -2, b = 1$

23. $m = 1, b = -4$

25. $m = \frac{5}{2}, b = -20$

27. $m = -\frac{3}{5}, b = \frac{3}{8}$

29.

31.

33.

35.

37. 6

39.

41.

Exercises 5.3, page 145

1. $x = 1.5, y = 1.4$ **3.** $x = 3.0, y = 1.0$

5. $x = 3.0, y = 0.0$ **7.** $x = 2.2, y = -0.3$

9. $x = -0.9, y = -2.3$ **11.** $s = 1.1, t = -1.7$

13. $x = 0.0, y = 3.0$ **15.** $x = -14.0, y = -5.0$

17. $r_1 = 4.0, r_2 = 7.5$ **19.** $x = 2.1, y = -0.4$

21. $x = -3.600, y = -1.400$ **23.** $x = 1.111, y = 0.841$

25. Dependent **27.** $t = -1.887, v = -3.179$

29. $x = 1.500, y = 4.500$ **31.** Inconsistent

33. 50 N, 47 N **35.** 8.5 ft, 11 ft **37.** 40 kg, 80 kg

Exercises 5.4, page 151

1. $x = -3, y = -3$ **3.** $x = \dfrac{16}{13}, y = -\dfrac{2}{13}$

5. $x = 1, y = -2$ **7.** $V = 7, p = 3$ **9.** $x = -1, y = -4$

11. $x = \dfrac{1}{2}, y = 2$ **13.** $x = 1, y = \dfrac{1}{2}$ **15.** $x = 3, y = 1$

17. $x = -1, y = -2$ **19.** $t = -\dfrac{1}{3}, y = 2$

21. Inconsistent **23.** $x = -\dfrac{14}{5}, y = -\dfrac{16}{5}$

25. $x = \dfrac{1}{2}, y = -4$ **27.** $t = -\dfrac{2}{3}, y = 0$

29. $x = \dfrac{3}{5}, y = \dfrac{1}{5}$ **31.** $C = -1, V = -2$

33. $A = -1, B = 3$ **35.** $a = \dfrac{1}{3}, b = -7$

37. $x = 2.38, y = 0.45$ **39.** $s = 3.9, t = 3.2$

41. $V_1 = 9.0\,V, V_2 = 6.0\,V$ **43.** $x = 6250\,L, y = 3750\,L$

45. 54 **47.** $W_r = 8120\,N, W_f = 9580\,N$

49. $t_1 = 32\,s, t_2 = 20\,s$

51. 34 at \$900/month, 20 at \$1250/month

53. 7.00 kW, 9.39 kW

55. Incorrect conclusion or error in sales figures; system of equations is inconsistent.

57. $a \neq b$ **59.** $(-3, -4), (9, 0)$

Exercises 5.5, Page 157

1. 86 **3.** $x = -13, y = -27$ **5.** -10 **7.** 29

9. 32 **11.** 9300 **13.** 1.083 **15.** 96 **17.** $x = 3, y = 1$

19. $x = -1, y = -2$ **21.** $t = -\dfrac{1}{3}, y = 2$

23. Inconsistent **25.** $x = -\dfrac{14}{5}, y = -\dfrac{16}{5}$

27. $x = 2.38, y = 0.45$ **29.** $s = 3.9, t = 3.2$

31. $x = -11.2, y = -9.26$ **33.** $x = -1.0, y = -2.0$

35. 0 **37.** 0 **39.** $F_1 = 15\,lb, F_2 = 6.0\,lb$

41. $x = 18.4\,gal, y = 17.6\,gal$

43. 160 3-bedroom homes, 80 4-bedroom homes

45. 210 phones, 110 detectors **47.** \$2000, 6%

49. 2.5 h, 2.1 h **51.** $V = 4.5i - 3.2$

Exercises 5.6, page 161

1. $x = 1, y = -4, z = \dfrac{1}{3}$ **3.** $x = 2, y = -1, z = 1$

5. $x = 4, y = -3, z = 3$ **7.** $l = \dfrac{1}{2}, w = \dfrac{2}{3}, h = \dfrac{1}{6}$

9. $x = \dfrac{2}{3}, y = -\dfrac{1}{3}, z = 1$ **11.** $x = \dfrac{4}{15}, y = -\dfrac{3}{5}, z = \dfrac{1}{3}$

13. $x = \dfrac{3}{4}, y = 1, z = -\dfrac{1}{2}$ **15.** $r = 0, s = 0, t = 0, u = -1$

17. $A = 2, B = 1, C = 1$

19. $P = 800\,h, M = 125\,h, I = 225\,h$

21. $F_1 = 9.43\,N, F_2 = 8.33\,N, F_3 = 1.67\,N$

23. $A = 22.5°, B = 45.0°, C = 112.5°$

25. $\theta = 0.295t^3 - 5.53t^2 + 24.2t$

27. 70 lb, 100 lb, 30 lb

29. Unlimited: $x = -10, y = -6, z = 0$ **31.** No solution

Exercises 5.7, page 167

1. -38 **3.** 122 **5.** 651 **7.** -439 **9.** 202

11. 29,440 **13.** 0.128 **15.** $x = -1, y = 2, z = 0$

17. $x = 2, y = -1, z = 1$ **19.** $x = 4, y = -3, z = 3$

21. $l = \dfrac{1}{2}, w = \dfrac{2}{3}, h = \dfrac{1}{6}$ **23.** $x = \dfrac{2}{3}, y = -\dfrac{1}{3}, z = 1$

25. $x = \dfrac{4}{15}, y = -\dfrac{3}{5}, z = \dfrac{1}{3}$ **27.** $p = -2, q = \dfrac{2}{3}, r = \dfrac{1}{3}$

29. $x = \dfrac{3}{4}, y = 1, z = -\dfrac{1}{2}$ **31.** -19 (sign changes)

33. 19 (no change) **35.** $A = 125\,N, B = 60\,N, F = 75\,N$

37. $s_0 = 2\,ft, v_0 = 5\,ft/s, a = 4\,ft/s^2$

39. $V = 5.0 + 0.5T + 0.1T^2$ **41.** 79% Ni, 16% Fe, 5% Mo

43. 30.9 mi/h, 45.9 mi/h, 551 mi/h

Review Exercises for Chapter 5, page 169

1. -17 **3.** -1485 **5.** -4 **7.** $\dfrac{2}{7}$

9. $m = -2, b = 4$ **11.** $m = 4, b = -\dfrac{5}{2}$

13. $x = 2.0, y = 0.0$ **15.** $A = 2.2, B = 2.7$

17. $x = 1.5, y = -1.9$ **19.** $M = 1.5, N = 0.4$

21. $x = 1, y = 2$ **23.** $x = \dfrac{1}{2}, y = -2$ **25.** $i = 2, v = -\dfrac{1}{3}$

27. $x = -\dfrac{6}{19}, y = \dfrac{36}{19}$ **29.** $x = 1.10, y = 0.54$

31. $x = 1, y = 2$ **33.** $x = \dfrac{1}{2}, y = -2$

35. $i = 2, v = -\dfrac{1}{3}$ **37.** $x = -\dfrac{6}{19}, y = \dfrac{36}{19}$

39. $x = 1.10, y = 0.54$ **41.** Best choices: 33, 34

43. Best choices: 39, 40   **45.** -115 **47.** 230.08

49. $x = 2, y = -1, z = 1$ **51.** $r = 3, s = -1, t = \dfrac{3}{2}$

53. $x = -0.17, y = 0.16, z = 2.4$ **55.** $x = 2, y = -1, z = 1$

57. $r = 3, s = -1, t = \dfrac{3}{2}$ **59.** $x = -0.17, y = 0.16, z = 2.4$

61. 4   **63.** $-\dfrac{15}{7}$ **65.** $x = \dfrac{8}{3}, y = -8$ **67.** $x = 1, y = 3$

69. -6 **71.** $-\dfrac{4}{3}$ **73.** $F_1 = 230\,\text{lb}, F_2 = 24\,\text{lb}, F_3 = 200\,\text{lb}$

75. $p_1 = 42\%, p_2 = 58\%$ **77.** \$40,000, \$4000

79. 28 tons of 6% copper, 14 tons of 2.4% copper

81. $a = 440\,\text{m}\cdot{}^\circ\text{C}, b = 9.6{}^\circ\text{C}$ **83.** 22,800 km/h, 1400 km/h

85. $R_1 = 0.50\,\Omega, R_2 = 1.5\,\Omega$ **87.** $L = 10\,\text{lb}, w = 40\,\text{lb}$

89. $A = 75^\circ, B = 65^\circ, C = 40^\circ$ **91.** 68 MB, 24 MB, 48 MB

93. 4.3 N, 1.7 N

Exercises 6.1, page 176

1. $9r^2 - 4s^2$ **3.** $x^2 + 2xy + y^2 - 4x - 4y + 4$

5. $40x - 40y$ **7.** $2x^3 - 8x^2$ **9.** $T^2 - 36$ **11.** $9v^2 - 4$

13. $16x^2 - 25y^2$ **15.** $144 - 25a^2b^2$ **17.** $25f^2 + 40f + 16$

19. $4x^2 + 68x + 289$ **21.** $L^4 - 2L^2 + 1$

23. $16a^2 + 56axy + 49x^2y^2$ **25.** $0.36s^2 - 1.2st + t^2$

27. $x^2 + 6x + 5$ **29.** $C^4 + 9C^2 + 18$

31. $20x^2 - 21x - 5$ **33.** $40v^2 + 138v - 45$

35. $600x^2 - 130xy - 63y^2$ **37.** $2x^2 - 8$ **39.** $8a^3 - 2a$

41. $6ax^2 + 24abx + 24ab^2$ **43.** $20n^4 + 100n^3 + 125n^2$

45. $16R^4 - 72R^2r^2 + 81r^4$

47. $x^2 + y^2 + 2xy + 2x + 2y + 1$

49. $x^2 + y^2 + 2xy - 6x - 6y + 9$ **51.** $125 - 75t + 15t^2 - t^3$

53. $27L^3 + 189L^2R + 441LR^2 + 343R^3$

55. $w^2 + 2wh + h^2 - 1$ **57.** $x^3 + 8$ **59.** $64 - 27x^3$

61. $x^4 - 2x^2y^2 + y^4$ **63.** $P_0P_1c + P_1G$

65. $4p^2 + 8pDA + 4D^2A^2$ **67.** $\tfrac{1}{2}\pi R^2 - \tfrac{1}{2}\pi r^2$

69. $\dfrac{L}{6}x^3 - \dfrac{L}{2}ax^2 + \dfrac{L}{2}a^2x - \dfrac{L}{6}a^3$

71. $L_0 + aL_0T - aL_0T_0$

73. $(49)(51) = (50 - 1)(50 + 1) = 50^2 - 1^2 = 2499$

75. $4x^2 - 9$ **77.** $(x + y)^2 = x^2 + 2xy + y^2$ (Eq. 6.3)

79. $4x^2 - y^2 - 4y - 4$

Exercises 6.2, page 181

1. $2ax(2x - 1)$ **3.** $5(x + 3)(x - 3)$ **5.** $6(x + y)$

7. $5(a - 1)$ **9.** $3x(x - 3)$ **11.** $7b(bh - 4)$

13. $24n(12n + 1)$ **15.** $2(x + 2y - 4z)$

17. $3ab(b - 2 + 4b^2)$ **19.** $4pq(3q - 2 - 7q^2)$

21. $2(a^2 - b^2 + 2c^2 - 3d^2)$ **23.** $(x + 2)(x - 2)$

25. $(10 + 3A)(10 - 3A)$ **27.** $36a^4 + 1$ (prime)

29. $2(9s + 5t)(9s - 5t)$ **31.** $(12n - 13p^2)(12n + 13p^2)$

33. $(x + y + 3)(x + y - 3)$ **35.** $2(x + 2)(x - 2)$

37. $300(x + 3z)(x - 3z)$ **39.** $2(I - 1)(I - 5)$

41. $(x^2 + 4)(x + 2)(x - 2)$

43. $(x^4 + 1)(x^2 + 1)(x + 1)(x - 1)$ **45.** $\dfrac{3 + b}{2 - b}$

47. $\dfrac{3}{2(t - 1)}$ **49.** $\tfrac{5}{2}$ **51.** $(3 + b)(x - y)$

53. $(a - b)(a + x)$ **55.** $(x + 2)(x - 2)(x + 3)$

57. $(x - y)(x + y + 1)$ **59.** $8^8 = 16,777,216$

61. $n^2 + n = n(n + 1)$; product of two consecutive positive integers is a positive integer.

63. $2\pi r(h + r)$ **65.** $Rv(1 + v + v^2)$

67. $r(R - r)(R + r)$ **69.** $r^2(\pi - 2)$ **71.** $\dfrac{i_2R_2}{R_1 + R_2}$

73. $\dfrac{5Y}{3(3S - Y)}$ **75.** $\dfrac{ER}{A(T_0 - T_1)}$

Exercises 6.3, page 187

1. $(x + 3)(x + 1)$ **3.** $(2x - 1)(x - 5)$

5. $2(x + 6)(x - 3)$ **7.** $(x + 1)(x + 4)$

9. $(s - 7)(s + 6)$ **11.** $(t + 8)(t - 3)$

13. $(x + 1)^2$ **15.** $(L - 2K)^2$ **17.** $(3x + 1)(x - 2)$

19. $4(3y + 1)(y - 3)$ **21.** $(2s + 11)(s + 1)$

23. $(3f^2 - 1)(f^2 - 5)$ **25.** $(2t - 3)(t + 5)$

27. $(3t - 4u)(t - u)$ **29.** $(4x - 7)(x + 1)$

31. $(9x - 2y)(x + y)$ **33.** $(2m + 5)^2$ **35.** $2(2x - 3)^2$

37. $(3t - 4)(3t - 1)$ **39.** $(8b^3 - 1)(b^3 + 4)$

41. $(4p - q)(p - 6q)$ **43.** $(12x - y)(x + 4y)$

45. $2(x - 1)(x - 6)$ **47.** $2x(2x^2 - 1)(x^2 + 4)$

49. $ax(x + 6a)(x - 2a)$ **51.** $(a + b + 2)(a + b - 2)$

53. $(5a + 5x + y)(5a - 5x - y)$ **55.** $(4x^n - 3)(x^n + 4)$

57. $16(t - 1)(t - 4)$ **59.** $4(s + 1)(s + 3)$

61. $100(2n + 3)(n - 12)$ **63.** $(V - nB)^2$

65. $wx^2(x - 2L)(x - 3L)$ **67.** $Ad(3u - v)(u - v)$

69. $k = -1, (2x + 1)^2$ **71.** $(x^2 + 2 + 2x)(x^2 + 2 - 2x)$

73. $9(4x^2 + 1)$

Exercises 6.4, page 189

1. $(x - 2)(x^2 + 2x + 4)$ **3.** $(x + 1)(x^2 - x + 1)$

5. $(2 - t)(4 + 2t + t^2)$ **7.** $(3x - 2a^2)(9x^2 + 6a^2x + 4a^4)$

9. $4(x + 2)(x^2 - 2x + 4)$ **11.** $6A^3(A - 1)(A^2 + A + 1)$

13. $6x^3y(9 - y^3)$ **15.** $x^3y^3(x + y)(x^2 - xy + y^2)$

17. $3a^2(a^2 + 1)(a + 1)(a - 1)$

19. $0.001(R - 4r)(R^2 + 4Rr + 16r^2)$

21. $27L^3(L + 2)(L^2 - 2L + 4)$

23. $(a + b + 4)(a^2 + 2ab + b^2 - 4a - 4b + 16)$

25. $(2 - x)(2 + x)(16 + 4x^2 + x^4)$

27. $2(x + 5)(x^2 - 5x + 25)$

29. $D(D - d)(D^2 + Dd + d^2)$

31. $QH(H + Q)(H^2 - HQ + Q^2)$

33. $x^5 - y^5 = (x - y)(x^4 + x^3y + x^2y^2 + xy^3 + y^4)$

$x^7 - y^7 = (x - y)(x^6 + x^5y + x^4y^2 + x^3y^3 + x^2y^4 + xy^5 + y^6)$

35. $(x + y)(x - y)(x^4 + x^2y^2 + y^4)$

37. $(n^3 + 1) = (n + 1)(n^2 - n + 1)$; If $n > 1$, both factors are integers and greater than 1.

Exercises 6.5, page 193

1. $\dfrac{3c}{4b}$ **3.** $\dfrac{x + 2}{x - 2}$ **5.** $\dfrac{14}{21}$ **7.** $\dfrac{2ax^2}{2xy}$ **9.** $\dfrac{2x - 4}{x^2 + x - 6}$

11. $\dfrac{ax^2 - ay^2}{x^2 - xy - 2y^2}$ **13.** $\dfrac{7}{11}$ **15.** $\dfrac{2xy}{4y^2}$ **17.** $\dfrac{2}{R + 1}$

19. $\dfrac{s - 5}{2s - 1}$ **21.** $9xy$ **23.** $a^2 - 25$ **25.** $2x$ **27.** 1

29. $\dfrac{1}{4}$ **31.** $\dfrac{3x}{4}$ **33.** $\dfrac{1}{5a}$ **35.** $\dfrac{2a - 2b}{2a - b}$

37. $\dfrac{4x^2 + 1}{(2x + 1)(2x - 1)}$ (cannot be reduced)   **39.** $3x$

41. $\dfrac{1}{2y^2}$ **43.** $\dfrac{x - 4}{x + 4}$ **45.** $\dfrac{2w^2 - 1}{w^2 + 8}$ **47.** $\dfrac{5x + 4}{x(x + 3)}$

49. $\dfrac{(N + 2)(N^2 + 4)}{8}$ **51.** $\dfrac{1}{2t + 1}$ **53.** $\dfrac{x + 3}{x - 3}$

55. $-\dfrac{x + y}{2}$ **57.** $\dfrac{x + 1}{x - 1}$ **59.** $\dfrac{(x + 5)(x - 3)}{(5 - x)(x + 3)}$

61. $\dfrac{x^2 - xy + y^2}{2}$ **63.** $\dfrac{2x}{9x^2 - 3x + 1}$

65. (a) $\dfrac{x^2(x + 2)}{x^2 + 4}$ (b) $\dfrac{x^2}{x^2 - 4}$ Numerator and denominator have no common factor. In each, x^2 is not a factor of the denominator.

67. (a) $\dfrac{(x - 2)(x + 1)}{x(x - 1)}$ (b) $\dfrac{x - 2}{x}$ Numerator and denominator have no common factor. In each, x is not a factor of the numerator.

69. $u + v$ **71.** $\dfrac{E^2(R - r)}{(R + r)^3}$

Exercises 6.6, page 197

1. $\dfrac{2(x + y)}{3(x - y)}$ **3.** $3(x + 2y)$ **5.** $\dfrac{3}{28}$ **7.** $6xy$ **9.** $\dfrac{7}{18}$

11. $\dfrac{xy^2}{bz^2}$ **13.** $4t$ **15.** $3(u + v)(u - v)$ **17.** $\dfrac{50}{3(a + 4)}$

19. $\dfrac{x^2 - 3}{x^2(x^2 + 3)}$ **21.** $\dfrac{3x}{5a}$ **23.** $\dfrac{(x + 1)(x - 1)(x - 4)}{4(x + 2)}$

25. $\dfrac{x^2}{a + x}$ **27.** $\dfrac{15}{4}$ **29.** $\dfrac{3}{4x + 3}$ **31.** $\dfrac{2(3T + N)(5V - 6)}{(V - 7)(4T + N)}$

33. $\dfrac{7x^4}{3a^4}$ **35.** $\dfrac{4t(2t - 1)(t + 5)}{(2t + 1)^2}$ **37.** $\dfrac{x + y}{2}$

39. $(x + y)(3p + 7q)$ **41.** $\dfrac{x - 2}{6x}$ **43.** $\dfrac{2(2x - 1)}{5(2 - x)}$

45. $\dfrac{2v_1v_2}{v_1 + v_2}$ **47.** $\dfrac{\pi}{2}$

Exercises 6.7, page 202

1. $12a^2b^3$ **3.** $\dfrac{x^3 - 10x^2 + 36x - 90}{2(x - 4)^2(x + 3)}$ **5.** $\dfrac{9}{5}$ **7.** $\dfrac{8}{x}$

9. $\dfrac{5}{4}$ **11.** $\dfrac{3 + 7ax + 8x}{4x}$ **13.** $\dfrac{ax - b}{x^2}$ **15.** $\dfrac{30 + ax^2}{25x^3}$

17. $\dfrac{14 - a^2}{10a}$ **19.** $\dfrac{xy + 7x + 2y - 8}{4xy}$ **21.** $\dfrac{7}{2(2x - 1)}$

23. $\dfrac{5 - 3x}{2x(x + 1)}$ **25.** $\dfrac{-3}{4(s - 3)}$ **27.** $\dfrac{7R + 6}{3(R + 3)(R - 3)}$

29. $\dfrac{2x - 5}{(x - 4)^2}$ **31.** $\dfrac{-4}{(v + 1)(v - 3)}$ **33.** $\dfrac{9x^2 + x - 2}{(3x - 1)(x - 4)}$

35. $\dfrac{-2t^3 + 3t^2 + 15t}{(t - 3)(t + 2)(t + 3)^2}$ **37.** $\dfrac{-2w^3 + w^2 - w}{(w + 1)(w^2 - w + 1)}$

39. $\dfrac{1}{x - 1}$ **41.** $\dfrac{x - y}{y}$ **43.** $-\dfrac{(3x + 4)(x - 1)}{2x}$

45. $\dfrac{h}{(x + 1)(x + h + 1)}$ **47.** $\dfrac{-2hx - h^2}{x^2(x + h)^2}$

49. $\dfrac{y^2 - rx + r^2}{r^2}$ **51.** $\dfrac{2a - 1}{a^2}$ **53.** $\dfrac{a^2 + 2a - 1}{a + 1}$

55. $\dfrac{a + b}{\frac{1}{a} + \frac{1}{b}} = ab$ **57.** $\dfrac{2mn}{m^2 + n^2}$ **59.** $\dfrac{3(H - H_0)}{4\pi H}$

61. $\dfrac{n(2n + 1)}{2(n + 2)(n - 1)}$ **63.** $\dfrac{P^2(9x^2 + L^2)}{4L^4}$

65. $\dfrac{sL + R}{s^2LC + sRC + 1}$

67. $\dfrac{\omega^2L^2 + \omega^4R^2C^2L^2 - 2\omega^2R^2CL + R^2}{\omega^2R^2L^2}$

Exercises 6.8, page 207

1. $\dfrac{2}{b - 1}$ **3.** $\dfrac{arv^2}{2aM - rM}$ **5.** 4 **7.** -3 **9.** $\dfrac{13}{2}$

11. $\dfrac{16}{21}$ **13.** -9 **15.** $-\dfrac{2}{13}$ **17.** $\dfrac{5}{3}$ **19.** -2

21. $-\dfrac{1}{2}$ **23.** $\dfrac{37}{6}$ **25.** 7 **27.** $\dfrac{63}{8}$ **29.** No solution

31. $\dfrac{2}{3}$ **33.** $\dfrac{3b}{1 - 2b}$ **35.** $\dfrac{(2b - 1)(b + 6)}{2(b - 1)}$

37. $\dfrac{2s - 2s_0 - v_0t}{t}$ **39.** $\dfrac{40V - 240}{78 - 5.0V}$ **41.** $\dfrac{jX}{1 - g_mz}$

43. $\dfrac{PV^3 - bPV^2 + aV - ab}{RV^2}$ **45.** $\dfrac{CC_2 + CC_3 - C_2C_3}{C_2 - C}$

47. $\dfrac{fnR_2 - fR_2}{R_2 + f - fn}$ **49.** 3.1 h   **51.** 3.2 min   **53.** 220 m

55. 80 km/h   **57.** 2.2 V   **59.** $A = 3, B = -2$

Review Exercises for Chapter 6, page 209

1. $12ax + 15a^2$ **3.** $4a^2 - 49b^2$ **5.** $4a^2 + 4a + 1$

7. $b^2 + 3b - 28$ **9.** $2x^2 - 13x - 45$ **11.** $4c^{2a} - d^2$

13. $3(s + 3t)$ **15.** $a^2(x^2 + 1)$ **17.** $b^x(Wb + 12)(Wb - 12)$

19. $(4x + 8 + t^2)(4x + 8 - t^2)$ **21.** $4(3t - 1)^2$

23. $(5t + 1)^2$ **25.** $(x + 8)(x - 7)$

27. $(t + 3)(t - 3)(t^2 + 4)$ **29.** $(2k - 9)(k + 4)$

31. $2(2x + 5)(2x - 7)$ **33.** $(5b - 1)(2b + 5)$

35. $4(x + 4y)(x - 4y)$ **37.** $2(5 - 2y^2)(25 + 10y^2 + 4y^4)$

39. $(2x + 3)(4x^2 - 6x + 9)$ **41.** $(a - 3)(b^2 + 1)$

43. $(x + 5)(n - x + 5)$ **45.** $\dfrac{16x^2}{3a^2}$ **47.** $\dfrac{3x + 1}{2x - 1}$

49. $\dfrac{16}{5x(x - y)}$ **51.** $-\dfrac{6}{L - 5}$ **53.** $\dfrac{x + 2}{2x(7x - 1)}$

55. $\dfrac{1}{x - 1}$ **57.** $\dfrac{16x - 15}{36x^2}$ **59.** $\dfrac{5y + 6}{2xy}$ **61.** $\dfrac{-2(2a + 3)}{a(a + 2)}$

63. $\dfrac{4x^2 - x + 1}{2x(x + 3)(x - 1)}$ **65.** $\dfrac{12x^2 - 7x - 4}{2(x - 1)(x + 1)(4x - 1)}$

67. $\dfrac{x^3 + 6x^2 - 2x + 2}{x(x - 1)(x + 3)}$

69.   **71.**

73. $(x + \sqrt{5})(x - \sqrt{5})$ **75.** $x\left(1 + \dfrac{y}{x}\right)$ **77.** 2

79. $-\dfrac{(a - 1)^2}{2a}$ **81.** $\dfrac{10}{3}$ **83.** -6

85. (a) Sign changes (b) No change in sign

87. $\dfrac{1}{4}[(x + y)^2 - (x - y)^2]$

$= \dfrac{1}{4}(x^2 + 2xy + y^2 - x^2 + 2xy - y^2)$

$= \dfrac{1}{4}(4xy) = xy$

89. $2zS^2 + 2zS$ **91.** $4b^2 + 4bn\lambda - 4b\lambda + n^2\lambda^2 - 2n\lambda^2 + \lambda^2$

93. $(c + R)(T_2 - T_1)$ **95.** $R(3R - 4r)$

97. $8n^6 + 36n^5 + 66n^4 + 63n^3 + 33n^2 + 9n + 1$

99. $10aT - 10at + aT^2 - 2aTt + at^2$

101. $4(3x^2 + 12x + 16)$ **103.** $\dfrac{12\pi^2 wv^2 D}{gn^2 t}$

105. $\dfrac{R^2 + r^2}{2}$ **107.** $\dfrac{120 - 60d^2 + 5d^4 - d^6}{120}$

109. $\dfrac{4k^2 + k - 2}{4k(k - 1)}$ **111.** $\dfrac{4r^3 - 3ar^2 - a^3}{4r^3}$

113. $\dfrac{c^2u^2 - 2gc^2x}{1 - c^2u^2 + 2gc^2x}$ **115.** $\dfrac{W}{g(h_2 - h_1)}$

117. $\dfrac{RHw}{w - RH}$ **119.** $\dfrac{p^2}{2E - 2V_0 - (m + M)V^2}$

121. $\dfrac{-mb^2s^2 - kL^2}{b^2s}$ **123.** $\dfrac{i}{sV - V_0}$ **125.** 3.4 h

127. 600 Hz   **129.** 11.3   **131.** 18 Ω

Exercises 7.1, page 216

1. $-\dfrac{1}{2}, 4$ **3.** $a = 1, b = -2, c = -4$

5. Not quadratic, no x^2-term   **7.** $a = 1, b = -1, c = 0$

9. $2, -2$ **11.** $\dfrac{3}{2}, -\dfrac{3}{2}$ **13.** $-1, 9$ **15.** $3, 4$ **17.** $0, \dfrac{5}{2}$

19. $\dfrac{1}{3}, -\dfrac{1}{3}$ **21.** $\dfrac{1}{3}, 4$ **23.** $-1, \dfrac{4}{7}$ **25.** $\dfrac{2}{3}, \dfrac{3}{2}$ **27.** $\dfrac{1}{2}, -\dfrac{3}{2}$

29. $-\dfrac{2b}{3}, \dfrac{b}{2}$ **31.** $\dfrac{5}{2}, -\dfrac{9}{2}$ **33.** $0, -2$ **35.** $b - a, -b - a$

37. $-a - b, -a + b$ **39.** $3 + \dfrac{1}{2} = \dfrac{-(-7)}{2} = \dfrac{7}{2}$

41. 2 A, -6 A (if current is negative)   **43.** 2 A.M., 10 A.M.

45. $-1, 0, 1$ **47.** $\dfrac{3}{2}, 4$ **49.** $-2, \dfrac{1}{2}$ **51.** 3 N/cm, 6 N/cm

53. 30 km/h, 40 km/h

Exercises 7.2, page 219

1. $-3 \pm \sqrt{17}$ **3.** $-5, 5$ **5.** $-\sqrt{7}, \sqrt{7}$ **7.** $-3, 7$

9. $-3 \pm \sqrt{7}$ **11.** $2, -4$ **13.** $-2, -1$ **15.** $2 \pm \sqrt{2}$

17. $-5, 3$ **19.** $-3, \dfrac{1}{2}$ **21.** $\dfrac{1}{6}(3 \pm \sqrt{33})$

23. $\dfrac{1}{4}(1 \pm \sqrt{17})$ **25.** $\dfrac{1}{5}(5 \pm \sqrt{5})$ **27.** $-\dfrac{1}{3}, -\dfrac{1}{3}$

29. $-b \pm \sqrt{b^2 - c}$ **31.** $5°C, 15°C$ **33.** 16 ft

Exercises 7.3, page 222

1. $-3, -2$ **3.** $\dfrac{4}{3}$ (double root)   **5.** $2, -4$ **7.** $-2, -1$

9. $2 \pm \sqrt{2}$ **11.** $-5, 3$ **13.** $-3, \dfrac{1}{2}$ **15.** $\dfrac{1}{6}(3 \pm \sqrt{33})$

17. $\dfrac{1}{4}(1 \pm \sqrt{17})$ **19.** $-\dfrac{8}{5}, \dfrac{5}{6}$ **21.** $\dfrac{1}{16}(-61 \pm \sqrt{-119})$

23. $\dfrac{1}{6}(1 \pm \sqrt{109})$ **25.** $\pm\dfrac{11}{5}$ **27.** $\dfrac{3}{4}, -\dfrac{5}{8}$ **29.** $-0.54, 0.74$

31. $-0.26, 2.43$ **33.** $-c \pm \sqrt{c^2 + 1}$

35. $\dfrac{b + 1 \pm \sqrt{-3b^2 + 2b + 4b^2a + 1}}{2b^2}$ **37.** Imaginary, unequal

39. Real, irrational, unequal   **41.** 4   **43.** $-2, -1, 1, 2$

45. 4.376 cm   **47.** $\dfrac{1}{2}(1 + \sqrt{5}) = 1.618$

49. $\dfrac{-R \pm \sqrt{R^2 - 4L/C}}{2L}$ **51.** 1.8 cm   **53.** 11.0 m by 23.8 m

55. 2.6 ft   **57.** 70 mi/h, 55 mi/h

Exercises 7.4, page 227

1.   **3.**

5.

7.

9.

11.

13.

15.

17. $-1.2, 1.2$ **19.** $0.5, 3.1$ **21.** No real roots

23. $-2.4, 1.2$

25.

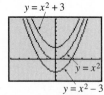

(a) The parabola $y = x^2 + 3$ is 3 units above, and the parabola $y = x^3 - 3$ is 3 units below.

(b) All parabolas open up.

27.

(a) Parabola (b) is 2 units to the right and 3 units up.
Parabola (c) is 2 units to the left and 3 units down.

(b) All parabolas open up.

29.

The parabola $y = 3x^2$ rises more quickly, and the parabola $y = \frac{1}{3}x^2$ rises more slowly.

31. $x = 7$ **33.** -2

35.

37.

39.

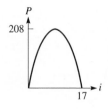

41. (a) After 16.2 s
(b) 1130 ft
(c) 3.3 s, 12.3 s

43. 4.53 cm   **45.** 173 ft by 116 ft, or 77 ft by 259 ft

Review Exercises for Chapter 7, page 228

1. $-4, 1$ **3.** $2, 8$ **5.** $\frac{1}{3}, -4$ **7.** $\frac{1}{2}, \frac{5}{3}$ **9.** $0, \frac{25}{6}$

11. $-\frac{3}{2}, \frac{7}{2}$ **13.** $-10, 11$ **15.** $-1 \pm \sqrt{7}$ **17.** $-4, \frac{9}{2}$

19. $\frac{1}{8}(3 \pm \sqrt{41})$ **21.** $\frac{1}{42}(-23 \pm \sqrt{-4091})$

23. $\frac{1}{6}(-2 \pm \sqrt{58})$ **25.** $-2 \pm 2\sqrt{2}$ **27.** $\frac{1}{3}(-4 \pm \sqrt{10})$

29. $-1, \frac{5}{4}$ **31.** $\frac{1}{4}(-3 \pm \sqrt{-47})$ **33.** $\dfrac{-1 \pm \sqrt{-1}}{a}$

35. $\dfrac{-3 \pm \sqrt{9 + 4a^2}}{2a}$ **37.** $-5, 6$ **39.** $\frac{1}{4}(1 \pm \sqrt{33})$

41. $3 \pm \sqrt{7}$ **43.** 0 (3 is not a solution)

45.

47.

49. $-1.7, 1.2$ **51.** No real roots   **53.** $x = -4$ **55.** $0, L$

57. 17   **59.** $202°F$ **61.** 0.6 s, 1.9 s   **63.** 8000

65. $\dfrac{-\pi h \pm \sqrt{\pi^2 h^2 + 2\pi A}}{2\pi}$ **67.** $\dfrac{1 + r \pm \sqrt{(1 + r)^2 - 4rp_2}}{2r}$

69.

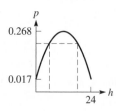

71. 61 m, 81 m   **73.** 9.9 cm

75. 970 m, 670 m; 570 m, 570 m   **77.** 16.0 in., 21.7 in.   **79.** 25

81.   6 h, 18 h

Exercises 8.1, page 234

1. (a) $-, +, -, -, +, -$ (b) $+, -, +, +, +, -$ **3.** $+, -$

5. $+, +$ **7.** $-, +$ **9.** $+, -$ **11.** $-, -$ **13.** $-, +$

15. $\sin \theta = \dfrac{1}{\sqrt{5}}$, $\cos \theta = \dfrac{2}{\sqrt{5}}$, $\tan \theta = \dfrac{1}{2}$, $\cot \theta = 2$,

$\sec \theta = \dfrac{1}{2}\sqrt{5}$, $\csc \theta = \sqrt{5}$

17. $\sin\theta = -\dfrac{3}{\sqrt{13}}, \cos\theta = -\dfrac{2}{\sqrt{13}}, \tan\theta = \dfrac{3}{2}, \cot\theta = \dfrac{2}{3},$

$\sec\theta = -\dfrac{1}{2}\sqrt{13}, \csc\theta = -\dfrac{1}{3}\sqrt{13}$

19. $\sin\theta = \dfrac{12}{13}, \cos\theta = -\dfrac{5}{13}, \tan\theta = -\dfrac{12}{5}, \cot\theta = -\dfrac{5}{12},$

$\sec\theta = -\dfrac{13}{5}, \csc\theta = \dfrac{13}{12}$

21. $\sin\theta = -\dfrac{2}{\sqrt{29}}, \cos\theta = \dfrac{5}{\sqrt{29}}, \tan\theta = -\dfrac{2}{5}, \cot\theta = -\dfrac{5}{2},$

$\sec\theta = \dfrac{1}{5}\sqrt{29}, \csc\theta = -\dfrac{1}{2}\sqrt{29}$

23. I, II   **25.** I, III   **27.** II, III   **29.** II   **31.** II
33. IV   **35.** III   **37.** II   **39.** −   **41.** −

Exercises 8.2, page 239

1. $\sin 200° = -\sin 20° = -0.3420,$
$\tan 150° = -\tan 30° = -0.5774,$
$\cos 265° = -\cos 85° = -0.0872,$
$\cot 300° = -\cot 60° = -0.5774, \quad \sec 344° = \sec 16° = 1.040,$
$\sin 397° = \sin 37° = 0.6018$

3. 296.00°   **5.** $\sin 20°; -\cos 40°$ **7.** $-\tan 75°; -\csc 58°$

9. $\cos 40°; -\tan 40°$ **11.** $-\sin 15° = -0.26$

13. $-\cos 73.7° = -0.281$ **15.** $\sec 31.67° = 1.175$

17. $-\tan 31.5° = -0.613$ **19.** −0.523   **21.** −0.7620

23. −3.910   **25.** 0.2989   **27.** 237.99°, 302.01°

29. 66.40°, 293.60°   **31.** 91°, 271°   **33.** 119.5°   **35.** 263°

37. 306.21°   **39.** 299.24°   **41.** −0.7003   **43.** −0.777

45. <   **47.** =   **49.** −0.9659   **51.** $\cot\theta$ **53.** 0.0183 A

55. 12.6 in.

Exercises 8.3, page 243

1. 160°   **3.** 1.082, 2.060   **5.** $\dfrac{\pi}{12}, \dfrac{5\pi}{6}$ **7.** $\dfrac{5\pi}{12}, \dfrac{11\pi}{6}$

9. $\dfrac{7\pi}{6}, \dfrac{3\pi}{20}$ **11.** $4\pi, -\dfrac{\pi}{20}$ **13.** 72°, 270°   **15.** 10°, 315°

17. 70°, 300°   **19.** −20°, 27°   **21.** 0.401   **23.** 4.40

25. 5.821   **27.** 8.351   **29.** 43.0°   **31.** 195.2°   **33.** 710°

35. −940.8°   **37.** 0.7071   **39.** 3.732   **41.** −0.8660

43. −8.327   **45.** 0.9056   **47.** −0.89   **49.** −2.1

51. 0.149   **53.** 0.3141, 2.827   **55.** 2.932, 6.074

57. 0.8309, 5.452   **59.** 2.442, 3.841   **61.** 1.25

63. −0.3827   **65.** $-\cot\theta$ **67.** 612 mil   **69.** 11.0 rad

71. 0.030 ft · lb   **73.** 2900 m

Exercises 8.4, page 248

1. 2.36 in.   **3.** 14.4 m/s   **5.** 3.46 in.   **7.** 426 mm

9. 0.1849 mi²   **11.** 67 cm²   **13.** 0.0647 ft   **15.** 8.21 m

17. 0.0431 mi   **19.** 1.81, 14.5 cm²   **21.** 32.73 min past noon

23. 4240 ft²   **25.** 0.52 rad/s   **27.** 34.73 m²   **29.** 2.30 ft

31. 22.6 m²   **33.** 369 m³   **35.** 0.4 km   **37.** 1070 ft/min

39. 5370 in./min   **41.** 3.18×10^5 mi/h   **43.** 72.0 r/min

45. 25.7 ft   **47.** 1400 in./min   **49.** 0.433 rad/s

51. 8800 ft/min   **53.** 250 rad   **55.** 940 m/s   **57.** 14.9 m³

59. Ratios become very close to 1   **61.** 7.13×10^7 mi

Review Exercises for Chapter 8, page 251

1. $\sin\theta = \dfrac{4}{5}, \cos\theta = \dfrac{3}{5}, \tan\theta = \dfrac{4}{3}, \cot\theta = \dfrac{3}{4}, \sec\theta = \dfrac{5}{3},$

$\csc\theta = \dfrac{5}{4}$

3. $\sin\theta = -\dfrac{2}{\sqrt{53}}, \cos\theta = \dfrac{7}{\sqrt{53}}, \tan\theta = -\dfrac{2}{7}, \cot\theta = -\dfrac{7}{2},$

$\sec\theta = \dfrac{\sqrt{53}}{7}, \csc\theta = -\dfrac{\sqrt{53}}{2}$

5. $-\cos 48°, \tan 14°$ **7.** $-\sin 71°, \sec 15°$ **9.** $\dfrac{2\pi}{9}, \dfrac{17\pi}{20}$

11. $\dfrac{34\pi}{15}, \dfrac{9\pi}{8}$ **13.** 252°, 130°   **15.** 12°, 330°   **17.** 32.1°

19. 2067°   **21.** 1.78   **23.** 0.3534   **25.** 4.5736

27. −11.10   **29.** −0.415   **31.** −0.47   **33.** −1.080

35. −0.4264   **37.** −1.64   **39.** 4.140   **41.** −0.5878

43. −0.8660   **45.** 0.5569   **47.** 1.197   **49.** 10.30°, 190.30°

51. 118.23°, 241.77°   **53.** 0.5759, 5.707   **55.** 4.187, 5.238

57. 223.76°   **59.** 246.78°   **61.** 10.8 in.

63. 3.23 = 185.3°   **65.** 14.2 m   **67.** 2120 in.²

69. $A = \dfrac{1}{2}r^2\theta - \dfrac{1}{2}r(r\sin\theta) = \dfrac{1}{2}r^2(\theta - \sin\theta)$ **71.** 1040 m²

73. 0.0562 W   **75.** 0.800 = 45.8°   **77.** 20.4 r/min

79. 2200 mi/h

81. (a) 4150 mi (b) 6220 mi; great circle route shortest of all routes

83. 4710 cm/s   **85.** 24.4 ft²   **87.** 1.81×10^6 cm/s

89. 138 m²   **91.** 3.58×10^5 km

Exercises 9.1, page 258

1. **3.**

5. (a) Scalar since no direction is given. (b) Vector since direction is given.

7. (a) Vector: magnitude and direction are specified; (b) scalar: only magnitude is specified.

9. **11.**

13.

15. 5.6, 50° **17.** 4.3, 156°

19. **21.** **23.** **25.**

27. **29.** **31.**

33. 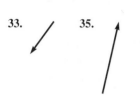 **35.** **37.** **39.** ←

41. **43.** 50 m/min 29° from horizontal

45. **47.**

Exercises 9.2, page 261

1. $V_x = -11.6, V_y = -8.46$ **3.** $A_x = 0, A_y = -375.4$
5. 662, 352 **7.** −349, −664 **9.** −750, 0
11. 3.22 lb, 7.97 lb **13.** −62.9 m/s, 44.1 m/s
15. 0, −9040 ft/s² **17.** −2.53 mN, −0.788 mN
19. −0.8088 ft, 0.3296 ft **21.** 23.9 km/h, 7.43 km/h
23. 2780 lb, 170 lb **25.** 116 km/h **27.** 0.692 ft, 3.43 ft
29. 89 N, −190 N **31.** 75 lb
33. 0.57 (km/h)/m, 0.48 (km/h)/m

Exercises 9.3, page 267

1. $R = 1650, \theta = 73.8°$ **3.** $R = 24.2, \theta = 52.6°$ with A
5. $R = 7.781, \theta = 66.63°$ with A **7.** $R = 10.0, \theta = 58.8°$
9. $R = 2.74, \theta = 111.0°$ **11.** $R = 2130, \theta = 107.7°$
13. $R = 7052, \theta = 349.82°$ **15.** $R = 29.2, \theta = 10.8°$
17. $R = 5920, \theta = 88.4°$ **19.** $R = 27.27, \theta = 33.14°$
21. $R = 50.2, \theta = 50.3°$ **23.** $R = 0.242, \theta = 285.9°$

25. $R = 532, \theta = 95.7°$ **27.** $R = 235, \theta = 121.7°$
29. 15,000 lb, 15° from 5500-lb force
31. 40 N **33.** 24.3 kg·m/s

Exercises 9.4, page 270

1. 47.10 mi, 10.27° N of E
3. 6.60 lb, 29.5° from the 5.75-lb force
5. 11,000 N, 44° above horizontal
7. 3070 ft, 17.8° S of W **9.** 229.4 ft, 72.82° N of E
11. 940 lb, 34° from 780 lb force
13. 25.3 km/h, 29.6° S of E **15.** 9.6 m/s², 7.0° from vertical
17. 175 lb, 9.3° above horizontal
19. 540 km/h, 6° from direction of plane
21. 18,130 mi/h, 0.03° from direction of shuttle
23. 27.0 mi, bearing 192.4° **25.** 9.6 km/h **27.** 910 N
29. 184,000 in./min², $\phi = 89.6°$ **31.** 138 mi, 65.0° N of E
33. 79.0 m/s, 11.3° from direction of plane, 75.6° from vertical
35. 4.05 A/m, 11.5° with magnet

Exercises 9.5, page 278

1. $b = 76.01, c = 50.01, C = 40.77°$
3. $b = 38.1, C = 66.0°, c = 46.1$
5. $a = 2800, b = 2620, C = 108.0°$
7. $B = 12.20°, C = 149.57°, c = 7.448$
9. $a = 11,050, A = 149.70°, C = 9.57°$
11. $A = 125.6°, a = 0.0776, c = 0.00566$
13. $A = 99.4°, b = 55.1, c = 24.4$
15. $A = 68.01°, a = 5520, c = 5376$
17. $A_1 = 61.36°, C_1 = 70.51°, c_1 = 5.628; A_2 = 118.64°,$
$C_2 = 13.23°, c_2 = 1.366$
19. $A_1 = 107.3°, a_1 = 5280, C_1 = 41.3°; A_2 = 9.9°, a_2 = 952,$
$C_2 = 138.7°$
21. No solution **23.** 25.3 cm **25.** 9.3° **27.** 1490 ft
29. 880 N **31.** 0.88 mi **33.** 13.94 cm **35.** 27,300 km
37. 77.3° with bank downstream **39.** 51°

Exercises 9.6, page 283

1. $A = 14.0°, B = 21.0°, c = 107$
3. $A = 50.3°, B = 75.7°, c = 6.31$
5. $A = 70.9°, B = 11.1°, c = 4750$
7. $A = 34.72°, B = 40.67°, C = 104.61°$
9. $A = 18.21°, B = 22.28°, C = 139.51°$
11. $A = 6.0°, B = 16.0°, c = 1150$
13. $A = 82.3°, b = 2160, C = 11.4°$
15. $A = 36.24°, B = 39.09°, a = 97.22$
17. $A = 46.94°, B = 61.82°, C = 71.24°$
19. $A = 137.9°, B = 33.7°, C = 8.4°$
21. $b = 3700, C = 25°, c = 2400$

23. (Intermediate steps) $a^2 = b^2 + c^2 - 2bc \cos A$,
$2bc \cos A = (b^2 + c^2 + 2bc) - a^2 - 2bc$,
$2bc(1 + \cos A) = (b + c)^2 - a^2$

25. Case 3—No; Case 4—Yes. One side may be longer than the sum of the other two sides.   **27.** 69.4 mi   **29.** 47.8°

31. 57.3°, 141.7°   **33.** 16.5 mm

35. 5.09 km/h   **37.** 17.8 mi   **39.** 2.05 km, 2.48 km

Review Exercises for Chapter 9, page 285

1. $A_x = 57.4, A_y = 30.5$ **3.** $A_x = -0.7485, A_y = -0.5357$

5. $R = 602, \theta = 57.1°$ with A **7.** $R = 5960, \theta = 33.60°$ with A

9. $R = 965, \theta = 8.6°$ **11.** $R = 26.12, \theta = 146.03°$

13. $R = 71.93, \theta = 336.50°$ **15.** $R = 0.9942, \theta = 359.57°$

17. $b = 181, C = 64.0°, c = 175$

19. $A = 21.2°, b = 34.8, c = 51.5$

21. $a = 17{,}340, b = 24{,}660, C = 7.99°$

23. $A = 39.88°, a = 5194, C = 30.03°$

25. $A_1 = 54.8°, a_1 = 12.7, B_1 = 68.6°; A_2 = 12.0°, a_2 = 3.24, B_2 = 111.4°$

27. $A = 32.3°, b = 267, C = 17.7°$

29. $A = 176.4°, B = 1.1°, c = 5.41$

31. $a = 1782, b = 1920, C = 16.00°$

33. $A = 37°, B = 25°, C = 118°$

35. $A = 20.6°, B = 35.6°, C = 123.8°$

37. Add the 3 forms of law of cosines together; simplify; divide by $2abc$.

39. 130 cm   **41.** $A_t = \frac{1}{2}ab; b = \dfrac{a \sin B}{\sin A}$; substitute

43. -155.7 lb, 81.14 lb   **45.** 2100 ft/s   **47.** 9.8 km

49. $R = 2700$ lb, $\theta = 107°$ **51.** 6.1 m/s, 35° with horizontal

53. 0.11 N   **55.** 1.52 pm   **57.** 2.30 m, 2.49 m

59. 0.039 km   **61.** 32,900 mi   **63.** 2.65 km   **65.** 190 mi

67. 2510 ft or 3370 ft (ambiguous)

69. 810 N, 36° N of E

Exercises 10.1, page 291

1. $y = 3 \cos x$

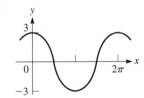

3. 0, -0.7, -1, -0.7, 0, 0.7, 1, 0.7, 0, -0.7, -1, -0.7, 0, 0.7, 1, 0.7, 0

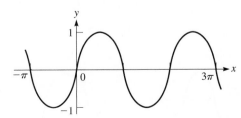

5. $-3, -2.1, 0, 2.1, 3, 2.1, 0, -2.1, -3, -2.1, 0, 2.1, 3, 2.1, 0, -2.1, -3$

7.

9.

11.

13.

15.

17.

19.

21.

23. 0, 0.84, 0.91, 0.14, -0.76, -0.96, -0.28, 0.66

25.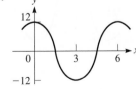

27. $y = -2 \sin x$

29. $y = -2\cos x$

31.

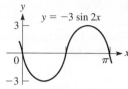

43. $y = \sin 6x$ **45.** $y = \sin 6\pi x$

47.

49.

51. 2π

33. $y = 4\sin x$ **35.** $y = -1.5\cos x$

37. $y = -2.50\sin x$ **39.** $y = -2.50\cos x$

53. $y = -2\sin 2x$ **55.** $y = 2\cos\frac{1}{2}x$

Exercises 10.2, page 294

1. $y = 3\sin 6x$

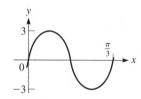

3. $\frac{\pi}{3}$ **5.** $\frac{\pi}{4}$ **7.** $\frac{\pi}{6}$ **9.** $\frac{\pi}{8}$ **11.** 1   **13.** $\frac{1}{2}$

15. 6π **17.** 3π **19.** 3   **21.** $\frac{2}{\pi}$

23.

25.

57.

59.

61. $y = \frac{1}{2}\cos 2x$ **63.** $y = -4\sin\pi x$

27.

29.

Exercises 10.3, page 298

1. $y = -\cos\left(2x - \frac{\pi}{6}\right)$ **3.** $1, 2\pi, \frac{\pi}{6}$

31.

33.

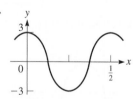

5. $1, 2\pi, -\frac{\pi}{6}$ **7.** $0.2, \pi, -\frac{\pi}{4}$

35.

37.

9. $1, \pi, \frac{\pi}{2}$ **11.** $\frac{1}{2}, 4\pi, \frac{\pi}{2}$

39.

41.

13. $30, 6\pi, -\pi$

15. $1, 2, -\frac{1}{8}$

17. $0.08, \frac{1}{2}, \frac{1}{20}$

19. $0.6, 1, \frac{1}{2\pi}$

21. $40, \frac{2}{3}, -\frac{2}{3\pi}$

23. $1, \frac{2}{\pi}, \frac{1}{\pi}$

25. $\frac{3}{2}, 2, -\frac{\pi}{6}$

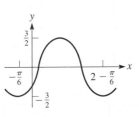

27. $y = 4\sin\left(\frac{2}{3}x + \frac{\pi}{6}\right)$

29. $y = 12\cos\left(4\pi x - \frac{\pi}{2}\right)$

31.

33.

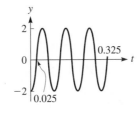

35. $y = 2\sin\left(2x + \frac{\pi}{4}\right)$

37.

39.

41. $y = 5\sin\left(\frac{\pi}{8}x + \frac{\pi}{8}\right)$. As shown: amplitude $= 5$;
period $= \frac{2\pi}{b} = 16, b = \frac{\pi}{8}$; displacement $= -\frac{c}{b} = -1, c = \frac{\pi}{8}$

43. $y = -0.8\cos 2x$. As shown: amplitude $= |-0.8| = 0.8$;
period $= \frac{2\pi}{b} = \pi, b = 2$; displacement $= -\frac{c}{b} = 0, c = 0$

Exercises 10.4, page 301

1. $y = 5\cot 2x$

3. Undef., $-1.7, -1, -0.58, 0, 0.58, 1, 1.7$, undef., $-1.7, -1, -0.58, 0$

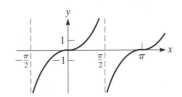

5. Undef., $2, 1.4, 1.2, 1, 1.2, 1.4, 2$, undef., $-2, -1.4, -1.2, -1$

7.

9.

11.

13.

15.

17.

19.

21.

23.

25. $y = -3\sec(x/2)$

21. $y = 12\sin 36\pi t$

27.

29.

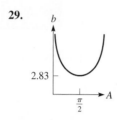

Exercises 10.6, page 307

1.

3.

Exercises 10.5, page 303

1. $d = R\cos \omega t$

3.

5.

7.

5.

7.

9.

11.

9.

11.

13.

15.

13.

15.

17.

19.

17.

21.

23.

19. (a)

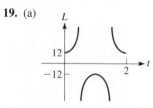

(b) $L > 0$, above x-axis

25.

27.

29.

31.

33.

35.

37. $p = 100 + 20\cos 2\pi t$

39.

41.

43.
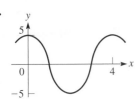

Review Exercises for Chapter 10, page 309

1.

3.

5.

7.

9.

11.

13.

15.

17.

19.

21.

23.

25.

27.

29.

31.

33.

35.

37.

39.

41. $y = 2\sin\left(2x + \frac{\pi}{2}\right)$ **43.** $y = \cos\left(\frac{\pi}{4}x - \frac{3\pi}{4}\right)$

45.

47.

49.
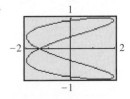

51.

53. 4π

55. $y = 3\sin x$ **57.** $y = 3\cos 3x$

59. $y = 3\sin(\pi x + 0.25\pi)$ **61.**

63.   **65.**

67.   **69.**

71.   **73.**

75.   **77.**

79.   **81.**

Exercises 11.1, page 316

1. $\dfrac{y^2}{4x^2}$ **3.** $\dfrac{3}{416}$ **5.** x^3 **7.** $\dfrac{2}{a^4}$ **9.** $\dfrac{1}{125}$ **11.** $\dfrac{4\pi^2}{x^2}$

13. $\dfrac{2n^2}{5a}$ **15.** 1   **17.** -7 **19.** $\dfrac{3}{x^2}$ **21.** $\dfrac{a^3}{343x^3}$

23. $\dfrac{n^9}{8}$ **25.** $\dfrac{3}{a^3b^6}$ **27.** $\dfrac{1}{a+b}$ **29.** $\dfrac{2x^2 + 3y^2}{x^2y^2}$

31. $\dfrac{8}{3a^n}$ **33.** $\dfrac{b^3}{432a}$ **35.** $\dfrac{4}{t^4V^4}$ **37.** $\dfrac{2a^6 + 16}{a^8}$ **39.** $\dfrac{10}{9}$

41. $\dfrac{R_1R_2}{R_1 + R_2}$ **43.** $\dfrac{4n^2 - 4n + 1}{n^4}$ **45.** $\dfrac{8}{99}$ **47.** $\dfrac{x+y}{xy}$

49. $\dfrac{t^2 + t + 2}{t^2}$ **51.** $\dfrac{2D}{D^2 - 1}$ **53.** No

55. (a) 4^5 (b) 2^{10}

57. (a) $\left(\dfrac{a}{b}\right)^{-n} = \dfrac{1}{\left(\dfrac{a}{b}\right)^n} = \dfrac{1}{\dfrac{a^n}{b^n}} = \dfrac{b^n}{a^n} = \left(\dfrac{b}{a}\right)^n$

     (b) $303.55182 = 303.55182$

59. $n = 3$ **61.** 7   **63.** $\text{N}\cdot\text{m}$ **65.** J/s^3 **67.** 1

69. $\dfrac{p\left[(1 + i)^n - 1\right]}{i(1 + i)^n}$

Exercises 11.2, page 320

1. 16   **3.**   **5.** 5   **7.** 3

9. 10^{25} **11.** $\dfrac{1}{2}$ **13.** $\dfrac{1}{16}$ **15.** 25   **17.** 81   **19.** -200

21. $\dfrac{3}{5}$ **23.** -2 **25.** $\dfrac{39}{1000}$ **27.** $\dfrac{3}{5}$ **29.** 2.059

31. 0.53891   **33.** $B^{7/6}$ **35.** $\dfrac{1}{-y^{9/10}}$ **37.** $\dfrac{1}{x^{3/2}}$ **39.** $2ab^2$

41. $\dfrac{1}{8a^3b^{9/4}}$ **43.** $a^{1/12}$ **45.** $\dfrac{4x}{(4x^2 + 1)^{1/2}}$ **47.** $-y$

49. $\dfrac{T}{(T + 2)^{1/2}}$ **51.** $\dfrac{a^2 + 1}{a^4}$ **53.** $\dfrac{a + 1}{a^{1/2}}$ **55.** $\dfrac{5x^2 - 2x}{(2x - 1)^{1/2}}$

57.   **59.**   **61.** $\sqrt[3]{x^2}$

63. If $(A/S)^{-1/4} = 0.5 = 1/2$, then $(A/S)^{1/4} = 2$. Raise each to the fourth power and get $A/S = 16$.

65. $R = \dfrac{T^{2/3}}{k^{1/3}} - d$ **67.** 1.91 mA

Exercises 11.3, page 324

1. $ab^2\sqrt{a}$ 3. $\dfrac{\sqrt[4]{54}}{3}$ 5. $2\sqrt{6}$ 7. $6\sqrt{3}$ 9. $xy^2\sqrt{y}$

11. $xy^2z\sqrt{z}$ 13. $3R^2V^2\sqrt{2RT}$ 15. $2\sqrt[3]{2}$ 17. $2\sqrt[5]{3}$

19. $2\sqrt[3]{a^2}$ 21. $2st\sqrt[4]{4r^3t}$ 23. 2   25. $P\sqrt[3]{V}$

27. $\dfrac{1}{2}\sqrt{6}$ 29. $\dfrac{1}{2}\sqrt[3]{6}$ 31. $\dfrac{1}{3}\sqrt[5]{27}$ 33. $2\sqrt{5}$ 35. 2

37. 200   39. 2000   41. $\sqrt{2a}$ 43. $\dfrac{1}{2}\sqrt{2}$ 45. $\sqrt[3]{2}$

47. $\sqrt[8]{n}$ 49. $\dfrac{2u\sqrt{7uv}}{v^3}$ 51. $4\sqrt{13}$ 53. $\dfrac{\sqrt{6x}}{3c^2}$

55. $\dfrac{3}{4}\sqrt{2}$ 57. $\dfrac{\sqrt{xy(x^2+y^2)}}{xy}$ 59. $\dfrac{\sqrt{C^2-4}}{C+2}$

61. $\sqrt{a^2+b^2}$ 63. $3x-1$ 65. $\sqrt[6]{a^3};\ \sqrt[6]{b^2};\ \sqrt[6]{c}$

67.

69. $\sqrt{2ag}$ 71. $\dfrac{8Af\sqrt{f^2+f_0^2}}{\pi^2(f^2+f_0^2)}$

43. $-\dfrac{\sqrt{x^2-y^2}+\sqrt{x^2+xy}}{y}$ 45. $a-1+\sqrt{a(a-2)}$

47. $-1-\sqrt{66}=-9.1240384$

49. $-\dfrac{16+5\sqrt{30}}{26}=-1.6686972$ 51. $\dfrac{2x+1}{\sqrt{x}}$

53. $\dfrac{5x^2+2x}{\sqrt{2x+1}}$ 55. $\dfrac{-1}{5\sqrt{2}-4\sqrt{5}}$ 57. $\dfrac{1}{\sqrt{x+h}+\sqrt{x}}$

59. $(1-\sqrt{2})^2-2(1-\sqrt{2})-1$
$$=1-2\sqrt{2}+2-2+2\sqrt{2}-1=0$$

61. $a=c$ 63. $\dfrac{\sqrt[3]{x}-1}{x-1}$

65. $\left[\dfrac{1}{2}(\sqrt{b^2-4k^2}-b)\right]^2+b\left[\dfrac{1}{2}(\sqrt{b^2-4k^2}-b)\right]+k^2$
$$=\dfrac{1}{4}(b^2-4k^2)-\dfrac{b}{2}\sqrt{b^2-4k^2}+\dfrac{1}{4}b^2$$
$$+\dfrac{b}{2}\sqrt{b^2-4k^2}-\dfrac{1}{2}b^2+k^2=0$$

67. $\dfrac{2500-50\sqrt{V}}{2500-V}$ 69. $2Q\sqrt{\sqrt{2}+1}$ 71. $\dfrac{\sqrt{LC-R^2C^2}}{LC}$

Exercises 11.4, page 326

1. $16\sqrt{5}$ 3. $7\sqrt{3}$ 5. $\sqrt{5}-\sqrt{7}$ 7. $3\sqrt{5}$

9. $-4t\sqrt{3}$ 11. $-2\sqrt{2a}$ 13. $19\sqrt{7}$ 15. $-20\sqrt{2}$

17. $23\sqrt{3R}-6\sqrt{2R}$ 19. $\dfrac{7}{3}\sqrt{15}$ 21. $-9\sqrt{2}$

23. $13\sqrt[3]{3}$ 25. $\sqrt[4]{2}$ 27. $(5a-2b^2)\sqrt{ab}$

29. $(3-2a)\sqrt{10}$ 31. $(2b-a)\sqrt[3]{3a^2b}$

33. $\dfrac{(a^2-c^3)\sqrt{ac}}{a^2c^3}$ 35. $\dfrac{(a-2b)\sqrt[3]{ab^2}}{ab}$

37. $\dfrac{-2V\sqrt{T^2-V^2}}{T^2-V^2}$ 39. $15\sqrt{3}-11\sqrt{5}=1.3840144$

41. $\dfrac{1}{6}\sqrt{6}=0.4082483$ 43. $\dfrac{7}{2}\sqrt[3]{2}=4.4097237$

45. $3\sqrt{3}$ 47. Positive, $\sqrt{1000}=10\sqrt{10},\ \sqrt{11}>\sqrt{10}$

49. $6\sqrt{2}+2\sqrt{6}$ 51. $18+3\sqrt{2}=22.2$ ft

Exercises 11.5, page 329

1. $3\sqrt{10}-16$ 3. $\sqrt{3}-\sqrt{2}$ 5. $\sqrt{30}$ 7. $2\sqrt{3}$

9. 2   11. 50   13. $2\sqrt{5}$ 15. $\sqrt{6}-\sqrt{15}$ 17. -1

19. $498+27\sqrt{10}$ 21. $66+13\sqrt{11x}-5x$

23. $a\sqrt{b}+c\sqrt{ac}$ 25. $\dfrac{2-\sqrt{6}}{2}$ 27. $2a-3b+2\sqrt{2ab}$

29. $\sqrt[6]{72}$ 31. $\dfrac{1}{4}(\sqrt{7}-\sqrt{3})$

33. $\dfrac{1}{11}(\sqrt{7}+3\sqrt{2}-6-\sqrt{14})$ 35. $\dfrac{1}{17}(-56+9\sqrt{15})$

37. $\dfrac{2x+2\sqrt{5x}}{x-5}$ 39. 1   41. $\dfrac{2-R-R^2}{R}$

Review Exercises for Chapter 11, page 331

1. $\dfrac{2}{a^2}$ 3. $\dfrac{9d^3}{c}$ 5. 375   7. $\dfrac{1}{8000}$ 9. $\dfrac{t^4}{9}$ 11. -28

13. $64a^2b^5$ 15. $-8m^9n^6$ 17. $\dfrac{2C-4L^2}{CL^2}$ 19. $\dfrac{2y}{x+2y}$

21. $\dfrac{b}{ab-3}$ 23. $\dfrac{(x^3y^3-1)^{1/3}}{y}$ 25. $\dfrac{1}{W+H}$

27. $\dfrac{-2(x+1)}{(x-1)^3}$ 29. $2\sqrt{17}$ 31. $b^2c\sqrt{ab}$ 33. $3ab^2\sqrt{a}$

35. $\dfrac{2t\sqrt{21st}}{u}$ 37. $\dfrac{5\sqrt{2s}}{2s}$ 39. $\dfrac{1}{9}\sqrt{33}$ 41. $mn^2\sqrt[4]{8m^2n}$

43. $\sqrt{2}$ 45. 0   47. $-7\sqrt{7}$ 49. $3ax\sqrt{2x}$

51. $(2a+b)\sqrt[3]{a}$ 53. $150-25\sqrt{7}$ 55. $4\sqrt{3}-4\sqrt{5}$

57. $6-51B-7\sqrt{17B}$ 59. $42-7\sqrt{7a}-3a$

61. $\dfrac{6x+\sqrt{3xy}}{12x-y}$ 63. $-\dfrac{8+\sqrt{6}}{29}$ 65. $\dfrac{13-2\sqrt{35}}{29}$

67. $\dfrac{6x-13a\sqrt{x}+5a^2}{9x-25a^2}$ 69. $\sqrt{4b^2+1}$ 71. $9x-4x^2$

73. $1+6^{1/2}$ 75. $\dfrac{15-2\sqrt{15}}{4}$ 77. $\dfrac{3+n+\sqrt{n^2+3n}}{3}$

79. $51-7\sqrt{105}=-20.728655$

81. $\sqrt{\sqrt{2}-1}(\sqrt{2}+1)=\sqrt{(\sqrt{2}-1)(\sqrt{2}+1)^2}$
$$=\sqrt{(\sqrt{2}-1)(3+2\sqrt{2})}=\sqrt{\sqrt{2}+1}=1.5537740$$

83. $12\sqrt{2}$ 85. $\dfrac{33}{4}-2\sqrt{3}$ 87. 2.784%

89. (a) $v=k(P/W)^{1/3}$ (b) $v=\dfrac{k\sqrt[3]{PW^2}}{W}$ 91. $\dfrac{100(1+2\sqrt{x})}{x}$

93. $\dfrac{c}{(c^2 - v^2)^{1/2}}$ **95.** $\dfrac{1}{\sqrt{A + h} + \sqrt{A}}$ **97.** $6\sqrt{2}$ cm

99. $\dfrac{\sqrt{LC_1C_2(C_1 + C_2)}}{2\pi LC_1C_2}$

Exercises 12.1, page 336

1. 6   **3.** −1   **5.** 9j **7.** −2j **9.** 0.6j **11.** $8j\sqrt{2}$

13. $\frac{1}{2}j\sqrt{7}$ **15.** −2ej **17.** (a) −7 (b) 7   **19.** (a) 4 (b) −4

21. $-\dfrac{3}{5}$ **23.** 10j **25.** (a) 1 (b) −1   **27.** 0   **29.** −2j

31. −j **33.** 2 + 3j **35.** −7j **37.** 2 + 2j

39. −2 + 3j **41.** $3\sqrt{2} - 2j\sqrt{2}$ **43.** −1

45. (a) 6 + 7j (b) 8 − j **47.** (a) −2j (b) −4

49. $x = 2, y = -2$ **51.** $x = 10, y = -6$

53. $x = -2, y = 3$ **55.** (a) No change (b) Sign changes

57. Yes   **59.** 0   **61.** Yes; imag. part is zero.

63. Each is x.

Exercises 12.2, page 339

1. 1 − 5j **3.** $\frac{1}{25}(29 + 22j)$ **5.** 5 − 8j **7.** −25 + 10j

9. −0.23 + 0.86j **11.** −36 + 21j **13.** 7 + 49j

15. 22 + 3j **17.** $-18j\sqrt{2}$ **19.** −28j **21.** $3\sqrt{7} + 3j$

23. −40 − 42j **25.** −2 − 2j **27.** $\frac{1}{29}(-30 + 12j)$

29. $-\frac{1}{3}(1 + j)$ **31.** $\frac{1}{11}(-13 + 8j\sqrt{2})$ **33.** $\dfrac{-1 + 3j}{5}$

35. $\dfrac{-45}{13} + \dfrac{48j}{13}$ **37.** −8j **39.** −80j **41.** $-\frac{7}{8} + 4j$

43. $(-1 - j)^2 + 2(-1 - j) + 2 = 1 + 2j - 1 - 2 - 2j + 2 = 0$

45. 10   **47.** $\frac{3}{10} + \frac{1}{10}j$ **49.** −1 + j **51.** −4

53. $\frac{1}{10}(11 + 27j)$ **55.** 281 + 35.2j volts

57. 0.016 + 0.038j amperes   **59.** (a) Real (b) Pure imaginary

61. Product is sum of squares of two real numbers
$(a + bj)(a - bj) = a^2 + b^2$

Exercises 12.3, page 341

1. 3 − j **3.**   **5.**

7. −3j **9.** 5 + 4j **11.** 8 + j

13. −1 + 9j **15.** −3j **17.** −1 + 4j

19. −180 + 150j **21.** 4.5 + 2.0j **23.** 4 + 2j

25. −2j **27.** −13 + j **29.**
Neg.   Conj.

31. 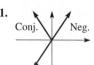 Conj. Neg.   **33.** −9 + 3j 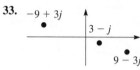 3 − j
9 − 3j

35. A complex number is on the opposite side of the real axis from its conjugate.
Conj.

37. Subtracting the conjugate from the complex number results in an imaginary number.
−Conj.   Conj.

39. 90 − 15j lb

Exercises 12.4, page 344

1. $5(\cos 126.9° + j \sin 126.9°)$

3. $10(\cos 36.9° + j \sin 36.9°)$

5. $50(\cos 306.9° + j \sin 306.9°)$

7. $3.61(\cos 123.7° + j \sin 123.7°)$

9. $0.60(\cos 204° + j \sin 204°)$

11. $2(\cos 60° + j \sin 60°)$

13. $8.062(\cos 295.84° + j\sin 295.84°)$

15. $3(\cos 180° + j\sin 180°)$ **17.** $9(\cos 90° + j\sin 90°)$

19. $2.94 + 4.05j$ **21.** $-139 + 80.0j$ **23.** $-1.85 - 2.36j$

25. 0.08 **27.** $-120j$

29. $-4.71 + 0.595j$ **31.** $-0.6052 - 0.7096j$ **33.** $7.32j$

35. $-6.961 + 86.14j$ **37.** $180°$

39. conj. of $r(\cos\theta + j\sin\theta) = r(\cos\theta - j\sin\theta)$
$$= r(\cos(-\theta) + j\sin(-\theta))$$

41. $3.03\underline{/339.5°}$ kV   **43.** $2.51 + 12.1j$ V/m

Exercises 12.5, page 346

1. $8.50e^{3.95j}$

3. $2.00(\cos 217.7° + j\sin 217.7°) = -1.58 - 1.22j$

5. $3.00e^{1.05j}$ **7.** $0.450e^{4.93j}$ **9.** $375.5e^{4.617j}$ **11.** $0.515e^{3.46j}$

13. $4.06e^{-1.07j} = 4.06e^{5.21j}$ **15.** $9245e^{5.172j}$ **17.** $5.00e^{5.36j}$

19. $36.1e^{2.55j}$ **21.** $6.37e^{0.386j}$ **23.** $825.7e^{3.836j}$

25. $3.00\underline{/28.6°}; 2.63 + 1.44j$

27. $464\underline{/106.0°}, -128 + 446j$

29. $3.20\underline{/50.0°} = 2.06 + 2.45j$

31. $0.1724\underline{/136.99°}; -0.1261 + 0.1176j$

33. $-18.2 + 9.95j, 20.7\underline{/151.3°}$

35. $-89.1 - 2750j; 2750\underline{/268.1°}$ **37.** $6.43e^{0.952j}$

39. $391e^{0.285j}$ ohms; $391\,\Omega$ **41.** $3.17 \times 10^{-4}e^{0.478j}$ $1/\Omega$

Exercises 12.6, page 352

1. $5.09(\cos 101.3° + j\sin 101.3°)$

3. $613(\cos 281.5° + j\sin 281.5°)$ **5.** $8(\cos 80° + j\sin 80°)$

7. $3(\cos 250° + j\sin 250°)$ **9.** $2(\cos 35° + j\sin 35°)$

11. $2.4\underline{/170°}$ **13.** $0.008(\cos 105° + j\sin 105°)$

15. $256(\cos 0° + j\sin 0°)$ **17.** $0.8\underline{/273°}$ **19.** $5\underline{/87°}$

21. $1.73\underline{/79.8°}$ **23.** $11{,}750\underline{/115.91°}$

25. $65.0(\cos 345.7° + j\sin 345.7°) = 63 - 16j$

27. $61.4(\cos 343.9° + j\sin 343.9°); 59 - 17j$

29. $2.21(\cos 71.6° + j\sin 71.6°); \dfrac{7}{10} + \dfrac{21}{10}j$

31. $3.85(\cos 120.5° + j\sin 120.5°) = \dfrac{10}{169}(-33 + 56j)$

33. $625(\cos 212.5° + j\sin 212.5°) = -527 - 336j$

35. $609(\cos 281.5° + j\sin 281.5°); 122 - 597j$

37. $2(\cos 30° + j\sin 30°); 2(\cos 210° + j\sin 210°)$

39. $-0.364 + 1.67j, -1.26 - 1.15j, 1.63 - 0.520j$

41. $1.10 + 0.455j, -1.10 - 0.455j$

43. $1, -1, j, -j$

45. $3j, -\dfrac{3}{2}(\sqrt{3} + j), \dfrac{3}{2}(\sqrt{3} - j)$

47. $1.62 + 1.18j, -0.618 + 1.90j, -2, -0.618 - 0.190j,$
$1.62 - 1.18j$

49. $-5, \dfrac{5}{2} + j\dfrac{5\sqrt{3}}{2}, \dfrac{5}{2} - j\dfrac{5\sqrt{3}}{2}$

51. $\left[\dfrac{1}{2}(1 - j\sqrt{3})\right]^3 = \dfrac{1}{8}[1 - 3(j\sqrt{3}) + 3(j\sqrt{3})^2$
$- (j\sqrt{3})^3] = \dfrac{1}{8}[1 - 3j\sqrt{3} - 9 + 3j\sqrt{3}]$
$= \dfrac{1}{8}(-8) = -1$

53. $-1, \dfrac{1}{2} + j\dfrac{\sqrt{3}}{2}, \dfrac{1}{2} - j\dfrac{\sqrt{3}}{2}$ **55.** $p = 0.479\underline{/40.5°}$ watts

57. $43.3\underline{/165°}, 43.3$ V

Exercises 12.7, page 358

1. $V_R = 24.0$ V, $V_L = 32.0$ V, $V_{RL} = 40.0$ V, $\theta = 53.1°$ (voltage
leads current)   **3.** 12.9 V   **5.** (a) $2850\,\Omega$ (b) $37.9°$

(c) 16.4 V   **7.** (a) $14.6\,\Omega$ (b) $-90.0°$ **9.** (a) $47.8\,\Omega$ (b) $19.8°$

11. 38.0 V   **13.** $54.5\,\Omega, -62.3°$ **15.** 0.682 H

17. 2.08×10^5 Hz   **19.** 1.30×10^{-11} F $= 13.0$ pF

21. 1.02 mW   **23.** $21.4 - 33.9j$

Review Exercises for Chapter 12, page 360

1. $10 - j$ **3.** $6 + 2j$ **5.** $9 + 2j$ **7.** $-12 + 66j$

9. $\dfrac{1}{85}(21 + 18j)$ **11.** $-2 - 3j$ **13.** $\dfrac{1}{10}(-12 + 9j)$

15. $\dfrac{1}{5}(13 + 11j)$ **17.** $x = -3, y = -2$ **19.** $x = \dfrac{10}{13}, y = \dfrac{11}{13}$

21. $3 + 11j$ **23.** $4 + 8j$

25. $1.41(\cos 315° + j\sin 315°) = 1.41e^{5.50j}$

27. $80.1(\cos 254.1° + j\sin 254.1°) = 80.1e^{4.43j}$

29. $4.67(\cos 76.8° + j \sin 76.8°); 4.67e^{1.34j}$

31. $5000\underline{/0°}, 5000e^{0j}$ **33.** $-1.41 - 1.41j$

35. $-2.789 + 4.162j$ **37.** $0.19 - 0.59j$ **39.** $26.31 - 6.427j$

41. $1.94 + 0.495j$ **43.** $-728.1 + 1017j$

45. $15(\cos 84° + j \sin 84°)$ **47.** $20\underline{/263°}$

49. $8(\cos 59° + j \sin 59°)$ **51.** $14.29\underline{/133.61°}$

53. $1.26\underline{/59.7°}$ **55.** $9682\underline{/249.52°}$

57. $1024(\cos 160° + j \sin 160°)$ **59.** $343\underline{/331.5°}$

61. $32(\cos 270° + j \sin 270°) = -32j$

63. $\dfrac{625}{2}(\cos 270° + j \sin 270°) = -\dfrac{625}{2}j$

65. $1 + j\sqrt{3}, -2, 1 - j\sqrt{3}$

67. $0.383 + 0.924j, -0.924 + 0.383j, -0.383 - 0.924j,$
$0.924 - 0.383j$ **69.** $40 + 9j, 41(\cos 12.7° + j \sin 12.7°)$

71. $-15.0 - 10.9j, 18.5(\cos 216.0° + j \sin 216.0°)$

73. $15 - 16j$ **75.** $-j, -2j$ **77.** $x^2 - 4x + 5 = 0$

79. $1 - j$, Yes; $-1 - j$, No   **81.** $\dfrac{1}{2}$ **83.** $2 + \dfrac{9}{2}j$

85. $5.83e^{5.74j}$ **87.** 60 V   **89.** $-21.6°$ **91.** 22.9 Hz

93. $5500(\cos 53.1° + j \sin 53.1°)$ N   **95.** $\dfrac{u - j\omega n}{u^2 + \omega^2 n^2}$

97. $e^{j\pi} = \cos \pi + j \sin \pi = -1$

Exercises 13.1, page 364

1. $-\dfrac{1}{4}$ **3.** (a) Yes (b) Yes

5. (a) No (base cannot be negative) (b) Yes   **7.** 3

9. $\dfrac{1}{81}$ **11.** $\dfrac{1}{27}$

13.   **15.**

17.   **19.**

21.   **23.**

25. 4   **27.** $f(a + b) = b^{a+b} = b^a b^b = f(a) \cdot f(b)$

29. 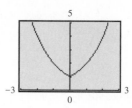   **31.** $-0.7667, 2, 4$

33. \$303.88   **35.** 0.0012 mA   **37.**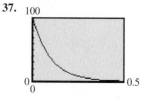

Exercises 13.2, page 368

1. $32^{4/5} = 16$ in logarithmic form is $\dfrac{4}{5} = \log_{32} 16$.

3. If $x = 16, y = \log_4 16$ means $y = 2$, since $4^2 = 16$. If $x = \dfrac{1}{16}$, $y = \log_4\left(\dfrac{1}{16}\right)$ means $y = -2$, since $4^{-2} = \dfrac{1}{16}$.

5. $\log_3 27 = 3$ **7.** $\log_4 256 = 4$ **9.** $\log_7\left(\dfrac{1}{49}\right) = -2$

11. $\log_2\left(\dfrac{1}{64}\right) = -6$ **13.** $\log_8 2 = \dfrac{1}{3}$

15. $\log_{1/4}\left(\dfrac{1}{16}\right) = 2$ **17.** $81 = 3^4$ **19.** $9 = 9^1$

21. $5 = 25^{1/2}$ **23.** $3 = 243^{0.2}$ **25.** $0.1 = 10^{-1}$

27. $16 = (0.5)^{-4}$ **29.** 2   **31.** -2 **33.** 343

35. $\dfrac{9}{4}$ **37.** $\sqrt{5}$ **39.** $\dfrac{1}{64}$ **41.** 0.2   **43.** -4

45.

47.

49.

51.

53.

55. (a) -3
(b) No value (x cannot be negative)

57. (a) $\dfrac{1}{2}$
(b) Not defined

59.

61. 0.0102, 2.376

63. $b_2 = b_1 10^{0.4(m_1 - m_2)}$

65. $N = N_0 e^{-kt}$

67. $V_1 = V_2 e^{-W/k}$

69.
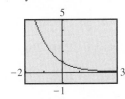

71. They are the same.

73.

75.
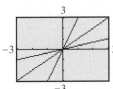

Exercises 13.3, page 373

1. $\log_4 3 + \log_4 7$ **3.** $\log_4\left(\dfrac{5}{x}\right)$ **5.** 3   **7.** $\dfrac{1}{2}$

9. $\log_5 3 + \log_5 11$ **11.** $\log_7 5 - \log_7 3$ **13.** $3\log_2 a$

15. $\log_6 a + \log_6 b + \log_6 c$ **17.** $2\log_5 y$

19. $\dfrac{1}{2}\log_2 x - 2\log_2 a$ **21.** $\log_b ac$ **23.** $\log_5 3$

25. $\log_b x^{3/2}$ **27.** $\log_e 4\pi^3$ **29.** -5 **31.** 2.5   **33.** 3

35. $\dfrac{3}{4}$ **37.** $2 + \log_3 2$ **39.** $-1 - \log_2 3$

41. $\dfrac{1}{2}(1 + \log_3 2)$ **43.** $3 + \log_{10} 3$ **45.** $y = 2x$

47. $y = \dfrac{3x}{5}$ **49.** $y = \dfrac{49}{x^3}$ **51.** $y = 2(2ax)^{1/5}$ **53.** $y = \dfrac{2}{x}$

55. $y = \left(\dfrac{x^2}{9}\right)^{1/\log_5 3}$ **57.** $\log_{10} x + \log_{10} 3 = \log_{10} 3x$

59. $y = \log_e(e^2 x) = 2\log_e e + \log_e x = 2 + \log_e x$ **61.** 8

63. $y_1 = y_2$ **65.** $D = ae^{cr^2 - br}$
67. $T = 90.0e^{-0.23t}$

Exercises 13.4, page 376

1. -0.4372 **3.** 2.754   **5.** 6.966   **7.** -0.2787

9. -0.0104 **11.** 1.219   **13.** 27,400   **15.** 0.04960

17. 2000.4   **19.** 0.0057882   **21.** 85.5   **23.** 7.37×10^{101}

25. $1.1461 - 0.3010 = 0.8451$ **27.** $1.9085 = 1.9085$

29. 8.9542   **31.** -13.886 **33.** $10^8\,\text{K}$ **35.** 0.45

37. $\log_b x$ **39.** 0.23   **41.** 7.9   **43.** $2^{400} = 2.58 \times 10^{120}$

Exercises 13.5, page 379

1. 5.298   **3.** 3.258   **5.** 0.4460   **7.** -4.91623

9. 1.92   **11.** 4.806   **13.** 1.795   **15.** 3.940   **17.** 0.3322

19. -0.008335 **21.** -17.39066 **23.** 1.6549

25. -0.16413 **27.** 8.935   **29.** 1.0085   **31.** 0.4757

33. 6.1993×10^{-11}

35.

37.
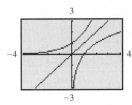

39. $1.6094 + 2.0794 = 3.6889$ **41.** $4.3944 = 4.3944$ **43.** 3

45. 10   **47.** $80 = e^{2x+y}$ **49.** $2.45 \times 10^9\,\text{Hz}$

51. 8.155%   **53.** 0.384 s   **55.** 21.7 s

Exercises 13.6, page 382

1. -0.535 **3.** 4   **5.** -0.748 **7.** 0.587   **9.** 1.43

11. 1.85   **13.** $0, \dfrac{\ln 0.6}{\ln 2} = -0.737$ **15.** $\dfrac{1}{4}$ **17.** 1 or 100

19. 3   **21.** -0.162 **23.** 250   **25.** 4   **27.** 2

29. 1.42   **31.** -0.1042 **33.** 0.2031   **35.** 0.9739

37. 10.87   **39.** ± 0.9624 **41.** 0.0025   **43.** -1.21

45. 28.0   **47.** Noon of previous day (36 h earlier)

49. 1.72×10^{-5} **51.** $10^{8.3} = 2.0 \times 10^8$

53. $c = 15e^{-0.20t}$ **55.** $P = P_0(0.999)^t$

57.

$x = 3.353$ **59.** $\pm 3.7\,\text{m}$

Exercises 13.7, page 386

1.
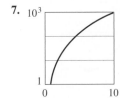

3.

5.

7.

9.

11.

13.

15.

Review Exercises for Chapter 13, page 388

1. 10,000   **3.** $\dfrac{1}{5}$ **5.** 6   **7.** $\dfrac{5}{3}$ **9.** 6   **11.** 100

13. $\log_3 2 + \log_3 x$ **15.** $2\log_3 t$ **17.** $2 + \log_2 7$

19. $1 + \dfrac{1}{2}\log_4 3$ **21.** $2 - \log_3 x$ **23.** $3 + 4\log_{10} x$

25. $y = \dfrac{4}{x}$ **27.** $y = e^{2/3}x$ **29.** $y = \sqrt{7x}$ **31.** $y = \dfrac{15}{x}$

33. $y = 8x^3$ **35.** $y = \dfrac{x}{\ln 2} = 1.44x$

17.

19. Semilog paper

37.

39.

21. Log-log paper

23. Semilog paper

41.

43.

25. Log-log paper

27. (a)
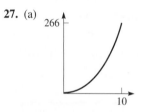

45. 2.18   **47.** -0.1301 **49.** 1.8188   **51.** -2.7657

53. 0.805   **55.** 4.30   **57.** 2   **59.** -0.2

61.

63.

27. (b)

29.
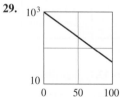

65. 4   **67.** 12   **69.** $0.9542 - 0.7782 = 0.1761$ **71.** 0

73. 2   **75.** 0.706   **77.** 1.28×10^{-5} or 3.925

79. $V = 1000(1.03)^{2t}, \; t = \dfrac{\log(V/1000)}{2\log 1.03}$ **81.** $I = I_0 e^{-\beta h}$

83. $4.60 \times 10^6 \, \text{N} \cdot \text{m}$

31.

33.

85.
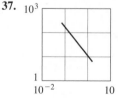

87.

89. $\sin \theta = \dfrac{l\omega^2}{3g}$ **91.** $R = 2^{C/B} - 1$ **93.** 910 times brighter

35.

37.

95. 1.17 min   **97.** 0.813 cm   **99.** $-\dfrac{1}{k}\ln\left(\dfrac{T - T_1}{T_0 - T_1}\right)$

39.

101. $n = 20e^{-0.04t}$ **103.**

Exercises 14.1, page 394

1. $y = 3x^2 + 6x$ **3.** $x = -1.2, y = 0.3; x = 0.8, y = 2.3$

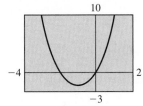

5. $x = 1.8, y = 3.6; x = -1.8, y = -3.6$

7. $x = 0.0, y = -2.0; x = 2.7, y = -0.7$

9. $x = 1.5, y = 0.2$ **11.** $x = 2.8, y = 10.7$

13. $x = 1.1, y = 2.8; x = -1.1, y = 2.8;$
$x = 2.4, y = -1.8; x = -2.4, y = -1.8$

15. No real solutions

17. $x = -2.8, y = -1.0; x = 2.8, y = 1.0;$
$x = 2.8, y = -1.0; x = -2.8, y = 1.0$

19. $x = -0.81, y = 2.52; x = 2.56, y = 0.66$

21. $x = 0.0, y = 0.0; x = 0.9, y = 0.8$

23. $x = -1.1, y = 3.1; x = 1.8, y = 0.2$

25. $x = 16.3, y = 16.1$

27. $x = 3.6, y = 1.0$

29. $x = -2.06, y = 4.23; x = 1.06, y = 1.12$

31.

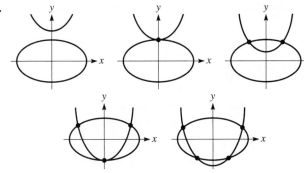

33. 4.9 mi N, 1.6 mi E   **35.** 2.2 A, 0.9 A   **37.** No

Exercises 14.2, page 397

1. $x = 2, y = 0; x = \dfrac{10}{3}, y = -\dfrac{8}{3}$

3. $x = -\sqrt{6}, y = -\sqrt{3}; x = -\sqrt{6}, y = \sqrt{3};$
$x = \sqrt{6}, y = -\sqrt{3}; x = \sqrt{6}, y = \sqrt{3}$

5. $x = 0, y = 1; x = 1, y = 2$

7. $x = -\dfrac{19}{5}, y = \dfrac{17}{5}; x = 5, y = -1$ **9.** $x = 1, y = 0$

11. $x = \dfrac{2}{7}(3 + \sqrt{2}), y = \dfrac{2}{7}(-1 + 2\sqrt{2});$

$x = \dfrac{2}{7}(3 - \sqrt{2}), y = \dfrac{2}{7}(-1 - 2\sqrt{2})$

13. $w = 1, h = 1$ **15.** $x = -\dfrac{3}{2}, y = -\dfrac{8}{3}; x = 2, y = 2$

17. $x = -5, y = 25; x = 5, y = 25$

19. $x = 1, y = 2; x = -1, y = 2$

21. $D = 1, R = 0; D = -1, R = 0; D = \dfrac{1}{2}\sqrt{6}, R = \dfrac{1}{2};$

$D = -\dfrac{1}{2}\sqrt{6}, R = \dfrac{1}{2}$

23. $x = \sqrt{19}, y = \sqrt{6}; x = \sqrt{19}, y = -\sqrt{6};$
$x = -\sqrt{19}, y = \sqrt{6}; x = -\sqrt{19}, y = -\sqrt{6}$

25. $x = -5, y = -2; x = -5, y = 2; x = 5, y = -2; x = 5, y = 2$

27. $x = -3, y = 2; x = -1, y = -2$ **29.** $x = a, y = b$

31. $x = 50 \text{ mi}, h = 25 \text{ mi}$ **33.** 2.00 cm, 4.00 cm

35. 1.5 cm, 1.4 cm   **37.** 2.19 m, 0.21 m   **39.** 12, 10

41. 12 in., 18 in.   **43.** 80 mi/h

Exercises 14.3, page 401

1. $\pm\dfrac{1}{2}j\sqrt{2}, \pm 2$ **3.** $-3, -2, 2, 3$ **5.** $-\dfrac{1}{2}, \dfrac{1}{4}$

7. $-\dfrac{1}{2}, \dfrac{1}{2}, \dfrac{1}{6}j\sqrt{6}, -\dfrac{1}{6}j\sqrt{6}$ **9.** $1, \dfrac{25}{4}$ **11.** $\dfrac{64}{729}, 1$

13. $-27, 125$ **15.** 81   **17.** 26   **19.** $-2, -1, 3, 4$

21. 18   **23.** $1, -1, j\sqrt{2}, -j\sqrt{2}$ **25.** 0   **27.** $\pm 2, \pm 4$

29. $1, 4$ **31.** $10, 100$ **33.** $-2, -1, 1, 2$

35. $R_1 = 2.62\ \Omega, R_2 = 1.62\ \Omega$ **37.** 0.610

39. 15.9 in., 21.8 in.

Exercises 14.4, page 405

1. $\dfrac{13}{12}$ **3.** 10   **5.** 12   **7.** 2   **9.** $\dfrac{2}{3}$ **11.** -1

13. 32   **15.** $4, 9$ **17.** ± 5

19. 12 (Extraneous root introduced in squaring both sides of $\sqrt{x + 4} = x - 8$.)

21. 16   **23.** $\dfrac{1}{2}$ **25.** $7, -1$ **27.** 0   **29.** 5   **31.** 25

33. 258   **35.** 4   **37.** 4   **39.** 16

41. $x = 2$; For $x = 5$, squaring $3 - x$ is squaring a negative number. A different extraneous root is introduced.

43. $-r \pm \sqrt{r^2 + R^2}$ **45.** $r_1^2 = (kC + A - \sqrt{R_1^2 - R_2^2})^2 + r_2^2$

47. $9.6\ \text{m}^2$ **49.** 9.2 km   **51.** 1.6 mi

Review Exercises for Chapter 14, page 406

1. $x = -0.9, y = 3.5; x = 0.8, y = 2.6$

3. $x = 2.0, y = 0.0; x = 1.6, y = 0.6$

5. $x = -0.563, y = 1.32; x = 0.563, y = 1.32$

7. $x = -2, y = 7; x = 2, y = 7$

9. $x = 0.0, y = 0.0; x = 2.4, y = 0.9$

11. $x = 0, y = 0; x = 2, y = 16$

13. $L = \sqrt{2}, R = 1; L = -\sqrt{2}, R = 1$

15. $u = \dfrac{1}{12}(1 + \sqrt{97}), v = \dfrac{1}{18}(5 - \sqrt{97});$

$u = \dfrac{1}{12}(1 - \sqrt{97}), v = \dfrac{1}{18}(5 + \sqrt{97})$

17. $x = 7, y = 5; x = 7, y = -5; x = -7, y = 5;$
$x = -7, y = -5$

19. $x = -2, y = 2; x = \dfrac{2}{3}, y = \dfrac{10}{9}$

21. $-4, -2, 2, 4$ **23.** $1, 16$ **25.** $\dfrac{1}{3}, -\dfrac{1}{7}$ **27.** $1, 1.65$

29. $\sqrt{3}, -\sqrt{3}, \dfrac{1}{2}j\sqrt{3}, -\dfrac{1}{2}j\sqrt{3}$ **31.** 6 **33.** 8 **35.** $\dfrac{9}{16}$

37. $\dfrac{1}{3}(11 - 4\sqrt{15})$ **39.** 2 **41.** 4 **43.** 5 **45.** $-1, 2$

47. $x = a + 1, y = a$ **49.** 25

51. $x = -2, y = -3; x = 3, y = 2$

53. $l = \dfrac{1}{2}(-1 + \sqrt{1 + 16\pi^2 L^2/h^2})$

55. $m = \dfrac{1}{2}(-y \pm \sqrt{2s^2 - y^2})$ **57.** 0.75 s, 1.5 s   **59.** 230 ft

61. 20 in., 26 in., 26 in.,   or   27.6 in., 22.2 in., 22.2 in.

63. 3.0 mm, 1.0 mm   **65.** 34 mm, 52 mm

67. 0.353 cm, 6.29 cm   **69.** 26 mi/h, 32 mi/h

Exercises 15.1, page 413

1. -25 **3.** Coefficients $1\ 0\ 0\ -4\ 11, R = -26$ **5.** 0

7. -40 **9.** -43 **11.** 51 **13.** -28 **15.** 14 **17.** Yes

19. No   **21.** No   **23.** $x^2 + 3x + 2, R = 0$

25. $x^2 + x - 4, R = 8$

27. $p^5 + 2p^4 + 4p^3 + 2p^2 + 2p + 4, R = 2$

29. $x^6 + 2x^5 + 4x^4 + 8x^3 + 16x^2 + 32x + 64, R = 0$

31. $x^3 + x^2 + 2x + 1, R = 0$ **33.** No   **35.** No   **37.** Yes

39. No   **41.** Yes   **43.** Yes   **45.** $2x^2 - 5x + 1 - \dfrac{8}{x + 4}$

47. $(4x^3 + 8x^2 - x - 2) \div (2x - 1) = 2x^2 + 5x + 2$; no, because the coefficient of x in $2x - 1$ is 2, not 1.

49. -7 **51.** $x^2 - (3 + j)x + 3j, R = 0$

53. Yes, because both functions cross the x-axis at the same values.

55. (a) No (b) Yes   **57.** Yes

Exercises 15.2, page 418

(Note: Unknown roots listed)

1. $1, 1, 1, -1, -1$ **3.** $-3, 2, 3$ **5.** $-3, -3, 2j, -2j$

7. $-2, 3$ **9.** $-2, -2$ **11.** $-j, -\dfrac{2}{3}$ **13.** $2j, -2j$

15. $-1, \dfrac{1}{2}$ **17.** $-2, 1$ **19.** $1 - j, \dfrac{1}{2}, -2$ **21.** $j, -j$

23. $-j, -1, 1$ **25.** $-2j, 3, -3$

27. If j is a root, $-j$ must also be a root, and $-j$ is not one of the roots.

29. $x^3 - 2x^2 + 2x$

Exercises 15.3, page 424

1. No more than two positive roots and three negative roots

3. $1, -1, -2$ **5.** $2, -1, -3$

7. $1, \dfrac{1}{6}(-3 + j\sqrt{15}), \dfrac{1}{6}(-3 - j\sqrt{15})$ **9.** $\dfrac{1}{3}, -3, -1$

11. $-2, -2, 2 \pm \sqrt{3}$ **13.** $-2, \dfrac{2}{5}, 1 + j\sqrt{3}, 1 - j\sqrt{3}$

15. $-3, -\dfrac{2}{3}, -\dfrac{1}{2}, \dfrac{1}{2}$ **17.** $2, 2, -1, -1, -3$

19. $2, 2, 2, \dfrac{1}{2}j, -\dfrac{1}{2}j$ **21.** $-1.86, 0.68, 3.18$

23. $-3.01, -1.49, -0.334, 0.334$ **25.** 0.59 **27.** -0.77

29. $-2, -2, 2 - \sqrt{3}, 2 + \sqrt{3}$ **31.** $-\dfrac{3}{4}, \sqrt{5}, -\sqrt{5}$

33. 1.8 s, 4.5 s   **35.** $0, L$ **37.** 0.9 s, 4.1 s

39. 1.23 cm or 2.14 cm   **41.** 3.0 mm, 4.0 mm; 5.0 mm, 6.0 mm

43. $2800\ \Omega, 4200\ \Omega, 8400\ \Omega$

45. Two positive, one negative; zero positive, one negative, two nonreal complex

Review Exercises for Chapter 15, page 425

1. 1 **3.** -107 **5.** Yes   **7.** No

9. $x^2 + 4x + 10, R = 11$ **11.** $2x^2 - 7x + 10, R = -17$

13. $x^3 - 3x^2 - 2x + 10, R = -4$

15. $2m^4 + 10m^3 + 4m^2 + 21m + 105, R = 516$ **17.** No

19. Yes   **21.** (unlisted roots) $-2, 1$ **23.** (unlisted roots) $3, 4$

25. (unlisted roots) $-1 + j\sqrt{2}, -1 - j\sqrt{2}$

27. (unlisted roots) $-j, \dfrac{1}{2}, -\dfrac{3}{2}$ **29.** (unlisted roots) $2, -2$

31. (unlisted roots) $-2 - j, \dfrac{1}{2}(-1 \pm j\sqrt{3})$ **33.** $1, 2, -4$

35. $-\dfrac{1}{2}, 1, 1$ **37.** $-1, -\dfrac{1}{2}, \dfrac{5}{3}$ **39.** $\dfrac{1}{2}, \dfrac{3}{2}, -1 + \sqrt{2}, -1 - \sqrt{2}$

41. $1, 3, 5$ **43.** 2 or 4   **45.** Use k in synthetic division, $k = 4$.

47. $x^3 - 5x^2 + x - 5$ **49.** $x = -1, y = -2; x = 2, y = 1$

51. 1.91 **53.** April   **55.** 0.75 cm   **57.** 2 in.

59. 5.1 m, 8.3 m   **61.** $h = 6.75$ ft, $w = 3.60$ ft

Exercises 16.1, page 431

1. $\begin{bmatrix} 5 & 7 & -1 & 9 \\ 6 & 4 & 1 & 12 \end{bmatrix}$ **3.** $a = 1, b = -3, c = 4, d = 7$

5. $x = -2, y = 5, z = -9, r = 48, s = 4, t = -1$

7. $C = 3, D = 2, E = -2$

9. Elements cannot be equated; different number of rows.

11. $\begin{bmatrix} 1 & 10 \\ 0 & 2 \end{bmatrix}$ **13.** $\begin{bmatrix} -5 & 0 \\ 11 & 71 \\ 11 & -5 \end{bmatrix}$ **15.** $\begin{bmatrix} 6 & 13 & -13 \\ 6 & -7 & 3 \end{bmatrix}$

17. Cannot be added   **19.** $\begin{bmatrix} 5 & 17 & -20 \\ 8 & -13 & 14 \end{bmatrix}$

21. $\begin{bmatrix} -15 & -14 & 5 \\ -6 & -4 & 27 \end{bmatrix}$ **23.** $\begin{bmatrix} 4 & -16 & 28 \\ -8 & 24 & -44 \end{bmatrix}$

25. Elements cannot be combined.

27. $\begin{bmatrix} -3 & 12 & -21 \\ 6 & -18 & 33 \end{bmatrix}$ **29.** $\begin{bmatrix} 20 & 31 & -25 \\ 14 & -9 & -13 \end{bmatrix}$

31. Cannot be subtracted

33. $\begin{bmatrix} -43 & -50 & 29 \\ -22 & 0 & 59 \end{bmatrix}$

35. $A + B = B + A = \begin{bmatrix} 3 & 1 & 0 & 7 \\ 5 & -3 & -2 & 5 \\ 10 & 10 & 8 & 0 \end{bmatrix}$

37. $-(A - B) = B - A = \begin{bmatrix} 5 & -3 & -6 & -7 \\ 5 & 3 & 0 & -3 \\ -8 & 12 & 8 & 4 \end{bmatrix}$

39. $v_w = 31.0 \text{ km/h}, v_p = 249 \text{ km/h}$

41. $\begin{bmatrix} 288 & 225 & 0 & 0 \\ 186 & 132 & 72 & 0 \\ 0 & 105 & 204 & 234 \end{bmatrix}$

43. $B = \begin{bmatrix} 15 & 25 & 10 & 25 \\ 10 & 10 & 10 & 45 \end{bmatrix}, J = \begin{bmatrix} 15 & 30 & 3 & 3 \\ 0 & 100 & 2 & 2 \end{bmatrix}$

Exercises 16.2, page 435

1. $\begin{bmatrix} 5 & 6 & 7 & -10 \\ -9 & 0 & -3 & 12 \\ 1 & 12 & 11 & -8 \end{bmatrix}$ **3.** $[-8 \quad -12]$ **5.** $\begin{bmatrix} 305 \\ -325 \end{bmatrix}$

7. $\begin{bmatrix} -\frac{1}{2} & \frac{111}{4} \\ -\frac{121}{8} & -\frac{101}{2} \\ \frac{1}{10} & 22 \end{bmatrix}$ **9.** $\begin{bmatrix} 33 & -22 \\ 31 & -12 \\ 15 & 13 \\ 50 & -41 \end{bmatrix}$ **11.** $\begin{bmatrix} -15 & 15 & -26 \\ 8 & 5 & -13 \end{bmatrix}$

13. $\begin{bmatrix} -49.43 & 55.20 \\ -53.02 & 79.16 \end{bmatrix}$

15. $AB = [40], BA = \begin{bmatrix} -1 & 3 & -8 \\ 5 & -15 & 40 \\ 7 & -21 & 56 \end{bmatrix}$

17. $AB = \begin{bmatrix} 45 \\ 327 \end{bmatrix}, BA$ not defined   **19.** $AI = IA = A$

21. $AI = IA = A$ **23.** $B = A^{-1}$ **25.** $B = A^{-1}$ **27.** Yes

29. No   **31.** $A^2 = A$ **33.** $B^3 = B$

35. $\begin{bmatrix} ac + bd & ad + bc \\ bc + ad & bd + ac \end{bmatrix}$

37. $\begin{bmatrix} -1 & 0 \\ 0 & -1 \end{bmatrix}\begin{bmatrix} -1 & 0 \\ 0 & -1 \end{bmatrix} = \begin{bmatrix} 1 & 0 \\ 0 & 1 \end{bmatrix}$

39. $A^2 - I = (A + I)(A - I) = \begin{bmatrix} 15 & 28 \\ 21 & 36 \end{bmatrix}$

41. $\begin{bmatrix} 0 & -j \\ j & 0 \end{bmatrix}\begin{bmatrix} 0 & -j \\ j & 0 \end{bmatrix} = \begin{bmatrix} 1 & 0 \\ 0 & 1 \end{bmatrix}$

43. $v_2 = v_1, i_2 = -v_1/R + i_1$ **45.** Ellipse

Exercises 16.3, page 440

1. $\begin{bmatrix} -\frac{5}{2} & \frac{3}{2} \\ -2 & 1 \end{bmatrix}$ **3.** $\begin{bmatrix} -2 & -\frac{5}{2} \\ -1 & -1 \end{bmatrix}$ **5.** $\begin{bmatrix} -\frac{1}{3} & \frac{1}{6} \\ \frac{2}{15} & \frac{1}{30} \end{bmatrix}$

7. $\begin{bmatrix} \frac{3}{4} & \frac{1}{2} \\ -\frac{1}{4} & 0 \end{bmatrix}$ **9.** $\begin{bmatrix} -\frac{8}{283} & -\frac{9}{566} \\ \frac{13}{1415} & \frac{5}{283} \end{bmatrix}$ **11.** $\begin{bmatrix} -3 & 2 \\ 2 & -1 \end{bmatrix}$

13. $\begin{bmatrix} -\frac{1}{2} & -2 \\ \frac{1}{2} & 1 \end{bmatrix}$ **15.** $\begin{bmatrix} \frac{2}{9} & -\frac{5}{9} \\ \frac{1}{9} & \frac{2}{9} \end{bmatrix}$ **17.** $\begin{bmatrix} \frac{7}{25} & \frac{3}{5} \\ -\frac{1}{5} & -\frac{1}{2} \end{bmatrix}$

19. $\begin{bmatrix} -18 & -7 & 5 \\ -3 & -1 & 1 \\ -5 & -2 & 1 \end{bmatrix}$ **21.** $\begin{bmatrix} 2 & 4 & \frac{7}{2} \\ -1 & -2 & -\frac{3}{2} \\ 1 & 1 & \frac{1}{2} \end{bmatrix}$

23. $\begin{bmatrix} \frac{5}{2} & -2 & -2 \\ -1 & 1 & 1 \\ \frac{7}{4} & -\frac{3}{2} & -1 \end{bmatrix}$ **25.** $\begin{bmatrix} 0.3 & -0.4 \\ 0.05 & 0.1 \end{bmatrix}$

27. $\begin{bmatrix} 2 & 4 & 3.5 \\ -1 & -2 & -1.5 \\ 1 & 1 & 0.5 \end{bmatrix}$ **29.** $\begin{bmatrix} 2.5 & -2 & -2 \\ -1 & 1 & 1 \\ 1.75 & -1.5 & -1 \end{bmatrix}$

31. $\begin{bmatrix} 1 & 2 & 3 & 1 \\ 1 & 3 & 3 & 2 \\ 2 & 4 & 3 & 3 \\ 1 & 1 & 1 & 1 \end{bmatrix}$

33. $\begin{bmatrix} 2.537 & -0.950 & 0.159 & 0.470 \\ -0.213 & 0.687 & 0.290 & -0.272 \\ -1.006 & 0.870 & 0.496 & -0.532 \\ 0.113 & -0.123 & -0.033 & 0.385 \end{bmatrix}$

35. $\begin{vmatrix} 1 & 1 \\ 1 & 1 \end{vmatrix} = 0$ means the inverse does not exist.

37. $\frac{1}{ad - bc}\begin{bmatrix} ad - bc & -ba + ab \\ cd - dc & -bc + ad \end{bmatrix} = \begin{bmatrix} 1 & 0 \\ 0 & 1 \end{bmatrix}$

39. $BC = CB = \begin{bmatrix} 0 & -1 \\ -\frac{1}{4} & \frac{1}{4} \end{bmatrix}$

41. $v_1 = (a_{22}i_1 - a_{12}i_2)/(a_{11}a_{22} - a_{12}a_{21})$

$v_2 = (-a_{21}i_1 + a_{11}i_2)/(a_{11}a_{22} - a_{12}a_{21})$

Exercises 16.4, page 445

1. $x = 2, y = -3$ **3.** $x = \frac{1}{2}, y = 3$ **5.** $x = 1, y = 3$

7. $x = 2, y = -2$ **9.** $x = -1, y = 0, z = 3$

11. $x = -\frac{3}{2}, y = -2$ **13.** $x = 1.6, y = -2.5$

15. $x = 2, y = -4, z = 1$ **17.** $x = 2, y = -\frac{1}{2}, z = 3$

19. $x = 1, y = -2, z = -3$ **21.** $u = 2, v = -5, w = 4$

23. $x = 2, y = 3, z = -2, t = 1$

25. $v = 2, w = -1, x = \frac{1}{2}, y = \frac{3}{2}, z = -3$

27. $x = \pm2, y = -2$

29. $x = 1, y = -2$; The three lines meet in the point $(1, -2)$.

31. $A = 1180 \text{ N}, B = 1860 \text{ N}$ **33.** 10 V, 8 V

35. 40 mL, 8 mL   **37.** 6.4 L, 1.6 L, 2.0 L

Exercise 16.5, page 449

1. $x = \frac{11}{7}, y = \frac{6}{7}$ **3.** $x = 2, y = 1$ **5.** $x = \frac{17}{14}, y = \frac{19}{14}$

7. Inconsistent   **9.** $x = -2, y = -\frac{2}{3}, z = \frac{1}{3}$

11. $w = 1, x = 0, y = -2, z = 3$

13. Unlimited: $x = -3, y = -1, z = 1; x = 12, y = 0, z = -10$

15. Inconsistent   **17.** $x = 4, y = \frac{2}{3}, z = -2$

19. $x = \dfrac{1}{2}, y = -\dfrac{3}{2}$ **21.** $x = \dfrac{1}{6}, y = -\dfrac{1}{2}$ **23.** Inconsistent

25. Inconsistent **27.** $r = 0, s = 0, t = 0, u = -1$

29. $x = \dfrac{c_1b_2 - c_2b_1}{a_1b_2 - a_2b_1}, y = \dfrac{a_1c_2 - a_2c_1}{a_1b_2 - a_2b_1}$

31. $x = 57 \times 10^6$ calc/s, $y = 68 \times 10^6$ calc/s

33. 250 parts/h, 220 parts/h, 180 parts/h

Exercises 16.6, page 454

1. $\begin{vmatrix} 3 & 0 & 0 \\ 1 & 1 & 0 \\ 2 & 1 & 3 \end{vmatrix} = 9, \begin{vmatrix} 0 & 0 & 3 \\ 0 & 1 & 1 \\ 3 & 1 & 2 \end{vmatrix} = -9$

3. -60 **5.** 0 **7.** -40 **9.** -40 **11.** 39 **13.** 57

15. -13 **17.** -72 **19.** 0 **21.** 39 **23.** 57

25. -13 **27.** -72 **29.** 0

31. $x = -1, y = 0, z = 2, t = 1$

33. $x = 1, y = 2, z = -1, t = 3$

35. $x = 2, y = -1, z = -1, t = 3$

37. $D = 1, E = 2, F = -1, G = -2$ **39.** 0 **41.** 8

43. $\dfrac{33}{16}A, \dfrac{11}{8}A, -\dfrac{5}{8}A, -\dfrac{15}{8}A, -\dfrac{15}{16}A$ **45.** 0.618, 1.618

47. ppm SO_2: 0.5, NO: 0.3, NO_2: 0.2, CO: 5.0

Review Exercises for Chapter 16, page 456

1. $a = 4, b = -1$

3. $x = 2, y = -3, z = \dfrac{5}{2}, a = -1, b = -\dfrac{7}{2}, c = \dfrac{1}{2}$

5. $x = -1, y = \dfrac{1}{2}, a = -\dfrac{1}{2}, b = -\dfrac{3}{2}$ **7.** $\begin{bmatrix} 1 & -3 \\ 8 & -5 \\ -8 & -2 \\ 3 & -10 \end{bmatrix}$

9. $\begin{bmatrix} -3 & 3 \\ 0 & -7 \\ 2 & -2 \\ -1 & -4 \end{bmatrix}$ **11.** $\begin{bmatrix} 7 & -6 \\ -4 & 20 \\ -1 & 6 \\ 1 & 15 \end{bmatrix}$ **13.** $\begin{bmatrix} 0 & 0 \\ 0 & 0 \end{bmatrix}$

15. $\begin{bmatrix} 0.34 & 0.11 & -0.05 \\ 0.02 & -0.08 & 0.10 \\ -0.01 & -0.17 & 0.20 \end{bmatrix}$ **17.** $\begin{bmatrix} -2 & \frac{5}{2} \\ -1 & 1 \end{bmatrix}$

19. $\begin{bmatrix} \frac{40}{3} & \frac{5}{3} \\ -\frac{20}{3} & \frac{35}{3} \end{bmatrix}$ **21.** $\begin{bmatrix} 11 & 10 & 3 \\ -4 & -4 & -1 \\ 3 & 3 & 1 \end{bmatrix}$

23. $\begin{bmatrix} \frac{1}{2} & -\frac{1}{2} & -1 \\ -3 & 2 & 1 \\ -4 & 3 & 2 \end{bmatrix}$ **25.** $x = -3, y = 1$

27. $x = 10, y = -15$ **29.** $u = -1, v = -3, w = 0$

31. $x = 1, y = \dfrac{1}{2}, z = -\dfrac{1}{3}$ **33.** $x = -3, y = 1$

35. $u = -1, v = -3, w = 0$ **37.** $x = 1, y = \dfrac{1}{2}, z = -\dfrac{1}{3}$

39. Unlimited number of solutions

41. $u = -1, v = -3, w = 0$ **43.** $x = 1, y = \dfrac{1}{2}, z = -\dfrac{1}{3}$

45. $x = 3, y = 1, z = -1$ **47.** $x = 1, y = 2, z = -3, t = 1$

49. $x = -\dfrac{1}{3}, y = 3, z = \dfrac{2}{3}, t = -4$

51. $r = \dfrac{1}{2}, s = -7, t = -\dfrac{3}{4}, u = 5, v = \dfrac{3}{2}$

53. $\begin{bmatrix} 1 & 0 \\ 15 & 16 \end{bmatrix}, \begin{bmatrix} 1 & 0 \\ 63 & 64 \end{bmatrix}, \begin{bmatrix} 1 & 0 \\ 255 & 256 \end{bmatrix}$

55. $B^3 = \begin{bmatrix} 1 & 0 & 0 \\ 0 & 1 & 0 \\ 0 & 0 & 1 \end{bmatrix}$ **57.** 77 **59.** 323 **61.** 77

63. 323 **65.** $N^{-1} = -N = \begin{bmatrix} 0 & 1 \\ -1 & 0 \end{bmatrix}$

67. $\begin{bmatrix} n & 1+n \\ 1-n & -n \end{bmatrix}\begin{bmatrix} n & 1+n \\ 1-n & -n \end{bmatrix}$

$= \begin{bmatrix} n^2 + (1-n^2) & (n+n^2) - (n+n^2) \\ (n-n^2) - (n-n^2) & (1-n^2) + n^2 \end{bmatrix}$

69. $(A + B)(A - B) = \begin{bmatrix} -6 & 2 \\ 4 & 2 \end{bmatrix}, A^2 - B^2 = \begin{bmatrix} -10 & -4 \\ 8 & 6 \end{bmatrix}$

71. $\begin{bmatrix} \frac{1}{2} & \frac{1}{3} \\ 0 & \frac{1}{6} \end{bmatrix} = \dfrac{1}{2}\begin{bmatrix} 1 & \frac{2}{3} \\ 0 & \frac{1}{3} \end{bmatrix}$ **73.** $R_1 = 4\,\Omega, R_2 = 6\,\Omega$

75. $F = 303$ lb, $T = 175$ lb **77.** $R_1 = 4\Omega, R_2 = 6\Omega$

79. $F = 303$ lb, $T = 175$ lb

81. 0.22 h after police pass intersection **83.** 30 g, 50 g, 20 g

85. $\begin{bmatrix} 12{,}000 & 24{,}000 & 4{,}000 \\ 15{,}000 & 8{,}000 & 30{,}000 \end{bmatrix} + \begin{bmatrix} 15{,}000 & 12{,}000 & 2{,}000 \\ 20{,}000 & 3{,}000 & 22{,}000 \end{bmatrix}$

$= \begin{bmatrix} 27{,}000 & 36{,}000 & 6{,}000 \\ 35{,}000 & 11{,}000 & 52{,}000 \end{bmatrix}$

87. $(R_1 + R_2)i_1 - R_2i_2 = 6$
$-R_2i_1 + (R_1 + R_2)i_2 = 0$

Exercises 17.1, page 464

1. $x + 1 < 0$ is true for all values of x less than -1; $x < -1$.

3. $-2 > -4$, multiplied by -3 gives $6 < 12$; divided by -2 gives $1 < 2$.

5. $7 < 12$ **7.** $20 < 45$ **9.** $-4 > -9$ **11.** $16 < 81$

13. $x > -2$ **15.** $x \le 45$ **17.** $1 < x < 7$

19. $x < -9$ or $x \ge -4$ **21.** $x < 1$ or $3 < x \le 5$

23. $-2 < x < 2$ or $3 \le x < 4$

25. x is greater than 0 and less than or equal to 2.

27. x is less than -10, or greater than or equal to 10 and less than 20.

29.   **31.**

33.   **35.**

37.   **39.**

41.   **43.**

45. Absolute inequality   **47.** No, $a - b < 0$

49. $|x| + |y| > |x + y|$

51. From step (4) to step (5), both sides divided by log 0.5, which is negative. Sign in step (5) should be $>$.

53. $2000 \le M \le 1{,}000{,}000$ **55.** $18{,}000 < v < 25{,}000 \text{ mi/h}$

57. $0 < n \le 2565 \text{ steps}$ **59.** $E = 0 \text{ for } 0 < r < a$

$E = k/r^2 \text{ for } r \ge a$

Exercises 17.2, page 469

1. $x \le 3$ **3.** $1 < x < \dfrac{9}{2}$ **5.** $x > -1$

7. $x < 64$ **9.** $x \le -2$

11. $y < -2$ **13.** $x \le \dfrac{5}{2}$

15. $x < -\dfrac{177}{20}$ **17.** $x > -1.80$

19. $L > -\dfrac{7}{9}$ **21.** $-1 < x < 1$

23. $2 < x \le 5$ **25.** $-3 \le x < -1$

27. No values   **29.** $x < \dfrac{5}{2}$

31. $x \ge -1$ **33.** $-2 < t < 2$

35. $x > -\dfrac{1}{2}$ **37.** $x \ge 5$

39. $-6 < k < 6$ **41.** $n > 35 \text{ h}$

43. $50° < F < 68°$ **45.** $0.54 \text{ m} < w < 0.81 \text{ m}$

47. $0.4 < t < 2.6 \text{ h}$ **49.** $0 \le x \le 500$

$200 \le y \le 700$

51. $2 \text{ min} \le \text{stop times} \le 4 \text{ min}$

Exercises 17.3, page 473

1. $x < 1 \text{ or } x > 3$ **3.** $-3 < x < 2 \text{ or } x > 4$

5. $-4 < x < 4$ **7.** $0 \le x \le 2$

9. $-4 \le x \le \dfrac{3}{2}$ **11.** $x = -2$

13. All R **15.** $-2 < x < 0, x > 1$

17. $-2 \le s \le -1, s \ge 1$ **19.** $-6 < x \le \dfrac{3}{2}$

21. $-5 < x < -1, x > 7$ **23.** $x < -1$

25. $x \le -2, x \ge \dfrac{1}{3}$ **27.** $T < 3, T \ge 8$

29. $-1 < x < \dfrac{3}{4}, x \geq 6$

31. $2 < x < 4, 5 < x < 9$

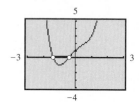

33. $x \leq -2, x \geq 1$

35. $-1 \leq x \leq 0$

37. $x > 1.52$

39. $-1.39 < x < -0.43$

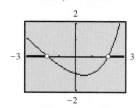

41. $x < -1.69, x > 2.00$

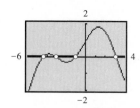

43. $x < -4.43, -3.11 < x < -1.08,$
$x > 3.15$

45. No; not true if $0 \leq x \leq 1$ **47.** $x^2 - 3x - 4 > 0$

49. $x < 3.113, x < \dfrac{3 \log 3 + 2 \log 2}{2 \log 3 - \log 2}$

51. If $b > a, a < x < b$; If $a > b, b < x < a$

53. $0.5 < i < 1\,\text{A}$ **55.** $x \leq 12$ **57.** $C_1 > \dfrac{4}{3}\,\mu\text{F}$

59. $h > 2640\,\text{km}$ **61.** $3.0 \leq w < 5.0\,\text{mm}$

63. $0 \leq t < 0.92\,\text{h}$

Exercises 17.4, page 479

1. $-2 < x < 3$

3. $3 < x < 5$

5. $x < -2, x > \dfrac{2}{5}$

7. $\dfrac{1}{6} \leq x \leq \dfrac{3}{2}$

9. $x < 0, x > \dfrac{3}{2}$

11. $-26 < x < 24$

13. $-6.4 \leq x \leq -2.1$

15. $x < -18, x > 66$

17. $1 < x < 2$

19. $x \leq \dfrac{1}{10}, x \geq \dfrac{7}{10}$

21. $-15 < R < \dfrac{35}{3}$

23. $x \leq 8.4, x \geq 17.6$

25. $1 < x < 4$

27. $x \leq 6, x \geq 10$

29. $x < -3, -2 < x < 1, x > 2$

31. $-3 < x < -2, 1 < x < 2$

33. $x = 0$ valid only for $a = 0$

35. $x < \dfrac{c - a}{b}$ or $x > \dfrac{a - c}{b}$ **37.** 4 km, 50 km

39. $|p - 2{,}000{,}000| \leq 200{,}000$; production is at least 1,800,000 barrels, but not greater than 2,200,000 barrels.

41. $|d - 3.765| \leq 0.002$ **43.** $0.020 < i < 0.040\,\text{A}$

Exercises 17.5, page 482

1. $y < 3 - x$

3.

5.

7.

9.

11.

13.

15.

17.

19.

21.

23.

25.

27.

29.

31.

33.

35.

37.

39.

41.

43. $y < 2x + \dfrac{5}{2}$

45. Below

47.

49.

51. p

53.

Exercises 17.6, page 486

1. Max $F = 18$ at $(0, 6)$

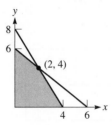

3. Max $F = 58$ at $(10, 7)$
    Min $F = 0$ at $(0, 0)$

5. Max $P = 30$ at $(0, 6)$

7. Max $P = 30$ at $(6, 0)$

9. Min $C = 24$ at $(3, 2)$

11. Min $F = 7$ at $(1, 2)$

Max $F = \dfrac{47}{3}$ at $\left(\dfrac{5}{3}, \dfrac{14}{3}\right)$

13. Max $P = 27$ at $(3, 0)$ **15.** Min $C = 40$ at $(4, 4)$

17. $6000 at 6%, $3000 at 5%

19. 40 business models
    60 graphing models

21. $4\dfrac{2}{7}$ oz of A
    $2\dfrac{6}{7}$ oz of B

Review Exercises for Chapter 17, page 487

1. $x > 6$

3. $\dfrac{5}{2} < x < 6$

5. $-2 < x < \dfrac{1}{5}$

7. $n < -\dfrac{7}{3}$ or $n > \dfrac{5}{2}$

9. $x < -4, \dfrac{1}{2} < x < 3$

11. $x < 0, x > 4$

13. $-2 \le x \le \dfrac{2}{3}$

15. $x < -\dfrac{4}{5}, x > 2$

17. $x > \dfrac{5}{3}$

19. $2 \le n < \dfrac{11}{4}$

21. $R < -\dfrac{1}{2}, R \ge 8$

23. $x < -18, x > 78$

25. $x < -0.68$

27. $x < 0.69$

29.

31.

33.

35.

37.

39.

41.

43.

45.

47.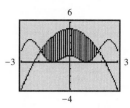

49. $x \le 3$ **51.** $x \le -4, x \ge 0$ **53.** Max $P = 29$ at $(1, 3)$

55. Min $C = \dfrac{65}{8}$ at $\left(\dfrac{3}{8}, \dfrac{7}{4}\right)$

57. a and b must have different signs.

59. $a + \dfrac{1}{a} - 2 > 0, a^2 + 1 - 2a > 0, (a - 1)^2 > 0$

61. $x^2 - 3x - 10 < 0$ **63.**

65. $f(x) > 0$ for $x < 2$ or $x > 3$
$f(x) < 0$ for $2 < x < 3$
$f(x) = 0$ for $x = 2, x = 3$
$f(0) = 6, f(5) = 6$

67. $0 < x \le 8.0$ cm   **69.** Between \$5 and \$12.50

71. $d > 39.5$ m   **73.** $168{,}000 \le B \le 400{,}000$ Btu

75. $0.456 < i < 0.816$ A   **77.** $r > 5.7$

79.

81. 300 regular
150 deluxe

Exercises 18.1, page 493

1. 56 mi   **3.** 6   **5.** $\dfrac{4}{3}$ **7.** 2   **9.** 40

11. $0.27 = 27\%$ **13.** 0.41   **15.** $0.025\,\mu$F   **17.** 5.25

19. 7.80   **21.** 3.5%   **23.** 863 kg   **25.** 0.0194

27. 144 Ω   **29.** 5.63 in.^{2} **31.** 0.335 hp   **33.** 1210 km

35. 12.5 m/s   **37.** 23,400 cm^3 **39.** 8.8 m   **41.** 6.0 ft, 9.0 ft

43. 19.67 kg   **45.** 17,500 chips   **47.** 44.3 in.

Exercises 18.2, page 498

1. $C = kd, k = \pi$ **3.** 667 Hz   **5.** $v = kr$

7. $R = \dfrac{k}{d^2}$ **9.** $P = \dfrac{k}{\sqrt{A}}$ **11.** $S = kwd^3$

13. A varies directly as the square of r.

15. f varies directly as L and inversely as the square root of m.

17. $V = \dfrac{H^2}{2048}$ **19.** $p = \dfrac{16q}{r^3}$ **21.** 25   **23.** 0.033

25. 180   **27.** 2.56×10^5

29. $A = k_1 x, B = k_2 x; A + B = (k_1 + k_2)x$ **31.** 10.2 kB/s

33. 209 kJ   **35.** 2.40 in.   **37.** 560 kJ   **39.** 1.4 h

41. (a) Inverse (b) $a = 60/m$

43. 10,200 hp   **45.** $F = 0.00523Av^2$ **47.** 0.536 N

49. 480 m/s   **51.** $R = \dfrac{4.44 \times 10^{-5}l}{A}$ **53.** 80.0 W

55. $G = \dfrac{5.9d^2}{\lambda^2}$ **57.** $-6.57 \, \text{ft/s}^2$ **59.** $0.288 \, \text{W/m}^2$

Review Exercises for Chapter 18, page 501

1. 200   **3.** 1.5   **5.** 3.1417   **7.** 5.6   **9.** $7.39 \, \text{lb/in.}^2$

11. 2260 J/g   **13.** 5.5%

15. $\dfrac{a}{b} + \dfrac{b}{b} = \dfrac{c}{d} + \dfrac{d}{d} = \dfrac{a}{b} + \dfrac{c}{d}$ **17.** 92 mi

19. $2.82 \times 10^6 \, \text{J} = 2.82 \, \text{MJ}$ **21.** 240   **23.** 310 mL

25. 71.9 ft   **27.** 4500, 7500   **29.** 30.2 kg

31. 115 bolts, 207 bolts   **33.** $y = 3x^2$ **35.** $v = \dfrac{128x}{y^3}$

37. 11.7 in.   **39.** 10 cm   **41.** $R = 0.94\,A$

43. 1500 bacteria/h   **45.** $F = \dfrac{5500}{L}$ **47.** 2.22 s

49. $4.3 \, \mu\text{C}$ **51.** 22.5 hp   **53.** 144 ft   **55.** 1.4

57. 48.7 Hz   **59.** $2.99 \times 10^8 \, \text{m/s}$ **61.** 5630 lb

63. 3.26 cm   **65.** $5.73 \times 10^4 \, \text{m}$ **67.** 150 ft   **69.** 150%

71. $125.00, $600.13   **73.** 4800 Btu   **75.** $0.023 \, \text{W/m}^2$

Exercises 19.1, page 509

1. -3 **3.** 125,250   **5.** 4, 6, 8, 10, 12

7. 2.5, 1.5, 0.5, -0.5, -1.5 **9.** 22   **11.** $-\dfrac{9\pi}{4}$ **13.** 30.9

15. $49b$ **17.** 440   **19.** $-\dfrac{85}{2}$ **21.** $n = 6, S_6 = 150$

23. $d = -\dfrac{2}{19}, a_{20} = -\dfrac{1}{3}$ **25.** $a_1 = 19, a_{30} = 106$

27. No solution (n cannot be negative)

29. $n = 13, a_{13} = 11k$ **31.** $n = 8, d = \dfrac{1}{14}(b + 2c)$

33. $a_1 = 360, d = 40, S_{10} = 5400$

35. Yes, $d = \ln 2; a_5 = \ln 48$ **37.** $a_n = 2n + 1$

39. $d = \dfrac{a + b}{2}$ **41.** $2b - c, b, c, 2c - b, 3c - 2b$

43. 5050   **45.** -8

47. $2700 \, \text{m}^2$ **49.** 195 logs   **51.** 10 rows

53. 13 years, $11,700   **55.** 185 m

57. $S_n = \dfrac{1}{2}n[2a_1 + (n - 1)d]$ **59.** $a_1 = 1, a_n = n$

Exercises 19.2, page 513

1. 243   **3.** 6400, 1600, 400, 100, 25   **5.** $\dfrac{1}{6}, \dfrac{1}{2}, \dfrac{3}{2}, \dfrac{9}{2}, \dfrac{27}{2}$

7. 128   **9.** $\dfrac{1}{125}$ **11.** $\dfrac{100}{729}$ **13.** 1   **15.** $\dfrac{341}{8}$ **17.** 762

19. $\dfrac{3(2^{10} - k^{10})}{16(2 + k)}$ **21.** $a_6 = 64, S_6 = \dfrac{1365}{16}$

23. $a_1 = 16, a_5 = 81$ **25.** $a_1 = 1, r = 3$

27. $n = 7, S_n = \dfrac{58,593}{625}$ **29.** Yes; $r = 3^x; a_{20} = 3^{19x+1}$

31. $r = \sqrt{\dfrac{b}{a}}$ **33.** 5   **35.** (a) a_1, (b) 0   **37.** 2 or -3

39. 1.4%   **41.** 1.09 mA   **43.** $443.96   **45.** 5800°C

47. 43%   **49.** 21.1°C   **51.** $671,088.64

53. $S_n = \dfrac{a_1 - ra_n}{1 - r}$ **55.** Yes, the ratio is squared.

57. 8, 1, -6; 8, 16, 24; 1, -6, 36; 16, 24, 36

Exercises 19.3, page 517

1. $\dfrac{32}{7}$ **3.** 8   **5.** 0.2   **7.** $\dfrac{400}{21}$ **9.** 8   **11.** $\dfrac{10,000}{9999}$

13. $\dfrac{1}{2}(5 + 3\sqrt{3})$ **15.** $\dfrac{1}{3}$ **17.** 0.5   **19.** $\dfrac{40}{99}$ **21.** $\dfrac{2}{11}$

23. $\dfrac{91}{3330}$ **25.** $\dfrac{11}{30}$ **27.** $\dfrac{100,741}{999,000}$ **29.** $\dfrac{125}{2}, \dfrac{125}{3}$

31. 350 gal   **33.** 346 g   **35.** 100 m   **37.** $-\dfrac{1}{4}$

Exercises 19.4, page 522

1. $32x^5 + 240x^4 + 720x^3 + 1080x^2 + 810x + 243$

3. $t^3 + 12t^2 + 48t + 64$ **5.** $16x^4 - 32x^3 + 24x^2 - 8x + 1$

7. 8445.96301

9. $n^5 + 10\pi n^4 + 40\pi^2 n^3 + 80\pi^3 n^2 + 80\pi^4 n + 32\pi^5$

11. $64a^6 - 192a^5 b^2 + 240a^4 b^4 - 160a^3 b^6 + 60a^2 b^8 - 12ab^{10} + b^{12}$

13. $625x^4 - 1500x^3 + 1350x^2 - 540x + 81$

15. $64a^6 + 192a^5 + 240a^4 + 160a^3 + 60a^2 + 12a + 1$

17. $x^{10} + 20x^9 + 180x^8 + 960x^7 + \cdots$

19. $128a^7 - 448a^6 + 672a^5 - 560a^4 + \cdots$

21. $x^6 - 48x^{11/2}y + 1056x^5 y^2 - 14,080x^{9/2}y^3 + \cdots$

23. $b^{40} + 10b^{37} + \dfrac{95}{2}b^{34} + \dfrac{285}{2}b^{31} + \cdots$

25. 1.338   **27.** 1.015   **29.** $1 + 8x + 28x^2 + 56x^3 + \cdots$

31. $1 + 6x + 27x^2 + 108x^3 + \cdots$

33. $1 + \dfrac{1}{2}x - \dfrac{1}{8}x^2 + \dfrac{1}{16}x^3 - \cdots$

35. $\dfrac{1}{3}\left[1 + \dfrac{1}{2}x + \dfrac{3}{8}x^2 + \dfrac{5}{16}x^3 + \cdots\right]$

37. (a) 3.557×10^{14} (b) 5.109×10^{19} (c) 8.536×10^{15}
(d) 2.480×10^{96}

39. $n! = n(n - 1)(n - 2)(\cdots)(2)(1) = n \times (n - 1)!$; for
$n = 1, 1! = 1 \times 0!$ Since $1! = 1, 0!$ must $= 1$.

41. $56a^3b^5$ **43.** $10,264,320x^8b^4$

45. $n!$ contains factors 2 and 5.

47. $n = 1, 1 - 1 = 0; n = 2, 1 - 2 + 1 = 0, n = 3,$
$1 - 3 + 3 - 1 = 0$, etc.

49. 2.45   **51.** $V = A(1 - 5r + 10r^2 - 10r^3 + 5r^4 - r^5)$

53. $1 - \dfrac{x}{a} + \dfrac{x^3}{2a^3} - \cdots$ **55.** $-\dfrac{2hk}{r^3} + \dfrac{3kh^2}{r^4}$

Review Exercises for Chapter 19, page 523

1. 81   **3.** 1.28×10^{-3} **5.** 66.5   **7.** $\dfrac{16}{243}$ **9.** $\dfrac{195}{2}$

11. $\dfrac{16,383}{1536}$ **13.** 81   **15.** 32   **17.** -15 **19.** -0.25

21. 186   **23.** $\dfrac{455}{2}$ (as), 127 (gs), or 43 (gs)   **25.** 2.7

27. 51   **29.** $\dfrac{1}{33}$ **31.** $\dfrac{4}{55}$

33. $x^4 - 8x^3 + 24x^2 - 32x + 16$

35. $x^{10} + 20x^8 + 160x^6 + 640x^4 + 1280x^2 + 1024$

37. $a^{10} + 20a^9e + 180a^8e^2 + 960a^7e^3 + \cdots$

39. $p^{18} - \dfrac{3}{2}p^{16}q + p^{14}q^2 - \dfrac{7}{18}p^{12}q^3 + \cdots$

41. $1 + 12x + 66x^2 + 220x^3 + \cdots$

43. $1 + \dfrac{1}{2}x^2 - \dfrac{1}{8}x^4 + \dfrac{1}{16}x^6 - \cdots$

45. $1 - \dfrac{1}{2}a^2 - \dfrac{1}{8}a^4 - \dfrac{1}{16}a^6 - \cdots$

47. $\dfrac{1}{8} + \dfrac{3}{4}x + 3x^2 + 10x^3 + \cdots$ **49.** 1,001,000

51. $4b - 3a$ **53.** No   **55.** $a_n = -3n - 2$ **57.** 0.7156

59. 5.475   **61.** 11th   **63.** 12.6 mm   **65.** 302.9 in.

67. \$4700   **69.** 4.4×10^9 in. $= 69,000$ mi   **71.** 624 ft

73. 100 cm   **75.** \$47,340.80   **77.** \$6.40   **79.** 4.0°C

81. $1 + \dfrac{1}{2}am^2 + \dfrac{1}{8}am^4$ **83.** 21 years   **85.** Five applications

87. 22 years   **89.** $S = \dfrac{n}{2}(a_1 + a_n); a_{mid} = \dfrac{a_1 + a_n}{2} = \dfrac{2S}{2n} = \dfrac{S}{n}$

91. Yes; term to term ratios are equal.

Exercises 20.1, page 531

(*Note:* "Answers" to trigonometric identities are intermediate steps of suggested reductions of the left member.)

1. $\sin x = \dfrac{\tan x}{\sec x} = \dfrac{\frac{\sin x}{\cos x}}{\frac{1}{\cos x}} = \dfrac{\sin x}{\cos x} \cdot \dfrac{\cos x}{1} = \sin x$

3. $1.483 = \dfrac{1}{0.6745}$ **5.** $\left(-\dfrac{1}{2}\sqrt{3}\right)^2 + \left(-\dfrac{1}{2}\right)^2 = \dfrac{3}{4} + \dfrac{1}{4} = 1$

7. $\sin x - 1$ **9.** $\cot \theta - 2 \cos \theta$ **11.** $\sec^2 u$

13. $\tan x \sec x$ **15.** $\sin t \cos t$ **17.** $\cot^2 y(\csc^2 y + 1)$

19. $\dfrac{\sin x}{\frac{\sin x}{\cos x}} = \sin x\left(\dfrac{\cos x}{1}\right)\left(\dfrac{1}{\sin x}\right)$ **21.** $\sin x\left(\dfrac{1}{\cos x}\right)$

23. $\csc^2 x(\sin^2 x)$

25. $\sin x(\csc^2 x) = (\sin x)(\csc x)(\csc x)$
$= \sin x\left(\dfrac{1}{\sin x}\right)\csc x$

27. $\cos \theta\left(\dfrac{\cos \theta}{\sin \theta}\right) + \sin \theta = \dfrac{\cos^2 \theta + \sin^2 \theta}{\sin \theta} = \dfrac{1}{\sin \theta}$

29. $\cot \theta(\sec^2 \theta - 1) = \cot \theta \tan^2 \theta = (\cot \theta \tan \theta)\tan \theta$

31. $\dfrac{\sin x}{\cos x} + \dfrac{\cos x}{\sin x} = \dfrac{\sin^2 x + \cos^2 x}{\cos x \sin x} = \dfrac{1}{\cos x \sin x}$

33. $(1 - \sin^2 x) - \sin^2 x$

35. $\dfrac{\sin \theta}{\frac{1}{\sin \theta}} + \dfrac{\cos \theta}{\frac{1}{\cos \theta}} = \sin^2 \theta + \cos^2 \theta$

37. $(2 \sin^2 x - 1)(\sin^2 x - 1)$

39. $\dfrac{\sin \pi t}{2}\left(\dfrac{\sin^2 \pi t + (1 - \cos \pi t)^2}{(1 - \cos \pi t) \sin \pi t}\right)$
$= \dfrac{\sin^2 \pi t + 1 - 2 \cos \pi t + \cos^2 \pi t}{2(1 - \cos \pi t)} = \dfrac{2(1 - \cos \pi t)}{2(1 - \cos \pi t)}$

41. Infinite series: $\dfrac{1}{1 - \sin^2 x} = \dfrac{1}{\cos^2 x}$ **43.** $\cot x$ **45.** $\sin x$

47. $\sec x$ **49.** $\cos x$

51.   **53.**

55. Yes   **57.** No

59. $0 = \cos A \cos B \cos C + \sin A \sin B,$
$\cos C = -\dfrac{\sin A \sin B}{\cos A \cos B}$

61. $l = a \csc \theta + a \sec \theta = a\left(\dfrac{1}{\sin \theta} + \dfrac{\tan \theta}{\sin \theta}\right)$

63. Replacing $\tan \theta$ with $\sin \theta/\cos \theta$ gives $\sin^2 \theta/\cos \theta + \cos \theta$. Using LCD of $\cos \theta$ gives $(\sin^2 \theta + \cos^2 \theta)/\cos \theta$ which equals $\sec \theta$.

65. $\sin^2 x - \sin^2 x \sec^2 x + \cos^2 x + \cos^2 x \sec^4 x$
$= \sin^2 x - \tan^2 x + \cos^2 x + \sec^2 x = 2$

67. $\left(\dfrac{r}{x}\right)^2 + \left(\dfrac{r}{y}\right)^2 = \dfrac{r^2(x^2 + y^2)}{x^2y^2}$

69. $\sqrt{1 - \cos^2 \theta} = \sqrt{\sin^2 \theta}$

71. $\sqrt{4 + 4\tan^2 \theta} = 2\sqrt{1 + \tan^2 \theta}$

Exercises 20.2, page 536

1. $\dfrac{16}{65}$

3. $\sin 105° = \sin 60° \cos 45° + \cos 60° \sin 45°$
$= \dfrac{\sqrt{3}}{2}\dfrac{\sqrt{2}}{2} + \dfrac{1}{2}\dfrac{\sqrt{2}}{2} = 0.9659$

5. $\cos 15° = \cos(60° - 45°)$
$= \cos 60° \cos 45° + \sin 60° \sin 45°$
$= \left(\dfrac{1}{2}\right)\left(\dfrac{1}{2}\sqrt{2}\right) + \left(\dfrac{1}{2}\sqrt{3}\right)\left(\dfrac{1}{2}\sqrt{2}\right)$
$= \dfrac{1}{4}\sqrt{2} + \dfrac{1}{4}\sqrt{6} = \dfrac{1}{4}(\sqrt{2} + \sqrt{6}) = 0.9659$

7. $-\dfrac{33}{65}$ **9.** $-\dfrac{56}{65}$ **11.** $\sin 3x$ **13.** $-\cos x$ **15.** $\sin x$

17. $\tan x$ **19.** 0 **21.** 1 **23.** 0

25. $(\sin x \cos y + \cos x \sin y)(\sin x \cos y - \cos x \sin y)$
$= \sin^2 x \cos^2 y - \cos^2 x \sin^2 y$
$= \sin^2 x(1 - \sin^2 y) - (1 - \sin^2 x)\sin^2 y$

27. $(\cos\alpha \cos\beta - \sin\alpha \sin\beta)$
$+ (\cos\alpha \cos\beta + \sin\alpha \sin\beta)$

29.   **31.**

33, 35, 37, 39. Use the indicated method.

41. $\dfrac{\sin(x + x)}{\sin x} = \dfrac{\sin x \cos x + \cos x \sin x}{\sin x}$
$= \dfrac{2\sin x \cos x}{\sin x}$

43. Using Eq. (20.9), $\sin 75° = \sin(30° + 45°)$. Using Eq. (20.11), $\sin 75° = \sin(135° - 60°)$. (Other angles are also possible.)

45. $I(\cos\theta \cos 30° - \sin\theta \sin 30°)$
$+ I(\cos\theta \cos 150° - \sin\theta \sin 150°)$
$+ I(\cos\theta \cos 270° - \sin\theta \sin 270°)$
$= I\left[\dfrac{1}{2}\sqrt{3}\cos\theta - \dfrac{1}{2}\sin\theta - \dfrac{1}{2}\sqrt{3}\cos\theta - \dfrac{1}{2}\sin\theta\right.$
$\left. + 0 - (-\sin\theta)\right]$

47.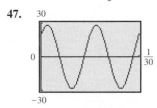

49. $i_0 \sin(\omega t + \alpha) = i_0(\sin\omega t \cos\alpha + \cos\omega t \sin\alpha)$

51. $\tan\alpha(R + \cos\beta) = \sin\beta, R = \dfrac{\sin\beta - \tan\alpha\cos\beta}{\tan\alpha}$
$= \dfrac{\sin\beta\cos\alpha - \cos\beta\sin\alpha}{\cos\alpha\tan\alpha}$

Exercises 20.3, page 539

1. $-\sqrt{3}$ **3.** $-\dfrac{24}{25}$

5. $\sin 60° = \sin 2(30°) = 2\sin 30° \cos 30°$
$= 2\left(\dfrac{1}{2}\right)\left(\dfrac{1}{2}\sqrt{3}\right) = \dfrac{1}{2}\sqrt{3}$

7. $\tan 120° = \dfrac{2\tan 60°}{1 - \tan^2 60°} = \dfrac{2\sqrt{3}}{1 - (\sqrt{3})^2} = -\sqrt{3}$

9. $\sin 258° = 2\sin 129° \cos 129° = -0.9781476$

11. $\cos 96° = \cos^2 48° - \sin^2 48° = -0.1045285$ **13.** 3.0777

15. $\dfrac{24}{25}$ **17.** $-\sqrt{3}$ **19.** $2\sin 8x$ **21.** $\cos 8x$ **23.** $\cos x$

25. $-2\cos 4x$ **27.** $2\cos 2\theta$ **29.** 2

31. $\cos^2\alpha - (1 - \cos^2\alpha)$

33. $\dfrac{\cos x - (\sin x/\cos x)\sin x}{1/\cos x} = \cos^2 x - \sin^2 x$

35. $\dfrac{2\sin\theta\cos\theta}{1 + 2\cos^2\theta - 1} = \dfrac{\sin\theta}{\cos\theta}$

37. $1 - (1 - 2\sin^2 2\theta) = \dfrac{2}{\csc^2\theta}$

39. $\ln\dfrac{1 - \cos 2x}{1 + \cos 2x} = \ln\dfrac{2\sin^2 x}{2\cos^2 x} = \ln\tan^2 x$

41.   **43.**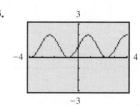

45. $3\sin x - 4\sin^3 x$ **47.** $8\cos^4 x - 8\cos^2 x + 1$

49. $1 - 2\sin^2 x + \dfrac{2\sin x \cos x \sin x}{\cos x} = 1$

51. amp. = 2, per. = π; write equation as
$y = 2(2\sin x \cos x) = 2\sin 2x$.

53. $\sqrt{\sin^2 x + 2\sin x \cos x + \cos^2 x} = \sqrt{1 + \sin^2 x}$

55. 474 m **57.** $R = v\left(\dfrac{2v\sin\alpha}{g}\right)\cos\alpha = \dfrac{v^2(2\sin\alpha\cos\alpha)}{g}$

59. $vi\sin\omega t\sin\left(\omega t - \dfrac{\pi}{2}\right)$
$= vi\sin\omega t\left(\sin\omega t\cos\dfrac{\pi}{2} - \cos\omega t\sin\dfrac{\pi}{2}\right)$
$= vi\sin\omega t[-(\cos\omega t)(1)] = -\dfrac{1}{2}vi(2\sin\omega t\cos\omega t)$

Exercises 20.4, page 543

1. $\cos 57°$

3. $\cos 15° = \cos\frac{1}{2}(30°) = \sqrt{\dfrac{1 + \cos 30°}{2}} = \sqrt{\dfrac{1.8660}{2}}$
$= 0.9659$

5. $\sin 105° = \sin\frac{1}{2}(210°) = \sqrt{\dfrac{1 - \cos 210°}{2}}$
$= \sqrt{\dfrac{1.8660}{2}} = 0.9659$

7. -0.9239 **9.** $\sin 118° = 0.8829476$

11. $\sqrt{2\left(\dfrac{1 + \cos 164°}{2}\right)} = \sqrt{2}\cos 82° = 0.1968205$

13. $\sin 3x$ **15.** $4\cos 2x$ **17.** $2\sqrt{2}\sin 5\theta$ **19.** 1

21. $\dfrac{1}{26}\sqrt{26}$ **23.** 0.1414 **25.** $\pm\sqrt{\dfrac{2\sec x}{\sec x - 1}}$

27. $\tan\frac{1}{2}\alpha = \dfrac{1 - \cos\alpha}{\sin\alpha} = \dfrac{\sin\alpha}{1 + \cos\alpha}$

29. $\dfrac{1 - \cos\alpha}{2\sin\frac{1}{2}\alpha} = \dfrac{1 - \cos\alpha}{2\sqrt{\frac{1}{2}(1 - \cos\alpha)}} = \sqrt{\dfrac{1 - \cos\alpha}{2}}$

31. $2\sqrt{2\dfrac{(1 + \cos x)(1 + \cos x)}{2(2)(1 + \cos x)}} = \dfrac{2(1 + \cos x)}{2}\cdot\sqrt{\dfrac{2}{1 + \cos x}}$

33.   **35.**

37. $\pm\dfrac{24}{7}$ **39.** $4\sin\dfrac{\theta}{2}$

41. $\sin\omega t = \pm\sqrt{\dfrac{1 - \cos 2\omega t}{2}}$,
$\sin^2\omega t = \dfrac{1}{2}(1 - \cos 2\omega t)$

43. $\dfrac{\sqrt{\dfrac{1 - \cos(A + \phi)}{2}}}{\sqrt{\dfrac{1 - \cos A}{2}}} = \sqrt{\dfrac{1 - \cos(A + \phi)}{1 - \cos A}}$

Exercises 20.5, page 547

1. $\dfrac{\pi}{4}, \dfrac{5\pi}{4}$ **3.** $0, \dfrac{\pi}{3}, \dfrac{5\pi}{3}$ **5.** $\dfrac{\pi}{2}$ **7.** $\dfrac{\pi}{2}, \dfrac{3\pi}{2}$

9. $\dfrac{\pi}{3}, \dfrac{2\pi}{3}, \dfrac{4\pi}{3}, \dfrac{5\pi}{3}$ **11.** $0, \dfrac{\pi}{6}, \dfrac{5\pi}{6}, \pi$ **13.** $\dfrac{\pi}{2}, \dfrac{3\pi}{2}$

15. $\dfrac{\pi}{4}, \dfrac{3\pi}{4}, \dfrac{5\pi}{4}, \dfrac{7\pi}{4}$ **17.** $0.2618, 1.309, 3.403, 4.451$

19. $0, \pi$ **21.** $\dfrac{3\pi}{4} = 2.36, \dfrac{7\pi}{4} = 5.50$

23. $1.9823, 4.3009$ **25.** $\dfrac{\pi}{3} = 1.05, \dfrac{2\pi}{3} = 2.09,$
$\dfrac{4\pi}{3} = 4.19, \dfrac{5\pi}{3} = 5.24$

27. $\dfrac{\pi}{12} = 0.26, \dfrac{\pi}{4} = 0.79, \dfrac{5\pi}{12} = 1.31, \dfrac{3\pi}{4} = 2.36,$
$\dfrac{13\pi}{12} = 3.40, \dfrac{5\pi}{4} = 3.93, \dfrac{17\pi}{12} = 4.45, \dfrac{7\pi}{4} = 5.50$

29. $0 = 0.00, \dfrac{\pi}{3} = 1.05, \pi = 3.14, \dfrac{5\pi}{3} = 5.24$

31. $3.569, 5.856$

33. $0.7854, 1.249, 3.927, 4.391$

35. $\dfrac{3\pi}{8} = 1.18, \dfrac{7\pi}{8} = 2.75, \dfrac{11\pi}{8} = 4.32, \dfrac{15\pi}{8} = 5.89$

37. $0, \dfrac{\pi}{2}, \pi, \dfrac{3\pi}{2}$ **39.** $0, \dfrac{\pi}{2}, \pi, \dfrac{3\pi}{2}$

41. No; $\cot\theta > 1, \sec\theta > 1, \csc\theta > 1$ for $0 < \theta < \dfrac{\pi}{2}$

43. $\theta = 0, r = 0; \theta = \dfrac{\pi}{3}, r = \dfrac{\sqrt{3}}{2}; \theta = \pi, r = 0; \theta = \dfrac{5\pi}{3},$
$r = -\dfrac{\sqrt{3}}{2}$

45. $37.8°$ **47.** 10.2 s, 15.7 s, 21.2 s, 47.1 s   **49.** 6.56×10^{-4}

51. 300.0 N, 400.0 N

53. $-2.28, 0.00, 2.28$ **55.** 0.29, 0.95

57. 2.10   **59.** 1.08

Exercises 20.6, page 553

1. y is the angle whose tangent is $3A$.   **3.** 0

5. y is the angle whose cotangent is $3x$.

7. y is twice the angle whose sine is x.

9. y is five times the angle whose cosine is $2x - 1$.

11. $\dfrac{\pi}{3}$ **13.** $\dfrac{\pi}{4}$ **15.** $-\dfrac{\pi}{3}$ **17.** No value   **19.** $-\dfrac{\pi}{4}$

21. $\dfrac{1}{2}\sqrt{3}$ **23.** $\dfrac{\pi}{4}$ **25.** 0.1961   **27.** -1 **29.** $2/\sqrt{21}$

31. $2/\sqrt{3}$ **33.** -1.3090 **35.** 0.0219   **37.** -1.2389

39. -0.2239 **41.** $x = \dfrac{1}{3}\sin^{-1} y$ **43.** $x = 4\tan y$

45. $x = 1 - \cos(1 - y)$ **47.** $\dfrac{x}{\sqrt{1 - x^2}}$

49. $xy + \sqrt{1 - x^2 - y^2 + x^2y^2}$ **51.** $\dfrac{3x}{\sqrt{9x^2 - 1}}$

53. $2x\sqrt{1 - x^2}$

55. No; $\sin^{-1}(\sin x) = x$ for $-\dfrac{\pi}{2} \le x \le \dfrac{\pi}{2}$.

57. $t = \dfrac{1}{2\omega}\cos^{-1}\dfrac{y}{A} - \dfrac{\phi}{\omega}$

59. $t = \dfrac{1}{\omega}\left(\sin^{-1}\dfrac{i}{I_m} - \alpha - \phi\right)$

61. $\sin\left(\sin^{-1}\dfrac{3}{5} + \sin^{-1}\dfrac{5}{13}\right) = \dfrac{3}{5}\cdot\dfrac{12}{13} + \dfrac{4}{5}\cdot\dfrac{5}{13} = \dfrac{56}{65}$

63. $\dfrac{\pi}{2}$ **65.** 0   **67.** $\sin^{-1}\left(\dfrac{a}{c}\right)$ **69.** $\tan^{-1}\left(\dfrac{b\tan B}{a}\right)$

71. Let y = height to top of pedestal; $\tan\alpha = \dfrac{151 + y}{d}$,

$\tan\beta = \dfrac{y}{d}; \tan\alpha = \dfrac{151 + d\tan\beta}{d}$

73. $\theta = \tan^{-1}\left(\dfrac{y + 50}{x}\right) - \tan^{-1}\left(\dfrac{y}{x}\right)$

Review Exercises for Chapter 20, page 555

1. $\sin(90° + 30°) = \sin 90°\cos 30° + \cos 90°\sin 30°$
$= (1)\left(\dfrac{1}{2}\sqrt{3}\right) + (0)\left(\dfrac{1}{2}\right) = \dfrac{1}{2}\sqrt{3}$

3. $\sin(360° - 45°) = \sin 360°\cos 45° - \cos 360°\sin 45°$
$= 0\left(\dfrac{1}{2}\sqrt{2}\right) - 1\left(\dfrac{1}{2}\sqrt{2}\right) = -\dfrac{1}{2}\sqrt{2}$

5. $\cos\left(2\dfrac{\pi}{2}\right) = \cos^2\dfrac{\pi}{2} - \sin^2\dfrac{\pi}{2} = 0 - 1^2 = -1$

7. $\tan 2(30°) = \dfrac{2\tan 30°}{1 - \tan^2 30°} = \dfrac{2(\sqrt{3}/3)}{1 - (\sqrt{3}/3)^2} = \sqrt{3}$

9. $\sin 52° = 0.7880108$ **11.** $\dfrac{1}{2}$ **13.** $\cos 215° = -0.8192$

15. $2\tan 24° = 0.8904574$ **17.** $\sin 5x$ **19.** $4\sin 12x$

21. $2\cos 12x$ **23.** $2\cos x$ **25.** $-\dfrac{\pi}{2}$ **27.** 0.2619

29. $-\dfrac{1}{3}\sqrt{3}$ **31.** $-\dfrac{\pi}{6}$ **33.** $\sec^2 y - \tan^2 y = 1$

35. $\sin x\csc x - \sin^2 x = 1 - \sin^2 x = \cos^2 x$

37. $\dfrac{(\sec^2 x - 1)(\sec^2 x + 1)}{\tan^2 x} = \sec^2 x + 1$

39. $2\left(\dfrac{1}{\sin 2x}\right)\left(\dfrac{\cos x}{\sin x}\right) = 2\left(\dfrac{1}{2\sin x\cos x}\right)\left(\dfrac{\cos x}{\sin x}\right)$
$= \dfrac{1}{\sin^2 x}$

41. $\dfrac{\cos^2\theta}{\sin^2\theta}$ **43.** $\dfrac{1}{2}\left(2\sin\dfrac{\theta}{2}\cos\dfrac{\theta}{2}\right)$ **45.** $\cot x$ **47.** $\sec x$

49. $\sin x$ **51.** $\sin x$

53.

55.

57.

59.

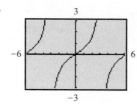

61. $x = \dfrac{1}{2}\cos^{-1}\dfrac{1}{2}y$ **63.** $x = \dfrac{1}{5}\sin\dfrac{1}{3}\left(\dfrac{1}{4}\pi - y\right)$

65. 1.2925, 4.4341   **67.** $\dfrac{\pi}{6}, \dfrac{5\pi}{6}, \dfrac{7\pi}{6}, \dfrac{11\pi}{6}$ **69.** $0, \dfrac{\pi}{3}, \dfrac{5\pi}{3}$

71. $0, \dfrac{2\pi}{3}$ **73.** 0.3142, 1.571, 2.827, 4.084, 4.712, 5.341

75. 0

77. Identity:
$$\tan x + \frac{1}{\tan x} = \frac{\tan^2 x + 1}{\tan x} = \sec^2 x(\cot x) = \frac{\sec^2 x \cos x}{\sin x}$$

79. $\dfrac{\pi}{2}, \pi$ **81.** 1.56, 2.16, 3.46   **83.** $-2.31, 1.14$

85. $\dfrac{1}{x}$ **87.** $2x\sqrt{1 - x^2}$ **89.** $\dfrac{\sqrt{1 - x^2} - xy}{\sqrt{1 + y^2}}$

91. $2\sqrt{1 - \cos^2\theta}$ **93.** $\dfrac{\tan\theta}{\sqrt{1 + \tan^2\theta}} = \dfrac{\tan\theta}{\sec\theta}$

95. $(\cos\theta + j\sin\theta)^2 = (\cos^2\theta - \sin^2\theta) + j(2\sin\theta\cos\theta)$

97. $\dfrac{\pi}{2} < x < \dfrac{3\pi}{2}$

99. $C\left(\dfrac{A}{C}\sin 2t + \dfrac{B}{C}\cos 2t\right)$
$= C(\cos\alpha\sin 2t + \sin\alpha\cos 2t)$

101. $R = \sqrt{(A\cos\theta - B\sin\theta)^2 + (A\sin\theta + B\cos\theta)^2}$
$= \sqrt{A^2(\cos^2\theta + \sin^2\theta) + B^2(\sin^2\theta + \cos^2\theta)}$

103. $\dfrac{k}{2} \cdot \dfrac{1}{\sin^2\frac{\theta}{2}} = \dfrac{k}{2} \cdot \dfrac{1}{\frac{1 - \cos\theta}{2}}$ **105.** $\theta = a + R\sin\omega t$

107. $\dfrac{\cos 2\alpha}{2\cos^2\alpha}$ **109.** $p = VI\cos\omega t[\cos(\phi + \omega t)]$

111. 13.6 ft   **113.** 54.7°

Exercises 21.1, page 562

1. $\sqrt{61}$ **3.** 150°   **5.** $2\sqrt{29}$ **7.** 3   **9.** 55

11. $2\sqrt{53}$ **13.** 2.86   **15.** $\dfrac{5}{2}$ **17.** Undefined   **19.** $-\dfrac{3}{4}$

21. $-\dfrac{5}{9}$ **23.** 0.747   **25.** $\dfrac{1}{3}\sqrt{3}$ **27.** -0.311 **29.** 20.0°

31. 98.50°   **33.** Parallel   **35.** Perpendicular   **37.** $8, -2$

39. -3 **41.** Two sides equal $2\sqrt{10}$.   **43.** $m_1 = \dfrac{5}{12}, m_2 = \dfrac{4}{3}$

45. 10   **47.** $4\sqrt{10} + 4\sqrt{2} = 18.3$ **49.** $(1, 5)$

51. $(-2.8, 4.2)$ **53.** $x^2 + y^2 = 9$ **55.** $m_1 = 1, m_2 = -1$

57. $\left(-\dfrac{11}{3}, 0\right)$ **59.** $\left(0, -\dfrac{7}{9}\right)$ **61.** $0, \sqrt{3}, -\sqrt{3}$

Exercises 21.2, page 567

1. $y + 2x - 7 = 0$ **3.** $2, \left(0, \dfrac{5}{2}\right)$

5. $4x - y + 20 = 0$ **7.** $7x - 2y - 24 = 0$

9. $x - y + 19 = 0$ **11.** $y = -2.7$

13. $x = -3$ **15.** $x - 3y - 7 = 0$

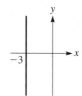

17. $x + y - 7 = 0$ **19.** $3x + y - 18 = 0$

21. $y = 4x - 8; m = 4, (0, -8)$

23. $y = -\dfrac{3}{5}x + 2; m = -\dfrac{3}{5}, (0, 2)$

25. $y = \dfrac{3}{2}x - \dfrac{1}{2}; m = \dfrac{3}{2}, \left(0, -\dfrac{1}{2}\right)$

27. $y = 3.5x + 0.5; m = 3.5, (0, 0.5)$ **29.** Parallel

31. Perpendicular   **33.** Neither   **35.** Perpendicular

37. -2

39. The slope of the first line is 3. A line perpendicular to it has a slope of $-\dfrac{1}{3}$. The slope of the second line is $-\dfrac{k}{3}$, so $k = 1$.

41. $b - a$ **43.** 5

45. $m_1 = -\dfrac{4}{5}, m_2 = \dfrac{2}{3}, m_3 = \dfrac{2}{3}; m_4 = -\dfrac{4}{5}; m_1 = m_4, m_2 = m_3$

47. $m_1 = -\dfrac{a}{b}, m_2 = -\dfrac{a}{b}, m_1 = m_2; m_3 = -\dfrac{a}{b}, m_4 = \dfrac{b}{a}, m_3 = 1/m_4$

49. $F = \dfrac{9}{4}R + 32$ **51.** $v = 0.607T + 331$

53. $T = \dfrac{4}{3}x + 3$ **55.** $P = 3.0 - 0.025T_S$

57. $y = 10^{-5}(2.4 - 5.6x)$

59. $n = \dfrac{7}{6}t + 10$; at 6:30, $n = 10$; at 8.30, $n = 150$

61.   **63.**

65.   **67.**

1. $C(1, -1), r = 4$ **3.** $C(3, 4), r = 7$

 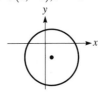

5. $(2, 1), r = 5$ **7.** $C(-1, 0), r = \dfrac{11}{2}$ **9.** $x^2 + y^2 = 9$

11. $(x - 2)^2 + (y - 2)^2 = 16$, or $x^2 + y^2 - 4x - 4y - 8 = 0$

13. $(x - 12)^2 + (y + 15)^2 = 324$, or $x^2 + y^2 - 24x + 30y + 45 = 0$

15. $(x + 3)^2 + (y - 4)^2 = 25$ **17.** $(x - 2)^2 + (y - 1)^2 = 8$

19. $(x + 3)^2 + (y - 5)^2 = 25$, or $x^2 + y^2 + 6x - 10y + 9 = 0$

21. $(x + 2)^2 + (y - 2)^2 = 4$ or $x^2 + y^2 + 4x - 4y + 4 = 0$ **23.** $x^2 + y^2 = 2$

25. $(0, 3), r = 2$ **27.** $(-1, 5), r = \dfrac{9}{2}$

 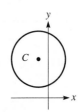

29. $(1, 0), r = 3$ **31.** $(-2.1, 1.3), r = 3.1$

 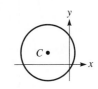

33. $(0, 2), r = \dfrac{5}{2}$ **35.** $(1, 2), r = \dfrac{1}{2}\sqrt{22}$

37. Symmetric to both axes and origin

39. Symmetric to y-axis   **41.** $(7, 0), (-1, 0)$

43. $3x^2 + 3y^2 + 4x + 8y - 20 = 0$, circle

45.

47. (a) Semicircle
(b) Semicircle
(c) Yes, there is only one value of y for each x in the domain.

49. Outside

51. $p > 0$, circle; $p = 0$, point; $p < 0$, does not exist

53. 0.0912 in.   **55.** 2.82 in.   **57.** $x^2 + y^2 = 0.0100$

59. $x^2 + (y + 7)^2 = 1$

61. $(x - 500 \times 10^{-6})^2 + y^2 = 0.16 \times 10^{-6}$

Exercises 21.4, page 576

1. $F(5, 0), x = -5$ **3.** $F\left(0, -\dfrac{3}{2}\right), y = \dfrac{3}{2}$

5. $F(1, 0), x = -1$ **7.** $F(-1, 0), x = 1$

9. $F(0, 18), y = -18$ **11.** $F(0, -1), y = 1$

13. $F\left(\dfrac{5}{8}, 0\right), x = -\dfrac{5}{8}$ **15.** $F\left(0, \dfrac{25}{48}\right), y = -\dfrac{25}{48}$

17. $y^2 = 12x$ **19.** $x^2 = -2y$ **21.** $x^2 = 0.64y$

23. $x^2 = 336y$ **25.** $x^2 = \dfrac{1}{8}y$ **27.** $y^2 = \dfrac{25}{3}x$

29. $y^2 = 3x$

31.   **33.** $(0, 0), (8, -4)$

35. $y^2 - 2y - 12x + 37 = 0$ **37.**

39.   **41.** $4p$ **43.** $y^2 = 2x$

45. $x^2 = 14{,}700y$ **47.** H **49.** 57.6 m

51. No (a course correction is necessary)   **53.** 2.16 cm from vertex

55.   **57.** $y^2 = 8x$ or $x^2 = 8y$ with vertex midway between island and shore

Exercises 21.5, page 582

1. $V(0, 6), V(0, -6)$, ends minor axis $(5, 0), (-5, 0)$ foci $(0, \sqrt{11}), (0, -\sqrt{11})$

3. $V(2, 0), V(-2, 0)$, $F(\sqrt{3}, 0), F(-\sqrt{3}, 0)$

5. $V(0, 12), V(0, -12)$, $F(0, \sqrt{119}), F(0, -\sqrt{119})$

7. $V\left(\dfrac{5}{2}, 0\right), V\left(-\dfrac{5}{2}, 0\right)$, $F\left(\dfrac{3}{2}, 0\right), F\left(-\dfrac{3}{2}, 0\right)$

9. $V(9, 0), V(-9, 0)$, $F(3\sqrt{5}, 0), F(-3\sqrt{5}, 0)$

11. $V(0, 7), V(0, -7)$, $F(0, \sqrt{45}), F(0, -\sqrt{45})$

13. $V(0, 4), V(0, -4)$, $F(0, \sqrt{14}), F(0, -\sqrt{14})$

15. $V(0.25, 0), V(-0.25, 0)$, $F(0.23, 0), F(-0.23, 0)$

17. $\dfrac{x^2}{225} + \dfrac{y^2}{144} = 1$, or $144x^2 + 225y^2 = 32{,}400$

19. $\dfrac{x^2}{225} + \dfrac{y^2}{289} = 1$, or $289x^2 + 225y^2 = 65{,}025$

21. $\dfrac{x^2}{208} + \dfrac{y^2}{144} = 1$, or $144x^2 + 208y^2 = 29{,}952$

23. $\dfrac{x^2}{64} + \dfrac{15y^2}{144} = 1$, or $3x^2 + 20y^2 = 192$

25. $\dfrac{x^2}{5} + \dfrac{y^2}{20} = 1$, or $4x^2 + y^2 = 20$

27. $\dfrac{x^2}{100} + \dfrac{y^2}{64} = 1$, or $16x^2 + 25y^2 = 1600$

29. $(1, -2), (1, 2)$

31. $16x^2 + 25y^2 - 32x - 50y - 359 = 0$

33.

35.

37. Write equation as $\dfrac{x^2}{1} + \dfrac{y^2}{1/k} = 1$. Thus, $\sqrt{\dfrac{1}{k}} > 1$, or $k < 1$,

39. $2x^2 + 3y^2 - 8x - 4 = 2x^2 + 3(-y)^2 - 8x - 4$

41.

43. $\sin^2 t + \cos^2 t = \dfrac{x^2}{4} + \dfrac{y^2}{9} = 1$

45. $i_1^2 + 4i_2^2 = 32$ **47.** 0.44   **49.** $7x^2 + 16y^2 = 112$

51. 27.5 m   **53.** 13 ft
55. 843 ft^3

Exercises 21.6, page 587

1. $V(0, -4), V(0, 4)$,
conj. axis $(-2, 0), (2, 0)$
$F(0, -2\sqrt{5}), F(0, 2\sqrt{5})$

3. $V(5, 0), V(-5, 0)$,
$F(13, 0), F(-13, 0)$

5. $V(0, 3), V(0, -3)$,
$F(0, \sqrt{10}), F(0, -\sqrt{10})$

7. $V\left(-\dfrac{5}{2}, 0\right), V\left(\dfrac{5}{2}, 0\right)$,
$F\left(-\dfrac{1}{2}\sqrt{41}, 0\right), F\left(\dfrac{1}{2}\sqrt{41}, 0\right)$

9. $V(1, 0), V(-1, 0)$,
$F(\sqrt{5}, 0), F(-\sqrt{5}, 0)$

11. $V(0, \sqrt{5}), V(0, -\sqrt{5})$,
$F(0, \sqrt{7}), F(0, -\sqrt{7})$

13. $V(0, 2), V(0, -2)$,
$F(0, \sqrt{5}), F(0, -\sqrt{5})$

15. $V(0.4, 0), V(-0.4, 0)$,
$F(0.9, 0), F(-0.9, 0)$

17. $\dfrac{x^2}{9} - \dfrac{y^2}{16} = 1$, or $16x^2 - 9y^2 = 144$

19. $\dfrac{y^2}{100} - \dfrac{x^2}{576} = 1$, or $144y^2 - 25x^2 = 14{,}400$

21. $\dfrac{x^2}{1} - \dfrac{y^2}{3} = 1$, or $3x^2 - y^2 = 3$

23. $\dfrac{x^2}{5} - \dfrac{y^2}{4} = 1$, or $4x^2 - 5y^2 = 20$

25. $x^2 - \dfrac{y^2}{4} = 1$, or $4x^2 - y^2 = 4$

27. $\dfrac{x^2}{36} - \dfrac{y^2}{64} = 1$, or $16x^2 - 9y^2 = 576$

29.

31. $\sec^2 t - \tan^2 t = x^2 - y^2 = 1$

33. $(-2, -3), (-2, 3), (2, -3), (2, 3)$

35. $9x^2 - 16y^2 - 108x + 64y + 116 = 0$

37.

39.

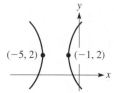

41. $x^2 - 2y^2 = 2$

43.

45. $9x^2 - 16y^2 = 144$

47. $3y^2 - x^2 = 27$

49.

51. $i = 6.00/R$

53.

Exercises 21.7, page 591

1. Hyperbola, $C(3, 2)$
transverse axis parallel to x-axis
$a = 5, b = 3$

3. Parabola, $(-1, 2)$

5. Hyperbola, $(1, 2)$

7. Ellipse, $(-1, 0)$

9. Parabola, $(-3, 1)$

11. $(y - 3)^2 = 16(x + 1)$, or $y^2 - 6y - 16x - 7 = 0$

13. $y^2 = 24(x - 6)$

15. $\dfrac{(x + 2)^2}{25} + \dfrac{(y - 2)^2}{16} = 1$, or
$16x^2 + 25y^2 + 64x - 100y - 236 = 0$

17. $\dfrac{(y - 1)^2}{16} + \dfrac{(x + 2)^2}{4} = 1$, or $4x^2 + y^2 + 16x - 2y + 1 = 0$

19. $\dfrac{(y - 2)^2}{1} - \dfrac{(x + 1)^2}{3} = 1$, or $x^2 - 3y^2 + 2x + 12y - 8 = 0$

21. $\dfrac{(x + 1)^2}{9} - \dfrac{(y - 1)^2}{16} = 1$, or
$16x^2 - 9y^2 + 32x + 18y - 137 = 0$

23. Parabola, $(-1, -1)$

25. Parabola, $(0, 6)$

27. Ellipse, $(-3, 0)$

29. Hyperbola, $(0, 4)$

31. Hyperbola, $(-2, 1)$

33. Hyperbola, $(-4, 5)$

35. Ellipse, $(1, -2)$

37. Hyperbola, $(4, 0)$

39. Circle, $\left(\dfrac{1}{3}, \dfrac{4}{3}\right)$

41. $x^2 - y^2 + 4x - 2y - 22 = 0$

43. $y^2 + 4x - 4 = 0$ **45.** $y^2 = 4p(x - h)$

47. $i' = \sin 2\pi t'$

49. $(x - 95)^2 = -\dfrac{95^2}{60}(y - 60)$

51. $\dfrac{x^2}{9.0} + \dfrac{y^2}{16} = 1, \dfrac{(x - 7.0)^2}{16} + \dfrac{y^2}{9.0} = 1$

Exercises 21.8, page 594

1. Hyperbola   **3.** Ellipse   **5.** Hyperbola   **7.** Circle

9. Parabola   **11.** Hyperbola   **13.** Circle

15. None (straight line)   **17.** Hyperbola   **19.** Ellipse

21. None (point at origin)   **23.** Ellipse

25. Parabola: $V(-4, 0); F(-4, 2)$

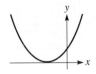

27. Hyperbola; $C(1, -2);$
$V(1, -2 \pm \sqrt{2})$

29. Ellipse; $C(5, 0);$
$V(5, \pm 2\sqrt{2})$

31. Ellipse

33. Parabola

35. Hyperbola

Enter both functions (from quadratic formula).

37. (a) Circle (b) Hyperbola (c) Ellipse   **39.** Point at the origin

41. Straight line   **43.** Parabola

45. Circle if light beam is perpendicular to floor; otherwise, an ellipse

47. One branch of a hyperbola

Exercises 21.9, page 598

1. Hyperbola; $2x'y' + 25 = 0$ **3.** Ellipse; $4x'^2 + 9y'^2 = 36$

5. Hyperbola   **7.** Parabola   **9.** Ellipse

11. Parabola; $x'^2 + \sqrt{2}y' = 0$ **13.** Hyperbola; $4x'^2 - y'^2 = 4$

15. Ellipse; $x'^2 - 2y'^2 = 2$ **17.** Parabola
$y'^2 - 4x' + 16 = 0$
$y''^2 = 4x''$

19. (a) Hyperbola   (b) Hyperbola

Exercises 21.10, page 602

1. $\left(3, \dfrac{-5\pi}{3}\right), \left(3, \dfrac{4\pi}{3}\right)$ **3.** (5.83, 2.11)

5.   **7.**   **9.**

11.   **13.**   **15.**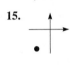

17. $\left(2, \dfrac{\pi}{6}\right)$ **19.** $\left(1, \dfrac{7\pi}{6}\right)$ **21.** $\left(-4, -4\sqrt{3}\right)$

23. $(2.76, -1.17)$ **25.** $r = 3 \sec \theta$

27. $r = \dfrac{-3}{\cos \theta + 2 \sin \theta}$ **29.** $r = 4 \sin \theta$

31. $r^2 = \dfrac{4}{1 + 3 \sin^2 \theta}$ **33.** $r = 6 \sin \theta$

35. $r = \dfrac{2 \sin 2\theta}{\cos^3 \theta + \sin^3 \theta}$ **37.** $x^2 + y^2 - y = 0$, circle

39. $x = 4$, straight line   **41.** $x - 3y - 2 = 0$, straight line

43. $x^2 + y^2 - 4x - 2y = 0$, circle

45. $x^4 + y^4 - 4x^3 + 2x^2y^2 - 4xy^2 - 4y^2 = 0$

47. $(x^2 + y^2)^2 = 2xy$ **49.** Yes $\left[\text{as } \left(-2, \dfrac{7\pi}{4}\right)\right]$

51. $x^2 + y^2 - bx - ay = 0$

53. $(2, 0), \left(2, \dfrac{\pi}{3}\right), \left(2, \dfrac{2\pi}{3}\right), \left(2, \dfrac{4\pi}{3}\right), \left(2, \dfrac{5\pi}{3}\right)$

55. $B_x = -\dfrac{k \sin \theta}{r}, B_y = \dfrac{k \cos \theta}{r}$

57. $x^4 + y^4 + 2x^2y^2 + 2x^2y + 2y^3 - 9x^2 - 8y^2 = 0$

59. 13.8 km

Exercises 21.11, page 605

1. $\theta = \dfrac{5\pi}{6}$

3. $r = 2 \sin 2\theta$

5.

7.

9.

11.

13.

15.

17.

19.

21.

23.

25.

27.

29.

31.

33.

18

−20   20

−18

35.

3

−3   7

−3

37.

3

−3   3

−3

39.

4

−2   4

−1

41.

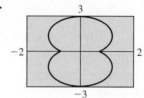

3

−2   2

−3

43. Straight line

45.

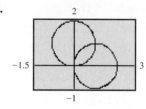

2

−1.5   3

−1

47.

49.

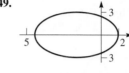

3

5   2

3

51.

53. If n is odd, there are n loops. If n is even, there are $2n$ loops.

Review Exercises for Chapter 21, page 608

1. $4x - y - 11 = 0$

3. $2x + 3y + 3 = 0$

$-\dfrac{3}{2}$ -1

5. $x^2 - 6x + y^2 - 1 = 0$

7. $y^2 = 12x$

9. $9x^2 + 25y^2 = 900$

11. $144y^2 - 169x^2 = 24{,}336$

13. $(-3, 0), r = 4$

15. $(0, -5), y = 5$

17. $V(0,1), V(0,-1), F\left(0,\frac{1}{2}\sqrt{3}\right), F\left(0,-\frac{1}{2}\sqrt{3}\right)$

19. $V\left(\frac{1}{2\sqrt{2}},0\right), V\left(-\frac{1}{2\sqrt{2}},0\right), F\left(\frac{\sqrt{70}}{20},0\right), F\left(-\frac{\sqrt{70}}{20},0\right)$

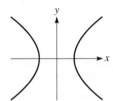

21. $V(4,-8), F(4,-7)$ **23.** $(2,-1)$

25. $(0,0)$ **27.**

29.   **31.**

33.   **35.** $\theta = \tan^{-1} 2 = 1.11$

37. $r^2 = \dfrac{2}{1 + \sin\theta\cos\theta}$ **39.** $(x^2 + y^2)^3 = 16x^2y^2$

41. $3x^2 + 4y^2 - 8x - 16 = 0$ **43.** 4   **45.** 2   **47.** 2

49.   **51.**

53.   **55.**

57.   **59.** $x^2 + y^2 - 6x + 8y + 9 = 0$

61. $\dfrac{(x-4)^2}{16} + \dfrac{(y+3)^2}{7} = 1$ or
$7x^2 + 16y^2 - 56x + 96y + 144 = 0$

63. Circle   **65.** 1   **67.** 6.71

69. $m_1 = -\dfrac{12}{5}, m_2 = \dfrac{5}{12}; d_1^2 = 169, d_2^2 = 169, d_3^2 = 338$

71.   **73.** Hyperbola   **75.** 8

77. $x^2 - 6x - 8y + 1 = 0$ **79.** $R_T = R + 2.5$

81. $v = 5.75 + 2.32t$ **83.** $y = 100.5T - 10050$

85. 11,000 ft^2 **87.** $y = -\dfrac{1}{80}x^2$ **89.** $y^2 = 32x$

91.   **93.**

95. 7500 ft^2 **97.** 18 cm, 8 cm   **99.** 37.8 ft

101. Dist. from rifle to $P -$ dist. from target to $P =$ constant (related to dist. from rifle to target).

103.   **105.**

Exercises 22.1, page 616

1.

Est. hours	0–5	6–11	12–17	18–23	24–29
f	5	12	19	9	5

3.

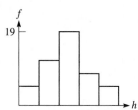

5.

No.	53	54	55	56	57	58	59	60	61	62	63
f	1	3	1	3	2	4	3	1	1	0	1

7.

No.	51–53	54–56	57–59	60–62	63–65
f	1	7	9	2	1

9.

11.

No.	<54	<57	<60	<63	<66
Cum. f	1	8	17	19	20

13.

No. inst.	18	19	20	21	22	23	24	25
f	1	3	2	4	3	1	0	1

15.

17.

19.

Time (s)	<2.22	<2.23	<2.24	<2.25	<2.26	<2.27	<2.28	<2.29	<2.30
Cum. f (%)	2	9	27	68	124	156	164	167	170

21.

Dist. (ft)	155–159	160–164	165–169	170–174	175–179	180–184	185–189
f (%)	1.7	12.5	26.7	30.0	20.0	8.3	0.8

23.

25.

27.

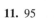

29.

31. The greatest frequency should be at 2 and the least at 0 and 4.

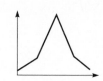
(Graphs will generally have approximately the shape shown.)

Exercises 22.2, page 620

1. 5 **3.** 5.5 **5.** 4 **7.** 0.495 **9.** 4.6 **11.** 0.505
13. 4 **15.** 0.49, 0.53 **17.** 57.5 mi/gal **19.** 58 mi/gal
21. 20.9 **23.** 2.25 s **25.** 172 ft **27.** 4.237 mR
29. 4.36 mR **31.** 31 h **33.** 0.00595 mm **35.** $625, $700
37. 862 kW·h **39.** 0.195, 0.18 **41.** $663
43. $725, $748, $800; if each value is increased by the same amount, the median, mean, and mode are also increased by this amount.
45. $884; an outlier can make the mean a poor measure of the center of the distribution.

Exercises 22.3, page 625

1. 2.2 **3.** 1.55 **5.** 0.035 **7.** 1.55 **9.** 0.035
11. 4.6, 1.55 **13.** 0.505, 0.035 **15.** 2.6 mi/gal **17.** 1.8
19. 2.6 h **21.** 0.014 s **23.** 0.047 **25.** 0.00022 mm

Exercises 22.4, page 630

1. The peak would be as high as for the left curve, and it would be centered as for the right curve.
3. 0.2072
5.

7.

9. 136 **11.** 95 **13.** 341 **15.** 78.81%
17. 2166 **19.** 179
21. 141; about 68% of samples have mean lifetime from 99,859 km to 100,141 km.
23. 2.28% **25.** −0.7 **27.** −0.6
29. 76% (normal dist. is 68%) **31.** 68% (normal dist. is 68%)

Exercises 22.5, page 635

1. $UCL(\bar{x}) = 504.7$ mg, $LCL(\bar{x}) = 494.7$ mg
3. The first point would be below the $\bar{x}$ line at 499.7. The UCL and LCL lines would be 0.2 unit lower.
5. $\bar{x} = = 361.8$ N·m, UCL = 368.9 N·m, LCL = 354.6 N·m

7.

9. $\bar{x} = 8.986$ V, UCL $= 9.081$ V, LCL $= 8.891$ V

11.

13. $\mu = 2.725$ in., UCL $= 2.729$ in., LCL $= 2.721$ in.
15. $\mu = 5.57$ mL, UCL $= 11.17$ mL, LCL $= 0.00$ mL
17. $\bar{p} = 0.0369$, UCL $= 0.0548$, LCL $= 0.0190$
19. $\bar{p} = 0.0580$, UCL $= 0.0894$, LCL $= 0.0266$

Exercises 22.6, page 640

1. $y = 2x + 2$ **3.** $y = -1.77x + 191$

5. $y = -0.308t + 9.07$ **7.** $V = -0.590i + 11.3$

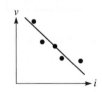

9. $h = 2.24x + 5.2$ **11.** $p = -0.200x + 649$

13. $V = 4.32 \times 10^{-15} f - 2.03$
  $f_0 = 0.470$ PHz

15. 0.953 **17.** -0.901

Exercises 22.7, page 644

1. $y = 2.59x^2 + 1.07$ **3.** $P = \dfrac{2390}{V}$

5. $y = 5.97t^2 + 0.38$ **7.** $p = 0.00967T^2 + 29.4$

9. $P = \dfrac{1343}{S}$ **11.** $y = 6.20e^{-t} - 0.05$

Review Exercises for Chapter 22, page 645

1. 76.5   **3.** 76.2

5.

No.	67–70	71–74	75–78	79–82	83–86
f	3	4	8	3	2

7.

9.

No.	<71	<75	<79	<83	<87
f	3	7	15	18	20

11. 0.264 Pa$\cdot$s   **13.** 0.014 Pa$\cdot$s

15.   **17.** 700 W   **19.** 700 W
21. 17.3 W

23.

Power (W)	<660	<670	<680	<690
Cum. f	3	5	12	24

Power (W)	<700	<710	<720	<730	<740
Cum. f	51	85	100	116	121

25. 4   **27.**   **29.** 66.2 mi/h
31. 9.0 mi/h

33. $\bar{p} = 0.0540$, UCL $= 0.0843$, LCL $= 0.0237$

35.

37. 322   **39.** 495   **41.** $R = 0.0983T + 25.0$

43. $s = 0.123t + 0.887$ **45.** $s = -4.90t^2 + 3000$

47. $i = 3.99(1 - e^{-0.500t})$, 3.99 A   **49.** $y = 0.0002x^2 + 15$

51. $y = 0.997x + 0.581$ **53.** $y = 1.39x^{0.878}$

55. 2.7   **57.** 9.2 ppm

59. Substitute expressions for m, b, and $\bar{x}$. Simplify to $\bar{y} = \bar{y}$.

Exercises for Appendix B, page A.7

1. MHz, 1 MHZ $= 10^6$ Hz   **3.** mm, 1 mm $= 10^{-3}$ m

5. kilovolt, 1 kV $= 10^3$ V   **7.** milliampere, 1 mA $= 10^{-3}$ A

9. 10^5 cm   **11.** 63,360 in.   **13.** 13.3 cm   **15.** 14.9 L

17. 144 in.^{2} **19.** 2.50×10^{-4} m^2 **21.** 20,040 g

23. 13.6 L   **25.** 1.68×10^6 N·cm/s   **27.** 90 MJ

29. 0.0101 hp   **31.** 52.6 min   **33.** 9.45×10^{15} m/year

35. 3.25 tons   **37.** 15 gal   **39.** 15,200 mi/h

41. 111 cm^2 **43.** 1240 km/h   **45.** 37 mi/h

47. 25,000 mg/dL   **49.** 101,000 Pa   **51.** 1.2×10^5 mA/cm^2

53. 3.73 h   **55.** 0.135 km/L

Exercises for C.2, page A.11

(Most answers have been rounded off to four significant digits.)

1. 56.02   **3.** 4162.1   **5.** 18.65   **7.** 0.3954

9. 14.14   **11.** 0.5251   **13.** 13.35   **15.** 944.6

17. 0.7349   **19.** -0.7594 **21.** -1.337 **23.** 1.015

25. 41.35°   **27.** -1.182 **29.** 0.5862   **31.** 6.695

33. 3.508   **35.** 0.005685   **37.** 2.053   **39.** 5.765

41. 4.501×10^{10} **43.** 497.2   **45.** 6.648   **47.** 401.2

49. 8.841   **51.** 2.523   **53.** 10.08   **55.** 22.36

57. 20.3°   **59.** 4729   **61.** 3.301×10^4

63. 1.056   **65.** 55.5°   **67.** 3.277   **69.** 8.125

71. 1.000   **73.** 1.000   **75.** 12.90   **77.** 8.001

79. 8.053   **81.** 0.04259   **83.** 0.4219   **85.** 0.7822

87. 2.0736465   **89.** 124.3   **91.** 252

93.   **95.**

97.   **99.**

101.   **103.**

105.   **107.**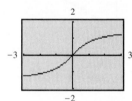

Exercises for C.3, page A.15

(Program names are suggested.)

1. CYLINDER

```
:Prompt R,H
:Disp "TOTAL AREA=",2πR(R+H)
:Disp "VOLUME=",πR²H
```

3. PRODBINO

```
:Prompt A,B,C,D
:Disp "E=",A*C
:Disp "F=",A*D+B*C
:Disp "G=",B*D
```

5. SLVTRASA

```
:Input "θA=",A
:Input "θB=",B
:Input "SC=",F
:180−A−B → C
:Disp "θC=",C
:Disp "SA=",Fsin(A)/sin(C)
:Disp "SB=",Fsin(B)/sin(C)
```

7. CNSTPROP

```
:Prompt Y,R,A,S,B,T,C
:Disp "K="
:Disp YT^C/((R^A)(S^B))
```

SOLUTIONS TO PRACTICE TEST PROBLEMS

Chapter 1

1. $\sqrt{9 + 16} = \sqrt{25} = 5$

2. $\dfrac{(7)(-3)(-2)}{(-6)(0)}$ is undefined (division by zero).

3. $\dfrac{3.372 \times 10^{-3}}{7.526 \times 10^{12}} = 4.480 \times 10^{-16}$

```
3.372E-3/7.526E1
2
     4.48046771E-16
```

4. $\dfrac{(+6)(-2) - 3(-1)}{5 - 2} = \dfrac{-12 - (-3)}{3} = \dfrac{-12 + 3}{3}$

$$= \dfrac{-9}{3} = -3$$

5. $\dfrac{346.4 - 23.5}{287.7} - \dfrac{0.944^3}{(3.46)(0.109)} = -1.108$

```
(346.4-23.5)/287
.7-.944^3/(3.46*
.109)
      -1.10820764
```

6. $(2a^0 b^{-2} c^3)^{-3} = 2^{-3} a^{0(-3)} b^{(-2)(-3)} c^{3(-3)} = 2^{-3} a^0 b^6 c^{-9}$

$$= \dfrac{b^6}{8c^9}$$

7. $(2x + 3)^2 = (2x + 3)(2x + 3)$

$$= 2x(2x) + 2x(3) + 3(2x) + 3(3)$$
$$= 4x^2 + 6x + 6x + 9$$
$$= 4x^2 + 12x + 9$$

8. $3m^2(am - 2m^3) = 3m^2(am) + 3m^2(-2m^3)$

$$= 3am^3 - 6m^5$$

9. $\dfrac{8a^3 x^2 - 4a^2 x^4}{-2ax^2} = \dfrac{8a^3 x^2}{-2ax^2} - \dfrac{4a^2 x^4}{-2ax^2} = -4a^2 - (-2ax^2)$

$$= -4a^2 + 2ax^2$$

10.
$$
\begin{array}{r}
3x - 5 \quad \text{(quotient)} \\
2x - 1 \overline{)\, 6x^2 - 13x + 7} \\
\underline{6x^2 - 3x} \\
-10x + 7 \\
\underline{-10x + 5} \\
2 \quad \text{(remainder)}
\end{array}
$$

11. $(2x - 3)(x + 7)$

$$= 2x(x) + 2x(7) + (-3)(x) + (-3)(7)$$
$$= 2x^2 + 14x - 3x - 21 = 2x^2 + 11x - 21$$

12. $3x - [4x - (3 - 2x)] = 3x - [4x - 3 + 2x]$

$$= 3x - [6x - 3]$$
$$= 3x - 6x + 3 = -3x + 3$$

13. $5y - 2(y - 4) = 7$

$$5y - 2y + 8 = 7$$
$$3y + 8 - 8 = 7 - 8$$
$$3y = -1$$
$$y = -1/3$$

14.
$$3(x - 3) = x - (2 - 3d)$$
$$3x - 9 = x - 2 + 3d$$
$$3x - 9 - x + 9 = x - 2 + 3d - x + 9$$
$$2x = 3d + 7$$
$$x = \dfrac{3d + 7}{2}$$

15. $0.000\,003\,6 = 3.6 \times 10^{-6}$ (six places to right)

16.

$-\pi$	-3	0.3	$\sqrt{2}$	$\lvert -4 \rvert$	(order)
-3.14	-3	0.3	1.41	4	(value)

17. $3(5 + 8) = 3(5) + 3(8)$ illustrates distributive law.

18. (a) 5, (b) 3.0 (zero is significant)

19. Evaluation:
$$1000(1 + 0.05/2)^{2(3)} = 1000(1.025)^6 = \$1159.69$$

20. $8(100 - x)^2 + x^2 = 8(100 - x)(100 - x) + x^2$

$$= 8(10{,}000 - 200x + x^2) + x^2$$
$$= 80{,}000 - 1600x + 8x^2 + x^2$$
$$= 80{,}000 - 1600x + 9x^2$$

21. $L = L_0[1 + \alpha(t_2 - t_1)]$

$$= L_0[1 + \alpha t_2 - \alpha t_1]$$
$$= L_0 + \alpha L_0 t_2 - \alpha L_0 t_1$$
$$L - L_0 + \alpha L_0 t_1 = \alpha L_0 t_2$$
$$t_2 = \dfrac{L - L_0 + \alpha L_0 t_1}{\alpha L_0}$$

22. Let n = number of pounds of second alloy

$$0.3(20) + 0.8n = 0.6(n + 20)$$
$$6 + 0.8n = 0.6n + 12$$
$$0.2n = 6$$
$$n = 30 \text{ lb}$$

Chapter 2

1. $\angle 1 + \angle 3 + 90° = 180°$ (sum of angles of a triangle)

$$\angle 3 = 52° \quad \text{(vertical angles)}$$
$$\angle 1 = 180° - 90° - 52° = 38°$$

2. $\angle 2 + \angle 4 = 180°$ (straight angle)

$$\angle 4 = 52° \quad \text{(corresponding angles)}$$
$$\angle 2 + 52° = 180°$$
$$\angle 2 = 128°$$

$AB \parallel CD$

3.

$$\frac{x}{8.0} = \frac{25.0}{10.0} \qquad x = \frac{8.0(25.0)}{10.0} = 20.0 \text{ ft}$$

4. Use Hero's formula:

$$s = \tfrac{1}{2}(2.46 + 3.65 + 4.07) = 5.09 \text{ cm}$$
$$A = \sqrt{5.09(5.09 - 2.46)(5.09 - 3.65)(5.09 - 4.07)}$$
$$= 4.43 \text{ cm}^2$$

5. $d^2 = 125^2 + 170$
$\quad d = \sqrt{125^2 + 170^2}$
$\quad\ = 211 \text{ ft}$

6. $A = \tfrac{1}{2}h(b_1 + b_2)$
$\quad\ = \tfrac{1}{2}(13.8)(49.8 + 23.5) = 506 \text{ mm}^2$

7. (a) $A = 6e^2 = 6(4.50)^2 = 122 \text{ cm}^2$
$\quad$ (b) $V = e^3 = 4.50^3 = 91.1 \text{ cm}^3$

8. $\quad c = 2\pi r \qquad A = 4\pi r^2$

$$21.0 = 2\pi r \qquad\ = 4\pi\left(\frac{10.5}{\pi}\right)^2 = \frac{4(10.5^2)}{\pi}$$

$$r = \frac{10.5}{\pi} \qquad\qquad = 140 \text{ cm}^2$$

9. $V = \tfrac{1}{3}\pi r^2 h = \tfrac{1}{3}\pi(2.08^2)(1.78)$
$\quad\ = 8.06 \text{ m}^3$

10. $\angle ACO + 64° = 90°$ (tangent perpendicular to radius)
$\qquad\ \angle ACO = 26°$
$\qquad \angle A = \angle ACO = 26°$ (isosceles triangle)
$\qquad \tfrac{1}{2}\overset{\frown}{CD} = 26°$ (intercepted arc)
$\qquad\quad \overset{\frown}{CD} = 52°$
$\qquad\quad\ \angle 1 = 52°$ (central angle)

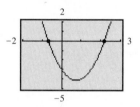

11. $\angle CBO + \angle 1 + 90° = 180°$ (sum of angles of triangle)
$\quad \angle CBO + 52° + 90° = 180°$
$\qquad\qquad \angle CBO = 180° - 52° - 90° = 38°$
$\qquad \angle 2 + \angle CBO = 180°$ (straight angle)
$\qquad\quad\ \angle 2 + 38° = 180°$
$\qquad\qquad\quad \angle 2 = 142°$

12. $r = \tfrac{1}{2}(2.25)$ cm
$\quad p = 3(2.25) + \tfrac{1}{2}(2\pi)\left[\tfrac{1}{2}(2.25)\right]$
$\quad\ = 10.3$ cm

2.25 cm

13. $A = 2.25^2 - \tfrac{1}{2}\pi\left[\tfrac{1}{2}(2.25)\right]^2 = 3.07 \text{ cm}^2$

14. $A = \tfrac{1}{2}(50)[0 + 2(90) + 2(145) + 2(260)$
$\qquad\quad + 2(205) + 2(110) + 20]$
$\quad\ = 41{,}000 \text{ ft}^2$

Chapter 3

1. $\quad f(x) = 2x - x^2 + \dfrac{8}{x}$

$$f(-4) = 2(-4) - (-4)^2 + \frac{8}{-4}$$
$$= -8 - 16 - 2$$
$$= -26$$

$$f(2.385) = 2(2.385) - (2.385)^2 + \frac{8}{2.385} = 2.436$$

2. $w = 2000 - 10t$

3. $f(x) = 4 - 2x$
$\quad y = 4 - 2x$
$\quad y = 4 - 2(-1) = 6$
$\quad y = 4 - 2(0) = 4$
$\quad y = 4 - 2(1) = 2$
$\quad y = 4 - 2(2) = 0$
$\quad y = 4 - 2(3) = -2$
$\quad y = 4 - 2(4) = -4$

x	y
-1	6
0	4
1	2
2	0
3	-2
4	-4

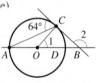

4. $y = 2x^2 - 3x - 3$

$x = -0.7$ and $x = 2.2$

5. $y = \sqrt{4 + 2x}$
$\quad y = \sqrt{4 + 2(-2)} = 0$
$\quad y = \sqrt{4 + 2(-1)} = 1.4$
$\quad y = \sqrt{4 + 2(0)} = 2$
$\quad y = \sqrt{4 + 2(1)} = 2.4$
$\quad y = \sqrt{4 + 2(2)} = 2.8$
$\quad y = \sqrt{4 + 2(4)} = 3.5$

x	y
-2	0
-1	1.4
0	2
1	2.4
2	2.8
4	3.5

6. On negative x-axis

7. $f(x) = \sqrt{6 - x}$
Domain: $x \le 6$; x cannot be greater than 6 to have real values of $f(x)$.
Range: $f(x) \ge 0$; $\sqrt{6 - x}$ is the principal square root of $6 - x$ and cannot be negative.

8. Shifting $y = 2x^2 - 3$ to the right 1 and up 3 gives
$y = 2(x - 1)^2 - 3 + 3 = 2(x - 1)^2 = 2x^2 - 4x + 2$.

9. Range: $y < -8.9$ or $y > 0.9$

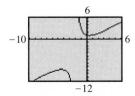

10. Let r = radius of circular part

square   semicircle

$A = (2r)(2r) + \frac{1}{2}(\pi r^2)$
$= 4r^2 + \frac{1}{2}\pi r^2$

11.

Voltage V	10.0	20.0	30.0	40.0	50.0	60.0
Current i	145	188	220	255	285	315

$f(45.0) = 270$ mA

12. $V = 20.0$ V for $i = 188$ mA.
$V = 30.0$ V for $i = 220$ mA.

$\dfrac{x}{10} = \dfrac{12}{32}$, $x = 3.8$ (rounded off)

$V = 20.0 + 3.8 = 23.8$ V
for $i = 200$ mA

$10\left[x\left[\begin{array}{cc} V & i \\ 20.0 & 188 \\ & 200 \\ 30.0 & 220 \end{array} \right]12 \right]32$

Chapter 4

1. $39' = \left(\dfrac{39}{60}\right)^\circ = 0.65^\circ$
$37°39' = 37.65^\circ$

2. $\tan 73.8^\circ = 3.44$

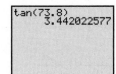

3. $\cos \theta = 0.3726$; $\theta = 68.12^\circ$

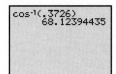

4. Let x = distance from course to east

$\dfrac{x}{22.62} = \sin 4.05^\circ$

$x = 22.62 \sin 4.05^\circ$
$= 1.598$ km

5. $\sin \theta = \dfrac{2}{3}$

$x = \sqrt{3^2 - 2^2} = \sqrt{5}$

$\tan \theta = \dfrac{2}{\sqrt{5}}$

6. $\tan \theta = 1.294$; $\csc \theta = 1.264$

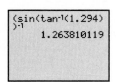

7. $B = 90^\circ - 37.4^\circ = 52.6^\circ$ $\dfrac{52.8}{c} = \cos 37.4^\circ$

$\dfrac{a}{52.8} = \tan 37.4^\circ$ $c = \dfrac{52.8}{\cos 37.4^\circ}$

$a = 52.8 \tan 37.4^\circ$ $= 66.5$
$= 40.4$

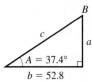

$b = 52.8$

8. $2.49^2 + b^2 = 3.88^2$
$b = \sqrt{3.88^2 - 2.49^2} = 2.98$

$\sin A = \dfrac{2.49}{3.88}$

$A = 39.9^\circ$ $B = 50.1^\circ$

9.

$\dfrac{s/2}{12.0} = \cos 42.0^\circ$

$s = 24.0 \cos 42.0^\circ = 17.8$

10.

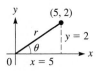

$h = \sqrt{9^2 + 40^2} = 41$

$\sin\theta = \dfrac{9}{41}, \quad \cos\theta = \dfrac{40}{41}$

$\dfrac{\sin\theta}{\cos\theta} = \dfrac{9/41}{40/41} = \dfrac{9}{40}$

11. $\lambda = d\sin\theta$

$= 30.05 \sin 1.167°$

$= 0.6120 \ \mu\text{m}$

12. $r = \sqrt{5^2 + 2^2} = \sqrt{29}$

$\sin\theta = \dfrac{2}{\sqrt{29}} = 0.3714 \qquad \csc\theta = \dfrac{\sqrt{29}}{2} = 2.693$

$\cos\theta = \dfrac{5}{\sqrt{29}} = 0.9285 \qquad \sec\theta = \dfrac{\sqrt{29}}{5} = 1.077$

$\tan\theta = \dfrac{2}{5} = 0.4000 \qquad \cot\theta = \dfrac{5}{2} = 2.500$

13. $\sin\theta = 2.5/9.5$

$\theta = 15°$

14. Distance between points is $x - y$.

$\dfrac{18.525}{x} = \tan 13.500° \qquad \dfrac{18.525}{y} = \tan 21.375°$

$x - y = \dfrac{18.525}{\tan 13.500°} - \dfrac{18.525}{\tan 21.375°} = 29.831 \text{ m}$

Chapter 5

1. Points $(2, -5)$ and $(-1, 4)$

$m = \dfrac{4 - (-5)}{-1 - 2} = \dfrac{9}{-3} = -3$

2. $2x + 5y = 11 \qquad 2(-2) + 5(3) = 11 \quad$ OK

$\dfrac{y - 5x = 12}{x = -2, y = 3} \qquad 3 - 5(-2) = 13 \qquad$ No

Not a solution. Values do not satisfy second equation.

3. $\begin{vmatrix} -1 & 3 & -2 \\ 4 & -3 & 0 \\ 5 & -4 & 2 \end{vmatrix} \begin{matrix} -1 & 3 \\ 4 & -3 \\ 5 & -4 \end{matrix} = 6 + 0 + 32 - 30 - 0 - 24 = -16$

4. $\begin{matrix} x + 2y = 5 \\ 4y = 3 - 2x \\ \overline{x = 5 - 2y} \end{matrix} \qquad \begin{matrix} 4y = 3 - 2(5 - 2y) \\ 4y = 3 - 10 + 4y \\ 0 = -7 \end{matrix}$

Inconsistent

5. $3x - 2y = 4$

$2x + 5y = -1$

$x = \dfrac{\begin{vmatrix} 4 & -2 \\ -1 & 5 \end{vmatrix}}{\begin{vmatrix} 3 & -2 \\ 2 & 5 \end{vmatrix}} = \dfrac{20 - 2}{15 - (-4)} = \dfrac{18}{19}$

$y = \dfrac{\begin{vmatrix} 3 & 4 \\ 2 & -1 \end{vmatrix}}{19} = \dfrac{-3 - 8}{19} = -\dfrac{11}{19}$

6. $2x + y = 4$

$y = -2x + 4$

$m = -2 \qquad b = 4$

7. $2l + 2w = 24 \quad$ (perimeter)

$\dfrac{l = w + 6.0}{l + w = 12}$

$\dfrac{l - w = 6.0}{2l = 18}$

$l = 9 \text{ km}$

$9 = w + 6$

$w = 3 \text{ km}$

8. $\begin{matrix} 6N - 2P = 13 \\ 4N + 3P = -13 \\ \overline{18N - 6P = 39} \\ 8N + 6P = -26 \\ \overline{26N \qquad = 13} \end{matrix}$

$N = \dfrac{1}{2}$

$6\left(\dfrac{1}{2}\right) - 2P = 13$

$-2P = 10$

$P = -5$

9. $C = 310 - 2d$

10. $\begin{matrix} 2x - 3y = 6 \\ x = 0: y = -2 \\ y = 0: x = 3 \end{matrix} \qquad \begin{matrix} 4x + y = 4 \\ x = 0: y = 4 \\ y = 0: x = 1 \end{matrix}$

Int: $(0, -2), (3, 0)$ Int: $(0, 4), (1, 0)$

$x = 1.3 \qquad y = -1.1$

11. Let x = vol. of first alloy
  y = vol. of second alloy
  z = vol. of third alloy

$$
\begin{aligned}
x + y + z &= 100 \quad \text{total vol.}\\
0.6x + 0.5y + 0.3z &= 40 \quad \text{copper}\\
\underline{0.3x + 0.3y \qquad\quad} &= \underline{15} \quad \text{zinc}\\
x + y + z &= 100\\
6x + 5y + 3z &= 400\\
\underline{3x + 3y \qquad} &= \underline{150}\\
3x + 2y \qquad &= 100\\
\underline{3x + 3y \qquad} &= \underline{150}\\
y &= 50 \text{ cm}^3\\
x &= 0 \text{ cm}^3\\
z &= 50 \text{ cm}^3
\end{aligned}
$$

12.
$$
\begin{aligned}
3x + 2y - z &= 4\\
2x - y + 3z &= -2\\
x \qquad + 4z &= 5
\end{aligned}
$$

$$
y = \frac{\begin{vmatrix} 3 & 4 & -1\\ 2 & -2 & 3\\ 1 & 5 & 4 \end{vmatrix}}{\begin{vmatrix} 3 & 2 & -1\\ 2 & -1 & 3\\ 1 & 0 & 4 \end{vmatrix}}
$$

$$
= \frac{-24 + 12 - 10 - 2 - 45 - 32}{-12 + 6 + 0 - 1 - 0 - 16}
$$

$$
= \frac{-101}{-23} = \frac{101}{23}
$$

Chapter 6

1. $2x(2x - 3)^2 = 2x[(2x)^2 - 2(2x)(3) + 3^2]$
  $= 2x(4x^2 - 12x + 9)$
  $= 8x^3 - 24x^2 + 18x$

2. $\dfrac{1}{R} = \dfrac{1}{R_1 + r} + \dfrac{1}{R_2}$

$\dfrac{RR_2(R_1 + r)}{R} = \dfrac{RR_2(R_1 + r)}{R_1 + r} + \dfrac{RR_2(R_1 + r)}{R_2}$

$R_2(R_1 + r) = RR_2 + R(R_1 + r)$

$R_1R_2 + rR_2 = RR_2 + RR_1 + rR$

$R_1R_2 - RR_1 = RR_2 + rR - rR_2$

$R_1(R_2 - R) = RR_2 + rR - rR_2$

$R_1 = \dfrac{RR_2 + rR - rR_2}{R_2 - R}$

3. $\dfrac{2x^2 + 5x - 3}{2x^2 + 12x + 18} = \dfrac{(2x - 1)(x + 3)}{2(x + 3)^2} = \dfrac{2x - 1}{2(x + 3)}$

4. $4x^2 - 16y^2 = 4(x^2 - 4y^2) = 4(x + 2y)(x - 2y)$

5. $pb^3 + 8a^3p = p(b^3 + 8a^3)$
  $= p(b + 2a)(b^2 - 2ab + 4a^2)$

6. $2a - 4T - ba + 2bT = 2(a - 2T) - b(a - 2T)$
  $= (2 - b)(a - 2T)$

7. $36x^2 + 14x - 16 = 2(18x^2 + 7x - 8)$
  $= 2(9x + 8)(2x - 1)$

8. $\dfrac{3}{4x^2} - \dfrac{2}{x^2 - x} - \dfrac{x}{2x - 2} = \dfrac{3}{4x^2} - \dfrac{2}{x(x - 1)} - \dfrac{x}{2(x - 1)}$

$= \dfrac{3(x - 1) - 2(4x) - x(2x^2)}{4x^2(x - 1)}$

$= \dfrac{3x - 3 - 8x - 2x^3}{4x^2(x - 1)}$

$= \dfrac{-2x^3 - 5x - 3}{4x^2(x - 1)}$

9. $\dfrac{x^2 + x}{2 - x} \div \dfrac{x^2}{x^2 - 4x + 4} = \dfrac{x^2 + x}{2 - x} \times \dfrac{x^2 - 4x + 4}{x^2}$

$= \dfrac{x(x + 1)}{2 - x} \times \dfrac{(x - 2)^2}{x^2}$

$= -\dfrac{x(x + 1)(x - 2)^2}{(x - 2)(x^2)}$

$= -\dfrac{(x + 1)(x - 2)}{x}$

10. $\dfrac{1 - \dfrac{3}{2x + 2}}{\dfrac{x}{5} - \dfrac{1}{2}} = \dfrac{\dfrac{2(x + 1) - 3}{2(x + 1)}}{\dfrac{2x - 5}{10}}$

$= \dfrac{2x + 2 - 3}{2(x + 1)} \times \dfrac{10}{2x - 5}$

$= \dfrac{5(2x - 1)}{(x + 1)(2x - 5)}$

11. Let t = time working together $\dfrac{48t}{12} + \dfrac{48t}{16} = 48$

$\dfrac{t}{12} + \dfrac{t}{16} = 1$ $4t + 3t = 48$

LCD of 12 and 16 is 48. $t = \dfrac{48}{7} = 6.9$ days

12. $\dfrac{3}{2x^2 - 3x} + \dfrac{1}{x} = \dfrac{3}{2x - 3}$ LCD $= x(2x - 3)$

$\dfrac{3x(2x - 3)}{x(2x - 3)} + \dfrac{x(2x - 3)}{x} = \dfrac{3x(2x - 3)}{2x - 3}$

$3 + (2x - 3) = 3x$

$x = 0$

No solution due to division by zero in first two terms

Chapter 7

1. $2x^2 + 5x = 12$
  $2x^2 + 5x - 12 = 0$
  $(2x - 3)(x + 4) = 0$
  $2x - 3 = 0 \quad x + 4 = 0$
  $x = \dfrac{3}{2} \quad$ or $\quad x = -4$

2. $x^2 = 3x + 5$
  $x^2 - 3x - 5 = 0$
  $a = 1, \quad b = -3, \quad c = -5$
  $x = \dfrac{-(-3) \pm \sqrt{(-3)^2 - 4(1)(-5)}}{2(1)}$
  $= \dfrac{3 \pm \sqrt{29}}{2}$

3. $4x^2 - 5x - 3 = 0$
On calculator
$Y_1 = 4X^2 - 5X - 3$
From zero feature
$x = -0.44$ or $x = 1.69$

4. $2x^2 - x = 6 - 2x(3 - x)$
$2x^2 - x = 6 - 6x + 2x^2$
$5x = 6$ (not quadratic)
$x = \dfrac{6}{5}$

5. $\dfrac{3}{x} - \dfrac{2}{x + 2} = 1$
$\dfrac{3x(x + 2)}{x} - \dfrac{2x(x + 2)}{x + 2} = x(x + 2)$
$3(x + 2) - 2x = x^2 + 2x$
$3x + 6 - 2x = x^2 + 2x$
$0 = x^2 + x - 6$
$(x + 3)(x - 2) = 0$
$x = -3, 2$

6. $y = 2x^2 + 8x + 5$
$\dfrac{-b}{2a} = \dfrac{-8}{2(2)} = -2$
$y = 2(-2)^2 + 8(-2) + 5 = -3$
Min. pt. $(a > 0)$ is $(-2, -3)$,
$c = 5$, y-intercept is $(0, 5)$.

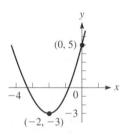

7. $P = EI - RI^2$
$RI^2 - EI + P = 0$
$I = \dfrac{-(-E) \pm \sqrt{(-E)^2 - 4RP}}{2R}$
$= \dfrac{E \pm \sqrt{E^2 - 4RP}}{2R}$

8. $x^2 - 6x - 9 = 0$
$x^2 - 6x \qquad = 9$
$x^2 - 6x + 9 = 9 + 9$
$(x - 3)^2 = 18$
$x - 3 = \pm\sqrt{18}$
$x = 3 \pm 3\sqrt{2}$

9. Let $w = $ width of window
$h = $ height of window
$2w + 2h = 8.4$ (perimeter)
$w + h = 4.2$ $h = 4.2 - w$
$w(4.2 - w) = 3.8$ (area)
$-w^2 + 4.2w = 3.8$
$w^2 - 4.2w + 3.8 = 0$
$w = \dfrac{-(-4.2) \pm \sqrt{(-4.2)^2 - 4(3.8)}}{2}$
$= 2.88, 1.32$
$w = 1.3$ m, $h = 2.9$ m   or
$w = 2.9$ m, $h = 1.3$ m

10. $y = x^2 - 8x + 8$
$a = 1, b = -8$
$\dfrac{-b}{2a} = \dfrac{-(-8)}{2(1)} = 4$
$y = 4^2 - 8(4) + 8 = -8$
$V(4, -8)$, $c = 8$, y-int $(0, 8)$
For $x = 8$, $y = 8$
$y = 0$ for $x = 1.2, 6.8$

Chapter 8

1. $150° = \left(\dfrac{\pi}{180}\right)(150) = \dfrac{5\pi}{6}$

2. (a) +   (b) −   (c) +

3. $\sin 205° = -\sin(205° - 180°) = -\sin 25°$

4. $x = -9$, $y = 12$
$r = \sqrt{(-9)^2 + 12^2} = 15$
$\sin \theta = \dfrac{12}{15} = \dfrac{4}{5}$
$\sec \theta = \dfrac{15}{-9} = -\dfrac{5}{3}$

5. $r = 2.80$ ft
$\omega = 2200$ r/min $= (2200$ r/min$)(2\pi$ rad/r$)$
$= 4400\pi$ rad/min
$v = \omega r = (4400\pi)(2.80) = 39{,}000$ ft/min

6. $3.572 = 3.572\left(\dfrac{180°}{\pi}\right) = 204.7°$

7. $\tan \theta = 0.2396$
$\theta_{ref} = 13.47°$
$\theta = 13.47°$ or
$\theta = 180° + 13.47° = 193.47°$

8. $\cos \theta = -0.8244$, $\csc \theta < 0$;
$\cos \theta$ negative, $\csc \theta$ negative,
θ in third quadrant
$\theta_{ref} = 0.6017$, $\theta = 3.7432$

9. $s = r\theta$, $32.0 = 8.50\theta$
$\theta = \dfrac{32.0}{8.50} = 3.76$ rad
$A = \dfrac{1}{2}\theta r^2 = \dfrac{1}{2}(3.76)(8.50)^2$
$= 136$ ft^2

$s = 32.0$ ft

$r = 8.50$ ft

10. $A = \dfrac{1}{2}\theta r^2$

$A = 38.5 \text{ cm}^2, d = 12.2 \text{ cm}, r = 6.10 \text{ cm}$

$38.5 = \dfrac{1}{2}\theta(6.10)^2, \theta = \dfrac{2(38.5)}{6.10^2} = 2.07$

$s = r\theta, s = 6.10(2.07) = 12.6 \text{ cm}$

Chapter 9

1.

2. $b^2 = a^2 + c^2 - 2ac \cos B$

$b = \sqrt{22.5^2 + 30.9^2 - 2(22.5)(30.9)\cos 78.6°}$

$= 34.4$

3. $x^2 = 36.50^2 + 21.38^2 - 2(36.50)(21.38)\cos 45.00°$

$x = 26.19 \text{ m}$

$\dfrac{21.38}{\sin \alpha} = \dfrac{26.19}{\sin 45.00°}$

$\sin \alpha = \dfrac{21.38 \sin 45.00°}{26.19}, \quad \alpha = 35.26°$

$\theta = 45.00° - 35.26° = 9.74°$

Displacement is 26.19 m,

9.74° N of E.

4. $C = 180° - (18.9° + 104.2°) = 56.9°$

$\dfrac{c}{\sin C} = \dfrac{a}{\sin A}$

$c = \dfrac{426 \sin 56.9°}{\sin 18.9°} = 1100$

5. Since a is longest side, find A first.

$a^2 = b^2 + c^2 - 2bc \cos A$

$\cos A = \dfrac{b^2 + c^2 - a^2}{2bc} = \dfrac{3.29^2 + 8.44^2 - 9.84^2}{2(3.29)(8.44)}$

$A = 105.4°$

$\dfrac{b}{\sin B} = \dfrac{a}{\sin A}$

$\sin B = \dfrac{b \sin A}{a} = \dfrac{3.29 \sin 105.4°}{9.84}$

$B = 18.8°$

$C = 180° - (105.4° + 18.8°) = 55.8°$

6. $R_x = -235, R_y = 152$

$R = \sqrt{(-235)^2 + 152^2} = 280$

$\tan \theta_{\text{ref}} = \left|\dfrac{152}{-235}\right| = \dfrac{152}{235}, \quad \theta_{\text{ref}} = 32.9°$

Quad. II: $\theta = 180° - 32.9° = 147.1°$

7. $A_x = 871 \cos 284.3° = 215$

$A_y = 871 \sin 284.3° = -844$

8. $\dfrac{63.0}{\sin 148.5°} = \dfrac{42.0}{\sin A} \qquad \dfrac{x}{\sin 11.1°} = \dfrac{63.0}{\sin 148.5°}$

$\sin A = \dfrac{42.0 \sin 148.5°}{63.0} \qquad x = \dfrac{63.0 \sin 11.1°}{\sin 148.5°}$

$A = 20.4° \qquad\qquad\qquad\quad = 23.2 \text{ mi}$

$C = 180° - (148.5° + 20.4°) = 11.1°$

9. $A_x = 449 \cos 74.2° \qquad B_x = 285 \cos 208.9°$

$A_y = 449 \sin 74.2° \qquad B_y = 285 \sin 208.9°$

$R_x = A_x + B_x = 449 \cos 74.2° + 285 \cos 208.9°$

$= -127.3$

$R_y = A_y + B_y = 449 \sin 74.2° + 285 \sin 208.9°$

$= 294.3$

$R = \sqrt{(-127.3)^2 + 294.3^3} = 321$

$\tan \theta_{\text{ref}} = \dfrac{294.3}{127.3}, \qquad \theta_{\text{ref}} = 66.6°, \qquad \theta = 113.4°$

θ is in second quadrant, since R_x is negative and R_y is positive.

10. $29.6 \sin 36.5° = 17.6$

$17.6 < 22.3 < 29.6$ means two solutions.

$\dfrac{29.6}{\sin B} = \dfrac{22.3}{\sin 36.5°}, \qquad \sin B = \dfrac{29.6 \sin 36.5°}{22.3}$

$B_1 = 52.1° \qquad C_1 = 180° - 36.5° - 52.1° = 91.4°$

$B_2 = 180° - 52.1° = 127.9°,$

$C_2 = 180° - 36.5° - 127.9° = 15.6°$

$\dfrac{c_1}{\sin 91.4°} = \dfrac{22.3}{\sin 36.5°}, \qquad c_1 = \dfrac{22.3 \sin 91.4°}{\sin 36.5°} = 37.5$

$\dfrac{c_2}{\sin 15.6°} = \dfrac{22.3}{\sin 36.5°}, \qquad c_2 = \dfrac{22.3 \sin 15.6°}{\sin 36.5°} = 10.1$

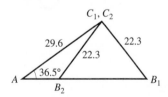

Chapter 10

1. $y = -3\sin(4\pi x - \pi/3)$

$a = -3, b = 4\pi, c = -\pi/3$ $\quad$ amplitude $= |-3| = 3$

period $= \dfrac{2\pi}{4\pi} = \dfrac{1}{2}$ $\quad$ displacement $= -\dfrac{-\frac{\pi}{3}}{4\pi} = \dfrac{1}{12}$

2. $y = 0.5\cos\dfrac{\pi}{2}x$

Amp. $= 0.5$, disp. $= 0$,

per. $= \dfrac{2\pi}{\pi/2} = 4$

x	0	1	2	3	4
y	0.5	0	−0.5	0	0.5

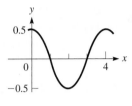

3. $y = 2 + 3\sin x$

For $y_1 = 3\sin x$,

amp. $= 3$, per. $= 2\pi$,

disp. $= 0$

x	0	$\frac{\pi}{2}$	π	$\frac{3\pi}{2}$	2π
$y_1 = 3\sin x$	0	3	0	−3	0
$y = 2 + 3\sin x$	2	5	2	−1	2

4. $y = 3\sec x$

$\sec x = \dfrac{1}{\cos x}$

5. $y = 2\sin\left(2x - \dfrac{\pi}{3}\right)$

Amp. $= 2$, per. $= \dfrac{2\pi}{2} = \pi$

disp. $= -\dfrac{-\pi/3}{2} = \dfrac{\pi}{6}$

x	$\frac{\pi}{6}$	$\frac{5\pi}{12}$	$\frac{2\pi}{3}$	$\frac{11\pi}{12}$	$\frac{7\pi}{6}$
y	0	2	0	−2	0

$\frac{\pi}{6} + \frac{\pi}{4} = \frac{5\pi}{12}, \frac{\pi}{6} + \frac{\pi}{2} = \frac{2\pi}{3}, \frac{\pi}{6} + \frac{3\pi}{4} = \frac{11\pi}{12}$

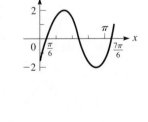

6. $y = A\cos\dfrac{2\pi}{T}t$

$A = 0.200$ in., $T = 0.100$ s,

$y = 0.200\cos 20\pi t$,

amp. $= 0.200$ in., per. $= 0.100$ s

7. $y = 2\sin x + \cos 2x$

For $y_1 = 2\sin x$,

amp. $= 2$, per. $= 2\pi$, disp. $= 0$

For $y_2 = \cos 2x$,

amp. $= 1$, per. $= \dfrac{2\pi}{2} = \pi$, disp. $= 0$

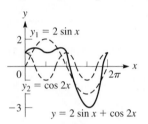

8. $x = \sin \pi t, y = 2\cos 2\pi t$

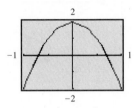

9. $d = R\sin\left(\omega t + \dfrac{\pi}{6}\right)$

$\omega = 2.00$ rad/s

$d = R\sin\left(2.00t + \dfrac{\pi}{6}\right)$

Amp. $= R$, per. $= \dfrac{2\pi}{2.00} = \pi$ s $= 3.14$ s,

disp. $= \dfrac{-\pi/6}{2.00} = -\dfrac{\pi}{12}$ s $= -0.26$ s

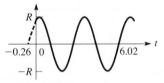

10. $y = 2\sin bx$

$\left(\dfrac{\pi}{3}, 2\right)$, is first max. with $x > 0$

period $= \dfrac{2\pi}{b}, \dfrac{1}{4}\left(\dfrac{2\pi}{b}\right) = \dfrac{\pi}{3}, \dfrac{\pi}{2b} = \dfrac{\pi}{3}, b = \dfrac{3}{2}$

$y = 2\sin\dfrac{3x}{2}$

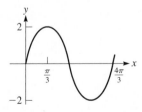

Chapter 11

1. $-5y^0 = -5(1) = -5$

2. $(3\pi x^{-4})^{-2} = (3\pi)^{-2}(x^{-4})^{-2} = \dfrac{x^8}{(3\pi)^2} = \dfrac{x^8}{9\pi^2}$

3. $\dfrac{100^{3/2}}{8^{-2/3}} = (100^{3/2})(8^{2/3}) = [(100^{1/2})^3][(8^{1/3})^2]$
$= (10^3)(2^2) = 4000$

4. $(as^{-1/3}t^{3/4})^{12} = a^{12}s^{(-1/3)(12)}t^{(3/4)(12)} = a^{12}s^{-4}t^9 = \dfrac{a^{12}t^9}{s^4}$

5. $2\sqrt{20} - \sqrt{125} = 2\sqrt{4 \times 5} - \sqrt{25 \times 5}$
$= 2(2\sqrt{5}) - 5\sqrt{5}$
$= 4\sqrt{5} - 5\sqrt{5} = -\sqrt{5}$

6. $(2x^{-1} + y^{-2})^{-1} = \dfrac{1}{2x^{-1} + y^{-2}} = \dfrac{1}{\dfrac{2}{x} + \dfrac{1}{y^2}} = \dfrac{1}{\dfrac{2y^2 + x}{xy^2}}$
$= \dfrac{xy^2}{2y^2 + x}$

7. $(\sqrt{2x} - 3\sqrt{y})^2 = (\sqrt{2x})^2 - 2\sqrt{2x}(3\sqrt{y}) + (3\sqrt{y})^2$
$= 2x - 6\sqrt{2xy} + 9y$

8. $\sqrt[3]{\sqrt[4]{4}} = \sqrt[12]{4} = \sqrt[12]{2^2} = 2^{2/12} = 2^{1/6} = \sqrt[6]{2}$

9. $\dfrac{3 - 2\sqrt{2}}{2\sqrt{x}} = \dfrac{3 - 2\sqrt{2}}{2\sqrt{x}} \times \dfrac{\sqrt{x}}{\sqrt{x}} = \dfrac{3\sqrt{x} - 2\sqrt{2x}}{2x}$

10. $\sqrt{27a^4b^3} = \sqrt{9 \times 3 \times (a^2)^2(b^2)(b)} = 3a^2b\sqrt{3b}$

11. $2\sqrt{2}(3\sqrt{10} - \sqrt{6}) = 2\sqrt{2}(3\sqrt{10}) - 2\sqrt{2}(\sqrt{6})$
$= 6\sqrt{20} - 2\sqrt{12}$
$= 6(2\sqrt{5}) - 2(2\sqrt{3})$
$= 12\sqrt{5} - 4\sqrt{3}$

12. $(2x + 3)^{1/2} + (x + 1)(2x + 3)^{-1/2}$
$= (2x + 3)^{1/2} + \dfrac{x + 1}{(2x + 3)^{1/2}}$
$= \dfrac{(2x + 3)^{1/2}(2x + 3)^{1/2} + x + 1}{(2x + 3)^{1/2}} = \dfrac{2x + 3 + x + 1}{(2x + 3)^{1/2}}$
$= \dfrac{3x + 4}{(2x + 3)^{1/2}}$

13. $\left(\dfrac{4a^{-1/2}b^{3/4}}{b^{-2}}\right)\left(\dfrac{b^{-1}}{2a}\right) = \dfrac{4a^{-1/2}b^{3/4-1}}{2ab^{-2}} = \dfrac{2b^{2-1/4}}{a^{1+1/2}} = \dfrac{2b^{7/4}}{a^{3/2}}$

14. $\dfrac{2\sqrt{15} + \sqrt{3}}{\sqrt{15} - 2\sqrt{3}} = \dfrac{2\sqrt{15} + \sqrt{3}}{\sqrt{15} - 2\sqrt{3}} \times \dfrac{\sqrt{15} + 2\sqrt{3}}{\sqrt{15} + 2\sqrt{3}}$
$= \dfrac{2(15) + 4\sqrt{45} + \sqrt{45} + 2(3)}{15 - 4(3)}$
$= \dfrac{36 + 15\sqrt{5}}{3}$
$= 12 + 5\sqrt{5}$

15. $\dfrac{3^{-1/2}}{2} = \dfrac{1}{2 \times 3^{1/2}} = \dfrac{1}{2\sqrt{3}}$
$= \dfrac{\sqrt{3}}{2\sqrt{3}\sqrt{3}} = \dfrac{\sqrt{3}}{6}$

16. $0.220N^{-1/6}$ $N = 64 \times 10^6$
$0.220(64 \times 10^6)^{-1/6} = \dfrac{0.220}{(64 \times 10^6)^{1/6}} = \dfrac{0.220}{2 \times 10} = 0.011$

Chapter 12

1. $(3 - \sqrt{-4}) + (5\sqrt{-9} - 1) = (3 - 2j) + [5(3j) - 1]$
$= 3 - 2j + 15j - 1$
$= 2 + 13j$

2. $(2\angle 130°)(3\angle 45°) = (2)(3)\angle 130° + 45° = 6\angle 175°$

3. $2 - 7j$: $r = \sqrt{2^2 + (-7)^2} = \sqrt{53} = 7.28$
$\tan\theta = \dfrac{-7}{2} = -3.500,$ $\theta_{ref} = 74.1°,$ $\theta = 285.9°$
$2 - 7j = 7.28(\cos 285.9° + j\sin 285.9°)$
$= 7.28\angle 285.9°$

4. (a) $-\sqrt{-64} = -(8j) = -8j$
(b) $-j^{15} = -j^{12}j^3 = (-1)(-j) = j$

5. $(4 - 3j) + (-1 + 4j) = 3 + j$

6. $\dfrac{2 - 4j}{5 + 3j} = \dfrac{(2 - 4j)(5 - 3j)}{(5 + 3j)(5 - 3j)} = \dfrac{10 - 26j + 12j^2}{25 - 9j^2}$
$= \dfrac{10 - 26j - 12}{25 - 9(-1)} = \dfrac{-2 - 26j}{34} = -\dfrac{1 + 13j}{17}$

7. $2.56(\cos 125.2° + j\sin 125.2°) = 2.56e^{2.185j}$
$125.2° = \dfrac{125.2\pi}{180} = 2.185$ rad

8. $R = 3.50\ \Omega, X_L = 6.20\ \Omega, X_C = 7.35\ \Omega$
$|Z| = \sqrt{R^2 + (X_L - X_C)^2}$
$= \sqrt{3.50^2 + (6.20 - 7.35)^2} = 3.68\ \Omega$
$\tan\theta = \dfrac{X_L - X_C}{R} = \dfrac{6.20 - 7.35}{3.50} = \dfrac{-1.15}{3.50}$
$\theta = -18.2°$

9. $3.47 - 2.81j = 4.47e^{5.60j}$
$R = \sqrt{3.47^2 + (-2.81)^2} = 4.47$
$\tan\theta = \dfrac{-2.81}{3.47},$ $\theta_{ref} = 39.0°$
$\theta = 321.0° = 5.60$ rad

10. $x + 2j - y = yj - 3xj$
$x - y + 3xj - yj = -2j$
$(x - y) + (3x - y)j = 0 - 2j$
$x - y = \ \ 0$
$\underline{3x - y = -2}$
$2x = -2$
$x = -1,$ $y = -1$

11. $L = 8.75 \text{ mH} = 8.75 \times 10^{-3} \text{ H}$
$f = 600 \text{ kHz} = 6.00 \times 10^5 \text{ Hz}$

$$2\pi f L = \frac{1}{2\pi f C}$$

$$C = \frac{1}{(2\pi f)^2 L} = \frac{1}{(2\pi)^2 (6.00 \times 10^5)^2 (8.75 \times 10^{-3})}$$
$$= 8.04 \times 10^{-12} = 8.04 \text{ pF}$$

12. $j = 1(\cos 90° + j \sin 90°)$

$$j^{1/3} = 1^{1/3}\left(\cos \frac{90°}{3} + j \sin \frac{90°}{3}\right)$$
$$= \cos 30° + j \sin 30° = 0.8660 + 0.5000j$$
$$= 1^{1/3}\left(\cos \frac{90° + 360°}{3} + j \sin \frac{90° + 360°}{3}\right)$$
$$= \cos 150° + j \sin 150° = -0.8660 + 0.5000j$$
$$= 1^{1/3}\left(\cos \frac{90° + 720°}{3} + j \sin \frac{90° + 720°}{3}\right)$$
$$= \cos 270° + j \sin 270° = -j$$

Cube roots of j: $0.8660 + 0.5000j$
$-0.8660 + 0.5000j$
$-j$

Chapter 13

1. $\log_9 x = -\frac{1}{2}$

$x = 9^{-1/2}$

$= \frac{1}{9^{1/2}} = \frac{1}{3}$

2. $\log_3 x - \log_3 2 = 2$

$\log_3 \frac{x}{2} = 2$

$\frac{x}{2} = 3^2$

$x = 2(3^2) = 18$

3. $\log_x 64 = 3$

$64 = x^3$

$4^3 = x^3$

$x = 4$

4. $3^{3x+1} = 8$

$(3x + 1)\log 3 = \log 8$

$3x + 1 = \frac{\log 8}{\log 3}$

$x = \frac{1}{3}\left(\frac{\log 8}{\log 3} - 1\right) = 0.298$

5. $y = 2 \log_4 x$

x	$\frac{1}{4}$	1	4	16
y	-2	0	2	4

$\log_4 \frac{1}{4} = -1$, $\log_4 16 = 2$

6. $y = 2(3^x)$

x	-1	0	1	2
y	0.7	2	6	18

x	3	4	5
y	54	162	486

7. $\log_5\left(\frac{4a^3}{7}\right) = \log_5 4a^3 - \log_5 7$

$= \log_5 4 + \log_5 a^3 - \log_5 7$

$= \log_5 4 + 3 \log_5 a - \log_5 7$

$= 2 \log_5 2 + 3 \log_5 a - \log_5 7$

8. $3 \log_7 x - \log_7 y = 2$

$\log_7 x^3 - \log_7 y = 2$

$\log_7 \frac{x^3}{y} = 2 \qquad \frac{x^3}{y} = 7^2$

$x^3 = 49y \qquad y = \frac{1}{49}x^3$

9. $\ln i - \ln I = -t/RC$

$\ln \frac{i}{I} = -t/RC$

$\frac{i}{I} = e^{-t/RC} \qquad i = Ie^{-t/RC}$

10. $\dfrac{2 \ln 0.9523}{\log 6066} = -0.02584$

11. $\log_b x = \dfrac{\log_a x}{\log_a b}$

$\log_5 732 = \dfrac{\log 732}{\log 5} = 4.098$

12. $A = A_0 e^{0.08t} \qquad A = 2A_0$

$2A_0 = A_0 e^{0.08t} \qquad 2 = e^{0.08t}$

$\ln 2 = \ln e^{0.08t} = 0.08t \qquad t = \dfrac{\ln 2}{0.08} = 8.66 \text{ years}$

Chapter 14

1. $x^{1/2} - 2x^{1/4} = 3$

Let $y = x^{1/4}$

$y^2 - 2y - 3 = 0$

$(y - 3)(y + 1) = 0$

$y = 3, -1$

$x^{1/4} \neq -1$

$x^{1/4} = 3, \qquad x = 81$

Check: $81^{1/2} - 2(81^{1/4}) = 3$

$9 - 6 = 3$

Solution: $x = 81$

2. $3\sqrt{x-2} - \sqrt{x+1} = 1$

$3\sqrt{x-2} = 1 + \sqrt{x+1}$

$9(x-2) = 1 + 2\sqrt{x+1} + (x+1)$

$8x - 20 = 2\sqrt{x+1}$

$4x - 10 = \sqrt{x+1}$

$16x^2 - 80x + 100 = x + 1$

$16x^2 - 81x + 99 = 0$

$x = \dfrac{81 \pm \sqrt{81^2 - 4(16)(99)}}{32}$

$= \dfrac{81 \pm 15}{32} = 3, \dfrac{33}{16}$

Check: $x = 3$: $3\sqrt{3-2} - \sqrt{3+1} \overset{?}{=} 1$

$3 - 2 = 1$

$x = \frac{33}{16}$: $3\sqrt{\frac{33}{16} - 2} - \sqrt{\frac{33}{16} + 1} \overset{?}{=} 1$

$\frac{3}{4} - \frac{7}{4} \neq 1$

Solution: $x = 3$

3. $x^4 - 17x^2 + 16 = 0$

Let $y = x^2$

$y^2 - 17y + 16 = 0$

$(y-1)(y-16) = 0$

$y = 1, 16$

$x^2 = 1, 16$

$x = -1, 1, -4, 4$

All values check.

4. $x^2 - 2y = 5$

$\underline{2x + 6y = 1}$

$y = \frac{1}{2}(x^2 - 5)$

$2x + 6\left(\frac{1}{2}\right)(x^2 - 5) = 1$

$3x^2 + 2x - 16 = 0$

$(3x + 8)(x - 2) = 0$

$x = -\frac{8}{3}, 2$

$x = -\frac{8}{3}$: $y = \frac{1}{2}\left(\frac{64}{9} - 5\right) = \frac{19}{18}$

$x = 2$: $y = \frac{1}{2}(4 - 5) = -\frac{1}{2}$

$x = -\frac{8}{3}, y = \frac{19}{18}$

or $x = 2, y = -\frac{1}{2}$

5. $\sqrt[3]{2x + 5} = 5$

$2x + 5 = 125$

$2x = 120$

$x = 60$

6. $v = \sqrt{v_0^2 + 2gh}$

$v^2 = v_0^2 + 2gh$

$2gh = v^2 - v_0^2$

$h = \dfrac{v^2 - v_0^2}{2g}$

7. $x^2 - y^2 = 4$ $xy = 2$

$y = \pm\sqrt{x^2 - 4}$ $y = \dfrac{2}{x}$

x	y
± 2	0
± 3	± 2.2
± 4	± 3.5

x	y
-4	$-\frac{1}{2}$
-2	-1
-1	-2
0	—
1	2
2	1
4	$\frac{1}{2}$

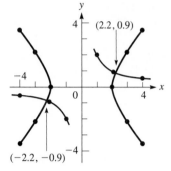

$x = 2.2, \quad y = 0.9; \quad x = -2.2, \quad y = -0.9$

8. Let l = length, w = width

$2l + 2w = 14$

$\underline{lw = 10}$

$l + w = 7$

$l = 7 - w$

$(7 - w)w = 10$

$7w - w^2 = 10$

$w^2 - 7w + 10 = 0$

$(w - 5)(w - 2) = 0$

$w = 5, 2$

$l = 5.00$ ft, $w = 2.00$ ft

(or $l = 2.00$ ft, $w = 5.00$ ft)

Chapter 15

1. $f(x) = 2x^3 + 3x^2 + 7x - 6$

$f(-3) = 2(-3)^3 + 3(-3)^2 + 7(-3) - 6$

$= -54 + 27 - 21 - 6 = -54$

$f(-3) \neq 0, \qquad -3$ is not a zero

2. $x^4 - 2x^3 - 7x^2 + 20x - 12 = 0$

1	-2	-7	20	-12	$\lfloor 2$
	2	0	-14	12	
1	0	-7	6		$\lfloor 2$
	2	4	-6		
1	2	-3			

$x^2 + 2x - 3 = (x + 3)(x - 1)$

Other roots: $x = -3, 1$

3. $(x^3 - 5x^2 + 4x - 9) \div (x - 3)$

1	-5	4	-9	$\lfloor 3$
	3	-6	-6	
1	-2	-2	-15	

Quotient: $x^2 - 2x - 2$

Remainder $= -15$

4. $f(x) = 2x^4 + 15x^3 + 23x^2 - 16$

$2x + 1 = 2\left(x + \frac{1}{2}\right)$

$$
\begin{array}{rrrrr|l}
2 & 15 & 23 & 0 & -16 & \underline{-\frac{1}{2}} \\
 & -1 & -7 & -8 & 4 & \\
\hline
2 & 14 & 16 & -8 & -12 &
\end{array}
$$

Remainder is not zero;
$2x + 1$ is not a factor.

5. $(x^3 + 4x^2 + 7x - 9) \div (x + 4)$

$f(x) = x^3 + 4x^2 + 7x - 9$

$f(-4) = (-4)^3 + 4(-4)^2 + 7(-4) - 9$

$\quad = -64 + 64 - 28 - 9 = -37$

Remainder $= -37$

6. $2x^4 - x^3 + 5x^2 - 4x - 12 = 0$

$f(x) = 2x^4 - x^3 + 5x^2 - 4x - 12$

$f(-x) = 2x^4 + x^3 + 5x^2 + 4x - 12$

$n = 4$; 4 roots

No more than 3 positive roots

One negative root

Rational roots: factors of 12 divided by factors of 2

Possible rational roots: $\pm 1, \pm 2, \pm 3, \pm 4, \pm 6, \pm 12, \pm\frac{1}{2}, \pm\frac{3}{2}$

$$
\begin{array}{rrrrr|l}
2 & -1 & 5 & -4 & -12 & \underline{2} \\
 & 4 & 6 & 22 & 36 & \\
\hline
2 & 3 & 11 & 18 & 24 &
\end{array}
$$

2 is too large.

$$
\begin{array}{rrrrr|l}
2 & -1 & 5 & -4 & -12 & \underline{\frac{3}{2}} \\
 & 3 & 3 & 12 & 12 & \\
\hline
2 & 2 & 8 & 8 & 0 & \underline{-1} \\
 & -2 & 0 & -8 & & \\
\hline
2 & 0 & 8 & 0 & &
\end{array}
$$

$2x^2 + 8 = 0, \quad x^2 + 4 = 0, \quad x = \pm 2j$

roots: $\frac{3}{2}, -1, 2j, -2j$

7. $y = kx^2(x^3 + 436x - 4000)$

$y = 0, kx^2(x^3 + 436x - 4000) = 0$

$x = 0, x^3 + 436x - 4000 = 0$

$$
\begin{array}{rrrr|l}
1 & 0 & 436 & -4000 & \underline{8} \\
 & 8 & 64 & 4000 & \\
\hline
1 & 8 & 500 & 0 &
\end{array}
$$

$y = 0$ for $x = 0$ ft, $x = 8$ ft

8. Let $x =$ length of edge.

$V = x^3, \quad 2V = (x + 1)^3$

$2x^3 = x^3 + 3x^2 + 3x + 1$

$x^3 - 3x^2 - 3x - 1 = 0$

$y = x^3 - 3x^2 - 3x - 1$

$x = 3.8$ mm

Chapter 16

1. $A = \begin{bmatrix} 3 & -1 & 4 \\ 2 & 0 & -2 \end{bmatrix}$ $B = \begin{bmatrix} 1 & 4 & 5 \\ -1 & -2 & 3 \end{bmatrix}$

$2B = \begin{bmatrix} 2 & 8 & 10 \\ -2 & -4 & 6 \end{bmatrix}$

$A - 2B = \begin{bmatrix} 3 - 2 & -1 - 8 & 4 - 10 \\ 2 + 2 & 0 + 4 & -2 - 6 \end{bmatrix}$

$\quad = \begin{bmatrix} 1 & -9 & -6 \\ 4 & 4 & -8 \end{bmatrix}$

2. $\begin{bmatrix} 2x & x - y & z \\ x + z & 2y & y + z \end{bmatrix} = \begin{bmatrix} 6 & -2 & 4 \\ a & b & c \end{bmatrix}$

$2x = 6 \qquad x - y = -2 \qquad z = 4$

$x = 3 \qquad 3 - y = -2$

$\qquad\qquad y = 5$

$x + z = a \qquad 2y = b \qquad y + z = c$

$3 + 4 = a \qquad 2(5) = b \qquad 5 + 4 = c$

$a = 7 \qquad b = 10 \qquad c = 9$

3. $CD = \begin{bmatrix} 1 & 0 & 4 \\ 2 & -2 & 1 \\ -1 & 3 & 2 \end{bmatrix} \begin{bmatrix} 2 & -2 \\ 4 & -5 \\ 6 & 1 \end{bmatrix}$

$= \begin{bmatrix} 2 + 0 + 24 & -2 + 0 + 4 \\ 4 - 8 + 6 & -4 + 10 + 1 \\ -2 + 12 + 12 & 2 - 15 + 2 \end{bmatrix}$

$= \begin{bmatrix} 26 & 2 \\ 2 & 7 \\ 22 & -11 \end{bmatrix}$

$DC = \begin{bmatrix} 2 & -2 \\ 4 & -5 \\ 6 & 1 \end{bmatrix} \begin{bmatrix} 1 & 0 & 4 \\ 2 & -2 & 1 \\ -1 & 3 & 2 \end{bmatrix}$ not defined, since D has 2 columns and C has 3 rows

4. $A = \begin{bmatrix} 2 & -5 \\ 1 & -2 \end{bmatrix}$ $B = \begin{bmatrix} -2 & 5 \\ -1 & 2 \end{bmatrix}$

$AB = \begin{bmatrix} 2 & -5 \\ 1 & -2 \end{bmatrix} \begin{bmatrix} -2 & 5 \\ -1 & 2 \end{bmatrix} = \begin{bmatrix} -4 + 5 & 10 - 10 \\ -2 + 2 & 5 - 4 \end{bmatrix}$

$= \begin{bmatrix} 1 & 0 \\ 0 & 1 \end{bmatrix} = I$

$BA = \begin{bmatrix} -2 & 5 \\ -1 & 2 \end{bmatrix} \begin{bmatrix} 2 & -5 \\ 1 & -2 \end{bmatrix} = \begin{bmatrix} -4 + 5 & 10 - 10 \\ -2 + 2 & 5 - 4 \end{bmatrix}$

$= \begin{bmatrix} 1 & 0 \\ 0 & 1 \end{bmatrix} = I$

$AB = BA = I, B = A^{-1}$

5. $\begin{bmatrix} 1 & 0 & 4 & | & 1 & 0 & 0 \\ 2 & -2 & 1 & | & 0 & 1 & 0 \\ -1 & 3 & 2 & | & 0 & 0 & 1 \end{bmatrix} \rightarrow \begin{bmatrix} 1 & 0 & 4 & | & 1 & 0 & 0 \\ 0 & -2 & -7 & | & -2 & 1 & 0 \\ 0 & 3 & 6 & | & 1 & 0 & 1 \end{bmatrix} \rightarrow$

$\begin{bmatrix} 1 & 0 & 4 & | & 1 & 0 & 0 \\ 0 & 1 & \frac{7}{2} & | & 1 & -\frac{1}{2} & 0 \\ 0 & 3 & 6 & | & 1 & 0 & 1 \end{bmatrix} \rightarrow \begin{bmatrix} 1 & 0 & 4 & | & 1 & 0 & 0 \\ 0 & 1 & \frac{7}{2} & | & 1 & -\frac{1}{2} & 0 \\ 0 & 0 & -\frac{9}{2} & | & -2 & \frac{3}{2} & 1 \end{bmatrix} \rightarrow$

$\begin{bmatrix} 1 & 0 & 4 & | & 1 & 0 & 0 \\ 0 & 1 & \frac{7}{2} & | & 1 & -\frac{1}{2} & 0 \\ 0 & 0 & 1 & | & \frac{4}{9} & -\frac{1}{3} & -\frac{2}{9} \end{bmatrix} \rightarrow \begin{bmatrix} 1 & 0 & 0 & | & -\frac{7}{9} & \frac{4}{3} & \frac{8}{9} \\ 0 & 1 & 0 & | & -\frac{5}{9} & \frac{2}{3} & \frac{7}{9} \\ 0 & 0 & 1 & | & \frac{4}{9} & -\frac{1}{3} & -\frac{2}{9} \end{bmatrix}$

$C^{-1} = \begin{bmatrix} -\frac{7}{9} & \frac{4}{3} & \frac{8}{9} \\ -\frac{5}{9} & \frac{2}{3} & \frac{7}{9} \\ \frac{4}{9} & -\frac{1}{3} & -\frac{2}{9} \end{bmatrix}$

6. $2x - 3y = 11$
$ x + 2y = 2$

$A = \begin{bmatrix} 2 & -3 \\ 1 & 2 \end{bmatrix}$ $A^{-1} = \begin{bmatrix} \frac{2}{7} & \frac{3}{7} \\ -\frac{1}{7} & \frac{2}{7} \end{bmatrix}$ $C = \begin{bmatrix} 11 \\ 2 \end{bmatrix}$

$A^{-1}C = \begin{bmatrix} \frac{2}{7} & \frac{3}{7} \\ -\frac{1}{7} & \frac{2}{7} \end{bmatrix}\begin{bmatrix} 11 \\ 2 \end{bmatrix} = \begin{bmatrix} \frac{22}{7} + \frac{6}{7} \\ -\frac{11}{7} + \frac{4}{7} \end{bmatrix} = \begin{bmatrix} 4 \\ -1 \end{bmatrix}$

$x = 4$ $y = -1$

7. $2x - 3y = 11$
$ \underline{x + 2y = 2}$
$ x + 2y = 2$ interchange equations
$ \underline{2x - 3y = 11}$
$ x + 2y = 2$ subtract 2 times first equation
$ \underline{-7y = 7}$ from second equation
$ x + 2y = 2$ divide second equation by -7
$ \underline{y = -1}$
$ x + 2(-1) = 2$ substitute $y = -1$ in first equation
$ x = 4$

8. $\begin{vmatrix} 2 & -4 & -3 \\ -3 & 6 & 2 \\ 5 & -1 & 5 \end{vmatrix} = 2\begin{vmatrix} 6 & 2 \\ -1 & 5 \end{vmatrix} - (-4)\begin{vmatrix} -3 & 2 \\ 5 & 5 \end{vmatrix} + (-3)\begin{vmatrix} -3 & 6 \\ 5 & -1 \end{vmatrix}$

$ = 2[30 - (-2)] + 4[-15 - 10] - 3[3 - 30]$
$ = 2(32) + 4(-25) - 3(-27) = 64 - 100 + 81 = 45$

9. $\begin{vmatrix} 2 & -4 & -3 \\ -3 & 6 & 2 \\ 5 & -1 & 5 \end{vmatrix} = \begin{vmatrix} 2 & 0 & -3 \\ -3 & 0 & 2 \\ 5 & 9 & 5 \end{vmatrix} = -\begin{vmatrix} 5 & 9 & 5 \\ -3 & 0 & 2 \\ 2 & 0 & -3 \end{vmatrix} = \begin{vmatrix} 9 & 5 & 5 \\ 0 & -3 & 2 \\ 0 & 2 & -3 \end{vmatrix}$

$ = 9\begin{vmatrix} -3 & 2 \\ 2 & -3 \end{vmatrix} = 9(9 - 4) = 45$

(1) $2 \times$ col. 1 added to col. 2; (2) rows 1 and 3 interchanged;
(3) cols. 1 and 2 interchanged; (4) expanded by col. 1

10. Let A = number of shares of stock A
$ B$ = number of shares of stock B
$ 50A + 30B = 2600$
$ \underline{30A + 40B = 2000}$
$ 5A + 3B = 260$
$ \underline{3A + 4B = 200}$
Let C = coefficient matrix

$C = \begin{bmatrix} 5 & 3 \\ 3 & 4 \end{bmatrix}$, $\begin{vmatrix} 5 & 3 \\ 3 & 4 \end{vmatrix} = 20 - 9 = 11$

$C^{-1} = \frac{1}{11}\begin{bmatrix} 4 & -3 \\ -3 & 5 \end{bmatrix} = \begin{bmatrix} \frac{4}{11} & -\frac{3}{11} \\ -\frac{3}{11} & \frac{5}{11} \end{bmatrix}$

$\begin{bmatrix} A \\ B \end{bmatrix} = \begin{bmatrix} \frac{4}{11} & -\frac{3}{11} \\ -\frac{3}{11} & \frac{5}{11} \end{bmatrix}\begin{bmatrix} 260 \\ 200 \end{bmatrix} = \begin{bmatrix} \frac{4}{11}(260) - \frac{3}{11}(200) \\ -\frac{3}{11}(260) + \frac{5}{11}(200) \end{bmatrix}$

$\phantom{\begin{bmatrix} A \\ B \end{bmatrix}} = \begin{bmatrix} 40 \\ 20 \end{bmatrix}$

$A = 40$ shares $B = 20$ shares

11. $A = \begin{bmatrix} 7 & -2 & 1 \\ 2 & 3 & -4 \\ 4 & -5 & 2 \end{bmatrix}$ $C = \begin{bmatrix} 6 \\ 6 \\ 10 \end{bmatrix}$

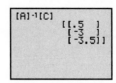
```
[A]⁻¹[C]
        [[.5  ]
         [-3  ]
         [-3.5]]
```

$x = 0.5$ $y = -3$ $z = -3.5$

Chapter 17

1. $x < 0, y > 0$ **2.** $\dfrac{-x}{2} \geq 3$ **3.** $3x + 1 < -5$
$\phantom{1. x < 0, y > 0 \quad 2.} -x \geq 6$ $ 3x < -6$
$\phantom{1. x < 0, y > 0 \quad 2.} x \leq -6$ $ x < -2$

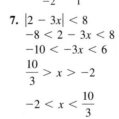

4. $-1 < 1 - 2x < 5$ **5.** $\dfrac{x^2 + x}{x - 2} \leq 0, \dfrac{x(x + 1)}{x - 2} \leq 0$
$ -2 < -2x < 4$
$ 1 > x > -2$
$ -2 < x < 1$

Interval	$\dfrac{x(x + 1)}{x - 2}$	Sign
$x < -1$	$\dfrac{- \ -}{-}$	$-$
$-1 < x < 0$	$\dfrac{- \ +}{-}$	$+$
$0 < x < 2$	$\dfrac{+ \ +}{-}$	$-$
$x > 2$	$\dfrac{+ \ +}{+}$	$+$

Solution: $x \leq -1$ or $0 \leq x < 2$

(x cannot equal 2)

6. $|2x + 1| \geq 3$
$ 2x + 1 \geq 3,$ $2x + 1 \leq -3$
$ 2x \geq 2$ $ 2x \leq -4$
$ x \geq 1$ or $x \leq -2$

7. $|2 - 3x| < 8$
$ -8 < 2 - 3x < 8$
$ -10 < -3x < 6$
$ \dfrac{10}{3} > x > -2$
$ -2 < x < \dfrac{10}{3}$

8.

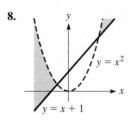

9. If $\sqrt{x^2 - x - 6}$ is real,
then $x^2 - x - 6 \geq 0$.
$(x - 3)(x + 2) \geq 0$
$x \leq -2$ or $x \geq 3$

10. Let $w = $ width, $l = $ length
$l = w + 20$ $w^2 + 20w - 4800 \geq 0$
$wl \geq 4800$ $(w + 80)(w - 60) \geq 0$
$w(w + 20) \geq 4800$ $w \geq 60$ m

11. Let $A = $ length of type A wire
$B = $ length of type B wire
$0.10A + 0.20B < 5.00$
$A + 2B < 50$

12. $|\lambda - 550 \text{ nm}| < 150 \text{ nm}$ (within 150 nm of $\lambda = 550$ nm)

13. $x^2 > 12 - x$

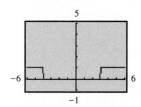

$x < -4, x > 3$

14. $P = 5x + 3y$
$x \geq 0, y \geq 0$
$2x + 3y \leq 12$
$4x + y \leq 8$

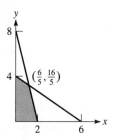

Vertices: $(0,0)$ $(2,0)$ $\left(\dfrac{6}{5}, \dfrac{16}{5}\right)$ $(0,4)$

$P = 5x + 3y$ 0 10 15.6 12

Max. value of $P = 15.6$ at $\left(\dfrac{6}{5}, \dfrac{16}{5}\right)$

Chapter 18

1. $\dfrac{180 \text{ s}}{4 \text{ min}} = \dfrac{180 \text{ s}}{240 \text{ s}} = \dfrac{3}{4}$ **2.** $\dfrac{1.8 \text{ m}}{20.0 \text{ mm}} = \dfrac{x}{14.5 \text{ mm}}$
$20.0x = 1.8(14.5)$
$x = 1.3$ m

3. $L = kT$
$L = 2.7$ cm   for $T = 150°C$
$2.7 = k(150)$
$k = 0.018$ cm/°C
$L = 0.018T$

4. $\dfrac{1.00 \text{ in.}}{2.54 \text{ cm}} = \dfrac{x}{7.24 \text{ cm}}$

$x = \dfrac{7.24}{2.54} = 2.85$ in.

5. $2l + 2w = 210.0$
$l = 105.0 - w$
$\dfrac{l}{w} = \dfrac{7}{3}$
$\dfrac{105.0 - w}{w} = \dfrac{7}{3}$
$315.0 - 3w = 7w$
$10w = 315.0$
$w = 31.5$ in.
$l = 105.0 - 31.5$
$= 73.5$ in.

6. $p = kdh$
Using values of water,
$1.96 = k(1000)(0.200)$
$k = 0.00980$ kPa $\cdot$ m^2/kg
For alcohol,
$p = 0.00980(800)(0.300)$
$= 2.35$ kPa

7. Let $L_1 = $ crushing load of first pillar
$L_2 = $ crushing load of second pillar
$$L_2 = \frac{kr_2^4}{l_2^2} \qquad L_1 = \frac{k(2r_2)^4}{(3l_2)^2}$$

$$\frac{L_1}{L_2} = \frac{\dfrac{k(2r_2)^4}{(3l_2)^2}}{\dfrac{kr_2^4}{l_2^2}} = \frac{16kr_2^4}{9l_2^2} \times \frac{l_2^2}{kr_2^4} = \frac{16}{9}$$

Chapter 19

1. (a) $a_1 = 8, d = -1/2$; arith. seq. 8, 15/2, 7, 13/2, 6, ...
(b) $a_1 = 8, r = -1/2$; geom. seq. 8, -4, 2, -1, 1/2, ...

2. $6, -2, \frac{2}{3}, \dots$;
geometric sequence
$$a_1 = 6 \qquad r = \frac{-2}{6} = -\frac{1}{3}$$

$$S_7 = \frac{6\left[1 - \left(-\frac{1}{3}\right)^7\right]}{1 - \left(-\frac{1}{3}\right)}$$

$$= \frac{6\left(1 + \dfrac{1}{3^7}\right)}{\dfrac{4}{3}} = \frac{1094}{243}$$

3. $a_1 = 6, d = 4, s_n = 126$;
arithmetic sequence
$126 = \frac{n}{2}(6 + a_n)$
$a_n = 6 + (n - 1)4 = 2 + 4n$
$126 = \frac{n}{2}[6 + (2 + 4n)]$
$252 = n(8 + 4n) = 4n^2 + 8n$
$n^2 + 2n - 63 = 0$
$(n + 9)(n - 7) = 0$
$n = 7$

4. $0.454545\ldots = 0.45 + 0.0045 + 0.000045 + \cdots$
$a = 0.45$ $r = 0.01$
$$S = \frac{0.45}{1 - 0.01} = \frac{0.45}{0.99} = \frac{5}{11}$$

5. $\sqrt{1-4x} = (1-4x)^{1/2}$

$$= 1 + \left(\tfrac{1}{2}\right)(-4x) + \frac{\tfrac{1}{2}\left(\tfrac{1}{2}-1\right)}{2}(-4x)^2 + \cdots$$

$$= 1 - 2x - 2x^2 + \cdots$$

6. $(2x-y)^5 = (2x)^5 + 5(2x)^4(-y) + \frac{5(4)}{2}(2x)^3(-y)^2$

$$+ \frac{5(4)(3)}{2(3)}(2x)^2(-y)^3 + \frac{5(4)(3)(2)}{2(3)(4)}(2x)(-y)^4 + (-y)^5$$

$$= 32x^5 - 80x^4y + 80x^3y^2 - 40x^2y^3 + 10xy^4 - y^5$$

7. $5\% = 0.05$

Value after 1 year is

$2500 + 2500(0.05) = 2500(1.05)$

$V_{20} = 2500(1.05)^{20}$

$\quad = \$6633.24$

8. $2 + 4 + \cdots + 200$

$a_1 = 2, \quad a_{100} = 200, \quad n = 100$

$S_{100} = \frac{100}{2}(2 + 200) = 10{,}100$

9. Ball falls 8.00 ft, rises 4.00 ft, falls 4.00 ft, etc.

Distance $= 8.00 + (4.00 + 4.00) + (2.00 + 2.00) + \cdots$

$\quad = 8.00 + 8.00 + 4.00 + 2.00 + \cdots$

$\quad = 8.00 + \dfrac{8.00}{1 - 0.5} = 8.00 + 16.0 = 24.0$ ft

Chapter 20

1. $\sec\theta - \dfrac{\tan\theta}{\csc\theta} = \dfrac{1}{\cos\theta} - \dfrac{\dfrac{\sin\theta}{\cos\theta}}{\dfrac{1}{\sin\theta}} = \dfrac{1}{\cos\theta} - \dfrac{\sin^2\theta}{\cos\theta}$

$$= \dfrac{1-\sin^2\theta}{\cos\theta} = \dfrac{\cos^2\theta}{\cos\theta} = \cos\theta$$

$$\cos\theta = \cos\theta$$

2. $\sin 2x + \sin x = 0$

$2\sin x\cos x + \sin x = 0$

$\sin x(2\cos x + 1) = 0$

$\sin x = 0 \qquad \cos x = -\dfrac{1}{2}$

$x = 0, \pi, \frac{2\pi}{3}, \frac{4\pi}{3}$

3. $\theta = \sin^{-1} x$

$\cos\theta = \cos(\sin^{-1} x)$

$\quad = \sqrt{1 - x^2}$

4. $i = 8.00e^{-20t}(1.73\cos 10.0t - \sin 10.0t)$

Since $8.00e^{-20t}$ will not equal zero, $i = 0$ if

$1.73\cos 10.0t - \sin 10.0t = 0$

$1.73 = \tan 10.0t$

$10.0t = \tan^{-1} 1.73$

$\quad t = 0.105$ s

5. $\dfrac{\tan\alpha + \tan\beta}{\tan\alpha - \tan\beta} = \dfrac{\dfrac{\sin\alpha}{\cos\alpha} + \dfrac{\sin\beta}{\cos\beta}}{\dfrac{\sin\alpha}{\cos\alpha} - \dfrac{\sin\beta}{\cos\beta}} = \dfrac{\sin\alpha\cos\beta + \sin\beta\cos\alpha}{\sin\alpha\cos\beta - \sin\beta\cos\alpha}$

$$= \dfrac{\sin(\alpha+\beta)}{\sin(\alpha-\beta)}$$

6. $\cot^2 x - \cos^2 x = \dfrac{\cos^2 x}{\sin^2 x} - \cos^2 x =$

$\cos^2 x(\csc^2 x - 1) = \cos^2 x\cot^2 x;$

$\cos^2 x\cot^2 x = \cos^2 x\cot^2 x$

7. $\sin x = -\frac{3}{5}$

$\cos x = \frac{4}{5}$

$\cos\dfrac{x}{2} = -\sqrt{\dfrac{1 + \left(\frac{4}{5}\right)}{2}} = -\sqrt{\dfrac{9}{10}} = -0.9487$

since $270° < x < 360°$, $135° < \frac{x}{2} < 180°$;

$\frac{x}{2}$ is in second quadrant, where $\cos\frac{x}{2}$ is negative.

8. $I = I_0 \sin 2\theta \cos 2\theta$

$\dfrac{2I}{I_0} = 2\sin 2\theta \cos 2\theta = \sin 4\theta$

$4\theta = \sin^{-1}\dfrac{2I}{I_0}$

$\theta = \dfrac{1}{4}\sin^{-1}\dfrac{2I}{I_0}$

9. $y = x - 2\cos x - 5$

$x = 3.02, 4.42, 6.77$

10. $\cos^{-1} x = -\tan^{-1}(-1)$

$\tan^{-1}(-1) = -\frac{\pi}{4}$

$\cos^{-1} x = -\left(-\frac{\pi}{4}\right) = \frac{\pi}{4}$

$x = \cos\frac{\pi}{4} = \frac{1}{2}\sqrt{2}$

Chapter 21

1. Points: $(4, -1), (6, 3)$

(a) $d = \sqrt{(6-4)^2 + [3-(-1)]^2} = \sqrt{4 + 16} = 2\sqrt{5}$

(b) Between points $m = \dfrac{3-(-1)}{6-4} = 2$; perp. line, $m = -\dfrac{1}{2}$

2. $\quad 2(x^2 + x) = 1 - y^2$

$2x^2 + y^2 + 2x - 1 = 0$

$A \neq C$ (same sign)

$B = 0$: ellipse

3. $4x - 2y + 5 = 0$

$\quad y = 2x + \frac{5}{2}$

$\quad m = 2 \qquad b = \frac{5}{2}$

4. $\quad x^2 = 2x - y^2$

$x^2 + y^2 = 2x$

$r^2 = 2r\cos\theta$

$r = 2\cos\theta$

5. $x^2 = -12y$

$4p = -12, \qquad p = -3$

$V(0,0) \qquad F(0,-3)$

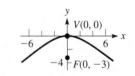

6. Center $(-1, 2)$; $h = -1$, $k = 2$
$$r = \sqrt{(2+1)^2 + (3-2)^2}$$
$$= \sqrt{10}$$
$$(x+1)^2 + (y-2)^2 = 10$$
or
$$x^2 + y^2 + 2x - 4y - 5 = 0$$

7. $m = \dfrac{-2-1}{2+4} = -\dfrac{1}{2}$
$$y - 1 = -\tfrac{1}{2}(x+4)$$
$$2y - 2 = -x - 4$$
$$x + 2y + 2 = 0$$

8. $x^2 = 4py$
$$(6.00)^2 = 4p(4.00)$$
$$p = 2.25$$
Focus is 2.25 cm
from vertex.

9. $a = 8 \qquad b = 4$

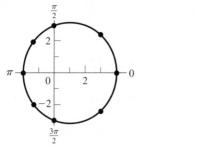

$$\frac{x^2}{64} + \frac{y^2}{16} = 1$$
Find y for $x = 4$ ft.
$$\frac{16}{64} + \frac{y^2}{16} = 1$$
$$y^2 = 12 \qquad y = 3.5\text{ ft}$$
$$h = 10.0 + 3.5 = 13.5\text{ ft}$$

10. $r = 3 + \cos\theta$

r	0	$\frac{\pi}{4}$	$\frac{\pi}{2}$	$\frac{3\pi}{4}$	π	$\frac{5\pi}{4}$	$\frac{3\pi}{2}$	$\frac{7\pi}{4}$	2π
θ	4.0	3.7	3.0	2.3	2.0	2.3	3.0	3.7	4.0

11. $4y^2 - x^2 - 4x - 8y - 4 = 0$
$$4(y^2 - 2y \quad) - (x^2 + 4x \quad) = 4$$
$$4(y^2 - 2y + 1) - (x^2 + 4x + 4) = 4 + 4 - 4$$
$$\frac{(y-1)^2}{1^2} - \frac{(x+2)^2}{2^2} = 1$$
$$C(-2, 1) \qquad V(-2, 0) \qquad V(-2, 2)$$

12. Equation of curve: $8x^2 - 4xy + 5y^2 = 36$; $A = 8$, $B = -4$,
$C = 5$
(a) $B^2 - 4AC = (-4)^2 - 4(8)(5) = 16 - 160 < 0$; ellipse
(b) $\tan 2\theta = \dfrac{B}{A-C} = \dfrac{-4}{8-5} = -\dfrac{4}{3}$; $\theta = \dfrac{1}{2}\tan^{-1}\left(-\dfrac{4}{3}\right) = 63.4°$

Chapter 22

1.

Number	1	2	3	4	5	6	7	8	9	10
Frequency	1	0	1	2	3	2	1	2	2	1

$\Sigma f = 15$
Median is eighth number; median $= 6$.

2. 5 appears three times, and no other number appears more than twice; mode $= 5$.

3.

Number	1–2	3–4	5–6	7–8	9–10
Frequency	1	3	5	3	3

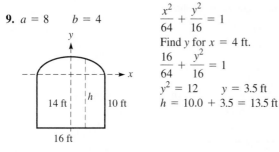

4. Let $t = $ thickness
$$\bar{t} = \frac{3(0.90) + 9(0.91) + 31(0.92) + 38(0.93) + 12(0.94) + 5(0.95) + 2(0.96)}{3 + 9 + 31 + 38 + 12 + 5 + 2}$$
$$= \frac{92.7}{100} = 0.927\text{ in.}$$

5. $\Sigma x^2 = 3(0.90)^2 + 9(0.91)^2 + 31(0.92)^2 + 38(0.93)^2$
$$+ 12(0.94)^2 + 5(0.95)^2 + 2(0.96)^2 = 85.9464$$
$$(\Sigma x)^2 = 92.7^2$$
$$s = \sqrt{\frac{100(85.9464) - 92.7^2}{100(99)}} = 0.117\text{ in.}$$

6.

7.

t (in.)	0.90	0.91	0.92	0.93	0.94	0.95	0.96
%	3	9	31	38	12	5	2

8.

t (in.)	<0.905	<0.915	<0.925	<0.935
cum f	3	12	43	81

t (in.)	<0.945	<0.955	<0.965
cum f	93	98	100

9. $0.5000 + 0.2257 = 0.7257$ (total area to the left)

$\qquad\qquad\qquad\quad = 72.57\%$

10. Find the range of each subgroup (subtract the lowest value from the largest value). Find the mean R of these ranges (divide their sum by 20). This is the value of the central line. Multiply R by the appropriate control chart factors to obtain the upper control limit (UCL) and the lower control limit (LCL). Draw the central line, the UCL line, and the LCL line on a graph. Plot the points for each R corresponding to its subgroup and join successive points by straight-line segments.

11.

x	y	xy	x^2
1	5	5	1
3	11	33	9
5	17	85	25
7	20	140	49
9	27	243	81
25	80	506	165

$n = 5$

$m = \dfrac{5(506) - (25)(80)}{5(165) - 25^2}$

$\quad = 2.65$

$b = \dfrac{165(80) - (506)(25)}{5(165) - 25^2}$

$\quad = 2.75$

$y = 2.65x + 2.75$

12.

x	$\sqrt{x}$	y	$\sqrt{x}\,y$	$(\sqrt{x})^2 = x$
1.00	1.000	1.10	1.1000	1.000
3.00	1.732	1.90	3.2908	3.000
5.00	2.236	2.50	5.5900	5.000
7.00	2.646	2.90	7.6734	7.000
9.00	3.000	3.30	9.9000	9.000
	10.614	11.70	27.5542	25.000

$n = 5$

$m = \dfrac{5(27.5542) - (10.614)(11.70)}{5(25.000) - (10.614)^2} = 1.10$

$b = \dfrac{(25.000)(11.70) - (27.5542)(10.614)}{5(25.000) - (10.614)^2} = 0.00$ (rounded off)

$y = 1.10\sqrt{x}$

INDEX OF APPLICATIONS

INDEX OF WRITING EXERCISES

The final exercise in each chapter is a writing exercise. Each of these 22 exercises will require at least a paragraph to provide a good explanation of the problem presented. Also, there are about 300 other exercises throughout the text marked with (W) before the exercise number or instructions that require at least one complete sentence (up to a paragraph) to provide the explanation needed.

Following is a listing of these writing exercises. The first number shown is the page number, and the numbers in parentheses are the exercise numbers. The * denotes the final review exercise of the chapter.

INDEX

Abscissa, 89
Absolute inequality, 461
Absolute value, 3; of complex numbers, 342; in inequalities, 477
Accuracy of number, 12, 117
Addition: of algebraic expressions, 27; of complex number, 337; of fractions, 198; of matrices, 429; of ordinates, 305; of radicals, 324; of signed numbers, 7; of vectors, 256, 263
Adjacent angles, 51
Algebraic expressions, 26
Alternate-exterior angles, 51
Alternate-interior, angles, 51
Alternating current, 303, 353
Altitude, 54
Ambiguous case, 277
Ampere, 356
Amplitude of sine curve, 290
Analytic geometry, 558
Angle, 50,109; acute, 50; central, 64; of depression, 124; of elevation, 124; inscribed, 64; negative, 109, 239; obtuse, 50; phase, 290; quadrantal, 110, 238; reference, 234; right, 50; of rotation, 596; standard position, 110; straight, 50
Angular velocity, 246
Antilogarithm, 374
Approximate numbers, 12, 114
Arc, 64
Arc length, 245
Area: of circular sector, 246; of geometric figures, 55; lateral, 72; under a curve, 743
Argument of complex number, 342
Aristotle, 80
Arithmetic mean, 618
Arithmetic sequence, 506
Array, 613
Associative law, 6, 430
Asymptote, 93, 300, 584
Attribute, 634
Axes, coordinate, 89
Axis: of ellipse, 579; of hyperbola, 584; of parabola, 574; polar, 599

Babbage, Charles, 408
Bar, 8, 28
Base: of exponents, 16; of logarithms, 365, 374, 377; of solid, 71; of trapezoid, 60; units, A.2
Binomial, 27
Binomial formula, 520
Binomial series, 521
Binomial theorem, 518
Bit, 6
Bode diagram, 387
Braces, 28

Brackets, 28
Briggs, Henry, 362
Byte, 6

Calculator, 11, 96, 116, 236, 363, 439, 468
Calculator operations (complete listing of operations), A.10
Cancellation, 190
Capacitance, 353
Cardioid, 603
Catenary, 383
Cauchy, Augustin-Louis, 391
Cayley, Arthur, 427
Central angle, 64
Central line, 632
Central tendency, 617
Centroid, 54
Character of roots, 221
Checking answers, 197
Chord, 63
Circle, 63, 392, 569
Circumference, 63
Class, 613
Class mark, 614
Coefficient, 27
Cofunction, 120
Column of determinant, 153
Common logarithms, 374
Commutative law, 6, 430
Complementary angles, 50, 119
Completing the square, 217, 225, 571, 590
Complex fraction, 202
Complex number, 2, 335
Complex plane, 340
Complex roots, 415, 422
Components of vector, 259
Composite trigonometric curves, 304
Computer, 460
Conditional equation, 35
Conditional inequality, 461
Cone, 71
Congruent triangles, 57
Conic sections, 392, 594
Conjugate: axis of hyperbola, 584; of complex number, 336
Constant, 4; of proportionality, 495, 496
Constraint, 483
Contradiction, 35
Control charts, 631
Conversion of units, 109, 241, A.5
Coordinates: polar, 599; rectangular, 89
Correlation, 641
Corresponding angles, 51, 56, 112
Corresponding segments, 51
Cosecant: of angle, 113; graph of, 299
Cosine: of angle, 113; of double angle, 537; graph of, 289; half angle, 541; of sum of two angles, 533

Cosines, law of, 280
Cotangent: of angle, 113; graph of, 299
Coterminal angles, 109
Coulomb, 356
Cramer, Gabriel, 134, 154
Cramer's rule, 154, 164, 453
Critical value, 471
Cube, 71
Cumulative frequency, 615
Current, 353
Cycle of sine curve, 297
Cylinder, 71

d'Alembert, Jean, 254
Dantzig, George, 460
Decibel, 381
Decimal, repeating, 2, 516
Degree: as measure of angle, 50, 109; of polynomial, 27
DeMoivre's theorem, 350
Denominate number, 4, 314, A.2
Denominator: common, 149; rationalizing of, 322, 329
Dependent system of equations, 145, 150, 157
Dependent variable, 81
Descartes, René, 89, 312, 335, 391, 420, 558
Descartes' rule of signs, 420
Determinant, 153; higher-order, 450; properties of, 451; second-order, 153; third-order, 163, 450
Deviation, 638; standard, 621
Diagonal, 60
Diagonal of determinant, 153
Diameter, 63
Difference, common, 506
Diocles, 212
Direct variation, 495
Directrix of parabola, 574
Displacement, 258, 761; of sine curve, 296
Distance formula, 559
Distributive law, 6
Dividend, 33
Division: of algebraic expressions, 32; of complex numbers, 337, 348; of fractions, 194; of radicals, 328; of signed numbers, 8; synthetic, 410; by zero, 9
Divisor, 33
Domain, 84, 290
Double-angle formulas, 537

e (irrational number), 344, 377
Eccentricity, 582, 588
Edison, Thomas, 288, 353, 526
Einstein, Albert, 38
Electric circuits, 353
Element: of determinant, 153; of matrix, 428

NOTES

NOTES

NOTES

ALGEBRA

Exponents and Radicals

$$a^m \cdot a^n = a^{m+n}$$

$$\frac{a^m}{a^n} = a^{m-n} \quad a \neq 0$$

$$(a^m)^n = a^{mn}$$

$$(ab)^n = a^n b^n$$

$$\left(\frac{a}{b}\right)^n = \frac{a^n}{b^n} \quad b \neq 0$$

$$a^0 = 1 \quad a \neq 0$$

$$a^{-n} = \frac{1}{a^n} \quad a \neq 0$$

$$a^{m/n} = \sqrt[n]{a^m} = (\sqrt[n]{a})^m$$

$$\sqrt{ab} = \sqrt{a}\,\sqrt{b}$$

Special Products

$$a(x + y) = ax + ay$$

$$(x + y)(x - y) = x^2 - y^2$$

$$(x + y)^2 = x^2 + 2xy + y^2$$

$$(x - y)^2 = x^2 - 2xy + y^2$$

Quadratic Equation and Formula

$$ax^2 + bx + c = 0$$

$$x = \frac{-b \pm \sqrt{b^2 - 4ac}}{2a}$$

Properties of Logarithms

$$\log_b x + \log_b y = \log_b xy$$

$$\log_b x - \log_b y = \log_b\left(\frac{x}{y}\right)$$

$$n \log_b x = \log_b(x^n)$$

Complex Numbers

$$\sqrt{-a} = j\sqrt{a} \quad (a > 0)$$

$$x + yj = r(\cos\theta + j\sin\theta)$$

$$= re^{j\theta} = r\underline{/\theta}$$

Variation

Direct variation: $y = kx$

Inverse variation: $y = k/x$

TRIGONOMETRY

$$\sin\theta = \frac{y}{r} \qquad \cos\theta = \frac{x}{r} \qquad \tan\theta = \frac{y}{x} \qquad \cot\theta = \frac{x}{y} \qquad \sec\theta = \frac{r}{x} \qquad \csc\theta = \frac{r}{y}$$

Law of Sines: $\dfrac{a}{\sin A} = \dfrac{b}{\sin B} = \dfrac{c}{\sin C}$

Law of Cosines: $a^2 = b^2 + c^2 - 2bc\cos A$

$\pi \text{ rad} = 180°$

Basic Identities

$$\sin\theta = \frac{1}{\csc\theta} \qquad \cos\theta = \frac{1}{\sec\theta} \qquad \tan\theta = \frac{1}{\cot\theta} \qquad \tan\theta = \frac{\sin\theta}{\cos\theta} \qquad \cot\theta = \frac{\cos\theta}{\sin\theta}$$

$$\sin^2\theta + \cos^2\theta = 1 \qquad 1 + \tan^2\theta = \sec^2\theta \qquad 1 + \cot^2\theta = \csc^2\theta$$

$$\sin 2\alpha = 2\sin\alpha\cos\alpha \qquad \cos 2\alpha = \cos^2\alpha - \sin^2\alpha = 2\cos^2\alpha - 1 = 1 - 2\sin^2\alpha$$

$$\sin\frac{\alpha}{2} = \pm\sqrt{\frac{1 - \cos\alpha}{2}} \qquad \cos\frac{\alpha}{2} = \pm\sqrt{\frac{1 + \cos\alpha}{2}}$$